BIOLOGY

concepts and applications

FOURTH EDITION

CECIE STARR

Brooks/Cole
Thomson Learning

Australia • Canada • Denmark • Japan • Mexico
New Zealand • Philippines • Puerto Rico • Singapore • South Africa
Spain • United Kingdom • United States

BIOLOGY PUBLISHER: Jack C. Carey

PROJECT DEVELOPMENT EDITOR: Kristin Milotich

MEDIA PROJECT MANAGER: Pat Waldo

EDITORIAL ASSISTANT: Susan Lussier

DEVELOPMENTAL EDITOR: Mary Arbogast

MARKETING MANAGER: Tami Cueny

MARKETING ASSISTANT: Kelly Fielding

PROJECT EDITOR: Sandra Craig

PRINT BUYER: Karen Hunt

PRODUCTION: Mary Douglas, Rogue Valley Publications

TEXT AND COVER DESIGN, ART DIRECTION: Gary Head,
Gary Head Design

ART COORDINATORS: Myrna Engler, Mary Douglas

EDITORIAL PRODUCTION: Karen Stough, Mary Wininger

PRIMARY ARTISTS: Lisa Starr; Raychel Ciemma; Precision
Graphics, Preface Inc. (Jan Flessner)

ADDITIONAL ARTISTS: Robert Demarest, Darwen Hennings,
Vally Hennings, Betsy Palay, Nadine Sokol, Kevin
Somerville, Lloyd Townsend

PHOTO RESEARCH, PERMISSIONS: Roberta Broyer

COVER PHOTOGRAPH: © 1999 John Warden/AlaskaStock.com

COMPOSITION: Preface, Inc. (Beth Morrison, Angela Harris)

COLOR PROCESSING: H&S Graphics (Tom Anderson,
Michelle Kessel, Rich Stanley, and John Deady)

PRINTING AND BINDING: R. R. Donnelley & Sons, Inc.

BOOKS IN THE BROOKS/COLE BIOLOGY SERIES

Biology: The Unity and Diversity of Life, Eighth, Starr/Taggart
Biology: Concepts and Applications, Fourth, Starr
Laboratory Manual for Biology, Perry and Morton
Human Biology, Third, Starr/McMillan
Perspectives in Human Biology, Knapp
Human Physiology, Third, Sherwood
Fundamentals of Physiology, Second, Sherwood
Dimensions of Cancer, Kupchella
Psychobiology: The Neuron and Behavior, Hoyenga/Hoyenga

Introduction to Cell and Molecular Biology, Wolfe
Molecular and Cellular Biology, Wolfe
Cell Ultrastructure, Wolfe

Introduction to Microbiology, Second, Ingraham/Ingraham

Genetics: The Continuity of Life, Fairbanks
Human Heredity, Fifth, Cummings
Introduction to Biotechnology, Barnum
Evolution: Process and Product, Third, Dodson/Dodson
Sex, Evolution, and Behavior, Second, Daly/Wilson

Plant Biology, Rost et al.
Plant Physiology, Fourth, Salisbury/Ross
Plant Physiology Laboratory Manual, Ross
Plants: An Evolutionary Survey, Second, Scagel et al.

General Ecology, Krohne
Living in the Environment, Eleventh, Miller
Environmental Science, Seventh, Miller
Sustaining the Earth, Fourth, Miller
Environment: Problems and Solutions, Miller
Environmental Science, Fifth, Chiras

Oceanography: An Invitation to Marine Science, Third, Garrison
Essentials of Oceanography, Garrison
Introduction to Ocean Sciences, Segar
Oceanography: An Introduction, Fifth, Ingmanson/Wallace
Marine Life and the Sea, Milne

Library of Congress Cataloging-in-Publication Data
Starr, Cecie.
    Biology : concepts and applications / Cecie Starr. — 4th ed.
      p. cm.
    Includes bibliographical references
    ISBN 0-534-56322-8 (hardback)
    1. Biology.   I. Title.
  QH307.2.S73 1999
  570—dc21
                                                    99-29085
                                                         CIP

Student Paperback Edition: ISBN 0-534-56325-2
Annotated Instructor's Edition: 0-534-56326-0

For more information, contact: BROOKS/COLE, THOMSON LEARNING,
511 Forest Lodge Road, Pacific Grove, CA 93950, USA
or electronically at http://www.brookscole.com

International Headquarters
Thomson Learning
290 Harbor Drive, 2nd Floor
Stamford, CT 06902-7477, USA

UK/Europe/Middle East
Thomson Learning
Berkshire House
168-173 High Holborn
London WC1V 7AA, United Kingdom

Australia
Nelson/Thomson Learning
102 Dodds Street
South Melbourne 3205, Victoria, Australia

Asia
Thomson Learning
60 Albert Street #15-01
Albert Complex, Singapore 189969

Canada
Nelson/Thomson Learning
1120 Birchmount Road
Scarborough, Ontario M1K 5G4, Canada

Japan
Thomson Learning
Hirakawacho Kyowa Building, 3F
2-2-1 Hirakawacho, Chiyoda-ku, Tokyo 102, Japan

Southern Africa
Thomson Learning
Building 18, Constantia Park
240 Old Pretoria Road
Halfway House, 1685 South Africa

CONTENTS IN BRIEF

DETAILED CONTENTS

MEERKATS AT SUNRISE

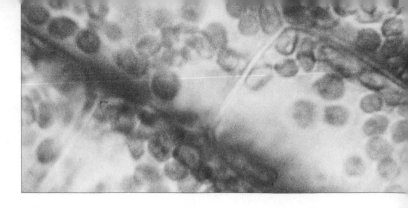

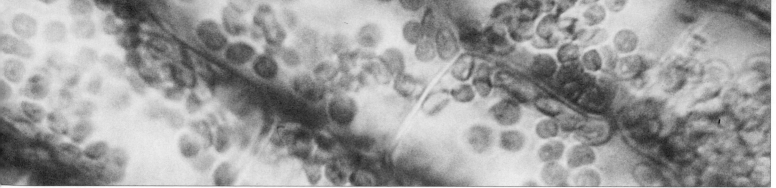

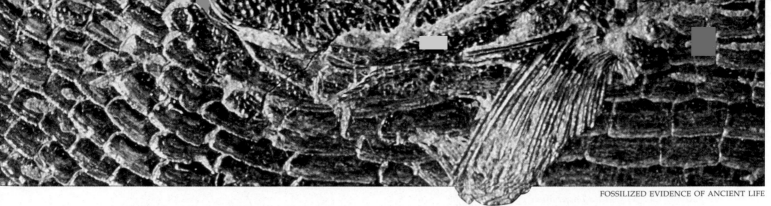
FOSSILIZED EVIDENCE OF ANCIENT LIFE

A SAMPLING OF PLANTS AND LICHENS

VI ANIMAL STRUCTURE AND FUNCTION

FLOWERING PLANT BUSILY ATTRACTING POLLINATORS

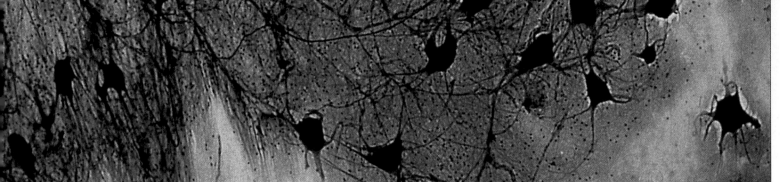

A SAMPLING OF MOTOR NEURONS

SOLDIER TERMITES DEFENDING THE COLONY

PREFACE

Teachers of introductory biology know all about the Red Queen effect, whereby one runs as fast as one can to stay in the same place. New and modified information from hundreds of fields of inquiry piles up daily, and somehow teachers are expected to distill it into Biology Lite, a one-course zip through the high points.

This book is a coherent account of the sweep of life's diversity and its underlying unity, although we structured it in such a way that teachers can easily and selectively assign chapters from it. The book highlights concepts, current understandings, and research trends for all major fields of inquiry. Through examples of problem solving and experiments, it shows the power of thinking critically. It explains the structure and function of a broad sampling of organisms in enough detail so students can develop a working vocabulary about life's parts and processes.

The book starts with an overview of basic concepts and methods. Three units on the principles of biochemistry, inheritance, and evolution follow; they are the conceptual framework for exploring life's unity and diversity, starting with an evolutionary survey of each kingdom. Units on comparative anatomy and physiology of plants, then of animals, are next. The last unit focuses on patterns and consequences of organisms interacting with one another and with the environment. This conceptual organization parallels the levels of biological organization.

CONCEPT SPREADS We keep the story line in focus for students by subscribing to the question "How do you eat an elephant?" and its answer, "One bite at a time." In this book students will find descriptions, art, and supporting evidence for each concept organized on two facing pages, at most. Each "concept spread" starts with a numbered tab and ends with boldface, summary statements of key points (*see below*). Students can use the statements to check whether they understand the concepts of one spread before starting another.

Writing that rambles quickly kills interest in any topic. Restricting the space available for each concept forced us to avoid the clutter of superfluous detail. Also, ongoing feedback from teachers of more than three million students guided our decisions about when to leave core material alone and when to loosen it up with applications. Within each spread, headings and subheadings help students track the hierarchy of information. Carefully crafted transitions between spreads help them keep the greater story in focus

To keep readers focused, we present each concept on one or two facing pages, starting with a numbered tab . . .

Pregnancy lasts an average of thirty-eight weeks from the time of fertilization. It takes about two weeks for a blastocyst to form. The time span from the third to the end of the eighth week is the *embryonic* period, when the major organ systems form. When it ends, the new individual has distinctly human features and is called a **fetus**. In the *fetal* period, from the start of the ninth week until birth, organs enlarge and become specialized.

d DAY 5. A blastocoel (a fluid-filled cavity) forms in the morula as a result of secretions from the surface cells. By the thirty-two-cell stage, cells of an inner cell mass are already differentiating. They will give rise to the embryo proper. This embryonic stage is called a blastocyst.

c DAY 4. By 96 hours, there is a ball of sixteen to thirty-two cells that is shaped like a mulberry. It is a morula (after *morum*, Latin for mulberry). Cells of the surface layer will function in implantation and will give rise to a membrane, the chorion.

b DAY 3. After the third cleavage, the cells suddenly huddle together into a compacted ball, which becomes stabilized by numerous tight junctions among the outer cells. Gap junctions form among the interior cells and enhance intercellular communication.

a DAYS 1–2. Cleavage begins within 24 hours after fertilization. The first cleavage furrow extends between the two polar bodies. Subsequent cuts are rotational, so the resulting cells are not symmetrically arranged (compare Section 38.3). Until the eight-cell stage forms, the cells are loosely arranged, with considerable space between them.

inner cell mass

FERTILIZATION

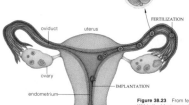
oviduct
uterus
ovary
IMPLANTATION
endometrium

Figure 38.23 From fertilization through implantation. Cleavage produces a blastocyst. Within the blastocyst, the inner cell mass gives rise to the disk-shaped early embryo. Three of the extraembryonic membranes (amnion, chorion, and yolk sac) start forming. The fourth extraembryonic membrane (allantois) forms after the blastocyst is implanted.

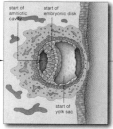

endometrium
trophoblast (surface layer of cells of the blastocyst)
blastocoel
inner cell mass
uterine cavity

e DAYS 6–7. Surface cells of the blastocyst attach to the endometrium and start to burrow into it. Implantation is under way.

actual size

We typically call the first three months of pregnancy the *first* trimester. The *second* trimester extends from the start of the fourth month to the end of the sixth. The *third* trimester extends from the seventh month until birth. Beginning with Figure 38.23, the next series of illustrations shows the characteristic features of the new individual at progressive stages of development.

Cleavage and Implantation

Three to four days after fertilization, the zygote is already undergoing cleavage as it tumbles through the oviduct. Genes are already being expressed; the early divisions depend on their products. At the eight-cell stage, the cells huddle into a compact ball. By the fifth day, there is a surface layer of cells (a trophoblast), a cavity filled with their secretions (a blastocoel), and a tiny cluster of interior cells (an inner cell mass). These are the defining features of a human blastocyst (Figure 38.23*d*).

Six or seven days after fertilization, **implantation** is under way. By this process, the blastocyst adheres to the uterine lining, some of its cells send out projections that invade the mother's tissues, and connections start forming that will metabolically support the developing embryo through the months ahead. While the invasion is proceeding, the inner cell mass develops into two

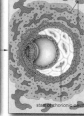

start of amniotic cavity
start of embryonic disk
start of yolk sac

f DAYS 10–11. The yolk sac, embryonic disk, and amniotic cavity have started to form from parts of the blastocyst.

actual size

cell layers of a flattened and somewhat circular shape. The two layers make up the embryonic disk—and in short order they will give rise to the embryo proper.

Extraembryonic Membranes

As implantation progresses, membranes start to form outside the embryo. First a fluid-filled *amniotic* cavity opens up between the embryonic disk and part of the blastocyst's surface (Figure 38.23*f*). Then cells migrate around the wall of the cavity and form the **amnion**, a membrane that will enclose the embryo. Fluid inside the cavity will function as a buoyant cradle where the embryo can grow, move freely, and be protected from abrupt temperature changes and mechanical impacts.

While the amnion forms, other cells migrate around the inner wall of the blastocyst's first cavity. They form a lining that becomes the **yolk sac**. This extraembryonic membrane speaks of the evolutionary heritage of land vertebrates (Sections 24.7 and 24.8). For most animals that produce shelled eggs, the sac holds nutritive yolk. In humans, part of the yolk sac becomes a site of blood cell formation, and part will give rise to germ cells, the forerunners of gametes.

Before the blastocyst is fully implanted, spaces open in maternal tissues and fill with blood seeping in from ruptured capillaries. Inside the blastocyst, another cavity opens around the amnion and yolk sac. Now fingerlike projections start to form on the cavity's lining, which is the **chorion**. This new membrane will become part of a spongy, blood-engorged tissue called the placenta.

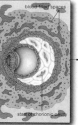

blood-filled spaces

g DAY 12. Blood-filled spaces form in maternal tissue. The chorionic cavity starts to form.

actual size

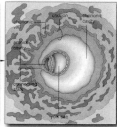

chorion
chorionic cavity
chorionic villi
amniotic cavity
connecting stalk
yolk sac

h DAY 14. A connecting stalk has formed between the embryonic disk and chorion. Chorionic villi, which will be features of a placenta, start to form.

actual size

After the blastocyst is finally implanted, another extraembryonic membrane will form as an outpouching of the yolk sac. This third membrane will become the **allantois**. An allantois functions differently in different animal groups. Among reptiles, birds, and some of the mammals, it has roles in respiration and in the storage of metabolic wastes. In humans, the urinary bladder as well as blood vessels for the placenta form from it.

One more point should be made here. Cells of the blastocyst secrete the hormone HCG (Human Chorionic Gonadotropin), which stimulates the corpus luteum to keep on secreting progesterone and estrogen. Thus the blastocyst itself prevents menstrual flow and works to avoid being sloughed off until the placenta takes over the task, some eleven weeks later. By the start of the third week, HCG can be detected in the mother's blood or urine. At-home *pregnancy tests* use a treated "dip-stick" that changes color when HCG is present in urine.

A human blastocyst is composed of a surface layer of cells around a fluid-filled cavity (blastocoel) and an inner cell mass, which will give rise to the embryo proper.

Six or seven days after fertilization, the blastocyst implants itself in the endometrium. Now projections from its surface invade maternal tissues, and connections start to form that in time will metabolically support the developing embryo.

Some parts of the blastocyst give rise to an amnion, yolk sac, chorion, and allantois. These extraembryonic membranes serve different functions. Together they are vital for the structural and functional development of the embryo.

. . . and ending with one or more summary statements.

and discourage memorization for its own sake. Also, for interested students, we integrate details that could disrupt the text's conceptual flow into optional illustrations.

Based on extensive user feedback, we know that our clearly defined organization helps students find assigned topics easily and translates into improved test scores.

BALANCING CONCEPTS WITH APPLICATIONS   Each chapter starts with a lively or sobering application and an adjoining list of key concepts (the chapter's advance organizer). Strategically placed examples of applications parallel core material, not so many as to be distracting, but enough to keep minds perking along with the conceptual development. Many brief applications are integrated in the text. Others are in focus essays that afford more depth on medical, environmental, and social issues for interested students but do not interrupt the text. The book's last four pages separately index all applications for quick reference.

FOUNDATIONS FOR CRITICAL THINKING   To help students increase their capacity for critical thinking, we walk them through experiments that yielded evidence in favor of or against hypotheses. The main index lists all experimental tests and observational tests (see the entries *Experiment* and *Test, observational*).

We use certain chapter introductions as well as entire chapters to show students some of the productive results of critical thinking. The introductions to Mendelian genetics (Chapter 10), DNA structure and function (12), speciation (17), immunology (34), and behavior (44) are examples.

Each chapter concludes with a set of *Critical Thinking* questions. Katherine Denniston developed most of those thought-provoking questions. Daniel Fairbanks developed many of the *Genetics Problems*, which help students grasp the principles of inheritance (Chapters 10 and 11).

VISUAL OVERVIEWS OF CONCEPTS   We simultaneously develop the text and art as inseparable parts of the same story. We also give visual learners a means to work their way through a visual overview of a major process before reading the corresponding text, which thereby becomes less intimidating. Students repeatedly let us know how much they appreciate this approach.

Our overview illustrations have step-by-step, written descriptions of biological parts and processes. Instead of "wordless" diagrams, we break down information into a series of illustrated steps. For example, the descriptions of Figure 13.10, integrated with the art, walk students step at a time through the stages by which a mature mRNA transcript becomes translated.

Many anatomical drawings are integrated overviews of structure and function. Students need not jump back and forth from text, to tables, to illustrations, and back again in order to comprehend how an organ system is put together and what its parts do. Individual descriptions of parts are hierarchically arranged to reflect the structural and functional organization of the system.

ZOOM SEQUENCES   Many illustrations progress from macroscopic to microscopic views of the same subject. Figure 7.2 is an example. It shows where the reactions of photosynthesis proceed, starting with a plant leaf. Figures 32.15 and 32.16 start with a human arm and move down through levels of skeletal muscle contraction.

COLOR CODING   Visual consistency throughout the book helps students track complex parts and processes. Our line illustrations consistently use the same colors for each kind of molecule and cell structure:

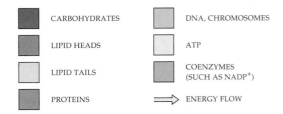

■	CARBOHYDRATES	■	DNA, CHROMOSOMES
■	LIPID HEADS	■	ATP
■	LIPID TAILS	■	COENZYMES (SUCH AS NADP⁺)
■	PROTEINS	⟹	ENERGY FLOW

ICONS   Small diagrams next to an illustration help students relate a topic to the big picture. For instance, a simple representation of a cell subtly reminds students of the location of the plasma membrane relative to the cytoplasm. Other icons serve as reminders of the location of cellular reactions and processes, and of how they relate to one another. Still other icons remind students of the evolutionary relationships among groups of organisms, as in Chapters 23 and 24.

A multimedia icon directs students to art in a CD-ROM enclosed in their book. Another icon directs them to the supplemental material on the Web and a third, to InfoTrac:

CD-ROM ICON: WEB ICON: *WWW*   INFOTRAC ICON:

CONTENT REVISIONS   For this edition we refined text and art in each concept spread and improved transitions between spreads. Overall, we tightened the writing but expanded certain topics that can be confusing if presented in insufficient detail. Chapters have crisper evolutionary story lines, a more balanced text-to-art ratio, and updated applications. Adjustments in the overall framework for the book reflect research trends. For example, in abidance with an emerging consensus, we subscribe to a six-kingdom classification scheme.

INTRODUCTION   The Chapter 1 conceptual overview for the book is more focused, starting with introductions to the molecular trinity (DNA to RNA to protein), energy, and levels of biological organization. An early look at biodiversity reflects the six-kingdom model. A new, simple explanation of evolution by natural selection, a dominant theme for the book, follows. Chapter 1 now introduces scientific methods and the value of thinking critically in more depth and with more straightforward definitions.

To highlight the power of scientific inquiry, we have a new section on experimental tests of an alternative to antibiotics. This example builds on the chapter's earlier description of natural selection for antibiotic resistance. New *Critical Thinking* questions expand on the chapter discussion of scientific methods.

UNIT I. PRINCIPLES OF CELLULAR LIFE Basic chemistry is divided into two more manageable chapters. The first starts with a vignette on elements that have accumulated in the environment and are targets of phytoremediation, as at Chernobyl. We moved the radiometric dating portion of an essay on radioisotopes to the evolution unit. We revised the section on acids, bases, and buffers. Table 2.2 summarizes terms the student will use often. Chapter 3 starts with a carbon story (changes in atmospheric CO_2), and has a new essay on pesticides and better coverage of proteins. Chapter 4 has updated material on the cytoskeleton and cell surface specializations.

Chapter 5 starts with a vignette on an application of bioluminescence. Simpler text and livelier illustrations make bioenergetics (a generally unloved topic) less intimidating. The chapter includes a refined treatment of membrane transport mechanisms.

New Chapter 6 art clearly shows the flow of hydrogen ions between the light-dependent and light-independent reactions. It has new icons, a new section on the properties of light and photosynthetic pigments, a visual comparison of carbon-fixing adaptations, and a new essay on light at Chapter 7 has new art and a livelier, updated treatment of alternative energy sources in humans and other mammals.

UNIT II. PRINCIPLES OF INHERITANCE New section 8.5 introduces chromosome structure earlier in the book and expands on the cell cycle to set the stage for later chapters. Chapter 9 has new art on crossing over and gametogenesis. Chapter 10 has cleaner definitions of many genetics terms and rewrites on pleiotropy and phenotypic variation. New to Chapter 11 is a vignette that uses a new karyotyping method to introduce the concept of genetic analysis (the progeria story moved to a Focus essay). The sections on karyotyping, linkage, and crossing over are rewritten. Sections on inheritance patterns as well as changes in chromosome number and chromosome structure contain new art and updated examples. More realistic art in the human genetics essay gets across the risk potential of amniocentesis and fetoscopy.

Chapter 12 has a new essay on Rosalind Franklin, a closer look at DNA replication, and a new essay on animal cloning. Sections on chromosome organization and on cancer moved to Chapters 9 and 14, respectively. Chapter 13 has a new tRNA model and icon for the art on protein synthesis. More art and a few more text examples make the mutation section easier to understand.

Sections on gene control are now in a separate chapter (14) that starts with a vignette on cancer. Text descriptions and definitions are crisper, particularly with respect to eukaryotic controls. The concluding essay that provides a closer look at cancer includes material on apoptosis.

Chapter 15 is reorganized, rewritten, and illustrated with striking new art. The changes include new material on DNA fingerprinting, automated DNA sequencing, gene isolation, mapping the human genome, and safety issues.

UNIT III. EVOLUTION Chapter 16 clarifies the influence of theories of catastrophism and uniformity on evolutionary thought. Chapter 17 has a tighter section on the biological species concept and reproductive isolating mechanisms. It has new art and text on allopatric speciation on archipelagos and other speciation routes, as by polyploidy. The extinction section is revised.

Chapter 18 is reorganized to separate the material on biological and geologic change. The geologic time scale and the comparative biochemistry section are revised. The five-kingdom and six-kingdom classification schemes are compared. A new section on plate tectonics and an essay on radiometric dating set the stage for the story of Earth's early evolution in next unit.

UNIT IV. EVOLUTION AND DIVERSITY Besides being a conceptual and chronological framework for the diversity chapters, survey Chapter 19 may help students sense their place in nature. Its tree of life reflects the six-kingdom scheme. Ediacaran and Cambrian forms are defined.

Chapter 20 has new material on the nature of bacterial growth and reproduction by prokaryotic fission. Detailed data in the old table on bacterial classification is now in Appendix I. The generalized viral multiplication cycle is revised for clarity. There is an essay on viroids and prions. Protistans get a more informative introduction. The essay on the nature of infectious diseases is expanded, with an evolutionary thrust. The algae are allocated a bit more text and new art (including a red algal life cycle).

Fungi now have their own chapter (21), which starts and ends with the concept of mutualism. The adaptive value of producing staggering numbers of spores is more clearly explained. The Chapter 22 section on evolutionary trends has new art and revised text. There is a new photo and description for peat bogs, a bit more on Carboniferous swamp forests, and more on seed-bearing plants (such as an essay on deforestation and a new concluding essay on seed plants and people).

Chapter 23 has new anatomicals for cnidarians and crustaceans, a section on cephalopod evolution, and a critical thinking question on rotifers. Its introduction to protostomes and deuterostomes is simply stated. Chapter 24 includes new illustrations (of a coelacanth, *Maiasaura*, chimpanzee big toe, and new diagrams (of the evolution of swim bladders and lungs; crocodile body plan; amniote egg, placenta, and hominid evolution). It includes a bit more on mammalian origins and convergent evolution on separate continents. The sections on human origins, early evolution, and dispersals are updated, as in a new Focus on Science essay.

UNIT V. PLANT STRUCTURE AND FUNCTION This unit has major changes. Chapter 25 has new art on meristems, plant tissues, stem and root primary growth patterns, secondary growth and wood characteristics. Chapter 26 has a more inviting introduction and a new section on soil properties, plant nutrients, leaching, and erosion. Water conservation (e.g., at stomata) follows the section on water transport. A new micrograph of tracheids and an experiment on wilting are included.

Chapter 27 includes a revised section on seed and fruit formation and dispersal, new art on patterns of plant growth and development, and a simple explanation of the difference between growth and development. It has the classic photograph of gibberelin-inspired cabbage plants and new art for experiments on flowering responses.

UNIT VI. ANIMAL STRUCTURE AND FUNCTION Chapter 28 is a proven, workable introduction to animal tissues, so we left it alone. The rest of the unit has a stronger systems-integration approach, and now starts with chapters on the nervous, sensory, and endocrine systems (29, 30, 31).

In Chapter 29, neuron structure and functioning are treated first, then invertebrate and vertebrate nervous systems. There is new art on action potential propagation along sheathed neurons and better descriptions of the evolution of the vertebrate nervous system and its functional divisions. The section on the human brain is revised; it now has descriptions of cerebrospinal fluid and the blood-brain barrier. The section on memory is updated. The summary includes a new table on features of the vertebrate brain and spinal cord.

The Chapter 30 introduction employs a python and a bat as lead-ins to the concept of receptor specificity. The chapter includes revised sections on somatic sensations, hearing, balance, and vision; and new sections on sensory pathways and chemical senses.

Chapter 31 has revised text and art on feedback control of adrenal hormones. Chapter 32 uses polar huskies to introduce systems responsible for the body's superficial features, shape, and movements. It has a revised sections on skin, bone, and ATP formation in muscle cells, and a critical thinking question on anabolic steroids. New to Chapter 33 are a simple analogy for flow velocity in a closed system, a diagram of flow distribution of cardiac output, a revised section on blood pressure and its control, and a separate section on capillary function.

Chapter 34 has new art and photographs throughout, a bit more on inflammation's chemical mediators and immunotherapy, and updates on HIV and AIDS. Chapter 35 has new art on gill function and on the evolution of complex lungs, more on vocal cords and speech, and a new section on breathing mechanisms. The section on gas exchange and transport is rewritten. Two of the critical thinking questions deal with deep-sea diving. Chapter 36 has new photographs and new art on intestinal structure, and the BMI formula for estimating weight.

Chapter 38 has a more informative overview of sexual vs. asexual reproduction, revised sections on cleavage, cell differentiation, morphogenesis; and a new section on pattern formation that reflects current research. It has revised sections on the emergence of the vertebrate body plan and on birth, aging, and birth control options.

UNIT VII. ECOLOGY AND BEHAVIOR Examples from conservation biology thread through the unit. Chapter 39 has a new essay on sampling population density, updates on human population growth, and age structure diagrams for baby boomers. Chapter 40 has the text on keystone species moved to the section on competition, a bit more on parasitism, new material on restoration ecology, and new sections on exotic and endangered species.

Chapter 41 has new art on the hydrologic, phosphorus, carbon, and nitrogen cycles; and updates on greenhouse gases and global warming. Eutrophication is described in the section on the phosphorus cycle. Critical thinking questions address endangered ecosystems and retreating polar ice shelves. Chapter 42 has a simpler diagram of the atmosphere, new art on climate zones and ocean currents, rain shadows, and biome distribution, and a photograph conveying the vulnerability of coniferous forests to fire. It has more on stream ecosystems and an update on newly discovered mid-ocean biodiversity. A new essay focuses on Rita Colwell's cholera research to reinforce a unifying concept, that the atmosphere, ocean, and land connect in ways that profoundly influence the world of life.

New to Chapter 43 is a look at Wangari Maathais work on reforestation, a more thoughtful essay on tropical forest destruction, a new map on aquifer depletion, new maps on energy consumption, and an update on Chernobyl's aftermath. Chapter 44 includes refined descriptions of instinctive behavior, learned behavior, and communication signals. Reorganization puts descriptions of the benefits and costs of social groups in separate sections. The section on altruism, the essay on self-sacrificing behavior, and the section on human social behavior are rewritten.

SUPPLEMENTS

Just inside the cover of instructors' examination copies is a foldout brochure describing the comprehensive package of supplements to this book.

A COMMUNITY EFFORT

After being responsible for two leading textbooks for more than twenty years, you'd think I would be full of myself. But I still wake up terrified in the middle of the night about possible flaws in some paragraph or diagram, about somehow LETTING DOWN STUDENTS. Given such compulsivity, I would have given up long ago if it were not for the ongoing guidance of my special advisors and contributors. Over the years, they have worked closely with me to evaluate my manuscripts and the reviews from an educational network of more than 2,000 teachers and researchers. On the next two pages, I acknowledge those individuals whose contributions continue to shape our thinking. There is no way to describe their thoughtful assistance. I can only salute their commitment to quality in education.

For this edition, Daniel Fairbanks was far more than advisor for the genetics unit. He wrote the new chapter on recombinant DNA technology and genetic engineering.

Our student readers will benefit from the years of research Dr. Fairbanks conducted while writing his own textbook, *Genetics: The Continuity of Life*. Stephen Wolfe, author of the leading textbook *Molecular and Cellular Biology*, gave line-by-line advice on the biochemistry, genetics, and evolution chapters. Paul Hertz and Tyler Miller carefully checked the evolution and ecology units, and Paul did sketches for the new biogeochemical cycles The highly respected physiology author Lauralee Sherwood rewrote critical passages in the unit on animals. Ron Hoham remains keeper of the pages on protistans, and Eugene Kozloff as keeper of the pages on invertebrates. Nancy Dengler, Richard Falk, John Jackson, and Thomas Rost helped improve the chapters on plants and Elizabeth Moore-Landecker guided us through the fungi. Alan Mann helped update the pages on human evolution. Linda Barham refined and carefully compiled answers for all end-of-chapter Self-Quizzes.

The revision started with evaluations and manuscript reviews from the fourth-edition advisors and contributers listed to the right. Bruce Reid, Michael Renfroe, and Robin Tyser were especially helpful with their detailed challenges. Tom Garrison and Mattie Roig also brought out the sun on dark days. David Goodin advised us on structural biology. Lisa Starr has a background in biochemistry and immunological research, and a passion for computers and art. Nearly all of the stunning new art in this edition is her work.

Most errors creep into textbooks during the crunched schedules of production. This edition is remarkably free of them, thanks to the professionalism of Mary Douglas and Gary Head, who worked with me every step of the way on its page-by-page design and production. Gary also designs our covers and creates such photographic gems as Fred-and-Ginger, the streetwise snail (Figure 16.1). Mary also keeps me focused. They are the best of the best, and I can't do books without them.

Pat Waldo, force of nature, and her superb colleagues Chris Evers, Amanda Kaufman, Veronica Oliva, and Steve Bolinger keep us at the creative forefront of multimedia. Kristin Milotich continues her efficient oversight of the vast supplements program and, amazingly, keeps smiling. Mary Arbogast continues as developmental editor in the wings. Sandra Craig helps us through tight spots. Karen Stough has a talent for zeroing in on amazing goofs that everyone else misses in the manuscripts. Myrna Engler, Bobbie Broyer, Susan Lussier, and Carole Lawson give us fine support. Beth Morrison, Angela Harris, and Michelle Kessell kept art and pages flowing. Tami Cueny and Kelly Fielding are exceptionally good at keeping an eye on the book's content. By giving me their trust, Tim McEwen, Susan Badger, Geoffrey Burn, Gary Carlson, and Kathie Head have contributed to the high quality of this edition.

So many years have passed since Jack Carey convinced me to write my first book. He remains close counselor and abiding friend. And nothing, in all that time, has shaken our shared belief in the intrinsic capacity of biology to enrich the lives of each new generation of students.

GENERAL ADVISORS/CONTRIBUTORS

JOHN ALCOCK *Arizona State University*

AARON BAUER *University of Chicago*

ROBERT COLWELL *University of Connecticut*

JERRY COYNE *University of Chicago*

GEORGE COX *San Diego State University*

KATHERINE DENNISTON *Towson State University*

DANIEL FAIRBANKS *Brigham Young University*

PAUL HERTZ *Barnard College*

JOHN JACKSON *North Hennipin Community College*

RONALD HOHAM *Colgate University*

EUGENE KOZLOFF *University of Washington*

ROBERT LAPEN *Central Washington University*

WILLIAM PARSON *University of Washington*

CLEON ROSS *Colorado State University*

LAURALEE SHERWOOD *West Virginia University*

STEPHEN WOLFE *University of California, Davis*

FOURTH-EDITION ADVISORS/CONTRIBUTORS

LINDA BARHAM *Meridian Community College*

A. KENT CHRISTIANSEN *University of Michigan*

FRED DELCOMYN *University of Illinois, Urbana*

NANCY DENGLER *University of California, Davis*

CATHY DONALD-WHITNEY *Collin County Community College*

RICHARD FALK *University of California, Davis*

TOM GARRISON *Orange Coast College*

DAVID GOODIN *The Scripps Research Institute*

ANN LUMSDEN *Florida State University*

ANNE MCNABB *Virginia Polytechnic Institute and State University*

ALAN MANN *University of Pennsylvania*

G. TYLER MILLER, JR. *Wilmington, North Carolina*

ELIZABETH MOORE-LANDECKER *Glassboro State University*

ALLISON MORRISON-SHETLER *Georgia State University*

DAVID MORTON *Frostburg State University*

RICHARD MURPHY *University of Virginia Medical School*

DAVID NORRIS *University of Colorado*

MICHAEL RENFROE *James Madison University*

BRUCE REID *Kean College of New Jersey*

DAVID REZNICK *University of California, Riverside*

MATTIE ROIG *Broward Community College*

THOMAS ROST *University of California, Davis*

JUDY SCHNEIDEWENT *Milwaukee Area Technical College*

ROGER SLOBODA *Darthmouth College*

LISA STARR *San Diego, California*

IAN TIZARD *Texas A&M University*

ROBIN TYSER *University of Wisconsin, LaCrosse*

MARK WEISS *Wayne State University*

CONTRIBUTORS OF INFLUENTIAL REVIEWS

ALDRIDGE, DAVID, *North Carolina Agricultural and Technical State University*
ARMSTRONG, PETER, *University of California at Davis*
BAJER, ANDREW, *University of Oregon*
BAKKEN, AIMEE, *University of Washington*
BARBOUR, MICHAEL, *University of California, Davis*
BARKWORTH, MARY, *Utah State University*
BELL, ROBERT A., *University of Wisconsin, Stevens Point*
BINKLEY, DAN, *Colorado State University*
BLEEKMAN, GEORGE, *American River College*
BRENGELMANN, GEORGE, *University of Washington*
BRINSON, MARK, *East Carolina University*
BROWN, ARTHUR, *University of Arkansas*
BUCKNER, VIRGINIA, *Johnson County Community College*
CALVIN, CLYDE, *Portland State University*
CASE, CHRISTINE, *Skyline College*
CASE, TED, *University of California, San Diego*
COLAVITO, MARY, *Santa Monica College*
CONKEY, JIM, *Truckee Meadows Community College*
CROWCROFT, PETER, *University of Texas at Austin*
DANIELS, JUDY, *Washtenau Community College*
DAVIS, JERRY, *University of Wisconsin, La Crosse*
DEMMANS, DANA, *Finger Lakes Community College*
DEMPSEY, JEROME, *University of Wisconsin*
DENGLER, NANCY, *University of Toronto*
DeSAIX, JEAN, *University of North Carolina*
DETHIER, MEGAN, *University of Washington*
DeWALT, R. EDWARD, *Louisiana State University*
DiBARTOLOMEIS, SUSAN, *Millersville University of Pennsylvania*
DIEHL, FRED, *University of Virginia*
DLUZEN, DEAN, *Northeastern Ohio Universities College of Medicine*
DOYLE, PATRICK, *Middle Tennessee State University*
DUKE, STANLEY H., *University of Wisconsin, Madison*
DYER, BETSEY, *Wheaton College*
EDLIN, GORDON, *University of Hawaii, Manoa*
EDWARDS, JOAN, *Williams College*
ELMORE, HAROLD W., *Marshall University*
ENDLER, JOHN, *University of California, Santa Barbara*
ENGLISH, DARREL, *Northern Arizona University*
ERWIN, CINDY, *City College of San Francisco*
EWALD, PAUL, *Amherst College*
FISHER, DAVID, *University of Hawaii, Manoa*
FISHER, DONALD, *Washington State University*
FLESSA, KARL, *University of Arizona*
FONDACARO, JOSEPH, *Hoechst Marion Roussel, Inc.*
FRAILEY, CARL, *Johnson County Community College*
FRISBIE, MALCOLM, *Eastern Kentucky University*
FROEHLICH, JEFFREY, *University of New Mexico*
FULCHER, THERESA, *Pellissippi State Technical Community College*
GAGLIARDI, GRACE S., *Bucks County Community College*
GENUTH, SAUL M., *Mt. Sinai Medical Center*
GHOLZ, HENRY, *University of Florida*
GIBSON, THOMAS, *San Diego State University*
GOODMAN, H. MAURICE, *University of Massachusetts Medical School*
GOSZ, JAMES, *University of New Mexico*
GREGG, KATHERINE, *West Virginia Wesleyan College*
HARRIS, JAMES, *Utah Valley Community College*
HARTNEY, KRISTINE BEHRENTS, *California State University, Fullerton*
HASSAN, ASLAM, *Univeristy of Illinois College of Veterinary Medicine*
HELGESON, JEAN, *Collin County Community College*
HESS, WILFORD M., *Brigham Young University*
HUFFMAN, DAVID, *Southwest Texas State University*
INGRAHAM, JOHN L., *University of California, Davis*
JENSEN, STEVEN, *Southwest Missouri State University*
JOHNSON, LEONARD R., *University of Tennessee College of Medicine*
JUILLERAT, FLORENCE, *Indiana University, Purdue University*
KAREIVA, PETER, *University of Washington*
KAUFMAN, JUDY, *Monroe Community College*
KAYE, GORDON I., *Albany Medical College*
KAYNE, MARLENE, *Trenton State College*
KENDRICK, BRYCE, *University of Waterloo*
KEYES, JACK L., *Linfield College, Portland Campus*
KILLIAN, JOELLA C., *Mary Washington College*
KIRKPATRICK, LEE A., *Glendale Community College*
KREBS, CHARLES, *University of British Columbia*
KREBS, JULIA E., *Francis Marion University*

KUTCHAI, HOWARD, *University of Virginia Medical School*
LANZA, JANET, *University of Arkansas, Little Rock*
LASSITER, WILLIAM, *University of North Carolina*
LEVY, MATTHEW, *Mt. Sinai Medical Center*
LEWIS, LARRY, *Bradford University*
LITTLE, ROBERT, *Medical College of Georgia*
LOHMEIER, LYNNE, *Mississippi Gulf Coast Community College*
LOPO, ALINA C., *University of California-Los Angeles Medical Center*
MACKLIN, MONICA, *Northeastern State University*
MANN, ALAN, *University of Pennsylvania*
MARTIN, JAMES, *Reynolds Community College*
MARTIN, TERRY, *Kishwaukee College*
MATSON, RONALD, *Kennesaw State College*
MATTHEWS, ROBERT, *University of Georgia*
MAXWELL, JOYCE, *California State University, Northridge*
McCLURE, JERRY, *Miami University*
McEDWARD, LARRY, *University of Florida*
McKEAN, HEATHER, *Eastern Washington State University*
McKEE, DOROTHY, *Auburn University, Montgomery*
McNABB, ANNE, *Virginia Polytechnic Institute & State University*
MILLER, TYLER, *St. Andrews Presbyterian College*
MOISES, HYLAN C., *University of Michigan Medical School*
MORTON, DAVID, *Frostburg State University*
MYRES, BRIAN, *Cypress College*
NAGARKATTI, PRAKASH, *Virginia Polytechnic Institute & State University*
NELSON, RILEY, *University of Texas at Austin*
PEARCE, FRANK, *West Valley College*
PECHENIK, JAN, *Tufts University*
PECK, JAMES H., *University of Arkansas, Little Rock*
PERRY, JAMES, *University of Wisconsin, Center-Fox Valley*
PETERSON, GARY, *South Dakota State University*
PIPERBERG, JOEL, *Millersville University*
PLETT, HAROLD, *Fullerton College*
REEVE, MARIAN, *Emeritus, Merritt Community College*
RICKETT, JOHN, *University of Arkansas, Little Rock*
ROBBINS, ROBERT, *National Science Foundation*
ROSE, GREIG, *West Valley College*
RUIBAL, RODOLFO, *University of California, Riverside*
SALISBURY, FRANK, *Utah State University*
SCHAPIRO, HARRIET, *San Diego State University*
SCHLESINGER, WILLIAM, *Duke University*
SCHNERMANN, JURGEN, *University of Michigan School of Medicine*
SHEGAL, PREM, *East Carolina University*
SHONTZ, NANCY, *Grand Valley State University*
SELLERS, LARRY, *Louisiana Tech University*
SHOPPER, MARILYN, *Johnson County Community College*
SMITH, JERRY, *St. Petersburg Junior College, Clearwater Campus*
SMITH, MICHAEL E., *Valdosta State College*
SMITH, ROBERT L., *West Virginia University*
SOLOMON, NANCY, *Miami University*
STEARNS, DONALD, *Rutgers University*
STEELE, KELLY P., *Appalachian State University*
STEINERT, KATHLEEN, *Bellevue Community College*
SUMMERS, GERALD, *University of Missouri*
SUNDBERG, MARSHALL D., *Louisiana State University*
SWANSON, ROBERT, *North Hennepin Community College*
SWEET, SAMUEL, *University of California, Santa Barbara*
TAYLOR, JANE, *Northern Virginia Community College*
TERHUNE, JERRY, *Jefferson Community College, University of Kentucky*
TIZARD, IAN, *Texas A & M University*
TROUT, RICHARD E., *Oklahoma City Community College*
WAALAND, ROBERT, *University of Washington*
WAHLERT, JOHN, *City University of New York, Baruch College*
WALSH, BRUCE, *University of Arizona*
WARING, RICHARD, *Oregon State University*
WARNER, MARGARET R., *Purdue University*
WEBB, JACQUELINE F., *Villanova University*
WEIGL, ANN, *Winston-Salem State University*
WEISS, MARK, *Wayne State University*
WELKIE, GEORGE W., *Utah State University*
WENDEROTH, MARY PAT, *University of Washington*
WHITE, EVELYN, *Alabama State University*
WHITENBERG, DAVID, *Southwest Texas State University*
WINICUR, SANDRA, *Indiana University, South Bend*
YONENAKA, SHANNA, *San Francisco State University*

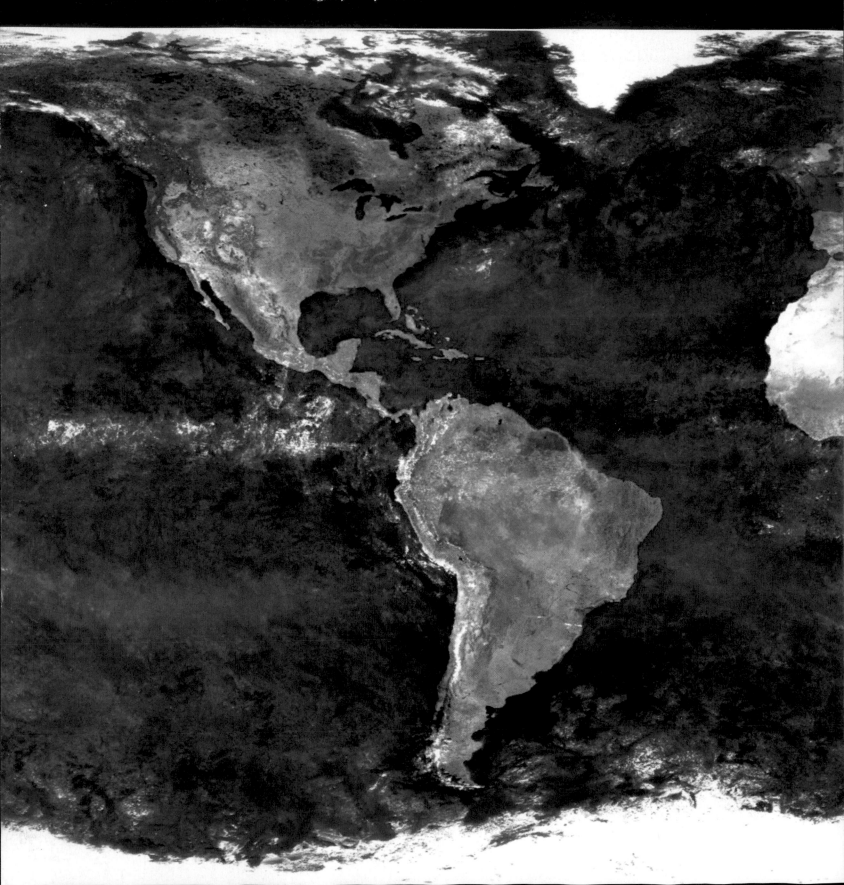

Current configurations of Earth's oceans and landmasses —the geologic stage upon which life's drama continues to unfold. Thousands of separate images were pieced together to create this remarkable, true-color image of our planet.

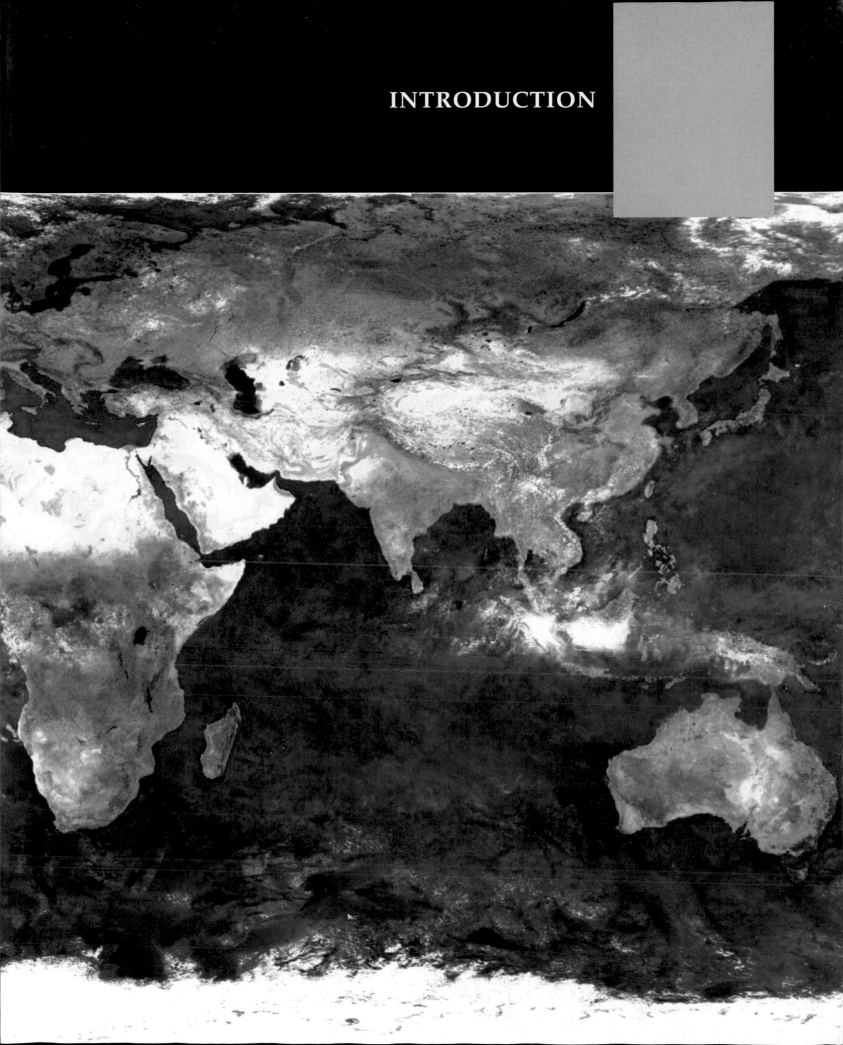

INTRODUCTION

CONCEPTS AND METHODS IN BIOLOGY

Biology Revisited

Buried somewhere in that mass of tissue just above and behind your eyes are memories of first encounters with the living world. In that brain are early memories of discovering your hands and feet, your family, friends, the change of seasons, the scent of rain-drenched earth and grass. Still in residence are memories of your early introductions to a great disorganized parade of plants and animals, mostly living, sometimes dead. There, too, are memories of questions—"*What is life?*" and, inevitably, "*What is death?*" There also are memories of answers, some satisfying, others less so.

By making observations, asking questions, and accumulating answers, you have gradually built up a store of knowledge about life. Experience and education have been refining the questions, and no doubt some answers are difficult to come by.

Think of the world's forests—once vast, now astonishingly diminished by logging, conversion for agriculture, urban expansion, and other activities that help keep many people sheltered, warm, fed, and alive. This year the human population surpassed 6 billion and is still growing. With few forests remaining, where will we find more usable land and forest products?

Think of a college student, twenty years old, whose motorcycle skidded into a truck. Now he is comatose, and doctors say his brain is functionally dead. His breathing, heart rate, and other basic functions will continue only as long as he stays hooked up to

Figure 1.1 Think back on all you have ever known and seen. This is a foundation for your deeper probes into life.

a respirator and other artificial support systems. Would you say the unfortunate student is still "alive"?

Or think of a human egg, recently penetrated by a sperm inside a woman's body. At first the fertilized egg does not get bigger, but a series of programmed cuts divides it into a cluster of a few dozen tiny cells. Would you call the microscopically small mass a *human* life? If you learn more about how embryos actually develop, will the knowledge influence your thoughts about such incendiary issues as birth control and abortion?

If questions like these have ever crossed your mind, your thoughts about life obviously run deep. And you can approach this course in **biology**—the scientific study of life—with confidence, because you have been studying life ever since information started penetrating your brain. You simply are *revisiting* biology, in ways that might help carry your thoughts to deeper, more organized levels of understanding.

Return to the question, *What is life?* Offhandedly, you might respond that you know it when you see it. However, as you will see later in the book, the question opens up a story that has been unfolding in countless directions for 3.8 billion years!

From the biological perspective, "life" is an outcome of ancient events by which nonliving matter—atoms and molecules—became assembled into the first living cells. "Life" is a way of capturing and using energy and raw materials. "Life" is a way of sensing and responding to changes in the environment. "Life" is a capacity to reproduce, grow, and develop. And "life" evolves, meaning that the traits characterizing the individuals of a population can change through the generations. Even so, this short list only hints at the meaning of life. Deeper insight requires wide-ranging study of life's characteristics.

Throughout this book, you will come across many diverse examples of how organisms are constructed, how they function, where they live, and what they do. The examples support certain concepts which, when taken together, will give you a sense of what "life" is.

This chapter introduces the basic concepts. It also sets the stage for forthcoming descriptions of scientific observations, experiments, and tests that help show how you can develop, modify, and refine your views of life. As you continue with your reading, you may find it useful to return occasionally to this simple overview as a way to reinforce your grasp of the details.

KEY CONCEPTS

1. Unity underlies the world of life, for all organisms are alike in key respects. They consist of one or more cells made of the same kinds of substances, put together in the same basic ways. Their activities require inputs of energy, which they must get from their surroundings. All organisms sense and respond to changing conditions in their environment. They all grow and reproduce, based on instructions contained in their DNA.

2. The world of life shows immense diversity. Many millions of different kinds of organisms, or species, now inhabit Earth, and many millions more lived in the past. Each one of those species is unique in some of its traits—that is, in some aspects of body plan, body functioning, and behavior.

3. Theories of evolution, especially a theory of natural selection as first formulated by Charles Darwin, help explain the meaning of life's diversity.

4. Biology, like other branches of science, is based on systematic observations, hypotheses, predictions, and observational and experimental tests. The external world, not internal conviction, is the testing ground for scientific theories.

Nothing Lives Without DNA

DNA AND THE MOLECULES OF LIFE Picture a frog on a rock, busily croaking. Without even thinking about it, you know the frog is alive and the rock is not. Would you be able to explain why? At a fundamental level, both are no more than concentrations of the same units of matter, called protons, electrons, and neutrons. The units are building blocks of atoms, which are building blocks of larger bits of matter called molecules. And it is at the molecular level that differences between living and nonliving things start to emerge.

You will never, ever find a rock made of nucleic acids, proteins, carbohydrates, and lipids. In nature, only cells build these molecules, which you will read about later. Cells are the smallest units of matter having a capacity for life; all living things consist of one or more of them. The signature molecule of cells is a nucleic acid known as **DNA**. No chunk of granite or quartz has it.

Encoded in DNA's structure are the instructions for assembling a dazzling array of proteins from a limited number of smaller building blocks, the amino acids. By analogy, if you follow suitable instructions and invest energy in the task, you can organize a heap of a few kinds of ceramic tiles (representing amino acids) into diverse patterns (representing proteins), as in Figure 1.2.

Among the proteins are enzymes. When these worker molecules get an energy boost, they swiftly build, split, and rearrange the molecules of life. Some enzymes work with a class of nucleic acids called RNAs in carrying out DNA's protein-building instructions. Think of this as a flow of information, from *DNA to RNA to protein*. As you will read in Chapter 13, this molecular trinity is central to our understanding of life.

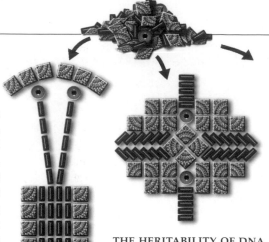
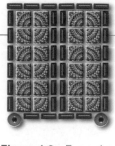

Figure 1.2 Examples of objects built from the same materials but with different assembly instructions.

THE HERITABILITY OF DNA We humans tend to think we enter the world rather abruptly and then leave it the same way, but we are much more than this. *We and all other organisms are part of a journey that began about 3.8 billion years ago, with the emergence of the first living cells.*

Under present-day conditions on Earth, cells arise only from cells that already exist. They do so by **reproduction**, one of the key defining features of life. By this process, parents transmit DNA instructions for duplicating their traits to offspring. Why do baby storks look like storks and not pelicans? They inherited stork DNA, which is not exactly the same as pelican DNA.

For many organisms, reproduction starts with a single cell that contains DNA from their parents. For example, think of a cell that forms when a sperm (a single cell) fuses with an egg (another single cell). The fertilized egg would not even exist if that sperm and egg had not formed earlier, according to DNA instructions passed on from cell to cell through countless generations.

For frogs and humans and other large organisms, DNA also guides **development**, the transformation of a fertilized egg into a multicelled adult with cells, tissues, and organs specialized for certain tasks. Development proceeds through a series of stages. As one example, a moth is only the adult stage of a winged insect (Figure 1.3). First a fertilized egg develops into an immature larval stage—a caterpillar that eats leaves and grows

Figure 1.3 "The insect"—a series of stages of development. Different adaptive properties emerge at each stage. Shown here, a silkworm moth, from the egg (**a**), to a larval stage (**b**), to a pupal stage (**c**), and on to the winged form of the adult (**d**,**e**).

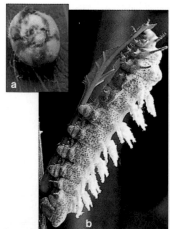
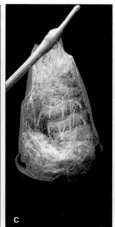

Figure 1.4 Response to signals from pain receptors, activated by a lion cub flirting with disaster.

rapidly until an internal alarm clock goes off. Then its tissues are remodeled into a different stage—a pupa. In time, an adult emerges that is adapted to reproduce. It produces sperm or eggs. Its wing color, patterns, and fluttering frequency are adapted for attracting a mate.

And so "the insect" is a series of organized stages. Its development from egg to adult will not proceed properly unless each stage is completed before the next begins. Instructions for each stage were written into moth DNA long before each moment of reproduction—and so the ancient moth story continues.

Nothing Lives Without Energy

ENERGY DEFINED Everything in the entire universe has some amount of **energy**, which is most simply defined as a capacity to do work. And nothing—absolutely nothing—happens in the universe without a *transfer* of energy. For instance, a lone, undisturbed atom can do nothing except vibrate incessantly with its own energy. Suppose it absorbs extra energy from the sun and starts vibrating faster. Now some energy is on the move. That energy can do work by getting transferred elsewhere. If by chance the atom collides with a neighboring atom, one may give up, grab, or share energy with the other. Molecules form, become rearranged, and are split by such energy transfers. When this kind of molecular work is done, cells stay alive, grow, and reproduce.

METABOLISM DEFINED Each living cell has the capacity to (1) obtain and convert energy from its surroundings and (2) use energy to maintain itself, grow, and make more cells. We call this capacity **metabolism**. Think of a cell in a leaf that produces food by photosynthesis. It intercepts sunlight energy and converts it to chemical energy, in the form of ATP molecules. ATP is an energy carrier that helps drive hundreds of activities. It easily transfers some energy to metabolic workers—in this case, enzymes that assemble sugar molecules. ATP also forms by aerobic respiration. This process can release energy that cells have tucked away in sugars and other kinds of molecules.

SENSING AND RESPONDING TO ENERGY It is often said that only organisms respond to the environment. Yet even a rock shows responsiveness, as when it yields to the force of gravity and tumbles down a hill or changes its shape slowly under the repeated battering of wind, rain, or tides. The difference is this: *Organisms sense changes in their surroundings, then they make controlled, compensatory responses to them.* How? Each organism has **receptors**, which are molecules and structures that detect stimuli. A **stimulus** is a specific form of energy detected by receptors. Examples are sunlight energy, heat energy, a hormone molecule's chemical energy, and the mechanical energy of a bite (Figure 1.4).

Cells adjust metabolic activities in response to signals from receptors. Each cell (and organism) can withstand only so much heat or cold. It must rid itself of harmful substances. It requires certain foods, in certain amounts. Yet temperatures do shift, harmful substances might be encountered, and food is sometimes plentiful or scarce.

For example, after you finish a snack, simple sugars leave your gut and enter your blood. Blood is part of an *internal* environment (the other part is tissue fluid that bathes your cells). Over the long term, too much or too little sugar in the blood causes problems, such as diabetes. When the sugar level rises, a glandular organ, the pancreas, normally steps up its secretion of insulin. Most of your cells have receptors for this hormone, which stimulates cells to take up sugar. When enough cells do this, the sugar level in blood returns to normal.

Organisms respond so exquisitely to energy changes that their internal operating conditions remain within tolerable limits. We call this a state of **homeostasis**, and it is one of the key defining features of life.

All organisms consist of one or more cells, the smallest units of life. Under present-day conditions, new cells form only through the reproduction of cells that already exist.

DNA, the molecule of inheritance, encodes protein-building instructions, which RNAs help carry out. Many proteins are enzymes, and these metabolic workers are necessary to construct DNA and all other complex molecules of life.

Cells live only for as long as they engage in metabolism. They acquire and transfer energy that is used to assemble, break down, stockpile, and dispose of materials in ways that promote survival and reproduction.

Single cells and multicelled organisms sense and respond to environmental conditions in ways that help maintain their internal operating conditions.

ENERGY AND LIFE'S ORGANIZATION

Levels of Biological Organization

Taken as a whole, the metabolic activities of single cells and multicelled organisms maintain the great pattern of organization in nature, as sketched out in Figure 1.5. Consider the hierarchy. Life's properties emerge when DNA and other molecules become organized into cells. The **cell** is the smallest unit of organization having a capacity to survive and reproduce on its own, given DNA instructions, suitable conditions, building blocks, and energy inputs. Free-living, single cells, such as an amoeba, fit the definition. Does the definition of cells hold for **multicelled organisms**, which typically consist of specialized, interdependent cells organized as tissues and organs? Yes. You might find this a strange answer. After all, your own cells could never live alone in nature, because body fluids must continually bathe them. Yet even isolated human cells stay alive under controlled conditions in laboratories around the world. Investigators routinely maintain isolated human cells for use in important experiments, as in cancer studies.

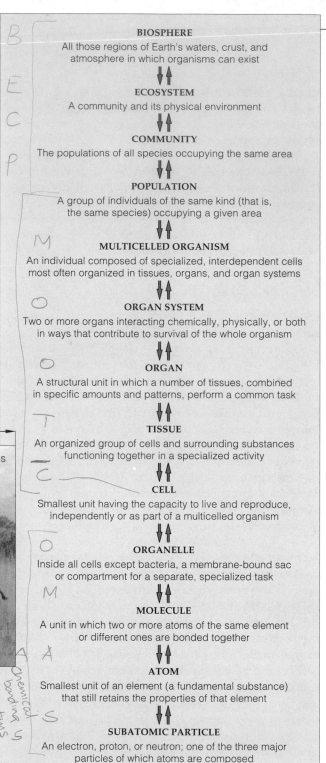

BIOSPHERE
All those regions of Earth's waters, crust, and atmosphere in which organisms can exist

⇩⇧

ECOSYSTEM
A community and its physical environment

⇩⇧

COMMUNITY
The populations of all species occupying the same area

⇩⇧

POPULATION
A group of individuals of the same kind (that is, the same species) occupying a given area

⇩⇧

MULTICELLED ORGANISM
An individual composed of specialized, interdependent cells most often organized in tissues, organs, and organ systems

⇩⇧

ORGAN SYSTEM
Two or more organs interacting chemically, physically, or both in ways that contribute to survival of the whole organism

⇩⇧

ORGAN
A structural unit in which a number of tissues, combined in specific amounts and patterns, perform a common task

⇩⇧

TISSUE
An organized group of cells and surrounding substances functioning together in a specialized activity

⇩⇧

CELL
Smallest unit having the capacity to live and reproduce, independently or as part of a multicelled organism

⇩⇧

ORGANELLE
Inside all cells except bacteria, a membrane-bound sac or compartment for a separate, specialized task

⇩⇧

MOLECULE
A unit in which two or more atoms of the same element or different ones are bonded together

⇩⇧

ATOM
Smallest unit of an element (a fundamental substance) that still retains the properties of that element

⇩⇧

SUBATOMIC PARTICLE
An electron, proton, or neutron; one of the three major particles of which atoms are composed

ecosystem (community together with its physical environment)

populations of shrubs and trees

community (all of the populations in this area)

populations of grasses

population of zebras

multicelled individual

former individual, about to revert to molecules and atoms

Figure 1.5 Levels of organization in nature.

You typically find cells and multicelled organisms as part of a **population**, defined as a group of organisms of the same kind, such as a herd of zebras. The next level of organization is the **community**—all the populations of all species living in the same area (such as the African savanna's bacteria, grasses, trees, zebras, lions, and so on). The next level, the **ecosystem**, is the community *and*

b

c

Figure 1.6 An example of the one-way energy flow and the cycling of materials through the biosphere.

(**a**) Plants of a warm, dry grassland called the African savanna capture energy from the sun and use it to build plant parts. Some of the energy ends up inside plant-eating organisms, including this adult male elephant. He eats huge quantities of plants to maintain his eight-ton self and produces great piles of solid wastes—dung—that still contain some unused nutrients. Although most organisms might not recognize it as such, elephant dung is an exploitable food source.

(**b**) And so we next have little dung beetles rushing to the scene almost simultaneously with the uplifting of an elephant tail. Working rapidly, they carve fragments of moist dung into round balls, which they roll off and bury in burrows. In the balls the beetles lay eggs—a reproductive behavior that will help assure their forthcoming offspring (**c**) of a compact food supply.

Thanks to beetles, dung does not pile up and dry out into rock-hard mounds in the intense heat of the day. Instead, the surface of the land is tidied up, the beetle offspring get fed, and the leftover dung accumulates in beetle burrows—there to enrich the soil that nourishes the plants that sustain (among others) the elephants.

Producers trap, convert, and use or store some energy from the sun.

PRODUCERS

NUTRIENT CYCLING

CONSUMERS, DECOMPOSERS

ONE-WAY FLOW OF ENERGY

Energy is transferred from one organism to another; in time, all flows back to the environment.

a

its physical and chemical environment. The **biosphere** encompasses all regions of Earth's atmosphere, waters, and crust in which organisms live. Astoundingly, *this globe-spanning organization begins with the convergence of energy, certain materials, and DNA in tiny, individual cells.*

Interdependencies Among Organisms

A great flow of energy into the world of life starts with the **producers**—plants and other organisms that make their own food. Animals are **consumers**. Directly or indirectly, they depend on energy that became stored in the tissues of producers. For example, some energy is transferred to zebras after they browse on grasses. It gets transferred again when lions devour a baby zebra that wandered away from its herd. And it is transferred again when fungal and bacterial decomposers feed on the tissues and remains of lions, elephants, or any other organism. **Decomposers** break down sugars and other biological molecules to simpler materials—which may

be cycled back to producers. In time, all of the energy that producers initially captured from the sun's rays returns to the environment, but that's another story.

For now, keep in mind that organisms connect with one another by a one-way flow of energy *through* them and a cycling of materials *among* them, as in Figure 1.6. Their interconnectedness affects the structure, size, and composition of populations and communities. It affects ecosystems, even the biosphere. Understand the extent of their interactions and you will gain insight into the environmental effects of acid rain, amplification of the greenhouse effect, and other modern-day problems.

Levels of organization exist in nature. The characteristics of life emerge at the level of single cells and extend through populations, communities, ecosystems, and the biosphere.

A one-way flow of energy through organisms and a cycling of materials among them organizes life in the biosphere. In nearly all cases, energy flow starts with energy from the sun.

So far, we have focused on life's unity, on the characteristics that all living things have in common. ① Think of it! They are put together from the same "lifeless" materials. ② They remain alive by metabolism —by ongoing energy transfers at ③ the cellular level. They interact in their requirements for energy and ④ for raw materials. They have the capacity to sense and respond to the environment in highly specific ⑤ ways. They all have a capacity to reproduce, based on instructions ⑥ encoded in DNA. And they all inherited their molecules of DNA from individuals of a preceding generation.

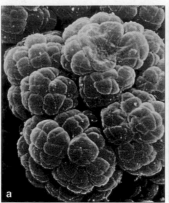
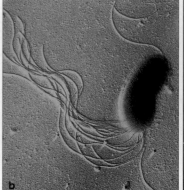
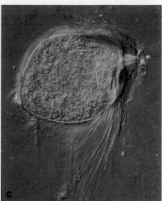

Superimposed on the common heritage is immense diversity. You share the planet with many millions of different kinds of organisms, or **species**. Many millions more preceded you during the past 3.8 billion years, but their lineages vanished; they are extinct. Centuries ago, scholars tried to make sense of the confounding diversity. One of them, Carolus Linneaus, came up with a classification scheme that assigns a two-part name to each newly identified species. The first part designates the **genus** (plural, genera). Each genus encompasses all species that seem closely related, by way of their recent descent from a common ancestor. The second part of the name designates a particular species within that genus.

For example, *Quercus alba* is the scientific name for the white oak, and *Q. rubra* is the name for the red oak. As this example suggests, once you spell out the name of the genus in a document, you may abbreviate the name wherever else it appears in that document.

Biologists also classify life's diversity by assigning species to groups at more encompassing levels. Among other things, they group genera that apparently share a common ancestor into the same *family*, related families into the same *order*, related orders into the same *class*, and related classes into the same *phylum* (plural, phyla). At a higher level, they assign related phyla to the same *kingdom*. They are still refining the groups. For instance, scholars once recognized only two kingdoms (animals and plants). Later, most biologists came to accept five kingdoms. Today, compelling evidence suggests there should be six: the **Archaebacteria**, **Eubacteria**, **Protista**, **Fungi**, **Plantae**, and **Animalia** (Figure 1.7). We use a six-kingdom scheme for this book. At this point in your reading, it is enough simply to become familiar with a few of the defining features of their members.

All of the world's bacteria (singular, bacterium) are single cells, of a type called *prokaryotic*. The word means they do not have a nucleus, a membrane-bound sac that otherwise would keep their DNA separate from the

Figure 1.7 A few representatives of life's diversity.

KINGDOM ARCHAEBACTERIA. (**a**) From the muck of an anaerobic (oxygen-free) habitat, a colony of cells (*Methanosarcina*).

KINGDOM EUBACTERIA. (**b**) A eubacterium sporting a number of bacterial flagella. Flagella are used for motility.

KINGDOM PROTISTA. (**c**) A trichomonad that lives in a termite's gut. Compared to bacteria, most protistans are much larger and have greater internal complexity. They range from microscopically small single cells, such as this one, to giant seaweeds.

KINGDOM FUNGI. (**d**) A stinkhorn fungus. Some species of fungi are parasites and some cause diseases, but the vast majority are decomposers. Without decomposers, communities would gradually become buried in their own wastes.

KINGDOM PLANTAE. (**e**) Trunk of a redwood growing near the coast of California. Like nearly all plants, redwoods produce their own food by photosynthesis. (**f**) Flower of a plant from the family of composites. Its colors and patterning guide bees to nectar. The bees get food. The plant gets help reproducing. Like many other organisms, they interact in a mutually beneficial way.

KINGDOM ANIMALIA. (**g**,**h**) Male bighorn sheep competing for females and so displaying a characteristic of the kingdom— they actively move about in their environment.

rest of the cell interior. Different bacterial species are producers, consumers, or decomposers. Of all organisms, bacteria show the greatest metabolic diversity.

We find diverse archaebacteria in extreme environments, much like the forms that are thought to have prevailed at the time life originated. Eubacteria are much more successful in their world distribution. Different kinds live in about every place you might imagine. Eubacteria inside your gut and on your skin outnumber the trillions of cells making up your body.

Generalizing about protistans is not easy. Like plants, fungi, and animals, they are *eukaryotic*, meaning their DNA is located inside a nucleus. Most types are larger and show far more internal complexity than bacteria. A spectacular variety of microscopically small, single-celled producers and consumers are protistans. But so are multicelled brown algae and other "seaweeds," some of which are giants of the underwater world.

Most fungi, including the common field mushrooms sold in grocery stores, are multicelled. These eukaryotic decomposers and consumers feed in a distinctive way. They secrete enzymes that digest food outside of the fungal body, then their cells absorb the digested bits.

Plants and animals include many familiar species and an astounding number of obscure ones. Members of both kingdoms are eukaryotic. Plants are multicelled producers that collectively represent the food base for nearly all communities of life. Nearly all plant species are photosynthetic. Animals are multicelled consumers that ingest other organisms. They include plant eaters, parasites, scavengers, and meat eaters. Unlike plants, they move about during at least some stage of their life.

Pulling this all together, you start to get a sense of what it means when someone says life shows unity *and* diversity.

Unity threads through the world of life, for all organisms are alike in important ways. They are composed of the same substances, which are assembled in the same basic ways. They engage in metabolism, and they sense and respond to their environment. They all have a capacity to reproduce, based on heritable instructions encoded in their DNA.

Immense diversity threads through the world of life, for organisms differ enormously in body form, in the functions of their body parts, and in their behavior.

To make the study of life's diversity more manageable, we group organisms into six kingdoms—the archaebacteria, eubacteria, protistans, fungi, plants, and animals.

WWW

AN EVOLUTIONARY VIEW OF DIVERSITY

Given that organisms are so much alike, what could account for their great diversity? One key explanation is called evolution by means of natural selection. A few simple examples will be enough to introduce you to its premises, which build on the simple observation that variation in traits exists in all populations.

Figure 1.9 Some of the 300+ varieties of domesticated pigeons produced by artificial selection practices. Breeders began with variant forms of traits in captive populations of wild rock doves (**a**).

Mutation—Original Source of Variation

DNA has two striking qualities. Its instructions work to ensure that offspring will resemble their parents, yet they also permit variations in the details of most traits. As an example, having five fingers on each hand is a human trait. Yet some humans are born with six fingers on each hand instead of five. This is an outcome of a **mutation**, a molecular change in the DNA. Mutations are the original source of variations in heritable traits.

Many mutations are harmful. A change in even a bit of DNA may be enough to sabotage the body's growth, development, or functioning. One such mutation causes *hemophilia A*. If a person affected by this blood-clotting disorder gets even a small cut or bruise, an abnormally lengthy time passes before a clot forms and stops the bleeding. Yet some variations are harmless or beneficial. A classic case is a mutation in light-colored moths that results in dark offspring. Moths fly at night and rest in the day, when birds that eat them are active. Birds tend to miss light-colored moths that rest on light tree trunks. Those moths are camouflaged; they are "hiding in the open," as shown in Figure 1.8.

Suppose that people build coal-burning factories nearby. In time, smoke laden with soot darkens the tree trunks. Now the dark moths are much less conspicuous to bird predators, so they have a better chance to survive and reproduce. Where soot prevails, the variant (darker) form of the trait is more adaptive than the common form. An **adaptive trait** is any form of a trait that gives the individual an advantage, in terms of surviving and reproducing, under a given set of environmental conditions.

Figure 1.8 Example of different forms of the same trait (body surface coloration), adaptive to two different environmental conditions. (**a**) On a light-colored tree trunk, light-colored moths (*Biston betularia*) are hidden from predators, but dark ones stand out. (**b**) The dark coloration is more adaptive in places where tree trunks are darkened with soot.

Evolution Defined

Now imagine a population of light-colored moths in a sooty forest. At some point, a DNA mutation arose in the population, and it resulted in a moth of a darker color. When the mutated individual reproduced, some of its offspring inherited the trait. Birds saw and ate many light-colored moths, but most of the dark ones escaped detection and lived long enough to reproduce. So did their dark offspring, and so did *their* offspring. Over the generations, the frequency of the dark form of the trait increased and that of the light form decreased. As more time passes, the dark form of the trait might even become the more common, and people might end up referring to "the population of dark-colored moths." **Evolution** is under way. In biology, the word means genetically based change in a line of descent over time. Like moths, individuals of most populations typically show different forms of many (or most) of their traits. And the frequencies of those different forms relative to one another can change over successive generations.

Natural Selection Defined

Long ago, the naturalist Charles Darwin used pigeons to explain a conceptual connection between evolution and variation in traits. Domesticated pigeons display great variation in size, feather color, and other traits (Figure 1.9). As Darwin knew, pigeon breeders select certain forms of traits. For example, if breeders prefer tail feathers that are black with curly edges, they will allow only those individual pigeons having the most black and the most curl in their tail feathers to mate and produce offspring. Over time, "black" and "curly" will become the most common forms of tail feathers in the captive population, and different forms of the two traits will become less common or will be eliminated.

Pigeon breeding is a case of **artificial selection**, for the selection among different forms of a trait is taking place in an artificial environment, under contrived, manipulated conditions. Yet Darwin saw the practice as a simple model for *natural* selection, a favoring of some forms of traits over others in nature. Whereas breeders are "selective agents" that promote the reproduction of some individuals over others in captive populations, a pigeon-eating peregrine falcon is one of many selective agents that operate across the range of variation among

pigeons in the wild. Generation after generation, the swifter or more effectively camouflaged pigeons have a better chance of living long enough to reproduce than the not-so-swift or too-conspicuous ones among them. What Darwin identified as **natural selection** is simply a difference in which individuals of a generation survive and reproduce, the difference being an outcome of which ones have adaptive forms of traits.

Unless you happen to be a pigeon breeder, Darwin's pigeon example may not be firing rockets through your imagination. So think about an example closer to home. Certain bacteria and fungi make *antibiotics*, metabolic products that kill bacterial competitors for nutrients in soil. Starting in the 1940s, we learned to use antibiotics to control diseases that result after bacteria invade the body and use its tissues for nutrients. Doctors routinely prescribed these "wonder drugs" for mild infections as well as serious ones. Some manufacturers even added an antibiotic to toothpaste and chewing gum.

As it turned out, antibiotics are powerful agents of natural selection. Over time, some bacteria have mutated in ways that help them resist the antibiotics produced by their bacterial neighbors. For example, streptomycin is an antibiotic that binds with some essential bacterial proteins and inhibits their activity. In certain variant strains of bacteria, mutations slightly changed the form of the proteins, so streptomycin is unable to bind with them. The mutant bacteria escape streptomycin's effects.

In infected patients, an antibiotic acts against bacteria that are susceptible to its action—but it actually favors variant strains that have resistance to it! Presently, such antibiotic-resistant strains are making it difficult to treat typhoid, tuberculosis, gonorrhea, staph (*Staphylococcus*) infections, and some other bacterial diseases. In a few patients, the "superbugs" responsible for tuberculosis cannot be successfully eliminated.

When resistance to antibiotics evolves by selection processes, the antibiotics must also evolve if they are to overcome the defenses. For example, drug companies have now modified parts of the streptomycin molecule. Such molecular changes in the laboratory produce more effective antibiotics—at least until new generations of even more resistant superbugs enter the deadly evolutionary competition for nutrients.

Later in the book, we will consider the mechanisms by which populations of moths, pigeons, bacteria, and all other organisms evolve. Meanwhile, keep in mind the following points about natural selection. They are central to biological inquiry, for they have consistently proved useful in explaining a great deal about nature.

1. Individuals of a population vary in form, function, and behavior. Much of the variation is heritable; it can be transmitted from parents to offspring.

2. Some forms of heritable traits are adaptive to the prevailing environmental conditions. They improve an individual's chance of surviving and reproducing, as by helping it secure food, a mate, hiding places, and so on.

3. Natural selection is the outcome of differences in the survival and reproduction of individuals that show variation in one or more traits.

4. Natural selection leads to a better fit with prevailing environmental conditions. Adaptive forms of traits tend to become more common and other forms less so. The population changes in its characteristics; it evolves.

In short, in the evolutionary view, *life's diversity is the sum total of variations in traits that have accumulated in different lines of descent generation after generation, as by natural selection and other processes of change.*

Mutations in DNA introduce variations in heritable traits.

Although many mutations are harmful, some give rise to variations in form, function, or behavior that are adaptive under prevailing environmental conditions.

Natural selection is a result of differences in survival and reproduction among individuals of a population that vary in one or more heritable traits. The process helps explain evolution—changes in lines of descent over the generations.

THE NATURE OF BIOLOGICAL INQUIRY

The preceding sections sketched out major concepts in biology. Now consider approaching this or any other collection of "facts" with a critical attitude. *"Why should I accept that they have merit?"* The answer requires insight into how biologists make inferences about observations and then test the predictive power of their inferences against actual experiences in nature or the laboratory.

Observations, Hypotheses, and Tests

To get a sense of "how to do science," start by following some practices that are pervasive in scientific research:

1. Observe some aspect of nature, carefully check what others have found out about it, and then frame a question or identify a problem related to your observation.

2. Develop **hypotheses**, or educated guesses, about possible answers to questions or solutions to problems.

3. Using hypotheses as a guide, make a **prediction**—that is, a statement of what you should observe in the natural world if you were to go looking for it. This is often called the "if-then" process. (*If* gravity does not pull objects toward Earth, *then* it should be possible to observe apples falling up, not down, from a tree.)

4. Devise ways to **test** the accuracy of your predictions, as by making systematic observations, building models, and conducting experiments. (**Models** are theoretical, detailed descriptions or analogies that help us visualize something that hasn't yet been directly observed.)

5. If the tests do not confirm the prediction, check to see what might have gone wrong. For example, maybe you overlooked a factor that influenced the test results. Or maybe the hypothesis is not a good one.

6. Repeat the tests or devise new ones—the more the better, for hypotheses that withstand many tests are likely to have a higher probability of being useful.

7. Objectively analyze and report the test results and the conclusions you have drawn from them.

You might hear someone refer to these practices as "the scientific method," as if all scientists march to the drumbeat of an absolute, fixed procedure. They do not. Many observe, describe, and report on some subject, then leave it to others to hypothesize about it. Some are lucky; they stumble onto information they are not even looking for, although chance does favor the prepared mind. It is not one single method they have in common. It is a critical attitude about being shown rather than told, and taking a logical approach to problem solving.

Logic encompasses thought patterns by which an individual draws a conclusion that does not contradict the evidence used to support it. Lick a cut lemon and you notice it is mouth-puckeringly sour. Lick ten more.

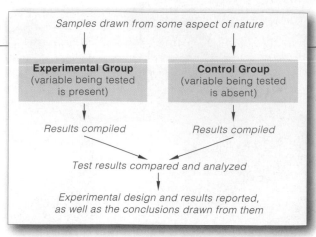

Figure 1.10 Generalized sequence of steps involved in an experimental test of a prediction based on a hypothesis.

You notice the same thing each time, so you conclude all lemons are mouth-puckeringly sour. You correlated one specific (lemon) with another (sour). By this pattern of thinking, called *inductive* logic, an individual derives a general statement from specific observations.

Express the generalization in "if-then" terms, and you have a hypothesis: "If you lick any lemon, then you will get an extremely sour taste in your mouth." By this pattern of thinking, called *deductive* logic, an individual makes inferences about specific consequences or specific predictions that must follow from a hypothesis.

You decide to test the hypothesis by tracking down and sampling all the varieties of lemons in the vicinity. One variety, the Meyer lemon, is actually mellow, for a lemon. You also discover that some people cannot taste anything. So you must modify the original hypothesis: "If most people lick any lemon *except* the Meyer lemon, they will get an extremely sour taste in their mouth." Suppose, after sampling all the known lemon varieties in the world, you conclude the modified hypothesis is a good one. You can never prove it beyond all shadow of a doubt, because there might be lemon trees growing in places people don't even know about. You *can* say the hypothesis has a high probability of not being wrong.

Comprehensive observations are a logical means to test the predictions that flow from hypotheses. So are **experiments**. These tests simplify observation in nature or the laboratory by manipulating and controlling the conditions under which observations are made. When suitably designed, observational and experimental tests allow you to predict that something will happen if a hypothesis isn't wrong (or won't happen if it *is* wrong). Figure 1.10 gives a general idea of the steps involved.

AN ASSUMPTION OF CAUSE AND EFFECT Experiments start from the premise that any aspect of nature has one or more underlying causes. With this premise, science is distinct from faith in the supernatural (meaning "beyond nature"). Experiments deal with potentially falsifiable hypotheses. Hypotheses of this sort can be tested in the natural world in ways that might disprove them.

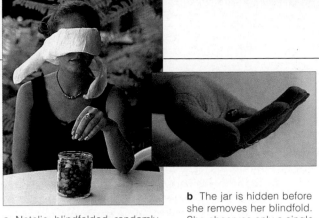

a Natalie, blindfolded, randomly plucks a jellybean from a jar of 120 green and 280 black jellybeans. That is a ratio of 30 to 70 percent.

b The jar is hidden before she removes her blindfold. She observes only a single green jellybean in her hand and assumes the jar holds only green jellybeans.

c Still blindfolded, Natalie randomly plucks 50 jellybeans from the jar and ends up with 10 green and 40 black ones.

d The larger sample leads her to assume one-fifth of the jar's jellybeans are green and four-fifths are black (a ratio of 20 to 80). Her larger sample more closely approximates the jar's green-to-black ratio. The more times Natalie repeats the sampling, the greater the chance she will come close to knowing the actual ratio.

Figure 1.11 A simple demonstration of sampling error.

EXPERIMENTAL DESIGN To get conclusive test results, experimenters rely on certain practices. They refine test designs by searching the literature for information that may relate to their inquiry. They design experiments to test one prediction of a hypothesis at a time. Each time, they set up a **control group**: a standard for comparison with one or more experimental groups. Ideally, their control group is identical with an experimental group in all respects *except* for the one variable being studied. **Variables** are specific aspects of objects or events that may differ or change over time and among individuals.

Section 1.6 shows the design of a recent experiment. As you will see, the experimenters directly manipulated a single variable in an attempt to support or disprove a prediction, and they also tried to hold constant any other variables that could influence the results.

SAMPLING ERROR Rarely can experimenters observe *all* individuals of a group. Rather, they use large-enough samples to avoid risking tests with groups that are not representative of the whole. They usually must rely on samples (or subsets) of populations, events, and other aspects of nature. In general, the larger their sampling, the less likely it will be that any differences among the individuals will distort results (Figure 1.11).

About the Word "Theory"

Suppose no one has disproved a hypothesis after years of rigorous tests. Suppose scientists use it to explain more data or observations, which could involve more hypotheses. When a hypothesis meets these criteria, it may become accepted as a **scientific theory**.

You may hear someone apply the word "theory" to a speculative idea, as in the expression "It's only a theory." However, a scientific theory differs from speculation for a simple reason: *Many researchers have tested its predictive power many times, in many ways, and have yet to find evidence that disproves it*. This is why Darwin's

view of natural selection is a respected theory. We use it successfully to explain many diverse issues, such as the origin of life, the relationship between plant toxins and plant-eating animals, the sexual advantages of strongly colored or patterned wings or feathers, the reason that certain cancers run in families, or why antibiotics that doctors often prescribe may no longer be effective. By yielding reasoned evidence that life evolved in the past, the theory even influenced views of Earth history.

An exhaustively tested theory might be as close to the truth as scientists can get with the evidence at hand. For example, Darwin's theory stands, with only minor modification, after more than a century's worth of many thousands of different tests. We cannot show that the theory holds under all possible conditions; an infinite number of tests would be required to do so. As for any theory, we can only say it has a *high or low probability* of being a good one. So far, biologists have not found any evidence that refutes Darwin's theory. Yet they still keep their eyes open for any new information and new ways of testing that might disprove its premises.

And this point gets us back to the value of thinking critically. Scientists must keep asking themselves: *Will observations or experiments show that a hypothesis is false?* They expect one another to put aside pride or bias by testing ideas, even in ways that may prove them wrong. Even if an individual doesn't or won't do this, others will—for science proceeds as a community that is both cooperative and competitive. Ideally, its practitioners share their ideas, with the understanding that it is just as important to expose errors as it is to applaud insights. Individuals can and often do change their minds when presented with contradictory evidence. As you will see, this is a strength of science, not a weakness.

A scientific approach to studying nature is based on asking questions, formulating hypotheses, making predictions, devising tests, and objectively reporting the results.

A scientific theory is a testable explanation about the cause or causes of a broad range of related phenomena. It remains open to tests, revision, and tentative acceptance or rejection.

WWW

1.6   THE POWER OF EXPERIMENTAL TESTS

BIOLOGICAL THERAPY EXPERIMENTS If you worry about the increasing frequencies of strains of pathogenic (disease-causing) bacteria that resist antibiotics, you are not alone. Bruce Levin of Emory University and Jim Bull of the University of Texas are among the investigators who search for possible alternatives to antibiotic therapy. They had been studying literature on a *biological therapy* that enlists bacteriophages—the "bacteria eaters"—to fight infections. Bacteriophages, a class of viruses, attack a narrow range of bacterial strains. When they contact a target, they typically inject a few enzymes and genetic material into it. What happens next is a hostile takeover of the cell's metabolic machinery, and the cell itself makes many new viral particles. The cell dies after viral enzymes

rupture its outer membrane. Virus particles slip through the ruptured membrane and typically infect new cells.

The biologists wondered, as others had decades ago, whether injections of bacteriophages could help people resist or fight off bacterial infections. The discovery of antibiotics in the 1940s had diverted attention away from that idea, at least in Western countries. Now, with so many lethal pathogens breaching the antibiotic arsenal, the bacteria eaters were starting to look good again.

Levin and Bull focused on promising phage therapy experiments that were conducted in 1982 by two British researchers, H. Williams Smith and Michael Huggins. They started their research by testing whether Smith and Huggins's results could be duplicated.

They selected 018:K1:H, a strain of *Escherichia coli* originally isolated from a human patient with meningitis. Like a harmless *E. coli* strain that lives in the intestines of humans and other mammals (Figure 1.12), this one departs from the body in feces. Bacteriophages used in laboratory work typically ignore 018:K1:H7, so Levin and Bull looked for some *E. coli* killers in samples from an Atlanta sewage treatment plant. They successfully isolated two kinds of bacteriophages and named them H (for Hero, an effective killer) and W (for the less effective Wimp).

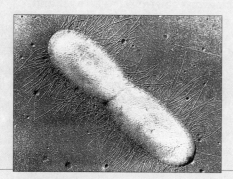

Figure 1.12 Two experiments to compare the effectiveness of bacteriophage injections against antibiotics for treating a bacterial infection. To the left, a micrograph shows a harmless *Escherichia coli* cell in the process of dividing in two.

Hypothesis: If bacteriophages specifically target and destroy cells of *E. coli* 018:K1:H7 in petri dishes, then they will do the same in laboratory mice that have been infected by that strain.

Prediction: Laboratory mice injected with a preparation that contains more than 10^7 particles of *H* bacteriophage will not die following an injection of *E. coli* strain 018:K1:H7.

Experimental test of the prediction:

Researchers establish large populations of *H* bacteriophage and of *E. coli* 018:K1:H7 from which to draw their samples, and select a specific strain of laboratory mice.

EXPERIMENTAL GROUP	CONTROL GROUP
E. coli injected into right thigh of 15 of the mice; bacteriophage injected into their left thigh.	*E. coli injected into 15 other mice; no bacteriophage injected into this sampling.*

Test results:

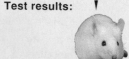

All mice survive.	*All mice die within 32 hours.*

a

Another prediction derived from the same hypothesis: *H* bacteriophage that target and destroy 018:K1:H7 will be more effective than single doses of streptomycin in treating mice infected with 018:K1:H7.

Experimental test of the new prediction:

Researchers inject 48 laboratory mice with 018:K1:H7, then divide them into four groups of 12 each. Eight hours later...

Test results:

EXPERIMENTAL GROUP 1 ...*12 mice receive a single injection of H bacteriophage.*	→	*11 of 12 mice survive.*
EXPERIMENTAL GROUP 2 ...*12 mice receive a single dose of 60 micrograms/gram steptomycin.*	→	*5 of 12 mice survive.*
EXPERIMENTAL GROUP 3 ...*12 mice receive a single dose of 100 micrograms/gram steptomycin.*	→	*3 of 12 mice survive.*
CONTROL GROUP ...*12 mice receive an injection of saline solution only.*	→	*All control mice die.*

b

With graduate student Terry DeRouin and laboratory technician Nina Moore Walker, the biologists grew a large population of 018:K1:H7 in a culture flask. They selected a specific strain of laboratory mice, all females of the same age. Fifteen mice of one experimental group were each injected with 018:K1:H7 in one thigh and more than 10^7 particles of H bacteriophage in the other thigh. All mice survived. Fifteen mice of a control group were injected with 018:K1:H7 only. None survived. Figure 1.12a outlines this experimental test. Figure 1.12b outlines a second test of another prediction derived from the same hypothesis. The results reinforced Smith and Huggins's conclusion that certain bacteriophages can be as good as or better than antibiotics at stopping specific bacterial infections.

Each type of bacteriophage targets specific strains of one or at most a few bacterial species. At present, it takes too long for clinicians to discover which specific bacterium is infecting a patient, so the right bacteriophage might not be enlisted until it is too late to do the patient any good. On the bright side, procedures are now being developed that can dramatically accelerate the identification process.

IDENTIFYING IMPORTANT VARIABLES In nature, many factors can influence the outcome of an infection. For example, genetic differences among infected individuals lead to differences in how their immune system responds to the invasion. Age, nutrition, and health at the time of infection influence the outcome. Some pathogens are deadlier than others. And so on. That is why researchers try to simplify and control variables in their experiments. Variables, recall, are specific aspects of objects or events that may differ over time and among individuals. All of the *E. coli* cells in Levin and Bull's experiments were descended from the same parent cell, raised on the same nutrients at the same temperature in a flask, and thus could be expected to respond in the same way to the bacteriophage attack. All the mice were the same age and sex, and were raised under identical laboratory conditions. Each mouse in each experimental group received the same amount of bacteriophages or streptomycin; each mouse in a given control group got the same injection of saline solution. Thus the focus was on *one variable*—a specific bacteriophage versus a specific antibiotic—in a simple, controlled, artificial situation.

BIAS IN REPORTING THE RESULTS Whether intentional or not, experimenters run the risk of interpreting data in terms of what they want to prove or dismiss. A few have even been known to fake measurements or nudge findings to reinforce their bias. That is why science emphasizes presenting test results in quantitative terms—that is, with actual counts or some other precise form. Doing so allows other experimenters to check or test the results readily and systematically, as Levin and Bull did. At this writing, they are assembling a detailed report of their own experiments, for publication in a science journal.

The call for objective testing strengthens the theories that emerge from scientific studies. It also puts limits on the kinds of studies that can be carried out. Beyond the realm of science, some events remain unexplained. Why do we exist, for what purpose? Why does any one of us have to die at a particular moment? Such questions lead to *subjective* answers, which come from within, as an outcome of all the experiences and mental connections shaping human consciousness. Because people differ vastly in this regard, subjective answers do not readily lend themselves to scientific analysis and experiments.

This is not to say subjective answers are without value. No human society can function for long unless its members share a commitment to certain standards for making judgments, even subjective ones. The moral, aesthetic, philosophical, and economic standards vary from one society to the next. But they all guide their members in deciding what is important and good, and what is not. All attempt to give meaning to what we do.

Every so often, scientists stir up controversy when they happen to explain some part of the world that was considered to be beyond natural explanation—that is, as belonging to the "supernatural." This is often the case when a society's moral codes are interwoven with religious narratives. Exploring some longstanding view of the world from a scientific perspective might be misinterpreted as questioning morality, even though the two are not the same thing.

As one example, centuries ago in Europe, Nicolaus Copernicus studied the planets and concluded the Earth circles the sun. Today this seems obvious enough. Back then it was heresy. The prevailing belief was that the Creator made the Earth (and, by extension, humans) the immovable center of the universe. Later a respected scholar, Galileo Galilei, studied the Copernican model of the solar system, thought it was a good one, and said so. He was forced to retract his statement publicly, on his knees, and put the Earth back as the fixed center of things. (Word has it that when he stood up he muttered, "Even so, it *does* move.") Later still, Darwin's theory of evolution ran up against the same prevailing belief.

Today, as then, society has sets of standards. Those standards might be questioned when some new, natural explanation runs counter to supernatural beliefs. This doesn't mean that the scientists who raise questions are less moral, less lawful, less sensitive, or less caring than anyone else. It simply means one more standard guides their work: *The external world, not internal conviction, must be the testing ground for scientific beliefs.*

Systematic observations, hypotheses, predictions, tests— in all these ways, science differs from systems of belief that are based on faith, force, or simple consensus.

---

1. There is unity in the living world, for all organisms have these characteristics in common: They all consist of one or more cells. They are assembled from the same kinds of atoms and molecules according to the same laws of energy. They survive by metabolism, and by sensing and responding to specific conditions in the environment. Organisms have the capacity for growth, development, and reproduction, based on the heritable instructions encoded in the molecular structure of their DNA. Table 1.1 lists these characteristics.

2. The characteristics of life extend from cells, through multicelled organisms, then populations, communities, ecosystems, and the biosphere.

3. Many millions of species (kinds of organisms) exist. Many millions more were alive in the past and became extinct. Classification schemes place all known species in ever more inclusive groupings, from genus on up through family, order, class, phylum, and kingdom.

4. Life's diversity arises through mutations (changes in the structure of DNA molecules). These molecular changes are the basis for variation in heritable traits. These are traits that parents bestow on their offspring, including most details of body form and functioning.

5. Darwin's theory of evolution by natural selection is a cornerstone of biological inquiry. Its key premises are:

   a. Individuals of a population differ in the details of their shared heritable traits. Variant forms of traits may affect the ability to survive and reproduce.

   b. Natural selection is the outcome of differences in survival and reproduction among individuals that differ in one or more traits. Adaptive forms of a trait tend to become more common; less adaptive ones become less common or disappear. Thus a population's defining traits may change over the generations; the population can evolve.

6. There are diverse methods of scientific inquiry. The following terms are important to all of them:

   a. Theory: Explanation of a broad range of related phenomena, supported by many tests. An example is Darwin's theory of evolution by natural selection.

   b. Hypothesis: A proposed explanation of a specific phenomenon. Sometimes called an educated guess.

   c. Prediction: A claim about what can be expected in nature, based on premises of a theory or hypothesis.

   d. Test: An attempt to produce actual observations that match predicted or expected observations.

   e. Conclusion: A statement about whether a theory or hypothesis should be accepted, modified, or rejected, based on tests of the predictions derived from it.

7. Logic is a pattern of thought by which an individual draws a conclusion that does not contradict evidence used to support the conclusion. Inductive logic means an individual derives a general statement from specific observations. Deductive logic means individuals make inferences about particular consequences or predictions that must follow from a hypothesis. Such a pattern of thinking is often expressed in "if-then" terms.

8. Predictions that flow from hypotheses can be tested by comprehensive observations or by experiments in nature or in the laboratory.

9. Experimental tests simplify observations in nature or the laboratory because conditions under which the observations are made are manipulated and controlled. They are based on the premise that any aspect of nature has one or more underlying causes, whether obvious or not. With such an assumption of cause and effect, only those hypotheses that can be tested in ways that might disprove them are scientific.

10. A control group is a standard against which one or more experimental groups (test groups) are compared. Ideally, it is the same as each experimental group in all variables except the one variable being investigated.

11. A variable is a specific aspect of an object or event that might differ over time and between individuals. Experimenters directly manipulate the variable they are studying to support or disprove their prediction.

12. Test results might be distorted by sampling error —chance differences between a population, event, or some other aspect of nature and the samples chosen to represent it. Test results are less likely to be distorted when samplings are large and when they are repeated.

13. Systematic observations, hypotheses, predictions, and experimental tests are the foundation of scientific theories. The external world, not internal conviction, is the testing ground for those theories.

Table 1.1   Summary of Life's Key Characteristics

SHARED CHARACTERISTICS THAT REFLECT LIFE'S UNITY:

1. Organisms consist of one or more cells.

2. Organisms are constructed of the same kinds of atoms and molecules according to the same laws of energy.

3. Organisms engage in metabolism; they acquire and use energy and materials to grow, maintain themselves, and reproduce.

4. Organisms sense and make controlled responses to internal and external conditions.

5. Heritable instructions encoded in DNA give organisms their capacity to grow and reproduce. DNA instructions also guide the development of large, complex organisms.

FOUNDATIONS FOR LIFE'S DIVERSITY:

1. Mutations (changes in the molecular structure of DNA) give rise to variation in heritable traits, including most details of body form, functioning, and behavior.

2. Diversity is the sum total of variations that accumulated in different lines of descent over the past 3.8 billion years, as by natural selection and other processes of change.

Review Questions

1. Why is it difficult to formulate a simple definition of life? *CI* For this chapter and subsequent chapters, *italics* after a review question identify the section where you can find answers. They include section numbers and *CI* (for Chapter Introduction).

2. Name the molecule of inheritance in cells. *1.1*

3. Write out simple definitions of the following terms: *1.1*
   a. cell
   b. energy
   c. metabolism
   d. photosynthesis
   e. ATP
   f. aerobic respiration

4. How do organisms sense changes in their surroundings? *1.1*

5. Study Figure 1.5. Then, on your own, arrange and define the levels of biological organization. *1.2*

6. Study Figure 1.6. Then, on your own, make a sketch of the one-way flow of energy and the cycling of materials through the biosphere. To the side of the sketch, write out definitions of producers, consumers, and decomposers. *1.2*

7. List the shared characteristics of life. *CI, 1.3*

8. What are the two parts of the scientific name for each kind of organism? *1.3*

9. List the six kingdoms of species as outlined in this chapter, and name some of their general characteristics. *1.3*

10. Define mutation and adaptive trait. Explain the connection between mutations and the immense diversity of life. *1.4*

11. Write brief definitions of evolution, artificial selection, and natural selection. *1.4*

12. Define and distinguish between: *1.5*
   a. hypothesis or speculation and scientific theory
   b. observational test and experimental test
   c. inductive and deductive logic

13. With respect to experimental tests, define variable, control group, and experimental group. *1.5*

14. What does sampling error mean? *1.5*

Self-Quiz *(Answers in Appendix III)*

1. _____ is the capacity of cells to extract energy from sources in their environment, and to transform and use energy to grow, maintain themselves, and reproduce.

2. _____ is a state in which the internal environment is being maintained within tolerable limits.

3. The _____ is the smallest unit of life.

4. If a form of a trait improves chances for surviving and reproducing in a given environment, it is a(n) _____ trait.

5. The capacity to evolve is based on variations in heritable traits, which originally arise through _____ .

6. You have some number of traits that also were present in your great-great-great-great-grandmothers and -grandfathers. This is an example of _____ .
   a. metabolism
   b. homeostasis
   c. a control group
   d. inheritance

7. DNA molecules _____ .
   a. contain instructions for traits
   b. undergo mutation
   c. are transmitted from parents to offspring
   d. all of the above

8. For many years in a row, a dairy farmer allowed his best milk-producing cows but not the poor producers to mate.

Figure 1.13 A spider (*Dolomedes*) and its prey: a tiny minnow. The spider delivered paralyzing venom as well as digestive enzymes into its captive, and is now sucking predigested juices from it. Such spiders can move about below the water's surface. Hairs on their body trap oxygen (for aerobic respiration) during their hunting expeditions.

Over many generations, milk production increased. This outcome is an example of _____ .
   a. natural selection
   b. artificial selection
   c. evolution
   d. both b and c

9. A control group is _____ .
   a. a standard against which experimental groups are compared
   b. identical to experimental groups except for one variable
   c. a standard with one variable against which an experimental group is compared
   d. both a and b are correct

10. A specific aspect of an object or event that may change over time or change among individuals is a _____ .
   a. control group
   b. experimental group
   c. variable
   d. sampling error

11. The fewer the individuals from a population that are chosen at random for an experimental group, _____ .
   a. the greater the chance of sampling error
   b. the smaller the chance of sampling error
   c. the less likely differences among them will distort the test results

12. Match the terms with the most suitable descriptions.
   ___ adaptive trait
   ___ natural selection
   ___ theory
   ___ hypothesis
   ___ prediction

   a. statement of what you should find in nature if you were to go looking for it
   b. educated guess
   c. improves chance of surviving and reproducing in environment
   d. related set of hypotheses that form a broad, applicable, testable explanation
   e. outcome of differences in survival and reproduction among individuals that differ in details of one or more traits

Critical Thinking

1. Some spiders (*Dolomedes*) that feed on insects around ponds occasionally capture tadpoles and small fishes, as in Figure 1.13, and isn't that fun to think about? While they are immature, the female spiders confine themselves to a small patch of vegetation next to the pond. When sexually mature, they mate and store sperm that fertilize the eggs. Only then do they move out and occupy larger areas around the pond. Develop hypotheses to explain what might cause the spiders to live in different places at different times. Design an experiment to test each hypothesis.

2. A scientific theory about some aspect of nature rests upon inductive logic. The assumption is that, because an outcome of some event has been observed to happen with great regularity, it will happen again. However, we cannot know this for certain, because there is no way to account for all possible variables that may affect the outcome. To illustrate this point, Garvin McCain and Erwin Segal offer a parable:

> Once there was a highly intelligent turkey. It lived in a pen, attended by a kind, thoughtful master, and had nothing to do but reflect on the world's wonders and regularities. It observed some major regularities. Morning always began with the sky getting light, followed by the clop, clop, clop of its master's friendly footsteps, then by the appearance of delicious food. Other things varied—sometimes the morning was warm and sometimes cold—but food always followed footsteps. The sequence of events was so predictable, it became the basis of the turkey's theory about the goodness of the world. One morning, after more than one hundred confirmations of the theory, the turkey listened for the clop, clop, clop, heard it, and had its head chopped off.

The turkey learned the hard way that explanations about the world only have a high or low probability of not being wrong. Today, some people take this uncertainty to mean that "facts are irrelevant—facts change." If that is so, should we just stop doing scientific research? Why or why not?

3. Witnesses in a court of law are asked to "swear to tell the truth, the whole truth, and nothing but the truth." What are some of the problems inherent in the question? Can you think of a better alternative?

4. Many popular magazines publish an astounding number of articles on diet, exercise, and other health-related topics. Some authors recommend a specific diet or dietary supplement. What kinds of evidence do you think the articles should include so that you can decide whether to accept their recommendations?

5. Although scientific information often is used when making a decision, it cannot tell an individual what is "right" or "wrong." Give an example of this from your own experience.

6. As the old saying goes, Everybody complains about the weather, but nobody does anything about it. Maybe you can start thinking about how we humans might change at least one aspect of it. Two researchers at Arizona State University found that, at least for their investigation of the northeast coast of North America, it really does rain more on weekends, when we'd rather be out having fun!

As R. Cerveny and R. Balling, Jr., reasoned, if the amount of rain falling each day of the week is a random event, then rules of probability should apply. (*Probability* means the chance that each possible outcome of an event will occur is proportional to the number of ways it can be reached.) Thus, each day of the week should get one-seventh (14.3 percent) of the total rainfall for the week. But it turns out Monday is driest (13.1 percent). Days of the week are wetter, and Saturday is the wettest of all (with 16 percent of the total). Sunday is a bit above average.

What causes this effect? As the researchers hypothesized, if *air pollution* is greater during weekdays than on weekends, then cyclic human activities may influence regional patterns of rainfall. Monday through Friday, coal-burning factories and gas-powered vehicles release quantities of tiny particles into the air. The particles can promote updrafts and act as "platforms" for the formation of water droplets. Air pollution must build up during the week and carry over into Saturday. Over the weekend, fewer factories operate and fewer commuters are on the road, so by Monday, the air must be cleaner—and drier.

What kind of evidence do you suppose Cerveny and Balling gathered to test the hypothesis? Jot down a few ideas and check them against the researchers' article in the 6 August 1998 issue of *Nature*. Whether or not you continue in biology, this exercise will be good practice for searching through scientific literature for actual data that you can evaluate on topics that interest you.

7. Scientists devised experiments to shed light on whether different fish species of the same genus compete in their natural habitat. They constructed twelve ponds, identical in chemical composition and physical characteristics. Then they released the following individuals in each pond:

Ponds 1, 2, 3:	Species A	(300 individuals each pond)
Ponds 4, 5, 6:	Species B	(300 individuals each pond)
Ponds 7, 8, 9:	Species C	(300 individuals each pond)
Ponds 10, 11, 12:	Species A, B, C	(300 individuals of each species in each pond)

Does this experimental design take into consideration all factors that can affect the outcome? If not, how would you modify it?

Selected Key Terms

For this chapter and subsequent chapters, these are the **boldface** terms that appear in the text, in the sections indicated here by *italic* numbers (or CI, short for Chapter Introduction). As a study aid, make a list of the terms, write a definition for each, and check it against the one in the text. You will use these terms later on. Becoming familiar with each one will help give you a foundation for understanding the material in later chapters.

adaptive trait *1.4*	ecosystem *1.2*	mutation *1.4*
Animalia *1.3*	energy *1.1*	natural selection *1.4*
Archaebacteria *1.3*	Eubacteria *1.3*	Plantae *1.3*
artificial	evolution *1.4*	population *1.2*
selection *1.4*	experiment *1.5*	prediction *1.5*
biology *CI*	Fungi *1.3*	producer *1.2*
biosphere *1.2*	genus *1.3*	Protista *1.3*
cell *1.2*	homeostasis *1.1*	receptor *1.1*
community *1.2*	hypothesis *1.5*	reproduction *1.1*
consumer *1.2*	logic *1.5*	species *1.3*
control group *1.5*	metabolism *1.1*	stimulus *1.1*
decomposer *1.2*	model *1.5*	test, scientific *1.5*
development *1.1*	multicelled	theory, scientific *1.5*
DNA *1.1*	organism *1.2*	variable *1.5*

Readings *See also www.infotrac-college.com*

Carey, S. 1994. *A Beginner's Guide to the Scientific Method.* Belmont, California: Wadsworth. Paperback.

Committee on the Conduct of Science. 1989. *On Being a Scientist.* Washington, D.C.: National Academy of Sciences. Paperback.

McCain, G., and E. Segal. 1988. *The Game of Science.* Fifth edition. Pacific Grove, California: Brooks/Cole. Paperback.

Moore, J. 1993. *Science as a Way of Knowing—The Foundations of Modern Biology.* Cambridge, Massachusetts: Harvard University Press.

WWW *http://www.brookscole.com/biology*

Practice quiz questions, hypercontents, BioUpdates, and critical thinking. The Brooks/Cole Biology Resource Center provides a wealth of information fully organized and integrated by chapter.

FACING PAGE: *Living cells of a green plant (Elodea), as seen with the aid of a microscope. Each rectangular cell contains efficient chemical factories called chloroplasts (the round, bright green parts inside).*

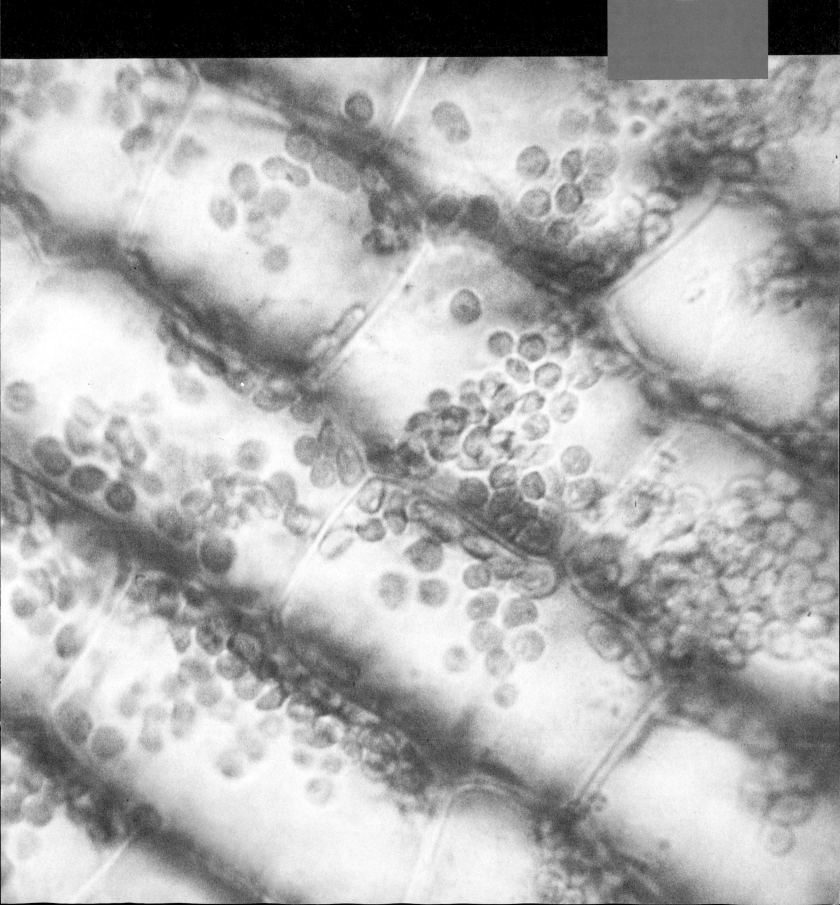

CHEMICAL FOUNDATIONS FOR CELLS

Checking Out Leafy Clean-Up Crews

Right now you are breathing in oxygen. You would die without it. Two centuries ago, no one had a clue to what oxygen is, where it comes from, and how it helps keep people alive. Then researchers started unlocking the secrets of this chemical substance and others. As their knowledge of chemistry deepened, they began to conjure up such amazing things as nuclear power, synthetic fertilizers, nylons, lipsticks, fabric cleaners, aspirin, antibiotics, and plastic parts of refrigerators, computers, television sets, jet planes, and cars.

Today, our chemical "magic" brings us benefits *and* problems. For example, the scale of agriculture required to sustain the human population, which now surpasses 6 billion, is astounding. Synthetic fertilizers boost crop yields by providing plants with nitrogen, phosphorus, and other growth-enhancing nutrients. The upside is that fertilizer-fed crops help keep more people than you might imagine from starving to death. The downside is that plants do not take up every last bit of fertilizer. Nutrients in runoff from the fields enter lakes, rivers, and seas—where they "feed" such organisms as the protistans that cause huge fish kills (Section 20.12).

What about diverting the runoff to holding ponds, where water can evaporate? Farmers commonly do this. However, evaporation ponds are like magnets that pull selenium from the soil. High concentrations of selenium are toxic to grazing animals and waterfowl.

In 1996, Norman Terry and his coworkers grew cattails and other grasses in ten experimental plots that had become contaminated by runoff from agricultural fields (Figure 2.1). As they knew, plants can incorporate selenium into their tissues and can convert some of it to dimethyl selenide—a gas about 600 times less toxic than selenium. They measured how much selenium settled in the mud and how much became incorporated into plant tissues or escaped into the air. As they predicted, before the runoff trickled away from the plots, the plants significantly reduced the selenium levels.

Terry's research is an example of *phytoremediation*—the use of living plants to withdraw harmful substances from the environment. Another example is the use of sunflowers to clean a pond at Chernobyl, in the Ukraine. Following a total meltdown at a nuclear power plant, strontium 90, cesium 137, and other extremely nasty radioactive elements contaminated the surroundings, including the pond (Section 43.8).

Fertilizers, wetlands, people, pumpkins, the air you breathe—*everything in and around you is "chemistry."* Every solid, liquid, or gaseous substance you care to think about is a collection of one or more elements. Think of the **elements** as fundamental forms of matter that have mass and take up space. Here on Earth, we can't break an element apart into something else except in physics laboratories. Break a chunk of copper into smaller and smaller bits, and the smallest bit you end up with will still be copper.

Ninety-two elements occur naturally on Earth. Four of these—oxygen, hydrogen, carbon, and nitrogen—are the most abundant elements in your body, just as they are in all other organisms (Figure 2.2). Your body also contains some phosphorus, potassium, sulfur, calcium, sodium, and chlorine, plus a number of trace elements such as iodine. A *trace* element simply is one that represents less than 0.01 percent of body weight.

The structure and function of each organism depend on the availability of certain kinds

Figure 2.1 At experimental wetlands in Corcoran, California, cattails, bulrushes, *Spartina*, and other grasses help clean up the water from agricultural runoff when they take up selenium.

and amounts of elements. For example, without daily intakes of magnesium, your muscles will become sore and weak, and your brain won't work properly. Without magnesium, older leaves of plants droop, turn yellow, then die. The herbicide 2,4-D stimulates weeds to grow faster than vital elements can be absorbed, so the weeds literally grow themselves to death.

As you might deduce from this story, safeguarding the environment, our food supplies, and our health depends on knowledge of chemistry. So do efforts to minimize side effects of its applications on biological systems. You owe it to yourself and others to gain insight into the structure and behavior of chemical substances. By demystifying chemistry's "magic," you will be better equipped to assess its benefits and risks.

KEY CONCEPTS

1. All substances consist of one or more elements, such as hydrogen, oxygen, and carbon. Each element is composed of atoms, which are the smallest units of matter that still display the element's properties. An element's atoms are composed of protons, electrons, and (except for hydrogen) neutrons.

2. The atoms that make up each kind of element have the same number of protons and electrons, but they may vary slightly from one another in their number of neutrons. Variant forms of an element's atoms are called isotopes.

3. Atoms have no overall electric charge unless they become ionized—that is, unless they lose electrons or acquire more of them. An ion is an atom or molecule that has lost or gained one or more electrons and thereby has acquired an overall positive or negative charge.

4. Whether a given atom will interact with other atoms depends on how many electrons it has and how they are structurally arranged within the atom. When energetic interactions unite two or more atoms, this is a chemical bond.

5. The molecular organization and activities of living things arise largely from ionic, covalent, and hydrogen bonds.

6. Life probably originated in water and is exquisitely adapted to its properties. Foremost among those properties are water's temperature-stabilizing effects, cohesiveness, and capacity to dissolve or repel a variety of substances.

7. Life depends on the controlled formation, use, and disposal of hydrogen ions (H^+). The pH scale is a measure of the concentration of these ions in solutions.

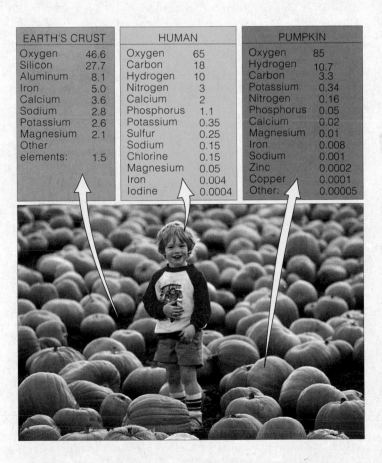

EARTH'S CRUST	
Oxygen	46.6
Silicon	27.7
Aluminum	8.1
Iron	5.0
Calcium	3.6
Sodium	2.8
Potassium	2.6
Magnesium	2.1
Other elements:	1.5

HUMAN	
Oxygen	65
Carbon	18
Hydrogen	10
Nitrogen	3
Calcium	2
Phosphorus	1.1
Potassium	0.35
Sulfur	0.25
Sodium	0.15
Chlorine	0.15
Magnesium	0.05
Iron	0.004
Iodine	0.0004

PUMPKIN	
Oxygen	85
Hydrogen	10.7
Carbon	3.3
Potassium	0.34
Nitrogen	0.16
Phosphorus	0.05
Calcium	0.02
Magnesium	0.01
Iron	0.008
Sodium	0.001
Zinc	0.0002
Copper	0.0001
Other:	0.00005

Figure 2.2 Proportions of elements in a human body and the fruit of pumpkin plants, compared to proportions of elements in the materials of the Earth's crust. How are the proportions similar? How do they differ?

REGARDING THE ATOMS

Structure of Atoms

What are the smallest particles that retain the properties of an element? **Atoms**. A line of about a million of them would fit in the period ending this sentence. Small as atoms are, physicists have split them into more than a hundred kinds of smaller particles. The only subatomic particles you will need to consider in this book are the ones called **protons**, **electrons**, and **neutrons**.

All atoms have one or more protons, which carry a positive electric charge (p^+). Except for hydrogen, atoms also have one or more neutrons, which are uncharged. Protons and neutrons make up the atom's core region, or atomic nucleus (Figure 2.3). Zipping about the nucleus and occupying most of the atom's volume are one or more electrons, which carry a negative charge (e^-). Each atom has just as many electrons as protons. This means that an atom carries no net charge, overall.

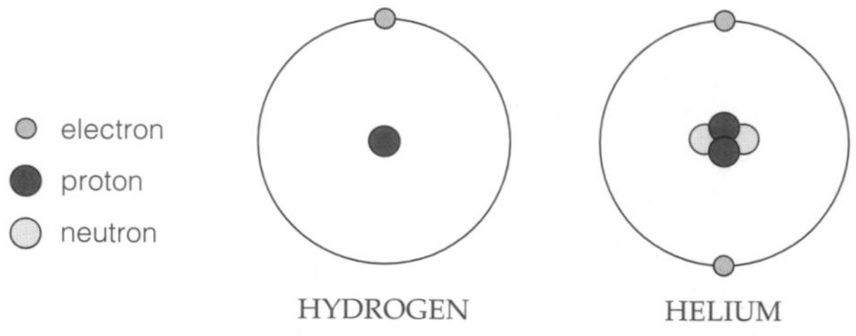

Each element has a unique *atomic* number, which refers to the number of protons in its atoms. To give examples, the atomic number is 1 for the hydrogen atom, which has a single proton. And it is 6 for the carbon atom, which has six protons (Table 2.1).

Also, each element has a *mass* number, which is the combined number of protons and neutrons in the atomic nucleus. A carbon atom, with six protons and six neutrons, has a mass number of 12.

Why bother with atomic numbers and mass numbers? They can give you an idea of whether and how substances will interact. *That knowledge may help you predict how substances might behave in individual cells, in multicelled organisms, and in the environment, under many conditions.*

Isotopes—Variant Forms of Atoms

Samples of most naturally occurring elements contain two or more **isotopes**. The word indicates that, even though all atoms of an element have the same number of protons, they don't all have the same number of neutrons. Carbon has three isotopes, nitrogen has two, and so on. If the element's symbol has a superscript number to the left of it, this signifies which isotope is being discussed. For example, carbon's three isotopes are abbreviated ^{12}C (six protons and six neutrons, this being the most common form), ^{13}C (six protons, seven

electron
proton
neutron

HYDROGEN   HELIUM

Figure 2.3 The hydrogen atom and helium atom according to one model of atomic structure. The model is highly simplified; the nucleus of these two representative atoms really would be an invisible speck at the scale employed here.

neutrons), and ^{14}C (six protons, eight neutrons). We also write these out as carbon 12, carbon 13, and carbon 14.

All isotopes of an element interact with other atoms in the same way. As you will see, this means cells are able to use any isotope of an element for metabolic activities.

Have you heard of radioactive isotopes (radioisotopes)? A physicist, Henri Becquerel, discovered them in 1896, after he had placed a heavily wrapped rock on top of an unexposed photographic plate located in a desk drawer. The rock contained isotopes of uranium, which emit energy. A few days after the plate was exposed to those energetic emissions, a faint image of that rock appeared on it. Marie Curie, Becquerel's coworker, gave the name "radioactivity" to the substance's chemical behavior.

As we now know, a **radioisotope** is an isotope that has an unstable nucleus and that stabilizes itself by spontaneously emitting energy and particles. (Those particles are much smaller than protons, electrons, and neutrons.) This process, radioactive decay, transforms a radioisotope into an atom of a different element at a known rate. For instance, over a predictable time span, carbon 14 becomes nitrogen 14. You will look at some uses of this radioisotope and others in Unit III.

Table 2.1 Atomic Number and Mass Number of Elements Common in Living Things

Element	Symbol	Atomic Number	Most Common Mass Number
Hydrogen	H	1	1
Carbon	C	6	12
Nitrogen	N	7	14
Oxygen	O	8	16
Sodium	Na	11	23
Magnesium	Mg	12	24
Phosphorus	P	15	31
Sulfur	S	16	32
Chlorine	Cl	17	35
Potassium	K	19	39
Calcium	Ca	20	40
Iron	Fe	26	56
Iodine	I	53	127

Elements are forms of matter that occupy space, have mass, and cannot be degraded to something else by ordinary means.

Atoms, the smallest particles that are unique to each element, have one or more positively charged protons, negatively charged electrons, and (except for hydrogen) neutrons.

Most elements have two or more isotopes (atoms that differ in the number of neutrons). A radioisotope has an unstable nucleus and stabilizes itself by spontaneously emitting particles and energy at a known rate.

WWW

2.2 USING RADIOISOTOPES TO TRACK CHEMICALS AND SAVE LIVES

Radioisotopes make splendid tracers. A **tracer** is any substance with a radioisotope attached to it, rather like a shipping label, that researchers can track after they deliver it into a cell, a body, an ecosystem, or some other system. Laboratory devices can detect emissions from the tracer and precisely follow its movement through a pathway or pinpoint its final destination.

Melvin Calvin and some other botanists gave us a classic example. They used tracers to figure out the steps by which plants synthesize carbohydrates during photosynthesis. As they knew, all isotopes of an element have the same number of electrons, so all the isotopes must interact with other atoms the same way. Plant cells, they hypothesized, should be able to use any isotope of carbon when they build carbon compounds. By putting plant cells in a medium enriched with a tracer (^{14}C instead of ^{12}C), they were able to track the uptake of carbon through each reaction step leading to the formation of sugars and starches.

As another example, botanists use a radioisotope of phosphorus (^{32}P) as a tracer to identify how plants take up and use soil nutrients and synthetic fertilizers. The findings may help improve crop yields.

Besides being research tools, radioisotopes also are diagnostic tools in medicine. For safety considerations, clinicians use only the kinds that rapidly decay into harmless elements. Consider how clinicians analyze a human thyroid. This gland of ours, located in front of the windpipe, is the only one that takes up iodine. Iodine is a building block for thyroid hormones, which have great influence over growth and metabolism. If a patient's symptoms point to abnormal outputs of the thyroid hormones, clinicians may inject a trace amount of a carbon radioisotope (^{123}I) into the patient's blood. Then they use a photographic imaging device to scan the gland. Figure 2.4 shows examples of the images.

Radioisotopes also have uses in *PET* (for *Positron-Emission Tomography*). With this device, clinicians obtain images of particular body tissues. Suppose clinicians attach a tracer to glucose (or some other molecule). They inject the labeled glucose into the patient, who is then moved into a PET scanner (Figure 2.5a). Because all cells require glucose, cells throughout the patient take up the labeled glucose. Uptake is greater in some tissues than in others, depending on the tissue's metabolic activity at the time of examination. Laboratory devices can detect the radioactive emissions and use them to form an image, such as the one in Figure 2.5b,c. Such images can reveal variations or abnormalities in metabolic activity.

Radioisotopes that are energetic enough to destroy cells have uses in some therapies. For example, emissions from plutonium 238 drive artificial pacemakers, which smooth out irregular heartbeats. The radioisotope is sealed in a case before the pacemaker is inserted into the body so that its dangerous emissions won't damage body cells.

By contrast, with *radiation therapy*, the idea is to allow radioisotopes to destroy or impair the activity of targeted living cells that are not functioning properly. For example, bombardment with emissions from a source of radium 226 or cobalt 60 may destroy small, localized cancers. *WWW*

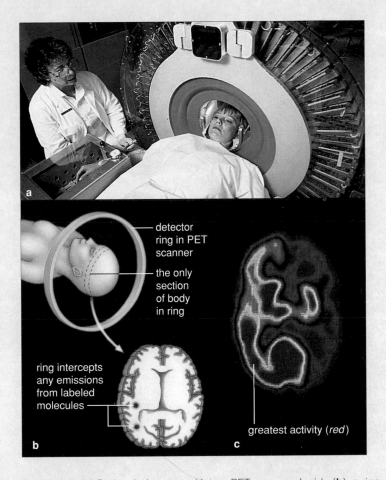

detector ring in PET scanner

the only section of body in ring

ring intercepts any emissions from labeled molecules

greatest activity (*red*)

Figure 2.5 (**a**) Patient being moved into a PET scanner. Inside (**b**), a ring of detectors intercepts the radioactive emissions from labeled molecules that were injected into the patient. Computers analyze and color-code the number of emissions from each location in the scanned body region.

(**c**) Brain scan of a child who has a neurological disorder. Different colors in a brain scan signify differences in metabolic activity. The right half of this brain shows very little activity. By comparison, cells of the left half absorbed and used the labeled molecule at expected rates.

normal   *enlarged*   *cancerous*

Figure 2.4 Scans of the thyroid gland from three patients.

WHAT HAPPENS WHEN ATOM BONDS WITH ATOM?

Electrons and Energy Levels

In a given chemical reaction, atoms may acquire extra electrons, share them, or donate them to another atom. The atoms of certain elements do this rather easily, but others do not. What determines whether one atom will interact with another atom in such ways? *The outcome depends on the number and arrangement of their electrons.*

Tinker with magnets and you can get a sense of the attractive force between unlike charges (+ −) as well as the repulsive force between like charges (++ or − −). Electrons carry a negative charge. In an atom, they repel each other but are attracted to the positive charge of protons. They spend as much time as possible near the protons and far away from each other by moving about in different orbitals. Think of orbitals as volumes of space around the atomic nucleus in which electrons are likely to be at any instant. As a rough analogy, picture three preschoolers circling a cookie jar, not yet expert in the art of sharing. Each is drawn inexorably to the cookies but dreads being shoved away by the others. Two of them might bob and weave about on opposite sides of the jar to avoid a direct hit. But all three never, ever will occupy the same space at the same time.

An orbital can house one or at most two electrons. Because atoms differ in their number of electrons, they also differ in how many occupied orbitals they have.

Hydrogen is the simplest atom. It has a lone electron in a spherical orbital, closest to the nucleus. The orbital corresponds to the *lowest available energy level*. In every other atom, two electrons fill that first orbital. And two more electrons are occupying a second spherical orbital

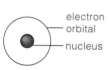

electron orbital

nucleus

HYDROGEN ATOM

around the first one. Larger atoms have even more electrons. These are occupying orbitals farther from the nucleus, at *higher energy levels*.

The **shell model** is a simple although not quite accurate way to think about how electrons are distributed in atoms. By this model, a series of "shells" enclose all orbitals available to an atom's electrons (Figure 2.6). The first shell comprises the first orbital, which is spherical. A second shell (at a higher energy level) surrounds the first shell. The next four available orbitals fit inside it. More orbitals fit inside a third shell, a fourth shell, and so on up through the large, complex atoms of heavier elements, as listed in Appendix VI.

The Nature of Chemical Bonds

Each **chemical bond** is a union between the electron structures of atoms. Take a moment to study Figure 2.7. It summarizes some of the conventions that are used to describe chemical (and metabolic) reactions.

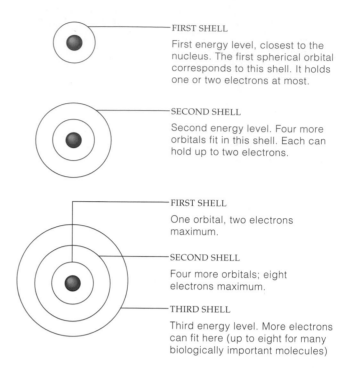

FIRST SHELL

First energy level, closest to the nucleus. The first spherical orbital corresponds to this shell. It holds one or two electrons at most.

SECOND SHELL

Second energy level. Four more orbitals fit in this shell. Each can hold up to two electrons.

FIRST SHELL

One orbital, two electrons maximum.

SECOND SHELL

Four more orbitals; eight electrons maximum.

THIRD SHELL

Third energy level. More electrons can fit here (up to eight for many biologically important molecules)

Figure 2.6 Shell model of electron distribution in atoms. Only three of the many possible energy levels (shells) are shown.

Figure 2.7 Chemical bookkeeping. We use symbols for elements when writing *formulas*, which identify the composition of compounds. For example, water has the formula H_2O. The subscript indicates two hydrogen (H) atoms are present for every oxygen (O) atom. We use such symbols and formulas when writing *chemical equations*, which are representations of the reactions among atoms and molecules. The substances entering a reaction (reactants) are to the left of the reaction arrow, and the products are to the right, as shown by the following chemical equation for photosynthesis:

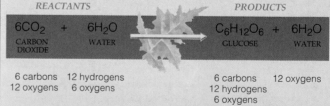

REACTANTS   *PRODUCTS*

$6CO_2$ + $6H_2O$ $C_6H_{12}O_6$ + $6H_2O$

CARBON DIOXIDE   WATER   GLUCOSE   WATER

6 carbons   12 hydrogens
12 oxygens   6 oxygens

6 carbons   12 oxygens
12 hydrogens
6 oxygens

You may read about reactions for which reactants and products are expressed in moles. A *mole* is a certain number of atoms or molecules of any substance, just as "a dozen" can refer to any twelve cats, roses, and so forth. Molar weight, in grams, equals the total atomic weight of all the atoms making up that substance.

For example, the atomic weight of carbon is 12, so one mole of carbon weighs 12 grams. A mole of oxygen (atomic weight 16) weighs 16 grams. Can you state why a mole of water (H_2O) weighs 18 grams, and why a mole of glucose ($C_6H_{12}O_6$) weighs 180 grams?

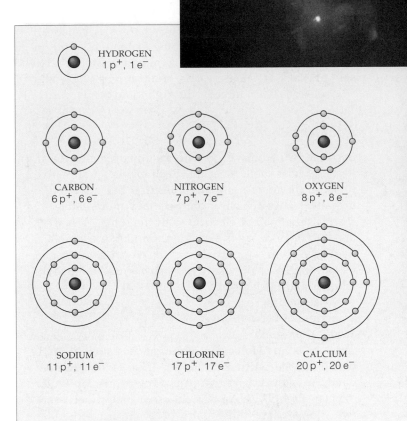

			Distribution of Electrons			
Element	Symbol	Atomic Number*	First Shell	Second Shell	Third Shell	Fourth Shell
Hydrogen	H	1	1	—	—	—
Helium	He	2	2	—	—	—
Carbon	C	6	2	4	—	—
Nitrogen	N	7	2	5	—	—
Oxygen	O	8	2	6	—	—
Neon	Ne	10	2	8	—	—
Sodium	Na	11	2	8	1	—
Magnesium	Mg	12	2	8	2	—
Phosphorus	P	15	2	8	5	—
Sulfur	S	16	2	8	6	—
Chlorine	Cl	17	2	8	7	—
Calcium	Ca	20	2	8	8	2

* The number of protons in the nucleus.

Figure 2.8 Examples of the shell model of the distribution of electrons in atoms. Hydrogen, carbon, and other atoms having electron vacancies (unfilled orbitals) in their outermost shell tend to give up, accept, or share electrons. Helium, neon, and other atoms with no electron vacancies in their outermost shell show no such tendency. The photograph lets us see how even a lone electron can make things happen. Physicists confined electrons in a bubble of liquid helium, then used fluctuating sound waves to pop the bubble. When an electron escaped from the bubble, it made the white-centered red flash shown here.

How can you predict whether two atoms will interact? Check for electron vacancies in their outermost shell. Vacancies mean an atom might give up, gain, or share electrons under suitable conditions, which of course will change the distribution or number of its electrons. You can use the shell model to visualize what goes on here.

In Figure 2.8, which shows the electron distribution for several atoms, electrons are assigned to an energy level (a circle, or shell). Count the electron vacancies—that is, one or more unfilled orbitals in the outermost shell of each atom. As the table shows, helium and neon have no vacancies. They are among the *inert* atoms, which show little tendency to enter chemical reactions.

Now look again at Figure 2.2, which lists the most abundant of the elements making up a typical organism (a human). These elements include hydrogen, oxygen, carbon, and nitrogen. Atoms of all four elements have electron vacancies. Because of this, *they tend to fill the vacancies by forming bonds with other atoms.*

From Atoms to Molecules

When two or more atoms bond together, the outcome is a **molecule**. Some molecules consist of one element only. Molecular nitrogen (N_2), which has two nitrogen atoms, is like this. So is molecular oxygen (O_2).

The molecules of **compounds** consist of two or more different elements in proportions that never vary. Water is a prime example of a compound. In each molecule of water you find one oxygen atom chemically bonded to two hydrogen atoms. Molecules of water in rainclouds, the ocean, a Siberian lake, your bathtub, the petals of a leaf or flower, or anywhere else always have twice as many hydrogen as oxygen atoms.

By contrast, in a **mixture**, two or more elements are simply intermingling in proportions that can vary (and usually do). Swirl together some water and the sugar sucrose, which is a compound of carbon, hydrogen, and oxygen, and you get a mixture.

In an atom, electrons occupy orbitals, which are volumes of space around the nucleus. By a simplified model, orbitals are arranged as a series of shells that surround the nucleus. The successive shells correspond to levels of energy, which become greater with distance from the nucleus.

One or two electrons at most occupy any orbital. Atoms with unfilled orbitals in their outermost shell tend to interact with other atoms; those with no vacancies do not.

In molecules of an element, all of the atoms are of the same kind. In molecules of a compound, atoms of two or more elements are bonded together, in unvarying proportions.

WWW

IMPORTANT BONDS IN BIOLOGICAL MOLECULES

Eat your peas! Drink your milk! Probably for longer than you care to remember, somebody has been telling you to eat foods that are rich in carbohydrates, proteins, and other "biological molecules." Only living organisms put together and use these molecules, which consist of a few kinds of atoms held together by only a few kinds of bonds. Foremost among the molecular interactions are the ionic, covalent, and hydrogen bonds.

Ion Formation and Ionic Bonding

An atom, recall, has just as many electrons as protons, so it carries no net charge. That balance can change for atoms having a vacancy—an unfilled orbital—in their outermost shell. For example, a chlorine atom has such a vacancy and can acquire another electron. A sodium atom has a lone electron in an orbital in its outermost shell, and that electron can be knocked out of or pulled away from the orbital. Any atom that has either lost or gained one or more electrons is an **ion**. The balance between its protons and its electrons has shifted, so the atom has become ionized; it has become positively or negatively charged (Figure 2.9a).

In living cells, neighboring atoms commonly accept or donate electrons among one another. When one atom loses an electron and one gains, both become ionized. Depending on cellular conditions, the two ions may not separate; they may remain together as a result of the mutual attraction of opposite charges. An association of two ions that have opposing charges is called an **ionic bond**. You see one outcome of ionic bonding in Figure 2.9b, which shows a portion of a crystal of table salt, or NaCl. In such crystals, sodium ions (Na^+) and chloride ions (Cl^-) interact through ionic bonds.

Covalent Bonding

Suppose two atoms, each with an unpaired electron in its outermost shell, meet up. Each exerts an attractive force on the other's unpaired electron but not enough to yank it away. Each atom becomes more stable by *sharing* its unpaired electron with the other. A sharing of a pair of electrons is a **covalent bond**. For example, a hydrogen atom can partially fill the electron vacancy in its outermost shell when it is covalently bonded to another hydrogen atom.

In structural formulas, a single line between two atoms represents a *single* covalent bond. Molecular hydrogen has such a bond, which can be written as H—H. In a *double* covalent bond, two atoms share two pairs of electrons. Molecular oxygen (O=O) is like this. In a *triple* covalent bond, two atoms share three pairs of electrons. Molecular nitrogen (N≡N) is like this. All three examples happen to be gaseous molecules. Each time you breathe in some air, you draw a stupendous number of H_2, O_2, and N_2 molecules into your nose.

Covalent bonds are nonpolar or polar. In a *nonpolar* covalent bond, participating atoms exert the same pull on the electrons and both share them equally. "Nonpolar" implies that there is no difference in charge

MOLECULAR HYDROGEN (H_2)

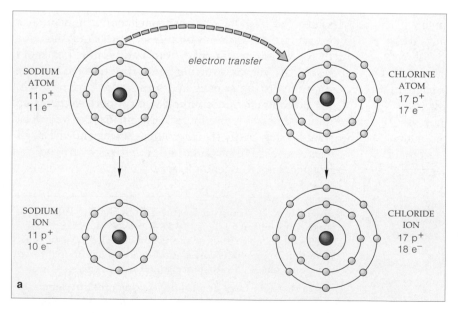

SODIUM ATOM
11 p$^+$
11 e$^-$

electron transfer

CHLORINE ATOM
17 p$^+$
17 e$^-$

SODIUM ION
11 p$^+$
10 e$^-$

CHLORIDE ION
17 p$^+$
18 e$^-$

a

1 mm

Figure 2.9 (a) Ionization by way of an electron transfer. In this case, a sodium atom donates the lone electron in its outermost shell to a chlorine atom, which has an unfilled orbital in *its* outermost shell. A sodium ion (Na^+) and a chloride ion (Cl^-) are the outcome of this interaction. (b) In each crystal of table salt, or NaCl, many sodium and chloride ions remain together because of the mutual attraction of their opposite charges. Their interaction is an example of ionic bonding.

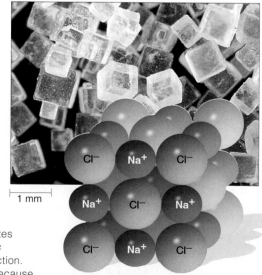

b Crystals of sodium chloride (NaCl)

between two ends of the bond (that is, at its two poles). Molecular hydrogen is a simple example of a nonpolar bond. Its two H atoms, each with one proton, attract the shared electrons equally.

In a *polar* covalent bond, atoms of different elements (which have different numbers of protons) do not exert the same pull on shared electrons. The more attractive atom ends up with a slight negative charge; the atom is "electronegative." Its effect is balanced out by the other atom, which ends up with a slight positive charge. In other words, taken together, the atoms interacting in a polar covalent bond have no *net* charge—but the charge is distributed unevenly between the bond's two ends.

As an example, a water molecule has two polar covalent bonds: H—O—H. In this molecule, electrons are less attracted to the hydrogens than to the oxygen, which has more protons. A water molecule carries no *net* charge, but you will see shortly that its polarity can weakly attract neighboring polar molecules and ions.

Hydrogen Bonding

The patterns of electron sharing in covalent bonds hold atoms together in specific arrangements in molecules. Some of the patterns also give rise to weak attractions and repulsions between charged functional groups of molecules, as well as between molecules and ions. Like interacting skydivers, such interactions break and form easily (Figure 2.10). Yet they have important roles in the structure and functioning of biological molecules.

For example, a **hydrogen bond** is a weak attraction between an electronegative atom (such as an oxygen or nitrogen atom taking part in a polar covalent bond) and

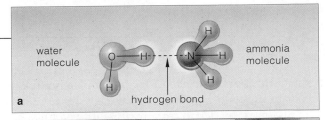
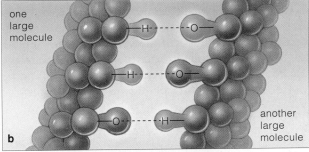

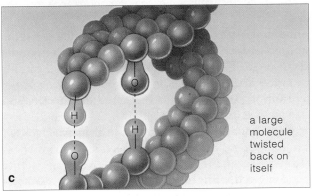

Figure 2.11 Three examples of hydrogen bonds. Compared with a covalent bond, a hydrogen bond is easier to break. Collectively, however, extensive hydrogen bonding is important in water, DNA, proteins, and many other substances.

a hydrogen atom taking part in a second polar covalent bond. Hydrogen's slight positive charge weakly attracts the atom with the slight negative charge (Figure 2.11).

Hydrogen bonds may form between two or more molecules. They also may form in different parts of the same molecule, where it twists and folds back on itself. For example, many such bonds form between the two strands of a DNA molecule. Individually, the hydrogen bonds break easily. Collectively, they stabilize DNA's structure. Similarly, hydrogen bonds form between the molecules that make up water. As you will read next, they contribute to water's life-sustaining properties.

In an ionic bond, two ions of opposite charge attract each other and stay together. Ions form when atoms gain or lose electrons and so acquire a net positive or negative charge.

In a covalent bond, atoms share a pair of electrons. If the atoms share the electrons equally, the bond is nonpolar. If the sharing is not equal, the bond itself is polar—slightly positive at one end, slightly negative at the other.

In a hydrogen bond, a covalently bound atom showing a slight negative charge weakly interacts with a covalently bound hydrogen atom showing a slight positive charge.

Figure 2.10 Like skydivers who briefly clasp hands to form an orderly pattern, weak attractions within and between molecules (and between ions and molecules) can form and break easily.

WWW

PROPERTIES OF WATER

No sprint through basic chemistry is complete unless it leads us to the collection of molecules called water. Life originated in water. Many organisms still live in it. The ones that don't cart water around with them, in cells and tissue spaces. Many metabolic reactions require water as a reactant. Cell shape and internal structure depend on it. These topics will repeatedly occupy our attention in the book, so you may find it useful to become familiar with the following points about water's properties.

Polarity of the Water Molecule

A water molecule, remember, has no net charge, but the charges that it does carry are unevenly distributed. As a result of its electron arrangements and bond angles, the water molecule's oxygen "end" is a bit negative and its hydrogen end is a bit positive. Figure 2.12a is a simple way to think about this charge distribution. Because of the resulting polarity, one water molecule attracts and hydrogen-bonds with others (Figures 2.12b and 2.13).

The polarity of the water molecule also attracts other polar molecules, including the sugars. We call these **hydrophilic** (water-loving) **substances**, for they readily hydrogen-bond with water. By contrast, water's polarity repels nonpolar molecules, including oils. We call these **hydrophobic** (water-dreading) **substances**. Observe this for yourself by shaking a bottle that contains water and salad oil, then setting it on a table. Not long afterward, new hydrogen bonds replace the ones that broke apart when you shook the bottle. As the water molecules reunite, they push molecules of oil aside, forcing them to cluster as droplets or as a film at the water's surface.

Life depends on hydrophobic interactions. For example, a thin, oily membrane separates the cell's watery surroundings and watery interior. Membrane organization starts with countless hydrophobic interactions (Section 4.1).

Figure 2.13 Hydrogen bonding pattern of ice, which in vast quantities blankets the habitat of choice for polar bears. Below 0°C, each water molecule is hydrogen-bonded to four others in a three-dimensional lattice. The molecules are spaced farther apart than they would be in liquid water at room temperature, when molecular motion is greater and not as many bonds form. That is why ice floats on water; it has fewer molecules than the same volume of liquid water (the lattice is less dense).

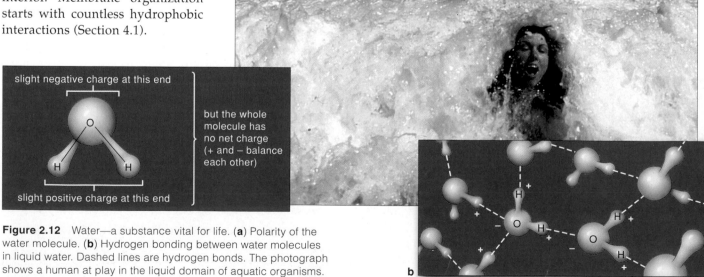

slight negative charge at this end

but the whole molecule has no net charge (+ and − balance each other)

slight positive charge at this end

a

Figure 2.12 Water—a substance vital for life. (**a**) Polarity of the water molecule. (**b**) Hydrogen bonding between water molecules in liquid water. Dashed lines are hydrogen bonds. The photograph shows a human at play in the liquid domain of aquatic organisms.

b

Figure 2.14 Water's cohesion. (**a**) Water strider (*Gerris*). This long-legged bug feeds on insects that land or fall on water. Because of its fine, water-resistant leg hairs and water's high surface tension, the bug scoots across the surface. (**b**) Water's cohesion, combined with evaporation from leaves, pulls water up to the tops of trees.

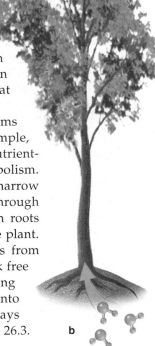

Water's Temperature-Stabilizing Effects

Cells consist mostly of water, and they release a great deal of heat energy during metabolism. If it were not for hydrogen bonds in liquid water, cells might cook in their own juices. To see why this is so, start with these observations: Every molecule vibrates incessantly, and its motion increases when it absorbs heat. **Temperature** is simply a measure of the molecular motion of a given substance. Compared to most other fluids, water can absorb more heat energy before its temperature rises measurably. Why? Much of the added energy disrupts hydrogen bonding *between* neighboring molecules of water; it does not cause an increase in the motion of individual molecules. In liquid water, the stupendous number of hydrogen bonds can buffer large swings in temperature. Such bonds help stabilize the temperature of multicelled organisms and of aquatic habitats.

Even when the temperature of liquid water is not shifting much, hydrogen bonds are constantly breaking, but they also are forming again just as fast. By contrast, a large energy input can increase molecular motion so much that hydrogen bonds *stay* broken, and individual molecules at the water's surface escape into the air. By this process, called **evaporation**, heat energy converts liquid water to the gaseous state. As large numbers of molecules break free and depart, they carry away some energy and lower the water's surface temperature.

Evaporative water loss can help cool you and some other mammals when you work up a sweat on hot, dry days. Under such conditions, sweat—which is about 99 percent water—evaporates from your skin.

Below 0°C, hydrogen bonds resist breaking and lock water molecules in the latticelike bonding pattern of ice (Figure 2.13). Ice is less dense than water. During winter freezes, ice sheets may form near the surface of ponds, lakes, and streams. Like a blanket, ice "insulates" the liquid water beneath it and helps protect many fishes, frogs, and other aquatic organisms against freezing.

Water's Cohesion

Life also depends on water's cohesion. **Cohesion** means something has a capacity to resist rupturing when placed under tension—that is, stretched—as by the weight of a bug's legs (Figure 2.14*a*). Think of a lake, pool, or some other body of liquid water. Uncountable episodes of hydrogen bonding exert a continual inward pulling on water molecules at or near the surface. The hydrogen bonding results in a high surface tension. That tension

grabs your interest when you swim in a lake on a summer night—when too many night-flying insects splat against water and float on it.

Cohesion works inside organisms as well as on the outside. For example, trees and other plants require nutrient-laden water for growth and metabolism. Largely because of the cohesion, narrow columns of liquid water move through pipelines of vascular tissues from roots to leaves and all other parts of the plant. On sunny days, water evaporates from leaves; individual molecules break free (Figure 12.14*b*). Hydrogen bonding "pulls up" more water molecules into leaf cells as replacements, in ways that you will read about in Section 26.3.

Water's Solvent Properties

Finally, water is a fine solvent; ions and polar molecules easily dissolve in it. Any dissolved substance is called a **solute**. In general, we say a substance is *dissolved* after water molecules cluster around its ions or molecules and keep them dispersed in a fluid. (Such clusters are "spheres of hydration.") This is what happens to solutes in cellular fluid, in the sap of maple trees, in blood, in fluid traveling through your gut, and in all other fluids associated with life.

Watch this happen when you pour some table salt (NaCl) into a cup of water. After a while, the salt crystals separate into Na^+ and Cl^-. Each Na^+ attracts the negative end of some of the water molecules at the same time Cl^- attracts the positive end of others (Figure 2.15). The spheres of hydration keep the ions dispersed in the fluid.

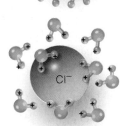

Figure 2.15 Two spheres of hydration.

A water molecule has no net charge, yet it shows polarity. The polarity allows water molecules to hydrogen-bond with one another and with other polar (hydrophilic) substances. Water molecules tend to repel nonpolar (hydrophobic) substances.

Water has temperature-stabilizing effects, internal cohesion, and a capacity to dissolve many substances. These properties influence the structure and functioning of organisms.

𝓦𝓦𝓦

ACIDS, BASES, AND BUFFERS

A great variety of ions dissolved in the fluids inside and outside cells influence cell structure and functioning. Among the most influential are **hydrogen ions**, or H^+. Hydrogen ions are the same thing as free (unbound) protons. They have far-reaching effects largely because they are chemically active and there are so many of them.

The pH Scale

At any instant in liquid water, some water molecules break apart into hydrogen ions and **hydroxide ions** (OH^-). This ionization of water is the basis of the **pH scale**, as in Figure 2.16. Biologists use this scale when measuring the H^+ concentration of seawater, tree sap, blood, and other fluids. Pure water (not rainwater or tapwater) always contains just as many H^+ as OH^- ions. This condition also may occur in other fluids, and it signifies neutrality. We assign neutrality a value of 7 at the midpoint of the pH scale, which ranges from 0 (the highest H^+ concentration) to 14 (the lowest). *The greater the H^+ concentration, the lower the pH.*

Starting at neutrality, each change by one unit of the pH scale corresponds to a tenfold increase or decrease in H^+ concentration. An easy way to sense the differences is to dissolve a bit of baking soda (pH 9) on your tongue, then water (7), then lemon juice (2.3).

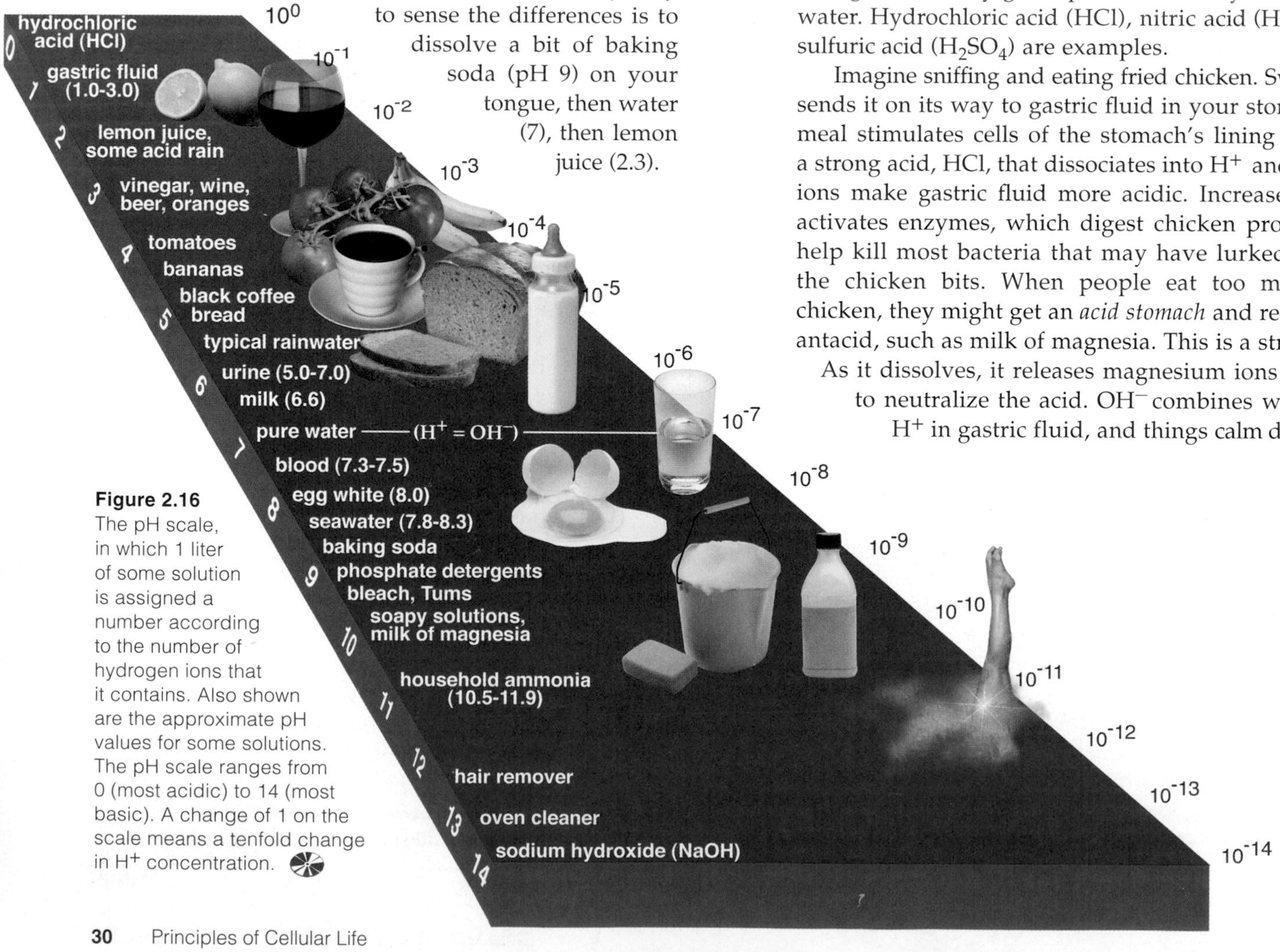

How Do Acids Differ From Bases?

When they dissolve in water, substances categorized as **acids** *donate* protons (H^+) to other solutes or to water molecules. By contrast, the substances we categorize as **bases** *accept* H^+ when dissolved in water, and OH^- forms directly or indirectly after they do this. *Acidic* solutions, such as lemon juice, gastric fluid, and coffee, release more H^+ than OH^-; their pH is below 7. *Basic* solutions, such as seawater, baking soda, and egg white, release more OH^- than H^+. Such solutions are also called "alkaline" fluids; they have a pH above 7.

The fluid inside most human cells is about 7 on the pH scale. The pH values of most fluids bathing those cells are slightly higher; they range between 7.3 and 7.5. This also is the case for the fluid portion of human blood. By contrast, seawater is more alkaline than body fluids of organisms that live in it.

Think of most acids as being either weak or strong. Weak ones such as carbonic acid (H_2CO_3) are reluctant H^+ donors. Depending on the pH, they just as easily accept H^+ after giving it up, so they alternate between acting as an acid and acting as a base. By contrast, strong acids totally give up H^+ when they dissociate in water. Hydrochloric acid (HCl), nitric acid (HNO_3), and sulfuric acid (H_2SO_4) are examples.

Imagine sniffing and eating fried chicken. Swallowing sends it on its way to gastric fluid in your stomach. The meal stimulates cells of the stomach's lining to secrete a strong acid, HCl, that dissociates into H^+ and Cl^-. The ions make gastric fluid more acidic. Increased acidity activates enzymes, which digest chicken proteins and help kill most bacteria that may have lurked in or on the chicken bits. When people eat too much fried chicken, they might get an *acid stomach* and reach for an antacid, such as milk of magnesia. This is a strong base. As it dissolves, it releases magnesium ions and OH^- to neutralize the acid. OH^- combines with excess H^+ in gastric fluid, and things calm down.

10^0
10^{-1}
10^{-2}
10^{-3}
10^{-4}
10^{-5}
10^{-6}
10^{-7}
10^{-8}
10^{-9}
10^{-10}
10^{-11}
10^{-12}
10^{-13}
10^{-14}

0 hydrochloric acid (HCl)
1 gastric fluid (1.0–3.0)
2 lemon juice, some acid rain
3 vinegar, wine, beer, oranges
4 tomatoes bananas
5 black coffee bread typical rainwater
6 urine (5.0–7.0) milk (6.6)
7 pure water —— ($H^+ = OH^-$) ——
blood (7.3–7.5)
8 egg white (8.0) seawater (7.8–8.3) baking soda
9 phosphate detergents bleach, Tums
10 soapy solutions, milk of magnesia
11 household ammonia (10.5–11.9)
12 hair remover
13 oven cleaner
14 sodium hydroxide (NaOH)

Figure 2.16 The pH scale, in which 1 liter of some solution is assigned a number according to the number of hydrogen ions that it contains. Also shown are the approximate pH values for some solutions. The pH scale ranges from 0 (most acidic) to 14 (most basic). A change of 1 on the scale means a tenfold change in H^+ concentration.

Figure 2.17 Sulfur dioxide emissions from a coal-burning power plant. Camera lens filters revealed the otherwise invisible emissions. Sulfur dioxide and other airborne pollutants dissolve in water vapor to form acidic solutions. They are a major component of acid rain.

High concentrations of strong acids or bases also can disrupt the external environment and pose dangers to life. Read the labels on bottles of ammonia, drain cleaner, and other products that are often stored in households. Many can cause severe *chemical burns*. So can sulfuric acid in car batteries. Smoke from fossil fuels, exhaust from motor vehicles, and nitrogen fertilizers release strong acids, which alter the pH of rain (Figure 2.17). Some regions are sensitive to the pH of this *acid rain*, owing to their soil type and vegetation cover. The altered chemistry of habitats in such regions drastically affects the functioning of organisms. We will return to this topic in Section 43.1.

Buffers Against Shifts in pH

Metabolic reactions are sensitive to even slight shifts in pII, for H^+ and OH^- can combine with many different molecules and alter their functions. Normally, control mechanisms minimize unsuitable shifts in pH, as they do when HCl enters the stomach in response to a meal. Many of the controls involve buffer systems.

A **buffer system** is a partnership between a weak acid and the base that forms when the acid dissolves in water. The two work as a pair to counter slight shifts in pH. Remember, when a strong base enters a fluid, the OH^- level rises. But a weak acid neutralizes part of the added OH^- by combining with it. By this interaction, the weak acid's partner forms. Later, if a strong acid floods in, the base will accept H^+ and thereby become its partner in the system.

Bear in mind, the action of a buffer system cannot make new hydrogen ions or eliminate ones that already are present. It can only bind or release them.

In all complex, multicelled organisms, diverse buffer systems operate in the internal environment—in blood and tissue fluids. For example, metabolic reactions in the vertebrate lungs and kidneys help control the acid-base balance of this environment, at levels suitable for life (Sections 35.4 and 37.4). For now, simply think of what happens when the blood level of H^+ decreases and the blood is not as acidic as it should be. At such times, carbonic acid that is dissolved in blood releases H^+ and so becomes the partner base, bicarbonate:

$$H_2CO_3 \longrightarrow HCO_3^- + H^+$$
CARBONIC ACID   BICARBONATE

When blood becomes more acidic, more H^+ becomes bound to the base, thus forming the partner acid:

$$HCO_3^- + H^+ \longrightarrow H_2CO_3$$
BICARBONATE   CARBONIC ACID

Uncontrolled shifts in pH have drastic outcomes. If blood's pH (7.3–7.5) declines even to 7, an individual will enter into a *coma*, a sometimes irreversible state of unconsciousness. An increase to 7.8 can lead to *tetany*, a potentially lethal stage in which the body's skeletal muscles enter a state of uncontrollable contraction. In *acidosis*, carbon dioxide builds up in blood, too much carbonic acid forms, and blood pH severely decreases. *Alkalosis* is an uncorrected increase in blood pH. Both conditions weaken the body and can be lethal.

Salts

Salts are compounds that release ions *other than* H^+ and OH^- in solutions. Salts and water often form when a strong acid and strong base interact. Depending on a solution's pH value, salts can form and dissolve easily. Consider how sodium chloride forms, then dissolves:

$$HCl \ (acid) + NaOH \ (base) \longrightarrow NaCl \ (salt) + H_2O$$
HYDROCHLORIC   SODIUM   SODIUM CHLORIDE
ACID   HYDROXIDE

$$Na^+ \quad Cl^- \ (ionization)$$

Many salts dissolve into ions that serve key functions in cells. For example, nerve cell activity depends on ions of sodium, potassium, and calcium; muscles contract with the help of calcium ions; and water absorption by plant cells depends largely on potassium ions.

Hydrogen ions (H^+) and other ions dissolved in the fluids inside and outside cells affect cell structure and function.

When dissolved in water, acidic substances release H^+, and basic (alkaline) substances accept them. Certain acid-base interactions, as in buffer systems, help maintain the pH value of a fluid—that is, its H^+ concentration.

A buffer system counters slight shifts in pH by releasing hydrogen ions when their concentration is too low or by combining with them when the concentration is too high.

Salts are compounds that release ions other than H^+ and OH^-, and many of those ions have key roles in cell functions.

WWW

1. Chemistry helps us understand the nature of all the substances that make up cells, organisms, and the Earth, its waters, and the atmosphere. Each substance consists of one or more elements. Of the ninety-two naturally occurring elements, the most common in organisms are oxygen, carbon, hydrogen, and nitrogen. Organisms also contain lesser and varying amounts of other elements, such as calcium, phosphorus, potassium, and sulfur.

2. Table 2.2 summarizes some key chemical terms that you will encounter throughout this book.

3. Elements consist of atoms. An atom has one or more positively charged protons, an equal number of negatively charged electrons, and (except for hydrogen atoms) one or more uncharged neutrons. The protons and neutrons occupy the core region, the atomic nucleus. Most elements have isotopes: two or more forms of atoms with the same number of protons but different numbers of neutrons.

4. An atom has no net charge. But it may lose or gain one or more electrons and so become an ion, which has an overall positive or negative charge.

5. Whether an atom interacts with others depends on the number and arrangement of its electrons, which occupy orbitals (volumes of space) inside a series of shells around the atomic nucleus. Atoms with unfilled orbitals in the outermost shell tend to bond with other elements.

6. Generally, a chemical bond is a union between the electron structures of atoms.

   a. In an ionic bond, a positive ion and negative ion stay together because of a mutual attraction of their opposite charges.

   b. Various atoms often share one or more pairs of electrons in single, double, or triple covalent bonds. Such electron sharing is equal in nonpolar covalent bonds, and it is unequal in polar covalent bonds. The interacting atoms have no net charge, but the bond is slightly negative at one end and slightly positive at the other.

   c. In a hydrogen bond, one covalently bonded atom (such as oxygen or hydrogen) that shows a slight negative charge is weakly attracted to the slight positive charge of a hydrogen atom that is taking part in a different polar covalent bond.

7. By the pH scale, a solution is assigned a number that reflects its H^+ concentration. This ranges from 0 (highest concentration) to 14 (lowest). At pH 7, the H^+ and OH^- concentrations are equal. Acids release H^+ in water, and bases combine with them. Buffer systems help maintain pH values of blood, tissue fluids, and the fluid inside cells.

8. Polar covalent bonds join together the three atoms of water molecules (two hydrogens, one oxygen). The polarity of individual water molecules invites hydrogen bonding between molecules. Such bonding is the basis of liquid water's ability to resist temperature changes more than other fluids do, to display internal cohesion, and to easily dissolve polar or ionic substances. These properties profoundly influence the metabolic activity, shape, and internal organization of cells.

Table 2.2 Summary of Key Players in the Chemical Basis of Life	
ELEMENT	Fundamental form of matter that occupies space, has mass, and cannot be broken apart into a different form of matter by ordinary physical or chemical means.
ATOM	Smallest unit of an element that still retains the characteristic properties of that element.
Proton (p^+)	Positively charged particle of the atomic nucleus. All atoms of an element have the same number of protons, which is the atomic number. A proton without an electron zipping around it is a hydrogen ion (H^+).
Electron (e^-)	Negatively charged particle that can occupy a volume of space (orbital) around an atomic nucleus. All atoms of an element have the same number of electrons. Electrons can be shared or transferred among atoms.
Neutron	Uncharged particle of the nucleus of all atoms except hydrogen. For a given element, the mass number is the number of protons and neutrons in the nucleus.
MOLECULE	Unit of matter in which two or more atoms of the same element, or different ones, are bonded together.
Compound	Molecule composed of two or more different elements in unvarying proportions. Water is an example.
Mixture	Intermingling of two or more elements in proportions that can and usually do vary.
ISOTOPE	One of two or more forms of atoms of an element that differ in their number of neutrons.
Radioisotope	Unstable isotope, having an unbalanced number of protons and neutrons, that emits particles and energy.
Tracer	Molecule of a substance to which a radioisotope is attached. In conjunction with tracking devices, it can be used to follow the movement or destination of that substance in a metabolic pathway, the body, etc.
ION	Atom that has gained or lost one or more electrons, thus becoming positively or negatively charged.
SOLUTE	Any molecule or ion dissolved in some solvent.
Hydrophilic substance	Polar molecule or molecular region that can readily dissolve in water.
Hydrophobic substance	Nonpolar molecule or molecular region that strongly resists dissolving in water.
ACID	Substance that donates H^+ when dissolved in water.
BASE	Substance that accepts H^+ when dissolved in water; OH^- forms directly or indirectly afterward.
SALT	Compound that releases ions other than H^+ or OH^- when dissolved in water.

Review Questions

1. What is an element? Name four elements (and their symbols) that make up more than 95 percent of the body weight of all living organisms. *CI*

2. Define atom, isotope, and radioisotope. *2.1*

3. How many electrons can occupy each orbital around an atomic nucleus? Using the shell model, explain how the orbitals available to electrons are distributed in an atom. *2.3*

4. Define molecule, compound, and mixture. *2.3*

5. Distinguish between:
   a. ionic and hydrogen bonds *2.4*
   b. polar and nonpolar covalent bonds *2.4*
   c. hydrophilic and hydrophobic interactions *2.5*

6. If a water molecule has no net charge, then why does it attract polar molecules and repel nonpolar ones? *2.5*

7. Label the atoms in each water molecule in the sketch below. Indicate which parts of each molecule carry a slight positive charge (+) and which carry a slight negative charge (−). *2.5*

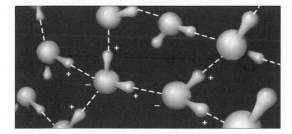

8. Define acid and base. Then describe the behavior of a weak acid in solutions having a high or low pH value. *2.6*

Self-Quiz *(Answers in Appendix III)*

1. Electrons carry (a) _____ charge.
   a. positive   b. negative   c. zero

2. Atoms share electrons unequally in a(n) _____ bond.
   a. ionic   c. polar covalent
   b. nonpolar covalent   d. hydrogen

3. A water molecule shows _____ .
   a. polarity   d. solvency
   b. hydrogen-bonding capacity   e. a and b
   c. heat resistance   f. all of the above

4. In liquid water, spheres of hydration form around _____ .
   a. nonpolar molecules   d. solvents
   b. polar molecules   e. b and c
   c. ions   f. all of the above

5. Hydrogen ions (H^+) are _____ .
   a. the basis of pH values   d. dissolved in blood
   b. unbound protons   e. both a and b
   c. targets of certain buffers f. all of the above

6. When dissolved in water, a(n) _____ donates H^+; however, a(n) _____ accepts H^+.

7. Match the terms with the most suitable descriptions.
   _____ trace element
   _____ buffer system
   _____ chemical bond
   _____ temperature
      a. weak acid and its partner base work as a pair to counter pH shifts
      b. union between electron structures of two atoms
      c. less than 0.01% of body weight
      d. measure of molecular motion in some defined region

Critical Thinking

1. An ionic compound forms when calcium combines with chlorine. Referring to Figure 2.8, give the compound's formula. (Hint: Be sure the outermost shell of each atom is filled.)

2. David, an inquisitive three-year-old, poked his fingers in the water inside a metal pan on the stove and discovered that it was warm. Then he touched the pan itself and got a nasty burn. Devise a hypothesis to explain why water in a metal pan heats up far more slowly than the pan itself.

3. When molecules absorb microwaves, which are a form of electromagnetic radiation, they move more rapidly. Explain why a microwave oven can heat foods.

4. From what you know about cohesion, devise a hypothesis to explain why water forms droplets.

5. Edward is trying to study a chemical reaction that an enzyme catalyzes (speeds up). H^+ forms in the reaction, but the enzyme is destroyed at low pH. What can he include in his reaction mix to protect the enzyme while he studies the reaction? Explain how your suggestion might solve the problem.

6. Many reactions occur on molecular parts of enzymes and other proteins. Cells must have access to those parts. Through interactions with water and ions, a soluble protein can become dispersed in cellular fluid rather than settling against some cell structure. An electrically charged cushion around it makes this happen. Using the sketch below as a guide, explain the chemical interactions by which such a cushion forms, starting with major bonds in the protein itself.

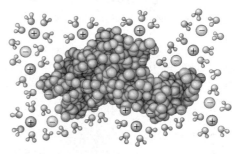

Selected Key Terms

acid *2.6*
atom *2.1*
base *2.6*
buffer system *2.6*
chemical bond *2.3*
cohesion *2.5*
compound *2.3*
covalent bond *2.4*
electron *2.1*
element *CI*
evaporation *2.5*

hydrogen bond *2.4*
hydrogen
  ion (H^+) *2.6*
hydrophilic
  substance *2.5*
hydrophobic
  substance *2.5*
hydroxide
  ion (OH^-) *2.6*
ion *2.4*
ionic bond *2.4*
isotope *2.1*

mixture *2.3*
molecule *2.3*
neutron *2.1*
pH scale *2.6*
proton *2.1*
radioisotope *2.1*
salt *2.6*
shell model *2.3*
solute *2.5*
temperature *2.5*
tracer *2.2*

Readings *See also www.infotrac-college.com*

Ritter, P. 1996. *Biochemistry: A Foundation.* Pacific Grove, California: Brooks/Cole.

WWW *http://www.brookscole.com/biology*

Practice quiz questions, hypercontents, BioUpdates, and critical thinking. The Brooks/Cole Biology Resource Center provides a wealth of information fully organized and integrated by chapter.

3 CARBON COMPOUNDS IN CELLS

Carbon, Carbon, in the Sky—Are You Swinging Low and High?

High in the mountains of the Pacific Northwest, vast forests of conifers have endured another murderously cold winter (Figure 3.1). Like all other organisms, those evergreen, cone-bearing trees cannot grow or reproduce in the absence of liquid water. Yet through the winter months, water in the surroundings is locked away from them, in the form of snow and ice.

The trees get by anyway. Their needle-shaped leaves have an epidermis—a layer of interconnected cells with thick, waxy walls that restricts water loss. About the only way precious water can escape is through small gaps across the epidermis that can close or open in response to changing conditions. During the cool, dry days of autumn, the trees enter dormancy. Metabolic activities idle and growth ceases, but water conserved inside them is enough to keep their cells alive.

With the arrival of spring, rising temperatures and water from melting snow stimulate renewed growth. Tree roots soak up mineral-laden water. And carbon dioxide, a gaseous molecule of one carbon atom and two oxygen atoms, moves in from the air, through gaps in the leaf epidermis. With their photosynthetic magic, the conifers turn those simple materials into sugars, starches, and other carbon-based compounds. They are premier producers of the northern forests.

Producers, recall, are organisms that use self-made organic compounds as their structural materials and as packets of energy. One way or another, compounds made by producers of forests and other ecosystems all over the world nourish every consumer and decomposer.

Plants of the great prevailing forests and plains at northern latitudes of Canada, the United States, and other parts of the Northern Hemisphere are now breaking dormancy sooner than they did just two decades ago. Why? No one knows for sure, but this change in their life cycles may be one outcome of long-term change in the global climate.

Figure 3.1 Conifers beneath the first snows of winter on Silver Star Mountain, Washington. As is the case for all other organisms, the structure, activities, and very survival of these trees start with the carbon atom and its diverse molecular partners in organic compounds.

Researchers have looked at atmospheric concentrations of carbon dioxide in the air around us since the early 1950s. Among other things, they found that the concentration shifts with the seasons. It declines during spring and summer, when great numbers of photosynthesizers take up stupendous amounts of the gas from their surroundings. It rises at other times of year, when huge populations of decomposers that release the gas as a metabolic by-product undergo rapid increases in their population size.

According to the researchers' measurements, the spring decline starts a full week earlier than it did in the mid-1970s. Besides this, the seasonal swings are becoming more pronounced—by as much as 20 percent in Hawaii and a whopping 40 percent in Alaska.

Swings in the atmospheric concentration of carbon dioxide were greatest in 1981 and 1990. Intriguingly, temperatures of the lower atmosphere are rising also, and in 1981 and 1990, they were uncommonly high. Are we in the midst of a long-term, worldwide rise in atmospheric temperature? And is this *global warming* promoting a longer growing season, hence the wider seasonal swings?

The picture gets more intricate. We humans burn great quantities of coal, gasoline, and other fossil fuels for energy. Fossil fuels are rich in carbon, which is released (as carbon dioxide) by the burning processes. The released carbon may be contributing to the global warming, in ways you will read about in later chapters.

For now, the point to keep in mind is this: *Carbon permeates the world of life—from the energy-requiring activities and structural organization of individual cells, to physical and chemical conditions that span the globe and that influence life everywhere.*

With this chapter, we turn to life-giving properties that emerge out of the molecular structure of carbon-rich compounds. Study the chapter well, especially the summary in Table 3.1. It will serve as your foundation for understanding how different organisms put such compounds together, how they use them, and how the effects of those uses ripple through the biosphere.

KEY CONCEPTS

1. Organic compounds have a backbone of one or more carbon atoms to which hydrogen, oxygen, nitrogen, and other atoms are attached. We define cells partly by their capacity to assemble the organic compounds known as carbohydrates, lipids, proteins, and nucleic acids.

2. Cells put together large biological molecules from their pools of smaller organic compounds, which include simple sugars, fatty acids, amino acids, and nucleotides.

3. Glucose and other simple sugars are carbohydrates. So are organic compounds composed of two or more sugar units, of one or more types, that are covalently bonded together. The most complex carbohydrates that cells assemble are polysaccharides, many of which consist of hundreds or thousands of sugar units.

4. Lipids are greasy or oily compounds that show little tendency to dissolve in water but dissolve in nonpolar compounds, including other lipids. They include the neutral fats (triglycerides), phospholipids, waxes, and sterols.

5. Cells use carbohydrates and lipids as building blocks and as their major sources of energy.

6. Proteins have truly diverse roles. Many are structural materials. Many are enzymes, a type of molecule that enormously increases the rate of specific metabolic reactions. Other kinds transport cell substances, contribute to cell movements, trigger changes in cell activities, and defend the body against injury and disease.

7. For living organisms, ATP and other nucleotides are crucial players in metabolism. DNA and RNA, strandlike nucleic acids assembled from nucleotide subunits, are the basis of inheritance and reproduction.

PROPERTIES OF ORGANIC COMPOUNDS

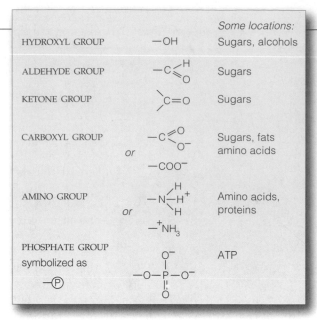

Figure 3.2 A few examples of functional groups.

The Molecules of Life

Under present-day conditions in nature, *only living cells synthesize carbohydrates, lipids, proteins, and nucleic acids.* These are the molecules characteristic of life. Different classes of biological molecules serve as the cells' packets of instantly available energy, energy stores, structural materials, metabolic workers, libraries of hereditary information, and cell-to-cell signals.

The molecules of life are **organic compounds**, with hydrogen and often other elements covalently bonded to carbon atoms. The term is a holdover from a time when chemists thought "organic" substances were the ones they got from animals and vegetables, as opposed to "inorganic" substances they got from minerals. The term persists, even though researchers now synthesize organic compounds in laboratories. And it persists even though there are reasons to believe organic compounds were present on Earth *before* organisms were.

Carbon's Bonding Behavior

By far, organisms consist mainly of oxygen, hydrogen, and carbon (Figure 2.3). Much of the oxygen and the hydrogen is in the form of water. Remove the water, and carbon makes up more than half of what's left.

Carbon's importance in life arises from its versatile bonding behavior. As shown in the sketch below, *each carbon atom can share pairs of electrons with as many as four other atoms.* Each covalent bond formed this way is relatively stable. Such bonds link carbon atoms together in chains. These form a backbone to which hydrogen, oxygen, nitrogen, and many other elements become attached.

The bonding arrangement is the start of wonderful three-dimensional shapes of organic compounds. A chain of carbon atoms, bonded covalently one after another, forms a backbone from which other atoms can project:

branching from backbone

carbon backbone

The backbone can coil back on itself in a ring structure, which we can diagram in such ways as:

or

carbon rings

Functional Groups

A carbon backbone with only hydrogen atoms bonded covalently to it is a hydrocarbon, which is a very stable structure. Besides hydrogen atoms, biological molecules also have **functional groups**: various kinds of atoms or clusters of them covalently bonded to the backbone.

To get a sense of their importance, consider a few of the functional groups shown in Figure 3.2. Sugars and other organic compounds classified as **alcohols** have one or more hydroxyl groups (—OH). Alcohols dissolve quickly in water, because water molecules easily form hydrogen bonds with —OH groups. The backbone of a protein forms by reactions between amino groups and carboxyl groups. As you will see, the backbone is the start of bonding patterns that produce the protein's three-dimensional structure. As other examples, amino groups can combine with H^+ and act as buffers against decreases in pH. And the functional groups shown in Figure 3.3 are a molecular starting point for differences between males and females of many species.

How Do Cells Build Organic Compounds?

Using carbon (from carbon dioxide), water, and sunlight (as an energy source), the photosynthetic cells you read about earlier put together simple sugar molecules. Like all living cells, they also use such molecules as a starting point for assembling other small molecules, especially the fatty acids, amino acids, and nucleotides. As you will read shortly, cells use some assortment of these four classes of small organic compounds as subunits for building all the organic compounds they require for their structure and functioning.

How do they do it? It will take more than one chapter to sketch out answers (and best guesses) to the question.

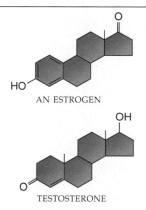

AN ESTROGEN

TESTOSTERONE

Figure 3.3 Notable differences in traits between male and female wood ducks (*Aix sponsa*). Two different sex hormones have key roles in the development of feather color and other traits that help the two ducks recognize each other (and therefore influence reproductive success). Both of the hormones, testosterone and one of the estrogens, have the same carbon ring structure. As you can see, however, the ring structures have different functional groups attached to them.

Figure 3.4 Examples of metabolic reactions by which most biological molecules are put together, rearranged, and broken apart.

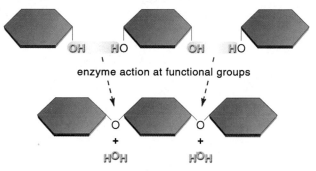

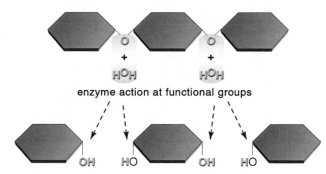

a Two condensation reactions. Enzymes remove an —OH group and H atom from two molecules, which covalently bond as a larger molecule. Two water molecules form.

b Hydrolysis, a water-requiring cleavage reaction. Enzyme action splits a molecule into three parts, then attaches an H atom and an —OH group derived from a water molecule to each exposed site.

At this point in your reading, simply become aware that the reactions by which cells build organic compounds, and even rearrange them and break them apart, require more than an energy input. They also require the class of proteins called **enzymes**, which make specific metabolic reactions proceed faster than they would on their own. Different enzymes mediate different kinds of reactions. Most of the reactions fall into five categories, which you will encounter in chapters to come:

1. *Functional-group transfer.* One molecule gives up a functional group, which another molecule accepts.

2. *Electron transfer.* One or more electrons stripped from one molecule are donated to another molecule.

3. *Rearrangement.* A juggling of internal bonds converts one type of organic compound into another.

4. *Condensation.* Through covalent bonding, two molecules combine to form a larger molecule.

5. *Cleavage.* A molecule splits into two smaller ones.

To get a sense of what goes on, consider two examples of these events. First, in many **condensation reactions**, enzymes remove a hydroxyl group from one molecule and an H atom from another, then speed the formation of a covalent bond between the two molecules at their exposed sites (Figure 3.4*a*). As a typical but incidental outcome of the reaction, the discarded atoms join to form

a molecule of water. A series of condensation reactions can produce starches and other polymers. A polymer is a large molecule having three to millions of subunits, which may or may not be identical. Biologists call the subunits monomers, as in the sugar monomers of starch.

As the second example, a cleavage reaction called **hydrolysis** is like condensation in reverse (Figure 3.4*b*). Enzymes that recognize specific functional groups split molecules into two or more parts, then attach an —OH group and a hydrogen atom derived from a molecule of water to the exposed sites. With hydrolysis, cells can cleave large polymers such as starch into smaller units when these are required for building blocks or energy.

Organic compounds have diverse, three-dimensional shapes and functions. The diversity starts at their carbon backbones and with bonding arrangements that arise from it.

Functional groups covalently bonded to the carbon backbone of organic compounds add enormously to the structural and functional diversity.

Carbohydrates, lipids, proteins, and nucleic acids are the main biological molecules, the organic compounds that only living cells can assemble under conditions that now occur in nature.

Cells assemble, rearrange, and degrade organic compounds mainly through enzyme-mediated reactions involving the transfer of functional groups or electrons, rearrangement of internal bonds, and a combining or splitting of molecules.

CARBOHYDRATES

Consider first the carbohydrates—the most abundant of all biological molecules, which cells use as structural materials and transportable or storage forms of energy. Most **carbohydrates** consist of carbon, hydrogen, and oxygen in a 1:2:1 ratio, which also may be written as $(CH_2O)_n$. Three classes are called the **monosaccharides**, **oligosaccharides**, and **polysaccharides**.

The Simple Sugars

"Saccharide" comes from a Greek word meaning sugar. A *mono*saccharide, meaning "one monomer of sugar," is the simplest carbohydrate. It has at least two —OH groups joined to the carbon backbone plus an aldehyde or a ketone group. Most monosaccharides are sweet tasting and readily dissolve in water. The most common have a backbone of five or six carbon atoms that tends to form a ring structure when dissolved in cells or body fluids. Ribose and deoxyribose, the sugar components of RNA and DNA, respectively, have five carbon atoms. Glucose has six (Figure 3.5a). Besides being the main energy source for most organisms, glucose is a precursor (parent molecule) of many compounds and a building block for larger carbohydrates. Vitamin C (a sugar acid) and glycerol (an alcohol with three —OH groups) are other examples of compounds having sugar monomers.

Short-Chain Carbohydrates

Unlike the simple sugars, an *oligo*saccharide is a short chain of two or more sugar monomers that are bonded covalently. (*Oligo*— means a few.) The type known as *di*saccharides consists of just two sugar units. Lactose, sucrose, and maltose are examples. Lactose (a glucose and a galactose unit) is a milk sugar. Sucrose, the most plentiful sugar in nature, consists of one glucose and one fructose unit (Figure 3.5c). Plants convert their stores of carbohydrates to sucrose, which can be easily transported through their leaves, stems, and roots. Table sugar is sucrose crystallized from sugarcane and sugar beets. Proteins and other large molecules often have oligosaccharides attached as side chains to the carbon backbone. Some chains take part in the body's defenses against disease, others in cell membrane functions.

Complex Carbohydrates

The "complex" carbohydrates, or *poly*saccharides, are straight or branched chains of many sugar monomers (often hundreds or thousands) of the same or different types. Cellulose, starch, and glycogen, the most common polysaccharides, consist only of glucose. In cell walls of plants, cellulose is a structural material (Figure 3.6a). Like steel rods in reinforced concrete, fibers of cellulose

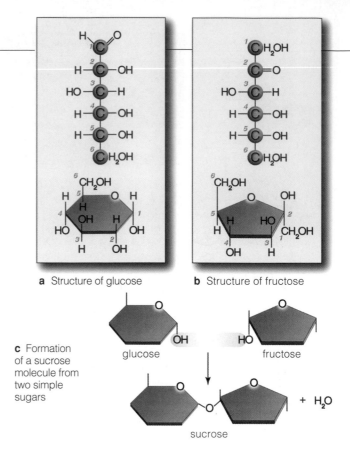

a Structure of glucose   **b** Structure of fructose

c Formation of a sucrose molecule from two simple sugars

glucose   fructose

sucrose   + H_2O

Figure 3.5 Straight-chain and ring forms of (**a**) glucose and (**b**) fructose. For reference purposes, carbon atoms of simple sugars are commonly numbered in sequence, starting at the end closest to the molecule's aldehyde or ketone group. (**c**) Condensation of two monosaccharides into a disaccharide.

are tough and insoluble; they withstand considerable weight and stress. Plants store their photosynthetically produced sugars as large starch molecules (Figure 3.6b). Enzymes can readily hydrolyze starch to glucose units.

If both cellulose and starch consist of glucose, why are their properties so different? The answer starts with differences in covalent bonding patterns between their monomers, which in both are bonded together in chains.

In cellulose, many glucose chains stretch out side by side and hydrogen-bond to one another at —OH groups (Figure 3.7a). This bonding arrangement stabilizes the chains into a tightly bundled pattern that resists being digested, at least by most enzymes.

In starch, the pattern of covalent bonding positions each sugar monomer at an angle relative to the next monomer in line. The chain ends up coiling like a spiral staircase (Figure 3.7b). The coils are not particularly stable. In the kinds of starches with branched chains, they are even less stable. A great many —OH groups project outward from the coiled chains, and this makes them readily accessible to enzymes.

In animals, glycogen is the sugar-storage equivalent of starch in plants. Muscle and liver cells house large stores of it. When the level of sugar in blood decreases, liver cells degrade glycogen, so glucose is released and

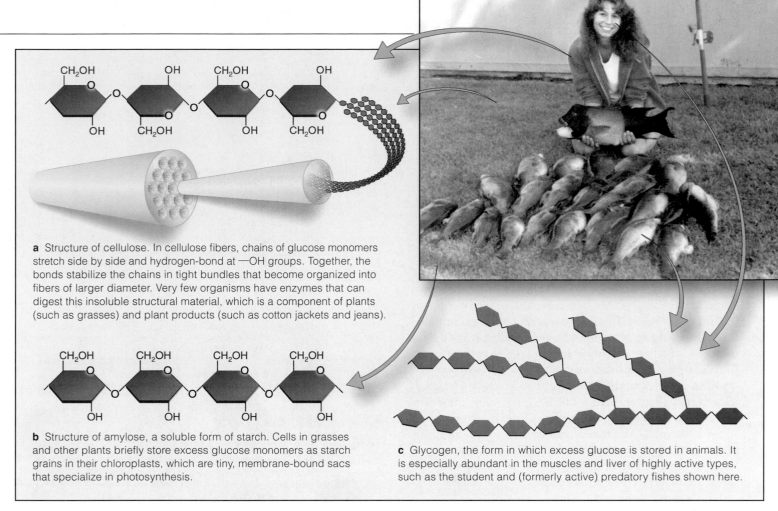

a Structure of cellulose. In cellulose fibers, chains of glucose monomers stretch side by side and hydrogen-bond at —OH groups. Together, the bonds stabilize the chains in tight bundles that become organized into fibers of larger diameter. Very few organisms have enzymes that can digest this insoluble structural material, which is a component of plants (such as grasses) and plant products (such as cotton jackets and jeans).

b Structure of amylose, a soluble form of starch. Cells in grasses and other plants briefly store excess glucose monomers as starch grains in their chloroplasts, which are tiny, membrane-bound sacs that specialize in photosynthesis.

c Glycogen, the form in which excess glucose is stored in animals. It is especially abundant in the muscles and liver of highly active types, such as the student and (formerly active) predatory fishes shown here.

Figure 3.6 Molecular structure of starch, cellulose, and glycogen, and their typical locations in a few organisms. All three carbohydrates consist only of glucose monomers.

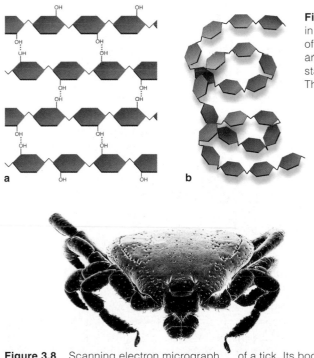

Figure 3.7 Comparison of bonding patterns between glucose monomers in starch and in cellulose. (**a**) In cellulose, bonds form between monomers of neighboring glucose chains. This bonding pattern stabilizes the chains and allows them to become tightly bundled. (**b**) In amylose, a form of starch, the bonding pattern between monomers causes the chains to coil. The coiling orients the bonds in a way that is accessible to enzymes.

Figure 3.8 Scanning electron micrograph of a tick. Its body covering is a protective cuticle reinforced with chitin.

enters the bloodstream. When you exercise strenuously but briefly, your muscle cells tap their glycogen stores for a rapid burst of energy. Figure 3.6*c* represents a few of glycogen's many branchings.

The polysaccharide chitin has nitrogen-containing groups attached to its glucose monomers. It is the main structural material for the external skeletons and other hard body parts of crabs, earthworms, insects, ticks, and many other animals (Figure 3.8). Chitin also is the main structural material in the cell walls of many fungi.

The simple sugars (such as glucose), oligosaccharides, and polysaccharides (such as starch) are carbohydrates. Every cell requires carbohydrates as structural materials, stored forms of energy, and transportable packets of energy.

𝒲𝒲𝒲

LIPIDS

Being mostly hydrocarbon, **lipids** show little tendency to dissolve in water, although they readily dissolve in nonpolar substances. All are greasy or oily to the touch. Cells use different kinds as their main energy reservoirs, as structural materials (for example, in cell membranes and surface coatings), and as signaling molecules. Let's look first at the kinds with fatty acid components—the fats, phospholipids, and waxes. We also will consider the sterols, each with a backbone of four carbon rings.

Fats and Fatty Acids

Lipids called **fats** have one, two, or three fatty acids attached to glycerol. Each **fatty acid** has a backbone of as many as thirty-six carbon atoms, a carboxyl group (—COOH) at one end, and hydrogen atoms occupying most or all of the remaining bonding sites. It typically stretches out like a flexible tail. *Unsaturated* tails incorporate one or more double bonds. *Saturated* tails contain single bonds only. Figure 3.9a shows examples.

Most animal fats have many saturated fatty acids, which pack together by weak interactions. They are solid at temperatures at which most plant fats remain liquid. The packing interactions in plant fats are not as stable because of rigid kinks in their fatty acid tails. And that is why "vegetable oils" flow freely.

Butter, lard, vegetable oils, and other natural fats consist mostly of **triglycerides**: neutral fats having three fatty acid tails attached to glycerol (Figure 3.9b). Triglycerides are the body's most abundant lipids and its richest energy source. Gram for gram, they yield more than twice as much energy when broken down, compared to starches and other complex carbohydrates. Notable quantities of triglycerides are stored as droplets in the cells of body fat (that is, adipose tissue) in vertebrates. A thick layer of triglycerides under the skin has survival value for some animals, including the penguins shown in Figure 3.10. It helps insulate their body against near-freezing temperatures of their habitats.

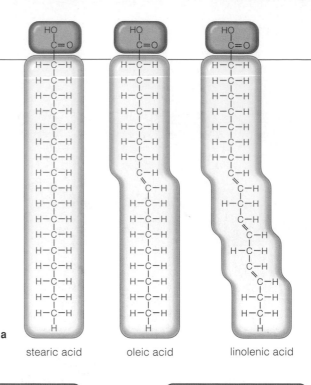

a

stearic acid   oleic acid   linolenic acid

Figure 3.9 (**a**) Structural formulas for three fatty acids. In stearic acid, the carbon backbone is fully saturated with hydrogen atoms. Oleic acid, with a double bond in its backbone, is unsaturated. Linolenic acid, which has three double bonds, is a "polyunsaturated" fatty acid. (**b**) Condensation of fatty acids and glycerol into a triglyceride.

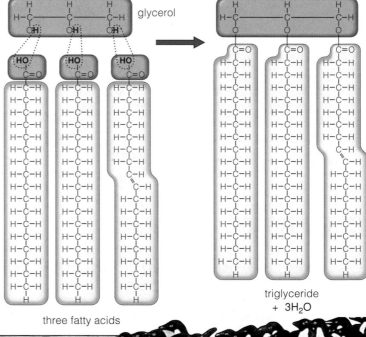

glycerol

three fatty acids

b

triglyceride + 3H$_2$O

Figure 3.10 Triglyceride-protected penguins taking the plunge.

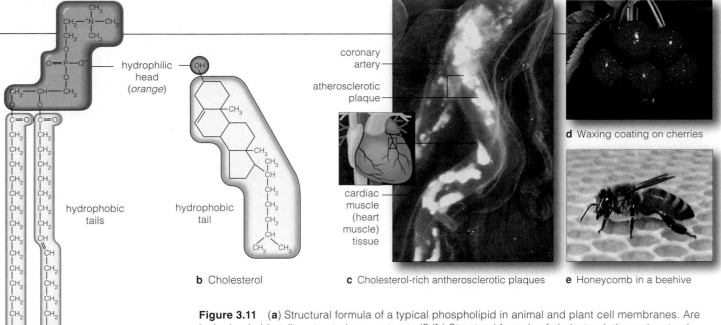

b Cholesterol

c Cholesterol-rich antherosclerotic plaques

d Waxing coating on cherries

e Honeycomb in a beehive

hydrophilic head (*orange*)

hydrophobic tails

hydrophobic tail

a One of the phospholipids

coronary artery

atherosclerotic plaque

cardiac muscle (heart muscle) tissue

Figure 3.11 (**a**) Structural formula of a typical phospholipid in animal and plant cell membranes. Are its hydrophobic tails saturated or unsaturated? (**b**) Structural formula of cholesterol, the major sterol of animal tissues. (**c**) Your liver synthesizes enough cholesterol for the body. A fat-rich diet may result in excessively high cholesterol levels in blood and the formation of abnormal masses of material in certain blood vessels (arteries). Such *atherosclerotic plaques* may clog the arteries that deliver blood to the heart. (**d**) Demonstration of the water-repelling attribute of a cherry cuticle. (**e**) Honeycomb, a structural material constructed of a firm, water-repellent, waxy secretion called beeswax.

One fatty acid is a precursor of eicosanoids, a class of local signaling molecules that include prostaglandins. Different eicosanoids bind to receptors on cells and help regulate many physiological processes, such as muscle contraction, message transmission through the nervous system, inflammation, and immune responses.

Phospholipids

A **phospholipid** has a glycerol backbone, two fatty acid tails, and a hydrophilic "head" with a phosphate group and another polar group (Figure 3.11*a*). Phospholipids are the main materials of cell membranes, which have two layers of lipids. Heads of one layer are dissolved in the cell's fluid interior, and heads of the other layer are dissolved in the surroundings. Sandwiched between the two are all the fatty acid tails, which are hydrophobic.

Sterols and Their Derivatives

Sterols are among the many lipids that have no fatty acids. Sterols differ in the number, position, and type of their functional groups, but all have a rigid backbone of four fused-together carbon rings:

sterol backbone

Sterols occur in eukaryotic cell membranes. Cholesterol is the most common type in tissues of animals (Figure 3.11*b*,*c*). Cells also remodel cholesterol into compounds

such as vitamin D (required for good bones and teeth), steroids, and bile salts. Steroids include sex hormones, such as estrogen and testosterone, that govern sexual traits and gamete formation. Bile salts have roles in the digestion of fats in the small intestine.

Waxes

The lipids called **waxes** have long-chain fatty acids tightly packed and linked to long-chain alcohols or to carbon rings. They have a firm consistency and repel water. Waxes and another lipid, cutin, make up most of the cuticle that covers the aboveground plant parts. The covering helps plants conserve water and fend off some parasites. A waxy cherry cuticle is an example (Figure 3.11*d*). In many animals, waxy secretions from cells are incorporated in coatings that protect, lubricate, and impart pliability to skin or hair. Among waterfowl and other birds, wax secretions help keep feathers dry. As another example, beeswax is the material of choice when bees construct their honeycombs (Figure 3.11*e*).

Being largely hydrocarbon, lipids can dissolve in nonpolar substances but tend not to dissolve in water.

Triglycerides (neutral fats), which have a glycerol head and three fatty acid tails, are the body's main energy reservoirs. Phospholipids are the main components of cell membranes.

Sterols such as cholesterol are membrane components as well as precursors of steroid hormones and other compounds. Waxes are firm yet pliable components of water-repelling and lubricating substances.

WWW

AMINO ACIDS AND THE PRIMARY STRUCTURE OF PROTEINS

Have you ever wondered how permanent waves work (Figure 3.12)? The characteristics of many mammalian body parts—including hair—start with the structure of proteins—just as they do for all other organisms.

Of all the large biological molecules, **proteins** are the most diverse. Proteins of the class called enzymes make metabolic events proceed much faster than they otherwise would. Structural proteins are the stuff of spider webs, butterfly wings, feathers, cartilage and bone, and a dizzying array of other body parts and products. Transport proteins move molecules and ions across cell membranes and cart them about through body fluids. Nutritious proteins abound in milk, eggs, and many seeds. Protein hormones and other regulatory types are signals for change in cell activities. Many proteins act as weapons against disease-causing bacteria and other invaders. Amazingly, cells build diverse proteins from their pools of only twenty kinds of amino acids!

Structure of Amino Acids

Every **amino acid** is a small organic compound that consists of an amino group, a carboxyl group (an acid), a hydrogen atom, and one or more atoms known as its R group. As you can see from the structural formula in Figure 3.13, these parts generally are covalently bonded to the same carbon atom. Figure 3.14 shows a number of amino acids that we will consider later in the book.

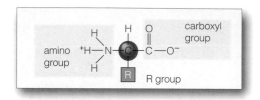

Figure 3.13 Generalized structural formula for amino acids.

Primary Structure of Proteins

When a cell synthesizes a protein, amino acids become linked, one after the other, by peptide bonds. As Figure 3.15 shows, this is the type of covalent bond that forms between one amino acid's amino group (NH_3^+) and the carboxyl group ($—COO^-$) of the next amino acid.

When peptide bonds join two amino acids together, we have a dipeptide. When they join three or more, we have a **polypeptide chain**. In such chains, the carbon

Figure 3.12 Appearance of the hair of actress Nicole Kidman (**a**) before and (**b**) after a permanent wave. The difference, as you will see from this section and the next, starts with strings of amino acids in polypeptide chains.

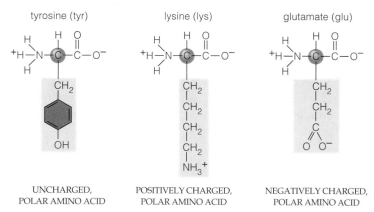

Figure 3.14 Structural formulas for eight of the twenty common amino acids. *Green* boxes highlight the R groups, which are side chains with functional groups. Each type of side chain contributes in a major way to the distinctive properties of each amino acid.

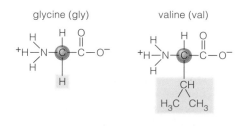
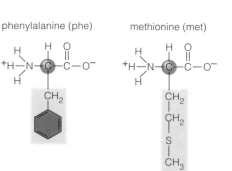

FIVE OF THE NONPOLAR AMINO ACIDS

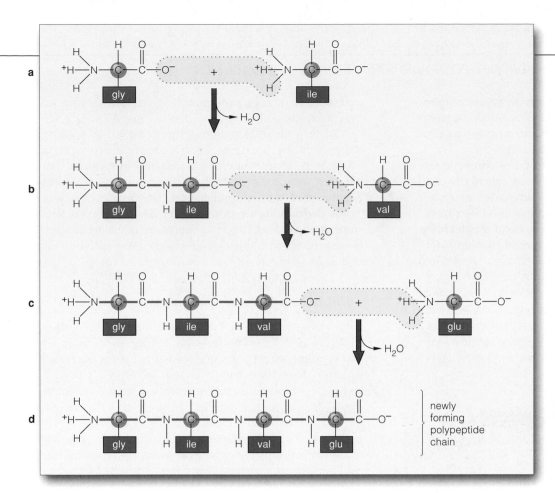

Figure 3.15 Peptide bond formation during protein synthesis.

(**a**) The first two amino acids shown are glycine (gly) and isoleucine (ile). They are at the start of the sequence for one of two polypeptide chains that make up the protein insulin in cattle.

(**b**) Through a condensation reaction, the isoleucine becomes joined to the glycine by a peptide bond. A water molecule forms as a by-product of the reaction.

(**c**) A peptide bond forms between the isoleucine and valine (val), another amino acid, and water again forms.

(**d**) Remember, DNA specifies the order in which the different kinds of amino acids follow one another in a growing polypeptide chain. In this case, glutamate (glu) is the fourth amino acid specified. Ultimately, the result is one of the two polypeptide chains of an insulin molecule, as in Figure 3.16. Chapter 13 has more details on the steps that lead from DNA instructions to proteins.

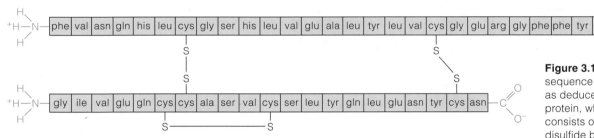

Figure 3.16 Diagram of the amino acid sequence for an insulin molecule (in cattle), as deduced by Frederick Sanger. This protein, which functions as a hormone, consists of two polypeptide chains. Two disulfide bridges (—S—S—), each formed by a condensation reaction at two R groups (sulfhydryl groups), link the chains together.

backbone has nitrogen atoms positioned in this regular pattern: —N—C—C—N—C—C—.

For each particular kind of protein, different amino acid units are selected one at a time from the twenty kinds available. Their orderly progression is prescribed by the cell's DNA. Overall, the resulting sequence of amino acids is unique for each kind of protein, and it is the protein's *primary* structure (Figures 3.15 and 3.16).

Now consider this: Different cells make thousands of different proteins. Many of the proteins are fibrous, with polypeptide chains organized as strands or sheets. Collectively, many such molecules contribute to the shape and internal organization of cells. Other kinds of proteins are globular, with one or more polypeptide chains folded in compact, rather rounded shapes. Most enzymes are globular proteins. So are actin and other proteins that contribute to cell movement.

Regardless of the type of protein, its shape and its function arise from the primary structure—that is, from information built into its amino acid sequence. As you will see next, that information dictates which parts of a polypeptide chain will coil, bend, or interact with other chains nearby. And the type and arrangement of atoms in coiled, stretched-out, or folded regions determine whether that protein will function as, say, an enzyme, a transporter, a receptor, or even an inadvertent target for a bacterium or virus.

A protein consists of one or more polypeptide chains, in which amino acids are strung together. The amino acid sequence (which kind of amino acid follows another in the chain) is unique for each kind of protein and gives rise to its unique structure, chemical behavior, and function.

WWW

HOW DOES A PROTEIN'S THREE-DIMENSIONAL STRUCTURE EMERGE?

The preceding section gave you a sense of how amino acids are strung together in a polypeptide chain, which is a protein's primary structure. Now consider just a few examples of the protein shapes that emerge.

For the most part, the primary structure gives rise to a protein's shape in two ways. First, it allows hydrogen bonds to form between different amino acids along the length of a polypeptide chain. Second, it puts R groups into positions that allow them to interact. Through their interactions, the chain is forced to bend and twist.

Second Level of Protein Structure

Hydrogen bonds form at regular, short intervals along a new polypeptide chain, and they give rise to a coiled or extended pattern known as the secondary structure of a protein. Think of a polypeptide chain as a set of rigid playing cards joined by links that can swivel a bit. Each "card" is a peptide group (Figure 3.17a). Atoms on either side of it can rotate slightly around their covalent bonds and form bonds with neighboring atoms. For instance, in many chains, hydrogen bonds readily form between every third amino acid. The bonding pattern forces the peptide groups to coil helically, like a spiral staircase (Figure 3.17b). In other proteins, a hydrogen-bonding pattern holds two or more chains side by side in an extended, sheetlike array (Figure 3.17c).

Third Level of Protein Structure

Most polypeptide chains that have a coiled secondary structure undergo more folding, owing to the number and location of certain amino acids (including the bulky proline) along their length. These amino acids bend a chain at certain angles and in certain directions to make the chain loop out. R groups far apart along its length interact and hold the loops in characteristic positions. A polypeptide chain folded as an outcome of its bend-producing amino acids and its R-group interactions has reached a third structural level (tertiary structure).

Figure 3.18a shows how a polypeptide chain became folded into the compact, tertiary structure of the protein globin. Hydrogen bonds between particular functional groups along the chain's length caused the folding.

Fourth Level of Protein Structure

Imagine that bonds form between *four* molecules of globin and that an iron-containing functional group, a heme group, is positioned near the center of each. The outcome is **hemoglobin**, an oxygen-transporting protein

a One peptide group

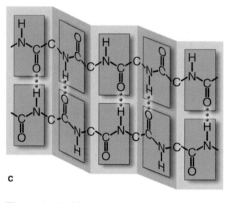

c

Figure 3.17 Major bonding patterns between (**a**) peptide groups of polypeptide chains. Extensive hydrogen bonding (the *dotted* lines) can bring about (**b**) coiling or (**c**) sheetlike formations.

b

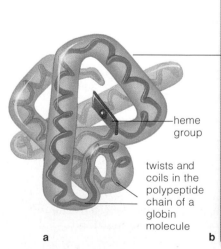

heme group

twists and coils in the polypeptide chain of a globin molecule

a

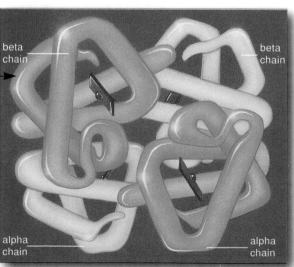

beta chain

beta chain

alpha chain

alpha chain

b

Figure 3.18 (**a**) Globin molecule. This coiled polypeptide chain associates with a heme group, an iron-containing group that strongly attracts oxygen. There is a whole family of globin molecules, identical in most parts of their amino acid sequences but unique in other parts. In this diagram, the artist drew a transparent "green noodle" around each chain to help you visualize how it folds in three dimensions.

(**b**) Hemoglobin, a protein that transports oxygen in blood, consists of four polypeptide chains (globin molecules) and four heme groups. Two chains, designated *alpha*, differ slightly from the other two (the *beta* chains) in their amino acid sequence. To keep this sketch from looking like a tangle of noodles, the chains in the background are tinted differently from those in the foreground.

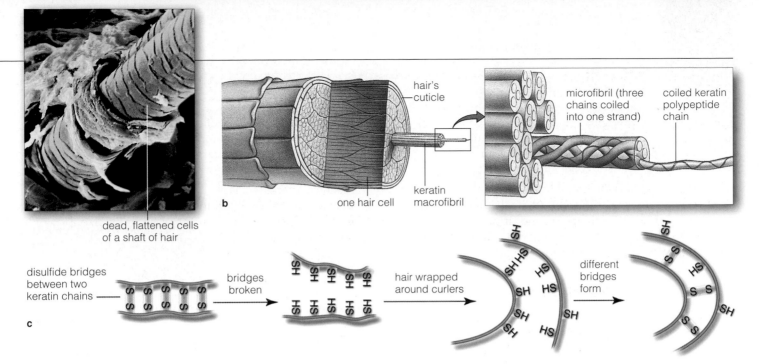

dead, flattened cells
of a shaft of hair

hair's
cuticle

microfibril (three
chains coiled
into one strand)

coiled keratin
polypeptide
chain

one hair cell

keratin
macrofibril

b

disulfide bridges
between two
keratin chains

bridges
broken

hair wrapped
around curlers

different
bridges
form

c

(Figure 3.18b). As you read this, each of your mature red blood cells is transporting a billion molecules of oxygen, bound to 250 million hemoglobin molecules.

Hemoglobin is a fine example of the fourth level of protein structure (quaternary structure). In all proteins at this level of organization, two or more polypeptide chains have become joined together by numerous weak interactions (such as hydrogen bonds) and sometimes by covalent bonds between sulfur atoms of R groups. Hemoglobin has four polypeptide chains.

Like most enzymes, hemoglobin is globular. Many other proteins having quaternary structure are fibrous. Keratin, a structural protein of hair, is like this (Figure 3.19). So is collagen, the most common animal protein. Skin, bone, corneas, arteries, and many other body parts depend on the strength inherent in collagen.

Glycoproteins and Lipoproteins

Some proteins have other organic compounds attached to their polypeptide chains. For example, lipoproteins form when certain proteins circulating in blood combine with cholesterol, triglycerides, and phospholipids that were absorbed from the gut after a meal. Similarly, most glycoproteins have linear or branched oligosaccharides bonded to them. Nearly all the proteins at the surface of animal cells are glycoproteins. So are most protein secretions from cells and many proteins in blood.

Structural Changes by Denaturation

Breaking weak bonds of a protein or any other large molecule disrupts its three-dimensional shape, an event called **denaturation**. For example, weak hydrogen bonds are sensitive to increases or decreases in temperature and pH. If the temperature or pH exceeds a protein's range of tolerance, its polypeptide chains will unwind

Figure 3.19 Structure of hair. (**a,b**) Hair cells develop from modified skin cells. They synthesize polypeptide chains of the protein keratin. Disulfide bridges link three chains together as fine fibers, which are bundled into larger, cablelike fibers. The cable-like fibers almost fill the cells, which eventually die off. (**a**) Dead, flattened cells form a tubelike cuticle around the developing hair shaft.

(**c**) For a permanent wave, hair is exposed to chemicals that break the disulfide bridges. When hairs wrap around curlers, their polypeptide chains are held in new positions. Now exposure to a different chemical causes new disulfide bridges to form. But they form between different sulfur-bearing amino acids than before. The displaced bonding locks the hair in curled positions (compare Figure 3.16).

or change shape, and the protein will lose its function. Consider the protein albumin, concentrated in the "egg white" of uncooked chicken eggs. When you cook eggs, the heat does not disrupt the strong covalent bonds of albumin's primary structure. But it destroys weaker bonds contributing to the three-dimensional shape. For some proteins, denaturation might be reversed when normal conditions are restored—but albumin isn't one of them. There is no way to uncook a cooked egg.

Proteins have a primary structure, which is the sequence of different kinds of amino acids along a polypeptide chain. An individual's DNA specifies that sequence.

Proteins have a secondary structure, a coiled pattern or an extended, sheetlike pattern that arises by hydrogen bonding at short, regular intervals along a polypeptide chain.

At the third level of protein structure, bend-producing amino acids make a coiled chain loop at certain angles and in certain directions, and interactions among R groups in the chain hold the loops in characteristic positions.

At the fourth level of protein structure, numerous hydrogen bonds and other interactions join together two or more polypeptide chains. Many proteins, including hemoglobin, are this structurally complex.

3.6 FOOD PRODUCTION AND A CHEMICAL ARMS RACE

The next time you shop for groceries, reflect on what it takes to provide you with your daily supply of organic compounds. For example, those heads of lettuce typically grew in fertilized cropland. Possibly they competed with weeds, which didn't know the nutrients in the fertilizer were meant for the lettuce. Leaf-chewing insects didn't know the lettuce wasn't meant to be their salad bar. Each

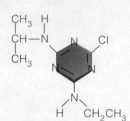

DDT, a nerve cell poison that breaks down after 2 to 15 years

Malathion, a nerve cell poison; lasts 1 to 12 weeks; some kinds persist several years

Atrazine, a herbicide sprayed directly on foliage; it kills weeds within a few days but does not affect corn and other crop plants

Figure 3.20 Lettuce, a common crop plant, and a low-flying crop duster with its rain of pesticides.

year, those food pirates and others ruin or gobble up nearly half of what people all over the world try to grow.

Most plants aren't entirely defenseless. They evolved under intense selection pressure from attacks by insects, fungi, and other organisms, and in natural settings they often can repel the attackers with toxins. A **toxin** is an organic compound, a normal metabolic product of one species, but its chemical effects can harm or kill individuals of a different species that come in contact with it. Humans, too, encounter traces of natural toxins in most of what they eat, even in such familiar edibles as hot peppers, potatoes, figs, celery, rhubarb, and alfalfa sprouts. Still, we do not die in droves from these natural toxins, so apparently our bodies have chemical defenses against them.

In 1945, we took a cue from plants and started using newly developed, synthetic toxins to protect crop yields, food stores, freshwater supplies, and even our health, pets, and ornamental plants. The *herbicides* kill weeds by disrupting metabolism and growth. Most *insecticides* clog a target insect's airways, disrupt functioning of its nerves and muscles, or block its reproduction. *Fungicides* work against harmful fungi, including a mold that makes aflatoxin, one of the deadliest poisons. By 1995, people in the United States were spraying or spreading more than 1.25 billion pounds of these toxins through their fields, gardens, homes, and industrial sites (Figure 3.20).

There are downsides to pesticide applications. Right along with the pests, some of the toxic organic compounds kill birds and other predators which, in natural settings, help control pest population sizes. Besides, targeted pests have been developing resistance to the chemical arsenal, for reasons that you read about earlier, in Chapter 1.

Also, pesticides cannot be released haphazardly into the environment, for people might inhale them, ingest them with food, or absorb them through the skin. Different types stay active for weeks or years (Figure 3.20). Some trigger rashes, hives, sickening headaches, asthma, and joint pain in millions of people. Some can trigger life-threatening allergic reactions in a number of abnormally sensitive individuals.

Presently, the long-lived pesticides are banned in the United States. Even rapidly degradable ones, such as malathion and other organophosphates, are subjected to rigorous application standards and continuing safety tests. However, with so many pests around us, the search for effective countermeasures goes on.

Nucleotides are small organic compounds with a sugar, at least one phosphate group, and a base. The sugar is either ribose or deoxyribose. Both sugars have a five-carbon ring structure. The only difference is that ribose has an oxygen atom attached to carbon 2 in the ring and deoxyribose does not. The bases have a single or double carbon ring structure that incorporates nitrogen.

One nucleotide, **ATP** (for adenosine triphosphate), has three phosphate groups attached to its sugar:

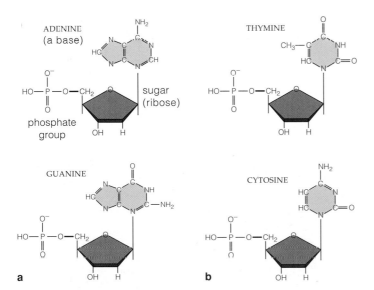

ATP molecules can readily transfer a phosphate group to many other molecules inside the cell, whereupon the acceptor molecules become energized enough to enter into a reaction. Thus ATP is central to metabolism.

Different nucleotides are building blocks for single- or double-stranded molecules classified as **nucleic acids**. In the backbone of such strands, each nucleotide's sugar component is covalently bonded to a phosphate group of the adjacent nucleotides (Figures 3.21 and 3.22). Most likely, you have heard about **DNA** (deoxyribonucleic acid). This molecule is composed of two strands of nucleotides, twisted together helically (Figure 3.22*b*). Hydrogen bonds between the nucleotide bases join the strands. Heritable (genetic) information is encoded in the base sequences. In some parts of the molecule, the

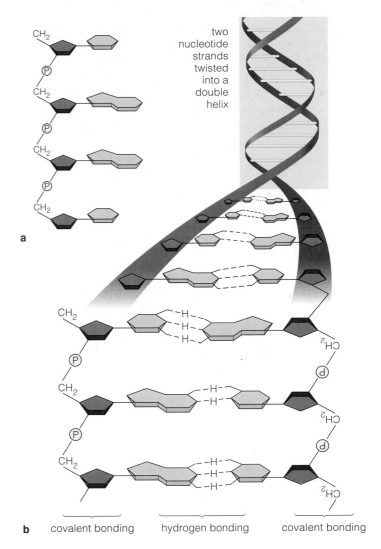

a

b   covalent bonding   hydrogen bonding   covalent bonding

Figure 3.22   (**a**) How nucleotides are joined in each strand of DNA or RNA. (**b**) Bonding patterns within the double-stranded DNA molecule. Hydrogen bonds connect bases (*blue*) of one strand with bases of the other strand. The base pairs are like rungs of a ladder. The sugar-phosphate backbones are like the ladder's posts. The "ladder" twists in the shape of a double helix.

Figure 3.21   The four kinds of nucleotides that serve as the building blocks for DNA molecules. Two nucleotide bases, adenine and thymine, have a double ring structure. The other two, thymine and cytosine, have a single ring structure.

sequence of particular bases is unique to each species. Unlike DNA, the **RNAs** (ribonucleic acids) usually are single strands of nucleotides. Ever since living cells first appeared, RNAs have functioned in processes by which genetic information is used to build proteins.

Nucleotides are small, nitrogen- and phosphorus-containing organic compounds. They are building blocks for DNA and RNA. Some, including ATP, have central roles in metabolism.

DNA, a double-stranded nucleic acid, has genetic information encoded in its sequence of nucleotide bases. The RNAs are single-stranded nucleic acids with roles in the processes by which DNA's genetic information is used to build proteins.

SUMMARY

1. Organic compounds consist of one or more elements covalently bonded to carbon atoms that commonly form linear and ring-shaped backbones. Different functional groups attached to the carbon backbone impart diverse properties to organic compounds.

2. In nature, only living cells can assemble the large organic compounds called biological molecules—the complex carbohydrates and lipids, proteins, and nucleic acids. To do so, they draw from small pools of organic compounds, which include simple sugars, fatty acids, amino acids, and nucleotides. Table 3.1 summarizes the major categories of these compounds.

3. Cells utilize the simple sugars (such as glucose) and oligosaccharides (such as the disaccharide sucrose) as building blocks and transportable forms of energy.

4. Cells use complex carbohydrates (including cellulose and starch) as structural materials and as forms of energy storage. They use lipids (including triglycerides) as energy storage forms, as structural components of cell membranes, and as precursors of other compounds.

5. Proteins, built of amino acids, are the most diverse biological molecules. They function in enzyme activity, structural support, transport, cell movements, changes in cell shape and activities, and defense against disease. The nucleic acids DNA and RNA, built of nucleotides, are the basis of inheritance and reproduction.

Table 3.1 Summary of the Main Organic Compounds in Living Things

Category	Main Subcategories	Some Examples and Their Functions	
CARBOHYDRATES . . . contain an aldehyde or a ketone group, and one or more hydroxyl groups	**Monosaccharides** (simple sugars)	Glucose	Energy source
	Oligosaccharides	Sucrose (a disaccharide)	Form of sugar transported in plants
	Polysaccharides (complex carbohydrates)	Starch, cellulose, glycogen	Energy storage Structural roles
LIPIDS . . . are largely hydrocarbon; generally do not dissolve in water but dissolve in nonpolar substances	**Lipids with fatty acids:** *Glycerides:* one, two, or three fatty acid tails attached to glycerol backbone	Fats (e.g., butter) Oils (e.g., corn oil)	Energy storage
	Phospholipids: phosphate group, one other polar group, and (often) two fatty acids attached to glycerol backbone	Phosphatidylcholine	Key component of cell membranes
	Waxes: long-chain fatty acid tails attached to alcohol	Waxes in cutin	Water retention by plants
	Lipids with no fatty acids: *Sterols:* four carbon rings; the number, position, and type of functional groups differ among various sterols	Cholesterol	Component of animal cell membranes; precursor of many steroids and of vitamin D
PROTEINS . . . are polypeptides (up to several thousand amino acids, covalently linked)	**Fibrous proteins:** Individual polypeptide chains, often linked into tough, water-insoluble molecules	Keratin Collagen	Structural element of hair, nails Structural element of bone, cartilage
	Globular proteins: One or more polypeptide chains folded and linked into globular shapes; many roles in cell activities	Enzymes Hemoglobin Insulin Antibodies	Increase in rates of reactions Oxygen transport Control of glucose metabolism Tissue defense
NUCLEIC ACIDS (AND NUCLEOTIDES) . . . are chains of units (or individual units) that each consist of a five-carbon sugar, phosphate, and a nitrogen-containing base	**Adenosine phosphates**	ATP cAMP (Section 31.2)	Energy carrier Messenger in hormone regulation
	Nucleotide coenzymes	NAD^+, $NADP^+$, FAD	Transport of protons (H^+), electrons from one reaction site to another
	Nucleic acids: Chains of thousands to millions of nucleotides	DNA, RNAs	Storage, transmission, translation of genetic information

Review Questions

1. Define organic compound. Name the type of chemical bond that predominates in the backbone of such a compound. *3.1*

2. Name the "molecules of life." Do they break apart most easily at their hydrocarbon portion or at functional groups? *3.1*

   Select one of the carbohydrates, lipids, proteins, or nucleic ids described in this chapter. Speculate on how its functional oups and bonds between the carbon atoms in its backbone contribute to its final shape and function. *3.2 through 3.7*

4. Which item listed includes all of the other items listed? *3.3*
   - a. triglyceride
   - b. fatty acid
   - c. wax
   - d. sterol
   - e. lipid
   - f. phospholipid

5. Explain how a hemoglobin molecule's three-dimensional shape arises, starting with its primary structure. *3.5*

Self-Quiz *(Answers in Appendix III)*

1. Each carbon atom can share pairs of electrons with as many as _____ other atoms.
   - a. one
   - b. two
   - c. three
   - d. four

2. Hydrolysis is a _____ reaction.
   - a. functional group transfer
   - b. electron transfer
   - c. rearrangement
   - d. condensation
   - e. cleavage
   - f. both b and d

3. _____ are simple sugars (monosaccharides).
   - a. glucose
   - b. sucrose
   - c. ribose
   - d. chitin
   - e. both a and b
   - f. both a and c

4. In unsaturated fats, fatty acid tails have one or more _____
   - a. single covalent bonds
   - b. double covalent bonds

5. _____ are to proteins as _____ are to nucleic acids.
   - a. sugars; lipids
   - b. sugars; proteins
   - c. amino acids; hydrogen bonds
   - d. amino acids; nucleotides

6. Nucleotides include _____ .
   - a. ATP
   - b. DNA subunits
   - c. RNA subunits
   - d. all of the above

7. A denatured protein or DNA molecule has lost its _____ .
   - a. hydrogen bonds
   - b. shape
   - c. function
   - d. all of the above

8. Match each molecule with the most suitable description.
   - (1) _____ long sequence of amino acids
   - (2) _____ the main energy carrier
   - (3) _____ glycerol, fatty acids, phosphate
   - (4) _____ two strands of nucleotides
   - (5) _____ one or more sugar monomers
   - a. carbohydrate
   - b. phospholipid
   - c. protein
   - d. DNA
   - e. ATP

Critical Thinking

1. Jack decided to celebrate summer by making a crab salad and a peach pie for some friends. He had such a good time, he forgot to put the cap on the bottle of olive oil after making the salad. He also didn't tighten the lid on the shortening can after making the pie crust. A few weeks later, the oil had turned rancid but the shortening still smelled okay. Both substances are fats, both were stored in a dark cupboard, at room temperature; yet one spoiled and the other stayed fresh. Speculate on which molecular bonds in these substances were the starting point for the difference.

2. In the following list, identify which is the carbohydrate, fatty acid, amino acid, and polypeptide:
   - a. $^+NH_3$—CHR—COO$^-$
   - b. $C_6H_{12}O_6$
   - c. (glycine)$_{20}$
   - d. $CH_3(CH_2)_{16}COOH$

3. A clerk in a health-food store tells you that certain "natural" vitamin C tablets extracted from rose hips are better for you than synthetic vitamin C tablets. Given your understanding of the structure of organic compounds, what would be your response? Design an experiment to test whether these vitamins differ.

4. Rabbits that eat green, leafy vegetables containing xanthophyll accumulate this yellow pigment molecule in body fat but not in muscles. What chemical properties of the molecule might cause this selective accumulation?

5. Cows can digest grasses, but people cannot. The cow's four-chambered stomach houses populations of certain microorganisms that are not normal residents of the human stomach. What kind of metabolic reactions do you think the microorganisms carry out? What do you think might happen to a cow undergoing treatment with an antibiotic that killed the microorganisms?

6. A Gary Larsen cartoon states that sheep with steel wool have no natural enemies. You might laugh at the thought of such a preposterous animal—but what exactly is wool? It is the keratin-rich, soft undercoat of sheep, Angora goats, llamas, and other hairy mammals. Keratin, a protein, is a long polypeptide chain. Along its length, many hydrogen bonds hold it in a helical shape. Visualize three such chains coiled together, with many disulfide bridges in the end regions stabilizing the trio in a tight, ropelike array. Each hair has linear aggregates of the ropelike fibers, which resist stretching. Given this structure for keratin, speculate why, when you run a wool sweater through the hot cycle of a dryer, it shrinks, pathetically and permanently.

7. Reflect on the component atoms of the different classes of organic compounds. Now look at the structural formulas for the pesticides shown in Figure 3.20. Which one is a glycoprotein? An organophosphate? A chlorinated hydrocarbon?

Selected Key Terms

alcohol *3.1*	functional group *3.1*	polypeptide chain *3.4*
amino acid *3.4*	hemoglobin *3.5*	polysaccharide *3.2*
ATP *3.7*	hydrolysis *3.1*	protein *3.4*
carbohydrate *3.2*	lipid *3.3*	RNAs *3.7*
condensation reaction *3.1*	monosaccharide *3.2*	sterol *3.3*
denaturation *3.3*	nucleic acid *3.7*	toxin *3.6*
DNA *3.7*	nucleotide *3.7*	triglyceride *3.3*
enzyme *3.1*	oligosaccharide *3.2*	wax *3.3*
fat *3.3*	organic compound *3.1*	
fatty acid *3.3*	phospholipid *3.3*	

Readings   *See also www.infotrac-college.com*

Atkins, P. 1987. *Molecules.* New York: Scientific American Library. A molecular "glossary" includes the formula, three-dimensional stucture, and action of many organic compounds.

"The Molecules of Life." October 1985. *Scientific American.*

Ritter, P. 1996. *Biochemistry: An Introduction.* Pacific Grove, California: Brooks/Cole. Chockful of human applications.

Wolfe, S. 1995. *Introduction to Molecular and Cellular Biology.* Belmont, California: Wadsworth.

WWW   *http://www.brookscole.com/biology*

Practice quiz questions, hypercontents, BioUpdates, and critical thinking. The Brooks/Cole Biology Resource Center provides a wealth of information fully organized and integrated by chapter.

CELL STRUCTURE AND FUNCTION

Animalcules and Cells Fill'd With Juices

Early in the seventeenth century, a scholar by the name of Galileo Galilei arranged two glass lenses within a cylinder. With this instrument he happened to look at an insect, and later he described the stunning geometric patterns of its tiny eyes. Thus Galileo, who was not a biologist, was among the first to record a biological observation made through a microscope. The study of the cellular basis of life was about to begin. First in Italy, then in France and England, scholars set out to explore a world whose existence had not even been suspected.

At midcentury Robert Hooke, Curator of Instruments for the Royal Society of England, was at the forefront of these studies. When Hooke first turned a microscope to thinly sliced cork from a mature tree, he observed tiny compartments (Figure 4.1c). He gave them the Latin name *cellulae*, meaning small rooms—hence the origin of the biological term "cell." They

actually were the interconnecting walls of dead plant cells, which is what cork is made of, but Hooke did not think of them as being dead because neither he nor anyone else at the time knew that cells could be alive. In still other plant tissues, he observed cells "fill'd with juices" but could not imagine what they represented.

Given the simplicity of their instruments, it is just amazing that the pioneers in microscopy observed as much as they did. Antony van Leeuwenhoek, a Dutch shopkeeper, had exceptional skill in constructing lenses and possibly the keenest vision of all (Figure 4.1a). By the late 1600s, he was discovering natural wonders everywhere, including "many very small animalcules, the motions of which were very pleasing to behold," in scrapings of tartar from his own teeth. Elsewhere he observed a variety of protistans, sperm, and even a bacterium—an organism so small that it would not be seen again for another two centuries!

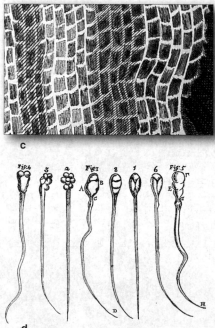

Figure 4.1 Early glimpses into the world of cells. (**a**) Antony van Leeuwenhoek, with microscope in hand. (**b**) Robert Hooke's compound microscope and (**c**) his drawing of cell walls from cork tissue. (**d**) One of van Leeuwenhoek's early sketches of sperm cells. (**e**) Cartoon evidence of the startling impact of microscopic observations on nineteenth-century London.

In the 1820s, improvements in lenses brought cells into sharper focus. Robert Brown, a botanist, was noticing an opaque spot in a variety of cells. He called it a nucleus. In 1838 still another botanist, Matthias Schleiden, wondered if the nucleus had something to do with a cell's development. As he hypothesized, each plant cell must develop as an independent unit even though it is part of the plant.

By 1839, after years of studying animal tissues, the zoologist Theodor Schwann had this to say: Animals as well as plants consist of cells and cell products—and even though the cells are part of a whole organism, to some extent they have an individual life of their own.

A decade later a question remained: Where do cells come from? Rudolf Virchow, a physiologist, completed his own studies of a cell's growth and reproduction— that is, its division into two daughter cells. Every cell, he reasoned, comes from a cell that already exists.

And so, by the middle of the nineteenth century, microscopic analysis had yielded three generalizations, which together constitute the **cell theory**. *First, every organism is composed of one or more cells. Second, the cell is the smallest unit having the properties of life. Third, the continuity of life arises directly from the growth and division of single cells.* All three insights still hold true.

This chapter is not meant for memorization. Read it simply to gain an overview of current understandings of cell structure and function. In later chapters, you can refer back to it as a road map through the details. With its images from microscopy, this chapter and others in the book can transport you into spectacular worlds of juice-fill'd cells and animalcules.

KEY CONCEPTS

1. All organisms consist of one or more cells. The cell is the smallest unit that still retains the characteristics of life. And each new cell arises from a preexisting cell. These are the three generalizations of the cell theory.

2. All cells have an outermost, double-layered membrane. This plasma membrane separates the cell interior from the surroundings. In addition, cells contain cytoplasm, an organized internal region where energy conversions, protein synthesis, movements, and other activities necessary for survival proceed. In eukaryotic cells only, DNA is enclosed within a membrane-bound nucleus. In prokaryotic cells (bacteria), DNA is concentrated in part of the cell interior.

3. The plasma membrane and internal cell membranes consist largely of phospholipid and protein molecules. The phospholipids form two adjacent layers that give the membrane its basic structure and prevent water-soluble substances from freely crossing it. Proteins embedded in cell membranes or positioned at their surfaces perform most membrane functions.

4. The nucleus is one of many organelles. Organelles are membrane-bound compartments inside eukaryotic cells. They physically separate different metabolic reactions and allow them to proceed in orderly fashion. Bacteria do not have comparable organelles.

5. When cells grow, they increase faster in volume than in surface area. This physical constraint on increases in size influences cell size and shape.

6. Different microscopes modify light rays or accelerated beams of electrons in ways that allow us to form images of incredibly small specimens. They are the foundation for our current understanding of cell structure and function.

e

BASIC ASPECTS OF CELL STRUCTURE AND FUNCTION

Inside your body and at its moist surfaces, trillions of cells live in interdependency. In northern forests, four-cell structures—pollen grains—escape from pine trees. In scummy pondwater, a single-celled amoeba moves freely, thriving on its own. For humans, pines, amoebas, and all other organisms, the **cell** is the smallest unit that retains the properties of life (Figure 4.2). Each living cell can survive on its own or has the potential to do so. Its structure is highly organized for metabolism. It senses and responds to its environment. And, based upon inherited instructions in its DNA, it has the potential to reproduce.

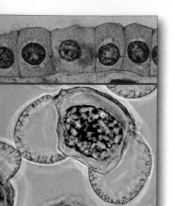

Figure 4.2 *Above:* Cross-section of cells lining a tiny tube inside a kidney. *Below:* A pollen grain. One of its cells will mature into a sperm that may meet up with an egg and help start a new pine tree.

Structural Organization of Cells

Cells differ enormously in size, shape, and activities, as you might gather by comparing a tiny bacterium with one of your relatively giant liver cells. Yet they are alike in three respects. All cells start out life with a plasma membrane, a region of DNA, and a region of cytoplasm:

1. **Plasma membrane.** This thin, outermost membrane maintains the cell as a distinct entity. By doing so, it allows metabolic events to proceed apart from random events in the environment. A plasma membrane does not isolate the cell interior. Substances and signals continually move across it in highly controlled ways.

2. **DNA-containing region.** DNA, with its heritable instructions, occupies part of the cell interior, along with molecules that can copy or read the instructions.

3. **Cytoplasm.** Cytoplasm is everything between the plasma membrane and region of DNA. It consists of a semifluid matrix and other components, such as **ribosomes** (structures on which proteins are built).

This chapter introduces two fundamentally different kinds of cells. The cytoplasm of **eukaryotic cells** includes organelles, which are tiny sacs and other compartments bounded by membranes. One sac, the nucleus, houses the DNA and is the defining feature of eukaryotic cells.

By contrast, **prokaryotic cells** do not have a nucleus; that is, no membranes intervene between their region of DNA and the cytoplasm around it. Bacteria are the only prokaryotic cells. Beyond the bacterial realm, all other organisms—from the amoebas to peach trees and puffball mushrooms to zebras—are eukaryotes.

Fluid Mosaic Model of Cell Membranes

Fluid bathes the two surfaces of a cell membrane and is vital for its functioning. The membrane itself has a fluid quality; it is not a solid, rigid barrier between the cytoplasmic and extracellular fluids. Why is this so? For the answer, start with the phospholipids.

A phospholipid, recall, has a phosphate-containing head and two fatty acid tails, all attached to a glycerol backbone (Figure 4.3a). The hydrophilic head dissolves in water. The tails are hydrophobic; water repels them. Immerse many phospholipids in water, and they interact with the water molecules and one another until they spontaneously cluster in a sheet or film at the surface. Their jostlings may even force them below the water's surface, where they become organized in two layers—with all the hydrophobic tails sandwiched between all the heads. This **lipid bilayer** arrangement, sketched in Figure 4.3b, is the structural basis of cell membranes.

Figure 4.4 shows a bit of membrane corresponding to the **fluid mosaic model**. By this model, membranes have a mixed composition—a "mosaic"—of diverse phospholipids, glycolipids, sterols, and proteins. Phospholipids are the most abundant of the components. Cholesterol is the most common sterol in animal cell membranes. In plants, phytosterols are their equivalent. Also by this model, a cell membrane has a "fluid" quality because of the motions

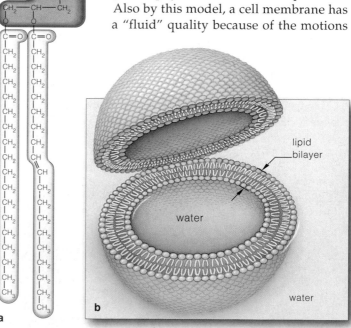

Figure 4.3 (a) Structural formula of phosphatidylcholine, a phospholipid that is one of the most common molecules of animal cell membranes. *Orange* indicates its hydrophilic head; *yellow* indicates its hydrophobic tails. (b) Diagram showing how lipids that are placed in liquid water may spontaneously organize themselves into a bilayer structure.

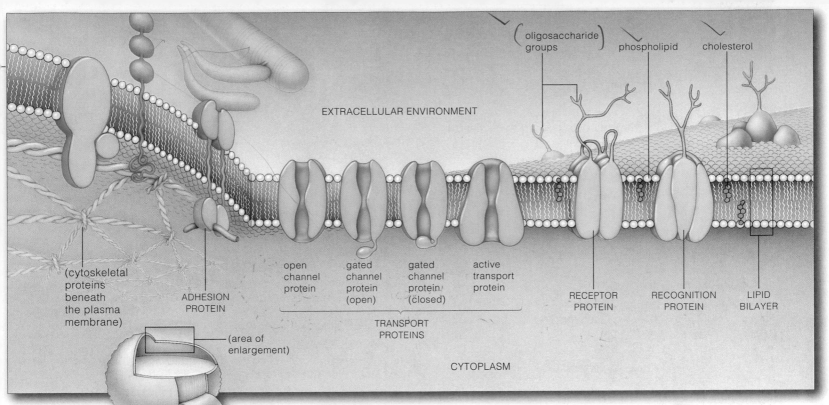

oligosaccharide groups   phospholipid   cholesterol

EXTRACELLULAR ENVIRONMENT

(cytoskeletal proteins beneath the plasma membrane)

ADHESION PROTEIN

open channel protein

gated channel protein (open)

gated channel protein (closed)

active transport protein

RECEPTOR PROTEIN

RECOGNITION PROTEIN

LIPID BILAYER

TRANSPORT PROTEINS

(area of enlargement)

CYTOPLASM

PLASMA MEMBRANE

Figure 4.4 Cutaway view of part of a plasma membrane, based on the fluid mosaic model. Besides the specialized proteins shown, enzymes also are associated with cell membranes.

and interactions of its component parts. Hydrophobic interactions that give rise to most of the membrane's structure are weaker than covalent bonds. This means most phospholipids and some proteins are free to drift sideways. Also, the phospholipids can spin about their long axis and flex their tails, which keeps neighboring molecules from packing together in a solid layer. Short or kinked (unsaturated) hydrophobic tails contribute to the membrane fluidity.

The fluid mosaic model is a good starting point for exploring membranes. But bear in mind, cell membranes differ in composition and molecular arrangements. They are not even the same on both surfaces of their bilayer. For example, oligosaccharides and other carbohydrates are covalently bonded to protein and lipid components of a plasma membrane, but only on its outward-facing surface (Figure 4.4). Moreover, they differ in number and kind from one species to the next, even among the different cells of the same individual.

Overview of Membrane Proteins

The proteins embedded in a lipid bilayer or attached to one of its surfaces carry out most membrane functions. Many are enzyme components of metabolic machinery. Others are *transport* proteins that span the bilayer and allow water-soluble substances to move through their interior. They bind molecules or ions on one side of the membrane, then release them on the other side.

The *receptor* proteins bind extracellular substances, such as hormones, that trigger changes in cell activities. For example, certain enzymes that crank up machinery

for cell growth and division become switched on when somatotropin, a hormone, binds with receptors for it. Different cells have different combinations of receptors.

Diverse *recognition* proteins at the cell surface are like molecular fingerprints; their oligosaccharide chains identify a cell as being of a specific type. For example, the plasma membrane of your cells bristles with "self" proteins. Certain white blood cells chemically recognize the proteins and leave your cells alone, but they attack invading bacterial cells bearing "nonself" proteins at their surface. Finally, *adhesion* proteins of multicelled organisms help cells of the same type locate and stick to one another and stay positioned in the proper tissues. They are glycoproteins, with oligosaccharides attached. After tissues form, the sites of adhesion may become a type of cell junction, as described in Section 4.10.

All cells have an outermost plasma membrane, an internal region of cytoplasm, and an internal region of DNA.

Besides the plasma membrane, eukaryotic cells have internal membrane-bound compartments, including a nucleus.

Each membrane has a bilayer structure, consisting largely of phospholipids. The hydrophobic parts of its lipid molecules are sandwiched between the hydrophilic parts, which are dissolved in the fluid surroundings.

A lipid bilayer imparts structure to the cell membrane and serves as a barrier to water-soluble substances.

Proteins associated with the bilayer carry out most membrane functions. Many are enzymes, transporters of substances across the bilayer, or receptors for extracellular substances. Other types function in cell-to-cell recognition or adhesion.

WWW

CELL SIZE AND CELL SHAPE

You may be wondering how small cells really are. Can any be observed with the unaided human eye? There are a few, including the "yolks" of bird eggs, cells in the red part of watermelons, and the fish eggs we call caviar. However, most cells cannot be observed without microscopes. To give you a sense of cell sizes, red blood cells are about 8 millionths of a meter across. You could fit a string of 2,000 of them across your thumbnail!

Why are most cells so small? The **surface-to-volume ratio** constrains increases in their size. By this physical relationship, an object's volume increases with the cube of the diameter, but its surface area increases only with the square. Simply put, *as the diameter of a growing cell expands, the cell's volume increases faster than its surface area does.* Here is an example:

diameter (cm):	0.5	1.0	1.5
surface area (cm^2):	0.79	3.14	7.07
volume (cm^3):	0.06	0.52	1.77
surface-to-volume ratio:	13.17:1	6.04:1	3.99:1

Suppose you figure out a way to make a round cell grow four times wider than it normally would grow. Its volume increases 64 times (4^3), but its surface area only increases 16 times (4^2). As a result, each unit of the cell's plasma membrane must now serve four times as much cytoplasm as it did previously! Moreover, past a certain point, the inward flow of nutrients and outward flow of wastes will not be fast enough, and the cell will die.

A large, round cell also would have trouble moving materials *through* its cytoplasm. In small cells, random, tiny motions of molecules easily distribute materials. If a cell is not small, you usually can expect it to be long and quite thin or to have outfoldings and infoldings that increase its surface area relative to its volume. *The smaller or thinner or more frilly-surfaced the cell, the more efficiently materials cross its surface and become distributed through the interior.*

We see evidence of surface-to-volume constraints on the body plans of multicelled organisms, also. For example, cells attach end to end in strandlike algae, and each one interacts directly with its surroundings. Your skeletal muscle cells are thin, but each one is as long as a biceps or some other muscle of which it is part.

When cells grow, they increase faster in volume than in surface area. The surface-to-volume ratio is a physical constraint on increases in cell size. It influences cell size, cell shape, and the body plans of multicelled organisms.

Focus on Science

MICROSCOPES: GATEWAYS TO CELLS

Modern microscopes are gateways to astounding worlds. Some even afford glimpses into the structure of molecules. Figure 4.5 gives you an idea of the range of magnifications possible. The micrographs in Figures 4.5 and 4.6 hint at the details now being observed. A **micrograph** simply is a photograph of an image that came into view with the aid of a microscope.

LIGHT MICROSCOPES Picture a series of waves moving across an ocean. Each **wavelength** is the distance from one wave's peak to the peak of the wave behind it. Light also travels as waves from sources such as the sun and illuminated specimens. In a *compound light microscope*, two or more sets of glass lenses bend light emanating from a cell or some other specimen in ways that form an enlarged image of it.

A living cell must be small or thin enough for light to pass through. It would help if cell parts differed in color and density from the surroundings, but most are nearly colorless and appear uniformly dense. To get around this problem, microscopists stain cells (expose them to dyes that react with some parts of a specimen but not others). Staining may alter and kill cells. Dead cells break down fast, so they typically are pickled or preserved before being stained.

Suppose you use the best glass lens system. When you magnify the diameter of a specimen by 2,000 times or more, you discover that cell parts appear larger but are not clearer. To understand why, think about the distance between the two crests of a wavelength of light. To give two examples, that distance is about 750 nanometers for red light and 400 nanometers for violet (wavelengths of all other colors fall in between). If a cell structure is less than one-half of a wavelength long, that structure will not be able to disturb the rays of light streaming past, and it will not be visible.

ELECTRON MICROSCOPES Better resolution of very fine details is possible with the assistance of electrons. Remember, electrons are particles, but they also behave like waves. Electron microscopy is based on accelerating the flow of streams of electrons to wavelengths of about 0.005 nanometer—about 100,000 times shorter than those of visible light. The electrons cannot pass through glass lenses, but they can be bent from their paths and focused by a magnetic field.

In a *transmission electron microscope*, a magnetic field is the "lens." Accelerated electrons are directed through a specimen, focused into an image, and magnified. With *scanning electron microscopes*, a narrow beam of electrons moves back and forth across a specimen that has been thinly coated with metal. The metal responds by emitting some of its own electrons. A detector that is connected to electronic circuitry transforms the electron energy into an image of the specimen's surface on a television screen. Most of the scanning images have fantastic depth.

WWW

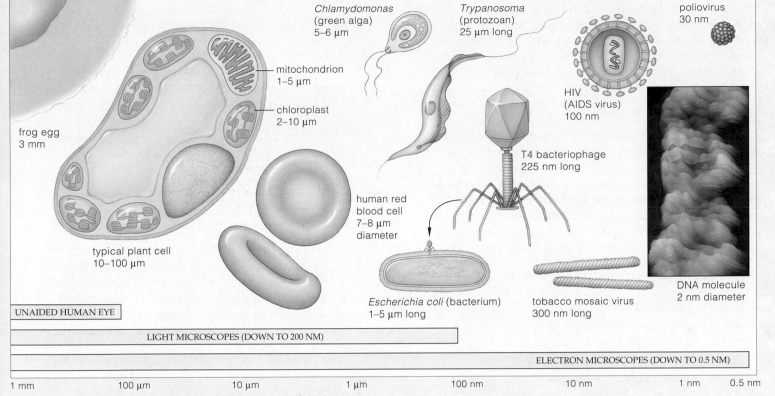

Figure 4.5 Units of measure used in microscopy. A *scanning tunneling microscope*, which can provide magnifications up to 100 million, gave us the photomicrograph of part of a DNA molecule. The scope's needlelike probe has a single atom at its tip. When voltage is applied between the tip and an atom at a specimen's surface, it tunnels into the electron orbitals. As the tip moves over a specimen's contours, a computer analyzes the tunneling motion and creates a three-dimensional view of the surface atoms.

1 centimeter (cm) = 1/100 meter, or 0.4 inch
1 millimeter (mm) = 1/1,000 meter
1 micrometer (μm) = 1/1,000,000 meter
1 nanometer (nm) = 1/1,000,000,000 meter

1 meter = 10^2 cm = 10^3 mm = 10^6 μm = 10^9 nm

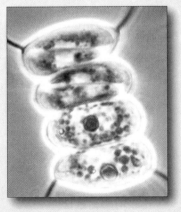

a Light micrograph (phase-contrast process)

b Light micrograph (Nomarski process)

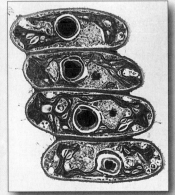

c Transmission electron micrograph, thin section

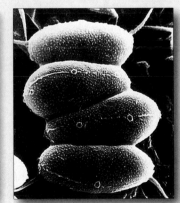

d Scanning electron micrograph

10 μm

Figure 4.6 How different microscopes can reveal different aspects of the same organism—in this case, a green alga (*Scenedesmus*). The images of all four specimens are at the same magnification. The phase-contrast and Nomarski processes mentioned in (**a**) and (**b**) can create optical contrasts without staining the cells. Both processes enhance the usefulness of light micrographs. As for other micrographs in the book, the short horizontal bar below the micrograph in (**d**) provides you with a visual reference for size. A micrometer (μm) is 1/1,000,000 of a meter. Using the scale bar, can you estimate the length and width of *Scenedesmus*?

THE DEFINING FEATURES OF EUKARYOTIC CELLS

We turn now to organelles and other structural features that are typical of the cells of plants, animals, fungi, and protistans. We define an **organelle** as an internal, membrane-bound sac or compartment that serves one or more specialized functions inside eukaryotic cells.

Major Cellular Components

Observe some micrographs of a typical eukaryotic cell, such as the ones included in the preceding section, and you probably will quickly notice that the nucleus is one of the most conspicuous features. Remember, any cell that starts out life with a nucleus is a eukaryotic cell; it contains a "true nucleus." Many other organelles and structures also are typical of these cells, although the numbers and kinds differ from one cell type to the next. Table 4.1 lists the most common features.

Table 4.1 Common Features of Eukaryotic Cells	
ORGANELLES AND THEIR MAIN FUNCTIONS:	
Nucleus	Localizing the cell's DNA
Endoplasmic reticulum	Routing and modifying the newly formed polypeptide chains; also, synthesizing lipids
Golgi body	Modifying polypeptide chains into mature proteins; sorting and shipping proteins and lipids for secretion or for use inside cell
Various vesicles	Transporting or storing a variety of substances; digesting substances and structures in the cell; other functions
Mitochondria	Producing many ATP molecules in highly efficient fashion
NON-MEMBRANOUS STRUCTURES AND THEIR FUNCTIONS:	
Ribosomes	Assembling polypeptide chains
Cytoskeleton	Imparting overall shape and internal organization to cell; moving the cell and its internal structures

Think about the list, and you might well find yourself asking: What is the advantage of partitioning the cell interior with such organelles? *The compartmentalization allows a large number of activities to occur simultaneously in very limited space.* Consider a photosynthetic cell in a leaf. It can put together starch molecules by one set of reactions *and* break them apart by another set. Yet the cell would get nothing if the synthesis and breakdown reactions proceeded at the exact same time on the same starch molecule. Without membranes of organelles, the balance of diverse chemical activities that helps keep eukaryotic cells alive would spiral out of control.

Figure 4.7 *Facing page*: Generalized sketches showing some of the features that are typical of (**a**) many plant cells and (**b**) many animal cells.

Organelle membranes have another function besides physically separating incompatible reactions. They allow compatible and interconnected reactions to proceed at different times. For instance, a plant's photosynthetic cells produce starch molecules in an organelle called a chloroplast, then store and later release starch for use in different reactions in the same organelle.

Which Organelles Are Typical of Plants?

Figure 4.7a can start you thinking about the location of organelles in a typical plant cell. Bear in mind, calling a cell "typical" is like calling a cactus or a water lily or an elm tree a "typical" plant. As is true of animal cells, variations on the basic plan are mind-boggling. With this qualification in mind, also take a close look at the micrograph in Figure 4.8, on the subsequent page. It shows the locations of organelles and structures you are likely to observe in many specialized plant cells.

Which Organelles Are Typical of Animals?

Next, start thinking about the organelles of a typical animal cell, such as the one shown in Figures 4.7b and 4.9. Like the plant cell, it contains a nucleus, numerous mitochondria, and the other components listed in Table 4.1. *The structural similarities point to basic functions that are necessary for survival, regardless of cell type.* We will reflect on this concept throughout the book.

Comparing Figures 4.7 through 4.9 also will give you an initial idea of how plant and animal cells differ in their structure. For example, you will never observe an animal cell surrounded by a cell wall. (You might see many different kinds of fungal and protistan cells with one, however.) What other differences can you identify?

Eukaryotic cells contain a number of organelles, which are internal, membrane-bound sacs and compartments that serve specific metabolic functions.

Organelles physically separate chemical reactions, many of which are incompatible.

Organelles separate different reactions in time, as when certain molecules are put together, stored, then used later in other reaction sequences.

All eukaryotic cells contain certain organelles (such as the nucleus) and structures (such as ribosomes) that perform functions essential for survival. Specialized cells also may incorporate other kinds of organelles and structures.

WWW

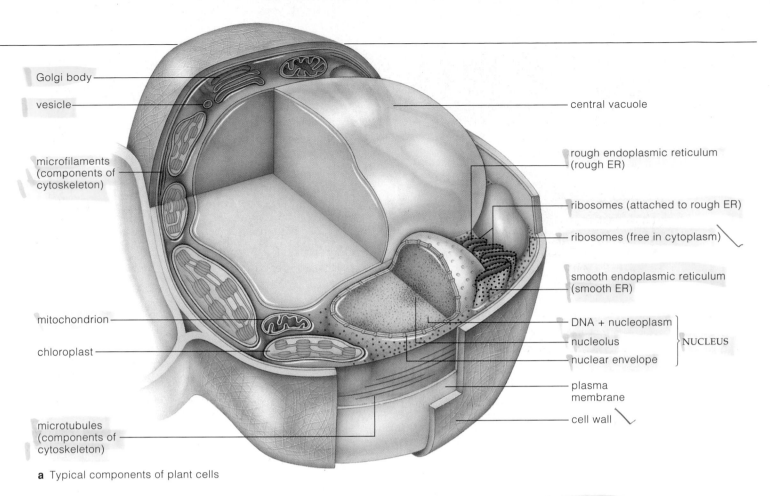

Golgi body

vesicle

microfilaments
(components of
cytoskeleton)

mitochondrion

chloroplast

microtubules
(components of
cytoskeleton)

central vacuole

rough endoplasmic reticulum
(rough ER)

ribosomes (attached to rough ER)

ribosomes (free in cytoplasm)

smooth endoplasmic reticulum
(smooth ER)

DNA + nucleoplasm

nucleolus   NUCLEUS

nuclear envelope

plasma
membrane

cell wall

a Typical components of plant cells

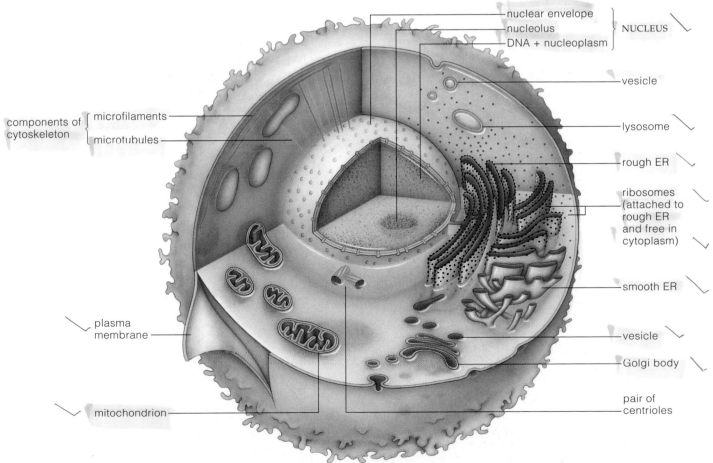

nuclear envelope

nucleolus   NUCLEUS

DNA + nucleoplasm

vesicle

lysosome

rough ER

ribosomes
(attached to
rough ER
and free in
cytoplasm)

smooth ER

vesicle

Golgi body

pair of
centrioles

components of
cytoskeleton
microfilaments

microtubules

plasma
membrane

mitochondrion

b Typical components of animal cells

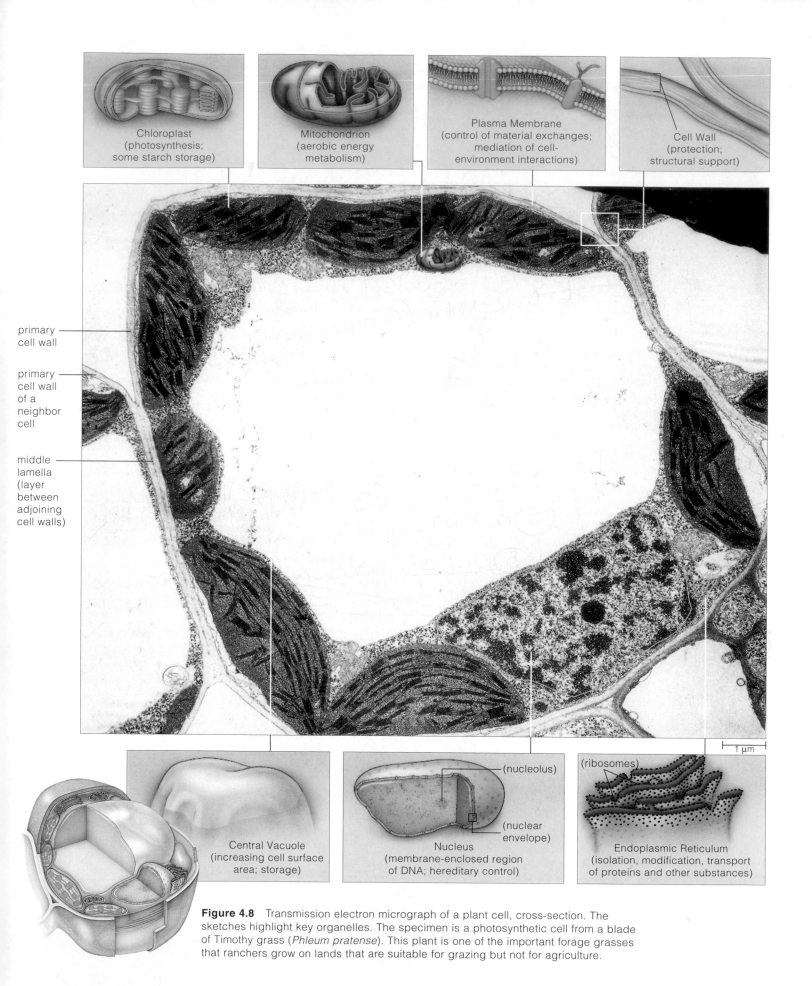

Chloroplast (photosynthesis; some starch storage)

Mitochondrion (aerobic energy metabolism)

Plasma Membrane (control of material exchanges; mediation of cell-environment interactions)

Cell Wall (protection; structural support)

primary cell wall

primary cell wall of a neighbor cell

middle lamella (layer between adjoining cell walls)

1 µm

Central Vacuole (increasing cell surface area; storage)

Nucleus (membrane-enclosed region of DNA; hereditary control)
(nucleolus)
(nuclear envelope)

Endoplasmic Reticulum (isolation, modification, transport of proteins and other substances)
(ribosomes)

Figure 4.8 Transmission electron micrograph of a plant cell, cross-section. The sketches highlight key organelles. The specimen is a photosynthetic cell from a blade of Timothy grass (*Phleum pratense*). This plant is one of the important forage grasses that ranchers grow on lands that are suitable for grazing but not for agriculture.

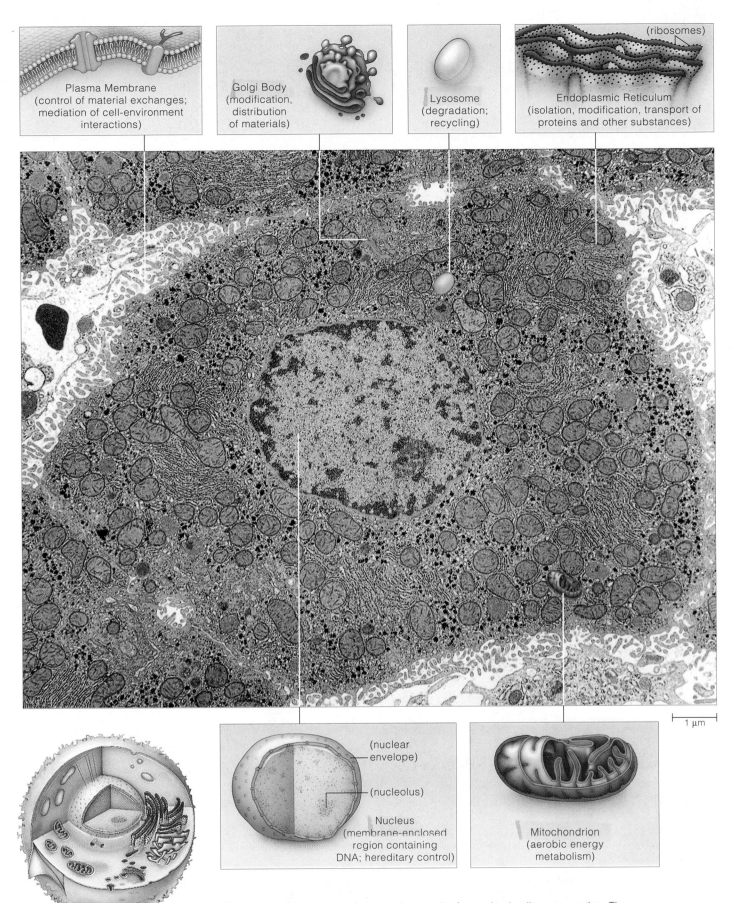

Plasma Membrane (control of material exchanges; mediation of cell-environment interactions)

Golgi Body (modification, distribution of materials)

Lysosome (degradation; recycling)

(ribosomes)

Endoplasmic Reticulum (isolation, modification, transport of proteins and other substances)

1 μm

(nuclear envelope)

(nucleolus)

Nucleus (membrane-enclosed region containing DNA; hereditary control)

Mitochondrion (aerobic energy metabolism)

Figure 4.9 Transmission electron micrograph of an animal cell, cross-section. The sketches highlight key organelles. The specimen is a cell from the liver of a rat.

Constructing, operating, and reproducing cells simply cannot be done without carbohydrates, lipids, proteins, and nucleic acids. It takes a class of proteins—enzymes —to build and use these molecules. Said another way, a cell's structure and function begin with proteins. *And instructions for building proteins are contained in DNA.*

Unlike bacteria, eukaryotic cells have their genetic instructions distributed among several to many DNA molecules of different lengths. For example, each human body cell has forty-six molecules of DNA. Stretch them end to end, and they would be about a meter long. Similarly, stretch the twenty-six DNA molecules in a frog cell end to end and they would extend ten meters. Compared to the lone DNA molecule in a bacterial cell, that's a lot of DNA!

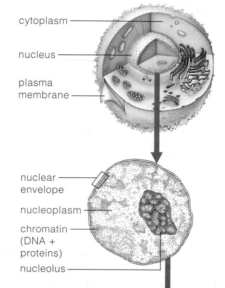

cytoplasm

nucleus

plasma membrane

Figure 4.10 The nucleus of a cell taken from a pancreas. Small arrows on this transmission electron micrograph point to pores where control systems operate to restrict or permit the passage of specific substances across the nuclear envelope.

nuclear envelope

nucleoplasm

chromatin (DNA + proteins)

nucleolus

Eukaryotic cells protect their DNA inside a **nucleus.** Figure 4.10 shows the distinctive structure of this type of organelle, which has two functions. *First*, the nucleus physically tucks away all the DNA molecules, apart from the complex metabolic machinery of the cytoplasm. This localization of DNA makes it easier for parent cells to copy their genetic instructions before the time comes for them to divide. DNA molecules can be sorted out into parcels—one for each daughter cell that forms. *Second*, outer membranes of the nucleus form a boundary where cells can control the passage of substances and signals to and from the cytoplasm.

Nuclear Envelope

Unlike the cell itself, a nucleus has two outer membranes, one wrapped around the other. This double-membrane system is called a **nuclear envelope**. It consists of two lipid bilayers in which numerous protein molecules are embedded (Figure 4.11). It surrounds the fluid portion of the nucleus (the nucleoplasm).

Stitched on the innermost surface of the nuclear envelope are attachment sites for protein filaments. These anchor the molecules of DNA to the envelope and help keep them organized. On the outer surface of the envelope is a profusion of ribosomes. All proteins are built either on such membrane-bound ribosomes or on ribosomes located in the cytoplasm.

As is the case for all cell membranes, a nuclear envelope's lipid bilayers keep water-soluble substances from moving freely into and out of the nucleus. But many pores, each composed of clusters of proteins, span both bilayers. Ions and small, water-soluble molecules cross the nuclear envelope at the pores. As you will see, large molecules (including the subunits of ribosomes) cross the bilayers at pores in highly controlled ways.

Nucleolus

Take a look at the nucleus in Figure 4.10. Inside is a rather round, dense mass of material. What is it? While eukaryotic cells are growing, one or more of these masses form inside its nucleus. Each is a **nucleolus** (plural, nucleoli). Here, a great number of protein and RNA molecules

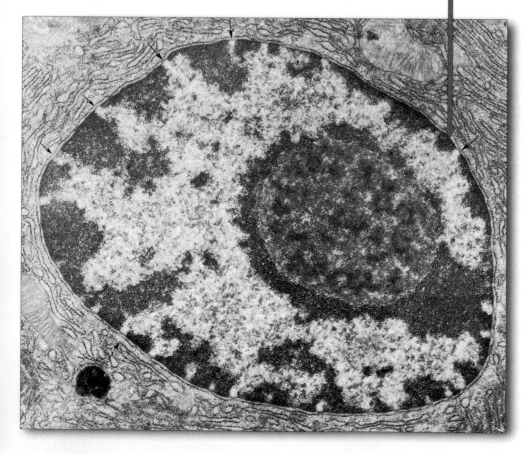

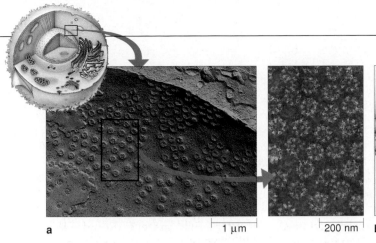

Figure 4.11 (a) Part of the outer surface of a nuclear envelope. *Left*: This specimen was fractured to reveal the intimate layering of its two lipid bilayers. *Right*: Closer view of some nuclear pores. Each pore across the envelope is an organized array of membrane proteins. It permits the selective transport of substances into and out of the nucleus. (b) Sketch of the nuclear envelope's structure.

are being constructed. These particular materials are subunits from which ribosomes are built. The subunits pass through nuclear pores and reach the cytoplasm. In times of protein synthesis, intact ribosomes form (each from two subunits) in the cytoplasm.

Chromosomes

When eukaryotic cells are not busy dividing, their DNA looks like thin threads inside the nucleus. Many protein molecules are attached to the threads, a bit like beads on a string. Except at extreme magnification, the beaded threads have a grainy appearance, as in Figure 4.10. However, when the cell is preparing to divide, it duplicates its DNA molecules so each of its daughter cells will get all of the required hereditary instructions. In addition, each DNA molecule becomes folded and twisted into a condensed structure, proteins and all.

Early microscopists bestowed the name *chromatin* on the seemingly grainy substance and *chromosomes* on the condensed structures. We now define **chromatin** as

Table 4.2 Summary of Components of the Nucleus

Nuclear envelope	Pore-riddled double-membrane system that selectively controls the passage of various substances into and out of the nucleus
Nucleoplasm	Fluid interior portion of the nucleus
Nucleolus	Dense cluster of RNA and proteins that will be assembled into subunits of ribosomes
Chromosome	One DNA molecule and many proteins that are intimately associated with it
Chromatin	Total collection of all DNA molecules and their associated proteins in the nucleus

the cell's collection of DNA, together with all proteins associated with it (Table 4.2). Each **chromosome** is one DNA molecule and its associated proteins, regardless of whether it is in threadlike or condensed form:

one chromosome (one threadlike DNA molecule + proteins; not duplicated)

one chromosome (threadlike but now duplicated; two DNA molecules + proteins)

one chromosome (duplicated and also condensed)

In other words, the appearance of "the chromosome" changes over the life of a eukaryotic cell. In chapters to come, you will look at different aspects of chromosomes, and you may find it useful to remember this point.

What Happens to the Proteins Specified by DNA?

Outside the nucleus, polypeptide chains for proteins are assembled on ribosomes. What happens to the new chains? Many become stockpiled in the cytoplasm or get used at once. Many others enter a cytomembrane system. As described in the next section, this system consists of different organelles, including endoplasmic reticulum, Golgi bodies, and vesicles.

Thanks to DNA's instructions, many proteins take on particular, final forms in the cytomembrane system. Lipids also are packaged and assembled in the system by enzymes and other proteins (which also were built according to DNA's instructions). As you will see next, vesicles deliver the proteins and lipids to specific sites within the cell or to the plasma membrane, for export.

The nucleus, an organelle with two outer membranes, keeps the DNA molecules of eukaryotic cells separated from the metabolic machinery of the cytoplasm.

The localization makes it easier to organize the DNA and to copy it, before a parent cell divides into daughter cells.

Pores across the nuclear envelope help control the passage of many substances between the nucleus and cytoplasm.

THE CYTOMEMBRANE SYSTEM

The **cytomembrane system** is a series of organelles in which lipids are assembled and new polypeptide chains are modified into final proteins. Its products are sorted and shipped to different destinations. Figure 4.12 shows how its organelles—the ER, Golgi bodies, and various vesicles—functionally interconnect with one another.

Endoplasmic Reticulum

The functions of the cytomembrane system begin with **endoplasmic reticulum**, or **ER**. In animal cells, the ER is continuous with the nuclear envelope and extends through cytoplasm. Its membrane regions appear rough or smooth, depending mainly on whether ribosomes are attached to the membrane facing the cytoplasm.

We typically observe *rough* ER arranged into stacks of flattened sacs with many ribosomes attached (Figure 4.13*a*). Every new polypeptide chain is synthesized on ribosomes. But only the newly forming chains having a built-in signal can enter the space within rough ER or become incorporated into ER membranes. (The signal is a string of fifteen to twenty specific amino acids.) Once the chains are in rough ER, enzymes may attach oligosaccharides and other side chains to them. Many specialized cells secrete the final proteins. Rough ER is abundant in such cells. For example, in your pancreas, ER-rich gland cells make and secrete enzymes that end up in the small intestine and help digest your meals.

Smooth ER is free of ribosomes and curves through cytoplasm like connecting pipes (Figure 4.13*b*). Many cells assemble most lipids inside the pipes. Smooth ER is well developed in seeds. In liver cells, some drugs and toxic metabolic wastes are inactivated in it. Sarcoplasmic reticulum, a type of smooth ER in skeletal muscle cells, functions in muscle contraction.

Golgi Bodies

In **Golgi bodies**, enzymes put the finishing touches on proteins and lipids, sort them out, and package them inside vesicles for shipment to specific locations. For example, an enzyme in one Golgi region might attach a phosphate group to a new protein, thereby giving it a mailing tag to its proper destination.

Commonly, a Golgi body looks vaguely like a stack of pancakes; it is composed of a series of flattened membrane sacs (Figure 4.14). In functional terms, the final portion of a Golgi body corresponds to the top pancake. Here, vesicles form as patches of the membrane bulge out, then break away into the cytoplasm.

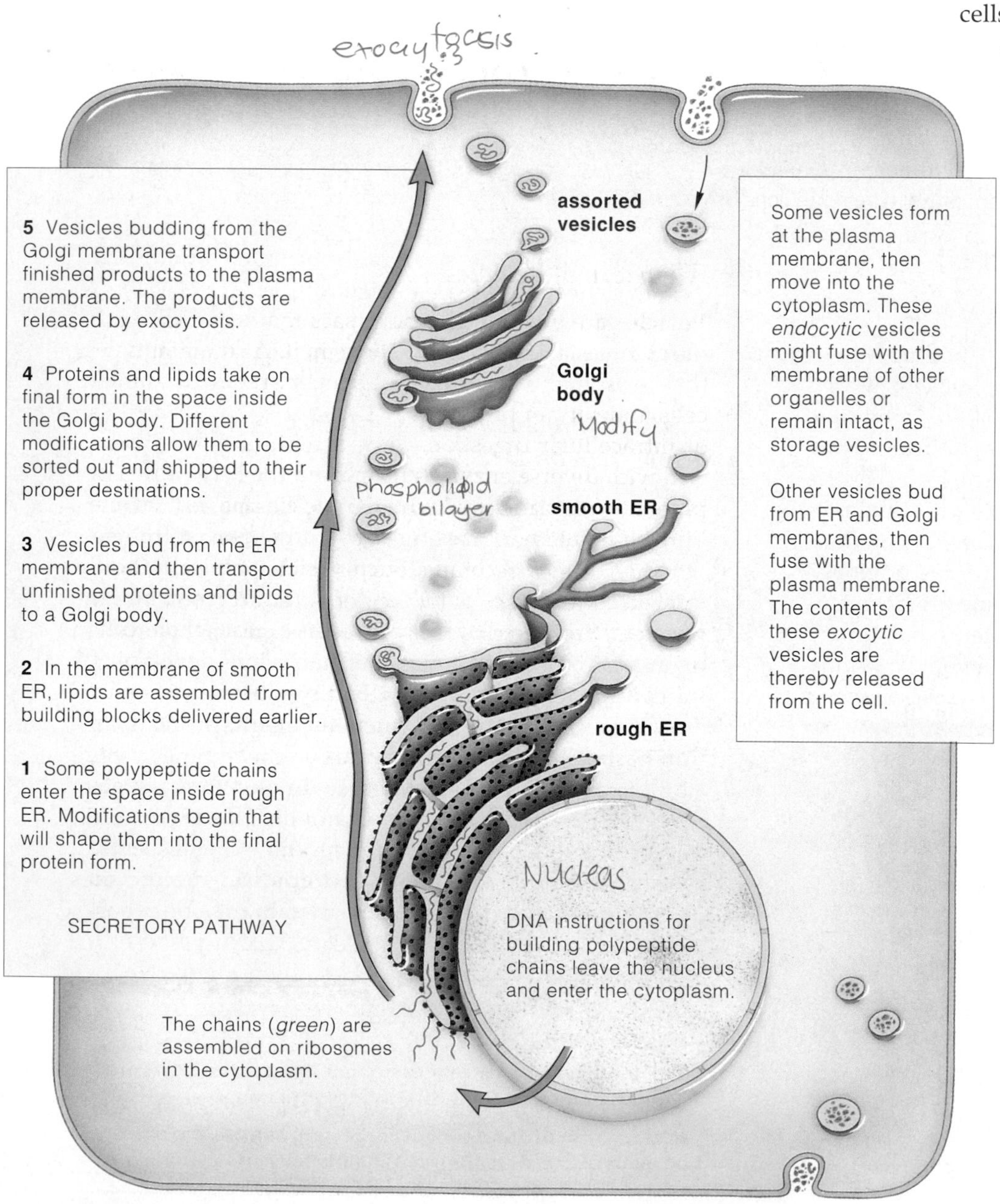

5 Vesicles budding from the Golgi membrane transport finished products to the plasma membrane. The products are released by exocytosis.

4 Proteins and lipids take on final form in the space inside the Golgi body. Different modifications allow them to be sorted out and shipped to their proper destinations.

3 Vesicles bud from the ER membrane and then transport unfinished proteins and lipids to a Golgi body.

2 In the membrane of smooth ER, lipids are assembled from building blocks delivered earlier.

1 Some polypeptide chains enter the space inside rough ER. Modifications begin that will shape them into the final protein form.

SECRETORY PATHWAY

assorted vesicles

Golgi body

smooth ER

rough ER

Nucleus
DNA instructions for building polypeptide chains leave the nucleus and enter the cytoplasm.

The chains (*green*) are assembled on ribosomes in the cytoplasm.

Some vesicles form at the plasma membrane, then move into the cytoplasm. These *endocytic* vesicles might fuse with the membrane of other organelles or remain intact, as storage vesicles.

Other vesicles bud from ER and Golgi membranes, then fuse with the plasma membrane. The contents of these *exocytic* vesicles are thereby released from the cell.

Figure 4.12 Cytomembrane system, a membrane system in the cytoplasm that assembles, modifies, packages, and ships proteins and lipids. *Green* arrows highlight a secretory pathway by which certain proteins and lipids are packaged and then released from many types of cells, including gland cells that secrete mucus, sweat, and digestive enzymes.

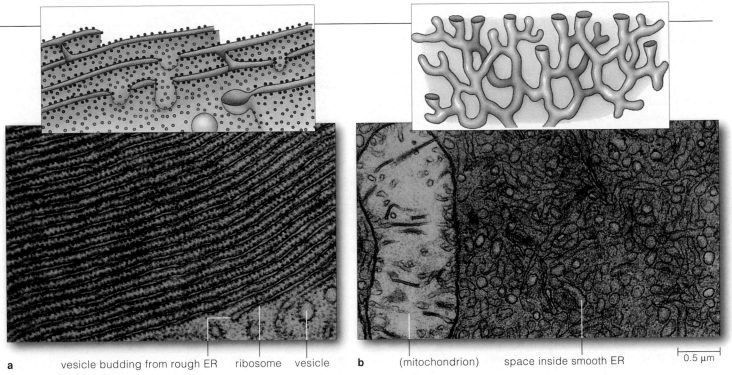

a   vesicle budding from rough ER   ribosome   vesicle   b   (mitochondrion)   space inside smooth ER   ⊢ 0.5 µm ⊣

Figure 4.13 Transmission electron micrographs and sketches of endoplasmic reticulum. (**a**) Many ribosomes dot the flattened surfaces of rough ER that face the cytoplasm. (**b**) This section reveals the diameters of the many interconnected, pipelike regions of smooth ER.

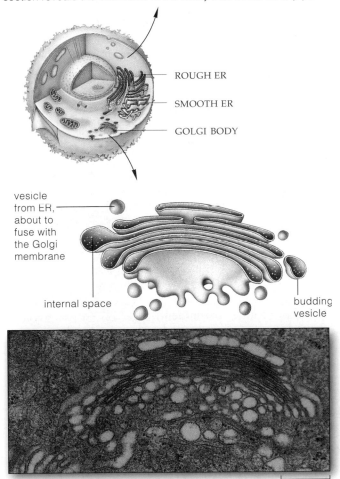

ROUGH ER

SMOOTH ER

GOLGI BODY

vesicle from ER, about to fuse with the Golgi membrane

internal space

budding vesicle

⊢ 0.25 µm ⊣

Figure 4.14 Sketch and micrograph of a Golgi body from an animal cell.

A Variety of Vesicles

Vesicles are tiny, membranous sacs that move through the cytoplasm or take up positions in it. A common type, the lysosome, buds from Golgi membranes of animal cells and certain fungal cells. Lysosomes are organelles of intracellular digestion. They contain a potent brew, rich with diverse enzymes that speed the breakdown of proteins, complex carbohydrates, nucleic acids, and some lipids. Often, lysosomes fuse with vesicles that formed at the plasma membrane. Such vesicles typically hold molecules, bacteria, or other items that docked at the plasma membrane. Lysosomes even digest whole cells or cell parts. For example, as a tadpole is developing into an adult frog, its tail slowly disappears. Lysosomal enzymes respond to developmental signals and help destroy cells that make up the tail.

Peroxisomes, another type, are tiny sacs of enzymes that break down fatty acids and amino acids. Hydrogen peroxide, a potentially harmful product, forms during the reactions. Enzyme action converts it to water and oxygen or channels it into reactions that break down alcohol. After someone drinks alcohol, nearly half of it is degraded in peroxisomes of liver and kidney cells.

In the ER and Golgi bodies of the cytomembrane system, many proteins take on final form and lipids are synthesized.

Lipids, proteins (such as enzymes), and other items become packaged in vesicles destined for export, storage, membrane building, intracellular digestion, and other cell activities.

MITOCHONDRIA

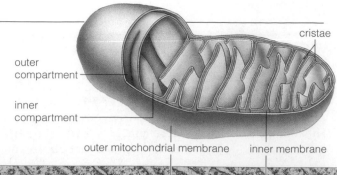

Recall, from Section 3.7, that ATP molecules are premier energy carriers. Energy associated with their phosphate groups can be delivered to nearly all reaction sites and drives nearly all cell activities. Many ATP molecules form when organic compounds are completely broken down to carbon dioxide and water in a **mitochondrion** (plural, mitochondria).

Only eukaryotic cells contain these organelles. The example in Figure 4.15 will give you an idea of their structure. The kind of ATP-forming reactions that proceed in mitochondria extract far more energy from organic compounds than can be done by any other means. They cannot run to completion without plenty of oxygen. Like all other land-dwelling vertebrates, every time you breathe in, you take in oxygen mainly for mitochondria in cells—in your case, many trillions of individual cells.

Each mitochondrion has a double-membrane system. As you can tell from Figure 4.7, the outermost membrane faces the cytoplasm. Most commonly, the inner membrane repeatedly folds back on itself. Each inner fold is a crista (plural, cristae).

What is the function of the intricate membrane system? It forms two distinct compartments within a mitochondrion. Enzymes and other proteins stockpile hydrogen ions in the outer compartment. Electron transfers drive the stockpiling, and oxygen helps keep the machinery running by binding and thus removing the spent electrons. Hydrogen ions flow out of the compartment in controlled ways. Energy inherent in the flow drives ATP formation, as Chapter 7 describes.

All eukaryotic cells contain one or more mitochondria. You may find only one in a single-celled yeast. You might find a thousand or more in energy-demanding cells, such as those of muscles. Take a look at the profusion of mitochondria in Figure 4.9, which is a micrograph of merely one thin slice from a liver cell. It alone tells you that the liver is an exceptionally active, energy-demanding organ.

In terms of size and biochemistry, mitochondria resemble bacteria. They even have their own DNA and some ribosomes, and they divide on their own. Perhaps they evolved from ancient bacteria that were engulfed by a predatory, amoebalike cell yet managed to escape digestion. Perhaps they were able to reproduce inside the cell and continued doing so in the descendant cells.

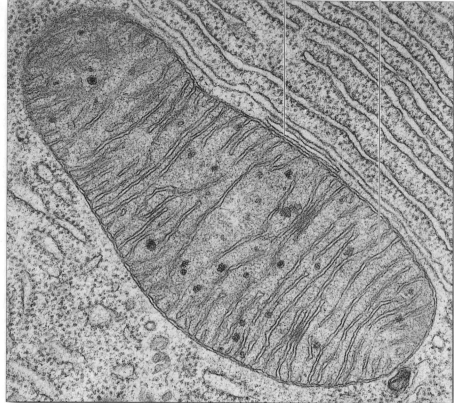

Figure 4.15 Sketch and transmission electron micrograph, thin section, of a typical mitochondrion. This organelle specializes in the production of large quantities of ATP, the main energy-carrying molecule between different reaction sites in cells. The process that produces ATP in mitochondria cannot proceed without free oxygen.

If they became permanent, protected residents, they might have lost structures and functions required for independent life while they were becoming mitochondria. We return to this topic in Section 19.4.

The organelles called mitochondria are the ATP-producing powerhouses of all eukaryotic cells.

Energy-releasing reactions proceed at the compartmented, internal membrane system of mitochondria. The reactions, which require oxygen, produce far more ATP than can be made by any other cellular reactions.

WWW

Chloroplasts and Other Plastids

Many plant cells contain plastids, a general category of organelles that specialize in photosynthesis or function in storage. Three types are common in different parts of plants. They are the chloroplasts, chromoplasts, and amyloplasts.

Of all eukaryotic cells, only the photosynthetic ones have **chloroplasts**. These organelles convert sunlight energy into the chemical energy of ATP, which is used to make sugars and other organic compounds. Chloroplasts commonly are oval or disk-shaped. Their semifluid interior, the stroma, is enclosed by two outer membrane layers. In the stroma is a third membrane called the thylakoid membrane. It is folded into a system of interconnecting, disk-shaped compartments. In many chloroplasts, these compartments stack, one atop the other, as in Figure 4.16. Each stack is a granum (plural, grana).

The first stage of photosynthesis starts and ends at a thylakoid membrane. Many light-trapping pigments, enzymes, and other proteins carry out the reactions. They work together to absorb light energy and "store" it in the form of ATP. Then, in the stroma, ATP energy is used to make sugars, then starch and other organic compounds, from carbon dioxide and water. Clusters of new starch molecules (starch grains) may briefly accumulate in the stroma.

The most abundant photosynthetic pigments are chlorophylls, which reflect or transmit green light. Others include carotenoids, which reflect or transmit yellow, orange, and red light. The relative abundances of the different pigments influences the colors of plant parts.

In many ways, chloroplasts resemble photosynthetic bacteria. Like mitochondria, they might have evolved from bacteria that were engulfed by predatory cells, yet escaped digestion and became permanent residents in them. We return to this idea in Section 19.4.

Unlike chloroplasts, chromoplasts lack chlorophylls but have an abundance of carotenoids. They are the source of red-to-yellow colors of many flowers, autumn leaves, ripening fruits, and carrots and other roots. The pigment colors commonly attract animals that pollinate plants or disperse seeds. Amyloplasts lack pigments. Often they store starch grains and are abundant in cells of stems, potato tubers (underground stems), and seeds.

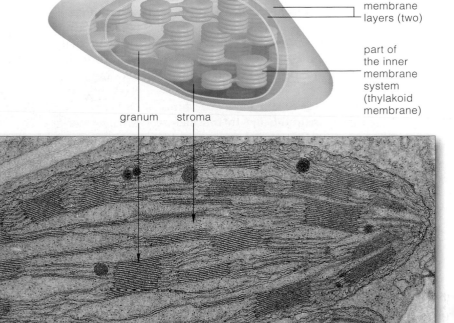

outermost membrane layers (two)

part of the inner membrane system (thylakoid membrane)

granum   stroma

Figure 4.16 Generalized sketch of a chloroplast, the key defining feature of every photosynthetic eukaryotic cell. The transmission electron micrograph shows a thin section of a chloroplast from a photosynthetic corn cell.

0.5μm

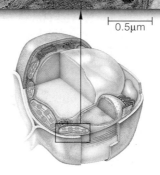

Central Vacuole

Many mature, living plant cells have a **central vacuole** (Figure 4.8). This fluid-filled organelle stores amino acids, sugars, ions, and toxic wastes. As it enlarges, it causes fluid pressure to build up inside the cell and so forces the cell's still-pliable cell wall to enlarge. Hence the cell itself enlarges. And as its surface area increases, so does the rate at which water and other substances can be absorbed across the plasma membrane.

In most cases, the central vacuole increases so much in volume that it takes up 50 to 90 percent of the cell's interior. The cytoplasm ends up as a very narrow zone between the central vacuole and plasma membrane.

Photosynthetic eukaryotic cells contain chloroplasts and other plastids that function in food production and storage.

Many plant cells have a central vacuole. When this storage vacuole enlarges during growth, cells are forced to enlarge, and this increases the surface area available for absorption.

WWW

THE CYTOSKELETON

The Main Components

The **cytoskeleton** is a system of interconnected fibers, threads, and lattices in the fluid part of the cytoplasm of eukaryotic cells. The system gives cells their internal organization, shape, and capacity to move. Figure 4.17 shows an animal cell, isolated from its home tissue, as it stretched out from left to right across a glass slide. Its internal mesh of filaments only hints at the extent of this cell's cytoskeleton. In this case, researchers were interested in its **microtubules** and its **microfilaments**. These two classes of cytoskeletal elements underlie nearly all movements of eukaryotic cells. Microtubules consist of protein subunits, called tubulins, that form a hollow cylinder. Microfilaments consist of two chains of actin subunits, twisted together. Some animal cells also have **intermediate filaments**, a class of ropelike cytoskeletal elements that impart mechanical strength to cells and tissues. Figure 4.18 gives examples of the basic structure of all three cytoskeletal elements.

Tubulin and actin have been highly conserved over evolutionary time; they are subunits for microtubules and microfilaments in all eukaryotic species. Yet they can serve diverse functions because different, *accessory* proteins can be joined to them. For instance, subunits of a "motor protein," such as myosin or dynein, typically project from microtubules or microfilaments with roles in cell movement. The myosin is part of the machinery by which muscle cells contract. "Crosslinking proteins" splice microtubules or microfilaments together. Certain kinds take part in forming the cell cortex, an extensive, three-dimensional mesh of microfilaments and other proteins underneath the plasma membrane. The cortex provides structural reinforcement for the cell surface. It also has roles in cell movements and changes in shape.

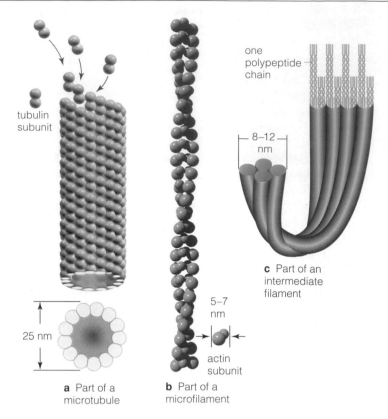

a Part of a microtubule **b** Part of a microfilament **c** Part of an intermediate filament

tubulin subunit · 25 nm · actin subunit · 5–7 nm · one polypeptide chain · 8–12 nm

Figure 4.18 Structural organization of (**a**) a microtubule, (**b**) a microfilament, and (**c**) an intermediate filament.

The Structural Basis of Cell Movements

Microfilaments, microtubules, or both take part in most aspects of motility. They do so by three mechanisms.

First, *the length of a microtubule or microfilament can grow or diminish by the controlled assembly or disassembly of its subunits.* When either lengthens or shortens at one end, a chromosome or some other structure attached to the other end is pushed or dragged through cytoplasm.

Some cells crawl about on protrusions of the body surface that form by microfilament assembly. *Amoeba proteus*, a soft-bodied protistan, has **pseudopods** ("false feet"): temporary, lobelike protrusions from the body. Inside each lobe, microfilaments grew rapidly in length, and the attached plasma membrane was dragged along with them. Similarly, the animal cell in Figure 4.17 was migrating on sheetlike extensions of microfilaments.

Second, *parallel rows of microfilaments or microtubules actively slide in specific directions.* For instance, a muscle cell has a series of contractile units. Microfilaments of actin are attached to one end or the other of each unit, and many myosin strands lie parallel with them. ATP energizes the myosin, which has oarlike projections that repeatedly bind and release microfilament neighbors. The short, repeated strokes make the microfilaments slide over the myosin strands, toward the center of the contractile unit—which shortens (contracts) as a result.

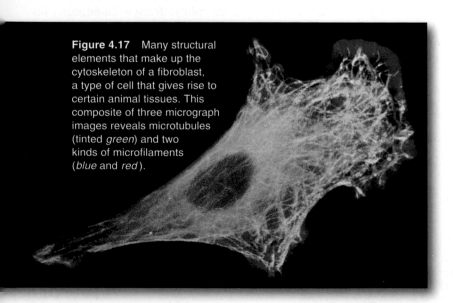

Figure 4.17 Many structural elements that make up the cytoskeleton of a fibroblast, a type of cell that gives rise to certain animal tissues. This composite of three micrograph images reveals microtubules (tinted *green*) and two kinds of microfilaments (*blue* and *red*).

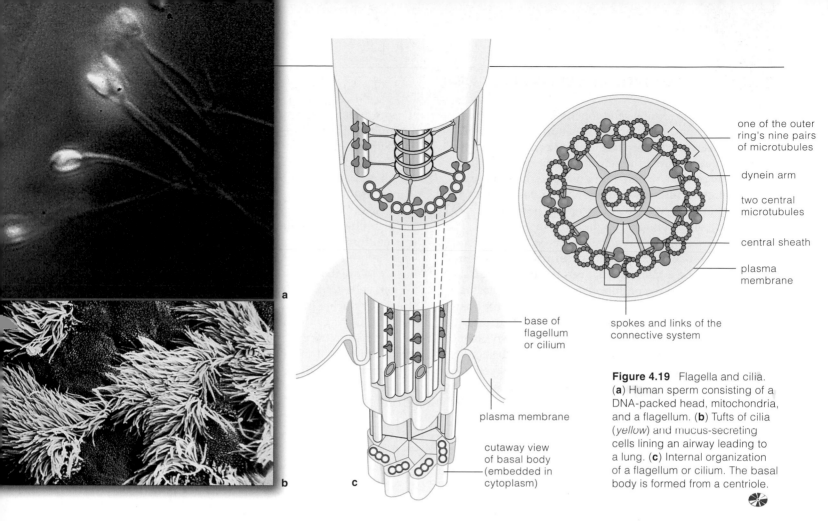

Figure 4.19 Flagella and cilia. (**a**) Human sperm consisting of a DNA-packed head, mitochondria, and a flagellum. (**b**) Tufts of cilia (*yellow*) and mucus-secreting cells lining an airway leading to a lung. (**c**) Internal organization of a flagellum or cilium. The basal body is formed from a centriole.

Labels in figure: one of the outer ring's nine pairs of microtubules; dynein arm; two central microtubules; central sheath; plasma membrane; spokes and links of the connective system; base of flagellum or cilium; plasma membrane; cutaway view of basal body (embedded in cytoplasm)

A similar sliding mechanism may be operating as cells crawl about. If the microfilament network beneath the plasma membrane at a cell's trailing end contracts, then some cytoplasmic gel would be squeezed into the leading end, which would bulge forward in response.

Third, *microtubules or microfilaments shunt organelles or parts of the cytoplasm from one location to another*. For example, chloroplasts move to new light-intercepting positions in response to the changing angle of the sun's overhead position. How? They are attached to myosin monomers that are "walking" over microfilaments and carrying chloroplasts with them. This shunt mechanism also causes cytoplasmic streaming: a dynamic and often rapid flowing of certain components in the cytoplasmic gel. Observe a living plant cell with a light microscope, and you may see pronounced streaming.

Flagella and Cilia

To biologists, the **flagellum** (plural, flagella) and **cilium** (plural, cilia) are classic examples of structures for cell motility. Both motile structures have a ring of nine pairs of microtubules and a central pair. A system of spokes and links stabilizes this "9 + 2 array." The array arises from a **centriole**, a barrel-shaped structure that is a type of microtubule-producing center. A centriole remains at the base of the completed array, where it is often called a **basal body** (Figure 4.19).

How do flagella and cilia differ? Flagella typically are longer and less profuse than cilia. Sperm and many other free-living cells use flagella as whiplike tails for swimming (Figure 4.19*a*). In multicelled species, some ciliated epithelial cells stir air or fluid. Thousands line airways in your chest (Figure 4.19*b*). As they beat, they direct mucus-trapped particles away from the lungs.

Flagella and cilia beat by a sliding mechanism. Short arms of motor proteins (dynein) extend from each pair of microtubules in the outer ring. When energized by ATP, the arms attach to the pair in front of them, tilt in a short, downward-directed stroke, then release their hold. Repeated strokes make the pairs slide down in sequence. Because spokes and links connect all the pairs, the sliding is converted into a bending motion.

Each eukaryotic cell has a cytoskeleton, the diverse elements of which are the basis of its shape, its internal structure, and its capacity for movements.

Microtubules are key organizers of the cytoskeleton, and they help move certain cell structures. Microfilaments take part in diverse movements and in forming and maintaining cell shape. Intermediate filaments structurally reinforce cells and internal cell structures.

Different mechanisms cause the cytoskeletal elements to assemble or disassemble, slide past one another, and shunt structures to new locations.

WWW

CELL SURFACE SPECIALIZATIONS

This survey of eukaryotic cells concludes with a look at cell walls and some other specialized surface structures. Many of these architectural marvels are constructed of various secretions from the cells themselves. Others are cytoplasmic bridges or sets of membrane proteins that connect neighboring cells and allow them to interact.

Eukaryotic Cell Walls

Single-celled eukaryotic species are directly exposed to their surroundings. Many have a **cell wall**, a structural component that wraps continuously around the plasma membrane. A cell wall protects and physically supports its owner. The wall is porous, so water and solutes can easily move to and from the plasma membrane. A great variety of protistans have a cell wall. (Figure 4.20 shows one.) So do plant cells and many types of fungal cells.

For instance, young plant cells in actively growing regions secrete gluelike polysaccharides (such as pectin), glycoproteins, and cellulose. The cellulose molecules join together

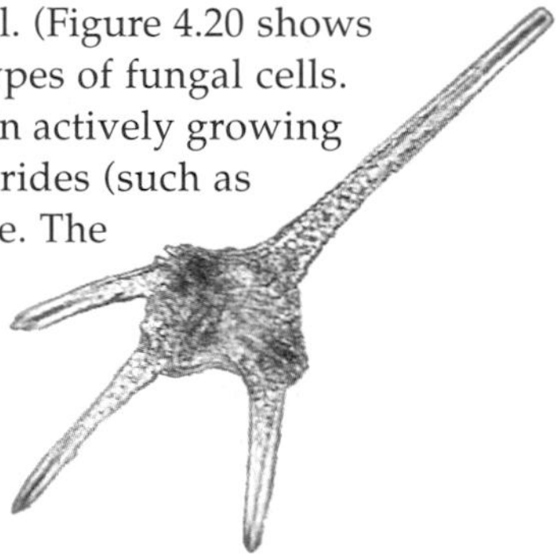

Figure 4.20 From a freshwater habitat, one of the single-celled, walled protistans (*Ceratium*, a dinoflagellate).

into ropelike strands, which become embedded in the gluey matrix. The secretions combine as a **primary wall** (Figure 4.21*a*). Primary walls are quite sticky, and they cement adjacent cells together. They also are thin and pliable, so the cell surface area can continue to enlarge under the pressure of incoming water.

At cell surfaces exposed to air, waxes and other cell secretions accumulate. The deposits form a cuticle. This semitransparent, protective surface covering restricts evaporative water loss from plants (Figure 4.22*a*).

Many plant cells develop only a thin wall. These are the cells that retain the capacity to divide or change shape during growth and development. As other plant cells mature, they stop enlarging and start secreting material on the primary wall's inner surface. The deposits combine to form a rigid, **secondary wall** that reinforces cell shape (Figure 4.21*e*). Whereas cellulose makes up less than 25 percent of the primary wall, the additional deposits now contribute more to structural support.

In woody plants, up to 25 percent of the secondary wall consists of lignin. Lignin is a complex molecule; it has a six-carbon ring structure to which a three-carbon chain and an oxygen atom are attached. Lignin makes plant parts stronger, more waterproof, and less inviting to insects and other plant-attacking organisms.

Figure 4.21 Examples of cell walls in flax plants. (**a**) Primary cell wall of young cells in flower petals. Cell secretions form the middle lamella, a layer between the walls of adjoining cells. The layer is thickest in adjoining corners. (**b**) Plasmodesmata, membrane-lined channels across the adjacent walls, connect the cytoplasm of neighboring cells. (**c,d**) Part of the lustrous fibers in a cell from a flax stem. We make linen from such fibers, which are three times stronger than cotton fibers. (**e**) In flax fibers, as in many other types of plant cells, more layers become deposited inside the primary wall. These layers stiffen the wall and help maintain its shape. Later, the cell dies, leaving the stiffened walls behind. This also happens in water-conducting pipelines that thread through most plant tissues. Interconnected, stiffened walls of dead cells form the tubes.

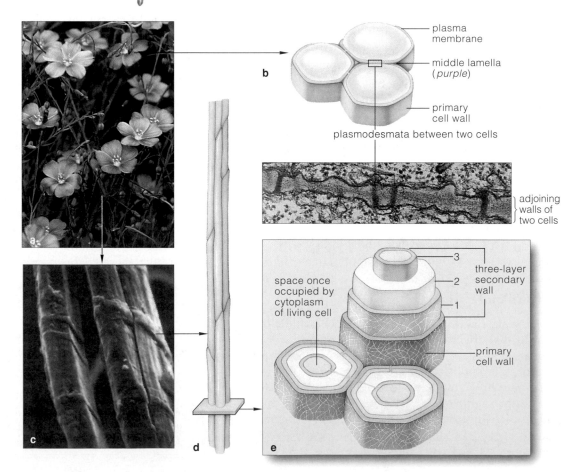

plasma membrane

middle lamella (*purple*)

primary cell wall

plasmodesmata between two cells

adjoining walls of two cells

space once occupied by cytoplasm of living cell

three-layer secondary wall

primary cell wall

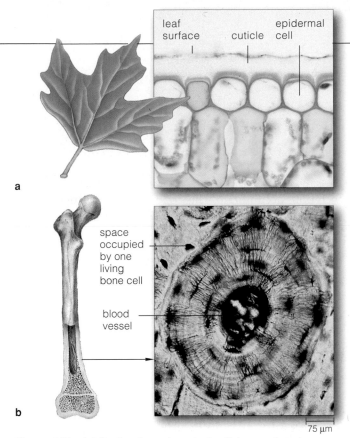

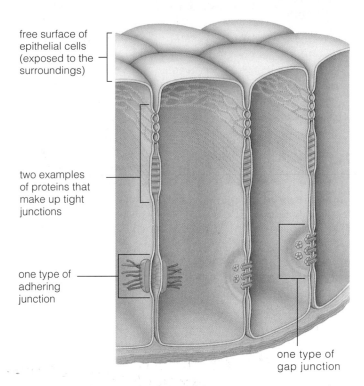

Figure 4.22 (**a**) Section through a plant cuticle, a surface layer composed of cell secretions. (**b**) Section through compact bone tissue, stained for microscopy.

Figure 4.23 Composite drawing of the most common types of cell junctions in animals. These are cells from epithelial tissue.

Matrixes Between Animal Cells

Although animal cells have no walls, diverse matrixes composed of cell secretions and even materials drawn from the surroundings intervene between many of them. Think of cartilage at the knobby ends of your leg bones. Cartilage consists of scattered cells and their secretions, which form collagen or elastin fibers embedded in a "ground substance" of modified polysaccharides. As another example, an extensive matrix widely separates the living bone cells in bone tissue (Figure 4.22*b*).

Cell-to-Cell Junctions

Even when a wall or some other structure imprisons a cell in its own secretions, the only contact that cell has with the outside world is *through* its plasma membrane. In multicelled species, membrane components project into adjacent cells as well as the surrounding medium. Among the components are junctions where the cell sends and receives diverse signals and materials, where it recognizes and cements itself to cells of the same type.

In plants, for instance, many tiny channels cross the adjacent primary walls of living cells and interconnect their cytoplasm. Figure 4.21*b* shows a few. Each channel is a plasmodesma (plural, plasmodesmata). The plasma membranes of adjoining cells have merged and fully line the channels, so there can be an uninterrupted flow

of substances between cells. Thus, all living cells in the plant body have the potential to exchange substances.

In most animal tissues, three categories of cell-to-cell junctions are common (Figure 4.23). *Tight* junctions link the cells of epithelial tissues, which line the body's outer surface and internal cavities and organs. They seal adjoining cells together; water-soluble substances can't leak between them. Thus gastric fluid cannot leak across the stomach's lining and damage surrounding tissues. *Adhering* junctions join cells in tissues of the skin, heart, and other organs subjected to stretching. *Gap* junctions link the cytoplasm of neighboring cells. They are open channels for a rapid flow of signals and substances.

We will be returning to the cell walls, intercellular substances, and cell-to-cell interactions in later chapters. For now, these are the points to remember:

A variety of protistan, plant, and fungal cells have a porous but protective wall that surrounds the plasma membrane. The cells themselves secrete the wall-forming materials.

Secretions from some cells form the cuticle at the surface of plants and some animals, extracellular matrixes in tissues, and other specialized structures.

In multicelled organisms, coordinated cell activities depend on cell-to-cell junctions, which are protein complexes or cytoplasmic bridges that serve as physical links and sites of communication between cells.

PROKARYOTIC CELLS—THE BACTERIA

We turn now to bacteria. Unlike the cells you have considered so far, all bacterial cells are prokaryotic; their DNA is *not* enclosed in a nucleus. *Prokaryotic* means "before the nucleus." Biologists selected the word as a reminder that bacterial cells already had appeared on the Earth before the nucleus evolved in the forerunners of eukaryotic cells.

As a group, the bacteria are the smallest cells, although a rare exception was recently discovered. Most are not much more than a micrometer wide; even the rod-shaped species are only a few micrometers in length. In terms of their structure, bacteria are the simplest kinds of cells to think about. Most species have a semirigid or rigid cell wall that wraps around the plasma membrane, structurally supports the cell, and imparts shape to it (Figure 4.24*a*). Dissolved substances can move freely to and from the plasma membrane because the wall is permeable. Often, sticky polysaccharides surround the cell wall. They help the bacterium attach to interesting surfaces, such as river rocks, teeth, and the vagina. In many of the disease-causing (pathogenic) bacteria, the polysaccharides form a thick, jellylike capsule that surrounds and helps protect the wall.

Like eukaryotic cells, bacteria have a plasma membrane that helps to control the movement of substances to and from the cytoplasm. A bacterial plasma membrane, too, has proteins that serve as channels, transporters, and receptors for signals and substances. It incorporates built-in machinery for metabolic reactions, such as the breakdown of energy-rich compounds. In photosynthetic types, clusters of some membrane proteins harness light energy and convert it to chemical energy in ATP.

Bacterial cells are too small to contain more than a small volume of cytoplasm, but they have many ribosomes upon which polypeptide chains are assembled. Apparently, these cells are small enough and so internally simple that they do not require a cytoskeleton.

The cytoplasm of a bacterium is distinct from that of eukaryotic cells. It is continuous with an irregularly shaped region of DNA. Membranes do not surround the region, which is named a **nucleoid** (Figure 4.24*c*). A circular molecule of DNA, also called the bacterial chromosome, occupies this region.

Extending from the surface of many bacterial cells are one or more threadlike motile structures known as bacterial flagella (singular, flagellum). These are not the same as eukaryotic flagella, because the 9 + 2 array of microtubules is absent. Bacterial flagella help a cell move rapidly in its fluid surroundings. Other surface

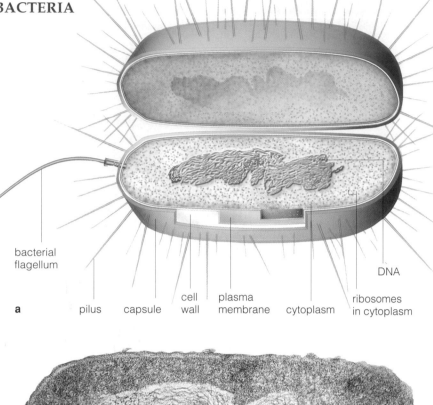

bacterial flagellum

a   pilus   capsule   cell wall   plasma membrane   cytoplasm   DNA   ribosomes in cytoplasm

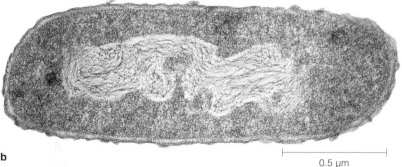

b   0.5 μm

Figure 4.24 (**a**) Generalized sketch of a typical prokaryotic body plan. (**b**) Micrograph of the bacterium *Escherichia coli.*

Your own gut is home to a large population of a normally harmless strain of *E. coli.* A harmful strain has contaminated large quantities of meat sold to restaurants. The same strain also has contaminated hard apple cider sold at a few roadside stands. Cooking the meat thoroughly or boiling the cider would have killed the bacterial cells. Where this was not done, people who ate the meat or drank the cider became quite sick. Some died.

Facing page: (**c**) Researchers manipulated this *E. coli* cell to release its single, circular molecule of DNA. (**d**) Cells of different bacterial species are shaped like balls, rods, or corkscrews. The ball-shaped cells of *Nostoc*, a photosynthetic bacterium, stick together inside a thick, gelatinlike sheath of their own secretions. Chapter 20 gives other fine examples. (**e**) Like this *Pseudomonas marginalis* cell, many species have one or more bacterial flagella, motile structures that propel the cell through fluid environments.

projections include pili (singular, pilus). These are the main protein filaments that help many kinds of bacteria attach to various surfaces, even to one another.

There are two kingdoms of prokaryotic cells: the Archaebacteria and Eubacteria (Section 1.3). Together, they contain the most metabolically diverse organisms. Different kinds have managed to exploit energy and raw materials in just about every kind of environment.

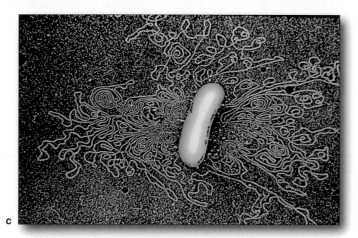

c

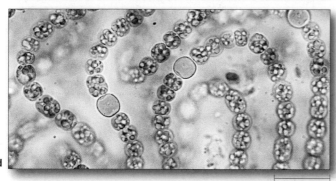

d

1 μm

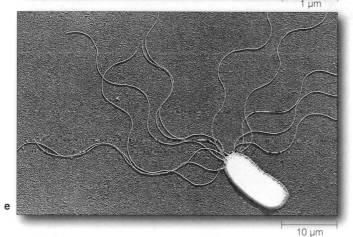

e

10 μm

In addition, ancient prokaryotic cells gave rise to all the protistans, plants, fungi, and animals ever to appear on Earth. The evolution, structure, and functions of these remarkable cells are topics of later chapters.

Bacteria alone are prokaryotic cells; their DNA is not housed inside a nucleus. Most species have a cell wall around the plasma membrane. Generally, their cytoplasm does not have any organelles comparable to those of eukaryotic cells.

Bacteria are the simplest cells, but as a group they show the most metabolic diversity. Their metabolic activities proceed at the plasma membrane and within the cytoplasm.

WWW

1. Three generalizations constitute the cell theory:
   a. All living things are composed of one or more cells.
   b. The cell is the smallest entity that still retains the properties of life. That is, it either lives independently or has a built-in, genetic capacity to do so.
   c. New cells arise only from cells that already exist.

2. At the minimum, a newly formed cell has a plasma membrane, a region of cytoplasm, and a region of DNA.
   a. The plasma membrane (a thin, outer membrane) maintains the cell as a distinct, separate entity. It allows metabolic events to proceed apart from random events in the environment. Many substances and signals are continually moving across it, in highly controlled ways.
   b. Cytoplasm is all the fluids, ribosomes, structural elements, and (in eukaryotic cells) organelles between the plasma membrane and the region of DNA.

3. Membranes are vital to cell structure and function. They consist of lipids (phospholipids, for the most part) and proteins. The lipids are arrayed as two layers, with all the hydrophobic tails of both layers sandwiched in between all the hydrophilic heads. This lipid bilayer imparts structure to the membrane and bars passage of water-soluble substances across it. Diverse proteins are embedded in the bilayer or attached to its surfaces.

4. Proteins carry out most cell membrane functions. For example, many serve as channels or pumps that allow or promote passage of water-soluble substances across the lipid bilayer. Others are receptors for extracellular substances that trigger changes in cell activities.

5. Cell membranes divide the cytoplasm of eukaryotic cells into functional compartments called organelles. Prokaryotic cells do not have comparable organelles.

6. Organelle membranes separate metabolic reactions in the space of the cytoplasm and allow different kinds to proceed in orderly fashion. (In bacteria, many similar reactions proceed at the plasma membrane.)
   a. The nuclear envelope functionally separates the DNA from the metabolic machinery of the cytoplasm.
   b. The cytomembrane system includes the ER, Golgi bodies, and vesicles. Many new proteins are modified into final form and lipids are assembled in this system. Finished products are packaged and then shipped off to destinations inside or outside the cell.
   c. Mitochondria are specialists in oxygen-requiring reactions that produce many ATP molecules.
   d. Chloroplasts trap sunlight energy and produce organic compounds in photosynthetic eukaryotic cells.

7. The cytoskeleton of eukaryotic cells functions in cell shape, internal organization, and movements.

8. Table 4.3, on the next page, summarizes the defining features of both prokaryotic and eukaryotic cells.

Table 4.3 Summary of Typical Components of Prokaryotic and Eukaryotic Cells

Cell Component	Function	PROKARYOTIC Archaebacteria, Eubacteria	EUKARYOTIC Protistans	Fungi	Plants	Animals
Cell wall	Protection, structural support	✓*	✓*	✓	✓	None
Plasma membrane	Control of substances moving into and out of cell	✓	✓	✓	✓	✓
Nucleus	Physical separation and organization of DNA	None	✓	✓	✓	✓
DNA	Encoding of hereditary information	✓	✓	✓	✓	✓
RNA	Transcription, translation of DNA messages into polypeptide chains of specific proteins	✓	✓	✓	✓	✓
Nucleolus	Assembly of subunits of ribosomes	None	✓	✓	✓	✓
Ribosome	Protein synthesis	✓	✓	✓	✓	✓
Endoplasmic reticulum (ER)	Initial modification of many of the newly forming polypeptide chains of proteins; lipid synthesis	None	✓	✓	✓	✓
Golgi body	Final modification of proteins, lipids; sorting and packaging them for use inside cell or for export	None	✓	✓	✓	✓
Lysosome	Intracellular digestion	None	✓	✓*	✓*	✓
Mitochondrion	ATP formation	**	✓	✓	✓	✓
Photosynthetic pigment	Light–energy conversion	✓*	✓*	None	✓	None
Chloroplast	Photosynthesis; some starch storage	None	✓*	None	✓	None
Central vacuole	Increasing cell surface area; storage	None	None	✓*	✓	None
Bacterial flagellum	Locomotion through fluid surroundings	✓*	None	None	None	None
Flagellum or cilium with 9 + 2 microtubular array	Locomotion through or motion within fluid surroundings	None	✓*	✓*	✓*	✓
Cytoskeleton	Cell shape; internal organization; basis of cell movement and, in many cells, locomotion	None	✓*	✓*	✓*	✓

* Known to be present in cells of at least some groups.

** Oxygen-requiring (aerobic) pathways of ATP formation do occur in many groups, but mitochondria are not involved.

Review Questions

1. Label the organelles in this diagram of a plant cell. *4.4*

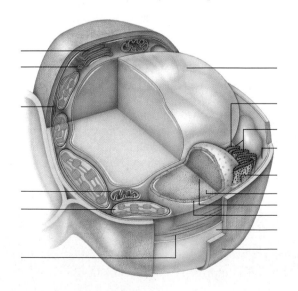

2. Label the organelles in this diagram of an animal cell. *4.4*

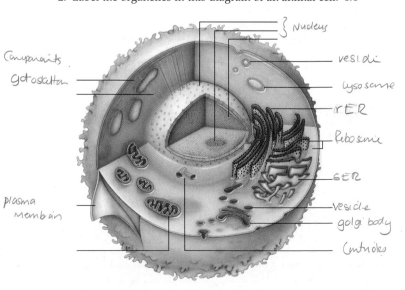

3. State the three key points of the cell theory. *CI*

4. Describe three features that all cells have in common. After reviewing Table 4.3, write a paragraph on the key differences between prokaryotic and eukaryotic cells. *4.1, 4.4, 4.11*

5. Suppose you want to observe the three-dimensional surface of an insect's eye. Would you benefit most by using a compound light microscope, transmission electron microscope, or scanning electron microscope? *4.3*

6. Briefly characterize the structure and function of the cell nucleus, the nuclear envelope, and the nucleolus. *4.5*

7. Define chromosome and chromatin. Do chromosomes always have the same appearance during a cell's life? *4.5*

8. Which organelles are part of the cytomembrane system? *4.6*

9. Is this statement true or false: Plant cells have chloroplasts, but not mitochondria. Explain your answer. *4.7, 4.8*

10. What are the functions of the central vacuole in mature, living plant cells? *4.8*

11. Define cytoskeleton. How does it aid in cell functioning? *4.9*

12. Are all components of the cytoskeleton permanent? *4.8, 4.9*

13. What gives rise to the 9 + 2 array of cilia and flagella? *4.9*

14. Cell walls are typical of which organisms: bacteria, protistans, fungi, plants, animals? Are the walls porous or nonporous? *4.10*

15. In certain plant cells, is a secondary wall deposited inside or outside the surface of the primary wall? *4.10*

16. In multicelled organisms, coordinated interactions depend on linkages and communications between cells. What types of junctions occur between adjacent animal cells? Plant cells? *4.10*

Self-Quiz (*Answers in Appendix III*)

1. Cell membranes consist mainly of a _____ .
   a. carbohydrate bilayer and proteins
   b. protein bilayer and phospholipids
   c. lipid bilayer and proteins
   d. none of the above

2. Organelles _____ .
   a. are membrane-bound compartments
   b. are typical of eukaryotic cells, not prokaryotic cells
   c. separate chemical reactions in time and space
   d. all of the above are features of the organelles

3. Cells of many protistans, plants, and fungi, but not animals, commonly have _____ .
   a. mitochondria   c. ribosomes
   b. a plasma membrane d. a cell wall

4. Is this statement true or false: All cells contain microtubules, microfilaments, and intermediate filaments.

5. Is this statement true or false: The plasma membrane is the outermost component of all cells. Explain your answer.

6. Unlike eukaryotic cells, prokaryotic cells _____ .
   a. lack a plasma membrane   c. do not have a nucleus
   b. have RNA, not DNA   d. all of the above

7. Match each cell component with its function.
   __e__ mitochondrion   a. synthesis of polypeptide chains
   __a__ chloroplast   b. initial modification of new
   __d__ ribosome   polypeptide chains
   __b__ rough ER   c. final modification of proteins; lipid
   __c__ Golgi body   synthesis; sorting, shipping tasks
                         d. photosynthesis
                         e. site of oxygen-requiring pathway
                            of ATP formation

Critical Thinking

1. Why is it likely that you will never encounter a predatory two-ton living cell on the sidewalk?

2. In compound light microscopes having blue filters, the lens transmits only blue light. Think about the spectrum of visible light (as in Figure 6.4), and then speculate on why blue light is efficient for viewing objects at high magnification.

3. Your professor shows you an electron micrograph of a cell with large numbers of mitochondria and Golgi bodies. You notice that this particular cell also contains a great deal of rough endoplasmic reticulum. What kinds of cellular activities would require such an abundance of the three kinds of organelles?

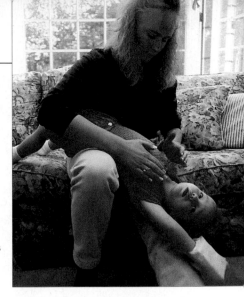

Figure 4.25 A daily chest thumping for a child with cystic fibrosis.

4. *Cystic fibrosis* is a fatal genetic disorder. Affected glands secrete far more than they should, with far-reaching effects. In time, digestive enzymes clog a duct between the pancreas and small intestine, food can't be digested properly, and even if food intake increases, malnutrition results. Cysts form in the pancreas, which degenerates and becomes fibrous (hence the disorder's name). Thick mucus builds up in the respiratory tract; affected people have trouble expelling airborne bacteria and particles that enter lungs (Figure 4.25). The disorder may arise from a defective protein in the plasma membrane of gland cells that secrete mucus, digestive enzymes, and sweat. Review Section 4.6, then name the organelles involved in the secretory pathway in those cells.

Selected Key Terms

basal body *4.9*	ER (endoplasmic	nucleoid *4.11*
cell *4.1*	reticulum) *4.6*	nucleolus *4.5*
cell theory *CI*	eukaryotic cell *4.1*	nucleus *4.5*
cell wall *4.10*	flagellum *4.9*	organelle *4.4*
central	fluid mosaic	plasma
vacuole *4.8*	model *4.1*	membrane *4.1*
centriole *4.9*	Golgi body *4.6*	primary wall *4.10*
chloroplast *4.8*	intermediate	prokaryotic cell *4.1*
chromatin *4.5*	filament *4.9*	pseudopod *4.9*
chromosome *4.4*	lipid bilayer *4.1*	ribosome *4.1*
cilium *4.9*	microfilament *4.9*	secondary wall *4.10*
cytomembrane	micrograph *4.3*	surface-to-volume
system *4.6*	microtubule *4.9*	ratio *4.2*
cytoplasm *4.1*	mitochondrion *4.7*	vesicle *4.6*
cytoskeleton *4.9*	nuclear envelope *4.5*	wavelength *4.3*

Readings *See also www.infotrac-college.com*

deDuve, C. 1985. *A Guided Tour of the Living Cell*. New York: Freeman. Beautiful introduction to the cell; two short volumes.

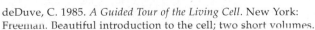

GROUND RULES OF METABOLISM

You Light Up My Life

Find yourself out and about at night or snorkeling in the ocean, and you might see fireflies and other insects, squids, or certain other organisms emitting light. Figure 5.1a shows fireflies (actually a type of beetle) engaging in communal flashings in a tropical forest. Different varieties, known locally as kittyboos, light up the night with green, yellow-green, yellow, or orange flashes.

Kittyboos emit light when enzymes—luciferases—convert chemical energy to light energy. The reactions get under way when ATP transfers a phosphate group to luciferin, a highly fluorescent substance. Luciferases, with a little help from oxygen, convert the activated luciferin to a different molecule. Their action boosts electrons of the molecule to a higher energy level. As the excited electrons quickly return to a lower energy level, they release energy as *fluorescent* light. Fluorescent light is emitted when a destabilized molecule reverts to a stable configuration. When organisms flash with it, this is called **bioluminescence**.

Imaginative biologists learned how to borrow light from such flashers to make *bioluminescent gene transfers*. They now insert copies of the genes for bioluminescence into bacteria, plants, and other organisms (Figure 5.1b)! Besides being fun to think about, these transfers have practical applications.

Each year, for example, 3 million people die from a lung disease caused by *Mycobacterium tuberculosis*. No one antibiotic is effective against all the different strains of this bacterium, so an infected person cannot receive effective treatment until the strain causing the infection is identified. A fast way to do this is to expose bacterial cells in samples taken from a patient to luciferase genes. The genes typically slip into the bacterial DNA of some cells. Clinicians isolate those cells, then expose colonies of the descendants to different antibiotics. If a particular antibiotic doesn't work, the colonies glow (their cells have churned out gene products—including luciferase). If the colonies don't glow, the antibiotic works.

Christopher and Pamela Contag, two postdoctoral students at Stanford University, wanted to light up bacteria that cause *Salmonella* infections in laboratory mice. Why? Researchers of viral or bacterial diseases typically infect dozens to hundreds of laboratory mice for experiments. The only option has been to kill infected mice and examine tissues to find out whether infection occurred—a costly, tediously painstaking practice that also happens to offend animal rights activists.

The Contags approached David Benaron, a medical imaging researcher at Stanford, with this hypothesis: If live, infectious bacteria were made bioluminescent, then flashes would shine through tissues of *live*, infected animals. In a preliminary test of this novel idea, the researchers put glowing *Salmonella* cells into a thawed chicken breast from a market. A glow showed through.

A kittyboo's bioluminescent organs, where luciferin and luciferases are kept separated until signals from the nervous system command them to mix it up

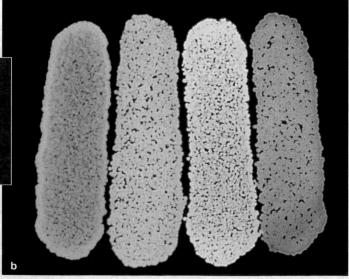

Figure 5.1 (a) Jamaican fireflies, known locally as kittyboos (*Pyrophorus noctilucus*). This type of beetle lights up the night sky with bioluminescent flashes, which help potential mates find each other in the dark. (b) Micrograph of four colonies of bacterial cells. Each colony started with a parent bacterial cell that had taken up a kittyboo gene for a glowing color.

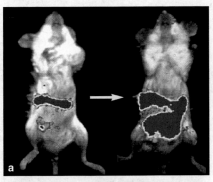

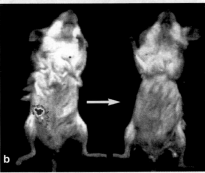

Figure 5.2 Using bioluminescent bacterial cells to chart the location of infectious bacteria inside living laboratory mice and their spread through body tissues. (**a**) False-color images in this pair of photographs show how the infection spread in a control group that had not been given a dose of antibiotics. (**b**) This pair shows how antibiotics had killed most of the infectious bacterial cells.

Next the Contags transferred bioluminescence genes into three strains of *Salmonella*. Then they injected the strains into mice of three experimental groups and used a digital imaging camera to track the infection in each group. The first strain was weak; the mice were able to fight off the infection in less than six days and did not glow. The second strain was not as weak but could not spread through the mouse body; it remained localized. The third strain was dangerous; it spread very rapidly through the mouse gut—and the entire gut glowed.

Thus bioluminescent gene transfer, combined with imaging of enzyme activity, can be used to track the course of infection and to evaluate the effectiveness of drugs in living organisms (Figure 5.2). It may also have uses in gene therapy, whereby copies of functional genes replace defective or cancer-causing genes in patients.

Why use bioluminescent organisms to introduce a chapter? They give us visible signs of **metabolism**—of the cell's capacity to acquire energy and use it to build, break apart, store, and release substances in controlled ways. Each flash reminds us that living cells are taking in energy-rich solutes, building membranes, storing things, replenishing enzymes, and checking out their DNA. A constant supply of energy drives all of these activities. The story of metabolism starts with ways in which cells get energy and channel it into the reactions by which they stay alive, grow, and reproduce.

KEY CONCEPTS

1. Cells engage in metabolism—that is, they use energy to build, stockpile, break apart, and eliminate substances in ways that help them survive and reproduce.

2. With each metabolic reaction, energy escapes into the environment. To stay alive, cells must balance their energy losses with energy gains. Yet they cannot create energy from scratch. They can only draw on existing sources, such as light energy from the sun and chemical energy in the bonds of glucose and other substances.

3. Metabolic pathways maintain, increase, or decrease the relative amounts of various substances in cells. Typically, the pathways couple reactions that release usable energy from substances to other reactions that require energy.

4. In cells, virtually all energy conversions start with electron transfers from one substance to another.

5. Chemical energy carried by ATP drives nearly all metabolic reactions. ATP activates substances such as glucose—it primes them for chemical change—when it donates a phosphate group to them.

6. Chemical reactions proceed far too slowly on their own to sustain life. In living organisms, the action of specific enzymes greatly increases the rate of specific reactions.

7. Compared to their environment, cells maintain greater or lesser amounts of certain substances, as required for metabolism. They do so even though the molecules or ions of any substance have a natural tendency to diffuse into regions where they are less concentrated. Membrane transport proteins work with or against this tendency.

ENERGY AND THE UNDERLYING ORGANIZATION OF LIFE

Defining Energy

If you have ever watched a house cat stalking a mouse, you know it can "freeze" its position to avoid detection before springing at its unsuspecting prey. Like anything else in the universe that is stationary, the cat has a store of **potential energy**—a capacity to do work, simply owing to its position in space and the arrangement of its parts. As a cat springs, some of its potential energy is transformed into **kinetic energy**, the energy of motion.

Energy on the move does work when it imparts motion to other things. In skeletal muscle cells inside the cat, ATP gave up some of its potential energy to molecules of contractile units and set them in motion. The combined motions in many muscle cells resulted in the movement of whole muscles. The transfer of energy from ATP also resulted in the release of another form of kinetic energy called **heat**, or *thermal* energy.

The potential energy of molecules has its own name: **chemical energy**. It is measurable, as in kilocalories. A kilocalorie is the same thing as 1,000 calories, which is the amount of energy it takes to heat 1,000 grams of water from 14.5°C to 15.5°C at standard pressure.

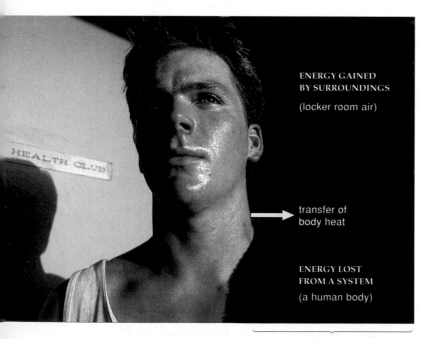

ENERGY GAINED
BY SURROUNDINGS

(locker room air)

transfer of
body heat

ENERGY LOST
FROM A SYSTEM

(a human body)

NET ENERGY CHANGE = 0

Figure 5.3 Example of how the total energy content of any system *together with its surroundings* remains constant.

"System" means all matter in a specific region, such as a human body, a plant, a DNA molecule, or a galaxy. "Surroundings" can be a small region in contact with the system or as vast as the entire universe. The system shown (a human male) is giving off heat to the surroundings (a locker room) by evaporative water loss from sweat. What one region loses, the other region gains, so the total energy content of both does not change.

What Can Cells Do With Energy?

All organisms have specific adaptations for securing energy from their environment. Some harness energy from the sun, and others extract energy from inorganic or organic substances in the environment. Regardless of the source, energy inputs become *coupled* to thousands of energy-requiring processes in cells. Cells use energy for *chemical* work, to stockpile, build, rearrange, and break apart substances. They channel it into *mechanical* work—to move flagella and other cell structures and (in multicelled species) the whole body or portions of it. They channel it into *electrochemical* work—to move charged substances into or out of the cytoplasm or an organelle compartment.

How Much Energy Is Available?

Like single cells, we cannot create energy from scratch; we must get it from someplace else. Why? According to the **first law of thermodynamics**, the total amount of energy in the universe remains constant. More energy cannot be created; existing energy cannot vanish. It can only be converted from one form to some other form.

Think about what the law means. The universe has only so much energy, distributed in a variety of forms. One form can be converted to another, as when corn plants absorb energy from the sun and convert it to the chemical energy of starch. After you eat and digest corn, your cells extract energy from starch and convert it to other forms, such as kinetic energy for moving about.

With each metabolic conversion, some of the energy escapes to the surroundings, as heat. Even when you "do nothing," your body gives off about as much heat as a 100-watt lightbulb because of conversions in your cells. The energy being released is transferred to atoms and molecules that make up the air, and in this way it "heats up" the surroundings (Figure 5.3). In the air, the kinetic energy increases the number of ongoing, random collisions among molecules. A bit more energy is released as heat with each collision. However, none of the energy ever vanishes.

The One-Way Flow of Energy

Energy available for conversions in cells resides mainly in covalent bonds. Glucose, glycogen, starches, fatty acids, and other organic compounds have organized arrangements of many of these bonds and are said to have a high energy content. When the compounds enter metabolic reactions, specific bonds break or become rearranged. During that molecular commotion, some amount of heat energy is lost to the surroundings. In general, cells cannot recapture energy lost as heat.

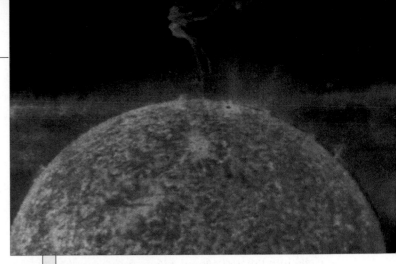

Figure 5.4 An example of the one-way flow of energy into the world of life that compensates for the one-way flow of energy out of it. The sun continuously loses energy, much of it in the form of wavelengths of light (Section 6.2). Living cells intercept some of the energy and convert it to useful forms of energy, stored in bonds of organic compounds. Each time a metabolic reaction proceeds in cells, stored energy is released— and some inevitably is lost to the surroundings, mostly as heat.

The lower photograph shows green, water-dwelling, photosynthetic cells (*Volvox*). They live in tiny, spherical colonies. The orange cells function in reproduction. They form new colonies inside the parent sphere.

ENERGY LOST
one-way flow of energy from sun to Earth's environment

ENERGY GAINED
one-way flow of energy from environment to organisms

For example, your cells release usable energy from glucose by breaking all of its covalent bonds. After many steps, six molecules of carbon dioxide and six of water remain. Compared with glucose, these leftovers have more stable arrangements of atoms, but chemical energy in their bonds is much less than the total chemical energy of glucose. Why? *Some energy was lost at each breakdown step leading to their formation.* Said another way, a glucose molecule is a better source of usable energy.

What about the heat that was transferred from cells to their surroundings when carbon dioxide formed? It is very low quality; cells cannot convert it to other forms, so it cannot be used to do work.

Bad news for cells of the remote future: The amount of "low-quality" energy in the universe is increasing. No energy conversion can ever be 100 percent efficient—even highly efficient ones lose heat—so the total amount of energy in the universe is spontaneously flowing from forms rich in energy to forms having less and less of it. That, basically, is the point to remember about the **second law of thermodynamics**.

Without energy inputs to maintain it, any organized system tends to get disorganized over time. **Entropy** is a measure of the degree of a system's disorder. Think of the Egyptian pyramids—originally organized, presently crumbling, and many thousands of years from now, dust. The ultimate destination of those pyramids and everything else in the universe is a state of maximum entropy. Billions of years from now, all of the energy available for conversions will be dissipated.

Can life be one glorious pocket of resistance to the depressing flow toward maximum entropy? After all, in each new organism, new bonds form and hold atoms together in precise arrays. So molecules become more organized and have a richer store of energy, not poorer!

Yet a simple example will show that the second law does indeed apply to life on Earth. The primary energy source for life is the sun, which has been losing energy

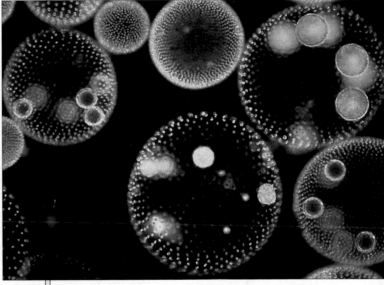

Producer organisms harness sun's energy, use it to build organic compounds from simple raw materials available in their environment.

All organisms tap potential energy stored in organic compounds to drive energy conversions that keep them alive. Some energy is lost with each conversion.

ENERGY LOST
one-way flow of energy from organisms back to the environment

since it first formed. Plants capture sunlight energy, convert it to other forms, then lose energy to other organisms that feed, directly or indirectly, on plants. At each energy transfer, some energy is lost as heat that joins the universal pool. *Overall, energy still flows in one direction.* The world of life maintains its high degree of organization only because it is being resupplied with energy that is being lost from someplace else (Figure 5.4).

The amount of energy in the universe remains constant. Energy can undergo conversions from one form to another, but it cannot be created out of nothing or destroyed.

The total amount of energy in the universe is spontaneously flowing from forms of higher to lower quality.

A steady flow of sunlight energy into the interconnected web of life compensates for the steady flow of energy leaving it.

DOING CELLULAR WORK

When cells convert one form of energy to another, there is a change in the amount of potential energy that is available to them. The greater the initial amount of potential energy, the larger the energy change will be and the more work can be done.

Imagine a Martian who is not happy that the JPL Rover is inching around her planet. She decides to push it to the top of a rocky hill (Figure 5.5a). To do this, she converts some potential energy stored in her muscles to kinetic energy. Once the Rover is precariously perched on the hill, it has potential energy (owing to its position) and it tends to roll down on its own, spontaneously. The higher up the Rover relative to its final position at the base of the hill, the greater the energy change and the more work done—in this case, a bigger impact and more broken parts (Figure 5.5b).

The same ground rules apply to metabolism. For instance, glucose ($C_6H_{12}O_6$) is built from carbon dioxide ($6CO_2$) and water ($12H_2O$). Each substance has potential energy in chemical bonds. The bond energy of glucose exceeds those of the other two substances combined. Thus the assembly of glucose from carbon dioxide and water is not something that tends to proceed on its own. The carbon dioxide and water are at the base of an energy hill. On their own, they don't have enough energy for an uphill run. In photosynthetic cells, *energy inputs* from the sun drive the synthesis reactions. The outcome is a net increase in energy (Figure 5.5c). Said another way, the reaction sequence by which glucose forms is **endergonic** (meaning energy in).

Now picture the reactions running in reverse, from glucose (at the top of the energy hill) to carbon dioxide and water (at the base). Energetically, a downhill run is favorable. It proceeds spontaneously and ends with a net loss in energy. So the reaction sequence that breaks down glucose is **exergonic** (meaning energy outward).

ATP—The Cell's Energy Currency

In cells, ATP couples energy released from exergonic reactions with energy-requiring ones. **ATP**, recall, is short for adenosine triphosphate, one of the organic compounds called nucleotides (Section 3.7). Each ATP molecule consists of the five-carbon sugar ribose to which adenine (a nucleotide base) and three phosphate groups are attached (Figure 5.6a). Its triphosphate tail

Energy input required to push Rover uphill

a

Potential energy released by downhill run used for mechanical work (for wrecking the Rover)

b

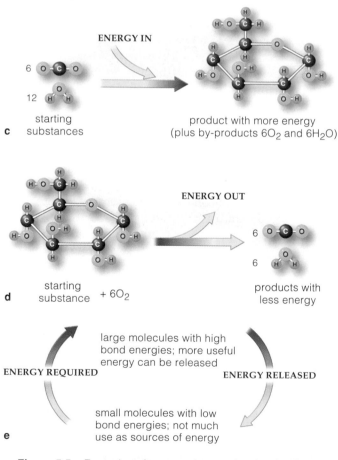

ENERGY IN

6 $O-C-O$

12 $H-O-H$

starting substances

c

product with more energy (plus by-products $6O_2$ and $6H_2O$)

ENERGY OUT

starting substance $+ 6O_2$

d

6 $O-C-O$

6 $H-H$

products with less energy

ENERGY REQUIRED

large molecules with high bond energies; more useful energy can be released

ENERGY RELEASED

small molecules with low bond energies; not much use as sources of energy

e

Figure 5.5 Examples of energy changes involved in (**a,b**) mechanical work and (**c,d**) chemical work. (**e**) Cells couple energy-requiring reactions with energy-releasing reactions.

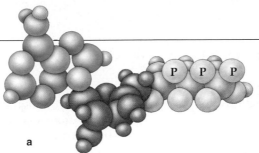
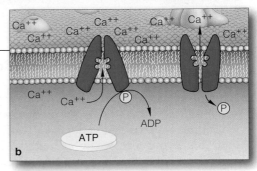

Figure 5.6 (**a**) Model of ATP showing its component atoms. (**b**) One example of cellular work. In a contractile unit of a muscle cell, ATP transfers a phosphate group to a transport protein. Energy associated with the transfer makes the protein's shape change in a way that pumps calcium ions out of the cell.

is where the action is, so to speak. Many hundreds of different enzymes can readily cleave the covalent bond between the two outermost phosphate groups of the molecule's tail, then attach it to another substance.

Any transfer of a phosphate group to a molecule is a **phosphorylation**. The energy transferred during such events is sufficient to activate hundreds of different molecules and drive hundreds of cellular activities. As examples, it can drive the synthesis and breakdown of organic compounds, contraction of muscle cells, and the pumping of substances across cell membranes in energetically unfavorable directions (Figure 5.6b).

Cells renew their ATP supplies. At certain steps of many metabolic processes, such as aerobic respiration, an unbound phosphate atom (P_i) or a phosphate group that an enzyme cleaves from some substance becomes attached to adenosine diphosphate, or ADP. The result is an ATP molecule. When ATP transfers a phosphate group elsewhere, it reverts to ADP, thereby completing the steps of the ATP/ADP cycle:

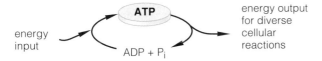

In short, phosphate-group transfers from ATP are a renewable, rapid, and near-universal mechanism for coupling energy-releasing and energy-requiring events. Its role is like currency in an economy: ATP is earned in exergonic reactions and spent in endergonic ones. That is why a cartoon "coin" is often used to symbolize ATP.

Electron Transfers

Cells can release energy from a substance by a series of small steps, each involving a transfer of electrons. You may hear someone refer to an electron transfer as an **oxidation-reduction reaction**, but this is only a technical way of saying the same thing. A molecule that gives up (donates) electrons is said to be "oxidized." A molecule that accepts electrons is "reduced."

By analogy, imagine trying to release energy from glucose by throwing some of it into a wood fire. Atoms of glucose molecules quickly let go of one another, then combine with oxygen in the air to form CO_2 and H_2O, and all the released energy is lost as heat. Cells release energy from glucose more efficiently by electrochemical work, which entails transferring electrons from glucose

through electron transport systems of cell membranes. Later chapters describe the components and functions of such systems. Often, energy released as an outcome of electron transfers through them drives ATP formation.

Metabolic Pathways

Energy inputs drive reactions that involve thousands of substances within the confines of a cell. Most of the reactions occur in orderly, enzyme-mediated sequences called **metabolic pathways**. In biosynthetic pathways, small molecules are assembled into molecules of higher bond energies, such as complex carbohydrates, lipids, and proteins. In degradative pathways, large molecules are broken down to products of lower bond energies. The participants of such pathways go by these names:

Substrates are substances that enter a reaction. They are also called reactants or precursors. Any substance that forms between the start and end of a pathway is an *intermediate*. Those remaining at the end of a reaction or a pathway are the *end products*. The *energy carriers* are ATP and a few other compounds that can activate substances by delivering energy (by way of functional groups) to them. Most *enzymes* are proteins that speed specific reactions. (A few RNAs also display enzyme activity.) *Cofactors* include coenzymes (certain organic compounds, such as NAD^+) and metal ions that assist enzymes or pick up electrons, atoms, or functional groups at one reaction site and taxi them to a different site. *Transport proteins* are proteins that span a cell membrane and let substances cross it in controlled ways. They help adjust concentrations of substances on both sides of the membrane and so influence metabolic reactions.

Many of the pathways advance step by step in linear fashion, from substrates to end products. Many others are cyclic; steps proceed in a circle, with end products serving as reactants to start things over. As you will see soon enough, the intermediates or end products of one pathway also can enter different metabolic pathways.

ATP is the main, renewable energy carrier between sites of metabolic reactions in cells. Its deliveries couple energy-releasing reactions with energy-requiring reactions.

Much of the cellular work driven by energy from ATP involves electron transfers from one substance to another.

Metabolic pathways are enzyme-mediated reaction sequences from substrates and intermediates to end products.

WWW

ENZYME STRUCTURE AND FUNCTION

Without enzymes, the dynamic, steady state called "you" would quickly cease to exist. Reactions simply would not proceed fast enough for the body to process food, build and tear down hemoglobin and other vital molecules, send signals to and from brain cells, make muscles contract, and do everything else to stay alive.

To see how enzymes work, start with this concept: *Cells control their internal concentrations of substances with respect to the surroundings, and eukaryotic cells also control the concentrations across membranes of organelles.* Remember, molecules or ions of any substance are in constant, random motion that puts them on collision courses. The more concentrated they are, the more often they collide. And energy associated with the collisions might be enough to cause a metabolic reaction—that is, to make a molecule combine with something else, split into smaller parts, or change its shape.

Nearly all metabolic reactions are reversible. They might start out in the "forward" direction, from starting substances to products. But they also run in "reverse," with products converted back to starting substances. Which way such a reaction runs depends partly on the ratio of reactant to product. A high concentration of reactant molecules is an energetically favorable state, so the reaction will run spontaneously and strongly in the forward direction. When the product concentration is high enough, more molecules or ions of the product are available to revert spontaneously to reactants.

Any reversible reaction tends to run spontaneously toward **chemical equilibrium**, the time at which it will be running at about the same pace in both directions.

Four Features of Enzymes

By definition, **enzymes** are catalytic molecules; *they speed the rate at which reactions approach equilibrium.* Again, nearly all enzymes are proteins. All share four features. First, enzymes do not make anything happen that could not happen on its own, but they usually make it happen hundreds to millions of times faster. Second, reactions do not permanently alter or use up enzyme molecules; the same one may act repeatedly. Third, the same type of enzyme usually works for the forward and reverse directions of a reaction. Fourth, each type of enzyme is very picky about its substrates. Its substrates are specific substances that it can chemically recognize, bind, and modify in certain ways. For example, thrombin is one of the enzymes required to clot blood. It only recognizes a side-by-side arrangement of arginine and glycine, two amino acids, in a protein molecule. When it does so, it cleaves the peptide bond between them.

Enzyme-Substrate Interactions

Take a look at Figure 5.7. A metabolic reaction occurs when participating molecules collide—provided that they collide with some minimum amount of energy called the **activation energy**.

activation energy without enzyme

activation energy with enzyme

starting substance

energy released by the reaction

6

products

a   direction of reaction ⟶

b   ENZYME!

Figure 5.7 Activation energy. (**a**) Before reactants enter a metabolic reaction, they must by activated by an energy input; only then will they spontaneously proceed to products. (**b**) An enzyme enhances the reaction rate by lowering the amount of activation energy required to boost the reactants to the crest (transition state) of the energy barrier.

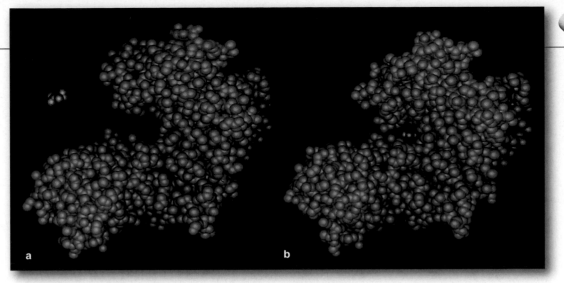

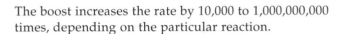

two substrate molecules

substrates contacting active site of enzyme

active site

TRANSITION STATE
(tightest binding but least stable)

end product

enzyme unchanged by the reaction

c

Figure 5.8 Model of an enzyme, hexokinase, at work. Hexokinase catalyzes phosphorylation of glucose. (**a**) A glucose molecule (color-coded *red*) is heading toward the active site, a cleft in the enzyme (*green*). (**b**) When the glucose molecule makes contact with the site, parts of the enzyme briefly close in around it and prod the molecule to enter into the reaction.

(**c**) Induced-fit model of enzyme-substrate interactions. Only when the substrate is bound in place is an enzyme's active site complementary to it. The fit is most precise during the transition state of a reaction. An enzyme-substrate complex is short-lived, for the attractive forces holding it together are usually weak.

Collision can be spontaneous or enzymes can promote it; this makes no difference. Activation energy is like a hill, an energy barrier, that must be surmounted one way or another before a reaction will proceed.

Enzymes can make the energy barrier smaller, so to speak. How? An enzyme has one or more **active sites**. At these surface crevices, an enzyme interacts with its substrates and catalyzes a reaction. Figure 5.8 shows the active site of one enzyme, which activates glucose by catalyzing the attachment of a phosphate group to it.

According to Daniel Koshland's **induced-fit model**, a surface region of each substrate has chemical groups that are almost but not quite complementary to chemical groups in an active site. When substrates first settle into the site, the contact strains some of their bonds. Strained bonds are easier to break, so they promote the formation of new bonds (in the products). Also in the active site, interactions among charged or polar groups often shift the electric charge in substrates, and that redistribution primes substrates for conversion to an activated state.

When substrates fit most precisely in the active site of an enzyme, they are in an activated, *transition* state and will now react spontaneously (Figure 5.8c). And so the reaction must proceed, just as NASA's Rover must roll spontaneously down the Martian hill if something pushes it over the crest (Figure 5.5).

What induces the transition state or gets substrates over the energy barrier once the state is reached? The following are among the mechanisms involved:

1. *Helping substrates get together.* Substrate molecules rarely collide if their concentrations are low. Binding at an active site is like a localized boost in concentration.

The boost increases the rate by 10,000 to 1,000,000,000 times, depending on the particular reaction.

2. *Orienting substrates in positions favoring reaction.* On their own, substrates collide from random directions. By contrast, weak but extensive bonding at an active site puts their chemical groups on precise collision courses much more frequently.

3. *Promoting acid-base reactions.* In many active sites, acidic or basic side groups of amino acids are poised to donate or accept hydrogen atoms from substrates. The loss or addition destabilizes covalent bonds in a substrate and makes them easier to break. Hydrolysis and stepwise electron transfers work this way.

4. *Shutting out water.* Some active sites bind substrates so tightly that some or all of the water molecules that bathed the site are shut out. A nonpolar environment lowers the activation energy for certain reactions, such as the attachment of a carboxyl group ($-COO^-$) to a molecule, by as much as 500,000 times.

Depending on the enzyme, such mechanisms work alone or in combination to bring about the straining and warping that convert substrates to the transition state.

Enzymes catalyze (speed) the rate at which specific reactions reach equilibrium. They do so by lowering the amount of activation energy necessary to make substrates react.

Enzymes change the rate, not the outcome, of a reaction. They only act on specific substrates. And they may catalyze the same reaction repeatedly, as long as substrates are available.

WWW

You probably don't get much done when you feel too hot or cold or out of sorts because you ate too many sour plums or salty potato chips. When the cupboard is bare, you focus on food. Maybe you call a friend to go shopping with you, and if you drive too fast to the grocery store, police tend to slow you down. In such respects, you have a lot in common with the enzymes. They, too, respond to shifts in temperature, pH, and salinity, and to the relative abundances of particular substances. Many even engage helpers for specific tasks. And all normal enzymes respond to metabolic police.

Enzymes and the Environment

Temperature, recall, is a measure of molecular motion. You may think increases in temperature must increase the rate of enzyme-mediated reactions by making the substrates collide more frequently with active sites. That is so, but only until some point on the temperature scale. Past that point—which differs among enzymes—the increased molecular motion disrupts weak bonds holding the enzyme in its three-dimensional shape. Substrates can no longer bind to the active site, so the reaction rate declines sharply, as in Figure 5.9a.

Expose an organism to temperatures that are far higher than it normally encounters and its enzymes will unravel, thereby throwing its metabolic activities into turmoil. This is what happens when sick people develop dangerously high fevers. They usually will die if their internal temperature reaches 44°C (112°F).

Similarly, pH values that rise or sink beyond each enzyme's range of tolerance disrupt enzyme structure and functioning (Figure 5.9c). Nearly all enzymes work best in the pH range between 6 and 8. For example, trypsin is active in a mammal's small intestine, where pH is around 8. Pepsin, a protein-digesting enzyme, is one of the exceptions. It functions in gastric fluid, even though this highly acidic fluid denatures most enzymes.

Enzyme activity also will suffer if the environment gets far saltier than is normally encountered. Extremely high ion concentrations disrupt interactions that help hold most enzymes in their three-dimensional shape.

How Is Enzyme Action Controlled?

Each cell controls its enzyme activity. By coordinating control mechanisms, it maintains, lowers, or raises the concentrations of substances. Controls that adjust how fast enzyme molecules are synthesized affect how many are available for a metabolic pathway. Other controls boost or slow the action of enzyme molecules that were synthesized earlier. For example, enzymes are activated or inhibited by way of allosteric control when a specific substance combines with them at a binding site *other than* the active site. (*Allo*– means different, *steric* means structure, or state.) Figure 5.10 shows two models of the binding, which is reversible.

Picture a bacterium, busily synthesizing tryptophan (and other amino acids) used to construct its proteins. After a bit, protein synthesis slows, so tryptophan is no longer required. But the tryptophan pathway is in full swing. The concentration of its end product (tryptophan) continues to rise. Now **feedback inhibition** kicks in: a cellular change, caused by a specific activity, *shuts down the activity that brought it about*. In this case, a feedback loop starts and ends with a key allosteric enzyme in the pathway. Unused tryptophan molecules bind with the allosteric site. This shuts down the enzyme and blocks the pathway. By contrast, if few tryptophan molecules are available when the demand for them increases, the enzyme remains free of inhibition, and so tryptophan production rises. Such feedback loops can quickly adjust the concentrations of many substances (Figure 5.11).

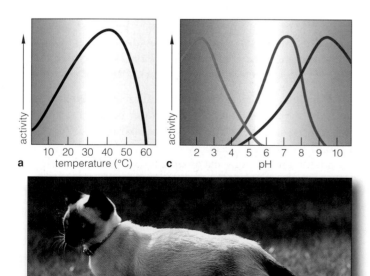

a   temperature (°C)   10 20 30 40 50 60
activity

c   pH   2 3 4 5 6 7 8 9 10
activity

b

Figure 5.9 (**a**) Example of how temperatures that fall outside the range of tolerance for one enzyme influence its activity. (**b**) Siamese cats show observable effects of such changes. Fur on the ears and paws contains more of a dark brown pigment, melanin, than the rest of the body does. A heat-sensitive enzyme controlling melanin production is less active in warmer parts of the body, which end up with lighter fur. (**c**) Diagram showing how the activity of three different enzymes is influenced by pH. One enzyme (the *brown* line) functions best in neutral solutions. Another enzyme (*red* line) functions best in basic solutions, and another (*purple* line) in acidic solutions. ✹

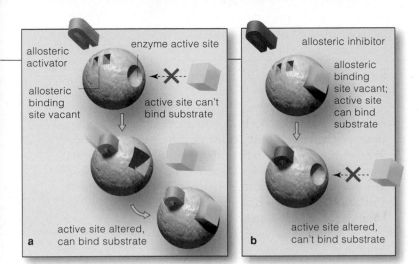

Figure 5.10 Allosteric control. (**a**) Activation of an allosteric enzyme. (**b**) Inhibition of an allosteric enzyme.

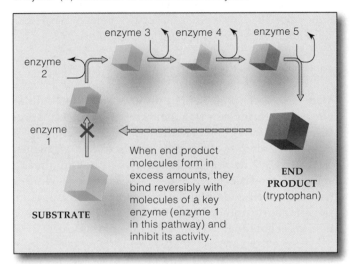

Figure 5.11 Example of feedback inhibition of a metabolic pathway. Five kinds of enzymes act in sequence to convert a substrate to an end product (tryptophan). When end product accumulates, some of the excess molecules bind to molecules of the first enzyme and block the entire pathway.

In humans and other multicelled organisms, control of enzyme activity is just amazing. Cells not only work to keep themselves alive, they work with other cells in ways that benefit the whole body! For example, this vast enterprise relies on hormones, a type of signaling agent. Specialized cells release hormones to the blood. Any cell with receptors for a given hormone responds to it, then its program for building a protein or some other activity changes. The hormone trips controls into action—and the activities of specific enzymes change.

Enzymes function best when the cellular environment stays within limited ranges of temperature, pH, and salinity. The range of tolerance differs among enzymes.

Control mechanisms govern the synthesis of new enzymes and stimulate or inhibit the activity of existing enzymes. By controlling enzymes, cells control the concentrations and kinds of substances available to them.

WWW

Reactant and product molecules contain energy in their chemical bonds. *Making that chemical energy available for metabolic reactions depends on cell membranes.* These help concentrate specific reactant and product molecules in the amounts required for specific reactions. Remember, substances move into and out of cells across the plasma membrane and across organelle membranes.

So picture a cell membrane, with water bathing both sides of its bilayer. Plenty of substances are dissolved in the water, but the kinds and amounts are not the same on the two sides. The membrane itself helps establish and maintain those differences, which are essential for cell functioning. How does it accomplish this feat? Like every cell membrane, it shows selective permeability. *Because of its molecular structure, it allows some substances but not others to cross it in certain ways, at certain times.*

Carbon dioxide, molecular oxygen, and other small, nonpolar molecules readily cross a membrane's lipid bilayer. Although water molecules are polar, some can move through temporary gaps that form in the bilayer when hydrocarbon chains of the lipids flex and bend:

O_2, CO_2, other small, nonpolar molecules, as well as H_2O

By contrast, large, polar molecules such as glucose almost never move freely across the lipid bilayer of a cell membrane. Neither do ions:

$C_6H_{12}O_6$, other large, polar, water-soluble molecules, ions (such as H^+, Na^+, K^+, Ca^{++}, Cl^-)

Such water-soluble substances cross the bilayer through the interior of transport proteins, and water in which they are dissolved crosses with them. As you will see, they do so by mechanisms of passive and active transport.

Cells also change the concentrations of substances by importing and exporting substances in bulk across the plasma membrane. These membrane crossings are called exocytosis and endocytosis.

At any time in a cell's life these mechanisms are operating simultaneously. They are the means by which cells increase, decrease, or maintain concentrations of the molecules and ions that are crucial for metabolism.

Metabolic reactions depend on the chemical energy inherent in concentrated amounts of molecules and ions. Cells have mechanisms for increasing or decreasing those concentrations across the plasma membrane and internal cell membranes.

By now, you probably sense that reactant and product concentrations affect the availability of chemical energy. *But how do cells actually work with those concentrations?* The answer starts with concentration gradients. In this context, "concentration" refers to the number of ions or molecules of a substance in some specified region, such as a volume of fluid or air. And "gradient" means the number in one region is not the same as it is in another. Therefore, a **concentration gradient** is a difference in the number of molecules or ions of a given substance in two adjoining regions.

In the absence of other forces, a substance moves from a region where it is more concentrated to a region where it is less concentrated. *The energy inherent in its individual molecules, which keeps them in constant motion, drives the directional movement.* Although the molecules collide randomly and career back and forth, millions of times a second, the *net* movement is away from the place of greater concentration (and the most collisions).

Diffusion is the name for the net movement of like molecules or ions down a concentration gradient. It is a key factor in the movement of substances across cell membranes and through cytoplasmic fluid. In multicelled organisms, diffusion moves substances to and from cells, the fluids bathing them, and the environment. For example, when oxygen builds up in photosynthetic leaf cells, it diffuses across their plasma membrane, into the air inside the leaf, then into air outside the leaf, where the oxygen concentration is lowest.

Like other substances, oxygen tends to diffuse in a direction established by its *own* concentration gradient, not those of other substances dissolved in the same fluid. You see the outcome of this tendency when you squeeze a drop of dye into a bowl of water. Molecules of dye diffuse to the region where they are less concentrated. And the water molecules move to the region where *they* are not as concentrated (Figure 5.12).

Diffusion is faster when the gradient is steep. In the region of greatest concentration, more molecules are moving outward, compared to the number that are moving in. As the gradient decreases, the difference in the number of molecules moving either way declines. When the gradient finally disappears, individual molecules are still in motion. However, now the total number moving one way or the other during a specified interval is about the same. When the net distribution of molecules becomes nearly uniform throughout two adjoining regions, we call this "dynamic equilibrium."

For charged molecules, transport is influenced by both the concentration gradient and the *electric* gradient (that is, a difference in electric charge across the cell membrane). The diffusion of ions down concentration and electric gradients is vital for cell functions. And remember, like large polar molecules, ions move across cell membranes only at transport proteins.

Passive Transport

Many transport proteins permit ions and other solutes to freely diffuse across a membrane. **Passive transport** is the name for a flow of solutes through the interior of transport proteins, down their concentration gradients. Energetically, the flow costs only what the cell already spent to produce or maintain the gradients; passive transport itself adds nothing more to the energy cost.

Transport proteins span the lipid bilayer, and their interior is able to open on both sides of it (Figure 5.13).

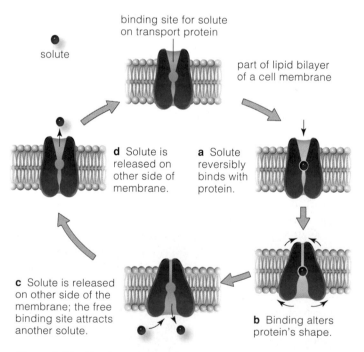

d Solute is released on other side of membrane.

a Solute reversibly binds with protein.

c Solute is released on other side of the membrane; the free binding site attracts another solute.

b Binding alters protein's shape.

binding site for solute on transport protein

solute

part of lipid bilayer of a cell membrane

Figure 5.13 Passive transport across a cell membrane. A solute can move in both directions through transport proteins. In passive transport, the *net* movement will be down its concentration gradient (from higher to lower concentration) until its concentrations are the same on both sides of the membrane.

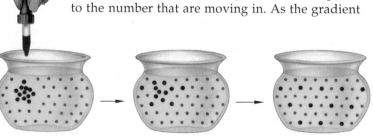

Figure 5.12 Example of diffusion. After a drop of dye enters a bowl of water, the dye molecules *and* the water molecules slowly become evenly dispersed, because each substance shows a net movement down its own concentration gradient.

Figure 5.14 Active transport across a cell membrane. ATP transfers a phosphate group to a transport protein. The transfer sets in motion reversible changes in the protein's shape that result in a greater net movement of solute particles against the concentration gradient.

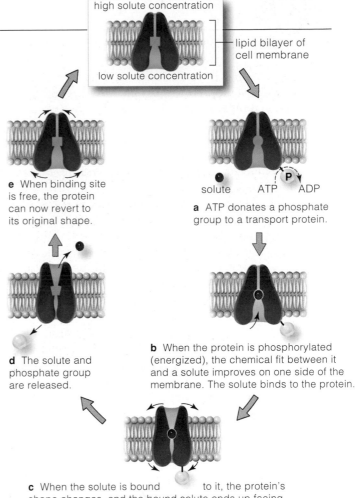

high solute concentration

low solute concentration

lipid bilayer of cell membrane

e When binding site is free, the protein can now revert to its original shape.

a ATP donates a phosphate group to a transport protein.

solute   ATP   ADP

d The solute and phosphate group are released.

b When the protein is phosphorylated (energized), the chemical fit between it and a solute improves on one side of the membrane. The solute binds to the protein.

c When the solute is bound to it, the protein's shape changes, and the bound solute ends up facing the opposite side of the membrane. There, the binding site reverts to the less attractive configuration.

To understand how they work, you have to know they are not rigid blobs of atoms. When the protein interacts with a solute, it changes from one shape to another, then back again. Those changes start when a solute weakly binds to a specific site on the protein's surface. Part of the protein closes in behind the bound solute, and part opens to the other side of the membrane. On that side, the solute dissociates (separates) from the binding site. Think of it as hopping onto the transport protein on one side of the membrane and hopping off on the other side.

Transport proteins allow solutes to move both ways across a cell membrane. In cases of passive transport, the *net* direction of movement during a given interval depends on how many molecules or ions of the solute are making random contact with vacant binding sites in the interior of the proteins (Figure 5.13). The binding and transport simply proceed more often on the side of the membrane where the solute is more concentrated. Because there are more molecules around, the random encounters with the binding sites are more frequent than they are on the other side of the membrane.

If nothing else were going on, the passive two-way transport would proceed until a solute's concentrations were equal across the membrane. But other processes usually affect the outcome. For example, blood delivers glucose to all body tissues, where nearly all cells use it as an energy source and as a building block. Cells can rapidly take up glucose when its blood concentration is high and still maintain the gradient. How? As fast as some glucose molecules diffuse into a cell, other glucose molecules are entering metabolic reactions. By quickly using glucose, then, the cells maintain a concentration gradient that favors the uptake of *more* glucose.

Active Transport

Only in dead cells do solute concentrations become equal on both sides of membranes. Living cells never stop expending energy to pump potassium and other solutes to and from their interior. With **active transport**, energy-driven mechanisms called "membrane pumps" make solutes cross membranes *against* concentration gradients. ATP provides most of the energy to do this.

When ATP gives up one of its phosphate groups to a transport protein, the chemical fit between the solute and the protein binding site improves on one side of the membrane (Figure 5.14*a,b*). After a solute particle binds at the site, the protein's folded shape changes in such a way that the bound solute becomes exposed to fluid bathing the other side of the membrane (Figure

5.14*c*). Now the binding site reverts to its less attractive state, and the solute is released. A less attractive site means fewer molecules or ions of the solute can make the return trip. The *net* movement is to the side of the membrane where the solute is more concentrated.

Operation of such transport systems helps maintain gradients for cell activities, such as muscle contraction, information flow in nervous systems, and the selective movement of dissolved solutes into plant roots.

Molecules or ions of a substance constantly collide because of their inherent energy of motion. The collisions result in diffusion, a net outward movement of a substance from one region into an adjoining region where it is less concentrated.

A concentration gradient is a form of energy that can drive the directional movement of a substance across membranes.

In passive transport, a solute diffuses through the interior of a transport protein that spans the cell membrane. Its net movement is down its concentration gradient.

In active transport, the net diffusion of a solute is against its concentration gradient. The transporting protein must be activated, as by ATP energy, to counter the force of the chemical energy inherent in the gradient.

MOVEMENT OF WATER ACROSS MEMBRANES

By far, more water diffuses across cell membranes than any other substance, so the key factors that influence its directional movement deserve special attention.

Osmosis

Turn on a faucet or watch a waterfall, and the moving water provides a demonstration of bulk flow. **Bulk flow** is the mass movement of one or more substances in response to pressure, gravity, or some other external force. It accounts for some movement of water through complex plants and animals. With each beat, your heart creates fluid pressure that drives a volume of blood, which is mainly water, through interconnected blood vessels. Sap runs inside conducting tissues that thread through maple trees, and this, too, is bulk flow.

What about the movement of water into and out of cells or organelles? A membrane intervening between two regions allows the small, polar water molecules to cross but restricts the passage of ions and large polar molecules. **Osmosis** is the name for the diffusion of water molecules in response to a water concentration gradient between two regions that are separated by a selectively permeable membrane.

Osmotic movement depends on the concentrations of solutes in water on both sides of a membrane. *The side with more solute particles has a lower concentration of water.* To see why this is so, dissolve a small amount of glucose in water. Compared to an equivalent volume of water, the glucose solution has fewer water molecules —because each molecule of glucose occupies some of the space formerly occupied by molecules of water.

It is mainly the *total number* of molecules or ions, not the type of solute, that dictates the concentration of water. Dissolve one mole of an amino acid or urea in 1 liter of water, and the water concentration decreases about as much as it did in the glucose solution. Add one mole of sodium chloride (NaCl) to 1 liter of water, and it dissociates into equal numbers of sodium ions and chloride ions. There are now two moles of solute particles—twice as many as in the glucose solution—so the water concentration has decreased proportionately.

Effects of Tonicity

Given that water molecules tend to move osmotically to a region where water is less concentrated, the direction of their movement tends to be toward a region where solutes are more concentrated. Figure 5.15 illustrates this tendency. Suppose you decide to make a simple observational test of this statement. You construct three sacs out of a membrane that water but not sucrose can cross, and you fill each with a 2M sucrose solution.

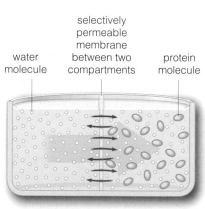

Figure 5.15 Effect of a solute concentration gradient on osmotic movement. Start with a container divided by a membrane that water but not proteins can cross. Pour water into the left compartment. Pour the same volume of a protein-rich solution into the right compartment. There, proteins occupy some of the space. The net diffusion of water in this example is from left to right (large *blue* arrow).

water molecule
selectively permeable membrane between two compartments
protein molecule

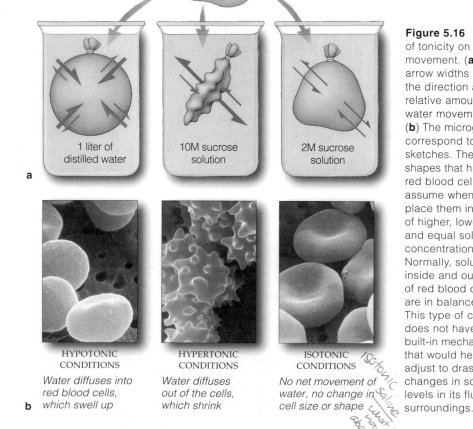

a

1 liter of distilled water

10M sucrose solution

2M sucrose solution

2M sucrose solution

HYPOTONIC CONDITIONS

Water diffuses into red blood cells, which swell up

HYPERTONIC CONDITIONS

Water diffuses out of the cells, which shrink

ISOTONIC CONDITIONS

No net movement of water, no change in cell size or shape

b

Figure 5.16 Effect of tonicity on water movement. (**a**) The arrow widths show the direction and relative amounts of water movement. (**b**) The micrographs correspond to the sketches. They show shapes that human red blood cells assume when you place them in fluids of higher, lower, and equal solute concentrations. Normally, solutions inside and outside of red blood cells are in balance. This type of cell does not have any built-in mechanisms that would help it adjust to drastic changes in solute levels in its fluid surroundings.

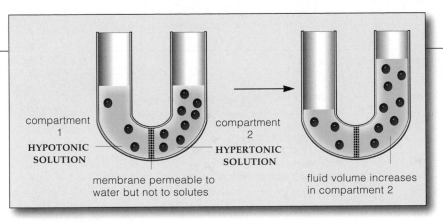

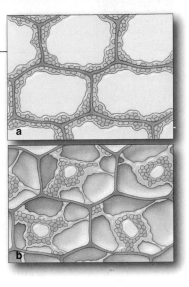

Figure 5.18
An osmotically induced loss of internal fluid pressure (called plasmolysis) in young plant cells. The central vacuole and the cytoplasm shrink; the plasma membrane moves away from the wall.

Figure 5.17 Increase in fluid volume owing to osmosis. In time, the net diffusion across a membrane separating two compartments is equal. Then, the fluid volume in compartment 2 is greater because the membrane is impermeable to solutes.

(*M* stands for *M*olarity, the number of moles of a solute in 1 liter of fluid.) Next you immerse one of the sacs in 1 liter of distilled water (which has no solutes), one in a 10M sucrose solution, and the third in a 2M sucrose solution. In each case, the extent and direction of water movement are dictated by tonicity (Figure 5.16).

Tonicity refers to the relative solute concentrations of two fluids. When two fluids on opposing sides of a membrane differ in solute concentration, the one having fewer solutes is called the **hypotonic solution**, and the one with more is the **hypertonic solution**. Water tends to diffuse from hypotonic to hypertonic fluids. **Isotonic solutions** have the same solute concentrations, so water shows no net osmotic movement from one to the other.

Normally, the fluid inside your cells and the tissue fluid bathing them are isotonic. If a tissue fluid became drastically hypotonic, so much water would diffuse into the cells that they would burst. If the fluid became too hypertonic, an outward diffusion of water would shrivel the cells. Most cells have built-in mechanisms that adjust to shifts in tonicity. Red blood cells do not; Figure 5.16 shows what happened to them during a demonstration of the effects of tonicity differences. That is why patients who are severely dehydrated are given infusions of a solution isotonic with blood. Such solutions move by bulk flow from a bottle positioned above the patient, through a tube, and directly into an incised vein.

Effects of Fluid Pressure

Animal cells generally can avoid bursting by engaging in the ongoing selective transport of solutes across the plasma membrane. Cells of plants and many protistans, fungi, and bacteria also avoid that unpleasant prospect with the help of pressure exerted on their cell walls.

Pressure differences as well as solute concentrations influence the osmotic movement of water. Take a look at Figure 5.17. It shows how water continues to diffuse from a hypotonic solution to a hypertonic one until its concentration becomes the same on both sides of the membrane between them. As you can see, the *volume* of

the formerly hypertonic solution has increased (because its solutes can't diffuse out). Any volume of fluid exerts **hydrostatic pressure**, or a force directed against a wall, membrane, or some other structure that encloses the fluid. The greater the solute concentration of the fluid, the greater will be the hydrostatic pressure it exerts.

As you know, living cells cannot increase in volume indefinitely (Section 4.2). At some point, the hydrostatic pressure that develops in the cell counters the inward diffusion of water. That point is the *osmotic* pressure, the amount of force which prevents any further increase in the volume of a solution.

Think of a young plant cell, with its pliable, primary wall. As it grows, water diffuses into it and hydrostatic pressure increases against the wall. The wall expands, and the cell volume increases. The thin walls are strong enough for the cell's internal fluid pressure to develop to the point where it counterbalances water uptake.

Plant cells are vulnerable to water loss, which can happen when soil dries or becomes too salty. Water stops diffusing in, the cells lose water, and internal fluid pressure drops. Such osmotically induced shrinkage of cytoplasm is called plasmolysis (Figure 5.18). Plants can adjust somewhat to the loss of pressure, as when they actively take up potassium ions against a concentration gradient by mechanisms outlined in the next section.

As you will read in Chapters 33 and 35, hydrostatic and osmotic pressure also influence the distribution of water in the blood, tissue fluid, and cells of animals.

Osmosis is the net diffusion of water between two solutions that differ in water concentration and that are separated by a selectively permeable membrane. The greater the number of molecules and ions dissolved in a solution, the lower its water concentration will be.

Water tends to move osmotically to regions of greater solute concentration (from hypotonic to hypertonic solutions). There is no net diffusion between isotonic solutions.

The fluid pressure that a solution exerts against a membrane or wall also influences the osmotic movement of water.

WWW

EXOCYTOSIS AND ENDOCYTOSIS

Transport proteins can only move small molecules and ions into or out of cells. When it comes to taking in or expelling large molecules or particles, cells use vesicles that form through exocytosis and endocytosis.

By **exocytosis**, a vesicle moves to the cell surface, and the protein-studded lipid bilayer of its membrane fuses with the plasma membrane. While this exocytic vesicle is losing its identity, its contents are released to the surroundings (Figure 5.19*a*).

By three pathways of **endocytosis**, a cell takes in substances next to its surface. In all three cases, a small indentation forms at the plasma membrane, balloons inward, and pinches off. The resulting endocytic vesicle transports its contents or stores them in the cytoplasm (Figure 5.19*b*). By *receptor-mediated* endocytosis, the first pathway, membrane receptors chemically recognize and bind specific substances, such as lipoproteins, vitamins, iron, peptide hormones, growth factors, and antibodies. The receptors become concentrated in tiny indentations

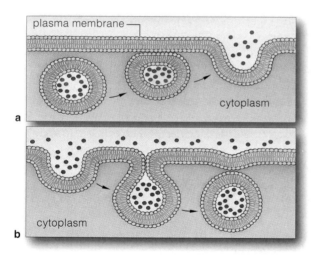

plasma membrane

cytoplasm

a

cytoplasm

b

Figure 5.19 (**a**) Exocytosis. Cells release substances when an exocytic vesicle's membrane fuses with the plasma membrane. (**b**) Endocytosis. A bit of plasma membrane balloons inward beneath water and solutes outside, then it pinches off as an endocytic vesicle that moves into the cytoplasm.

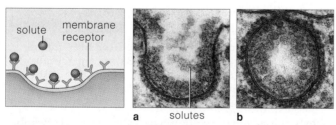

solute

membrane receptor

a solutes **b**

Figure 5.20 Receptor-mediated endocytosis. (**a**) This shallow indentation in the plasma membrane is a coated pit; its side facing the cytoplasm has a basketlike array of protein filaments. Receptor proteins at the outer surface bind solutes (in this case, lipoprotein particles). (**b**) The pit deepens and becomes an endocytic vesicle. The cell will use or store the lipoproteins.

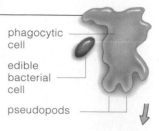

phagocytic cell

edible bacterial cell

pseudopods

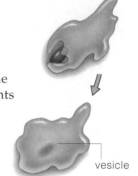

vesicle

Figure 5.21 Phagocytosis, a process by which amoebas, macrophages, and some other cells engulf their targets. A vesicle forms around the target and moves into the cytoplasm, then fuses with lysosomes.

in the plasma membrane (Figure 5.20). Each pit looks like a woven basket on its cytoplasmic side. The basket is made of protein filaments (clathrin) interlocked into stable, geometric patterns. When the pit sinks in the cytoplasm, the basket closes back on itself and becomes the vesicle's structural framework.

The second pathway, *bulk-phase* endocytosis, is less selective. An endocytic vesicle forms around a small volume of extracellular fluid regardless of what kinds of substances happen to be dissolved in it. Bulk-phase endocytosis operates at a fairly constant rate in nearly all eukaryotic cells. By continually pulling patches of plasma membrane into the cytoplasm, this pathway compensates for membrane that steadily departs from the cytoplasm in the form of exocytic vesicles.

The third pathway, **phagocytosis**, is an active form of endocytosis by which a cell engulfs microorganisms, large edible particles, and cellular debris. (Phagocytosis literally means "cell eating.") Amoebas and some other protistans get food this way. In multicelled organisms, macrophages and some other white blood cells engage in phagocytosis when they defend the body against harmful viruses, bacteria, and other threats to health.

A phagocytic cell gets busy after a target binds with certain receptors that bristle from its plasma membrane. Binding sends signals into the cell. The signals trigger a directional assembly and crosslinking of microfilaments into a dynamic, ATP-requiring network just beneath the plasma membrane. The network contracts in ways that squeeze some cytoplasm toward the cell margins, thus forming lobes called pseudopods (Figure 5.21). The pseudopods flow over the target and fuse at their tips. The result is a phagocytic vesicle, which sinks into the cytoplasm. There it fuses with lysosomes, the organelles of intracellular digestion in which trapped items are digested to fragments and smaller, reusable molecules.

By exocytosis, a cytoplasmic vesicle fuses with the plasma membrane, so that its contents are released outside the cell. By endocytosis, a small patch of the plasma membrane sinks inward and seals back on itself, forming a vesicle inside the cytoplasm. Membrane receptors often mediate this process.

WWW

5.9 SUMMARY

1. Cells store, break down, and dispose of substances by acquiring and using energy and raw materials from outside sources. Metabolism, the sum of these energy-driven activities, underlies the survival of organisms.

2. Two laws of thermodynamics affect life. *First*, energy undergoes conversion from one form to another, but its total amount never increases or decreases as a result of conversions. Thus, the total amount of energy in the universe holds constant. *Second*, energy spontaneously flows in one direction, from forms of higher quality to forms of lower quality.

   a. All matter has some amount of potential energy (as measured by the capacity to do work) by virtue of its position in space and the arrangement of its parts.

   b. Potential energy may be transformed into kinetic energy, the energy of motion. Mechanical movements and heat, which corresponds to the degree of molecular motion, are two common forms of kinetic energy.

   c. Chemical energy (the potential energy inherent in molecular bonds) is often measured in kilocalories.

3. Like all organized systems, the cell tends to become disorganized without energy. It inevitably loses some of its chemical potential energy during every metabolic reaction, mainly in the form of heat. It stays organized and alive as long as it counters its energy expenditures (outputs) with energy replacements (inputs).

4. The sun is life's primary energy source. In plants and other photosynthetic organisms, cells trap energy from the sun and convert it to chemical bond energy of organic compounds. Plants, then organisms that feed on plants and one another, use energy that was stored in those organic compounds to do cellular work.

5. Exergonic reactions (energy out) end with a net loss in energy. Endergonic reactions (energy in) end with a net gain in energy. Cells conserve energy by coupling energy-releasing reactions with energy-requiring ones.

6. ATP is the main energy carrier in cells. ATP forms when a phosphate group or inorganic phosphate is attached to ADP. It gives up energy at many reaction sites when it phosphorylates reactants or intermediates, which thus become primed to enter specific reactions.

7. Much of the cellular work triggered by ATP involves a transfer of electrons from one substance to another. We also call the transfers oxidation-reduction reactions.

8. Metabolic pathways are orderly, stepwise sequences of enzyme-mediated reactions. Table 5.1 summarizes their key participants. Energy-rich organic compounds are assembled from smaller molecules of lower energy content in the biosynthetic pathways. By contrast, energy-rich molecules are broken down to smaller ones of lower energy content in the degradative pathways.

Table 5.1 Main Participants in Metabolic Pathways

SUBSTRATE	Substance that enters a metabolic reaction or pathway; also called a reactant
INTERMEDIATE	Substance formed between reactants and end products of a pathway
END PRODUCT	Substance remaining at end of reaction or pathway
ENZYME	Usually a protein that enhances reaction rates
COFACTOR	Coenzyme (such as NAD$^+$) or metal ion; assists enzymes or taxis electrons, hydrogen, or functional groups between reaction sites
ATP	Main energy carrier in cells; couples energy-releasing reactions with energy requiring ones
TRANSPORT PROTEIN	Protein that passively assists substances across a cell membrane or actively pumps them across

9. Enzymes are catalysts; they greatly enhance the *rate* of a reaction involving specific substrates but do not alter the outcome. Nearly all enzymes are proteins, although some RNAs also show catalytic activity.

   a. Enzymes lower the activation energy necessary to start a reaction. They bind substrates at an active site, strain its bonds, and make bonds easier to break.

   b. Each kind of enzyme operates best within limited ranges of temperature, pH, and salinity.

   c. Controls stimulate or inhibit enzyme activity at key steps in metabolic pathways. They help coordinate the kinds and amounts of substances available.

10. Molecules or ions of a substance tend to move from a region of higher to lower concentration. Movement in response to a concentration gradient is called diffusion.

   a. Diffusion rates are influenced by the steepness of the concentration gradient, temperature, and molecular size, as well as by gradients in electrical charge and pressure that may occur between two regions.

   b. Cells have built-in mechanisms that work with or against gradients to move solutes across membranes.

   c. Metabolism requires chemical energy inherent in concentration and electric gradients across cell membranes.

11. Oxygen, carbon dioxide, and other small nonpolar molecules diffuse across a membrane's lipid bilayer. Ions, glucose, and other large, polar molecules cross it with the passive or active help of transport proteins. Water moves through proteins and through the bilayer.

12. Transport proteins bind specific solutes on one side of a cell membrane, and they shunt solutes to the other side through reversible changes in their shape.

   a. Passive transport does not require energy inputs; the protein allows a solute simply to diffuse through its interior, in the direction of the concentration gradient.

   b. Active transport requires energy boosts, as from ATP. The protein pumps a solute across the membrane against its concentration gradient.

13. Osmosis is the diffusion of water across a selectively permeable membrane in response to a concentration gradient.

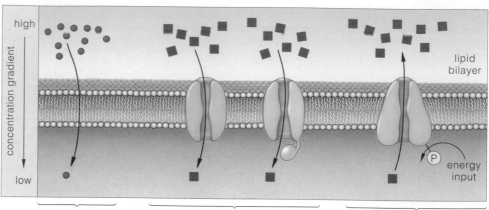

Figure 5.22 Summary of the major mechanisms by which solutes cross cell membranes. (Exocytosis and endocytosis proceed only at the plasma membrane.)

high

concentration gradient

low

lipid bilayer

P energy input

DIFFUSION ACROSS LIPID BILAYER
Lipid-soluble substances as well as water diffuse across.

PASSIVE TRANSPORT
Water-soluble substances, and water, diffuse through interior of transport proteins. No energy boost required. Also called facilitated diffusion.

ACTIVE TRANSPORT
Specific solutes are pumped through interior of transport proteins. Requires energy boost.

a. We can predict the direction in which molecules of water will move between two solutions by comparing their tonicity. Tonicity is a measure of the concentration of solutes in any solution relative to the concentration of solutes in another solution. Extracellular fluid compared with cytoplasmic fluid is an example.

b. Water tends to move from a hypotonic solution (one with the lower solute concentration) into a hypertonic solution (one with the higher solute concentration).

c. When two fluids are isotonic, their solute concentrations are equal, and water will show no net movement in either direction.

14. Cells acquire and get rid of raw materials in bulk by the processes of exocytosis and endocytosis.

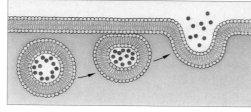

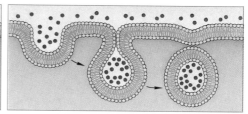

EXOCYTOSIS
Vesicle in cytoplasm moves to plasma membrane, fuses with it; contents released to the outside.

ENDOCYTOSIS
Vesicle forms from a patch of inward-sinking plasma membrane, enters cytoplasm.

a. By exocytosis, a cytoplasmic vesicle moves to the plasma membrane, and then its membrane fuses with it. The contents of the vesicle are automatically released to the outside.

b. By endocytosis, a small indentation forms on the plasma membrane, then it sinks into the cytoplasm and seals back on itself as a vesicle. Receptor-mediated endocytosis depends on recognition of specific solutes. Bulk-phase endocytosis is the indiscriminate uptake of extracellular fluid, and phagocytosis is the engulfment of large particles, cell parts, or whole cells.

15. Figure 5.22 summarizes the routes by which cells acquire and rid themselves of materials for metabolism.

Review Questions

1. State the first and second laws of thermodynamics. Does life violate the second law? *5.1*

2. Define and give examples of potential energy and kinetic energy. What is the potential energy of molecules called? *5.1*

3. Define mechanical work, chemical work, and electrical work as accomplished by cells. *5.1*

4. Give examples of a change in potential energy involved in (a) mechanical work and (b) chemical work. *5.1*

5. Make a simple diagram of the ATP molecule. Highlight which part of the molecule can be transferred to another molecule and later replaced. *5.2*

6. What is an oxidation-reduction reaction, and what is its functional connection to ATP? *5.2*

7. Define and describe the four main features of enzymes. *5.3*

8. Define activation energy, then state four ways in which enzymes may lower it. *5.3*

9. Briefly describe the induced-fit model of enzyme-substrate interactions. *5.3*

10. Define feedback inhibition as it relates to the activity of an allosteric enzyme. *5.4*

11. Define and give examples of the selective permeability of cell membranes. *5.5*

12. Define diffusion. Does diffusion occur in response to a solute concentration gradient, an electric gradient, a pressure gradient, or some combination of these? *5.6*

13. If all transport proteins can shunt substances across cell membranes by changing their shape, then how do the passive transporters differ from the active transporters? *5.6*

14. Define osmosis. Briefly explain how solute concentrations of two solutions on either side of a membrane can influence the osmotic movement of water down the water concentration gradient. *5.7*

15. Distinguish among hypertonic, hypotonic, and isotonic solutions. Does each term refer to a property inherent in a given type of solution? Or are the terms used only when comparing one solution to another? *5.7*

16. Define exocytosis and endocytosis. Describe the main features of the three pathways of endocytosis. *5.8*

1. _____ is the primary source of energy for life on Earth.
   a. Food   d. The sun
   b. Water   e. The supermarket
   c. ATP   f. Electron flow

2. An _____ reaction is an uphill run.
   a. endergonic   c. ATP-assisted
   b. exergonic   d. both a and c

3. Phosphate-group transfers from ATP to another molecule are a _____ mechanism for delivering energy.
   a. rapid   c. near-universal
   b. renewable   d. all of the above

4. Enzymes are _____ .
   a. enhancers of reaction rates   d. not influenced by salinity
   b. influenced by temperature   e. a through c
   c. influenced by pH   f. all of the above

5. Immerse a living cell in a hypotonic solution, and water will tend to _____ .
   a. move into the cell   c. show no net movement
   b. move out of the cell   d. move in by endocytosis

6. _____ can readily diffuse across a lipid bilayer.
   a. Glucose   c. Carbon dioxide
   b. Oxygen   d. b and c

7. Potassium ions cross a membrane through transport proteins that receive an energy boost. This is an example of _____ .
   a. passive transport   c. facilitated diffusion
   b. active transport   d. a and c

8. Match each substance with the most suitable description.
   ____ mainly ATP   a. reactant or
   ____ adjusts gradients at membrane   substrate
   ____ substance entering a reaction   b. enzyme
   ____ substance formed while   c. coenzyme NAD⁺
       a reaction is proceeding   d. intermediate
   ____ substance at end of reaction   e. product
   ____ enhances reaction rate   f. energy carrier
   ____ taxis electrons, atoms, or   g. transport protein
       functional groups to new site

Critical Thinking

1. *Pyrococcus furiosus* thrives at 100°C, the boiling point of water. This species of bacterium was discovered growing in a volcanic vent in Italy. Enzymes isolated from *P. furiosus* cells do not function well below 100°C. What is it about the structure of these enzymes that allows them to remain stable and active at such high temperatures? (Hint: Review Section 3.5, which summarizes the interactions that maintain protein structure.)

2. The bacterium *Vibrio cholerae* causes the disease *cholera*. Infected people have severe diarrhea and may lose up to twenty liters of fluid in a day. The bacterium enters the body if someone drinks contaminated water, then it adheres to the lining of the intestines. It secretes a metabolic product that is toxic to cells of the lining, and they in turn start secreting chloride ions (Cl⁻). Sodium ions (Na⁺) follow the chloride ions into the fluid in the intestines. Explain how the sequence of ion movements causes the massive fluid loss.

3. Many cultivated fields in California require heavy irrigation. Over the years, most of the water has evaporated from the soil, leaving behind all of the irrigation water's solutes. What kinds of problems might the altered soil conditions cause for plants?

4. Imagine that you are a juvenile shrimp living in an *estuary*, where freshwater draining from the land mixes with saltwater

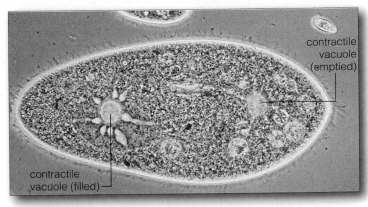

Figure 5.23 Contractile vacuoles of *Paramecium*, a protistan.

from the sea. Many people who own homes around a large lake want boat access to the sea, so they ask their city for permission to build a canal between the lake and estuary. If they succeed, what might happen to you and other estuary inhabitants?

5. Water moves osmotically into *Paramecium*, a single-celled protistan of aquatic habitats. If unchecked, the influx would bloat the cell and rupture its plasma membrane. An energy-requiring mechanism involving contractile vacuoles expels the excess (Figure 5.23). Water enters tubelike extensions of this organelle and collects in a central space in the vacuole. When full, the vacuole contracts and squirts excess water out of a pore that opens to the outside. Are the fluid surroundings hypotonic, hypertonic, or isotonic relative to *Paramecium*'s cytoplasm?

Selected Key Terms

activation energy *5.3*
active site *5.3*
active transport *5.6*
ATP *5.2*
bioluminescence *CI*
bulk flow *5.7*
chemical energy *5.1*
chemical equilibrium *5.3*
concentration gradient *5.6*
diffusion *5.6*
endergonic reaction *5.2*
endocytosis *5.8*
entropy *5.1*
enzyme *5.3*
exergonic reaction *5.2*
exocytosis *5.8*
feedback inhibition *5.4*

first law of thermodynamics *5.1*
heat *5.1*
hydrostatic pressure *5.7*
hypertonic solution *5.7*
hypotonic solution *5.7*
induced-fit model *5.3*
isotonic solution *5.7*
kinetic energy *5.1*
metabolic pathway *5.2*
metabolism *CI*
osmosis *5.7*
oxidation-reduction reaction *5.2*
passive transport *5.6*
phagocytosis *5.8*
phosphorylation *5.2*
potential energy *5.1*
second law of thermodynamics *5.1*

Readings See also www.infotrac-college.com

Adams, S. 22 October 1994. "No Way Back." *New Scientist*. A refreshing look at the second law of thermodynamics.

Ritter, P. 1996. *Biochemistry: A Foundation*. Brooks/Cole: Pacific Grove, California. Chapter 8 is about as simple an introduction to bioenergetics as you will get.

WWW *http://www.brookscole.com/biology*

Practice quiz questions, hypercontents, BioUpdates, and critical thinking. The Brooks/Cole Biology Resource Center provides a wealth of information fully organized and integrated by chapter.

HOW CELLS ACQUIRE ENERGY

Sunlight and Survival

Think about the last time you were hungry and craved a bit of apple, maybe, or lettuce, chicken, pizza, bread, or any other kind of food. Where did it come from? For the answer, look past the refrigerator, the market or restaurant, or even the farm. Look to individual plants—the starting point for nearly all of the food you put into your mouth. Plants use environmental sources of energy and raw materials to build glucose and all other organic compounds necessary for survival. Organic compounds, recall, are built on a framework of carbon atoms. So the questions become these:

1. *Where does the carbon come from in the first place?*

2. *Where does the energy come from to drive the synthesis of carbon-based compounds?*

The answers vary according to an organism's mode of nutrition.

Plants generally are "self-nourishing" organisms, or **autotrophs**. As their carbon source, they use carbon dioxide (CO_2), a gaseous compound present in the air and dissolved in aquatic habitats. Plants, some bacteria, and many protistans are *photo*autotrophs, which means they capture sunlight energy to drive a metabolic process called **photosynthesis**. By this process, energy from the sun is converted to chemical bond energy of ATP, then ATP gives up energy at sites where glucose and other organic compounds are synthesized. An enzyme helper, the coenzyme $NADP^+$, typically picks up electrons and hydrogen, then delivers them to those same sites.

Many other organisms are **heterotrophs**, meaning they must feed on autotrophs, one another, and organic wastes. (*Hetero–* means other, as in "being nourished by other organisms.") That is how most bacteria, many protistans, and all fungi and animals stay alive. Unlike plants, heterotrophs cannot nourish themselves with sunlight and raw materials from their environment.

How do we know such things? We didn't have a clue until observational and experimental tests began in the mid-seventeenth century. Before then, most people assumed that plants got the raw materials they needed to make food from the soil they grew in. By 1882 a few chemists had an inkling that plants use sunlight, water, and something in the air to make food. T. Englemann, a botanist, was curious: What parts of sunlight do plants favor? As he already knew, when plants and algae are photosynthesizing, they release oxygen. He also knew that, like many other organisms, certain free-living bacterial cells require oxygen for aerobic respiration. Those cells move toward places where conditions favor their activities and away from unfavorable conditions.

Englemann hypothesized: If bacterial cells require oxygen, then they will move toward places where photosynthesis is proceeding most effectively. He put a strand of the green alga *Spirogyra* in a water droplet that contained such bacteria. He mounted the alga on a microscope slide, then used a crystal prism to break up a beam of sunlight and to cast a spectrum of colors across it. As Figure 6.1 shows, bacterial cells became concentrated primarily where violet and red light illuminated the algal strand. Englemann concluded that light of those colors is the most effective for oxygen production—hence for photosynthesis.

Such observations yielded insight into the process of photosynthesis. Ultimately, they also helped reveal a great pattern in nature, because nearly all organisms depend on photosynthesis, *the main pathway by which carbon and energy enter the web of life.* Once a cell makes or takes up organic compounds, it uses or stores them. *All* autotrophs and heterotrophs store energy in organic compounds, and that energy can be released by other processes. Aerobic respiration, the most common energy-releasing process, requires oxygen to run to completion. Use Figure 6.2 as your preview of the chemical links between photosynthesis and aerobic respiration—the focus of this chapter and the next.

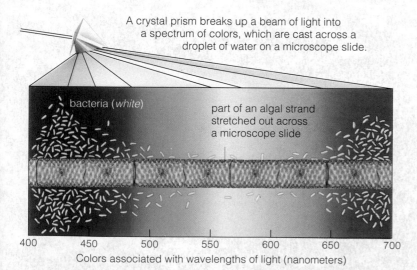

A crystal prism breaks up a beam of light into a spectrum of colors, which are cast across a droplet of water on a microscope slide.

bacteria (*white*)

part of an algal strand stretched out across a microscope slide

400   450   500   550   600   650   700

Colors associated with wavelengths of light (nanometers)

Figure 6.1 Results from T. Englemann's observational test that correlated portions of visible light with photosynthesis in *Spirogyra*, a strandlike green alga. Large numbers of oxygen-requiring bacteria moved to the colors where algal cells were releasing the most oxygen, which is a by-product of their photosynthetic activity.

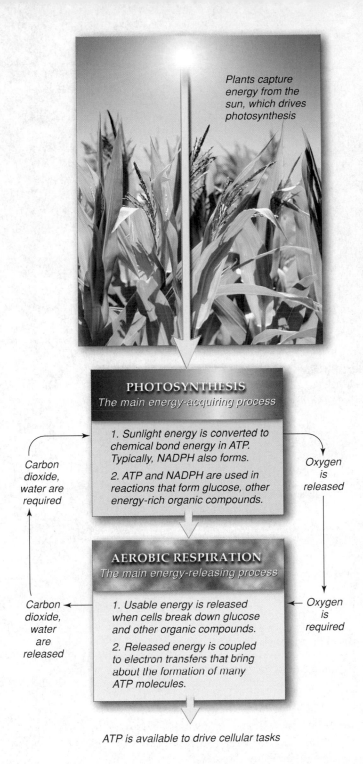

Plants capture energy from the sun, which drives photosynthesis

PHOTOSYNTHESIS
The main energy-acquiring process

1. Sunlight energy is converted to chemical bond energy in ATP. Typically, NADPH also forms.

2. ATP and NADPH are used in reactions that form glucose, other energy-rich organic compounds.

Carbon dioxide, water are required

Oxygen is released

AEROBIC RESPIRATION
The main energy-releasing process

1. Usable energy is released when cells break down glucose and other organic compounds.

2. Released energy is coupled to electron transfers that bring about the formation of many ATP molecules.

Carbon dioxide, water are released

Oxygen is required

ATP is available to drive cellular tasks

Figure 6.2 Links between photosynthesis (the main energy-requiring process) and aerobic respiration (the main energy-releasing process) in the world of life.

At times, such processes might seem far removed from your daily interests. But remember this: The food that nourishes you and most other organisms cannot be produced or used without them. You will be returning to this point in later chapters. It provides perspective on many important issues, including nutrition and dieting, agriculture and human population growth, genetic engineering, as well as the impact of pollution on our sources of food—hence on our survival.

KEY CONCEPTS

1. Plants, some bacteria, and many protistans use sunlight energy, carbon dioxide, and water to produce glucose and other organic compounds, which have a backbone of carbon atoms. The metabolic process by which they accomplish this is called photosynthesis.

2. Photosynthesis is the main route by which carbon and energy enter the web of life.

3. In the first stage of photosynthesis, sunlight energy is trapped and converted to chemical bond energy in ATP molecules. Typically, water molecules are split, their electrons and hydrogen atoms are picked up by $NADP^+$ to form NADPH, and their oxygen atoms are released as a by-product.

4. In the second stage of photosynthesis, ATP delivers energy to reaction sites where glucose is synthesized. Carbon dioxide provides carbon and oxygen for the reactions, and NADPH provides the electrons and hydrogen.

5. In the photosynthetic cells of plants and in many protistans, both stages of the reactions proceed inside organelles called chloroplasts.

6. Often we summarize photosynthesis this way:

$$12H_2O + 6CO_2 \xrightarrow{\text{LIGHT ENERGY}} 6O_2 + C_6H_{12}O_6 + 6H_2O$$

WATER   CARBON DIOXIDE   OXYGEN   GLUCOSE   WATER

Where the Reactions Take Place

Let's start with a look at **chloroplasts**, the organelles of photosynthesis in plants and in the protistans called algae. Chloroplasts, recall, specialize in producing and briefly storing food (Section 4.8). Figure 6.3 shows the structure and functional zones of one type. All chloroplasts have two outer membranes, which enclose a largely fluid interior, the **stroma**. Still another membrane weaves through the stroma; we call it a thylakoid membrane system. In many species, parts of the system are often folded repeatedly into disk-shaped sacs, or **thylakoids**, stacked one on top of the other as in Figure 6.3d,e. Each stack of disks is a granum (plural, grana). Membranous channels interconnect the stacks.

Two stages of photosynthesis—the *light-dependent* and *light-independent* reactions—proceed in the interior of a chloroplast. The first stage occurs at the thylakoid membrane system (Figure 6.3e,f). The interconnecting spaces within all thylakoids and channels form a single compartment in which hydrogen ions (H⁺) accumulate. ATP forms when the ions flow across the membrane. The second stage of photosynthesis, the set of reactions by which sugars are assembled, occurs in the stroma.

Energy and Materials for the Reactions

In the light-dependent reactions, absorption of energy from the sun drives the formation of ATP from the chloroplast's pool of ADP and inorganic phosphate (P_i). In this way, energy from the sun becomes temporarily stored as chemical bond energy of ATP. Typically, water molecules are split and a coenzyme, $NADP^+$, picks up their electrons and hydrogen atoms to form NADPH.

In the light-independent reactions, the ATP donates energy to sites where carbon, hydrogen, and oxygen are assembled into glucose ($C_6H_{12}O_6$). NADPH provides the electrons and hydrogen atoms. And carbon dioxide (CO_2) provides the carbon and oxygen atoms.

Overall, the reactions of photosynthesis are often summarized as a simple equation:

$$12H_2O + 6CO_2 \xrightarrow{\text{LIGHT ENERGY}} 6O_2 + C_6H_{12}O_6 + 6H_2O$$

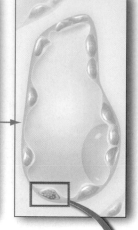

upper surface of leaf   photosynthetic cells

a Part of a leaf

b Cutaway view of a small section from the leaf. Its upper and lower surfaces enclose many photosynthetic cells.

c One of the photosynthetic cells

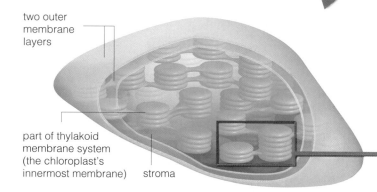

two outer membrane layers

part of thylakoid membrane system (the chloroplast's innermost membrane)   stroma

d Cutaway view of one of the chloroplasts inside the photosynthetic cell shown in (**c**).

Figure 6.3 Zooming in on the sites of photosynthesis inside one of the leaves of a typical plant.

Remember the description of tracers in Section 2.2? By attaching radioisotopes to atoms of the two kinds of reactants given in the summary equation, researchers showed where each atom ends up in the three products:

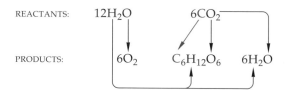

REACTANTS: $12H_2O$ $6CO_2$

PRODUCTS: $6O_2$ $C_6H_{12}O_6$ $6H_2O$

In the summary equation we show *glucose* as the carbon-rich end product of the reactions to keep the chemical bookkeeping simple. But you will find very little glucose in a chloroplast. Each newly formed glucose molecule has a phosphate group attached; it is primed to react

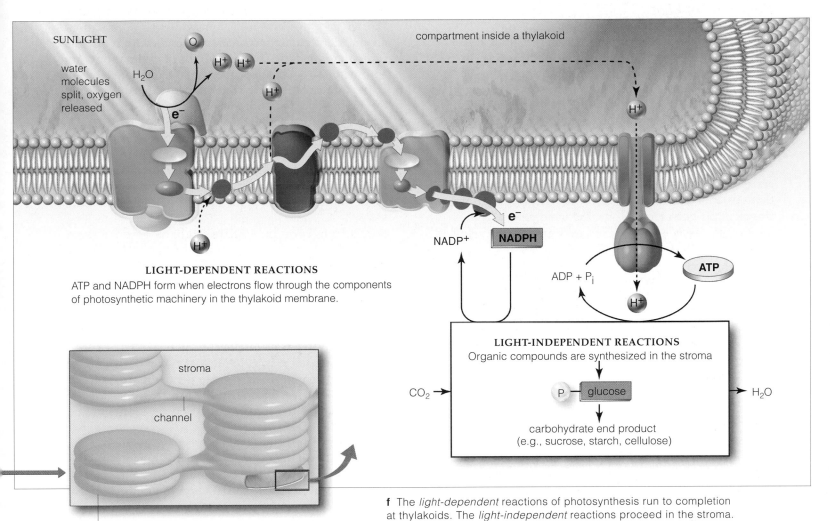

SUNLIGHT

water molecules split, oxygen released

H_2O

O

H^+ H^+

H^+

H^+

e^-

H^+

compartment inside a thylakoid

H^+

e^-

$NADP^+$

NADPH

ADP + P$_i$

H^+

ATP

LIGHT-DEPENDENT REACTIONS

ATP and NADPH form when electrons flow through the components of photosynthetic machinery in the thylakoid membrane.

stroma

channel

LIGHT-INDEPENDENT REACTIONS

Organic compounds are synthesized in the stroma

CO_2

P glucose

H_2O

carbohydrate end product (e.g., sucrose, starch, cellulose)

f The *light-dependent* reactions of photosynthesis run to completion at thylakoids. The *light-independent* reactions proceed in the stroma.

e One granum (a stack of disks, or thylakoids, that are part of the thylakoid membrane system)

with something else. Almost always, it is used at once in reactions that form sucrose, starch, and cellulose. Just keep in mind that these are the main end products of photosynthesis, even though we "stop" with glucose.

The diagram below is a simplified version of Figure 6.3. Where you see it repeated in sections to follow, use it as a reminder to refer back to Figure 6.3 to reinforce your grasp of how the details fit into the big picture:

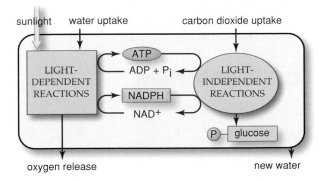

sunlight   water uptake   carbon dioxide uptake

LIGHT-DEPENDENT REACTIONS

ATP

ADP + P$_i$

NADPH

NAD$^+$

LIGHT-INDEPENDENT REACTIONS

P glucose

oxygen release   new water

We turn next to the details of photosynthesis. Before we do, think about this: Two thousand chloroplasts, lined up single file, would be no wider than a dime. Imagine all of the chloroplasts in just one corn or rice plant—each a tiny factory for producing sugars and starch—and you may get an inkling of the magnitude of the metabolic events required to feed you and every other organism on this planet.

Chloroplasts, organelles of photosynthesis in plants and many protistans, specialize in food production.

In the first stage of photosynthesis, sunlight energy drives the formation of ATP and NADPH, and oxygen is released. In chloroplasts, this stage occurs at thylakoids.

The second stage occurs in the stroma, where energy from ATP drives glucose formation. For these synthesis reactions, carbon dioxide provides the carbon and oxygen atoms, and NADPH provides the electrons and hydrogen atoms.

WWW

SUNLIGHT AS AN ENERGY SOURCE

Properties of Light

Photosynthesis starts with energy from the sun that has radiated across space in undulating motion, a bit like waves crossing a sea. The horizontal distance between crests of every two successive waves is a **wavelength** (Figure 6.4a). There are many different wavelengths of radiant energy. The entire range of all the wavelengths represents the electromagnetic spectrum (Figure 6.4b).

Photoautotrophs intercept only about 1 percent of the wavelengths that reach the Earth's surface. Yet they are entry points for a grand, one-way flow of energy that sustains nearly all webs of life. They can absorb wavelengths between 380 and 750 nanometers. That is the range of *visible* light, which we humans and many other organisms perceive as various colors. Wavelengths shorter than this, including ultraviolet (UV) radiation, are energetic enough to break the bonds within organic compounds and kill cells. Hundreds of millions of years ago, before a layer of ozone (O_3) accumulated in the upper atmosphere, only water shielded Earth's surface from ultraviolet radiation. The lethal bombardment kept photoautotrophs below the surface of the seas.

When absorbed by matter, energy of visible light can be measured as if it were organized in packets, which we call photons. Each type of photon has a fixed amount of energy. Those having the most energy travel as the shortest wavelengths, which correspond to blue-violet light. Those having the least energy travel as long wavelengths corresponding to red light (Figure 6.4b).

The Rainbow Catchers

Collectively, wavelengths of visible light look white to us. A crystal prism intercepting white light can sort it out into its component colors by bending the different wavelengths by different degrees. In the 1800s, recall, T. Englemann used prisms to identify which wavelengths drive photosynthesis in a green alga (Figure 6.1). But molecular biology was far in the future, so he was not able to identify the actual bridge between sunlight and photosynthetic activity.

Pigments, molecules that absorb wavelengths of light, are the bridge. Organisms synthesize pigments and put them to many uses. Most pigments absorb only certain incoming wavelengths. The rest bounce off the pigment molecule or travel through it (they are either reflected or transmitted). A few pigments can absorb so many different wavelengths that they appear dark or black.

An absorption spectrum is a graph of how effectively a particular pigment absorbs different wavelengths of visible light. Think about the **chlorophylls**. *Chlorophylls, the main photosynthetic pigments, absorb the wavelengths*

400-nanometer
wavelength

750-nanometer
wavelength

a

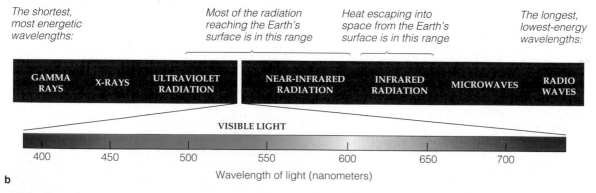

The shortest, most energetic wavelengths:

Most of the radiation reaching the Earth's surface is in this range

Heat escaping into space from the Earth's surface is in this range

The longest, lowest-energy wavelengths:

GAMMA RAYS	X-RAYS	ULTRAVIOLET RADIATION	NEAR-INFRARED RADIATION	INFRARED RADIATION	MICROWAVES	RADIO WAVES

VISIBLE LIGHT

400   450   500   550   600   650   700

Wavelength of light (nanometers)

b

Figure 6.4 (**a**) Examples of wavelengths, the horizontal distance between crests of successive waves. (**b**) The electromagnetic spectrum. Visible light is one of many forms of electromagnetic radiation in the spectrum.

Figure 6.5 (a) Two absorption spectra that indicate the efficiency with which chlorophylls *a* and *b* respond to different wavelengths. (b) Absorption spectra for a carotenoid (beta-carotene) and for a phycobilin.

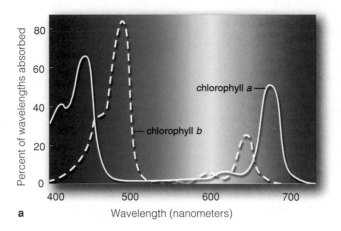

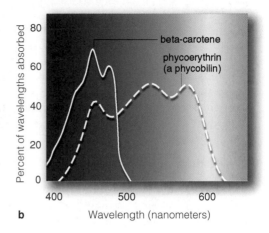

that are most efficient at driving photosynthesis. As Figure 6.5a indicates, chlorophyll molecules can absorb every wavelength of visible light except a few yellow-green to green ones. They reflect or transmit the unabsorbed wavelengths, which is why chlorophyll-rich plant parts appear green to us.

Chlorophyll *a*, a grass-green pigment that absorbs blue-violet and red wavelengths, is a key player in the light-dependent reactions. Chlorophyll *b* absorbs blue and red-orange wavelengths. In chloroplasts, it is one of several *accessory* pigments, which harvest wavelengths that chlorophyll *a* misses. The **carotenoids** are accessory pigments that absorb light of blue-violet and blue-green wavelengths, and reflect yellow, orange, and red ones. Carotenoids color many flowers, vegetables, and fruits. They are less abundant than chlorophylls in green leaves, but in many plant species, they are visible in autumn (Figure 6.6). Each autumn, tourists visit New England and spend about a billion dollars to observe the demise of chlorophyll in great forests of deciduous trees.

Other accessory pigments include the phycobilins, the signature pigments of red algae and cyanobacteria.

Why Aren't All Pigments Black?

If trapping energy is so vital to life, why doesn't each photosynthetic pigment go after the whole rainbow? In other words, why isn't each one black? Remember, the earliest photoautotrophs evolved in the seas—and so did their pigments. Ultraviolet and blue light does not penetrate water as deeply as green and red light.

Did natural selection favor the evolution of different pigments at different depths? Maybe. Red algae evolved in deep, dimly lit waters, and the kinds that still live there *are* nearly black. Green algae evolved in shallow water, and the main pigments of the kinds still living there absorb red light. Their accessory pigments help harvest photon energy in the sunlit waters—and some even function as shields against ultraviolet radiation.

Figure 6.6 Leaf color. In intensely green leaves, chlorophyll molecules are continually synthesized and broken down, and they mask the presence of carotenoids and other accessory pigments. In autumn, however, chlorophyll synthesis lags behind chlorophyll breakdown in many species. The other leaf pigments are unmasked, and more colors show through.

Also in autumn, water-soluble pigments called anthocyanins accumulate in leaf cells. They appear red if fluids moving through plants are slightly acidic, blue if the fluids are basic (alkaline), or colors in between if fluids are of intermediate pH. Soil conditions contribute to pH.

Radiation from the sun travels in waves that differ in length and energy content. We perceive the different wavelengths of visible light as different colors. We measure their energy as distinct packets called photons.

Pigment molecules are the molecular bridge between the sun's energy and photosynthesis. Each kind absorbs certain wavelengths and reflects or transmits the rest.

Chlorophyll *a* is the main photosynthetic pigment. Other, accessory pigments include chlorophyll *b* and carotenoids.

WWW

THE LIGHT-DEPENDENT REACTIONS

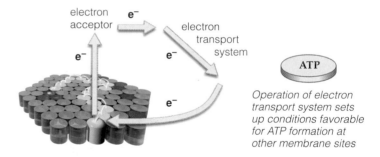

We are now ready to look more closely at the first stage of photosynthesis, the **light-dependent reactions**, by tracking three events that proceed in the chloroplast. *First*, pigments absorb light energy and give up excited electrons, which enter electron transport systems. *Second*, water molecules are split, ATP and NADPH form, and oxygen is released. *Third*, the pigments that gave up electrons in the first place get electron replacements.

What Happens to the Absorbed Energy?

Only certain atoms of a photosynthetic pigment absorb photons. Absorption destabilizes the way electrons are distributed around the atomic nucleus, and these move to a higher energy level, as outlined in Section 2.2. The excited electrons quickly return to a lower energy level. As they do, they emit extra energy as light and heat. When a destabilized part of any molecule emits light as it reverts to stable form, we call this a "fluorescence."

All of the energy emitted from an excited electron would escape as fluorescent light and heat if nothing else were around to intercept it. But the photosynthetic pigments are not isolated. They have neighbors.

In the membrane of the chloroplast's thylakoids are **photosystems**—many thousands of them, in some plants. Each photosystem is a cluster of proteins and 200 to 300 pigment molecules. Most of those pigments only harvest photon energy. Instead of releasing excitation energy as a fluorescent afterglow, a harvester directly transfers it to a neighbor, which randomly passes it to another neighbor, and so on (Figure 6.7).

Each time a harvester makes a transfer, some energy is lost (as heat). In no time at all, the energy remaining corresponds to a wavelength that only a specialized

chlorophyll *a* at the photosystem's reaction center can trap. That chlorophyll accepts excitation energy but does not pass it on to another pigment. A suitably activated reaction center donates electrons to a nearby acceptor molecule, poised next to an electron transport system.

Electron transport systems are organized arrays of enzymes, coenzymes, and other proteins embedded in or attached to a cell membrane. They transfer electrons step-by-step through the array. As you will see, some energy escapes at each transfer, but much of it is used to drive machinery that produces ATP and NADPH.

Cyclic and Noncyclic Electron Flow

Photosynthesis started out in ancient bacteria as a way to produce ATP, not to synthesize organic compounds. (Those bacteria got enough electrons to make NADPH by stripping them from simple inorganic compounds in their environment.) They converted sunlight energy to chemical bond energy of ATP by cycling electrons from a photosystem, to a transport system, then back to the photosystem. This *cyclic* pathway still operates in all photoautotrophs. It requires a *type I* photosystem, which has a reaction center designated P700:

electron acceptor   e⁻   electron transport system

e⁻   e⁻

e⁻

ATP

Operation of electron transport system sets up conditions favorable for ATP formation at other membrane sites

The electron flow from P700 does not pack enough punch to make NADPH also. But more than 2 billion years ago, it seems that the energy for electron transfers increased as if two batteries had been hooked together. Another photosystem, *type II*, entered the picture. And the two photosystems operating together could harness enough energy to strip electrons from water molecules.

Take a look at Figure 6.8*a*. As you see, the newer pathway of electron flow is *noncyclic*. There is a linear flow of electrons from water to photosystem II, through a transport system to photosystem I, then on through a transport system that delivers the electrons to NADP⁺.

The machinery works when photosystem II absorbs photon energy and P680 (a chlorophyll *a* molecule) at its reaction center gives up electrons. The energy input also triggers **photolysis**, a reaction sequence by which molecules of water split into oxygen, hydrogen ions, and electrons. When P680 releases excited electrons, new electrons released from water replace them (Figure 6.8*a*).

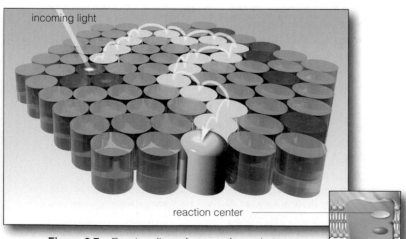

incoming light

reaction center

PHOTOSYSTEM

Figure 6.7 Random flow of energy from photons among a sampling of harvester pigments of a photosystem. The energy of excitation quickly reaches a reaction center which, when suitably activated, releases electrons for photosynthesis.

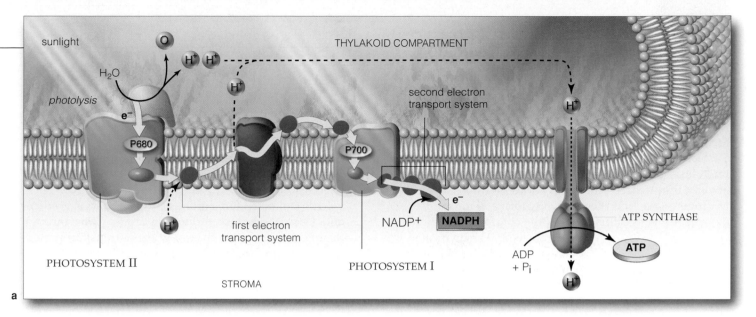

a

Figure 6.8 (**a**) Filling in the details for Figure 6.3—the noncyclic pathway of photosynthesis, by which ATP and NADPH form. *Yellow* arrows show electron flow. Electrons released from water molecules split by photolysis move through two photosystems (*light green*) and two transport systems (*dark green*). The joint operation of the two photosystems boosts electrons to an energy level high enough to drive NADPH formation. Also, as you will read in Section 6.4, hydrogen ions move across the thylakoid membrane in ways that lead to ATP formation at proteins called ATP synthases.

(**b**) If you were to diagram the energy changes associated with the noncyclic pathway, you would end up with this "Z scheme."

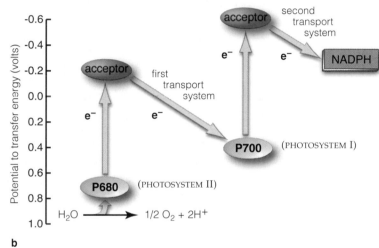

b

The excited electrons move on to a transport system, then to P700 of photosystem I. They still have a bit of extra energy. And they get more upon their arrival, for photons are also bombarding photosystem I. The energy boost puts electrons at a higher energy level that allows them to enter a transport system. There, NADP⁺ accepts two electrons and a hydrogen ion to become NADPH.

Today, both the cyclic and noncyclic pathways of electron flow operate in all photosynthetic organisms. Which one dominates at a given time depends on the organism's metabolic demands for ATP and NADPH.

The Legacy—A New Atmosphere

On sunny days, on the surfaces of aquatic plants, you see bubbles of oxygen, a by-product of the noncyclic pathway (Figure 6.9). When the pathway first evolved, the released oxygen dissolved in seas, lakes, mud, and other habitats of bacteria. By about 1.5 billion years ago, great

Figure 6.9 Visible evidence of photosynthesis—bubbles of oxygen escaping from the leaves of *Elodea*, a plant of freshwater habitats.

quantities of dissolved oxygen were escaping to what had been an oxygen-free atmosphere. The accumulation of oxygen changed the atmosphere forever. That vast, global change made possible aerobic respiration, which became the most efficient pathway for releasing usable energy from organic compounds. The emergence of the noncyclic pathway ultimately allowed you and every other animal to be around today, breathing the oxygen that helps keep your cells alive.

In the light-dependent reactions, sunlight energy drives the release of electrons from photosystems in thylakoid membranes. The electron flow through nearby transport systems results in the formation of ATP, NADPH, or both.

ATP can form when electrons cycle from and back to a type I photosystem. Both ATP and NADPH can form by a noncyclic pathway in which electrons flow from water, through two photosystems (types II and I), and finally to NADP⁺.

All photosynthetic species employ one or both pathways in response to the cell's changing needs for ATP and NADPH.

Oxygen, a by-product of the noncyclic pathway, changed the early atmosphere and made aerobic respiration possible.

𝒲𝒲𝒲

A CLOSER LOOK AT ATP FORMATION IN CHLOROPLASTS

As you probably have noticed, we saved the trickiest question for last. Exactly how does ATP form during the noncyclic pathway of photosynthesis? To arrive at the answer, let's walk through Figure 6.10.

Inside the chloroplast, photon absorption triggers photolysis, the pathway's first step. At this step, enzyme activity repeatedly splits water molecules into oxygen, hydrogen ions, and electrons. O_2 forms from the free oxygen atoms. It diffuses out of the chloroplast and out of the cell. The hydrogen ions (H^+) are left behind. And they accumulate in the fluid inside the thylakoid compartment (Figure 6.10a).

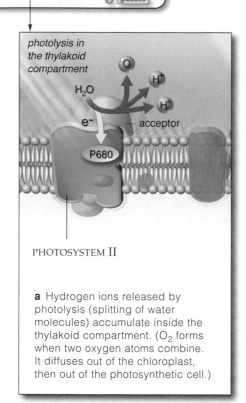

Therefore, *by a combination of photolysis and electron transport*, the concentration of hydrogen ions becomes greater inside the thylakoid compartment, compared to the stroma. The unequal distribution of these positively charged ions also creates a difference in electric charge across the membrane. An electric gradient, as well as a concentration gradient, has become established.

The combined force of the H^+ concentration gradient and electric gradient propels hydrogen ions through the interior of ATP synthases, a type of transport protein that spans the thylakoid membrane (Figure 6.10c). In other words, the ions flow out from the compartment, through ATP synthases, into the stroma. ATP synthases have built-in enzymatic machinery. And the flow of ions

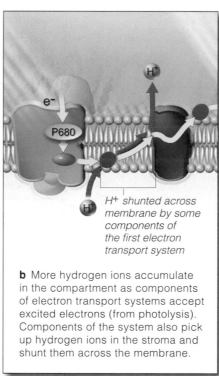

a Hydrogen ions released by photolysis (splitting of water molecules) accumulate inside the thylakoid compartment. (O_2 forms when two oxygen atoms combine. It diffuses out of the chloroplast, then out of the photosynthetic cell.)

b More hydrogen ions accumulate in the compartment as components of electron transport systems accept excited electrons (from photolysis). Components of the system also pick up hydrogen ions in the stroma and shunt them across the membrane.

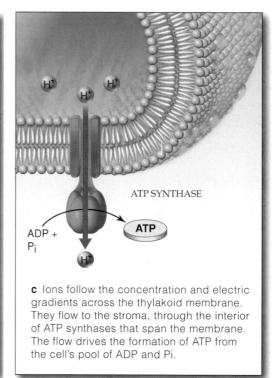

c Ions follow the concentration and electric gradients across the thylakoid membrane. They flow to the stroma, through the interior of ATP synthases that span the membrane. The flow drives the formation of ATP from the cell's pool of ADP and Pi.

Figure 6.10 How ATP forms in chloroplasts during the noncyclic pathway of photosynthesis.

The released electrons are picked up by a primary acceptor molecule, which transfers them to the electron transport system positioned next to photosystem I in the thylakoid membrane. More hydrogen ions flow into the thylakoid compartment while the electron transport system is operating. This is the case for both the cyclic and noncyclic pathways. The thylakoid membrane has many photosystems and many transport systems. At the same time that certain components of the transport systems accept electrons, they also pick up hydrogen ions from the stroma. They immediately shunt those ions across the membrane, as shown in Figure 6.10b.

through them drives the machinery, which catalyzes the attachment of unbound phosphate to a molecule of ADP. In this way, ATP forms.

The sequence of events just described is called the chemiosmotic model of ATP formation in chloroplasts. As you will see in the next chapter, the same model applies to ATP formation in mitochondria.

In chloroplasts, H^+ concentration and electric gradients form across the thylakoid membrane. The flow of ions from the thylakoid compartment to the stroma drives ATP formation.

THE LIGHT-INDEPENDENT REACTIONS

The **light-independent reactions** are the "synthesis" part of photosynthesis. They occur as a cyclic pathway called the **Calvin-Benson cycle**. The reactions require energy from ATP, hydrogen and electrons from NADPH, and carbon and oxygen from carbon dioxide (CO_2), which is present in the air (or water) that surrounds photosynthetic cells. We say the reactions are light-independent because they don't depend directly on sunlight. They can proceed just as well in the dark, as long as ATP and NADPH are available.

How Do Plants Capture Carbon?

Let's track a CO_2 molecule that diffuses into air spaces inside a leaf and ends up next to a photosynthetic cell. It diffuses into the cell, then into a chloroplast's stroma. An enzyme attaches the carbon atom of CO_2 to **RuBP** (ribulose bisphosphate), a compound with a backbone of five carbon atoms. **Rubisco** (for RuBP carboxylase) is the enzyme's name. Its action produces an unstable six-carbon intermediate that splits into two molecules of **PGA** (phosphoglycerate). PGA is a stable molecule with a three-carbon backbone. Incorporating a carbon atom from CO_2 into a stable organic compound is called **carbon fixation**. Quite simply, food can't be produced without this first step of the Calvin-Benson cycle.

The cycle yields phosphorylated glucose and it also regenerates the RuBP. For our purposes, we can focus on the carbon atoms of the substrates, intermediates, and end products, as Figure 6.11 shows.

How Do Plants Build Glucose?

Each PGA accepts a phosphate group from ATP, then hydrogen and electrons from NADPH. The resulting intermediate is called **PGAL** (phosphoglyceraldehyde). To build *one* six-carbon sugar phosphate, carbon atoms from six CO_2 must be fixed and twelve PGAL must form. Most of the PGAL becomes rearranged to form new RuBP, which can be used to fix more carbon. But two of the PGAL combine, thus forming glucose with a phosphate group attached to its six-carbon backbone.

When phosphorylated this way, glucose is primed to enter other reactions. Plants use it as a building block for their main carbohydrates—sucrose, cellulose, and starch. *The synthesis of these organic compounds by other pathways marks the end of the light-independent reactions.*

With six turns of the cycle, enough RuBP molecules form to replace the ones used in carbon fixation. The ADP, NADP+, and phosphate remaining diffuse through the stroma, to sites of the light-dependent reactions. There they are converted back to NADPH and ATP.

Photosynthetic cells convert phosphorylated glucose to sucrose or starch during daylight hours. Of all plant

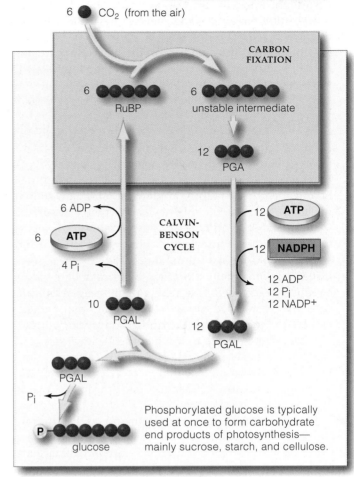

Figure 6.11 Summary of the light-independent reactions of photosynthesis. *Red* circles are carbon atoms of key molecules. All intermediates have one or two phosphate groups attached, but to keep things simple, we show only the phosphate group on the resulting glucose. Also, many water molecules that formed in the light-dependent reactions enter this pathway, and six remain at its conclusion. Appendix V (Figure C) shows more details.

carbohydrates, sucrose is the most easily transported. Starch is the most common storage form. The cells also convert excess PGAL to starch. They briefly store starch (as starch grains) inside the stroma. After the sun goes down, they convert starch to sucrose, for export to other living cells in leaves, stems, and roots. Ultimately, the products and intermediates of photosynthesis end up as energy sources and building blocks for all of the lipids, amino acids, and other organic compounds that are required for growth, survival, and reproduction.

In the light-independent reactions (the Calvin-Benson cycle), carbon is "captured" from carbon dioxide, glucose forms during reactions that require ATP and NADPH, and RuBP (necessary to capture the carbon) is regenerated.

WWW

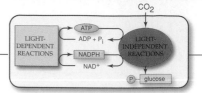

If sunlight intensity, air temperature, rainfall, and soil composition were the same everywhere, photosynthesis might proceed the same way in every plant. However, environments differ—and so do photosynthetic details. A brief comparison of two carbon-fixing adaptations to stressful environments will illustrate this point.

C4 Plants

All plants must take up CO_2 for growth. But CO_2 is not always plentiful inside leaves, which have a waxy cover that helps plants conserve water. The CO_2 must diffuse in and the O_2 out mainly at **stomata** (singular, stoma), which are microscopic openings across the leaf surface.

Stomata close on hot, dry days. Water is conserved but CO_2 can't get into leaves. Photosynthetic cells are still busy, so oxygen accumulates. A high O_2 level inside leaves triggers *photorespiration,* a process that wastes fixed CO_2 and thereby reduces a plant's sugar-building capacity. Remember rubisco, the enzyme that attaches carbon to RuBP in the Calvin-Benson cycle? Rubisco also can attach *oxygen* to it when the O_2 level rises and the CO_2 level falls. Only one (not two) PGA forms, along with a glycolate molecule that is later degraded to CO_2.

Think of Kentucky bluegrass, one of many **C3 plants**. "C3" refers to *three*-carbon PGA, the first intermediate of its carbon-fixing pathway (Figure 6.12*a*). By contrast, the first intermediate that forms when corn and many other plants fix carbon is the *four*-carbon oxaloacetate. Hence *their* name, **C4 plants**.

When C4 plants close stomata on hot, dry days, they still maintain enough CO_2 in leaves by *fixing carbon twice,* in two types of photosynthetic cells. Mesophyll cells, the first type, use CO_2 to form oxaloacetate—which is transferred to bundle-sheath cells around leaf veins (Figure 6.12*b*). There, the CO_2 is released and its local concentration rises, so rubisco must use oxygen. In the mesophyll cells CO_2 is fixed again during the Calvin-Benson cycle. With this pathway, C4 plants get by with tinier stomata, lose less water, and make more glucose than C3 plants can when conditions are hot, bright, and dry.

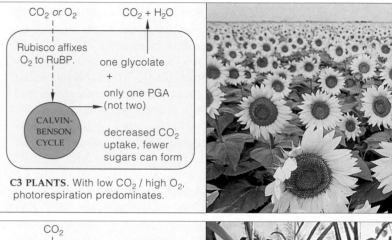

a

C3 PLANTS. With low CO_2 / high O_2, photorespiration predominates.

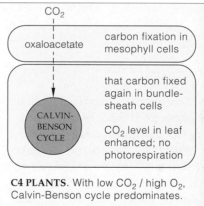

b

C4 PLANTS. With low CO_2 / high O_2, Calvin-Benson cycle predominates.

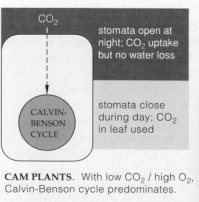

c

CAM PLANTS. With low CO_2 / high O_2, Calvin-Benson cycle predominates.

cross-section of leaf from a C4 plant

upper leaf surface   vein   mesophyll cell

lower leaf surface   bundle-sheath cell CO_2 moves through stoma, into air spaces in leaf

Figure 6.12 Three ways of fixing carbon in hot, dry weather, when the CO_2 level is low and the O_2 level is high in leaves. (**a**) The C3 pathway is common among evergreen trees and shrubs as well as many nonwoody plants of temperate zones, such as sunflowers. (**b**) The C4 pathway is common among grasses and other plants that evolved in the tropics and fix CO_2 twice. Corn, crabgrass, and the sugarcane shown here are examples. (**c**) CAM plants open stomata and fix carbon at night. They include pineapple, cacti and many other succulents (plants having a low surface-to-volume ratio), and orchids.

Especially in hot weather, photorespiration greatly lowers the photosynthetic efficiency of many C3 crop plants, including tomatoes, rice, wheat, soybeans, and potatoes. (In some experiments, hothouse tomatoes were grown at CO_2 concentrations high enough to eliminate photorespiration. Growth rates increased by as much as five times.) If photorespiration is so wasteful, then why hasn't natural selection eliminated it? The answer may lie with rubisco, the switch-hitting enzyme that sets the whole process in motion. The enzyme evolved long ago, when atmospheric levels of oxygen were still low and carbon dioxide levels high. Maybe a gene specifying its structure cannot mutate without adversely affecting the vital carbon-fixing activity. Or maybe the pathway that degrades glycolate to carbon dioxide has proved so adaptive it cannot be eliminated. Glycolate can be toxic at high concentrations.

The C4 pathway evolved separately in many lineages of flowering plants over the past 50 to 60 million years. Before then, atmospheric CO_2 levels were higher, and they gave C3 plants a selective advantage.

Which pathway will be the most adaptive in years to come? Atmospheric CO_2 levels have been increasing for decades. Some ecologists predict that they will double over the next fifty years. If that happens, photorespiration will again decline—and many vital crops might benefit.

CAM Plants

We see a carbon-fixing adaptation to desert conditions in cacti. A cactus is one of the succulents; it has juicy, water-storing tissues and thick surface layers that restrict water loss. It cannot open stomata on hot days without losing precious water. It opens them and fixes CO_2 *at night*. Its cells store the resulting intermediate in their central vacuoles, then use it for photosynthesis the next day, when the stomata close. Many plants are adapted this way. They are **CAM plants** (short for *Crassulacean Acid Metabolism*). Unlike C4 species, CAM plants do not fix carbon twice in different types of cells. They fix it in the same cells, but at different times (Figure 6.12c).

Many plants die during prolonged droughts, but some CAM plants survive by keeping stomata closed even at night. They repeatedly fix the CO_2 that forms by aerobic respiration. Not much forms, but it is enough to allow these plants to maintain very low rates of metabolism. CAM plants grow slowly. Try growing a cactus plant in Seattle or some other place with a mild climate, and it will compete poorly with C3 and C4 plants.

C4 plants and CAM plants both have modified ways of fixing carbon for photosynthesis. The modifications counter the stress imposed by hot, dry conditions in their environments.

WWW

LIGHT IN THE DEEP DARK SEA?

In the late 1980s a graduate student, Cindy Lee Van Dover, was puzzling over an eyeless shrimp. Divers had discovered it at a **hydrothermal vent** deep in the Atlantic Ocean. Such vents are fissures in the seafloor where molten rock rises and mixes with cold seawater, so that mineral-rich water, superheated by 650 degrees, spews into the perpetually dark surroundings. The sides of the vents teem with life—including shrimp and bacteria.

When Van Dover saw videotapes of shrimps clinging to the sides of the vent, she noticed a pair of bright strips on their back. By examining specimens in the lab, she saw that the strips connected to a nerve. The eyeless shrimp, it seemed, were equipped with a novel sensory organ.

Steven Chamberlain, a neuroscientist, confirmed that the strips are a sensory organ having photoreceptors—which are light-absorbing pigmented cells. Ete Szuts, an expert on pigments, isolated the photoreceptor pigment. Its absorption spectrum is the same as that for rhodopsin, a visual pigment in eyes as complex as yours.

Van Dover had a hunch: If the strips serve as an eye, then we should find *light* on the seafloor. Special cameras used during subsequent dives confirmed her hypothesis. Like coils in a toaster oven, vents release heat (infrared radiation). But they also release faint radiation at the low end of the visible spectrum—light up to nineteen times brighter than expected for infrared radiation. A billion billion photons per square inch per second strike a typical sunlit tree. A trillion photons per square inch per second reach photoautotrophic bacteria living 72 meters (240 feet) below the surface of the Black Sea. *Just as many photons are available to organisms at hydrothermal vents.*

About this time Van Dover casually asked a colleague, "Hey, what if there's enough light for photosynthesis?" The response was, "What a stupid idea."

Euan Nisbet, who studies ancient environments, thought about this. The first cells arose 3.8 billion years ago. As Nisbet and Van Doren hypothesized, What if they arose at hydrothermal vents? They could have used inorganic compounds such as hydrogen sulfide (for their hydrogen and electrons) and carbon dioxide (for carbon). Survival probably depended on being able to detect light from vents and move away from it (and so avoid being boiled alive). Millions of years later, some bacterial descendants were evolving in hot springs near the ocean's surface. In time they used their light-detecting machinery to absorb light from the sun. Those cells were now adapted to a new energy source; they had become photosynthetic.

Some observations support the intriguing hypothesis that light-sensing machinery of deep-sea bacteria became modified for shallow-water photosynthesis. For example, absorption spectra for the chlorophyll in evolutionarily ancient photosynthetic bacteria happen to correspond to the light measured at hydrothermal vents. In addition, photosynthetic machinery incorporates iron, sulfur, manganese, and other minerals—which happen to be abundant at hydrothermal vents.

6.8 AUTOTROPHS, HUMANS, AND THE BIOSPHERE

We conclude this chapter with a story that reinforces how photosynthesizers and other autotrophs fit in the world of living things. It is a story of mind-boggling numbers of single-celled and multicelled species on land and in the sunlit waters of the Earth.

Each spring, you sense their renewed growth on land when leaves unfurl and lawns and fields turn green. You might not be aware that uncountable numbers of single-celled autotrophs also drift through the surface waters of the world ocean. You can't see them without a microscope; a row of 7 million cells of one aquatic species would be less than a quarter-inch long. In some regions, a cupful of seawater might hold 24 million cells of one species; and that wouldn't include cells of all the other aquatic species.

Nearly all of the drifters are photoautotrophic bacteria and protistans. Together they are the "pastures of the seas," the producers that ultimately feed all other marine organisms. The pastures "bloom" in the spring, when nutrient inputs sustain rapid reproduction. At that time seawater becomes warmer and enriched with nutrients churned up from the deep by winter currents.

Until NASA gathered data from satellites in space, we had no idea of their numbers and distribution. Figure 6.13*a* shows visual evidence of their activities one winter in the surface waters of the North Atlantic Ocean. Figure 6.13*b* shows a springtime bloom stretching from North Carolina all the way past Spain!

Collectively, the cells help shape the global climate, for they deal in staggering numbers of reactant and product molecules. For instance, they sponge up nearly half of the carbon dioxide that we humans release each year, as when we burn fossil fuels or burn vast tracts of forests to clear land for farming. Without aquatic photoautotrophs, atmospheric carbon dioxide would accumulate more rapidly and possibly contribute to global warming, as described in Section 41.8. If the atmosphere warms by only a few degrees, sea levels will rise and all lowlands along the coasts of continents and islands will be submerged.

Even though such global change is a real possibility, tons of industrial wastes, raw sewage, and fertilizers in runoff from croplands drain into the ocean each day. The pollutants seriously alter the chemical composition of seawater. How long will marine photoautotrophs be able to function in this new chemical brew? The answer will have impact on your own life in more ways than one.

Other autotrophs also influence your life in ways you might not expect. At hydrothermal vents, in hot springs, even in waste heaps of coal mines are diverse bacteria classified as *chemo*autotrophs. They, too, use carbon dioxide as a carbon source. But they use inorganic compounds in their environment as energy sources.

For example, some of those bacteria living on the sides of hydrothermal vents strip hydrogen and electrons from hydrogen sulfide (Section 6.7). Other chemoautotrophs living in soil obtain energy from nitrogen-containing wastes and remains of animals and other organisms.

Although they, too, are microscopically small, the chemoautotrophs also exist in monumental numbers. They influence the cycling of nitrogen, phosphorus, and other elements through the biosphere. We will be returning to their impact on the environment. In this unit, we turn next to pathways by which cells release the chemical bond energy of glucose and other biological molecules —the chemical legacy of autotrophs everywhere.

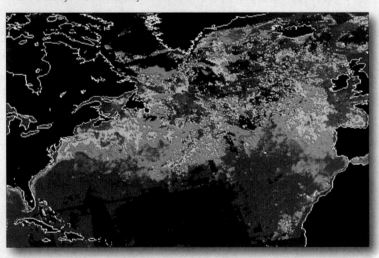

a Photosynthetic activity in winter.

b Photosynthetic activity in spring.

Figure 6.13 Two satellite images that help convey the magnitude of photosynthetic activity during springtime in the North Atlantic portion of the world ocean. In these color-enhanced images, *red-orange* shows where chlorophyll is most concentrated.

WWW

SUMMARY

1. Cell structure and functions are based on organic compounds, the synthesis of which depends on sources of carbon and energy. Plants and other autotrophs use carbon dioxide as their source of carbon. Some types use sunlight as the energy source; others use inorganic compounds. Animals and other heterotrophs are not self-nourishing; they must get carbon and energy from organic compounds already synthesized by autotrophs.

2. Photosynthesis is the main process by which carbon and energy enter the web of life. It has two stages: the light-dependent and light-independent reactions. Key reactants, intermediates, and products are summarized in Figure 6.14 and in this simplified equation:

$$12H_2O + 6CO_2 \xrightarrow{\text{LIGHT ENERGY}} 6O_2 + C_6H_{12}O_6 + 6H_2O$$

WATER CARBON OXYGEN GLUCOSE WATER
DIOXIDE

3. In plants and algae, photosynthesis proceeds inside the organelles called chloroplasts.

a. Two outer membranes enclose the chloroplast's semifluid interior. A third membrane extends through the stroma as a system of interconnected channels and disks (which are often arranged in stacks).

b. Light-dependent reactions proceed at thylakoids: disk-shaped sacs of the inner membrane system. ATP and NADPH form. Oxygen is released as a by-product.

c. Light-independent reactions occur in the stroma. Glucose molecules form. Each has a phosphate group attached and typically is used at once for the assembly of sucrose, starch, and cellulose. Those three organic compounds are the main end products of photosynthesis.

4. The light-dependent reactions start after pigments absorb wavelengths of visible light from the sun.

a. Chlorophyll a, the main photosynthetic pigment, absorbs wavelengths that are most efficient for driving photosynthesis. Like other chlorophylls, it absorbs nearly all wavelengths of visible light except for some yellow-green and green ones, which it reflects or transmits.

b. Accessory pigments can absorb wavelengths that chlorophyll a cannot absorb. They include chlorophyll b and the carotenoids (such as beta-carotene).

5. In chloroplasts, photosynthetic pigments are part of photosystems. Each photosystem has a cluster of 200 to 300 pigments, most of which are harvesters of energy. Light absorption raises some electrons of the harvester pigments to a higher energy level. Then the energy of excitation passes randomly among pigments until it reaches a chlorophyll a that acts as the reaction center. When suitably activated, that chlorophyll alone releases excited electrons to a nearby acceptor molecule, which is poised at the start of an electron transport system.

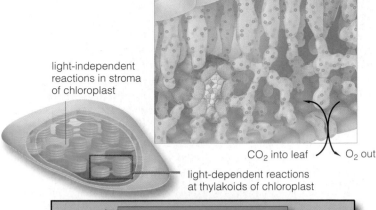

light-independent reactions in stroma of chloroplast

CO₂ into leaf / O₂ out

light-dependent reactions at thylakoids of chloroplast

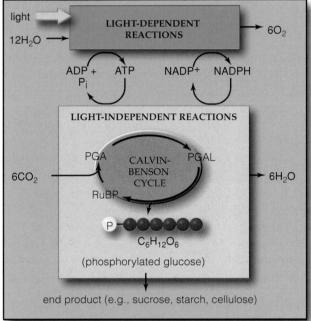

Figure 6.14 Summary of photosynthesis. With the light-dependent reactions, sunlight energy is converted to chemical bond energy of ATP. With a noncyclic pathway, water molecules are split (by photolysis). The coenzyme NADP⁺ accepts the released electrons and hydrogen to form NADPH. The oxygen is released as a by-product.

With the light-independent reactions, a phosphorylated glucose molecule forms in the Calvin-Benson cycle, RuBP on which the cycle turns is regenerated, and new water molecules form. *Each turn* of the cycle requires one CO₂, three ATP, and two NADPH. Because each glucose molecule has a backbone of six carbon atoms, its formation requires six turns of the cycle.

6. A thylakoid membrane incorporates photosystems of two types: I and II.

a. An ancient pathway of electron flow leading to ATP formation uses photosystem I alone. A chlorophyll a, called P700, is at this photosystem's reaction center.

b. Photosystems I and II both operate in a noncyclic pathway by which ATP and NADPH form. The reaction center of photosystem II has a chlorophyll a called P680.

7. In the cyclic pathway, sunlight energy is converted to chemical bond energy of ATP when excited electrons cycle from P700, through a transport system, then back to photosystem I. New electrons are not required; the electrons released from P700 simply return to P700.

8. Noncyclic electron flow has these features:

a. There is a *linear* flow of excited electrons from water to P680 (photosystem II), into a transport system to P700 (photosystem I), then on through a transport system that delivers electrons to NADP+.

b. Water molecules give up the electrons when they are split into oxygen and hydrogen (photolysis).

c. By 1.5 billion years ago, oxygen released by the noncyclic pathway had accumulated in the atmosphere. It ultimately made possible aerobic respiration.

9. With the light-dependent reactions, ATP forms from the cell's pools of ADP and inorganic phosphate.

a. While components of the transport systems accept electrons, they also pick up hydrogen ions (H+) from the stroma and shunt them into a thylakoid compartment.

b. H+ from water molecules split during photolysis also collects in the compartment.

c. The activity sets up H+ concentration and electric gradients across the thylakoid membrane. So H+ flows out through ATP synthases (into the stroma), and the energy behind this flow drives the formation of ATP.

10. Here are the key points about the light-independent reactions, which proceed in the stroma:

a. The reactions require ATP and NADPH. The ATP delivers energy (by phosphate group transfers), and the NADPH delivers hydrogen and electrons to reaction sites. Glucose forms and is used in the assembly of starch, cellulose, and other end products of photosynthesis.

b. The reactions are steps of the Calvin-Benson cycle. They start when an enzyme, rubisco, affixes carbon from CO_2 to five-carbon RuBP. The intermediate formed splits into two PGA. ATP phosphorylates each one. NADPH gives electrons and H+ to the formation of two PGAL.

c. For every six carbon atoms that enter the cycle by way of carbon fixation, twelve PGAL form. Two PGAL are used to produce a six-carbon sugar phosphate. The remainder are used to regenerate the RuBP.

11. Rubisco, a carbon-fixing enzyme, evolved when the atmosphere had far more CO_2 and far less oxygen. Today, when the O_2 level is higher than the CO_2 level in leaves, it attaches oxygen rather than carbon to RuBP. This results in the formation of only one PGA (not two) and glycolate, a compound that cannot be used to form sugars but instead is broken down to carbon dioxide and water. This wasteful process is photorespiration.

12. Photorespiration predominates in C3 plants, such as sunflowers, during hot, dry conditions. Then, stomata close and O_2 from photosynthesis builds up inside the leaf to levels higher than the CO_2 levels. Sugarcane and other C4 plants can raise the CO_2 level by fixing carbon twice, in two cell types. CAM plants such as cacti raise it by fixing carbon at night, when stomata are open.

Review Questions

1. A cat eats a bird, which earlier ate a caterpillar that chewed on a weed. Which organisms are autotrophs? Heterotrophs? *CI*

2. Summarize the photosynthesis reactions as an equation. Name where each stage takes place inside a chloroplast. *6.1*

3. Which of the following pigments are most visible in a maple leaf in summer? Which become the most visible in autumn? *6.2*
   a. chlorophylls   c. anthocyanins
   b. phycobilins   d. carotenoids

4. Fill in Figure 6.15 for the light-dependent reactions. *6.3*

5. How does chlorophyll *a* differ in function from accessory pigments during the light-dependent reactions? *6.3*

6. With respect to the light-dependent reactions, how do the cyclic and noncyclic pathways of electron flow differ? *6.3*

7. Which substance does *no*t take part in Calvin-Benson cycle: ATP, NADPH, RuBP, carotenoids, O_2, CO_2, or enzymes? *6.5*

8. Fill in the blanks for Figure 6.16. Which substances are the original sources of carbon atoms and hydrogen atoms used in the synthesis of glucose in the Calvin-Benson cycle? *6.1, 6.5*

9. How many carbon atoms from CO_2 must enter the Calvin-Benson cycle to produce one glucose molecule? Why? *6.5*

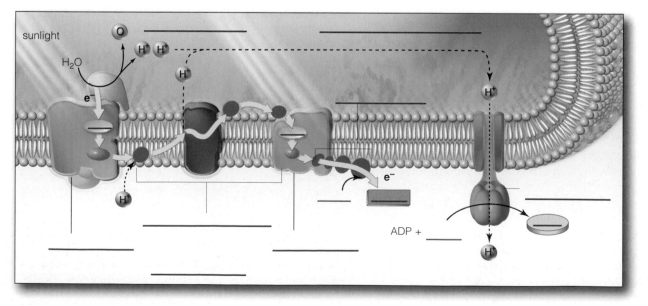

sunlight

H_2O

ADP +

Figure 6.15
Review Figure 6.8 on the light-dependent reactions. Then, on your own, fill in the blanks (*red* lines) with the names of key components and activities.

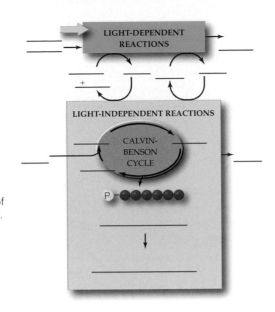

Figure 6.16
Review Figure
6.14. Then,
on your own,
fill in the
blanks (*red*
lines) with
the names
of the key
reactants,
intermediates,
and products of
photosynthesis.

Self-Quiz *(Answers in Appendix III)*

1. Photosynthetic autotrophs use _____ from the air as a carbon source and _____ as their energy source.

2. In plants, light-dependent reactions proceed at the _____ .
   a. cytoplasm
   b. plasma membrane
   c. stroma
   d. thylakoid membrane

3. In the light-*dependent* reactions, _____ .
   a. carbon dioxide is fixed
   b. ATP and NADPH form
   c. CO_2 accepts electrons
   d. sugar phosphates form

4. Identify which of the following substances accumulates inside the thylakoid compartment of chloroplasts during the light-dependent reactions:
   a. glucose
   b. carotenoids
   c. chlorophyll
   d. fatty acids
   e. hydrogen ions

5. When a photosystem absorbs light, _____ .
   a. sugar phosphates are produced
   b. electrons are transferred to ATP
   c. RuBP accepts electrons
   d. light-dependent reactions begin

6. The light-*independent* reactions proceed in the _____ .
   a. cytoplasm
   b. stroma
   c. plasma membrane
   d. grana

7. The Calvin-Benson cycle starts when _____ .
   a. light is available
   b. light is not available
   c. carbon dioxide is attached to RuBP
   d. electrons leave a photosystem

8. In the light-independent reactions, ATP phosphorylates:
   a. RuBP
   b. $NADP^+$
   c. PGA
   d. PGAL

9. Match each event with its most suitable description.
   _____ RuBP used; PGA forms
   _____ ATP and NADPH used
   _____ NADPH forms
   _____ ATP and NADPH form
   _____ only ATP forms
   _____ energy matches amount needed to excite electrons

   a. photon absorption
   b. noncyclic pathway
   c. CO_2 fixation
   d. PGAL forms
   e. H^+ and e^- to $NADP^+$
   f. cyclic pathway

Critical Thinking

1. About 200 years ago, Jan Baptista van Helmont performed an experiment on the nature of photosynthesis. He wanted to know where growing plants acquire the raw materials necessary to increase in size. For his experiment, he planted a tree seedling weighing 5 pounds in a barrel filled with 200 pounds of soil. He watered the tree regularly. After five years passed, van Helmont again weighed the tree and the soil. At that time the tree weighed 169 pounds, 3 ounces. The soil weighed 199 pounds, 14 ounces. Because the tree's weight had increased so much and the soil's weight had decreased so little, he concluded the tree had gained weight after absorbing the water he had added to the barrel.

Given what you know about the composition of biological molecules, why was van Helmont's conclusion misguided? Reflect on the current model of photosynthesis and then give a more plausible explanation of his results.

2. Like other accessory pigments, carotenoids extend the effective range of light absorption beyond chlorophyll *a*, the main pigment of photosynthesis. They also protect plants from *photo-oxidation*. This destructive process begins when the excitation energy of chlorophylls drives the conversion of oxygen into free radicals.

Oxygen normally picks up electrons from certain reactions. But sometimes it picks up only one—which is not enough to complete the reactions, but is enough to give the oxygen a negative charge (O_2^-). An unbound molecular fragment having the wrong number of electrons is a *free radical*. Free radicals are so reactive, they even attach to molecules that usually will not enter any reaction, including DNA. An attack by free radicals can disrupt DNA's structure and destroy its function, damage organic compounds, and kill cells.

Plants that have been damaged by photo-oxidation cannot produce carotenoids. Grow them in light and they bleach white and die. Given this observation, which molecules in the plant cells are among the first to go?

3. Suppose a garden in your neighborhood is filled with red, white, and blue petunias. Explain the floral colors in terms of which wavelengths of light they are absorbing and reflecting.

4. A busily photosynthesizing plant takes up molecules of CO_2 that have incorporated radioactively labeled carbon atoms ($^{14}CO_2$). Identify the compound in which the labeled carbon will appear first: NADPH, PGAL, pyruvate, or PGA.

Selected Key Terms

autotroph *CI*
C3 plant *6.6*
C4 plant *6.6*
Calvin-Benson cycle (light-independent reactions) *6.5*
CAM plant *6.6*
carbon fixation *6.5*
carotenoid *6.2*
chlorophyll *6.2*
chloroplast *6.1*
electron transport system *6.3*
heterotroph *CI*
hydrothermal vent *6.7*
light-dependent reactions *6.3*
PGA *6.5*
PGAL *6.5*
photolysis *6.3*
photosynthesis *CI*
photosystem *6.3*
rubisco *6.5*
RuBP *6.5*
stoma (stomata) *6.6*
stroma *6.1*
thylakoid *6.1*
wavelength *6.2*

Readings *See also www.infotrac-college.com*

Bazzaz, F., and E. Jajer. January 1992. "Plant Life in a CO_2-rich World." *Scientific American*, 266:68–74.

Hendry, George. May 1990. "Making, Breaking, and Remaking Chlorophyll." *Natural History*, 36–41.

Zimmer, C. November 1996. "The Light at the Bottom of the Sea." *Discover*, 63–73.

WWW *http://www.brookscole.com/biology*

Practice quiz questions, hypercontents, BioUpdates, and critical thinking. The Brooks/Cole Biology Resource Center provides a wealth of information fully organized and integrated by chapter.

HOW CELLS RELEASE STORED ENERGY

The Killers Are Coming! The Killers Are Coming!

In 1990, thanks to selective breeding experiments gone wrong, descendants of "killer" bees that flew out of South America a few decades earlier buzzed across the border between Mexico and Texas. By 1995, they had invaded 13,287 square kilometers of southern California and were busily setting up colonies. By1998, when nectar-rich desert flowers bloomed profusely after heavy El Niño storms, the invasion extended even farther west and north than scientists had predicted.

When provoked, the bees behave in a terrifying way. For example, thousands flew into action simply because a construction worker started up a tractor a few hundred yards away from their hive. Agitated bees entered a nearby subway station and started stinging passengers on the platform and in trains. They killed one person and injured a hundred others.

Where did these bees come from? Some queen bees were shipped from Africa to Brazil for breeding experiments in the 1950s. Why? As it happens, honeybees are big business. Besides being a source of nutritious honey, bees are rented to commercial orchards—where their collective pollinating activities may significantly enhance fruit production. For example, enclose a blossoming orchard tree in a pollinator-excluding cage, and less than 1 percent of the tree's flowers will set fruit. Yet, put a hive of honeybees inside the cage with the tree and 40 percent of the flowers will set fruit. Compared to their relatives in Africa, bees in Brazil are rather sluggish pollinators and honey producers. By cross-breeding the two, researchers thought they might come up with a strain of mild-mannered but zippier bees. So they put local bees and imported ones together in netted enclosures, complete with artificial hives. Then they let nature take its course.

Figure 7.1 One of the mild-mannered honeybees buzzing in for a landing on a flower, wings beating with energy provided by ATP. If this were one of its Africanized relatives protecting a hive, possibly you would not stay around to watch the landing. Both kinds of bees look alike. How can we tell them apart? From our own biased perspective, Africanized bees are the ones with an attitude problem.

Twenty-six African queen bees escaped. That was bad enough. Then beekeepers got wind of preliminary experimental results. After learning that the first few generations of offspring were more energetic but not overly aggressive, they imported hundreds of African queens and encouraged them to mate with the locals. And they set off a genetic time bomb.

Before long, African bees became established in commercial hives—and in wild bee populations. Their traits became dominant. The "Africanized" bees do everything other bees do, but they do more of it faster. Their eggs develop into adults more quickly. Adults fly more rapidly, outcompete other bees for nectar, and even die sooner.

When something disturbs their hives or swarms, Africanized bees become extremely agitated. They can remain that way for as long as eight hours. Whereas a mild-mannered honeybee might chase an intruding animal fifty yards or so, a squadron of Africanized bees will chase it a quarter of a mile. If they catch up to it, they collectively can sting it to death.

Doing things faster means having a continuous supply of energy and efficient ways of using it. An Africanized bee's stomach can hold thirty milligrams of sugar-rich nectar—which is enough fuel to fly sixty kilometers. That's more than thirty-five miles! Besides this, compared to other kinds of bees, the flight muscle cells of an Africanized bee have larger mitochondria. These organelles specialize in releasing a great deal of energy from sugars and other organic compounds, then converting it to the energy of ATP.

Whenever they tap into the stored energy of organic compounds, Africanized bees reveal their biochemical connection with other organisms. Study a primrose or puppy, a mold growing on stale bread, an amoeba in pondwater, or a bacterium living on your skin, and you will discover that their energy-releasing pathways differ in some details. But all of the pathways require characteristic starting materials. They yield predictable products and by-products. And they yield the universal energy currency of life—ATP.

In fact, throughout the biosphere, organisms put energy and raw materials to use in amazingly similar ways. *At the biochemical level, we find undeniable unity among all forms of life.* We will return to this idea in the concluding section of the chapter.

KEY CONCEPTS

1. All organisms can release energy stored in glucose and other organic compounds, then use it in ATP production. The energy-releasing pathways differ from one another. But the main types all start with the breakdown of glucose to pyruvate.

2. The initial breakdown reactions, known as glycolysis, can proceed in the presence of oxygen or in its absence. Said another way, these reactions can be the first stage of either aerobic or anaerobic pathways.

3. Two kinds of energy-releasing pathways are completely anaerobic, from start to finish. We call them fermentation and anaerobic electron transport. They proceed only in the cytoplasm, and none yields more than a small amount of ATP for each glucose molecule metabolized.

4. Another pathway, aerobic respiration, also starts in the cytoplasm. But this one runs to completion in organelles called mitochondria. Compared with the other pathways, aerobic respiration releases far more energy from glucose.

5. Aerobic respiration has three stages. First, pyruvate forms from glucose (through glycolysis). Second, different reactions break down the pyruvate to carbon dioxide. These reactions liberate electrons and hydrogen, which coenzymes deliver to an electron transport system. Third, stepwise electron transfers through the system help set up the conditions that favor ATP formation. Free oxygen accepts the electrons at the end of the line and combines with hydrogen, thereby forming water.

6. Over evolutionary time, photosynthesis and aerobic respiration became linked on a global scale. The oxygen-rich atmosphere, a long-term outcome of photosynthetic activity, sustains aerobic respiration, which has become the dominant energy-releasing pathway. And most kinds of photosynthesizers use carbon dioxide and water from aerobic respiration as raw materials when they synthesize organic compounds:

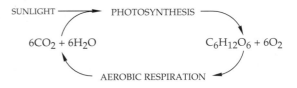

HOW DO CELLS MAKE ATP?

Organisms stay alive by taking in energy. Plants and all other organisms that engage in photosynthesis get energy from the sun. Animals get energy secondhand, thirdhand, and so on, by eating plants and one another. Regardless of its source, energy must be in a form that can drive thousands of life-sustaining reactions. Energy that becomes converted into the chemical bond energy of adenosine triphosphate—ATP—serves that function.

Plants make ATP during photosynthesis, which they then use to produce glucose and other carbohydrates. Plants and all other organisms also can make ATP by breaking down carbohydrates (glucose especially), fats, and proteins. During the breakdown reactions, electrons are stripped from intermediates, then energy associated with the liberated electrons drives the formation of ATP. Electron transfers of the sort outlined in Section 5.2 are central to these energy-releasing pathways.

Comparison of the Main Types of Energy-Releasing Pathways

The first energy-releasing pathways evolved about 3.8 billion years ago, when conditions were very different on Earth. Because the atmosphere had little free oxygen, the pathways must have been *anaerobic*, which means they could run to completion without utilizing oxygen. Many bacteria and protistans still live in places where oxygen is absent or not always available. They make ATP by anaerobic routes, mainly fermentation pathways and anaerobic electron transport. Some cells in your own body can use an anaerobic route for short periods, but only when they are not receiving enough oxygen. Your cells, like most others, mainly use **aerobic respiration**, an oxygen-dependent pathway of ATP formation. With each breath you take, you are providing your actively respiring cells with a fresh supply of oxygen.

Make note of this point: *The main energy-releasing pathways all start with the same reactions in the cytoplasm.* During this initial stage of reactions, called **glycolysis**, enzymes cleave and rearrange each glucose molecule into two molecules of **pyruvate**, each with a backbone of three carbon atoms. The energy-releasing pathways differ once this stage is completed. Most importantly, the aerobic pathway continues within a mitochondrion

a

(Figure 7.2). There, oxygen serves as the final acceptor of electrons used during the reactions. The anaerobic pathways start and end in the cytoplasm. A substance other than oxygen is the final electron acceptor.

As you examine the energy-releasing pathways in sections to follow, keep in mind that the reaction steps do not proceed by themselves. Enzymes catalyze each step, and intermediate molecules formed at one step serve as substrates for the next enzyme in the pathway.

Overview of Aerobic Respiration

Of all energy-releasing pathways, aerobic respiration gets the most ATP for each glucose molecule. Whereas anaerobic routes typically have a net yield of two ATP molecules, the aerobic route commonly yields thirty-six or more. If you were a bacterium, you wouldn't require much ATP. Being far larger, more complex, and highly active, you rely absolutely on the aerobic route's high yield. When a glucose molecule is the starting material, aerobic respiration can be summarized this way:

$$C_6H_{12}O_6 \ + \ 6O_2 \ \longrightarrow \ 6CO_2 \ + \ 6H_2O$$

GLUCOSE   OXYGEN   CARBON DIOXIDE   WATER

However, as you can see, the summary equation only tells us what the substances are at the start and finish of the pathway. In between are three reaction stages.

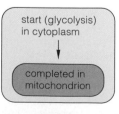

start (glycolysis) in cytoplasm

↓

completed in mitochondrion

AEROBIC RESPIRATION

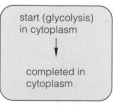

start (glycolysis) in cytoplasm

↓

completed in cytoplasm

ANAEROBIC ENERGY-RELEASING PATHWAYS

Figure 7.2 Where the aerobic and anaerobic pathways of ATP formation start and finish.

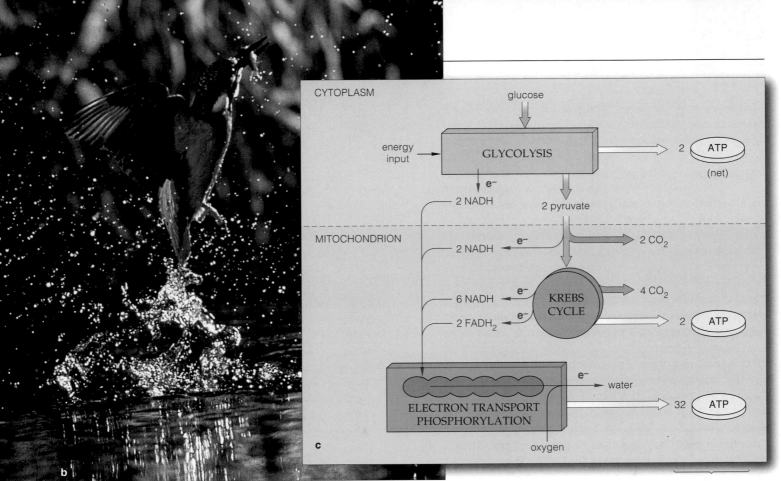

Figure 7.3 Overview of aerobic respiration, which has three stages. From start to finish, the typical net energy yield from each glucose molecule is thirty-six ATP. Only this pathway delivers enough ATP to construct and maintain giant redwoods and all other large, multicelled organisms. It alone delivers enough ATP for highly active animals, such as bees, humans (**a**), and kingfishers (**b**) and other birds.

(**c**) Glucose partially breaks down to pyruvate in the first stage (glycolysis). In the second stage, which is mainly the Krebs cycle, pyruvate breaks down completely to carbon dioxide. The coenzymes NAD$^+$ and FAD pick up electrons and hydrogen stripped from intermediates at both stages. In the final stage (electron transport phosphorylation), the loaded-down coenzymes (NADH and FADH$_2$) give up the electrons and hydrogen to a transport system. Energy released during the flow of electrons through the system drives ATP formation. Oxygen accepts the electrons at the end of the third stage.

Let's use Figure 7.3 and the following descriptions as a brief overview of these reactions. The initial stage, again, is glycolysis. During the second stage, mainly a cyclic pathway of reactions known as the **Krebs cycle**, enzymes degrade pyruvate to carbon dioxide and water.

NAD$^+$ (nicotinamide adenine dinucleotide) and **FAD** (flavin adenine dinucleotide) take part in glycolysis and the Krebs cycle. These organic compounds, derived from vitamins, function as *coenzymes*. They assist enzymes by accepting electrons (e$^-$) and hydrogen removed from intermediates, then transferring the electrons elsewhere. Unbound hydrogen atoms, recall, are naked protons, or hydrogen ions (H$^+$). So they tag along with the oppositely charged electrons. When carrying electrons and H$^+$, the two coenzymes are abbreviated NADH and FADH$_2$.

Not much ATP forms during glycolysis or the Krebs cycle. The large energy harvest comes *after* coenzymes deliver their cargo to an electron transport system.

In the third stage, the transport system functions as machinery for **electron transport phosphorylation**. It sets up H$^+$ concentration and electric gradients that drive ATP formation at nearby membrane proteins. It is during this final stage that so many ATP molecules are produced. As it ends, oxygen inside the mitochondrion accepts the "spent" electrons from the last component of the transport system. Oxygen picks up hydrogen at the same time and thereby forms water.

Cells drive nearly all metabolic activities by releasing energy from glucose and other organic compounds and converting it to chemical bond energy of ATP.

All of the main energy-releasing pathways start inside the cytoplasm with glycolysis, a stage of reactions that break down glucose to pyruvate.

The most common anaerobic pathways, which include the fermentation routes, end in the cytoplasm. Each has a net energy yield of two ATP.

Aerobic respiration, an oxygen-dependent pathway, runs to completion in the mitochondrion. From start (glycolysis) to finish, it commonly has a net energy yield of thirty-six ATP.

WWW

GLYCOLYSIS: FIRST STAGE OF ENERGY-RELEASING PATHWAYS

Let's track what happens to a glucose molecule in the first stage of aerobic respiration. Remember, the same things happen to glucose in the anaerobic pathways.

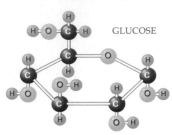
GLUCOSE

As described earlier, in Section 3.2, glucose is one of the simple sugars. Each molecule of it contains six carbon, twelve hydrogen, and six oxygen atoms, all joined by covalent bonds. The carbons make up the backbone. With glycolysis, glucose or some other carbohydrate in the cytoplasm is partially broken down, so that two molecules of the three-carbon compound pyruvate form:

glucose ⟶ glucose-6-phosphate ⟶ 2 pyruvate

The first steps of glycolysis are *energy-requiring*. As Figure 7.4 indicates, they proceed only when two ATP molecules each transfer a phosphate group to glucose and so donate energy to it. Such transfers, recall, are "phosphorylations." In this case, they raise the energy content of glucose to a level that is high enough to allow entry into the *energy-releasing* steps of glycolysis.

The first energy-releasing step cleaves the activated glucose into two molecules, which we can call PGAL (phosphoglyceraldehyde). Both PGALs are converted to an unstable intermediate, both of which allow ATP to form by giving up a phosphate group to ADP. The next intermediate in the sequence does the same thing.

Thus, a total of four ATP form by **substrate-level phosphorylation**. We define this metabolic event as the direct transfer of a phosphate group from a substrate of a reaction to some other molecule, such as ADP. Remember, though, two ATP were invested to start the reactions. So the *net* energy yield is only two ATP.

Meanwhile, the coenzyme NAD⁺ picks up electrons and hydrogen atoms liberated from each PGAL, thereby becoming NADH. When the NADH gives up its cargo at a different reaction site, it reverts to NAD⁺. Said another way, like other coenzymes, NAD⁺ is reusable.

In sum, glycolysis converts energy stored in glucose to a transportable form of energy, in ATP. NAD⁺ picks up electrons and hydrogen stripped from glucose. These have roles in the next stage of reactions. So do the end products of glycolysis—the two molecules of pyruvate.

Glycolysis is an energy-releasing stage of reactions in which glucose or some other carbohydrate is partially broken down to two molecules of pyruvate.

Two NADH and four ATP form. However, when we subtract the two ATP required to start the reactions, the *net* energy yield of glycolysis is two ATP per glucose molecule.

𝒲𝒲𝒲

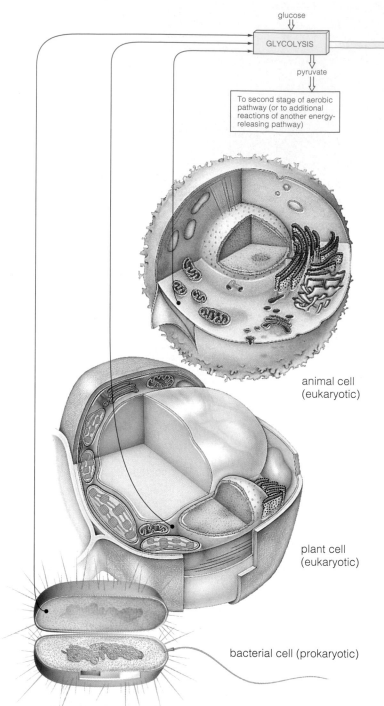

animal cell (eukaryotic)

plant cell (eukaryotic)

bacterial cell (prokaryotic)

Figure 7.4 Glycolysis, first stage of the main energy-releasing pathways. The reaction steps proceed inside the cytoplasm of every living prokaryotic and eukaryotic cell. In this example, glucose is the starting material. By the time the reactions end, two pyruvate, two NADH, and four ATP have been produced. Cells invest two ATP to start glycolysis, however, so the *net* energy yield of glycolysis is two ATP. For an expanded picture of the reactions, refer to Appendix V (Figure A).

Depending on the type of cell and on environmental conditions, the pyruvate may be used in the second set of reactions of the aerobic pathway, which includes the Krebs cycle. Or it may be used in other reactions, such as those of fermentation pathways.

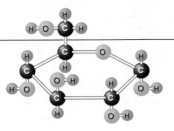

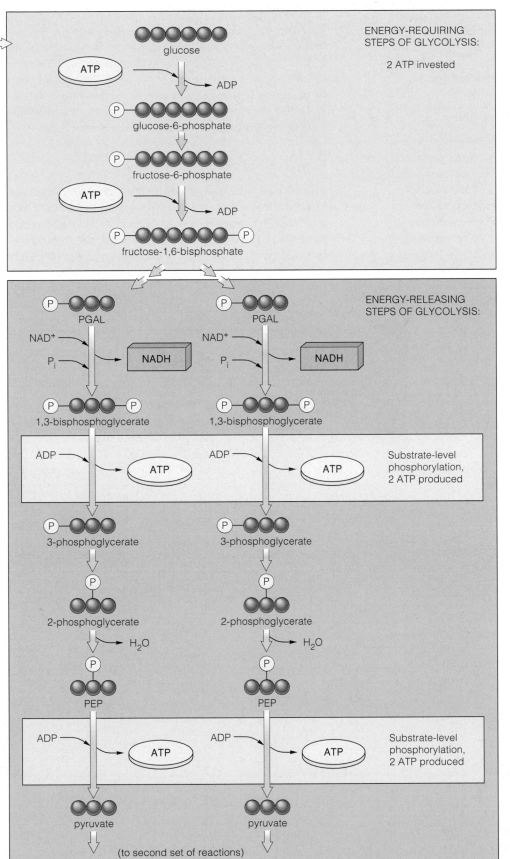

ENERGY-REQUIRING
STEPS OF GLYCOLYSIS:

2 ATP invested

ENERGY-RELEASING
STEPS OF GLYCOLYSIS:

Substrate-level
phosphorylation,
2 ATP produced

Substrate-level
phosphorylation,
2 ATP produced

NET ENERGY YIELD: 2 ATP

a This diagram tracks the fate of the six carbon atoms (*red* spheres) of the glucose molecule shown above. Glycolysis starts with an energy investment of two ATP. First, enzyme action promotes transfer of a phosphate group from ATP to glucose. With this transfer, atoms in the molecule undergo rearrangements.

b Enzyme action promotes the transfer of a phosphate group from another ATP to the rearranged molecule.

c The resulting fructose-1,6-bisphosphate molecule splits at once into two molecules, each with a three-carbon backbone. We can call these two PGAL.

d During enzyme-mediated reactions, two NADH form after each PGAL gives up two electrons and a hydrogen atom to NAD^+. Each PGAL also combines with inorganic phosphate (P_i) present in the cytoplasm, then donates a phosphate group to ADP.

e *Thus two ATP have formed by the direct transfer of phosphate from two intermediate molecules that serve as substrates in the reactions.* With this formation of two ATP molecules, the original energy investment of two ATP is paid off.

f In the next two enzyme-mediated reactions, each of the two intermediate molecules releases a hydrogen atom and an —OH group, which combine to form water.

g The resulting intermediates (two molecules of 3-phosphoenolpyruvate, or PEP) are rather unstable. Each gives up a phosphate group to ADP. *Once again, two ATP have formed by substrate-level phosphorylation.*

h Thus the net energy yield from glycolysis is two ATP for each glucose molecule entering the reactions. The end products of glycolysis are two molecules of pyruvate, each with a three-carbon backbone.

SECOND STAGE OF THE AEROBIC PATHWAY

Suppose two pyruvate molecules, formed by glycolysis, leave the cytoplasm and enter a **mitochondrion** (plural, mitochondria). In this organelle, both the second and third stages of the aerobic pathway run to completion. Figure 7.5 shows its structure and functional zones.

Preparatory Steps and the Krebs Cycle

During the second stage, a bit more ATP forms. Carbon atoms depart from the pyruvate, in the form of carbon dioxide. And coenzymes latch onto the electrons and hydrogen stripped from intermediates of the reactions.

In a few preparatory reactions, an enzyme removes a carbon atom from each pyruvate molecule. An enzyme helper, coenzyme A, becomes **acetyl-CoA** by combining with the remaining two-carbon fragment. The fragment is transferred to **oxaloacetate**, the entry point for the Krebs cycle. The name of this cyclic pathway honors Hans Krebs, who began working out its details in the 1930s. (It also is called the citric acid cycle.) Notice, in Figure 7.6, that *six* carbon atoms enter the second stage of reactions (three in each pyruvate backbone). Notice also that *six* depart, in six carbon dioxide molecules, during the preparatory steps and the Krebs cycle.

Functions of the Second Stage

The second stage of reactions serves three functions. First, it loads electrons and hydrogen onto both NAD^+ and FAD, which results in NADH and $FADH_2$. Second, through substrate-level phosphorylations, it produces two ATP molecules. And third, it rearranges the Krebs

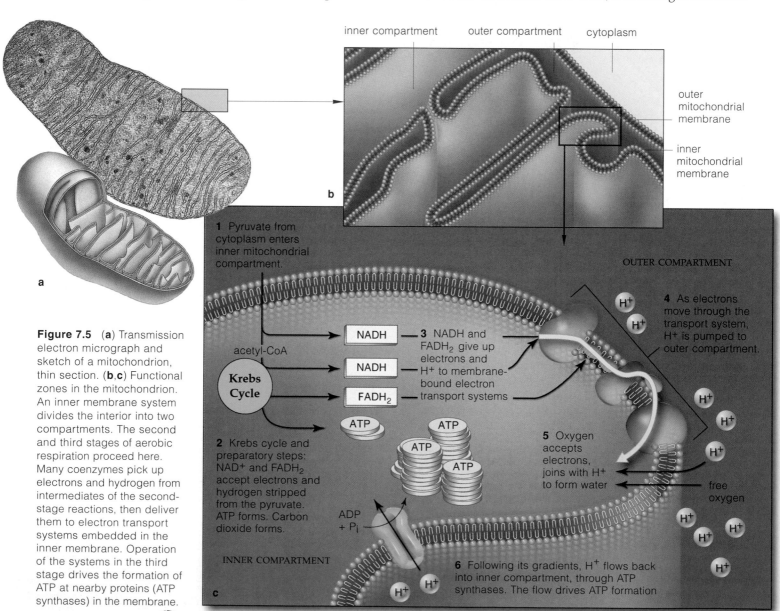

Figure 7.5 (**a**) Transmission electron micrograph and sketch of a mitochondrion, thin section. (**b,c**) Functional zones in the mitochondrion. An inner membrane system divides the interior into two compartments. The second and third stages of aerobic respiration proceed here. Many coenzymes pick up electrons and hydrogen from intermediates of the second-stage reactions, then deliver them to electron transport systems embedded in the inner membrane. Operation of the systems in the third stage drives the formation of ATP at nearby proteins (ATP synthases) in the membrane.

inner compartment   outer compartment   cytoplasm

outer mitochondrial membrane

inner mitochondrial membrane

b

1 Pyruvate from cytoplasm enters inner mitochondrial compartment.

OUTER COMPARTMENT

acetyl-CoA

NADH

Krebs Cycle

NADH

FADH₂

ATP

3 NADH and $FADH_2$ give up electrons and H^+ to membrane-bound electron transport systems

4 As electrons move through the transport system, H^+ is pumped to outer compartment.

2 Krebs cycle and preparatory steps: NAD^+ and $FADH_2$ accept electrons and hydrogen stripped from the pyruvate. ATP forms. Carbon dioxide forms.

ATP
ATP
ATP
ATP

5 Oxygen accepts electrons, joins with H^+ to form water

free oxygen

ADP + Pᵢ

INNER COMPARTMENT

6 Following its gradients, H^+ flows back into inner compartment, through ATP synthases. The flow drives ATP formation

c

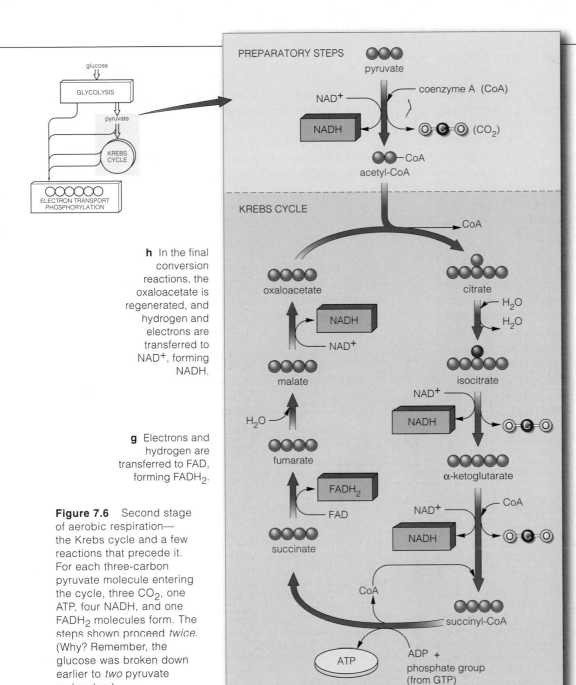

PREPARATORY STEPS

KREBS CYCLE

a Pyruvate from glucose enters a mitochondrion, where it undergoes preparatory conversions before entering cyclic reactions (the Krebs cycle).

b Pyruvate is stripped of a functional group (COO^-), which departs as CO_2. It gives up hydrogen and electrons to NAD^+, forming NADH. A coenzyme joins with the remaining two-carbon fragment, forming acetyl-CoA.

c The acetyl-CoA transfers its two-carbon group to oxaloacetate, a four-carbon compound that is the point of entry into the Krebs cycle. The result is citrate, with a six-carbon backbone. The addition and removal of H_2O changes citrate to isocitrate.

d Isocitrate enters conversion reactions in which a COO^- group departs (as CO_2). Hydrogen and electrons are transferred to NAD^+, forming NADH.

e Another COO^- group departs (as CO_2) and another NADH forms. The resulting intermediate attaches to a coenzyme A molecule, forming succinyl-CoA. *At this point, three carbon atoms have been released, balancing out the three that entered the mitochondrion (in pyruvate).*

f The attached coenzyme is replaced by a phosphate group (donated by a substrate known as GTP). Then that phosphate group gets attached to ADP. Thus, for each turn of the cycle, one ATP forms by substrate-level phosphorylation.

h In the final conversion reactions, the oxaloacetate is regenerated, and hydrogen and electrons are transferred to NAD^+, forming NADH.

g Electrons and hydrogen are transferred to FAD, forming $FADH_2$.

Figure 7.6 Second stage of aerobic respiration—the Krebs cycle and a few reactions that precede it. For each three-carbon pyruvate molecule entering the cycle, three CO_2, one ATP, four NADH, and one $FADH_2$ molecules form. The steps shown proceed *twice*. (Why? Remember, the glucose was broken down earlier to *two* pyruvate molecules.)

cycle intermediates into oxaloacetate. Cells have only so much oxaloacetate, which must be regenerated to keep the cyclic reactions going.

The two ATP that form in the second stage do not add much to the small yield from glycolysis. However, for each glucose molecule, *many* coenzymes pick up electrons and hydrogen for transport to the sites of the third and final stage of the aerobic pathway:

Glycolysis:	2 NADH
Pyruvate conversion before Krebs cycle:	2 NADH
Krebs cycle:	2 $FADH_2$ + 6 NADH
Coenzymes sent to third stage:	2 $FADH_2$ + 10 NADH

Overall, these are the key points to remember about the second stage of aerobic respiration:

During the second stage of aerobic respiration, two pyruvate molecules from glycolysis enter a mitochondrion.

Each pyruvate molecule gives up a carbon atom, then its two-carbon remnant enters the Krebs cycle. All of the carbon atoms of pyruvate eventually end up in carbon dioxide.

The second stage yields only two ATP. But the reactions regenerate oxaloacetate, the entry point for the Krebs cycle. And many coenzymes pick up electrons and hydrogen that were stripped from substrates, for delivery to the final stage of the pathway.

WWW

THIRD STAGE OF THE AEROBIC PATHWAY

ATP production goes into high gear in the third stage of the aerobic pathway. Electron transport systems and neighboring proteins called ATP synthases serve as the production machinery. They are embedded in the inner membrane that divides the mitochondrion into two compartments (Figure 7.7). They interact with electrons and unbound hydrogen—that is, H$^+$ ions. Remember, coenzymes deliver this bounty from reaction sites of the first two stages of the aerobic pathway.

Electron Transport Phosphorylation

Briefly, during the final stage, electrons get transferred from one molecule of each transport system to the next in line. When certain molecules accept and then donate the electrons, they also pick up hydrogen ions in the inner compartment. Shortly afterward, they release them to the outer compartment. Their shuttling action sets up H$^+$ concentration and electric gradients across the inner mitochondrial membrane. Nearby in that membrane, the ions follow the gradients and flow back to the inner

compartment, through the interior of ATP synthases. The H$^+$ flow through these transport proteins drives formation of ATP from ADP and unbound phosphate. Free oxygen keeps ATP production going. It withdraws spent electrons at the end of the transport systems and then combines with H$^+$. Water is the result.

Summary of the Energy Harvest

In many types of cells, thirty-two ATP form during the third stage of aerobic respiration. Add these to the net yield from the preceding stages, and the total harvest is thirty-six ATP from one glucose molecule (Figure 7.8). That's a lot! An anaerobic pathway may use eighteen glucose molecules to produce the same amount of ATP.

Think of thirty-six ATP as a typical yield only. The actual amount depends on cellular conditions, as when cells require a given intermediate elsewhere and pull it out of the reaction sequence.

The yield also depends on how particular cells use the NADH that formed during glycolysis. Any NADH produced in the cytoplasm can't enter a mitochondrion. It can only deliver electrons and hydrogen *to* the outer mitochondrial membrane, where proteins shuttle them across to NAD$^+$ or FAD molecules already inside this organelle. Then both coenzymes deliver the electrons to transport systems of the inner membrane. However, FAD puts them at a *lower* entry point in the transport system, so *its* deliveries produce less ATP (Figure 7.8).

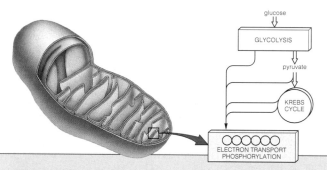

Figure 7.7 Electron transport phosphorylation, the third and final stage of aerobic respiration. The reactions proceed at electron transport systems and at ATP synthases, a type of transport protein, in the inner mitochondrial membrane. Each electron transport system consists of specific enzymes, cytochromes, and other proteins that act in sequence.

The inner membrane functionally divides the mitochondrion into two compartments. The third-stage reactions start in the inner compartment, when NADH and FADH$_2$ give up electrons and hydrogen to transport systems. Electrons are transferred *through* the system, but electron transport proteins drive unbound hydrogen (H$^+$) *to*   the outer compartment:

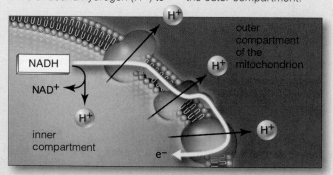

In short order, there is a higher concentration of H$^+$ in the outer compartment compared to the inner one. Concentration and electric gradients now exist across the membrane. The ions follow the gradients and flow across the membrane, through the interior of the ATP synthases. Energy associated with the flow drives the formation of ATP from ADP and unbound phosphate (P$_i$). Hence the name, electron transport *phosphorylation*:

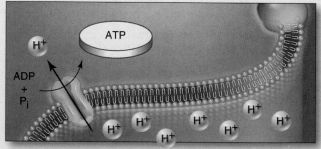

Do these metabolic events sound familiar? They should. As you may recall from Section 6.4, ATP forms in much the same way inside chloroplasts. According to the *chemiosmotic* model, H$^+$ concentration and electric gradients across a cell membrane drive ATP formation. The model applies also to mitochondria, although the ions flow in the opposite direction compared to chloroplasts.

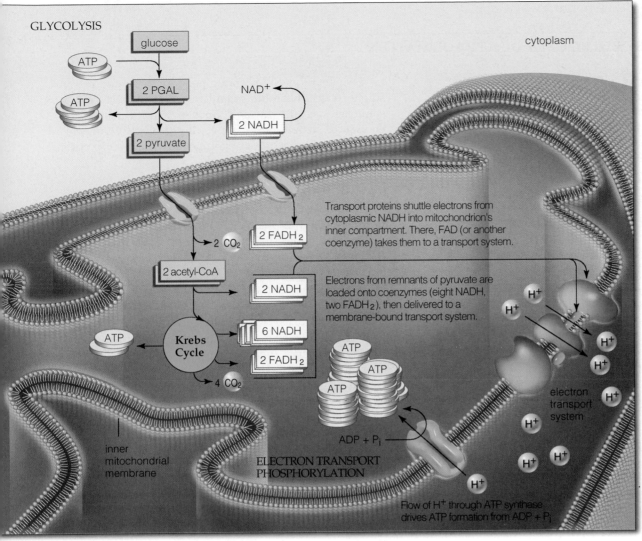

GLYCOLYSIS

cytoplasm

a Two ATP formed in first stage in cytoplasm (during glycolysis, by *substrate-level* phosphorylations).

b NADH that formed in cytoplasm during first stage delivers electrons and hydrogen that help drive the formation of four ATP during third stage at the inner mitochondrial membrane (by *electron transport* phosphorylations).

Transport proteins shuttle electrons from cytoplasmic NADH into mitochondrion's inner compartment. There, FAD (or another coenzyme) takes them to a transport system.

Electrons from remnants of pyruvate are loaded onto coenzymes (eight NADH, two FADH₂), then delivered to a membrane-bound transport system.

c Two ATP form at second stage in mitochondrion (by *substrate-level* phosphorylations of Krebs cycle).

d Coenzymes from Krebs cycle and its preparatory steps deliver electrons and hydrogen that drive formation of twenty-eight ATP at third stage (by *electron transport* phosphorylations at the inner mitochondrial membrane).

36 ATP

TYPICAL NET
ENERGY YIELD

ELECTRON TRANSPORT
PHOSPHORYLATION

electron transport system

Flow of H⁺ through ATP synthase drives ATP formation from ADP + Pᵢ

Figure 7.8 Summary of the harvest from the energy-releasing pathway of aerobic respiration. Commonly, thirty-six ATP form for each glucose molecule that enters the pathway. However, the net yield varies according to shifting concentrations of reactants, intermediates, and end products of the reactions. It also varies among different types of cells.

As you read earlier, cells differ in how they use the NADH from glycolysis, which cannot enter mitochondria. These NADH only deliver their cargo of electrons and hydrogen to certain transport proteins of the outer mitochondrial membrane. The proteins shuttle the electrons and hydrogen across the membrane, to NAD⁺ or FAD already inside the mitochondrion, to form NADH or FADH₂.

Any NADH inside the mitochondrion delivers electrons to the highest possible point of entry into a transport system. When it does, enough H⁺ can be pumped across the inner membrane to produce *three* ATP. By contrast, any FADH₂ delivers them to a lower entry point in the transport system. Fewer hydrogen ions can be pumped, so only *two* ATP can be produced.

In liver, heart, and kidney cells, for example, electrons and hydrogen from glycolysis enter the highest entry point of transport systems, so the energy harvest is thirty-eight ATP. More commonly, as in skeletal muscle and brain cells, they are transferred to FAD—so the harvest is thirty-six ATP.

One final point. Glucose, recall, has more energy (stored in more covalent bonds) than carbon dioxide or water. About 686 kilocalories of energy are released when glucose is broken down to those more stable end products. Much of this energy escapes (as heat), but about 7.5 kilocalories are conserved in each mole of ATP. So when 36 ATP form through breakdown of a glucose molecule, the energy-conserving efficiency of aerobic respiration is 36 × 7.5 /686 × 100, or 39 percent.

In the final stage of the aerobic pathway, coenzymes deliver electrons to transport systems of the inner mitochondrial membrane. As electrons move through the system, they set up H⁺ gradients that drive ATP formation at nearby proteins in the membrane. Oxygen is the final electron acceptor.

Again, from start (glycolysis in the cytoplasm) to finish (in mitochondria), the pathway commonly has a net yield of thirty-six ATP for every glucose molecule metabolized.

WWW

So far, we have tracked the fate of a glucose molecule through the pathway of aerobic respiration. We turn now to its use as a substrate for fermentation pathways. Remember, these are anaerobic pathways; they do *not* use oxygen as the final acceptor of the electrons that ultimately drive the ATP-forming machinery.

Fermentation Pathways

Diverse kinds of organisms use fermentation pathways. Many are bacteria and protistans that make their homes in marshes, bogs, mud, deep-sea sediments, the animal gut, canned foods, sewage treatment ponds, and other oxygen-free settings. Some kinds of fermenters actually die if exposed to oxygen. The bacteria responsible for many diseases, including botulism and tetanus, are like this. Other kinds of fermenters, including the bacterial "employees" of yogurt manufacturers, are indifferent to the presence of oxygen. Still others can use oxygen, but they also might use a fermentation pathway when oxygen becomes scarce. Even your muscle cells do this.

As is true of aerobic respiration, glycolysis serves as the first stage of the fermentation pathways. Here also, enzymes split glucose and rearrange the fragments into two pyruvate molecules. Here again, two NADH form, and the net energy yield is two ATP.

However, fermentation reactions do not completely break down glucose to carbon dioxide and water, and they produce no more ATP beyond the tiny yield from glycolysis. *The final steps of fermentation serve only to regenerate NAD⁺, the coenzyme with a central role in the breakdown reactions.*

Fermentation yields enough energy to sustain many single-celled anaerobic organisms. It even helps carry some aerobic cells through times of stress. But it is not enough to sustain large, active, multicelled organisms, this being one reason why you never will meet up with an anaerobic elephant.

LACTATE FERMENTATION With these points in mind, take a look at Figure 7.9, which tracks the main steps of **lactate fermentation**. During this anaerobic pathway, the *pyruvate* molecules from the first stage of reactions (glycolysis) accept the hydrogen and electrons from NADH. The transfer regenerates the NAD⁺ and, at the same time, converts each pyruvate to a three-carbon compound called lactate. You may hear people refer to this compound as "lactic acid." However, its ionized form (lactate) is far more common in cellular fluids.

Some bacteria, such as *Lactobacillus*, rely exclusively on this anaerobic pathway. Left to their own devices, their fermentation activities often spoil food. Yet certain fermenters have commercial uses, as when they break down glucose in huge vats where cheeses, yogurt, and sauerkraut are produced.

In humans, rabbits, and many other animals, some types of cells also may switch to lactate fermentation for a quick fix of ATP. When your own demands for energy are intense but brief—say, during a short race —muscle cells use this pathway. They cannot do so for long; they would throw away too much of glucose's stored energy for too little ATP. When their stores of glucose become depleted, muscles fatigue and lose their ability to contract.

ALCOHOLIC FERMENTATION In the anaerobic route called **alcoholic fermentation**, enzymes convert each pyruvate molecule that formed during glycolysis to an intermediate form: acetaldehyde. The NADH transfers electrons and hydrogen to this form and so converts it to an alcoholic end product—ethanol (Figure 7.10).

Certain species of single-celled fungi called yeasts are renowned for their use of this pathway. One type, *Saccharomyces cerevisiae*, makes bread dough rise. Bakers mix the yeast with sugar, then blend both into dough. When yeast cells degrade the sugar, they release carbon dioxide. Bubbles of the gas expand the dough (make it rise). Oven heat forces the gas out of the dough, and a porous product remains.

Beer and wine producers use yeasts on a large scale. Vintners use wild yeasts living on grapes and cultivated strains of *S. ellipsoideus*, which remain active until the alcohol concentration in wine vats exceeds 14 percent. (Wild yeast cells die when the concentration exceeds 4 percent.) That is why some birds get drunk on naturally fermenting berries. That is why landscapers don't plant

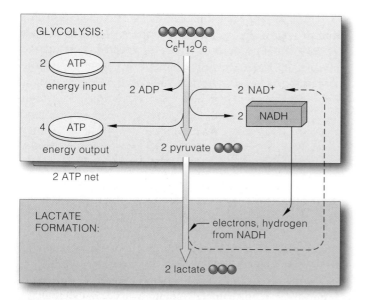

GLYCOLYSIS:

$C_6H_{12}O_6$

2 ATP
energy input

2 ADP

2 NAD⁺

2 NADH

4 ATP
energy output

2 pyruvate

2 ATP net

LACTATE FORMATION:

electrons, hydrogen from NADH

2 lactate

Figure 7.9 Lactate fermentation. In this anaerobic pathway, electrons end up in lactate, the reaction product.

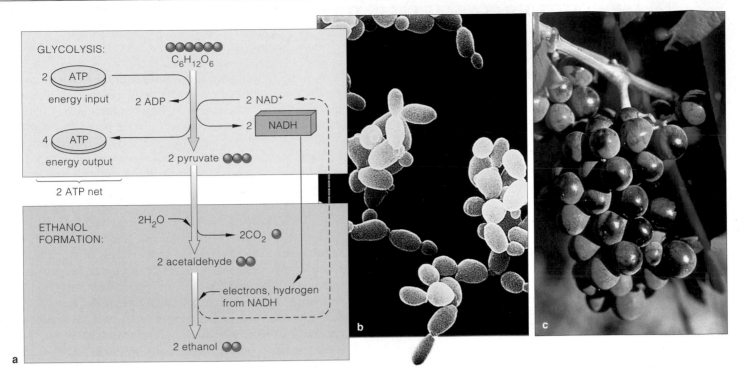

GLYCOLYSIS:

$C_6H_{12}O_6$

2 ATP — energy input

2 ADP

2 NAD^+

2 NADH

4 ATP — energy output

2 ATP net

2 pyruvate

ETHANOL FORMATION:

$2H_2O$

$2CO_2$

2 acetaldehyde

electrons, hydrogen from NADH

2 ethanol

a

b

c

Figure 7.10 (**a**) Alcoholic fermentation. In this anaerobic pathway, acetaldehyde, an intermediate of the reactions, is the final acceptor of electrons. Ethanol is the end product. Yeasts, single-celled organisms, use this pathway. (**b**) Activity of one species of *Saccharomyces* makes bread dough rise. Another (**c**) lives on sugar-rich tissues of ripened grapes.

prodigious berry-producing shrubbery near highways; drunk birds doodle into windshields (Figure 7.11). Wild turkeys similarly have been known to get tipsy when they gobble fermenting apples in untended orchards.

Anaerobic Electron Transport

Especially among the bacteria, we find less common energy-releasing pathways, some of which are topics of later chapters in the book. For example, many bacterial species have key roles in the global cycling of sulfur, nitrogen, and other crucial elements. Collectively, their metabolic activities influence nutrient availability for organisms everywhere.

For example, certain bacteria use **anaerobic electron transport**. Electrons stripped from some type of organic compound move on through transport systems of their plasma membrane. Commonly, an inorganic compound in the environment serves as the final electron acceptor. The net energy yield varies, but it is always small.

Even as you read this, some anaerobic bacteria that live in waterlogged soil are stripping electrons from a variety of compounds. They dump electrons on sulfate. Hydrogen sulfide, a putrid-smelling gas, is the result. The sulfate-reducing bacteria also live in many aquatic habitats that are enriched with decomposed organic material. They even live on the deep ocean floor, near hydrothermal vents. As described in Section 42.11, they form the food production base for unique communities.

In fermentation pathways, an organic substance that forms during the reactions serves as the final acceptor of electrons from glycolysis. The reactions regenerate NAD^+, which is required to keep the pathway operational.

In anaerobic electron transport, an inorganic substance (but not oxygen) usually serves as the final electron acceptor.

For each glucose molecule metabolized, anaerobic pathways typically have a net energy yield of two ATP, which only form during glycolysis.

Figure 7.11 Robin feasting on the fermented berries of a pyracantha bush.

WWW

ALTERNATIVE ENERGY SOURCES IN THE HUMAN BODY

So far, you have looked at what happens after a lone glucose molecule enters an energy-releasing pathway. Now you can start thinking about what cells do when they have too many or too few molecules of glucose.

Carbohydrate Breakdown in Perspective

THE FATE OF GLUCOSE AT MEALTIME   Consider what happens to you or any other mammal during a meal. Glucose and certain other small organic molecules are being absorbed across the gut lining, then the blood transports them through the body. A rise in the glucose level in blood prompts the pancreas to release insulin, a hormone that stimulates cells to take up glucose at a faster rate. The cells convert the windfall of glucose to glucose-6-phosphate and so "trap" it in the cytoplasm. (When phosphorylated, glucose cannot be transported back out, across the plasma membrane.) Look again at Figure 7.4, and you see that glucose-6-phosphate is the first activated intermediate of glycolysis.

If your glucose intake exceeds cellular demands for energy, ATP-producing machinery goes into high gear. Unless a cell is rapidly using ATP, its concentration of ATP can rise to a high level. Then, glucose-6-phosphate is diverted into a biosynthesis pathway that assembles glucose units into glycogen, a storage polysaccharide (Section 3.2). This is especially the case for muscle and liver cells, which maintain the largest glycogen stores.

THE FATE OF GLUCOSE BETWEEN MEALS When you are not eating, glucose is not entering your bloodstream and its level in the blood declines. If the decline were not countered, that would be bad news for the brain, your body's glucose hog. The brain constantly takes up more than two-thirds of the freely circulating glucose because its many hundreds of millions of cells simply use this sugar alone as their preferred energy source.

The pancreas responds to the decline by secreting glucagon, a hormone that prompts liver cells to convert glycogen back to glucose and send it back to the blood. Only liver cells do this; muscle cells won't give it up. The blood glucose level rises, and brain cells keep on functioning. Thus, *hormones control whether the body's cells use free glucose as an energy source or tuck it away.*

A word of caution: Don't let the preceding examples lead you to believe cells squirrel away large amounts of glycogen. In adult humans, glycogen makes up merely 1 percent or so of the body's total energy reserves, the energy equivalent of two cups of cooked pasta. Unless you eat on a regular basis, you will deplete the liver's small glycogen stores in less than twelve hours. Of the total energy reserves in, say, a typical adult American, 78 percent (about 10,000 kilocalories) is concentrated in body fat and 21 percent in proteins.

Energy From Fats

The question becomes this: How does the body access its huge reservoir of fats? A fat molecule, recall, has a glycerol "head" and one, two, or three fatty acid "tails." Most fats that become stored in your body are in the form of triglycerides, with three tails each. Triglycerides accumulate in fat cells of adipose tissue, which forms at the buttocks and other strategic places beneath the skin.

When blood glucose levels decline, triglycerides can be tapped as an energy alternative. Then, enzymes in fat cells cleave the bonds holding the glycerol and fatty acids together, and the breakdown products enter the bloodstream. Afterward, enzymes in the liver convert the glycerol to PGAL—an intermediate of glycolysis. Nearly all cells can take up the circulating fatty acids. Enzymes cleave the carbon backbone of the fatty acid tails and convert the fragments to acetyl-CoA—which can enter the Krebs cycle (Figures 7.6 and 7.12).

Each fatty acid tail has many more carbon-bound hydrogen atoms than glucose, so its breakdown yields much more ATP. In between meals or during sustained exercise, fatty acid conversions supply about half of the ATP that muscle, liver, and kidney cells require.

What happens if you eat too many carbohydrates? Exceed the glycogen-storing capacity of your liver and muscle cells, and the excess gets converted to fats. *Too much glucose ends up as excess fat.* Worse yet, 25 percent of the people in the United States are blessed with a combination of genes that allows them to eat as much as they like without gaining weight—but a diet far too rich in carbohydrates keeps the other 75 percent fat. For them, insulin levels remain elevated, which "tells" the body to store fat rather than use it for energy. We will return to this topic in Sections 31.7 and 36.10.

Energy From Proteins

Eat more proteins than your body requires to grow and maintain itself, and its cells won't store them. Enzymes split these proteins into amino acid units. Then they remove the amino group ($-NH_3^+$) from each unit, and ammonia (NH_3) forms. What happens to the leftover carbon backbones? Depending on conditions in the cell, the outcome varies. The backbones can be converted to carbohydrates or fats. Or they may enter the Krebs cycle, as shown in Figure 7.12, where coenzymes can pick up hydrogen and electrons stripped away from the carbon atoms. The ammonia that forms undergoes conversions to become urea. This nitrogen-containing waste product would be toxic if it accumulated to high concentrations. Normally your body excretes urea, in urine.

As this brief discussion makes clear, maintaining and accessing the body's energy reserves is complicated

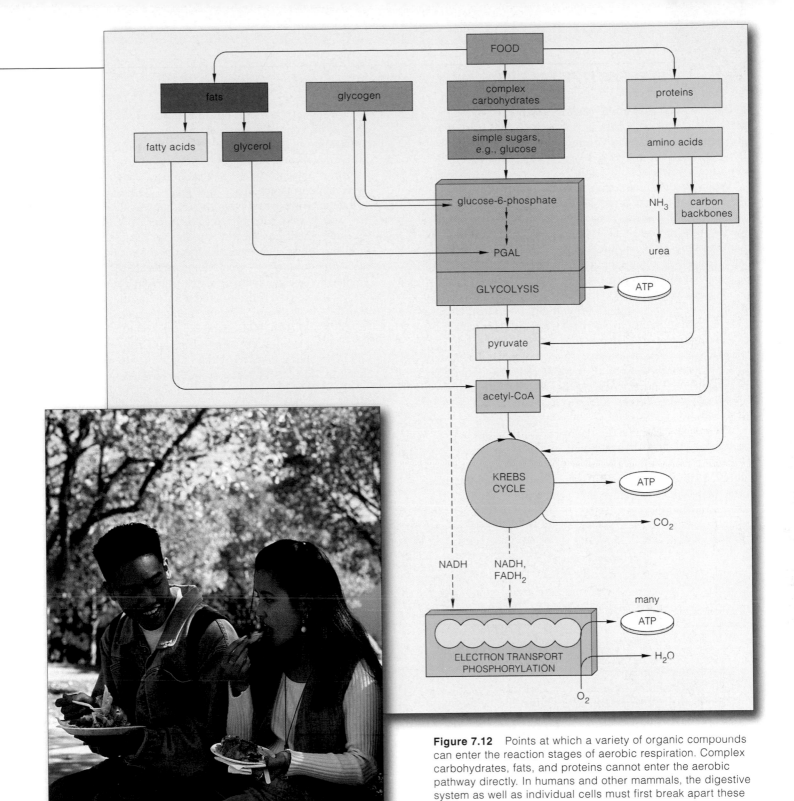

Figure 7.12 Points at which a variety of organic compounds can enter the reaction stages of aerobic respiration. Complex carbohydrates, fats, and proteins cannot enter the aerobic pathway directly. In humans and other mammals, the digestive system as well as individual cells must first break apart these large molecules into simpler, degradable subunits.

business. Hormonal controls over the disposition of glucose are special only because glucose is the fuel of choice for the all-important brain. However, as you will see in later chapters, providing all of your cells, organs, and organ systems with energy starts with the kinds and proportions of food you put in your mouth.

This concludes our look at aerobic respiration and other energy-releasing pathways. The section to follow may help you get a sense of how they fit into the larger picture of life's evolution and interconnectedness.

In humans and other mammals, the entrance of glucose or other organic compounds into an energy-releasing pathway depends on the kinds and proportions of carbohydrates, fats, and proteins in the diet as well as on the type of cell.

In this unit, you read about photosynthesis and aerobic respiration—the main pathways by which cells trap, store, and release energy. What you might not know is that the two pathways became linked, on a grand scale, over evolutionary time.

When life originated more than 3.8 billion years ago, the Earth's atmosphere had little free oxygen. The earliest single-celled organisms probably used reactions similar to glycolysis to make ATP. Without oxygen, fermentation pathways must have dominated. By about 1.5 billion years later, oxygen-producing photosynthetic cells had emerged. They irrevocably changed the course of evolution.

Oxygen, a by-product of the noncyclic pathway of photosynthesis, began to accumulate in the atmosphere. Probably through mutations that affected the proteins of electron transport systems, some cells started using oxygen as an electron acceptor. At some point in the past, descendants of those fledgling aerobic cells abandoned photosynthesis entirely. Among them were the forerunners of animals and all other organisms that engage in aerobic respiration.

With aerobic respiration, the flow of carbon, hydrogen, and oxygen through the metabolic pathways of living organisms came full circle. For the final products of this aerobic pathway—carbon dioxide and water—are the same materials that are necessary to build organic compounds in photosynthesis:

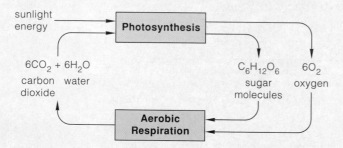

sunlight energy → **Photosynthesis**

$6CO_2 + 6H_2O$
carbon dioxide   water

$C_6H_{12}O_6$
sugar molecules

$6O_2$
oxygen

Aerobic Respiration

Perhaps you have difficulty fathoming the connection between yourself—an intelligent being—and such remote-sounding events as energy flow and the cycling of carbon, hydrogen, and oxygen. Is this really the stuff of humanity?

Think back, for a moment, on the structure of a water molecule. Two hydrogen atoms sharing electrons with an oxygen atom may not seem close to your daily life. And yet, through that sharing, water molecules show polarity—and they hydrogen-bond with one another. Their chemical behavior is a beginning for the organization of lifeless matter that leads to the organization of all living things.

For now you can imagine other molecules interspersed through water. The nonpolar kinds resist interaction with water; the polar kinds dissolve in it. On their own, the phospholipids among them assemble into a two-layered film. Such lipid bilayers, remember, serve as the very framework of all cell membranes, hence all cells. From the beginning, the cell has been the fundamental *living* unit.

The essence of life is not some mysterious force. It is metabolic control. With a cell membrane to contain them, reactions *can* be controlled. With mechanisms built into their membranes, cells can respond to energy changes and shifting concentrations of substances in the environment. The response mechanisms operate by "telling" proteins—enzymes—when and what to build or tear down.

And it is not some mysterious force that creates the proteins themselves. DNA, the slender double-stranded treasurehouse of inheritance, has the chemical structure—*the chemical message*—that allows molecule to reproduce molecule, one generation after the next. In your own body, DNA strands tell trillions of cells how countless molecules must be built or torn apart for their stored energy.

So yes, carbon, hydrogen, oxygen, and other atoms of organic molecules represent the stuff of you, and us, and all of life. But it takes more than molecules to complete the picture. Life exists as long as an unbroken flow of energy sustains its organization. Molecules are assembled into cells, cells into organisms, organisms into communities, and so on up through the biosphere. It takes energy inputs from the sun to maintain these levels of organization. And energy flows through time in one direction—from organized to less organized forms. Only as long as energy continues to flow into the web of life can life continue in all its rich diversity.

In short, life is no more *and no less* than a marvelously complex system of prolonging order. Sustained by energy transfusions from the sun, life continues onward, through its capacity for self-reproduction. For with the hereditary instructions contained in DNA, energy and materials can be organized, generation after generation. Even with the death of individuals, life elsewhere is prolonged. With each death, molecules are released and may be cycled once more, as raw materials for new generations.

In this flow of energy and cycling of material through time, each birth is affirmation of our ongoing capacity for organization, each death a renewal.

SUMMARY

1. Energy inherent in the organization of phosphate groups in ATP and other substances drives nearly all metabolic reactions. Aerobic respiration, fermentation, and other pathways that release chemical energy from organic compounds, such as glucose, produce ATP.

2. After glucose enters these pathways, enzymes strip electrons and hydrogen from intermediates that form along the way. Coenzymes pick these up and deliver them to other reaction sites, where the pathway ends. NAD$^+$ is the main coenzyme; the aerobic route also uses FAD. When loaded with electrons and hydrogen, these coenzymes are designated NADH and FADH$_2$.

3. The main energy-releasing pathways all start with glycolysis, a stage of reactions that begin and end in the cytoplasm. The glycolytic reactions can be completed either in the presence of oxygen or in its absence.
   a. During glycolysis, enzymes break down a glucose molecule to two pyruvate molecules. Two NADH and four ATP form during the reactions.
   b. The net energy yield is two ATP (because two ATP had to be invested up front to get the reactions going).

4. Aerobic respiration continues on through two more stages: (1) the Krebs cycle and a few steps preceding it, and (2) electron transport phosphorylation. The stages proceed only inside the organelles called mitochondria, which occur only in eukaryotic cells.

5. The second stage of the aerobic pathway starts when an enzyme strips a carbon atom from each pyruvate. Coenzyme A binds the remaining two-carbon fragment (to form acetyl-CoA), then transfers it to oxaloacctate, the entry point of the Krebs cycle. The cyclic reactions, along with the steps immediately preceding them, load up ten coenzymes with electrons and hydrogen (eight NADH and two FADH$_2$). Two ATP form. Three carbon dioxide molecules are released for each pyruvate that entered this second stage.

6. The third stage of the aerobic pathway proceeds at a membrane that divides the interior of a mitochondrion into two compartments. Electron transport systems and ATP synthases are embedded in this inner membrane.
   a. Coenzymes deliver electrons from the first two stages to transport systems. In the outer compartment, hydrogen ions accumulate; therefore, concentration and electric gradients form across the membrane.
   b. Hydrogen ions follow the gradients and flow from the outer to the inner compartment, through the interior of ATP synthases. Energy released during the ion flow drives the formation of ATP from ADP and unbound phosphate.
   c. Oxygen withdraws electrons from the transport system and, at the same time, combines with hydrogen ions to form water molecules. The oxygen is the final acceptor of electrons that initially resided in glucose.

7. Aerobic respiration has a typical net energy yield of thirty-six ATP for each glucose molecule metabolized. Yields vary, according to cell type and cell conditions.

8. The fermentation pathways and anaerobic electron transport also start with glycolysis, but they do not use oxygen. They are anaerobic, start to finish.
   a. Lactate fermentation has a net energy yield of two ATP, which form in glycolysis. The remaining reaction regenerates NAD$^+$. The two NADH from glycolysis transfer electrons and hydrogen to two pyruvate from glycolysis. Two lactate molecules are the end products.
   b. Alcoholic fermentation has a net energy yield of two ATP from glycolysis, and its remaining reactions serve to regenerate NAD$^+$. Enzymes convert pyruvate from glycolysis to acetaldehyde, and carbon dioxide is released. The NADH from glycolysis transfer electrons and hydrogen to the two acetaldehyde molecules, thus forming two ethanol molecules, the end products.
   c. Certain bacteria use anaerobic electron transport. Electrons are stripped from various organic compounds and travel through transport systems in the bacterial cell's plasma membrane. An inorganic compound in the environment often serves as the final electron acceptor.

9. In humans and other mammals, simple sugars such as glucose from carbohydrates, glycerol and fatty acids from fats, and carbon backbones of amino acids from proteins can enter ATP-producing pathways.

Review Questions

1. Is this true or false: Aerobic respiration occurs in animals but not plants, which make ATP only by photosynthesis. *7.1*

2. For this diagram of the aerobic pathway, fill in all blanks and write in the number of molecules of pyruvate, coenzymes, and end products. Also write in the net ATP formed in each stage, and the net ATP formed from start (glycolysis) to finish. *7.1*

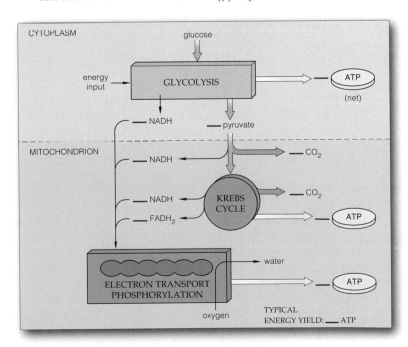

3. Is glycolysis energy-*requiring* or energy-*releasing*? Or do both kinds of reactions occur during glycolysis? *7.2*

4. In what respect does *electron transport* phosphorylation differ from *substrate-level* phosphorylation? *7.2, Figure 7.7*

5. Sketch the double-membrane system of the mitochondrion and show where transport systems and ATP synthases are located. *7.3*

6. Name the compound that is the entry point for the Krebs cycle, and state whether it directly accepts the pyruvate from glycolysis. For each glucose molecule, how many carbon atoms enter the Krebs cycle? How many depart from it, and in what form? *7.3*

7. Is this statement true or false: Muscle cells cannot contract at all when deprived of oxygen. If true, explain why. If false, name the alternative(s) available to them. *7.5*

Self-Quiz *(Answers in Appendix III)*

1. Glycolysis starts and ends in the _____ .
   a. nucleus   c. plasma membrane
   b. mitochondrion   d. cytoplasm

2. Which of the following does *not* form during glycolysis?
   a. NADH   c. FADH$_2$
   b. pyruvate   d. ATP

3. Aerobic respiration is completed in the _____ .
   a. nucleus   c. plasma membrane
   b. mitochondrion   d. cytoplasm

4. In the last stage of aerobic respiration, _____ is the final acceptor of electrons that originally resided in glucose.
   a. water   c. oxygen
   b. hydrogen   d. NADH

5. _____ engage in lactate fermentation.
   a. *Lactobacillus* cells   c. Sulfate-reducing bacteria
   b. Muscle cells   d. a and b

6. In alcoholic fermentation, _____ is the final acceptor of the electrons stripped from glucose.
   a. oxygen   c. acetaldehyde
   b. pyruvate   d. sulfate

7. The fermentation pathways produce no more ATP beyond the small yield from glycolysis, but the remaining reactions _____ .
   a. regenerate ADP   c. dump electrons on an
   b. regenerate NAD$^+$ inorganic substance (not oxygen)

8. In certain organisms and under certain conditions, _____ can be used as an energy alternative to glucose.
   a. fatty acids   c. amino acids
   b. glycerol   d. all of the above

9. Match the event with its most suitable metabolic description.
   ____ glycolysis   a. ATP, NADH, FADH$_2$, CO$_2$,
   ____ fermentation   and water form
   ____ Krebs cycle   b. glucose to two pyruvate
   ____ electron transport   c. NAD$^+$ regenerated, two ATP net
        phosphorylation   d. H$^+$ flows through ATP synthases

Critical Thinking

1. Diana suspects that a visit to her family doctor is in order. After eating carbohydrate-rich food, she always experiences sensations of being intoxicated and becomes nearly incapacitated, as if she had been drinking alcohol. She even wakes up with a hangover the next day. Having completed a course in freshman biology, Diana has an idea that something is affecting the way her body is metabolizing glucose. Explain why.

2. The cells of your body absolutely do not use nucleic acids as alternative energy sources. Suggest why.

3. The body's energy needs and its programs for growth depend on balancing the levels of amino acids in blood with proteins in cells. Cells of the liver, kidneys, and intestinal lining are especially important in this balancing act. When the levels of amino acids in blood decline, lysozymes in cells can rapidly digest some of their proteins (structural and contractile proteins are spared, except in cases of malnutrition). The amino acids released this way enter the blood and thereby help maintain the required levels.

Suppose you embark on a body-building program. You already eat plenty of carbohydrates, but a nutritionist advises a protein-rich diet that includes protein supplements. Speculate on how the extra dietary proteins will be put to use, and in which tissues.

4. Each year, Canada geese lift off in precise formation from their northern breeding grounds. They head south to spend the winter months in warmer climates, and then they make the return trip in spring. As is true of other migratory birds, their flight muscle cells are efficient at using fatty acids as an energy source. (Remember, the carbon backbone of fatty acids can be cleaved into fragments that can be converted to acetyl-CoA for entry into the Krebs cycle.)

Suppose a lesser Canada goose from Alaska's Point Barrow has been steadily flapping along for three thousand kilometers and is nearing Klamath Falls, Oregon. It looks down and notices a rabbit sprinting like the wind from a coyote with a taste for rabbit. With a stunning burst of speed, the rabbit reaches the safety of its burrow.

Which energy-releasing pathway predominated in muscle cells in the rabbit's legs? Why was the Canada goose relying on a different pathway for most of its journey? And why wouldn't the pathway of choice in goose flight muscle cells be much good for a rabbit making a mad dash from its enemy?

5. Reflect on this chapter's introduction, then question 4. Now speculate on which energy-releasing pathway is predominating in an agitated Africanized bee chasing a farmer in a cornfield.

Selected Key Terms

acetyl-CoA *7.3*
aerobic respiration *7.1*
alcoholic fermentation *7.5*
anaerobic electron transport *7.5*
electron transport
   phosphorylation *7.1*
FAD *7.1*
glycolysis *7.1*

Krebs cycle *7.1*
lactate fermentation *7.5*
mitochondrion *7.3*
NAD$^+$ *7.1*
oxaloacetate *7.3*
pyruvate *7.1*
substrate-level
   phosphorylation *7.2*

Readings   *See also www.infotrac-college.com*

Levi, P. October 1984. "Travels with C." *The Sciences.* Journey of a carbon atom through the world of life.

Wolfe, S. 1995. *An Introduction to Molecular and Cellular Biology.* Belmont, California: Wadsworth. Exceptional reference text.

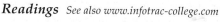
WWW   *http://www.brookscole.com/biology*

Practice quiz questions, hypercontents, BioUpdates, and critical thinking. The Brooks/Cole Biology Resource Center provides a wealth of information fully organized and integrated by chapter.

FACING PAGE: *Human sperm, one of which will penetrate this mature egg and so set the stage for the development of a new individual in the image of its parents.*

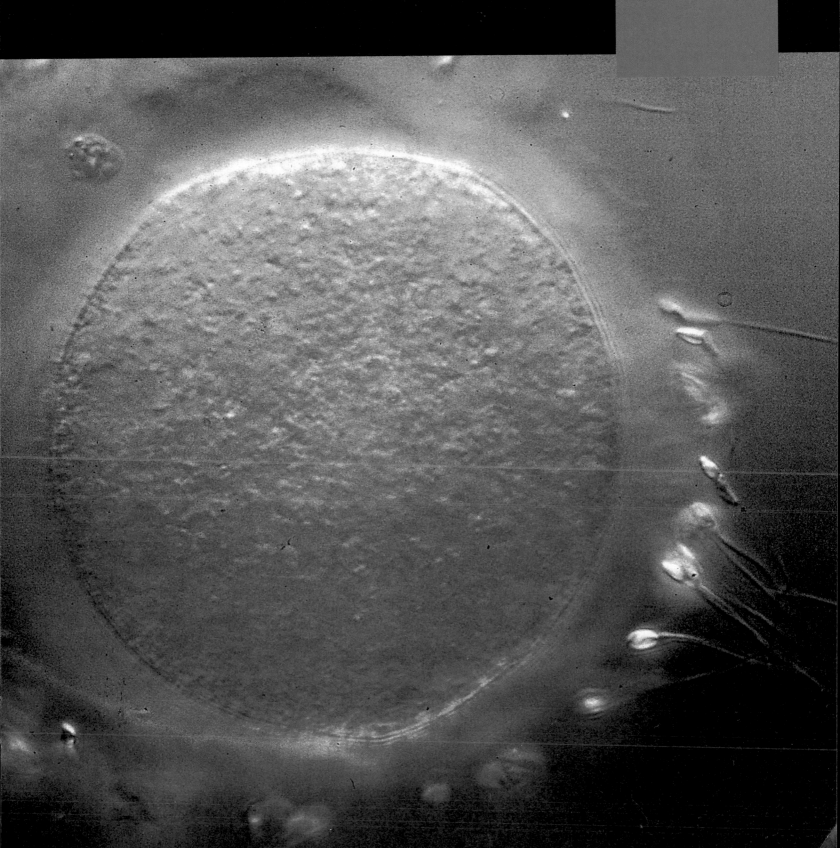

CELL DIVISION AND MITOSIS

Silver In the Stream of Time

Five o'clock, and the first rays from the sun dance over the wild Alagnak River of the Alaskan tundra. It is September, and life is ending and beginning in the clear, cold waters. By the thousands, mature silver salmon have returned from the open ocean to spawn in their shallow native home. The females are tinged with red, the color of spawners, and they are dying.

This morning you observe a female salmon releasing translucent pink eggs into a shallow "nest," which her fins hollowed out in the gravel riverbed (Figure 8.1). Within moments a male sheds a cloud of sperm, and fertilization follows. Trout and other predators eat most of the eggs. But if you wait around, you will find that some eggs survive and give rise to a new generation.

Within three years, the pea-size eggs have become streamlined salmon, fashioned from billions of cells. A few of their cells will develop into eggs or sperm. In time, on some September morning, they will take part in an ongoing story of birth, growth, death, and rebirth.

For you, as for salmon and every other multicelled species, growth as well as reproduction depends on *cell division*. Inside your mother, a fertilized egg divided in two, then the two into four, and so on until billions of cells were growing, developing in specialized ways, and dividing at different times to produce all of your genetically prescribed body parts. Your body now has roughly 65 trillion living cells—and many of them are still dividing. Every five days, for instance, divisions replace the tissue that lines your small intestine.

Understanding cell division—and, ultimately, how new individuals are put together in the image of their parents—begins with answers to three questions. *First,* what instructions are necessary for inheritance? *Second,* how are those instructions duplicated for distribution into daughter cells? *Third,* by what mechanisms are the duplicated instructions parceled out to daughter cells? We will require more than one chapter to consider the nature of cell reproduction and other mechanisms of inheritance. Even so, the points made early in this chapter can help you keep the overall picture in focus.

Begin with the word **reproduction**. In biology, this means that parents produce a new generation of cells or multicelled individuals like themselves. The process starts in cells that are programmed to divide. And the

sexually mature
female salmon

Figure 8.1   The last of one generation and the first of the next in Alaska's Alagnak River.

ground rule for division is this: *Parent cells must provide their daughter cells with hereditary instructions, encoded in DNA, and enough metabolic machinery to start up their own operation.*

DNA, recall, contains instructions for synthesizing proteins. Some proteins are structural materials. Many are enzymes that speed the assembly of specific organic compounds, such as the lipids that cells use as building blocks and sources of energy. Unless a daughter cell receives the necessary instructions for making proteins, it simply will not be able to grow or function properly.

Also, the cytoplasm of a parent cell already contains enzymes, organelles, and other operating machinery. When a daughter cell inherits what looks like a blob of cytoplasm, it really is getting start-up machinery—which will keep that cell operating until it can use its inherited DNA for growing and developing on its own.

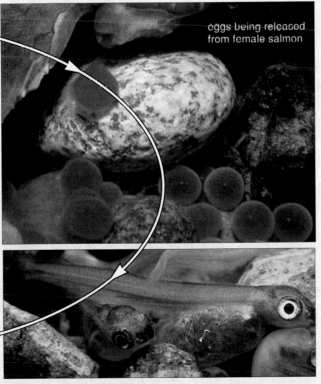

eggs being released from female salmon

fingerlings—young fishes growing, from fertilized eggs, by way of mitotic cell divisions

KEY CONCEPTS

1. The continuity of life depends on reproduction, by which parents produce a new generation of cells or multicelled individuals like themselves. Cell division is the bridge between generations.

2. When a cell divides, its two daughter cells must each receive a required number of DNA molecules and some cytoplasm. For eukaryotic cells, a division mechanism called mitosis sorts out the DNA into two new nuclei. A separate mechanism divides the cytoplasm.

3. Mitosis is one part of the cell cycle. The other part is interphase, an interval when each new cell formed by mitosis and cytoplasmic division increases in mass, increases the number of its components, then duplicates its DNA. The cycle ends when that cell divides.

4. In eukaryotic cells, many proteins with structural and functional roles are attached to the DNA. Each DNA molecule, with its attached proteins, is a chromosome.

5. Members of the same species have a characteristic number of chromosomes in their cells. The chromosomes differ from one another in length and shape, and they carry different portions of the hereditary instructions.

6. The body cells of humans and many other organisms have a diploid chromosome number; they contain two of each type of chromosome characteristic of the species.

7. Mitosis keeps the chromosome number constant, from one cell generation to the next. Therefore, if a parent cell is diploid, the daughter cells also will be diploid.

8. Mitotic cell division is the basis of growth and tissue repair in multicelled eukaryotes. It also is the means by which single-celled eukaryotes and many multicelled eukaryotes reproduce asexually.

DIVIDING CELLS: THE BRIDGE BETWEEN GENERATIONS

Overview of Division Mechanisms

In plants, animals, and all other eukaryotic organisms, hereditary instructions are distributed among a number of DNA molecules. Before the cells of such organisms reproduce, they must undergo *nuclear* division. **Mitosis** and **meiosis** are two nuclear division mechanisms. Both sort out and then package a parent cell's DNA into new nuclei, for the forthcoming daughter cells. A separate mechanism splits the cytoplasm into daughter cells.

Multicelled organisms grow, replace dead or worn-out cells, and repair tissues by way of mitosis and the cytoplasmic division of body cells, which are called **somatic cells**. (Cut yourself peeling a potato and mitotic cell divisions will replace the cells that the knife sliced away.) Also, many protistans, fungi, plants, and some animals reproduce asexually by mitotic cell division.

By contrast, meiosis occurs only in **germ cells**, a cell lineage set aside for the formation of gametes (such as sperm and eggs) and sexual reproduction. As you will read in the next chapter, meiosis has much in common with mitosis, but the end result is different.

Prokaryotic cells, or bacteria, reproduce asexually by a different mechanism, called prokaryotic fission. We will consider the bacteria later, in Section 20.3.

Some Key Points About Chromosomes

In a nondividing cell, the DNA molecules are stretched out like thin threads, with many proteins attached to them. Each DNA molecule, with its attached proteins, is a **chromosome**. When a cell prepares for mitosis, each

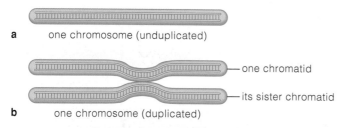

a   one chromosome (unduplicated)

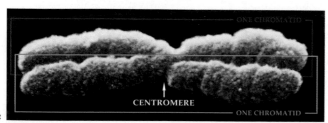

— one chromatid

— its sister chromatid

b   one chromosome (duplicated)

Figure 8.2   (**a**,**b**) Sketches of a chromosome in the unduplicated and duplicated states. Chromosomes are duplicated before cell division. (**c**) This scanning electron micrograph shows a human chromosome in the duplicated state; it consists of two sister chromatids attached at the centromere.

threadlike chromosome is duplicated. It now consists of two DNA molecules, which will stay together until late in mitosis. As long as they remain attached, the two are called **sister chromatids** of the chromosome.

Figure 8.2 illustrates a eukaryotic chromosome in the unduplicated and duplicated states. Notice how the duplicated chromosome narrows down in one region along its length. This is the **centromere**, a small region with attachment sites for microtubules that move the chromosome during nuclear division. Bear in mind, the sketch in Figure 8.2 is highly simplified. For example, the centromere's location differs among chromosomes. And the two strands of a DNA molecule do not look like a ladder; they twist rather like a spiral staircase and are much longer than can be shown here.

Mitosis and the Chromosome Number

Each species has a characteristic **chromosome number**, which refers to the sum total of chromosomes in cells of a given type. Human somatic cells have 46, those of gorillas have 48, and those of pea plants have 14.

Actually, your 46 chromosomes are like volumes of two sets of books. Each set is numbered, from 1 to 23. For example, you have two "volumes" of chromosome 22—that is, *a pair of them*. Generally, both members of each pair have the same length and shape, and they carry the same portion of hereditary instructions for the same traits. Think of them as two sets of books on how to build a house. Your father gave you one set. Your mother had her own ideas about storage, plumbing, and so on, so she gave you an alternate edition. Her set covers the same topics but says slightly different things about many of them.

We say the chromosome number is **diploid**, or $2n$, if a cell has two of each type of chromosome characteristic of the species. The body cells of humans, gorillas, pea plants, and a great many other organisms are like this. (By contrast, as explained in Chapter 9, eggs and sperm of these organisms have a *haploid* chromosome number, or n. This means that they have only one of each type of chromosome characteristic of the species.)

With mitosis, a diploid parent cell can produce two diploid daughter cells. This doesn't mean each merely gets forty-six or forty-eight or fourteen chromosomes. If only the total mattered, one cell might get, say, two pairs of chromosome 22 and no pairs whatsoever of chromosome 9. Neither would be able to function like the parent cell *without two of each type of chromosome*.

Mitosis keeps the chromosome number constant, division after division, from one cell generation to the next. Thus, if a parent cell is diploid, its daughter cells will be diploid also.

THE CELL CYCLE

Mitosis is only one phase of the **cell cycle**. Such cycles start each time new cells form, and they end when those cells complete their own division. The cycle starts again for each new daughter cell, as in Figure 8.3. Usually, the longest phase of the cell cycle is **interphase**, which has three parts. During interphase, a cell increases its mass, roughly doubles its number of cytoplasmic components, then duplicates its DNA. The three parts of a cell cycle are often abbreviated this way:

G1 Of interphase, a "*Gap*" (interval) of cell growth and functioning before the onset of DNA replication

S   Of interphase, a time of "*Synthesis*" (DNA replication)

G2 Of interphase, a second "*Gap*" (interval) following DNA replication, when the cell prepares for division

M   *Mitosis*; nuclear division only, usually followed by cytoplasmic division

The cell cycle lasts about the same length of time for cells of the same type. Its duration differs among cells of different types. As examples, all neurons (nerve cells) in your brain are arrested at interphase and usually do not divide again. All red blood cells form and replace your worn-out ones at an average rate of 2–3 million each second. During development, the cells of a sea urchin embryo may double in number every two hours.

Adverse conditions may disrupt a cell cycle. When deprived of a vital nutrient, for instance, the free-living cells called amoebas do not leave interphase. Even so, if any cell proceeds past a certain point in interphase,

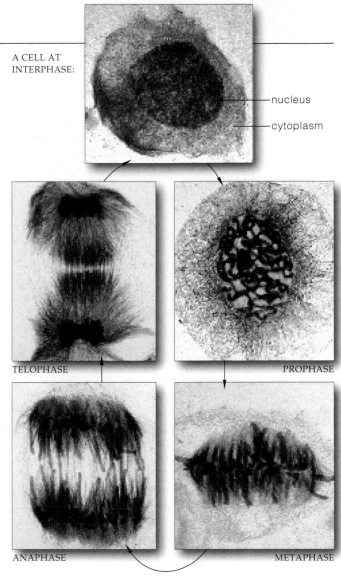

A CELL AT INTERPHASE:
— nucleus
— cytoplasm

TELOPHASE

PROPHASE

ANAPHASE

METAPHASE

Figure 8.4 Mitosis in a cell from the African blood lily, *Haemanthus*. The chromosomes are stained *blue*, and the many microtubules are stained *red*. Before reading further, take a moment to become familiar with the labels on the micrographs.

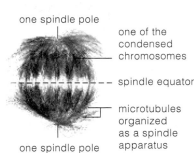

one spindle pole

one of the condensed chromosomes

spindle equator

microtubules organized as a spindle apparatus

one spindle pole

the cycle normally will continue regardless of outside conditions owing to built-in controls over its duration.

We turn now to mitosis and to how it maintains the chromosome number through turn after turn of the cell cycle. Figure 8.4 only hints at the divisional ballet that begins when a cell leaves interphase.

A cell cycle starts at interphase, when a new cell (formed by mitosis and cytoplasmic division) increases its mass and the number of its cytoplasmic components, then duplicates its chromosomes. The cycle ends when the cell divides.

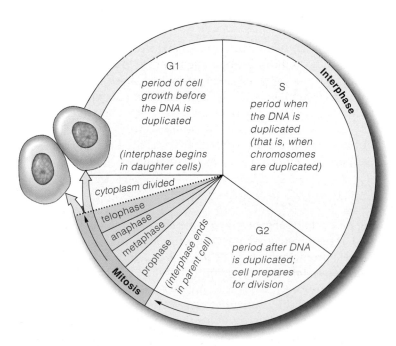

G1
period of cell growth before the DNA is duplicated

(interphase begins in daughter cells)

cytoplasm divided
telophase
anaphase
metaphase
prophase
(interphase ends in parent cell)

Interphase

S
period when the DNA is duplicated (that is, when chromosomes are duplicated)

G2
period after DNA is duplicated; cell prepares for division

Mitosis

Figure 8.3 Eukaryotic cell cycle, generalized. The length of each part differs among different cell types.

THE STAGES OF MITOSIS—AN OVERVIEW

When a cell makes the transition from interphase to mitosis, it has stopped constructing new cell parts, and the DNA has been replicated. Within that cell, major changes will now proceed smoothly, one after the other, through four stages. The sequential stages of mitosis are **prophase**, **metaphase**, **anaphase**, and **telophase**.

Figure 8.5 illustrates mitosis in an animal cell. By comparing the series of photographs against those of the plant cell in Figure 8.4, we clearly see that the chromosomes in those cells are changing positions. They are not doing this on their own. A **spindle apparatus** is moving them.

A spindle consists of microtubules organized into two sets. Each set extends from one of the two poles (end points) of the spindle. The two sets overlap each other a bit at the spindle equator, or midway between the two poles. The formation of this bipolar, microtubular spindle establishes what will be the final destinations of chromosomes during mitosis, as you will see shortly.

Figure 8.5 Mitosis in a generalized animal cell. This nuclear division mechanism ensures that every daughter cell will have the same chromosome number as the parent cell. For clarity, the diagram shows only two pairs of chromosomes from a diploid (2n) animal cell. With only rare exceptions, the picture is more complicated than this, as indicated by the micrographs of mitosis in a whitefish cell (*facing page*).

MITOSIS

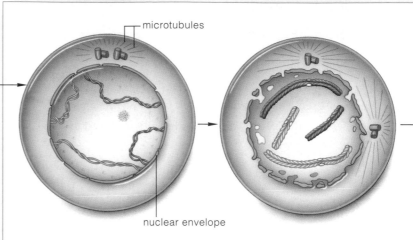

nucleus   uncondensed DNA

microtubules

plasma membrane   pair of centrioles

nuclear envelope

CELL AT INTERPHASE
The cell duplicates its DNA; then it prepares for nuclear division.

EARLY PROPHASE
The DNA and its associated proteins have started to condense. The two chromosomes colored *purple* were inherited from the female parent. The other two (*blue*) are their counterparts, inherited from the male parent.

LATE PROPHASE
Chromosomes continue to condense. New microtubules are assembled; they move one of the two centriole pairs to the opposite end of the cell. The nuclear envelope starts to break up.

Prophase: Mitosis Begins

We know a cell is in prophase when its chromosomes become visible in the light microscope as threadlike forms. ("Mitosis" comes from the Greek *mitos*, meaning thread.) Each chromosome was duplicated earlier, during interphase. In other words, each now consists of two sister chromatids, joined at the centromere. Early on, the two sister chromatids twist and fold into a more compact form. By late prophase, all the chromosomes will be condensed into thicker, rod-shaped forms.

Meanwhile, in the cytoplasm, most microtubules of the cytoskeleton are breaking apart into their tubulin subunits (Section 4.9). Near the nucleus, the subunits reassemble as *new* microtubules of the spindle. Many of the new microtubules will extend from one spindle pole or the other to the centromere of a chromosome. The remainder will not interact at all with the chromosomes. They will extend from the poles and overlap each other.

While new microtubules are assembling, the nuclear envelope physically prevents them from interacting with chromosomes inside the nucleus. However, the nuclear envelope starts breaking up as prophase ends.

Many cells have two barrel-shaped **centrioles**. Each centriole started duplicating itself during interphase, so there are two pairs of them when prophase is under way. Microtubules start moving one pair to the opposite pole of the newly forming spindle. Centrioles, recall, give rise to flagella or cilia. If you observe them in cells of an organism, you can bet that flagellated cells (such as sperm) or ciliated cells develop during its life cycle.

Transition to Metaphase

So much happens between prophase and metaphase that researchers give this transitional period its own name, "prometaphase." The nuclear envelope breaks up completely into numerous tiny, flattened vesicles. Now the chromosomes are free to interact with microtubules that are extending toward them, from the poles of the forming spindle. Microtubules from both poles harness each chromosome and start pulling on it. The two-way pulling orients the chromosome's two sister chromatids toward opposite poles. Meanwhile, overlapping spindle

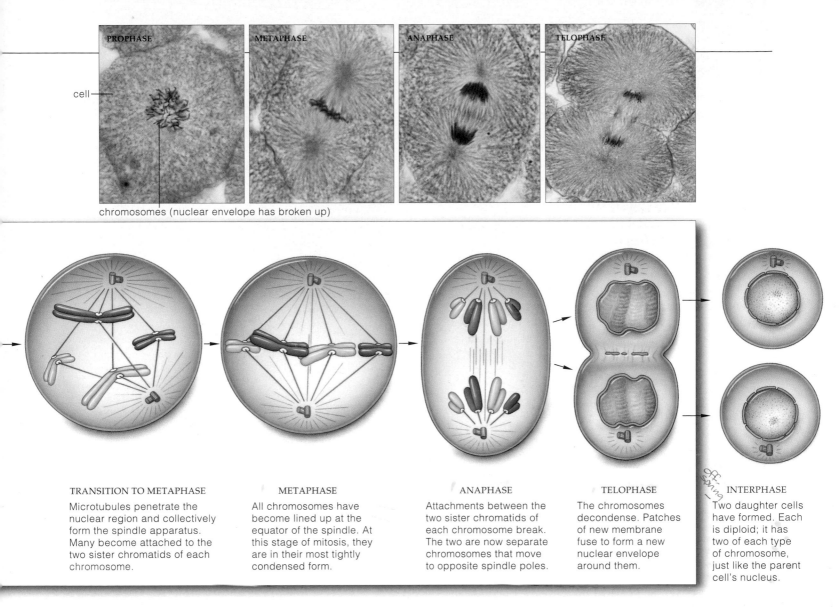

PROPHASE

METAPHASE

ANAPHASE

TELOPHASE

cell

chromosomes (nuclear envelope has broken up)

TRANSITION TO METAPHASE
Microtubules penetrate the nuclear region and collectively form the spindle apparatus. Many become attached to the two sister chromatids of each chromosome.

METAPHASE
All chromosomes have become lined up at the equator of the spindle. At this stage of mitosis, they are in their most tightly condensed form.

ANAPHASE
Attachments between the two sister chromatids of each chromosome break. The two are now separate chromosomes that move to opposite spindle poles.

TELOPHASE
The chromosomes decondense. Patches of new membrane fuse to form a new nuclear envelope around them.

INTERPHASE
Two daughter cells have formed. Each is diploid; it has two of each type of chromosome, just like the parent cell's nucleus.

microtubules ratchet past each other and push the poles of the spindle apart. The push–pull forces are balanced when the chromosomes reach the spindle's midpoint.

When all the duplicated chromosomes are aligned midway between the poles of a completed spindle, we call this metaphase (*meta*— means "midway between"). The alignment is crucial for the next stage of mitosis.

From Anaphase Through Telophase

At anaphase, sister chromatids of each chromosome separate from each other and move to opposite poles by two mechanisms. First, microtubules attached to the centromere regions shorten and *pull* the chromosomes to the poles. Second, the spindle elongates as overlapping microtubules continue to ratchet past each other and *push* the two spindle poles even farther apart. Once each chromatid is separated from its sister, it has a new name. It is a separate chromosome in its own right.

Telophase gets under way as soon as each of two clusters of chromosomes arrives at a spindle pole. The chromosomes, no longer harnessed to the microtubules, return to threadlike form. Vesicles derived from the old nuclear envelope fuse and form patches of membrane around the chromosomes. Patch joins with patch, and soon a new nuclear envelope separates each cluster of chromosomes from the cytoplasm. If the parent cell was diploid, each cluster contains two chromosomes of each type. With mitosis, remember, each new nucleus has the same chromosome number as the parent nucleus. Once two nuclei form, telophase is over—and so is mitosis.

Prior to mitosis, each chromosome in a cell's nucleus is duplicated, so that it consists of two sister chromatids.

Mitosis proceeds through four consecutive stages, called prophase, metaphase, anaphase, and telophase.

A microtubular spindle moves sister chromatids of every chromosome apart, to opposite spindle poles. Around each of two clusters of chromosomes, new nuclear envelope forms. Both daughter nuclei formed this way have the same chromosome number as the parent cell's nucleus.

WWW

DIVISION OF THE CYTOPLASM

The cytoplasm usually divides at some time between late anaphase and the end of telophase. As you might gather by comparing Figures 8.6 and 8.7, the actual mechanism of **cytoplasmic division** (or cytokinesis, as it is often called) differs among organisms.

Cell Plate Formation in Plants

As described in Section 4.10, most plant cells are walled, which means their cytoplasm cannot be pinched in two. Cytoplasmic division of such cells involves **cell plate formation**, as shown in Figure 8.6. By this mechanism, vesicles chockfull of wall-building materials fuse with one another and with remnants from the microtubular spindle. Together, they form a disklike structure—a cell plate. At this location, deposits of cellulose accumulate.

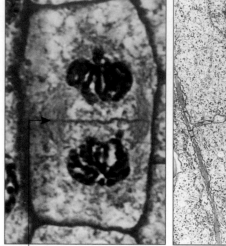
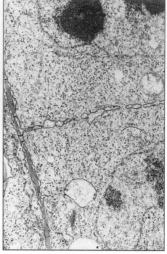

light micrograph and transmission electron micrograph showing cell plate formation in a dividing plant cell

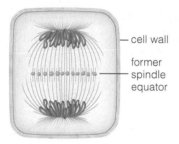
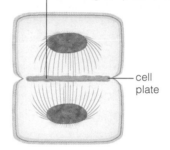
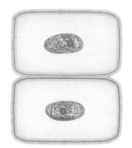

— cell wall

former spindle equator

vesicles converging

cell plate

a As mitosis ends, vesicles converge at the spindle equator. They contain cementing materials and structural materials for a new primary cell wall.

b A cell plate starts forming as membranes of the vesicles fuse. Materials inside the vesicles get sandwiched between two new membranes that elongate along the plane of the cell plate.

c Cellulose is deposited on the inside of the "sandwich." (In time, deposits will form two cell walls. Other deposits will form a middle lamella and cement the walls together; refer to Section 4.10.)

d A cell plate grows at its margins until it fuses with the parent cell's plasma membrane. During growth, when new plant cells expand and their walls are still thin, new material is deposited on the old primary wall.

Figure 8.6 Cytoplasmic division of a plant cell, as brought about by cell plate formation.

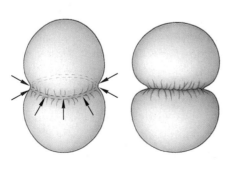

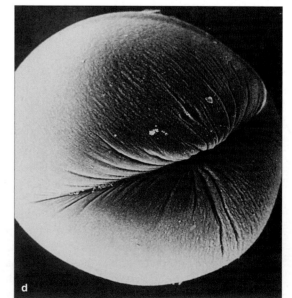

a Mitosis is over, and the spindle is now disassembling.

b Just beneath the plasma membrane, a band of microfilaments at the former spindle equator contract, so that their diameter closes all around the cell.

c The contractions continue and cut the cell in two.

Figure 8.7 (**a–c**) Cytoplasmic division of an animal cell. (**d**) Scanning electron micrograph of the cleavage furrow at the plane of the former spindle's equator. Beneath it, a band of microfilaments attached to the plasma membrane contracts and pulls the surface inward. It continues to deepen until the egg is cut in two.

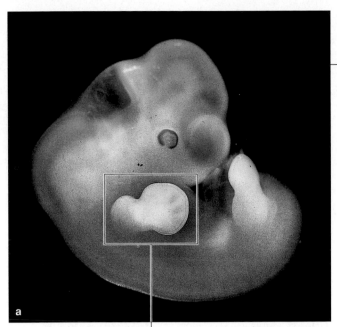

future arm and hand of embryo, five weeks old

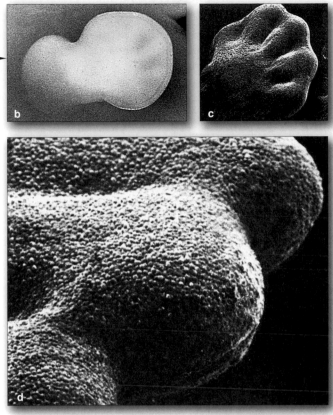

In time, the cellulose deposits are thick enough to form a crosswall. The new crosswall bridges the cytoplasm and divides the parent cell into two daughter cells.

Cytoplasmic Division of Animal Cells

Unlike plant cells, an animal cell is not confined within a cell wall, and its cytoplasm typically "pinches in two." Look at Figure 8.7, a surface view of a newly fertilized animal egg. An indentation is forming about midway between the egg's two poles ("ends"). The indentation in its plasma membrane is a **cleavage furrow**. Such a furrow is the first visible sign that an animal cell is undergoing cytoplasmic division. It will extend around the cell and continue to deepen along the plane of the former spindle's midpoint until the cell is cut in two.

A band of microfilaments beneath the cell's plasma membrane generates the force for cytoplasmic division. Microfilaments, remember, are threadlike cytoskeletal elements. These particular ones are organized so that they slide past one another (Section 4.9). When they do, they pull the plasma membrane inward until the two daughter nuclei are cut off in separate cells, each with its own cytoplasm and plasma membrane.

This concludes our picture of mitotic cell division. Look now at your two hands and try to visualize all of the cells in your palms, thumbs, and fingers. Imagine the divisions that produced all the generations of cells that preceded them when you were developing early on, inside your mother (Figure 8.8). And be grateful for the astonishing precision of the mechanisms that led to their formation at certain times, in certain numbers, for the alternatives can be terrible indeed. Why? Good

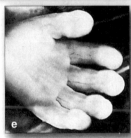

Figure 8.8 Moments in the development of the human hand by way of mitosis, cytoplasmic divisions, and other processes. Many individual cells that were produced by mitotic cell divisions are visible in (**d**). The photograph in (**e**) shows the digits at a later stage of development.

health, and survival itself, depends absolutely on the proper timing and the completion of cell cycle events, including mitosis. Some genetic disorders arise from mistakes in the duplication or distribution of just one chromosome. Also, when normal controls that prevent cells from dividing are lost, unchecked cell divisions may destroy the surrounding tissues and, ultimately, the organism. Section 8.6 touches on a landmark case of such losses, which are further explored in Section 14.5.

Following mitosis, a separate mechanism cuts the cytoplasm into two daughter cells, each with a daughter nucleus.

In plants, cytoplasmic division often involves the formation of a cell plate and a crosswall in between the adjoining, new plasma membranes of daughter cells.

Cytoplasmic division in animals may involve cleavage. Rings of microfilaments around a parent cell's midsection slide past one another in a way that pinches the cytoplasm in two.

A CLOSER LOOK AT THE CELL CYCLE

Close this book for a moment. Be sure you have a clear picture of the flow of events in interphase, mitosis, and on through cytoplasmic division. How easily you will get through many later chapters depends on how well you understand the cell division story. If parts of the picture are still not clear, it may be worth your time to read the preceding sections once again before getting into the details presented here.

On Chromosomes and Spindles

The precision with which mitosis parcels out DNA for forthcoming daughter cells is impressive. That precision depends on chromosome organization and interactions among microtubules and motor proteins.

ORGANIZATION OF METAPHASE CHROMOSOMES Even during interphase, eukaryotic DNA has many proteins bound tightly to it. **Histones** are among them. Many histones are like spools for winding up small stretches of DNA. Each histone-DNA spool is a single structural unit called a **nucleosome**. Other histones stabilize the spools. During mitosis (and meiosis also), interactions between histones and DNA make the chromosome coil back on itself repeatedly. The coiling greatly increases the chromosome's diameter. Other proteins besides the histones form a structural scaffold when the DNA folds even more, possibly into a series of loops (Figure 8.9).

A chromosome acquires its distinct shape and size late in prophase, when condensation is nearly complete. By then, each of its sister chromatids has at least one constricted region, the most prominent of which is the centromere. As you will read next, small, disk-shaped structures at the surface of centromeres serve as the docking sites for spindle microtubules (Figure 8.9a). We call them **kinetochores**.

SPINDLES COME, SPINDLES GO All through anaphase, the spindle microtubules attached to chromosomes do not change position, yet the distance shrinks between those microtubules and the spindle poles. How? When a kinetochore slides over them, the microtubules shorten by disassembling. A kinetochore is like a train chugging along a railroad track—except the track falls apart after the "train" has passed. It contains two motor proteins, dynein and kinesin, that may drive the sliding motion.

Motor proteins, remember, are accessory proteins for cytoskeletal elements. As described in Section 4.9, they are attached to and project from microtubules and microfilaments that take part in cell movements.

What about spindle microtubules extending from both poles but not attached to kinetochores? They, too, incorporate dynein and kinesin. They actively slide past one another where they overlap in a spindle.

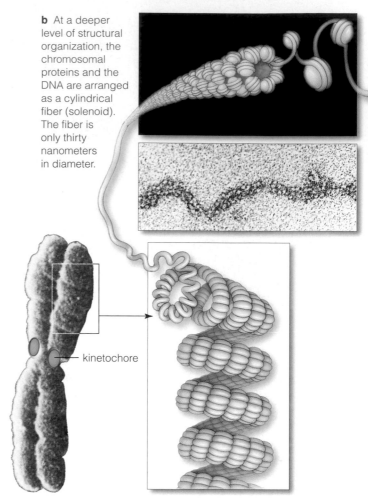

b At a deeper level of structural organization, the chromosomal proteins and the DNA are arranged as a cylindrical fiber (solenoid). The fiber is only thirty nanometers in diameter.

kinetochore

a A duplicated human chromosome at metaphase, at its most condensed. Interactions among some chromosomal proteins may keep its loops of DNA tightly packed in a "supercoiled" array.

Figure 8.9 One model of the levels of organization in a human metaphase chromosome.

A final point about spindle structure and function: Over evolutionary time, spindle microtubules became targets in the chemical warfare between certain edible plants and animals that browse on them. For instance, plants of the genus *Colchicum* make a poison that blocks assembly and promotes disassembly of microtubules. The poison, colchicine, is a favorite of those who study cancer and other expressions of cell division. (Critical Thinking question 4 at the end of the chapter describes another.) Spindles in cells disassemble within seconds or minutes after exposure to such microtubule poisons.

The Wonder of Interphase

If you could coax the DNA molecules from just one of your somatic cells to stretch in a single line, one after another, that line would extend past the fingertips of

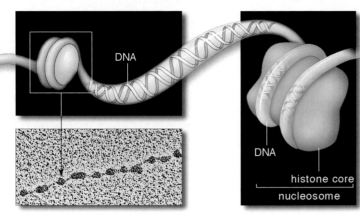

c Immerse a chromosome in saltwater and it loosens up to a beads-on-a-string organization. The "string" is one molecule of DNA. Each "bead" on the string is a nucleosome.

d A nucleosome consists of a double loop of DNA around a core of eight histones. Other histones stabilize the structural array.

your outstretched arms. Salamander DNA is even more amazing. A single line of it would extend ten meters! The wonder is, enzymes and other proteins in the cell selectively scan all the DNA, switch protein-building instructions on and off, and make base-by-base copies of each DNA molecule—all during interphase.

G1, S, and G2 of interphase have distinct patterns of biosynthesis. Most of your cells remain in G1, when they assemble most of the carbohydrates, lipids, and proteins they use or export. Those destined to divide enter S, when they copy the DNA and the histones and other proteins associated with it. During G2, those cells produce proteins that will drive mitosis to completion.

Once S begins, events normally proceed at about the same rate in all cells of a species and continue through mitosis. Given this observation, you might well assume the cycle has built-in, molecular brakes. Apply brakes that operate in G1, and the cycle stalls in G1. Lift the brakes and the cycle runs to completion. Said another way, *control mechanisms govern the rate of cell division.*

Imagine a car losing its brakes just as it starts down a steep mountain road. As you will read later in the book, that is how cancer starts. Controls over division are lost, and the cell cycle cannot stop turning.

The condensed form of metaphase chromosomes arises by interactions among DNA and structural proteins, including histones, that associate with DNA throughout the cell cycle.

During mitosis, motor proteins act on microtubules of the spindle to produce chromosomal movements.

Once the S stage of interphase begins, the cell cycle proceeds at about the same rate in all cells of a given type, all the way through mitosis. Molecular mechanisms control whether a cell enters S and thereby control the rate of cell division.

HENRIETTA'S IMMORTAL CELLS

Each human starts out as a single fertilized egg. By the time of birth, mitotic cell divisions and other processes have resulted in a human body of about a trillion cells. Even in an adult, billions of cells are still dividing. For example, cells of the stomach's lining divide every day. Liver cells usually do not divide, but if part of the liver becomes injured or diseased, repeated cell divisions will keep on producing more new cells until the damaged part is finally replaced.

In 1951, George and Margaret Gey of Johns Hopkins University were trying to develop a way to keep human cells dividing *outside* the body. With such isolated cells, these researchers and others could investigate basic life processes. They also could conduct studies of cancer and other diseases without having to experiment directly on patients and thereby gamble with human lives.

The Geys had samples of normal and diseased human cells, which local physicians had sent to them. But they just couldn't stop the descendants of those precious cells from dying out within a few weeks.

Mary Kubicek, a laboratory assistant, worked with the Geys in their efforts to start a self-perpetuating lineage of cultured human cells. She was about to give up after dozens of failed attempts. Even so, in 1951 she decided to prepare one more sample of cancer cells for culture. She gave the sample the code name **HeLa cells**, for the first two letters of the patient's first and last names.

The HeLa cells began to divide. And divide. And divide again! By the fourth day there were so many cells that Kubicek had to subdivide them into more culture tubes. As months passed, the culture continued to thrive.

Unfortunately, the tumor cells inside the patient's body were just as vigorous. Six months after the patient was first diagnosed as having cancer, tumor cells had spread to tissues throughout her body. Only eight months after the diagnosis, Henrietta Lacks, a young woman from Baltimore, was dead.

Although Henrietta passed away, some of her cells lived on in the Geys' laboratory as the first successful human cell culture. HeLa cells were soon shipped to other researchers, who passed cells on to others, and so on. HeLa cells came to live in laboratories all over the world. Some of the cultured cells even journeyed far into space, as components of experiments to be carried out aboard the *Discoverer XVII* satellite. Every year, hundreds of scientific papers describe research that is based on work with HeLa cells.

Henrietta was only thirty-one years old when runaway cell divisions quickly killed her. Yet now, many decades later, her legacy is still benefiting humans everywhere—in cellular descendants that are still alive and dividing, day after day after day.

WWW

SUMMARY

1. Through specific division mechanisms, summarized in Table 8.1, a parent cell provides each daughter cell with the hereditary instructions (DNA) and cytoplasmic machinery necessary to start up its own operation.

a. In eukaryotic cells, the nucleus divides by mitosis or meiosis. Cytoplasmic division typically follows.

b. Prokaryotic cells divide by prokaryotic fission.

2. Each eukaryotic chromosome is one DNA molecule with numerous proteins attached. The chromosomes in a given cell differ in length, shape, and which part of the hereditary instructions they carry.

a. "Chromosome number" refers to the sum total of chromosomes in cells of a given type. Cells having a diploid chromosome number (2*n*) contain two of each kind of chromosome.

b. Mitosis divides the nucleus into two equivalent nuclei, each having the same chromosome number as the parent cell. Therefore, it maintains the chromosome number from one cell generation to the next.

c. Mitosis is the basis of growth, tissue repair, and cell replacements among multicelled eukaryotes. Often it is the basis of asexual reproduction among single-celled eukaryotes. (Meiosis occurs only in germ cells.)

3. A cell cycle starts when a new cell forms. It proceeds through interphase and ends when the cell reproduces by mitosis and cytoplasmic division. In interphase, a cell carries out its functions. If it is to divide again, the cell increases in its mass and cytoplasmic components, and duplicates its chromosomes in preparation for division.

4. A duplicated chromosome has two DNA molecules attached at the centromere. While the two stay attached to each other, they are called sister chromatids.

5. Mitosis proceeds through four continuous stages:

a. Prophase. Duplicated chromosomes are in their threadlike form and start to condense. Near the nucleus, new microtubules start to assemble into the spindle apparatus. The nuclear envelope starts to break up. In the *transition* to metaphase (prometaphase), the nuclear envelope breaks apart, and its remnants form vesicles. The microtubules from opposite poles of the developing spindle attach to only one of the two sister chromatids of each chromosome.

b. Metaphase. *At* metaphase, all the chromosomes are aligned at the spindle equator.

c. Anaphase. Microtubules pull sister chromatids of each chromosome away from each other, to opposite spindle poles. Now each type of parental chromosome is represented by a daughter chromosome at both poles:

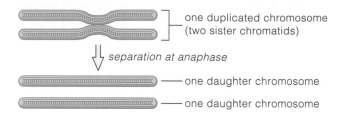

one duplicated chromosome
(two sister chromatids)

separation at anaphase

—— one daughter chromosome

—— one daughter chromosome

d. Telophase. Chromosomes now decondense to a threadlike form. A new nuclear envelope forms around them. Each nucleus has the same chromosome number as the parent cell. Mitosis is completed.

6. Different mechanisms divide the cytoplasm near the end of nuclear division or afterward. In plant cells, a cell plate, new plasma membrane, then a crosswall form. Animal cells pinch in two, as by cleavage.

Review Questions

1. Define mitosis and meiosis, two mechanisms that operate in eukaryotic cells. Does either one divide the cytoplasm? 8.1

2. Define somatic cell and germ cell. 8.1

3. What is a chromosome called when it is in the unduplicated state? In the duplicated state (with two sister chromatids)? 8.1

4. Describe the microtubular spindle and its functions. 8.3

5. Using Figure 8.10 as a guide, name and describe the key features of the stages of mitosis. 8.3

6. Briefly explain how cytoplasmic division differs in typical plant and animal cells. 8.4

Self-Quiz (*Answers in Appendix III*)

1. A somatic cell having two of each type of chromosome typical of the species has a(n) _____ chromosome number.
   a. diploid   b. haploid   c. tetraploid d. abnormal

2. A duplicated chromosome has _____ chromatid(s).
   a. one   b. two   c. three   d. four

3. In a chromosome, a _____ is a constricted region with attachment sites for microtubules.
   a. chromatid   c. cell plate
   b. centromere d. cleavage furrow

4. Interphase is the part of the cell cycle when _____ .
   a. a cell ceases to function
   b. a germ cell forms its spindle apparatus
   c. a cell grows and duplicates its DNA
   d. mitosis proceeds

Table 8.1	Summary of Cell Division Mechanisms
Mechanisms	Functions
MITOSIS, CYTOPLASMIC DIVISION	In all multicelled eukaryotes, the basis of increases in body size during growth, tissue. repair, and cell replacements. Also the basis of *asexual* reproduction in single-celled and many multicelled eukaryotes.
MEIOSIS, CYTOPLASMIC DIVISION	In single-celled and multicelled eukaryotes, the basis of gamete formation and of *sexual* reproduction.
PROKARYOTIC FISSION	In bacterial cells only, the basis of *asexual* reproduction.

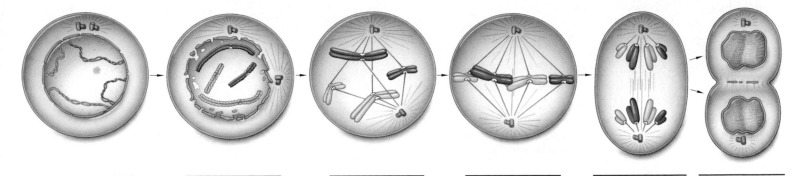

Figure 8.10 Name and briefly describe the key features of the stages of mitosis.

5. After mitosis, the chromosome number of a daughter cell is _____ the parent cell's.
   a. the same as   c. rearranged compared to
   b. one-half   d. doubled compared to

6. Mitosis and cytoplasmic division function in _____ .
   a. asexual reproduction of single-celled eukaryotes
   b. growth, tissue repair, and sometimes asexual reproduction in many multicelled eukaryotes
   c. gamete formation in prokaryotes
   d. both a and b

7. Only _____ is not a stage of mitosis.
   a. prophase   b. interphase   c. metaphase   d. anaphase

8. Match each stage with the events listed.
   _____ metaphase   a. sister chromatids move apart
   _____ prophase   b. chromosomes start to condense
   _____ telophase   c. chromosomes decondense and daughter nuclei form
   _____ anaphase   d. all duplicated chromosomes are aligned at spindle equator

Critical Thinking

1. Suppose you have a way to measure the amount of DNA in a single cell during the cell cycle. You first measure the amount at the G1 phase. At what points during the rest of the cell cycle would you predict changes in the amount of DNA per cell?

2. A cell from a tissue culture has 38 chromosomes. After mitosis and cytoplasmic division, one daughter cell has 39 chromosomes. The other daughter cell has 37. Speculate on what might have caused the abnormal chromosome numbers. Generally speaking, how might the abnormality affect the structure, function, or both of the tissue descendants of the abnormal cells?

3. Each year, 250,000 to 1 million women in the United States are at risk to develop *cervical cancer.* The cervix is a narrowed part of the uterus, an organ in which embryos grow and develop. A screening procedure called the *Pap smear* can detect the earliest stages of cervical cancer, when chances for survival are greatest.

   For this procedure, a doctor scrapes a few living cells from the cervix and sends them out for laboratory analysis. Precancerous and cancerous cervical cells have an altered appearance. If cells are precancerous, freezing them or hitting them with a laser beam will kill them. A hysterectomy (removal of the uterus) can remove the cancerous cervical cells still localized in the uterus. If caught early, treatment is over 90 percent effective. If cancer spreads from the uterus and invades other parts of the body, the chance of effective treatment plummets to less than 9 percent.

   Most cervical cancers develop slowly. Past age eighteen (or earlier, if sexual activity starts sooner), all women should have Pap smears annually (or every other year if results are negative

three years in a row). Engaging in unsafe sex increases the risk (Section 38.19). Why? A major risk factor is infection by human papillomaviruses that cause genital warts. In 93 percent of all cases, viral genes coding for tumor-inducing proteins had been inserted into the DNA of previously normal cervical cells.

   Not all women request Pap smears. Many wrongly believe the procedure is costly (low-cost screening is available). Many do not recognize the importance of abstinence or "safe" sex. Many others can't bear the thought of having cancer and do not want to know whether they do. Knowing what you have learned so far about the cell cycle and cancer, what would you say to a woman who falls in one or more of these groups?

4. Pacific yews (*Taxus brevifolius*) face extinction. People started stripping its bark and killing the trees when they heard that *taxol,* a chemical extracted from the bark, might be useful for treating breast and ovarian cancer. (Synthesizing taxol in the laboratory may save the species.) Taxol can prevent microtubules from disassembling into tubulin subunits. What does this tell you about its potential as an anticancer drug?

5. X-rays and gamma rays emitted from some radioisotopes chemically damage DNA, especially in cells engaged in DNA replication. High-level exposure can result in *radiation poisoning.* Hair loss and damage to the lining of the stomach and intestines are two early symptoms. Speculate why. Also speculate on why highly focused radiation therapy is used against some cancers.

Selected Key Terms

anaphase 8.3	cytoplasmic	metaphase 8.3
cell cycle 8.2	division 8.4	mitosis 8.1
cell plate	diploid (chromosome	motor protein 8.5
formation 8.4	number) 8.1	nucleosome 8.5
centriole 8.3	germ cell 8.1	prophase 8.3
centromere 8.1	HeLa cell 8.6	reproduction CI
chromosome 8.1	histone 8.5	sister chromatid 8.1
chromosome	interphase 8.2	somatic cell 8.1
number 8.1	kinetochore 8.5	spindle apparatus 8.3
cleavage 8.4	meiosis 8.1	telophase 8.3

Readings See also *www.infotrac-college.com*

Murray, A., and M. Kirschner. March 1991. "What Controls the Cell Cycle?" *Scientific American* 264(3): 56–63.

WWW *http://www.brookscole.com/biology*

Practice quiz questions, hypercontents, BioUpdates, and critical thinking. The Brooks/Cole Biology Resource Center provides a wealth of information fully organized and integrated by chapter.

9 MEIOSIS

Octopus Sex and Other Stories

The couple clearly are interested in each other. First he caresses her with one tentacle, then another—and then another and another. She reciprocates with a hug here, a squeeze there. This goes on for hours. Finally the male reaches under his mantle, a fold of tissue that drapes around most of his body. He removes a packet of sperm from a reproductive organ and inserts it into an egg chamber underneath the female's mantle. For every sperm that successfully fertilizes an egg, a new octopus may grow and develop.

Unlike the one-to-one coupling between a male and female octopus, sex for the slipper limpet is a group activity. Slipper limpets are marine animals, relatives of those familiar snails on land. Before becoming transformed into a sexually mature adult, a slipper limpet must pass on through a free-living stage of development called a larva. When a limpet larva is about to undergo transformation, it settles on a rock or pebble or shell. If it settles down all by itself, it will become a female. Then, if another larva settles and develops on the first limpet, the second limpet will function right off as a male. However, if *another* limpet develops into a male on top of *it*, that second male limpet will gradually become a female. The third also will become a female if a fourth male develops on top of *it*—and so on amongst ten or more limpets!

Figure 9.1 Variations in reproductive modes among eukaryotic organisms.

(**a**) Slipper limpets, busily perpetuating the species through group participation in sexual reproduction. The tiny crab in front of the topmost limpet in the stack is merely a passerby. (**b**) Live birth of an aphid, a type of insect that reproduces sexually in autumn but can switch to an asexual mode in summer.

With this chapter, we turn to the kinds of cells that serve as the bridge between generations of organisms. For many eukaryotic species, specialized phases of reproduction and development, including asexual episodes, loop out from the basic life cycle. Regardless of their specialized details, all of the life cycles turn on two basic events: *gamete formation* and *fertilization*.

Slipper limpets typically live in such piles, with the bottom one always being the oldest female and the uppermost one being the youngest male (Figure 9.1a). Until they make the gender switch, the males release sperm, which fertilize a female's eggs, which grow up to become males and then, most likely, females—and so it goes, from one limpet generation to the next.

Limpets are not alone in having unusual variations in their mode of reproduction. For example, sexual reproduction is common in many life cycles—and so are asexual episodes that are based on mitotic cell divisions. Orchids, dandelions, and many other plants reproduce very well with or without engaging in sex. Aquatic animals called flatworms can engage in sex or can split their small body into two roughly equivalent parts, which each grow into a new flatworm.

And what about aphids! In summertime, nearly all aphids are females, which produce more females from *unfertilized* egg cells (Figure 9.1b). Only when autumn approaches do male aphids finally develop and do their part in the sexual phase of the life cycle. Even then, females that manage to survive through the winter can do without males. Come summer, the females begin another round of producing offspring all by themselves.

These examples only hint at the immense variation in reproductive modes among eukaryotic organisms. And yet, despite the variation, *sexual* reproduction dominates nearly all of the life cycles, and it always involves certain events. Briefly, before the time of cell division, chromosomes are duplicated in reproductive cells. For instance, immature reproductive cells called **germ cells** develop in male and female animals. Germ cells undergo meiosis and cytoplasmic division. In time, the cellular descendants of germ cells mature and become **gametes**, or sex cells. When gametes manage to get together at fertilization, they form the first cell of a new individual.

Meiosis, the formation of gametes, and fertilization are the hallmarks of sexual reproduction. As you will see, these interconnected events have contributed to the diversity of life.

KEY CONCEPTS

1. Sexual reproduction proceeds through three events: meiosis, gamete formation, and fertilization. Sperm and eggs are familiar gametes.

2. Meiosis, a nuclear division mechanism, occurs only in cells that are set aside for sexual reproduction. The immature germ cells of male and female animals are examples. Meiosis sorts out a germ cell's chromosomes into four new nuclei. After meiosis, gametes form by way of cytoplasmic division and other events.

3. Cells with a diploid chromosome number contain two of each type of chromosome characteristic of the species. The two function as a pair during meiosis. Commonly, one chromosome of the pair is maternal, with hereditary instructions from a female parent. The other is paternal, with the same categories of hereditary instructions from a male parent.

4. Meiosis divides the chromosome number by half for each forthcoming gamete. Thus, if both parents have a diploid chromosome number ($2n$), the gametes that form will be haploid (n). Later, union of two gametes at fertilization restores the diploid number in the new individual ($n + n = 2n$).

5. During meiosis, each pair of chromosomes may swap segments. Each time they do, they exchange hereditary information about certain traits. Also, meiosis assigns one of each pair of chromosomes to a forthcoming gamete—but *which* gamete is its destination is a matter of chance. Hereditary instructions are further shuffled at fertilization. All three reproductive events lead to variations in traits among offspring.

6. In most plants, spore formation and other events intervene between meiosis and gamete formation.

FROM GAMETES TO OFFSPRING

The gametes that form following meiosis are not all the same in their details. For example, human sperm have one tail, opossum sperm have two, and roundworm sperm have none. Crayfish sperm look like pinwheels. Most eggs are microscopic in size, yet an ostrich egg tucked inside its shell is as large as a baseball. From its appearance alone, you might not believe that a plant gamete is even remotely like an animal's.

Later chapters contain details of how gametes form in the life cycles of representative organisms, including humans. Figure 9.7 and the following points may help you keep the details in perspective.

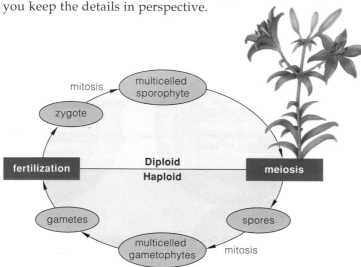

a Generalized life cycle for most kinds of plants

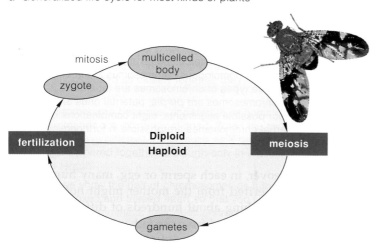

b Generalized life cycle for animals

Figure 9.7   Generalized life cycles for (**a**) most plants and (**b**) animals. The zygote is the first cell that forms when the nuclei of two gametes fuse together at fertilization.

For plants, a sporophyte (spore-producing body) develops, by way of mitotic cell divisions, from the zygote. After meiosis, gametophytes (gamete-producing bodies) form. A lily plant is a sporophyte. Gametophytes form in parts of its flowers.

Chapters 20 through 23 as well as Chapter 27 include many examples of the life cycles of representative organisms.

Gamete Formation in Plants

For pine trees, apple trees, roses, dandelions, corn, and nearly all other familiar plants, certain events intervene between the times of meiosis and gamete formation. Among other things, spores form.

Spores are haploid resting cells, often walled, that are good at resisting drought, cold, and other adverse environmental conditions. When favorable conditions return, spores germinate (resume growth) and develop into a haploid body or structure that produces gametes. So *gamete*-producing bodies and *spore*-producing bodies develop during the life cycle of most kinds of plants. Figure 9.7*a* is a generalized diagram of these events.

Gamete Formation in Animals

In male animals, gametes form by a process known as spermatogenesis. Inside the male reproductive system, a diploid germ cell grows in size. It becomes a primary spermatocyte, a large immature cell that enters meiosis and cytoplasmic divisions. Four haploid cells result and develop into spermatids (Figure 9.8). These immature cells change in form and develop a tail, thus becoming a **sperm**, a common type of mature male gamete.

In female animals, gametes form by a process that is called oogenesis. In human females, for example, a diploid germ cell develops into an **oocyte**, or immature egg. Unlike a sperm cell, an oocyte accumulates many cytoplasmic components. Also, as Figure 9.9 indicates, its daughter cells differ in size and function. When the oocyte undergoes cytoplasmic division after meiosis I, one cell—the secondary oocyte—gets nearly all of the cytoplasm. The other cell, called the first polar body, is quite small. Some time afterward, both cells undergo meiosis II, then cytoplasmic division. One daughter cell of the secondary oocyte develops into a second polar body. The other gets most of the cytoplasm and develops into the gamete. A mature female gamete is called an ovum (plural, ova) or, more commonly, an **egg**.

Thus, one egg and three polar bodies have formed. Polar bodies don't have much in the way of nutrients or metabolic machinery and don't function as gametes. In time they degenerate. But polar body formation allows the egg to end up with a suitable (haploid) number of chromosomes. Also, by getting most of the cytoplasm, the egg receives enough start-up machinery to support the new individual right after fertilization.

More Shufflings at Fertilization

The chromosome number characteristic of the parents is restored at **fertilization**, the time when a male gamete unites with a female gamete and their haploid nuclei

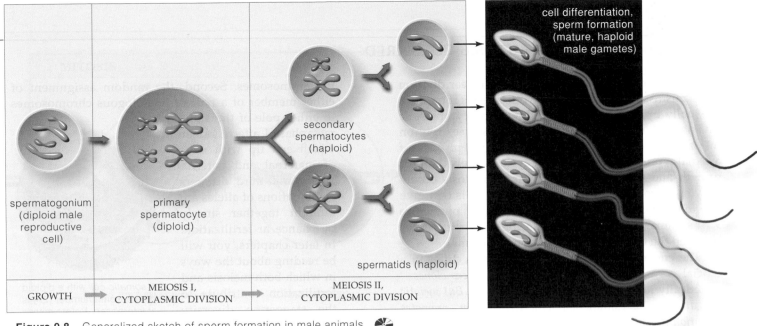

Figure 9.8 Generalized sketch of sperm formation in male animals.

Labels in figure: cell differentiation, sperm formation (mature, haploid male gametes); secondary spermatocytes (haploid); spermatids (haploid); spermatogonium (diploid male reproductive cell); primary spermatocyte (diploid); GROWTH; MEIOSIS I, CYTOPLASMIC DIVISION; MEIOSIS II, CYTOPLASMIC DIVISION

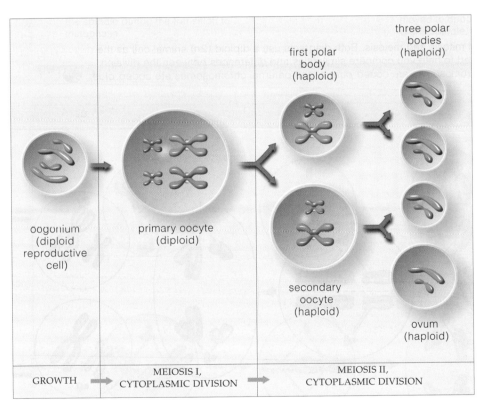

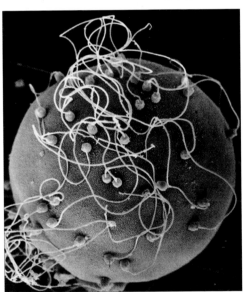

Figure 9.9 Egg formation in female animals. The generalized sketch is not at the same scale as Figure 9.8. Eggs are far larger than sperm, as the micrograph of sea urchin gametes shows. Also, the three polar bodies that form in meiosis are extremely small compared to the egg.

Labels in figure: three polar bodies (haploid); first polar body (haploid); oogonium (diploid reproductive cell); primary oocyte (diploid); secondary oocyte (haploid); ovum (haploid); GROWTH; MEIOSIS I, CYTOPLASMIC DIVISION; MEIOSIS II, CYTOPLASMIC DIVISION

fuse. If meiosis did not precede it, fertilization would double the chromosome number each new generation. Such changes in chromosome number would disrupt the hereditary instructions encoded in chromosomes, usually for the worse, for those instructions operate as a complex, fine-tuned package in each individual.

Fertilization contributes to variation in offspring. Reflect on the possibilities for humans alone. During prophase I, an average of two or three crossovers takes place in each human chromosome. Even without the crossovers, the random positioning of pairs of paternal and maternal chromosomes at metaphase I results in one of millions of possible chromosome combinations in each gamete. And of all the male and female gametes that are produced, which two actually get together is a matter of chance. The sheer number of combinations that can exist at fertilization is staggering!

Cumulatively, crossing over, the distribution of random mixes of homologous chromosomes into gametes, and fertilization contribute to variation in the traits of offspring.

SUMMARY

1. The life cycle of each sexually reproducing species includes meiosis, gamete formation, and fertilization.

   a. Meiosis, a nuclear division mechanism, reduces the chromosome number of a parent germ cell by half. It precedes the formation of haploid gametes (typically, sperm in males, eggs in females).

   b. At fertilization, a sperm and egg nuclei fuse. This event restores the chromosome number (Figure 9.11).

2. A germ cell with a diploid chromosome number ($2n$) has *two* of each type of chromosome characteristic of its species. Commonly, one of each pair of chromosomes is maternal (inherited from a female parent) and the other is paternal (from a male parent).

3. Each pair of maternal and paternal chromosomes shows homology, meaning the two are alike. Generally the two have the same length, same shape, and same sequence of genes. And they interact during meiosis.

4. Chromosomes are duplicated in interphase. Each consists of two DNA molecules that will stay attached, as sister chromatids, during mitosis.

5. Meiosis consists of two consecutive divisions that both require a microtubular spindle apparatus.

   a. In meiosis I, kinetochores (of chromosomes) and motor proteins (projecting from microtubules) interact to move each duplicated chromosome away from its partner, the homologous chromosome.

   b. In meiosis II, similar interactions move the sister chromatids of each chromosome away from each other.

6. The following events characterize meiosis I, the first nuclear division:

   a. Crossing over occurs in prophase I. Two nonsister chromatids of each pair of homologous chromosomes commonly break at corresponding sites and exchange segments. And this puts new combinations of alleles together. Alleles (slightly different molecular forms of the same gene) specify different forms of the same trait.

   b. Different combinations of alleles lead to variation in the details of a given trait among offspring.

   c. Also in prophase I, a microtubular spindle forms outside the nucleus, and the nuclear envelope starts to break up. In cells with duplicated pairs of centrioles, one pair starts moving to the opposite spindle pole.

   d. All the pairs of homologous chromosomes have become positioned at the spindle equator at metaphase I. For each pair, either the maternal chromosome or its homologue can be oriented toward either pole.

   e. In anaphase I, spindle microtubules interact with the kinetochores to move each duplicated chromosome away from its homologue, to opposite spindle poles.

7. The following key events characterize meiosis II, the second nuclear division:

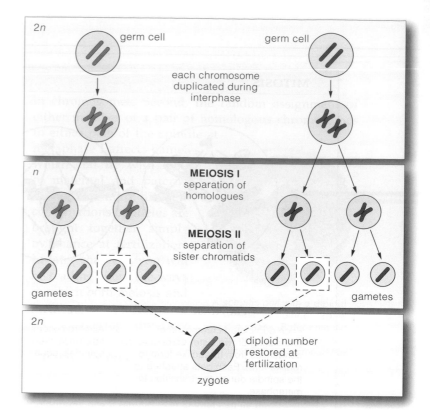

Figure 9.11 Summary of changes in the chromosome number at different stages of sexual reproduction, using diploid ($2n$) germ cells as the example. Meiosis reduces the chromosome number by half (n). Then the union of haploid nuclei of two gametes at fertilization restores the diploid number.

   a. At metaphase II, all the duplicated chromosomes are positioned at the spindle equator.

   b. Sister chromatids are moved apart in anaphase II. Each is now a separate, unduplicated chromosome.

   c. By the end of telophase II, four nuclei that have a haploid chromosome number (n) have been formed.

8. When the cytoplasm divides, there are four haploid cells. One or all of these may function as gametes (or as plant spores that give rise to gamete-producing bodies).

9. Crossing over, the chance allocation of different mixes of pairs of maternal and paternal chromosomes to different gametes, and the chance of any two gametes meeting at fertilization all contribute to the immense variation in the details of traits among offspring.

Review Questions

1. The diploid chromosome numbers for the somatic cells of a few kinds of organisms are listed at right. How many chromosomes will end up in gametes of each of these organisms? *9.2*

Fruit fly, *Drosophila melanogaster*	8
Garden pea, *Pisum sativum*	14
Corn, *Zea mays*	20
Frog, *Rana pipiens*	26
Earthworm, *Lumbricus terrestris*	36
Human, *Homo sapiens*	46
Chimpanzee, *Pan troglodytes*	48
Amoeba, *Amoeba*	50
Horsetail, *Equisetum*	216

2. A diploid germ cell has four pairs of homologous chromosomes, designated AA, BB, CC, and DD. How would the chromosomes of the gametes be designated? *9.2*

The cell is at anaphase ____ rather than anaphase ____ . I know this because:

_____
_____
_____

3. Look at the chromosomes in the germ cell in the diagram at the right. Is this cell at anaphase I or anaphase II? *9.3, 9.6*

4. Define meiosis and describe its main stages. In what key respects is meiosis *not* like mitosis? *9.3, 9.6*

5. From each of his parents, actor Michael Douglas (Figure 9.12*a*) inherited a gene that influences the chin dimple trait. One form of the gene called for a dimple and the other didn't, but one is all it takes for this particular trait. Figure 9.12*b* shows what the chin of Mr. Douglas might have looked like if he had inherited two ordinary forms of the gene instead. What is the name for the alternative forms of the same gene? *9.1*

6. Outline the main steps by which gametes form in plants. Do the same for gamete formation in animals. *9.5*

7. Genetically speaking, what is the key difference between the outcomes of sexual and asexual reproduction? *9.1, 9.6*

Self-Quiz (Answers in Appendix III)

1. Sexual reproduction requires _____ .
   a. meiosis   c. fertilization
   b. gamete formation d. all of the above

2. Meiosis is a division mechanism that produces _____ .
   a. two cells   c. four cells
   b. two nuclei d. four nuclei

3. An animal cell having two rather than one of each type of chromosome has a _____ chromosome number.
   a. diploid   c. normal gamete
   b. haploid   d. both b and c

4. Meiosis _____ the parental chromosome number.
   a. doubles   c. maintains
   b. reduces   d. corrupts

5. Generally, a pair of homologous chromosomes _____ .
   a. carry the same genes   c. interact at meiosis
   b. are the same length, shape d. all of the above

6. Before the onset of meiosis, all chromosomes are _____ .
   a. condensed   c. duplicated
   b. released from protein d. both b and c

7. Each chromosome moves away from its homologue and ends up at the opposite spindle pole during _____ .
   a. prophase I   c. anaphase I
   b. prophase II d. anaphase II

8. Sister chromatids of each chromosome move apart and end up at opposite spindle poles during _____ .
   a. prophase I   c. anaphase I
   b. prophase II d. anaphase II

9. Match each term with its description.
   ____ chromosome   a. different molecular forms
        number   of the same gene
   ____ alleles   b. none between meiosis I and II
   ____ metaphase I c. pairs of homologous
   ____ interphase   chromosomes are now aligned
                        at the spindle equator
                     d. sum total of chromosomes
                        in all cells of a given type

Figure 9.12 Example of the chin dimple trait (actually a fissure in the chin surface).

Critical Thinking

1. Assume you can measure the amount of DNA in a primary oocyte, then in a primary spermatocyte, which gives you a mass *m*. What mass of DNA would you expect to find in each mature gamete (egg and sperm) that forms after meiosis? What mass of DNA would you expect to find (1) in an egg fertilized by one of the sperm and (2) in that egg after the first DNA duplication?

2. Adam has a pair of alleles that influence whether a person is right- or left-handed. One allele says "left," and its partner says "right." Visualize one of his germ cells, in which chromosomes are being duplicated prior to meiosis. Visualize what happens to the chromosomes during anaphase I and II. (It might help to use index cards as models of the sister chromatids of each chromosome.) What fraction of Adam's sperm will carry the gene for right-handedness? For left-handedness?

3. Adam also has one allele for long eyelashes, and a partner allele (on the homologous chromosome) for short eyelashes. What fraction of his sperm will have these gene combinations:
   right-handed, long eyelashes   left-handed, long eyelashes
   right-handed, short eyelashes   left-handed, short eyelashes

---

Selected Key Terms

allele *9.1*	fertilization *9.5*	oocyte *9.5*
asexual reproduction *9.1*	gamete *CI*	sexual reproduction *9.1*
chromosome number *9.1*	gene *9.1*	sister chromatid *9.2*
diploidy *9.2*	germ cell *CI*	sperm *9.5*
egg (ovum) *9.5*	haploidy *9.2*	spore *9.5*
	homologous chromosome *9.2*	
	meiosis *9.2*	

Readings   *See also www.infotrac-college.com*

Klug, W., and M. Cummings. 1994. *Concepts of Genetics.* Fourth edition. New York: Macmillan.

Wolfe, S. 1995. *Introduction to Molecular and Cellular Biology.* Belmont, California: Wadsworth.

WWW   *http://www.brookscole.com/biology*

Practice quiz questions, hypercontents, BioUpdates, and critical thinking. The Brooks/Cole Biology Resource Center provides a wealth of information fully organized and integrated by chapter.

OBSERVABLE PATTERNS OF INHERITANCE

A Smorgasbord of Ears and Other Traits

Basketball ace Charles Barkley has them. So does actor Tom Cruise. Actress Joan Chen doesn't, and neither did a monk named Gregor Mendel. To see how you fit in with these folks, use a mirror to check out your ears. Is the fleshy lobe at the base of each ear attached to the side of your head? If so, you and Barkley and Cruise have something in common. Or is the fleshy lobe not attached, so that you can flap it back and forth? If so, you are like Chen and Mendel (Figure 10.1).

Charles Barkley

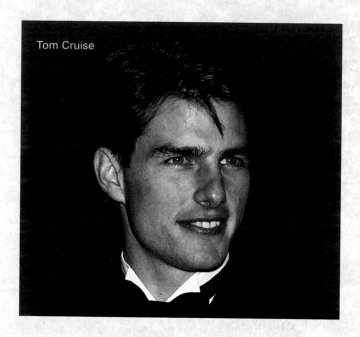

Tom Cruise

Whether a person is born with attached or detached earlobes depends on a single kind of gene. That gene comes in slightly different molecular forms—alleles. Only one form has information about detached lobes. The information is put to use while a human body is developing inside the mother. It calls for a death signal, which is sent to all the cells positioned between the newly forming lobes and the head. Without the signal, the cells don't die and earlobes don't detach.

We all have genes for thousands of traits, including earlobes, cheeks, lashes, and eyeballs. Most traits vary in their details from one person to the next. Remember, we inherit pairs of genes, on pairs of chromosomes. In some pairs, one allele has strong effects and overwhelms the other allele's contribution. The outgunned allele is said to be recessive to the dominant one. If you have detached earlobes, dimpled cheeks, long lashes, or large

Figure 10.1 Attached and detached earlobes of a few representative humans. This sampling provides observable evidence of a trait governed by a certain gene, which exists in different molecular forms in the human population. Do you have one or the other version of the trait? It depends on which molecular forms of the gene you inherited from your mother and father. As Gregor Mendel perceived, such easily observable traits can be used to identify patterns of inheritance that exist from one generation to the next.

eyeballs, you have at least one and quite possibly two dominant alleles that affect the trait in a particular way.

When both alleles of a pair are recessive, nothing masks their effect on a trait. You get *attached* earlobes with one pair of recessive alleles (and *flat* feet with another, a *straight* nose with another, and so on).

How did we discover such remarkable things about our genes? It started with Mendel. By analyzing garden

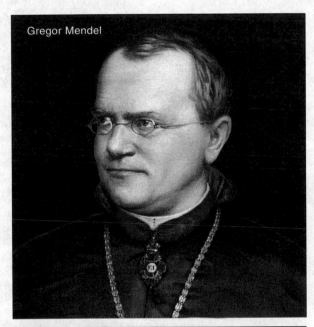

Gregor Mendel

Joan Chen

pea plants generation after generation, Mendel found indirect but *observable* evidence of how parents bestow units of hereditary information—genes—on offspring. This chapter focuses on both the methods and the results of Mendel's experiments. They remain a classic example of how a scientific approach can pry open important secrets about the natural world. And to this day, they serve as the foundation for modern genetics.

KEY CONCEPTS

1. Genes are units of information about heritable traits. Alleles, which are slightly different molecular forms of a gene, specify different versions of the same trait.

2. Each gene has a particular location on a particular chromosome of a species. Humans, pea plants, and other organisms with a diploid chromosome number inherit *pairs* of genes, at equivalent locations on pairs of homologous chromosomes.

3. During meiosis, when the two members of a pair of homologous chromosomes are moved apart, their pairs of genes are moved apart also and end up in different gametes. Gregor Mendel found indirect evidence of this gene segregation when he crossbred pea plants showing different versions of the same trait, such as purple or white flowers.

4. Each pair of homologous chromosomes in a germ cell is sorted out for distribution into one gamete or another independently of how the other pairs of homologous chromosomes are assorted. Mendel discovered indirect evidence of this when he tracked many plants having observable differences in two traits, such as flower color and plant height.

5. The contrasting traits that Mendel happened to study were specified by nonidentical alleles. One allele was dominant, in that its effect on a trait masked the effect of a recessive allele paired with it.

6. Not all traits have such clearly dominant or recessive forms. One allele of a pair may be fully or partially dominant over its partner or codominant with it. Two or more gene pairs often influence the same trait, and some single genes influence many traits. Besides this, environmental factors induce further variation in traits.

MENDEL'S INSIGHT INTO PATTERNS OF INHERITANCE

More than a century ago, people wondered about the basis of inheritance. It was common knowledge that sperm and eggs both transmit information about traits to offspring. But few suspected that the information is organized in units (genes). Instead, the idea was that a father's blob of information "blended" with the mother's blob, like cream into coffee, at fertilization.

However, carried to its logical conclusion, blending would slowly dilute a population's shared pool of hereditary information until there was only a single version left of each trait. If that were so, why did, say, freckles keep showing up among the children of nonfreckled parents over the generations? Why weren't all the descendants of a herd of white stallions and black mares gray? The blending theory could scarcely explain the obvious variation in traits that people could observe with their own eyes. Nevertheless, few disputed the theory.

Charles Darwin was among the scholarly dissidents. According to a key premise of his theory of natural selection, individuals of a population show variation in heritable traits. Through the generations, the variations that improve the chance of surviving and reproducing occur with greater frequency than those that do not. The less advantageous variations might persist among fewer individuals or may even disappear. It is not that separate versions of a given trait are "blended out" of the population. Rather, *each version of a trait may persist in a population, at frequencies that may change over time.*

Just before Darwin presented his theory, someone was gathering evidence that eventually would support his key premise. A monk, Gregor Mendel, had already guessed that sperm and eggs carry distinct "units" of information about heritable traits. By carefully analyzing traits of pea plants generation after generation, Mendel found indirect but *observable* evidence of how parents transmit genes to offspring.

Mendel's Experimental Approach

Mendel spent most of his adult life in a monastery in Brno, a city near Vienna that has since become part of the Czech Republic. However, Mendel was not a man of narrow interests who accidentally stumbled onto principles of great import. The monastery of St. Thomas was close to the European capitals, which were then the centers of scientific inquiry.

Having been raised on a farm, Mendel was aware of agricultural principles and their applications. He kept abreast of the breeding experiments and developments described in the available literature. He was a member of the regional agricultural society. He also won several awards for developing improved varieties of vegetables and fruits. Shortly after entering the monastery, he spent two years studying mathematics, physics, and botany at the University of Vienna. Few scholars of his time had combined talents in plant breeding and mathematics.

Figure 10.2 The garden pea plant (*Pisum sativum*), the organism Mendel chose for experimental tests of his ideas about inheritance. (**a**) This flower has been sectioned to show the location of its stamens and carpel. Sperm-producing pollen grains form in stamens. Eggs develop, fertilization takes place, and seeds mature inside the carpel.

a
carpel   stamen

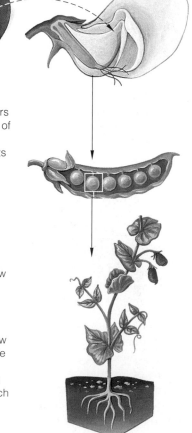

b Pollen from a plant that breeds true for purple flowers is brushed onto a floral bud of a plant that breeds true for white flowers and that had its own stamens snipped off. This is one way to assure cross-fertilization of plants.

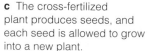

c The cross-fertilized plant produces seeds, and each seed is allowed to grow into a new plant.

d The flower color of the new plants can be used as visible evidence of patterns in how hereditary material might be transmitted to them from each parent plant.

Shortly after his university training, Mendel began experimenting with the garden pea plant, *Pisum sativum* (Figure 10.2). This plant is self-fertilizing. Male as well as female gametes (call them sperm and eggs) develop in different parts of the same flower, where fertilization takes place. Nearly all the plants breed true for certain traits. In other words, successive generations are just like their parents in one or more traits, as when all of the offspring grown from seeds of self-fertilized, white-flowered parent plants have white flowers.

As Mendel knew, we can cross-fertilize pea plants by transferring pollen from one plant's flower to the flower of another plant. For his experimental studies, he could open flower buds of a plant that bred true for a trait—say, white flowers—and snip out the stamens. (Stamens bear pollen grains in which sperm develop.) Then he could brush buds without stamens with pollen from a plant that bred true for a *different* version of the same trait—purple flowers. As Mendel hypothesized, he could use such clearly observable differences to track a given trait through many generations. If there were patterns to the trait's inheritance, *those patterns might tell him something about hereditary itself.*

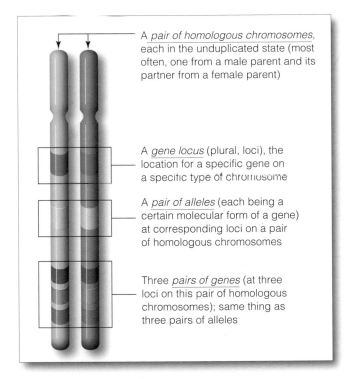

A *pair of homologous chromosomes*, each in the unduplicated state (most often, one from a male parent and its partner from a female parent)

A *gene locus* (plural, loci), the location for a specific gene on a specific type of chromosome

A *pair of alleles* (each being a certain molecular form of a gene) at corresponding loci on a pair of homologous chromosomes

Three *pairs of genes* (at three loci on this pair of homologous chromosomes); same thing as three pairs of alleles

Figure 10.3 A few genetic terms illustrated. Diploid organisms have pairs of genes, on pairs of homologous chromosomes. For example, you inherited one chromosome of each pair from your mother and the other, homologous chromosome from your father.

Most genes can have slightly different molecular forms, called alleles. Different alleles specify different versions of the same trait. An allele at one location on a chromosome may or may not be identical to its partner on the homologous chromosome.

Some Terms Used in Genetics

Having read the chapter on meiosis, you already have insight into the mechanisms of sexual reproduction, which is more than Mendel had. Neither he nor anyone else of his era knew about chromosomes. So he could not have known that a chromosome number is reduced by half in gametes, then restored when gametes meet at fertilization. Even so, Mendel sensed what was going on. As we follow his thinking, let's simplify the story by substituting a few of the modern terms used in studies of inheritance (see also Figure 10.3):

1. **Genes** are units of information about specific traits, and they are passed from parents to offspring. Each gene has a specific location (locus) on a chromosome.

2. Cells with a diploid chromosome number ($2n$) have pairs of genes, on pairs of homologous chromosomes.

3. Mutation can change a gene's molecular structure and thus its information about a trait (as when the gene for flower color specifies purple and a mutated version specifies white). All the different molecular forms of the same gene are called **alleles**.

4. When offspring of genetic crosses inherit a pair of *identical* alleles for a trait, generation after generation, they are a **true-breeding lineage**. By contrast, when offspring of a genetic cross inherit a pair of *nonidentical* alleles for a trait, they are **hybrid offspring**.

5. When both alleles of a pair are identical, this is a *homozygous* condition. When the two are not identical, this is a *heterozygous* condition.

6. An allele is said to be *dominant* when its effect on a trait masks that of any *recessive* allele paired with it. We use capital letters for dominant alleles and lowercase letters for recessive ones; for instance, *A* and *a*.

7. Putting this all together, a **homozygous dominant** individual has a pair of dominant alleles (*AA*) for a trait under study. A **homozygous recessive** individual has a pair of recessive alleles (*aa*). And a **heterozygous** individual has a pair of nonidentical alleles (*Aa*).

8. Two terms help keep the distinction clear between genes and the traits they specify. **Genotype** refers to the particular genes an individual carries. **Phenotype** refers to an individual's observable traits.

9. When tracking the inheritance of traits through generations of offspring, these abbreviations apply:

P	parental generation
F_1	first-generation offspring
F_2	second-generation offspring

WWW

MENDEL'S THEORY OF SEGREGATION

Mendel had an idea that in every generation, a plant inherits two "units" (genes) of information about a trait, one from each parent. To test this idea, he performed what is now known as **monohybrid crosses**. These are experimental crosses between two parents that breed true (are homozygous) for different versions of a single trait. The F_1 offspring are hybrids; each inherits a pair of nonidentical alleles (is heterozygous) for that trait.

Predicting Outcomes of Monohybrid Crosses

Mendel tracked many individual traits through two generations. For instance, in one series of experiments, he crossed true-breeding purple-flowered plants and true-breeding white-flowered ones. All plants grown from the seeds that resulted from this cross had purple flowers. Mendel allowed these plants to self-fertilize. Some plants grown from the seeds had white flowers!

If Mendel's hypothesis were correct—if each plant had inherited two units of information about flower color—then the unit for "purple" had to be dominant, because it had masked the unit for "white" in F_1 plants.

Let's rephrase his thinking. Germ cells of pea plants are diploid, with pairs of homologous chromosomes. Assume one parent is homozygous dominant (AA) and the other is homozygous recessive (aa) for flower color. Following meiosis, a sperm or egg carries one allele for flower color (Figure 10.4). Thus, when a sperm fertilizes an egg, only one outcome is possible: $A + a = Aa$.

Before continuing, you should know Mendel crossed hundreds of plants and tracked thousands of offspring. Besides this, he counted and recorded the number of plants showing dominance or recessiveness. As you can see from Figure 10.5, an intriguing ratio emerged. On average, three of every four F_2 plants had the dominant phenotype, and one had the recessive phenotype.

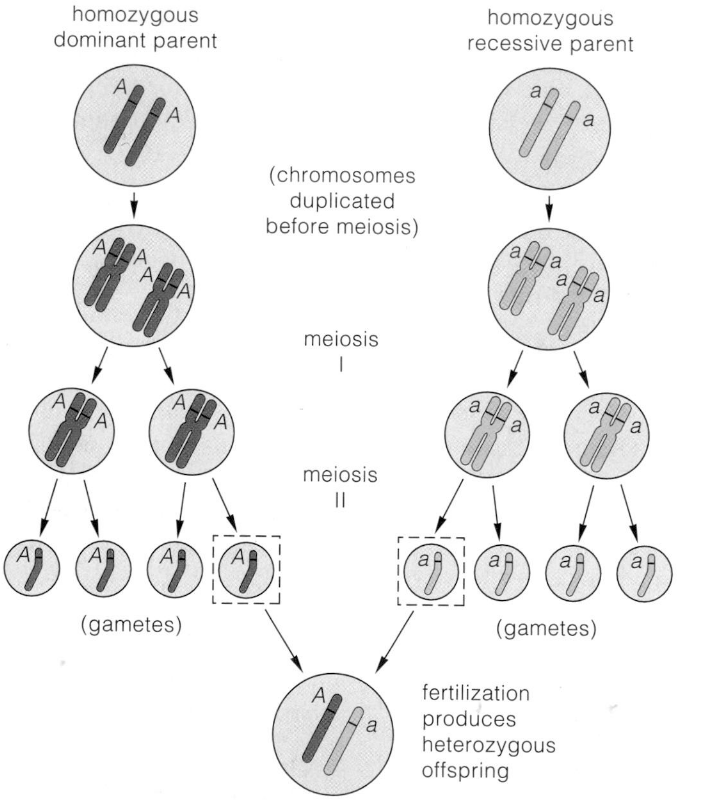
homozygous dominant parent
homozygous recessive parent
(chromosomes duplicated before meiosis)
meiosis I
meiosis II
(gametes)
(gametes)
fertilization produces heterozygous offspring

Figure 10.4 A monohybrid cross, showing how one gene of a pair segregates from the other gene. Two parents that breed true for two different versions of a trait can produce only heterozygous offspring.

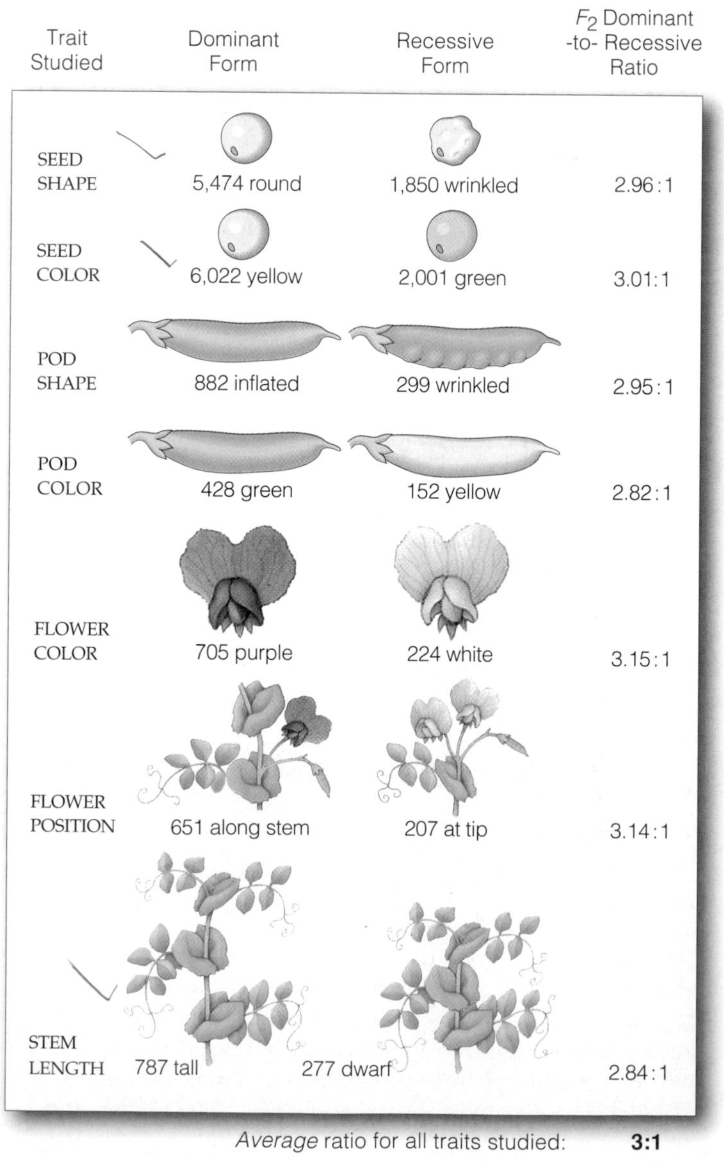
Trait Studied
Dominant Form
Recessive Form
F2 Dominant -to- Recessive Ratio
SEED SHAPE
5,474 round
1,850 wrinkled
2.96 : 1
SEED COLOR
6,022 yellow
2,001 green
3.01 : 1
POD SHAPE
882 inflated
299 wrinkled
2.95 : 1
POD COLOR
428 green
152 yellow
2.82 : 1
FLOWER COLOR
705 purple
224 white
3.15 : 1
FLOWER POSITION
651 along stem
207 at tip
3.14 : 1
STEM LENGTH
787 tall
277 dwarf
2.84 : 1
Average ratio for all traits studied: 3:1

Figure 10.5 *Right*: Results from Mendel's monohybrid cross experiments with the garden pea plant (*P. sativum*). Numbers given are his counts of the F_2 plants that carried dominant or recessive hereditary "units" (alleles) for the trait. On average, the dominant-to-recessive ratio was 3 : 1.

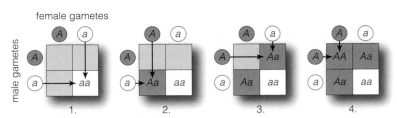

female gametes

male gametes

1.   2.   3.   4.

Figure 10.6 Punnett-square method of predicting the probable outcome of a genetic cross. Circles represent gametes. *Italic* letters on the gametes represent dominant or recessive alleles. The different squares show the different genotypes possible among offspring. In this case, gametes are from a self-fertilizing heterozygous (*Aa*) plant.

Figure 10.7 *Right*: Results from one of Mendel's monohybrid crosses. On average, the dominant-to-recessive ratio among the second-generation (F_2) plants was 3 : 1.

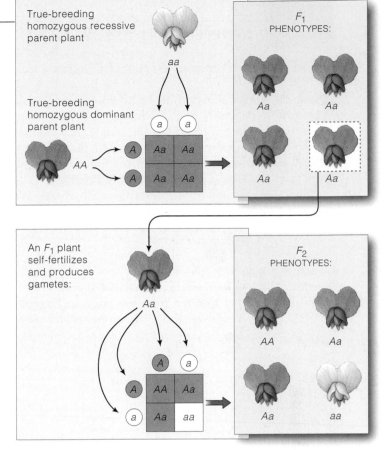

To Mendel, the ratio suggested that fertilization is a chance event, with a number of possible outcomes. And he had an understanding of probability, which applies to chance events *and therefore could help him predict the possible outcomes of crosses*. **Probability** simply means this: The chance that each outcome of a given event will occur is proportional to the number of ways in which that event can be reached.

The **Punnett-square method**, explained in Figure 10.6 and applied in Figure 10.7, may help you visualize the possibilities. As you can see, if half of a plant's sperm (or eggs) were *a* and half were *A*, then four outcomes would be possible each time a sperm fertilized an egg:

POSSIBLE EVENT:	PROBABLE OUTCOME:
sperm *A* meets egg *A*	1/4 *AA* offspring
sperm *A* meets egg *a*	1/4 *Aa*
sperm *a* meets egg *A*	1/4 *Aa*
sperm *a* meets egg *a*	1/4 *aa*

By this prediction, an F_2 plant has three chances in four of getting at least one dominant allele (purple flowers). It has one chance in four of getting two recessive alleles (white flowers). That is a probable phenotypic ratio of three purple to one white, or 3:1.

Mendel's observed ratios weren't *exactly* 3:1. You can see this for yourself by looking at the numerical results listed in Figure 10.5. Why did Mendel put aside the deviations? To understand why, flip a coin several times. As we all know, a coin is just as likely to end up heads as tails. But often a coin ends up heads, or tails, several times in a row. So if you flip the coin only a few times, the observed ratio might differ greatly from the predicted ratio of 1:1. Flip the coin many, many times, and you are more likely to come close to the predicted ratio. Mendel understood the rules of probability—and performed a large number of crosses. Almost certainly, this kept him from being confused by minor deviations from the predicted results of the experimental crosses.

Testcrosses

By running **testcrosses**, Mendel gained support for his prediction. In this type of experimental test, an organism shows dominance for a specified trait but its genotype is unknown, so it is crossed to a known homozygous recessive individual. Test results may reveal whether the organism is homozygous dominant or heterozygous.

Regarding the monohybrid crosses just described, Mendel tested his prediction that the purple-flowered F_1 offspring were heterozygous by crossing them with true-breeding, white-flowered plants. If they were all homozygous dominant, then all the F_2 offspring would show the dominant form of the trait. If heterozygous, there would be about as many dominant as recessive plants. Sure enough, when old enough to flower, about half the F_2 plants had purple flowers (*Aa*) and half had white (*aa*). Can you construct two Punnett squares that show the possible outcomes of this testcross?

The results from Mendel's monohybrid crosses and testcrosses became the basis of a theory of **segregation**, which we state here in modern terms:

MENDEL'S THEORY OF SEGREGATION. **Diploid cells have pairs of genes, on pairs of homologous chromosomes. During meiosis, the two genes of each pair segregate from each other. As a result, they end up in different gametes.**

WWW

INDEPENDENT ASSORTMENT

Predicting Outcomes of Dihybrid Crosses

By another series of experiments, Mendel attempted to explain how *two* pairs of genes might be assorted into gametes. He selected true-breeding plants that differed in two traits, including flower color and plant height. In such **dihybrid crosses**, F_1 offspring inherit two gene pairs, each consisting of two nonidentical alleles.

Let's diagram one of Mendel's dihybrid crosses. We can use *A* for flower color and *B* for height as dominant alleles, and as their recessive counterparts, *a* and *b*:

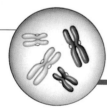

TRUE-BREEDING PARENTS: purple flowers, tall × white flowers, dwarf

AABB **aabb**

GAMETES: AB AB ab ab

F_1 HYBRID OFFSPRING: **AaBb**

As Mendel would have predicted, the F_1 offspring from this cross are all purple-flowered and tall (*AaBb*).

When F_1 plants mature and reproduce, how will the two gene pairs be assorted into gametes? The answer partly depends on the chromosomal locations of the gene pairs. Assume one pair of homologous chromosomes carries the *Aa* alleles and a *different* pair carries the *Bb* alleles. Next, think of how all chromosomes become positioned at the spindle equator during metaphase I of meiosis (Figures 8.4 and 10.8). The chromosome with the *A* allele might be positioned to move to either spindle pole (and then into one of four gametes). The same is true of its homologue. And the same is true of the chromosomes with the *B* and *b* alleles. Thus, after meiosis and gamete formation, four combinations of alleles are possible in the sperm or eggs: 1/4 *AB*, 1/4 *Ab*, 1/4 *aB*, and 1/4 *ab*.

Given the alternative alignments of chromosomes at metaphase I, several allelic combinations are possible at fertilization. Simple multiplication (four kinds of sperm times four kinds of eggs) tells us sixteen combinations of gametes are possible in the F_2 offspring of a dihybrid cross. Use the Punnett-square method to diagram the probabilities (Figure 10.9). Now add up all the possible phenotypes and you get 9/16 tall purple-flowered, 3/16 dwarf purple-flowered, 3/16 tall white-flowered, and 1/16 dwarf white-flowered plants. That is a probable phenotypic ratio of 9:3:3:1. Results from one dihybrid cross that Mendel described were close to this ratio.

Nucleus of a diploid (2n) reproductive cell with only two pairs of homologous chromosomes

Figure 10.8 Independent assortment. This example tracks two pairs of homologous chromosomes. An allele at one locus on a chromosome may or may not be identical with its partner allele on the homologous chromosome. At meiosis, either chromosome of a pair may become attached to either pole of the spindle. Thus, in this case, two different lineups are possible at metaphase I.

OR

Possible alignments of the homologous chromosomes at metaphase I of meiosis, as shown by two diagrams:

The resulting alignments of chromosomes at metaphase II:

The combinations of alleles possible in the forthcoming gametes:

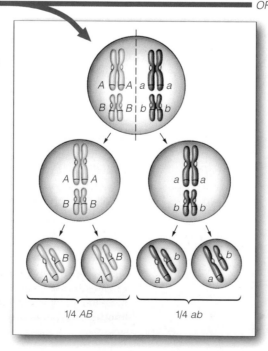

1/4 AB   1/4 ab

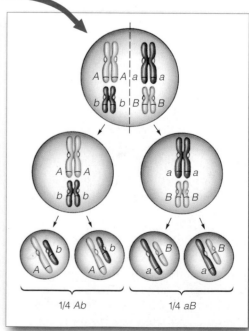

1/4 Ab   1/4 aB

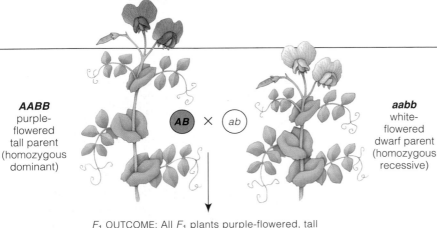

AABB
purple-
flowered
tall parent
(homozygous
dominant)

AB × **ab**

aabb
white-
flowered
dwarf parent
(homozygous
recessive)

F_1 OUTCOME: All F_1 plants purple-flowered, tall
(**AaBb** heterozygotes)

AaBb

AaBb

meiosis,
gamete formation

meiosis,
gamete formation

	1/4 **AB**	1/4 **Ab**	1/4 **aB**	1/4 **ab**
1/4 **AB**	1/16 **AABB**	1/16 **AABb**	1/16 **AaBB**	1/16 **AaBb**
1/4 **Ab**	1/16 **AABb**	1/16 **AAbb**	1/16 **AaBb**	1/16 **Aabb**
1/4 **aB**	1/16 **AaBB**	1/16 **AaBb**	1/16 **aaBB**	1/16 **aaBb**
1/4 **ab**	1/16 **AaBb**	1/16 **Aabb**	1/16 **aaBb**	1/16 **aabb**

Possible outcomes of cross-fertilization

ADDING UP THE F_2 COMBINATIONS POSSIBLE:

- 9/16 or 9 purple-flowered, tall
- 3/16 or 3 purple-flowered, dwarf
- 3/16 or 3 white-flowered, tall
- 1/16 or 1 white-flowered, dwarf

Figure 10.9 Results from Mendel's dihybrid cross between parent plants that bred true for different versions of two traits: flower color and plant height. *A* and *a* represent dominant and recessive alleles for flower color. *B* and *b* represent dominant and recessive alleles for plant height. As the Punnett square indicates, the probabilities of certain combinations of phenotypes among F_2 offspring occur in a 9:3:3:1 ratio, on the average.

The Theory in Modern Form

Mendel could do no more than analyze the numerical results from his dihybrid crosses, because he didn't know that seven pairs of homologous chromosomes carry the pea plant's "units" of inheritance. It just seemed to him that the two units for the first trait he was tracking had been assorted into gametes independently of the two units for the other trait. In time his interpretation became known as the theory of **independent assortment**, which we state here in modern terms: By the end of meiosis, each pair of homologous chromosomes—and the genes they carry—have been sorted for shipment into gametes independently of how the other pairs were sorted out.

Independent assortment and hybrid crossing lead to staggering genetic variation. In a monohybrid cross that involves only a single gene pair, three genotypes are possible: *AA*, *Aa*, and *aa*. We can represent this as 3^n, where *n* is the number of gene pairs. When we consider more gene pairs, the number of possible combinations increases dramatically. Even if parents differ in merely ten pairs of genes, nearly 60,000 genotypes are possible among their offspring. If they differ in twenty pairs of genes, the number approaches 3.5 billion!

In 1865 Mendel presented his ideas to the Brünn Natural History Society. His ideas had little impact. The next year his paper was published, and apparently it was read by few and understood by no one. In 1871 he became abbot of the monastery, and his experiments gradually gave way to administrative tasks. He died in 1884, never to know that his work would be the starting point for the development of modern genetics.

Today, Mendel's theory of segregation still stands. Hereditary material is indeed organized in units (genes) that retain their identity and are segregated from each other for distribution into different gametes. But the theory of independent assortment does not apply to all gene combinations, as you will see in the next chapter.

MENDEL'S THEORY OF INDEPENDENT ASSORTMENT. **By the end of meiosis, the genes on pairs of homologous chromosomes have been sorted out for distribution into one gamete or another independently of gene pairs of other chromosomes.**

WWW

For the most part, Mendel studied traits having clearly dominant or recessive forms. As the remaining sections of this chapter will make clear, however, the expression of other traits is not as straightforward.

Incomplete Dominance

In **incomplete dominance**, one allele of a pair isn't fully dominant over its partner, so a heterozygous phenotype *somewhere in between* the two homozygous phenotypes emerges. Cross a true-breeding red snapdragon and a true-breeding white one. All F_1 offspring will have pink flowers. Cross two F_1 plants and expect red, pink, and white snapdragons in a predictable ratio (Figure 10.10). What causes this inheritance pattern? Red snapdragons have two alleles that allow them to make an abundance of red pigment molecules. White snapdragons have two mutated alleles that render them pigment-free. The pink ones are heterozygous. Their one red allele can specify enough pigment to make flowers pink but not red.

ABO Blood Types: A Case of Codominance

In **codominance**, a pair of nonidentical alleles specify two phenotypes, which are both expressed at the same time in heterozygotes. As an example, think about one of the glycolipids projecting from the plasma membrane of your red blood cells. It helps give red blood cells their unique identity, although it comes in slightly different molecular forms. A method of analysis known as *ABO blood typing* reveals which form a person has.

In humans, an enzyme dictates the glycolipid's final structure. The gene for that enzyme has three alleles. Two alleles, I^A and I^B, are codominant when paired. The third, i, is recessive; a pairing with either I^A or I^B masks its effect. Together, they are a **multiple allele system**, which we define as the presence of three or more alleles of a gene among individuals of a population.

Before each glycolipid molecule became positioned at the cell surface, it was modified in the cytomembrane system (Section 4.6). First an oligosaccharide chain was attached to a lipid molecule, then a sugar was attached to the end of the chain. Alleles I^A and I^B code for two slightly different versions of the enzyme catalyzing that final step. The two attach different sugars, which give the glycolipid its identity—either A or B.

Which alleles do you have? With either $I^A I^A$ or $I^A i$, you have type A blood. With $I^B I^B$ or $I^B i$, your blood is type B. With codominant alleles $I^A I^B$, it is AB—meaning you have both versions of the sugar-attaching enzyme. But if you are homozygous recessive (ii), the molecules never did get an additional sugar attached to them. Your blood type is neither A nor B; that is what type "O" means. Figure 10.11 summarizes the possibilities.

homozygous parent × homozygous parent

All F_1 offspring are heterozygous for flower color:

Cross two of the F_1 plants, and the F_2 offspring will show three phenotypes in a 1:2:1 ratio:

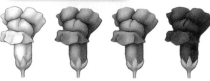

Figure 10.10 Visible evidence of incomplete dominance in heterozygous snapdragons, in which an allele for red pigment is paired with a "white" allele.

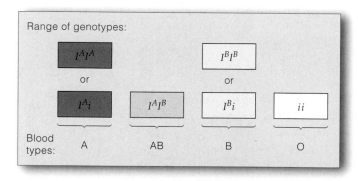

Figure 10.11 Allelic combinations related to ABO blood typing.

During *transfusions*, the blood of two people mixes. Unless their blood is compatible, the recipient's immune system perceives the red blood cells from the donor as "nonself." It will act against those cells and may cause death (Sections 33.4 and 34.4).

One allele may be fully dominant, incompletely dominant, or codominant with its partner on the homologous chromosome.

WWW

MULTIPLE EFFECTS OF SINGLE GENES

Expression of the alleles at just a single location on a chromosome may have positive or negative effects on two or more traits. This phenotypic outcome of a single gene's activity is known as **pleiotropy** (after the Greek *pleio–*, meaning more, and *–tropic*, meaning to change).

The genetic disorder *sickle-cell anemia* is a classic example of how the alleles at a single locus can have pleiotropic effects. The disorder arises from a mutated gene for beta-globin, one of two kinds of polypeptide chains in the hemoglobin molecule. Hemoglobin, recall, is the oxygen-transporting protein in red blood cells. We designate the mutated allele as Hb^S instead of Hb^A. Heterozygotes (Hb^A/Hb^S) usually show few symptoms of the disorder. Their red blood cells are able to produce enough normal hemoglobin molecules to compensate for the abnormal ones. In homozygotes (Hb^S/Hb^S), the red blood cells can only produce abnormal hemoglobin. This one abnormality may have drastic repercussions throughout the body, for it disrupts the concentration of oxygen in the bloodstream.

Humans, like most organisms, depend on the intake of oxygen for aerobic respiration. Oxygen in air flows into the lungs, then diffuses into the blood. There it binds to hemoglobin, which transports it through arteries, arterioles, and then small-diameter, thin-walled capillaries threading past all living cells in the body. A steep oxygen concentration gradient exists between blood and the fluid within the surrounding tissues, so most of the oxygen diffuses into the tissues, and on into cells. With this extensive cellular uptake, the concentration of oxygen in blood declines. The decline is most pronounced at high altitudes and during strenuous activity.

In the red blood cells of people who bear the mutant gene, abnormal hemoglobin molecules stick together in rod-shaped arrangements. The rods distort cells into a sickle shape, as in Figure 10.12*b*. (A sickle is a farm tool having a crescent-shaped blade.) The distorted cells rupture easily, and then their remnants clog and rupture capillaries. When these oxygen transporters are rapidly destroyed, the cells in affected tissues become starved for oxygen. Also, their clumping effect leads to local failures in the capacity of the circulatory system to deliver oxygen and to carry away carbon dioxide and other metabolic wastes.

Over time, ongoing expression of the mutant gene can damage tissues and organs throughout the body. Figure 10.12*c* tracks how the successive alterations in

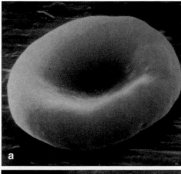

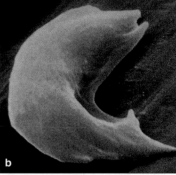

Figure 10.12 (**a**) A red blood cell from a person affected by sickle-cell anemia, which is a genetic disorder. This scanning electron micrograph shows the surface appearance of the affected individual's red blood cells when the blood is adequately oxygenated. (**b**) This scanning electron micrograph shows the sickle shape that red blood cells assume when the concentration of oxygen in blood is low. (**c**) This diagram summarizes the range of symptoms that are characteristic of an individual who is homozygous recessive for the disorder.

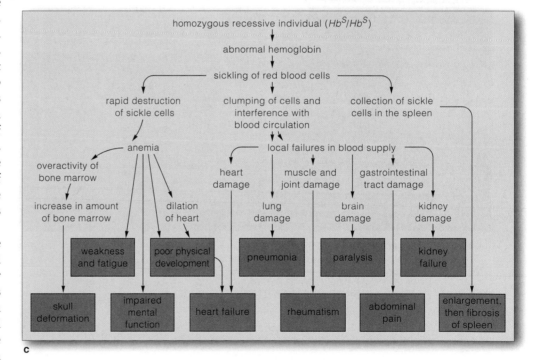

phenotype may proceed. In chapters to come, you will read about other aspects of this genetic disorder.

The alleles at a single gene location may have positive or negative effects on two or more traits.

The effects may not be simultaneous. Rather, they may have repercussions over time. The gene may lead to an alteration in one trait. That change may alter another trait, and so on.

INTERACTIONS BETWEEN GENE PAIRS

Often a trait results from interactions among products of two or more gene pairs. For example, two alleles of one gene may mask expression of another gene's alleles, so some expected phenotypes may not appear at all. Such interactions between pairs of genes are called **epistasis** (meaning the act of stopping).

Hair Color in Mammals

Epistasis is common among the gene pairs responsible for the coloration of fur or skin in mammals. Consider the black, brown, or yellow fur of Labrador retrievers (Figure 10.13). The different colors arise from variations in the amount and distribution of melanin, a brownish black pigment. Enzymes and other products of many gene pairs influence different steps in the production of melanin and its deposition in certain body regions.

The alleles of one gene specify an enzyme required to produce melanin. Expression of allele B (black) has a more pronounced effect and is dominant to b (brown). Alleles of a different gene control the extent to which molecules of melanin will be deposited in a retriever's hairs. Allele E permits full deposition. Two recessive alleles (ee) reduce deposition, and fur will be yellow.

In some individuals, those two gene pairs are not able to interact, owing to a certain allelic combination at still another gene locus. There, a gene (C) calls for tyrosinase, the first of several enzymes in a melanin-producing pathway. An individual bearing one or two dominant alleles (CC or Cc) can make the functional enzyme. An individual bearing two recessive alleles (cc) cannot. When the biosynthetic pathway for melanin production is blocked, then *albinism*—the absence of melanin—is the resulting phenotype (Figure 10.14).

a BLACK LABRADOR **b** YELLOW LABRADOR **c** CHOCOLATE LABRADOR

Figure 10.13 The heritable basis of coat color among Labrador retrievers. The trait arises through interactions among the alleles of two pairs of genes.

One kind of gene is involved in melanin production. Allele B (black) of this gene is dominant to allele b (brown). A different kind of gene influences the deposition of melanin pigment in individual hairs. Allele E of this gene promotes deposition, but a pairing of recessive alleles (ee) of the gene blocks deposition, and a yellow coat results.

F_1 offspring of a dihybrid cross produce F_2 offspring in a 9:3:4 ratio, as the Punnett-square diagram to the right indicates.

The yellow Labrador in photograph (**b**) probably has genotype BBee, because it can produce melanin but cannot deposit pigment in hairs. After studying the photograph, can you say why?

HOMOZYGOUS PARENTS: $BBEE \times bbee$

F_1 PUPPIES: $BbEe$

ALLELIC COMBINATIONS POSSIBLE AMONG F_2 PUPPIES:

	BE	Be	bE	be
BE	BBEE	BBEe	BbEE	BbEe
Be	BBEe	BBee	BbEe	Bbee
bE	BbEE	BbEe	bbEE	bbEe
be	BbEe	Bbee	bbEe	bbee

RESULTING PHENOTYPES:

☐ 9/16 or 9 black
▨ 3/16 or 3 brown
☐ 4/16 or 4 yellow

Figure 10.14 A rare albino rattlesnake. Like other animals that cannot produce melanin, it has pink eyes and its body surface is white, overall. (Eyes look pink because the absence of melanin from a tissue layer in the eyeball allows red light to be reflected from blood vessels in the eyes.) In birds and mammals, surface coloration results from pigments in feathers, fur, or skin. In fishes, amphibians, and reptiles, color-bearing cells give skin its surface coloration. Some of the cells contain melanin pigments or yellow-to-red pigments. Others contain crystals that reflect light and alter the effect of other pigments present.

The mutation affecting melanin production in the snake shown here had no effect on the production of yellow-to-red pigments and of light-reflecting crystals. Therefore, the snake's skin appears to be iridescent yellow as well as white.

a WALNUT COMB **b** ROSE COMB **c** PEA COMB **d** SINGLE COMB

Figure 10.15 Interaction between two genes that affect the same trait in domestic breeds of chickens. The initial cross is between a Wyandotte (with a rose comb, **b**, on the crest of its head) and a brahma (pea comb, **c**). With complete dominance at the locus for pea comb and at the locus for rose comb, products of the two gene pairs interact and give rise to a walnut comb (**a**). With complete recessiveness at both gene loci, products interact and give rise to a single comb (**d**).

Comb Shape in Poultry

In some cases, interaction between two gene pairs results in a phenotype that neither pair can produce alone. The geneticists W. Bateson and R. Punnett identified two interacting gene pairs (R and P) that affect comb shape in chickens. Allelic combinations of rr at one gene locus and pp at the other locus result in the least common phenotype, the single comb. The presence of dominant alleles (R, P, or both) results in varied phenotypes.

Take a look at Figure 10.15. The diagram shows the combinations of alleles that interact to specify the rose, pea, and walnut combs shown in the photographs.

Genes often interact, as when alleles of one gene mask the expression of another gene, and some expected phenotypes may not appear at all.

WWW

HOW CAN WE EXPLAIN LESS PREDICTABLE VARIATIONS?

Regarding the Unexpected Phenotype

As Mendel demonstrated, the phenotypic effects of one or two pairs of certain genes show up in predictable ratios when you track them from one generation to the next. Besides this, certain interactions among two or more gene pairs can produce phenotypes in predictable ratios, as the example of Labrador coat color clearly demonstrated in Section 10.6. However, even when you decide to track a single gene through the generations, you may discover that the resulting phenotypes are not quite what you had expected.

Consider *camptodactyly*—a rare genetic abnormality that affects both the shape and the movement of fingers. Certain people who carry the mutant allele for this heritable trait develop immobile, bent fingers on both hands. Other people develop immobile, bent fingers on the left or right hand only. Others who carry the mutant allele develop fingers that are not affected either way.

What is the source of such confounding variation? Recall that most organic compounds are synthesized by a series of metabolic steps, *and different enzymes—each a gene product—regulate different steps.* Maybe one gene has mutated in one of a number of different ways. Maybe the product of another gene blocks the pathway or causes it to run nonstop, or not long enough. Or maybe poor nutrition or another factor that is variable in the individual's environment affects a crucial enzyme in the pathway. These are the sorts of variable factors that commonly introduce far less predictable variations in the phenotypes resulting from gene expression.

Continuous Variation in Populations

Generally, the individuals of a population display a range of small differences in most traits.

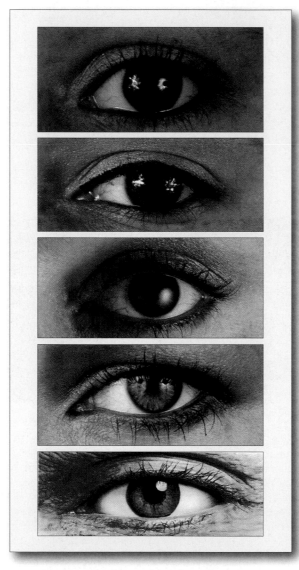

Figure 10.16 Samples from a range of continuous variation in human eye color. Different pairs of genes interact to produce and deposit melanin. Among other things, this pigment helps color the eye's iris. Different combinations of alleles result in small differences in eye color. Thus the frequency distribution for the eye-color trait appears to be continuous over a range from black to light blue.

This characteristic of populations, which is known as **continuous variation**, is primarily an outcome of the number of genes affecting a trait and the number of environmental factors that influence their expression. In most cases, the greater the number of genes and environmental factors, the more continuous will be the expected distribution of all the versions of that trait.

Think about your own eye color. The colored part is the iris, a doughnut-shaped, pigmented structure just beneath the cornea. As is true of all humans, the color of your iris is the cumulative outcome of a number of gene products. Some products take part in the stepwise production and distribution of melanin, the same light-absorbing pigment that influences coat color in mammals. Dark eyes that seem to be almost black have abundant deposits of melanin molecules in the iris—so much so that most of the light that enters gets absorbed. Dark brown eyes don't have as many deposits of melanin, and some light that is not absorbed is reflected out from the iris. Light brown or hazel eyes have even less (Figure 10.16).

Green, gray, or blue eyes do not contain green, gray, or blue pigments. In these cases, the iris does incorporate different amounts of melanin, but not very much of it. As a result, many or most of the blue wavelengths of light that do enter the eye are reflected out.

How might you describe the continuous variation of some trait within a group, such as the college students in Figure 10.17a? They range from very short to very tall, with average heights much more common than either extreme. You can start out by dividing the full range of the different phenotypes into measurable categories. Next, count all the individual students in each category. This will give you the relative frequencies of all phenotypes distributed across the range of measurable values.

The bar chart in Figure 10.17c plots the proportion of students in each category against the range of the measured phenotypes. The vertical bars that are the shortest

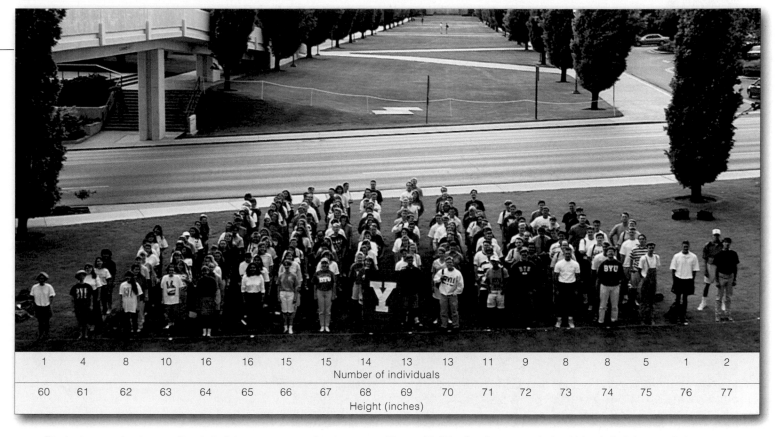

1	4	8	10	16	16	15	15	14	13	13	11	9	8	8	5	1	2

Number of individuals

60	61	62	63	64	65	66	67	68	69	70	71	72	73	74	75	76	77

Height (inches)

a Students organized according to height, as an example of continuous variation

Figure 10.17 Continuous variation in body height, a trait that is one of the characteristics of the human population.

(**a**) Suppose you want to find the frequency distribution for height in a group of 169 biology students at Brigham Young University. You decide on how finely the range of possible heights should be divided. Then you measure each student and assign her or him to the appropriate category. Finally, you divide the number in each category by the total number of all students in all categories.

(**b**) Often a bar graph is used to depict continuous variation in a population. In such graphs, the proportion of individuals in each category is plotted against the range of measured phenotypes. Notice the curved line above the bars. It is an idealized example of the kind of "bell-shaped" curve that emerges for populations showing continuous variation in a trait. The bell-shaped curve in (**c**) is a specific example of this type of diagram.

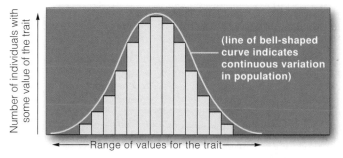

b Idealized bell-shaped curve for a population that displays continuous variation in some trait

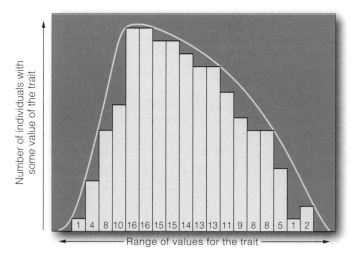

c Specific bell-shaped curve corresponding to the distribution of the trait (height) illustrated by the photograph in (**a**)

represent categories with the least number of students. The bar that is tallest represents the category with the greatest number of students. Finally, draw a graph line around all of the bars and you end up with a "bell-shaped" curve. Such curves are typical of populations that show continuous variation in a trait.

Enzymes and other products of genes regulate each step of most metabolic pathways. Mutations, gene interactions, and environmental conditions may affect one or more of the steps, and this leads to variations in phenotypes.

For most traits, the individuals of a population display continuous variation—that is, a range of small differences.

The greater the number of genes and environmental factors that can influence a trait, the more continuous will be the expected distribution of all the versions of that trait.

EXAMPLES OF ENVIRONMENTAL EFFECTS ON PHENOTYPE

We have mentioned, in passing, that environmental conditions often contribute to variable gene expression among individuals of a population. Before leaving this chapter, consider just two examples of environmentally induced variations in phenotype.

Possibly you have observed a Himalayan rabbit and probably a Siamese cat. Both of these furry mammals carry an allele that specifies a heat-sensitive version of one of the enzymes necessary for melanin production. At the surface of warm body regions, the enzyme is less active. Fur growing there is lighter in color than fur in cooler regions, which include the ears and other body parts that project away from the main body mass. The experiment shown in Figure 10.18 provided observable evidence of an environmental effect on gene expression.

The environment also influences genes that govern phenotypes of plants. You may have observed the color variation in the floral clusters of *Hydrangea macrophylla*, a species widely favored in home gardens (Figure 10.19). In this type of plant, the action of genes responsible for floral color produces different phenotypes, depending on the acidity of the soil in which the plant is growing.

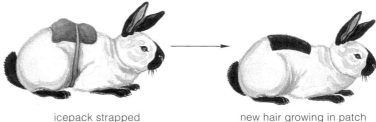

icepack strapped to hair-free patch

new hair growing in patch exposed to cold is black

Figure 10.18 An observable effect of differences in environmental conditions on gene expression in animals. A Himalayan rabbit normally has black hair only on its long ears, nose, tail, and lower leg limbs. For one experiment, a patch of a rabbit's white fur was plucked clean, and then an icepack was secured over the hairless patch. Where the colder temperature had been maintained, the hairs that grew back were black.

Himalayan rabbits are homozygous for the *ch* allele of a gene that codes for tyrosinase, an enzyme required to produce melanin. The allele specifies a heat-sensitive version of the enzyme, which is able to function only when the air temperature is below about 33°C. When cells that give rise to hairs grow under warmer conditions, they cannot produce melanin and their hairs appear light. This happens in body regions that are massive enough to conserve a fair amount of metabolic heat. Ears and other slender extremities are cooler because they tend to lose metabolic heat more rapidly.

Figure 10.19 Effect of environmental conditions on gene expression in a favorite garden plant (*Hydrangea macrophylla*). Even plants that carry the same alleles have floral colors ranging from pink to blue, depending on the acidity of the soil in which they happen to be growing.

And so we conclude this chapter, which has dealt with heritable and environmental factors that give rise to variations in phenotype. What is the take-home lesson? Simply this: An individual's phenotype is an outcome of complex interactions among genes, enzymes and other gene products, and environmental factors.

Owing to gene mutations, cumulative gene interactions, and environmental effects on genes, individuals of a population show degrees of variation for many traits.

10.9 SUMMARY

1. A gene is a unit of information about a heritable trait. Alleles of a gene are different molecular versions of that information. Through experimental crosses with pea plants, Mendel gathered indirect evidence that diploid organisms have two genes for each trait and that genes retain their identity when transmitted to offspring.

2. *Monohybrid* crosses are experimental crosses between two individuals that are homozygous (true-breeding) for different versions of a trait; offspring are heterozygous (they inherit a pair of nonidentical alleles) for the trait. Mendel's monohybrid crosses with garden pea plants provided indirect evidence that some forms of a gene may be dominant over other, recessive forms.

3. A homozygous dominant individual has inherited two dominant alleles (AA) for the trait being studied. A homozygous recessive has two recessive alleles (aa), and a heterozygote has two nonidentical alleles (Aa).

4. In Mendel's monohybrid crosses ($AA \times aa$), all of the F_1 offspring were Aa. Then crosses between F_1 plants resulted in these combinations of alleles in F_2 offspring:

	A	a
A	AA	Aa
a	Aa	aa

AA (dominant)
Aa (dominant)
Aa (dominant)
aa (recessive)
} the expected phenotypic ratio of 3:1

5. Results from Mendel's monohybrid crosses led to the formulation of a theory of segregation. In modern terms, diploid organisms have pairs of genes, on pairs of homologous chromosomes. The two genes of each pair segregate from each other at meiosis, so that each gamete formed ends up with one or the other gene.

6. *Dihybrid* crosses are experimental crosses between two individuals that breed true (are homozygous) for different versions of two traits; so the F_1 offspring are heterozygous (inherit two nonidentical alleles) for both traits. The phenotypes of the F_2 offspring from Mendel's dihybrid crosses were close to a 9:3:3:1 ratio:

> 9 dominant for both traits
> 3 dominant for A, recessive for b
> 3 dominant for B, recessive for a
> 1 recessive for both traits

7. Mendel's dihybrid crosses led to the formulation of a theory of independent assortment. In modern terms, by the end of meiosis, the gene pairs of two homologous chromosomes have been sorted out for distribution into one gamete or another, independently of how the gene pairs of other chromosomes were sorted out.

8. Four factors commonly influence gene expression:
   a. Degrees of dominance may occur between some pairs of genes.
   b. The products of pairs of genes may interact to influence the same trait.
   c. One gene may have positive or negative effects on two or more traits, a condition called pleiotropy.
   d. Environmental conditions to which an individual is subjected may affect gene expression.

Review Questions

1. Define the difference between these terms: *10.1*
   a. gene and allele
   b. dominant allele and recessive allele
   c. homozygote and heterozygote
   d. genotype and phenotype

2. Define a true-breeding lineage. What is a hybrid? *10.1*

3. Distinguish between monohybrid and dihybrid crosses. What is a testcross, and why is it useful in genetic analysis? *10.2, 10.3*

4. Do segregation and independent assortment proceed during mitosis, meiosis, or both? *10.2, 10.3*

5. What do the vertical and horizontal arrows of the diagram at right represent? What do the bars and the curved line represent? *10.7*

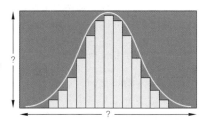

Self-Quiz (*Answers in Appendix III*)

1. Alleles are _____ .
   a. different molecular forms of a gene
   b. different phenotypes
   c. self-fertilizing, true-breeding homozygotes

2. A heterozygote has a _____ for the trait being studied.
   a. pair of identical alleles
   b. pair of nonidentical alleles
   c. haploid condition, in genetic terms
   d. a and c

3. The observable traits of an organism are its _____ .
   a. phenotype   c. genotype
   b. sociobiology   d. pedigree

4. F_1 offspring of the monohybrid cross $AA \times aa$ are _____ .
   a. all AA c. all Aa
   b. all aa d. 1/2 AA and 1/2 aa

5. Second-generation offspring from a cross are the _____ .
   a. F_1 generation   c. hybrid generation
   b. F_2 generation   d. none of the above

6. Assuming complete dominance will occur, the offspring of the cross $Aa \times Aa$ will show a phenotypic ratio of _____ .
   a. 3:1   c. 1:2:1
   b. 9:1   d. 9:3:3:1

7. Crosses between F_1 individuals resulting from the cross $AABB \times aabb$ lead to F_2 phenotypic ratios close to _____ .
   a. 1:2:1   c. 1:1:1:1
   b. 3:1   d. 9:3:3:1

8. Match each example with the most suitable description.
   _____ dihybrid cross   a. bb
   _____ monohybrid cross   b. $AaBb \times AaBb$
   _____ homozygous condition   c. Aa
   _____ heterozygous condition d. $Aa \times Aa$

Critical Thinking—Genetics Problems
(*Answers in Appendix IV*)

1. One gene has alleles *A* and *a*. Another has alleles *B* and *b*. For each genotype listed, what type(s) of gametes will be produced? (Assume independent assortment occurs before gametes form.)
   a. *AABB*   c. *Aabb*
   b. *AaBB*   d. *AaBb*

2. Still referring to Problem 1, what will be the genotypes of the offspring from the following matings? Indicate the frequencies of each genotype among them.
   a. *AABB* × *aaBB*   c. *AaBb* × *aabb*
   b. *AaBB* × *AABb*   d. *AaBb* × *AaBb*

3. In one experiment, Mendel crossed a pea plant that bred true for green pods with one that bred true for yellow pods. All the F_1 plants had green pods. Which form of the trait (green or yellow pods) is recessive? Explain how you arrived at your conclusion.

4. Return to Problem 1, and assume you now study a third gene having alleles *C* and *c*. For each genotype listed, what type(s) of gametes will be produced?
   a. *AABBCC*   c. *AaBBCc*
   b. *AaBBcc*   d. *AaBbCc*

5. Mendel crossed a true-breeding tall, purple-flowered pea plant with a true-breeding dwarf, white-flowered plant. All F_1 plants were tall and had purple flowers. If an F_1 plant self-fertilizes, then what is the probability that a randomly selected F_2 offspring will be heterozygous for the genes specifying height and flower color?

6. At a gene location on a human chromosome, a dominant allele controls *tongue rolling*, an ability to curl up the two sides of the tongue (Figure 10.20). People who are homozygous for a recessive allele at that locus can't roll the tongue. At a different gene locus, a dominant allele controls whether the earlobes will be attached or detached (refer to Figure 10.1). These two pairs of genes assort independently. Suppose a tongue-rolling, detached-earlobed woman marries a man who has attached earlobes and cannot roll his tongue. Their first child has the phenotype of the father. Given this outcome,
   a. What are the genotypes of the mother, father, and child?
   b. What is the probability that a second child of theirs will have detached earlobes and won't be a tongue roller?

7. Bill and his wife, Marie, are hoping to have children. Both have notably flat feet and long eyelashes, and they tend to sneeze a lot (hence the name of the *achoo syndrome*). Dominant alleles give rise to these traits: *A* (foot arch), *E* (eyelash length), and *S* (chronic sneezing). Bill is heterozygous for all three dominant alleles and Marie is homozygous.
   a. What is Bill's genotype? What is Marie's genotype?
   b. Marie becomes pregnant four times. What is the probability each child will have the dominant phenotypes for all three traits?
   c. What is the probability that each child will have short lashes, high arches, and no chronic tendency to sneeze?

8. *DNA fingerprinting* is a method of identifying individuals based on locating unique base sequences in their DNA (Section 15.3). Before researchers refined the method, attorneys often relied on the ABO blood-typing system to settle disputes over paternity. Suppose that you, as a geneticist, are asked to testify during a paternity case in which the mother has type A blood, the child has type O blood, and the alleged father has type B blood. How would you respond to the following statements?
   a. Attorney of the alleged father: "The mother's blood is type A, so the child's type O blood must have come from the father. Because my client has type B blood, he could not be the father."
   b. Mother's attorney: "Because further tests prove this man is heterozygous, he must be the father."

9. Suppose you identify a new gene in mice. One of its alleles specifies white fur color. A second allele specifies brown fur color. You want to determine whether the relationship between the two alleles is one of simple dominance or incomplete dominance. What sorts of genetic crosses would give you the answer? On what types of observations would you base your conclusions?

10. Your sister moves away and gives you her purebred Labrador retriever, a female named Dandelion. Suppose you decide to breed Dandelion and sell puppies to help pay for your college tuition. Then you discover that two of her four brothers and sisters show *hip dysplasia*, a heritable disorder arising from a number of gene interactions. If Dandelion mates with a male Labrador known to be free of the harmful genes, can you guarantee to a buyer that puppies will not develop the disorder? Explain your answer.

11. A dominant allele *W* confers black fur on guinea pigs. If a guinea pig is homozygous recessive (*ww*), it has white fur. Fred would like to know whether his pet black-furred guinea pig is homozygous dominant (*WW*) or heterozygous (*Ww*). How might he determine his pet's genotype?

12. Red-flowering snapdragons are homozygous for allele R^1. White-flowering snapdragons are homozygous for a different allele (R^2). Heterozygous plants (R^1R^2) bear pink flowers. What phenotypes should appear among F_1 offspring of the crosses listed? What are the expected proportions for each phenotype?
   a. $R^1R^1 \times R^1R^2$ c. $R^1R^2 \times R^1R^2$
   b. $R^1R^1 \times R^2R^2$ d. $R^1R^2 \times R^2R^2$

Notice, in Problem 12, that in cases of incomplete dominance it is inappropriate to refer to either allele of a pair as dominant or recessive. When the phenotype of a heterozygous individual is halfway between those of the two homozygotes, then there is no dominance. Such alleles are usually designated by superscript numerals, as shown here, rather than by uppercase letters for dominance and lowercase letters for recessiveness.

Figure 10.20 A student at San Diego State University exhibiting the tongue-rolling trait for the benefit of a tongue-roll-challenged student.

13. Two pairs of genes affect comb type in chickens (Figure 10.15). When both genes are recessive, a chicken has a single comb. A dominant allele of one gene, *P*, gives rise to a pea comb. Yet a dominant allele of the other (*R*) gives rise to a rose comb. An epistatic interaction occurs when a chicken has at least one of both dominants, *P— R —* , which gives rise to a walnut comb. Predict the F_1 ratios resulting from a cross between two walnut-combed chickens that are heterozygous for both genes (*PpRr*).

14. As described in Section 10.5, a single mutated allele gives rise to an abnormal form of hemoglobin (Hb^S instead of Hb^A). Homozygotes ($Hb^S Hb^S$) develop sickle-cell anemia. But the heterozygotes ($Hb^A Hb^S$) show few outward symptoms.

Suppose a woman's mother is homozygous for the Hb^A allele and her father is homozygous for the Hb^S allele. She marries a male who is heterozygous for the allele, and they plan to have children. For *each* of her pregnancies, state the probability that this couple will have a child who is:

   a. homozygous for the Hb^S allele

   b. homozygous for the Hb^A allele

   c. heterozygous $Hb^A Hb^S$

15. Certain dominant alleles are so vital for normal development that an individual who is homozygous recessive for a mutant recessive form of the allele cannot survive. Such recessive, *lethal alleles* can be perpetuated by heterozygotes. Consider the Manx allele (M^L) in cats. Homozygous cats ($M^L M^L$) die when they are still embryos inside the mother cat. In heterozygotes ($M^L M$), the spine develops abnormally, and the cats end up with no tail whatsoever (Figure 10.21).

Two $M^L M$ cats mate. Among their *surviving* progeny, what is the probability that any one kitten will be heterozygous?

16. A recessive allele *a* is responsible for *albinism*, an inability to produce or deposit melanin in tissues. Humans and some other organisms can have this phenotype (Figure 10.22). In each of the following cases, what are the possible genotypes of the father, of the mother, and of their children?

   a. Both parents have normal phenotypes; some of their children are albino and others are unaffected.

   b. Both parents are albino and have only albino children.

   c. The woman is unaffected, the man is albino, and they have one albino child and three unaffected children.

Figure 10.22 An albino male in India.

17. Kernel color in wheat plants is determined by two pairs of genes. Alleles of one pair show incomplete dominance over alleles of the other pair. For the gene pair at one locus on the chromosome, allele A^1 imparts one dose of red color to the kernel, whereas allele A^2 does not. At the second locus, allele B^1 gives one dose of red color to the kernel, whereas allele B^2 does not. One kernel with genotype $A^1 A^1 B^1 B^1$ is dark red. A different kernel with genotype $A^2 A^2 B^2 B^2$ is white. All other genotypes have kernel colors in between these two extremes.

   a. Suppose you cross a plant grown from a dark-red kernel with a plant grown from a white kernel. What genotypes and phenotypes would you expect among the offspring?

   b. If a plant with genotype $A^1 A^2 B^1 B^2$ self-fertilizes, what genotypes and what phenotypes would be expected among the offspring? In what proportions?

Selected Key Terms

allele *10.1*

codominance *10.4*

continuous variation *10.7*

dihybrid cross *10.3*

epistasis *10.6*

F_1 *10.1*

F_2 *10.1*

gene *10.1*

genotype *10.1*

heterozygous *10.1*

homozygous dominant *10.1*

homozygous recessive *10.1*

hybrid offspring *10.1*

incomplete dominance *10.4*

independent assortment *10.3*

monohybrid cross *10.2*

multiple allele system *10.4*

phenotype *10.1*

pleiotropy *10.5*

probability *10.2*

Punnett-square method *10.2*

segregation *10.2*

testcross *10.2*

true-breeding lineage *10.1*

Readings See also *www.infotrac-college.com*

Fairbanks, D. J., and W. R. Andersen. 1999. *Genetics: The Continuity of Life*. Monterey, California: Brooks-Cole.

Orel, V. 1996. *Gregor Mendel: The First Geneticist*. New York: Oxford University Press.

WWW   *http://www.brookscole.com/biology*

Practice quiz questions, hypercontents, BioUpdates, and critical thinking. The Brooks/Cole Biology Resource Center provides a wealth of information fully organized and integrated by chapter.

Figure 10.21 Manx cat.

11

CHROMOSOMES AND HUMAN GENETICS

The Philadelphia Story

Positioned at strategic locations in chromosomes are genes that work together to bring about orderly cell growth and division. Some of the genes specify the enzymes and other proteins that perform these tasks, and neighboring genes control when, whether, and how fast the tasks proceed. If something disturbs the neighborhood, cell growth and division can spiral out of control and lead to cancer.

The first abnormal chromosome to be associated with cancer was named the *Philadelphia chromosome* after the city in which it was discovered. It shows up in cells of humans affected by a type of leukemia.

Leukemias arise when stem cells in bone marrow overproduce the white blood cells necessary for the body's housekeeping and defense. (All stem cells are unspecialized and retain the capacity for mitotic cell division. Their descendants also divide, and a portion become specialized cell types.) Leukemic cells infiltrate bone marrow and often crowd out stem cells that are giving rise to red blood cells and platelets. Without red blood cells, anemia develops. Without platelets, the body starts bleeding internally. Leukemic cells also infiltrate the blood and organs, including the lymph nodes, spleen, and liver, and skew their functioning.

Untreated patients with acute leukemia often die within weeks or months. Those with chronic leukemia live longer but often die after infections overwhelm their body's weakened defenses. The Philadelphia chromosome is associated with a chronic leukemia

No one knew about the Philadelphia chromosome until microscopists learned how to identify its physical appearance. Chromosomes, remember, are most highly condensed at metaphase of mitosis. In that state their size, length, and centromere location are easiest to identify. Chromosomes of many species also have distinct patterns of bands when stained a certain way (Figure 9.3). A preparation of metaphase chromosomes based on their defining features is a **karyotype**. Section 11.2 shows you how to construct a karyotype diagram from a photograph of metaphase chromosomes, just like microscopists do. The idea is to line up all of the chromosomes, from largest to smallest, and position the sex chromosomes last in the line-up.

As it turned out, the Philadelphia chromosome is physically longer than its normal counterpart, human chromosome 9. The extra length is actually a piece of chromosome 22! What happened?

By chance, both chromosomes broke in a stem cell in bone marrow. Then, in an instance of reciprocal translocation, each broken piece reattached to the wrong chromosome—and a gene located at the end of chromosome 9 fused with a gene in chromosome 22. The altered gene specifies an abnormal protein. In some way, that protein stimulates the unrestrained division of white blood cells.

A Philadelphia chromosome is not easy to identify in standard karyotypes. Things should change with *spectral* karyotyping, a newer research and diagnostic tool that artificially colors human chromosomes in an unambiguous way. Figure 11.1 gives an example.

You began this unit of the book by looking at cell division, the starting point of inheritance. You thought about how chromosomes—and the genes they carry—are shuffled during meiosis and at fertilization. In this chapter you will delve more deeply into patterns of chromosomal inheritance. As the Philadelphia story suggests, the described methods of analysis are not remote from the world of your interests. An inherited collection of bits of information in chromosomal DNA gives rise to traits that, for better or worse, define each organism, young and old alike (Figure 11.2).

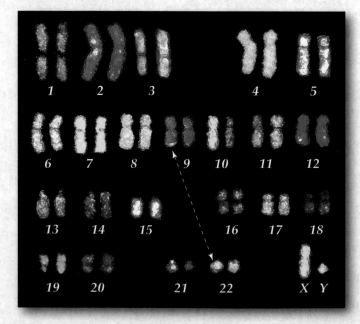

Figure 11.1 Pinpointing a killer—how a reciprocal translocation between human chromosome 9 and chromosome 22 might appear with the help of a new imaging technique called spectral karyotyping. One outcome of this translocation, the Philadelphia chromosome, is a risk factor in chronic myelogenous leukemia, a type of cancer.

Figure 11.2 The DNA icon for an Internet site, *Rare Genetic Diseases in Children*, which allows interested individuals to seek information about specific genetic diseases and disorders. At this writing, the Philadelphia chromosome is on the site's topic board.

Insights into the chromosomal basis of inheritance started gathering force with the rediscovery of Gregor Mendel's work. In 1884 Mendel had just passed away, and his paper on pea plants had been gathering dust in a hundred libraries for nearly two decades. Then improvements in the resolving power of microscopes rekindled efforts to find the hereditary material in cells. By 1882, Walther Flemming had seen threadlike bodies—chromosomes—inside the nuclear region of dividing cells. By 1884, a question was taking shape: Could chromosomes be the hereditary material?

Microscopists soon realized each gamete has half the number of chromosomes of a fertilized egg. In 1887, August Weismann hypothesized that a special division process must reduce the chromosome number by half before gametes form. Sure enough, in that same year meiosis was discovered. Weismann began to promote this theory of heredity: The chromosome number is halved during meiosis, then restored at fertilization; so a cell's hereditary material is half paternal and half maternal in origin. His theory was hotly debated, and it prompted a flurry of experimental crosses—just like the ones Mendel had carried out.

Finally, in 1900, researchers came across Mendel's paper while checking literature related to their own genetic crosses. To their surprise, their experimental results confirmed what Mendel's results had already suggested: Diploid cells have two units (genes) for each heritable trait, and the two units are segregated from each other before gametes form.

In the decades that followed, researchers learned a great deal more about chromosomes. Turn now to a few high points of their work, which will be our starting point for exploring human inheritance.

KEY CONCEPTS

1. Cells of many organisms contain pairs of homologous chromosomes that interact during meiosis. Typically, one chromosome of each pair is maternal in origin, and its homologue is paternal in origin.

2. One gene follows another in sequence along the length of a chromosome. Each kind of gene has its own position, or locus, in that sequence. Different molecular forms of the gene (alleles) may occupy the locus.

3. The combination of alleles in a chromosome does not necessarily remain intact throughout meiosis and gamete formation. During an event called crossing over, some alleles in the sequence swap places with their partners on the homologous chromosome. The alleles that are swapped may or may not be identical.

4. Allelic recombinations contribute to variations in the phenotypes of offspring.

5. A chromosome may change structurally, as when a segment of it is deleted, duplicated, inverted, or moved to a new location. Also, the chromosome number of an individual's cells may change as a result of an improper separation of duplicated chromosomes during meiosis or mitosis.

6. Chromosome structure and the parental chromosome number rarely change. When changes do occur, they often give rise to genetic abnormalities or disorders.

Genes and Their Chromosome Locations

Earlier chapters provided you with a general sense of the structure of chromosomes and of what happens to them during meiosis. To refresh your memory and get a glimpse of where you are going from here with your reading, take a moment to study the following list:

1. **Genes** are units of information about heritable traits. The genes of eukaryotic species are distributed among a number of chromosomes, and each has its own location—a gene locus—in a particular type of chromosome.

2. A cell with a diploid chromosome number ($2n$) has inherited pairs of **homologous chromosomes**. All but one pair of homologous chromosomes are identical in their length, shape, and gene sequence. The single exception is a pairing of nonidentical sex chromosomes, such as X with Y. During meiosis, the two members of a pair of homologous chromosomes interact and then segregate from each other.

3. When comparing the gene at a given locus on one chromosome with its partner on a homologous chromosome, their molecular forms may be the same or slightly different. Although many different forms of that gene may have arisen in a population, each diploid cell has only a pair of them.

4. The different molecular forms of a gene that are possible at a given locus are called **alleles**. They arise through mutation.

5. A *wild-type* allele is the most common form of a gene, either in a natural population or in standard, laboratory-bred strains of a species. Any other form of the gene is called a mutated allele.

6. Genes on the same chromosome are physically linked together. The farther apart two linked genes are, the more vulnerable they are to **crossing over**, an event by which homologous chromosomes exchange corresponding segments. Crossing over results in **genetic recombination**: nonparental combinations of alleles in gametes, then in offspring.

7. The random alignment of each pair of homologous chromosomes at metaphase I of meiosis results in nonparental combinations of alleles in gametes, then in offspring (independent assortment).

8. Abnormal occurrences during meiosis or mitosis occasionally change the structure of chromosomes and the parental chromosome number.

Autosomes and Sex Chromosomes

In all but one case, a pair of homologous chromosomes are exactly like each other in their length, shape, and gene sequence. Microscopists discovered the exception in the late 1800s. For many organisms, a distinctive chromosome occurs in female *or* male individuals, but not both. For example, a diploid cell of a human male has one **X chromosome** and one **Y chromosome** (XY), and that of a human female has two X chromosomes (XX). This inheritance pattern is common among many organisms, including all mammals and fruit flies.

A different pattern prevails for butterflies, moths, some birds, and certain fishes. For them, inheriting two identical sex chromosomes results in a male. Inheriting two nonidentical sex chromosomes results in a female.

Figure 11.1 shows examples of the human X and Y chromosomes. Notice that they are physically different; one is much shorter than the other. Besides this, they do not carry the same genes. Despite the differences, the two can synapse (become zippered together briefly) in a small region along their length, and this allows them to function as homologues during meiosis.

The human X and Y chromosomes fall in the more general category of **sex chromosomes**. The term refers to distinctive types of chromosomes which, in certain combinations, determine a new individual's sex—that is, whether a male or a female will develop. All other chromosomes in an individual's cells are the same in both sexes; they are called **autosomes**.

Karyotype Analysis

Today, microscopists can routinely analyze the physical appearance of sex chromosomes and autosomes taken from a cell at metaphase of mitosis. A preparation of metaphase chromosomes can be sorted out by their defining features and used to construct a karyotype diagram. You read about a new karyotyping method of karyotyping in this chapter's introduction. You can use the next section to walk through a simple procedure for constructing your own karyotype diagrams.

Diploid cells have pairs of genes, on pairs of homologous chromosomes. At each gene locus, the alleles (alternative forms of a gene) may be identical or nonidentical.

As a result of crossing over and other events during meiosis, new combinations of alleles and parental chromosomes end up in offspring. Abnormal events during meiosis or mitosis also can change the structure and number of chromosomes.

Autosomes are the pairs of chromosomes that are the same in males and females of a species. One other pair, the sex chromosomes, govern the sex of a new individual.

WWW

11.2

KARYOTYPING MADE EASY

Karyotype diagrams help answer questions about an individual's chromosomes. Chromosomes are the most condensed and easiest to identify in cells that are going through metaphase of mitosis. Technicians don't count on finding a cell that happens to be dividing in the body when they go looking for it. Instead, they culture cells **in vitro** (literally, "in glass"). They put a sample of cells, usually from blood, into a glass container. The container holds a solution that stimulates cell growth and mitotic cell divisions.

Dividing cells can be arrested at metaphase by adding colchicine to a culture medium. As mentioned in Section 8.5, technicians and researchers use colchicine to block spindle formation. (Colchicine is an extract of autumn crocus and other plants of the genus *Colchicum*.) When a spindle can't form, the sister chromatids of duplicated chromosomes can't move to opposite spindle poles during nuclear division. As you can imagine, by using suitable colchicine concentrations and exposure times, technicians can stockpile metaphase cells, and this increases the chance of finding candidates for karyotype diagrams.

Following colchicine treatment, the culture medium is transferred to the tubes of a centrifuge, a motor-driven spinning device (Figure 11.3a). The cells have greater mass and density than the solution bathing them. As a result, the spinning force moves them farthest from the center of rotation, to the bottom of the attached tubes. Separation in response to a spinning force is called **centrifugation**.

Afterward, the cells are transferred to a saline solution. When immersed in this hypotonic fluid, they swell (by osmosis) and move apart. The metaphase chromosomes thus move apart also. The cells are ready to be mounted on a microscope slide, fixed as by air-drying, and stained.

Chromosomes take up some stains uniformly along their length, and this allows identification of chromosome size and shape. With other staining procedures, horizontal bands show up along the length of the chromosomes of certain species. If researchers direct a ray of ultraviolet light at the chromosomes, the bands will fluoresce. You can see an example of this in Figure 9.3.

In the final steps of karyotype preparation, the chromosomes are photographed through the microscope, and the image is enlarged. Next, the photograph itself is cut apart, one chromosome at a time, and individual cutouts are arranged according to size, shape, and length of arms. Then all the pairs of homologous chromosomes are horizontally aligned by centromeres. Figure 11.3f shows an example of a karyotype diagram prepared in this manner.

WWW

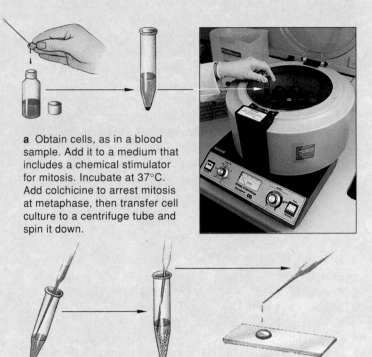

a Obtain cells, as in a blood sample. Add it to a medium that includes a chemical stimulator for mitosis. Incubate at 37°C. Add colchicine to arrest mitosis at metaphase, then transfer cell culture to a centrifuge tube and spin it down.

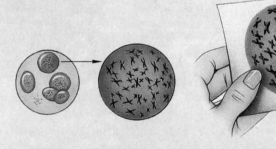

b Cells are now forced to bottom of the tube. Draw off culture medium. Add a dilute saline solution to the tube, then a fixative.

c Prepare and stain the cells for microscopy.

d Place cells on microscope slide. Observe.

e Photograph on cell through the microscope. Enlarge the image of its chromosomes. Cut the image apart. Arrange chromosomes as a set.

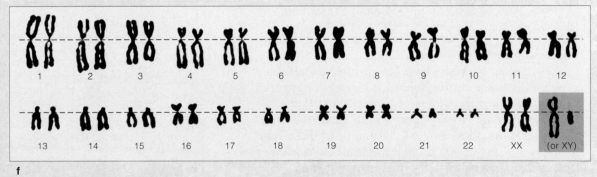

f

Figure 11.3 (**a**–**e**) Karyotype preparation. (**f**) Human karyotype. Human somatic cells have 22 pairs of autosomes and 1 pair of sex chromosomes (XX or XY). That is a diploid number of 46. These are metaphase chromosomes; each is in the duplicated state, with two sister chromatids joined at the centromere.

Analyzing human cells, as by karyotyping, has yielded evidence that each egg produced by a female carries one X chromosome. Half the sperm cells produced by a male carry an X chromosome and half carry a Y.

If an X-bearing sperm fertilizes an X-bearing egg, the new individual will develop into a female. If the sperm happens to carry a Y chromosome, the individual will develop into a male (Figure 11.4).

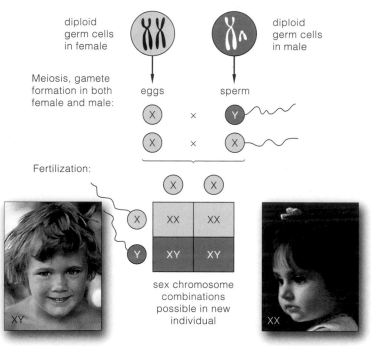

Figure 11.4   Pattern of sex determination in humans.

At this writing fewer than two dozen genes have been identified on the human Y chromosome. One of them is the master gene for male sex determination. Its expression leads to the formation of testes, the primary male reproductive organs (Figure 11.5). In the gene's absence, ovaries form automatically. Ovaries are the primary female reproductive organs. Testes and ovaries both produce important sex hormones that influence the development of particular sexual traits.

The human X chromosome carries more than 2,300 genes. Like other chromosomes, it carries some genes associated with sexual traits, such as the distribution of body fat and hair. However, most of its genes deal with *nonsexual* traits, such as blood-clotting functions. These genes can be expressed in males as well as in females (males, remember, also carry one X chromosome).

A certain gene on the human Y chromosome dictates that a new individual will develop into a male. In the absence of the Y chromosome (and the gene), a female develops.

WWW

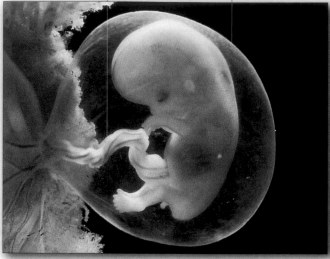

umbilical cord (lifeline between the embryo and the mother's tissues)

amnion (a protective, fluid-filled sac surrounding and cushioning the embryo)

a   A human embryo, eight weeks old and about an inch long. The mass at left is part of the placenta, an organ that forms from maternal and embryonic tissues.

Figure 11.5   Boys, girls, and the Y chromosome.

For about the first four weeks of its existence, a human embryo has neither male nor female traits, even though it normally carries either XY or XX chromosomes. Soon enough, however, ducts and other internal structures that can go either way start forming.

(**a–c**) In an XX embryo, the ovaries (primary female reproductive organs) start to form automatically—*in the absence of a Y chromosome.* By contrast, in an XY embryo, the testes (primary male reproductive organs) start to form during the next four to six weeks. Apparently, a gene region on the Y chromosome governs a fork in the developmental road that can lead to maleness.

The newly forming testes start to produce testosterone and other sex hormones. These hormones are crucial for the development of a male reproductive system. By contrast, in the XX embryo, the newly forming ovaries start to produce different kinds of sex hormones, including estrogens. These are crucial for the development of a female reproductive system.

The master gene for male sex determination is named *SRY* (short for the *S*ex-determining *R*egion of the *Y* chromosome). The same gene has been identified in DNA from male humans, chimpanzees, mice, rabbits, pigs, horses, cattle, and tigers, among others. None of the females tested had the gene. Tests with mice indicate that the gene region becomes active about the time that testes are starting to develop.

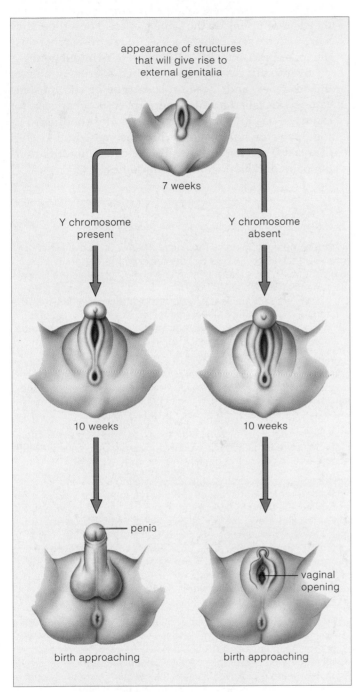

b External appearance of the newly forming reproductive organs in human embryos.

The *SRY* gene resembles DNA regions that specify regulatory proteins. As described in Section 14.1, such proteins bind with certain parts of DNA and thereby turn genes on and off. It seems that the *SRY* gene product regulates a cascade of reactions that are necessary for male sex determination.

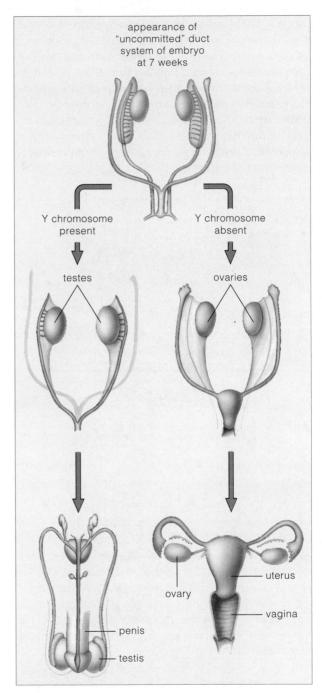

c A duct system that forms early on in the human embryo. The same internal system may develop into the male primary reproductive organs *or* the female primary reproductive organs. The presence or absence of the *SRY* gene in the embryo is the major factor that determines which developmental pathway will be followed.

EARLY QUESTIONS ABOUT GENE LOCATIONS

Linked Genes—Clues to Inheritance Patterns

By the early 1900s, researchers were suspecting that each gene has a specific location on a chromosome. Through hybridization experiments involving mutant fruit flies (*Drosophila melanogaster*), Thomas Hunt Morgan and his coworkers helped confirm this. For example, they found evidence that a gene for eye color and another gene for wing size are located on the *Drosophila* X chromosome. Figure 11.6 describes one series of their experiments.

It seemed, during the early *Drosophila* experiments, that two mutant genes on the X chromosome (*w* for white eyes and *m* for miniature wings) were "linked." That is, they traveled together during meiosis, so that they ended up inside the same gamete. For a time, they were called "sex-linked genes." Today, researchers use the more precise terms **X-linked** and **Y-linked genes**.

Eventually, researchers identified a large number of linked genes, and those on each type of chromosome came to be called a **linkage group**. *D. melanogaster*, for example, has four linkage groups, corresponding to its four pairs of homologous chromosomes. Indian corn (*Zea mays*) has ten linkage groups, corresponding to its ten pairs of homologous chromosomes.

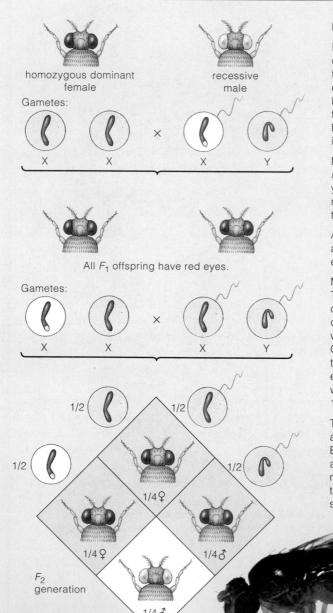

Figure 11.6 X-linked genes as clues to inheritance patterns.

In the early 1900s, the embryologist Thomas Morgan was studying patterns of inheritance. He and his coworkers discovered an apparent genetic basis for connections between sex determination and some nonsexual traits. For example, human males and females both have blood-clotting factors. Yet blood-clotting disorders (hemophilias) show up most often in males, not females, of a family lineage. Sex-specific outcomes were not like anything Mendel saw in his hybrid crosses of pea plants. With respect to phenotype, it made no difference which parent plant carried a recessive allele.

Morgan studied eye color and other nonsexual traits of fruit flies (*Drosophila melanogaster*). The flies can live in bottles on cornmeal, molasses, agar, and yeast. A female lays hundreds of eggs in a few days, and the offspring can reproduce in less than two weeks. So Morgan could track hereditary traits through nearly thirty generations of thousands of flies in a year's time.

At first, all flies were wild type for eye color; they had brick-red eyes. Then, as a result of an apparent mutation in a gene controlling eye color, a white-eyed male appeared in one of the bottles.

Morgan established true-breeding strains of white-eyed males and females. Then he did paired, **reciprocal crosses**. (In the first of such paired crosses, one parent displays the trait of interest. In the second, the other parent displays it.) For the first cross, Morgan allowed white-eyed males to mate with homozygous red-eyed females. All of the F_1 offspring had red eyes. Of the F_2 offspring, however, only some of the males had white eyes. In the second cross, white-eyed females were mated with true-breeding red-eyed males. Half of the F_1 offspring were red-eyed females and half were white-eyed males. And of the F_2 offspring, 1/4 were red-eyed females, 1/4 white-eyed females, 1/4 red-eyed males, and 1/4 white-eyed males!

The seemingly odd results implied a relationship between an eye-color gene and sex determination. Probably the gene locus was on a sex chromosome. But which one? Because females (XX) could be white-eyed, the recessive allele would have to be on one of their X chromosomes. Suppose white-eyed males (XY) also carry the recessive allele on their X chromosome. Suppose there is no corresponding eye-color allele on the Y chromosome. If that were so, then the males would have white eyes, for they have no dominant allele to mask the effect of the recessive one.

The diagram at left illustrates the expected results when Morgan's idea of an X-linked gene is combined with Mendel's concept of segregation. By proposing that one particular gene is located on an X chromosome but not on the Y, Morgan was able to explain the outcome of his reciprocal crosses. His experimental results matched the predicted outcomes.

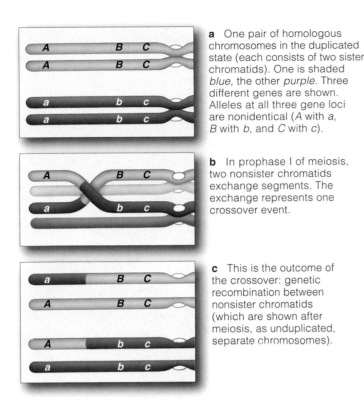

a One pair of homologous chromosomes in the duplicated state (each consists of two sister chromatids). One is shaded *blue*, the other *purple*. Three different genes are shown. Alleles at all three gene loci are nonidentical (*A* with *a*, *B* with *b*, and *C* with *c*).

b In prophase I of meiosis, two nonsister chromatids exchange segments. The exchange represents one crossover event.

c This is the outcome of the crossover: genetic recombination between nonsister chromatids (which are shown after meiosis, as unduplicated, separate chromosomes).

Figure 11.7 Simplified diagram of crossing over. This event occurs in prophase I of meiosis. (Compare Section 9.4.)

Crossing Over and Genetic Recombination

If linked genes always stayed together through meiosis, then a dihybrid cross between true-breeding parents should always have a predictable outcome. Specifically, the most frequent phenotypes among the F_2 offspring should be the same as the phenotypes of the original parents. However, a number of results from the early *Drosophila* experiments did not match this expectation.

As we now know, linked genes are vulnerable to crossing over (Figure 11.7). We also know that crossing over is not a rare event. In humans and most other eukaryotic species, meiosis cannot even be completed properly unless each pair of homologous chromosomes takes part in at least one crossover.

Imagine any two genes at two different locations on the same chromosome. *The probability that a crossover will disrupt their linkage is proportional to the distance that separates them.* Suppose genes *A* and *B* are twice as far apart as two other genes, *C* and *D*:

We would expect crossing over to disrupt the linkage between *A* and *B* much more often.

Two genes are very closely linked when the distance between them is small; they nearly always end up in

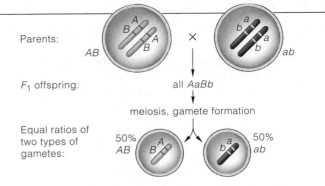

a Full linkage between two genes (no crossovers): half of the gametes have one parental genotype and half have the other.

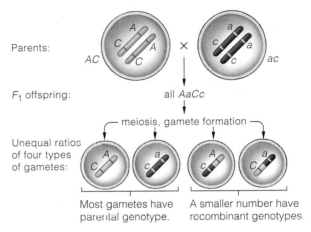

b Incomplete linkage; crossing over affected the outcome.

Figure 11.8 Examples of results from two hybrid crosses (**a**) when gene linkage is complete and (**b**) when crossing over affects the outcome.

the same gamete (Figure 11.8). Linkage is much more vulnerable to crossing over when the distance between genes is greater. When two genes are very far apart, crossing over disrupts linkage so often that those genes assort independently of each other into gametes.

As results from many experimental crosses tell us, genes undergo recombination in fairly regular patterns. The patterns have been used to locate relative positions of genes along chromosomes, an activity called linkage mapping. For example, of the several thousand known genes in the four types of *Drosophila* chromosomes, the positions of about a thousand have been mapped. What about the 50,000 to 100,000 genes on the twenty-three types of human chromosomes? These are being mapped by methods described in Chapter 15. Humans, unlike garden pea plants or fruit flies, do not lend themselves to experimental crosses. And with this jarring thought in your mind, you are ready to begin the next section.

The farther apart two genes are on a chromosome, the greater will be the frequency of crossing over and recombination.

WWW

HUMAN GENETIC ANALYSIS

Some organisms, including pea plants and fruit flies, are ideal for genetic analysis. They grow and reproduce rapidly in small spaces, under controlled conditions. It does not take very long to track a trait through many generations. Humans are another story. We live under variable conditions in diverse environments. We select our own mates and reproduce if and when we want to. Humans live as long as the geneticists who study them, so tracking traits through generations can be tedious. Most human families are so small, there are not enough offspring for easy inferences about inheritance.

Constructing Pedigrees

To get around the problems associated with analyzing human inheritance, geneticists put together pedigrees. A **pedigree** is a chart that shows genetic connections among individuals. Genetic researchers construct them by using standardized methods and definitions, as well as standardized symbols to represent individuals, as shown in Figure 11.9a.

When they analyze pedigrees, geneticists rely on their knowledge of probability and Mendelian inheritance patterns, which may yield clues to the genetic basis for a trait. For example, clues might suggest that an allele responsible for a disorder is dominant or recessive or that it is located on a particular autosome or sex chromosome.

Gathering a great many family pedigrees increases the numerical base for analysis. When any trait follows a simple Mendelian inheritance pattern, a geneticist has more confidence in predicting the probability of its occurrence among children of prospective parents. We will return to this topic later on.

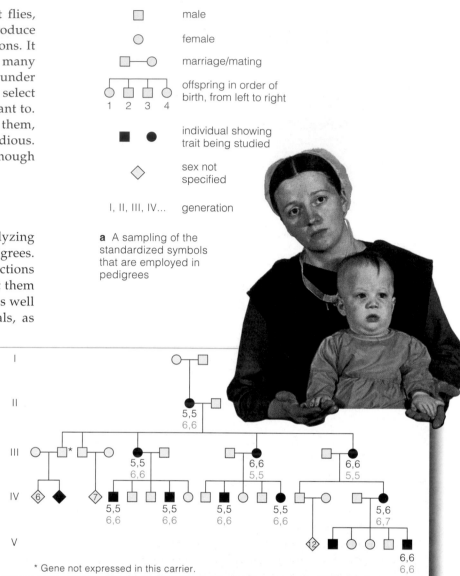

a A sampling of the standardized symbols that are employed in pedigrees

b Pedigree for a family in which polydactyly recurs

* Gene not expressed in this carrier.

Figure 11.9 (**a**) Some standardized symbols used in constructing pedigree diagrams. (**b**) Example of a pedigree for *polydactyly*, a condition in which an individual has extra fingers, extra toes, or both. Expression of the gene for this trait can vary from one individual to the next. *Black* numerals signify the number of fingers on each hand where data were available; *blue* numerals signify the number of toes on each foot.

(**c**) From the human genetic researcher Nancy Wexler, a pedigree for *Huntington disorder*, by which the nervous system progressively degenerates. Wexler and her team constructed an extended family tree for nearly 10,000 people in Venezuela. Analysis of affected and unaffected individuals revealed that one dominant allele on human chromosome 4 is the genetic culprit. Wexler has a special interest in Huntington disorder; she herself has a 50 percent chance of developing it.

c

Table 11.1 Examples of Human Genetic Disorders and Genetic Abnormalities

Disorder or Abnormality*	Main Consequences	Disorder or Abnormality*	Main Consequences
AUTOSOMAL RECESSIVE INHERITANCE		**X-LINKED RECESSIVE INHERITANCE**	
Albinism *10.6; 10.9 CT*	Absence of pigmentation	Color blindness *11.7*	Inability to distinguish among some or all colors
Blue offspring *16.12 CT*	Bright blue skin coloration	Fragile X syndrome *11.6*	Mental retardation
Cystic fibrosis *4.12 CT*	Excessive glandular secretions leading to tissue, organ damage	Hemophilia *11.6; 11.11*	Impaired blood-clotting ability
Ellis-van Creveld syndrome *16.11*	Extra fingers, toes, short limbs	Testicular feminization syndrome *31.2*	XY individual but having some female traits, sterility
Galactosemia *11.6*	Brain, liver, eye damage	X-linked anhidrotic dysplasia *14.3*	In human females, mosaic patches of skin, some with and others without sweat glands
Phenylketonuria (PKU *11.10*	Mental retardation		
Sickle-cell anemia *10.5; 13.4; 16.9*	Adverse plieiotropic effects on organs throughout body		
AUTOSOMAL DOMINANT INHERITANCE		**CHANGES IN CHROMOSOME NUMBER**	
Achondroplasia *11.6*	One form of dwarfism	Down syndrome *11.6; 11.9*	Mental retardation, heart defects
Achoo syndrome *10.9 CT*	Chronic sneezing	Klinefelter syndrome *11.9*	Sterility, retardation
Camptodactyly *10.7*	Rigid, bent little fingers	Turner syndrome *11.9*	Sterility; abnormal ovaries, abnormal sexual traits
Familial hypercholesterolemia *15 CI*	High cholesterol levels in blood; eventually clogged arteries	XYY condition *11.9*	Mild retardation or free of symptoms
Huntington disorder *11.5; 11.6*	Nervous system degenerates progressively, irreversibly	**CHANGES IN CHROMOSOME STRUCTURE**	
Marfan syndrome *11.11 CT*	Absence or abnormal formation of connective tissue	Chronic myelogenous leukemia *10 CI*	Excessive production of white blood cells in bone marrow, followed by malfunctioning of the liver, spleen, other organs
Polydactyly *11.5*	Extra fingers, toes, or both	Cri-du-chat syndrome *11.8*	Mental retardation, abnormally shaped larynx
Progeria *11.7*	Drastic premature aging		
Neurofibromatosis *13.5*	Tumors of nervous system, skin		

* *Italic* numbers indicate sections in which a disorder is described. *CI* signifies Chapter Introduction. *CT* signifies an end-of-chapter Critical Thinking question.

Regarding Human Genetic Disorders

Table 11.1 lists some heritable traits that have been studied in detail. A few are abnormalities, or deviations from the average condition. Said another way, a **genetic abnormality** is nothing more than a rare, uncommon version of a trait, as when a person is born with six toes on each foot instead of five. Whether an individual or society at large views an abnormal trait as disfiguring or merely interesting is subjective. As the classic novel *The Hunchback of Notre Dame* suggests, there is nothing inherently life-threatening or even ugly about it.

By comparison, a **genetic disorder** is an inherited condition that sooner or later will cause mild to severe medical problems. A **syndrome** is a recognized set of symptoms that characterize a given disorder.

Because alleles underlying severe genetic disorders put people at great risk, they are rare in populations. Why, then, don't they disappear entirely? There are two reasons. First, rare mutations introduce new copies of the alleles into the population. Second, in heterozygotes, the harmful allele is paired with a normal one that may cover its functions, so it still can be passed to offspring.

You may hear someone refer to a genetic disorder as a disease, but the terms are not always interchangeable.

A disease also is an abnormal alteration in the way the body functions, and it, too, is characterized by a set of symptoms. But **disease** is illness caused by infectious, dietary, or environmental factors, not by inheritance of mutant genes. If an individual's previously workable genes get altered in a way that disrupts body functions, the illness might be called a *genetic* disease.

With these qualifications in mind, we will turn next to examples of inheritance in the human population. As you will see, genetic analyses of family pedigrees have often revealed simple Mendelian inheritance patterns for certain traits. Researchers have traced many of the traits to dominant or recessive alleles on an autosome or X chromosome. They have traced others to changes in the structure or number of chromosomes.

For many genes, pedigree analysis might reveal simple Mendelian inheritance patterns that will allow inferences about the probability of their transmission to children.

A genetic abnormality is a rare or less common version of an inherited trait. A genetic disorder is an inherited condition that results in mild to severe medical problems.

WWW

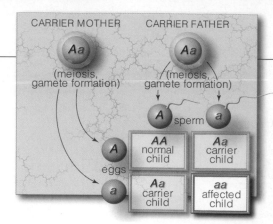

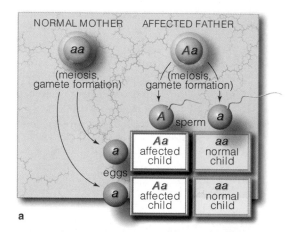

11.6 INHERITANCE PATTERNS

Autosomal Recessive Inheritance

For some traits, inheritance patterns reveal two clues that point to a recessive allele on an autosome. *First,* if both parents are heterozygous, any child of theirs will have a 50 percent chance of being heterozygous and a 25 percent chance of being homozygous recessive, as Figure 11.10 indicates. *Second,* if the parents are both homozygous recessive, any child of theirs will be, also.

On average, 1 in 100,000 newborns is homozygous for a recessive allele that causes *galactosemia.* It cannot produce functional molecules of an enzyme that stops a product of lactose breakdown from accumulating to toxic levels. Lactose normally is converted to glucose and galactose, then to glucose-1-phosphate (which is broken down by glycolysis or converted to glycogen). In affected persons, the full conversion is blocked:

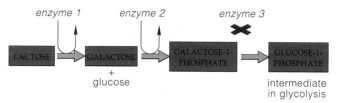

A high blood level of galactose can damage the eyes, liver, and brain. Malnutrition, diarrhea, and vomiting are early symptoms. A high galactose level, the telling symptom, can be detected in urine samples. Untreated galactosemics often die in childhood. But if affected individuals are placed on a restricted diet that excludes dairy products, they can grow up symptom-free.

Autosomal Dominant Inheritance

Two clues of a different sort indicate that an autosomal dominant allele is responsible for a trait. *First,* the trait typically appears in each generation, for the allele is usually expressed, even in heterozygotes. *Second,* if one parent is heterozygous and the other is homozygous recessive, there is a 50 percent chance that any child of theirs will be heterozygous (Figure 11.11*a*).

A few dominant alleles persist in populations even though they cause severe genetic disorders. Some persist by spontaneous mutations. For others, expression of the dominant allele may not interfere with reproduction or affected people reproduce before symptoms are severe.

For example, *Huntington disorder* is characterized by progressive involuntary movements and deterioration of the nervous system, and eventual death. Symptoms may not even start to show up until an affected person is past age thirty. Most people have already reproduced by then. Affected individuals usually die during their forties or fifties, before they might realize that they have transmitted the mutant allele to their children.

Figure 11.10 A pattern for autosomal recessive inheritance. In this case, both parents are heterozygous carriers of the recessive allele (shown in *red*).

Figure 11.11 (**a**) A pattern for autosomal dominant inheritance. In this case, the dominant allele (*red*) is fully expressed in the carriers. (**b**) Infanta Margarita Teresa of the Spanish court and her maids, including the achondroplasic woman at the far right.

As another example, *achondroplasia* affects about 1 in 10,000 people. Commonly, the homozygous dominant condition leads to stillbirth, but heterozygotes are able to reproduce. While they are young, the cartilage of bones forms improperly. At maturity, the affected individuals have abnormally short arms and legs relative to other body parts (Figure 11.11*b*). Adult achondroplasics are less than 4 feet, 4 inches tall. Often the dominant allele has no other phenotypic effects than this.

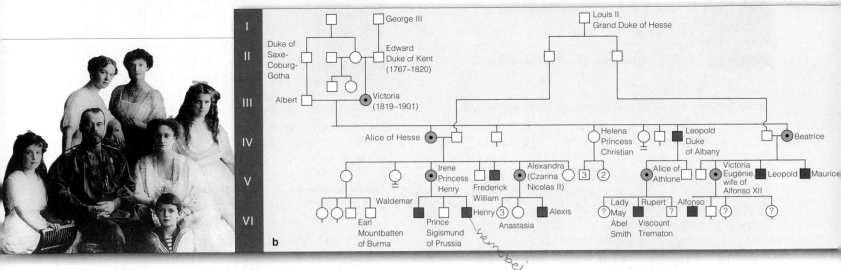

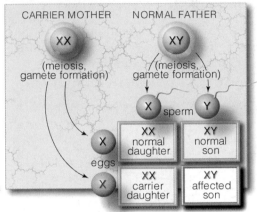

Figure 11.12 (**a**) One pattern for X-linked inheritance. In this example, assume the mother carries the recessive allele on one of her X chromosomes (*red*).

(**b**) *Above:* Partial pedigree for Queen Victoria's descendants, showing carriers and affected males who carried the X allele for hemophilia A. Of the Russian royal family members in the photograph, the mother was a carrier and Crown Prince Alexis was hemophilic. He was a focus of political intrigue that helped trigger the Russian Revolution of 1917.

Figure 11.13 Scanning electron micrograph of the fragile X chromosome from a cultured cell. The arrow points to the fragile site.

X-Linked Recessive Inheritance

Distinctive clues often show up when a recessive allele on an X chromosome causes a genetic disorder. First, males show the recessive phenotype more often than females. A recessive allele can be masked in females, who may inherit a dominant allele on their other X chromosome. The allele is not masked in males, who have only one X chromosome (Figure 11.12*a*). Second, a son cannot inherit the recessive allele from his father. A daughter can. If a daughter is heterozygous, there is a 50 percent chance each son of hers will inherit the allele.

Color blindness, an inability to distinguish among some or all colors, is a common X-linked recessive trait. For instance, in red-green color blindess, the individual lacks some or all of the sensory receptors that normally respond to visible light of red or green wavelengths.

Hemophilia A, a blood-clotting disorder, is a case of X-linked recessive inheritance. In most people, a blood-clotting mechanism quickly stops bleeding from minor injuries. Clot formation requires the products of several genes, some of which are on the X chromosome. If any of the X-linked genes is mutated in a male, the absence of its functional product prolongs bleeding. About 1 in 7,000 males inherits the mutant allele for hemophilia A. Clotting is close to normal in heterozygous females.

The frequency of hemophilia A was unusually high in royal families of nineteenth-century Europe. Queen Victoria of England was a carrier (Figure 11.12*b*). At one time, the recessive allele was present in eighteen of her sixty-nine descendants.

Fragile X syndrome, an X-linked recessive disorder that causes mental retardation, affects 1 in 1,500 males in the United States. In cultured cells, an X chromosome carrying the mutant allele is constricted near the end of its long arm (Figure 11.13). This constriction is called a fragile site because the end of the chromosome tends to break away. The term is misleading, for the end breaks away only in cultured cells, not in cells in the body.

A mutant gene, not breakage, causes the syndrome. It specifies a protein required for normal development of brain cells. Within that gene, a segment of DNA is repeated several times. Certain mutations result in the addition of many more repeats in the DNA (hence their name, expansion mutations). The outcome is a mutant allele that cannot function properly. Because that allele is recessive, males who inherit it and female who are homozygous for it have fragile X syndrome. Expansion mutations are now known to be the cause of Huntington disorder and other genetic disorders.

Genetic analyses of family pedigrees have revealed simple Mendelian inheritance patterns for certain traits, as well as for many genetic disorders that arise from expression of specific alleles on an autosome or X chromosome.

WWW

11.7 TOO YOUNG TO BE OLD

Imagine being ten years old with a mind trapped inside a body that is rapidly getting a bit more shriveled, more frail—*old*—with each passing day. You are just barely tall enough to peer over the top of the kitchen counter, and you do weigh less than thirty-five pounds. Already you are bald and have a crinkled nose. Possibly you have a few more years to live. Could you, like Mickey Hayes and Fransie Geringer, still laugh with your friends?

Of every 8 million newborn humans, one is destined to grow old far too soon. On one of its autosomes, that rare individual carries a mutated gene that gives rise to *Hutchinson-Gilford progeria syndrome*. Through hundreds, thousands, then many billions of DNA replications and mitotic cell divisions, terrible information encoded in that gene was systematically distributed to every cell in the growing embryo, and later in the newborn. Its legacy will be accelerated aging and a greatly reduced life span. The photograph of Mickie Hayes and Fransie Geringer in Figure 11.14 illustrates some of the symptoms.

The mutation causes gross disruptions in interactions among genes that bring about the body's growth and development. Observable symptoms start before age two. Skin that should be plump and resilient starts to thin. Skeletal muscles weaken. Tissues in limb bones that should lengthen and grow stronger start to soften. Hair loss is pronounced; extremely premature baldness is inevitable. There are no documented cases of progeria running in families, so it seems likely the gene mutates spontaneously, at random. It appears to be dominant over its normal partner on the homologous chromosome.

Most progeriacs can expect to die in their early teens as a result of strokes or heart attacks. These final insults are brought on by a hardening of the walls of arteries, a condition typical of advanced age. When Mickey Hayes turned eighteen, he was the oldest living progeriac. Fransie was seventeen when he died.

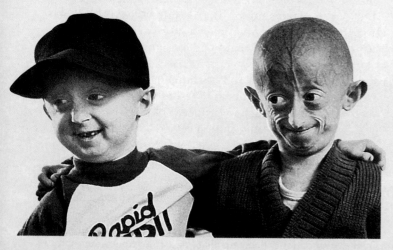

Figure 11.14 Two boys who met at a gathering of progeriacs at Disneyland, California, when they were not yet ten years old.

11.8 CHANGES IN CHROMOSOME STRUCTURE

On rare occasions, the physical structure of one or more chromosomes changes, and the outcome is a genetic disorder or abnormality. Such changes spontaneously occur in nature and are induced in research laboratories by exposure to chemicals or by irradiation. Either way, changes may be detected by microscopic examination and karyotype analysis of cells at mitosis or meiosis. Let's review four kinds of structural change. As you will see, some have serious or lethal consequences.

Major Categories of Structural Change

DUPLICATION Even normal chromosomes have gene sequences that are repeated several to many hundreds or thousands of times. These are **duplications**:

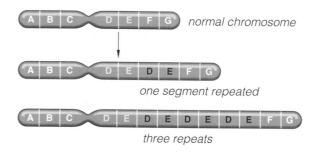

normal chromosome

one segment repeated

three repeats

INVERSION With an **inversion**, a linear stretch of DNA within the chromosome becomes oriented in the reverse direction, with no molecular loss:

inversion of segments GHI

TRANSLOCATION As the chapter introduction showed, the Philadelphia chromosome that has been linked to a form of leukemia results from **translocation**, whereby a broken part of a chromosome becomes attached to a *non*homologous chromosome. Most translocations are reciprocal (both chromosomes exchange broken parts):

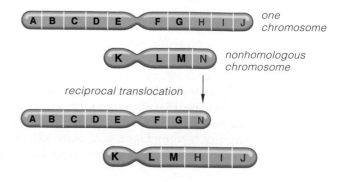

one chromosome

nonhomologous chromosome

reciprocal translocation

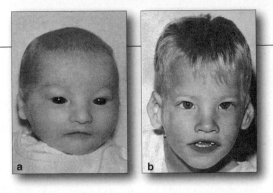

Figure 11.15 (**a**) A male infant who later developed cri-du-chat syndrome. The ears are positioned low on the side of the head relative to the eyes. (**b**) The same boy, four years later. By this age, affected humans no longer make mewing sounds typical of the syndrome.

DELETION   Viral attacks, irradiation (ionizing radiation especially), chemical assaults, or other environmental factors can trigger a **deletion**, the loss of some segment of a chromosome:

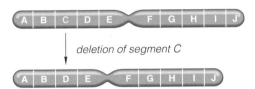

Do species with diploid cells have an advantage after deletions happen? That is, wouldn't genes on the homologous chromosome cover the loss? Maybe, but the sword of chance cuts both ways. If the remaining segment carries a harmful recessive allele, nothing will mask or compensate for *its* effects.

Most deletions are lethal or cause serious disorders in mammals, for they disrupt normal gene interactions that underlie a program of growth, development, and maintenance activities. For example, one deletion from human chromosome 5 results in mental retardation and the development of an abnormally shaped larynx. When affected infants cry, they produce sounds rather like a cat's meow. Hence *cri-du-chat* (cat-cry), the name of this disorder. Figure 11.15 shows an affected child.

Does Chromosome Structure Ever Evolve?

Changes in chromosome structure tend to be selected against rather than conserved over evolutionary time. Even so, for many species, we have interesting signs of changes that occurred in the past. To give one example, many duplications have done their bearers no harm. Although duplications are relatively rare events, they have accumulated over millions, even billions, of years and are now built into the DNA of all species.

Biologists speculate that duplicates of some gene sequences with neutral effects could have an adaptive advantage. In effect, a copy could free up a gene for chance mutations that might turn out to be useful; and

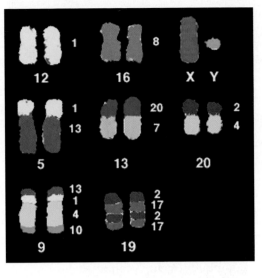

Gibbon and human chromosomes 12, 16, X, and Y are structurally identical (compare Figure 11.1).

Gibbon chromosomes 5, 13, and 20 show translocations that correspond to human chromosomes 1/13, 20/7, and 2/4.

Gibbon chromosome 9 corresponds to parts of human chromosomes; and chromosome 19 has duplications that correspond to human chromosomes 2 and 17.

Figure 11.16 Spectral karyotype of duplicated chromosomes from an ape. Colors identify which parts are structurally identical in human chromosomes.

the normal gene would still issue the required product. One or more duplicated gene sequences could become slightly modified, and then the products of those genes could function in slightly different or new ways.

Several kinds of duplicated, modified genes seem to have had pivotal roles in evolution. Consider the gene regions for the polypeptide chains of hemoglobin, as shown in Section 3.5. In humans and other primates, gene regions for the polypeptide chains of hemoglobin contain multiple sequences that are remarkably similar. The sequences specify whole families of chains that have slight structural differences. Each of the resulting hemoglobin molecules transports oxygen with slightly different efficiencies, depending on cellular conditions.

Certain duplications, inversions, and translocations probably helped put primate ancestors of humans on a unique evolutionary road. They apparently contributed to divergences that led to the modern apes and humans. Of our twenty-three pairs of chromosomes, eighteen are nearly identical to their counterparts in chimpanzees and gorillas. The other five pairs differ at inverted and translocated regions. You can observe such similarities for yourself by comparing chromosomes of a human with those of a gibbon, one of the apes (Figure 11.16).

On rare occasions, a segment of a chromosome may get lost, inverted, moved to a new location, or duplicated.

Most chromosome changes are harmful or lethal, especially when they alter gene interactions that underlie growth, development, and maintenance activities.

Over evolutionary time, many changes have been conserved; they confer adaptive advantages or have had neutral effects.

CHANGES IN CHROMOSOME NUMBER

Occasionally, abnormal events occur before or during cell division, then gametes and new individuals end up with the wrong chromosome number. Consequences range from minor to lethal physical changes.

Categories and Mechanisms of Change

With **aneuploidy**, individuals usually have one extra or one less chromosome. This condition is a major cause of human reproductive failure. Possibly it affects about one-half of all fertilized eggs. As autopsies reveal, most *miscarriages* (spontaneous aborting of embryos before pregnancy reaches full term) are aneuploids.

With **polyploidy**, individuals have three or more of each type of chromosome. About one-half of all species of flowering plants are polyploid (Section 17.3). Often, researchers can induce polyploidy in undifferentiated plant cells by exposing them to colchicine (Section 11.2). Some species of insects, fishes, and other animals are polyploids. However, polyploidy is lethal for humans. All but about 1 percent of human polyploids die before birth, and the rare newborns die soon after birth.

Chromosome numbers can change during mitotic or meiotic cell divisions. Suppose a cell cycle proceeds through DNA duplication and mitosis but is arrested before the cytoplasm divides. The cell is now *tetra*ploid, with four of each type of chromosome. Suppose one or more pairs of chromosomes fail to separate in mitosis or meiosis, an event called **nondisjunction**. Some or all of the forthcoming cells will have too many or too few chromosomes, as in the example in Figure 11.17.

The chromosome number also may get changed at fertilization. Visualize a normal gamete that unites by chance with an $n + 1$ gamete (one extra chromosome). The new individual will be "trisomic" ($2n + 1$); it will have three of one type of chromosome and two of every other type. What if an $n - 1$ gamete unites with a normal n gamete? Then, the new individual will turn out to be "monosomic" ($2n - 1$).

Change in the Number of Autosomes

Most changes in the number of autosomes arise by nondisjunction during gamete formation. Let's look at one of the most common of the resulting disorders. A trisomic 21 newborn, with three chromosomes 21, will develop *Down syndrome*. The symptoms vary greatly, but most affected individuals show moderate to severe mental impairment. About 40 percent develop heart defects. Abnormal development of the skeleton means older children have shortened body parts, loose joints, and poorly aligned bones of the hips, fingers, and toes. Muscles and muscle reflexes are weaker than normal, and speech and other motor skills develop slowly. With special training, trisomic 21 individuals often take part in normal activities; Figure 11.18 gives examples. As a group, they tend to be cheerful and affectionate, and they derive great pleasure from socializing.

Down syndrome is one of many disorders that can be detected before birth. Before detection procedures were widespread, about 1 in 700 newborns of all ethnic groups was trisomic 21. The number is now closer to 1 in 1,100, for some women elect to terminate pregnancy when the condition is detected (Section 11.11). The risk is greater if pregnant women are more than thirty-five years old, as indicated by the diagram in Figure 11.18b.

Change in the Number of Sex Chromosomes

Most sex chromosome abnormalities arise as a result of nondisjunction during meiosis and gamete formation. Let's look at a few phenotypic outcomes.

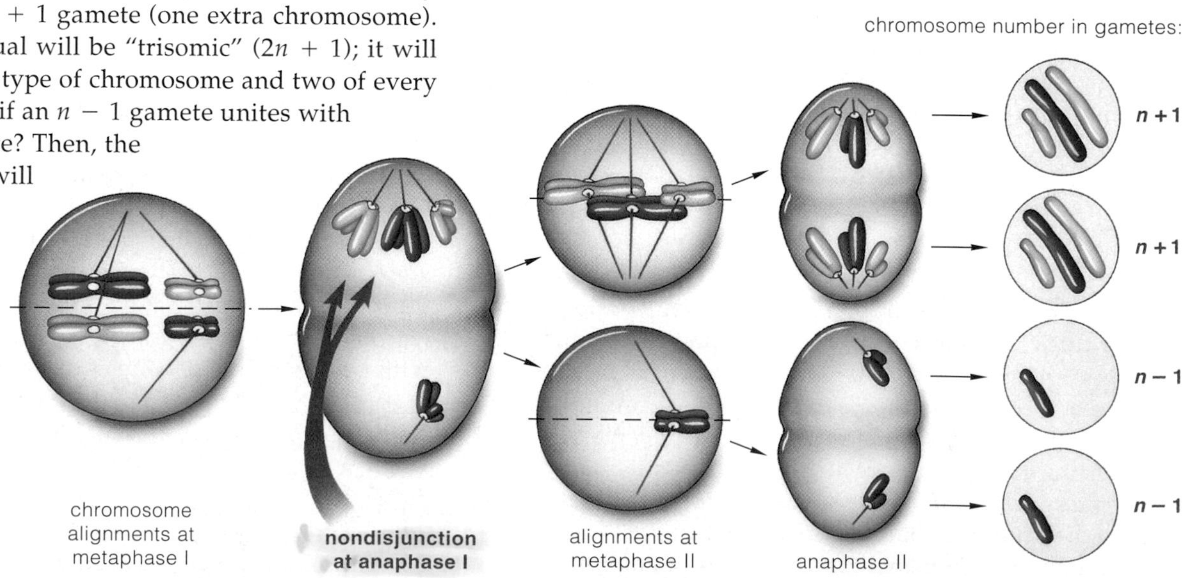

Figure 11.17 Example of nondisjunction. Of two pairs of homologous chromosomes shown, one pair fails to separate at anaphase I of meiosis. The chromosome number changes in the gametes. (Make a sketch of nondisjunction at anaphase II. What will the chromosome numbers be in gametes?)

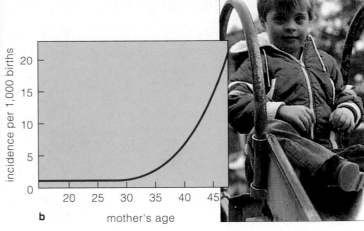

b   mother's age

Figure 11.18 Down syndrome. (**a**) Karyotype revealing a trisomic 21 condition (*arrows*). (**b**) Relationship between the frequency of Down syndrome and mother's age at the time of childbirth. Results are from a study of 1,119 affected children who were born in Victoria, Australia, between 1942 and 1957. The young lady above was a lively participant in the Special Olympics, held annually in San Mateo, California.

TURNER SYNDROME   Inheriting one X chromosome with no corresponding X or Y chromosome gives rise to *Turner syndrome*, which affects 1 in 2,500 to 10,000 or so newborn girls. Nondisjunction affecting sperm accounts for 75 percent of the cases. We see fewer people with Turner syndrome compared to people having other sex chromosome abnormalities. The likely reason is that at least 98 percent of all X0 zygotes spontaneously abort early in pregnancy. Approximately 20 percent of all the spontaneously aborted embryos in which chromosome abnormalities were detected have been X0.

Despite the near-lethality, X0 survivors are not as disadvantaged as other aneuploids. They grow up well proportioned, albeit short—4 feet, 8 inches tall, on the average. Generally, their behavior is normal during childhood. But most Turner females are infertile. They do not have functional ovaries and so cannot produce eggs or sex hormones. Without sex hormones, breast enlargement and the development of other secondary sexual traits are reduced. Possibly as a result of their arrested sexual development and small size, X0 females often become passive and are easily intimidated by peers during their teens. Some patients have benefited from hormone therapy and corrective surgery.

KLINEFELTER SYNDROME   Of every 500 to 2,000 liveborn males, one has inherited two X chromosomes and one Y chromosome. This XXY condition results mainly from nondisjunction in the mother (about 67 percent of the time, compared to 33 percent in the father). Symptoms of the resulting *Klinefelter syndrome* develop after the onset of puberty. XXY males are taller than average and are sterile or nearly so. Their testes usually are much

smaller than average; the penis and scrotum are not. Often facial hair is sparse, and the breasts somewhat enlarged. Injections of the hormone testosterone can reverse the feminized traits but not the low fertility. Some XXY males display mild mental impairment, but many fall within the normal range for intelligence. Except for their low fertility, many affected individuals show no outward symptoms at all.

XYY CONDITION   About 1 in 1,000 males has one X and two Y chromosomes. *XYY males* tend to be taller than average. Some may be mildly retarded, but most are phenotypically normal. At one time, XYY males were thought to be genetically predisposed to become criminals. The erroneous conclusion was based on small numbers of cases in highly selected groups, such as prison inmates. Investigators often knew who the XYY males were, and this may have biased their evaluations. There were no **double-blind studies**, by which different investigators gather data independently of one another and then match them up only after both sets of data are completed. In this case, the same investigators gathered the karyotypes and the personal histories. Fanning the stereotype was a sensationalized report in 1968 that a mass-murderer of young nurses was XYY. He wasn't.

In 1976, a Danish geneticist issued a report on a large-scale study based on records of 4,139 tall males, twenty-six years old, who had reported to their draft board. Besides giving results of physical examinations and intelligence testing, the records provided clues to the socioeconomic status, educational history, and any criminal convictions. Only twelve of the males were XYY, which left more than 4,000 for the control group. The only finding of significance was that tall, mentally impaired males who engage in criminal activity are just more likely to get caught—irrespective of karyotype.

Most changes in chromosome number arise as an outcome of nondisjunction during meiosis and gamete formation.

WWW

11.10 PROSPECTS IN HUMAN GENETICS

With the arrival of their newborn, parents typically ask, "Is our baby normal?" Quite naturally, they want their baby to be free of genetic disorders, and most of the time it is. But what are the options when it is not?

We do not approach heritable disorders and diseases the same way. We attack diseases with antibiotics, surgery, and other weapons. But how do we attack a heritable "enemy" that can be transmitted to offspring? Do we institute regional, national, or global programs to identify people who might be carrying harmful alleles? Do we tell them they are "defective" and run a risk of bestowing a disorder on their children? Who decides which alleles are "harmful"? Should society bear the cost of treating genetic disorders before and after birth? If so, should society also have a say in whether an affected embryo will be born at all, or whether it should be aborted? An **abortion** is the induced expulsion of a pre-term embryo from the uterus.

Questions such as these are only the tip of an ethical iceberg. And we do not have anwers that are universally acceptable throughout our society.

PHENOTYPIC TREATMENTS   Often, the symptoms of genetic disorders can be either minimized or suppressed by (1) exerting dietary controls, (2) making adjustments to specific environmental conditions, and (3) intervening surgically or by way of hormone replacement therapy.

For example, dietary control works in *phenylketonuria*, or PKU. A certain gene specifies an enzyme that converts one amino acid to another—phenylalanine to tyrosine. If an individual is homozygous recessive for a mutated form of the gene, the first of these amino acids accumulates inside the body. If excess amounts are diverted into other pathways, then phenylpyruvate and other compounds may form. High levels of phenylpyruvate in the blood can impair the functioning of the brain. When affected people restrict their intake of phenylalanine, they are not required to dispose of excess amounts, so they can lead normal lives. Among other things, they can avoid soft drinks and other food products that are sweetened with aspartame, a compound that contains phenylalanine.

Environmental adjustments help counter or minimize the symptoms of some disorders, as when albinos avoid exposure to direct sunlight. Surgical reconstructions also minimize many physical problems. For example, surgeons close up a form of *cleft lip* in which a vertical fissure cuts through the lip and extends into the roof of the mouth.

GENETIC SCREENING   Through large-scale screening programs in the general population, affected persons or carriers of a harmful allele often can be detected early enough to start preventive measures before symptoms develop. For example, most hospitals in the United States routinely screen newborns for PKU, so today it is less common to see people with symptoms of this disorder.

GENETIC COUNSELING   If a first child or close relative has a severe heritable problem, prospective parents may worry about their next child. They may request help in evaluating their options from a qualified professional counselor. *Genetic counseling* often includes diagnosis of parental genotypes, detailed pedigrees, and genetic testing for hundreds of known metabolic disorders. Geneticists, too, may be contacted to help predict risks for genetic disorders. During the genetic counseling, prospective parents also must be reminded that the same risk usually applies to each pregnancy.

PRENATAL DIAGNOSIS   Methods of *prenatal diagnosis* can be used to determine the sex as well as more than a hundred genetic conditions of the embryo or fetus. (*Prenatal* means "before birth." The term *embryo* applies until eight weeks after fertilization, after which the term *fetus* is appropriate.)

For example, suppose a woman who is forty-five years old becomes pregnant. She worries that her forthcoming child may develop Down syndrome. She might request prenatal diagnosis by *amniocentesis* (Figure 11.19). With this diagnostic procedure, a clinician withdraws a tiny sample of the fluid inside the amnion, the membranous sac surrounding the fetus. Some cells that the fetus has *delay the pregnancy 35- 40*

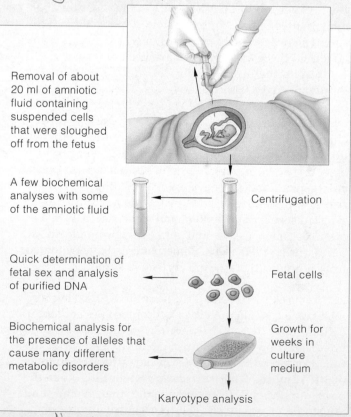

Removal of about 20 ml of amniotic fluid containing suspended cells that were sloughed off from the fetus

A few biochemical analyses with some of the amniotic fluid

Centrifugation

Quick determination of fetal sex and analysis of purified DNA

Fetal cells

Biochemical analysis for the presence of alleles that cause many different metabolic disorders

Growth for weeks in culture medium

Karyotype analysis

Figure 11.19 Steps in amniocentesis, a prenatal diagnostic tool.

#21

sloughed off are suspended in the sample. The cells are cultured and then analyzed.

The woman also might ask for *chorionic villi sampling* (CVS). By this diagnostic procedure, a clinician withdraws some cells from the chorion, a fluid-filled, membranous sac that surrounds the amnion. It is possible to perform CVS weeks before amniocentesis. It can yield results as early as the ninth week of pregnancy.

Direct visualization of the developing fetus is possible with *fetoscopy*. A fiber-optic device, an endoscope, uses pulsed sound waves to scan the uterus and visually locate particular parts of the fetus, umbilical cord, or placenta (Figure 11.20). Fetoscopy has been used to diagnose blood cell disorders, such as sickle-cell anemia and hemophilia.

All three procedures may accidentally cause infection or puncture the fetus. Occasionally the punctured amnion does not reseal itself quickly, and an excessive amount of amniotic fluid may leak out and cause problems for the fetus. A mother-to-be who requests amniocentesis runs a 1 to 2 percent greater risk of miscarriage. For CVS, she runs a 0.3 percent risk that her future child will have missing or underdeveloped fingers and toes. Fetoscopy increases the risk of a miscarriage by 2 to 10 percent.

Parents-to-be probably should seek counseling from their doctor to help them weigh the risks and benefits of such procedures as applied to their own circumstances. They may wish to (1) ask about the small overall risk of 3 percent that any child will have some kind of birth defect, (2) ask about the severity of genetic disorder that a child might be at risk of developing, and (3) consider how old the woman is at the time of pregnancy.

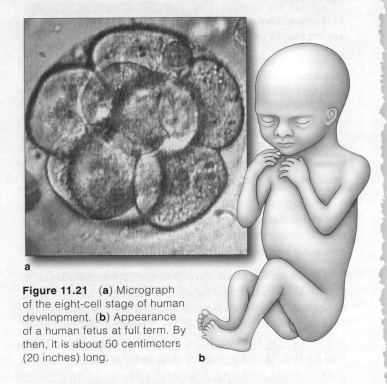

Figure 11.21 (**a**) Micrograph of the eight-cell stage of human development. (**b**) Appearance of a human fetus at full term. By then, it is about 50 centimeters (20 inches) long.

REGARDING ABORTION What happens if prenatal diagnosis does reveal a serious problem? Do prospective parents opt for induced abortion? We can only say here that they must weigh their awareness of the severity of the genetic disorder against ethical and religious beliefs. Worse, they must play out their personal tragedy on a larger stage, dominated by a nationwide battle between fiercely vocal "pro-life" and "pro-choice" factions. We return to this volatile issue in Section 38.18.

PREIMPLANTATION DIAGNOSIS Another procedure, *preimplantation diagnosis*, relies on **in-vitro fertilization**. In vitro literally means "in glass." Sperm and eggs taken from prospective parents are quickly transferred to an enriched medium in a glass petri dish. One or more eggs may become fertilized. In two days, mitotic cell divisions may convert a fertilized egg into a ball of eight cells, such as the one shown in Figure 11.21*a*.

According to one view, the tiny, free-floating ball is a *pre*-pregnancy stage. Like the unfertilized eggs discarded monthly from a woman, the ball is not attached to the uterus. All cells in the ball have the same genes and are not yet committed to giving rise to specialized cells of a heart, lungs, and other organs. Doctors take one of the undifferentiated cells and analyze its genes for suspected disorders. If the cell has no detectable genetic defects, the ball is inserted into the uterus.

Several couples who are at risk of passing on muscular dystrophy, cystic fibrosis, and other disorders have opted for the procedure. Many *"test-tube" babies* have been born in good health and are free of the mutant alleles. *WWW*

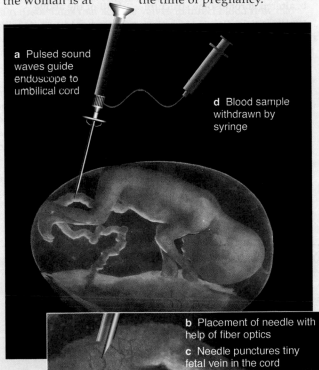

a Pulsed sound waves guide endoscope to umbilical cord

d Blood sample withdrawn by syringe

b Placement of needle with help of fiber optics

c Needle punctures tiny fetal vein in the cord

Figure 11.20 Fetoscopy for prenatal diagnosis.

12 DNA STRUCTURE AND FUNCTION

Cardboard Atoms and Bent-Wire Bonds

One might have wondered, in the spring of 1868, why Johann Friedrich Miescher was collecting cells from the pus of open wounds and, later, from the sperm of a fish. Miescher, a physician, wanted to identify the chemical composition of the nucleus. These particular cells have very little cytoplasm, which makes it easier to isolate the nuclear material for analysis.

Miescher finally succeeded in isolating an organic compound with the properties of an acid. Unlike other substances in cells, it incorporated a notable amount of phosphorus. Miescher called the substance nuclein. He had discovered what came to be known many years later as **deoxyribonucleic acid**, or **DNA**.

The discovery did not cause even a ripple through the scientific community. At the time, no one really knew much about the physical basis of inheritance—that is, *which chemical substance encodes the instructions for reproducing parental traits in offspring*. Few even suspected that the cell nucleus might hold the answer. For a time, researchers generally believed that hereditary instructions had to be encoded in the structure of some unknown class of proteins. After all, heritable traits are spectacularly diverse. Surely the molecules encoding information about those traits were structurally diverse also. Proteins are put together from potentially limitless combinations of twenty different amino acids, so the thinking was that they could function as the sentences (genes) in each cell's book of inheritance.

By the early 1950s, however, the results of many ingenious experiments clearly indicated that DNA was the substance of inheritance. Moreover, in 1951, Linus Pauling did something that no one had done before. Through his training in biochemistry, a talent for

model building, and a few great educated guesses, he deduced the three-dimensional structure of the protein collagen. Pauling's discovery was truly electrifying. If someone could pry open the secrets of proteins, then why not assume the same might be done for DNA? And once the structural details of the DNA molecule were understood, wouldn't they provide clues to its biological functions? *Who would go down in history as having discovered the very secrets of inheritance?*

Scientists around the world started scrambling after that ultimate prize. Among them were James Watson, a young post-doctoral student from Indiana University, and Francis Crick, an exuberant researcher working at Cambridge University. Exactly how could DNA, a molecule consisting of only four kinds of subunits, hold genetic information? Watson and Crick spent long hours arguing over everything they had read about the size, shape, and bonding requirements of the subunits of DNA. They fiddled with cardboard cutouts of the subunits. They even badgered chemists to help them identify any potential bonds they might have overlooked. Then they assembled models from bits of metal, held together with wire "bonds" bent at seemingly suitable angles.

In 1953, they put together a model that fit all of the pertinent biochemical rules and all of the facts about DNA they had gleaned from other sources (Figures 12.1 and 12.2). They had discovered the structure of DNA. And the breathtaking simplicity of that structure enabled them to solve another long-standing riddle—*how life can show unity at the molecular level and yet give rise to such spectacular diversity at the level of whole organisms.*

Figure 12.1 James Watson and Francis Crick posing in 1953 by their newly unveiled structural model of DNA.

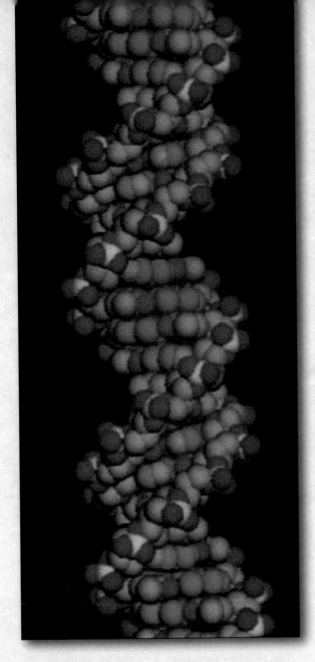

Figure 12.2 A more recent, computer-generated model of DNA. Although more sophisticated in appearance, it is much the same as the prototype Watson and Crick put together decades ago.

KEY CONCEPTS

1. In all living cells, DNA molecules are the storehouses of information about heritable traits.

2. In a DNA molecule, two strands of nucleotides twist together, like a spiral stairway. Each strand consists of four kinds of nucleotides that are the same except for one component—a nitrogen-containing base. The four bases are adenine, guanine, thymine, and cytosine.

3. Great numbers of nucleotides are arranged one after another in each strand of the DNA molecule. In at least some regions, the order in which one kind of nucleotide follows another is unique for each species. Hereditary information is encoded in that particular sequence.

4. Hydrogen bonds connect the bases of one strand of the DNA molecule to bases of the other strand. As a rule, adenine pairs (hydrogen-bonds) with thymine, and guanine with cytosine.

5. Before a cell divides, its DNA is replicated with the assistance of enzymes and other proteins. Each double-stranded DNA molecule starts unwinding. A new, complementary strand is assembled bit by bit on the exposed bases of each parent strand, according to the base-pairing rule.

With this chapter, we turn to investigations that led to our current understanding of DNA. The story is more than a march through details of its structure and function. *It also is revealing of how ideas are generated in science.* On the one hand, having a shot at fame and fortune quickens the pulse of men and women in any profession, and scientists are no exception. On the other hand, science proceeds as a community effort, with individuals sharing not only what they can explain but also what they do not understand. Even when an experiment fails to produce the anticipated results, it might turn up information that others can use or lead to questions that others can answer. Unexpected results, too, might be clues to something important about the natural world.

12.1 DISCOVERY OF DNA FUNCTION

Early and Puzzling Clues

The year was 1928. Frederick Griffith, an army medical officer, was attempting to develop a vaccine against *Streptococcus pneumoniae*, a bacterium that is one cause of the lung disease pneumonia. (When introduced into the body, vaccines mobilize internal defenses against a real attack. Many vaccines are preparations of either killed or weakened bacterial cells.) Griffith never did develop a vaccine. But his work unexpectedly opened a door to the molecular world of heredity.

Griffith isolated and cultured two different strains of the bacterium. He noticed that colonies of one strain had a rough surface appearance, but those of the other strain appeared smooth. He designated the two strains *R* and *S* and used them in a series of four experiments:

1. Laboratory mice were injected with live R cells. As Figure 12.3 indicates, they did not develop pneumonia. *The R strain was harmless.*

2. Mice were injected with live S cells. The mice died. Blood samples taken from them teemed with live S cells. *The S strain was pathogenic* (disease-causing).

3. S cells were killed by exposure to high temperature. Mice injected with these cells did not die.

4. Live R cells were mixed with heat-killed S cells and injected into mice. The mice died—and blood samples from them teemed with *live* S cells!

What was going on in the fourth experiment? Maybe heat-killed S cells in the mixture weren't really dead. But if that were true, then mice injected with heat-killed S cells alone (experiment 3) would have died. Maybe harmless R cells in the mixture had mutated into a killer form. But if that were true, then mice injected with the R cells alone (experiment 1) would have died.

The simplest explanation was as follows: *Heat killed the S cells but did not destroy their hereditary material— including the part that specified "how to cause infection."* Somehow, that material had been transferred from the dead S cells to living R cells, which put it to use.

Further experiments showed the harmless cells had indeed picked up information on causing infections and were permanently transformed into pathogens. After a few hundreds of generations, descendants of the transformed bacterial cells were still infectious!

The unexpected results of Griffith's experiments intrigued Oswald Avery and his fellow biochemists. In time they were even able to transform harmless bacterial cells with *extracts* of killed pathogenic cells. Finally in 1944, after rigorous chemical analyses, they felt confident in reporting that the hereditary substance in their extracts probably was DNA—not proteins, as was then widely believed. To give experimental evidence for their conclusion, they reported that they had added certain protein-digesting enzymes to some extracts, but cells exposed to those extracts were transformed anyway. To other extracts, they had added an enzyme that digests DNA but not proteins. Doing so blocked hereditary transformation.

Despite these impressive experimental results, many biochemists refused to give up on the proteins. Avery's findings, they said, probably applied only to bacteria.

Confirmation of DNA Function

By the early 1950s molecular detectives, including Max Delbrück, Alfred Hershey, Martha Chase, and Salvador Luria, were using viruses as experimental subjects. The viruses they had selected, called **bacteriophages**, infect *Escherichia coli* and other bacteria.

Viruses are as biochemically simple as you can get. Although they are not alive, they do contain hereditary information about building more new virus particles. At some point after a virus infects a host cell, viral enzymes take over a portion of the cell's metabolic machinery—which starts churning out substances that are necessary to construct new virus particles.

genetic material
viral coat
sheath
base plate
tail fiber

a

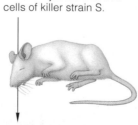

Figure 12.3 Results of Griffith's experiments with a harmless and a pathogenic strain of *Streptococcus pneumoniae*, as described in the text above.

1 Mice injected with live cells of harmless strain R.

Mice do not die. No live R cells in their blood.

2 Mice injected with live cells of killer strain S.

Mice die. Live S cells in their blood.

3 Mice injected with heat-killed S cells.

Mice do not die. No live S cells in their blood.

4 Mice injected with live R cells *plus* heat-killed S cells.

Mice die. Live S cells in their blood.

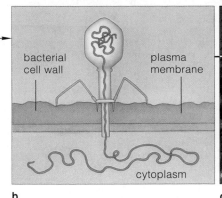

bacterial cell wall

plasma membrane

cytoplasm

b

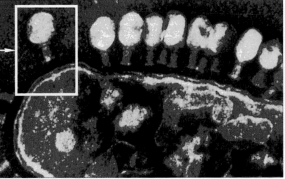

c

Figure 12.4 (**a,b**) *Far left:* Structural organization of a T4 bacteriophage. The diagram shows the genetic material of this type of virus being injected into the cytoplasm of a host cell. For this bacteriophage, it is DNA (the *blue,* threadlike strand). (**c**) Electron micrograph of T4 virus particles infecting a bacterium (*Escherichia coli*), which has just become an unwilling host.

Figure 12.5 Two examples of the landmark experiments that pointed to DNA as the substance of heredity. In the 1940s, Alfred Hershey and his colleague, Martha Chase, were studying the biochemical basis of inheritance. They were aware that certain bacteriophages consisted of proteins and DNA. *Did the proteins, DNA, or both contain the viral genetic information?*

To find a possible answer, Hershey and Chase designed two experiments, based on two known biochemical facts. First, the proteins of bacteriophages incorporate sulfur (S) but not phosphorus (P). Second, their DNA contains phosphorus but not sulfur.

(**a**) In one experiment, some bacterial cells were grown on a culture medium that included a radioisotope of sulfur, ^{35}S. Therefore, when the bacterial cells synthesized proteins, they would take up that radioisotope—which would serve as a tracer. (Here you may wish to review Section 2.2.) After the cells became labeled with the tracer, bacteriophages were allowed to infect them. As the infection ran its course, viral proteins were synthesized inside the host cells. These proteins also became labeled with ^{35}S. And so did the new generation of virus particles.

Next, the labeled bacteriophages were allowed to infect a new batch of unlabeled bacteria that were suspended in a fluid culture medium. Afterward, Hershey and Chase whirred the fluid in a kitchen blender. Whirring dislodged the viral protein coats from the cells. The particles became suspended in the fluid medium. Analysis revealed the presence of labeled protein in the fluid—but there was no evidence of labeled protein *inside* the bacterial cells.

(**b**) For the second experiment, Hershey and Chase cultured more bacterial cells. The phosphorus that was available to them for synthesizing DNA included the radioisotope ^{32}P. Later, bacteriophages were allowed to infect the cells.

As predicted, viral DNA synthesized inside the infected cells became labeled, as did the new generation of virus particles. Next, the labeled particles were allowed to infect bacteria that were suspended in a fluid medium. Then they were dislodged from the host cells. Analysis revealed that the labeled viral DNA was not in the fluid. DNA remained *inside* the host cells, where its hereditary information had to be put to use to make more virus particles. Here was strong evidence that DNA is the genetic material of this type of virus.

virus particle labeled with ^{35}S

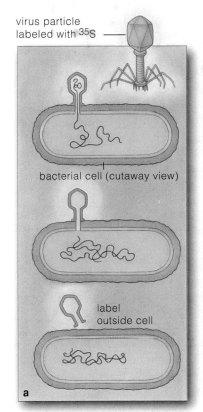

bacterial cell (cutaway view)

label outside cell

a

virus particle labeled with ^{32}P

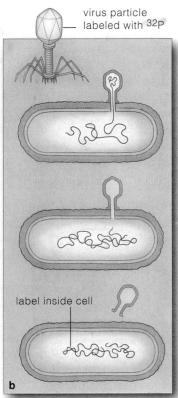

label inside cell

b

By 1952, researchers knew that some bacteriophages consist only of DNA and a protein coat. Also, electron micrographs revealed that the main part of the viruses remains *outside* the cells they are infecting (Figure 12.4). Possibly, such viruses were injecting genetic material alone *into* host cells. If that were true, was the material DNA, protein, or both? Through many experiments, researchers accumulated strong evidence that DNA, not proteins, serves as the molecule of inheritance. Figure 12.5 describes two of these landmark experiments.

Information for producing the heritable traits of single-celled and multicelled organisms is encoded in DNA.

WWW

What Are the Components of DNA?

Long before the bacteriophage studies were under way, biochemists knew that DNA contains only four types of nucleotides that are the building blocks of nucleic acids. Each **nucleotide** consists of a five-carbon sugar (which, in DNA, is deoxyribose), a phosphate group, and one of the following nitrogen-containing bases:

adenine	guanine	thymine	cytosine
(A)	(G)	(T)	(C)

As you can see from Figure 12.6, all four types of nucleotides in DNA have their component parts joined together in much the same way. However, T and C are pyrimidines, which are single-ring structures. A and G are purines, which are larger, bulkier molecules; they have double-ring structures.

By 1949, Erwin Chargaff, a biochemist, had shared two crucial insights into the composition of DNA with the scientific community. First, the amount of adenine relative to guanine differs from one species to the next. Second, the amount of adenine in DNA always equals that of thymine, and the amount of guanine always equals that of cytosine. We may show this as:

$$A = T \quad \text{and} \quad G = C$$

The proportions of those four kinds of nucleotides relative to each other were tantalizing clues. In some way, the proportions almost certainly were related to the arrangement of the nucleotides in a DNA molecule.

The first convincing evidence of that arrangement emerged from Maurice Wilkins's research laboratory in England. Rosalind Franklin, one of Wilkins's colleagues, had obtained especially good **x-ray diffraction images** of DNA fibers (Section 12.3). By this process, a beam of x-rays is directed at a molecule. The molecule scatters the beam in patterns that can be captured on film. The pattern itself consists only of dots and streaks; in itself, it doesn't reveal the molecular structure. However, the photographic images of those patterns can be used to calculate the positions of the molecule's atoms.

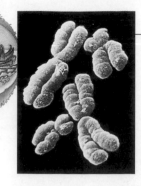

All chromosomes in a cell contain DNA. What does DNA contain? Four kinds of nucleotides, A, G, C, and T. Here are the structural formulas for those nucleotides:

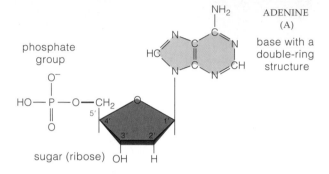

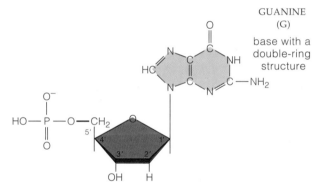

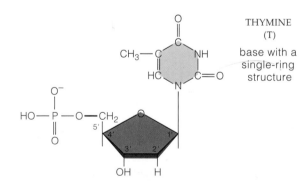

Figure 12.6 Four kinds of nucleotides that serve as building blocks for DNA. The small numerals on the structural formulas identify the carbon atoms to which other parts of the molecule are attached.

As you can see, each nucleotide in a DNA molecule has a five-carbon sugar (shaded *red*), which has a phosphate group attached to the fifth carbon atom of its carbon ring structure. A nucleotide also has one of four kinds of nitrogen-containing bases (*blue*), which is attached to the first carbon atom. The four kinds of nucleotides in DNA differ only in which base they have: adenine, guanine, thymine, or cytosine.

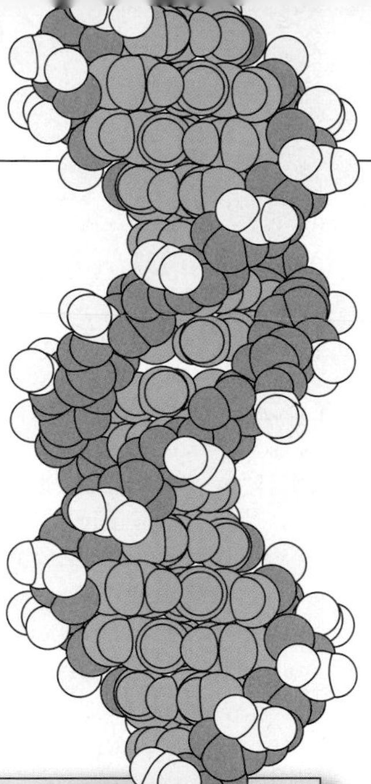

Figure 12.7 Representations of a DNA double helix. Notice how the two sugar-phosphate backbones run in *opposing* directions. (Think of the deoxyribose units of one strand as being upside down.) By comparing the numerals used to identify each carbon atom of the deoxyribose molecule (1′, 2′, 3′, and so on), you see that one strand runs in the 5′→3′ direction and the other runs in the 3′→5′ direction.

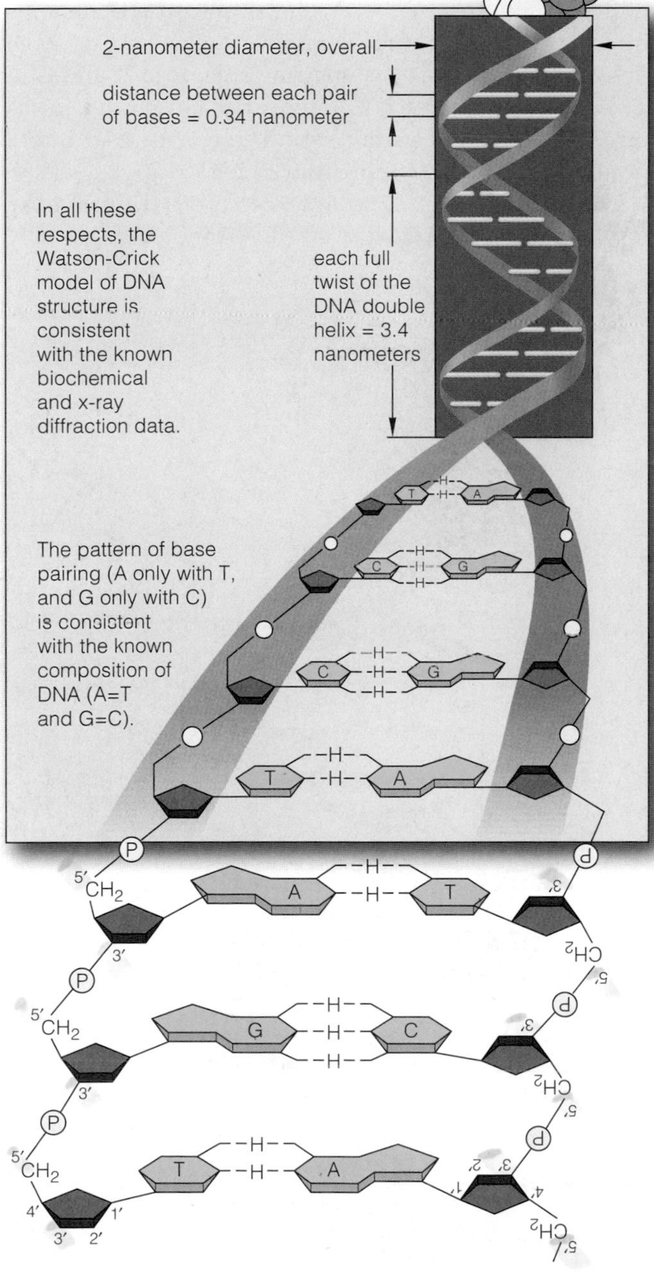

2-nanometer diameter, overall

distance between each pair of bases = 0.34 nanometer

In all these respects, the Watson-Crick model of DNA structure is consistent with the known biochemical and x-ray diffraction data.

each full twist of the DNA double helix = 3.4 nanometers

The pattern of base pairing (A only with T, and G only with C) is consistent with the known composition of DNA (A=T and G=C).

DNA does not readily lend itself to x-ray diffraction. However, researchers could rapidly spin a suspension of DNA molecules, spool them onto a rod, and gently pull them into gossamer fibers, like cotton candy. If the atoms in DNA were arranged in a regular order, x-rays directed at a fiber should scatter in a regular pattern that could be captured on film.

Calculations based on Franklin's images strongly indicated that the DNA molecule had to be long and thin, with a 2-nanometer diameter. Some molecular configuration was being repeated every 0.34 nanometer along its length, and another one, every 3.4 nanometers.

Could the sequence of nucleotide bases be twisting, like a circular stairway? Certainly Pauling thought so. After all, he discovered the helical shape of collagen. He and everybody else—including Wilkins, Watson, and Crick—were thinking "helix." Watson later wrote, "We thought, why not try it on DNA? We were worried that *Pauling* would say, why not try it on DNA? Certainly he was a very clever man. He was a hero of mine. But we beat him at his own game. I still can't figure out why."

Pauling, it turned out, made a big chemical mistake. His model had hydrogen bonds at phosphate groups holding DNA's structure together. That does happen in highly acidic solutions. It doesn't happen in cells.

Patterns of Base Pairing

As Watson and Crick perceived, DNA consists of *two* strands of nucleotides, held together at their bases by hydrogen bonds. The bonds form when the two strands run in opposing directions and twist together into a double helix (Figure 12.7). Two kinds of base pairings form along the length of the molecule: A—T and G—C. This bonding pattern permits variation in the order of bases in any given strand. For example, in even a tiny stretch of DNA from a rose, gorilla, human, or any other organism, the sequence might be:

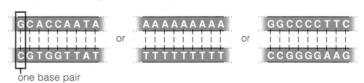

GCACCAATA   or   AAAAAAAAA   or   GGCCCCTTC
| | | | | | | | | | | | | | | | | | | | | | | | | | |
CGTGGTTAT   TTTTTTTTT   CCGGGGAAG

one base pair

In fact, even though all DNA molecules show the same bonding pattern, each species has unique base sequences in its DNA. *This molecular constancy and variation among species is the foundation for the unity and diversity of life.*

The pattern of base pairing between the two strands in DNA is constant for all species—A with T, and G with C. However, the DNA molecules of each species show unique differences in the sequence of base pairs along their length.

WWW

Focus on Bioethics

In 1951, Rosalind Franklin arrived at King's Laboratory of Cambridge University with impressive credentials. Earlier, in Paris, she had refined existing procedures for x-ray diffraction while studying the structure of coal. She also had devised a new mathematical approach to interpreting x-ray diffraction images and had built three-dimensional models of molecules, as Pauling had done. Now she had been asked to run an x-ray crystallography laboratory, which she would create with state-of-the-art equipment. Her assignment? Investigate the structure of DNA.

No one bothered to tell her that, down the hall, Maurice Wilkins was already working on the puzzle. Even the graduate student assigned to assist her failed to mention it. And no one bothered to tell Wilkins about Franklin's assignment, so he assumed she was just a technician hired to do his x-ray crystallography work because he could not

Figure 12.8   Rosalind Franklin.

do it himself. And so began a poisonous clash. To Franklin, Wilkins seemed inexplicably prickly; to Wilkins, Franklin displayed an appalling lack of deference that technicians usually show to researchers.

Wilkins had a prized cache of crystalline fibers of DNA, each having parallel arrays of hundreds of millions of DNA molecules—and these he gave to his "technician."

Five months later, Franklin gave a talk on what she had learned so far. DNA, she said, might have two, three, or four parallel chains twisted into a helix, with phosphate groups projecting outward. She had measured DNA's density and assigned DNA fibers to 1 of 230 categories of crystals, based on the symmetry of their parallel chains.

With his background in crystallography, Crick would have recognized the significance of that symmetry *if* he had been present. (To wit, *paired* chains running in opposite directions would look the same even if flipped 180°. Two paired chains? No. DNA's density ruled that out. But *one pair* of chains? Yes!) Watson was in the audience, but he didn't have a clue to what Franklin was talking about.

Later, Franklin created an outstanding x-ray diffraction image of wet DNA fibers that fairly screamed *Helix!* She also worked out DNA's length and diameter. But she had been working with dry fibers for so long, she did not dwell on her new data. Wilkins, however, did. In 1953, he let Watson see Franklin's exceptional x-ray diffraction image and reminded him of what she had reported fourteen months earlier. And when Watson and Crick finally did focus on her data, they had the final bits of information necessary to start building a DNA model—one with two helically twisted chains running in opposing directions.

Not until ten years after Franklin's untimely death in 1958 did Watson acknowledge her pivotal discoveries.

How Is a DNA Molecule Duplicated?

The discovery of DNA structure was a turning point in studies of inheritance. Until then, no one could explain **DNA replication**, or how the molecule of inheritance is duplicated before the cell divides. Once Watson and Crick had assembled their model, Crick understood at once how this might be done.

As he knew, enzymes can easily break the hydrogen bonds between the two nucleotide strands of a DNA molecule. When these enzymes and other proteins act on the molecule, one strand can unwind from the other, thereby exposing a stretch of nucleotide bases. Cells have stockpiles of free nucleotides, and these can pair with the exposed bases. Each parent strand remains intact, and a companion strand is assembled on each one according to this base-pairing rule: A to T, and G to C. As soon as a stretch of a new, partner strand forms on a stretch of the parent strand, the two twist into a double helix, as shown in Figure 12.9.

Because the parent strand is conserved during the replication process, each "new" DNA molecule is really

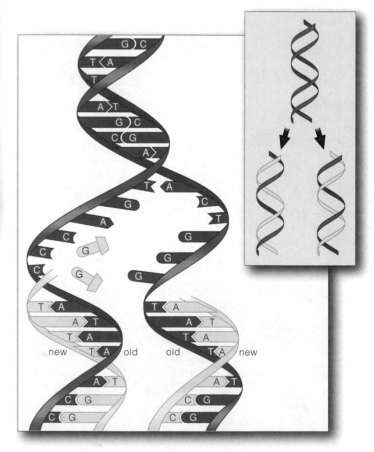

Figure 12.9   Overview of the semiconservative nature of DNA replication. The original two-stranded DNA molecule is shown in *blue*. Each parent strand remains intact, and a new strand (*yellow*) is assembled on each one.

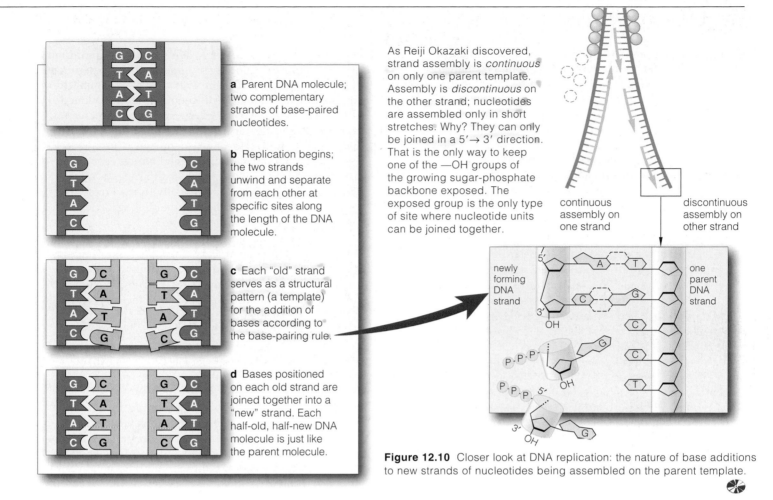

a Parent DNA molecule; two complementary strands of base-paired nucleotides.

b Replication begins; the two strands unwind and separate from each other at specific sites along the length of the DNA molecule.

c Each "old" strand serves as a structural pattern (a template) for the addition of bases according to the base-pairing rule.

d Bases positioned on each old strand are joined together into a "new" strand. Each half-old, half-new DNA molecule is just like the parent molecule.

As Reiji Okazaki discovered, strand assembly is *continuous* on only one parent template. Assembly is *discontinuous* on the other strand; nucleotides are assembled only in short stretches. Why? They can only be joined in a 5'→ 3' direction. That is the only way to keep one of the —OH groups of the growing sugar-phosphate backbone exposed. The exposed group is the only type of site where nucleotide units can be joined together.

continuous assembly on one strand

discontinuous assembly on other strand

newly forming DNA strand

one parent DNA strand

Figure 12.10 Closer look at DNA replication: the nature of base additions to new strands of nucleotides being assembled on the parent template.

half old, half new (Figure 12.9). That is why biologists refer to this process as *semiconservative* replication.

DNA replication uses a team of molecular workers. In response to cellular signals, the replication enzymes become active along the length of the DNA molecule. Together with other proteins, some enzymes unwind the strands in both directions and prevent them from rewinding. Enzyme action jump-starts the unwinding but is not required to unzip hydrogen bonds between the strands. (Hydrogen bonds are individually weak.)

Now enzymes called **DNA polymerases** can attach short stretches of free nucleotides to unwound portions of the parent template (Figure 12.10). Free nucleotides brought up for strand assembly supply the energy that drives the replication process. Each has three phosphate groups attached. DNA polymerase splits away two of the groups, and some energy released by the reaction is used to attach the nucleotide to a growing strand.

DNA ligases fill in tiny gaps between the new short stretches to form a continuous strand. Then enzymes wind the template strand and complementary strand together to form a DNA double helix.

As you will read in Section 15.2, some replication enzymes also have been put to use in recombinant DNA technology.

Monitoring and Fixing the DNA

DNA polymerases, DNA ligases, and other enzymes also engage in **DNA repair**. By this process, enzymes excise and repair altered parts of the base sequence in one strand of a double helix. Suppose a few bases are lost from one strand during replication. DNA polymerases can "read" the complementary sequence on the other strand. With assistance from other repair enzymes, they restore the original sequence. And remember crossing over, by which nonsister chromatids of homologous chromosomes break and exchange segments at meiosis? The same enzymes that bring about that recombination also repair double-strand breaks by initiating strand exchange with homologous DNA. Section 13.4 gives you an idea of what happens when mutations or some other factor compromise the excision-repair function.

DNA is replicated prior to cell division. Enzymes unwind its two strands. Each strand remains intact throughout the process—it is conserved—and enzymes assemble a new, complementary strand on each one.

Enzymes involved in replication also repair the DNA where base-pairing errors have crept into the nucleotide sequence.

WWW

12.5 DOLLY, DAISIES, AND DNA

Imagine the possibility of making a genetically identical copy—a clone—of yourself. Is the image that far-fetched? Consider this: Researchers have been cloning complex animals for more than a decade. For example, some use in vitro fertilization methods to grow cattle embryos in petri dishes. After cleavage is under way in a new embryo, they split the tiny cluster of embryonic cells, and the two tinier clusters continue to develop as two identical-twin cattle embryos. These are implanted in surrogate mothers, and they are born as cloned calves. Figure 12.11*a* shows cloned Holsteins that are prized for their milk production.

A researcher who clones farm animals derived from embryonic cells has to wait for the clones to grow up to see if they display a desired trait. Using a differentiated cell from a prized adult would be faster, for the prized genotype would already be demonstrated and could be maintained indefinitely. However, it seemed impossible to trick a differentiated cell into using its DNA in a new way, and so become the first cell of an embryo.

What does "differentiated" mean? When an embryo first grows from a fertilized egg, all of its cells have the same DNA and are pretty much alike. Then different embryonic cells start using different parts of their DNA. Their unique selections commit them to being liver cells, heart cells, brain cells and other specialists in structure, composition, and function (Section 14.3).

Then, in 1997 in Scotland, a research group led by Ian Wilmut coaxed a differentiated cell from a sheep udder into becoming an "uncommitted" first cell of an embryo. Thus a lamb named Dolly was born a clone of her mother without help from a father. Wilmut's group had managed to transfer nuclei from differentiated cells into enucleated unfertilized eggs, which they had removed earlier from a pregnant ewe. (*Enucleated* means that the nucleus was surgically removed from a cell with a microscopically small needle.) After hundreds of attempts, one egg proceeded to grow, by mitotic cell divisions, into a whole sheep. Dolly developed into a healthy adult and gave birth to a lamb of her own (Figure 12.11*b*).

In science, extraordinary claims call for extraordinary proof—in this case, successful repeats of the experiment. Resounding success came in 1998. Ryozo Yanagimachi and his coworkers at the University of Hawaii cloned three generations of mice (*Mus musculus*). They quickly transferred nuclei from mature cumulus cells from an ovary into unfertilized, enucleated eggs. (Cumulus cells provide nutritional support to neighboring eggs.) Shortly afterward, they chemically activated the eggs as a way to jump-start development into fully formed mice. The resting period may have helped the cells adjust to the unconventional mode of fertilization.

By the year's end, experimenters in Japan gave us a new way of looking at Daisy the Cow. Yukio Tsunoda and his coworkers slipped a cow's cumulus cells and oviduct cells into enucleated eggs. Of 240 transfers in the laboratory, 67 embryos formed. Ten of the survivors were transferred into cows that served as surrogate mothers. Eight cloned calves were born. Only four survived, yet the results suggest that cloning cows from differentiated cells may be at least as efficient as producing them by in vitro fertilization.

In short, individuals of three genera of mammals have now been cloned. More importantly, the efficiency of cloning techniques is increasing at a startling pace.

All of which raises some far-out questions. Is human cloning next? For example, will a human Daisy be able to reproduce copies of herself without involving men? Will men be able to pay to have their DNA inserted into an enucleated cell, have the cell implanted in a surrogate mother, and make a little repeat of themselves?

Recently Lee Silver, a molecular biologist at Princeton University, said he knows of two qualified specialists in human fertility who are interested in cloning humans. He suggested that anyone who thinks the technology will move slowly is being naive. Ethically speaking, should such an attempt be made at all? At this writing, scientists and nonscientists alike are actively debating the issue. *WWW*

Figure 12.11 (**a**) A clone of Holsteins resulting from in vitro fertilization. (**b**) Dolly, a cloned sheep. She started life as a differentiated cell that was extracted from an adult ewe, then induced to undergo mitotic cell divisions. She is shown with her first lamb; clearly, Dolly can breed normally and reproduce the old-fashioned way.

a

b

12.6 SUMMARY

1. The hereditary information of cells and multicelled organisms is encoded in DNA (deoxyribonucleic acid).

2. DNA consists of nucleotide subunits. Each of these has a five-carbon sugar (deoxyribose), one phosphate group, and one of four kinds of nitrogen-containing bases (adenine, thymine, guanine, or cytosine).

3. A DNA molecule consists of two nucleotide strands twisted together into a double helix. The bases of one strand pair (hydrogen-bond) with bases of the other.

4. The bases of the two strands in a DNA double helix pair in constant fashion. Adenine pairs with thymine (A to T), and guanine with cytosine (G to C). *Which* base pair follows the next (A—T, T—A, G—C, or C—G) varies along the length of the strands.

5. Overall, the DNA of one species includes a number of unique stretches of base pairs that set it apart from the DNA of all other species.

6. During DNA replication, enzymes unwind the two strands of a double helix and assemble a new strand of complementary sequence on each parent strand. Two double-stranded molecules result. One strand of each molecule is "old" (it is conserved); the other is "new."

7. Some of the enzymes involved in DNA replication also repair DNA where base-pairing errors have been introduced into the nucleotide sequence.

Review Questions

1. Name the three molecular parts of a nucleotide in DNA. Also name the four different bases in these nucleotides. *12.2*

2. What kind of bond joins two DNA strands in a double helix? Which nucleotide base-pairs with adenine? With guanine? *12.2*

3. Explain how DNA molecules can show both constancy and variation from one species to the next. *12.2*

Self-Quiz (*Answers in Appendix III*)

1. Which is *not* a nucleotide base in DNA?
   a. adenine   c. uracil   e. cytosine
   b. guanine   d. thymine

2. What are the base-pairing rules for DNA?
   a. A–G, T–C   c. A–U, C–G
   b. A–C, T–G   d. A–T, G–C

3. A DNA strand having the sequence C–G–A–T–T–G would be complementary to the sequence _____ .
   a. C–G–A–T–T–G   c. T–A–G–C–C–T
   b. G–C–T–A–A–G   d. G–C–T–A–A–C

4. One species' DNA differs from others in its _____ .
   a. sugars   c. base sequence
   b. phosphate groups   d. all of the above

5. When DNA replication begins, _____ .
   a. the two DNA strands unwind from each other
   b. the two DNA strands condense for base transfers
   c. two DNA molecules bond
   d. old strands move to find new strands

6. DNA replication requires _____ .
   a. free nucleotides   c. many enzymes
   b. new hydrogen bonds   d. all of the above

7.   Match the DNA terms appropriately.
   _____ DNA polymerase   a. two nucleotide strands that
   _____ constancy in   are twisted together
          base pairing   b. A with T, G with C
   _____ replication   c. hereditary material duplicated
   _____ DNA double helix   d. replication enzyme

Critical Thinking

1. Chargaff's data suggested that adenine pairs with thymine, and guanine pairs with cytosine. What other data available to Watson and Crick suggested that adenine-guanine and cytosine-thymine pairs normally do not form?

2. One of Matthew Meselson and Frank Stahl's experiments supported the semiconservative model of DNA replication. The researchers made "heavy" DNA by growing *Escherichia coli* in a medium enriched with ^{15}N, a heavy isotope of nitrogen. They prepared "light" DNA by growing *E. coli* in the presence of ^{14}N, the more common isotope. An available technique helped them identify which replicated molecules were heavy, light, or hybrid (one heavy strand, one light). Use two pencils of two different colors, one for heavy strands and one for light strands. Starting with a DNA molecule having two heavy strands, sketch the daughter molecules that would form after one replication in a ^{14}N-containing medium. Now sketch the four DNA molecules that would result if these daughter molecules were replicated a second time in the ^{14}N medium.

3. Mutations (permanent changes in base sequences of genes) are the original source of genetic variation. This variation is the raw material of evolution. Yet how can both statements be true, given that cells have efficient mechanisms to repair DNA before mutations can become established?

4. As indicated in Section 4.11, a pathogenic strain of *E. coli* has acquired an ability to produce a dangerous toxin that has caused medical problems and fatalities. This is especially true in young children who have ingested undercooked, contaminated beef. Develop a hypothesis to explain how a normally harmless bacterium such as *E. coli* can become a pathogen.

Selected Key Terms

adenine (A) *12.2*	DNA ligase *12.4*	nucleotide *12.2*
bacteriophage *12.1*	DNA polymerase *12.4*	thymine (T) *12.2*
cytosine (C) *12.2*	DNA repair *12.4*	x-ray diffraction
deoxyribonucleic	DNA replication *12.4*	image *12.2*
acid (DNA) *CI*	guanine (G) *12.2*	

Readings *See also www.infotrac-college.com*

Watson, J. 1978. *The Double Helix*. New York: Atheneum. Highly personal view of scientists and their methods, interwoven into an account of how DNA structure was discovered.

Wolfe, S. 1995. *Introduction to Molecular and Cellular Biology*. Belmont, California: Wadsworth. Comprehensive, current, and accessible.

WWW   *http://www.brookscole.com/biology*

Practice quiz questions, hypercontents, BioUpdates, and critical thinking. The Brooks/Cole Biology Resource Center provides a wealth of information fully organized and integrated by chapter.

13

FROM DNA TO PROTEINS

Beyond Byssus

Picture a mussel, of the sort shown in Figure 13.1. Hard-shelled but soft of body, it is using its muscular foot to probe a wave-scoured rock. At any moment, pounding waves can whack the mussel into the water, hurl it repeatedly against the rock with shell-shattering force, and so offer up a gooey lunch for gulls.

By chance, the mussel's foot comes across a crevice in the rock. The foot moves, broomlike, and sweeps the crevice clean. It presses down, forcing air out from underneath it, then arches up. The result is a vacuum-sealed chamber, rather like the one that forms when a plumber's rubber plunger is being squished down and up to unclog a drain. Into this vacuum chamber the mussel spews a fluid, consisting of keratin and other proteins, which bubbles into a sticky foam. Now, by curling its foot into a small tubular shape and pumping the foam through it, the mussel forms sticky threads about as wide as a human whisker. As a final touch, it varnishes the threads with another type of protein and ends up with an adhesive called byssus, which anchors the mussel to the rock.

Byssus is the world's premier underwater adhesive. Nothing that humans have manufactured even comes close. (Sooner or later, water chemically degrades or deforms synthetic adhesives.) Byssus truly fascinates biochemists, dentists, and surgeons looking for better ways to do tissue grafts and to rejoin severed nerves. Genetic engineers insert mussel DNA into yeast cells,

Figure 13.1 Mussels (*Mytilus californianus*) busily demonstrating the importance of proteins for survival. When mussels come across a suitable anchoring site, they use their muscular foot rather like a plumber's plunger to create a vacuum chamber. In this chamber they manufacture the world's best underwater adhesive from a mix of proteins. The adhesive anchors them to substrates in their wave-swept habitat.

which go on to reproduce in large numbers and serve as "factories" for translating mussel genes into useful quantities of proteins. This exciting work, like the mussel's own byssus-building efforts, starts with one of life's universal concepts: *Every protein is synthesized in accordance with instructions contained in DNA.*

You are about to trace the steps leading from DNA to proteins. Many enzymes are players in this pathway. So is another kind of nucleic acid besides DNA. The same steps produce *all* proteins, from mussel-inspired adhesives to the keratin in your hair and fingernails to the insect-digesting enzymes of a Venus flytrap.

Start out by thinking of each cell's DNA as a book of protein-building instructions. The alphabet used to create the book is simple enough: A, T, G, and C (for the nucleotide bases adenine, thymine, guanine, and cytosine). How do you get from that alphabet to a protein? The answer starts with DNA's structure.

DNA, recall, is a double-stranded molecule. Which kind of nucleotide base follows the next along the length of a strand—that is, the **base sequence**—differs from one kind of organism to the next. As you read in the preceding chapter, before a cell divides, its DNA is replicated as the two strands unwind entirely from each other. However, at other times in a cell's life, the two strands unwind only in certain regions to expose particular base sequences—genes. Most of those genes contain instructions for building proteins.

It takes two steps, **transcription** and **translation**, to carry out a gene's protein-building instructions. In eukaryotic cells, transcription proceeds in the nucleus. By this step, a selected base sequence in DNA serves as a structural pattern— a template—for assembling a strand of **ribonucleic acid** (RNA) from the cell's pool of free nucleotides. Afterward, the RNA moves into the cytoplasm, where translation proceeds. In this second step, RNA directs the assembly of amino acids into polypeptide chains. The newly formed chains become folded into the three-dimensional shapes of proteins.

In short, DNA guides the synthesis of RNA, then RNA guides the synthesis of proteins:

$$\text{DNA} \xrightarrow{\text{transcription}} \text{RNA} \xrightarrow{\text{translation}} \text{PROTEIN}$$

The newly synthesized proteins will play structural and functional roles in cells. And some even will have roles in synthesizing more DNA, RNA, and proteins.

KEY CONCEPTS

1. Life cannot exist without enzymes and other proteins. Proteins consist of polypeptide chains, which consist of amino acids. The sequence of amino acids corresponds to a gene, which is a sequence of nucleotide bases in a DNA molecule.

2. The path leading from genes to proteins consists of two steps, called transcription and translation.

3. In transcription, the double-stranded DNA molecule is unwound at a gene region, then an RNA molecule is assembled on the exposed bases of one of the strands.

4. In translation, a certain type of RNA directs the linkage of one amino acid after another, in the sequence required to produce the specified polypeptide chain.

5. With few exceptions, the genetic "code words" by which DNA's instructions are translated into proteins are the same in all species of organisms.

6. A mutation is a permanent alteration in a gene's base sequence. Such changes are the original source of genetic variation in populations.

7. Mutations give rise to alterations in protein structure, protein function, or both. The alterations may lead to small or large differences in traits among the individuals of a population.

HOW IS DNA TRANSCRIBED INTO RNA?

The Three Classes of RNA

Before turning to the details of protein synthesis, let's clarify one point. The chapter introduction might have left you with the impression that synthesis of proteins requires only one class of RNA molecules. Actually, it requires three. Transcription of most genes produces **messenger RNA**, or **mRNA**—the only class of RNA that carries *protein-building* instructions. Transcription of some other genes produces **ribosomal RNA**, or **rRNA**, a major component of ribosomes. Ribosomes, recall, are the structural units upon which polypeptide chains are assembled. Transcription of still other genes produces **transfer RNA**, or **tRNA**, which delivers amino acids one by one to a ribosome in the order specified by mRNA.

The Nature of Transcription

An RNA molecule is almost but not quite like a single strand of DNA. RNA, too, consists of only four types of nucleotides. Each nucleotide has a five-carbon sugar, ribose (not DNA's deoxyribose), a phosphate group, and a base. Three types of bases—adenine, cytosine, and guanine—are the same in RNA and DNA. But in RNA, the fourth type of base is **uracil**, not thymine (Figure 13.2). Like thymine, uracil can pair with adenine. This means a new RNA strand can be put together on a DNA region according to base-pairing rules (Figure 13.3).

Transcription resembles DNA replication in another respect. Enzymes add nucleotides to a growing RNA

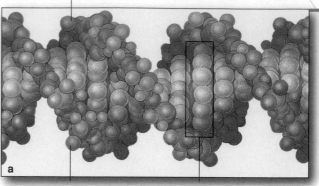

sugar-phosphate backbone of one strand of nucleotides in a DNA double helix

sugar-phosphate backbone of the other strand of nucleotides

part of the sequence of base pairs in DNA

Figure 13.4 The process of gene transcription, by which an RNA molecule is assembled on a DNA template. The sketch in (**a**) shows a gene region in part of a DNA double helix. In this region, the base sequence of one of the two nucleotide strands (not both) is about to be transcribed into an RNA molecule, in the manner shown in (**b**) through (**e**).

strand one at a time, in the 5′ → 3′ direction. (Here you might wish to refer to the simple explanation of strand assembly in Section 12.4, Figure 12.10.)

Transcription *differs* from DNA replication in three key respects. First, only a selected stretch of one DNA strand, not the whole molecule, serves as the template. Second, instead of DNA polymerase, a different enzyme, **RNA polymerase**, catalyzes the addition of nucleotides to the 3′ end of a growing strand of RNA. Third, transcription results in a single, free strand of RNA nucleotides, not in a double helix.

Transcription is initiated at a **promoter**, a base sequence in DNA that signals the start of a gene. Proteins position an RNA polymerase on DNA and thereby helps it bind to the promoter. The enzyme moves along the DNA strand, joining nucleotides one after another (Figure 13.4). When it reaches a particular base, the new RNA molecule is released as a free transcript.

Finishing Touches on mRNA Transcripts

In eukaryotic cells alone, a new mRNA molecule is unfinished. That "pre-mRNA" must be modified before its protein-building instructions can be put to use. Just as a dressmaker might snip off some threads or add bows on a dress before it leaves the shop, so do eukaryotic cells tailor their pre-mRNA.

For example, enzymes attach a cap to the 5′ end of a pre-mRNA molecule. The cap is a nucleotide

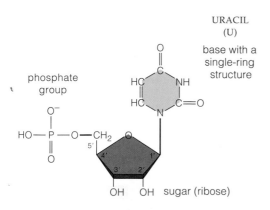

URACIL
(U)

base with a single-ring structure

phosphate group

OH   OH   sugar (ribose)

Figure 13.2 Structural formula for one of the four types of RNA nucleotides. The three others have a different base (adenine, guanine, or cytosine instead of the uracil shown here). Compare Section 12.2, which shows DNA's four nucleotides. Notice that the sugars of DNA and RNA differ at one group only (*yellow*).

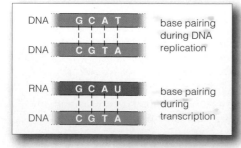

| DNA | G C A T | base pairing during DNA replication |
| DNA | C G T A | |

| RNA | G C A U | base pairing during transcription |
| DNA | C G T A | |

Figure 13.3 An example of base pairing of RNA with DNA during transcription, as compared to base pairing during DNA replication.

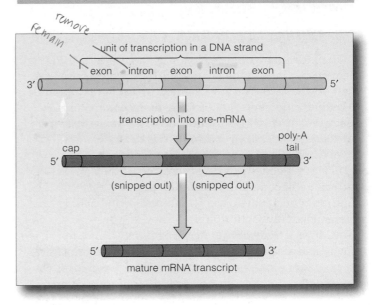

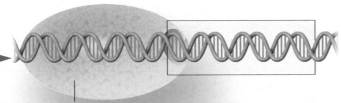

RNA Polymerase

b A molecule of RNA polymerase binds with a promoter region in the DNA. It will recognize the base sequence located downstream from that site as a template for linking together the nucleotides adenine, cytosine, guanine, and uracil into a strand of RNA.

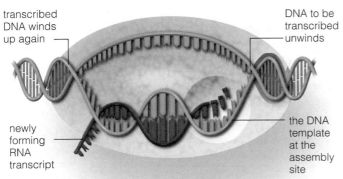

transcribed DNA winds up again

DNA to be transcribed unwinds

newly forming RNA transcript

the DNA template at the assembly site

c All through transcription, the DNA double helix is unwound in front of the RNA polymerase. Short lengths of the newly forming RNA strand briefly wind up with its DNA template strand. New stretches of RNA unwind from the template (and the two DNA strands wind up again).

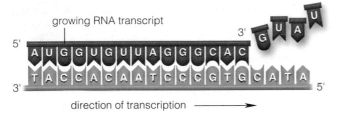

growing RNA transcript

direction of transcription ⟶

d What happened at the assembly site? RNA polymerase catalyzed the base-pairing of RNA nucleotides, one after another, with exposed bases on the DNA template strand.

e At the end of the gene region, the last stretch of the new mRNA transcript is unwound and released from the DNA.

that has a methyl group and phosphate groups bonded with it. Also, enzymes attach a tail of about 100 to 200 adenine-containing nucleotides to the 3′ end of most pre-mRNA transcripts. Hence the name, "poly-A tail" (for multiple adenine units). The tail gets wound up with proteins. Later, in the cytoplasm, the cap will assist the binding of mRNA to a ribosome. Also, enzymes will

Figure 13.5 Transcription and modification of new mRNA in the nucleus of eukaryotic cells. Its cap is a nucleotide with functional groups attached. Its tail is a string of adenine nucleotides.

gradually destroy the wound-up tail from the tip on back. Such tails "pace" the enzyme access to mRNA and thus dictate how long the mRNA lasts. Apparently they help keep protein-building messages intact for as long as the cell requires them.

Besides these alterations, the pre-mRNA itself gets modified. Most eukaryotic genes contain one or more **introns**, base sequences that must be removed before a pre-mRNA molecule can be translated. The introns intervene between **exons**, the parts that remain in the mRNA when it gets translated into protein. As Figure 13.5 shows, the introns are transcribed right along with the exons, but they are snipped out before the mRNA leaves the nucleus in mature form.

It could be that some introns are evolutionary junk, the leftovers of past mutations that led nowhere. Yet many introns are sites where instructions for building a particular protein can be snipped apart and spliced back together in various ways. The alternative splicing allows different cells in your body to use the same gene to make different versions of a pre-mRNA transcript, and therefore different versions of the resulting protein. We return to this topic in Section 14.4.

During gene transcription, a sequence of exposed bases in one of the two strands of a DNA molecule serves as the template for assembling a single strand of RNA. The assembly follows base-pairing rules (adenine only with uracil, cytosine only with guanine).

Before leaving the nucleus, each new mRNA transcript, or pre-mRNA, undergoes modification into final form.

www

What Is the Genetic Code?

Like a strand of DNA, an mRNA molecule is a linear sequence of nucleotides. What are the protein-building "words" encoded in that sequence? Gobind Khorana, Marshall Nirenberg, and other investigators came up with the answer. They deduced that ribosomes "read" nucleotide bases *three at a time*, as triplets. In an mRNA strand, such base triplets are called **codons**.

Figure 13.6 will give you an idea of how the order of different codons in an mRNA strand dictates the order in which particular amino acids will be added to a growing polypeptide chain.

Count the codons listed in Figure 13.7, and you see that there are sixty-four kinds. Notice how most of the twenty kinds of amino acids correspond to more than one codon. Glutamate corresponds to the code words GAA *and* GAG, for example. Also notice how AUG has dual functions. It codes for the amino acid methionine, and it also is an initiation codon, a START signal for translating an mRNA transcript at a ribosome. That is, "three-bases-at-a-time" selections start at a particular

a Base sequence of a gene region in DNA:

G C A C C A A T A A C C A T A

b Part of an mRNA strand, transcribed from that DNA:

C G U G G U U A U U A U

c What the amino acid sequence will be when the mRNA is translated into a polypeptide chain:

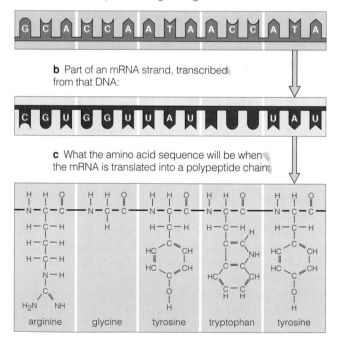

arginine   glycine   tyrosine   tryptophan   tyrosine

Figure 13.6 The steps from genes to proteins. (**a**) This diagram represents a region of a DNA double helix that was unwound during transcription. (**b**) The exposed bases on one DNA strand served as a template for assembling an mRNA strand. In the new mRNA transcript, every three nucleotide bases equaled one codon. Each codon calls for one amino acid in a polypeptide chain. (**c**) Referring to Figure 13.7, can you fill in the blank codon for tryptophan in the chain? ✿

FIRST BASE	SECOND BASE OF A CODON				THIRD BASE
	U	C	A	G	
U	phenylalanine	serine	tyrosine	cysteine	U
	phenylalanine	serine	tyrosine	cysteine	C
	leucine	serine	STOP	STOP	A
	leucine	serine	STOP	tryptophan	G
C	leucine	proline	histidine	arginine	U
	leucine	proline	histidine	arginine	C
	leucine	proline	glutamine	arginine	A
	leucine	proline	glutamine	arginine	G
A	isoleucine	threonine	asparagine	serine	U
	isoleucine	threonine	asparagine	serine	C
	isoleucine	threonine	lysine	arginine	A
	methionine (or START)	threonine	lysine	arginine	G
G	valine	alanine	aspartate	glycine	U
	valine	alanine	aspartate	glycine	C
	valine	alanine	glutamate	glycine	A
	valine	alanine	glutamate	glycine	G

Amino acids that correspond to base triplets:

Figure 13.7 The genetic code. The codons in mRNA are nucleotide bases, "read" in blocks of three. Sixty-one of the base triplets correspond to specific amino acids. Three others serve as signals that stop translation. The left column of the diagram shows the first of the three nucleotides in each codon in mRNA. The middle columns show the second nucleotide. The right column shows the third. Reading from left to right, for instance, the triplet U G G corresponds to tryptophan. Both U U U and U U C correspond to phenylalanine. ✿

AUG in the transcript's nucleotide sequence. Codons UAA, UAG, and UGA do not correspond to amino acids. They are the STOP signals that prevent the further addition of amino acids to a new polypeptide chain.

The set of sixty-four different codons is the **genetic code**. It is the basis of protein synthesis in all organisms.

Structure and Function of tRNA and rRNA

In a cell's cytoplasm are pools of free amino acids and free tRNA molecules. The tRNAs each have a molecular "hook," an attachment site for amino acids. They also have an **anticodon**, a nucleotide triplet that is able to base-pair with a codon (Figure 13.8). When tRNAs bind

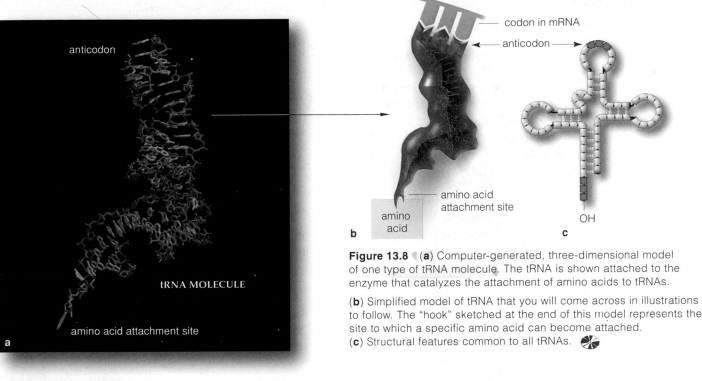

codon in mRNA

anticodon

amino acid attachment site

amino acid

OH

b

c

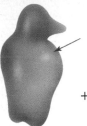

anticodon

tRNA MOLECULE

amino acid attachment site

a

Figure 13.8 (**a**) Computer-generated, three-dimensional model of one type of tRNA molecule. The tRNA is shown attached to the enzyme that catalyzes the attachment of amino acids to tRNAs.

(**b**) Simplified model of tRNA that you will come across in illustrations to follow. The "hook" sketched at the end of this model represents the site to which a specific amino acid can become attached. (**c**) Structural features common to all tRNAs.

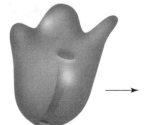

a A small ribosomal subunit. (The *red* arrow points to a platform for chain assembly on the surface)

+

b A large ribosomal subunit, with a tunnel through parts of its interior

c Side view of an intact ribosome, showing how the platform and tunnel are aligned.

Figure 13.9 Model of eukaryotic ribosomes. Polypeptide chains are assembled on the platform of the small ribosomal subunit. Newly forming chains may move through the tunnel of the large ribosomal subunit.

to the codons, they position their attached amino acids automatically, in the order specified by mRNA.

A cell has a cytoplasmic pool of sixty-four kinds of codons, but it is able to utilize fewer kinds of tRNAs. How do tRNAs match up with more than one type of codon? According to the base-pairing rules, adenine must pair with uracil, and cytosine with guanine. For codon-anticodon interactions, however, the rules loosen up for the first and third bases. To give one example, AUU, AUC, and AUA specify isoleucine. All three of these codons can pair with a single type of tRNA that carries isoleucine. Such freedom in codon-anticodon pairing at a base is known as the "wobble effect."

Even before anticodons interact with the codons of an mRNA strand, that strand must bind to specific sites on the surface of ribosomes. As shown in Figure 13.9, each ribosome has two subunits. These are assembled in the nucleus from rRNA and protein components, some of which show enzyme activity.

At some point the subunits are shipped separately to the cytoplasm. There, intact, functional ribosomes are put together, each from two subunits—but only when messages encoded in mRNA are to be translated.

The nucleotide sequence of both DNA and mRNA encodes protein-building instructions. The genetic code is a set of sixty-four base triplets (nucleotide bases, read in blocks of three). A codon is a base triplet in mRNA.

Different combinations of codons specify the amino acid sequence of different polypeptide chains, start to finish.

mRNAs are the only molecules that carry protein-building instructions from DNA to the cytoplasm.

tRNAs deliver amino acids to ribosomes and base-pair with codons in the order specified by mRNA. Their action translates mRNA into a sequence of amino acids.

rRNAs are components of ribosomes, the structures upon which amino acids are assembled into polypeptide chains.

WWW

HOW IS mRNA TRANSLATED?

Stages of Translation

The protein-building code built into mRNA transcripts of DNA becomes translated at intact ribosomes in the cytoplasm. Translation proceeds through three stages: initiation, elongation, and termination.

During the stage called *initiation*, a particular tRNA that can start transcription and an mRNA transcript are both loaded onto a ribosome. First, the initiator tRNA

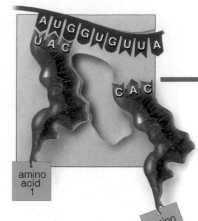

c As the final step of the initiation stage, a large ribosomal subunit joins with the small one. Once this initiation complex has formed, chain *elongation*—the second stage of translation—is about to get under way.

intact ribosome

b *Initiation*, the first stage of translating the mRNA transcript, is about to begin. An initiator tRNA (one that can start this stage) is loaded onto the platform of a small ribosomal subunit. The small subunit/tRNA complex attaches to the 5′ end of the mRNA. It moves along the mRNA and "scans" it for an AUG START codon.

INITIATION

a A mature mRNA transcript leaves the nucleus by passing through pores in the nuclear envelope. It enters the cytoplasm, which contains pools of many free amino acids, tRNAs, and ribosomal subunits.

mRNA transcript

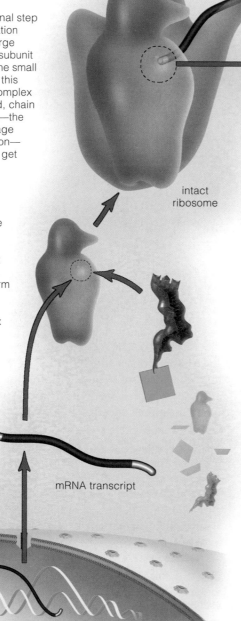

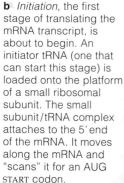

d This close-up of the small ribosomal subunit's platform shows the relative positions of binding sites for an mRNA transcript and for tRNAs that deliver amino acids to the intact ribosome.

ELONGATION

e The initiator tRNA has become positioned in the first tRNA binding site (designated *P*) on the ribosome platform. Its anticodon matches up with the START codon (AUG) of the mRNA, which also has become positioned in *its* binding site. Another tRNA is about to move into the platform's second tRNA binding site (designated *A*). It is one that can bind with the codon following the START codon.

amino acid 1

amino acid 2

binds with the small ribosomal subunit. AUG, the start codon for the transcript, matches up with this tRNA's anticodon. At the same time, the AUG binds with the small subunit. Second, a large ribosomal subunit binds with the small subunit. When joined together this way, the three form an initiation complex (Figure 13.10*b*). Now the next stage can begin.

In the *elongation* stage of translation, a polypeptide chain is assembled as the mRNA passes between two ribosomal subunits, like a thread being moved through the eye of a needle. Some components of the ribosomes are enzymes. They join individual amino acids together in the sequence dictated by the codon sequence in the mRNA molecule. Figure 13.10*f–i* shows how a peptide bond forms between the most recently attached amino acid of the growing polypeptide chain and the next amino acid delivered to the intact ribosome. (Here you might wish to refer to the description and sketch of peptide bond formation in Section 3.4.)

During the last stage of translation, *termination*, a STOP codon in the mRNA moves onto the platform, and no tRNA has a corresponding anticodon. Now proteins called release factors bind to the ribosome. They trigger enzyme activity that detaches the mRNA *and* the chain from the ribosome (Figure 13.10*j–l*).

Figure 13.10 Translation, the second step of protein synthesis.

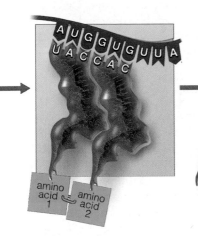

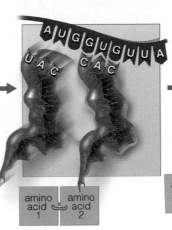

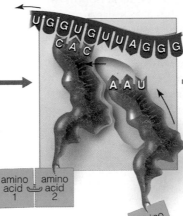

f Enzyme action breaks the bond between the initiator rRNA and the amino acid hooked to it. At the same time, enzyme action catalyzes the formation of a peptide bond between that amino acid and the one hooked to the second tRNA. Then the initiator tRNA is released from the ribosome.

g Now the first amino acid is attached only to the second one—which is still hooked to the second tRNA. This tRNA is about to move into the ribosomal platform's *P* site and slide the mRNA along with it by one codon. This will align the third codon in the *A* site.

h A third tRNA is about to move into the vacated *A* site. Its anticodon is able to base-pair with the third codon of the mRNA transcript. Next, through enzyme action, a peptide bond will form between amino acids 2 and 3.

i Steps (**f**) through (**g**) are repeated for as long as one codon after another becomes positioned above the *A* binding site on the ribosomal platform.

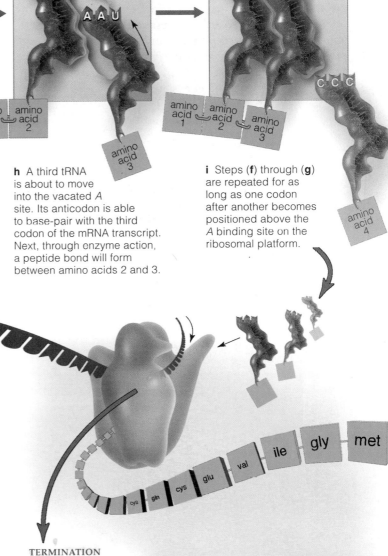

What Happens to the New Polypeptides?

Unfertilized eggs and other cells that will be called upon to rapidly synthesize many copies of different proteins typically stockpile mRNA transcripts in their cytoplasm. In cells that are rapidly using or secreting proteins, you often observe polysomes. Each polysome is a cluster of many ribosomes translating the same mRNA transcript at the same time. The transcript threads through all of them, one after another.

After the new polypeptide chains are synthesized, many join the cytoplasm's pool of free proteins. Many others start to enter the rough ER of the cytomembrane system, as described in Section 4.6. There they take on final form before they are shipped to their ultimate destinations inside or outside the cell.

Translation is initiated when a small ribosomal subunit and an initiator tRNA arrive at an mRNA's START codon and a large ribosomal subunit binds to them.

tRNAs deliver amino acids to the ribosome in the order dictated by the sequence of mRNA codons, to which the tRNA anticodons base-pair. A polypeptide chain grows as peptide bonds form between the amino acids.

Translation is over when a STOP codon triggers events that cause the chain and the mRNA to detach from the ribosome.

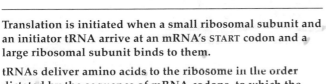

TERMINATION

j A STOP codon moves onto the ribosomal assembly platform. It is the signal to release the mRNA transcript from the ribosome.

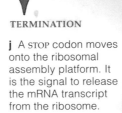

k The newly formed polypeptide chain also is released from the ribosome. It is free to join the pool of proteins in the cytoplasm or to enter rough ER of the cytomembrane system.

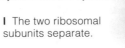

l The two ribosomal subunits separate.

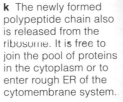

DO MUTATIONS AFFECT PROTEIN SYNTHESIS?

Whenever a cell puts its genetic code into action, it is making precisely those proteins that it requires for its structure and functions. If something changes a gene's code words, the resulting protein may also change. If the protein is central to cell architecture or metabolism, we can expect the outcome to be an abnormal cell.

Gene sequences do change. Sometimes one base gets substituted for another in the nucleotide sequence. At other times, an extra base is inserted into the sequence or a base is lost from it. Such small-scale changes in the nucleotide sequence of genes in a DNA molecule are **gene mutations**. There is some leeway here; remember, more than one codon may specify the same amino acid.

For example, if a mutation were to change UCU to UCC, it probably would not have dire effects, for both codons specify serine. More often, however, mutations give rise to proteins with altered or lost functions.

Common Gene Mutations and Their Sources

Figure 13.11*a* shows a common gene mutation. In this example, one base (adenine) is wrongly paired with another (cytosine) during DNA replication. Proofreading enzymes can recognize an error in a newly replicated strand and fix it. If they don't, a mutation will become established in one DNA molecule in the next round of replication. As a result of this mutation, a **base-pair substitution**, one amino acid can replace another during protein synthesis.

That happened in people with *sickle-cell anemia*; they carry a mutated gene for HbS hemoglobin. Hemoglobin, an oxygen carrier inside red blood cells, has four polypeptide chains. Two are alpha chains and two are beta chains (Section 3.5). An abnormal beta chain arises from a mutation that affected protein synthesis. Instead of glutamate (the sixth amino acid in the chain), valine was added. Glutamate carries an overall negative charge; however, valine has no net charge. Because of the difference, HbS hemoglobin has an abnormal hydrophobic patch that is "sticky." In the smallest blood vessels of the circulatory system, oxygen levels are at their lowest. This condition causes hemoglobin molecules to interact at their sticky patches and aggregate as rods, which distort the red blood cells housing them. The cells clump together and disrupt blood circulation. The adverse consequences for organs throughout the body are described in Section 10.5.

Figure 13.12 shows a different mutation. Here, an *extra* base was inserted into a gene

Figure 13.11
(**a**) Example of a base-pair substitution. (**b**) The single amino acid substitution that gives rise to the abnormal beta chain of the HbS molecule, the culprit behind sickle-cell anemia.

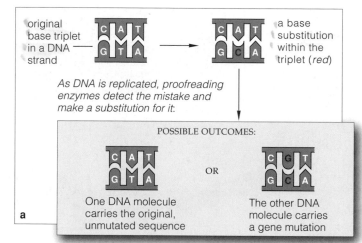

original base triplet in a DNA strand — a base substitution within the triplet (*red*)

As DNA is replicated, proofreading enzymes detect the mistake and make a substitution for it:

POSSIBLE OUTCOMES:

One DNA molecule carries the original, unmutated sequence — OR — The other DNA molecule carries a gene mutation

a

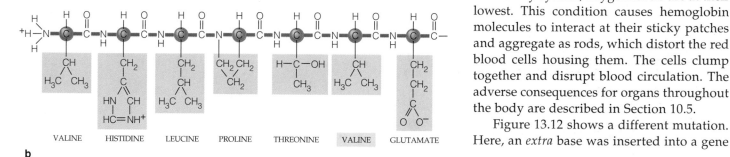

VALINE HISTIDINE LEUCINE PROLINE THREONINE VALINE GLUTAMATE

b

Figure 13.12 Example of an insertion, a mutation in which an extra base gets inserted into a gene region of DNA. This insertion has caused a *frameshift*; it has changed the reading frame for base triplets in the DNA and in the mRNA transcript of that region. As a result, the wrong amino acids will be called up when the mRNA transcript becomes translated into protein.

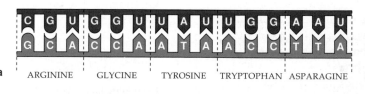

mRNA transcribed from the DNA

PART OF PARENTAL DNA TEMPLATE

a ARGININE GLYCINE TYROSINE TRYPTOPHAN ASPARAGINE

resulting amino acid sequence

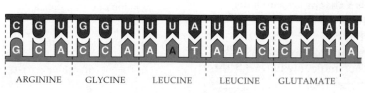

altered message in mRNA

A BASE INSERTION (RED) IN DNA

b ARGININE GLYCINE LEUCINE LEUCINE GLUTAMATE

the altered amino acid sequence

Figure 13.13 Barbara McClintock, who won a Nobel Prize for her insight that some DNA segments can move from one site to another in DNA molecules, as transposable elements. In her hands is an ear of Indian corn (*Zea mays*). Its kernels sent her on the road to discovery.

Each kernel is a seed, a potential new corn plant. All of its cells have the same pigment-coding genes. But some of the kernels are colorless or spottily colored. In the ancestor of the plant from which this ear of corn was plucked, a gene in a germ cell left its position in a DNA molecule, invaded another DNA molecule, and shut down a pigment-encoding gene. The plant inherited the mutation. As cell divisions proceeded in the growing plant, none of the mutated cell's descendants was able to synthesize pigment molecules. Each gave rise to colorless kernel tissue. Later, in some cells, the movable DNA slipped out of the pigment-encoding gene. And all the descendants of *those* cells produced pigment—and colored kernel tissue.

region. Polymerases, recall, read nucleotide sequences in blocks of three. In this example, the insertion shifts the three-at-a-time reading frame by one base; hence the name, "frameshift mutation." The gene now has a different message, and an altered version of the protein will be synthesized. Frameshift mutations fall within broader categories of gene mutation called **insertions** and **deletions**. In such cases, one to several base pairs are inserted into a DNA molecule or deleted from it.

As a final example, Barbara McClintock discovered that mutations can result when **transposable elements** are on the move. These are DNA segments that move spontaneously from one location to another in the same DNA molecule or to a different one. Often they may inactivate the genes into which they become inserted. As Figure 13.13 indicates, the unpredictability of such jumps can cause interesting variations in phenotype.

Causes of Gene Mutations

Many gene mutations arise spontaneously while DNA is being replicated. This should not come as a surprise, given the rapid pace of replication and the huge pools of free nucleotides concentrated around the growing strands. Proofreading and repair enzymes detect and fix most of them but, like most people, they are good but not perfect. A low number of mistakes do slip past the enzymes with predictable frequency.

Not all mutations are spontaneous. Many result after exposure to mutagens (mutation-causing agents in the environment). Ultraviolet radiation, especially the 260-nanometer wavelength in sunlight, is mutagenic. This is the wavelength DNA absorbs most strongly, and it may induce crosslinks to form between two pyrimidine neighbors on the same DNA strand. Skin cancers are one outcome. Other mutagens are gamma rays and x-rays. They can ionize water and other molecules around the DNA, so free radicals form. Free radicals are molecular

fragments having an unpaired electron—and they can attack DNA's structure. Such **ionizing radiation** causes base substitutions or breaks in one or both strands.

Natural and synthetic chemicals in the environment can accelerate the rate of spontaneous mutations. For example, substances called **alkylating agents** transfer methyl or ethyl groups to reactive sites on the bases or phosphate groups of DNA. At an alkylated site, DNA becomes more susceptible to base-pair disruptions that invite mutation. Many **carcinogens**, which are cancer-causing agents, operate by alkylating the DNA.

The Proof Is in the Protein

Spontaneous mutations are rare in terms of a human life. (The rate for eukaryotes ranges between 10^{-4} and 10^{-6} per gene per generation.) If one arises in a somatic cell, any good or bad consequences will not endure, for it cannot be passed on to offspring. If the mutation arises in a germ cell or gamete, however, it may enter the evolutionary arena. The same can happen with a mutation in an asexually reproducing organism or cell.

In all such cases, nature's test is this: *A protein that is specified by a heritable mutation may have harmful, neutral, or beneficial effects on an individual's ability to function in the prevailing environment.* As you will read in the next unit of the book, the outcomes of gene mutations can have powerful evolutionary consequences.

A gene mutation is an alteration in one to several bases in the nucleotide sequence of DNA.

Each gene has a spontaneous and characteristic mutation rate, which may be accelerated by exposure to harmful radiation and certain chemicals in the environment.

A protein specified by a mutated gene may have harmful, neutral, or beneficial effects on the ability of an individual to function in the prevailing environment.

WWW

SUMMARY

1. Cells, and multicelled organisms, cannot stay alive without enzymes and other proteins. A protein consists of one or more polypeptide chains, each of which is composed of a linear sequence of amino acids.

   a. The amino acid sequence of a polypeptide chain corresponds to a gene region in a double-stranded DNA molecule. Each gene is a sequence of nucleotide bases in one of the two strands. The bases are adenine, thymine, guanine, and cytosine (A, T, G, and C).

   a. The path from genes to proteins has two steps, called transcription and translation, as in Figure 13.14:

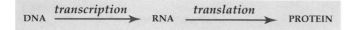

2. In transcription, double-stranded DNA is unwound at a gene region. Enzymes use the exposed bases as a template to assemble a strand of ribonucleic acid (RNA)

from the cell's pool of free nucleotides. Base-pairing rules govern its assembly. In RNA, guanine pairs with cytosine, and uracil (not thymine) pairs with adenine:

   a. Different gene regions in DNA serve as templates for assembling different RNA molecules.

   b. Messenger RNA (mRNA) is the only class of RNA that has protein-building instructions.

   c. Together with proteins, ribosomal RNA (rRNA) becomes a component of ribosomes, the physical units upon which polypeptide chains will be assembled.

   d. Transfer RNA (tRNA) is the vehicle of translation; many kinds latch onto free amino acids in cytoplasm and deliver them to ribosomes in a sequence dictated by the sequential message of mRNA.

   e. RNA transcripts of eukaryotic cells are processed into final form before being shipped from the nucleus.

3. In translation, the three classes of RNAs interact in the synthesis of polypeptide chains, which later twist, fold, and also may be modified into the final, three-dimensional shape of the protein.

   a. tRNAs and ribosomes interact with mRNA to link amino acids in the sequence that is required to produce a specific kind of polypeptide chain.

   b. Translation follows a genetic code, a set of sixty-four base triplets. Triplets are a series of nucleotide bases. Ribosomal proteins "read" the series of bases three at a time (hence the name, triplet).

   c. Each base triplet in an mRNA molecule is a codon. An anticodon is a complementary triplet in a tRNA molecule. Some particular combination of codons specifies what the amino acid sequence of a polypeptide chain will be, start to finish.

4. Translation proceeds through three stages:

   a. Initiation. One small ribosomal subunit and an initiator tRNA bind with mRNA and move along it until they encounter an AUG start codon. The small subunit binds with a large ribosomal subunit.

   b. Chain elongation. tRNAs deliver amino acids to an intact ribosome. Their anticodons base-pair with mRNA codons. The amino acids become linked by peptide bonds to form a new polypeptide chain.

   c. Chain termination. After an mRNA stop codon moves onto the ribosomal platform, the polypeptide chain and the mRNA detach from the ribosome.

5. Gene mutations are heritable, small-scale changes in DNA's base sequence. Many arise spontaneously during DNA replication. Other mutations arise after DNA is exposed to ultraviolet radiation, ionizing radiation, alkylating agents, or some other kind of mutagen in the environment.

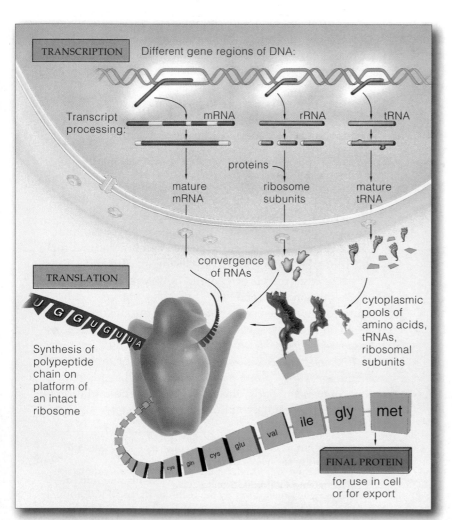

Figure 13.14 Summary of flow of genetic information from DNA to proteins, in eukaryotic cells. The DNA is transcribed into RNA in the nucleus; RNA is translated in the cytoplasm. Prokaryotic cells do not have a nucleus; transcription and translation proceed in their cytoplasm.

Review Questions

1. Are the polypeptide chains of proteins assembled on DNA? If so, state how. If not, state how they are assembled, and on which molecules. *CI, 13.1*

2. Name the three classes of RNA and state the function of each class in protein synthesis. *13.1, 13.2*

3. The pre-mRNA transcripts of eukaryotic cells contain both introns and exons. Are the introns or exons snipped out before the transcript leaves the nucleus? *13.1*

4. Distinguish between codon and anticodon. *13.2*

5. Name the three stages of translation, and briefly describe the key events of each stage. *13.3*

6. Define gene mutation. Give three examples of agents that cause mutations. *13.4*

7. Do all mutations arise spontaneously? Are environmental agents always the trigger for change? *13.4*

8. Define and then state the possible outcomes of the following types of mutation: a base-pair substitution, a base insertion, and finally an insertion of a transposable element at a new location in the DNA. *13.4*

Self-Quiz *(Answers in Appendix III)*

1. DNA contains many different genes that are transcribed into different _____ .
   a. proteins
   b. mRNAs only
   c. mRNAs, tRNAs, and rRNAs
   d. all are correct

2. An RNA molecule is _____ .
   a. a double helix
   b. usually single-stranded
   c. always double-stranded
   d. usually double-stranded

3. An mRNA molecule is produced by _____ .
   a. replication
   b. duplication
   c. transcription
   d. translation

4. Each codon calls for a specific _____ .
   a. protein
   b. polypeptide
   c. amino acid
   d. carbohydrate

5. Referring to Figure 13.7, use the genetic code to translate the mRNA sequence UAUCGCACCUCAGGAGACUAG. Notice that the first codon in the frame is UAU. Which amino acid sequence is being specified?

   TYR — ARG — THR — SER — GLY — ASP
   or
   TYR — ARG — THR — SER — GLY
   or
   TYR — ARG — TYR — SER — GLY — ASP

6. Anticodons pair with _____ .
   a. mRNA codons
   b. DNA codons
   c. tRNA anticodons
   d. amino acids

7. Match the terms with the suitable description.
   _____ alkylating agent
   _____ chain elongation
   _____ exons
   _____ genetic code
   _____ anticodon
   _____ intron
   _____ codon

   a. part remaining in mRNA transcript
   b. base triplet coding for amino acid
   c. second stage of translation
   d. Base triplet that pairs with codon
   e. One environmental agent that induces mutation in DNA
   f. set of sixty-four codons for mRNA
   g. part removed from a pre-mRNA transcript

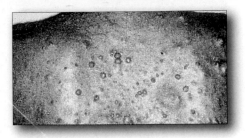

Figure 13.15 Soft skin tumors on a person with neurofibromatosis, an autosomal dominant disorder.

Critical Thinking

1. Sandra discovered a tRNA with a mutation in DNA that encodes the anticodon 3'-AAU instead of 3'-AUU. In cells with the mutated tRNA, what will be the effect on protein synthesis?

2. A DNA polymerase made an error during the replication of an important gene region of DNA. None of the DNA repair enzymes detected or repaired the damage. A portion of the DNA strand with the error is shown here:

   ..AATTCCGACTCCTATGG
   ..TTAAGGTTGAGGATACC

After the DNA molecule is replicated and two daughter cells have formed, one cell is carrying a mutation and the other cell is normal. Develop a hypothesis to explain this observation.

3. *Neurofibromatosis* is a human autosomal dominant disorder caused by mutations in the *NF1* gene. It is characterized by soft, fibrous tumors in the peripheral nervous system and skin as well as abnormalities in muscles, bones, and internal organs (Figure 13.15). Because the gene is dominant, an affected child usually has an affected parent. Yet in 1991, scientists reported on a boy who had neurofibromatosis yet whose parents did not. When they examined both copies of his *NF1* gene, they found the copy he had inherited from his father contained a transposable element. Neither the father nor the mother had a transposable element in any of the copies of their own *NF1* genes. Explain the cause of neurofibromatosis in the body and how it arose.

Selected Key Terms

alkylating agent *13.4*	ionizing radiation *13.4*
anticodon *13.2*	mRNA (messenger RNA) *13.1*
base sequence *CI*	mutation rate *13.5*
base-pair substitution *13.4*	polysome *13.4*
carcinogen *13.4*	promoter (RNA) *13.1*
codon *13.2*	ribonucleic acid (RNA) *CI*
deletion (of base) *13.4*	RNA polymerase *13.1*
exon *13.1*	rRNA (ribosomal RNA) *13.1*
gel electrophoresis *13.1*	transcription *CI*
gene mutation *13.4*	translation *CI*
genetic code *13.2*	transposable element *13.4*
insertion (of base) *13.4*	tRNA (transfer RNA) *13.1*
intron *13.1*	uracil *13.1*

Readings *See also www.infotrac-college.com*

Crick, F. 1988. *What Mad Pursuit: A Personal View of Scientific Discovery.* New York: HarperCollins. Crick's autobiography.

Fairbanks, F., and W. R. Anderson. 1999. *Genetics: The Continuity of Life.* Monterey, California: Brooks-Cole.

WWW *http://www.brookscole.com/biology*

Practice quiz questions, hypercontents, BioUpdates, and critical thinking. The Brooks/Cole Biology Resource Center provides a wealth of information fully organized and integrated by chapter.

When DNA Can't Be Fixed

1992 was an unforgettable year for Laurie Campbell. She finally turned eighteen. In that same year she just happened to notice a suspiciously black mole on her skin. It had an odd lumpiness about it, a ragged border, and an encrusted surface. No fool, Laurie quickly made an appointment with her family doctor, who quickly ordered a biopsy. The mole turned out to be a *malignant melanoma*, the deadliest form of skin cancer.

Laurie was lucky. She detected a cancer in its earliest stage, before it could spread through her body. Now she regularly checks out the appearance of other moles on her skin. She is aware of having become a statistic— one of 500,000 people in the United States who develop skin cancer in any given year and one of the 23,000 with malignant melanoma. She knows now that, of 7,500 who will die each year from skin cancer, 5,600 of them have malignant melanoma.

Laurie is smart. She plotted out the position of every mole on her body. Once a month, that body map is her guide for a quick but thorough self-examination. Figure 14.1 shows examples of what she looks for. Laurie also schedules a medical examination every six months.

Changes in DNA are triggers for skin cancer. The ultraviolet wavelengths in rays from the sun, tanning lamps, and other environmental sources can cause the bad molecular changes. For example, they can promote covalent bonding between two adjacent thymine bases in a nucleotide strand. The two nucleotides to which

the bases belong become an abnormal, bulky structure, a thymine dimer, within the DNA. Normally, at least seven gene products interact as a DNA repair mechanism to remove such bulky lesions. But a mutation in one or more of the genes can disrupt the repair machinery. Thymine dimers can accumulate in skin cells and trigger cancers and other lesions.

thymine dimer

The risk of skin cancer is greater for some than others. At one extreme is *xeroderma pigmentosum*. People affected by this genetic disorder cannot be exposed to sunlight, even briefly, without risking skin tumors and early death from cancer.

You are at risk if you have moles that are chronically irritated, as by shaving or abrasive clothing. You are at risk if your skin, including the lip surface, is chronically chapped, cracked, or sore. You are at risk if your family has a history of cancer or if you have had radiation

Figure 14.1 Examples of what can happen after repair enzymes have not been able to fix changes in the nucleotide sequence of DNA. (**a**) *Basal cell carcinoma*, the most common skin cancer. This slow-growing, raised lump may be uncolored, reddish-brown, or black. (**b**) *Squamous cell carcinoma*, the second most common skin cancer. The pink growths, firm to the touch, grow rapidly under the surface of skin exposed to the sun. (**c**) *Malignant melanoma*, which spreads most rapidly. The malignant cells form very dark, encrusted lumps. They may itch like an insect bite or bleed easily. (**d**) Laurie Campbell avoiding the sun—and melanoma.

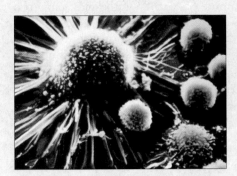

Figure 14.2 A patrol of white blood cells in an encounter with a body cell that has undergone cancerous transformation.

therapy. And, like Laurie, you are at risk if you burn easily. Damaged DNA is the reality for all of us, in spite of the ill-advised, socially promoted allure of tanning and staying out unprotected under the sun.

By definition, all **cancers** are malignant forms of tumors, which are tissue masses that arise through mutations in the genes that govern cell growth and division. Tumor cells don't respond to normal controls; they go on dividing as long as conditions for growth remain favorable. Cells of common skin moles and other *benign* tumors grow slowly, in an unprogrammed way. They still have surface recognition proteins that hold them together in their home tissue. Most benign tumors are left alone unless they become overly large or irritating to the patient.

In a *malignant* tumor, the abnormal cells grow and divide more rapidly, with destructive physical and metabolic effects on surrounding tissues. The cells are grossly disfigured (Figure 14.2). They cannot construct a normal cytoskeleton or plasma membrane, and they cannot synthesize normal recognition proteins. Also, malignant cells can break loose from their home tissue, enter lymph or blood vessels, travel along, then slip out and invade other parts of the body where they do not belong. And there they may start growing as new tumors. **Metastasis** (meh-TAH-stu-SIS) is the name for the process of abnormal cell migration and tissue invasion.

This chapter is your invitation to learn about the controls that govern when and how fast genes will be transcribed and translated, and whether gene products will be activated or shut down. By starting out with cancerous transformations, it invites you to reflect on how lucky we are when proper gene controls are in place and cells are operating as they should. Each year in the developed countries alone, 15 to 20 percent of all deaths result from cancer. It is not just a human problem. Cancer has also been observed in most of the animals that have been studied.

KEY CONCEPTS

1. In cells, a variety of controls govern when, how, and to what extent genes are expressed. The control elements operate in response to changing chemical conditions and to reception of external signals.

2. Control is exerted by way of regulatory proteins and other molecules that operate before, during, or after gene transcription. Different controls interact with DNA, with RNA that has been transcribed from the DNA, or with gene products—that is, with the resulting polypeptide chains or final proteins.

3. Prokaryotic cells depend on rapid control over short-term shifts in nutrient availability and other aspects of their surrounding environment. Commonly, they rely on regulatory proteins that have roles in the rapidly adjusting rates of transcription.

4. All eukaryotic cells depend on controls over short-term shifts in diet and levels of activity. In multicelled species, they also depend on controls over an intricate, long-term program of growth and development.

5. Controls over eukaryotic cells come into play when new cells contact one another in developing tissues. They also come into play when those cells start interacting with their neighbors by way of hormones and other signaling molecules.

6. Although all cells of a multicelled organism inherit the same genes, different cell types activate or suppress many of those genes in different ways. The controlled, selective use of genes leads to the synthesis of the proteins that give each type of cell its distinctive structure, function, and products.

At this very moment, bacteria are feeding on nutrients in your gut. Red blood cells are binding, transporting, or giving up oxygen, and great numbers of epithelial cells in your skin are busily synthesizing the protein keratin. Like cells everywhere, they are functioning by virtue of the protein products of genes.

Cells are selective about which gene products they make or require. Different kinds express certain genes just once, only at certain times, all the time, or not at all. *Which genes are being expressed depends on the type of cell, its moment-by-moment adjustments to changing chemical conditions, which external signals it happens to receive—and its built-in control systems.*

For example, availability of nutrients shifts rapidly and often for enteric bacteria (those living in intestines). Like other prokaryotes, they rapidly transcribe genes for nutrient-digesting enzymes and make many copies of those enzymes when nutrients are moving through the intestines. They cut back on the synthesis of those enzymes when nutrients are scarce. By comparison, your cells do not encounter drastic shifts in the solute concentrations and composition of the fluid that bathes them, and few exhibit rapid shifts in transcription.

The "systems" that control the expression of genes consist of molecules. For instance, **regulatory proteins** influence transcription, translation, and gene products. As you will see, such molecules exert their effects by interacting with DNA, RNA, new polypeptide chains, or final proteins (such as enzymes). Some components are activated or inhibited by signaling molecules, such as hormones. Others operate in response to changing concentrations of substances outside or inside the cell.

Negative control systems block a particular activity in the cell, and **positive control systems** promote it. For instance, one regulatory protein inhibits transcription of a gene when it binds with DNA, but the action of a different regulatory protein enhances its transcription. Usually, regulatory proteins do not act alone. To give an example, some types bind to DNA when required to do so, then release their grip on the binding site when a different control element interacts with them.

Summing up, these are the main concepts to keep in mind as you read through the rest of this chapter:

Cells exert selective control over when, how, and to what extent each of their genes is expressed.

The expression of a given gene depends on the type of cell and its functions, on chemical conditions, and on signals from the outside environment.

Many regulatory proteins and other molecules exert control over gene expression through their interactions with DNA or RNA. Others exert control through their interaction with gene products, either polypeptide chains or final proteins.

Let's first consider some examples of gene control in prokaryotic cells—that is, bacteria. When nutrients are plentiful and other environmental conditions also favor growth, bacteria tend to grow and divide indefinitely. Gene controls promote the rapid synthesis of enzymes that have roles in nutrient digestion and other growth-related activities. Transcription is rapid, and translation is initiated even before mRNA transcripts are finished. Remember, bacteria have no nucleus; nothing separates their DNA from ribosomes in the cytoplasm.

When a nutrient-degrading pathway utilizes several enzymes, all genes for those enzymes are transcribed, often as a continuous mRNA molecule. The genes are not transcribed when conditions turn unfavorable. The rest of this section gives two cases of how controls can adjust transcription rates downward or upward.

Negative Control of Transcription

The enteric bacterium *Escherichia coli* inhabits the gut of all mammals. It survives on glucose, lactose (a sugar in milk), and other ingested nutrients. Like other adult mammals, you probably do not drink milk around the clock. When you do, however, *E. coli* cells in your gut rapidly transcribe three genes for enzymes that have roles in breakdown reactions that begin with lactose.

A promoter precedes the three genes, which are next to one another. A **promoter**, recall, is a base sequence that signals the start of a gene. A different sequence, an **operator**, intervenes between a promoter and bacterial genes. It is a binding site for a **repressor**, a regulatory protein that can block transcription (Figure 14.3). An arrangement in which a promoter and operator service more than one gene is an **operon**. Elsewhere in *E. coli* DNA, a different gene codes for this repressor, which can bind with the operator *or* with a lactose molecule.

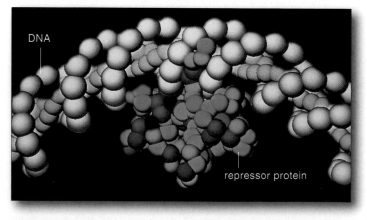

Figure 14.3 Computer model of one type of repressor protein binding to an operator at a site in a bacterial DNA molecule.

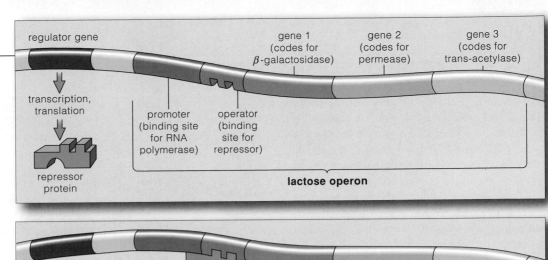

a A repressor protein exerts negative control over three genes of the lactose operon by binding to the operator and inhibiting transcription.

regulator gene

transcription, translation

repressor protein

gene 1 (codes for β-galactosidase) gene 2 (codes for permease) gene 3 (codes for trans-acetylase)

promoter (binding site for RNA polymerase) operator (binding site for repressor)

lactose operon

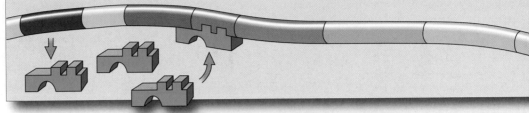

b When the concentration of lactose is low, the repressor is free to block transcription. Being bulky, it overlaps the promoter and prevents binding by RNA polymerase. The enzymes (not needed) are not produced.

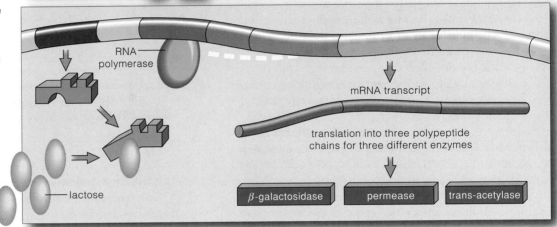

c At high concentration, lactose is an inducer of transcription. It binds to and distorts the shape of the repressor—which now cannot bind to the operator. The promoter is exposed and the genes can be transcribed.

RNA polymerase

mRNA transcript

translation into three polypeptide chains for three different enzymes

β-galactosidase permease trans-acetylase

lactose

Figure 14.4 Negative control of the lactose operon. The first gene of this operon codes for an enzyme that splits lactose, a disaccharide, into two subunits (glucose and galactose). The second codes for an enzyme that transports lactose into cells. The third enzyme functions in metabolizing certain sugars.

When the lactose concentration is low, a repressor binds with the operator, as in Figure 14.4. Being a large molecule, it overlaps the promoter, so transcription is blocked. Thus, *lactose-degrading enzymes are not built if they are not required.* When many lactose molecules are around, chances are greater that one of them will bind with the repressor. Binding alters the repressor's shape, so it cannot bind with the operator, so RNA polymerase can now transcribe the genes. Thus, *lactose-degrading enzymes are synthesized only when required.*

Positive Control of Transcription

E. coli cells pay far more attention to glucose than to lactose. They transcribe genes for glucose breakdown continually, at faster rates. Even if lactose is present, the lactose operon isn't used much—*unless glucose is absent.*

At such times, CAP, a regulatory protein, acts on the operon. CAP is one of the **activator proteins**, which are part of positive control systems. To understand how it works, you have to know the lactose operon's promoter is not good at binding RNA polymerase. It does better when CAP adheres to it first. But CAP won't do this until it is activated by a small molecule called cAMP.

Among other things, cAMP forms from ATP, which *E. coli* can make by glucose breakdown (by glycolysis). Not much cAMP is available in the cell when glucose is plentiful and glycolysis is running full bore. When such conditions prevail, CAP does not get primed to adhere to the promoter, so transcription of the lactose operon genes slows almost to a standstill.

Suppose glucose is scarce but lactose is available. cAMP accumulates, and CAP-cAMP complexes form. Now the lactose operon genes are transcribed rapidly, and the cell uses lactose as an alternative energy source.

Prokaryotic cells, which must respond rapidly to changing conditions, commonly rely on a small number of regulatory proteins that exert rapid, on-off control of transcription.

CONTROLS IN EUKARYOTIC CELLS

Like bacteria, eukaryotic cells control short-term shifts in diet and level of activity. If the cells are one of hundreds or trillions of cells in a multicelled organism, long-term controls also come into play. Gene activity changes as the new organism develops, as new cells contact one another in growing tissues and start interacting with particular neighbors by way of hormones and other signaling molecules.

Cell Differentiation and Selective Gene Expression

To get a sense of how intricate the controls become in eukaryotes, start with the idea that all cells in your body inherited the same genes (because they descended from the same fertilized egg). Many of those genes specify proteins that are absolutely vital for any cell's structure and everyday functioning. That is why the genes coding for those proteins are under controls that promote ongoing, low levels of transcription.

Even so, *nearly all of your cells became specialized in composition, structure, and function.* This process of **cell differentiation** proceeds during the development of all multicelled species. It arises as embryonic cells and cell lineages descended from them activate and suppress a small fraction of their genes in unique, selective ways. As examples, only immature red blood cells use the genes for making hemoglobin. Only certain white blood cells use the genes for making weapons called antibodies.

By some estimates, cells of a multicelled organism rarely use more than 5 to 10 percent of their genes at a given time. One way or another, controls repress most of the genes. Also, which genes are being expressed varies over time, depending on the stage of growth and development that the organism happens to be passing through. The following examples hint at the kinds of control mechanisms that work at different stages.

X Chromosome Inactivation

As you know, proteins impart structural organization to eukaryotic DNA. In female mammals, the structure of one of the two X chromosomes in each body cell is chemically modified in a way that completely closes off access to its genes.

A mammalian embryo destined to become female inherits one X chromosome from the mother and another from the father. One of the two condenses in each cell and is inactivated. Condensation is part of the program of development, but the outcome is random. *One or the other chromosome may be inactivated.* We observe such a condensed X chromosome in the interphase nucleus; it

Figure 14.5 (**a**) Inactivated X chromosome (*white arrow*) in a cell from a female mammal. (**b**) Why is this female calico cat "calico"? In her cells, one X chromosome carries a dominant allele for melanin, a dark pigment. The partner allele on her other X chromosome specifies yellow fur. When the cat was an early embryo, one of her two X chromosomes was inactivated at random in each cell that had formed by then. In all descendants of those cells, the same chromosome also became inactivated, leaving only one functional allele for the coat-color trait. We see patches of different colors, depending on which allele was inactivated in cells that formed a given tissue region. (The white patches result from a gene interaction involving the "spotting gene," which blocks melanin synthesis entirely.)

shows up as a dense spot (Figure 14.5*a*). The condensed X chromosome was named a **Barr body** (after Murray Barr, its discoverer).

Whenever a maternal (or paternal) X chromosome is inactivated in a cell, it also becomes inactivated in all descendants of that cell. Thus, when development ends, the female is a "mosaic" for those X chromosomes. *She bears patches of tissue where the genes of the maternal X chromosome are expressed—and patches of tissues in which the genes of the paternal X chromosome are expressed.* Any pair of alleles on those two chromosomes may or may not be identical. That is why tissues such as skin may have different features in different body regions.

Mary Lyon discovered the mosaic tissue effect that arises from random X chromosome inactivation. We see this effect in human females who are heterozygous for a recessive allele on the X chromosome that prevents sweat glands from forming. The lack of sweat glands is a symptom of *anhidrotic ectodermal dysplasia*. In affected females, the X chromosome with the dominant allele condensed. Genes on the chromosome with the mutant allele are being transcribed in patches of skin that lack sweat glands. Patches in which the dominant allele is active have sweat glands. The same effect is apparent in female calico cats, which are heterozygous for black and yellow coat-color alleles on their X chromosomes. The coat color in a given body region depends on which X chromosome's genes are transcribed (Figure 14.5*b*).

Examples of Signaling Mechanisms

Hormones, a major category of signaling molecules, can stimulate or inhibit gene activity in target cells. Any cell with receptors for a given hormone is a target. Animal cells secrete hormones into tissue fluid. Most hormone molecules are picked up by the bloodstream, which distributes them to cells some distance away. In plants, hormones do not travel far from cells that secrete them.

Certain hormones bind to membrane receptors at the surface of target cells. Others enter the cells, bind to regulatory proteins, and thus help initiate transcription. Often a hormone must activate an enhancer to exert an effect on cells. **Enhancers** are base sequences in DNA that act as binding sites for specific activator proteins. Some are located next to a gene's promoter; others are brought next to it when the DNA molecule loops back on itself in organized ways. RNA polymerase binds avidly to such promoter-enhancer complexes.

In vertebrates, certain hormones have widespread effects on gene expression because many types of cells have receptors for them. For instance, somatotropin (or growth hormone) is a signal from the pituitary gland. It stimulates the synthesis of proteins required for cell division and, ultimately, for the body's growth. Most cells have receptors for somatotropin.

Other vertebrate hormones signal only certain cells at certain times. Prolactin, another pituitary hormone, is like this. A few days after a female mammal gives birth, prolactin acts on mammary gland cells, which have receptors for it. It triggers activity of genes with roles in milk production. Liver and heart cells have the same genes, but they don't have prolactin receptors.

Explaining hormonal control of gene activity is like explaining a full symphony orchestra to someone who has never seen one or heard it perform. Many separate parts must be defined before their interactions can be understood! We will return to this topic, starting with Chapter 31 on the endocrine system. Studies of animal reproduction and development provide us with some of the most elegant examples of hormonal controls, as you will see in Chapter 38.

Other organisms also deploy signaling mechanisms. Plant a few seeds from a corn or bean plant in a pot that has moist, nutrient-rich soil. Let the seeds germinate, but keep them in total darkness. Eight days later, they have developed into pale, spindly seedlings; they have no chlorophyll (Figure 14.6). Now expose the seedlings to a single burst of dim light from a flashlight. Within ten minutes, they start converting stockpiles of certain molecules to the activated form of chlorophylls—the light-trapping protein molecules of photosynthesis.

Phytochrome, a blue-green pigment, is a signaling molecule that helps plants adapt over the short term

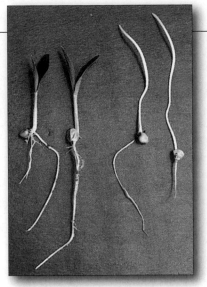

control group   experimental group

Figure 14.6 Sunlight as a signal for gene activity. The photograph shows the effect of the absence of light on corn seedlings. The two seedlings at left were the control group; they were grown in sunlight, inside a greenhouse. The two seedlings next to them were grown in the dark for eight days. The dark-grown plants could not convert stored precursors of chlorophyll molecules to active form. They never did green up.

to changes in light conditions. Section 27.10 takes a close look at phytochrome. For now, simply be aware that it alternates between active and inactive forms.

At sunset, during the night, or in the shade, when far-red wavelengths predominate, phytochrome in cells is inactive. It becomes activated at sunrise, when red wavelengths dominate the sky. The amount of red or far-red wavelengths a plant intercepts also varies from day to night and as the seasons change. Such variations are control signals over phytochrome activity—which influences the transcription of certain genes at certain times of day and year. Those genes specify enzymes and other proteins that help seeds to germinate, stems to lengthen and branch, and leaves to grow. The gene products also help flowers, fruits, and seeds to form.

Elaine Tobin and her coworkers at the University of California, Los Angeles, performed experiments that provided evidence in favor of phytochrome control. Using dark-grown seedlings of duckweed (*Lemna*), they discovered a marked increase in the number of certain mRNA transcripts after a one-minute exposure to red light. Exposure had enhanced transcription of the genes for proteins that bind chlorophylls and for rubisco, an enzyme that mediates carbon fixation (Sections 6.5 and 6.6). In the absence of the proteins, chloroplasts do not develop properly and they don't turn green.

In each multicelled organism, gene controls underlie basic, short-term housekeeping tasks in cells and more intricate, long-term patterns of bodily growth and development.

All cells in a multicelled organism inherit the same genes, yet most become specialized in their composition, structure, and function. This process of cell differentiation arises as different populations of cells activate and suppress some of their genes in highly selective, unique ways.

Hormones and other signals from the environment outside cells influence gene expression inside cells.

WWW

Our focus in this chapter is on transcriptional controls simply because they are the best understood. However, bear in mind that eukaryotic cells exert selective control of gene expression at other levels as well.

Many genes concerned with housekeeping tasks are under positive controls that promote continuous, low levels of transcription. For many other genes, however, we see fluctuating adjustments of transcription. Why? In multicelled organisms, the controls over the internal environment (that is, tissue fluids and blood) work to stabilize operating conditions for individual cells. Most often, the rates of transcription rise or fall by degrees only, in response to slight shifts in the concentrations of hormones and other signaling molecules, nutrients, and products in the internal environment.

Even *before* some gene sequences are transcribed, they might be chemically modified and shut down in programmed ways. These are not mutations; they are events that govern whether and how the genes will be expressed. Inactivation of one of the X chromosomes in the cells of female mammals is an example of this.

Control mechanisms work *after* gene transcription, also. Some govern transcript processing, transport of mature RNAs from the nucleus, and translation rates. For example, for some genes, introns can be removed and different exons spliced together in more than one way. In other words, a pre-mRNA transcript of a single gene may undergo alternative splicing.

Pre-mRNA transcribed from the gene that specifies a contractile protein, troponin-1, is like this. Enzymes snip out bits of the transcript, but different cuts are made in different cell types. And so, when the exons are spliced together, the protein-building message is a bit different from one cell type to the next. The resulting proteins are similar. But each is unique in a small part of its amino acid sequence, so it functions in a slightly unique way. Such alternative splicing might account for subtle variations in how different types of muscles function.

Other post-transcriptional controls modify the new polypeptide chains or activate, inhibit, or degrade the final proteins. Enzymes, proteins that catalyze nearly all metabolic reactions, are prime targets. Besides affecting gene transcription and translation for specific enzymes, control systems switch existing enzyme molecules on or off (Section 5.4). Just imagine the coordination required to govern which of the thousands of types of enzymes become stockpiled, deployed, or inactivated in a given interval. *That internal coordination governs all short-term and long-term aspects of cell structure and function.*

Eukaryotic gene controls operate in highly coordinated ways before, during, and after transcription and translation.

THE CELL CYCLE REVISITED   Every second, millions of cells in your skin, bone marrow, gut lining, liver, and elsewhere divide and replace their worn-out, dead, and dying predecessors. They do not divide willy-nilly. Controls govern the expression of genes that specify enzymes and other proteins required for cell growth, DNA replication, chromosome movements, and division of the cytoplasm. Diverse regulatory proteins control the synthesis and use of those gene products. And they control when the division machinery is put to rest.

Controls over the cell cycle are extensive. At various interrelated points, gene activity can be stepped up or slowed down, and gene products can advance, delay, or block the cycle. Some genes directly regulate passage through the cycle and are its primary controllers. Their products include **protein kinases**, a class of enzymes that attach phosphate groups to proteins. They operate at the boundary between G1 and S of the cell cycle, and during the transition from G2 to mitosis (Section 8.2).

Other genes encode proteins that modify the activity of the primary control genes or their products. Among them are genes for **growth factors**: transcriptional signals sent by one cell to trigger growth in other cells. As two examples, CSF stimulates certain white blood cells to grow, and EFG stimulates skin cells to grow and divide. Different genes specify cell receptors for growth factors.

Even more gene products exert indirect control over the cell cycle. Enzymes and other factors maintain DNA replication and stockpile weapons of cell suicide, the **ICE-like proteases** (Figure 14.7). If the genes specifying them become mutated, a cell may be deprived of crucial enzymes or proteins—and cancer may be the result.

CHARACTERISTICS OF CANCER   At the least, four features characterize all cancer cells. First, *their plasma membrane and cytoplasm change profoundly.* The membrane becomes more permeable; its proteins are lost or altered, and abnormal ones form. The cytoskeleton becomes disorganized, shrinks, or both. Enzyme action shifts, as in amplified reliance on glycolysis. Second, *cancer cells grow and divide abnormally.* Controls that prevent overcrowding in tissues are lost. Cell populations reach high densities, and different proteins trigger abnormal increases in small blood vessels that service the growing cell mass. Third, *cancer cells have a weakened capacity for adhesion.* Recognition proteins become altered or lost, so the cells cannot stay anchored in proper tissues. Fourth, *cancer cells are lethal.* Unless eradicated, they eventually kill the individual.

Any gene having the potential to induce a cancerous transformation is an **oncogene**. Oncogenes were first identified in retroviruses, which are a class of RNA viruses. They are altered forms of normal genes, now called **proto-oncogenes**, that specify certain proteins required for normal cell functioning. In other words, the *normal* expression of proto-oncogenes is vital, which helps explain why their *abnormal* expression is lethal.

signal
to die

ICE-like
protease

Figure 14.7 *Here's to suicidal cells!*

Transformation of any cell may start with mutations in control elements that govern proto-oncogenes or with genes themselves. This happens with certain insertions of viral DNA into cellular DNA. It happens after carcinogens (cancer-inducing agents) change the DNA. Ultraviolet radiation and ionizing radiation (x-rays and gamma rays) are common carcinogens. So are many natural and synthetic compounds, including asbestos and certain components of tobacco smoke.

Remember those chromosome alterations and gene mutations? Some cancers arise when base substitutions or deletions alter a proto-oncogene or one of the controls over its transcription. Others arise by translocations. You read about translocations earlier, in the introduction to Chapter 11 and in Section 11.8.

Yet cancer is a multistep process, involving more than one oncogene. Researchers identified three such genes with roles in nearly all colon cancers and a key gene in ovarian and breast cancers. From their work, it may be possible to diagnose carriers early and frequently. Cancer detected early enough may still be curable, as by surgery. But if cancer cells live long enough, they may end up with a number of mutations that allow them to break away and establish colonies in distant tissues (Figure 14.8).

HERE'S TO SUICIDAL CELLS! We conclude this chapter with a case study of a gene and its cancerous transformation.

For all multicelled organisms, the first cell of a new individual contains marching orders that will take its descendants along a program of growth, development, and reproduction, then on to death. As part of that program, many cells heed calls to self-destruct when they complete a prescribed function. If they become altered in ways that might pose a threat to the body as a whole, as by infection or cancerous transformation, they can execute themselves. **Apoptosis** (app-oh-TOE-sis) is the name for this form of cell death. It starts with molecular signals that activate and unleash lethal weapons of self-destruction that were stockpiled earlier within the cell itself.

A body cell in the act of suicide shrinks away from its neighbors. Its cytoplasm seems to roil, and its surface repeatedly bubbles outward and inward. No longer are its chromosomes extended through the nucleoplasm; they have bunched together near the nuclear envelope. The nucleus, then the cell itself, breaks apart. Phagocytic white blood cells patrol the body's tissues. If they encounter suicidal cells or remnants of them, the patrolling cells will swiftly engulf them.

The timing of cell death is predictable in some cells. Examples are pigment-packed keratinocytes. They form the densely packed sheets of dead cells that are continually sloughed off and replaced at the skin surface. They have a three-week life span, more or less. Yet keratinocytes and other body cells, even the kinds that are supposed to last a lifetime, can be induced to die ahead of schedule. All it takes is sensitivity to specific signals that can activate weapons of death. Those weapons are protein-cleaving enzymes, the ICE-like proteases mentioned earlier.

Think of them as folded pocketknives or lethal Ninja weapons. When popped open, the ICE-like proteases chop apart structural proteins, including building blocks of cytoskeletal elements and nucleosomes that organize the DNA (Figure 14.7). They do so, for example, when a cell is deprived of growth factors, loses contact with its neighbors, or receives abnormal signals about when to grow, divide, or cease dividing. The weapons also are unleashed in the presence of certain regulatory proteins that can induce apoptosis.

As you might suspect, the knives remain sheathed in cancer cells, which are supposed to—but don't—commit suicide on cue. Researchers already know that the gene coding for a protein designated p53 has been tampered with or shut down entirely in many masses of transformed cells. The normal form of that protein will induce apoptosis if a cell's DNA gets damaged. Intriguingly, in more than half of all cancer patients who have been investigated to date, molecules of p53 are either absent or malfunctioning.

Cancer cells break away from home tissue.

Metastasizing cells attach to and secrete digestive enzymes onto a blood or lymph vessel, then cross the breached wall.

Cancer cells creep or tumble along, then leave the same way they got in to found tumors in new tissue.

Figure 14.8 Steps in metastasis.

WWW

SUMMARY

1. Cells control gene expression; that is, which gene products appear, when, and in what amounts. When the control mechanisms come into play depends on the type of cell, on prevailing chemical conditions, and on signals from the outside environment that can change the cell's activities.

2. Regulatory proteins, enzymes, hormones, and other molecules are components of control systems. Various kinds interact with one another, with DNA and RNA, and with products of genes. Control systems operate before, during, and after transcription and translation.

3. In all cells, two of the most common categories of control systems operate in ways that block or enhance transcription. Their effects are reversible.

   a. In negative control systems, a regulatory protein binds at a specific DNA sequence and prevents one or more genes from being transcribed.

   b. In positive control systems, a regulatory protein binds to DNA and promotes transcription.

4. Most types of prokaryotic cells (bacteria) do not require many genes to live, grow, and reproduce. Most bacterial control systems affect transcription rates. The control of operons (groupings of related bacterial genes and control elements) is an example.

5. Promoters are binding sites in DNA that signal the start of a gene. In all cells, activator proteins bind next to promoters, and the complex promotes transcription. In bacteria alone, operators serve as binding sites for regulatory proteins that can inhibit transcription.

6. Eukaryotic cells—those of multicelled organisms especially—require more complex gene controls. Gene activity must change rapidly in response to short-term shifts in the surroundings, as in prokaryotic cells. But it also must be adjusted in intricate ways during long-term growth and development, when great numbers of cells multiply, make physical contact with one another in tissues, and interact chemically.

7. Being descended from the same cell, all of the cells in a multicelled organism inherit the same assortment of genes. However, different types of cells activate and suppress some fraction of the genes in different ways. This behavior is called selective gene expression.

8. Cell differentiation is one outcome of selective gene expression. By this process, different lineages of cells in multicelled organisms become specialized in function, appearance, and composition.

9. Cancer results when the normal controls over the cell cycle and mechanisms of cell death are lost. Some genes encode proteins that are the primary regulators of the cycle. Other genes encode proteins that modify the action of the primary controllers or their products.

Still other genes encode proteins that indirectly affect the cell cycle when their structure and function become modified in abnormal ways.

Review Questions

1. Describe the general characteristics of cells, then briefly explain the difference between a benign tumor and a malignant tumor. Can cancer arise in every type of multicelled organism? CI, 14.5

2. In what fundamental way do negative and positive controls of transcription differ? Is the effect of one or the other form of control (or both) reversible? 14.1

3. Distinguish between: 14.2, 14.3
   a. repressor protein and activator protein
   b. promoter and operator
   c. repressor and enhancer

4. Describe one type of control of transcription for the lactose operon in E. coli, a prokaryotic cell. 14.2

5. A plant, fungus, or animal consists of diverse cell types. How might the diversity arise, given that body cells in each organism inherit the same set of genetic instructions? As part of your answer, define cell differentiation and the general way that selective gene expression brings it about. 14.3

6. What is a Barr body? Does it appear in the cells of males, females, or both? Explain your answer. 14.3

7. Using the diagram on the facing page, define at least three types of gene controls in eukaryotic cells and indicate the levels at which they take effect. 14.4

Self-Quiz (Answers in Appendix III)

1. The expression of a given gene depends on the _____ .
   a. type of cell and its functions
   b. chemical conditions
   c. environmental signals
   d. all of the above

2. Regulatory proteins interact with _____ .
   a. DNA   c. gene products
   b. RNA   d. all of the above

3. In prokaryotic cells but not eukaryotic cells, a(n) _____ precedes the genes of an operon.
   a. lactose molecule   c. operator
   b. promoter   d. both b and c

4. A base sequence signaling the start of a gene is a(n) _____ .
   a. promoter   c. enhancer
   b. operator   d. activator protein

5. An operon most typically governs _____ .
   a. bacterial genes   c. genes of all types
   b. a eukaryotic gene   d. DNA replication

6. Prokaryotic cells rely most heavily on controls that assure
   _____ .
   a. slow transcription   c. rapid translation
   b. rapid transcription   d. both b and c

7. Cell differentiation _____ .
   a. occurs in all multicelled organisms
   b. requires different genes in different cells
   c. involves selective gene expression
   d. both a and c
   e. all of the above

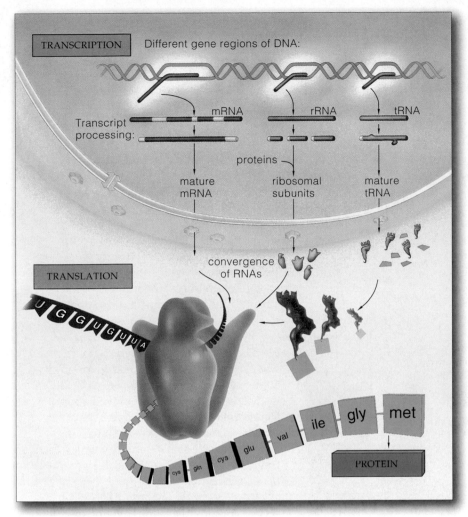

TRANSCRIPTION

Different gene regions of DNA:

mRNA rRNA tRNA

Transcript processing:

proteins

mature mRNA ribosomal subunits mature tRNA

convergence of RNAs

TRANSLATION

U G G U G U U A

cys gln cys glu val ile gly met

PROTEIN

transcription of the lactose operon when cells of this bacterial strain are subjected to these conditions:
   a. Lactose and glucose are available.
   b. Lactose is available but glucose is not.
   c. Both lactose and glucose are absent.

2. *Duchenne muscular dystrophy*, a genetic disorder, affects boys almost exclusively. Early in childhood, muscles begin to atrophy (waste away) in affected individuals, who typically die in their teens or early twenties as a result of respiratory failure. Muscle biopsies of women who carry a gene associated with the disorder reveal some regions of atrophied muscle tissue. Yet muscle tissue adjacent to these regions was normal or even larger and more chemically active, as if to compensate for the weakness of the adjoining region. How can you explain these observations?

3. Unlike most rodents, guinea pigs are already well developed at the time of birth. Within a few days, they are able to eat grass, vegetables, and other plant material. Suppose a breeder decides to separate the baby guinea pigs from their mothers after three weeks. He wants to keep the males and females in different cages, but it is quite difficult to identify the sex of guinea pigs when they are so young. Suggest a simple test that the breeder can perform to identify their sex.

4. Individuals affected by *pituitary dwarfism* cannot synthesize somatotropin (growth hormone). Children with this genetic abnormality will not grow to normal height. Develop a hypothesis to explain why therapy involving somatotropin injections may be effective.

5. Transcription is generally controlled by the binding of a protein to a DNA sequence "upstream" from a gene, rather than by modification of RNA polymerase. Develop a hypothesis to explain why this is so.

8. X chromosome inactivation in mammalian females may result in a _____ for some traits.
   a. male phenotype   c. rise in transcription rates
   b. mosaic tissue effect d. rise in translation rates

9. Hormones interact with _____ .
   a. membrane receptors   c. enhancers
   b. regulatory proteins   d. all of the above

10. Apoptosis is _____ .
   a. cell division after severe tissue damage
   b. cell death by suicide
   c. a popping sound in mutated toes

11. ICE-like proteases are _____ .
   a. regulatory proteins   c. environmental signals
   b. lethal weapons   d. low-temperature enzymes

12. Match the terms with their most suitable descriptions.
   ____ phytochrome   a. inhibits gene transcription
   ____ Barr body   b. gene that induces cancerous
   ____ oncogene   transformation
   ____ repressor   c. normally required, lethal
   ____ proto-oncogene   when mutated
                        d. helps plants adapt to daily
                           and seasonal changes in light
                        e. inactivated X chromosome

Critical Thinking

1. Geraldo isolated a strain of *E. coli* in which a mutation has affected the capacity of CAP to bind to a region of the lactose operon, as it would do normally. State how the mutation affects

Selected Key Terms

activator protein *14.2*
apoptosis *14.5*
Barr body *14.3*
cancer *CI*
cell differentiation *14.3*
enhancer *14.3*
growth factor *14.5*
hormone *14.3*
ICE-like protease *14.5*
metastasis *CI*
negative control system *14.1*

oncogene *14.5*
operator *14.2*
operon *14.2*
phytochrome *14.3*
positive control system *14.1*
promoter *14.2*
protein kinase *14.5*
proto-oncogene *14.5*
regulatory protein *14.1*
repressor *14.2*

Readings See also www.infotrac-college.com

Duke, R. D. Ojcius, and J. Ding-E Young. December 1996. "Cell Suicide in Health and Disease." *Scientific American*, 80–87.

Murray, A., and M. Kirschner. March 1991. "What Controls the Cell Cycle?" *Scientific American*, 264(3): 56–63.

Tijan, R. February 1995. "Molecular Machines That Control Genes." *Scientific American*, 54–61.

WWW   http://www.brookscole.com/biology

Practice quiz questions, hypercontents, BioUpdates, and critical thinking. The Brooks/Cole Biology Resource Center provides a wealth of information fully organized and integrated by chapter.

15

RECOMBINANT DNA AND GENETIC ENGINEERING

Mom, Dad, and Clogged Arteries

Butter! Bacon! Eggs! Ice cream! Cheesecake! Possibly you think of such foods as enticing, off-limits, or both. After all, who among us doesn't know about animal fats and the dreaded cholesterol?

Soon after you feast on those fatty foods, cholesterol enters the bloodstream. Cholesterol is important. It is a structural component of animal cell membranes, and without membranes, there would be no cells. Cells also remodel cholesterol into a variety of molecules, such as the vitamin D necessary for the development of good bones and teeth. Normally, however, the liver itself synthesizes enough cholesterol for your cells.

Some proteins circulating in blood combine with cholesterol and other substances to form lipoprotein particles. The high-density lipoproteins, or *HDLs*, transport cholesterol to the liver, where it can be metabolized. The low-density lipoproteins (*LDLs*) normally end up in cells that store or use cholesterol.

Sometimes too many LDLs form. The excess infiltrates the elastic walls of arteries, where it promotes the formation of atherosclerotic plaques (Figure 15.1). These abnormal masses interfere with blood flow and narrow the arterial diameter. If the plaques clog one of the tiny coronary arteries that deliver blood to the heart, chest pains or a heart attack may result.

How well you handle dietary cholesterol depends on genes you inherited from your parents. One gene codes for a protein that serves as the cell's receptor for LDLs. Inherit two "good" alleles of that gene, and your blood level of cholesterol will tend to remain so low that your arteries may never clog up, even with a high-fat diet. Inherit two copies of a certain mutated allele, however, and you are destined to develop a rare genetic disorder called *familial cholesterolemia*. Levels of cholesterol become abnormally high. Many affected people die of heart attacks in childhood or their teens.

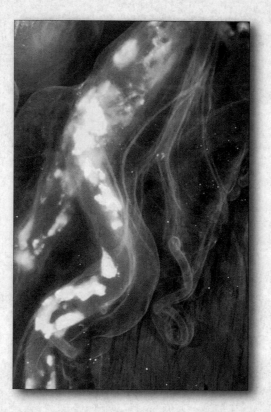

Figure 15.1 Life-threatening plaques (*bright yellow-white*) in one of the coronary arteries, small vessels that deliver blood to the heart. The plaques are the legacy of abnormally high levels of cholesterol.

In 1992 a woman from Quebec, Canada, became a milestone in the history of genetics. She was thirty years old. Like two of her younger brothers who had died from heart attacks in their early twenties, she inherited the mutant allele for the LDL receptor. She herself survived a heart attack when she was sixteen. At twenty-six, she had coronary bypass surgery.

At the time, people were hotly debating the risks and the promises of **gene therapy**—the transfer of one or more normal or modified genes into an individual's body cells to correct a genetic defect or boost resistance to disease. Even so, the woman opted for an untried, physically wrenching procedure that was designed to give her body working copies of the good gene.

Medical researchers removed about 15 percent of the woman's liver. They put cells from it in a nutrient-rich medium to promote cell growth and division. *And they spliced the functional allele for the LDL receptor into the genetic material of a harmless virus.* That modified virus served roughly the same function as a hypodermic needle. It was allowed to infect the cultured liver cells and insert copies of the good gene into them.

Later, the researchers infused about a billion of the modified cells into the woman's portal vein, the main blood vessel leading directly to the liver. There, some cells took up residence and started to produce the missing cholesterol receptor. Two years later, a fraction of the woman's liver cells were behaving normally and sponging up cholesterol from the blood. Her blood levels of LDLs had declined nearly 20 percent. Scans of her arteries showed no evidence of the progressive clogging that had nearly killed her.

Her cholesterol levels remain more than twice as high as normal, and it is too soon to know whether the gene therapy will prolong her life. Yet the intervention does provide strong evidence that the concept of gene therapy is sound.

As you might gather from that pioneering clinical application, recombinant DNA technology has great potential for medicine. It also has staggering potential

Viral injected genes into the corn less virus.

Figure 15.2 One outcome of genetic manipulation by way of artificial selection practices: a large kernel from a modern strain of corn next to the tiny kernels of an ancestral species, which was recovered from a prehistoric cave in Mexico.

for agriculture and industry. Think of it! For thousands of years, we humans have been changing genetically based traits of species. By artificial selection practices, we produced new crop plants and breeds of cattle, cats, dogs, and birds from wild ancestral stocks. We were selective agents for meatier turkeys, sweeter oranges, seedless watermelons, flamboyant ornamental roses, and big juicy corn kernels (Figure 15.2). We produced splendid hybrid organisms, including the mule (horse × donkey) and the plants that produce the tangelo (tangerine × grapefruit).

Of course, we have to remember we are newcomers on the evolutionary stage. During the 3.8 billion years before we even made our entrance, nature conducted uncountable numbers of genetic experiments by way of mutation, crossing over, and other events that introduce changes in genetic messages. Those countless changes are the source of life's rich diversity.

But the striking thing about human-directed changes is that the pace has picked up. Today, researchers use **recombinant DNA technology** to analyze genes. They cut and recombine DNA from different species, then insert it into bacterial, yeast, or mammalian cells that replicate their DNA and divide rapidly. The cells copy the foreign DNA as if it were their own and churn out useful quantities of recombinant DNA molecules. The technology also is the basis of **genetic engineering**, by which genes are isolated, modified, and inserted back into the same organism or into a different one. The protein products of those modified genes may cover the functions of their missing or malfunctioning counterparts.

The new technology does not come without risks. With this chapter, we consider its basic aspects. At the chapter's end, we also address some ecological, social, and ethical questions related to its application.

KEY CONCEPTS

1. Genetic experiments have been proceeding in nature for billions of years, through gene mutations, crossing over, recombination, and other natural events.

2. Humans are now purposefully bringing about genetic changes by way of recombinant DNA technology. Such enterprises are called genetic engineering.

3. With this technology, researchers isolate, cut, and splice together gene regions from different species, then greatly amplify the number of copies of the genes that interest them. The genes, and in some cases the proteins they specify, are produced in quantities that are large enough to use for research and for practical applications.

4. Three activities are at the heart of recombinant DNA technology. First, procedures based on specific types of enzymes are used to cut DNA molecules into fragments. Second, the fragments are inserted into cloning tools, such as plasmids. Third, the fragments containing the genes of interest are identified, then copied rapidly and repeatedly.

5. Genetic engineering involves isolating, modifying, and inserting genes back into the same organism or into a different one. The goal is to beneficially modify traits that the genes influence. Human gene therapy, which focuses on controlling or curing genetic disorders, is an example.

6. The new technology raises social, legal, ecological, and ethical questions regarding its benefits and risks.

Many years have passed since foreign DNA was first transferred into a plasmid. Yet that transfer started a debate that will continue into the next century. The issue is this: *Do the benefits of gene modifications and gene transfers outweigh the potential dangers?*

Genetically engineered bacteria are "designed" to stay confined to the laboratory. As an added precaution, "fail-safe" genes are included in foreign DNA in case they escape. Such genes stay silent unless the bacteria are exposed to environmental conditions—whereupon the genes become activated, with lethal results for their owner. Say the package includes a *hok* gene next to a promoter of the lactose operon (Section 14.2). Sugars are plentiful in the environment, and when they switch on the gene, the protein it specifies destroys membrane function and the wayward cell. But what about a worst-case scenario? Remember how retroviruses are being used to insert genes into cultured cells? If they escape from laboratory isolation and infect organisms, what might be the consequences?

Unlike modified species in laboratories, genetically engineered plants and animals are being released into the environment. Not long ago, Steven Lindlow thought about how many crop plants are vulnerable to frost. He found that a surface protein of a bacterium promotes formation of ice crystals. He excised the "ice-forming" gene from some cells. As he hypothesized, spraying his "ice-minus bacteria" on strawberry plants in an isolated field prior to a frost would make plants resist freezing. He actually had deleted a *harmful* gene from a species, and yet it triggered a bitter legal battle. The courts ruled in Lindlow's favor, and his coworkers sprayed a small strawberry patch. Nothing bad ever happened.

Then there was the potato plant designed to kill the insects that attacked it. It also was too toxic for people to eat. Or think of crop plants, which compete poorly with weeds for nutrients. Many have been designed to resist weedkillers, so farmers can spray for weeds and not worry about killing their crops. Herbicides happen to have toxic effects on more than just their targets (Section 3.6). If crop plants offer herbicide resistance, will farmers be less or more apt to spread them about?

And what if engineered plants or animals transfer modified genes with organisms in the wild? Think of how the advantage for acquisition of resources will tilt toward vigorous weeds that acquire herbicide-resistant genes. Such possibilities may be remote, but they are why standards for rigorous, extended safety testing must be in place *before* genetically modified organisms are released into the environment.

Rigorous safety tests are carried out before genetically modified organisms are released into the environment.

1. Uncountable gene mutations and other forms of genetic "experiments" have been proceeding in nature for at least 3 billion years.

2. Through artificial selection practices, humans have manipulated the genetic character of many species for many thousands of years. Today, recombinant DNA technology has enormously expanded our capacity to genetically modify organisms.

3. A genome is all the DNA in the haploid chromosome number for a species. By recombinant DNA technology, a genome can be cut into fragments, then the fragments can be amplified (copied over and over again) into useful quantities that permit analysis of the nucleotide sequence of the genome or a specific portion of it.

4. Researchers work with chromosomal DNA or cDNA, a strand of DNA transcribed by an enzyme (reverse transcriptase) from a mature mRNA transcript. The first choice is better for questions about DNA regions that control gene expression or that contain introns. cDNA is a better choice for questions about the amino acid sequence of a protein. Either way, researchers use restriction enzymes, which cut the DNA into fragments.

5. The use of plasmids or other cloning vectors is one way to amplify DNA fragments. Many bacteria carry plasmids, which are small circles of DNA with a few genes in addition to those of the bacterial chromosome.

a. Certain restriction enzymes make staggered cuts, so the fragments have single-stranded tails. Such tails base-pair with complementary tails of any other DNA cut by the same enzyme, such as plasmid DNA.

b. DNA ligase, a modification enzyme, seals base-pairing sites between plasmid DNA and foreign DNA. A plasmid modified to accept foreign DNA is a cloning vector; it can deliver foreign DNA into a bacterium or some other cell that can give rise to a population of rapidly dividing descendant cells. All the descendants have identical copies of the foreign DNA. Collectively, all of the identical copies are a DNA clone.

6. Currently, the polymerase chain reaction (PCR) is the fastest way to amplify fragments of chromosomal DNA or cDNA. Bacterial factories are not needed; the reactions occur in test tubes. The reactions use a supply of nucleotide building blocks and primers: synthetic, short nucleotide sequences that will base-pair with any complementary DNA sequence and that enzymes (DNA polymerases) recognize to start replication. Each round of replication doubles the number of fragments.

7. For sexually reproducing species, no two individuals have exactly the same DNA base sequence (except for identical twins). When restriction enzymes are used to cut an individual's DNA, the result is a unique array of restriction fragments called a DNA fingerprint.

a. The DNA in all humans is more than 99 percent identical except at tandem repeats: short stretches of repeated base sequences, such as TTTC. The number and combination of tandem repeats is unique in each individual and can be detected by gel electrophoresis.

b. DNA fingerprinting has uses in forensic science and in resolving paternity suits.

8. Automated DNA sequencing can rapidly reveal the base sequence of cloned DNA or PCR-amplified DNA fragments. The method labels each fragment with one of four modifed nucleotides, then separates them according to length (by gel electrophoresis). Each label fluoresces a certain color under a laser beam. A machine reads each as it peels off the gel and assembles the whole sequence.

9. A gene library is a mixed collection of bacterial cells that took up different cloned DNA or cDNA fragments. A gene may be isolated from the library with use of a probe, a very short stretch of radioactively labeled DNA that is known or suspected to be similar to or identical with part of that gene and can base-pair with it. A base pairing between nucleotide sequences from different sources is called nucleic acid hybridization.

10. In genetic engineering, specific genes are modified and inserted into the same organism or a different one. In gene therapy, copies of normal or modified genes are inserted into an individual in order to correct a genetic defect or boost resistance to disease.

11. Generally, recombinant DNA technology and genetic engineering have enormous potential for research and applications in medicine and agriculture, in the home and industry. As with any new technology, however, the potential benefits must be weighed against the potential risks, including ecological and social repercussions.

Review Questions

1. Distinguish between recombinant DNA technology and genetic engineering. CI

2. Distinguish these terms from one another: 15.1
   a. chromosomal (genomic) DNA and cDNA
   b. cloning vector and DNA clone

3. Define PCR. Can fragments of chromosomal DNA, cDNA, or both be amplified by PCR? 15.2

4. Define DNA fingerprint. Then describe which portions of the DNA are used in DNA fingerprinting. 15.3

5. Describe the steps of automated DNA sequencing. 15.4

6. Define cDNA library, then briefly explain how a gene can be isolated from it. Define probe and nucleic acid hybridization as part of your answer. 15.5

7. Give three examples of applications that can be derived from knowledge of an organism's genome. CI, 15.6–15.8

8. Name one of the ways in which modified genes have been inserted into mammalian cells. 15.8

9. Define gene therapy. Once the human genome has been fully sequenced, why will it be difficult to manipulate its genes to advantage? 15.8

Self-Quiz (Answers in Appendix III)

1. _____ is the transfer of normal genes into body cells to correct a genetic defect.
   a. Reverse transcription   c. Gene mutation
   b. Nucleic acid hybridization d. Gene therapy

2. DNA fragments result when _____ cut DNA molecules at specific sites.
   a. DNA polymerases   c. restriction enzymes
   b. DNA probes   d. RFLPs

3. _____ are small circles of bacterial DNA that are separate from the circular bacterial chromosome.

4. Foreign DNA that was inserted into a plasmid and then replicated many times in a population of bacteria is a _____ .
   a. DNA clone   c. DNA probe
   b. gene library   d. gene map

5. In reverse transcription, _____ is assembled on _____ .
   a. mRNA; DNA   c. DNA; enzymes
   b. cDNA; mRNA   d. DNA; agar

6. PCR stands for _____ .
   a. polymerase chain reaction
   b. polyploid chromosome restrictions
   c. polygraphed criminal rating
   d. politically correct research

7. By gel electrophoresis, fragments of a gene library can be separated according to _____ .
   a. shape   b. length   c. species

8. Automated DNA sequencing relies on _____ .
   a. supplies of standard and labeled nucleotides
   b. primers and DNA polymerases
   c. gel electrophoresis and a laser beam
   d. all of the above

9. Match the terms with the most suitable description.
   ____ DNA fingerprint   a. selecting "desirable" traits
   ____ Ti plasmid   b. deciphering 3.2 billion base pairs
   ____ nature's genetic   of 23 human chromosomes
        experiments   c. used in some gene transfers
   ____ nucleic acid   d. unique array of DNA fragments
        hybridization   inherited in Mendelian pattern
   ____ Human Genome   from each of two parents
        Initiative   e. base pairing of nucleotide
   ____ eugenic   sequences from different
        engineering   DNA or RNA sources
                             f. mutations, crossovers

Critical Thinking

1. In the following diagram, which restriction enyzme made the cuts (indicated in red) in part of a DNA molecule from two different organisms?

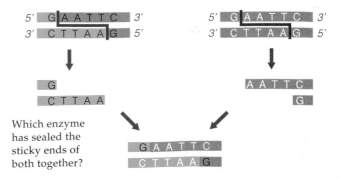

Which enzyme has sealed the sticky ends of both together?

2. Organisms of all kingdoms dabble in gene transfers and recombinations. For example, many bacteria transfer plasmid genes to a bacterial neighbor of the same species or a different one. We know this happens among *Salmonella, Streptococcus,* and *E. coli,* even between *E. coli* and yeast cells in the laboratory. Genes on the F (Fertility) plasmid confer the means to engage in *bacterial conjugation,* a mechanism by which a donor cell transfers plasmid DNA to a recipient cell. The F plasmid genes specify a sex pilus. This cell surface structure can hook onto a recipient cell and pull it right next to a donor. Shortly after two cells make contact, a conjugation tube develops between them. Plasmid DNA is transferred through the tube, in the manner shown in Figure 15.15.

Replication enzymes may even integrate the transferred plasmid into the bacterial chromosome of a recipient cell. A recombinant DNA molecule is the result. Given that gene transfers and recombinations occur so often in nature, why are some individuals so nervous about genetic engineering?

3. Ryan, a forensic scientist, obtained a very small DNA sample from a crime scene. In order to examine the sample by DNA fingerprinting, he must amplify the sample by the polymerase chain reaction. He estimates there are 50,000 copies of the DNA in his sample. Derive a simple formula and calculate the number of copies he will have after fifteen cycles of PCR.

4. A game warden in Africa confiscated eight ivory tusks from elephants. Some tissue is still attached to the tusks. Now he must determine whether the tusks were taken illegally from northern populations of endangered elephants, or from other elephants from populations to the south that can be hunted legally. How can he use DNA fingerprinting to find the answer?

5. Lunardi's Market put out a bin of tomatoes having splendid vine-ripened redness, flavor, and texture. The sign posted above the bin identified them as genetically engineered produce. Most shoppers selected unmodified tomatoes in the adjacent bin, even though those tomatoes were pale pink, mealy-textured, and tasteless. Which ones would you pick? Why?

6. Besides this chapter's examples, list what you believe might be some potential benefits of genetic engineering.

7. The Human Genome Initiative is not yet completed, yet knowledge about a number of the newly discovered genes is already being used to detect genetic disorders. Ask yourself: What will be done with genetic information about individuals? Will insurance companies and potential employers request it? At this writing, many women have already refused to take advantage of genetic screening for a gene associated with the development of breast cancer. Should medical records about people participating in genetic research and genetic clinical services be made available to other individuals? If not, how could such information be protected?

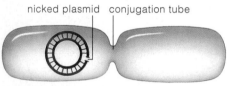

a A conjugation tube has already formed between a donor and a recipient cell. An enzyme has nicked the donor's plasmid.

b DNA replication starts on the nicked plasmid. The displaced DNA strand moves through the tube and enters the recipient cell

c In the recipient cell, replication starts on the transferred DNA.

d The cells separate from each other; the plasmids circularize.

Figure 15.15 Plasmid gene transfer during bacterial conjugation. The bacterial chromosome is not shown.

Selected Key Terms

automated DNA sequencing 15.4	nucleic acid
cDNA 15.1	hybridization 15.5
cloning vector 15.1	PCR 15.2
DNA clone 15.1	plasmid 15.1
DNA fingerprint 15.3	primer 15.2
DNA ligase 15.1	probe (nucleic acid) 15.5
gel electrophoresis 15.3	recombinant DNA
gene library 15.5	technology CI
gene therapy CI	restriction enzyme 15.1
genetic engineering CI	reverse transcriptase 15.1
genome 15.1	

Readings See also www.infotrac-college.com

Fairbanks, D., and W. R. Anderson. 1999. *Genetics: The Continuity of Life.* Monterey, California: Brooks-Cole.

Joyce, G. December 1992. "Directed Molecular Evolution." *Scientific American* 267(6): 90–97.

Shreve, J. May 1998. "The Code Breaker." *Discover,* 45–51. Craig Ventner's breakneck speed at decoding of the human genome.

Watson, J. D. et al., 1992. *Recombinant DNA.* Second edition. New York: Scientific American Books.

WWW *http://www.brookscole.com/biology*

Practice quiz questions, hypercontents, BioUpdates, and critical thinking. The Brooks/Cole Biology Resource Center provides a wealth of information fully organized and integrated by chapter.

FACING PAGE: *Millions of years ago, a bony fish died, and sediments gradually buried it. Today its fossilized remains are studied as one more piece of the evolutionary puzzle.*

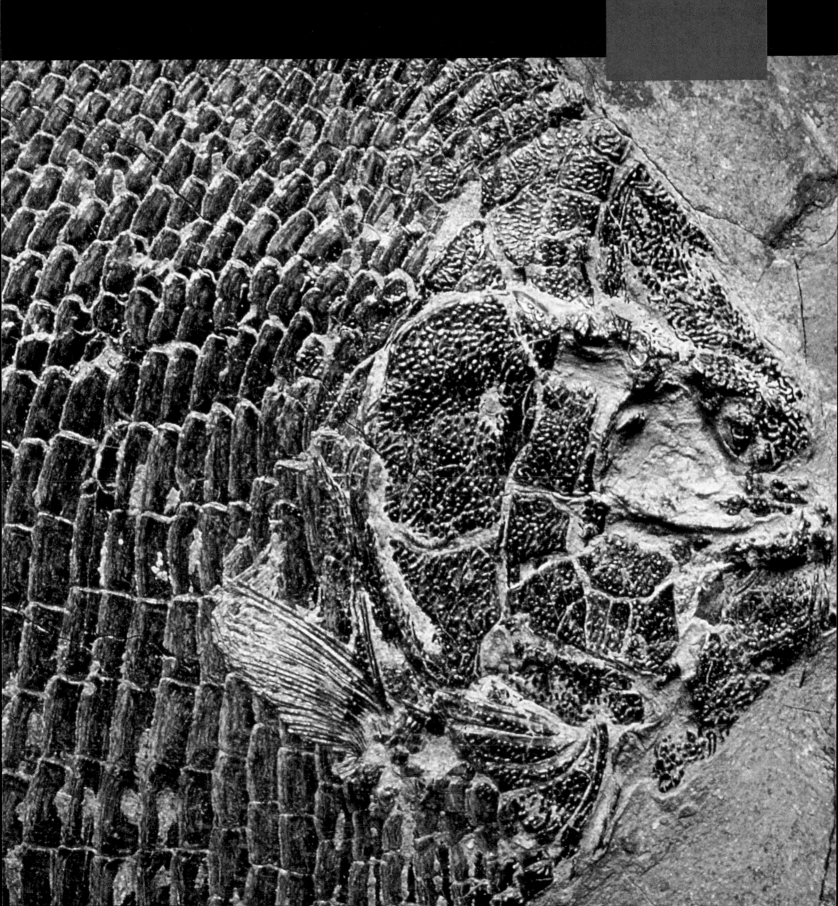

16 MICROEVOLUTION

Designer Dogs

With this unit, we turn to evolutionary theories and to the ways in which they can be used to interpret the past and present, even to predict possible futures for the natural world. Today the theories are routinely used to guide scientific investigations in many fields, and they are widely accepted in society at large. Quite simply, they help explain heritable changes in lines of descent over time—including changes that we humans imposed on a long and distinguished lineage. The lineage arose some 40 million years ago from small, tree-dwelling,

weasel-shaped carnivores. Their descendants evolved along branchings that now include weasels, badgers, otters, skunks, bears, pandas, raccoons—and dogs.

About 50,000 years ago, we began domesticating wild dogs. No doubt the advantages of doing so were important. Times were tough in the days before police protection and supermarkets. Dogs welcomed to the hearth could guard people and their possessions. They could corner, kill, eat, and thus dispose of big rats and other unwelcome vermin.

By 14,000 years ago, we started to develop different varieties (breeds) through artificial selection. Individual dogs having desired forms of traits were selected from each new litter and, later, encouraged to breed. Those having undesired forms of traits were passed over.

After favoring the pick of the litter over hundreds or thousands of generations, we ended up with sheep-herding collies, badger-hunting dachshunds, wily retrievers, and sled-pulling huskies. And at some point we began to delight in the odd, extraordinary dog. Imagine! In practically no time at all, evolutionarily speaking, we picked our way through the pool of variant dog alleles and came up with such extremes as Great Danes and chihuahuas (Figure 16.1).

Sometimes the canine designs exceeded the limits of biological common sense. For example, how long would a tiny, finicky-eating, nearly hairless, nearly defenseless chihuahua last in the wild? Not long. What about the English bulldog, bred for a very short snout and a compressed face? Long ago, breeders thought those particular traits would allow the dogs to get a better grip on the nose of bulls. (Why they wanted dogs to bite bulls is a story in itself.) So now the roof of the bulldog mouth is ridiculously wide and it is often flabby, so bulldogs have trouble breathing. They sometimes get so short of air, they pass out.

Through our centuries-old fascination with artificial selection, we produced thousands of varieties of crop plants, cats, cattle, and birds as well as dogs. With the currently available technologies of genetic engineering, we are now mixing the genes of many different species and producing incredible new varieties, including tobacco plants that produce hemoglobin and mustard plants that produce plastic.

So, when you hear someone wonder about whether "evolution" takes place, remind yourself that evolution simply means *genetic change in a line of descent through successive generations*. Selective breeding practices yield abundant, tangible evidence that heritable changes do, indeed, occur. The actual mechanisms of change are the focus of this chapter. Later chapters explain their role in the evolution of new species from parent species.

Figure 16.1 Two designer dogs. About 50,000 years ago, humans began domesticating wild dogs. From that ancestral stock, artificial selection produced diverse yet rather closely related breeds, such as the Great Dane (*legs, left*) and the chihuahua (*possibly fearful of being stepped on, right*).

KEY CONCEPTS

1. We define biological evolution as heritable changes in lineages, or lines of descent. Evidence of evolution comes from many different investigations that began nearly two centuries ago.

2. As Charles Darwin and Alfred Wallace perceived, populations can evolve by way of natural selection. In their view of this process, individuals of a population show differences in their shared traits. The forms of a trait that turn out to be most adaptive under prevailing environmental conditions tend to increase in frequency, and other forms do not. In other words, the traits that characterize a population can change through successive generations; the population can evolve.

3. In general, all individuals of a population have the same number and kinds of genes, which give rise to the same array of traits.

4. In the population as a whole, each gene may exist in two or more slightly different molecular forms, called alleles. Individuals may or may not inherit the same combinations of alleles, so they may not be exactly alike in the details of their traits.

5. Any allele at a given gene locus may become more or less common in the population, relative to the other kinds, or it may disappear. *Microevolution* means that changes have occurred in a population's allele frequencies over time.

6. Allele frequencies can change through mutation, gene flow, genetic drift, and natural selection. Mutation alone produces new alleles. Gene flow, genetic drift, and natural selection shuffle existing alleles into, through, or out of populations.

7. Natural selection is not a purposeful search through a population for the "best" individuals. It simply is the result of differences by which individuals of a population survive and reproduce, based on differences in their traits. Over time, it leads to continuing adaptation of a species to its environment.

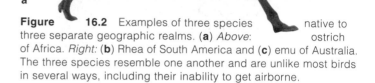

The Great Chain of Being

Our story begins more than two thousand years ago, when the seeds of biological inquiry were beginning to take hold among the ancient Greeks. At the time, popular belief held that supernatural beings intervened directly in human affairs. For example, people "knew" that angry gods inflicted a common ailment known as the sacred disease. And yet, from one physician of the school of Hippocrates, these thoughts come down to us:

It seems to me the disease called sacred . . . has a natural cause, as other diseases have. Men think it divine merely because they do not understand it. But if they called everything divine that they did not understand, there would be no end of divine things! . . . If you watch these fellows treating the disease, you see them use all kinds of incantations and magic—but they are also careful in regulating diet. Now if food makes the disease better or worse, how can they say it is the gods who do this? . . . It does not really matter whether you call such things divine or not. In Nature, all things are alike in this, in that they can be traced to preceding causes.

— ON THE SACRED DISEASE (400 B.C.)

Passages such as this reflect the early attempts to find natural explanations for observable events.

Aristotle was foremost among the early naturalists, and he described the world around him in great detail. He had no reference books or instruments to guide him, for biological science in the Western world began with the great thinkers of this age. Yet here was a man who was no mere collector of random tidbits of information. In his thoughtful descriptions, we find evidence of a mind perceiving connections between observations and attempting to explain the order of things.

Aristotle believed (as did others) that each kind of organism was distinct from all the rest. Nevertheless, he wondered about organisms that seemed to have rather blurred positions in nature. For example, even though some sponges look very much like plants, they do not make their own food, as plants do. They capture and digest it, as animals do. In time, Aristotle came to view nature as a continuum of organization, from lifeless matter through complex forms of plant and animal life.

By the fourteenth century, Aristotle's idea had been transformed into a rigid view of life. A Chain of Being was seen to extend from the "lowest" forms of life to humans and on up to spiritual beings. Each kind of being, or "species" as it was called, was a separate link in the great chain. All of the links were designed and forged at the same time, at the same center of creation, and had not changed since. Scholars thought that once they had discovered, named, and described all of the links, the meaning of life would be revealed to them.

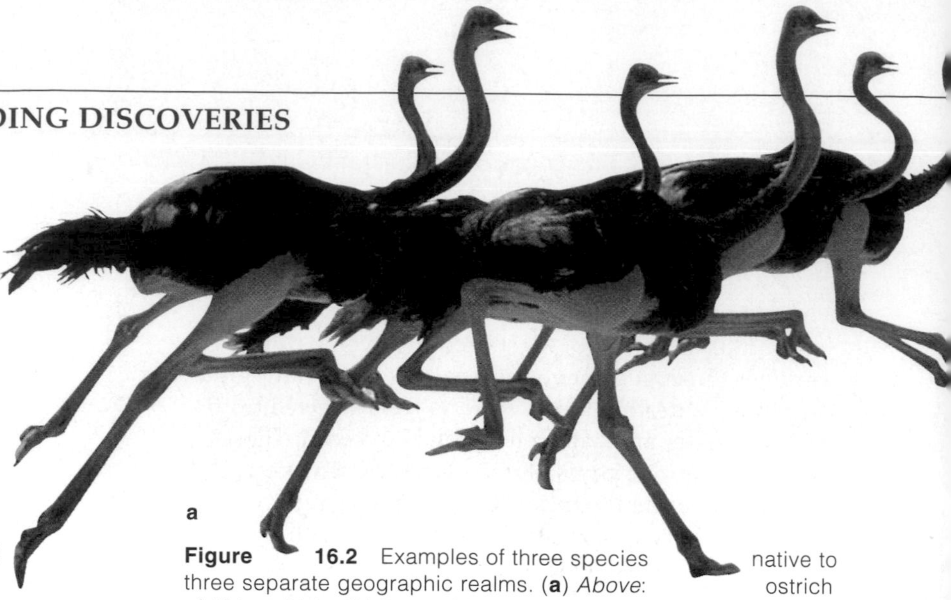

Figure 16.2 Examples of three species native to three separate geographic realms. (**a**) *Above:* ostrich of Africa. *Right:* (**b**) Rhea of South America and (**c**) emu of Australia. The three species resemble one another and are unlike most birds in several ways, including their inability to get airborne.

Questions From Biogeography

Until the fifteenth century, naturalists were not aware that the world is a great deal bigger than Europe, so the task of locating and describing all species seemed manageable. Then global explorations began, and the known world expanded enormously. Naturalists were soon overwhelmed by descriptions of tens of thousands of plants and animals that explorers were discovering in Asia, Africa, the Pacific Islands, and the New World.

In 1590 the naturalist Thomas Moufet attempted to sort through the bewildering array. He simply gave up and wrote such gems as this description of locusts and grasshoppers: "Some are green, some black, some blue. Some fly with one pair of wings; others with more; those that have no wings they leap; those that cannot fly or leap they walk; some have long shanks, some shorter. Some there are that sing, others are silent." It was not a work of subtle distinctions.

Even so, a few scholars began to examine the world distribution of organisms, a discipline now known as **biogeography** (see the Chapter 42 introduction). They soon realized that many plants and animals are unique to isolated places, such as remote oceanic islands. But certain species separated by great distances resemble one another (Figure 16.2).

How did so many species get from one center of creation to oceanic islands and other isolated locations? And what did the similarities and differences among them mean?

Questions From Comparative Anatomy

By the eighteenth century, many scholars were engaged in **comparative morphology**, the systematic study of similarities and differences in the body plans between major groups, such as different kinds of vertebrates. Think of the bones of a human arm, whale flipper, and

bat wing. They differ in size, shape, and function. Yet all have similar locations in the body (compare Section 18.2). They consist of the same tissues, arranged in the same overall patterns. They develop in similar ways in embryos. Comparative anatomists who discovered all of this wondered: Why are some animals that are so different in some features so much alike in others?

By one hypothesis, basic body plans were so perfect that there was no need to come up with a totally new one for each organism at the time of their creation. Yet if that were true, then how could there be parts with no function? For example, some snakes have bones that correspond to a pelvic girdle—a set of bones to which hind legs attach (Figure 16.3). Snakes don't have legs. Why the bones? Humans have bones that correspond to a few tailbones of many other mammals, but they don't have a tail. Why do they have parts of one?

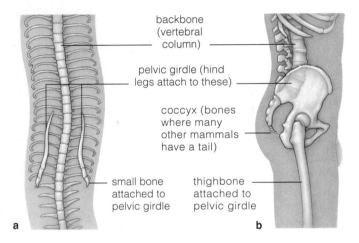

Figure 16.3 (a) Python bones corresponding to the pelvic girdle of other vertebrates, including humans (b). Small "hind limbs" protrude through the skin on the underside of the snake.

Questions About Fossils

From the late 1600s on, geologists added to the growing confusion. They began mapping layers of rocks at sites where erosion or quarrying had cut deep into the earth. They found similar layers around the world. (Such beds consist of distinct, multiple layers of sedimentary rock. Section 18.6 shows an example.) Most agreed that sand and other sediments had been deposited at different times and had slowly compacted into rock layers. They realized that certain layers contained distinctive **fossils**, which were known to be evidence of life in the past.

For example, deep layers contained fossils of simple marine organisms. Some fossils in the overlying layers were similar—but structurally more complex. Fossils in layers above these closely resembled living marine organisms. Were these *sequences* of fossils, separated in time? If so, what did increasing structural complexity among fossils of a given type of organism represent?

Besides this, many of the organisms represented in the fossil record are similar to living organisms, yet are

clearly different. What were these organisms, and how do we account for their similarity to existing species?

Taken as a whole, the findings from biogeography, comparative anatomy, and geology did not fit well with prevailing beliefs. Georges-Louis Leclerc de Buffon and a few others started to formulate novel hypotheses. If dispersal of all species from a center of creation was not possible, given the vast oceans and other barriers, *then perhaps species had originated in more than one place.* And if they were not created in a perfect state (and fossil sequences and the presence of "useless" body parts in certain organisms suggested they were not), *then perhaps species had been modified over time.* Awareness of change in lines of descent—**evolution**—was in the wind.

Awareness of biological evolution emerged over centuries, through the cumulative observations of many naturalists, biogeographers, comparative anatomists, and geologists.

WWW

A FLURRY OF NEW THEORIES

Squeezing New Evidence Into Old Beliefs

In the nineteenth century, naturalists tried to reconcile the growing evidence of change in lines of descent with a traditional conceptual framework that did not allow for change. Foremost among them was Georges Cuvier, a respected anatomist. For years he had compared body plans of fossils and living organisms. He acknowledged the abrupt changes in the fossil record, corresponding to discontinuities between certain layers of sedimentary beds. Was the record evidence of changing populations of organisms that lived in those ancient environments? Cuvier thought so. And he was right, as you will see from the evolutionary story in Chapter 19.

Based on that inference, Cuvier constructed his own theory of **catastrophism**. There was one time of creation, he said, that populated the world with all species. A global catastrophe destroyed many of them. Survivors repopulated the world. It was not that survivors were *new* species; naturalists simply hadn't yet found earlier fossils that would date to the time of creation. Later catastrophes wiped out more species, and repopulation by the survivors followed, as recorded by fossils.

Many scholars accepted his theory, but others kept at the puzzle. One hypothesis, inheritance of *acquired* characteristics, was pushed by Jean Lamarck. During each individual's life, said Lamarck, environmental pressures and internal "needs" bring about permanent changes in body form and functioning, and offspring inherit the needed changes. And so life, created long ago in a simple state, gradually improved. The force for change was a drive toward perfection, up the Chain of Being. The drive was centered in nerves that directed an unknown "fluida" to body parts in need of change.

Apply his hypothesis to modern giraffes. Say they had a short-necked ancestor. Pressed by a need to find food, it kept stretching its neck to browse on leaves beyond the reach of other animals. Stretching directed fluida to its neck, which lengthened permanently. The longer neck was inherited by offspring, which stretched their necks, also. So generations of animals desiring to reach loftier leaves led to the modern giraffe.

As Lamarck correctly inferred, the environment *is* a factor in evolution. His hypothesis, however, like others proposed at the time, has not been supported by any investigation carried out since then. No one has found evidence that environmental modifications of the traits of an existing individual can be passed to offspring.

Voyage of the Beagle

In 1831, in the midst of the confusion, Charles Darwin was twenty-two years old and wondering what to do with his life. Ever since he was eight, he had wanted to

Figure 16.4 (**a**) Charles Darwin and (**b**) a blue-footed booby, one of the species he observed during his five-year voyage around the world on the *Beagle*. A replica of this ship is shown in (**c**), sailing off the coast of South America. During stops along Argentina's coast, Darwin ventured into the Andes. He saw fossils of marine organisms embedded in rocks 3.6 kilometers above sea level. (**d,e**) The Galápagos Islands, isolated in the ocean far to the west of Ecuador. They arose through volcanic action about 5 million years ago, so organisms could not have originated there. Winds or ocean currents must have carried them to the new islands.

hunt, fish, collect shells, or simply watch insects and birds—anything but sit in school. Later, at his father's insistence, he did try to study medicine in college. The crude, painful procedures used on patients at that time sickened him. Then his father strongly urged him to become a clergyman, so Darwin packed for Cambridge. His grades were good enough to earn him a degree in theology. But he spent most of his time among faculty members with leanings toward natural history.

John Henslow, a botanist, perceived Darwin's real interests. He arranged for Darwin to function as ship's naturalist aboard H.M.S. *Beagle*. The *Beagle* was about to embark on a five-year voyage that would take Darwin around the world (Figure 16.4). And the young man who hated schoolwork and had no formal training as a naturalist suddenly began to work enthusiastically.

The *Beagle* sailed first to South America to complete work on mapping the coastline. During the Atlantic crossing, Darwin collected and examined marine life. He also read Henslow's parting gift, the first volume of Charles Lyell's *Principles of Geology*. During stops along the coast and at various islands, he observed diverse species in environments ranging from sandy shores to

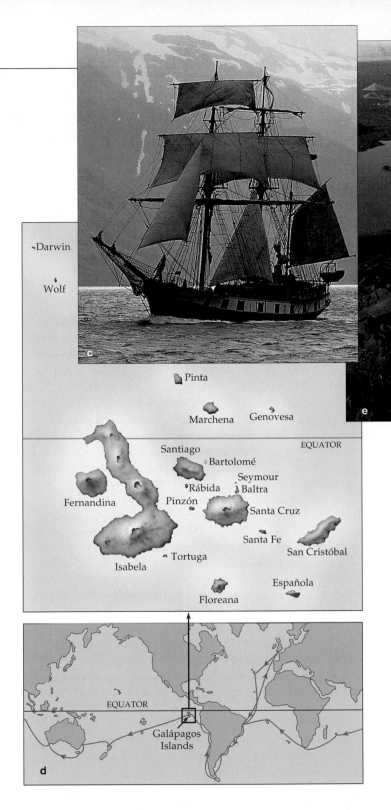

high mountains. And he started circling the question of evolving life, which was now on the minds of many respected individuals—including his own grandfather.

Darwin started mulling over a rather radical theory that Lyell was advancing in his book. Lyell and other geologists were arguing that catastrophes had no more and possibly less effect on Earth history than did subtle processes of change. They had thought about how long it takes for rains, the pounding surf, and other forces of nature to sculpt the landscape. For years geologists had chipped away at layers of sandstones, limestones, and other rocks, which form after sediments erode from the land and accumulate in the beds of rivers and seas. They thought about how many stacked layers occurred in many sedimentary beds. If, they hypothesized, the deposition had occurred as gradually in the past as it did in their own era, then surely it took many millions of years—not a few thousand—for such thick stacks to form. They even managed to incorporate earthquakes and other infrequent events into their view of Earth history. After all, major floods, more than a hundred great earthquakes, and twenty or so volcanic eruptions occur every year, so catastrophes are not unusual.

Their view of gradual, uniformly repetitive change became the **theory of uniformity**. It directly challenged prevailing views of the age of the Earth.

The theory bothered scholars who firmly believed the Earth was less than 6,000 years old. They thought people had recorded everything that happened during those thousands of years, and in all that time no one ever mentioned seeing a species evolve. Yet by Lyell's calculations, it must have taken millions of years to mold the present landscape. *Wasn't that enough time for species to evolve in diverse ways?* Later, Darwin thought so. But exactly *how* did they evolve? He would end up devoting the rest of his life to that burning question.

Prevailing beliefs can influence how we interpret clues to natural processes and their observable outcomes.

Darwin's observations during a global voyage helped him think about these processes in a novel way.

WWW

DARWIN'S THEORY TAKES FORM

Old Bones and Armadillos

After Darwin returned to England in 1836, he talked with other naturalists about possible evidence that life evolves. By carefully studying all the notes from his journey, he came up with some possibilities.

For example, he had observed fossils of glyptodonts in Argentina. Of all the animals on Earth, only living armadillos are like glyptodonts, which are extinct (Figure 16.5). Of all places on Earth, armadillos live only in the same places where glyptodonts once lived. If the two kinds of animals had been created at the same time, lived in the same place, and were so much alike, why is only one still alive? Wouldn't it be reasonable to assume glyptodonts were early relatives of armadillos? Many of their shared traits might have been retained through countless generations. Other traits might have been modified in the armadillo branch of a family tree. Descent with modification—it seemed possible. What, then, was the driving force for evolution?

A Key Insight—Variation in Traits

While Darwin assessed his notes, an influential essay by Thomas Malthus, a clergyman and economist, made him look closely at a topic of social interest. Malthus had correlated population size with famine, disease, and war. Humans, he claimed, run out of food, living space, and other resources because they reproduce too much. The larger a population gets, the more people there are to reproduce in each generation. Population size burgeons, resources dwindle, and the struggle to live intensifies. Many people starve, get sick, and engage in war and other forms of competition for the remaining resources.

After Darwin reflected on his personal observations, he suspected that any population has the capacity to produce more individuals than the environment is able to support. To give one example, a single sea star can release 2,500,000 eggs per year, but the seas obviously do not fill up with sea stars. In each generation, nearly all of the eggs and larvae end up inside the bellies of predators. Many of the survivors starve or succumb to disease or some other environmental insult.

Assume that the environment restricts the number of reproducing individuals. Which individuals will be the winners and losers? What influences the outcome? Darwin thought about the populations he had observed during his voyage. As he recalled, individuals were not alike down to the last detail. They showed variations in size, coloration, and other traits. *And it dawned on him that variations in traits might affect an individual's ability to secure resources—and therefore to survive and reproduce in particular environments.*

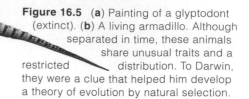

Figure 16.5 (a) Painting of a glyptodont (extinct). (b) A living armadillo. Although separated in time, these animals share unusual traits and a restricted distribution. To Darwin, they were a clue that helped him develop a theory of evolution by natural selection.

Did the Galápagos Islands show evidence of this? Between these volcanic islands and the South American coastline are 900 kilometers of open ocean. The islands offer diverse habitats along their rocky shores, deserts, and mountain flanks. Nearly all of their inhabitants live nowhere else—although they resemble species on the mainland. Were those remote islands colonized by species that flew, floated, or were blown over from the mainland? If so, then in the different island habitats, island-hopping descendants of colonizers might have changed over time in response to local conditions.

As Darwin learned from other naturalists, thirteen finch species were distributed among the Galápagos Islands. He himself had collected bird specimens, and now he attempted to correlate variations in their traits with environmental challenges. Imagine yourself in his place. The finches of one population, you notice, have a large, strong bill suitable for cracking seeds (Figure 16.6). Yet a few with a slightly stronger bill crack seeds that are too tough for other birds. If most of the seeds being produced in a particular habitat have hard coats, then a strong-billed bird will have a competitive edge. It will have a better chance of surviving and producing offspring. Assuming the trait is heritable, the same will be true of that bird's strong-billed descendants.

Take these thoughts one step further. If factors in the environment continue to "select" the most adaptive version of the trait, the population will become one of mostly strong-billed birds. *And a population is evolving if forms of heritable traits are changing over the generations.*

Recall, from Section 1.4, that Darwin offered pigeon breeding and other *artificial* selection practices as a way to explain *natural* selection of traits in the wild. When breeders favor pigeons with, say, black tail feathers, they encourage all new black-tailed pigeons to mate but not white-tailed ones. It was an easy way to show how selection could cause an increase in the frequency of one form of a trait in a captive population.

Anticipating that his ideas would be controversial, Darwin waited to announce it and searched for flaws in his reasoning. He waited too long. More than a decade

Figure 16.6 Four finch species from the Galápagos Islands. (**a**) *Geospiza conirostris* and (**b**) *G. scandens*, both with a bill adapted for eating cactus flowers and fruits. Other finches have thick, strong bills that crush cactus seeds. (**c**) *Certhidea olivacea*, a tree-dwelling finch, resembles warblers in song and behavior. It uses its slender beak to probe for insects. (**d**) *Camarhynchus pallidus* feeds on wood-boring insects such as termites. It has learned to break cactus spines and twigs to suitable lengths, then hold the "tools" and use them to probe bark for insects hidden from its view.

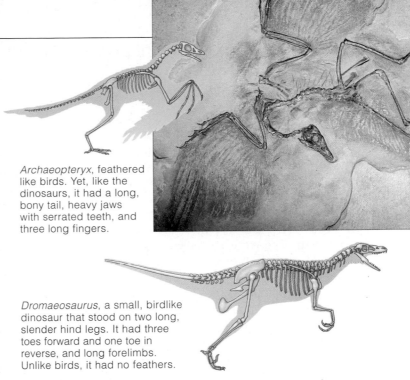

Archaeopteryx, feathered like birds. Yet, like the dinosaurs, it had a long, bony tail, heavy jaws with serrated teeth, and three long fingers.

Dromaeosaurus, a small, birdlike dinosaur that stood on two long, slender hind legs. It had three toes forward and one toe in reverse, and long forelimbs. Unlike birds, it had no feathers.

Figure 16.8 Comparison of the skeletons of *Archaeopteryx* and *Dromaeosaurus*, a dinosaur with traits closest to those of birds.

after he wrote up but never published an essay on his theory, another respected naturalist—Alfred Wallace— sent him a letter (Figure 16.7). Wallace had developed the same theory, then quickly wrote up and circulated his ideas. Colleagues insisted Darwin present a formal paper at the same time Wallace presented his. (Wallace also believed Darwin deserved most of the credit.) The next year, in 1859, Darwin did publish *On the Origin of Species*, his detailed evidence in support of the theory.

In that publication, Darwin gave examples of how artificial selection could mold traits of a population in no time at all, so he argued it was entirely possible for one species to evolve gradually into a separate species, with its own unique traits, over many generations. But if that were so, then where were the "missing links"? If

Figure 16.7 Alfred Wallace. Darwin worked out a theory of natural selection long before Wallace did. However, Wallace was first to report it; he circulated a brief letter describing his views about the process.

each species evolved from others, where were fossils of all the *transitional* forms? Where were the species with traits that bridged two major groups of organisms?

Ironically, a fossil of just such a transitional form had already been unearthed in southern Germany. At first the pigeon-size specimen was viewed as a theropod, one of the meat-eating dinosaurs. Its shape said "dinosaur." Like theropods, it had a long, bony tail, clawed fingers, and a heavy jaw with short, spiky teeth (Figure 16.8). Later, another fossil of the same type was unearthed. Still later, someone noticed the feathers. If those fossils were of dinosaurs, what were they doing with *feathers*? Close examination revealed that the feathers were like those of modern birds. The specimen type was named *Archaeopteryx*, meaning "ancient winged one."

Although you may have heard that Darwin's book fanned an intellectual firestorm, the idea that diversity is the product of evolution was accepted almost at once by most naturalists. But Darwin's specific explanation, of gradual evolution by natural selection, was fiercely debated. Nearly seventy years passed before advances in a new field, genetics, led to widespread acceptance of his explanation. Until that happened, people generally associated Darwin's name mainly with the premise that life evolves—something others had suggested before him.

DARWIN'S THEORY OF EVOLUTION BY NATURAL SELECTION. A population can evolve (change over time) when individuals differ in one or more heritable traits that are responsible for differences in the ability to survive and reproduce.

WWW

INDIVIDUALS DON'T EVOLVE—POPULATIONS DO

Examples of Variation in Populations

As Charles Darwin perceived, the individual does not evolve; populations do. By definition, a **population** is a group of individuals of the same species occupying a given area. To understand how a population evolves, start with the variation in traits among its individuals.

Certain features characterize a population. All of its members have the same body plan, as when jays have wings, feathers, feet with three toes forward and one toe back, and so on. These are *morphological* traits (*morpho—* means form). The cells and body parts of all individuals in the population operate much the same way during short-term metabolic tasks, growth, and reproduction. These are *physiological* traits, which relate to the body's functioning. Also, individuals respond the same way to basic stimuli, as when babies instinctively imitate adult facial expressions. These are *behavioral* traits.

Most traits differ in their details from one individual to the next, especially in sexually reproducing species. Pigeon feathers and snail shells differ in patterning or coloration in a given population (Section 1.4 and Figure 16.9). Some individuals of a frog population might be more sensitive to winter cold or better at attracting a mate than others. Humans differ in the distribution, color, texture, and amount of hair. And these examples only hint at the staggering variation within populations; almost every trait of every species is variable.

Many traits, such as those of the pea plants studied by Gregor Mendel, vary in *qualitatively different* ways. They come in two or more distinct forms (morphs) in a population, as when feathers are yellow or white. Such qualitative variation is known as **polymorphism**. Other traits, such as eye color and height, show continuous, *quantitatively different* variation. That is, individuals of the population show small, incremental differences in traits that can be quantified, as described in Section 10.7.

The "Gene Pool"

Information governing heritable traits resides in genes, which are specific regions of DNA molecules. In general, all individuals of a population have the same number and kinds of genes. We say "in general" because the males and females of sexually reproducing populations differ in some genes on the sex chromosomes.

Think of all the genes in the entire population as a **gene pool**—a pool of genetic resources that, in theory at least, is shared by all members of a population and passed on to the next generation. Each kind of gene in the pool usually exists in two or more slightly different molecular forms, called **alleles**.

Figure 16.9 *Facing page:* From islands of the Caribbean, variation in shell color and banding patterns in populations of the same snail species. Different individuals carry different alleles for the genes that specify most traits. The shells differ because their owners had different combinations of alleles at particular gene locations along their chromosomes.

Individuals inherit different combinations of alleles. This leads to variations in phenotype (to differences in the details of traits). For example, whether your hair is black, brown, red, or blond depends on which alleles of certain genes you inherited from your two parents. When studying the basis of evolution, also remember this: *Offspring inherit genes, not phenotypes.* What may appear to be a heritable trait might be the outcome of environmental effects on gene expression (Section 10.8).

Which alleles end up in a given gamete and, later, in a new individual? The outcome depends on five events, as described in earlier chapters and summarized here:

1. Gene mutation (produces new alleles)

2. Crossing over at meiosis I (puts novel combinations of alleles in chromosomes)

3. Independent assortment at meiosis I (puts mixes of maternal and paternal chromosomes in gametes)

4. Fertilization (combines alleles from two parents)

5. Change in chromosome number or structure (leads to the loss, duplication, or repositioning of genes)

Of the five events just listed, mutation alone *creates* new alleles. The other four only shuffle *existing* alleles into different combinations. But what a shuffle! Consider this: A human gamete will end up with one of 10^{600} possible combinations of alleles. Not even 10^{10} humans are alive today. So unless you are an identical twin, it is extremely unlikely that another person with your exact genetic makeup has ever lived, or ever will.

Stability and Change in Allele Frequencies

Imagine yourself in a large garden in summer, watching butterflies flitting about. They all look the same except in wing color. A few wings are white. Most are blue. Most likely, you muse, the "blue" allele must be more common than the "white" allele. With genetic analysis, you could identify **allele frequencies**, the abundance of each kind of allele in the population. You could track the rate of genetic change over the generations.

Suppose you start with the "Hardy-Weinberg rule," as given in Section 16.5, to set up a theoretical reference point for measuring patterns of change. At that point, called **genetic equilibrium**, the frequencies of alleles at a given gene locus remain stable, one generation after

the next. The population is *not* evolving with respect to that gene, for five conditions are being met. First, there have been no gene mutations. Second, the population is very large. Third, it is isolated from other populations of the species. Fourth, the gene has no effect at all on survival or reproduction. Finally, all mating is random.

Rarely, if ever, do all five conditions prevail at the same time in nature. Gene mutation is infrequent but inevitable. And three occurrences—*natural selection, gene flow,* and *genetic drift*—may drive the population away from genetic equilibrium, even in a few generations. The term **microevolution** refers to small-scale changes in allele frequencies, as brought about by mutation, natural selection, gene flow, and genetic drift.

Mutations Revisited

Mutations, remember, are heritable changes in DNA that typically give rise to altered gene products. They are the only source of new alleles. We cannot predict exactly when or in which individual they will appear. Yet each gene has its own **mutation rate**, which is the probability of its mutating during or in between DNA replications. On average, the rate is between 10^{-5} and 10^{-6} per gene locus per gamete, each generation. In a single reproductive season, merely 1 gamete in 100,000 to 1,000,000 has a new mutation at any given locus.

Mutations often give rise to structural, functional, or behavioral modifications that decrease the individual's chances of surviving and reproducing. Even a single biochemical change may have devastating effects.

For example, skin, bones, tendons, and many other vertebrate organs cannot develop without collagen, a structural protein. If the gene specifying the molecular form of collagen mutates, drastic changes in the lungs, arteries, skeleton, and other body parts may follow. As

you know, gene interactions underlie the phenotype of complex organisms. A mutation that has severe effects on phenotype usually will result in death; it is a **lethal mutation**.

By comparison, a **neutral mutation** does not help *or* harm an individual. Natural selection can neither increase nor decrease the frequency of neutral mutations in a population, for these do not have any discernible effect on the individual's chance of surviving or reproducing. For example, if you have a mutated gene that resulted in attached earlobes instead of detached ones, this alone should not stop you from surviving and reproducing just as well as anybody else.

Finally, every so often, a new gene mutation bestows an advantage. For instance, a product of a mutated gene that helps control growth might make a corn plant grow larger or faster and so give it the best access to sunlight and nutrients. Or maybe a neutral mutation turns out to be advantageous after environmental conditions change. Even if the advantage is small, chance events or natural selection might preserve the mutated gene and assure its representation in the next generation.

Mutations are so rare, they usually have little or no immediate effect on allele frequencies of a population. But beneficial mutations, and neutral ones, have been accumulating in different lineages for billions of years. Through all that time, they have been the raw material for evolutionary change—the basis for the staggering range of biological diversity, past and present.

In the evolutionary view, then, the reason you don't look like a bacterium or a grass plant or an earthworm or even neighbors down the street began with different mutations that originated at different times in the past, in different lines of descent.

Certain morphological, physiological, and behavioral traits characterize a population. The traits differ in their details from one individual to the next.

Differences in the combinations of alleles that individuals of the population carry give rise to variations in phenotype. By phenotypic variation, we mean differences in the details of structural, functional, and behavioral traits shared by those individuals.

For sexually reproducing species, individuals of a population represent a pool of genetic resources—that is, a gene pool.

Mutation alone creates *new* alleles. Natural selection, gene flow, and genetic drift change the *frequencies* of alleles in the gene pool. The evolutionary story begins with these changes.

Look again at Figure 10.1, which shows the earlobes of Tom Cruise, Gregor Mendel, and other celebrated people. Now ask this: Is the number of individuals with *detached* earlobes staying the same, one generation after the next, in the whole human population? What about the number of individuals who have *attached* earlobes? Stated more broadly, how do we know whether or not a population is evolving with respect to earlobes or any other trait?

Almost a hundred years ago, a mathematician and a doctor came up with the answer. Working independently, they thought about what it would take to maintain the frequencies of alleles of an idealized population. They came up with what is now called the **Hardy-Weinberg rule**, in their honor. They started out with this formula: In a population at genetic equilibrium, the proportions of genotypes at one gene locus, for which there are two kinds of alleles, are

$$p^2 \ AA + 2pq \ Aa + q^2 \ aa = 1$$

where *p* is the frequency of allele *A*, and *q* is the frequency of allele *a*. Their rule is this: *The allele frequencies will stay the same through the generations if there is no mutation, if the population is infinitely large and is isolated from other populations of the same species, if mating is random regarding the alleles, and if all individuals survive and reproduce equally.* Thus, the Hardy-Weinberg rule presents a hypothesis for determining whether any of the assumptions just listed currently apply to the population.

To test whether the allele frequencies of a population will remain constant from one generation to the next in the absence of evolutionary forces, you decide to track a pair of alleles through a population of butterflies. These sexually reproducing organisms have pairs of genes, on pairs of homologous chromosomes. You start by assuming the population is currently at genetic equilibrium. The pair of alleles you are interested in govern wing color. Allele *A* specifies dark-blue wings. Allele *a* is associated with white wings. The heterozygous (*Aa*) condition results in medium-blue wings.

In the population as a whole, the frequencies of *A* and *a* must add up to 1. For example, if *A* occupies half of all the loci for this gene in the population, then 0.5 + 0.5 = 1. If *A* occupies 90 percent of all the loci, then *a* must occupy the remaining 10 percent (0.9 + 0.1 = 1). No matter what the proportions of the two kinds of alleles,

$$p + q = 1$$

During meiosis in germ cells, each allele segregates from its partner, and the two end up in separate gametes. Therefore, *p* is also the proportion of gametes carrying the *A* allele, and *q* is the proportion with the *a* allele. To find the expected frequencies of the three genotypes (*AA*, *Aa*, and *aa*) that are possible in the next generation, you construct a Punnett square:

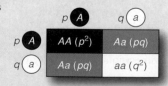

The frequencies of the genotypes add up to 1:

$$p^2 + 2pq + q^2 = 1$$

To see whether the allele frequencies and genotypic frequencies will remain the same through the generations, you work through an example. You find the population has 1,000 butterflies, each of which produces two gametes:

490 *AA* individuals produce 980 *A* gametes
420 *Aa* individuals produce 420 *A* and 420 *a* gametes
90 *aa* individuals produce 180 *a* gametes

You notice the frequency of *A* among the 2,000 gametes is

$$(980 + 420)/2,000 = 0.7$$

Also,

$$q = (420 + 180)/2,000 = 0.3$$

At fertilization, the gametes combine at random and give rise to the next generation, as given in the Punnett square. Assuming the population remains constant at 1,000 individuals, you now have

$$p^2 \ AA = 0.7 \times 0.7 = 0.49 \qquad 490 \ AA \text{ individuals}$$
$$2pq \ Aa = 2 \times 0.7 \times 0.3 = 0.42 \quad or \quad 420 \ Aa \text{ individuals}$$
$$q^2 \ aa = 0.3 \times 0.3 = 0.09 \qquad 90 \ aa \text{ individuals}$$

and

$$p^2 + 2pq + q^2 = 0.49 + 0.42 + 0.09 = 1$$

The allele frequencies have not changed:

$$A = \frac{2 \times 490 + 420}{2,000 \text{ alleles}} = \frac{1,400}{2,000} = 0.7 = p$$

$$a = \frac{2 \times 90 + 420}{2,000 \text{ alleles}} = \frac{600}{2,000} = 0.3 = p$$

You notice the genotype frequencies have not changed, either. And as long as the five assumptions of the Hardy-Weinberg rule are being met, frequencies should stay the same through the generations. To test this, you calculate allele frequencies in the gametes of the *next* generation:

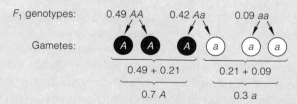

which is back where you started from. Because the allele frequencies for dark-blue, medium-blue, and white wings are the same as they were in the original gametes, they will yield the same phenotypic frequencies as you saw in the second generation.

You could do similar calculations for additional wing colors. You could go on with your calculations until you ran out of paper (or patience). However, as long as the five assumptions continue to hold, the allele frequencies and the range of values for the wing-color trait will not change, as you can see from Figure 16.10.

Therefore, when genotypes and phenotypes do *not* show up in the proportions that you predicted on the basis of the Hardy-Weinberg rule, this tells you that one or more conditions of the rule are being violated. And the hunt can begin for the specific evolutionary force, or forces, driving the change.

WWW

Figure 16.10 A hypothetical population of butterflies at genetic equilibrium.

Starting generation:

490 *AA* butterflies (dark-blue wings)

420 *Aa* butterflies (medium-blue wings)

90 *aa* butterflies (white wings)

The next generation:

490 *AA* butterflies

420 *Aa* butterflies

90 *aa* butterflies

(NO CHANGE)

The next generation:

490 *AA* butterflies

420 *Aa* butterflies

90 *aa* butterflies

(NO CHANGE)

We now turn from our idealized population that never changes to the real-world processes of change. Of those processes, natural selection might account for most of the morphological and physiological changes that have occurred throughout the history of life.

Darwin, recall, was able to explain natural selection after correlating his understanding of inheritance with certain features of populations and the environment. Before we consider the modes of natural selection, let's restate his correlations in modern terms:

1. *Observation*: All populations in nature have the reproductive capacity to increase in numbers over the generations.

2. *Observation*: No population is able to increase indefinitely, for its individuals will run out of food, living space, and other resources that sustain it.

3. *Inference*: Sooner or later, the individuals of a population will end up competing for resources.

4. *Observation*: All of the individuals have the same genes, which specify the same assortment of traits. Collectively, their genes represent a pool of heritable information.

5. *Observation*: Most, if not all, kinds of genes occur in slightly different molecular forms (alleles), which give rise to differences in phenotypic details.

6. *Inferences*: Some phenotypes are better than others at helping the individual compete for resources, and therefore to survive and reproduce. Thus, the alleles for those phenotypes increase in the population, and other alleles decrease. Over time, the genetic change leads to increased **fitness**—that is, to an increase in adaptation to the environment.

7. *Conclusions*: **Natural selection** simply is the result of differences in survival and reproduction among individuals that differ in heritable traits. Adaptation is one outcome of this microevolutionary process.

Evolutionary biologists have been documenting the results of natural selection in thousands of field studies of populations of all kinds of organisms. They find that this microevolutionary process has different results. As you will see in sections to follow, sometimes the result is a shift in the range of values for a given trait in some direction. At other times, the result may be stabilization or disruption of an existing range of values.

As Darwin perceived, natural selection is an outcome of differences in survival and reproduction among individuals that show variation in heritable traits. Over the generations, natural selection can lead to increased fitness—that is, to an increase in adaptation to the environment.

DIRECTIONAL CHANGE IN THE RANGE OF VARIATION

What Is Directional Selection?

In cases of **directional selection**, allele frequencies that underlie a range of variation in phenotypes shift in a consistent direction. The shifts occur in response to a directional change in the environment or to one or more new environmental conditions. They also occur when a mutation appears and proves to be adaptive. As Figure 16.11 indicates, the forms of a trait at one end of the range become more common than the midrange forms. Let's look at a few documented cases of this outcome.

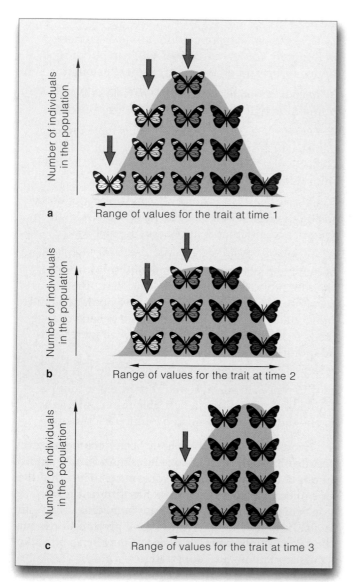

a Range of values for the trait at time 1

b Range of values for the trait at time 2

c Range of values for the trait at time 3

Figure 16.11 Directional selection, using phenotypic variation within a population of butterflies as the example. A bell-shaped curve (*brown*) represents the range of continuous variation in wing color. The most common forms (powder blue) are between extreme forms of the trait (white at one end of the curve, deep purple at the other). *Orange* arrows signify which forms are being selected against over time.

The Case of the Peppered Moths

In England, biologists tracked directional selection in about a hundred moth species, including the peppered moth (*Biston betularia*). Peppered moths feed and mate at night. During the day, they rest motionless on birches and other trees. Their wings and body have a mottled pattern, in shades that range from light gray to nearly black. Apparently, their behavior, coloration, and wing patterns help camouflage them from moth-eating birds, which actively hunt for food during the day.

The English industrial revolution began during the 1850s. Outpourings of sooty smoke altered conditions in many parts of the surrounding countryside. Before then, the light moths were the most common form, and a dark form was rare. Also before conditions changed, light-gray speckled lichens grew in profusion on tree trunks. Lichens can camouflage light moths that rest on them, but not dark ones (Figure 16.12*a*).

Lichens are sensitive to air pollution. Between 1848 and 1898, the soot and other pollutants from factories started to kill the lichens and darken tree trunks. Now the rare form of the moth was better camouflaged, as Figure 16.12*b* indicates. Moth collectors hypothesized that conditions had previously favored light moths—but that the changed conditions would favor dark ones.

In the 1950s, H. B. Kettlewell used a *mark-release-recapture* method to test the prediction. He bred both forms of the moth in captivity, then marked hundreds so that they could be identified. He released moths near heavily industrialized areas around Birmingham and in Dorset, an unpolluted area. After a time, he recaptured as many moths as he could. As Table 16.1 shows, more dark moths were recaptured in the polluted area and more light ones in the pollution-free area. By stationing observers in blinds near groups of moths that were tethered to trees, Kettlewell and his coworkers gathered direct evidence of birds capturing more of the light moths around Birmingham and more of the dark ones around Dorset. Directional selection was operating.

In 1952, strict pollution controls went into effect. Lichens made a comeback. Tree trunks became free of soot, for the most part. As you might have predicted, directional selection started to operate in the reverse direction. Where levels of pollution have declined, the frequency of dark moths has been declining, also.

Pesticide Resistance

Widespread use of chemical pesticides in agriculture has resulted in directional selection. Initial applications kill most of the insects, worms, or other pests, but some individuals usually manage to survive. Some aspect of their structure, physiology, or behavior allows them to

Figure 16.12 Individuals of a population of peppered moths (*Biston betularia*) that has undergone directional selection in response to changes in the environment. Light-winged and dark-winged individuals are resting on a lichen-covered tree trunk in (**a**) and on a soot-darkened tree trunk in (**b**).

Table 16.1 Marked *Biston betularia* Recaptured in Polluted Versus Nonpolluted Areas

	Near Birmingham (pollution high)	Near Dorset (pollution low)
LIGHT-GRAY MOTHS:		
Released	64	393
Recaptured	16 (25%)	54 (13.7%)
DARK-GRAY MOTHS:		
Released	154	406
Recaptured	82 (53%)	19 (4.7%)

Data after H. B. Kettlewell.

resist the chemical effects. When their resistance has a heritable basis, it becomes more common in the next generation, and the next, and the next. The chemicals are the agents of selection; they favor the most resistant forms! Today, 450 different species are resistant to one or more pesticides. Worse, the pesticides also kill the natural predators of the pests. When freed from natural constraints, the populations of resistant pests burgeon, and crop damage is greater than ever. This outcome of directional selection is called *pest resurgence*.

Maybe pesticide use will lessen in fields of plants that are genetically engineered for pesticide resistance. Those plants will not escape the coevolutionary arms race, but they might help keep our food supplies one step ahead of the pests. However, many consumers are leery of genetically engineered food.

What about using *biological controls*? By this practice, natural enemies of pests, including parasitic wasps and predatory beetles, are raised in commercial insectaries in great numbers and then released at preselected sites. The practice has an advantage, in that a control species can coevolve with the pests. But farmers must replace the ones that migrate from the fields or are destroyed at harvest time, and they pass on the cost to consumers.

Antibiotic Resistance

When your grandparents were young, up to one-fourth of the annual deaths in the United States alone were caused by bacterial agents of tuberculosis, pneumonia, and scarlet fever. In the 1940s, we started treating such bacterial-induced diseases with antibiotics.

Remember, certain microorganisms that live in soil produce diverse antibiotics, which can destroy bacterial competitors for nutrients (Section1.4). Streptomycins, for example, block protein synthesis in target cells. The penicillins disrupt the formation of covalent bonds that hold bacterial cell walls together. Penicillin derivatives cause the wall to weaken until it ruptures.

Antibiotics should be prescribed with restraint and care. Why? Besides performing their intended function, they commonly disrupt the balances among bacterial populations that normally compete for resources in the mammalian intestines and of yeast cells in the vaginal canal. Such disruptions lead to secondary infections.

Worse yet, antibiotics have been overprescribed in the human population. Too frequently they have been used for simple infections that many individuals could have overcome successfully on their own. Disturbingly, antibiotics have lost their punch. Over time, they did destroy the most susceptible cells of target populations. But they also favored their replacement by much more resistant cells. Millions of people around the world are now dying each year of tuberculosis, cholera, and other bacterial infections. Even vancomycin, held in reserve as "the antibiotic of last resort," is no longer effective against certain pathogenic strains of enteric bacteria. In 1996 the World Health Organization announced that, in the race for supremacy, pathogens are sprinting ahead.

With directional selection, allele frequencies underlying a range of variation tend to shift in a consistent direction in response to directional change in the environment.

WWW

SELECTION AGAINST OR IN FAVOR OF EXTREME PHENOTYPES

As you have seen, natural selection can bring about a directional shift in a population's range of phenotypic variation. Depending on the prevailing environmental conditions, the process also may favor either the most common or the most extreme phenotypes in that range.

Stabilizing Selection

In **stabilizing selection**, intermediate forms of a trait are favored and alleles that specify extreme forms are eliminated from a population (Figure 16.13). This mode of selection tends to counteract the effects of mutation, gene flow, and genetic drift and to preserve the most common phenotypes.

Consider the selection pressures on the gallmaking fly *Eurosta solidaginis*. After a fly larva develops from an egg, it bores into a stem of a tall goldenrod (*Solidago altissima*). There, it feeds upon plant tissues as it grows into an adult. In response, plant cells multiply rapidly and enclose the invader in a gall, a type of tumorous mass. As genetic analysis reveals, flies having different phenotypes induce formation of galls of different sizes.

A parasitic wasp, *Eurytoma gigantea*, can only attack larvae inside small galls (Figure 16.14), so it promotes selection in favor of flies that induce formation of *large* galls. The downy woodpecker and some other insect-eating birds focus on larvae inside large-diameter galls.

As researchers discovered after doing a number of studies in Pennsylvania, flies that induced formation of *intermediate-size* galls had the highest survival rate and fitness. Extreme phenotypes of the fly actually invited natural enemies and were selected against. So were the alleles responsible for the fly phenotypes that triggered the formation of small galls and large galls inside the goldenrod plants.

Thus, wasps work against one extreme in the range of phenotypes, birds work against the other—and the net result is an increase in intermediate phenotypes.

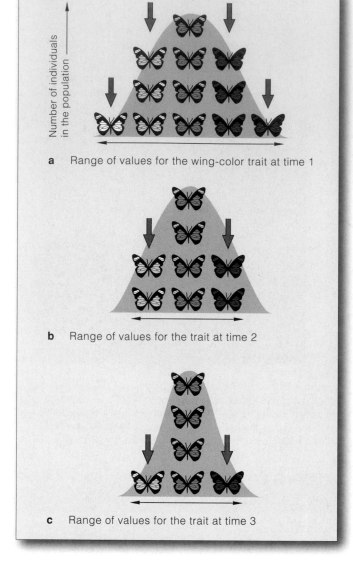

a   Range of values for the wing-color trait at time 1

b   Range of values for the trait at time 2

c   Range of values for the trait at time 3

Figure 16.13   Stabilizing selection, using phenotypic variation within a population of butterflies as the example.

Figure 16.14   Example of stabilizing selection. (**a**,**b**) Larvae of the fly *Eurosta solidaginis* induce formation of tumors (galls) on goldenrod stems. (**c**) Downy woodpeckers (*Dendrocopus pubescens*) and other birds prefer to chisel into large-size galls and eat the larvae. (**d**) The egg-laying device of the wasp *Eurytoma gigantea* can only penetrate the thin wall of small galls. Its eggs develop into larvae, then *its* larvae eat fly larvae.

Warren Abrahamson and his coworkers monitored twenty populations of *Eurosta* in Pennsylvania. They found that the larvae in small and large galls have low relative fitnesses, and larvae in intermediate-size galls have relatively high fitnesses. Thus, there is a stabilizing component to selection pressures created by the fly's natural enemies.

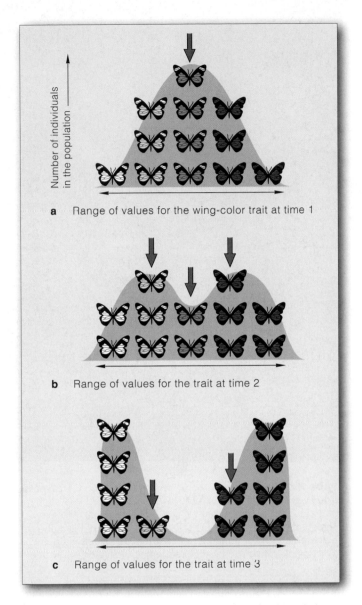

Figure 16.15 Disruptive selection, using phenotypic variation within a population of butterflies as the example.

a Range of values for the wing-color trait at time 1

b Range of values for the trait at time 2

c Range of values for the trait at time 3

Disruptive Selection

In **disruptive selection**, forms at both ends of the range of variation are favored and intermediate forms are selected against (Figure 16.15). Thomas Smith found a splendid example of this in a remote rain forest in Cameroon, West Africa. Smith had read about unusual variation in bill size in populations of the black-bellied seedcracker (*Pyrenestes ostrinus*). These African finches have large or small bills—but no sizes in between. The pattern holds for females and males, throughout their geographic range. (This is simply remarkable; imagine every person in Texas being four feet *or* six feet tall, with no intermediates.) If the bill pattern is unrelated to gender or geography, what causes it?

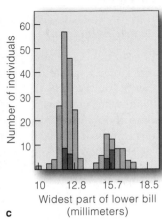

Figure 16.16 Disruptive selection among African finches. During feeding trials conducted by Thomas Smith, birds with large bills were more efficient users of hard seeds. Small-billed birds were most efficient at using soft seeds, not hard ones. (**a,b**) Two individuals displaying small and large bill sizes.

(**c**) Survival of juvenile birds during the dry season, when competition for resources is most intense. Individuals with *very* small, *very* large, or intermediate-size bills cannot feed efficiently on either type of seed; they survive poorly. For this graph, *tan* bars show the number of nestlings; *orange* bars show the only survivors among them. The findings are based on measurements of 2,700 netted individuals.

If, as Smith hypothesized, the persistence of only two sizes of bills relates to seed-cracking ability (which directly affects survival), then disruptive selection may be eliminating birds with intermediate-size bills. But what factors cause the disruptive selection pressure on feeding performance? Cameroon's swamp forests are flooded during the wet season, then lightning-sparked fires burn during the dry season. Two species of sedge (fire-resistant, grasslike plants) dominate these forests. One has hard seeds; the other has soft. When finches reproduce, both hard and soft seeds are abundant. All birds prefer soft seeds for as long as they can get them. But food supplies dwindle when the dry season peaks. Then, the youngest birds especially are at a competitive disadvantage. Many do not survive (Figure 16.16).

Smith also performed experimental crosses between large- and small-billed birds. All offspring had large *or* small bills. Along with other data, the crosses suggest two alleles at a single autosomal gene locus control bill size and feeding performance.

With stabilizing selection, intermediate phenotypes are favored and extreme phenotypes at both ends of the range of variation are eliminated.

With disruptive selection, intermediate forms are selected against; extreme forms in the range of variation are favored.

WWW

SPECIAL TYPES OF SELECTION

Sexual Selection

As you may have noticed, individuals of most sexually reproducing species have a distinctively male or female phenotype. We call this sexual dimorphism (*dimorphos* means "having two forms"). How does this condition come about, and what maintains it? Here again, natural selection is at work. In this case, it is **sexual selection**. The traits being favored are advantageous, with respect to survival and reproduction, simply because males or females prefer them. Through nonrandom mating, the alleles for preferred traits prevail over the generations.

Sexual dimorphism is particularly striking among many mammals and birds, including the pair in Figure 16.17. Often males are larger, have flashier coloration and patterning, and are much more aggressive than the females. Remember those male bighorn sheep butting heads (Section 1.3)? Such fights waste time and energy, and they might cause serious injuries. Why, then, do alleles that contribute to aggressive behavior persist in a population? The increased chance of mating offsets the costs. Male bighorn sheep fight only to control areas where receptive females gather during a winter rutting season. Winners mate often, with a number of females. Losers will not challenge a stronger, larger male.

Females are the main agents of selection. They exert direct control over reproductive success by choosing their mates. We return to this topic in Chapter 44.

Maintaining Two or More Alleles

Balancing selection includes all forms of selection that maintain two or more alleles for a trait in a population. When this type of genetic variation persists over time, we call it **balanced polymorphism** (after *polymorphos*, "having many forms"). A population is in a state of balanced polymorphism when nonidentical alleles for a trait are being maintained at frequencies greater than 1 percent. Frequencies may shift slightly, but over time they often will bounce back to the same values. Smith's work with African finches is only one example of the effect of balancing selection through the generations.

Sickle-Cell Anemia—Lesser of Two Evils?

We sometimes find cases of balanced polymorphism where environmental conditions favor heterozygotes (which carry nonidentical alleles for a specified trait), not the homozygotes (which carry identical alleles for the trait). Said another way, *the heterozygote has a higher fitness than either homozygote.*

Let's look at the environmental pressures that favor an Hb^A/Hb^S pairing in humans. Hb^S specifies a mutated form of hemoglobin, an oxygen-transporting protein

Figure 16.17 One outcome of sexual selection. This male bird of paradise (*Paradisaea raggiana*) is engaged in a flashy courtship display. He caught the eye (and, perhaps, the sexual interest) of the smaller, less colorful female. Males of this species compete fiercely for females, which serve as selective agents. (Why do you suppose drab-colored females have been favored?)

in the blood. Homozygotes (Hb^S/Hb^S) develop *sickle-cell anemia*, a genetic disorder that can have quite serious phenotypic outcomes (Section 10.5). The frequency of Hb^S is high in tropical and subtropical regions of Africa and Asia. Often, Hb^S/Hb^S homozygotes die in their early teens or early twenties. However, in those same regions, heterozygotes (Hb^A/Hb^S) make up nearly one-third of the human population! Why is this combination of alleles maintained at such high frequency?

The balancing act, an outcome of natural selection, is most pronounced in areas with the highest incidence of *malaria* (Figure 16.18). There a mosquito transmits *Plasmodium*, the parasite responsible for the disease, to humans. The parasite multiplies in the liver, then in red blood cells. The cells rupture and release new parasites during severe, recurring bouts of infection.

Who is far more likely to survive the recurrences? Hb^A/Hb^S heterozygotes. That allelic combination gives them two forms of hemoglobin, with interesting results.

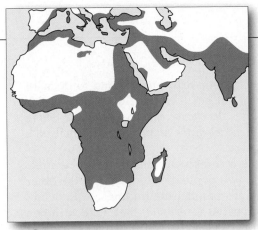

a

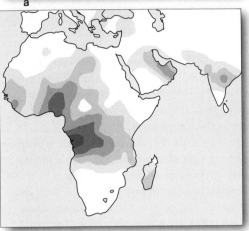

b

Figure 16.18
(**a**) Distribution of malaria cases in Africa, Asia, and the Middle East during the 1920s, before mosquito control programs were instituted. (**b**) For the same regions, the distribution and frequency of individuals with the sickle-cell trait. Notice the close correlation between the color-coded parts of the two maps.

☐ less than 1 in 1,600
☐ 1 in 400–1,600
☐ 1 in 180–400
☐ 1 in 100–180
☐ 1 in 64–100
☐ more than 1 in 64

They have enough nonmutated molecules to maintain body functions, even if at reduced levels. But the *altered* molecules distort the shape of red blood cells, and this interferes with blood circulation (the misshapen cells are destroyed in the spleen). The slowdown in blood flow is a factor, for it hampers the parasite's ability to rapidly infect new cells during an infective cycle.

Thus the persistence of the "harmful" Hb^S allele is a matter of relative evils. Natural selection has favored one allelic combination, Hb^A/Hb^S, because its bearers show greatest fitness in environments where malaria is the most prevalent. In those places, Hb^A/Hb^S has a higher fitness than either Hb^S/Hb^S or Hb^A/Hb^A.

In tropical and subtropical habitats of Asia and the Middle East, malaria has been a selective force for more than 2,000 years. Although sickle-cell anemia occurs at high frequencies in these regions, its symptoms are much less severe than they are in Central Africa, where the Hb^S allele became established much later in time. Most likely, other gene products might be modifying some of the Hb^S allele's widespread effects in ways that minimize symptoms of the disorder.

With sexual selection, some version of a trait simply gives the individual an advantage in reproductive success. Sexual dimorphism is one outcome of sexual selection.

Balanced polymorphism is a state in which natural selection is maintaining two or more alleles over the generations at frequencies greater than 1 percent.

WWW

Over time, individuals of the same species may move about, so that alleles move between populations. Alleles are lost from a population when individuals leave it permanently, an act called *emigration*. They are added when new individuals move to the population, an act called *immigration*.

The physical flow of alleles, or **gene flow**, tends to counter genetic differences that arise through mutation, natural selection, and genetic drift. And it helps keep separated populations genetically similar.

Think of the acorns that blue jays disperse when they store nuts for the winter. Each fall the jays may make hundreds of round trips from acorn-bearing oak trees to bury acorns in the soil of their home territories, which may be up to a mile away (Figure 16.19). The alleles flowing in with "immigrant acorns" help reduce genetic differences that might otherwise arise among neighboring stands of oaks. Gene flow apparently is operating among peppered moths, also, for the form that doesn't match the prevailing background color is being maintained at higher than expected frequencies.

Figure 16.19 Gene flow among oak populations, courtesy of feathered travel agents. Blue jays hoard acorns in their home territory, but they might shop at nut-bearing trees up to a mile away. Some acorns contribute to the allele pool of an oak population some distance away from the parent tree.

Or think of the millions of people from economically bankrupt, politically explosive countries who seek more stable homes. The sheer scale of their movements is unprecedented, but hardly unique. Throughout human history, immigrations may have minimized many of the genetic differences that otherwise would have built up among geographically separated groups of people.

Gene flow is the physical movement of alleles into and out of a population, through immigration and emigration.

Chance Events and Population Size

Genetic drift is a random change in allele frequencies over the generations, brought about by chance alone. The magnitude of its effect on genetic diversity and on the range of phenotypes relates to population size. Its impact tends to be minor or insignificant in very large populations but significant in small ones.

Sampling error, a rule of probability, helps explain the difference. By this rule, the fewer times a chance event occurs, the greater will be the variance from the expected outcome of that occurrence. You saw a simple demonstration of this in Section 1.5.

Also think back on the coin-flipping example given earlier, in Section 10.2. Each time you flip a coin, there is a 50 percent chance it will turn up heads or tails. With, say, only ten flips, the odds are great that the coin will turn up heads 7 times and tails 3 times. But with a thousand flips, the odds that it will turn up heads 700 times and tails 300 times are virtually nil. Similarly, sampling error applies every time random mating and fertilization take place in a population.

Bear in mind, genetic drift has nothing to do with how a population got small in the first place. *Genetic drift simply increases the chance of a given allele becoming more or less prevalent when the number of individuals in a population is small.*

Figure 16.20 is a computer simulation of the effect of genetic drift in two populations: one large, the other small. The outcomes parallel the findings of an actual experiment with beetles (*Tribolium*) that mate with one another at random. Researchers grouped 1,320 beetles into twelve populations of 10 beetles and twelve of 100. Which beetles ended up in which group was a matter of chance. At the start, the frequency of a wild-type allele (call it *A*) was 0.5. The researchers tracked allele *A* for twenty generations. Each time, they randomly removed some offspring to maintain each population's original size. At the end of the experiment, *A* was not the only allele left in the large groups, but it was fixed in seven of the small groups. **Fixation** means that only one kind of allele remains at a particular locus in the population; all individuals are homozygous for it.

Thus, *in the absence of other forces, random change in allele frequencies leads to the homozygous condition and a loss of genetic diversity over the generations.* This happens in all populations; it just happens faster in small ones. Once alleles inherited from an original population are fixed, their frequencies will not change again unless new alleles appear by mutation or gene flow.

Bottlenecks and the Founder Effect

Genetic drift is pronounced when very few individuals rebuild a population or found a new one. A **bottleneck** is a severe reduction in population size, as brought about by intense selection pressure or natural calamity. Suppose contagious disease, habitat loss, hunting, or a massive volcanic blast destroys much of a population. Even if a moderate number do survive the bottleneck, the allele frequencies will have been altered at random.

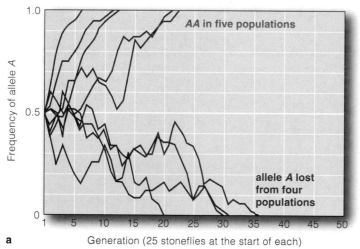

a Generation (25 stoneflies at the start of each)

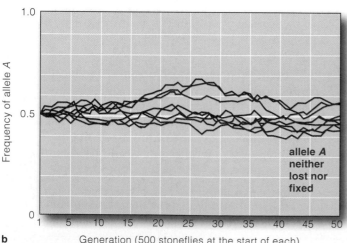

b Generation (500 stoneflies at the start of each)

Figure 16.20 Computer simulation of the effect of genetic drift on one allele's frequency in small populations and in large populations. Equal fitness is assumed for three simulations (*AA* = 1, *Aa* = 1, and *aa* = 1).

(**a**) The size of nine populations of a species (stoneflies, in this case) was maintained at 25 breeding individuals in each generation, through fifty generations. The five graph lines reaching the top of the diagram tell you that allele *A* became fixed in five of the small populations. The four lines plummeting off the bottom of the diagram tell you it was lost from four of them. As you can see, *alleles can be fixed or lost even in the absence of selection.*

(**b**) The size of nine other populations was maintained at 500 individuals in each generation, through fifty generations. Allele *A* did not become fixed in any of these large populations. The magnitude of genetic drift was much less in every generation than in the small populations.

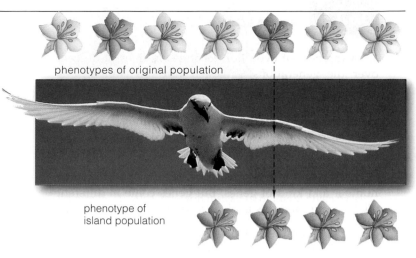

phenotypes of original population

phenotype of
island population

Figure 16.21 Example of the founder effect. A seabird carries a few seeds, stuck to its feathers, from the mainland to a remote oceanic island. By chance, most of the seeds carry an allele for orange flowers that was not common in the original population. In the absence of further gene flow or selection for flower color, genetic drift will fix the allele in the island population.

In the 1890s, hunters killed all but twenty of a large population of northern elephant seals. The population recovered. When its size passed 30,000, electrophoretic analysis of a sample of twenty-four genes revealed no variation in the population. A number of alleles had been lost in the bottleneck.

The genetic outcome can be similarly dicey after a few individuals leave a population and establish a new one elsewhere. This form of bottlenecking is called a **founder effect**. By chance, the allele frequencies of the founders may not be the same as those of the original population. In the absence of further gene flow, natural selection will influence those frequencies in drastically different ways, because of its interaction with genetic drift. As you might deduce, the founder effect is quite pronounced on isolated islands (Figure 16.21).

Genetic Drift and Inbred Populations

Inbreeding refers to nonrandom mating among closely related individuals, which have many identical alleles in common. Inbreeding is a form of genetic drift in a small population—that is, within the group of relatives that are preferentially interbreeding. Like genetic drift, inbreeding leads to the homozygous condition. It also can lower fitness when the alleles that are increasing in frequency are recessive and have harmful effects.

Most human societies forbid or discourage incest (inbreeding between parents and children or between siblings). But inbreeding among other close relatives is common in small communities that are geographically or culturally isolated from a larger population. The Old

Order Amish of Pennsylvania, for instance, is a highly inbred group having distinct genotypes. One outcome of their inbreeding is a high frequency of a recessive allele that causes *Ellis-van Creveld syndrome*. Affected individuals have extra fingers or toes and short limbs (Section 11.6 shows one such individual). The allele probably was rare when the small number of founders immigrated to Pennsylvania. At present, about 1 in 8 are heterozygous and 1 in 200 are homozygous for it.

Bottlenecks and inbreeding are a bad combination for *endangered species*, the populations of which have become small and vulnerable to extinction. Cheetahs, for instance, apparently survived a drastic bottleneck in the nineteenth century. Survivors mated with their own offspring when no other options were available.

Inbreeding among survivors and their descendants left the 20,000 existing cheetahs (Figure 16.22) with strikingly similar alleles. One, a mutated allele, affects fertility. Most male cheetahs have a low sperm count, and 70 percent of the sperm are abnormal. Other shared alleles result in much lower resistance to disease. Thus, infections that are seldom life-threatening to other cat species can be devastating to cheetahs. In one outbreak of *feline infectious peritonitis* in a wild animal park, the viral pathogen had very little effect on captive lions but killed many cheetahs. For them, infection triggered an uncontrollable inflammatory response. Fluid filled the body cavity housing the heart and other organs, and the cats died in agony. There is no vaccine.

The Florida panther also is highly inbred. Pressure from hunting and urban sprawl has helped put this cat on the endangered species list. Only fifty remain.

Figure 16.22 A few of the remaining cheetahs, which have some bad alleles that made it through a severe bottleneck.

Genetic drift is the random change in allele frequencies over the generations, brought about by chance alone. The magnitude of its effect is greatest in small populations, such as the ones that make it through a bottleneck.

Barring mutation, selection, and gene flow, the chance losses and increases of the various alleles at a given locus lead to the homozygous condition and a loss of genetic diversity.

WWW

17

SPECIATION

The Case of the Road-Killed Snails

If you happen to be a snail living in a garden in Bryan, Texas, it doesn't take much to keep your genes away from snails in a backyard across the street. By day, the sunbaked asphalt would be about as inviting to a snail as a desert would be to a catfish. Besides, day or night, a street-traversing snail is vulnerable to cars, trucks, skateboards, and bicycles (Figure 17.1*a*). Whatever else it might be, that strip of asphalt is a formidable barrier to the flow of genes between populations. For snails, that is.

Whether any physical barrier deters gene flow depends in large part on the organism's mode of locomotion or dispersal. It also depends on how fast and how long an organism *can* move in response to environmental factors or its own hormonal signals.

A snail is not swift, and it does not roam very far from its home population. Compare it with the wild black duck, leg-banded in Virginia in 1969, that turned up eight years later

in Korea. Compare it to the wandering albatross, one of the supreme barrier busters. After lifting off from Kerguelen Island in the Indian Ocean, one of those birds soared westward past the southern tip of Africa, across the Atlantic, and around South America's Cape Horn. After it had traveled thirteen thousand kilometers, it landed in Chile. With its lightweight body and wingspan of 3.65 meters (12 feet across), that bird was able to exploit the great prevailing winds of the Southern Hemisphere.

And yet, in 1859, some snails did cross an ocean. Humans had transported garden-variety snails (*Helix aspersa*) all the way from France to California, then turned them loose. The idea was that the snails would mate, multiply, and meet the demand for that French delicacy, *escargots aux fines herbes*. It was bad enough that the importers brought over a less tasty species by mistake. Worse, the exotic species exceeded expectations and became

Figure 17.1 (**a**) A snail (*Helix aspersa*) encountering a major barrier to gene flow. It only appears to be reading the warning label. (**b**) Results from a study of neighboring populations of snails, all descended from founders that ended up in a small town in Texas in the 1930s. For each population, a circle represents relative abundances of three alleles (color-coded *pink*, *orange*, or *brown*) for an enzyme, leucine aminopeptidase. Genetic variation is greater between populations living on opposite sides of Twenty-Second Street. For example, notice the higher frequencies of the allele coded *orange* in the block to the west of the street.

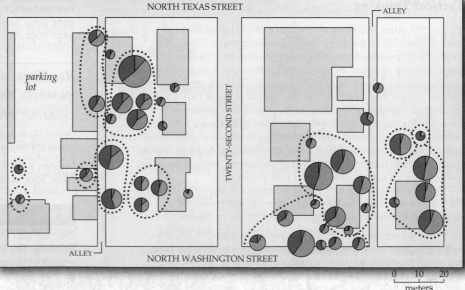

an absolute nuisance in gardens and nurseries through much of the southwestern United States.

By the 1930s, *H. aspersa* had hitched rides, possibly as eggs in the soil of plant containers, to Bryan, Texas. They founded small colonies in the local vegetation. Forty years later, Robert Selander, now of Pennsylvania State University, was down on his hands and knees with a few graduate students, scouring patches of vegetation on two adjacent city blocks. Why? Selander wanted to determine the effect of genetic drift on the introduced populations. He and his students collected every single snail—2,218 of them—from fourteen local populations. For each population, they determined the allele frequencies for five different genes.

For each of the five genes, results from their analysis pointed to some genetic variation among the colonies on the same block—and to major differences in allele frequencies *between* different blocks. Figure 17.1*b* shows the results for one of the genes studied.

Assume the genetic differences between populations will continue to increase, as through natural selection or genetic drift. Will the time come when snails from opposite sides of the street can no longer interbreed successfully, even if they do manage to get together? In other words, *will they become members of separate species*? Or will stabilizing selection work against significant genetic divergence by eliminating extreme phenotypes from the neighboring populations? After all, how much can a workable package of *H. aspersa* genes evolve in adjacent patches of vegetation—and under very similar environmental pressures—in a small town in Texas?

And with these questions, we arrive at the topic of **speciation**—that is, to changes in allele frequencies that are significant enough to mark the formation of daughter species from a parent species. As you read in the preceding chapter, Charles Darwin proposed long ago that natural selection might lead to speciation but was not sure how this actually happened. Obviously, no one was around to watch species originate during the past. No one lives long enough to know whether many existing populations are at intermediate stages leading to speciation. However, evidence is accumulating that supports certain models of speciation.

As you read through this chapter, it is important for you to recognize that speciation is not the same thing as natural selection. Evolutionary biologists now view speciation as a *potential consequence* of natural selection —and of other microevolutionary processes that may work in concert with natural selection.

KEY CONCEPTS

1. A species consists of one or more populations of individuals that can interbreed under natural conditions and produce fertile offspring, and that are reproductively isolated from other such populations. This definition is restricted to sexually reproducing species.

2. Populations of the same species have a shared genetic history, they are maintaining genetic contact over time, and they are evolving independently of other species.

3. Speciation is the process by which daughter species evolve from a parent species.

4. By one model, speciation starts when a geographic barrier arises between populations or subpopulations of a species. Thereafter, mutation, natural selection, and genetic drift operate independently in each population and lead to irreversible genetic divergence of one from the other.

5. Such populations may come to differ in certain alleles that affect morphological, physiological, or behavioral traits associated with reproduction. When the genetic differences affecting reproduction become great enough, the populations can no longer interbreed even if they later coexist in the same area. Speciation is completed.

6. The model of speciation just described, which applies to geographically separated populations, is known as allopatric speciation. This might be the way most species originate in nature. Sympatric speciation and parapatric speciation might be less prevalent routes.

7. By sympatric speciation, species form within the home range of their parent species. By parapatric speciation, adjacent populations become distinct species while still maintaining contact along a common border between their home ranges.

8. The timing, rate, and direction of speciation vary within and between lineages. The extinction of some number of species is inevitable for all lineages.

ON THE ROAD TO SPECIATION

What Is a Species?

"If it looks like a duck, walks like a duck, and quacks like a duck, probably it's a duck." Let's use this familiar saying as a starting point for defining what is and what is not a species. **Species** is a Latin word. Generally, it simply means "kind," as in "a particular kind of duck."

Even when early naturalists were defining species on the basis of morphological traits, common sense told them other factors also must be considered. Sometimes, for example, individuals of the same species look quite different because they respond to different conditions in the environment, as arrowhead plants do (Figure 17.2).

Morphological details vary so enormously, perhaps we should extend our consideration past those details and look to a basic function that unites the members of a species and isolates them from all other species.

For example, using *reproduction* as a basic, defining function is central to the **biological species concept**. The evolutionary biologist Ernst Mayr has phrased the concept this way: "Species are groups of interbreeding natural populations that are reproductively isolated from other such groups." No matter how extensive the phenotypic variation, populations belong to the same species for as long as their individuals have the form, physiology, and behavior that allow them to interbreed and produce fertile offspring. Their capacity to contribute to a shared gene pool is qualification for membership. Mayr's concept doesn't apply to asexually reproducing organisms, such as bacteria. But it is a useful guide for research into factors that define the vast majority of species, which do reproduce sexually.

If we subscribe to Mayr's concept, then speciation is the attainment of reproductive isolation. Bear in mind, reproductive isolation does not evolve purposefully to promote formation of a species or maintain its identity. Rather, *any structural, functional, and behavioral difference that favors reproductive isolation simply is a by-product of genetic change.*

Recall that genetic changes between populations of the same species can only be countered by **gene flow**, the movement of alleles into and out of populations by immigration and emigration. Gene flow can exert its homogenizing effect even when two populations are geographically separate, as long as some gene exchange continues between them. This microevolutionary process helps maintain their common reservoir of alleles.

What if some geographic barrier arises and prevents intermingling of genes between populations or even subpopulations of a species? Then, each will embark on its own evolutionary road. Through **genetic divergence**, differences will build up between the gene pools of the genetically separate populations, for mutation, natural selection, and genetic drift will now be free to operate

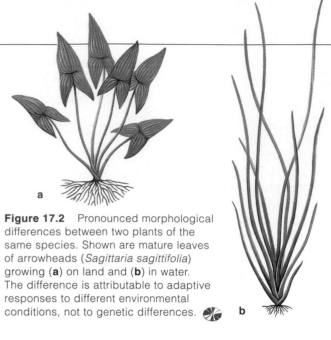

Figure 17.2 Pronounced morphological differences between two plants of the same species. Shown are mature leaves of arrowheads (*Sagittaria sagittifolia*) growing (**a**) on land and (**b**) in water. The difference is attributable to adaptive responses to different environmental conditions, not to genetic differences.

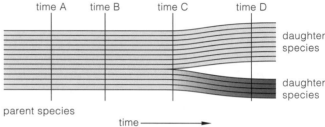

Figure 17.3 Simplified diagram of genetic divergence. Each horizontal line represents a different population.

independently in each one. Figure 17.3 is a simple way to think about genetic divergence. Each horizontal line of the sketch represents a population. At times A and B, gene flow is helping to keep them in genetic contact, and they belong to the same species. After a geographic barrier arose at time C, divergences were under way.

Most often, speciation occurs by genetic divergence. For some groups, it has proceeded far more rapidly by way of changes in chromosome number. Said another way, *the process of speciation can vary in its duration.*

Depending on how the affected populations interact and on their patterns of distribution, *speciation also may vary in its details.* The preceding paragraphs started you thinking about what is probably the main route, which is called allopatric speciation. Later in the chapter you will be taking a closer look at this route and two others.

Reproductive Isolating Mechanisms

Let's now define **reproductive isolating mechanisms** as any heritable feature of body form, functioning, or behavior that prevents interbreeding between one or more genetically divergent populations. Some prevent successful mating or pollination between individuals of the divergent populations so hybrid zygotes can't form. Let's look first at some pre-zygotic mechanisms.

a

b

Figure 17.4 (**a**) A sampling of courtship displays that precede sex between a male and a female albatross. Individuals of their species recognize the tactile, visual, and acoustical components of such displays. (**b**) Adult *Magicicada septendecim*, a periodical cicada that matures underground and then emerges to reproduce every 17 years. Often its populations overlap the habitats of its sibling species (*M. tredecim*), which reproduces every 13 years. Adults live only a few weeks. They are a taste thrill to birds, which often gorge themselves so frenziedly that they vomit; which makes one suspect why periodical cicadas stay underground for so long.

Behavioral isolation: Differences in behavior are key barriers to gene flow among related species living in the same territory. For instance, before male and female birds copulate, they often engage in intricate courtship rituals (Figure 17.4). A female is genetically equipped to recognize singing, head bobbing, wing spreading, or prancing of a male of the same species as an overture to sex. Females of different species usually ignore it.

Temporal isolation: What if individuals of divergent populations still have the potential to interbreed? If they reproduce at different times, this makes no difference. Most animals mate and most plants get pollinated quickly, sometimes in less than a day, so there is not much chance of overlap with others. Extreme temporal isolation occurs among periodical cicadas (*Magicicada*) of the eastern United States. These insects mature below the soil surface, where they feed on juices of tree roots. Three species differ in size, color, and song. They emerge and reproduce every 17 years (Figure 17.4*b*). Each of the three has a "sibling species," which is just about the same, morphologically. But its sibling species emerges every 13 years. Only once every 221 years do the two species release gametes at the same time!

Mechanical isolation: Incompatibility between body parts of potential mates or pollinators is an example of mechanical isolation. *Salvia apiana* and *S. mellifera* are two sage species. A nectar cup in their flowers attracts pollinators that use some petals as a landing platform. *S. apiana*'s pollen-bearing stamens extend from the cup, above a platform of petals large enough to hold large pollinators. *S. mellifera* has a smaller landing platform for small pollinators. When small bees visit *S. apiana*, the larger flower's stamens often do not brush against them, so they don't transport pollen to flowers of the other species. Similarly, large pollinators of *S. apiana* cannot land on *S. mellifera* and cross-pollinate it.

Ecological isolation: Populations adapted to different microenvironments in the same habitat may be isolated ecologically. In the seasonally dry foothills of the Sierra Nevada are populations of two manzanita species. One grows at elevations between 600 and 1,850 meters in open forests of conifers, the other at elevations between 750 and 3,350 meters. Where their ranges overlap, the two rarely hybridize. Like all manzanitas, both species have built-in, physiological mechanisms that enhance water conservation during the dry season. But one is adapted to the more sheltered sites, where water stress is not as intense as it is on more exposed (and drier) rocky hillsides—which the other species prefers. Those ecological differences make cross-pollination unlikely.

Gametic mortality: Gametes of different species may have evolved incompatibilities at the molecular level. For example, when the pollen of one species lands on a plant of a different species, it usually does not recognize the molecular signals that are supposed to trigger its growth down through the plant's tissues, to the egg.

What about the isolating mechanisms that take effect *after* zygotes form? Some post-zygotic mechanisms take effect after fertilization, when an embryo is developing. Unsuitable interactions among some of the embryo's genes or gene products lead to early death, sterility, or hybrids of low fitness. Hybrid offspring commonly are weak, with low survival rates. A few types are sturdy but sterile. Mules, which result from a cross between a female horse and a male donkey, are strong but sterile.

THE BIOLOGICAL SPECIES CONCEPT. **A species is one or more populations of individuals that (1) are interbreeding under natural conditions and producing fertile offspring, and (2) are reproductively isolated from other such populations.**

Speciation is the process by which a daughter species forms from a population or subpopulation of the parent species. The process can vary in its details and in the length of time it takes before reproductive isolation is complete.

By perhaps the most common mechanism of speciation, a geographic barrier separates populations of a species. Then differences build up in their gene pools because mutation, natural selection, and genetic drift operate independently in each. This outcome is called genetic divergence.

Reproductive isolating mechanisms may evolve simply as by-products of the genetic changes. They are heritable traits that, one way or another, prevent interbreeding.

WWW

Allopatric Speciation Defined

If we assume physical separation between populations promotes the genetic changes that are necessary for speciation, then allopatry may be the main speciation route. By the model for **allopatric speciation**, some type of physical barrier arises and prevents gene flow between populations or subpopulations of a species. (*Allo*– means different, and *patria* can be taken to mean homeland.) Reproductive isolating mechanisms arise in the genetically diverging populations. The process of speciation is completed when individuals of the two populations no longer will interbreed even if changing circumstances put them back together in the same area.

Whether a geographic barrier proves to be effective at blocking gene flow between populations depends on an organism's means of travel, how fast it can travel, and whether it is compelled or behaviorally inclined to disperse. You have only to reflect on the garden snails and the wandering albatross described at the start of this chapter to conclude that this is so.

The Pace of Geographic Isolation

Some measurable distance separates the populations of most species, so gene flow among them is more of an intermittent trickle than a steady stream. In addition, barriers can rapidly arise and shut off the trickles. For example, in the 1800s, a monstrous earthquake changed the course of the Mississippi River. As one outcome, the change separated populations of insects that could neither swim nor fly, and it effectively cut off gene flow between insects now living along opposite shores.

Geographic isolation also can proceed slowly, over great spans of time. We find evidence of such extended events in the fossil record, which affords glimpses into the breakup of formerly continuous environments.

For example, during past ice ages, glaciers advanced down through North America and Europe, and they gradually cut off populations from one another. When the glaciers retreated, the descendant plants, animals, and other organisms came in contact. Some groups that had descended from the same parent population were no longer reproductively compatible; they had evolved into separate species. In other groups, however, genetic divergences did not proceed far, and the descendants are still interbreeding. For them, reproductive isolation was not completed. Speciation did not occur.

As another example, as you might know, the Earth's crust is fractured into gigantic plates, somewhat like a cracked eggshell. In the past, imperceptibly slow but colossal movements of the plates caused land masses to collide and to break up. Such geologic movements uplifted part of the seafloor in the region now called the Isthmus of Panama. The uplifting divided an ancient ocean basin. It set the stage for allopatric speciation among populations of fishes and other marine species.

In the 1980s, John Graves compared four enzymes from the muscle cells of two related species of isthmus fishes (Figure 17.5). Both species are strong swimmers. As Graves knew, temperature has different effects on the activity of different enzymes. He also knew that seawater on the Pacific side of the isthmus is cooler by about 2°–3°C than it is on the Atlantic side, and that it varies more with the changing seasons. Graves found that all four "Pacific" enzymes function better at lower temperatures than the four "Atlantic" enzymes can do. Analysis by gel electrophoresis revealed slight differences in electric charge between two of the four pairs of enzyme molecules, owing to slight differences in their amino acid sequences.

Graves drew these tentative conclusions: First, selection pressures that are imposed by even small differences in environmental conditions may have caused divergences in the molecular structure of enzymes. Second, it appears that closely related populations of fishes on both sides of the isthmus are starting to diverge from each other. Why? The alternative molecular forms of the enzymes that Graves happened to study are already showing detectable differences in catalytic activity.

How might we interpret results from his study of geographically isolated populations? For the isthmus fishes, a gradual accumulation of mutations that are neutral or adaptive may be like a microevolutionary "foot in the door." In other words, genetic divergences here might be evidence of speciation in progress.

ISTHMUS OF PANAMA

Figure 17.5   (**a**) Blue-headed wrasse (*Thalassoma bifasciatum*) from the Atlantic Ocean near the Isthmus of Panama. (**b**) Cortez rainbow wrasse (*T. lucasanum*) from the Pacific Ocean near the isthmus. The fishes apparently are related by descent from a common ancestral population that split when geologic forces created the isthmus. As is common among reef fishes, these individuals of the same species differ in body coloration and patterning.

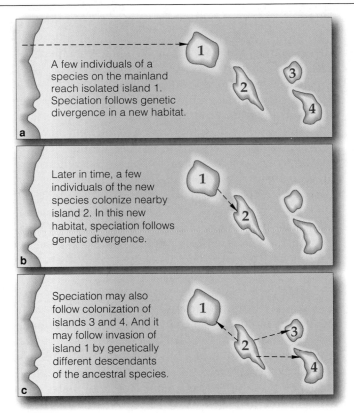

Figure 17.6 Allopatric speciation on an isolated archipelago. (Can you envision other possibilities?)

a A few individuals of a species on the mainland reach isolated island 1. Speciation follows genetic divergence in a new habitat.

b Later in time, a few individuals of the new species colonize nearby island 2. In this new habitat, speciation follows genetic divergence.

c Speciation may also follow colonization of islands 3 and 4. And it may follow invasion of island 1 by genetically different descendants of the ancestral species.

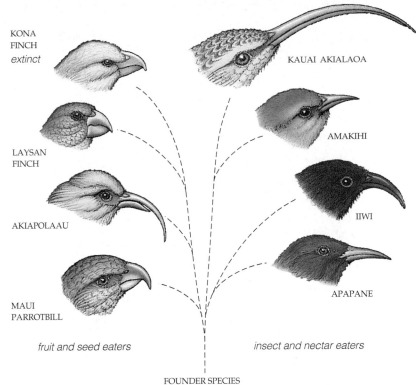

KONA FINCH *extinct*

KAUAI AKIALAOA

LAYSAN FINCH

AMAKIHI

AKIAPOLAAU

IIWI

MAUI PARROTBILL

APAPANE

fruit and seed eaters   *insect and nectar eaters*

FOUNDER SPECIES

Figure 17.7 A few Hawaiian honeycreepers, an example of how a new arrival in species-poor habitats on an isolated archipelago can be the start of a flurry of allopatric speciation.

Allopatric Speciation on Archipelagos

An **archipelago** is an island chain some distance away from a continent. Some, including the Florida Keys, are so near the mainland that gene flow remains more or less unimpeded, so that there is little, if any, speciation. Other archipelagos are so isolated that they are like laboratories for the study of evolution (Figure 17.6). Foremost among these are the Hawaiian Archipelago, nearly 4,000 kilometers from the California coast, and the Galápagos Islands, about 900 kilometers from the South American coast. Islands of both chains are only the tops of great volcanoes, some still active, that rise from a deep seafloor. When each volcano first broke the surface of the sea, it was devoid of life.

Remember those Galápagos finches (Section 16.3)? A few finches from the mainland apparently evolved in isolation on one Galápagos island. Later, some of their descendants reached other islands in the chain. In those new and unoccupied habitats, even in different parts of the same habitats, conditions varied, so that the island hoppers were subject to a variety of selection pressures. In time, genetic divergences within and between the islands paved the way for new episodes of allopatric speciation. Later on, some island-hopping new species even invaded the island of the ancestors. The distances between these islands are enough to foster divergence, but not enough to stop occasional invasions.

Bursts of speciation in the Hawaiian Archipelago have been so dramatic, they are a premier example of adaptive radiation in a species-poor environment, as described in Section 17.4. The youngest island, Hawaii, is less than a million years old. Here alone we find a great diversity of habitats, ranging from rain forests to alpine grasslands below high, snow-capped volcanoes. When the ancestors of Hawaiian honeycreepers arrived, they found a veritable buffet of fruits, seeds, nectars, and tasty insects—and not many competitors for them. The near-absence of competition from other bird species fanned allopatric speciations. Figure 17.7 only hints at the resulting variation that arose among the different species of Hawaiian honeycreepers. Today, these and thousands of other species of animals as well as plants that evolved in the archipelago are found nowhere else in the world. As a final example of the potential for speciation, the Hawaiian Islands represent less than 2 percent of the world's land mass, yet they are home to 40 percent of all species of *Drosophila*.

ALLOPATRIC SPECIATION. Some type of physical barrier intervenes between populations or subpopulations of a species and prevents gene flow among them. By doing so, it favors genetic divergence and speciation.

WWW

Geographic isolation might not always be required for reproductive isolation. By the models of sympatric and parapatric speciation, genetically divergent populations that are *not* separated by geographic barriers and are still in contact can become reproductively isolated, also.

Sympatric Speciation

By the model for **sympatric speciation**, species may form *within* the home range of an existing species, in the absence of a physical barrier. (*Sym—* means together with, as in "together with others in the homeland.")

EVIDENCE FROM CICHLIDS IN AFRICA   In 1994, Ulrich Schliewen and his coworkers found evidence in favor of the model of sympatric speciation. They had studied certain fishes, cichlids, that were living together in two lakes in Cameroon, West Africa. The lake basins are the collapsed cones of volcanoes. Each lake is quite small (Figure 17.8). Yet eleven cichlid species coexist in one, and nine coexist in the other.

The researchers analyzed mitochondrial DNA from all the species in one lake. They did the same for all the species in the other lake as a basis for comparison. In nucleotide sequences, the species of each lake are like each other and *not* like related species in nearby lakes and rivers. For example, nine species in one lake are the only ones that carry an unusual base-pair substitution in the gene that specifies the protein cytochrome *b*.

Physical and chemical conditions are too uniform in the lakes to foster geographic separation. For instance, the uniform shorelines do not present even small-scale topographic barriers. Therefore, allopatric populations could not have formed even on a small scale. The crater rim isolates the lakes from all but tiny creeks trickling in from higher elevations. Thus there has been no gene flow from outside.

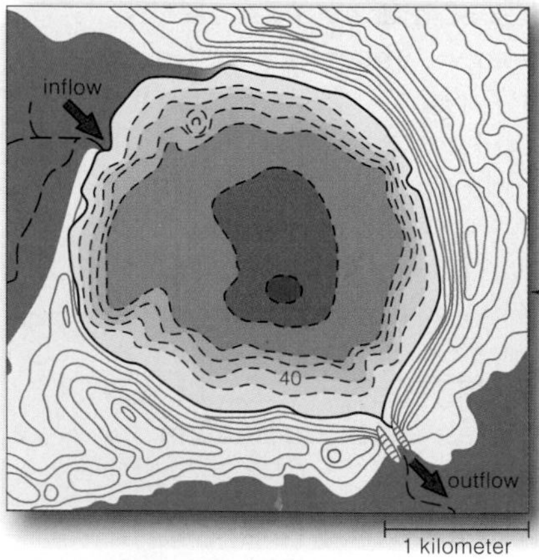

Figure 17.8 Topographical map of Lake Barombi Mbo in Cameroon, West Africa. Apparently, nine kinds of cichlids evolved by sympatric speciation in this small, isolated crater lake. Even microgeographic separation is absent from the lake. There is separation by feeding preferences, but all species breed near the lake bottom, in sympatry.

The lakes must have been colonized before connections with a river system on the outside were severed.

Cichlids are mobile, not sluggish, fishes. Besides this, it's safe to assume individuals of different species often encounter one another in such small crater lakes. Said another way, they all must be living in sympatry.

Identifying clusters of species that are more closely related to one another than to species anywhere else suggests they are all descended from a single ancestor. The restricted distribution suggests each formed in the same lake, but there is nothing about these lakes that can isolate different populations of fishes. Apparently, then, the ancestors of each present-day cichlid species were never isolated from one another while they were evolving in separate directions.

Inside each lake, species do show small degrees of *ecological* separation that arises through differences in feeding preferences. Some cichlid species feed in open waters and others at the lake bottom. Even so, they all *breed* close to the bottom, in sympatry. It may be that ecological separation on a small scale was enough to influence sexual selection among potential mates and, over generations, to provide the reproductive isolation that can lead to speciation.

SPECIATION BY WAY OF POLYPLOIDY   Quite possibly, sympatric speciation has been a prevalent evolutionary event among flowering plants. Consider that about half of all known species of flowering plants are polyploid. **Polyploidy**, remember, is a change in the chromosome number; offspring inherit three or more of each type of chromosome characteristic of the parental stock. Such changes arise when chromosomes separate improperly during meiosis or mitosis. They also arise when a germ cell replicates its DNA but fails to divide, then goes on to function as a gamete (Section 11.9).

Speciation may have been rapid for many flowering plants that engage in self-fertilization or some asexual reproductive mode. Suppose they produced polyploid offspring. If the extra chromosomes paired with each other during meiosis, maybe the extra set of genes did no harm. Common bread wheat is one of the species that might have arisen by polyploidy, although in this case cross-fertilization also was involved (Figure 17.9).

Polyploid animals are rare. Allen Orr has evidence that this is a result of a failed "dosage compensation," meaning genes on sex chromosomes are expressed at the same levels in females *and* males. For instance, X chromosome inactivation in female mammals means the cells of both males and females have a single active X chromosome (Section 14.3). Polyploidy skews dosage compensation, with bad or fatal effects. In plants, the mechanism is absent, so chromosome doublings are not as problematic as they are in animals.

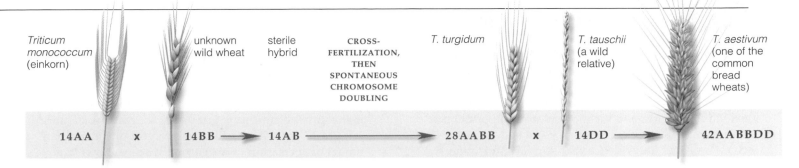

| 14AA | x | 14BB $\rightarrow$ 14AB | 28AABB | x | 14DD $\rightarrow$ | 42AABBDD |

a By 11,000 years ago, humans had started to cultivate wild wheats. Einkorn (*Triticum monococcum*) is still around. It has diploid chromosome number of 14 (two sets of 7 chromosomes, shown above as 14AA). Long ago, einkorn wheat probably hybridized with another species having the same chromosome number.

b If the AB hybrid offspring were sterile but self-fertilizing, an interbreeding population of AB plants could have arisen by asexual reproduction. About 8,000 years ago, polyploidy did indeed arise in such a population. Wild emmer (*T. turgidum*) plants are tetraploid (AABB), with a chromosome number of 28 (two sets of 14). They are fertile; at meiosis, the A chromosomes pair with each other, and B chromosomes pair with each other.

c Later, an AABB plant probably hybridized with *T. tauschii*, a wild relative of the wild emmer. Its diploid chromosome number must have been 14 (two sets of 7 DD). Populations of the hybrid descendants now include common bread wheats, such as *T. aestivum*, which have a chromosome number of 42 (six sets of 7 AABBDD).

Figure 17.9 Presumed sympatric speciation in wheat by polyploidy and hybridizations over time. Wheat grains 11,000 years old were found in the Near East. Diploid wild wheats still grow there.

BULLOCK'S ORIOLE   BALTIMORE ORIOLE

Figure 17.10 A possible setting for parapatric speciation? Hybrids arise along the common border of the ranges of two species of orioles. To the east are Baltimore orioles; to the west are Bullock's orioles. Hybridization was once so common, they were considered subspecies. In 1997, the American Ornithologist Union decided they are separate species.

Parapatric Speciation

By the model for **parapatric speciation**, neighboring populations become distinct species while maintaining contact along a common border. (*Para*— means near, as in "near another homeland.") Interbreeding individuals produce hybrid offspring in this region, which is called a **hybrid zone**. Evidence that might support parapatric speciation is sketchy. Why? In many cases, it is difficult to determine whether or not geographically separated populations are only *subspecies* (that is, geographically distinct populations of the same species).

For example, Bullock's orioles differ from Baltimore orioles in the color patterning of most of their feathers.

The male orioles have different territorial songs, and generally they mate with their own kind. As K. Corbin determined, these birds also differ in the frequencies of alleles for several enzymes. Their geographic ranges differ, but the ranges overlap in the American Midwest (Figure 17.10). Interbreeding was once common in the hybrid zone, although it is now becoming less frequent. Is this an example of parapatric speciation in progress? Maybe. But it might also be an example of "secondary contact." In other words, previously isolated subspecies that diverged recently from a common ancestor may have been getting together again.

SYMPATRIC SPECIATION. **Daughter species arise from a group of individuals within an existing population. This may have been a common speciation event among polyploid flowering plants.**

PARAPATRIC SPECIATION. **Adjacent populations evolve into distinct species even while maintaining contact along their common border.**

WWW

PATTERNS OF SPECIATION

Branching and Unbranched Evolution

All species, past and present, are related by descent. They have genetic connections through lineages that extend back in time to the molecular origin of the first prototypic cells, some 3.8 billion years ago. Subsequent chapters focus on evidence that supports this view. In anticipation of those chapters, let's start thinking about ways to interpret the large-scale histories of species.

The fossil record provides evidence of two patterns of evolutionary change in lineages—one branching and the other unbranched. The first is called **cladogenesis** (from the Greek *klados*, meaning branch; and *genesis*, meaning origin). This is the pattern by which a lineage splits, with populations becoming genetically isolated and then diverging in different evolutionary directions. This is the pattern of speciation described earlier.

By the second pattern, **anagenesis**, changes in allele frequencies and in morphology accumulate within an unbranched line of descent. (In this context, *ana*— means renewed.) Directional changes are confined to a single lineage, and gene flow never does cease among all the populations of the lineage. In time, the morphology and underlying allele frequencies change so much that we give the descendants their own name. Such changes are occurring in the moth populations you read about earlier.

Evolutionary Trees and Rates of Change

Evolutionary trees summarize information about the continuity of relationship among species. Figure 17.11 is a simple way to start thinking about how tree diagrams are constructed. Each *branch* represents a single line of descent from a common ancestor. And each *branch point* represents a time of genetic divergence and speciation, as brought about by microevolutionary processes.

Figure 17.11 also shows how an evolutionary tree diagram conveys rates of change, or the approximate length of time between two speciation events. Branches with slight angles imply the species emerged through many small changes in form over long spans of time (Figure 17.11a). This is the key premise of the **gradual model of speciation**, which actually fits well with many fossil sequences. As an example of this, in many layers of sedimentary rock we find sequences of intricately perforated shells of single-celled foraminiferans, a type of protistan. The observable characteristics of fossils in the sequences provide evidence of slow change.

Alternatively, evolutionary tree diagrams for some lineages are constructed with short, horizontal branches that abruptly make a 90-degree turn (Figure 17.11b). These diagrams are consistent with the **punctuation model of speciation**. According to this model, most morphological changes are compressed within a brief period when populations are starting to diverge—say, within merely hundreds or thousands of years. The idea is that bottlenecks, founder effects, strong directional selection, or some combination of these brings about rapid speciation. Daughter species recover quickly from the adaptive wrenching, then change little over the next 2 million to 6 million years or so. Said another way, reproductive cohesion seems to have prevailed for about 99 percent of the history of most lineages. This pattern seems to pervade many parts of the fossil record.

Apparently, changes in evolutionary trees have been gradual, abrupt, or both. Species originated at different times, and they differed in how long they persisted on the evolutionary stage. Remember, some lineages have endured without much change, producing a species here and losing a species there, often over millions of years. Other lineages have branched bushily, and sometimes spectacularly, during times of adaptive radiation.

Adaptive Radiations

An **adaptive radiation** is a burst of divergences from a single lineage that give rise to many new species, each adapted to an unoccupied or a new habitat or to using a novel resource. Figure 17.12 provides an example. In the past, lineages often diversified in such a way when the member species found themselves in vacant **adaptive zones**. Think of

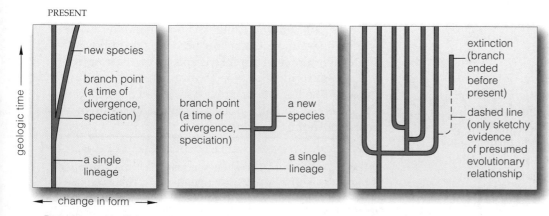

PRESENT

geologic time

← change in form →

a Branching with slight angle; new species formed through gradual changes in traits over geologic time.

b Horizontal branching; rapid change in traits around time of speciation. Vertical continuation of branch means traits of new species did not change much thereafter.

c Many branchings of the same lineage at or near the same point in geologic time; an adaptive radiation has taken place.

Figure 17.11 How to read evolutionary tree diagrams.

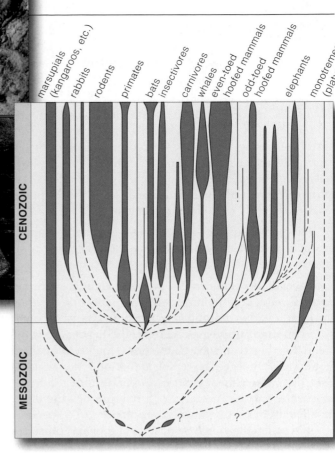

Figure 17.12 Example of an adaptive radiation, as shown by an evolutionary tree diagram. This one started about 65 million years ago, at the start of a geologic era called the Cenozoic, and it led to mammals as different as the opossum and walrus. Variations in the width of a given branch correspond to the range of species diversity (the number of groups) within the lineage as represented at different points in time. The wider the branch, the greater the diversity.

adaptive zones as ways of life, such as "burrowing in seafloor sediments" or "catching insects in the air at night." A lineage may radiate into such zones when it has physical, evolutionary, or ecological access to them.

Physical access means a lineage happens to be there when adaptive zones open up. For example, mammals were once distributed through uniform tropical regions of a single continent. That continent split into several land masses. Habitats and resources changed in many different ways on the separate masses and set the stage for independent radiations.

Evolutionary access means that modification of some structure or function will permit a lineage to exploit the environment in novel or more efficient ways. Such modifications are called **key innovations**. For example, the forelimbs of certain five-toed vertebrates evolved into wings. That innovation opened up new adaptive zones for the ancestors of birds and bats.

Ecological access means that the lineage can enter an unoccupied adaptive zone or displace resident species. We will say more about this in Chapter 40.

Extinctions—End of the Line

Extinction is the irrevocable loss of a species. The fossil record shows twenty or more **mass extinctions**—large, catastrophic events in which entire families or other major groups disappeared during the same phase or at the same point in geologic time. For example, 95 percent of all known species throughout the world abruptly vanished 250 million years ago. In other times, only certain groups or certain regions were extinguished.

By studying the statistical distribution of extinctions, David Raup found that most were clustered in time, even though they differed in size. Like George Simpson before him, he realized that times of reduced diversity follow extinction events, then new species arise and fill

vacant adaptive zones. Apparently, luck has a lot to do with it. For example, in the past, asteroids repeatedly hit the Earth and leveled the playing field. Some of the survivors of one impact radiated into adaptive zones occupied earlier by diverse dinosaurs. Among them were mammals—some of which gave rise to your early ancestors. As another example, the species that are not widely distributed tend to be hit hard by pronounced declines in global temperature. This is usually the case for species adapted to warm regions. Unlike species in cold regions, they have nowhere else to go. Asteroids, drifting continents, climatic change—as you will see in later chapters, these are some of the major factors that have contributed to a general pattern of extinctions.

Lineages have changed gradually, abruptly, or both. Their member species originated at different times and differ in how long they have persisted.

Adaptive radiations are bursts of divergences from a single lineage that gave rise to many new species, each adapted to a vacant or new habitat or to using a novel resource.

Repeated and often large extinctions have occurred in the past. After times of reduced diversity, new species formed and occupied the new or vacated adaptive zones.

Taken together, the persistence, branchings, and extinctions of species account for the full range of biological diversity at any point in geologic time.

SUMMARY

1. A species is a single kind of organism, recognized partly in terms of its morphology.

2. By the biological species concept, a species consists of one or more populations of individuals that are, under natural conditions, interbreeding and producing fertile offspring, and that are reproductively isolated from other such populations. The concept identifies a species primarily in terms of the portion of alleles that promote or maintain reproductive isolation. It applies only to sexually reproducing organisms.

3. The populations of a species have a shared genetic history, they are maintaining genetic contact over time, and they are evolving independently of other species.

4. Speciation is the process by which daughter species form from a population or subpopulation of a parent species. It is the attainment of reproductive isolation.

  a. As an example, if a geographic barrier prevents gene flow between such populations, they will undergo genetic divergence. The reason is that mutation, natural selection, and genetic drift can operate independently in each. Genetic divergence is a buildup of differences in allele frequencies between populations of a species.

  b. Among their effects, microevolutionary processes may simply by chance give rise to reproductive isolating mechanisms that prevent interbreeding. Isolation may then result in irreversible genetic differences between the populations.

  c. Speciation proceeds gradually, by way of genetic divergence. It also can proceed instantaneously, as by polyploidy or other changes in chromosome number.

5. Prezygotic isolating mechanisms prevent mating or pollination between individuals of populations. They include differences in reproductive timing or behavior, incompatibilities in reproductive structures or gametes,

and occupation of different microenvironments in the same area. Postzygotic mechanisms lead to early death, sterility, or unfit hybrid offspring. They take effect after fertilization, while the embryo develops.

6. There are three current models of speciation:

  a. Allopatric speciation. Geographic barriers prevent gene flow between the populations of a species, which genetically diverge in ways that interbreeding will not happen in nature even if their individuals later make contact with each other. Allopatric speciation might be the most prevalent speciation route.

  b. Sympatric speciation. Reproductive isolation of individuals that share the same home range leads to speciation. Speciation by polyploidy is an example.

  c. Parapatric speciation. Adjacent populations give rise to a new species while maintaining contact along a border between their home ranges.

7. Lineages differ in their time, rates, and direction of speciation. Speciation can proceed gradually, rapidly, or both. Extensive branching of a lineage (divergences of one or more of its populations) in the same geologic time span is an adaptive radiation. Often this occurs as member species of a lineage become adapted to new or unoccupied habitats or to using novel resources.

8. Evolutionary tree diagrams indicate the relationships among groups of species. Each branch of such trees represents a line of descent (lineage). Branch points are speciation events, as brought about by natural selection, genetic drift, and other microevolutionary processes.

9. Extinctions are losses of species. Most are clustered in time, and their size varies greatly. Asteroid impacts, volcanism, continental drift, and other environmental changes have apparently brought about most of them.

10. Persistences, branchings, and extinctions of species account for the full range of biological diversity through geologic time (Table 17.1).

Table 17.1 Summary of Processes and Patterns of Evolution		
MICROEVOLUTIONARY PROCESSES		
Mutation	Original source of alleles	Stability or change in a species is the outcome of balances or imbalances among all of these processes, the effects of which are influenced by population size and by the prevailing environmental conditions.
Gene flow	Preserves species cohesion	
Genetic drift	Erodes species cohesion	
Natural selection	Preserves or erodes species cohesion, depending on environmental pressures	
MACROEVOLUTIONARY PROCESSES		
Genetic persistence	Basis of the unity of life. The biochemical and molecular basis of inheritance extends from the origin of first cells through all subsequent lines of descent.	
Genetic divergence	Basis of life's diversity, brought about by adaptive shifts, branchings, and radiations. Rates and times of change have varied within and between lineages.	
Genetic disconnect	Extinction. End of the line for a species. Extinctions tend to be clustered in time as discrete events; many are simultaneous, mass extinctions of many species.	

Review Questions

1. How does the biological species concept differ from a definition of species based on morphological traits alone? *17.1*

2. Define speciation and describe a speciation model. *CI, 17.2–17.4*

3. Give examples of reproductive isolating mechanisms. *17.2*

4. Interpret the following features of evolutionary tree diagrams: *17.4*
  a. a single line
  b. soft-angled branching of line
  c. horizontal branching
  d. vertical continuation of branch
  e. many branchings of a line
  f. dashed line
  g. branch ending before present

1. Sexually reproducing individuals of a species _____ .
   a. can interbreed under natural conditions
   b. can produce fertile offspring
   c. have a shared genetic history
   d. all of the above

2. Reproductive isolating mechanisms _____ .
   a. prevent interbreeding   c. reinforce genetic divergence
   b. prevent gene flow   d. all of the above

3. When potential mates occupy overlapping ranges but reproduce at different times, this is a case of _____ isolation.
   a. postzygotic   c. temporal
   b. mechanical   d. gametic

4. In an evolutionary tree diagram, a branch point represents _____ , and a branch that ends represents _____ .
   a. a single species; incomplete data on lineage
   b. a single species; a time of extinction
   c. a time of divergence; extinction
   d. a time of divergence; speciation complete

5. An evolutionary tree diagram with horizontal branches that abruptly become vertical is consistent with the _____ .
   a. gradual model of speciation
   b. punctuation model of speciation
   c. idea of small changes in form over long spans of time
   d. both a and c

6. Match each term with its most suitable description.
   ____ cladogenesis   a. burst of microevolutionary
   ____ anagenesis   activity within a lineage
   ____ adaptive   b. catastrophic disappearance
        radiation   of major groups of organisms
   ____ extinction   c. branching lineages
   ____ mass extinction   d. a species lost from a lineage
                          e. genetic and morphological
                             change in unbranched lineage

Critical Thinking

1. You notice several duck species in the same lake habitat. The females of different species look very similar to one another, but the males of each particular species have feathers with distinctive patterns and colors. Speculate on which forms of reproductive isolation may be keeping each species distinct. How does the appearance of the male ducks provide a clue to the answer?

2. One family of mammals includes true horses (*Equus*), zebras (*Hippotrigris* and *Dolichohippus*), and donkeys and asses (*Asinus*). Zebroids are hybrid offspring of wild zebras and domesticated horses that were confined to the same pasture (Figure 17.13). The unnatural confinement breached the reproductive barriers between the two lineages. Those barriers have been in place since a divergence more than 3 million years ago. What does the breach say about the number of genetic changes required to attain reproductive isolation in nature?

3. Suppose a small population that was founded by only a few individuals is cut off from gene exchange with the main population of their species. By Mayr's view of the *founder effect*, the frequencies of alleles for a few gene products change as a result of genetic drift, and the change triggers a number of changes in other genes that are affected by those alleles. If that is so, then genetic changes in newly founded populations may be so great that speciation occurs rapidly—so rapidly that we will seldom be able to document it through the fossil record. Does this view correspond to the gradual model or punctuated model of speciation?

Figure 17.13 A mixed herd of zebroids and horses.

4. A key innovation, recall, is some modification in structure or function that permits a species to exploit the environment in a more efficient or novel way, compared to the ancestral species. Identify a key innovation of an existing species, such as humans. Then describe how that innovation might be the basis of an adaptive radiation in environments of the distant future.

5. Richard Lenski uses bacterial populations in culture tubes to develop model systems for studying evolution. He is fond of saying that such a population is the equivalent of the entire human population, that he can replicate it several times over, that bacteria produce several generations in a day, and that he can store them in the deep freeze, then bring them back to active form, unaltered, to directly compare ancestors and descendants. Are bacterial models relevant to evolutionary studies of sexually reproducing organisms? Before you answer, read a short article by P. Raine and M. Travisano entitled "Adaptive Radiation in a Heterogenous Environment" (*Nature*, 2 July 1998, 69–72).

Selected Key Terms

adaptive radiation *17.4*   hybrid zone *17.3*
adaptive zone *17.4*   key innovation *17.4*
allopatric speciation *17.2*   mass extinction *17.4*
anagenesis *17. 4*   parapatric speciation *17.3*
archipelago *17.2*   polyploidy *17.3*
biological species concept *17.1* punctuation model,
cladogenesis *17.4*   speciation *17.4*
evolutionary tree *17.4*   reproductive isolating
extinction *17.4*   mechanism *17.1*
gene flow *17.1*   speciation *CI*
genetic divergence *17.1*   species *17.1*
gradual model, speciation *17.4*   sympatric speciation *17.3*

Readings See also *www.infotrac-college.com*

Futuyma, D. 1998. *Evolutionary Biology*. Third edition. Sunderland, Massachusetts: Sinauer.

Mayr, E. 1976. *Evolution and the Diversity of Life*. Cambridge, Massachusetts: Belknap Press of Harvard University Press.

WWW *http://www.brookscole.com/biology*

Practice quiz questions, hypercontents, BioUpdates, and critical thinking. The Brooks/Cole Biology Resource Center provides a wealth of information fully organized and integrated by chapter.

THE MACROEVOLUTIONARY PUZZLE

Of Floods and Fossils

About 500 years ago, Leonardo da Vinci was brooding about seashells entombed in layered rocks of northern Italy's high mountains, hundreds of kilometers from the sea. How did they get there? If he accepted the traditional explanation, he would have to agree that stupendous floodwaters deposited those shells in the mountains during the Great Deluge (Figure 18.1). But many shells were thin and fragile—yet intact. Surely they would have been battered to bits if they had been swept across such distances, then up the mountains.

Da Vinci also brooded about the rocks. They were stacked like cake layers, and some contained shells but others had none. Then he remembered how large rivers deposit silt with each spring flood. *Did the layers slowly accumulate long ago, as a series of silt deposits where rivers emptied into the sea*? If so, then shells in the mountains would be evidence of a progression of communities of organisms that once lived in the seas! Da Vinci did not announce his novel idea, perhaps knowing it would be met with deafening silence, imprisonment, or worse.

By the 1700s, fossils were accepted as evidence of past life. They were still being interpreted through the prism of prevailing cultural beliefs, as when a Swiss naturalist excitedly unveiled the remains of a giant salamander and announced they were the skeleton of a man who had drowned in the Deluge.

By midcentury, however, scholars began to question such interpretations. Extensive mining, quarrying, and canal excavations were under way. The diggers were finding similar rock layers and similar fossil sequences in distant places, such as the cliffs on both sides of the English Channel. More than a few scholars began to view the findings as evidence of connections between Earth history and the history of life.

Ever since, fossils have been analyzed in increasingly refined ways. Together with biochemical studies and other modern sources of information, they yield good evidence of evolution through vast spans of time.

What does the evidence tell us? *Only populations that already exist can evolve.* As a result of mutations, natural selection, and genetic drift, each species is a mosaic of ancestral and novel traits that emerged earlier, along unbranched or branching evolutionary roads that began with the first populations of the very first species on

Earth. Thus, *all species that ever evolved are related to one another, by way of descent*. This principle of evolution guides efforts to make sense of often puzzling scraps of evidence of past life. It guides the task of identifying and sorting out the many lines of descent, or **lineages**, that connect all species, past and present.

Ultimately, then, life is a story of *species*—of how and when each kind of organism originated, whether its defining traits persisted or were modified, and whether it vanished or endured. Life also is a story of **macroevolution**—of large-scale patterns, trends, and rates of change among families and other more inclusive groups of species. We turn to this larger picture in the next unit of the book. Here we begin with the nature of the evidence that supports it.

Figure 18.1 Two interpretations of the past. *Left*: From the Sistine Chapel, Michelangelo's painting of the onset of the Great Deluge, a catastrophic flood that became integrated into biblical thought. *Below*: A modern-day photographer captures a pattern of fossilized shells of ammonites. About 65 million years ago, all of these marine animals abruptly perished, along with many other groups of organisms. That mass extinction is one intriguing piece of the evolutionary puzzle.

KEY CONCEPTS

1. In the evolutionary view, all species that have ever lived are related—some closely, others remotely so. Their relatedness is based on this premise: Each new species evolved from variant individuals of species that already existed, starting with the first living cells to appear on Earth.

2. The term macroevolution refers to the patterns, trends, and rates of change among lineages over geologic time.

3. The fossil record, the geologic record, and radiometric dating of rocks yield evidence of macroevolution.

4. Anatomical comparisons help us understand and reconstruct patterns of change through time. Among the most revealing aspects of anatomy are cases of similar, homologous structures in the adult forms or embryos of different lineages.

5. Homology implies descent from a common ancestor, with evolution in different lineages proceeding through conservative modifications to the shared body plan.

6. Biochemical comparisons within and between major lineages provide strong evidence of macroevolution.

7. Stunning diversity characterizes the distribution of species through time and through the global environment. Biological systematics attempts to discern patterns in life's diversity through taxonomy, phylogenetic reconstruction, and classification.

8. Taxonomy is concerned with identifying and naming new species. Phylogenetic reconstruction is concerned with working out evolutionary connections. Classification is an attempt to organize information about species into retrieval systems.

Fossilization

"Fossil" comes from a Latin word for something that has been "dug up." In general, **fossils** are recognizable, physical evidence of ancient life, as in Figure 18.2.

Most of the fossils discovered so far are bones, teeth, shells, seeds, capsules of spores, and other hard parts. (Soft parts usually are the first to decompose when an organism dies.) Imprints of leaves, stems, tracks, trails, burrows, and other *trace* fossils yield indirect evidence

Figure 18.2 (**a**) A fossil hunter's dream: a complete skeleton of a bat that lived 50 million years ago. Erosion and other forces of nature have left few ancient burial sites undisturbed, so intact fossils are rare. Even jumbled parts of the ducklike birds in (**b**) are a good find. It will take hours of preparation and analysis to identify the species. (**c**) Fossilized parts of the oldest known land plant (*Cooksonia*). Its stems were not even seven centimeters tall. (**d**) A complete fossil—skeletal remains of an ichthyosaur, a dolphin-like marine reptile that lived 200 million years ago.

of past life. Even fossilized feces, or coprolites, contain residual evidence of which species were being eaten—and therefore were present—in ancient environments.

Fossilization is a very slow process that starts when an organism, or traces of it, becomes buried in volcanic ash or in sediments at the bottom of some lake, lagoon, or sea. Sooner or later, water infiltrates the organic remains, which become infused with dissolved metal ions and other inorganic compounds. More and more sediments gradually accumulate above the burial site, and they exert ever increasing pressure on the remains. Over great spans of time, the pressure and the chemical changes transform those remains to stony hardness.

Preservation is favored when organisms are buried rapidly, in the absence of oxygen. Gentle entombment by volcanic ash or anaerobic mud is best. Preservation also is favored when a burial site stays undisturbed. Most often, however, erosion and other geologic insults crush, deform, break, or scatter the fossils.

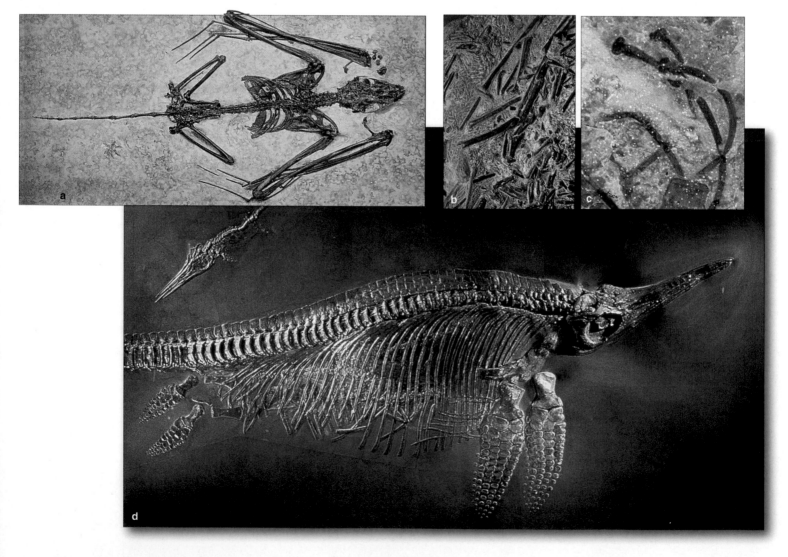

Figure 18.3 The geologic time scale. The major boundaries mark times of mass extinctions. Radiometric dating methods (Section 18.7) allowed researchers to assign absolute dates. Life originated during the Archean.

The time spans shown are not to scale. If they were, the Archean and Proterozoic portions would run off the page. Think of the spans as minutes on a clock that runs from midnight to noon. If we say that life originated at midnight, the Paleozoic began at 10:04 A.M., the Mesozoic at 11:09 A.M., and the Cenozoic at 11:47 A.M. The recent epoch of the Cenozoic started during the last 0.1 second before noon.

Eon	Era	Period	Epoch	Millions of Years Ago (mya)
PHANEROZOIC	CENOZOIC	QUATERNARY	Recent	0.01–
			Pleistocene	1.55
		TERTIARY	Pliocene	5
			Miocene	25
			Oligocene	38
			Eocene	54
			Paleocene	65
	MESOZOIC	CRETACEOUS	Late	100
			Early	138
		JURASSIC		205
		TRIASSIC		240
	PALEOZOIC	PERMIAN		290
		CARBONIFEROUS		360
		DEVONIAN		410
		SILURIAN		435
		ORDOVICIAN		505
		CAMBRIAN		570
PROTEROZOIC				2,500
ARCHEAN				4,600

Interpreting the Geologic Tombs

We find similar fossil-containing layers of sedimentary rock over vast areas, even on different continents. Such layers formed long ago by the gradual deposition of volcanic ash, silt, and other materials, one above the other. This layering of sedimentary deposits is called **stratification**. Generally, the deepest layers were the first to form. Layers closest to the surface formed last. Because particles tend to settle in response to gravity, we can assume that most sedimentary layers must have formed horizontally. Where they are tilted or ruptured, this is evidence of subsequent geologic disturbance.

Understand how rock layers form, and you realize that fossils within a particular layer are from a similar age in Earth history. *And the older the layer, the older the fossils.* Given that rock layers formed in sequence, then the fossil assemblages that they contain were unique to sequential ages. Therefore, fossils can be used to assign relative dates to the record of the rocks.

Early geologists identified four abrupt transitions in the sequence of fossil assemblages. They found them in the fossil record from regions around the world. They used those transitions as the boundaries between four great intervals in time. At the time, the oldest fossils were from the first interval, the Proterozoic. They were followed in time by fossils from the Paleozoic, Mesozoic, and finally the "modern" era, the Cenozoic.

We still use the boundaries between those intervals in a chronological chart of Earth history. Figure 18.3 is a modern version of this **geologic time scale**. As we now know, the boundaries correlate with extinction events. For example, the boundary between the Mesozoic and Cenozoic marks the mass extinction of the dinosaurs and other groups of reptiles, which was followed by a major adaptive radiation of mammals (Section 17.4).

Interpreting the Fossil Record

The known fossils for about 250,000 species are clues to evolutionary history. Judging from the current range of diversity, there must have been many, many millions of ancient, now-extinct species. We never will be able to recover fossils for most of them, so our "record" of past life is incomplete, with built-in biases. Why is this so?

Most importantly, colossal movements in the Earth's crust obliterated evidence from crucial intervals in the past. Besides this, most species of ancient communities simply weren't preserved. For example, bony fishes and hard-shelled mollusks are well represented in the fossil record. Soft-bodied worms and jellyfishes are not, even though they may have been just as common or more so. Population density and body size also skew the record. For example, a population of plants may have released millions of spores in a single growing season, whereas the earliest humans lived in small groups and produced few offspring. What are the chances of finding even one fossilized skeleton of an early human, as compared to finding spores of plants that lived at the same time?

Also, the fossil record is heavily biased toward some environments. Most species we know about lived either on land or in shallow seas which, as a result of geologic uplifting, became part of continents. We have recovered very few fossils from the Southern Hemisphere and from sediments beneath the ocean—which covers nearly three-fourths of the Earth's surface! Why? Less time has been spent searching for fossils in those environments.

Fossils, the stone-hard physical evidence of ancient life, are present in layers of sedimentary rock. The deeper the layers, the older the fossils. The geologic time scale is based on sequences of fossils in sedimentary rocks.

The completeness of the fossil record varies as a function of the kinds of organisms represented, where they lived, and the stability of their burial sites.

WWW

18.7 DATING PIECES OF THE MACROEVOLUTIONARY PUZZLE

For interested students, we leave this chapter with a look at a powerful tool for dating fossils and gaining insights into the history of life. Remember, probably for as long as they have been digging up the earth, people have been turning up fossilized leaves, shells, skeletons, and other stone-hard evidence of past life. Figure 18.18 shows two examples. *How do we know how old they are?*

Figure 18.18 (**a**) Fossilized frond of a tree fern that unfurled its leaves more than 250 million years ago. (**b**) A fossilized leaf that dropped from a sycamore tree 50 million years ago.

As you read in Section 18.6, some time ago, geologists hypothesized that if newly formed rocks accumulate on top of older rocks, then fossils in older rock layers must be more ancient than those in more recently deposited ones. They used this perception to count backwards through great numbers of rock layers and construct a chronology of Earth history—a geologic time scale. They defined the boundaries of successive spans by sequences of fossils and other clues in the rocks. However, *no one was able to assign firm dates along the length of the scale until the discovery of radioactive decay—which made it possible to convert relative ages into absolute ones.*

By a method called **radiometric dating**, researchers now measure the proportions of (1) a radioisotope in a mineral that became trapped long ago in a brand-new rock and (2) a daughter isotope that formed from it in the same rock. Remember, the atomic nucleus of each radioactive element consists of an *unstable* number of protons and neutrons. The nucleus spontaneously decays—it gives up energy and one or more particles of itself—until it reaches a more stable configuration.

Half-life is the time it takes for half of a given quantity of any radioisotope to decay into a different, daughter isotope that is less unstable (Figure 18.19). The rate of decay is constant; changes in pressure, temperature, or chemical state cannot alter it. For example, uranium 238 occurs in zircon, which is a mineral in most volcanic rocks. Its nuclei decay into thorium 234, a still-unstable daughter isotope that in turn decays into something else, and so on through a series of intermediate daughter isotopes to lead 206—the final, most stable configuration for this series (Table 18.1). By using uranium 238, with its half-life of 4.5 billion years, researchers realized that the Earth formed more than 4.6 billion years ago.

Radiometric dating works for volcanic rocks or ashes. However, most fossil-containing rocks formed by the compaction of sand and other sediments. Thus, the only way to date most fossil-containing rocks is to determine their position relative to volcanic rocks in the same area.

You may have heard of dating methods based on carbon 14 (^{14}C). The greatest numbers of this radioisotope form continually in the upper atmosphere, where they combine with oxygen to form carbon dioxide. Along with the more stable isotopes of carbon, small amounts of ^{14}C enter the web of life by photosynthesis. All organisms incorporate it. When organisms die, the amount of ^{14}C starts to decrease as an outcome of radioactive decay. ^{14}C has a half-life of 5,730 years. Thus, amounts in bone, wood, shells, and other organic substances are too small to detect by direct analysis. Instead, researchers employ scintillation counters to monitor emissions from a given sample. The older the sample, the fewer emissions (counts) will be recorded in a given period. *WWW*

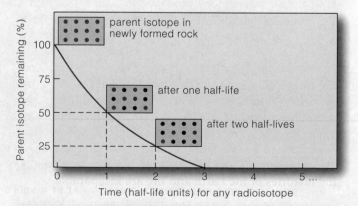

Figure 18.19 Generalized diagram of radioactive decay. Notice the proportion of the parent radioisotope in a rock sample over time, compared to the proportion of each decay product. During *each* half-life unit, the amount of the parent radioisotope will diminish by half.

Table 18.1 Radioisotopes Commonly Used in Radiometric Dating

Radioisotope (unstable)	More Stable Product	Half-Life (years)	Useful Range (years)
Samarium 147 →	Neodymium 143	106 billion	>100 million*
Rubidium 87 →	Strontium 87	48.8 billion	>100 million
Thorium 232 →	Lead 208	14 billion	>200 million
Uranium 238 →	Lead 206	4.5 billion	>100 million
Potassium 40 →	Argon 40	1.25 billion	>100,000
Uranium 235 →	Lead 207	700 million	>100 million
Carbon 14 →	Nitrogen 14	5,730	0–60,000

* The symbol > means greater than.

SUMMARY

1. All species, past and present, are related by way of descent from their common ancestors, starting with the origin of the first cells on Earth.

2. Macroevolution refers to patterns, trends, and rates of change among groupings of species over long spans of time. Those clues to the past provide us with insight into the sweeping continuity of relationship in nature.

3. The fossil record, the geologic record, comparative morphology, and comparative biochemistry have yielded extensive evidence of evolution. The evidence is based on similarities and differences in body form, functions, behavior, and biochemistry.

4. Fossils are recognizable, physical evidence of life in the distant past. They start to form after an organism or traces of it become buried in volcanic ash or sediments. The organic remains become infused with mineral-rich water. As sediments accumulate above the burial site, they impose increasing pressure which, with chemical changes, transforms the remains to stony hardness.

   a. Fossils occur in layers of sedimentary rock. The deepest layers accumulate first; they are the oldest. Thus, the older the layers, the older the fossils.

   b. Scientists use abrupt transitions in the sequence of fossil assemblages as boundaries for intervals of the geologic time scale. In the modern version of the scale, life originated during the Archean eon. Oldest to most recent fossils extend from the Proterozoic era through the Paleozoic, Mesozoic, and Cenozoic eras.

   c. The completeness of the fossil record is variable in terms of the species represented, where they lived, and the stability of their tombs since fossilization occurred.

5. Comparative morphology often reveals similarities that imply evolutionary relationship among groups.

   a. *Homology* refers to similarity in one or more body parts between different groups of organisms that imply descent from a shared ancestor. Homologous structures suggest morphological divergence: modification of the same body parts in different ways in different lines of descent from the common ancestor.

   b. By contrast, *analogy* refers to body parts that were once different in evolutionarily remote lineages, but that converged in structure and function because those lineages responded to similar environments. They are evidence of morphological convergence.

   c. Similarities show up in the patterns by which the embryos of plants and animals develop. Mutations that affected steps in a program of early development might have been enough to bring about major differences in the adult form of related lineages.

6. Comparative biochemistry can identify similarities and differences among species at the molecular level.

   a. Nucleic acid comparisons, as by hybridization of the DNA from different species, give strong evidence of evolutionary relationships. The strength with which single-stranded DNA or RNA from one species base-pairs with a single strand from another species is a rough measure of evolutionary distance between them.

   b. The DNA of all species contains an accumulation of neutral mutations in highly conserved genes. Such mutations are like ticks of a molecular clock; they help date divergences from a common ancestor.

7. Taxonomists identify, name, and classify species. By the Linnaean binomial system, each kind of organism can be assigned a two-part Latin name. The first part (genus) identifies species that have shared similarities and presumably were derived from common ancestors. In conjunction with the second part of the name (species epithet), it is the name of the particular species.

8. Classification systems use higher taxa, or ever more inclusive groupings, from species to genera, families, orders, classes, phyla, and kingdoms. Such groupings reflect phylogeny (presumed evolutionary relatedness).

9. This book uses a scheme that groups organisms into six great kingdoms: Archaebacteria, Eubacteria, Protista, Plantae, Fungi, and Animalia. It is based on perceived evolutionary relationships; it is a phylogenetic scheme.

10. Like life, the Earth has evolved. Some of its features change repetitively, as when mountains rise and erode slowly by the same processes. Other features changed irreversibly, as explained by plate tectonics theory:

    a. The Earth's crust is fractured into huge, thin, rigid plates that slowly split apart, drift, and collide with one another, rafting the land masses along with them.

    b. Great plumes of molten material welling up from Earth's interior drive the movements. At mid-oceanic ridges, material seeps out, cools, hardens, and laterally displaces older seafloor. Seafloor spreading has forced older crust down into great trenches. Great mountain ranges have gradually formed when one plate thrust under another, which uplifted as a result.

    c. Large-scale, long-term changes in the Earth's land masses have changed the ocean and atmosphere, and they have profoundly influenced the evolution of life.

Review Questions

1. Distinguish macroevolution from microevolution. *CI* (*Also compare 17.5*)

2. Will biologists ever be able to read a complete fossil record? Why or why not? *18.1*

3. Name the three eras of the geologic time scale. One of these, the Proterozoic, was so vast that it has since been divided into great eons. In which eon did life originate? *18.1*

4. Define the difference between: *18.2*
   a. homologous and analogous structures
   b. morphological divergence and convergence

5. Comparative morphology refers to anatomical comparisons of major lineages. This applies to adult forms and to embryonic forms. Give an example of such a comparison. *18.3*

6. Give reasons why two organisms that are quite different in outward appearances may belong to the same lineage. *18.2, 18.3*

7. Name a protein specified by a gene that has been highly conserved in organisms ranging from bacteria to humans. *18.4*

8. Why do evolutionary biologists apply heat energy to hybrid molecules that contain DNA from two species? *18.4*

9. What type of mutations are the basis of a molecular clock? What does the last tick of a molecular clock signify? *18.4*

10. What type of evidence favors putting archaebacteria in their own kingdom? *18.5*

11. What is the basis of continental drift? Of seafloor spreading? In general, how did such changes in the Earth influence the evolution of life? *18.6*

12. How do biologists determine the age of a fossil? *18.1, 18.7*

13. Define radiometric dating. What does half-life mean? *18.7*

Self-Quiz *(Answers in Appendix III)*

1. Mutations underlying evolutionary changes in morphology of plants and animals most likely occurred _____ .
   a. when embryos were first developing
   b. in nearly mature embryos
   c. in adults facing environmental challenges

2. Morphological convergences may lead to _____ .
   a. analogous structures   c. divergent structures
   b. homologous structures   d. both a and c

3. A classification system that is _____ is based on presumed evolutionary relationship.
   a. epigenetic   c. credited to Linnaeus
   b. phylogenetic   d. both b and c

4. *Pinus banksiana, Pinus strobus,* and *Pinus radiata* are _____ .
   a. three families of pine trees
   b. three different names for the same organism
   c. several species belonging to the same genus
   d. both a and c

5. Increasingly inclusive taxa range from _____ to _____ .
   a. kingdom; species   c. genera; kingdom
   b. kingdom; genera   d. species; kingdom

6. Match these terms suitably.
   ____ phylogeny   a. accumulation of neutral mutations
   ____ fossil   b. stone-hard evidence of life in past
   ____ stratification   c. similar body parts in different
   ____ homology   lineages owing to common descent
   ____ molecular clock d. e.g., shark fins, penguin flippers
   ____ analogy   e. evolutionary relationship among
                            species, ancestors to descendants
                         f. layers of sedimentary rock

Critical Thinking

1. At the end of your backbone are several small, fused bones, called the coccyx. Is the coccyx a vestigial structure—all that is left of a tail that was a feature of the evolutionarily distant vertebrate (and primate) ancestors of humans? Or is it the start of a newly evolving structure with an as-yet undetermined function? Take an educated guess, then describe some ways in which you might test whether your answer is plausible.

2. Protein comparisons and nucleic acid hybridization studies help us estimate evolutionary relationship and approximate times for divergences from ancestral stocks. DNA sequence comparisons yield even more accurate estimates. Reflect on the genetic code (Section 13.2), then suggest why this may be a more accurate measure of mutations, mutation rates, and biochemical relatedness.

3. Shannon thinks there are too many kingdoms and sees no good reason to make another one for something as small as archaebacteria. "Keep them with the other prokaryotes!" she says. Taxonomists would call her a "lumper." By contrast, Geoffrey is a "splitter." He sees no good reason to withhold kingdom status from archaebacteria simply because they are part of a microscopic world that not many people know about. Which may be the more useful choice: more or fewer boundaries between groups of organisms? Explain your answer.

4. When walking along a path cut into the side of a steep mountain, you pass rocky layers that are fractured and folded back on themselves in bizarre patterns. You look closely and discover a fossilized shell in the "highest" layer in the series. How would you use the theories of continental drift and plate tectonics to help you decide whether the fossil is of recent or ancient origin?

5. "Missing links" are undiscovered transitional forms between major groups of organisms. *Archaeopteryx*, described earlier in Section 16.3, is one example. Drawing information from Sections 18.1 and 18.6, what are some possible reasons for apparent gaps in the fossil record?

Selected Key Terms

analogy *18.2*	morphological
Animalia *18.5*	convergence *18.2*
Archaebacteria *18.5*	morphological
binomial system *18.5*	divergence *18.2*
classification scheme *18.5*	neutral mutation *18.4*
comparative morphology *18.2*	nucleic acid hybridization *18.4*
Eubacteria *18.5*	phylogeny *18.5*
fossil *18.1*	Plantae *18.5*
fossilization *18.1*	plate tectonics theory *18.6*
Fungi *18.5*	Protista *18.5*
genus *18.5*	radiometric dating *18.7*
geologic time scale *18.1*	six-kingdom classification
half-life *18.7*	scheme *18.5*
homology *18.2*	specific name *18.5*
lineage *CI*	stratification *18.1*
macroevolution *CI*	taxonomy *18.5*
molecular clock *18.4*	theory of uniformity *18.6*
Monera *18.5*	

Readings See also *www.infotrac-college.com*

Brooks, D. R., and D. A. McLennan. 1991. *Phylogeny, Ecology, and Behavior.* Chicago: University of Chicago Press.

Dott, R., Jr., and R. Batten. 1998. *Evolution of the Earth.* Fourth edition. New York: McGraw-Hill. Good historical perspective on correlations between biological and geologic evolution.

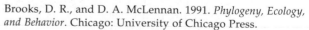

WWW   *http://www.brookscole.com/biology*

Practice quiz questions, hypercontents, BioUpdates, and critical thinking. The Brooks/Cole Biology Resource Center provides a wealth of information fully organized and integrated by chapter.

FACING PAGE: *Patterns of diversity in nature, here represented by different species of plants and fungi.*

THE ORIGIN AND EVOLUTION OF LIFE

In The Beginning . . .

Some clear evening, watch the moon as it rises from the horizon and think of the 380,000 kilometers between it and you. *Five billion trillion times* farther away from you are galaxies—systems of stars—at the boundary of the known universe. Wavelengths traveling through space move faster than anything else—millions of meters per second—yet long wavelengths that originated from faraway galaxies many billions of years ago are only now reaching the Earth.

If we are to accept every known measure, all the near and distant galaxies in the vast space of the universe are moving away from one another, which means the universe must be expanding. And the prevailing view of how the colossal expansion came about may account for every bit of matter in every living thing.

Think about how you rewind a videotape on a VCR, then imagine "rewinding" the universe. As you do this, the galaxies start moving back together. After 12 to 15 billion years of rewinding, all galaxies, all matter, and all of space are compressed into a hot, dense volume about the size of the sun. You have arrived at time zero.

That incredibly hot, dense state lasted only for an instant. What happened next is known as the **big bang**,

Figure 19.1 Part of the great Eagle nebula, a hotbed of star formation 7,000 light-years from Earth, in the constellation Serpens. (The Latin *nebula* means mist.) Not shown in this image are a few huge, young stars above the pillars. For the past few million years, intense ultraviolet radiation from the stars has been eroding the less dense surface of the pillars. (By analogy, visualize a strong prevailing wind blowing away sand in a desert and exposing rocks.) Globules of denser gases and dust that have resisted erosion are visible at the surface. Each one is wider than our solar system—more than 10 billion miles across! New stars are hatching from the protruding globules; some are shining brightly on the tips of gaseous streamers.

a stupendous, nearly instantaneous distribution of matter and energy throughout the known universe. About a minute later, temperatures dropped to a billion degrees. Fusion reactions produced most of the light elements, including helium, which are still the most abundant elements in the universe. Radio telescopes detect cooled, diluted background radiation that is a relic of the "big bang," left over from the beginning of time.

Over the next billion years, uncountable numbers of gaseous particles collided and condensed, under gravity's force, to become the first stars. When the stars became massive enough, nuclear reactions were ignited in their core, and they gave off tremendous light and heat. Massive stars continued to contract, and many became dense enough to promote the formation of heavier elements.

All stars have a life history, from birth to an often spectacularly explosive death. In what might be called the original stardust memories, the heavier elements released during the explosions became swept up in the gravitational contraction of new stars, and they became raw materials for the formation of even heavier elements. Even as you read this page, the Hubble space telescope is revealing astounding glimpses of such star-forming activity, as in the dust clouds of Orion, Serpens, and other constellations (Figure 19.1).

Now imagine a time long ago, when explosions of dying stars ripped through our galaxy and left behind a dense cloud of dust and gas that extended trillions of kilometers in space. As the cloud cooled, countless bits of matter gravitated toward one another. By 4.6 billion years ago, the cloud had flattened out into a slowly rotating disk. At the dense, hot center of that great disk, the shining star of our solar system—the sun—was born.

The remainder of this chapter is a sweeping slice through time, one that cuts back to the formation of the Earth and the chemical origins of life. It is the starting point for the next five chapters, which will take us along lines of descent that led to the present range of species diversity.

The story is not complete. Even so, the available evidence from many avenues of research points to a principle that can help us organize bits of information about the past: *Life is a magnificent continuation of the physical and chemical evolution of the universe, of galaxies and stars, and of the planet Earth.*

KEY CONCEPTS

1. We have evidence that life originated about 3.8 billion years ago. The origin and subsequent evolution of life have been correlated with the physical and chemical evolution of the universe, the stars, and the planet Earth.

2. All inorganic and organic compounds necessary for self-replication, membrane assembly, and metabolism—for the structure and functions of living cells—could have formed spontaneously under conditions that apparently prevailed on the early Earth.

3. The history of life, from its chemical beginnings to the present, spans five intervals of geologic time. It extends through two great eons—the Archean and Proterozoic—and the Paleozoic, Mesozoic, and Cenozoic eras.

4. Not long after life originated, divergences led to two great prokaryotic lineages called the archaebacteria and eubacteria. Soon afterward, the ancestors of eukaryotes diverged from the archaebacterial lineage.

5. Archaebacteria and eubacteria dominated the Archean and Proterozoic eons. Eukaryotic cells originated late in the Proterozoic era and became spectacularly diverse. A theory of endosymbiosis helps explain the profusion of specialized organelles that evolved in eukaryotic cells.

6. All six kingdoms of life have a history of persistences, extinctions, and radiations of different lineages.

7. Throughout the history of life, asteroid impacts, drifting and colliding continents, and other environmental insults have had profound impact on the direction of evolution.

CONDITIONS ON THE EARLY EARTH

Origin of the Earth

Figure 19.1 gave you a view of one of the vast clouds in the universe. Such clouds consist mostly of hydrogen gas, along with water, iron, silicates, hydrogen cyanide, ammonia, methane, formaldehyde, and other simple organic and inorganic substances. The contracting cloud that became our solar system probably was similar in composition. We assume the edges of that cloud cooled between 4.6 and 4.5 billion years ago. Mineral grains and ice orbiting the new sun started clumping together as a result of electrostatic attraction and gravity's pull (Figure 19.2). In time, larger, faster clumps collided and shattered. Some became more massive by sweeping up asteroids, meteorites, and the other rocky remnants of collisions, and gradually they evolved into planets.

Figure 19.2 Representation of the cloud of dust, gases, and clumps of rock and ice around the early sun.

As the Earth was forming, much of its inner rocky material melted. Asteroid impacts and the Earth's own internal compression and radioactive decay of minerals could have generated the heat necessary to do this. As rocks melted, nickel, iron, and other heavy materials moved to the Earth's interior; lighter ones floated to the surface. The process produced a crust, mantle, and core. The **crust** is an outer region of basalt, granite, and other low-density rocks. It envelops the intermediate-density rocks of the **mantle**, which wraps around a core of very high-density, partially molten nickel and iron.

Four billion years ago, the Earth was a thin-crusted inferno (Figure 19.3a). Within 200 million years, life had originated on its surface! We have no record of the event. As far as we know, movements in the mantle and crust, volcanic activity, and erosion obliterated all traces of it. Still, we can put together a plausible explanation of how life originated by considering three questions:

First, can we identify physical and chemical conditions that prevailed on the Earth when life originated?

Second, do known physical, chemical, and evolutionary principles support the hypothesis that large organic molecules formed spontaneously, then evolved into molecular systems displaying the fundamental properties of life?

Third, can we devise experiments to test the hypothesis that living systems emerged by chemical evolution?

The First Atmosphere

When the first patches of crust were forming, hot gases blanketed the Earth. Probably this first atmosphere was a mix of gaseous hydrogen (H_2), nitrogen (N_2), carbon monoxide (CO), and carbon dioxide (CO_2). Did it hold gaseous oxygen (O_2)? Probably not. Rocks subjected to intense heat, as happens during volcanic eruptions, do release oxygen, but not much. Also, free oxygen would have reacted at once with other elements. An oxygen atom has an electron vacancy in its outermost shell and tends to fill it by bonding with other atoms (Section 2.3).

If the early atmosphere had not been relatively free of oxygen, organic compounds required to assemble cells in the first place would not have been able to form spontaneously—*on their own*. Any oxygen would have attacked their structure and disrupted their functioning.

What about liquid water? Dense clouds blanketed the early Earth, but water reaching the molten surface would have evaporated at once. After the crust cooled and solidified, however, rains fell on the parched rocks. For millions of years, the runoff eroded mineral salts and other compounds from the rocks. Salt-laden waters collected in the depressions in the crust and formed the first seas. If liquid water had not accumulated, then cell membranes—which take on their bilayer organization only in water—could not have formed. No membrane, no cell. Life at its most basic level *is* the cell, which has a capacity to survive and reproduce on its own.

Synthesis of Organic Compounds

Reduce a cell to its lowest common denominator and all that remains are proteins, complex carbohydrates and lipids, and nucleic acids. Existing cells assemble these molecules from smaller organic compounds: the simple sugars, fatty acids, amino acids, and nucleotides. Energy from the environment drives the synthesis reactions. Were small organic compounds also present on the early Earth? Were there sources of energy that spontaneously drove their assembly into the large molecules of life?

Mars, meteorites, the Earth's moon, and the Earth all formed at the same time, from the same cosmic cloud. Rocks collected from Mars, meteorites, and the moon contain precursors of biological molecules, so the same precursors must have been present on the Earth. If this were the case, *then energy from sunlight, lightning, or*

Figure 19.3 (**a**) Representation of the Earth during its formation, when the moon's orbit was much closer than it is today. If the Earth had condensed into a planet of smaller diameter, its gravitational mass would not have been great enough to hold on to an atmosphere. If it had settled into an orbit closer to the sun, water would have evaporated from its hot surface. If the Earth's orbit had been more distant from the sun, its surface would have been far colder, and water would have been locked up as ice. Without liquid water, life as we know it never would have originated on Earth.

(**b**) Stanley Miller's experimental apparatus, used to study the synthesis of organic compounds under conditions that presumably existed on the early Earth. The condenser cools circulating steam so that water droplets form.

even heat escaping from the crust could have been enough to drive their combination into organic molecules.

Stanley Miller conducted the first experimental test of that prediction. First he mixed methane, hydrogen, ammonia, and water inside a reaction chamber of the sort depicted in Figure 19.3*b*. Then he kept the mixture circulating and bombarded it with a spark discharge to simulate lightning. In less than one week, amino acids and other small organic compounds had formed.

In other experiments that simulated conditions on the early Earth, glucose, ribose, deoxyribose, and other sugars formed spontaneously from formaldehyde, and adenine from hydrogen cyanide. Ribose and adenine occur in ATP, NAD$^+$, and other nucleotides vital to cells.

However, if *complex* organic compounds had formed directly in the seas, they would not have lasted long. The spontaneous direction of the necessary reactions would have been toward hydrolysis, not condensation, in water. How did more lasting bonds form?

By one hypothesis, clay in the rhythmically drained muck of tidal flats and estuaries served as templates (structural patterns) for the spontaneous assembly of proteins and other complex organic compounds. Clay consists of thin, stacked layers of aluminosilicates with metal ions at their surfaces, which attract amino acids. Expose amino acids to some clay, warm the clay with rays from the sun, then alternately moisten and dry it. Condensation reactions will proceed at its surfaces and yield proteins and other complex organic compounds.

By another hypothesis, complex organic compounds formed spontaneously near hydrothermal vents on the seafloor, where species of archaebacteria are thriving today. As experimental tests by Sidney Fox and others show, when amino acids are heated and then placed in water, they spontaneously order themselves into small protein-like molecules, which Fox calls "protenoids."

However the first proteins formed, their molecular structure dictated how they could interact with other compounds. Suppose some proteins had the structure to function as weak enzymes, and that they hastened bond formation between amino acids. Such enzyme-directed synthesis of proteins would have had selective advantage. In the chemical competition for available amino acids, protein configurations that could promote reactions would win. Proteins would have been favored in still another way—for proteins have the capacity to bind metal ions and other agents of metabolism.

As you will read next, the evolution of metabolism must have been based on such chemical modification. For now, simply reflect on the possibility that selection was at work before the origin of living cells, favoring the chemical evolution of enzymes and other complex organic compounds.

Many diverse experiments yield indirect evidence that the complex organic molecules characteristic of life could have formed under conditions that existed on the early Earth.

WWW

EMERGENCE OF THE FIRST LIVING CELLS

Origin of Agents of Metabolism

A defining characteristic of life is *metabolism*. The word refers to all the reactions by which cells harness energy and use it to drive their activities, such as biosynthesis. During the first 600 million years or so of Earth history, enzymes, ATP, and other organic compounds may have assembled spontaneously, perhaps in the same physical locations. If they did so, their close association would have naturally promoted chemical interactions and the beginning of metabolic pathways.

Imagine an ancient estuary, rich in clay deposits. It is a coastal region where seawater mixes with mineral-rich water being drained from the land. There, beneath the sun's rays, countless aggregations of organic molecules stick to the clay. At first there are quantities of an amino acid; call it *D*. Throughout the estuary, *D* molecules get incorporated into new proteins—until the supply of *D* dwindles. However, suppose a protein that is weakly catalytic also is present in the estuary. It can promote the formation of *D* by acting on an abundant, simpler substance—call it *C*. By chance, some aggregations of organic molecules include that particular enzyme-like protein, and so they have an advantage in the chemical competition for starting materials.

In time, *C* molecules become scarce. At that point, the advantage tilts to aggregations that can promote formation of *C* from even simpler substances *B* and *A*. Assume *B* and *A* are carbon dioxide and water. As you know, carbon dioxide and water occur in essentially unlimited amounts in the atmosphere and in the seas. Chemical selection has favored a synthetic pathway:

$$A + B \longrightarrow C \longrightarrow D$$

Finally, suppose some aggregations are better than others at absorbing and using energy. Which molecules could bestow such an advantage? Think of the energy-trapping pathway that now dominates the world of life: photosynthesis. It starts at pigments called chlorophylls. The portion of a chlorophyll molecule that absorbs light and gives up electrons is a porphyrin ring structure. Porphyrins also are present in cytochromes, which are part of electron transport systems in all photosynthetic and aerobically respiring cells. Porphyrin molecules can spontaneously assemble from formaldehyde—one of the molecular legacies of cosmic clouds (Figure 19.4). Was porphyrin a major electron transporter of some of the early metabolic pathways? Perhaps.

Origin of Self-Replicating Systems

Another defining characteristic of life is a capacity for reproduction, which now starts with protein-building instructions in DNA. The DNA molecule is fairly stable,

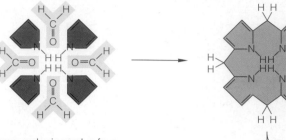

four pyrrole rings plus four
formaldehyde molecules

porphyrin ring system

Figure 19.4 One hypothetical sequence by which formaldehyde, an organic compound, underwent chemical evolution into porphyrin. Formaldehyde was present when the Earth formed. Porphyrin is the light-trapping and electron-donating component of all existing chlorophyll molecules. It also is a component of cytochrome, which is a protein component of electron transport systems that are part of many metabolic pathways.

chlorophyll *a*, one of the light-trapping pigments of photosynthetic plant cells

and it is easily replicated before each cell division. As you know from earlier chapters, arrays of enzymes and RNA molecules operate together to carry out DNA's encoded instructions.

Most existing enzymes get assistance from small organic molecules or metal ions called coenzymes. Intriguingly, some categories of coenzymes have a structure identical to that of RNA nucleotides. Another clue: Mix together and then heat up precursors of ribonucleotides and short chains of phosphate groups, and they self-assemble into single strands of RNA. On the early Earth, energy from the sun's rays or from geothermal events could have driven the spontaneous formation of RNA from such starting materials.

Very simple self-replicating systems of RNA, enzymes, and coenzymes have been created in some laboratories. Did RNA later become the information-

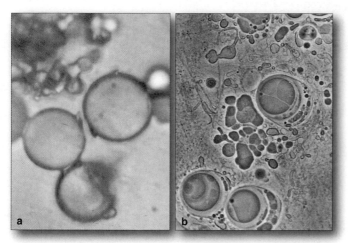

Figure 19.5 Microscopic spheres of (**a**) proteins and (**b**) lipids that self-assembled under abiotic conditions.

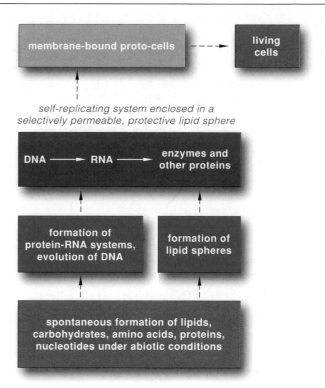

Figure 19.6 One possible sequence of events that led to the first self-replicating systems, then to the first living cells.

storing templates upon which proteins could be synthesized? Perhaps it did so initially. We can only speculate at present, because the RNA molecules that we know about are too chemically fragile to act as templates for protein synthesis.

Yet the use of RNA templates might have set up an **RNA world** that preceded DNA's dominance as the main informational molecule. Whatever the case, DNA eventually assumed this function, probably because it can form long nucleotide chains in more stable fashion.

We still don't know how DNA entered the picture. Until we identify the likely chemical ancestors of RNA and DNA, the story of life's origin will be incomplete. Filling in the details will require imaginative sleuthing. For instance, researchers ran a computer program that incorporated information about natural energy sources and simple inorganic compounds of the sort thought to have been present on the early Earth. They asked their advanced computer to subject the chosen compounds to random chemical competition and natural selection as might have occurred untold billions of times in the past. They ran the program repeatedly. And always the outcome was the same: Simple precursors inevitably evolved—and they organized themselves as interacting systems of large, complex molecules.

Origin of the First Plasma Membranes

Experimental tests are more revealing of the origin of the plasma membrane—the outermost membrane of all living cells. This cellular component consists of a lipid bilayer, studded with proteins that carry out diverse functions. Its main role is to control which substances move into and out of the cell. Without control, cells can neither exist nor reproduce.

Maybe one avenue of molecular evolution led to **proto-cells**: simple membrane sacs that surrounded and

protected information-storing templates and metabolic agents from the environment. We do know that simple membrane sacs can form spontaneously. For example, for one experiment, Fox heated amino acids until they formed protein-like chains, which he then placed in hot water. After cooling, the chains assembled into small, stable spheres (Figure 19.5a). Like membranes of cells, these proteinoid spheres were selectively permeable to various substances. In other experiments, the spheres picked up free lipids from their surroundings, and a lipid-protein film formed at their surface.

In still other experiments, by David Deamer and his coworkers, fatty acids and glycerol combined to form long-tail lipid molecules under laboratory conditions that simulated evaporating tidepools. The lipids self-assembled into small, water-filled sacs. Many were like cell membranes (Figure 19.5b).

In short, there are major gaps in the story of life's origins. But there also is strong experimental evidence that chemical evolution probably led to the molecules and structures that are characteristic of life. Figure 19.6 summarizes the milestones in that chemical evolution, which preceded the first cells.

Although the story is not yet complete, many laboratory experiments and computer simulations indirectly show that chemical and molecular evolution could have given rise to proto-cells.

ORIGIN OF PROKARYOTIC AND EUKARYOTIC CELLS

The first living cells originated in the **Archean** eon, which lasted from 3.9 billion to 2.5 billion years ago. Those cells emerged as molecular extensions of the evolving universe, of our solar system and the Earth. Maybe they originated in tidal flats or seafloor sediments (Section 6.7). Fossils indicate they were like the existing bacteria. Specifically, they were **prokaryotic cells**, without a nucleus. They may have been little more than membrane-bound, self-replicating sacs of DNA and other complex organic molecules. Given the absence of free oxygen, they must have secured energy through anaerobic pathways—fermentation, most likely. Energy was plentiful. Geologic processes had enriched the seas with organic compounds. So "food" was available, predators were absent, and cellular structures were free from oxygen attacks.

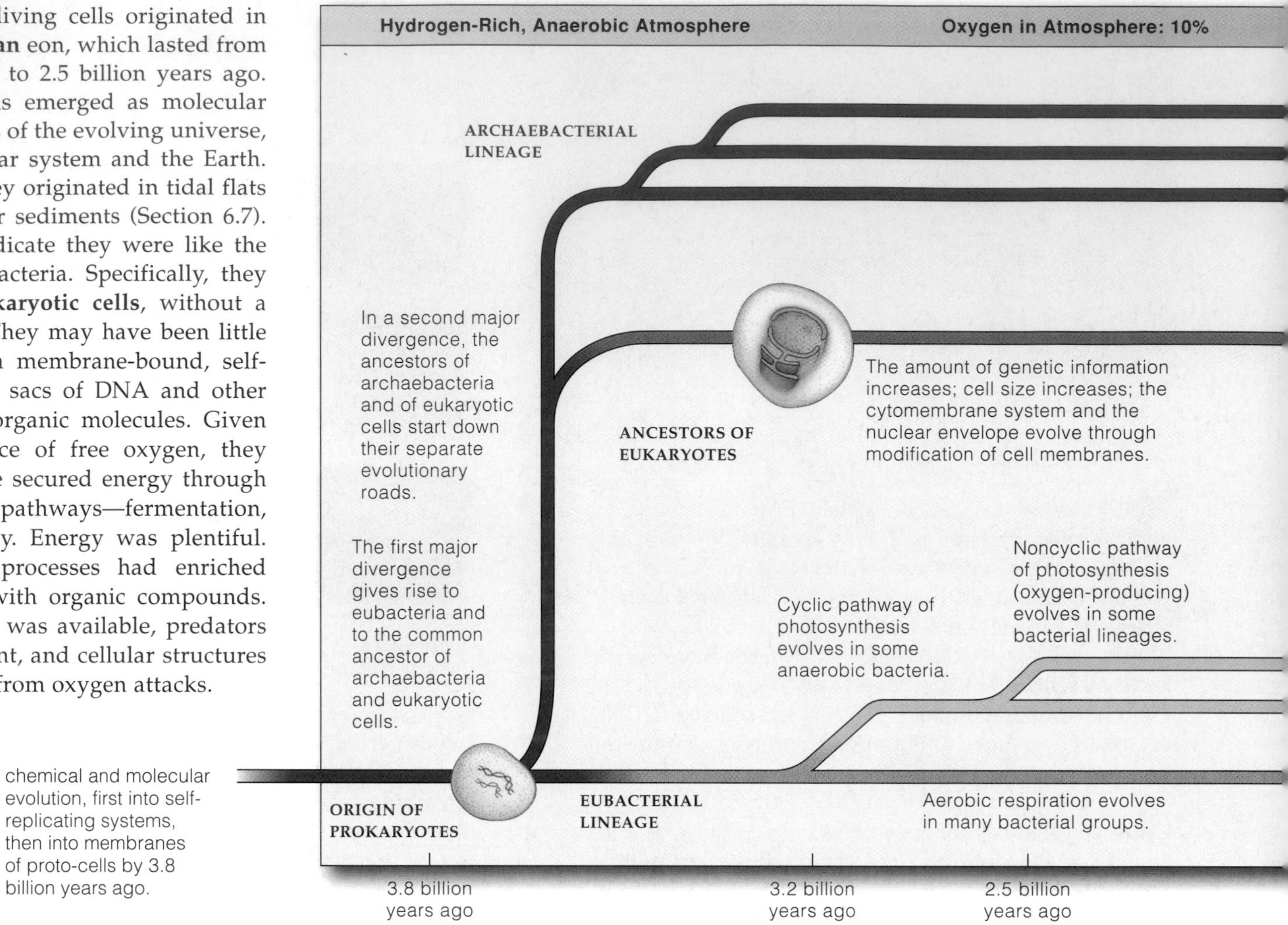

Hydrogen-Rich, Anaerobic Atmosphere	Oxygen in Atmosphere: 10%

ARCHAEBACTERIAL LINEAGE

In a second major divergence, the ancestors of archaebacteria and of eukaryotic cells start down their separate evolutionary roads.

ANCESTORS OF EUKARYOTES

The amount of genetic information increases; cell size increases; the cytomembrane system and the nuclear envelope evolve through modification of cell membranes.

The first major divergence gives rise to eubacteria and to the common ancestor of archaebacteria and eukaryotic cells.

Cyclic pathway of photosynthesis evolves in some anaerobic bacteria.

Noncyclic pathway of photosynthesis (oxygen-producing) evolves in some bacterial lineages.

chemical and molecular evolution, first into self-replicating systems, then into membranes of proto-cells by 3.8 billion years ago.

ORIGIN OF PROKARYOTES

EUBACTERIAL LINEAGE

Aerobic respiration evolves in many bacterial groups.

3.8 billion years ago

3.2 billion years ago

2.5 billion years ago

Some populations of those first prokaryotic cells apparently diverged in two major directions shortly after the time of origin. One lineage gave rise to the **eubacteria**. The other lineage gave rise to the common ancestor of **archaebacteria** and **eukaryotic cells** (Figure 19.7).

Between 3.5 and 3.2 billion years ago, light-trapping pigments, electron transport systems, and other metabolic machinery evolved in some of the anaerobic eubacteria, which became the earliest practitioners of the cyclic pathway of photosynthesis. (You read about this ATP-forming pathway in Section 6.3.) An unlimited source of energy—sunlight—had been tapped. For nearly 2 billion years, the photosynthetic descendants of those bacterial cells dominated the world of life. Their tiny but numerous populations formed very large mats in which sediments collected. The mats built up, one above the other. Calcium deposits hardened and preserved the mats, which came to be called **stromatolites** (Figure 19.8).

Figure 19.7 An evolutionary tree of life that reflects mainstream thinking about the connections among major lineages. The diagram incorporates ideas about the origins of some eukaryotic organelles.

Figure 19.8 Stromatolites exposed at low tide in western Australia's Shark Bay. These mounds started forming 2,000 years ago. Structurally, they are identical with stromatolites that formed more than 3 billion years ago.

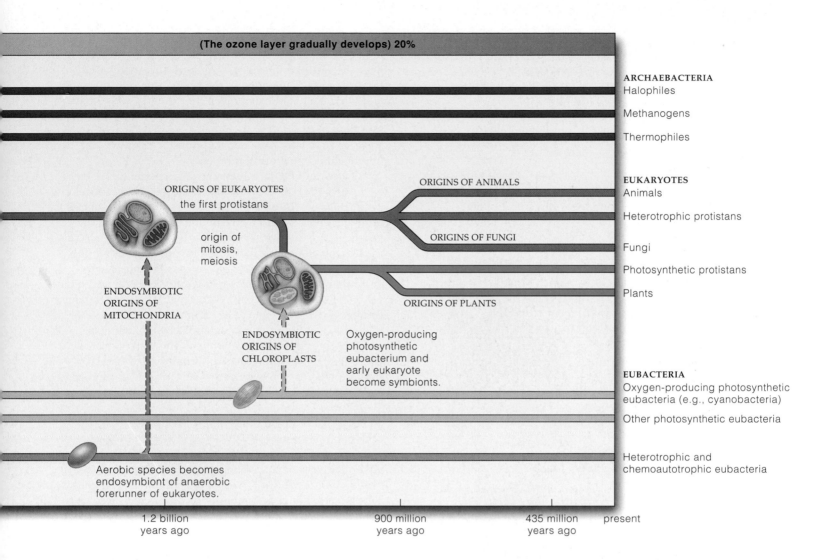

(The ozone layer gradually develops) 20%

ARCHAEBACTERIA
Halophiles

Methanogens

Thermophiles

ORIGINS OF EUKARYOTES
the first protistans

ORIGINS OF ANIMALS

EUKARYOTES
Animals

Heterotrophic protistans

origin of
mitosis,
meiosis

ORIGINS OF FUNGI

Fungi

Photosynthetic protistans

ENDOSYMBIOTIC
ORIGINS OF
MITOCHONDRIA

Plants

ORIGINS OF PLANTS

ENDOSYMBIOTIC
ORIGINS OF
CHLOROPLASTS

Oxygen-producing
photosynthetic
eubacterium and
early eukaryote
become symbionts.

EUBACTERIA
Oxygen-producing photosynthetic
eubacteria (e.g., cyanobacteria)

Other photosynthetic eubacteria

Heterotrophic and
chemoautotrophic eubacteria

Aerobic species becomes
endosymbiont of anaerobic
forerunner of eukaryotes.

| 1.2 billion years ago | 900 million years ago | 435 million years ago | present |

By the dawn of the **Proterozoic** eon, 2.5 billion years ago, photosynthetic machinery had become altered in some eubacterial species, and the noncyclic pathway of photosynthesis emerged. Oxygen, one of the pathway's by-products, started to accumulate. In time, this had two irreversible effects. First, *an oxygen-rich atmosphere stopped the further chemical origin of living cells.* Except in a few anaerobic habitats, complex organic compounds could no longer form spontaneously and resist attack. Second, *aerobic respiration became the dominant energy-releasing pathway.* In many prokaryotic lineages, selection favored metabolic equipment that "neutralized" oxygen by using it as an electron acceptor! This key innovation contributed to the rise of multicelled eukaryotes and their invasion of far-flung environments (Section 7.7).

Eukaryotic cells evolved in the Proterozoic, possibly before 1.2 billion years ago. We have fossils, 900 million years old, of well-developed algae, fungi, and plant spores. As you know, organelles are the key defining features of eukaryotic cells. *Where did they come from?* The next section describes a few plausible hypotheses.

By about 800 million years ago, stromatolites along the shores of Laurentia, an early supercontinent, were declining dramatically. Perhaps newly evolved, bacteria-eating animals were using them as a concentrated food source. Fossilized embryos found in China (clusters of cells no wider than a pin) give evidence that the first animals were soft-bodied forerunners of today's sponges and marine worms. Before 570 million years ago, in "Precambrian" times, some of their small descendants started the first adaptive radiation of animals.

The first cells evolved by about 3.8 billion years ago, during the Archean. All were prokaryotic cells, and most, if not all, probably made ATP by fermentation routes.

Early on, ancestors of archaebacteria and eukaryotic cells diverged from the lineage that led to modern eubacteria.

Oxygen-releasing photosynthetic bacteria evolved. In time, the oxygen-enriched atmosphere put an end to the further spontaneous chemical evolution of life. That atmosphere was a key selection pressure in the evolution of eukaryotic cells.

WWW

WHERE DID ORGANELLES COME FROM?

Thanks to Andrew Knoll, William Schopf, Jr., and other globe-hopping microfossil hunters, we have tantalizing evidence of early life. One fossil treasure is a strand of bacterial cells that lived 3.5 billion years ago, not long after life originated. Others are from the Proterozoic. They were eukaryotic cells that had a few membrane-bound organelles in the cytoplasm, as shown in Figure 19.9. Most of their living descendants have a profusion of organelles (Figure 19.10).

Where did eukaryotic organelles come from? Speculations abound. Some organelles probably evolved through gene mutations and natural selection. For others, researchers make a good case for evolution by way of endosymbiosis.

ORIGIN OF THE NUCLEUS AND ER Prokaryotic cells do not have an abundance of organelles, but some species have interesting infoldings of their plasma membrane (Figure 19.11). Embedded in that membrane are enzymes and other agents of metabolism. In the early forerunners of eukaryotic cells, similar infoldings may have extended into the cytoplasm and served as passageways to the surface. They may have evolved into ER channels and into an envelope around the DNA.

What would be the advantage of such membranous enclosures? Maybe they protected the genes and protein products from "foreigners." Remember, bacterial species often transfer plasmid DNA among themselves. So do yeasts, which are very simple eukaryotic cells. At first, a nuclear envelope may have been favored because it got the cell's genes, replication enzymes, and transcription enzymes out of the cytoplasm. It would have allowed vital genetic messages to be copied and read, free of metabolic competition from what could become an unmanageable hodgepodge of foreign genes. Similarly, ER channels might have kept important proteins and other organic compounds away from metabolically hungry "guests"— foreign cells that one way or another became permanent residents inside the host cell, as described next.

A THEORY OF ENDOSYMBIOSIS It appears likely that accidental partnerships between a variety of prokaryotic species arose countless times on the evolutionary road to eukaryotic cells. Perhaps some partnerships resulted in the origin of mitochondria, chloroplasts, and other organelles. This is a story of endosymbiosis, as developed in greatest

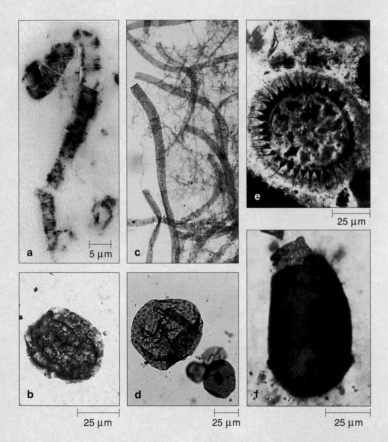

Figure 19.9 From Australia, (**a**) a strand of walled prokaryotic cells 3.5 billion years old and (**b**) one of the oldest known eukaryotes— a protistan 1.4 billion years old. From Siberia, (**c**) a multicelled alga 900 million to 1 billion years old and (**d**) eukaryotic microplankton 850 million years old. (**e**) From China, a eukaryotic cell that lived 560 million years ago. (**f**) From Spitsbergen, Norway, a protistan that was alive 50 million years before the dawn of the Cambrian.

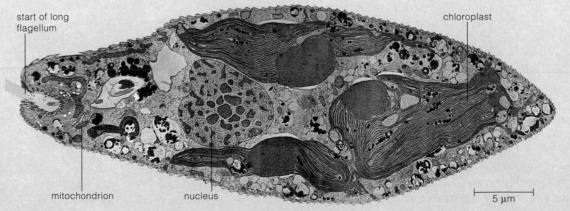

Figure 19.10
A fine example of the profusion of diverse organelles that are hallmarks of eukaryotic cells: *Euglena*, a single-celled protistan, sliced lengthwise for this micrograph. It also has a long flagellum, which could not fit in the image area at this magnification.

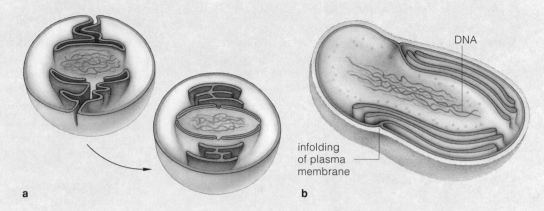

Figure 19.11 (a) Sketch of one idea concerning the origin of endoplasmic reticulum and the nuclear envelope. In prokaryotic ancestors of eukaryotic cells, infoldings of the plasma membrane might have given rise to both cell components. (b) Such infoldings are present in the cytoplasm of many kinds of existing bacteria, including *Nitrobacter*, sketched here in cutaway view.

a

DNA

infolding of plasma membrane

b

detail by Lynn Margulis. *Endo—* means within; *symbiosis* means living together. In cases of **endosymbiosis**, one species (the guest) lives permanently inside another species (the host), and the interaction benefits both.

According to one theory, eukaryotic cells evolved by endosymbiosis long after the noncyclic pathway of photosynthesis emerged and oxygen had accumulated to significant levels in the atmosphere. In some bacterial groups, certain electron transport systems in the plasma membrane had become modified and included "extra" cytochromes. The cytochromes could donate electrons to oxygen. Thus the bacteria could extract energy from organic compounds by aerobic respiration.

By 1.2 billion years ago, and possibly much earlier, the forerunners of eukaryotes were engulfing aerobic bacteria. Maybe they were like existing soft-bodied, amoebalike cells that weakly tolerate free oxygen. They would have entrapped food by sending out cytoplasmic extensions from the cell body. Endocytic vesicles could form around food and deliver it to the cytoplasm for digestion.

A key point of the theory is that some aerobic bacteria resisted digestion and thrived in the protected, nutrient-rich environment. In time, they were releasing extra ATP, which their hosts came to depend on for cell growth, increased activity, and assembly of hard parts and other structures. How did the guests benefit? They no longer had to search for food or duplicate metabolic functions that hosts performed for them. The anaerobic and aerobic cells were now incapable of independent life. The guests had become mitochondria, supreme suppliers of ATP.

EVIDENCE OF ENDOSYMBIOSIS Strong evidence favors Margulis's theory. There are plenty of examples of nature continuing to tinker with endosymbionts, including the cell in Figure 19.12. Its mitochondria are like bacteria in size and structure. The inner mitochondrial membrane is like a bacterial plasma membrane. Each mitochondrion replicates its own DNA and divides independently of the host cell's division. A few genetic code words in its DNA and mRNA have unique meanings. They are translated into a few proteins required for special mitochondrial tasks. Compared to the genetic code of living cells, the "mitochondrial code" has a few distinct differences.

Chloroplasts, too, may be stripped-down descendants of oxygen-evolving, photosynthetic bacteria. Perhaps predatory aerobic bacteria engulfed such photosynthetic cells, which escaped digestion, absorbed nutrients from their host's cytoplasm, and continued to function. By providing aerobically respiring hosts with oxygen, they would have promoted their endosymbiotic existence.

In their metabolism and overall nucleic acid sequence, chloroplasts resemble some eubacteria. Their DNA is self-replicating, and they divide independently of the cell's division. Chloroplasts vary in their shape and array of light-absorbing pigments, just as photosynthetic eubacteria do. They might have originated a number of times, in a number of different lineages.

Adding to the intrigue are ciliated protistans and marine slugs that "enslave" chloroplasts! The slugs eat algae but retain algal chloroplasts in their gut cells. The chloroplasts draw nutrients from the host tissues, and they continue to photosynthesize and release oxygen for weeks.

However they arose, new kinds of cells did appear on the evolutionary stage. They had become equipped with a nucleus, cytomembranes, and mitochondria, chloroplasts, or both. They were eukaryotic cells, the first **protistans**. With their efficient metabolic strategies, early protistans underwent rapid divergences and adaptive radiations. In no time at all, evolutionarily speaking, some of their descendants gave rise to the great kingdoms of plants, fungi, and animals, as sketched out earlier in Figure 19.7. *WWW*

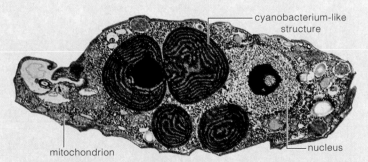

cyanobacterium-like structure

mitochondrion

nucleus

Figure 19.12 *Cyanophora paradoxa*, a protistan. Its mitochondria resemble bacteria. Its photosynthetic structures look like spherical cyanobacteria (which are photosynthetic) without the cell wall.

LIFE IN THE PALEOZOIC ERA

We divide the **Paleozoic** into the Cambrian, Ordovician, Silurian, Devonian, Carboniferous, and Permian periods. Before the dawn of that era, gradual rifting had split the supercontinent Laurentia apart. From the Cambrian on into the Silurian, the great fragments straddled the equator, and warm, shallow seas lapped their margins:

FRAGMENTS OF LAURENTIA

GONDWANA

The global conditions restricted pronounced seasonal changes in prevailing winds, ocean currents, and the upward churning of nutrient deposits on the seafloor. As a result, supplies of nutrients along shorelines at or near the equator were stable but limited.

Most major animal phyla had evolved earlier, in the Precambrian seas. Possibly some of their ancestors were among the **Ediacarans**, peculiar organisms shaped like fronds, disks, and blobs that nearly defy classification (Figure 19.13*a,b*). Like Ediacarans, the early Cambrian animals had flattened bodies, with a good surface-to-volume ratio for taking up nutrients (Figure 19.13*c*). Most existed on or in seafloor sediments, where dead organisms and organic debris settled. They ranged from sponges to simple vertebrates, and they were diverse.

How could so much diversity arise? Possibly genes governing early growth and development were far less intertwined than they are at present, so there might not have been as much selection against mutant alleles and

novel traits. Also, warm, shallow waters near the new coasts were vacant adaptive zones, with many enticing opportunities for new ways to secure food.

Entombed in sedimentary beds from the Cambrian are fossilized organisms that have punctures, missing chunks, and healed wounds. These are not artifacts of fossilization; the animals were injured while they were alive. Diverse predators and prey with armorlike shells, spines, mouths, and novel feeding structures evolved in short order. Things were starting to get lively!

Later in the Cambrian, temperatures in the shallow seas changed drastically. Trilobites (Figure 19.13*d*), one of the most common invertebrates, almost vanished. In the Ordovician, the supercontinent Gondwana had been drifting south, and parts became submerged in shallow seas. The emergence of vast new marine environments promoted adaptive radiations. Many new reef organisms evolved. Among them were the swift, shelled predators called nautiloids. Their surviving descendants include the chambered nautiluses (Section 42.12).

Later on, Gondwana straddled the South Pole, and huge glaciers formed across its surface. When enormous volumes of water were locked up as ice, shallow seas throughout the world were drained. This was the first ice age that we know about. It may have triggered the first global mass extinction. At the Ordovician-Silurian boundary, reef life everywhere collapsed.

Gondwana drifted north during the Silurian and on into the Devonian. This was a pivotal time of evolution. Reef communities recovered. Armor-plated fishes with massive jaws diversified. And in the wet lowlands, small stalked plants appeared (Figure 19.14*b,c*). So did fungi and many invertebrates, such as segmented worms. In

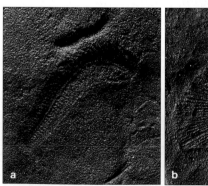

Figure 19.13 Representatives from the Precambrian and Cambrian seas. Two Ediacarans, about 600 million years old: (**a**) *Spriggina* and (**b**) *Dickensonia*. The oldest known Ediacarans lived 610 million years ago and the most recent in Cambrian times, 510 million years ago. (**c**) From British Columbia's Burgess Shale, a fossilized marine worm. (**d**) A beautifully preserved fossil of one of the earliest trilobites.

Figure 19.14 (**a**) Life in the Silurian seas. Some of those shelled animals (nautiloids) were twelve feet long. (**b**) A Silurian swamp, dominated by forerunners of modern ferns and club mosses. (**c**) Fossils of a Devonian plant (*Psilophyton*), possibly one of the earliest ancestors of conifers and other seed-bearing plants. (**d**) Reptiles (*Dimetrodon*) of a largely hotter and drier time—the Permian. Some fossils of those carnivores were found in Texas. Giant club mosses and horsetails had declined, and conifers, cycads, and other gymnosperms replaced them.

Devonian times, the fishes that would become ancestors to amphibians invaded the land. Those fishes had lobed fins, the forerunners of legs and other limbs. They had simple lungs. As described in Section 24.6, lobed fins and lungs were key innovations; they would prove most advantageous for life out of water, in dry land habitats.

Then, as they say, something bad happened. At the boundary between the Devonian and Carboniferous, sea levels swung catastrophically for unknown reasons, and caused another mass extinction. Afterward, plants and insects embarked on adaptive radiations on land.

Throughout Carboniferous times, land masses were gradually submerged and drained many times. Organic debris piled up. Then it was compacted and converted to coal, in the manner described in Section 22.3.

Insects, amphibians, and early reptiles flourished in vast swamp forests of Permian times (Figure 19.14*d*). Ancestors of the seed-producing plants called cycads, ginkgos, and conifers dominated those forests.

As the Permian drew to a close, something caused the greatest of all mass extinctions. More than 50 percent of all families disappeared, and only 5 percent of the known species survived. At the time, all land masses were colliding together. Eventually they formed Pangea, a supercontinent that extended all the way from the North to the South Pole. A single world ocean lapped its margins:

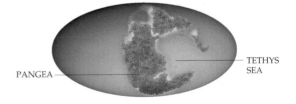

As you will see, the new distribution of oceans, land masses, and land elevations had catastrophic effects on global climates—and on the course of life's evolution.

Early in the Paleozoic era, organisms of all six kingdoms were flourishing in the seas. By the end of the era, many lineages had successfully invaded the land, including the wet lowlands of the supercontinent Pangea.

WWW

LIFE IN THE MESOZOIC ERA

Speciation on a Grand Scale

We divide the **Mesozoic** into the Triassic, Jurassic, and Cretaceous periods. It lasted about 175 million years. Early on in the Cretaceous, the supercontinent Pangea started to break up. Its huge fragments slowly drifted apart, and we can assume that the resulting geographic isolation favored divergences and speciation:

This was an era of spectacular expansion in the range of global diversity. In the seas, invertebrates and fishes underwent adaptive radiations. On land, conifers and other seed-bearing **gymnosperms** as well as insects and reptiles became the most visibly dominant lineages. Flowering plants, or **angiosperms**, originated before the end of the era. Within a mere 30 to 40 million years, they would displace conifers and related plants in most environments (Figure 19.15 and Section 22.8).

Rise of the Ruling Reptiles

Early in the Triassic, the first **dinosaurs** evolved from a reptilian lineage. They were not much bigger than a turkey. Possibly most species had high metabolic rates, and maybe they were warm-blooded. Many sprinted about on two legs. Dinosaurs weren't the dominant land animals. Instead, center stage belonged to *Lystrosaurus* and other plant-eating, mammal-like reptiles that were too large to be bothered by most predators.

Adaptive zones may have opened up for dinosaurs in a frightening way. There is a string of five craters in France, Quebec, Manitoba, and North Dakota; one is the size of Rhode Island. About 214 million years ago, according to radioisotope dating, fragments from an asteroid or a comet apparently fell one after another as the Earth spun beneath them, just as a string of huge fragments from comet Shoemaker-Levy 9 hit Jupiter in 1994. The resulting blast waves, global firestorm, lava flows, and earthquakes must have been stupendous. Most of the animals lucky enough to survive that time of mass extinction (as well as later ones) were smaller, had higher rates of metabolism, and were less vulnerable than others to drastic changes in outside temperatures.

Descendants of the surviving dinosaurs became the ruling reptiles; they endured for 140 million years. Some species reached monstrous proportions. Among them were the ultrasaurs, fifteen meters tall.

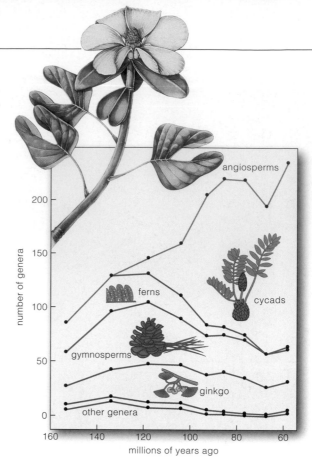

Figure 19.15 Range of diversity among vascular plants during the Jurassic and Cretaceous. Conifers and other gymnosperms were dominant before then. They were already declining when flowering plants began a spectacular radiation that continued into the Cenozoic. Also shown is a floral shoot of *Archaeanthus linnenbergeri*, a flowering plant of Cretaceous times. In many of its traits, this now-extinct species resembled living magnolias.

Many dinosaurs perished in another mass extinction at the end of the Jurassic, then in a pulse of extinctions in the Cretaceous. Perhaps plumes of molten material ruptured the crust and triggered changes in the global climate. Perhaps asteroids or comets inflicted the blows. Not all lineages survived those episodes. Yet some did recover, and new ones evolved. Duckbilled dinosaurs appeared in forests and swamps. Tanklike *Triceratops* and other plant eaters flourished in open habitats. They were prey for the fearsomely toothed, agile, and swift *Velociraptor* of motion picture fame.

About 120 million years ago, global temperatures skyrocketed 25 degrees. By one theory, plumes spread out beneath the crust and "greased" the crustal plates into moving twice as fast. A superplume or a rash of them broke through the crust. The crust in what is now the South Atlantic opened like a zipper. Basalt and lava poured from the fissures; volcanoes spewed nutrient-rich ashes. Simultaneously, the plumes released great amounts of carbon dioxide, one of the "greenhouse" gases that absorb some of the heat radiating from the Earth before it escapes into outer space. The nutrient-enriched planet warmed—and remained warm for 20

Figure 19.16 If dinosaurs of this sort had not disappeared at the end of the Cretaceous, would then-tiny mammals ever have ventured out from under the shrubbery? Would *you* even be here today?

million years. Photosynthetic organisms flourished on land and in shallow seas. Their remains were slowly buried and converted into the world's vast oil reserves.

About 65 million years ago, the last dinosaurs and many marine organisms vanished in a mass extinction (Figure 19.16). As described in the next section, their disappearance apparently coincided with a direct hit by an asteroid the size of Mount Everest. Over time, the impact site slowly drifted into a position that we now call the northern Yucatán peninsula.

The Mesozoic was a time of major adaptive radiations and of a mass extinction in which the last dinosaurs and many marine organisms disappeared.

www

HORRENDOUS END TO DOMINANCE

Figure 19.18 Artist's interpretation of what might have happened in the last few minutes of the Cretaceous.

It has only been about 55,000 years since the first fully modern humans walked the Earth. Since then, how many times have people, puffed up with self-importance, set out to conquer neighbors, the land, and the seas? Think about it. Then think about the dinosaurs. Were they good at reigning supreme? No question about it; their lineage dominated the land for 140 million years. In the end, did it matter? Not a bit. Sixty-five million years ago, at the Cretaceous-Tertiary (K-T) boundary, nearly all remaining members of their most excellent lineage perished. Why? Bad luck.

A thin layer of iridium-rich rock distributed around the world dates precisely to the K-T boundary. Iridium is rare here but common in asteroids. The **asteroids** are rocky, metallic bodies, 1,000 kilometers to a few meters in diameter, that are hurtling through space. When planets of our solar system were forming, their gravitational pull swept most asteroids from the sky. At least 6,000 still orbit the sun in a belt between Mars and Jupiter (Figure 19.17). The orbits of dozens of others take them across Earth's orbit, like Russian roulette on a cosmic scale.

By analyzing iridium levels in soils, gravity maps, and other evidence, Walter Alvarez and Luis Alvarez hypothesized that an asteroid impact caused the K—T mass extinction. Later, researchers identified the impact site. Massive movements in the crust transported the site to what is now the northern Yucatán peninsula of Mexico (Figure 19.18). The impact crater is 9.6 kilometers deep and 300 kilometers across—wider than Connecticut. This crater as well as other evidence strongly supports what is now called the **K-T asteroid impact theory**.

To make a crater that big, the asteroid had to hit the Earth at 160,000 kilometers (100,000 miles) per hour. At least 200,000 cubic kilometers of debris and dense gases were blasted skyward. The crust itself heaved violently. Monstrous waves, 120 meters high, raced across the ocean, obliterating life on islands and then slamming

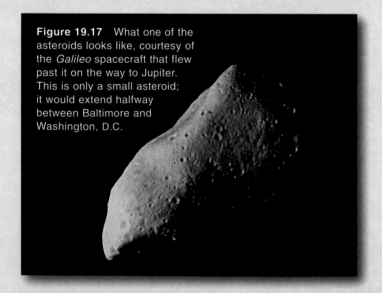

Figure 19.17 What one of the asteroids looks like, courtesy of the *Galileo* spacecraft that flew past it on the way to Jupiter. This is only a small asteroid; it would extend halfway between Baltimore and Washington, D.C.

into coasts of continents. Researchers long hypothesized that atmospheric debris blocked out sunlight for months. If so, plants and other producers that sustained the web of life would have withered and died; animals and other consumers would have starved to death. The hypothesis had problems. By some calculations, the volume of debris blasted aloft would not have been enough to have had such significant consequences.

Then comet Shoemaker-Levy 9 slammed into Jupiter. Particles blasted into the Jovian atmosphere triggered an intense heating of an area larger than the Earth. That event supports a **global broiling hypothesis**, proposed first by H. J. Melosh and his colleagues. Briefly stated, energy released at the K-T impact site was equivalent to detonating 100 million nuclear bombs. Trillions of tons of vaporized debris rose in a colossal fireball, then rapidly condensed into particles the size of sand grains. Seconds later, a cooler fireball made of steam, carbon dioxide, and unmelted rock formed. When the debris fell to the Earth, it raised the atmosphere's temperature by thousands of degrees. The sky above the planet must have glowed with heat ten times more intense than the noonday sun above Death Valley in summer. In one horrific hour, nearly all plants erupted in flames and every animal out in the open was broiled alive.

Things haven't settled down much since dinosaurs disappeared. For instance, about 2.3 million years ago, a huge chunk from outer space apparently hit the Pacific Ocean. At about the same time, vast ice sheets started forming abruptly in the Northern Hemisphere. Long-term shifts in climate may have been ushering in this most recent ice age, but the global impact would have accelerated it. Water vaporized by the impact could have contributed to a global cloud cover that limited the amount of sunlight reaching the surface. Ancestors of humans were around when all this happened. The extreme climate shift surely tested their adaptability.

Almost certainly, severe environmental tests await all existing lineages. When we become too smitten with our importance in the world of life, we would do well to step back, from time to time, and reflect on what is going on above and beneath the Earth's surface. Asteroids and superplumes do have a way of leveling the playing field, as they did for gigantic dinosaurs and tiny mammals.

WWW

19.8 LIFE IN THE CENOZOIC ERA

The breakup of Pangea triggered events that continued into the present era, the **Cenozoic**. At the dawn of the Cenozoic, land masses were on collision courses:

Coastlines fractured. The Cascades, Andes, Himalayas, and Alps rose through volcanic activity, uplifting, and other events at crustal rifts and plate boundaries. These geologic changes brought about major shifts in climate that influenced the further evolution of life.

During the Paleocene epoch, climates were warmer and wetter. Tropical and subtropical forests extended farther north and south than they do today. Woodlands and forests spread even into polar regions. Although their key traits developed before the dinosaurs left the scene, mammals now began their major radiation.

The global climate warmed even more in the Eocene epoch, and subtropical forests extended north into the polar regions. A variety of mammals, including primates, bats, rhinos, hippos, elephants, horses, and assorted carnivores, emerged. By the late Eocene, climates became cooler and drier, and seasonal changes became pronounced. Woodlands and dry grasslands dominated vast tracts of land. Patterns of vegetational growth changed so much that many mammals were driven to extinction.

From the Oligocene through the Pliocene, an abundance of grazing and browsing animals flourished in the woodlands and grasslands. Among them were camels and the "giraffe rhinoceros," along with some fearsome carnivores that stalked them (Figure 19.19).

Today the distribution of land masses favors species diversity. The richest ecosystems are the tropical forests of South America, Madagascar, and Southeast Asia, as well as the marine ecosystems of archipelagos in the tropical Pacific. Yet we are in the midst of what may be one of the greatest mass extinctions. About 50,000 years ago, nomadic humans followed migrating herds of wild animals around the Northern Hemisphere. By

Figure 19.19 Representatives from Cenozoic times. (**a**) In the early Paleocene, in what is now Wyoming, diverse mammals lived in dense forests of sequoia trees, laurel, and other plants. On the ground is the raccoonlike *Chriacus*, facing a tree-climbing rodent (*Ptilodus*). Higher in the tree is *Peradectes*, a marsupial. (**b**) From the late Eocene to the early Miocene, *Indricotherium* browsed on woodland trees. This "giraffe rhinoceros," the largest land mammal known, weighed 15 tons and was 5.5 meters (18 feet) at the shoulder. (**c**) From the Pleistocene, a small horse and the saber-tooth cat (*Smilodon*). Both mammals became extinct. Fossils of them have been found in a pitch pool at Rancho LaBrea, California.

a few thousand years later, major groups of mammals were extinct. The pace of extinction has since accelerated as humans hunt for food, fur, feathers, or fun, and as they destroy habitats to clear land for farm animals or crops. Chapter 43 focuses on the global repercussions.

Major geologic changes during the Cenozoic triggered shifts in climate. The great adaptive radiation of mammals began, first in tropical forests, then in woodlands and grasslands.

WWW

SUMMARY OF EARTH AND LIFE HISTORY

We conclude this overview chapter with an illustration that correlates milestones in the evolution of life with the evolution of the Earth. As you study Figure 19.20, keep in mind that it is only a generalized summary. For example, it charts five of the greatest mass extinctions, but there were many others in between. The illustration's diagram of the full range of biodiversity conveys the shrinking and expansion of species through time. It is a combined range for *all* of the major groups on land and in the seas. Remember, each major group has its own evolutionary history. Some of its member species may have persisted to the present; some or all may be extinct. Some of the groups may be represented by only one or a few species; others may be represented by hundreds or thousands or a million.

With these qualifications in mind, you are ready to turn to the next chapters in this unit. They will provide you with far richer detail of the history and the current range of biodiversity for all six kingdoms of life.

WWW

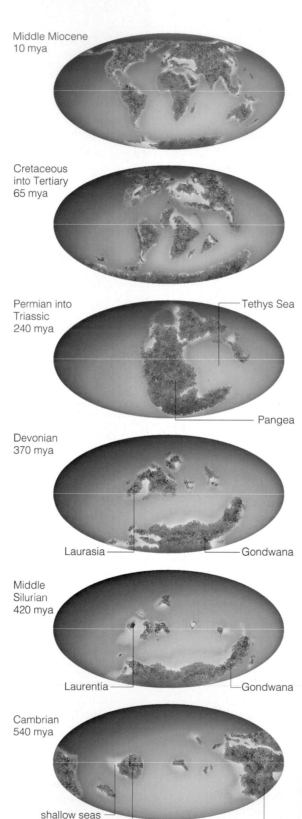

Middle Miocene
10 mya

Cretaceous
into Tertiary
65 mya

Permian into
Triassic
240 mya
— Tethys Sea
— Pangea

Devonian
370 mya

Laurasia — — Gondwana

Middle
Silurian
420 mya

Laurentia — — Gondwana

Cambrian
540 mya

shallow seas
(light blue)   Laurentia   Gondwana

Figure 19.20 Summary of major events in the evolution of the Earth and of life. As you read through the chapters to follow, you may wish to return to this illustration now and then to remind yourself of how the details fit into the greater evolutionary picture.

	Period	Epoch
CENOZOIC ERA	Quaternary	Recent
		Pleistocene
	Tertiary	Pliocene
		Miocene
		Oligocene
		Eocene
		Paleocene
MESOZOIC ERA	Cretaceous	Late
		Early
	Jurassic	
	Triassic	
PALEOZOIC ERA	Permian	
	Carboniferous	
	Devonian	
	Silurian	
	Ordovician	
	Cambrian	
PROTEROZOIC EON		
ARCHEAN EON AND EARLIER		

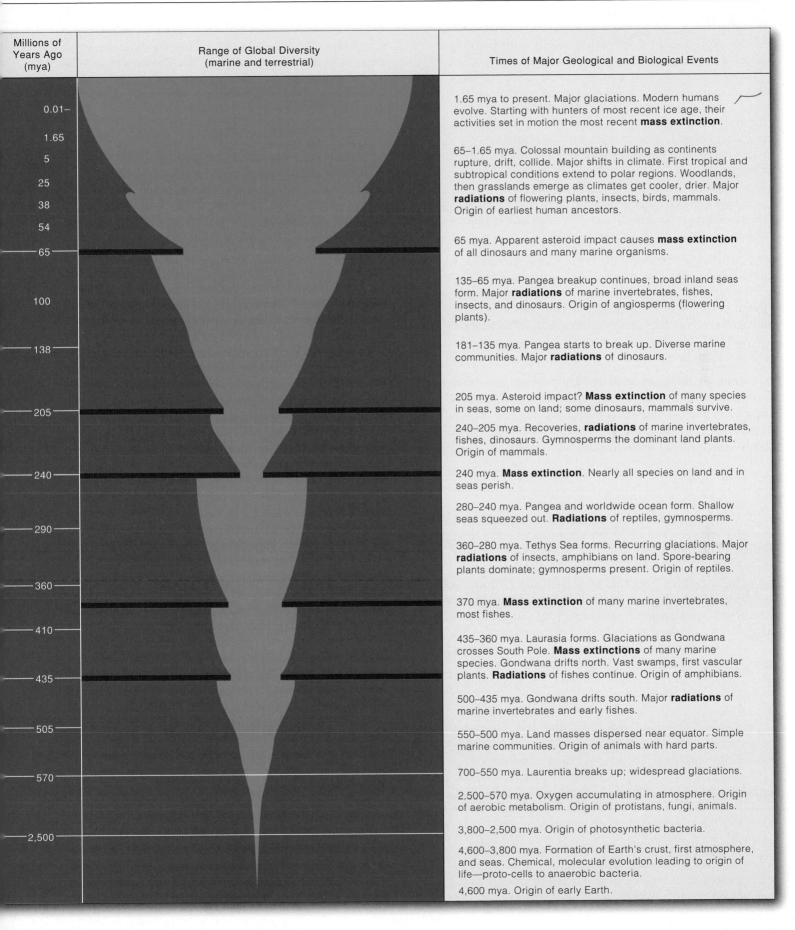

Millions of Years Ago (mya)	Range of Global Diversity (marine and terrestrial)	Times of Major Geological and Biological Events

0.01–
1.65
5
25
38
54

—65—

100

—138—

—205—

—240—

—290—

—360—

—410—

—435—

—505—

—570—

—2,500—

1.65 mya to present. Major glaciations. Modern humans evolve. Starting with hunters of most recent ice age, their activities set in motion the most recent **mass extinction**.

65–1.65 mya. Colossal mountain building as continents rupture, drift, collide. Major shifts in climate. First tropical and subtropical conditions extend to polar regions. Woodlands, then grasslands emerge as climates get cooler, drier. Major **radiations** of flowering plants, insects, birds, mammals. Origin of earliest human ancestors.

65 mya. Apparent asteroid impact causes **mass extinction** of all dinosaurs and many marine organisms.

135–65 mya. Pangea breakup continues, broad inland seas form. Major **radiations** of marine invertebrates, fishes, insects, and dinosaurs. Origin of angiosperms (flowering plants).

181–135 mya. Pangea starts to break up. Diverse marine communities. Major **radiations** of dinosaurs.

205 mya. Asteroid impact? **Mass extinction** of many species in seas, some on land; some dinosaurs, mammals survive.

240–205 mya. Recoveries, **radiations** of marine invertebrates, fishes, dinosaurs. Gymnosperms the dominant land plants. Origin of mammals.

240 mya. **Mass extinction**. Nearly all species on land and in seas perish.

280–240 mya. Pangea and worldwide ocean form. Shallow seas squeezed out. **Radiations** of reptiles, gymnosperms.

360–280 mya. Tethys Sea forms. Recurring glaciations. Major **radiations** of insects, amphibians on land. Spore-bearing plants dominate; gymnosperms present. Origin of reptiles.

370 mya. **Mass extinction** of many marine invertebrates, most fishes.

435–360 mya. Laurasia forms. Glaciations as Gondwana crosses South Pole. **Mass extinctions** of many marine species. Gondwana drifts north. Vast swamps, first vascular plants. **Radiations** of fishes continue. Origin of amphibians.

500–435 mya. Gondwana drifts south. Major **radiations** of marine invertebrates and early fishes.

550–500 mya. Land masses dispersed near equator. Simple marine communities. Origin of animals with hard parts.

700–550 mya. Laurentia breaks up; widespread glaciations.

2,500–570 mya. Oxygen accumulating in atmosphere. Origin of aerobic metabolism. Origin of protistans, fungi, animals.

3,800–2,500 mya. Origin of photosynthetic bacteria.

4,600–3,800 mya. Formation of Earth's crust, first atmosphere, and seas. Chemical, molecular evolution leading to origin of life—proto-cells to anaerobic bacteria.

4,600 mya. Origin of early Earth.

SUMMARY

1. The evolutionary story of life begins with the "big bang," a model for the origin of the universe.

   a. By this model, all matter and all of space were once compressed in a fleeting state of enormous heat and density. Time began with the near-instantaneous distribution of matter and energy through the known universe, which has been expanding ever since.

   b. Helium and the other light elements formed right after the big bang. Heavier elements originated during the formation, evolution, and death of stars.

   c. Every element of the solar system, the Earth, and life itself are products of the physical and chemical evolution of the universe and its stars.

2. Four billion years ago, the Earth was organized as a high-density core, a mantle of intermediate density, and a thin, extremely unstable crust of low-density rocks. Its first atmosphere probably contained gaseous hydrogen, nitrogen, carbon monoxide, and carbon dioxide. Free oxygen and liquid water could not have accumulated at the surface under the prevailing conditions.

3. Water accumulated after the crust cooled off. Over hundreds of millions of years, runoff from rains carried dissolved mineral salts and other compounds to crustal depressions, where the early seas formed. Life could not have originated without salty liquid water.

4. Many diverse studies and experiments have yielded indirect evidence that life originated under conditions that presumably existed on the early Earth.

   a. Comparative investigations of the composition of cosmic clouds, rocks from other planets, and rocks from the Earth's moon suggest that precursors of complex molecules associated with life were available.

   b. In laboratory tests that simulated the primordial conditions, including the absence of free oxygen, those precursors spontaneously assembled into sugars (such as glucose), amino acids, and other organic compounds.

   c. Known chemical principles as well as advanced computer simulations indicate that metabolic pathways could have evolved through chemical competition for limited supplies of organic molecules (which could have accumulated by natural geologic processes in the seas).

   d. Self-replicating systems of RNA, enzymes, and coenzymes have been synthesized in the laboratory. How DNA entered the picture is not yet understood.

   e. In laboratory simulations of conditions thought to have prevailed on the early Earth, lipids as well as lipid-protein membranes having some of the properties of cell membranes have formed spontaneously.

5. Life originated by 3.8 billion years ago. Since then, it has been influenced by major changes in the Earth's crust, atmosphere, and oceans. Forces of change have included plate movements, asteroid impacts, and the activities of organisms (including oxygen-producing photosynthesizers and, currently, the human species).

6. Abrupt discontinuities in the fossil record mark the times of global mass extinctions. They are the boundary markers for five great intervals in a geologic time scale. Radiometric dating has allowed us to assign absolute dates to this time scale:

   a. Archean: 3.9 billion to 2.5 billion years ago
   b. Proterozoic: 2.5 billion to 570 million years ago
   c. Paleozoic: 570 to 240 million years ago
   d. Mesozoic: 240 to 65 million years ago
   e. Cenozoic: 65 million years ago to the present

7. The first living cells were prokaryotic (bacteria). Not long after they arose in the Archean, the first divergence led to eubacteria, and to a shared prokaryotic ancestor of both archaebacteria and eukaryotes. Some eubacteria used a cyclic pathway of photosynthesis.

8. During the Proterozoic, the noncyclic pathway of photosynthesis evolved in some lineages of eubacteria. Oxygen, a by-product of the pathway, slowly started to accumulate in the atmosphere.

   a. Eventually, the atmospheric concentration of free oxygen blocked the further spontaneous formation of organic molecules. From that time on, the spontaneous origin of life was no longer possible on Earth.

   b. The abundance of atmospheric oxygen became a selective pressure that brought about the evolution of aerobic respiration. Aerobic respiration was a key step toward the origin of the first eukaryotic cells.

   c. Mitochondria and chloroplasts, both important eukaryotic organelles, probably evolved as an outcome of endosymbiosis between certain aerobic bacteria and the anaerobic bacterial forerunners of eukaryotes.

   d. The oxygen-rich atmosphere promoted formation of a layer of ozone (O_3). In time, that atmospheric shield against destructive ultraviolet radiation allowed some lineages to move out of the seas, into low wetlands.

9. By the early Paleozoic, diverse organisms of all six lineages had become established in the seas. By its end, the invasion of land was under way. From that time on, there have been pulses of mass extinctions and adaptive radiations. Asteroid impacts and other catastrophes triggered many of these events. So did plate movements that changed the distribution of oceans and land masses as well as the prevailing global and regional climates.

Review Questions

1. Compare the presumed chemical and physical conditions that are thought to have prevailed on the Earth 4 billion years ago with conditions that are now prevailing. *19.1*

2. Describe examples of the kinds of experimental evidence for the spontaneous origin of (a) large organic molecules, (b) the self-assembly of proteins, and (c) the formation of organic membranes and spheres, under laboratory conditions similar to those of the early Earth. *19.1, 19.2*

3. Summarize the key points of the theory of endosymbiotic origins for mitochondria and chloroplasts. Cite evidence that favors this theory. *19.4*

4. Describe the prevailing conditions that probably favored the Cambrian "explosion" of diversity among marine animals, as evidenced by the fossil record. *19.5*

5. During which geologic time spans did plants, fungi, and insects invade the land? What kind of vertebrates first invaded the land, and when? *19.5*

6. What were global conditions like when gymnosperms and dinosaurs originated? *19.6*

7. Briefly explain how an asteroid impact and "global broiling" may have caused the mass extinctions at the K–T boundary. *19.7*

8. Would you expect the Paleozoic, Mesozoic, or Cenozoic to be called "the age of mammals"? As part of your answer, explain the differences between global conditions in each era. *19.8*

Self-Quiz *(Answers in Appendix III)*

1. Life originated by _____ .
   a. 4.6 billion years ago   c. 3.8 billion years ago
   b. 2.8 million years ago   d. 3.8 million years ago

2. Through study of the geologic record, we know that the evolution of life has been profoundly influenced by _____ .
   a. tectonic movements of the Earth's crust
   b. bombardment of the Earth by celestial objects
   c. profound shifts in land masses, shorelines, and oceans
   d. physical and chemical evolution of the Earth
   e. all of the above

3. _____ was the first to obtain indirect evidence that organic molecules could have been formed on the early Earth.
   a. Darwin   c. Fox
   b. Miller   d. Margulis

4. An abundance of _____ was conspicuously absent from the Earth's atmosphere 4 billion years ago.
   a. hydrogen   c. carbon monoxide
   b. nitrogen   d. free oxygen

5. Which of the following statements is false?
   a. The first living cells were prokaryotes.
   b. The cyclic pathway of photosynthesis first appeared in some eubacterial species.
   c. Oxygen began accumulating in the atmosphere after the noncyclic pathway of photosynthesis evolved.
   d. In the Proterozoic, increasing levels of atmospheric oxygen enhanced the spontaneous formation of organic molecules.
   e. All are correct.

6. The first eukaryotic cells emerged during the _____ .
   a. Paleozoic   c. Archean   e. Cenozoic
   b. Mesozoic   d. Proterozoic

7. Match the geologic time interval with the events listed.
   ____ Archean   a. major radiations of dinosaurs, origin
   ____ Proterozoic   of flowering plants and mammals
   ____ Paleozoic   b. chemical evolution, origin of life
   ____ Mesozoic   c. major radiations of flowering plants,
   ____ Cenozoic   insects, birds, mammals; emergence
                           of human forms
                        d. oxygen present; origin of aerobic
                           metabolism, protistans, fungi,
                           animals
                        e. rise of early plants on land, origin
                           of amphibians, and origin of reptiles

Critical Thinking

1. Briefly explain, in terms of hydrophilic and hydrophobic interactions, how proto-cells might have formed in water from aggregations of lipids, proteins, and nucleic acids.

2. The Atlantic Ocean is gradually widening, and the Pacific Ocean and Indian Ocean are closing. Many millions of years from now, the continents will collide and form a second Pangea. Write a short essay on what environmental conditions might be like on that future supercontinent and on what types of species might survive there.

3. According to one estimate, there is a chance that about 10^{20} planets have formed in the universe that are capable of sustaining life—but there is only one chance at intelligent life per planet. Given your knowledge of molecular biology and evolutionary processes, do you agree with this estimate? If so, speculate on why the odds are so low.

4. We know of a number of large asteroids that may intersect Earth's orbit in the distant future. There probably are a number we don't know about. Would you use this as an excuse not to worry about polluting the environment, not to take care of your physical health (as by avoiding drugs), and not to care about our cultural evolution? Why or why not?

Selected Key Terms

angiosperm *19.6*	global broiling
archaebacterium *19.3*	hypothesis *19.7*
Archean *19.3*	gymnosperm *19.6*
asteroid *19.7*	K–T asteroid impact theory *19.7*
big bang *CI*	mantle, of Earth *19.1*
Cenozoic *19.8*	Mesozoic *19.6*
crust, of Earth *19.1*	Paleozoic *19.5*
dinosaur *19.6*	prokaryotic cell *19.3*
Ediacaran *19.5*	Proterozoic *19.3*
endosymbiosis	protistan *19.4*
(theory of) *19.4*	proto-cell *19.2*
eubacterium *19.3*	RNA world *19.2*
eukaryotic cell *19.3*	stromatolite *19.3*

Readings   *See also www.infotrac-college.com*

de Duve, C. September–October 1995. "The Beginnings of Life on Earth." *American Scientist* 83: 428–437.

———. April 1996. "The Birth of Complex Cells." *Scientific American* 274(4): 50–57.

Dobb, E. February 1992. "Hot Times in the Cretaceous." *Discover* 13: 11–13.

Dott, R., Jr., and R. Batten. 1988. *Evolution of the Earth.* Fourth edition. New York: McGraw-Hill.

Hartman, W., and Chris Impey. 1994. *Astronomy: The Cosmic Journey.* Fifth edition. Belmont, California: Wadsworth.

Horgan, J. February 1991. "Trends in Evolution: In the Beginning . . . ." *Scientific American* 264(2): 116–125.

Wright, K. March 1997. "When Life Was Odd." *Discover* 18(3): 52–61. Update on the bizarre Ediacarans.

WWW   *http://www.brookscole.com/biology*

Practice quiz questions, hypercontents, BioUpdates, and critical thinking. The Brooks/Cole Biology Resource Center provides a wealth of information fully organized and integrated by chapter.

BACTERIA, VIRUSES, AND PROTISTANS

The Unseen Multitudes

Did a friend ever mention that you are nearly 1/1,000 of a mile tall? Probably not. What would be the point of measuring people in units as big as miles? Even so, we think this way, in reverse, whenever we measure microorganisms. For the most part, **microorganisms** are single-celled organisms that are too small to be seen without the aid of a microscope.

The bacterial cells shown in Figure 20.1 are a case in point. To measure them, you would have to divide one meter into a thousand units, or millimeters. Next, you would have to divide one of the millimeters into a thousand smaller units, or micrometers. To give you a sense of how small that is, a single millimeter would be about as small as the dot of this "i." And a *thousand* bacteria would fit side by side on top of the dot!

To be sure, viruses are smaller still, as you might deduce after looking at Figure 20.1*d*. We measure them in nanometers—and one nanometer is only a billionth of a meter. But viruses are not alive. We consider them in this chapter because one kind or another infects just about every organism, starting with the bacteria.

Bacteria generally are the smallest microorganisms, but they vastly outnumber the individuals in all other kingdoms combined. Their reproductive potential is staggering. Under ideal conditions, some species divide about every twenty minutes. If that rate of reproduction were to hold constant, a single bacterium could have nearly a billion descendants in ten hours!

So why don't the unseen multitudes take over the world? Sooner or later, their burgeoning populations use up available nutrients and pollute the surroundings with their own metabolic wastes. In other words, the cells alter the very conditions that initially favored their reproduction. Besides this, certain kinds of viruses

Figure 20.1 (**a–c**) How small are bacteria? Shown here, *Bacillus* cells on the tip of a pin. The cells in (**c**) are magnified 14,000 times. (**d**) How small are viruses? Shown here, bacteriophage particles, each about 225 nanometers tall, infecting a bacterial cell. (**e**) Mealtime for *Didinium*, a single-celled, predatory protistan with a big mouth. *Paramecium*, another protistan, is poised at the mouth (*left*) and swallowed (*right*). Generally, protistans are much larger than bacteria, as you can tell by comparing the scale bars at the lower right corner of all five micrographs.

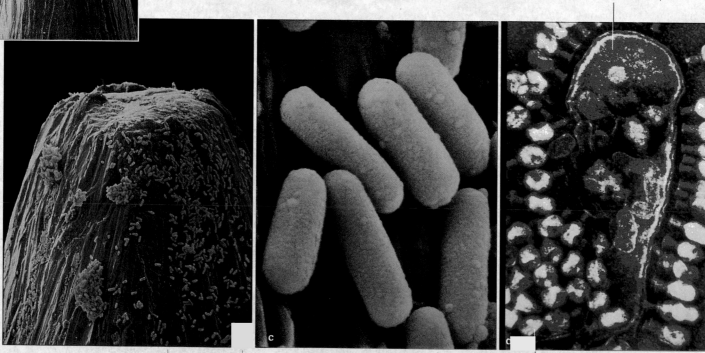

infected bacterial cell that ruptured

20 μm

0.5 μm

virus particle   0.75 μm

attack just about every species of bacterium and help keep their population sizes in check. So do seasonal changes in living conditions.

Of course, you probably do not find much comfort in this when you serve as an unwilling host for one of the pathogenic types. **Pathogens**, recall, are infectious, disease-causing agents that invade target organisms and multiply inside them or on them. Disease follows when metabolic products secreted from pathogenic cells damage tissues of the body and interfere with their normal functioning.

Certain pathogens can indeed make you suffer, but they should not give every microorganism a bad name. For example, think back on the uncountable numbers of photosynthetic bacteria and protistans in the seas (Section 6.8). Collectively, they help provide food and oxygen for entire communities, and they have major roles in the global cycling of carbon. Or think about the microorganisms that feed on organic debris. Together with other decomposers, they help cycle nutrients that sustain entire communities.

From the human perspective, microorganisms are good or bad, or even dangerous. Basically, however, they are simply surviving and reproducing like the rest of us, in ways that are topics of this chapter.

Paramecium, a predatory protistan that ended up as a meal

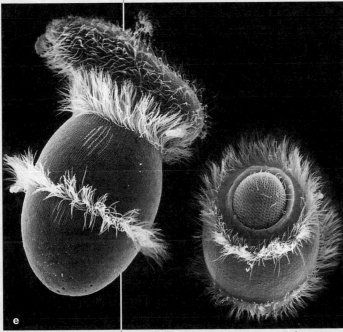

Didinium, another predatory protistan

50 µm

KEY CONCEPTS

1. Bacteria alone are prokaryotic cells. They do not have a profusion of internal, membrane-bound organelles, as eukaryotic cells do. As a group, bacteria show great metabolic diversity, and many species show complex behavior.

2. Most bacteria reproduce by prokaryotic fission. This cell division mechanism follows DNA replication and results in two genetically equivalent daughter cells.

3. Bacteria were the first living organisms on Earth. Not long after they originated, they diverged into lineages that gave rise to eubacteria and to the common ancestor of archaebacteria and eukaryotic cells.

4. A virus, a noncellular infectious particle, consists of nucleic acid (either DNA or RNA), a protein coat, and sometimes an outer envelope. It cannot replicate without pirating the metabolic machinery of a specific type of host cell.

5. Nearly all viral multiplication cycles proceed through five steps: attachment to a host cell, penetration of its plasma membrane, replication of viral DNA or RNA and synthesis of viral proteins, then assembly of new viral particles, and finally release from the infected cell.

6. Protistans are easily distinguishable from bacteria but are difficult to classify with respect to other eukaryotes. They differ enormously in morphology and life-styles.

7. Fungus-like protistans include parasitic and predatory molds that produce spores. Animal-like protistans include nonphotosynthetic flagellated protozoans, sporozoans, and ciliates. Plant-like protistans include single-celled, photosynthetic flagellates. They also include the red, brown, and green algae.

8. Most of us tend to judge microorganisms through the prism of human interests. Yet their lineages are the most ancient, their adaptations are diverse, and they are simply surviving and reproducing like the rest of us.

CHARACTERISTICS OF BACTERIA

Of all organisms, bacteria are the most abundant and far-flung. Thousands of species inhabit diverse places, such as deserts, hot springs, glaciers, and seas. Some have been carrying on for millions of years 2,780 meters (9,121 feet) below the Earth's surface! Billions may live in a handful of rich soil. The ones in your gut and on your skin outnumber your own cells. Bacteria have the longest evolutionary history. Trace any lineage back far enough and you come across bacterial ancestors. From *Escherichia coli* to amoebas, elephants, clams, and coast redwoods, all organisms interconnect, regardless of their differences in size, numbers, and evolutionary distance.

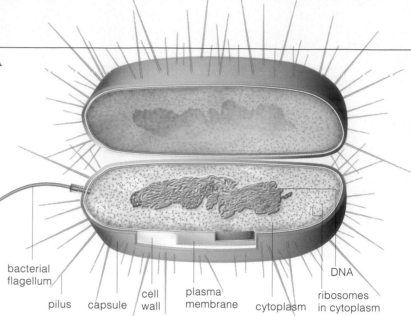

Figure 20.2 Generalized body plan of a bacterium.

bacterial flagellum / pilus / capsule / cell wall / plasma membrane / cytoplasm / DNA / ribosomes in cytoplasm

Bacterial Classification

Just a few decades ago, reconstructing the evolutionary history of bacteria appeared to be an impossible task. Except for stromatolites (Section 19.3), the most ancient groups of bacteria are not well represented in the fossil record. Most groups are not represented at all.

Given their elusive histories, the many thousands of known species of prokaryotic cells traditionally have been classified by **numerical taxonomy**. By this practice, traits of an unidentified cell are compared with those of a known bacterial group. Such traits typically include cell shape, motility, staining attributes of the cell wall, nutritional requirements, metabolic patterns, and the presence or absence of endospores. The greater the total number of traits that the cell has in common with the known group, the closer is their inferred relatedness.

Since the 1970s, nucleic acid hybridization studies, gene sequencing, and other methods of comparative biochemistry have been revealing compelling evidence of bacterial phylogenies (Section 19.3 and Appendix I). For example, it seems a key divergence began shortly after prokaryotic cells originated. One branch led to the **eubacteria**, the most common prokaryotic cells. (Here, *eu–* is meant to signify "typical.") The other branch led to the common ancestor of both **archaebacteria** and the first eukaryotic cells. Table 20.1 and Figure 20.2 show the general characteristics of both bacterial groups.

Table 20.1   Characteristics of Bacterial Cells

1. All bacterial cells are prokaryotic; they do not have a membrane-bound nucleus.

2. Bacterial cells in general have a single chromosome (a circular DNA molecule); many species also have plasmids.

3. Most bacteria have a cell wall of peptidoglycan.

4. Most bacteria reproduce by prokaryotic fission.

5. Collectively, bacteria show great metabolic diversity.

Splendid Metabolic Diversity

All organisms take in energy and carbon to meet their nutritional requirements. Compared with other species, however, bacteria display the greatest diversity in their means of securing resources.

Like plants, *photoautotrophic* bacteria build organic compounds by photosynthesis; they are "self-feeders." They tap sunlight for energy and use carbon dioxide as their carbon source. Their plasma membrane contains the photosynthetic machinery. Some photosynthesizers use electrons and hydrogen from water molecules for the synthesis reactions, and release oxygen. Anaerobic photosynthesizers (which die in the presence of oxygen) get electrons and hydrogen from inorganic compounds, such as gaseous hydrogen and hydrogen sulfide.

Chemoautotrophic bacteria are self-feeders. Carbon dioxide is the usual carbon source. Some species strip organic compounds for electrons and hydrogen. Others use inorganic substances, such as gaseous hydrogen, sulfur, nitrogen compounds, and a form of iron.

Photoheterotrophic bacteria are not self-feeders. They use energy from the sun for photosynthesis, but their carbon sources are fatty acids, complex carbohydrates, and other compounds that various organisms produce.

Chemoheterotrophic bacteria are parasites or saprobes, not self-feeders. The parasites live on or in a living host and draw glucose and other nutrients from it. Saprobic types get nutrients from the organic products, wastes, or remains of other organisms.

Bacterial Sizes and Shapes

So far, you have a general sense of the microscopically small sizes of bacteria. Typically, the width or length of these cells is between one and ten micrometers. Three basic shapes are common among bacteria. A spherical

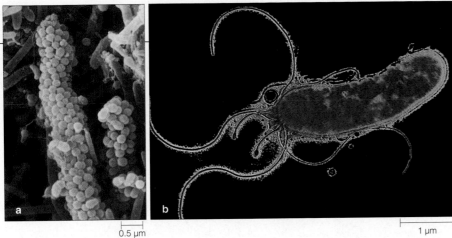

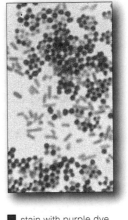

Figure 20.4 Gram staining. Cocci and rods smeared on a slide are stained with a purple dye (such as crystal violet), washed off, and then stained with iodine. All cells are now purple. The slide is washed with alcohol, which renders the Gram-negative cells colorless. Now the slide is counterstained (with safranin), washed, and dried. Gram-positive cells (*Staphylococcus aureus* in this case) remain purple. But the counterstain colors the Gram-negative cells (*Escherichia coli*) pink.

■ stain with purple dye
■ stain with iodine
□ wash with alcohol
■ counterstain with safranin

0.5 μm   1 μm

Figure 20.3 (**a**) Surface view of many bacilli and cocci attached to a human tooth; and doesn't that make you want to grab a toothbrush? (**b**) *Helicobacter pylori* cell, with its tuft of flagella. This pathogen can colonize the stomach lining and trigger inflammation. If untreated, infection can lead to gastritis, peptic ulcers, and possibly stomach cancer. *H. pylori* may contaminate water and food, especially unpasteurized milk. Currently, a combination of antibiotics and an antacid rids the body of the pathogen, after which ulcers heal.

shape is a coccus (plural, cocci; from a word that means berries). A rod shape is a bacillus (plural, bacilli, which means small staffs). A cell body that has one or more twists to it is a spirillum (plural, spirilla):

coccus   bacillus   spirillum

Don't let these simple categories fool you. Many cocci are oval or flattened. Bacilli can be skinny (like straws) or tapered (like cigars). Besides this, after the daughter cells divide, they often remain stuck together in chains, sheets, and other aggregations, as in Figure 20.3.

Structural Features

Bacteria alone are **prokaryotic cells**, meaning they were around before the evolution of nucleated cells (*pro—*, before; *karyon*, nucleus). Few have membrane-bound compartments of any sort for isolating metabolic events. Reactions take place in the cytoplasm or at the plasma membrane. For example, protein synthesis proceeds at ribosomes that are distributed through the cytoplasm or attached to the inner side of the plasma membrane. This does not mean bacteria are somehow inferior to the eukaryotic cells. Being tiny, fast reproducers, they do not require great internal complexity.

A wall usually surrounds the plasma membrane. A cell wall is a semirigid, permeable structure that helps the cell maintain its shape and resist rupturing when the internal fluid pressure increases (Section 5.7). The cell walls of eubacteria are composed of peptidoglycan molecules. In such molecules, peptide groups crosslink numerous polysaccharide strands to one another.

Clinicians identify many bacterial species on the basis of their wall structure and composition. Consider

the staining reaction known as a Gram stain. A sample of unknown bacterial cells is exposed to a purple dye, then to iodine, an alcohol wash, and a counterstain. The cell wall of a *Gram-positive* species remains purple. The wall of a *Gram-negative* species loses color after the wash, but the counterstain turns it pink (Figure 20.4).

A sticky mesh or slimy layer often surrounds the cell wall. It consists of polysaccharides, polypeptides, or both. The mesh helps a bacterial cell attach to teeth, mucous membranes, rocks in streams, and many other surfaces (Figure 20.3a). It also helps some encapsulated species avoid being eaten by the phagocytic, infection-fighting cells of a host organism.

Some bacteria move by use one or more bacterial flagella (singular, flagellum). These differ structurally from a eukaryotic flagellum and don't operate the same way. They move the cell with propeller-like rotations. Many bacteria display pili (singular, pilus). These short, filamentous proteins project above the cell wall (Figure 20.2). They help many species adhere to surfaces and, in some cases, to one another as a prelude to bacterial conjugation. Section 15.11 (*Critical Thinking*) highlights this mode of gene transfer between bacterial cells.

Prokaryotic cells are classified by numerical taxonomy (the total percentage of observable traits they have in common with a known bacterial group). They also are classified more directly by comparisons at the biochemical level.

Prokaryotic cells are now assigned to one of two lineages, the eubacteria and archaebacteria.

Bacteria are microscopic, prokaryotic cells. They generally have one circular bacterial chromosome, and often they have extra DNA in the form of plasmids.

Nearly all bacteria have a wall around the plasma membrane, and many have a capsule or slime layer around the cell wall.

WWW

Archaebacteria

We divide the new kingdom Archaebacteria into three main groups, the methanogens, halophiles, and extreme thermophiles. In many respects, the archaebacteria are unique in chemical composition, structure, metabolism, and nucleic acid sequences. They differ as much from other bacteria as they do from eukaryotic cells. Many investigators suspect that the existing archaebacteria, which can withstand conditions as hostile as those on the early Earth, resemble the first living cells. Hence the name of the kingdom (*archae*– means beginning).

The **methanogens**, or "methane makers," thrive in swamps, sewage, stockyards, the animal gut, and other oxygen-free habitats. They are strict anaerobes; they die in the presence of oxygen. They make ATP by anaerobic electron transport, a pathway that ends with methane (CH_4). As a group, methanogens produce 2 billion tons of methane each year. Their activities affect atmospheric carbon dioxide levels and the global cycling of carbon.

Long ago, methane accumulated on the ocean floor. There is a huge deposit (35 billion tons) 400 kilometers off South Carolina's coast. If ocean circulation changes, as it has in the past, all of that gas could move abruptly to the surface. Such a rapid release into the atmosphere could change the global climate (compare Section 41.8).

"Salt lovers," or **halophiles**, live in brackish ponds, in salt lakes, near hydrothermal vents (volcanic fissures on the ocean floor), and in other high-salinity settings (Figure 20.5*a*). Halophiles can spoil salted fish, animal hides, and commercial sea salt. Most produce ATP by aerobic pathways. When oxygen levels are low, some also produce ATP by a unique photosynthetic pathway.

"Heat lovers," or **extreme thermophiles**, inhabit such places as highly acidic soils, hot springs, even coal mine wastes. Some species are the basis of remarkable food webs in sediments around hydrothermal vents. Sometimes the water above the sediments is as hot as 110°C. Extreme thermophiles use the hydrogen sulfide spewing from the vents as a source of electrons for ATP formation. Their existence at vents is cited as evidence that life could have originated deep in the oceans.

Eubacteria

Most of the 400 genera of prokaryotes are eubacteria. Unlike other bacterial cells, eubacteria have fatty acids in their plasma membrane. Nearly all have a cell wall, which incorporates peptidoglycan. We still do not know enough about the evolutionary histories of eubacteria to move much beyond taxonomic classification. Here, we focus on their modes of nutrition.

Photoautotrophs. Cyanobacteria, or blue-green algae, are among the most common photoautotrophs. These

Figure 20.5 Archaebacteria. (**a**) Pinkish, saline water in Great Salt Lake, Utah—a sign of colonies of halophilic bacteria and algae that contain pinkish to red-orange carotenoid pigments. (**b**) Transmission electron micrograph of *Methanococcus jannaschii*, which had its DNA fully sequenced (Section 18.5).

aerobic cells release oxygen when photosynthesizing. Most live in ponds and other freshwater habitats. They may grow as mucus-sheathed chains of cells, which form dense, slimy mats near the surface of nutrient-enriched water. *Anabaena* and other types also convert nitrogen gas (N_2) to ammonia for biosynthesis. When nitrogen compounds are scarce, some cells develop into **heterocysts**. These modified cells make a nitrogen-fixing enzyme (Figure 20.6). They produce and share nitrogen compounds with the photosynthetic cells in the chains and receive carbohydrates in return.

Anaerobic photoautotrophs, such as green bacteria, get electrons from hydrogen sulfide or hydrogen gas, not from water. They may resemble anaerobic bacteria in which the cyclic pathway of photosynthesis evolved.

Chemoautotrophs. Many eubacteria in this category influence the global cycling of nitrogen, sulfur, and other nutrients. For example, as you know, nitrogen is a key building block for amino acids and proteins. Without it, there would be no life. Nitrifying bacteria in soil strip electrons from ammonia. Plants use the end product, nitrate, as a nitrogen source.

Chemoheterotrophs. Most of the eubacteria fall in this category. Many are decomposers; their enzymes break down organic compounds and pesticides in soil. Other "good" species are *Lactobacillus* (used in manufacturing pickles, sauerkraut, buttermilk, and yogurt) and the actinomycetes (sources of antibiotics). *E. coli*, an enteric bacterium, produces vitamin K and compounds that help humans digest fat. It also helps newborns digest milk, and its activities help prevent many food-borne pathogens from colonizing the gut. Sugarcane and corn benefit from a symbiont, the nitrogen-fixing spirochete *Azospirillum*. The plants take up some of the nitrogen initially fixed by this spirochete, and they give up some of their sugars to it. Beans, peas, and other legumes are mutualists with *Rhizobium*, a symbiont in their roots.

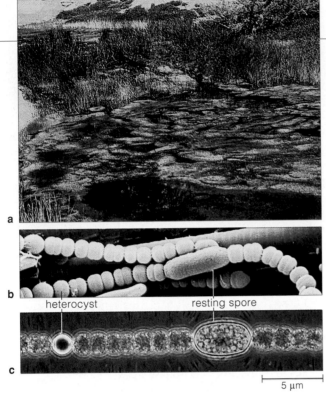

Figure 20.6 Cyanobacteria—common photoautotrophs. (**a**) A cyanobacterial population near the surface of a nutrient-enriched pond. (**b,c**) Resting spores form when conditions do not favor growth. A nitrogen-fixing heterocyst is also shown.

heterocyst   resting spore

5 µm

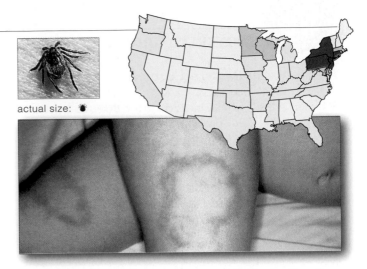

actual size: •

Figure 20.7 An extreme reaction to an infection by *Borrelia burgdorferi*, a spirochete. The rash is a symptom of what is now the most common tick-borne disease in the United States: Lyme disease. Tick bites deliver the spirochete to new hosts. The map shows statewide cases reported in 1996 (*tan*, fewer than 10; *yellow*, 11 to 99; *gold*, 100 to 600; *red*, more than 2,000).

Also in this category are most pathogenic bacteria. We admire pseudomonads as decomposers in soil, not when they grow on "our" soaps and other carbon-rich goods. They are especially dangerous because they can transfer plasmids with antibiotic-resistance genes.

Some *E. coli* strains cause a form of diarrhea that is the main cause of infant death in developing countries. *Clostridium botulinum* can taint fermented grain as well as food in improperly sterilized or sealed cans and jars. Its toxins cause *botulism*, a form of poisoning that can disrupt breathing and lead to death. *C. tetani*, one of its relatives, causes the disease *tetanus* (Section 29.5).

Like many other bacteria that commonly live in soil, *C. tetani* can make an **endospore**. This resting structure forms around a copy of the bacterial chromosome and part of the cytoplasm when depletion of nitrogen or some other nutrient arrests cell growth. It is released as a free spore when the plasma membrane ruptures. It resists heat, drying out, irradiation, acids, disinfectants, and boiling. When favorable conditions do return, the endospore develops into a single bacterial cell.

Many chemoautotrophic eubacteria taxi from host to host inside the gut of insects. For example, bites from blood-sucking ticks can transmit *Borrelia burgdorferi* from deer and some other wild animals to humans, who develop *Lyme disease*. A "bull's-eye" rash typically forms around the bite (Figure 20.7). Severe headaches, backaches, chills, and fatigue follow. Without prompt treatment, the condition worsens.

Regarding the "Simple" Bacteria

Bacteria are small. Their insides are not elaborate. *But bacteria are not simple.* A brief look at their behavior will reinforce this point. Bacteria move toward nutrient-rich regions. Aerobes move toward oxygen; anaerobes avoid it. Photosynthetic types move into light (or away from intense light). Many species tumble away from toxins. Such behaviors often start with membrane receptors that change shape when stimulated by light or chemical compounds. Receptors are stimulated differently when a bacterium alters its direction. This triggers a fleeting "memory," a changing biochemical condition that can be compared against that of the immediate past.

Some species even show collective behavior, as when millions of *Myxococcus xanthus* cells form a "predatory" colony. The cells secrete enzymes that digest "prey"— cyanobacteria and other microorganisms that get stuck to the colony—then they absorb breakdown products. What's more, the cells migrate, change direction, and move as a single unit toward what may be food!

Many myxobacteria produce fruiting bodies (spore-bearing structures). Under suitable conditions, some cells in a colony differentiate and form a slime stalk, others form branches, and still others form clusters of spores. Spores are dispersed when a cluster bursts open, and each may form a new colony. As you will see, certain eukaryotic organisms also form such structures.

The first prokaryotes diverged into three great lineages: archaebacteria, eubacteria, and the bacterial ancestors of eukaryotes. Archaebacteria are confined to extreme habitats and may resemble the first cells. Eubacteria are the most common prokaryotes. They occur in nearly all environments.

WWW

The Nature of Bacterial Growth

Between cell divisions, bacteria grow through increases in their component parts. We measure the growth of a large, multicelled organism in terms of increases in size, but doing this for a microscopically small bacterium would be a bit pointless. Instead, we measure bacterial growth as an increase in the number of cells in a given population. Under ideal conditions, each cell divides in two, the division of two cells results in four, four result in eight, and so forth. Many types can divide every half hour; a few can do so every ten minutes. Such rates of increase lead to large population sizes in short order.

parent cell. Most often, however, bacteria reproduce by a division mechanism called **prokaryotic fission**.

As prokaryotic fission gets under way, a parent cell replicates its DNA (Figure 20.8). Two DNA molecules result, and they are attached to the plasma membrane at adjacent sites. The cell synthesizes lipid and protein molecules that become incorporated into the plasma membrane between those attachment sites. Membrane growth moves the two DNA molecules apart, and new wall material is deposited above the membrane. The membrane and wall grow through the cell midsection and divide the cytoplasm. The result is two genetically equivalent daughter cells.

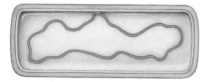

a A bacterial cell (cutaway view) before its DNA is copied. The DNA is attached to the plasma membrane.

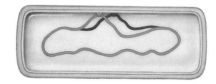

b Replication starts and proceeds in two directions, away from some point in the bacterial DNA molecule.

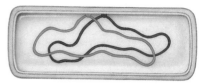

c The DNA copy is attached at a membrane site near the attachment site of the parent DNA molecule.

d Membrane growth proceeds between the two attachment sites and moves the two DNA molecules apart.

e New membrane and new wall material start forming through the cell midsection.

f The ongoing, organized deposition of membrane and wall material at the cell midsection divides the cytoplasma in two.

In nature, conditions that promote the growth of some bacterial populations might not sustain growth of others. Most species could not establish themselves in the most extreme environments—say, Antarctica, the Negev Desert, or deep inside the Earth. There, the few species that do endure, inside rocks, rarely reproduce. And few can live in the highly acidic wastewater from mining operations. K-12, a strain of *E. coli* originally isolated from the human gut, has been cultivated for such a long time in the laboratory that it no longer is able to grow when reintroduced into the gut. Through microevolutionary processes, it became adapted to the conditions in its artificial environment.

Prokaryotic Fission

When a bacterium has nearly doubled in size, it divides in two. Each daughter cell inherits a single bacterial chromosome, which is a circularized, double-stranded DNA molecule that has a few proteins attached to it. In some species, the daughter cell merely buds from the

new deposits at cell midsection

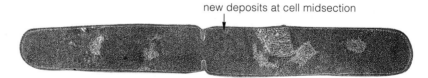

Figure 20.8 Bacterial reproduction by prokaryotic fission, a cell division mechanism. The micrograph shows the division of the cytoplasm of *Bacillus cereus*, as brought about by the formation of new membrane and wall material.

Many bacterial cells also may inherit one or more plasmids. A **plasmid**, recall, is a small circle of extra DNA. It gets replicated independently of the bacterial chromosome. One, the F (Fertility) plasmid, has genes that confer the means for bacterial conjugation.

Nearly all bacteria reproduce by prokaryotic fission, a cell division mechanism that follows DNA replication. Each daughter cell inherits a single bacterial chromosome (one DNA molecule). Many species also transfer plasmid DNA.

WWW

CHARACTERISTICS OF VIRUSES

In ancient Rome, *virus* meant "poison" or "venomous secretion." In the late 1800s, this rather nasty word was bestowed on newly discovered pathogens, smaller than the bacteria being studied by Louis Pasteur and others. Many viruses deserve the name. They attack humans, cats, cattle, birds, insects, plants, fungi, protistans, and bacteria. You name it, there are viruses that can infect it (Figures 20.9 and 20.10).

Today we define a **virus** as a noncellular infectious agent that has two characteristics. First, a viral particle consists of a protein coat surrounding a nucleic acid core—that is, around its genetic material. Second, a virus cannot reproduce itself. It can be reproduced only after its genetic material and a few enzymes enter a host cell and subvert the cell's biosynthetic machinery.

The genetic material of a virus is DNA *or* RNA. The coat consists of one or more types of protein subunits organized into a rodlike or many-sided shape (Figure 20.10). It protects the viral genetic material during the journey from one host cell to the next. It also contains proteins that can bind with specific receptors on host cells. Some coats are enclosed in an envelope of mostly membrane remnants from a previously infected cell.

The vertebrate immune system can detect certain viral proteins. The problem is, the genes for many viral proteins mutate so frequently that a virus may elude the immune fighters. For instance, health care providers often urge people susceptible to lung infections to get new "flu shots" each year because envelope spikes on influenza viruses (Figure 20.9a) keep changing.

Each kind of virus multiplies only in certain hosts. It cannot be studied easily unless investigators culture living host cells. This is why much of our knowledge of viruses comes from **bacteriophages**, a group of viruses that infect bacterial cells. Unlike cells of humans and other complex, multicelled species, bacterial cells can

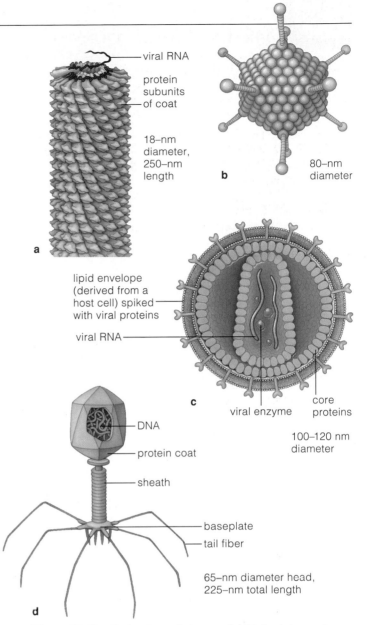

Figure 20.10   Body plans of viruses. (**a**) *Helical* viruses have a rod-shaped coat of protein subunits, coiled helically around their nucleic acid. The upper subunits of the coat have been removed from this portion of a tobacco mosaic virus to reveal the RNA. (**b**) *Polyhedral* viruses, including this adenovirus, have a many-sided coat. (**c**) *Enveloped* viruses, including HIV, have an envelope around a helical or polyhedral coat. (**d**) *Complex* viruses, such as T-even bacteriophages, have additional structures attached to the coat.

be cultured easily and rapidly. This is also why bacteria and bacteriophages were used in early experiments to determine DNA function (Section 12.1). They are still used as research tools in genetic engineering.

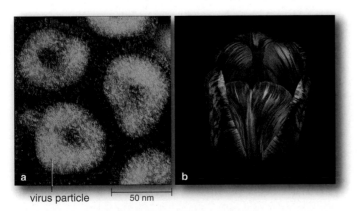

Figure 20.9   (**a**) Particles of an enveloped RNA virus that causes influenza in humans. Spikes project from the lipid envelope. (**b**) Streaking of a tulip blossom. A harmless virus infected pigment-forming cells in the colorless parts.

A virus is a nonliving infectious particle that consists of nucleic acid enclosed in a protein coat and sometimes an outer envelope. It cannot be replicated without pirating the metabolic machinery of a specific type of host cell.

WWW

VIRAL MULTIPLICATION CYCLES

Viruses multiply in a variety of ways. Even so, nearly all of their multiplication cycles proceed through five basic steps, as outlined here:

1. *Attachment.* The virus attaches to a host cell. Any cell is a suitable host if molecular groups on a virus particle are able to chemically recognize and lock onto specific molecular groups at the cell surface.

2. *Penetration.* Either the whole virus or its genetic material alone penetrates the cell's cytoplasm.

3. *Replication and synthesis.* In an act of molecular piracy, the viral DNA or RNA directs the host cell into producing many copies of viral nucleic acids and proteins, including enzymes.

4. *Assembly.* The viral nucleic acids and viral proteins are put together to form new infectious particles.

5. *Release.* New virus particles are released from the cell.

Consider this list with respect to some bacteriophages. Lytic and lysogenic pathways are common among their replication cycles. In a *lytic* pathway, steps 1 through 4 proceed rapidly, and new particles are released when a host cell undergoes lysis (Figure 20.11). **Lysis** means the plasma membrane, cell wall, or both are damaged, so that cytoplasm leaks out and the cell dies. Late into most lytic pathways, a viral enzyme that causes swift destruction of the bacterial cell wall is synthesized.

In a *lysogenic* pathway, a latent period extends the cycle. The virus does not kill its host outright. Instead, a viral enzyme cuts a host chromosome, then integrates viral genes into it. Thus, when an infected cell prepares to divide, the recombinant molecule is replicated. As a result of that single instance of genetic recombination, miniature time bombs get passed on to all of that cell's descendants. Later, a molecular signal or some other stimulus may reactivate the cycle.

Latency occurs in the multiplication cycles of many viruses, not just among bacteriophages. Type I *Herpes simplex*, which causes *cold sores*, is an example. Nearly everybody harbors this virus. It remains latent inside a ganglion (a cluster of cell bodies of neurons) in facial tissue. Stress factors, such as sunburn, can reactivate the virus. Then, virus particles move down the neurons to their tips near the skin. There they infect epithelial cells and cause painful skin eruptions.

Like other enveloped viruses, the herpesviruses can enter a host cell by their own version of endocytosis, then leave it by budding from the plasma membrane. Figure 20.12 shows how they accomplish this.

The multiplication cycle of the RNA viruses has an interesting twist to it. In the host cell's cytoplasm, their

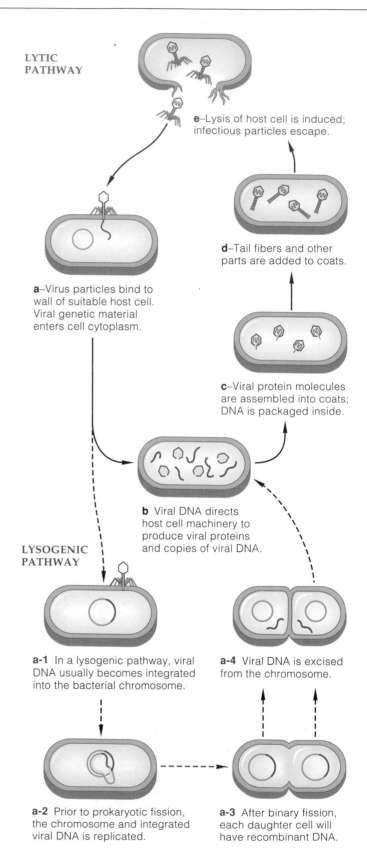

e–Lysis of host cell is induced; infectious particles escape.

a–Virus particles bind to wall of suitable host cell. Viral genetic material enters cell cytoplasm.

d–Tail fibers and other parts are added to coats.

c–Viral protein molecules are assembled into coats; DNA is packaged inside.

b Viral DNA directs host cell machinery to produce viral proteins and copies of viral DNA.

LYTIC PATHWAY

LYSOGENIC PATHWAY

a-1 In a lysogenic pathway, viral DNA usually becomes integrated into the bacterial chromosome.

a-4 Viral DNA is excised from the chromosome.

a-2 Prior to prokaryotic fission, the chromosome and integrated viral DNA is replicated.

a-3 After binary fission, each daughter cell will have recombinant DNA.

Figure 20.11 Generalized multiplication cycle for some bacteriophages. New viral particles may be produced and released by a lytic pathway. For certain viruses, the lytic pathway may be expanded to include a lysogenic pathway.

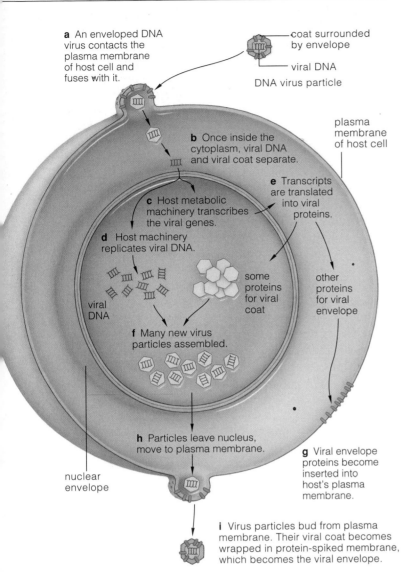

a An enveloped DNA virus contacts the plasma membrane of host cell and fuses with it.

coat surrounded by envelope

viral DNA

DNA virus particle

b Once inside the cytoplasm, viral DNA and viral coat separate.

plasma membrane of host cell

e Transcripts are translated into viral proteins.

c Host metabolic machinery transcribes the viral genes.

d Host machinery replicates viral DNA.

viral DNA

some proteins for viral coat

other proteins for viral envelope

f Many new virus particles assembled.

h Particles leave nucleus, move to plasma membrane.

nuclear envelope

g Viral envelope proteins become inserted into host's plasma membrane.

i Virus particles bud from plasma membrane. Their viral coat becomes wrapped in protein-spiked membrane, which becomes the viral envelope.

j The finished particle is equipped to infect a new potential host cell.

Figure 20.12 Multiplication cycle of one type of enveloped DNA virus infecting an animal cell.

RNA serves as a template for synthesizing either DNA or mRNA. For example, HIV, a retrovirus, carts its own enzymes into cells. These assemble DNA on viral RNA by reverse transcription (Sections 15.1 and 34.11).

The multiplication cycles of nearly all viruses include five basic steps: attachment to a suitable host cell, penetration into the cell, viral DNA or RNA replication and synthesis of viral proteins, assembly of new viral particles, and then release from the infected cell.

Bacteriophage replication cycles commonly follow a rapid, lytic pathway and an extended, lysogenic pathway.

Replication cycles of RNA viruses involve the use of viral RNA as a template for synthesizing DNA or mRNA.

WWW

INFECTIOUS PARTICLES TINIER THAN VIRUSES

VIROIDS You may have trouble visualizing this, but some infectious agents are more stripped down than viruses. **Viroids** are tightly folded strands or circles of RNA, smaller than anything in viruses. They resemble introns (noncoding portions of eukaryotic DNA) and may have evolved from them. Viroids have no protein coat, but their tight folding may help protect them from a host's enzymes. These bits of "naked" RNA are known to cause plant diseases. Each year, they destroy many millions of dollars' worth of potatoes, citrus, and other important crop plants.

PRIONS Eight rare, fatal degenerative diseases of the nervous system are linked to small proteins known as **prions**. The proteins are altered products of a gene that is present in unaffected as well as infected individuals. Unaltered and altered forms of the protein molecule are found at the surface of neurons, the communication cells of the nervous system. You may have heard of *kuru* and *Creutzfeldt-Jakob* diseases (CJD). They slowly destroy muscle coordination and brain function in humans. *Scrapie*, a disease of sheep, is so named because infected animals rub against trees or posts until they scrape off most of their wool.

In 1996, more than 150,000 cattle in Great Britain staggered about, drooled, and showed other symptoms of BSE (bovine spongiform encephalopathy), or *mad cow disease*. Prions had caused abnormal changes in proteins in neurons, and spongy holes and amyloid deposits had formed in the cattle brains. BSE is fatal.

How did it happen? Ground-up tissues of sheep that had died of scrapie had been used as a supplement in cattle feed. The practice was banned in 1988, but prions were already in the food chain.

Medical researchers already knew that 1 in 1 million people around the world develops Creutzfeldt-Jakob disease each year. Prions also are linked to this fatal condition. Symptoms, which include loss of vision and speech, rapid mental deterioration, and spastic paralysis, may take decades to develop. At the same time as the major outbreak of BSE in Great Britain, ten people were diagnosed with a variant of CJD; four had already died, probably as a result of eating meat of infected animals. Genetic analysis and interviews linked all the cases to exposure to BSE.

Ten other countries had reported cases of BSE, possibly as a result of importing cattle feed from Great Britain. The incidence of BSE in Great Britain is now declining since strict precautionary measures have been instituted. In the United States, importation of live cattle or meat products from BSE-affected countries has been banned since 1989.

WWW

By about 2.5 billion years ago, in tidal flats and shallow sediments, the noncyclic, oxygen-releasing pathway of photosynthesis was operating in bacterial populations. At first the oxygen combined with iron in rocks. When iron-rich deposits were rusted out (oxidized), oxygen started accumulating. With nothing else around to hold it, free oxygen accumulated in water, then in the air. This triggered rampant competition for resources. In the presence of so much oxygen, energy-rich organic compounds could no longer build up by geochemical processes. The organic compounds formed by living cells became the premier source of carbon and energy. Novel ways of acquiring and using organic compounds developed. Many bacterial species engaged in new kinds of partnerships, predations, and parasitic interactions. As outlined in Section 19.4, some of those evolutionary experiments gave rise to eukaryotic cells.

Of all existing organisms, **protistans** are most like the earliest, structurally simple eukaryotes. They differ from bacteria in several respects. At the least, they have a nucleus, large ribosomes, mitochondria, endoplasmic reticulum, and Golgi bodies. Their chromosomes are DNA molecules with a great many histones and other proteins. Protistans assemble microtubules for use in a cytoskeleton, in spindles that move chromosomes, and in the 9+2 core of flagella or cilia. Many species have chloroplasts. And depending on the species, protistans divide by way of mitosis, meiosis, or both.

The photosynthetic types range from microscopic single cells to giant seaweeds. The saprobes resemble some bacteria and fungi. Some of the predators and parasites resemble animals. Although the vast majority of protistans are single-celled, nearly every lineage also has multicelled forms. Some groups have notably close evolutionary ties to other kingdoms.

We now turn to the major lineages. The chytrids, water molds, slime molds, protozoans, and sporozoans are heterotrophs. Among the euglenoids, chrysophytes, and dinoflagellates are species that are photosynthetic, heterotrophic, or both. Most red, brown, and green algae are evolutionarily committed to photosynthesis.

Opinions differ on how to classify many of these confoundingly diverse organisms. For example, many biologists view the multicelled algae as protistans, and about as many view them as plants. Don't worry about memorizing the classification schemes. Simply become familiar with various groups. In time, gene sequencing studies will bring the evolutionary picture into focus.

Structural and functional traits distinguish protistans from plants, fungi, and animals. By a process of elimination, "protistans" are organisms that do not show these traits and that are not bacteria.

Three groups of protistans, the **chytrids**, **water molds**, and **slime molds**, have members that resemble fungi in some respects. Like fungi, all produce spore-bearing structures and are heterotrophs. Like fungi, many are saprobic decomposers or parasites. "Saprobes" secrete digestive enzymes that break down organic compounds made by other organisms, then absorb the breakdown products. Unlike fungi, the slime molds are phagocytic predators. Also, members of all three groups differ from fungi in producing motile cells during the life cycle.

Chytrids (Chytridiomycota) are common in marine and freshwater habitats, where they absorb nutrients from plant debris or from necrotic plant tissues. Most of the 575 species are saprobic decomposers, but some are parasites. In damp environments, chytrid populations become fuzzy, pale masses that resemble fungal molds. Water molds (Oomycota) may be distantly related to yellow-green and brown algae. Most of the 580 known types are saprobes in aquatic habitats. Some parasitize aquatic organisms. Those pale, cottony growths on some aquarium fish are absorptive structures (mycelia) of a parasitic type, *Saprolegnia* (Figure 20.13a).

Some water molds are major plant pathogens. For example, grapes with *downy mildew* have been infected by *Plasmopara viticola* (Figure 20.13b). Another pathogen adversely affected human affairs. Over a century ago, Irish peasants grew potatoes as their main food crop. Between 1845 and 1860, the growing seasons were cool and damp, year after year, and conditions encouraged the rapid spread of *Phytophthora infestans*. This water mold causes *late blight*, a rotting of potato (and tomato) plants. Its abundant spores were dispersed unimpeded through the watery film on the plants. Destruction was rampant. During a fifteen-year period, one-third of the population in Ireland starved to death, died during the outbreak of typhoid fever that followed as a secondary effect, or fled to the United States and other countries.

Slime molds produce free-living, amoebalike cells for part of the life cycle. The group includes cellular slime molds (Acrasiomycota) and the plasmodial slime

Figure 20.13 Effects of two water molds. (**a**) A parasitic water mold (*Saprolegnia*) has destroyed tissues of this aquarium fish. (**b**) *Plasmopara viticola* causes downy mildew in grapes. At times it has threatened large vineyards in Europe and North America.

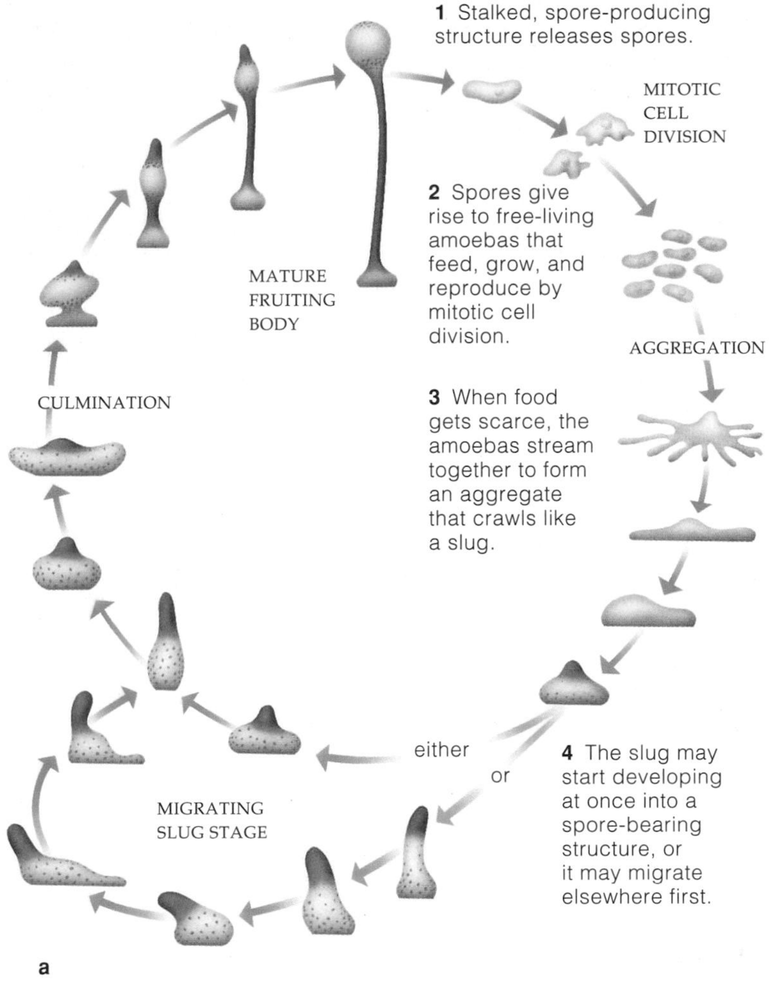

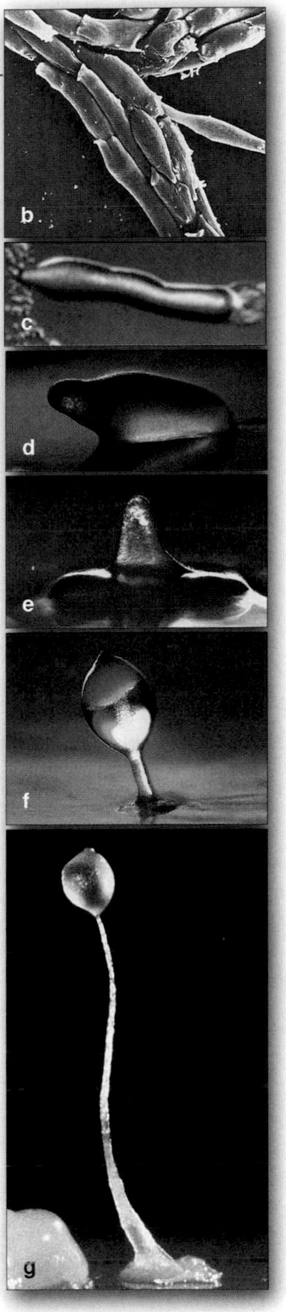

Figure 20.14 A cellular slime mold, *Dictyostelium discoideum*. (**a**) Its life cycle includes a spore-producing stage. Spores give rise to free-living amoebas, which grow and divide until food (soil bacteria) dwindles. In response to cyclic AMP, a chemical signal that they secrete, amoebas stream toward one another. As the amoebas aggregate, their plasma membranes become sticky and they adhere to one another. A cellulose sheath forms around the aggregation, which starts crawling like a slug (**b–d**). Some slugs may incorporate 100,000 amoebas.

As a slug migrates, amoebas develop and differentiate into prestalk (*red*), prespore (*white*), and anteriorlike cells (*brown dots*). Prestalk cells secrete ammonia in amounts that vary with temperature and light intensity. The slug moves most rapidly in response to intermediate levels of ammonia (not too little, not too much), which correspond to warm, moist conditions (not too cold or hot, not soppy or dry). Where such conditions occur at the surface of a substrate, prestalk cells and prespore cells differentiate and form a stalked, spore-bearing structure (**e–g**). The anteriorlike cells, which sort into two groups, may function in elevating the nonmotile spores for dispersal from the top of the spore-bearing structure.

1 Stalked, spore-producing structure releases spores.

MITOTIC CELL DIVISION

2 Spores give rise to free-living amoebas that feed, grow, and reproduce by mitotic cell division.

AGGREGATION

3 When food gets scarce, the amoebas stream together to form an aggregate that crawls like a slug.

4 The slug may start developing at once into a spore-bearing structure, or it may migrate elsewhere first.

either or

MATURE FRUITING BODY

CULMINATION

MIGRATING SLUG STAGE

a

molds (Myxomycota). Their cells crawl on rotting plant parts, such as decaying leaves and bark. Like the true amoebas, they are phagocytic predators. They engulf bacteria and yeasts, spores, and organic compounds. When nutrients are scarce and cells are starving, many of the cells aggregate and form a slimy mass that may migrate to a more favorable location. Later, they form a spore-bearing structure, as shown in Figure 20.14*a*. Slime molds also reproduce sexually. Figure 20.15 highlights one of the 500 species of plasmodial slime molds. When amoebalike cells of the members of that phylum aggregate, each cell's plasma membrane breaks down. Cytoplasm flows unimpeded and distributes nutrients and oxygen through the mass. A streaming mass of this sort (the plasmodium) often will occupy several square meters and migrate if food runs out. You might have seen one crossing a lawn or a road, or even climbing a tree.

Most chytrids and water molds are saprobic decomposers of aquatic habitats. Some are single cells. Like slime molds, many are free-living predators some of the time and simple experiments in multicellularity at other times.

During a slime mold life cycle, amoeboid cells aggregate to form a migrating mass. Cells in the mass differentiate, then they form reproductive structures and spores or gametes.

Figure 20.15 *Physarum*, a plasmodial slime mold. This aggregation of cells is migrating along a rotting log.

WWW

THE ANIMAL-LIKE PROTISTANS

About 65,000 named species of protistans are known informally as **protozoans** ("first animals"), because they may resemble single-celled, heterotrophic protistans that gave rise to animals. They include amoeboid protozoans, ciliated protozoans, and animal-like flagellates. Among their ranks are predators, parasites, and grazers that actively move about to secure a meal. Endosymbiotic algae inhabit certain species in all four groups

Free-living species live in damp soil; in ponds, lakes, and other freshwater habitats; and in marine habitats. The symbionts and parasites live in or on moist tissues of other organisms. Some species are major pathogens.

Protozoans reproduce either asexually or sexually, and many species can alternate between reproductive modes in response to environmental conditions. Most commonly, asexual reproduction is by **binary fission**, a process by which the individual's body divides in two. The division plane is random for amoebas, longitudinal for flagellates, and transverse for ciliates. Budding from the parent organism also occurs. Some species undergo multiple fission. More than two nuclei form, then each nucleus and a bit of cytoplasm separate as a daughter cell. Many parasitic types go through an encysted stage. The **cyst**, a resistant body covering made of their own secretions, helps them wait out stressful conditions.

Amoeboid Protozoans

Amoeboid protozoans (Sarcodina), are naked amoebas, foraminiferans, heliozoans, and radiolarians. Some of these soft-bodied single cells have hardened skeletal elements. They all use pseudopods for motility and for capturing prey (Figure 20.16*a*). In this group are gut-dwelling species and free-living species in damp soil, freshwater, and saltwater. Most species are free-living phagocytes that engulf other protozoans and bacteria. A few parasitize certain invertebrates or vertebrates.

Entamoeba histolytica, a pathogen, completes its life cycle in human intestines. It causes *amoebic dysentery*, an infectious disease most prevalent in subtropical and tropical areas where sanitation is poor. Worldwide, this is a leading cause of death among infants and toddlers, whose immune systems are not yet well developed.

Unlike the naked amoebas, foraminiferans have a highly perforated external skeleton of secretions that are hardened with calcium carbonate (Figure 20.16*b*). Thin pseudopods extend through the perforations and trap prey in their mucous covering. Most foraminiferans live on the seafloor. All but 1 percent of the named species became extinct. Fossilized shells of countless foraminiferans became compressed as a distinctive type of sedimentary rock, including England's famous white cliffs of Dover. We now mine extensive calcified remains for making cement and blackboard chalk.

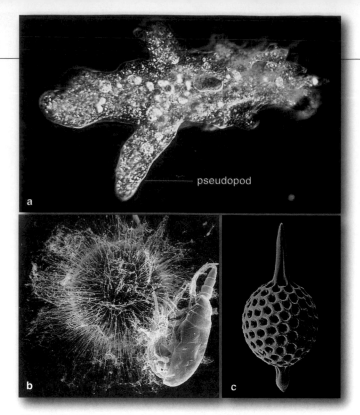

Figure 20.16 Amoeboid protozoans. (**a**) *Amoeba proteus.* (**b**) Foraminiferan with prey. (**c**) Radiolarian shell.

Radiolarians and heliozoans have many slender, reinforced pseudopods that radiate from the body. Like foraminiferans, they are well represented in the fossil record, for their silica-hardened parts resist degradation (Figure 20.16*c*). Many thin skeletal parts and stiffened pseudopods project outward. Almost all radiolarians drift with the ocean currents, as part of **plankton**. The word refers to aquatic communities of passive drifters and motile organisms, mostly microscopic. A few kinds of radiolarians form colonies in which numerous cells become cemented together by their own secretions.

Ciliated Protozoans

Ciliated protozoans (Ciliophora) typically have profuse arrays of cilia at their surface. You read about these fine motile structures in Section 4.9. Most ciliates use them for swimming through freshwater and marine habitats, where they prey on bacteria, algae, and one another, as Figure 20.1*e* so aptly suggests. Their cilia beat in such a synchronized pattern over the body surface, they call to mind a soft wind through a field of tall grasses.

Paramecium is typical of the group (Figure 20.17). A fully grown cell is about 150 to 200 micrometers long. Outer membranes form a pellicle, a body covering that may be rigid or quite flexible, depending on how those membranes are organized. In the gullet, which starts as an oral depression at the body's surface, cilia sweep food-laden water into the cell body. The food becomes enclosed in enzyme-filled vesicles and is digested. Like the amoeboid protozoans, *Paramecium*'s internal solute

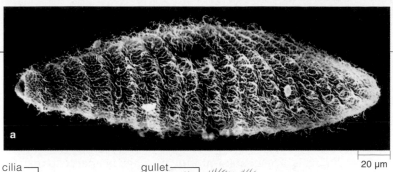

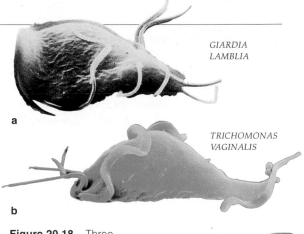

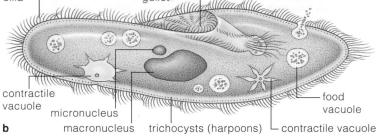

cilia — gullet

contractile vacuole

micronucleus

macronucleus trichocysts (harpoons) — contractile vacuole

food vacuole

b

20 µm

Figure 20.17 (**a**,**b**) Body plan and scanning electron micrograph of a ciliated protozoan, *Paramecium*. (**c**) From the Bahamas, one of the hypotrichs, the most animal-like ciliates. They run around on leglike tufts of cilia. Some have a "head" end with modified sensory cilia.

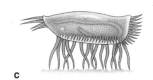

c

GIARDIA LAMBLIA

a

TRICHOMONAS VAGINALIS

b

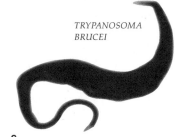

TRYPANOSOMA BRUCEI

Figure 20.18 Three animal-like flagellates. *Giardia lamblia* causes intestinal disturbances. *Trichomonas vaginalis* causes trichomoniasis, a prevalent sexually transmitted disease. *Trypanosoma brucei* causes African sleeping sickness.

c

concentrations are greater than that of the fluid outside the cell. It must continually counter water's tendency to diffuse inward by osmosis. Like amoebas, it depends on **contractile vacuoles**. Tubes extend from the center of these organelles and collect excess water that moves osmotically into the cell. A filled vacuole contracts and forces water through a pore to the outside (Section 5.9).

Like other protistan groups, ciliates show diversity in life-styles. About 65 percent of the known species are free-living and motile. Others attach permanently or temporarily to substrates, often by stalks. Some form colonies. About 30 percent live with other organisms as symbionts or parasites.

Ciliates can reproduce sexually and asexually, and things get interesting because each cell commonly has two types of nuclei. When a cell reproduces asexually by binary fission, a small, diploid *micro*nucleus divides by mitosis. And the large *macro*nucleus lengthens and splits in two a bit sloppily; some of the DNA may spill out. Most often, sexual reproduction is by a unique form of conjugation. Ciliates repeatedly divide their micronuclei, swap two daughter micronuclei with a partner ciliate, and let two other micronuclei fuse and form a diploid macronucleus to replace the one that disappears. And you probably thought that sex among single-celled critters was simple!

Animal-Like Flagellates

Animal-like flagellates (Mastigophora) are free-living predators and parasites bearing one to several flagella. Members of one group are equipped with flagella *and*

pseudopods. Free-living types thrive in freshwater and marine habitats. Parasitic types live in moist tissues of plants and animals, including humans. *Giardia lamblia*, an internal parasite, is a major cause of diarrhea (Figure 20.18*a*). It now contaminates surface waters of aquatic habitats and reservoirs. Foraging cattle, wild animals, and people often harbor the parasite and its cysts. It leaves a host encysted in feces, then infects people who drink water or ingest food tainted with the feces.

Many trichomonads also are parasites. *Trichomonas vaginalis*, a worldwide nuisance, can infect new human hosts during sexual intercourse (Figure 20.18*b*). Unless trichomonad infections are treated, they may damage the urinary and reproductive tracts.

Trypanosomes also are parasites. One, *Trypanosoma brucei*, causes *African sleeping sickness* (Figure 20.18*c*). Bites of the tsetse fly transmit it to hosts, where it may damage the nervous system. *T. cruzi*, prevalent in South America and Mexico, causes *Chagas disease*. Bugs pick up the parasite when they feed on infected humans and other animals. The parasite multiplies in the insect gut, then it may be excreted onto a host. Scratches in skin invite infection. The liver and spleen enlarge, eyelids and the face swell up, then the brain and heart become severely damaged. There is no cure for Chagas disease.

Protozoans are single-celled protistans that are animal-like in some respects, as in the way they actively move about to secure their food. They include predators, parasites, and grazers. Some species are pathogens that cause serious diseases in humans and other animals.

WWW

THE NOTORIOUS SPOROZOANS

Sporozoan is an informal name for parasitic protistans that complete part of the life cycle inside specific cells of host organisms. These parasites form sporozoites (a motile infective stage), and some have encysted stages. At one end of the body is a complex of structures that have roles in penetrating host cells. Many cause human diseases. *Cryptosporidium* contaminates most freshwater supplies. One strain causes serious intestinal disorders; an outbreak in 1993 hospitalized thousands of people in Milwaukee. The disease symptoms normally subside after ten days. People with weakened immune systems can suffer for months or years.

recur. In time, anemia may develop and the liver may become grossly enlarged.

Amazingly, the *Plasmodium* life cycle is sensitive to human and mosquito body temperature and oxygen levels. The gamete-producing stage is unable to mature in humans, who are warm-bodied and have little free oxygen in blood (most is bound to hemoglobin). They are induced to mature inside the mosquito, which has a lower body temperature and which incidentally slurps in oxygen from the air along with the infected blood.

Throughout human history, malaria has been most prevalent in tropical and subtropical parts of Africa.

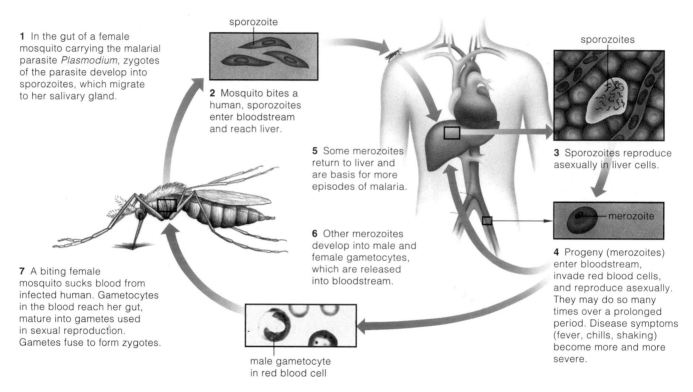

1 In the gut of a female mosquito carrying the malarial parasite *Plasmodium*, zygotes of the parasite develop into sporozoites, which migrate to her salivary gland.

sporozoite

2 Mosquito bites a human, sporozoites enter bloodstream and reach liver.

sporozoites

3 Sporozoites reproduce asexually in liver cells.

5 Some merozoites return to liver and are basis for more episodes of malaria.

merozoite

6 Other merozoites develop into male and female gametocytes, which are released into bloodstream.

4 Progeny (merozoites) enter bloodstream, invade red blood cells, and reproduce asexually. They may do so many times over a prolonged period. Disease symptoms (fever, chills, shaking) become more and more severe.

7 A biting female mosquito sucks blood from infected human. Gametocytes in the blood reach her gut, mature into gametes used in sexual reproduction. Gametes fuse to form zygotes.

male gametocyte in red blood cell

Figure 20.19 Life cycle of a sporozoan that causes malaria.

Four parasitic species of *Plasmodium* cause *malaria*. This long-lasting disease affects more than 100 million people and kills 2.7 million of them every year. Only female *Anopheles* mosquitoes transmit it to humans. A host's bloodstream transports a motile, infective stage to the liver, where the stage divides repeatedly. Some of the resulting cells (merozoites) penetrate and asexually reproduce inside red blood cells, which they destroy. Others will mature into gametes (Figure 20.19).

Disease symptoms start after infected cells rupture and release merozoites, metabolic wastes, and cellular debris into the bloodstream. Shaking, chills, a burning fever, and drenching sweats are classic signs. After one episode, they subside for a few weeks or months and infected people may feel all right. However, relapses

Today, cases reported in North America (Florida and New Jersey) and elsewhere are increasing dramatically, this being a result of hordes of globe-hopping travelers and unprecedented levels of human immigration.

Certain strains of *Plasmodium* are now resistant to antimalarial drugs. Vaccines (preparations that induce the body to build up resistance to a specific pathogen) have been difficult to develop, although some new types are entering clinical trials. Experimental vaccines for malaria are not equally effective against all stages that develop in sporozoan life cycles. This is generally the case for vaccines that researchers hope to develop against most parasites with complex life cycles.

Sporozoans are internal parasites that produce infective, motile stages and often form encysted stages.

WWW

20.11 THE NATURE OF INFECTIOUS DISEASES

CATEGORIES OF DISEASES Just by being human, you are a potential host for many pathogenic bacteria, viruses, fungi, protozoans, and parasitic worms. When a pathogen invades your body and is multiplying in host cells and tissues, this is an **infection**. Its outcome, **disease**, results if the body's defenses cannot be mobilized fast enough to prevent the pathogen's activities from interfering with normal body functions. With *contagious* diseases, the pathogen can be transmitted by direct contact with body fluids secreted or otherwise released (as by explosive wet sneezes) from infected individuals.

During an **epidemic**, a disease spreads rapidly through part of a population for a limited time, then the outbreak subsides. If an epidemic breaks out in several countries around the world at the same time, this a pandemic. AIDS, an incurable disease, is an example. Millions are infected by the agent, HIV (*Human Immunodeficiency Virus*).

Sporadic diseases such as whooping cough break out irregularly and affect few people. *Endemic* diseases pop up more or less continuously, but they don't spread far in large populations. Tuberculosis is like this. So is impetigo, a highly contagious bacterial infection that often spreads no farther than, say, a single day-care center.

AN EVOLUTIONARY VIEW Now look at disease in terms of the pathogen's prospects for survival. *A pathogen stays around only for as long as it has access to outside sources of energy* and *raw materials*. To a microscopic organism or virus, a human is a treasurehouse of both. With such bountiful resources, it can multiply or replicate itself to amazing population sizes. Evolutionarily speaking, the ones that leave the most descendants win.

There are two huge barriers to total world dominance by pathogens. First, any species with a history of being attacked by a particular pathogen has coevolved with it and has built-in defenses against it. A premier example is the vertebrate immune system, as described in Chapter 34. Second, if a pathogen is too good at rapidly killing an individual, it may vanish before having time to infect a new one. This is one reason why most have less-than-fatal effects on a host. After all, infected people who live longer can spread more germs and contribute to the pathogen's reproductive or replicative success. Usually, a host only dies if overwhelming numbers of the pathogen enter its body, if it is a novel host (one with no coevolved defenses against the pathogen), or if a mutant pathogenic strain emerges and can surmount the host's current defenses.

Being equipped with an evolutionary perspective, you probably can work out the numbers on your own. To wit: *The greater the population density of host individuals, the greater will be the kinds and frequencies of infectious diseases transmitted among them.* And this brings us to the bad news.

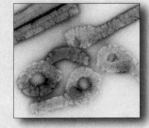

Figure 20.20 Electron micrograph of *Ebola* virus.

DRUG-RESISTANT STRAINS There is an old saying that, when you attack nature, it will come back at you with a pitchfork. At this point in the book, we have looked at antibiotic resistance in several contexts. Here we reinforce the point that infectious diseases are on the upsurge.

A mere twenty-five years ago, antibiotics and vaccines were considered invincible, and the Surgeon General of the United States announced we could "close the book on infectious diseases." There are now more than 5.9 billion of us. Most crowd in cities, and at any time, as many as 50 million are on the move within and between countries, in search of a better life (Section 39.7). Even putting aside the emerging pathogens, is it any wonder that pathogens responsible for cholera, tuberculosis, and other familiar diseases are striking with a vengeance?

For example, because of economic pressures over the past several decades, the number of working mothers skyrocketed in the United States. So did the number of preschoolers enrolled in day-care centers. In effect, each center is a population of host individuals whose immune systems are still developing and vulnerable to contagious diseases. Now think about *Streptococcus pneumoniae*. This bacterium causes pneumonia, meningitis, and middle-ear infections in people of all ages; 40,000 to 50,000 cases end in death each year. The risk of infection is 36 times greater in large day-care centers than it is for children cared for at home. As recently as 1988, drug-resistant strains of *S. pneumoniae* were practically unheard of in the United States. They are now the rule, not the exception.

EMERGING PATHOGENS Thanks to planes, trains, and automobiles, people now travel often and in droves to virgin tropical forests and other remote places where humans were infrequent (or nonexistent) opportunities for pathogens. Today, strange and dangerous pathogens encounter novel, two-legged treasurehouses of metabolic machinery and nutrients entering their habitats. Within a matter of hours, human travelers can become infected. They can taxi such pathogens far away, and back to home. The deadly, *emerging pathogens* may have been around for quite a long time and only now are taking advantage of the increased presence of novel hosts. Maybe some are newly mutated strains of existing species.

For example, it appears that *Ebola*, an RNA virus, may have coevolved with monkeys in African tropical forests (Figure 20.20). It is 70–90 percent lethal, and there are no treatments or vaccines. High fever and flulike aches mark the onset of disease. Within a few days, nausea, vomiting, and diarrhea begin. Cells making up the lining of blood vessels are destroyed. Blood seeps into the surrounding tissues and out through the body's orifices. Organs may turn to mush. Patients often become deranged and die of circulatory shock. During recent epidemics, government agencies around the world were mobilized. Only the fast implementation of quarantine procedures limited the spread of the disease. WWW

The rest of this chapter surveys the protistans, largely photosynthetic, that are informally called "the algae" (Table 20.2). Let's start with euglenoids, chrysophytes, and dinoflagellates. Single-celled species dominate all three groups, which is not the case for red, brown, and green algae. Most are members of *phyto*plankton, the communities of aquatic photosynthetic species, mainly microscopic, that drift or swim weakly through water. As food producers, they are the start of nearly all food webs in aquatic habitats. Diverse species occur in great numbers in the water provinces, including lakes, rivers, streams, ponds, and the ocean (Sections 42.10 and 42.11).

Euglenoids

Euglenoids (Euglenophyta) are a classic example of evolutionary experimentation. These flagellated, free-living cells abound in freshwater or stagnant ponds and lakes. Most of the 1,000 known species engage in photosynthesis. The rest are heterotrophs that subsist on organic compounds dissolved in the water.

Euglena has a profusion of organelles (Figure 20.21). Among these are chloroplasts with chlorophylls *a* and *b* and carotenoids, like chloroplasts in plants. *Euglena* has flagella (one long, one short) and a contractile vacuole,

as animal-like protozoans do. Like some protozoans, it has a pellicle, this one being a flexible cover with spiral strips of a translucent, protein-rich material. Pigments form an "eyespot" that partly shields a light-sensitive receptor. The cell moves where the amount of light is most suitable for its activities.

In general, "self-feeders" make their own vitamins, which they require for growth. But all photosynthetic euglenoids get vitamin B_{12} from the surroundings. Their chloroplasts probably originated with green algae that long ago entered an endosymbiotic relationship with a host cell, in the manner described in Section 19.4.

Chrysophytes

Chrysophytes (Chrysophyta) are mainly photosynthetic, free-living cells with chlorophylls a, c_1, and c_2. They include diatoms, golden algae, and yellow-green algae.

Most diatoms are photosynthetic. The 5,600 existing species have a silica "shell" of two perforated parts that overlap like a pillbox (Figure 20.22*a*). Substances move to and from the plasma membrane through perforations in the shell. For 100 million years, finely crushed shells of at least 35,000 now-extinct species accumulated on the bottom of lakes and seas. Many sediments now contain the deposits, which we use in insulation, abrasives, and filters. Each year, for example, more than 270,000 metric tons are quarried near Lompoc, California.

Among the 500 species of golden algae are many with no obvious cell wall; they have silica scales or skeletal elements. Their chlorophylls are masked by a golden-brown carotenoid, fucoxanthin. Except for their chloroplasts, some amoeboid species closely resemble

Table 20.2 Mostly Photosynthetic Protistans

EUGLENOPHYTA	Euglenoids. Free-living cells. Autotrophic *and* heterotrophic.
CHRYSOPHYTA	Free-living cells, mostly autotrophic.
	CHRYSOPHYTES (golden algae, yellow-green algae). Most photosynthetic; some heterotrophs.
	DIATOMS. Most photosynthetic; some symbionts with foraminiferans.
PYRRHOPHYTA	Dinoflagellates. Free-living cells. Autotrophic *and* heterotrophic.
RHODOPHYTA	Red algae. Most multicelled, photosynthetic.
PHAEOPHYTA	Brown algae. Multicelled, photosynthetic.
CHLOROPHYTA	Green algae. Most multicelled, photosynthetic

long flagellum
contractile vacuole   chloroplast
eyespot shielding a light-sensitive receptor
nucleus   Golgi body   mitochondrion   pellicle
ER

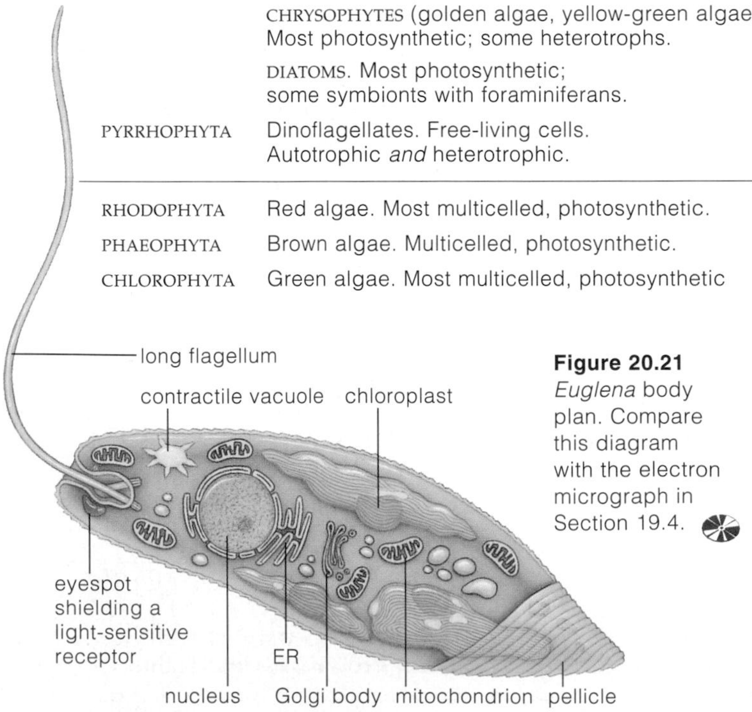

Figure 20.21 *Euglena* body plan. Compare this diagram with the electron micrograph in Section 19.4.

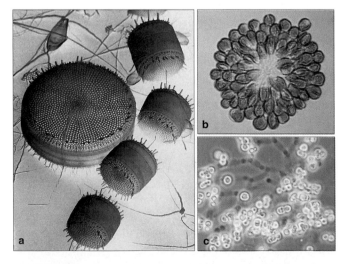

Figure 20.22 Representative chrysophytes. (**a**) Diatom shells. (**b**) *Synura*, a golden alga with a fishy odor. It forms colonies in phytoplankton. (**c**) Another member of phytoplankton: a colonial type of yellow-green alga, *Mischococcus*.

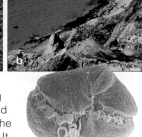

Figure 20.23 (**a**) Fish kill resulting from a dinoflagellate bloom. (**b**) Red tide in a central California bay. (**c**) The dinoflagellate *Gymnodinium breve.* It causes red tides along Florida's coast.

the true amoebas. Fucoxanthin is absent from the 600 known species of yellow-green algae. Most of these are nonmotile but have flagellated gametes. Yellow-green algae are common in many aquatic habitats.

Dinoflagellates

More than 1,200 species of dinoflagellates (Pyrrhophyta) live in marine phytoplankton. Most are photosynthetic; many are heterotrophs. Some have cellulose plates and flagella. They are yellow-green, green, blue, brown, or red, depending on their pigments and endosymbiotic history. **Red tides** form near coasts when some species undergo population explosions and the water appears red or brown (Figure 20.23). A few produce toxins that kill shellfish, fishes, and fish-eating migratory birds; in one year, 150,000 eared grebes and 5,000 brown pelicans died in California's Salton Sea. People who have eaten seafood contaminated with some of the toxins have impaired brain function, including memory loss.

Fertilizers in runoff and raw sewage fan algal blooms around the world. Since 1991, *Pfiesteria* has killed more than a billion fish along the coasts of North Carolina, Virginia, and Maryland. This "cell from hell" releases a neurotoxin, then releases amoeboid stages that feast on tissues sloughed from the dead fish. JoAnn Burkholder of North Carolina State University correlated its blooms with hundreds of millions of gallons of raw sewage from factory-like hog farms and chicken farms in the region. The nutrient-enriched waters enter rivers that drain into the poorly flushed estuaries of Chesapeake Bay. Ten years ago, *Pfiesteria* was not a problem. Today, water pollution has drastically disrupted the region's aquatic ecosystems.

Nearly all single-celled photosynthetic protistans, including most euglenoids, chrysophytes, and dinoflagellates, belong to phytoplankton, the food-producing base of aquatic habitats.

WWW

Of the 4,100 known species of **red algae** (Rhodophyta), nearly all are marine; only 200 live in freshwater. The life cycle of most species includes multicelled stages, but these lack tissues or organs. Reproductive modes are diverse, with complex asexual and sexual phases. Figure 20.24 shows one example of a life cycle.

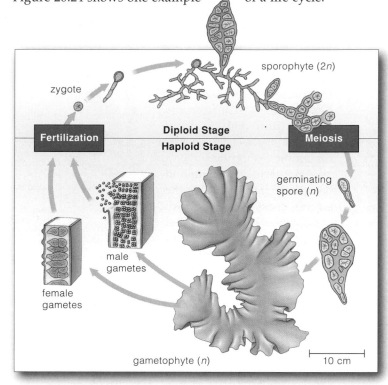

Figure 20.24 Life cycle of *Porphyra.*

Not all "red algae" are red. Diverse photosynthetic species are green, purple, or almost black. Phycobilins and other accessory pigments mask their chlorophyll *a* (Section 6.2). Parasitic species have little pigmentation.

Red algae have a flexible, slippery texture owing to mucous material in their cell walls. Agar is made from extracts of wall material of several species. This inert, gelatinous substance has uses as a moisture-preserving agent in baked goods and cosmetics, as a setting agent for jellies, and as a culture gel. Carrageenan, extracted from *Eucheuma,* is a stabilizer in paints, dairy products, and many other emulsions.

Humans find different species of *Porphyra* tasty as well as nutritious. You may know it from sushi bars as *nori,* an algal wrapping for rice and fish.

Red algae are photosynthetic protistans with phycobilins and other accessory pigments that mask their chlorophyll *a.* Usually, multicelled stages develop during the life cycles. Most species are aquatic. They show great diversity in size, morphology, life-styles, reproductive modes, and habitats.

WWW

You find most of the 1,500 or so species of **brown algae** (Phaeophyta) living in cool or temperate marine waters, from the intertidal zone on into the open ocean. Different kinds are dark brown, olive-green, or golden, depending on the pigment array. Brown algae have fucoxanthin and other carotenoids as well as the chlorophylls a, c_1, and c_2. They range in size from the giant kelps, twenty to thirty meters long, to microscopic, threadlike *Ectocarpus*.

Macrocystis, Laminaria, and the other giant kelps are the largest, most complex of all protistans. Their multicelled sporophytes consist of stipes (stemlike structures), blades (leaflike parts), and holdfasts (anchoring structures). Gas-filled bladders on the stipes impart buoyancy and keep blades upright (Figure 20.25). Within the stipes are tubelike arrangements of elongated cells. Through all the tubes, sugars and other dissolved products of photosynthesis quickly move to living cells throughout the body.

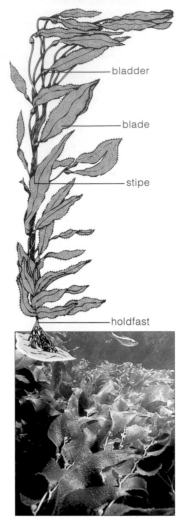

Figure 20.25   *Macrocystis.*

Giant kelp beds function as productive ecosystems. Think of them as underwater forests within which great numbers of diverse bacteria and protistans, as well as fishes and other animals, carry out their lives.

We commercially harvest *Macrocystis* and certain other brown algae. Extracts from them have uses in ice cream, pudding, jelly beans, salad dressings, canned foods, frozen foods, beer, cough syrup, toothpaste, floor polish, cosmetics, and paper. Alginic acid in the cell walls of some species is used to make algins, which are added to various products as thickening, emulsifying, and suspension agents. Many people, especially in the Far East, harvest kelps as sources of food and mineral salts, and as a fertilizer for crops.

The brown algae range in size from the microscopic to the largest of all of the multicelled protistans. Nearly all species live in marine habitats, ranging from the intertidal zone to the surface waters of the open ocean.

Of all protistans, the **green algae** (Chlorophyta) are structurally and biochemically most like the plants and are their closest relatives. Nearly all are photosynthetic. Like plants, they have chlorophylls a and b. They, too, temporarily store carbohydrates in starch grains inside their chloroplasts. In addition, the cell walls of some species are composed of cellulose, pectins, and other polysaccharides typical of plants.

Figure 20.26 is a sampling of the more than 7,000 known species of green algae. They include single-celled, colonial, filamentous, sheetlike, and tubular forms. You won't be able to see many species without the aid of a microscope. Most, including the *Micrasterias* cell shown in Figure 20.26b, live in freshwater. *Micrasterias* is one of thousands of desmid species, which are important food producers in nutrient-poor ponds and peat bogs. Green algae also grow at the ocean surface, in marine sediments, just below the surface of soil, and on rocks,

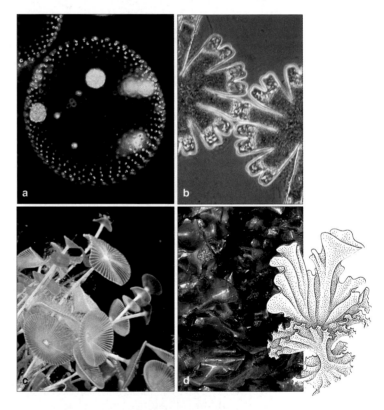

Figure 20.26   Representative green algae. (**a**) *Volvox*, a colony of interdependent green algal cells that bear resemblances to free-living, flagellated cells of the genus *Chlamydomonas*. (**b**) A *Micrasterias* cell reproducing itself. (**c**) From a marine habitat, *Acetabularia*, fancifully called the mermaid's wineglass. Each individual in the cluster is a multinucleate cell mass with a rootlike structure, stalk, and cap in which gametes form. (**d**) Sea lettuce (*Ulva*) grows in estuaries and attaches to kelps in the seas. Reproductive cells form within and are released from the margins of the sporophyte and gametophyte, both of which have the same form shown in the diagram.

WWW

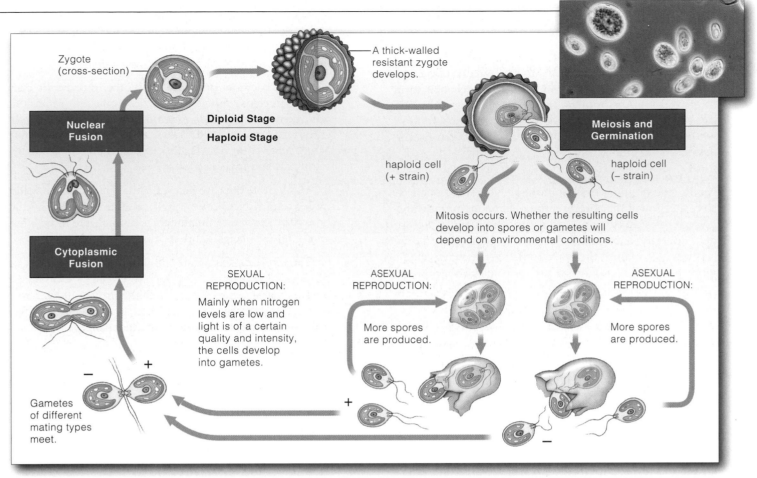

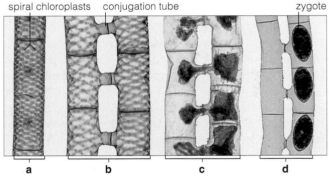

Figure 20.27 Life cycle of a species of *Chlamydomonas*, one of the most common green algae of freshwater habitats. This single-celled species reproduces asexually most often. It also reproduces sexually under certain environmental conditions.

tree bark, other organisms, and snow. Some types are symbionts with fungi, protozoans, and a few marine animals. A colonial form (*Volvox*) is a hollow whirling sphere of 500 to 60,000 flagellated cells. Those beautiful white, powdery beaches in tropical regions are largely the work of uncountable numbers of green algal cells (*Halimeda*) that formed calcified cell walls, then died and disintegrated. Some day, green algae might even accompany astronauts in space. They can grow in very small spaces, give off vital oxygen, and take up carbon dioxide exhaled by the aerobically respiring crew.

Green algae employ diverse modes of reproduction. *Chlamydomonas* provides a classic example. This single-celled alga is five to ten micrometers wide, on average. It is able to reproduce sexually, but most of the time it engages in asexual reproduction, with typically 16 to 32 daughter cells forming by mitotic cell division in the confines of the parent cell wall. Daughter cells may live at home for a while. But sooner or later they depart by secreting enzymes, which digest what's left of their parent. Figure 20.27 shows the life cycle of one species.

Figure 20.28 One mode of sexual reproduction in *Spirogyra*, or watersilk. (**a**) This green alga has spiral, ribbonlike chloroplasts. (**b**) A conjugation tube forms between cells of adjacent haploid filaments of different mating strains. (**c,d**) The cellular contents of one strain pass through the tubes into cells of the other strain, where zygotes form. Zygotes develop thick walls. They undergo meiosis when they germinate and give rise to haploid filaments.

To give a final example, Figure 20.28 shows *Spirogyra*, a filamentous green alga, reproducing sexually.

The green algae show great diversity in size, morphology, life-styles, and habitats. The structure and biochemistry of some groups indicate they have close evolutionary ties to the plant kingdom.

WWW

1. Bacteria are prokaryotic cells. None has a nucleus or the profusion of organelles typical of eukaryotic cells. Eubacteria are the most common types. Archaebacteria (methanogens, halophiles, and extreme thermophiles) live in extreme habitats. They differ from eubacteria in wall structure and other features, and they share some features with eukaryotic cells.

2. Three bacterial shapes are common: cocci (spheres), bacilli (rods), and spirilla (spirals). Nearly all bacteria have a cell wall that protects the plasma membrane and helps it resist rupturing. Wall composition and structure help identify a bacterial species. A capsule or slime layer may surround the wall. Some species have one or more bacterial flagella that function in motility. Many species have pili, filamentous proteins that help the cell adhere to a surface or facilitate bacterial conjugation.

3. Bacteria reproduce by prokaryotic fission. The single bacterial chromosome is replicated, and the parent cell divides into two genetically equivalent daughter cells. Many species have plasmids: small circles of extra DNA that are replicated independently of the chromosome.

4. Bacteria as a group show great metabolic diversity, as in their modes of acquiring energy and carbon. They also make behavioral responses to stimuli.

5. Viruses are nonliving, noncellular agents that infect particular species of nearly all organisms.

    a. A virus particle has a core of DNA or RNA and a protein coat that may be enclosed in a lipid envelope from which viral proteins project. Coats of complex viruses have tail fibers and other accessory structures.

    b. A virus particle cannot reproduce on its own. Its genetic material must enter a host cell and direct the cellular machinery to synthesize materials necessary to produce new virus particles.

6. Nearly all viral multiplication cycles have five steps: attachment to a suitable host cell, penetration of it, DNA or RNA replication and then synthesis of viral proteins, assembly of new viral particles, and release.

7. Unlike bacteria, protistans and other eukaryotic cells have a nucleus, mitochondria, endoplasmic reticulum with distinct ribosomes, microtubules, and at least two chromosomes (each with DNA and numerous proteins). Many have chloroplasts. Eukaryotic cells alone engage in mitosis and meiosis (Table 20.3).

8. Many protistans are heterotrophs. They include water molds, chytrids, slime molds (cellular and plasmodial types), protozoans (the amoebas, animal-like flagellates, and ciliated species), and sporozoans (Table 20.4).

9. The chytrids and water molds are decomposers or parasites. Like fungi, they secrete enzymes that digest organic matter, then absorb breakdown products. Slime molds are predatory (animal-like) phagocytes and also form spore-producing structures (like fungi).

10. Like animals, protozoans are predators, grazers, or parasites. The amoebas, foraminiferans, heliozoans, and radiolarians are amoeboid protozoans with or without skeletal elements. Animal-like protozoans are internal parasites or live freely in aquatic habitats; sporozoans are internal parasites. Some cause human diseases. The ciliated protozoans typically use cilia for motility.

11. Most of the euglenoids, chrysophytes (the diatoms, golden algae, and yellow-green algae), dinoflagellates, and red, brown, and green algae are photoautotrophs. Many are members of phytoplankton.

Review Questions

1. Describe key metabolic and structural features of bacteria. Make sketches of the three basic shapes of bacterial cells. *20.1*

2. How is bacterial growth measured? *20.3*

    3. Name a few photoautotrophic and chemoheterotrophic eubacteria. Describe one or two that are likely to give humans trouble, medically speaking. *20.2*

    4. Define and describe the characteristics of a virus. *20.4*

    5. Distinguish between epidemic and pandemic. *20.11*

    6. Review Table 20.4, cover it with a sheet of paper, then list the major categories of protistans. *20.9, Table 20.4*

    7. Select three protistan species, then briefly explain how they affect our affairs, such as human health. *20.8–20.11*

Self-Quiz *(Answers in Appendix III)*

1. _____ are confined to extreme environments, probably much like those that prevailed on the early Earth.
    a. Cyanobacteria   c. Archaebacteria
    b. Eubacteria   d. Protozoans

2. Which of the following is *not* an archaebacterium?
    a. Halophile   c. Thermophile
    b. Cyanobacterium   d. Methanogen

Table 20.3   **Comparison of Prokaryotes With Eukaryotes**

	PROKARYOTES	EUKARYOTES
Organism	Bacteria only	Protistans, fungi, plants, and animals
Level of organization	Single-celled	Single-celled (protistans, mostly), or multicelled, usually tissues, organs
Typical cell size	Small (1–10 μm)	Large (10–100 μm)
Cell wall	Nearly all are walled	Cellulose or chitin; none in animals
Organelles	Very rarely	Typically profuse
Metabolism	Anaerobic, aerobic	Aerobic modes predominate
Genetic material	Bacterial chromosome, sometimes plasmids	Complex chromosomes (DNA, many associated proteins) within a nucleus
Mode of division	Prokaryotic fission, mostly; also budding	Nuclear division (mitosis, meiosis, or both), and then cytoplasmic division

3. Bacteria reproduce by _____ .
   a. mitosis   c. prokaryotic fission
   b. meiosis   d. longitudinal fission

4. A nondividing bacterial cell contains
   _____ chromosome(s) and may have extra
   circles of _____ called plasmids.
   a. one; RNA   c. one; DNA
   b. two; RNA   d. two; DNA

5. Viruses have a _____ and a _____ .
   a. DNA core; carbohydrate coat
   b. DNA or RNA core; plasma membrane
   c. nculeus with DNA; lipid envelope
   d. DNA or RNA core; protein coat

6. The multiplication cycle of all viruses
   requires _____ .
   a. penetration   c. budding
   b. latency   d. none is correct

7. During a _____ life cycle, amoeboid
   cells aggregate and form a migrating mass.
   a. slime mold   c. protozoan
   b. water mold   d. chytrid

8. Amoebas, foraminiferans, radiolarians,
   and heliozoans are all _____ .
   a. parasites   c. protozoans
   b. ciliates   d. sporozoans

9. Trypanosomes, protozoans classified as
   animal-like flagellates, cause which disease(s)?
   a. Sleeping sickness d. toxoplasmosis
   b. Chagas disease   e. malaria
   c. amoebic dysentery f. both a and b

10. The euglenoids and chrysophytes are mostly _____ .
    a. photoautotrophs   c. heterotrophs
    b. chemoautotrophs   d. omnivorous

11. Single-celled photosynthetic protistans, which include most
    euglenoids, chrysophytes, and dinoflagellates, are members of
    _____ , the "pastures" of most aquatic habitats.
    a. zooplankton   c. brown algae
    b. red algae   d. phytoplankton

12. Match the terms with the most suitable descriptions.
    ____ archaebacteria   a. bacterial cell division mechanism
    ____ eubacteria   b. nonliving infectious particle with
    ____ viruses   nucleic acid core and protein coat
    ____ *Plasmodium*   c. agent of malaria
    ____ prokaryotic   d. methanogens, halophiles, and
         fission   extreme thermophiles
                           e. most common prokaryotic cells

Table 20.4 Summary of Major Protistan Groups

HETEROTROPHIC
(Different species are decomposers, predators, grazers, or parasites):

Chytrids
Water molds
Cellular slime molds
Plasmodial slime molds
Protozoans:
   Amoeboid protozoans (naked amoebas, foraminiferans, heliozoans, radiolarians)
   Animal-like flagellated protozoans
   Ciliated protozoans
Sporozoans

AUTOTROPHIC AND HETEROTROPHIC
Euglenoids
Dinoflagellates

MOSTLY AUTOTROPHIC
(Photosynthesizers, but some parasites):

Chrysophytes (golden algae, yellow-green algae, diatoms)
Red algae
Brown algae
Green algae

---

3. Suppose you vacation in a developing country where sanitation practices and standards of personal hygiene are poor. Having read about parasitic protozoans that can contaminate water and damp soil, what would you consider safe to drink? Which foods may be best to avoid or which food preparation methods might make them safe to eat?

4. As you read in this chapter, red tides are associated with "algal blooms." Algal blooms often follow enrichment of aquatic habitats with water that drains into them from heavily fertilized croplands or from concentrated sources of raw sewage. After thinking about it, do you accept the resulting destruction of aquatic species, birds, and other forms of wildlife as an unfortunate but necessary side effect of human actions? If you do not accept it, how would you stop the pollution? And if you could stop it, what sorts of measures would you take to feed the enormous human population, which is dependent on high-yield (and very heavily fertilized) crops? How would you propose to dispose of or prevent the accumulation of fecal matter and other wastes of close to 6 billion people?

Selected Key Terms

archaebacterium *20.1*
bacteriophage *20.4*
binary fission *20.9*
brown alga *20.14*
chrysophyte *20.12*
chytrid *20.8*
contractile vacuole *20.9*
disease *20.11*
endospore *20.2*
epidemic *20.11*
eubacterium *20.1*
euglenoid *20.12*

extreme thermophile *20.2*
green alga *20.15*
halophile *20.2*
heterocyst *20.2*
infection *20.11*
lysis *20.5*
methanogen *20.2*
microorganism *CI*
numerical taxonomy *20.1*
pathogen *CI*
plankton *20.9*
plasmid *20.3*

prion *20.6*
prokaryotic cell *20.3*
prokaryotic fission *20.1*
protistan *20.7*
protozoan *20.9*
red alga *20.13*
red tide *20.12*
slime mold *20.8*
sporozoan *20.10*
viroid *20.6*
virus *20.4*
water mold *20.8*

Critical Thinking

1. *Salmonella* bacteria cause a form of food poisoning. Often they live in poultry and eggs. Newly hatched chicks acquire them when they ingest bacteria-laden feces of healthy adult chickens. However, harmless bacteria ingested the same way colonize the surface of intestinal cells, leaving no place for *Salmonella* to take hold. Some farmers raise thousands of chicks in confined quarters with no adult chickens. Should they feed the chicks a known mixture of bacteria from a lab or a mixture of unknown bacteria from healthy adult chickens? Devise an experiment to test which approach may be more effective.

2. Reflect on the description of mad cow disease in Section 20.6. The FDA is moving to prohibit all protein supplements for sheep, cattle, and other ruminants. Investigate the extent of this practice in the United States and how it has been monitored.

Readings *See also www.infotrac-college.com*

Hively, W. May 1997. "Looking for Life in All the Wrong Places." *Discover*, 76–85. Account of microbes that live in the most extreme environments on and in the Earth.

Margulis, L. 1993. *Symbiosis in Cell Evolution*. Second edition. New York: Freeman. Paperback.

Satchell, M. 28 July 1997. "The Cell From Hell." *U.S. News and World Report*, 26–28.

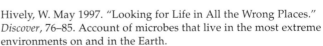

WWW   *http://www.brookscole.com/biology*

Practice quiz questions, hypercontents, BioUpdates, and critical thinking. The Brooks/Cole Biology Resource Center provides a wealth of information fully organized and integrated by chapter.

21

FUNGI

Ode to the Fungus Among Us

When push comes to shove, plants need certain fungi more than they need us. (In fact, they don't need us at all.) Fungi were there, as symbionts, when plants first invaded the land. **Symbiosis**, recall, refers to species that live closely together. In cases called **mutualism**, their interaction benefits both partners or does one of them no harm. Lichens and mycorrhizae are like this.

A **lichen** is a vegetative body in which a fungus has become intertwined with one or more photosynthetic organisms. From the Antarctic to the Arctic, you find lichens surviving in a variety of habitats that are just too hostile for most organisms.

Lichens absorb mineral ions from substrates; some absorb nitrogen from the air. They make antibiotics (against bacteria that can decompose the lichen body) and toxins (against invertebrate larvae that graze on it). Coincidentally, their metabolic products help form new soils or enrich existing ones. This is what happens when lichens colonize new habitats, such as bare soil exposed by a retreating glacier. Conditions improve so much, other species move in and replace the pioneers. Actually, this may have happened when plants first invaded the land. Cyanobacteria-containing lichens even help maintain ecosystems. They put captured nitrogen into a form that plants use. *Lobaria* provides old-growth forests in the Pacific Northwest with up to 20 percent of their required nitrogen (Figure 21.1).

Lichens serve as early warnings of deteriorating environmental conditions. They absorb toxins but can't get rid of them. From extensive studies in New York City and in England, we know that when lichens die around cities, air pollution is getting bad.

Fungi, too, enter into mutualistic interactions with young tree roots. Together they form a **mycorrhiza** (plural, mycorrhizae), which means "fungus-root." The fungus absorbs sugars from the plant, which absorbs minerals from its partner. The underground parts of a fungus thread through soil and afford a huge surface area for absorption. The fungus rapidly takes up many ions of phosphorus and other minerals when they are abundant and releases them to the plant when the ions are scarce. Many plants do not grow well at all in the absence of mycorrhizae.

Since the early 1900s, collectors have recorded data on wild mushroom populations in European forests. As the records tell us, the number and kinds of fungi are declining at alarming rates. Mushroom gatherers aren't to blame; inedible as well as edible species are vanishing. But the decline does correlate with rising air pollution. Vehicle exhaust, smoke from coal burning, and emissions from nitrogen fertilizers pump ozone, nitrogen oxides, and sulfur oxides into the air. As a tree ages, one species of mycorrhizal fungus normally gives way to another in predictable patterns. When fungi die, the tree loses this vital support system and becomes vulnerable to severe frost and drought. Are the North American forests at risk? Conditions there are deteriorating in comparable ways.

Plants depend on fungi in still another way, because many are premier **decomposers**. Figure 21.2 shows just a few of these. Like all other decomposers, fungi break down organic compounds in their surroundings. But few organisms besides fungi digest their dinner out on the table, so to speak. As fungi grow in or on organic matter, they secrete enzymes that digest it into bits that individual cells can absorb. Such a mode of nutrition is called **extracellular digestion and absorption**. Plants —the primary producers of nearly all ecosystems—also benefit because they can take up some of the released carbon and other nutrients.

Keep this global perspective in mind. Why? As you will see, the activities of some fungi do cause diseases in humans, pets and farm animals, ornamental plants, and important crop plants. Some species are notorious spoilers of food supplies. Others help us manufacture substances ranging from antibiotics to cheeses. We tend to assign "value" to fungi and other organisms in terms of their direct effect on our lives. There is nothing wrong with battling dangerous species and admiring beneficial ones—as long as we do not lose sight of the greater roles of fungi or any other kind of organism in nature.

Figure 21.1 *Lobaria oregana.*

a PURPLE CORAL FUNGUS *Clavaria*

b RUBBER CUP FUNGUS *Sarcosoma*

c BIG LAUGHING MUSHROOM *Gymnophilus*

d TRUMPET CHANTERELLE *Craterellus*   **e** SCARLET HOOD *Hygrophorus*

Figure 21.2 Fungal species from southeastern Virginia. This sampling hints at the rich diversity within the kingdom Fungi.

KEY CONCEPTS

1. Fungi are heterotrophs. Together with heterotrophic bacteria, they are the bisophere's decomposers. Saprobic types obtain nutrients from nonliving organic matter. Parasitic types obtain them from tissues of living hosts.

2. Fungi secrete enzymes that digest food outside their body, then fungal cells absorb breakdown products. Their metabolic activities release carbon dioxide to the atmosphere and return many nutrients to the soil, where they become available to producer organisms.

3. Most fungi are multicelled. A mycelium, which is the food-absorbing portion of a fungal body, develops during the fungal life cycle. Each mycelium is a mesh of hyphae, elongated filaments that grow and develop by repeated mitotic cell divisions.

4. Commonly, a portion of the fungal hyphae becomes modified and weaves together to form a reproductive structure in or upon which fungal spores develop. A "mushroom" is such a structure. Germinating spores grow and develop into a new mycelium.

5. Many fungi are symbionts with other organisms. Some are partners with algae and other organisms, in lichens. Many are part of mycorrhizae; they are locked in mutually beneficial relationships with young roots of land plants. Metabolic activities of cells making up the fungal hyphae provide the plants with nutrients, and the plants provide the fungi with carbohydrates.

6. We tend to assign value to plants and fungi in terms of their direct effect on our lives. Our battles with the "bad" ones and reliance on the "good" ones should start from a solid understanding of their long-established roles in nature.

Mode of Nutrition

Fungi are heterotrophs, meaning they require organic compounds that other organisms synthesize. Most are **saprobes**; they obtain nutrients from nonliving organic matter and so cause its decay. Others are **parasites**; they extract nutrients from tissues of a living host. When the cells of any species grow in or on organic matter, they secrete digestive enzymes and then absorb breakdown products. Again, their mode of extracellular digestion helps plants, which readily absorb some of the released nutrients and carbon dioxide by-products. Without the fungi and heterotrophic bacteria, communities would slowly become buried in their own garbage, nutrients would not be cycled, and life could not go on.

Major Groups

For many of us, "fungi" are drab mushrooms sold in grocery stores. These are simply fungal body parts, and they are produced by a fungus with stunningly diverse relatives. The few species in Figure 21.2 don't do justice to the 56,000 fungal species we know about—and there may be at least a million more we don't know about!

We know, from the fossil record, that fungi evolved before 900 million years ago. Some accompanied plants onto the land 430 million years ago. About 100 million years later, three major lineages were well established. We call them the **zygomycetes** (Zygomycota), **sac fungi** (Ascomycota), and **club fungi** (Basidiomycota). Other, puzzling kinds known as "imperfect fungi" are lumped together but are not a formal taxonomic group. The vast majority of species in all these groups are multicelled.

Key Features of Fungal Life Cycles

Fungi reproduce asexually quite often, but given the opportunity, they also reproduce sexually. They form great numbers of nonmotile spores. As in plants, their **spores** are reproductive cells or multicelled structures, often walled, that germinate after dispersal from the parent. In multicelled species, spores give rise to a mesh of branched filaments. The mesh, a **mycelium** (plural, mycelia), rapidly grows over or into organic matter and has a good surface-to-volume ratio for food absorption. Each filament in a mycelium is called a **hypha** (plural, hyphae). Hyphal cells commonly have chitin-reinforced walls. Their cytoplasm interconnects, so that nutrients flow unimpeded throughout the mycelium.

Fungi are major decomposers that engage in extracellular digestion and absorption of organic matter. Multicelled types form absorptive mycelia and spore-producing structures.

𝒲𝒲𝒲

A Sampling of Spectacular Diversity

Fungal life cycles and life-styles show dizzying variety. The most we can do here is to sample a few species, starting with the club fungi. The 25,000 or so club fungi include mushrooms, shelf fungi, coral fungi, puffballs, and stinkhorns. Figures 21.2 through 21.5 and 1.7*d* show examples. Some of the saprobic species are important decomposers of plant debris. As you will see, other species are symbionts that live in close association with the young roots of trees in forests. The fungal rusts and smuts can destroy fields of wheat, corn, and other crop plants. Cultivation of the common mushroom (*Agaricus brunnescens*) is a multimillion-dollar business. It is the mushroom of grocery-store and pizza-topping fame. Yet some of its relatives produce toxins that can kill you or any other organism that nibbles on them.

Have you ever wondered which organisms are the oldest and the largest? *Armillaria bulbosa* is one of them. The mycelium of one individual, discovered in a forest in northern Michigan, extends through fifteen hectares of soil. (Each hectare is the equivalent of 10,000 square meters.) By some estimates, this fungus weighs more than 10,000 kilograms and has been spreading beneath the forest floor for more than 1,500 years!

Figure 21.3   Two club fungi. (**a**) The light-red coral fungus *Ramaria*. (**b**) The shelf fungus *Polyporus*. With the exception of the rubber cup fungus, all of the fungal species shown in Figure 21.2 are club fungi.

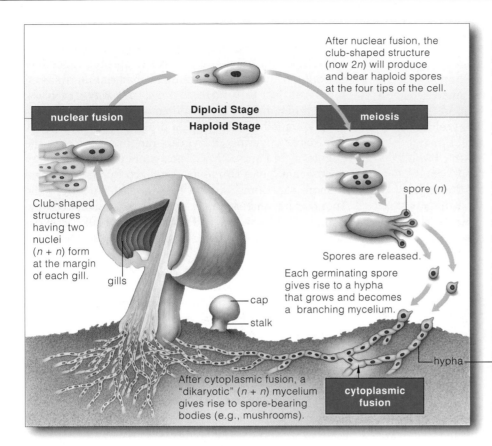

After nuclear fusion, the club-shaped structure (now 2n) will produce and bear haploid spores at the four tips of the cell.

nuclear fusion

Diploid Stage

Haploid Stage

meiosis

Club-shaped structures having two nuclei (n + n) form at the margin of each gill.

gills

cap

stalk

spore (n)

Spores are released.

Each germinating spore gives rise to a hypha that grows and becomes a branching mycelium.

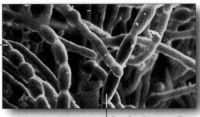

hypha

After cytoplasmic fusion, a "dikaryotic" (n + n) mycelium gives rise to spore-bearing bodies (e.g., mushrooms).

cytoplasmic fusion

hypha in mycelium

Figure 21.4 Generalized life cycle for many club fungi. When hyphal cells of two compatible mating strains grow together, their cytoplasm (but not nuclei) fuse. Cell divisions result in a dikaryotic mycelium (its cells each have two nuclei). Mushrooms form, with club-shaped, spore-bearing structures on their gill surface. The two nuclei in each structure fuse to form a diploid zygote, which starts the cycle anew.

When you see mushrooms or any other fungus growing outdoors, think twice before devouring them. Unlike cultivated species, such as the common mushroom, many are toxic, including the two species shown in Figure 21.5. *No one should eat any fungi that were gathered in the wild unless they have been accurately identified as edible.* As the saying goes, there are old mushroom hunters and bold mushroom hunters, but no old, bold mushroom hunters.

Figure 21.5 (**a**) Fly agaric mushroom (*Amanita muscaria*). It causes hallucinations when eaten. It was used to induce trances in ancient rituals in Central America, Russia, and India. (**b**) From California, *A. ocreata*. It can be fatal if eaten. A close relative, the death cap mushroom (*A. phalloides*), lives up to its name. Nibble as little as five milligrams of its toxin, and vomiting and diarrhea will begin eight to twenty-four hours later. Your liver and kidneys will degenerate; death may follow within a few days.

Example of a Fungal Life Cycle

Probably you already know what a common mushroom (*A. brunnescens*) looks like, so let's use it as our example of a fungal life cycle. Like most of the other club fungi, this species produces short-lived reproductive bodies—**mushrooms**—that are merely its aboveground parts; its living mycelium is buried in soil or decaying wood. A mushroom has a stalk and a cap. Fine tissue sheets, or gills, line the cap's inner surface. On these gills, club-

shaped, spore-bearing structures form. The spores that form here are **basidiospores**. When a spore dispersed from a particular strain of mushroom lands on a suitable site, it germinates and gives rise to a haploid mycelium.

Suppose hyphae of two compatible mating strains make contact. They may undergo cytoplasmic fusion, but their nuclei do not fuse at once. The fused part may be the start of a *dikaryotic* mycelium, in which hyphal cells have one nucleus of each mating type (Figure 21.4). An extensive mycelium forms. And when conditions are favorable, mushrooms form. Each spore-producing structure of the mushroom is initially dikaryotic, but then its two nuclei fuse and form a short-lived zygote. The zygote undergoes meiosis, haploid spores develop outside on small stalks, then air currents disperse them.

Club fungi, the fungal group with the greatest diversity, produce spores outside distinctive club-shaped structures.

WWW

SPORES AND MORE SPORES

A fungus has a thing about spores. It produces sexual spores, asexual spores, or both, depending on contact with a suitable hypha, food availability, and how cool or damp conditions are. Its spores are usually small and dry, and air currents disperse them. Each spore that germinates can be the start of a hypha and a mycelium. Stalked reproductive structures may develop on many of the hyphae and produce asexual spores. After these spores germinate, each may be the start of still *another* extensive mycelium. In no time at all, that one fungus and staggering numbers of its descendants are busily decomposing organic stuff or pirating nutrients from a host! Look at what can happen to a slice of stale bread:

Each fungal class produces unique sexual spores. Club fungi form basidiospores, zygomycetes form spores by way of zygosporangia, and sac fungi form ascospores.

Think of the zygomycetes. Parasitic species feed on houseflies and other insects. Most saprobic types live in soil, decaying plant or animal material, and stored food. You just saw what *Rhizopus stolonifer*, the black bread mold, does to bread. When it reproduces sexually, a thick-walled sexual spore, a diploid zygote, forms. A thin, clear covering (a zygosporangium) encloses the zygote, as in Figure 21.6a. The zygote proceeds through meiosis and gives rise to a specialized hypha that bears a spore sac. Some number of spores form inside the sporangium, and each may give rise to a new mycelium. Stalked hyphae grow out of such mycelia. Asexual spores form in a spore sac perched on top of each stalk (Figure 21.6b).

We know of more than 30,000 kinds of sac fungi. The vast majority are multicelled. Most form sexual spores called **ascospores** inside sac-shaped cells. They alone form these cells, which are called asci (singular, ascus). Reproductive structures that consist of tightly interwoven hyphae enclose asci of multicelled species. These structures resemble flasks, globes, and shallow cups. Figures 21.2b and 21.7 show some examples.

Sac fungi include highly prized, tasty truffles and morels, as in Figure 21.7b). They also include about 500 single-celled yeast species, although other yeasts are classified as club

Figure 21.6 Life cycle of the black bread mold *Rhizopus stolonifer*. Asexual phases are common. Different mating strains (+ and −) also reproduce sexually. Either way, haploid spores form and give rise to mycelia. Chemical attraction between a + hypha and a − hypha causes them to fuse. Two gametangia form, each with several haploid nuclei. Later their nuclei fuse to form a zygote. The zygote develops a thick wall, so becoming a zygospore, and may remain dormant for several months. Meiosis occurs as the zygospore germinates, and asexual spores form.

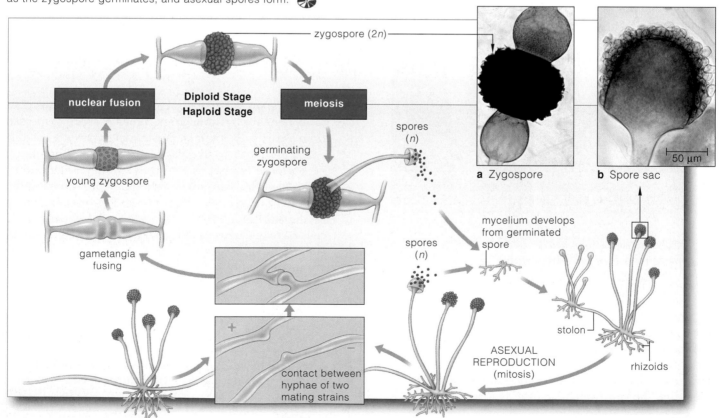

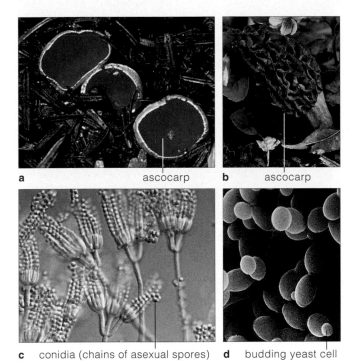

a ascocarp   **b** ascocarp

c conidia (chains of asexual spores)   **d** budding yeast cell

Figure 21.7 Sac fungi. (**a**) Saclike structures on the inner surface of this sac fungus produce sexual spores (ascospores) by meiosis. (**b**) Morel (*Morchella esculenta*). This edible kind has a poisonous relative. (**c**) From *Eupenicillium*, chains of asexual spores of a type called conidiospores. These drift from the chains, like dust, even after the slightest jiggling. "Conidia" means dust. (**d**) Cells of *Candida albicans*, agent of "yeast" infections of the vagina, mouth, intestines, and skin.

fungi. Yeasts reproduce sexually when two cells fuse and become a spore-producing sac. Some types live in the nectar of flowers and on fruits and leaves. Bakers and vintners put the fermenting by-products of great populations of yeasts to use (Section 7.5). Then again, *Candida albicans*, a notorious relative of "good" yeasts, causes vexing infections in humans (Figure 21.7*d*).

Sac fungi include some species of *Penicillium* that "flavor" Camembert and Roquefort cheeses as well as species that make the penicillins we use as antibiotics. They include some species of *Aspergillus*, which we use to make citric acid for candies and soft drinks, and to ferment soybeans for soy sauce. Opinions differ on how to classify red, bluish-green, and brown fungal molds that spoil stored food. Most are multicelled. One of them, *Neurospora crassa*, has uses in genetic research.

Through their exuberant and rapid production of asexual and sexual spores, fungi take quick advantage of available organic matter, whether it has been discarded or is part of a living or dead organism. Their penchant for producing spores underlies their success as decomposers and parasites.

WWW

You know you are a serious student of biology when you view organisms objectively in terms of their place in nature, not in terms of their impact on humans generally and you in particular. As a student you salute saprobic fungi as vital decomposers and praise parasitic fungi that help keep harmful insects and weeds in check. The true test is when you open the fridge to get a bowl of high-priced raspberries and discover that a fungus beat you to them. The true test is when a fungus starts feeding on warm, damp tissues between your toes and turns skin scaly, reddened, and cracked (Figure 21.8*a*).

Which home gardeners wax poetic about black spot or powdery mildew on roses? Which farmers happily give up millions of dollars each year to sac fungi that attack corn, wheat, peaches, and apples (Figure 21.8*b*)? Who cares if the sac fungus *Cryphonectria parasitica* turned most of eastern North America's chestnut trees into stubby versions of their former selves?

Who willingly inhales airborne spores of *Ajellomyces capsulatus*? When these dimorphic beasties alight on soil, they form mycelia. When they alight in moist lung tissue, they form populations of yeastlike cells that cause a respiratory disease, *histoplasmosis*. Macrophages defend infected tissues and usually engulf all of the invaders. Masses of them become calcified as the tissue heals. With heavy exposure to the spores, pneumonia develops. In rare cases, the calcified masses form in lymph nodes, the liver, and other organs. Progressive histoplasmosis usually ends in death. Where nitrogen-rich droppings of many birds or bats accumulate, the fungus thrives.

Fungi even thread through human history. One species, *Claviceps purpurea*, parasitizes rye and other cereal grains. We use some of its alkaloid products to treat migraine headaches and, after childbirth, to shrink the uterus to stop hemorrhaging. But its alkaloids are toxic when ingested in large amounts. Eat a lot of bread made with tainted rye flour and you end up with *ergotism*. Symptoms include vomiting, diarrhea, hallucinations, and convulsions. Untreated, the disease invites gangrene and death.

Ergotism epidemics were common in Europe in the Middle Ages, when rye was a key crop. Ergotism thwarted Peter the Great, a Russian czar obsessed with conquering ports along the Black Sea for his nearly landlocked empire. Soldiers laying siege to the ports ate mostly rye bread and fed rye to their horses. The former went into convulsions and the latter into "blind staggers." Possibly, outbreaks of ergotism were even used as an excuse to launch witch-hunts in colonial Massachusetts and elsewhere.

WWW

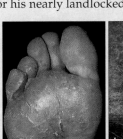
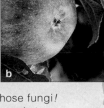

Figure 21.8 Love those fungi! (**a**) Thanks to *Epidermophyton floccosum*, athlete's foot. (**b**) Apple scab, hallmark of *Venturia inequalis*.

Recall, from the introduction, that symbiosis refers to species that live together in close ecological association. Often one is a parasite's victim, not a partner. In cases of mutualism, interaction benefits both partners or does one of them no harm. Here are more detailed examples.

Lichens

In the single vegetative body called a lichen, a fungus is intertwined with one or more photosynthetic species. The fungal part is the *myco*biont. The photosynthetic part is the *photo*biont. Of about 13,500 known types of lichens, nearly half incorporate sac fungi. Only 100 or so species serve as photobionts, and most often these are green algae and cyanobacteria.

Lichens typically colonize sites that are hostile for most organisms, including sunbaked or frozen rocks, fenceposts, gravestones, and plants, even the tops of giant Douglas firs. Almost always, the fungus is the largest component. Cyanobacteria typically reside in a separate structure inside or outside the main body. The fungus gets a long-term source of nutrients, which it absorbs from the photobiont cells. Nutrient withdrawals affect the photobiont's growth a bit, but the lichen may help shelter it. If more than one fungus is present in the

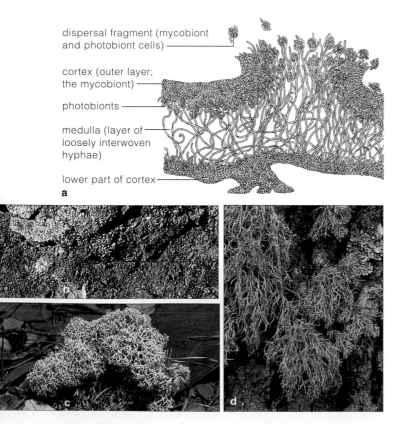

dispersal fragment (mycobiont and photobiont cells)

cortex (outer layer; the mycobiont)

photobionts

medulla (layer of loosely interwoven hyphae)

lower part of cortex

a

Figure 21.9 (**a**) Sketch of a stratified lichen, cross-section. (**b**) Encrusting lichens. (**c**) Erect, branching lichen, *Cladonia rangiferina*. (**d**) *Usnea* (old man's beard), a pendant lichen.

hyphal strands

small, young tree root

a

b   c

Figure 21.10 (**a**) Mycorrhiza from a hemlock tree. (**b**) Juniper seedlings, six months old, that were grown with a mycorrhizal fungus in sterilized, phosphorus-poor soil. (**c**) Juniper seedlings grown without the fungus.

lichen, it might be a mycobiont, a parasite, or even an opportunist that is using the lichen as a substrate.

A lichen forms after the tip of a fungal hypha binds with a suitable host cell. Both lose their wall, and their cytoplasm fuses or the hypha induces the host cell to cup around it. The mycobiont and the photobiont grow and multiply together. The lichen commonly has distinct layers. The overall pattern of growth may be leaflike), flattened, pendulous, or erect (Figures 21.1 and 21.9).

Mycorrhizae

Fungi, recall, also are mutualists with young tree roots, as mycorrhizae. Many plants do not grow as efficiently without mycorrhizae. In an *ecto*mycorrhiza, hyphae form a dense net around living cells in the roots but do not penetrate them (Figure 21.10*a*). Other hyphae form a velvety wrapping around the roots, with the mycelium radiating outward in soil. Ectomycorrhizae are common in temperate forests. They help trees withstand seasonal changes in temperature and rainfall. About 5,000 fungal species, mostly club fungi, enter into such associations.

The more common *endo*mycorrhizae form in about 80 percent of all vascular plants. These fungal hyphae penetrate plant cells, as they do in lichens. Fewer than 200 species of zygomycetes serve as the fungal partner. Their hyphae branch extensively, forming tree-shaped absorptive structures in plant cells. Hyphae also extend for several centimeters into the soil. Chapter 26 offers a closer look at these beneficial species.

Lichens and mycorrhizae are both symbiotic associations between fungi and other organisms, with mutual benefits.

WWW

SUMMARY

1. Fungi are heterotrophs and major decomposers. The saprobes feed on nonliving organic matter; the parasites feed on the tissues of living organisms. Some species of fungi are symbiotic partners with other organisms. The cells of all species secrete digestive enzymes that break down food into small molecules, which the cells absorb.

2. Nearly all fungi are multicelled. The food-absorbing portion, the mycelium, consists of a mesh of filaments (hyphae). Aboveground reproductive structures, such as mushrooms, form from tightly interwoven hyphae.

3. The major groups of fungi are the zygomycetes, the ascomycetes (sac fungi), and the basidiomycetes (club fungi). Each is characterized by distinctive sexual and asexual spores. When a sexual phase cannot be detected or is absent from the life cycle, a fungus is assigned to an informal category called the imperfect fungi.

4. Lichens are mutualistic associations of fungi with photosynthetic partners (green algae and cyanobacteria, mostly). Mycorrhizae are mutualistic associations of a fungus and the young roots of plants. Fungal hyphae provide nutrients for their symbiont, which provides the fungus with carbohydrates.

Review Questions

1. Describe the fungal mode of nutrition, and explain how the structure of mycelia facilitates this mode. *21.1*

2. How does a lichen differ from a mycorrhiza? *21.1, 21.5*

Self-Quiz (*Answers in Appendix III*)

1. New mycelia form after _____ germinate.
   a. hyphae   b. mycelia   c. spores   d. mushrooms

2. A "mushroom" is _____ .
   a. the food-absorbing part of a fungal body
   b. the part of the fungal body not constructed of hyphae
   c. a reproductive structure
   d. a nonessential part of the fungus

3. A mycorrhiza is a _____ .
   a. fungal disease of the foot   c. parasitic water mold
   b. fungus-plant relationship   d. fungus of barnyards

4. Parasitic fungi obtain nutrients from _____ .
   a. tissues of living hosts   c. only living animals
   b. nonliving organic matter   d. none of the above

5. Saprobic fungi derive nutrients from _____ .
   a. nonliving organic matter   c. root hairs
   b. living organisms   d. both b and c

6. Match the terms appropriately.
   ____ zygomycetes
   ____ conidia
   ____ hypha
   ____ club fungi
   ____ asci
   ____ sac fungi

   a. mushrooms, shelf fungi
   b. sac-shaped cells
   c. chains of asexual spores in *Eupenicillium*
   d. each filament in a mycelium
   e. black bread mold
   f. truffles, morels, some yeasts

Figure 21.11 Reproductive structures of *Pilobolus*, a name from a Greek word for "hat-thrower." The "hats" are spore sacs.

Critical Thinking

1. *Pilobolus* is a type of fungus that commonly dines on horse dung. Each morning, stalked reproductive hyphae emerge from irregularly spaced piles of dung. By early afternoon, they have dispersed spores to sunlit grasses where horses feed. The spores pass through the horse gut unharmed and exit with their own pile of dung. At the tip of each stalked hypha is a dark-walled, spore-containing sac (Figure 21.11). Just below the sac, the stalk is differentiated into a vesicle, swollen with a fluid-filled central vacuole. At the base of the vesicle is a ring of light-sensitive, pigmented cytoplasm. The stalk bends as it grows until its wall is parallel with the sun's rays and light strikes all of the ring. When that happens, turgor pressure builds up inside the central vacuole until the vesicle ruptures. The forceful blast can propel spore sacs two meters away—which is amazing, considering that the stalk is less than ten millimeters tall. Reflect on the examples of fungi discussed in this chapter. Would you say *Pilobolus* is a zygomycete, club fungus, or sac fungus?

2. Diana sees in the laboratory that the fungus *Trichoderma* grows well in distilled water. It continues to do so even after she rigorously treats the water and glassware to remove all traces of organic carbon. This fungus is not a photoautotroph. Suggest a metabolic life-style that lets it grow under these conditions.

3. *Trichoderma* is being tested as a natural pest control agent. Laboratory experiments demonstrated that some strains of this fungus combat other fungi that cause plant diseases. Some even promote seed germination and plant growth. During one set of twenty trials, workers increased lettuce yields by 54 percent. What concerns must be addressed before *Trichoderma* can be released into the environment for commercial applications?

Selected Key Terms

ascospore *21.3*	fungus *21.1*	parasite *21.1*
basidiospore *21.2*	hypha *21.1*	sac fungus
club fungus	lichen *CI*	(ascomycetes) *21.1*
(basidiomycetes) *21.1*	mushroom *21.2*	saprobe *21.1*
decomposer *CI*	mutualism *CI*	spore (fungal) *21.1*
extracellular digestion	mycelium *21.1*	symbiosis *CI*
and absorption *CI*	mycorrhiza *CI*	zygomycetes *21.1*

Readings   *See also www.infotrac-college.com*

Moore-Landecker, E. 1996. *Fundamentals of the Fungi.* Fourth edition. Englewood Cliffs, New Jersey: Prentice-Hall.

WWW   *http://www.brookscole.com/biology*

Practice quiz questions, hypercontents, BioUpdates, and critical thinking. The Brooks/Cole Biology Resource Center provides a wealth of information fully organized and integrated by chapter.

22 PLANTS

Pioneers In a New World

Seven hundred million years ago, no shorebirds stirred and noisily announced the dawn of a new day. There were no crabs to clack their claws together and skitter off to burrows. The only sounds were the rhythmic muffled thuds of waves in the distance, at the outer limits of another low tide.

More than 3 billion years before, life had its beginning somewhere in the waters of the Earth. And now, quietly, the invasion of the land was under way.

Why did it happen? Astronomical numbers of photosynthetic cells had come and gone, and the oxygen-producing types had slowly changed the atmosphere. High above the Earth, the sun's energy had converted much of the oxygen into a dense ozone layer. That layer became a shield against lethal doses of ultraviolet radiation, which had kept early organisms beneath the water's surface.

Were cyanobacteria the first to adapt to intertidal zones, where mud dried out with each retreating tide? Were they the first to spread into shallow, freshwater streams flowing down to the coasts? Probably so. From fossil evidence, we know that later in time, green algae and fungi made the same journey together.

Every plant around you today is a descendant of ancient species of green algae that lived near the water's edge or made it onto the land. Diverse fungi still associate with nearly all of them. Together, plants and fungi became the basis of communities in coastal lowlands, near the snow line of mountains, and in just about all places in between (Figure 22.1).

We have a few tantalizing fossils of the first pioneers. We also are learning about them through comparative biochemistry and studies of existing species. Today, as in the late Precambrian, cyanobacteria and green algae grow in mats in nearshore waters and on the banks of freshwater streams (Figure 22.1a). After a volcano erupts or a glacier retreats, cyanobacteria are the first to colonize the barren rocks. Symbiotic associations between green algae and fungi follow. Gradually their organic products and remains accumulate and create pockets of soil. And then mosses and other species of

Figure 22.1 (a) Filaments of a green alga, massed in a shallow stream. More than 400 million years ago, green algal species that might have been ancestral to all plants, past and present, lived in similar streams that meandered down to the shores of early continents. (b) One land-dwelling descendant of those ancestral forms— a Ponderosa pine high above the floor of Yosemite Valley in the Sierra Nevada of California. (c) Flowers of one of the most highly prized flowering plants—orchids— growing on a branch of a living tree in a tropical rain forest. With this chapter, we turn to the beginning—and end of the line—of some ancient lineages.

plants can become established in the newly forming soil and further enrich it.

With this chapter we turn to the plant kingdom. Nearly all plants are multicelled photoautotrophs. They absorb energy from the sun, carbon dioxide from the air, and some minerals dissolved in water to synthesize organic compounds. These metabolic wizards also can split water molecules. In doing so, they get stupendous numbers of the electrons and hydrogen atoms required for growth into multicellular forms as tall as the giant redwoods, as extensive as an aspen forest that is one continuous clone.

We know of at least 295,000 kinds of existing plants. Be glad their ancient ancestors left the water. Without them, we humans and other land-dwelling animals never would have made it onto the evolutionary stage.

1. With very few exceptions, the plant kingdom consists of multicelled photoautotrophs. From earlier chapters, you know that plants use chlorophylls *a* and *b* as their main photosynthetic pigments. In this respect they are like green algae, which are their closest relatives.

2. Unlike their algal ancestors, which were adapted to aquatic habitats, nearly all existing plants live on land.

3. In general, plants are structurally adapted to intercept sunlight, absorb water and mineral ions, and conserve water. Their lignin-reinforced tissues permit upright growth. Root systems mine the soil for water and ions, and internal tissues conduct water and solutes to all living cells in belowground and aboveground parts.

4. Land plants are reproductively adapted to withstand dry periods. During the life cycle, a sporophyte develops roots, stems, and leaves. It holds on to its developing gametes and supplies them with water and food resources. And it disperses the new generation in ways that are responsive to the conditions of specific habitats.

5. Early divergences gave rise to the bryophytes, then seedless vascular plants, and then seed-bearing vascular plants. Of these categories, the seed producers were the most successful in radiating into drier environments.

6. The seed-bearing vascular plants called gymnosperms include the cycads, ginkgo, gnetophytes, and conifers. The angiosperms, another group of vascular plants, bear flowers as well as seeds. There are two classes of flowering plants, informally called the dicots and monocots.

EVOLUTIONARY TRENDS AMONG PLANTS

Overview of the Plant Kingdom

The plant kingdom includes at least 295,000 species of photoautotrophs and a few heterotrophs. Most kinds are **vascular plants**, defined in part by internal tissues that conduct water and solutes through roots, stems, and leaves. Fewer than 19,000 species are *non*vascular plants called **bryophytes**. Plants, like photoautotrophic bacteria and protistans, are producers for ecosystems.

Liverworts, hornworts, and mosses are bryophytes. The whisk ferns, lycophytes, horsetails, and ferns are *seedless* vascular plants. Cycads, ginkgo, gnetophytes, and conifers belong to a group of *seed-bearing* vascular plants called **gymnosperms**. The **angiosperms**, another group of vascular plants, bear flowers and seeds. Dicots and monocots are two classes of flowering plants.

The ancestors of plants evolved in the seas by 700 million years ago. About 265 million more years passed before simple stalked plants appeared along coasts and streams. Evolutionarily, the pace picked up after that. Within 60 million years, plants radiated through much of the land. Long-term changes in their structure and reproduction explain how the diversity came about.

Evolution of Roots, Stems, and Leaves

Simple underground structures started to evolve when plants first colonized the land. In the lineages that led to vascular plants, they developed into root systems. Most **root systems** have many underground absorptive structures that collectively afford a large surface area. These rapidly take up soil water and dissolved mineral ions. In many species, root systems anchor the plant.

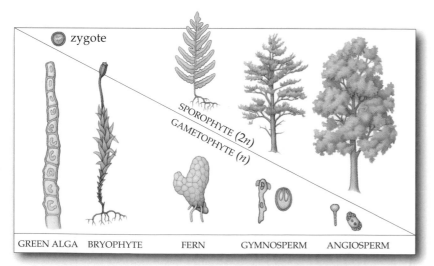

Figure 22.2 Evolutionary trend from gametophyte (haploid) dominance to sporophyte (diploid) dominance in the life cycle, represented here by existing species ranging from a green alga (*Ulothrix*) to a flowering plant. This trend occurred when early plants were colonizing habitats on land. See also Section 9.5.

Aboveground, **shoot systems** evolved. These have stems and leaves, which efficiently absorb energy from the sun and carbon dioxide from the air. Stems grew and branched extensively only after plants developed a biochemical capacity to synthesize and deposit **lignin**, an organic compound, in cell walls. Collectively, cells with lignified walls are strong enough to structurally support stems, which grow in patterns that increase the total light-intercepting surface of leaves.

Cellular pipelines for water and solutes evolved in many plants. The pipelines were major factors in the evolution of roots, stems, and leaves. They developed as components of xylem and phloem, which are two vascular tissues. **Xylem** distributes water and dissolved ions to all of the plant's living cells. **Phloem** distributes dissolved sugars and other photosynthetic products.

Life on land also depended on water conservation, which had not been a problem in most aquatic habitats. Shoots became protected by a **cuticle**, a waxy coat that helps conserve water on hot, dry days. Also, **stomata** (singular, stoma), tiny openings across the surfaces of leaves and some stems, helped control the absorption of carbon dioxide and restrict evaporative water loss. Later chapters describe these tissue specializations.

From Haploid to Diploid Dominance

As early plants radiated into higher, drier places, their life cycles changed. Think about the gametes of algae, which can get together only through liquid water. As shown in earlier chapters, a *haploid* (n) phase in the form of **gametophytes** (gamete-producing bodies), dominates their life cycles. The diploid ($2n$) phase is the zygote, which forms when gametes fuse at fertilization.

Now look at Figure 22.2. *In most plant life cycles, the diploid phase dominates.* After a diploid zygote forms at fertilization, mitotic cell divisions and cell enlargements transform it into a multicelled diploid body, of a type called a **sporophyte**. Pine trees are an example. In time, some cells of the sporophyte undergo meiosis and give rise to haploid cells of a type called **spores** (sporophyte means spore-producing body). Later, the spores divide by way of mitosis and give rise to the gametophytes.

The shift to diploid dominance was an adaptation to land habitats, most of which show seasonal changes in the availability of free water and dissolved nutrients. Long ago in those challenging habitats, natural selection must have favored sporophytes with well-developed root systems. Young roots of such systems interact with fungal symbionts (Section 21.5). The association, called a mycorrhiza, enhances the plant's uptake of water and scarce minerals, even during dry seasons.

Unlike algae and bryophytes, vascular plants have a sporophyte that is larger and structurally far more

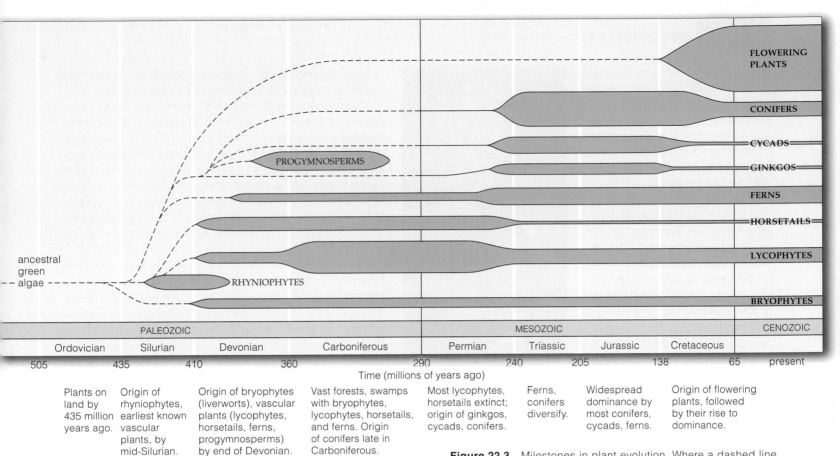

PALEOZOIC				MESOZOIC				CENOZOIC
Ordovician	Silurian	Devonian	Carboniferous	Permian	Triassic	Jurassic	Cretaceous	
505	435	410	360	290	240	205	138	65 present

Time (millions of years ago)

Plants on land by 435 million years ago.	Origin of rhyniophytes, earliest known vascular plants, by mid-Silurian.	Origin of bryophytes (liverworts), vascular plants (lycophytes, horsetails, ferns, progymnosperms) by end of Devonian.	Vast forests, swamps with bryophytes, lycophytes, horsetails, and ferns. Origin of conifers late in Carboniferous.	Most lycophytes, horsetails extinct; origin of ginkgos, cycads, conifers.	Ferns, conifers diversify.	Widespread dominance by most conifers, cycads, ferns.	Origin of flowering plants, followed by their rise to dominance.

Figure 22.3 Milestones in plant evolution. Where a dashed line turns solid, this indicates when the lineage originated. The width of each lineage indicates variations in the range of diversity over time.

complex than the gametophyte. As you will see, the gametophytes of the seedless vascular plants develop independently of the sporophyte that produces them. But they protect their gametes, and, after fertilization, they nourish the embryo sporophytes. The sporophyte became most dominant among the gymnosperms and, later, the angiosperms. It retains, nourishes, and protects developing gametophytes as well as young sporophytes. *And it does so until environmental conditions are suitable for fertilization and for dispersal of the new generation.*

Evolution of Pollen and Seeds

Like some seedless species, seed-bearing plants produce not one but two types of spores. We call this condition *hetero*spory, as opposed to *homo*spory (only one type). In both gymnosperms and angiosperms, one type of spore develops into female gametophytes, where eggs form and become fertilized. The other spore type gives rise to **pollen grains**, which develop into the mature, sperm-bearing male gametophytes. Pollen grains hitch rides on air currents, insects, birds, and so on; they do not require free-standing water to reach the eggs. In this respect they differ greatly from algae. The evolution of pollen grains contributed to the successful radiation of seed-bearing plants into high and dry habitats.

Seed production also was adaptive in drier habitats. Female gametophytes (and eggs) of seed-bearing plants form inside nutritive tissues and a jacket of cell layers. Each **seed** consists of an embryo sporophyte, nutritive tissues, and a protective coat (which develops from the jacket). Seeds withstand hostile conditions. It was no coincidence that seed plants rose to dominance during Permian times, when shifts in climate were extreme.

Before turning to the spectrum of diversity among plants, take a look at Figure 22.3. You can use it as a map for tracking the branching evolutionary roads.

The plant kingdom includes multicelled, photosynthetic species called bryophytes, seedless vascular plants, and seed-bearing vascular plants. Most of these live on land.

In most lineages, structural adaptations to life on land included root and shoot systems, waxy cuticles, stomata, vascular tissues, and lignin-reinforced tissues.

Sporophytes with well-developed roots, stems, and leaves came to dominate the life cycle of most land plants. Parts of these complex sporophytes nourish and protect fertilized eggs and embryos through unfavorable conditions.

Some plants started producing two types of spores, not one. This led to the evolution of male gametes adapted for dispersal without liquid water and to the evolution of seeds.

Today, the bryophyte lineage consists of about 18,600 species called **mosses**, **liverworts**, and **hornworts**. These nonvascular plants are mostly well adapted to grow in fully or seasonally moist habitats, although you will find some mosses growing in deserts and on windswept plateaus of Antarctica. Mosses especially are sensitive to air pollution. Where air quality is poor, mosses are few or absent.

All known bryophytes are less than twenty centimeters (eight inches) tall. They have leaflike, stemlike, and rootlike parts, but these don't contain xylem or phloem. Like lichens and some algae, the bryophytes can dry out, then revive after absorbing moisture. Most have rhizoids, which are elongated

cells or threadlike structures that attach gametophytes to the soil and serve as absorptive structures.

Bryophytes are the simplest plants to display three features that evolved early in land plants. *First*, a cuticle prevents water loss from aboveground parts. *Second*, a cellular jacket around the parts that produce sperm and eggs holds in moisture. *Third*, of all plants, bryophytes alone have large gametophytes that do not depend on sporophytes for nutrition. Instead, embryo sporophytes start to develop inside gametophyte tissues—and even at maturity, they are *attached to* the gamete-producing body and still gain some nutritional support from it.

With 10,000 species, mosses are the most common bryophytes. Gametophytes of some species grow in clusters and form low, cushiony mounds (Figure 22.4*a*). Those of others commonly grow in branched, feathery patterns on tree trunks and branches in humid climates. Eggs and sperm develop in tiny, jacketed vessels at the

Figure 22.4 (**a**) Moss-covered rocks near a small stream. (**b**,**c**) Photograph and life cycle of a common bryophyte, a moss (*Polytrichum*). The moss sporophyte remains attached to the gametophyte and depends on it for water and nutrients.

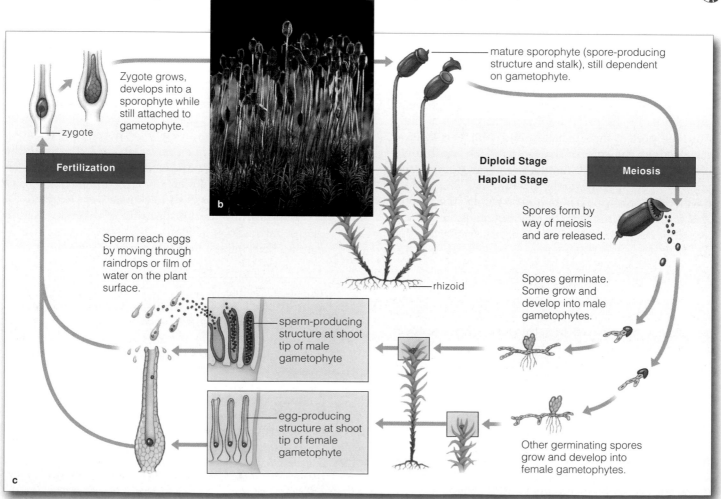

Zygote grows, develops into a sporophyte while still attached to gametophyte.

zygote

Fertilization

mature sporophyte (spore-producing structure and stalk), still dependent on gametophyte.

Diploid Stage

Haploid Stage

Meiosis

Spores form by way of meiosis and are released.

Sperm reach eggs by moving through raindrops or film of water on the plant surface.

rhizoid

Spores germinate. Some grow and develop into male gametophytes.

sperm-producing structure at shoot tip of male gametophyte

egg-producing structure at shoot tip of female gametophyte

Other germinating spores grow and develop into female gametophytes.

Figure 22.5 (**a**) A peat bog in Ireland. This family is cutting blocks of peat, and stacking them to dry, as a fuel source for their home. Nearly all of the peat harvested and dried in Ireland and elsewhere is burned to generate electricity in power plants. Compared with coal burning, peat fires generate fewer air pollutants. Every so often, peat harvesters come across well-preserved bodies of humans that lived 2,000 to 3,000 years ago; the high acidity prevented decomposition. (**b**) Gametophyte of a peat moss (*Sphagnum*) with a few sporophytes attached. The sporophytes are the brown, jacketed structures on white stalks.

shoot tips of gametophytes. Sperm swim through water on plant parts to reach eggs. After fertilization, zygotes give rise to sporophytes, each consisting of a stalk and a jacketed structure in which spores will develop.

Figure 22.5 shows one of 350 kinds of peat mosses (*Sphagnum*). Whereas most bryophytes grow slowly, the peat mosses grow fast enough to yield twelve metric tons of organic matter per hectare, which is about twice as much as corn plants yield. They soak up five times as much water as cotton does, owing to large, dead cells in leaflike parts. They also produce acids that inhibit growth of bacterial and fungal decomposers. Their remains accumulate as compressed, excessively moist mats known as **peat bogs**. In cold and temperate regions, peat bogs cover an area equal to one-half of the United States. Only the most acid-tolerant plants, such as cranberries, blueberries, and Venus flytraps, can grow in the bogs, which can be as acidic as vinegar.

Bryophytes are nonvascular plants with flagellated sperm that require liquid water to reach and fertilize the eggs.

A sporophyte of these plants develops within gametophyte tissues. It remains attached to the gametophyte and receives some nutritional support from it.

WWW

22.3 **ANCIENT CARBON TREASURES**

Three hundred million years ago, about halfway through the Carboniferous, mild climates prevailed and swamp forests carpeted the wet lowlands of the continents. The absence of pronounced seasonal swings in temperature favored plant growth through much of the year. The plants having lignin-reinforced tissues and well-developed root and shoot systems had the competitive edge under those growth conditions, and some of them evolved into giants. Massive-stemmed lycophyte trees—giant club mosses—topped out at nearly forty meters. Horsetails, including species of *Calamites*, were close to twenty meters tall. Often, stems that grew fast from spreading underground rhizomes formed dense thickets.

This was a period when sea levels rose and fell fifty times. When seas receded, swamp forests flourished. When seas moved back in, forest plants were submerged and became buried in sediments that protected them from decay. Gradually the sediments compressed the saturated, undecayed remains into what we now call peat. As more sediments accumulated, increased heat and pressure made the peat even more compact. It became **coal** (Figure 22.6).

Coal has a high percentage of carbon; it is energy-rich. It is a premier "fossil fuel." It took a fantastic amount of photosynthesis, burial, and compaction to form each large seam of coal. It has taken us only a few centuries to deplete many of the world's known coal deposits.

Often you will hear about annual "production rates" for coal or some other fossil fuel. But how much do we really produce each year? None. We simply *extract* it from the Earth. Coal is a nonrenewable source of energy.

Figure 22.6 Reconstruction of a Carboniferous swamp forest. The boxed inset shows part of a seam of coal.

EXISTING SEEDLESS VASCULAR PLANTS

Figure 22.7a shows one of the early seedless vascular plants. Descendants of certain lineages are still with us; we call them **whisk ferns**, **lycophytes**, **horsetails**, and **ferns**. Like their ancestors, they differ from bryophytes in three key respects. The sporophyte does not remain attached to a gametophyte; it has true vascular tissues; and it is the larger, longer lived phase of the life cycle.

Most seedless vascular plants live in wet, humid places, and their gametophytes lack vascular tissues. Water droplets clinging to the plants are the only means by which flagellated sperm can reach the eggs. The few species in dry habitats reproduce sexually during brief, seasonal pulses of heavy rains. In a sense, whisk ferns, lycophytes, horsetails, and ferns are the "amphibians" of the plant kingdom. They have not fully escaped the aquatic habitats of their ancestors.

Whisk Ferns

Whisk ferns (Psilophyta), which are not ferns, resemble whisk brooms. Florist suppliers commonly cultivate them in Hawaii, Texas, Louisiana, Florida, Puerto Rico, and other tropical or subtropical regions. One genus,

Psilotum, is a unique vascular plant, for its sporophytes have no roots or leaves. The photosynthetic, branched stems have scalelike projections and, internally, xylem and phloem (Figure 22.7b). Belowground are **rhizomes**, branching, short, mostly horizontal absorptive stems.

Lycophytes

About 350 million years ago, lycophytes (Lycophyta) included tree-sized members of swamp forests. About 1,100 far tinier species exist today. The most familiar are club mosses, members of communities in the Arctic, the tropics, and regions in between. Many form mats on forest floors. One type, called the resurrection plant, is common in Texas, New Mexico, and Mexico.

Sporophytes of most club mosses have leaves and a branching rhizome that gives rise to vascularized roots and stems. Some have nonphotosynthetic, cone-shaped leaf clusters with spore-producing structures (Figure 22.7c). Each cluster is a **strobilus** (plural, strobili). After spores disperse, they germinate and develop into small, free-living gametophytes. *Selaginella* is heterosporous; two kinds of spores develop in the same strobilus.

Horsetails

Tree-sized sphenophytes (Sphenophyta) flourished in ancient swamp forests. Twenty-five or so smaller species of one genus, *Equisetum*, made it to the present. These are the horsetails. Their body plan has changed little over the past 300 million years.

Horsetails thrive in streambank muds, vacant lots, roadsides, and other disrupted habitats. Figure 22.7d–f shows the vegetative, photosynthetic stems and fertile stems of one species. Its spores give rise to free-living

Figure 22.7 (**a**) *Cooksonia*, one of the earliest known vascular plants, no more than a few centimeters tall. It probably grew in mud flats. Its upright, branching stems had a cuticle. Its spores formed in sporangia at stem tips. Compare Figure 18.2c. (**b**) Sporophytes of a whisk fern (*Psilotum*), a seedless vascular plant. Pumpkin-shaped, spore-producing structures form at the ends of stubby branchlets. (**c**) Sporophyte of one of the lycophytes (*Lycopodium*). (**d**) Vegetative stem of *Equisetum*, which resembles a horsetail. (**e**) Fertile, nonphotosynthetic stems of *Equisetum*. (**f**) Closer look at the spore-producing structure of a fertile stem.

strobilus, an aggregation of spore-producing structures, at tip of a vegetative shoot of the horsetail sporophyte

f Each petal-shaped structure of a strobilus contains many spores that formed by way of meiotic cell divisions.

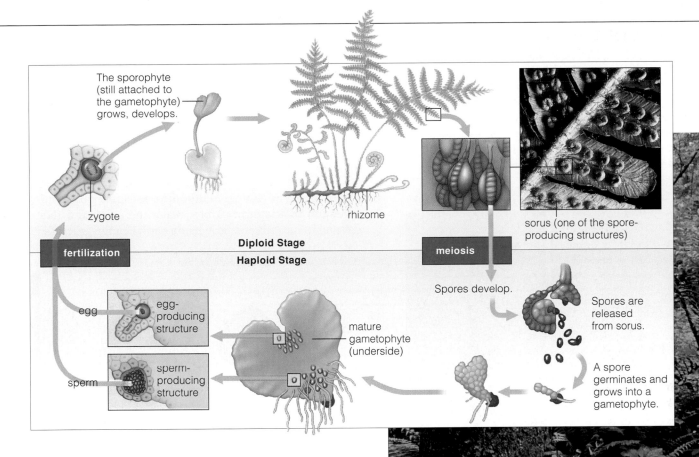

The sporophyte (still attached to the gametophyte) grows, develops.

zygote

rhizome

sorus (one of the spore-producing structures)

Diploid Stage

fertilization

Haploid Stage

meiosis

Spores develop.

egg

egg-producing structure

sperm

sperm-producing structure

mature gametophyte (underside)

Spores are released from sorus.

A spore germinates and grows into a gametophyte.

gametophytes 1 millimeter to 1 centimeter across. The sporophytes of most horsetails have rhizomes, hollow photosynthetic stems, and scale-shaped leaves. Strands of xylem and phloem are arrayed in a ring in the stems. Silica-reinforced ribs structurally support the stems and give them a texture like sandpaper. Pioneers of the American West, who did not have many places to store and wash towels, gathered horsetails on their westward journeys and used them as pot scrubbers.

Ferns

With 12,000 or so species, the ferns (Pterophyta) are the largest and most diverse group of seedless vascular plants. All but about 380 are native to the tropics, but they are popular houseplants all over the world. Their size range is stunning. Some floating species are less than 1 centimeter across. Some tropical tree ferns are 25 meters (82 feet) tall. One climbing fern has a modified leaf stalk about 30 meters long.

Most ferns have vascularized rhizomes that give rise to roots and leaves. Exceptions include tropical tree ferns and epiphytes. (*Epiphyte* refers to any aerial plant that grows attached to tree trunks or branches.) While they develop, young fern leaves are coiled, rather like a fiddlehead. At maturity, the leaves (fronds) commonly are divided into leaflets.

You may have noticed rust-colored patches on the lower surface of many fern fronds. Each patch, a cluster

Figure 22.8 Life cycle of a fern. The photograph shows ferns growing in a moist habitat in Indiana. Ferns with finely divided fronds are in the foreground.

of spore-producing structures, is called a sorus (plural, sori). At dispersal time, each sorus snaps open with such force that the released spores catapult through the air. Each spore that germinates develops into a small gametophyte. The green, heart-shaped gametophyte shown in Figure 22.8 is one example.

Seedless vascular plants (whisk ferns, lycophytes, horsetails, and ferns) have sporophytes adapted to conditions on land. Yet they have not entirely escaped their aquatic ancestry. When they reproduce sexually, their flagellated sperm cannot reach the eggs unless liquid water is clinging to the plant.

WWW

RISE OF THE SEED-BEARING PLANTS

About 360 million years ago, as the Devonian gave way to the Carboniferous, the first seed-bearing plants arose. In terms of diversity, numbers, and distribution, they would become the most successful groups of the plant kingdom. Seed ferns, gymnosperms, and (much later) angiosperms were the dominant groups. They differed from seedless vascular plants in three crucial respects.

First, all seed-bearing plants produce pollen grains, a type of sperm-bearing male gametophyte. Remember, these plants produce not one but two types of spores. As shown in Figure 22.9, their **microspores** give rise to pollen grains.

Like a suitcase, a pollen grain is a means of getting its contents (the sperm) to the eggs even during times of prolonged drought. Seedless vascular plants do not have such an advantage; without predictable rains and moisture, their sperm cannot reach the eggs, and this obviously has adverse effects on reproductive success.

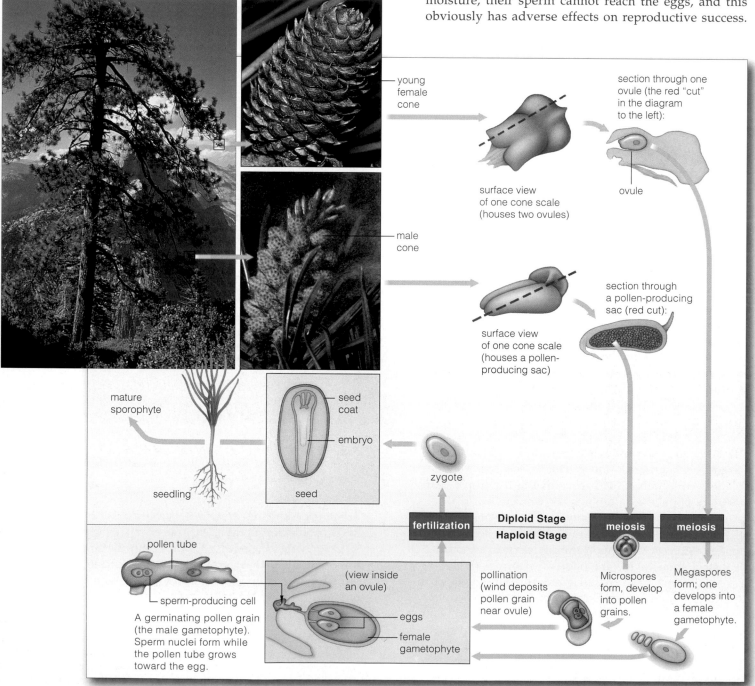

Figure 22.9 Life cycle of one of the conifers, the ponderosa pine (*Pinus ponderosa*).

By contrast, pollen grains of gymnosperms simply drift with air currents. Some angiosperm pollen grains do the same, but many others are loaded onto insects, birds, bats, and other animals that truck them to the eggs. **Pollination** is the name for the arrival of pollen grains on the female reproductive structures. By this process, seed-bearing plants escaped dependence on free water for fertilization.

Second, besides microspores, seed-bearing plants also produce **megaspores**. These develop within **ovules**, the female reproductive structures which, at maturity, are seeds (Figure 22.9). Each ovule consists of a female gametophyte (with its egg cell), nutrient-rich tissue, and a jacket of cell layers which, recall, develops into the seed coat. A zygote will form inside the ovule when a sperm reaches the egg and then fertilizes it. An embryo sporophyte will develop, and when the time comes to leave the parent plant, the coat around the seed will afford protection for the journey.

Third, compared with the seedless vascular plants, gymnosperms had water-conserving traits, including thick cuticles and stomata recessed below the surface of the leaf. These traits gave competitive advantages in drier, cooler environments. Such environments were ushered in after the Carboniferous period gave way to the Permian. Swamplands disappeared and the cycads, conifers, and other gymnosperms rose to dominance.

Who among us hasn't noticed the woody, shelflike scales of "a pine cone"? The scales are parts of a mature female cone, which bears ovules in which megaspores formed and developed into female gametophytes. Pine trees also produce male cones, in which microspores form and develop into pollen grains (Figure 22.9). Each spring, millions of pollen grains drift from male cones. Pollination is completed when some land on ovules of female cones. After each pollen grain germinates, cell divisions and enlargements transform it into a tubular structure. The germinating pollen grain is the sperm-bearing male gametophyte. It grows toward the egg in the female gametophyte. For pines, fertilization occurs months or a year after pollination.

As with other gymnosperms, seed formation begins at the ovule (Figure 22.9). An embryo sporophyte starts developing from the fertilized egg. The outer layers of the jacket around the female gametophyte and embryo mature into a hard coat. The seed coat will protect the embryo sporophyte after it is dispersed from the parent plant. Nutrients inside will help it through the critical time of germination, before its roots and shoots become fully functional.

Seed-bearing plants rely on pollen grains, ovules that mature into seeds, and tissue changes adapted to dry conditions.

WWW

22.6   GOOD-BYE, FORESTS

Conifers dominated many habitats on land during the Mesozoic, but their slow reproductive pace put them at a competitive disadvantage when the flowering plants began their great adaptive radiation (Section 19.6). Coniferous forests still predominate in the far north, at higher elevations, and in some parts of the Southern Hemisphere. However, existing conifers face more than competition with flowering plants for resources. Now they are vulnerable to **deforestation**: the removal of all trees from large tracts, as by clear-cutting (Figure 22.10). Conifers just have the bad luck to be premier sources of lumber, paper, and other wood products required in human societies. We return to this topic in Chapter 43.

For now, keep this in mind: Where flowering plants flourish, the conifers are at a competitive disadvantage, partly because they take so long to reproduce. Rampant deforestation isn't helping them one bit, either.

Figure 22.10 Clear-cut peaks in Alaska. This is not an isolated example. In the early 1980s, 400 million board feet of timber were being cut in Washington's Olympic Peninsula every year. In Arkansas, about one-third of the Ouachita National Forest has been clear-cut. Its once-diverse forest communities have been replaced by "tree farms" of a single species of pine. Throughout the world, huge tracts of land that were deforested years ago still show no signs of recovery.

WWW

GYMNOSPERM DIVERSITY

With a bit of history behind us, we turn now to a survey of some existing gymnosperms. Unlike the seeds of flowering plants, which are enclosed in a reproductive chamber (an ovary), gymnosperm seeds grow, in an exposed location, on top of a spore-producing structure. (*Gymnos* means naked; *sperma* is taken to mean seed.)

Conifers are woody trees and shrubs that produce scalelike or needlelike leaves and bear seeds exposed on cone scales. (Cones are clusters of modified leaves around spore-producing structures.) Most conifers shed some leaves all year but stay leafy, or *evergreen*. A few are *deciduous*; they shed their leaves all at once in cold, dry seasons. The most abundant trees (pines), the tallest (redwoods), and oldest (bristlecone pine, Figure 22.11*a*) are all conifers. One bristlecone pine sprouted when Egyptians built the Great Sphinx, 4,725 years ago. This group also includes firs, spruces, yews, and junipers.

The 185 or so existing **cycad** species form pollen-bearing and seed-bearing cones on separate plants. Insects or air currents transfer pollen from "male" to

Figure 22.11 (**a**) Bristlecone pine (*Pinus longaeva*) growing high in the Sierra Nevada. (**b**) A cycad's seed-bearing cone.

"female" plants. Cycad leaves resemble a palm tree's (Figure 22.11*b*), but palms are flowering plants. Most cycads live in tropical and subtropical regions. Two species (*Zamia*) grow wild in Florida and are widely planted as ornamentals. Cycad seeds and a flour made from the trunks are edible after their toxic alkaloids are removed. Many species are vulnerable to extinction.

Ginkgos were diverse in dinosaur times. The only surviving species is the maidenhair tree, *Ginkgo biloba*. Like a few gymnosperms, these plants are deciduous. Several thousand years ago, they were planted around temples in China. Natural populations nearly became extinct even though the trees seem hardier than many others. Perhaps too many were cut down for firewood. Today, male ginkgo trees are widely planted. They have attractive, fan-shaped leaves and are resistant to insects, disease, and air pollutants. Female trees are not favored. Their thick, fleshy seeds, which are the size of small plums, give off an awful stench (Figure 22.12).

At present, there are three genera of woody plants called **gnetophytes**. *Gnetum* trees and leathery leafed vines thrive in the humid tropics. The shrubby *Ephedra* lives in California deserts and some other arid regions (Figure 22.12*d*). Photosynthesis occurs in its green stems. *Welwitschia mirabilis* grows in hot deserts of south and west Africa. Its sporophyte is mainly a deep-reaching taproot. Its exposed part, a woody, disk-shaped stem, has cones and one or two strap-shaped leaves that split lengthwise repeatedly as the plant ages (Figure 22.12*e*).

Figure 22.12 (**a**) Ginkgo and (**b**) its fleshy-coated seeds. (**c**) Ginkgo leaf compared with a fossilized leaf, 65 million years old, from an ancestor. (**d**) *Ephedra*. Its striking branches are popular with florists. (**e**) *Welwitschia mirabilis*.

Gymnosperms include the conifers, cycads, ginkgo, and gnetophytes. Like their ancestors, they bear seeds on exposed surfaces of cones and other spore-producing structures.

WWW

ANGIOSPERMS—THE FLOWERING, SEED-BEARING PLANTS

Only the angiosperms produce specialized reproductive structures called **flowers** (Figure 22.13). *Angeion*, which means vessel, refers to the female reproductive parts at the center of a flower. The enlarged base of the "vessel" is the floral ovary, where ovules and seeds develop.

Most flowering plants coevolved with **pollinators**— insects, bats, birds, and other animals that withdraw nectar or pollen from a flower and, in so doing, transfer pollen to its female reproductive parts. The recruitment of animals as assistants in reproduction probably has contributed to the success of flowering plants, which have dominated the land for 100 million years.

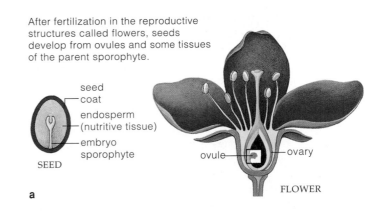

After fertilization in the reproductive structures called flowers, seeds develop from ovules and some tissues of the parent sporophyte.

seed coat
endosperm (nutritive tissue)
embryo sporophyte

SEED

ovule — ovary

a

FLOWER

Figure 22.13 (**a**) Unique to angiosperms— the flower, a reproductive structure that has roles in pollination and seed formation. (**b**) A hummingbird pollinator sipping nectar from a passion flower. Representing angiosperm diversity: (**c**) The water lily (*Nymphaea*), one of a few that live in water. (**d**) Dwarf mistletoe (*Arceuthobium*), a parasitic plant, limits the growth of forest trees in the western United States. (**e**) Indian pipe (*Monotropa uniflora*) one of the rare nonphotosynthetic species. It withdraws nutrients from mycorrhizae on the roots of photosynthetic plants.

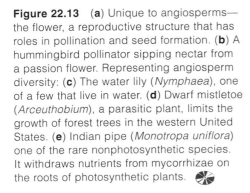

At least 260,000 species live in a variety of habitats. They range in size from tiny duckweeds (a millimeter or so long) to towering *Eucalyptus* trees (some are more than 100 meters tall). A few species, including mistletoes and Indian pipe, are not photosynthetic. They withdraw nutrients directly from other plants or from mycorrhizae.

There are two classes of flowering plants, called the **dicots** and **monocots** (more formally, the Dicotyledonae and Monocotyledonae). Among the 180,000 dicots are most herbaceous (nonwoody) plants, such as cabbages and daisies; most flowering shrubs and trees, such as oaks and apple trees; water lilies; and cacti. Among 80,000 or so species of monocots are the orchids, palms, lilies, and grasses, including rye, sugarcane, corn, rice, and wheat, as well as many other highly valued crop plants.

The next unit deals with the structure and function of flowering plants. For now, simply start thinking about how a large sporophyte dominates the life cycles. It retains and nourishes gametophytes; its sperm are dispersed within pollen grains. Endosperm, a nutritive tissue, surrounds embryo sporophytes inside the seeds of flowering plants. As the seeds develop, the ovaries (along with other structures) mature into fruits. Fruits protect and help disperse embryos. Figure 23.14, in the next section, is an overview of these life cycle events.

Angiosperms are the most successful plants in terms of their diversity, numbers, and distribution. They alone produce flowers. Most species coevolved with animal pollinators.

WWW

VISUAL OVERVIEW OF FLOWERING PLANT LIFE CYCLES

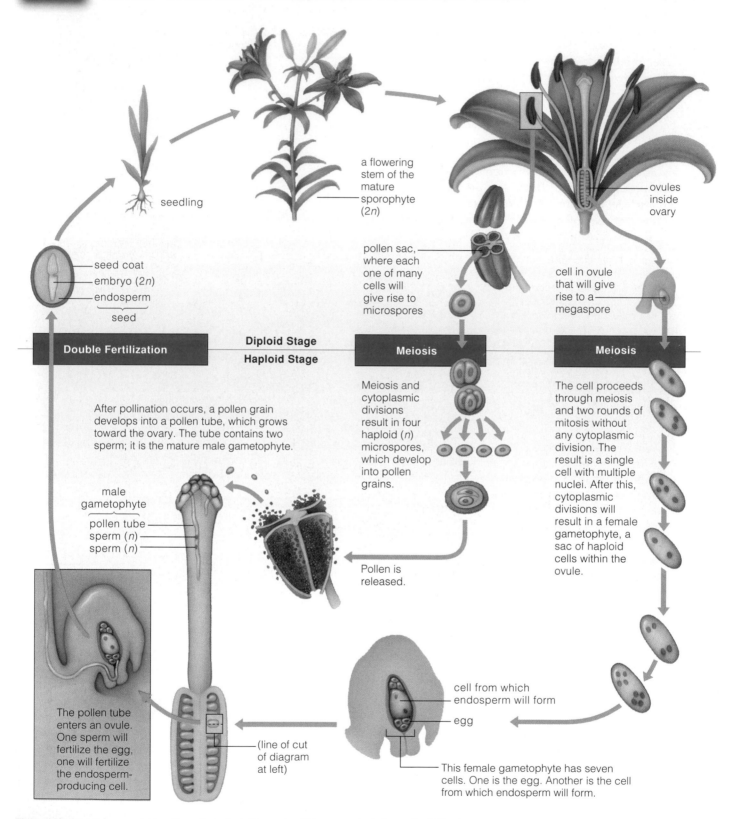

seedling

a flowering stem of the mature sporophyte (2n)

ovules inside ovary

pollen sac, where each one of many cells will give rise to microspores

cell in ovule that will give rise to a megaspore

seed coat
embryo (2n)
endosperm
seed

Double Fertilization	Diploid Stage	Meiosis	Meiosis
	Haploid Stage		

After pollination occurs, a pollen grain develops into a pollen tube, which grows toward the ovary. The tube contains two sperm; it is the mature male gametophyte.

Meiosis and cytoplasmic divisions result in four haploid (n) microspores, which develop into pollen grains.

The cell proceeds through meiosis and two rounds of mitosis without any cytoplasmic division. The result is a single cell with multiple nuclei. After this, cytoplasmic divisions will result in a female gametophyte, a sac of haploid cells within the ovule.

male gametophyte
pollen tube
sperm (n)
sperm (n)

Pollen is released.

The pollen tube enters an ovule. One sperm will fertilize the egg, one will fertilize the endosperm-producing cell.

(line of cut of diagram at left)

cell from which endosperm will form

egg

This female gametophyte has seven cells. One is the egg. Another is the cell from which endosperm will form.

Figure 22.14 Representative flowering plant life cycle. This example is for a lily (*Lilium*), one of the monocots. "Double" fertilization is a distinctive feature of flowering plant life cycles. A male gametophyte delivers two sperm to an ovule. One sperm fertilizes the egg, and the other fertilizes a cell that gives rise to endosperm, a tissue that will nourish the forthcoming embryo. Section 27.3 provides a closer look at flowering plant life cycles, using a dicot as the example.

22.10 SEED PLANTS AND PEOPLE

Which plants provide taste thrills and which kill? Starting with trials and errors of the earliest human species, we started collecting intimate knowledge of plants. Certainly by 300,000 years ago, *Homo erectus* clans in China were stashing pine nuts, walnuts, hazelnuts, and rose hips in caves and roasting seeds. At least by then, grown-ups were teaching children which plants are edible and which are toxic. Through language, youngsters learned about the plants that had evolved in their parts of the world.

About 11,000 years ago, people started to domesticate wheat, barley, and other plants, this being a way to count on having reliable quantities of food. Of an estimated 3,000 species that different populations recognized as food, only about 200 became *the* major crops (Figure 22.15).

Plant lore still threads through our lives in other ways. We learned to use fast-growing, soft-wooded conifers for lumber and paper—and slow-growing, hard-wooded flowering plants, such as cherry, maple, and mahogany, for fine furniture. We make twine and rope from century plant leaves (*Agave*), cords and textiles from the leaf fibers of Manila hemp, and thatched roofs from grasses and palm fronds. Insecticides derived from Mexican cockroach plants kill cockroaches, fleas, lice, and flies. Extracts of neem tree leaves kill nematodes, insects, and mites but not the natural predators of these common pests.

Flowers grace our homes and customs (Figure 22.16). Their oils impart scents to perfumes. Oils of eucalyptus and camphor are used medicinally. Digitalin extracted from foxglove (*Digitalis purpurea*) stabilizes the heartbeat and blood circulation. Juices from *Aloe vera* leaves soothe sun-damaged skin. Alkaloids from periwinkle leaves can slow the growth of some cancer cells.

And have we, in all this time, learned to abuse plants, also? You bet. Section 25.4 gives a few choice examples.

Figure 22.16 A bride tossing her bouquet to single women who attended her wedding, a ritual that supposedly reveals who will be married next. (A sixty-eight-year-old optimist caught this one.)

Figure 22.15 A few of the prized flowering plants. (**a**) Indonesians gathering tender shoots of tea plants, evergreen shrubs related to camellias. Plants growing on hillsides in moist, cool regions yield leaves with the best flavors. Only the terminal bud and two or three of the youngest leaves are picked for fine teas. (**b**) From the American Midwest, mechanized harvesting of a field of common bread wheat, *Triticum*. (**c**) Triticale, a popular hybrid grain; parental stocks are wheat and rye (*Secale*). It has wheat's high yield and rye's tolerance of harsh environmental conditions. (**d**) From Hawaii, sugarcane (*Saccharum officinarum*). Wild stock of this cultivated species may have evolved in New Guinea. Sap extracted from its cut stems is boiled down to make sucrose crystals (table sugar) or syrups. (**e**) Fruits of *Theobroma cacao*. Each one holds up to forty seeds, which are processed into cocoa butter or chocolate essences. The average American who buys 8–10 pounds of chocolate per year might not be happy that a pathogenic fungus (*Monilia*) may be driving *T. cacao* to extinction.

SUMMARY

1. Green algae probably gave rise to plants, which had invaded the land by 435 million years ago. Nearly all species of plants are multicelled photoautotrophs. Table 22.1 summarizes and compares the major phyla.

2. Several trends in plant evolution may be identified by comparing different lineages (see also Table 22.2):

    a. Structural adaptations to dry conditions, such as vascular tissues (xylem and phloem).

    b. A shift from haploid to diploid dominance during the life cycle. Complex sporophytes evolved; they hold on to, nourish, and protect spores and gametophytes.

    c. A shift from one to two spore types (homospory to heterospory) that, among gymnosperms and flowering plants, led to the evolution of pollen grains and seeds.

3. Mosses, liverworts, and hornworts are bryophytes, nonvascular plants that have no well-developed xylem or phloem and that require free water for fertilization.

Table 22.1	Comparison of Major Plant Groups

Nonvascular land plants. Fertilization requires free water. Haploid dominance. Cuticle, stomata present in some.

BRYOPHYTES	18,600 species. Moist, humid habitats.

Seedless vascular plants. Fertilization requires free water. Diploid dominance. Cuticle, stomata present.

WHISK FERNS	7 species, sporophytes with no obvious roots or leaves. *Psilotum.*
LYCOPHYTES	1,100 species with simple leaves. Mostly wet or shady habitats.
HORSETAILS	25 species of single genus. Swamps, disturbed habitats.
FERNS	12,000 species. Wet, humid habitats in mostly tropical, temperate regions.

Gymnosperms—vascular plants with "naked seeds." Free water not required for fertilization. Diploid dominance. Cuticle, stomata present.

CONIFERS	550 species, mostly evergreen, woody trees and shrubs having pollen- and seed-bearing cones. Widespread distribution.
CYCADS	185 slow-growing species. Tropics, subtropics.
GINKGO	1 species, a tree with fleshy-coated seeds.
GNETOPHYTES	70 species. Limited to some deserts, tropics.

Angiosperms—vascular plants with flowers and protected seeds. Free water not required for fertilization. Diploid dominance. Cuticle, stomata present.

FLOWERING PLANTS

Monocots	80,000 species. Floral parts often arranged in threes or in multiples of three; one seed leaf; parallel leaf veins common.
Dicots	At least 180,000 species. Floral parts often arranged in fours, fives, or multiples of these; two seed leaves; net-veined leaves common.

Table 22.2	Evolutionary Trends Among Plants

Bryophytes	Ferns	Gymnosperms	Angiosperms
Nonvascular ⟶	Vascular ——————————————⟶		
Haploid dominance ⟶	Diploid dominance ——————————⟶		
Spores of one type ——⟶	Spores of two types ——————⟶		
Motile gametes ———————————⟶		Nonmotile gametes* ⟶	
Seedless ——————————⟶	Seeds ——————————⟶		

* Require pollination by wind, insects, animals, etc.

4. Nearly all vascular plants are adapted to land. Their cuticle and stomata conserve water. Root systems mine soil for nutrients. Upright and branched growth patterns of shoot systems intercept sunlight and carbon dioxide. Tissues enclose and protect spores and gametes.

5. The seedless vascular plants include the whisk ferns, lycophytes, horsetails, and ferns. Their flagellated sperm require ample water to swim to the eggs.

6. Gymnosperms and flowering plants (angiosperms) are vascular plants. Both produce pollen grains (mature microspores that develop into male gametophytes) and megaspores (which give rise to female gametophytes).

    a. Megaspores develop within ovules: reproductive structures that contain (1) the egg-producing female gametophytes, (2) the precursor of nutritive tissue, and (3) a jacket of cell layers, the outer portion of which will develop into the seed coat. A mature ovule is a seed.

    b. The evolution of pollen grains freed these plants from dependence on water for fertilization. Their seeds are efficient means of dispersing new generations even during hostile conditions. Pollen grains and seeds were key adaptations in the move to high and dry habitats.

7. Only angiosperms produce flowers. Most coevolved with pollinators, such as insects, which enhance the transfer of pollen grains to female reproductive parts. Their seeds contain a nutritive tissue (endosperm) and are usually surrounded by fruit, which aids in dispersal.

Review Questions

1. Identify a few structural and reproductive modifications that helped plants invade and diversify in habitats on land. *22.1*

2. Does the haploid phase or diploid phase dominate the life cycles of most plants? *22.1*

3. Name representatives of the following groups of plants and then compare their main characteristics: (*also refer to Table 22.1*)

    a. Bryophytes and seedless vascular plants *22.2, 22.4*
    b. Gymnosperms and angiosperms *22.6–22.8*

4. Distinguish between:

    a. Root system and shoot system *22.1*
    b. Xylem and phloem *22.1*
    c. Sporophyte and gametophyte *22.1*
    d. Ovule and seed *22.1, 22.5*
    e. Microspore and megaspore *22.5*

Figure 22.17 Where many conifers end up.

Self-Quiz (*Answers in Appendix III*)

1. Which of the following statements is *not* true?
   a. Monocots and dicots are two classes of angiosperms.
   b. Bryophytes are nonvascular plants.
   c. Lycophytes and angiosperms are both vascular plants.
   d. Gymnosperms are the simplest vascular plants.

2. Of all land plants, bryophytes alone have independent
   _____ and attached, dependent _____ .
   a. sporophytes; gametophytes   c. rhizoids; zygotes
   b. gametophytes; sporophytes   d. rhizoids; stalked
                                       sporangia

3. Psilophytes, lycophytes, horsetails, and ferns are classified
   as _____ plants.
   a. multicelled aquatic   c. seedless vascular
   b. nonvascular seed   d. seed-bearing vascular

4. Which does *not* apply to gymnosperms and angiosperms?
   a. vascular tissues   c. single spore type
   b. diploid dominance   d. all of the above

5. A seed is _____ .
   a. a female gametophyte   c. a mature pollen tube
   b. a mature ovule   d. an immature embryo

6. Match the terms appropriately.
   ____ gymnosperm   a. produces haploid gametes
   ____ sporophyte   b. control water loss
   ____ seedless vascular c. "naked" seeds
        plant   d. protects, disperses embryo
   ____ ovary   sporophyte
   ____ bryophyte   e. produces haploid spores
   ____ gametophyte   f. nonvascular land plant
   ____ stomata   g. lycophyte
   ____ angiosperm seed   h. usually a fruit at maturity

Critical Thinking

1. Elliot Meyerowitz of the California Institute of Technology
has studied the genetic basis of flower formation in *Arabidopsis
thaliana*. By inducing mutations in seeds of this small weed, he
discovered three genes (call them *A*, *B*, and *C*) that interact in
different parts of a developing flower. Gene interactions lead to
the formation of different structures—sepals, petals, stamens,
and carpels—from the same mass of undifferentiated tissue.
From what you know of gene regulation, suggest ways in which
the *A*, *B*, and *C* genes might be controlling flower development.

2. Genes nearly identical to the *A*, *B*, and *C* genes of *A. thaliana*
also have been isolated from snapdragons and other flowering
plants. These plants originated by 150 million years ago and
quickly rose to dominance in nearly all land habitats.

Review Section 19.6, then speculate on how the spectacular
and rapid adaptive radiation of flowering plants came about.

3. Figure 22.17 shows a forest in the Nahmint Valley of British
Columbia, before and after logging. It also shows the wood
frames of homes that are in the process of being built. Reflect on
these photographs and Figure 22.10. To stop the loggers, would
you chain yourself to a tree in an old-growth forest scheduled
for clear-cutting? If your answer is yes, would you also give up
the chance of owning a wood-frame home (as most homes are
in developed countries)? What about forest products, including
newspapers, toilet tissue, and fireplace wood?

4. With respect to question 3, multiply each of your answers
by 6 billion (there are almost that many people in the world)
and describe what might happen when, inevitably, we run out
of trees. Also describe what you might consider to be some of
the pros and cons of tree farms of, say, a single species of pine.

Selected Key Terms

angiosperm *22.1*	hornwort *22.2*	pollinator *22.8*
bryophyte *22.1*	horsetail *22.4*	rhizome *22.4*
coal *22.3*	lignin *22.1*	root system *22.1*
conifer *22.7*	liverwort *22.2*	seed *22.1*
cuticle *22.1*	lycophyte *22.4*	shoot system *22.1*
cycad *22.7*	megaspore *22.5*	spore *22.1*
deforestation *22.6*	microspore *22.5*	sporophyte *22.1*
dicot *22.8*	monocot *22.8*	stoma (plural,
fern *22.4*	moss *22.2*	stomata) *22.1*
flower *22.8*	ovule *22.5*	strobilus *22.4*
gametophyte *22.1*	peat bog *22.2*	vascular plant *22.1*
ginkgo *22.7*	phloem *22.1*	whisk fern *22.4*
gnetophyte *22.7*	pollen grain *22.1*	xylem *22.1*
gymnosperm *22.1*	pollination *22.5*	

Readings *See also www.infotrac-college.com*

Gray, J., and W. Shear. September–October 1992. "Early Life
on Land." *American Scientist* 80: 444–456.

Moore, R., W. D. Clark, and K. Stern. 1995. *Botany*. Dubuque,
Iowa: W. C. Brown. Beautifully illustrated.

WWW *http://www.brookscole.com/biology*

Practice quiz questions, hypercontents, BioUpdates, and critical
thinking. The Brooks/Cole Biology Resource Center provides a
wealth of information fully organized and integrated by chapter.

23 ANIMALS: THE INVERTEBRATES

Madeleine's Limbs

In August of 1994, about 900 million years after the first animals appeared on Earth, Madeleine made *her* entrance. As they are wont to do, her grandmothers and aunts made a quick count on the sly—arms, legs, ears, and eyes, two of each; fully formed mouth and nose—just to be sure these were present and accounted for.

One grandmother, having been too long in the company of biologists, experienced an epiphany as she witnessed Madeleine's birth. In that profound instant she sensed ancestral connections, emerging from the distant past and through her, into the future.

Madeleine's body plan did not emerge out of thin air. Thirty-five thousand years ago, people just like us were having children just like Madeleine. And if we are interpreting the fossil record correctly, then five million years ago the offspring of individuals on the road to modern humans resembled her in some respects but not others. Sixty million years ago, primate ancestors of those individuals were giving birth precariously, up in the trees. Two hundred and fifty million years ago, mammalian ancestors of those primates were giving birth—and so on back in time to the very first animals, which had no limbs or eyes or noses at all.

We have very few clues to what those first animals looked like, but one thing is clear. By the dawn of the Cambrian, they had given rise to all major groups of invertebrates—animals without backbones—even to Madeleine's backboned but limbless ancestors.

And what stories those Cambrian animals tell! One bunch flourished 530 million years ago, in a submerged basin that had formed between a reef and the coast of an early continent. Protected from ocean currents, the sediments had piled up against the steep reef. About 500 feet below the surface, the water was oxygenated and clear. Small, well-developed animals lived in, on, and above the dimly lit, muddy sediments (Figure 23.1).

Like castles built from wet sand along a seashore, their living quarters were unstable. Part of the bank above the community slumped abruptly and obliterated it. The sediments from that underwater avalanche kept scavengers from reaching and removing all traces of the dead. Gradually through time, muddy silt rained down on the tomb. Increased pressure and chemical changes transformed the sediments into finely stratified shale, and the soft parts of the flattened animals became shimmering, mineralized films.

Sixty-five million years ago, part of the seafloor was plowing under the North American plate, and western Canada's mountain ranges were slowly rising. By 1909, the fossils had traveled high into the eastern mountains of British Columbia. In that year a fossil hunter tripped over a chunk of shale, which split apart into thin, fine layers—and so the Burgess Shale story came to light.

In this chapter and the next, you will be comparing body plans of different groups of animals. Such comparisons give insight into evolutionary relatedness and help us to construct family trees, such as the one in Figure 23.2. Don't assume that structurally simple animals of the most ancient lineages are somehow primitive or evolutionarily stunted. As you will see, they, too, are exquisitely adapted to their environment.

Figure 23.1 Reconstruction of a few Cambrian animals, known from fossils of the Burgess Shale, British Columbia.

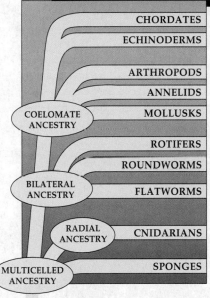

	CHORDATES
	ECHINODERMS
	ARTHROPODS
	ANNELIDS
COELOMATE ANCESTRY	MOLLUSKS
	ROTIFERS
	ROUNDWORMS
BILATERAL ANCESTRY	FLATWORMS
RADIAL ANCESTRY	CNIDARIANS
MULTICELLED ANCESTRY	SPONGES

SINGLE-CELLED, PROTISTAN-LIKE ANCESTORS

Figure 23.2
Evolutionary tree diagram showing the presumed relationship among major groups of animals. Take a moment to study this family tree. We will use it repeatedly as a road map through our discussions of each group, starting with the invertebrates and ending with the chordates, including young Madeleine (*above*).

As you poke through the branches of the animal family tree, keep the greater evolutionary story in mind. At each branch point, microevolutionary processes gave rise to workable changes in body plans. Madeleine's uniquely human traits, and yours, emerged through modification of certain traits that had evolved earlier in countless generations of vertebrates and, before them, in ancient invertebrate forms.

KEY CONCEPTS

1. All animals are multicelled, aerobic heterotrophs that ingest or parasitize other organisms. Nearly all kinds have tissues, organs, and organ systems, and most are motile during at least part of their life cycle. Animals reproduce sexually and often asexually, and their embryos develop in a series of continuous stages.

2. Animals originated approximately 900 million years ago. More than 2 million existing species have been identified. Of these, more than 1,950,000 are invertebrates (animals with no backbone). Fewer than 50,000 species are vertebrates (animals with a backbone).

3. Comparisons of the body plans of existing animals, in conjunction with the fossil record, reveal that there were several trends in the evolution of certain lineages. The most revealing aspects of an animal's body plan are its type of symmetry, gut, and cavity (if any) between the gut and body wall; whether it has a distinct head end; and whether it is divided into a series of segments.

4. The placozoans and sponges are structurally simple animals with no body symmetry. Both are at the cellular level of body construction. Cnidarians, which have radial symmetry, are at the tissue level of body construction.

5. Flatworms, roundworms, rotifers, and nearly all other animals that are more complex than the cnidarians show bilateral symmetry. They consist of tissues, organs, and organ systems.

6. Not long after the flatworms evolved, divergences gave rise to two major lineages. One evolutionary branching gave rise to the mollusks, annelids, and arthropods. The other gave rise to the echinoderms and chordates.

7. By biological measures, including diversity, sheer numbers, and distribution, the arthropods—especially insects—have been the most successful animal group.

OVERVIEW OF THE ANIMAL KINGDOM

General Characteristics of Animals

What, exactly, are **animals**? We can only define them by a list of general characteristics, not with a sentence or two. *First*, animals are multicelled, and in most cases their body cells form tissues that become arranged as organs and as organ systems. The body cells of nearly all species have a diploid chromosome number. *Second*, animals are heterotrophs that obtain carbon and energy by ingesting other organisms or by absorbing nutrients from them. *Third*, animals require oxygen, for use in aerobic respiration. *Fourth*, animals reproduce sexually and, in many cases, asexually. *Fifth*, most animals are motile during at least part of the life cycle. *Sixth*, their life cycles include stages of embryonic development. In brief, mitotic cell divisions transform the animal zygote into a multicelled embryo. The embryonic cells are the forerunners of **ectoderm**, **endoderm**, and, in nearly all species, **mesoderm**. These are the primary tissue layers which give rise to all tissues and organs of the adult, as described in Sections 28.6 and 38.2.

Diversity in Body Plans

Mammals, birds, reptiles, amphibians, and fishes are the most familiar animals. All are **vertebrates**, the only animals with a "backbone." And yet, of probably more than 2 million species of animals, fewer than 50,000 are vertebrates! What we call the **invertebrates** are animals with plenty of diverse features, but not a backbone.

We group the animals into more than thirty phyla. Table 23.1 lists the groups described in this book. The characteristics they share with one another arose early in time, before divergences from a common ancestor gave rise to separate lineages. Later, as morphological differences accumulated among them, the lineages took off in amazingly diverse directions. How might we get a conceptual handle on their modern-day descendants—on animals as different as flatworms, hummingbirds, spiders, toads, humans, and giraffes? We can compare their similarities and differences with respect to five basic features. These are body symmetry, cephalization, type of gut, type of body cavity, and segmentation.

BODY SYMMETRY AND CEPHALIZATION   With very few exceptions, animals are radial or bilateral. Those with **radial symmetry** have body parts arranged regularly around a central axis, like spokes of a bike wheel. Thus a cut down the center of a hydra (Figure 23.3*a*) divides it into equal halves; another cut at right angles to the first divides it into equal quarters. Radial animals live in water. Their body plan is adapted to intercepting food that is coming toward them from any direction.

Animals with **bilateral symmetry** have a right half and left half that are mirror images of each other. Most

Table 23.1	Animal Phyla Described in This Book	
Phylum	Some Representatives	Existing Species
PLACOZOA (*Trichoplax*)	Simplest animal; like a tiny plate, but with layers of cells	1
PORIFERA (poriferans)	Sponges	8,000
CNIDARIA (cnidarians)	Hydrozoans, jellyfishes, corals, sea anemones	11,000
PLATYHELMINTHES (flatworms)	Turbellarians, flukes, tapeworms	15,000
NEMATODA (roundworms)	Pinworms, hookworms	20,000
ROTIFERA (rotifers)	Tiny body with crown of cilia, great internal complexity	2,000
MOLLUSCA (mollusks)	Snails, slugs, clams, squids, octopuses	110,000
ANNELIDA (segmented worms)	Leeches, earthworms, polychaetes	15,000
ARTHROPODA (arthropods)	Crustaceans, spiders, insects	1,000,000+
ECHINODERMATA (echinoderms)	Sea stars, sea urchins	6,000
CHORDATA (chordates)	Invertebrate chordates: Tunicates, lancelets	2,100
	Vertebrates: Fishes	21,000
	Amphibians	3,900
	Reptiles	7,000
	Birds	8,600
	Mammals	4,500

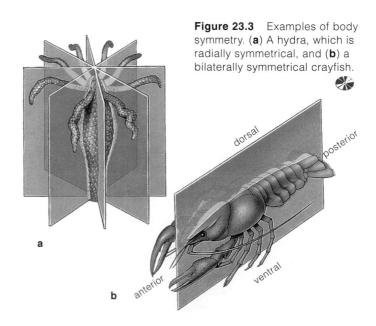

Figure 23.3 Examples of body symmetry. (**a**) A hydra, which is radially symmetrical, and (**b**) a bilaterally symmetrical crayfish.

dorsal

posterior

anterior

ventral

a

b

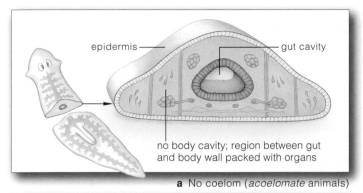

epidermis — — gut cavity

no body cavity; region between gut and body wall packed with organs

a No coelom (*acoelomate* animals)

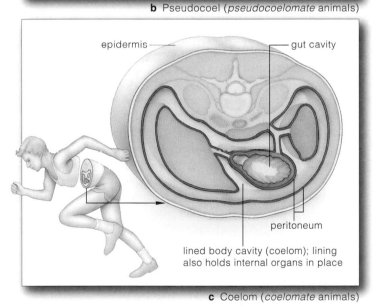

epidermis — — gut cavity

unlined body cavity (pseudocoel) around gut

b Pseudocoel (*pseudocoelomate* animals)

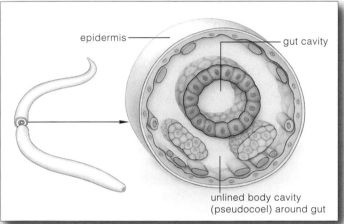

epidermis — — gut cavity

peritoneum

lined body cavity (coelom); lining also holds internal organs in place

c Coelom (*coelomate* animals)

Figure 23.4 Type of body cavity (if any) in animals.

have an *anterior* end (head) and an opposite, *posterior* end. They have a *dorsal* surface (a back) and an opposite, *ventral* surface (Figure 23.3*b*). As fossils show, this body plan evolved among the first forward-creeping species. Their forward end would have encountered food and other stimuli first, so there must have been selection for **cephalization**. With this evolutionary process, sensory

structures and nerve cells became concentrated in the head. The joint evolution of bilateral body plans and cephalization resulted in pairs of muscles and pairs of sensory structures, nerves, and brain regions.

TYPE OF GUT   The **gut** is a tubular or saclike region in the body in which food is digested, then absorbed into the internal environment. Saclike guts have one opening (a mouth) for taking in food and expelling residues. Other guts are part of a tubelike, "complete" digestive system with an opening at two ends (mouth and anus). Different parts of the system have specialized functions, such as preparing, digesting, and storing material. As more efficient digestive systems evolved, this helped pave the way for increases in body size and activity.

BODY CAVITIES   In between the gut and body wall of most bilateral animals is a body cavity (Figure 23.4). One type of cavity, a **coelom**, has a unique tissue lining called a peritoneum. This lining also encloses organs in the coelom and helps hold them in place. For example, your body has a coelom, and a sheetlike muscle called a diaphragm divides it into two smaller cavities. Your heart and lungs are positioned in the upper (thoracic) cavity, and your stomach, intestines, and other organs occupy the lower (abdominal) cavity.

Some invertebrates don't have a body cavity; tissues fill the region between their gut and body wall. Others have a pseudocoel ("false coelom"), a body cavity with no peritoneum. The coelom was a key innovation by which larger and more complex animals evolved from ancestral forms. It favored increases in size and activity by cushioning and protecting internal organs.

SEGMENTATION   Segmented animals have a repeating series of body units that may or may not be similar to one another. The many segments of earthworms have a similar outward appearance. Insect segments are fused into three units (head, thorax, and abdomen) and differ greatly from one another. Especially among the insects, diverse head parts, legs, wings, and other appendages evolved from less specialized segments.

Animals are multicelled, and most have tissues, organs, and organ systems. The body cells of most species are diploid.

Animals are aerobically respiring heterotrophs that ingest other organisms or absorb nutrients from them.

Animals reproduce sexually and, in many cases, asexually. They go through a period of embryonic development, and most are motile during at least part of the life cycle.

Body plans of animals differ with respect to five features: body symmetry, cephalization, type of gut, type of body cavity, and segmentation.

WWW

From telltale tracks and burrows left behind in marine sediments, we suspect that the first multicelled animals evolved by 900 million years ago. *Where did they come from?* They probably originated with protistan lineages, but we don't know which ones (Sections 19.3 and 19.4).

By one hypothesis, the forerunners of animals were ciliates, like *Paramecium*, and had multiple nuclei in a single-celled body. Over time, each nucleus became compartmentalized in individual cells of a multicelled body. However, we don't know of any existing animal that develops by compartmentalization.

By another hypothesis, multicelled animals arose from spherical colonies of a number of flagellated cells, maybe like the *Volvox* colonies shown in Section 20.15. In time, as a result of mutation, some cells in the colony became modified in ways that enhanced reproduction and other specialized tasks. And so began the division of labor that characterizes multicellularity.

Suppose such colonies became flattened and started creeping around on the seafloor. A creeping life-style could have favored the evolution of cell layers, such as those of *Trichoplax adhaerens*. This is the only known **placozoan** (Placozoa, after *plax*, meaning plate, and *zoon*, meaning animal). *Trichoplax* is a soft-bodied marine animal, shaped a bit like a tiny pita bread. It has no symmetry and no mouth. Its several thousand cells are arranged into two distinctly different layers. When it glides over food, its body briefly humps up (Figure 23.5). Gland cells in the lower layer secrete digestive enzymes onto the food, then individual cells absorb the breakdown products. Reproduction might be asexual (by budding or fission) or sexual, by mechanisms not yet understood. In sum, structurally and functionally, *Trichoplax* is as simple as animals get.

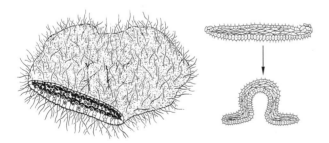

Figure 23.5 Cutaway view of *Trichoplax adhaerens*, an animal with a two-layer body measuring about three millimeters across.

Possibly the question of origins requires more than one answer. It may be that different lineages descended from more than one group of protistan-like ancestors.

Multicelled animals arose from protistan-like ancestors that may have resembled ciliates, colonial flagellates, or both.

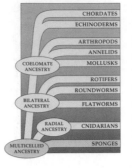

Sponges (Porifera) are animals with no symmetry, tissues, or organs, and yet they are one of nature's success stories. They have been abundant in the seas ever since the Precambrian, especially in waters off coasts and along reefs. Of 8,000 or so known species, about 100 live in freshwater. Small marine worms, shrimps, and other animals make their home in or on the sponge body.

Some sponges are large enough to sit in, and others are as small as a fingernail. Figures 23.6 and 23.7 show a few sprawling, flattened, lobed, compact, tubular, cuplike, and vaselike shapes. Regardless of the shape, the sponge body is not symmetrical. Flattened cells line its outer surface and inner cavities. But these linings are not much more specialized than the cell layers of *Trichoplax*, and they differ from tissues of other animals. Amoeboid cells live in a gelatin-like substance between the two linings (Figure 23.7b). Spicules, tough fibers of a protein (spongin), or both stiffen the sponge body and impart structure to it. The sharp, glasslike spicules are made of calcium carbonate or silica.

The skeletal elements may be a reason why sponges as a group have endured so long. Cleveland Hickman put it this way: Most potential predators discover that sampling a sponge is about as pleasant as eating a mouthful of glass splinters embedded in fibrous gelatin. Besides, chemically speaking, many sponges stink.

Water flows into the sponge body through many microscopic pores and chambers, then out through one or more large openings. It does so when thousands or millions of **collar cells** beat their flagella. The cells are components of the body's inner lining. Their "collars" are food-trapping structures called microvilli (Figure 23.7c). Bacteria and other food in the water get trapped in the collars, then are engulfed by phagocytosis. Some

Figure 23.6 A sprawling, red-orange sponge, one of many types that encrust underwater ledges in temperate seas.

Figure 23.7 (**a**,**b**) Body plan of a simple sponge. The outer lining consists of flattened cells. It also has some contractile cells, most of which are arranged around the large opening at the top. These cells contract slowly, independently of the other cell types, and so influence water flow through the body. Amoeboid cells inside the gelatin-like matrix between the inner and outer linings secrete materials from which the body's spicules and fibers are constructed. Other amoeboid cells digest and transport food. They do not lose the capacity to divide, and their cellular descendants can differentiate into any other type of sponge cell. They also have roles in asexual processes, such as gemmule formation.

(**c**) A great many flagellated cells line inner canals and chambers of the sponge body. Each of these phagocytic cells has a collar of food-trapping structures called microvilli. Fine filaments connect the microvilli to each other; they form a "sieve" that strains food particles from the water. At the base of the collar, the cell engulfs the trapped food.

(**d**) An example of skeletal elements—Venus's flower basket (*Euplectella*). This marine sponge has six-rayed spicules of silica fused in a rigid, elaborate network. A thin layer of cells stretches over all interconnecting spicules. At the base of the body is an anchoring tuft of spicules.

(**e**) A basket sponge of the Caribbean, releasing a cloud of sperm into the water.

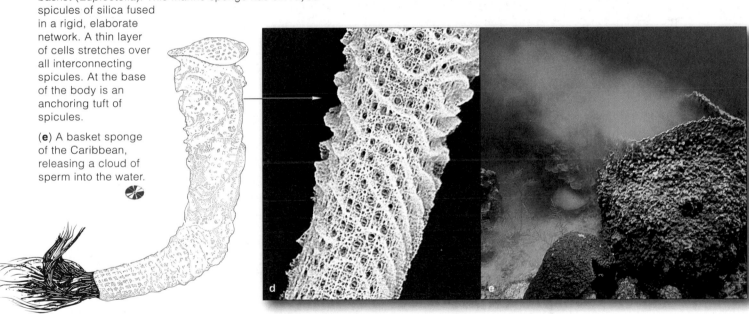

water out

central cavity

water in

a

glasslike structural elements

amoeboid cell

pore

semifluid matrix

flattened surface cells

b

flagellum   microvilli   nucleus

c Collar cell

d

e

of the food also is transferred to the amoebalike cells for further breakdown, storage, and distribution.

Sponges reproduce sexually. Most release sperm into the water (Figure 23.7e), but the eggs typically are retained until after they have been fertilized and the embryos start development. A young sponge proceeds through a microscopic, swimming larval stage. A **larva** (plural, larvae) is a sexually immature stage that grows and develops into an adult, the sexually mature form of the species. As you will see, the life cycle of many animal species include larval stages.

Some kinds of sponges also reproduce asexually by fragmentation; small fragments break away from the parent and grow into new sponges. Most freshwater species also reproduce asexually by way of gemmules. These are clusters of sponge cells, some of which form a hard covering around others. The clusters inside are protected from extreme cold or drying out. Later, when favorable conditions return, the gemmules germinate and establish a new colony of sponges.

Sponges have no symmetry, tissues, or organs; they are at the cellular level of construction. Yet they have successfully endured through time, possibly because most predators find their spicule-rich and often stinky bodies unappetizing.

WWW

CNIDARIANS—TISSUES EMERGE

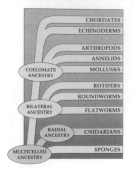

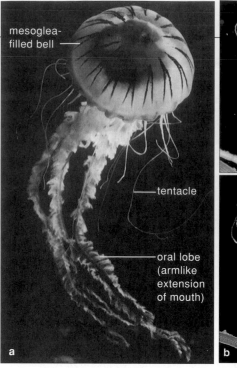

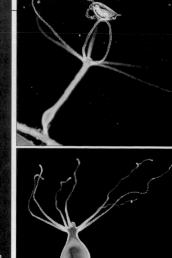

mesoglea-filled bell

tentacle

oral lobe (armlike extension of mouth)

a

b

Take a look at the jellyfish in Figure 23.8. It is one of the scyphozoans which, along with anthozoans (such as sea anemones) and hydrozoans (such as *Hydra*), are tentacled, radial animals of phylum **Cnidaria**. Most cnidarians live in the seas. Of 11,000 or so known species, fewer than 50 are adapted to freshwater habitats.

Cnidarians are at the tissue level of organization. They alone make **nematocysts**, capsules with dischargeable, tube-shaped threads. Many threads have prey-piercing barbs and an open tip for delivering toxin (Figure 23.9). Others, when discharged, ooze a sticky substance from the tip or entangle prey. Toxin-tipped threads of some species inflict a painful sting. Hence the name of the phylum Cnidaria, after the Greek word for nettle.

Cnidarian Body Plans

The **medusa** (plural, medusae) and **polyp** are the most common cnidarian body forms. Both have a saclike gut (Figure 23.8a,b). Medusae float. Some look like bells, others like upside-down saucers. The mouth, centered under the bell, may have extensions that assist in prey capture and feeding. Polyps have a tubelike body with a tentacle-fringed mouth at one end. Usually the other end is attached to a substrate. When *Trichoplax* is draped over and digesting food, it is rather like a gut on the run. By contrast, the cnidarian gut is a permanent food-processing chamber. It has a gastrodermis, a sheetlike lining with many glandular cells that secrete digestive enzymes. Epidermis lines the rest of the body's surfaces (Figure 23.8c,d). Each lining is an **epithelium** (plural, epithelia), a tissue with

Figure 23.8 (**a**) A jellyfish, the sea nettle (*Chrysaora*). (**b**) Hydrozoan polyp (*Hydra*) attached to a substrate, first capturing and then digesting its prey. The most common cnidarian body plans: (**c**) medusa and (**d**) polyp, sliced through the midsection.

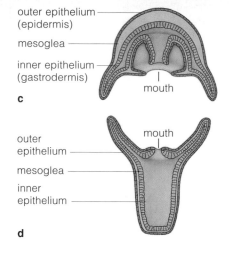

outer epithelium (epidermis)

mesoglea

inner epithelium (gastrodermis)

mouth

c

mouth

outer epithelium

mesoglea

inner epithelium

d

a free surface facing the environment or some fluid in the body. All animals more complex than sponges have epithelium. In a cnidarian, it incorporates **nerve cells**. Receptors detect external changes and signal nerve cells, which order **contractile cells** to carry out the responses. When stimulated, contractile cells *shorten*, then go back to their original length when the stimulation stops. The nerve cells interact as a "nerve net," a simple nervous tissue, to control movement and changes in shape.

Between the epidermis and gastrodermis is a layer of gelatinous secreted material, the mesoglea ("middle jelly"). The volume of mesoglea in jellyfishes imparts buoyancy and serves as a firm yet deformable skeleton against which contractile cells act. Visualize many cells contracting in a jellyfish bell. Their action narrows the bell and forces a jet of water out from underneath it. Cells contract each time the bell returns to its original position, and their coordinated contractions allow the animal to move slowly through the water.

capsule's lid at free surface of epidermal cell

exposure of barbs on discharged thread

trigger (modified cilium)

barbed thread inside capsule

nematocyst (capsule at free surface of epidermal cell)

Figure 23.9 One type of nematocyst before and after a prey organism (not shown) touched its trigger. The contact made the capsule more "leaky" to water. Water diffused inward, turgor pressure built up within the capsule, and the thread was forced to turn inside out. The thread's tip pierced the prey's body.

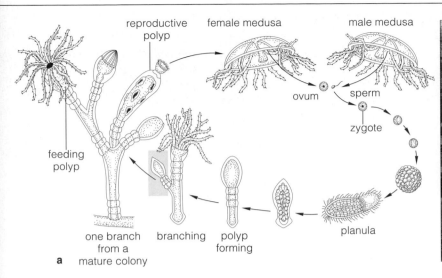

land barrier reef open ocean   interconnected
skeletons of polyps

Figure 23.10 (**a**) Life cycle of *Obelia*. (**b**) A coral reef. (**c**) One of the colonial corals. (**d**) Portuguese man-of-war (*Physalia*).

Any fluid-filled cavity or cell mass against which contractile cells can act is a **hydrostatic skeleton**. With coordinated contractions, the cavity's volume or mass does not change but rather is shunted about, so that the shape of the body changes. The contractile cells of most polyps, which have little mesoglea, act against water in their gut. As you will see, nearly all animals have some form of skeletal-muscular system of movement.

Stages in Cnidarian Life Cycles

Medusae, polyps, or both grow and develop during the life cycle of different cnidarian species. Let's use *Obelia* as an example (Figure 23.10*a*). Its medusa is a sexual stage. Embedded in part of its epithelium are **gonads**, the primary (gamete-producing) reproductive organs of animals. In this case, the gonads release gametes by rupturing. Most of the zygotes that form at fertilization develop into **planulas**, a type of swimming or creeping larva, usually with ciliated epidermal cells. In time, a mouth opens at one end of a planula, which becomes a polyp or medusa, and the cycle begins anew.

Reef-forming corals and other colonial anthozoans show variations on the basic cnidarian body plan. They thrive in clear, warm water. Secretions from polyps in a colony form external skeletons that interconnect with one another (Figure 23.10*b*,c). Skeletons accumulate over time and are key building material for reefs. One impressive accumulation, called the Great Barrier Reef, parallels Australia's coast for about 1,600 kilometers.

Reef-building corals feed on nutrients stirred up by tides. They also benefit from mutualists living in their tissues. Dinoflagellates (photosynthetic protistans) live as mutualists in coral tissues. They supply their hosts with oxygen, recycle mineral wastes, and adjust the pH of seawater in a way that enhances calcium deposition

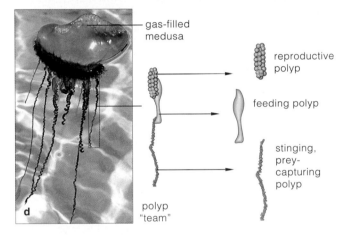

gas-filled medusa

reproductive polyp

feeding polyp

stinging, prey-capturing polyp

polyp "team"

for the host's skeleton. The corals give dinoflagellates a relatively safe, sunlit habitat, dissolved carbon dioxide, and mineral ions. In sunlit reefs, nutrients are cycled quickly, directly, and efficiently.

Physalia, the Portuguese man-of-war, also illustrates diversity in body plans. The nematocyst toxin of this infamous colonial hydrozoan poses a danger to bathers, fishermen, and fish prey. *Physalia* lives mainly in warm waters, but currents sometimes move it to the cooler Atlantic coastal waters of North America and Europe. A blue, gas-filled float that develops from the planula keeps the colony near the water's surface, where winds move it about (Figure 23.10*d*). Under the float, groups of polyps and medusae interact as "teams" in feeding, reproduction, defense, and other specialized tasks.

Cnidarians are radial animals with tentacles, a saclike gut, epithelia, a nerve net, and a hydrostatic skeleton. They alone produce nematocysts.

Cnidarians are at the tissue level of construction; compared with sponges, they have layers of cells interacting in more coordinated fashion in the performance of specific tasks.

WWW

ACOELOMATE ANIMALS—AND THE SIMPLEST ORGAN SYSTEMS

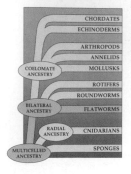

When we move beyond cnidarians in our survey, we find animals that range from flatworms to humans. Most show bilateral symmetry, and all have simple or complex organs. An **organ** is an association of one or more kinds of tissues, arranged in particular proportions and patterns. Most often, one organ interacts with others in helping the body function. By definition, each **organ system** is two or more organs that are interacting efficiently in the performance of some task. We turn now to the simplest animals at the organ-system level of construction. They are not what you would call breathtakingly complex. However, as you will see shortly, a few can make our lives breathtakingly miserable.

Flatworms

Among the 15,000 or so known species of **flatworms** (phylum Platyhelminthes) are turbellarians, flukes, and tapeworms. Most of these bilateral, cephalized animals have simple organ systems in a flattened body, as in Figure 23.11. Their digestive system, for example, has a pharynx (a muscular tube, which flatworms use for feeding) as well as a saclike, often branching gut. Their reproductive systems differ, but most flatworms are **hermaphrodites**. That is, an individual has female and male gonads. Two individuals can reproduce sexually through the mutual transfer of sperm. Each has a penis (a sperm-delivery structure) and glands that produce a protective capsule around fertilized eggs.

TURBELLARIANS Most turbellarians (class Turbellaria) live in the seas; only planarians and a few others live in freshwater. Some eat tiny animals or suck tissues from dead or wounded ones. A planarian can divide in half at its midsection, then each half can regenerate missing parts. Thus it reproduces asexually by *transverse* fission. Like you, a planarian can adjust the composition and volume of its body fluids. Its water-regulating system has one or more branched tubes called protonephridia (singular, protonephridium). These tubes extend from pores at the body surface to bulb-shaped flame cells in tissues. When excess water diffuses into the flame cells, a tuft of cilia "flickering" in the bulb drives the water through the tubes, to the outside (Figure 23.11b).

FLUKES Flukes (class Trematoda) are parasitic worms. A parasite, recall, lives in or on a living host and feeds on its tissues. Most do not kill the host, at least not until

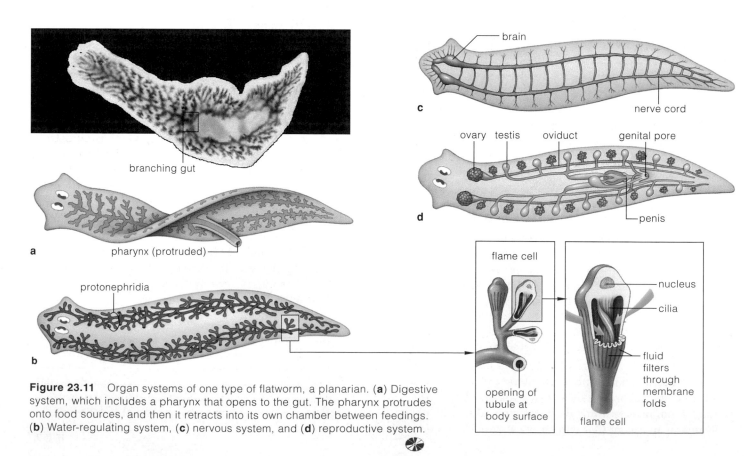

Figure 23.11 Organ systems of one type of flatworm, a planarian. (**a**) Digestive system, which includes a pharynx that opens to the gut. The pharynx protrudes onto food sources, and then it retracts into its own chamber between feedings. (**b**) Water-regulating system, (**c**) nervous system, and (**d**) reproductive system.

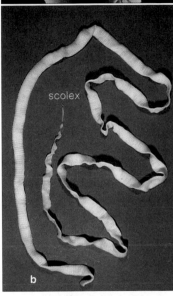

Figure 23.12 (**a**) Sheep tapeworm. (**b**) Tapeworm scolex. This one attaches to the intestinal lining of a shorebird, its definitive host.

after they reproduce. The life cycles have sexual and asexual phases—and two or more different kinds of hosts. As you will see in Section 23.7, the *definitive* host is one in which the parasite grows in size and reaches sexual maturity. In one or more *intermediate* hosts, the immature stages (such as larvae) develop or become encysted.

TAPEWORMS Tapeworms (class Cestoda) parasitize intestines of vertebrates. It seems likely that ancestral tapeworms had a gut but lost it as they evolved in the intestines of animals—which, for tapeworms, are habitats that are rich with predigested food. Existing descendants attach to the intestinal wall by a scolex, a structure equipped with suckers, hooks, or both. Figure 23.12*a* is a close-up of one of these formidable structures.

Proglottids, new units of a tapeworm body, bud just in back of the scolex. These are hermaphroditic units; they can mate and transfer sperm among one another. Older proglottids (the ones farthest from the scolex) store the fertilized eggs. They break off from the younger ones, then leave a host body in feces. Later, an intermediate host may meet up with their eggs. Section 23.7 includes a detailed look at proglottid formation in one tapeworm life cycle.

In some respects, the simplest turbellarians, larval flukes, and larval tapeworms resemble the planulas of cnidarians. The resemblance inspires speculation that bilateral animals evolved from planula-like ancestors, through increased cephalization and the development of tissues derived from mesoderm.

Flatworms are among the simplest bilateral, cephalized, acoelomate animals with organ systems.

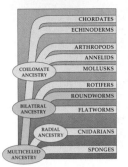

Roundworms (phylum Nematoda) are pseudocoelomate worms. They thrive nearly everywhere and may be the most abundant animals alive. In most sediments beneath shallow saltwater or freshwater, you often find a million roundworms in each square meter. Thousands can occupy a handful of rich topsoil, in which scavenging types make fast work of dead earthworms or rotting plants. We know of 20,000 species, but there may be a hundred times more. Figure 23.13 shows one of them.

A roundworm is bilateral. But its body is cylindrical, typically tapered at both ends, and covered by a protective cuticle. (Animal cuticles are tough, often flexible body coverings.) A roundworm is the simplest animal equipped with a complete digestive system. Between the gut and body wall is a false coelom, typically jam-packed with reproductive organs. The cells in every tissue absorb nutrients from the coelomic fluid and give up wastes to it.

Parasitic species can severely damage their hosts, which include humans, cats, dogs, cows, sheep, soybeans, potatoes, and other crop plants. Although thin, they can grow long; the roundworms in female sperm whales may be nine meters long, stretched out. Elephantiasis, one of the world's most rapidly spreading diseases, is the work of a few species. (See also Sections 23.7 and 40.4.)

Most roundworms are free-living types that are harmless or beneficial, as when they cycle nutrients for communities. One species, *Caenorhabditis elegans*, is used in research into inheritance, development, and aging. The *C. elegans* genome was the first of any multicelled organism to be fully sequenced (Section 38.5).

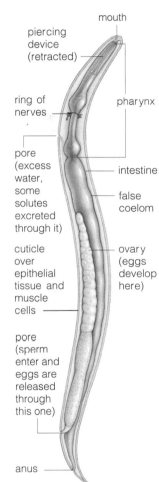

Figure 23.13 Body plan of a roundworm (*Paratylenchus*). This one parasitizes the roots of plants.

Roundworms are cylindrical, bilateral, cephalized animals with a false coelom and a complete digestive system. Most cycle nutrients in communities; the parasites are notorious.

A number of parasitic flatworms and roundworms call the human body home. In any year, for instance, 200 million people house blood flukes responsible for *schistosomiasis*. One of these, the Southeast Asian blood fluke *Schistosoma japonicum*, requires a human definitive host, standing water in which its larvae can swim, and an aquatic snail as an intermediate host. The flukes grow, become sexually mature, and mate inside a human host (Figure 23.14*a*). After being fertilized, a female's eggs (*b*) leave the human body in feces and hatch into ciliated, swimming larvae (*c*) that burrow into a snail (*d*) and multiply asexually. In time, many fork-tailed larvae develop (*e*). These leave the snail and swim about until they contact human skin (*f*). They bore in and migrate to thin-walled intestinal veins, then the cycle begins anew. In infected humans, white blood cells that defend the body attack the masses of fluke eggs, and grainy masses form in tissues. In time, the liver, spleen, bladder, and kidneys deteriorate.

Some tapeworms parasitize humans. Different species use pigs, freshwater fish, or cattle as intermediate hosts. Humans become infected when they eat pork, fish, or beef that is raw, improperly pickled, or insufficiently cooked—and contaminated with tapeworm larvae (Figure 23.15).

Or consider a parasitic roundworm that causes thin, serpentlike ridges in human skin. For several thousand years, healers have been extracting the "serpents" by winding them out slowly, painfully, around a stick. The roundworms called pinworms and hookworms cause other problems. *Enterobius vermicularis*, a pinworm of

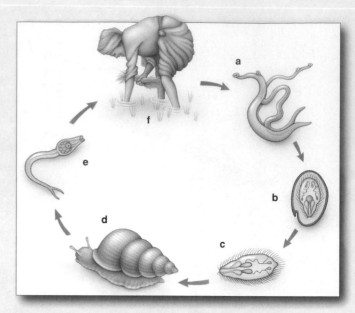

Figure 23.14 Life cycle of a dangerous blood fluke, *Schistosoma japonicum*.

temperate regions, parasitizes humans. It lives in the large intestine, but at night the centimeter-long females migrate to the anal region of the host and lay eggs. Their presence causes itching, and scratchings made in response transfer some eggs to hands, then to other objects. Newly laid eggs contain embryonic pinworms, but within a few hours they

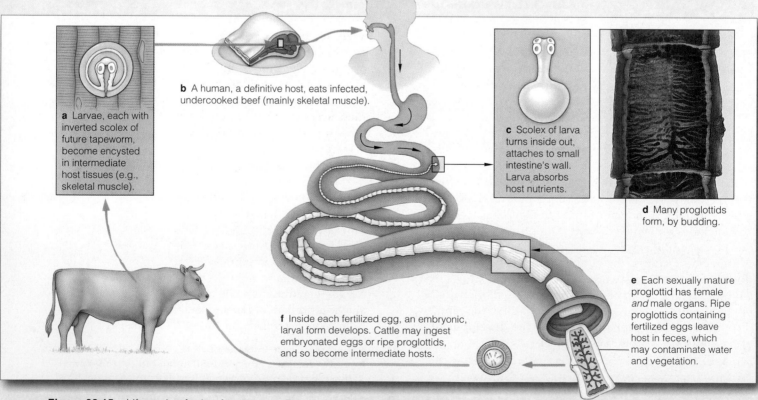

a Larvae, each with inverted scolex of future tapeworm, become encysted in intermediate host tissues (e.g., skeletal muscle).

b A human, a definitive host, eats infected, undercooked beef (mainly skeletal muscle).

c Scolex of larva turns inside out, attaches to small intestine's wall. Larva absorbs host nutrients.

d Many proglottids form, by budding.

e Each sexually mature proglottid has female *and* male organs. Ripe proglottids containing fertilized eggs leave host in feces, which may contaminate water and vegetation.

f Inside each fertilized egg, an embryonic, larval form develops. Cattle may ingest embryonated eggs or ripe proglottids, and so become intermediate hosts.

Figure 23.15 Life cycle of a beef tapeworm, *Taenia saginata*.

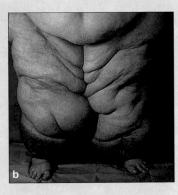

Figure 23.16 (**a**) Juveniles of a roundworm, *Trichinella spiralis*, inside the muscle tissue of a host animal. (**b**) Legs of a woman parasitized by the roundworm *Wuchereria bancrofti*.

have developed into juveniles and are ready to hatch if another human inadvertently ingests them.

Hookworms are especially serious in impoverished areas of the tropics and subtropics. Adult hookworms live in the small intestine. After the toothlike devices or sharp ridges bordering their mouth cut into the intestinal wall, they feed on blood and other tissues, then compete with their host for nutrients. Adult females, about a centimeter long, can release a thousand eggs daily. These leave the body in feces, then hatch into juveniles. When it meets up with bare skin of a host, a juvenile hookworms cuts its way inside. The parasite travels the bloodstream to the lungs. There it works its way into the air spaces. After moving up the windpipe, the parasite moves into the gut when the host swallows. Soon it is in the small intestine, where it may mature and live for several years.

Another roundworm, *Trichinella spiralis*, causes painful, sometimes fatal symptoms. Adults live in the lining of the small intestine. Females release juveniles (Figure 23.16a), which work their way into blood vessels and travel to muscles. There they become encysted; they secrete a covering around themselves and enter a resting stage. Humans become infected mainly by eating insufficiently cooked meat from pigs or some game animals. It is not easy to detect the encysted juveniles when fresh meat is being examined, even in a slaughterhouse.

Figure 23.16b shows the results of prolonged, repeated infections by *Wuchereria bancrofti*, another roundworm. Adult worms become lodged in lymph nodes, organs that filter lymph (excess tissue fluid) that normally flows into the bloodstream. There, the worms obstruct lymph flow. If an obstruction causes fluid to back up and accumulate in tissues, legs and other body regions may enlarge grossly. This condition is called *elephantiasis*.

A mosquito is *Wuchereria*'s intermediate host. Females of this parasite produce active young that travel about at night, in the bloodstream. If a mosquito sucks blood from an infected person, the juveniles may enter the insect's tissues. In time they move near the insect's sucking device and enter a new host when it draws blood again.

WWW

Bilateral animals not much more complex than modern flatworms evolved during Cambrian times. Shortly after this, some species gave rise to two lineages of animals equipped with a coelom (Figure 23.1b). We call these great lineages the **protostomes** and the **deuterostomes**. Mollusks, annelids, and arthropods are all protostomes. Echinoderms and chordates are deuterostomes.

As a result of mutations, embryos of the animals in each lineage develop differently from fertilized eggs. For instance, mitotic cell divisions cut the egg cytoplasm repeatedly, along prescribed planes, to form a tiny ball of cells (the early embryo). Protostomes undergo *spiral* cleavage, a pattern in which the earliest cuts are made at oblique angles relative to the genetically prescribed body axis. But deuterostomes undergo *radial* cleavage, a developmental pattern in which the earliest cuts are made parallel with and perpendicular to the axis:

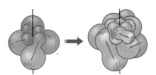

Early protostome embryo. Its four cells are undergoing cleavages *oblique to* the original body axis.

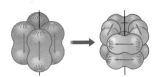

Early deuterostome embryo. Its four cells are undergoing cleavages *parallel with* and *perpendicular to* the original body axis:

As another example, the very first opening to form at the surface of a protostome embryo becomes the mouth; an anus forms elsewhere. In a deuterostome embryo, the first opening becomes the anus; the second becomes the mouth. As another example, a protostome coelom arises from spaces in the mesoderm, but a deuterostome coelom forms from outpouchings of the gut wall:

How a coelom forms in a protostome embryo: pouch will form mesoderm around coelom — developing gut

coelom

How a coelom forms in a deuterostome embryo: solid mass of mesoderm — developing gut

Such modifications to the embryonic stages of the two kinds of animals led to major differences in body plans.

Soon after the coelomate animals evolved in Cambrian times, two great lineages—the protostomes and deuterostomes—evolved through mutations that affected how their embryos develop, and this led to major differences in body plans.

WWW

A SAMPLING OF MOLLUSKS

Molluscan Diversity

Few of us have trouble recognizing a snail (Figure 23.17). Yet few know much about its 110,000 relatives in one of the largest animal phyla, the Mollusca. A **mollusk** is a bilateral animal having a small coelom and a fleshy soft body (*molluscus*, meaning soft). Most have a shell of calcium carbonate and protein, which were secreted from cells of a tissue that drapes like a skirt over the body mass. This tissue, the **mantle**, is unique to mollusks. Special gills with thin-walled leaflets serve in gas exchange. Most mollusks have a fleshy foot. Many have a radula, a tonguelike, toothed organ. (A radula, by its rhythmic protractions and retractions, rasps small algae or other tidbits from substrates and draws it into the mouth.) Mollusks with a well-developed head have eyes and tentacles, but not all have a head. Beyond these generalizations we have rampant diversity, ranging from tiny snails in treetops to gigantic predators of the seas. Here we sample four classes: chitons, gastropods, bivalves, and cephalopods.

The largest class has 90,000 species of snails and slugs (Figure 23.17a–c). These are the gastropods ("belly foots"), so named because their soft foot spreads out as they crawl. Many have spirally coiled or conical shells. Coiling compacts the organs into a mass that can be balanced above the body, rather like a backpack. Other species have a reduced shell or none at all.

Chitons are slow-moving or sedentary grazers with a dorsal shell divided into eight plates (Figure 23.17d). Bivalves, animals having a "two-valved shell," include clams, scallops, oysters, and mussels (Figure 23.17e). The shells of many are lined with iridescent mother-of-pearl. Some species are only a millimeter across. A few giant clams are a meter across and weigh 225 kilograms (close to 500 pounds). Humans have been eating one type of bivalve or another since prehistoric times.

All cephalopods are active predators of the seas. In this class we find the largest and swiftest invertebrates (certain squids, such as the one shown in Figure 23.17f) and the smartest (the octopuses). Show an octopus some object having a distinct shape and then give this cephalopod a mild electric shock and it will thereafter avoid the object. With respect to memory and a capacity for learning, the octopuses and some squids are the most complex invertebrates in the world.

Figure 23.17 Some of the gastropods: (**a**) aquatic snail, (**b**) land snail, and (**c**) sea slugs ("Mexican dancers"). Gastropods creep, swim, or float as they graze on, prey upon, or parasitize other organisms. (**d**) Chiton. With its broad foot, it creeps on and clings to rocks. (**e**) A scallop, one of the bivalves. Light-sensitive "eyes" (the tiny dark dots) fringe the two halves of its shell. (**f**) A squid (*Dosidiscus*) and a diver inspecting each other. Being highly active, cephalopods have great demands for oxygen. They are the only mollusks with a closed circulatory system. A main heart pumps blood to two gills, each with a booster (accessory) heart at its base that speeds blood flow, hence the uptake of oxygen (for muscle cells especially) and removal of carbon dioxide.

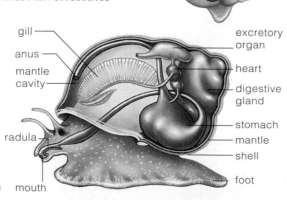

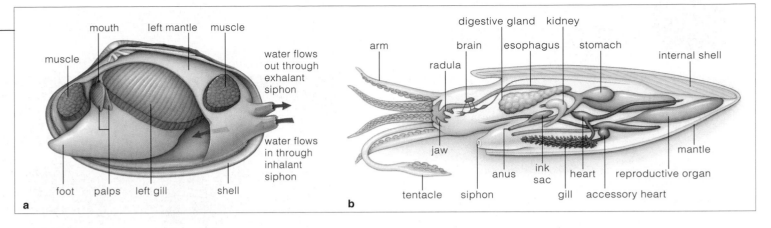

Evolutionary Experiments With Body Plans

Maybe it was their soft, fleshy bodies—so forgiving of chance evolutionary changes in morphology—that gave the ancestors of mollusks the potential to diversify in so many ways and radiate into so many habitats. Let's explore this idea by looking at a few species.

TWISTING AND DETWISTING OF SOFT BODIES Notice, in Figure 23.17a, how evolution put an odd twist in the soft snail body. The anus dumps wastes near the mouth! As a gastropod embryo develops, a cavity between the mantle and shell twists 180° counterclockwise. So does nearly all of the visceral mass (the gut, heart, gills, and other internal organs). This process of *torsion* operates only in gastropods. It must have created a distasteful sanitation problem in ancestral torsioned species.

We find evolutionary compensations for this state of affairs. Most gastropods now have enough cilia in this region to create currents that can sweep the wastes away. Also, torsion is not as pronounced as it once was in some lineages. Nudibranchs have even undergone a detorsion; the soft larval body twists, then it untwists.

HIDING OUT, ONE WAY OR ANOTHER If you were small, edible, and soft of body, an external shell would be a distinct advantage, as it is for chitons and clams. When a chiton is disturbed by predators or the surf or exposed by a receding tide, it hunkers under its shell. Foot muscles pull the body mass down, and the mantle's edge around the shell's rim presses like a suction cup against a rock. That eight-plated shell is flexible. Pull a chiton away from a rock, and it will roll up into a ball until it can unroll and reattach itself elsewhere.

Besides hiding in a shell, protection also can be had by hiding in sand or some other substrate. A bivalve's head is not much to speak of, but its foot is usually large and specialized for burrowing. Bivalves burrowed in sand or mud have a pair of siphons: extensions of the mantle edges, fused into tubes (Figure 23.18a). Water is drawn into the mantle cavity through one siphon and leaves through the other, carrying wastes. One wonders what preyed on the ancestors of geoducks of the Pacific Northwest, which have siphons more than a meter long.

Figure 23.18 (**a**) Body plan of a clam, with half of its shell removed. In nearly all bivalves, gills collect food and serve in respiration. As water moves through the mantle cavity, mucus on the gills traps food. Cilia move the mucus and food to palps, where final sorting occurs before suitable bits are driven to the mouth. (**b**) Generalized body plan of a cuttlefish, a cephalopod. Its tentacles, thinner than its arms, specialize in prey capture.

ON THE CEPHALOPOD NEED FOR SPEED Five hundred million years ago, cephalopods had buoyant, chambered shells and were the supreme predators of the Ordivician seas (Section 19.5). And yet, of a lineage that once had more than 7,000 species, the shells of all but one species that made it to the present are gone or reduced, as in Figure 23.18f. What happened? This evolutionary trend coincided with a great adaptive radiation of the bony fishes—which preyed upon cephalopods or were strong competitors for the same prey.

During what may have been a long-term race for speed and wits, cephalopods lost their thick external shell and became streamlined and highly active. Of all mollusks, they now have the largest brain relative to body size and show the most complex behavior. Nerves connect the brain to muscles that respond quickly to food or to danger. Blood circulation and respiration are highly efficient. And except for the chambered nautilus, cephalopods can discharge dark fluid from an ink sac, maybe to confuse predators.

Long ago, *jet propulsion* became the name of the game. Cephalopods force a jet of water out from their mantle cavity and a funnel-shaped siphon. As mantle muscles relax, water is drawn into the cavity. As they contract, a water jet is squeezed out. When the mantle's free edge closes down on the head at the same time, a jet shoots out through the siphon. The brain controls the siphon's activity and thus the direction of escape or pursuit.

Mollusks are bilateral, soft-bodied, coelomate animals that vary tremendously in body details, size, and life-styles.

Lively stories emerge when evolutionary theory is used to interpret the fossil record and the range of existing species diversity, as we have done for the mollusks.

WWW

ANNELIDS—SEGMENTS GALORE

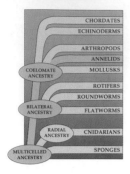

Maybe you've noticed earthworms after a downpour, as they wriggle out from their burrows to avoid drowning. They are among 15,000 or so species of bilateral, segmented animals called **annelids** (Annelida). Their relatives are a few kinds of leeches and the polychaetes, which are far more diverse but not nearly as well known (Figures 23.19 and 23.20). The phylum's name means "ringed forms." But the "rings" actually are a series of repeating body units. The segmentation is pronounced. Also, except for leeches, nearly all segments have pairs or clusters of chitin-reinforced bristles on each side of the body. The bristles are also called setae or chaetae, but these are just formal names for "bristles." When pushed into the soil, the bristles provide the traction required for crawling or burrowing. They have become broad paddles in some aquatic species. Earthworms, one of the oligochaetes, have only a few setae per body segment. The marine polychaete worms typically have many of them (*oligo–*, few; *poly–*, many).

Advantages of Segmentation

A segmented body has great evolutionary potential, for individual parts can undergo modification and become highly adapted for specialized tasks. Although most of an earthworm's segments are similar, the leeches have suckers at both ends, and polychaetes have an elaborate head and fleshy-lobed appendages known as parapods ("closely resembling feet"). By analyzing the existing species, we catch glimpses of developments that led to increases in size and to more complex internal organs.

Annelid Adaptations—A Case Study

Earthworms are usual textbook examples of annelids. As Figure 23.21 shows, partitions divide the body into a series of coelomic chambers. In most of the chambers we find repeats of muscles, blood vessels, branching nerves, and other

Figure 23.19 Leeches. Most leeches have sharp jaws and a blood-sucking device. This one is shown before and after gorging on human blood. For at least 2,000 years, *Hirudo medicinalis*, a freshwater leech, has been employed as a blood-letting tool to "cure" problems ranging from nosebleeds to obesity. Today leeches are still used, but more selectively. Thus, after surgeons reattach a severed ear, lip, or fingertip, leeches may be used to draw off pooled blood. A patient's body cannot do this on its own until the severed blood circulation routes are reestablished.

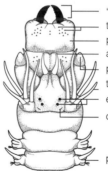

"jaws"
toothlike structures
pharynx (everted)
antenna
palp (food handling)
tentacle
eyes
chemical-sensing pit

parapod

c

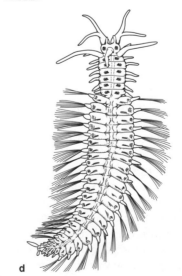

d

Figure 23.20 (a) A familiar annelid—one of the earthworms. (b–d) Polychaetes. These give a sense of the dizzying variety of modifications that have evolved in this group, starting with a segmented, coelomic body plan. The species in (b) is one of the tube-dwellers. Featherlike structures at its head end are coated with mucus. After the mucus has trapped bacteria and other bits of food, coordinated beating of cilia sweeps them to the mouth. Most polychaetes live in marine habitats. They actually are one of the most common types of animals along coasts. Many are predators or scavengers; others dine on algae.

Figure 23.21 Earthworm body plan. (**a**) Section through the midbody. (**b**) A nephridium, one of many functional units that help maintain the volume and composition of body fluids. (**c**) Portion of the closed circulatory system, which is functionally linked with nephridia. (**d**) Part of the digestive system, near the worm's head end. (**e**) Part of the nervous system.

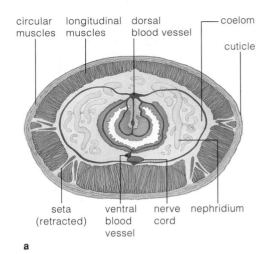

circular muscles · longitudinal muscles · dorsal blood vessel · coelom · cuticle

seta (retracted) · ventral blood vessel · nerve cord · nephridium

a

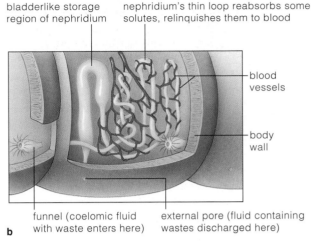

bladderlike storage region of nephridium · nephridium's thin loop reabsorbs some solutes, relinquishes them to blood

blood vessels

body wall

funnel (coelomic fluid with waste enters here) · external pore (fluid containing wastes discharged here)

b

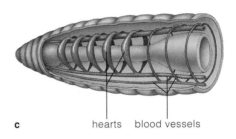

c · hearts · blood vessels

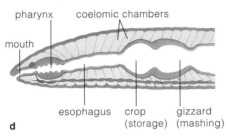

pharynx · coelomic chambers

mouth

esophagus · crop (storage) · gizzard (mashing)

d

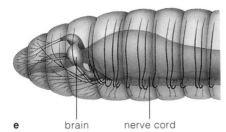

e · brain · nerve cord

organs. The gut extends through all the chambers, from mouth to anus. Like all annelids, an earthworm has a cuticle of secreted material that surrounds the body surface. It bends easily, and it is permeable to water as well as to gases, which is one reason why annelids are restricted to aquatic habitats or moist habitats on land.

Earthworms are scavengers. They ingest moist soil and mud that contains decomposing plant material and other organic matter. Each worm can eat its own weight every twenty-four hours. As numerous worms burrow and feed, they collectively aerate soil and lift nutrients to the surface, to the benefit of many plants.

As in other annelids, the fluid-cushioned coelomic chambers serve as a hydrostatic skeleton against which muscles act. Each segment's wall incorporates a layer of circular muscles (Figure 23.21*a*). When longitudinal muscles that span several segments contract, circular muscles relax, so segments shorten and fatten. When the pattern reverses, the segments lengthen. While this is going on, bristles on different segments protract and retract. When the first few segments lengthen and their setae are not touching the ground, the front part of the body is extended. Bristles of the segments behind them plunge into the ground and anchor the worm. Next, the first few segments plunge *their* bristles into the ground, and the segments toward the posterior end of the body retract their bristles and are pulled forward. The whole worm moves forward as alternating contractions and elongations proceed along the body's length.

Figure 23.21*b* shows part of a system of **nephridia** (singular, nephridium), units that regulate the volume and composition of body fluids. In many annelids, cells of the units are similar to flame cells, which implies an evolutionary link between the flatworms and annelids. More often, a nephridium starts out as a funnel that collects excess fluid from one coelomic chamber. The funnel connects with a tubular part of the nephridium, which delivers fluid to a pore at the surface in the body wall of the next coelomic chamber.

The worm's head end has a rudimentary **brain**, an aggregation of nerve cell bodies that integrate sensory input as well as commands for muscle responses for the whole body. Paired **nerve cords**, each a bundle of long extensions of nerve cell bodies, lead away from the brain. They are pathways for rapid communication. In each body segment, the paired nerve cords broaden into a ganglion (plural, ganglia), a cluster of nerve cell bodies that controls local activity.

Finally, as is the case for most annelids, earthworms have a closed circulatory system, with blood confined in hearts and muscularized blood vessels. Contractions keep blood circulating in one direction. Smaller blood vessels service the gut, nerve cord, and body wall.

Annelids are bilateral, coelomate, segmented worms that have complex organ systems. Some species show the degree of specialization possible with a segmented body plan.

WWW

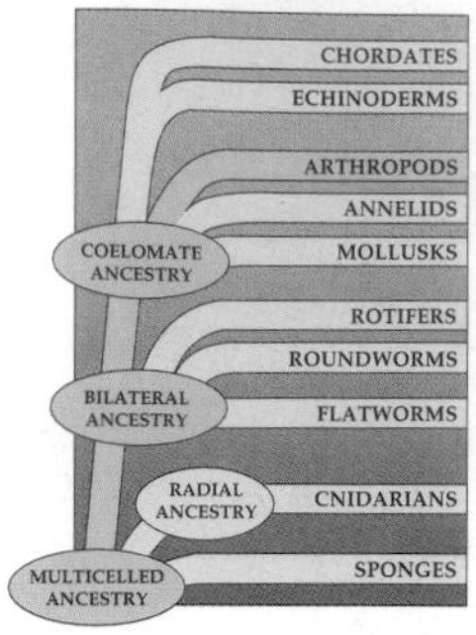

Arthropod Diversity

Evolutionarily speaking, "success" means having the greatest number of species, producing the greatest number of offspring, occupying the most habitats, effectively fending off predators and competitors, and having the capacity to exploit the greatest amounts and kinds of food. These are the features that come to mind when we attempt to characterize the **arthropods** (Arthropoda). Over a million species (mostly insects) are known, and new ones are being discovered weekly. Of four major lineages, trilobites are extinct (Section 19.5). The other three are chelicerates (spiders and their relatives), crustaceans (such as barnacles and crabs), and uniramians (centipedes, millipedes, and insects).

Adaptations of Insects and Other Arthropods

Six important adaptations contributed to the success of arthropods in general and the insects in particular: a hardened exoskeleton, jointed appendages, fused and modified segments, specialized respiratory structures, efficient nervous system and sensory organs, and often a division of labor in the life cycle.

HARDENED EXOSKELETONS   Arthropods have a cuticle of chitin, proteins, and surface waxes that may even be impregnated with calcium carbonate deposits. It acts as a rigid, protective external skeleton—an **exoskeleton**. Such cuticles might have evolved as defenses against predation. They took on added functions when some arthropods first invaded the land. They support a body deprived of water's buoyancy, and their waxy surface restricts evaporative water loss. Hard cuticles stop size increases, but arthropods grow in spurts by **molting**. At certain stages of their life cycle, they secrete a new, soft cuticle under the old one, which they shed (Figure 23.22). The body mass increases at first by uptake of air or water, then by repeated, rapid cell division before the new cuticle can harden.

JOINTED APPENDAGES   If an arthropod had a cuticle that was uniformly hardened, it wouldn't move much. However, the cuticle is thinned at *joints*, where two body

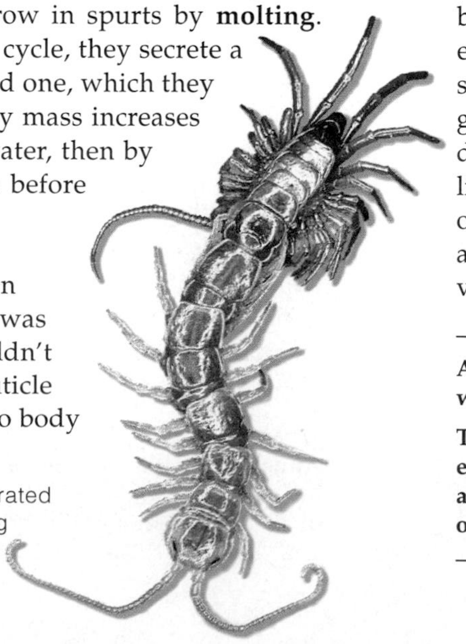

Figure 23.22   Molting, demonstrated by an orange centipede wriggling out of its old exoskeleton.

parts abut. Muscles near the joints make the thinned cuticle bend in specific directions and move attached body parts. A jointed exoskeleton was a key innovation that led to appendages as diverse as wings, antennae, and legs (*arthropod* means jointed foot).

FUSED AND MODIFIED SEGMENTS   The first arthropods were segmented, like the annelid stock that presumably gave rise to them. In most existing descendants, serial repeats of the body wall and organs are masked, for many fused-together, modified segments perform more specialized functions. For example, in the ancestors of insects, different segments fused to form three regions —head, thorax, and abdomen—which morphologically diverged from one another in astounding ways.

RESPIRATORY STRUCTURES   Many aquatic arthropods depend on gills for gas exchange. Air-conducting tubes evolved among insects and other land-dwellers. Insect tracheas begin as pores on the body surface and branch into tubes that deliver oxygen directly to tissues. They support energy-consuming activities, such as flight.

SPECIALIZED SENSORY STRUCTURES   Intricate eyes and other sensory organs contributed to arthropod success. Numerous species have a wide angle of vision and can process visual information from many directions.

DIVISION OF LABOR   Moths, butterflies, beetles, flies, and many other species divide the job of surviving and reproducing among different stages of development. For many species, the new individual is a *juvenile*—a miniaturized version of the adult that simply changes in size and proportion until reaching sexual maturity. Other species show **metamorphosis**, meaning the body form changes from embryo to adult. Under hormonal commands, their size increases, tissues reorganize, and body parts undergo remodeling. Section 31.8 includes examples of this transitional time. Typically, immature stages, such as caterpillars, specialize in feeding and growing in size. The adult stage specializes mainly in dispersal and reproduction. For such species, then, the life cycle turns on a *division of labor*: the different stages of development become specialized in ways that are adaptive to environmental conditions, such as seasonal variation in food resources and water supplies.

As a group, the arthropods are exceptionally abundant and widespread, and they have enormously different life-styles.

Their success arises largely from their hardened, jointed exoskeletons; fused, modified body segments; specialized appendages; specialized respiratory, nervous, and sensory organs; and often a division of labor in the life cycle.

WWW

A LOOK AT SPIDERS AND THEIR KIN

The chelicerates originated in shallow seas early in the Paleozoic. The only surviving marine species are a few mites, sea spiders, and horseshoe crabs (Figure 23.23). The familiar chelicerates on land—spiders, scorpions, ticks, and chigger mites—are all classified as arachnids. And we might say this about the whole group: Never have so many been loved by so few.

forebody
hindbody

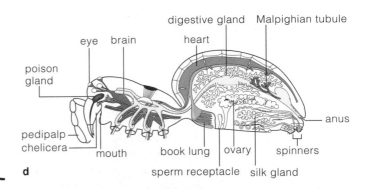
digestive gland   Malpighian tubule
eye   brain   heart
poison gland
pedipalp
chelicera   mouth   book lung   ovary   spinners
sperm receptacle   silk gland
anus
d

e

f

Figure 23.23 (**a,b**) A horseshoe crab. Its *five* pairs of legs, one of its defining features, are hidden beneath a hard, shieldlike cover. Spiders: (**c**) Brown recluse, with a violin-shaped mark on its forebody. Its bite can be severe to fatal. (**d**) A spider's internal organization. (**e**) Wolf spider. Like most spiders, it helps keep insect populations in check. Its bite is harmless to humans. (**f**) Female black widow. It prefers dark, dry habitats, as in a dusty garage cabinet or behind a clothes dryer. Its bite can be painful; sometimes it is dangerous.

Scorpions and spiders are efficient predators; they sting or bite and may subdue prey with venom. When they sting or bite us, the reaction might be painful, but it is rarely serious. Collectively, the spiders especially are beneficial in that they prey upon great numbers of pestiferous insects.

Bites of the blood-sucking ticks that parasitize deer, mice, and other vertebrates cause maddening itches and often serious diseases. For example, bites of some ticks transmit the bacterial agents of Rocky Mountain spotted fever or Lyme disease to humans (Section 20.2). Most mites are free-living scavengers.

Arachnids have segments fused into a forebody and hindbody. The forebody's jointed appendages include four pairs of legs, a pair of pedipalps with primarily sensory functions, and a pair of chelicerae that inflict wounds and discharge venom. The appendages of the hindbody spin out silk threads for webs and for egg cases. Most webs are netlike. One type of spider spins a vertical thread with a ball of sticky material at the end. It uses a leg to swing the ball at insects passing by!

Inside the body is an *open* circulatory system, with a heart that pumps blood into body tissues, then receives blood through small openings in its wall. Some of the slowly circulating blood travels through moist folds of book lungs. These respiratory organs resemble book pages, and they greatly increase the surface area that is available for gas exchange with the air. Figure 23.23*d* shows the arrangement of book lungs and other major organs inside the spider body.

The spiders, scorpions, and their relatives have a variety of appendages specialized for predatory or parasitic life-styles.

WWW

A LOOK AT THE CRUSTACEANS

Shrimps, lobsters, crabs, barnacles, pillbugs, and other crustaceans got their name because they have a hard yet flexible "crust" (an external skeleton)—but so do nearly all arthropods. Only some of the 35,000 species live in freshwater or on land. The vast majority live in marine habitats, where they are so abundant they have been dubbed the insects of the seas. Lobsters and crabs are "giants" of this subphylum; most crustaceans are less than a few centimeters long. All have major roles in food webs, and humans harvest many edible types.

By having many pairs of similar appendages along most of their length, the simplest crustaceans might resemble their annelid ancestors. In the more complex lineages, unspecialized appendages evolved into diverse structures of the sort shown in Figure 23.24. The strong claws of lobsters and crabs, for example, are used to collect food, intimidate other animals, and sometimes dig burrows. Feathery appendages of barnacles comb microscopic bits of food from the water.

Many crustaceans have sixteen to twenty segments; some have more than sixty. In crabs, lobsters, and some other crustaceans, the dorsal cuticle extends back from the head as a shieldlike cover over some or all of the segments. The head has two pairs of antennae, one pair of jawlike appendages (mandibles), and two pairs of appendages for handling food. Crayfish, crabs, lobsters,

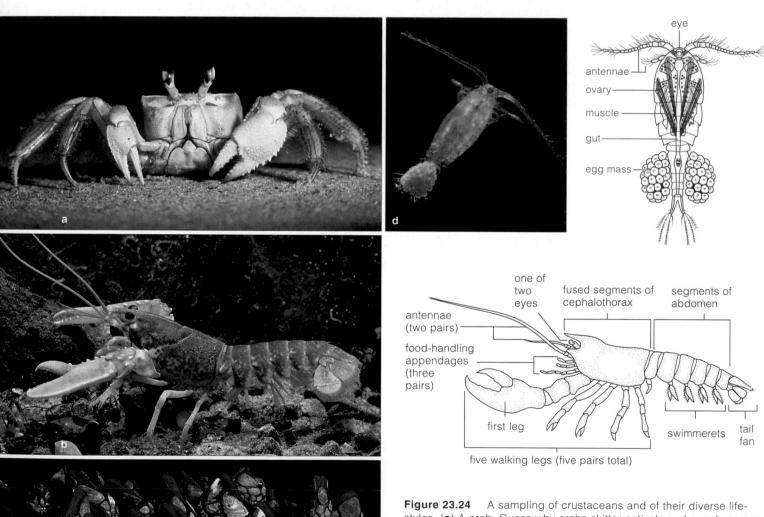

Figure 23.24 A sampling of crustaceans and of their diverse lifestyles. (**a**) A crab. Guess why crabs skitter actively and openly across sand and rocks at night but not during the day. (**b**) Photograph and body plan of a lobster. Lobsters are secretive for most of their lives. (**c**) Stalked barnacles. Adult barnacles cement themselves to one spot. You might mistake them for mollusks, but as soon as they open their hinged shell to filter-feed, jointed appendages (the hallmark of arthropods) are apparent. (**d**) Photograph and body plan of a copepod. Copepods are free-living filter feeders, predators, or parasites. This female has a pair of long antennae and is carrying her eggs with her.

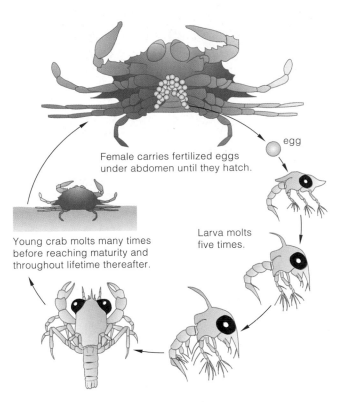

Figure 23.25 Life cycle of a crab. The larval and juvenile stages molt repeatedly and grow in size.

shrimps, and their various relatives have five pairs of walking legs.

Of all the arthropods, only barnacles have a calcified "shell," a modified external skeleton that protects them from predators, drying out, battering surf, and strong currents. Adult barnacles cement themselves to rocks, wharf pilings, and similar surfaces (Figure 23.24c). A few kinds attach themselves only to the skin of whales.

Figure 23.24d shows a copepod. Copepods are less than two millimeters long and are the most numerous animals in aquatic habitats, maybe even in the world. About 1,500 kinds parasitize various invertebrates and fishes. The majority—8,000 species—are consumers of phytoplankton, the "pastures" of aquatic habitats. Some also eat larval or small adult invertebrates, fish eggs, and fish larvae, which they grab with their pair of food-handling appendages. In turn, the copepods are food for different invertebrates, fishes, and baleen whales.

Like other arthropods, crustaceans molt repeatedly to shed the exoskeleton during their life cycle. Figure 23.25 shows a crab's larval stages and increases in size.

Crustaceans differ greatly in the number and kind of their appendages. As for arthropods generally, they repeatedly replace their external skeleton by molting.

HOW MANY LEGS?

The **millipedes** and **centipedes** have a long, segmented body with many legs. Of course, millipedes don't have "a thousand," as their name implies. Most have about 100 legs, although one exuberant individual grew 752. And centipedes have between 15 and 177 pairs of legs, not a nicely rounded number of "one hundred."

As the millipedes develop, pairs of segments fuse, so each segment in the cylindrical body ends up with two pairs of legs (Figure 23.26a). Millipedes scavenge decaying plant material in soil and forest litter.

Figure 23.26 (**a**) Millipede. (**b**) A Southeast Asian centipede.

Centipedes have a flattened body, and all but two segments have a pair of walking legs. All species are fast-moving, aggressive predators, outfitted with fangs and venom glands. They prey on insects, earthworms, and snails. The one in Figure 23.26b can subdue small lizards, toads, and frogs. A house centipede (*Scutigera*) often lurks in buildings, where it hunts cockroaches, flies, and other pests. Athough helpful in this respect, its vaguely terrifying body keeps it from being welcomed.

The mild-mannered, scavenging millipedes and aggressive, predatory centipedes do not lend themselves to leg counts as they walk by.

A LOOK AT INSECT DIVERSITY

As a group, insects share the adaptations listed earlier in Section 23.11. Here we expand the list a bit. Insects have a head, a thorax, and an abdomen. The head has paired sensory antennae and paired mouthparts that specialize in biting, chewing, sucking, or puncturing (Figure 23.27). Three pairs of legs and usually two pairs of wings project from the thorax. Most appendages of the abdomen are reproductive structures, such as egg-laying devices. An insect has a foregut, midgut (where most digestion proceeds), and hindgut (where water is reabsorbed). Insects get rid of waste material through **Malpighian tubules**, small tubes that connect with the midgut. When they break down proteins, the nitrogen-containing wastes diffuse from blood into the tubules and are converted into harmless crystals of uric acid. The crystals are eliminated with feces. This system lets land-dwelling insects get rid of potentially toxic wastes without losing precious water.

We've already catalogued more than 800,000 species of insects. If we use sheer numbers and distribution as the yardstick, the most successful insects are small in size and have a staggering reproductive capacity. For example, you may find some of these species growing and reproducing in great numbers on a single plant that might be only an appetizer for another animal. By one estimate, if all the progeny of a single female fly were to survive and reproduce through six more generations, that fly would have more than 5 trillion descendants!

Besides this, the most successful insect species are winged. In fact, they are the *only* winged invertebrates. They can move among food sources that are too widely scattered to be exploited by other kinds of animals. The capacity for flight contributed to their success on land.

Finally, insect life cycles commonly proceed through stages that allow exploitation of different resources at different times. As an insect embryo develops, organs required for feeding and other vital activities form and become functional. Before an insect becomes an adult (the sexually mature form of the species), it proceeds through immature, post-embryonic stages. **Nymphs** and **pupae**, as well as larvae, are examples of the stages.

Like human infants, nymphs of some insects are shaped like miniature adults. But unlike children, they undergo growth and molting (Figure 23.28a). Other insects go through post-embryonic stages of reactivated growth, tissue reorganization, and remodeling of body parts. Metamorphosis, remember, is the name for this resumption of growth and transformation into an adult form. Figure 23.28 shows how the transformation is more drastic in some species than in others.

The factors that contribute to insect success also make them our most aggressive competitors. Insects destroy crops, stored food, wool, paper, and timber. As they stealthily draw blood from us and from our pets,

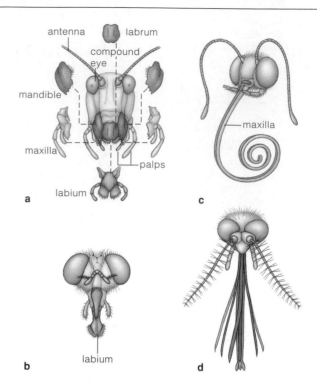

Figure 23.27 Examples of insect appendages. Headparts of (**a**) grasshoppers, a chewing insect; (**b**) flies, which sponge up nutrients with a specialized labium; (**c**) butterflies, which siphon up nectar with a specialized maxilla; and (**d**) mosquitoes, with piercing and sucking appendages.

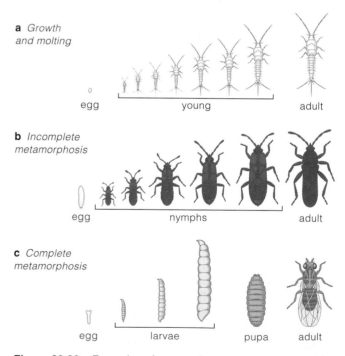

Figure 23.28 Examples of post-embryonic development. (**a**) Young silverfish are adults in miniature, changing little except in size and proportion as they mature into adults. (**b**) True bugs show *incomplete* metamorphosis, which involves gradual, partial change from the first immature form until the last molt. (**c**) Fruit flies show *complete* metamorphosis. Tissues of immature forms are destroyed and replaced before emergence of the adult.

Figure 23.29 Representative insects. (**a**) Duck louse (order Mallophaga). It eats bits of feathers and skin. (**b**) European earwig (order Dermaptera), a common household pest. (**c**) Flea (order Siphonaptera), with strong legs for jumping onto and off animal hosts. (**d**) Mediterranean fruit fly (order Diptera). Its larvae destroy citrus fruit and other crops.

(**e**) Stinkbugs (order Hemiptera), newly hatched. (**f**) At center, a honeybee (order Hymenoptera) attracting its hive mates with a dance, as described in Section 44.4. (**g**) Ladybird beetles (order Coleoptera) swarming. These beetles are commercially raised and released as biological controls of aphids and other pests. Also in this order, the scarab beetle (**h**). With more than 300,000 species, Coleoptera is the largest order of the animal kingdom. (**i**) Luna moth (order Lepidoptera) of North America. Microscopic scales cover the wings and body of most butterflies and moths, including this one. (**j**) One of the dragonflies (order Odonata). It swiftly captures and eats insects in midflight.

they often transmit pathogenic microorganisms. On the bright side, many insects pollinate flowering plants in general and crop plants in particular. And many "good" insects attack or parasitize the ones we would rather do without. We now leave our survey of insects and other arthropods with representatives from some of the major orders of insects, shown in Figure 23.29.

As a group, insects show immense variation on the basic arthropod body plan. Many have wings, and their life cycles have stages that allow exploitation of different and often widely scattered food sources. Many species produce great numbers of small individuals that pass through immature, post-embryonic stages before the adult form emerges.

WWW

THE PUZZLING ECHINODERMS

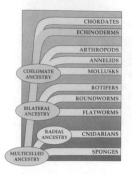

tube feet   spine

We turn, finally, to the second great lineage of coelomate animals, the deuterostomes. Major invertebrate members of this lineage are called **echinoderms** (Echinodermata). Sea stars as well as the sea urchins, sea cucumbers, brittle stars, and feather stars shown in Figure 23.30 belong to this phylum. So do the sea lilies (or crinoids), sea biscuits, and sand dollars. Sea lilies, which look a bit like stalked plants, flourished in Silurian times (Section 19.5). About 13,000 echinoderm species are known from the fossil record, but most became extinct. Nearly all of the 6,000 or so existing species live in marine habitats.

An echinoderm body wall bears a number of spines, spicules, or plates made rigid with calcium carbonate. (*Echinodermata* means spiny-skinned.) These structures serve defensive functions, as you might suspect if you have ever stepped barefoot on a sea urchin. Its spines trigger painful swelling if they break off under your skin. Most echinoderms have a well-developed internal skeleton, which is composed of calcium carbonate and other substances secreted from specialized cells.

Oddly, the adult echinoderms are radial with some bilateral features. Most species even produce bilateral larvae. Did bilateral invertebrates give rise to ancestors of echinoderms, in which some radial features evolved at a later time? Maybe.

Adult echinoderms have no brain. However, their decentralized nervous system allows them to respond to information about food, predators, and so forth that is coming from different directions. For instance, any

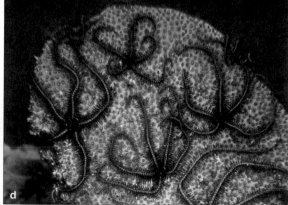

Figure 23.30 Representative echinoderms. (**a**) Feather star, with finely branched food-gathering appendages. (**b**) Sea urchin, which moves about on spines and tube feet. (**c**) Sea cucumber, with rows of tube feet along its body. (**d**) Brittle stars. Their arms (rays) make rapid, snakelike movements.

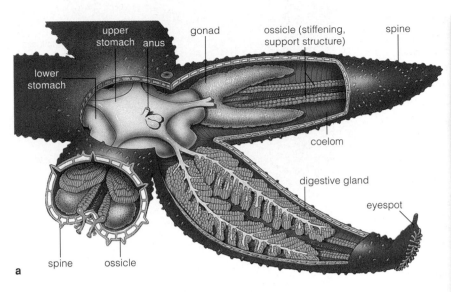

upper stomach — anus — gonad — ossicle (stiffening, support structure) — spine

lower stomach

coelom

digestive gland

eyespot

spine — ossicle

a

part of the water-vascular system

sieve plate

ring canal

ampulla

b

c

d

tube feet on the ventral surface of a sea star arm

Figure 23.31 (**a**) Some key aspects of the radial body plan of a sea star. The water-vascular system, in combination with great numbers of tube feet, is the basis of locomotion. (**b**–**d**) Five-armed sea star, with closer views of tube feet.

arm of a sea star that senses the shell of a tasty scallop can become the leader, directing the rest of the body to move in a direction suitable for prey capture.

Figures 23.30 and 23.31 show examples of tube feet. These fluid-filled, muscular structures have suckerlike adhesive disks. Sea stars use their tube feet for walking, burrowing, clinging to rocks, or gripping a clam or snail about to become a meal. Tube feet are components of a **water-vascular system** unique to echinoderms. In sea stars, that system includes a main canal in each arm. Short side canals extend from them and deliver water to the tube feet. Each tube foot has an ampulla, a fluid-filled, muscular structure shaped rather like the rubber bulb on a medicine dropper. As an ampulla contracts, it forces fluid into the foot and causes it to lengthen.

Tube feet change shape constantly as muscle action redistributes fluid through the water-vascular system. Hundreds of tube feet may move at a time. After being released, each one swings forward, reattaches to the substrate, then swings backward and is released before swinging forward again. Their motions are splendidly coordinated, so sea stars glide rather than lurch along.

On their ventral surface, sea stars have a formidable feeding apparatus, such as the one shown here:

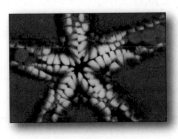

Some eager types simply swallow their prey whole. Others push part of their stomach outside the mouth and around their prey, then start digesting their meal even before swallowing it. Sea stars get rid of coarse, undigested residues through the mouth. They do have a small anus, but this is of no help in getting rid of an empty clam shell or snail shell.

With their curious traits, echinoderms are a suitable point of departure for this chapter. Even though we can identify broad trends in animal evolution, we should keep in mind that there are confounding exceptions to the perceived macroevolutionary patterns.

Echinoderms are coelomate animals with spines, spicules, or plates in the body wall. From the evolutionary perspective, they are a puzzling mix of bilateral and radial features.

WWW

SUMMARY

1. Animals are multicelled, aerobic heterotrophs that ingest or parasitize other organisms. Nearly all have diploid body cells organized into tissues, organs, and organ systems. Animals reproduce sexually and often asexually. They undergo embryonic development. Most are motile during at least part of the life cycle.

2. Animals range from structurally simple placozoans and sponges to vertebrates.

   a. By comparing major animal phyla and integrating the information with the fossil record, biologists have identified major evolutionary trends among them.

   b. Revealing aspects of body plans are the type of symmetry, gut, and cavity (if any) between the gut and body wall; whether there is a head end; and whether the body is divided into segments (Figure 23.32).

3. *Trichoplax*, the only known placozoan, is the simplest animal. It is composed of little more than two layers of cells, with a fluid matrix in between.

4. The sponge body has no symmetry and it is at the cellular level of construction. Although it consists of several kinds of cells, these are not organized into the epithelia and other tissues seen in complex animals.

5. Cnidarians include the jellyfishes, sea anemones, and hydras. They have radial symmetry and are at the tissue level of construction. Cnidarians alone produce nematocysts, capsules with dischargeable threads that are used mainly in prey capture.

6. Nearly all animals more complex than cnidarians show bilateral symmetry, and they form tissues, organs, and organ systems. Their gut may be saclike, as it is in flatworms, but it usually is complete, with an anus and a mouth. Most animals more complex than flatworms have a coelom or false coelom (cavities between the gut and body wall). A coelom's lining is the peritoneum.

7. Two major lineages diverged shortly after flatworms evolved. One (protostomes) gave rise to the mollusks, annelids, and arthropods. The other (deuterostomes) gave rise to echinoderms and chordates.

8. All mollusks have a fleshy, soft body and a mantle. Most have either a shell or a remnant of one. Mollusks vary greatly in size, body details, and life-styles.

9. The annelids (earthworms, polychaetes, and leeches) have a segmented body, complex organs, and a series of coelomic chambers.

10. Collectively, arthropods are the most successful of all groups in terms of diversity, numbers, distribution, defenses, and capacity to exploit food resources.

    a. All arthropods, including arachnids, crustaceans, and insects, have hardened, jointed exoskeletons. They also have modified segments, specialized appendages,

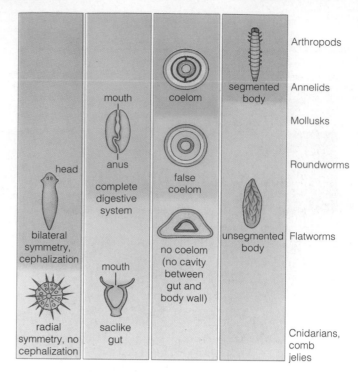

Figure 23.32 Summary of key trends in the evolution of animals, as identified by comparing body plans of major phyla. All of these features did not appear in every group.

highly specialized respiratory, nervous, and sensory organs, and (in insects only) wings.

   b. Arthropods develop by growth in size and molting or develop through a series of immature stages, such as larvae and nymphs. Many metamorphose; the tissues of immature forms undergo major reorganization and body parts are remodeled before the adult emerges.

11. Echinoderms have spines, spicules, or plates in their body wall. Evolutionarily, the phylum is puzzling. The larvae of most species form bilateral features, but they go on to develop into basically radial adults.

Review Questions

1. List the six main features that characterize animals. *23.1*

2. When attempting to discern evolutionary relationships among major groups of animals, which aspects of their body plans provide the most useful clues? *23.1*

3. What is a coelom? Why was it important in the evolution of certain animal lineages? *23.1*

4. Name some animals with a saclike gut. Evolutionarily, what advantages does a complete gut afford? *23.1, 23.4–23.6*

5. Choose a species of insect that lives in your neighborhood and describe some of the observable adaptations that underlie its success. *23.11, 23.15*

Self-Quiz (Answers in Appendix III)

1. Which is *not* a general characteristic of the animal kingdom?
   a. multicellularity; most have tissues, many form organs
   b. exclusive reliance on sexual reproduction
   c. motility at some stage of the life cycle
   d. embryonic development during the life cycle

2. Jellyfishes, sea anemones, and their relatives have _____ symmetry, and their cells form _____ .
   a. radial; mesoderm   c. radial; tissues
   b. bilateral; tissues   d. bilateral; mesoderm

3. In sheer numbers and distribution, _____ are the most successful animals.
   a. arthropods   c. snails and clams   e. vertebrates
   b. sponges   d. sea stars

4. Bilateral, segmented bodies and hardened exoskeletons occur among the _____ .
   a. arthropods   c. snails and clams   e. vertebrates
   b. sponges   d. sea stars

5. Which phylum contains members that are notorious for causing serious diseases in humans?
   a. cnidarians   c. segmented worms
   b. flatworms   d. chordates

6. Most animals that are more complex than cnidarians have _____ symmetry, and _____ forms in their embryos.
   a. radial; mesoderm   c. bilateral; mesoderm
   b. bilateral; endoderm   d. radial; endoderm

7. Match the terms with the appropriate groups.
   _____ sponges   a. spiny-skinned
   _____ cnidarians   b. vertebrates and kin
   _____ flatworms   c. flukes and tapeworms
   _____ roundworms   d. no tissue organization
   _____ rotifers   e. cilia crown; no males for some
   _____ mollusks   f. nematocysts, radial symmetry
   _____ annelids   g. hookworms, elephantiasis
   _____ arthropods   h. jointed exoskeleton
   _____ echinoderms   i. "belly-foots" and kin
   _____ chordates   j. segmented worms

Critical Thinking

1. A carnivorous sponge was discovered in an underwater cave in the Mediterranean Sea. Unlike other sponges, it has no pores and canals. Projecting from branching outgrowths of its body surface are hooklike spicules that act like Velcro to trap shrimp and other animals. The outgrowths surround prey, which sponge cells then digest. Which components of the sponge body had to evolve for such a feeding strategy?

2. Tapeworms are hermaphroditic. What selective advantages might this feature offer, and in what kinds of environments?

3. People who eat raw oysters or clams harvested from sewage-polluted waters develop mild to severe gastrointestinal ailments. Think about the feeding modes of these mollusks and develop a hypothesis to explain why mollusk eaters can get sick.

4. You are diving in calm, warm waters behind a tropical reef. You see a branching soft coral consisting of small individual polyps, each with a profusion of tiny tentacles. You will never see such a coral growing on reef surfaces exposed to the open sea. Propose two possible explanations for this.

5. A *rotifer* is a bilateral, cephalized animal with a crown of cilia that assists in swimming and in wafting food toward the mouth (Figure 23.33a). Nearly all rotifers live in freshwater, such as lakes, ponds, and films of water on mosses. A liter of pond-water typically holds 40 to 50 individuals; 5,000 were once recorded. They eat bacteria and microscopic algae. Most are not even a millimeter long, yet numerous organs are crammed into an unlined cavity between the gut and body wall. Among other things, rotifers have a pharynx, an esophagus, digestive glands, a stomach, protonephridia, usually an intestine and anus, and reproductive organs. Two "toes" exude substances that attach

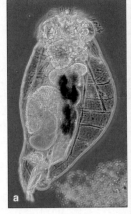

Figure 23.33 (a) A rotifer, releasing its eggs. (b) Go ahead, identify the mystery animal.

free-living species to substrates at feeding time. A cluster of nerve cell bodies in the head end integrates body activities. Some species have "eyes" (cells associated with light-sensitive pigments). Many have no males; females produce diploid eggs that become diploid females. In other species, males develop from haploid eggs that will develop into females if fertilized by males. Occasionally, dwarfed, short-lived males appear. Reflect on the characteristics just described. Then state which animals are closest to rotifers, evolutionarily speaking: (a) flatworms, (b) roundworms, or (c) mollusks.

6. Figure 23.33b shows a marine worm emerging from its sandy burrow. The worm has a segmented body. Most of its segments are similar to one another, and each has bristles on the sides that were used to dig into the sediments. To which phylum does this animal belong?

Selected Key Terms

animal 23.1	flatworm 23.5	nephridium 23.10
annelid 23.10	gonad 23.4	nerve cell 23.4
arthropod 23.11	gut 23.1	nerve cord 23.10
bilateral	hermaphrodite 23.5	nymph 23.15
symmetry 23.1	hydrostatic	organ 23.5
brain 23.10	skeleton 23.4	organ system 23.5
centipede 23.14	invertebrate 23.1	placozoan 23.2
cephalization 23.1	larva 23.3	planula 23.4
cnidarian 23.4	Malpighian	polyp 23.4
coelom 23.1	tubule 23.15	proglottid 23.5
collar cell 23.3	mantle 23.9	protostome 23.8
contractile cell 23.4	medusa 23.4	pupa 23.15
deuterostome 23.8	mesoderm 23.1	radial symmetry 23.1
echinoderm 23.16	metamorphosis 23.11	roundworm 23.6
ectoderm 23.1	millipede 23.14	sponge 23.3
endoderm 23.1	mollusk 23.9	vertebrate 23.1
epithelium 23.4	molting 23.11	water-vascular
exoskeleton 23.11	nematocyst 23.4	system 23.16

Readings   See also www.infotrac-college.com

Brusca, R., and G. Brusca. 1990. *Invertebrates.* Sunderland, Massachusetts: Sinauer. Fine reference book for serious students.

Pechenik, J. 1996. *Biology of Invertebrates.* Third edition. New York: McGraw-Hill. Paperback.

WWW   *http://www.brookscole.com/biology*

Practice quiz questions, hypercontents, BioUpdates, and critical thinking. The Brooks/Cole Biology Resource Center provides a wealth of information fully organized and integrated by chapter.

ANIMALS: THE VERTEBRATES

Making Do (Rather Well) With What You've Got

In 1798, a few naturalists were skeptically probing a specimen delivered to the British Museum in London, looking for signs that a prankster had slyly stitched the bill of an oversized duck onto the pelt of a small furry mammal. They didn't know it, but they were examining the remains of a platypus, a web-footed mammal about half the size of a housecat. Like the other mammals, the duck-billed platypus (*Ornithorhynchus anatinus*) has mammary glands and hair. However, like birds and reptiles, it has a cloaca, an enlarged duct through which gametes, feces, and excretions from the kidneys pass. It lays shelled eggs, as birds and most reptiles do. Its young hatch pink and unfinished, as embryonic stages too helpless to fend for themselves (Figure 24.1*a*). Its fleshy bill does look ducklike and its broad, flat, furry tail looks like the one on a beaver (Figure 24.1*b*).

With its unusual traits, the platypus invites us to challenge preconceived notions of what constitutes "an animal." This particular combination of traits started coming together when the supercontinent Pangea was breaking up. As you will see, the platypus's ancestors happened to be stuck on a huge fragment that remained isolated from the other continents for 150 million years; eventually it became Australia. The geographic separation was a springboard for genetic divergence, and unique mutations in combination with selection pressures gave rise to the platypus body. Even though that body plan is much ridiculed, we scarcely can call it a failure. It has endured far longer than the body plan we humans inherited.

The platypus inhabits streams and lagoons in remote regions of Australia and Tasmania. During the day it hides in underground burrows. At night it slips into the water, where it hunts for small invertebrates. Even when it stays in water all night, its dense fur

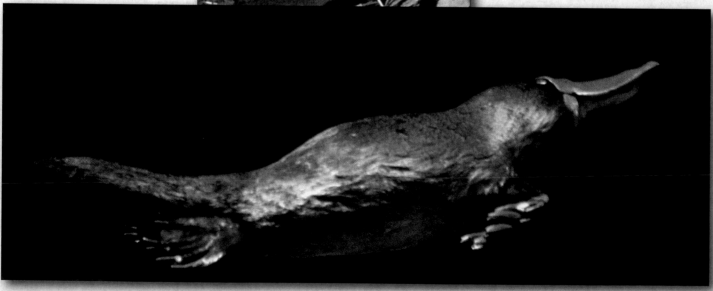

Figure 24.1 One of evolution's success stories—the platypus, (**a**) raising its incompletely formed offspring in a burrow and (**b**) underwater, eyes and ears shut, yet homing in on prey.

helps maintain body temperature, just as it does all day long when the platypus is resting in its cool burrow. The broad, thick tail is a most excellent rudder during underwater maneuvers and a fine storehouse for energy-rich reserves

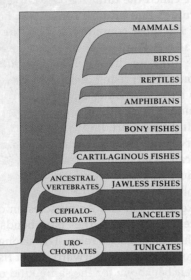

Figure 24.2 Family tree for the vertebrates and other chordates.

of fat. The broad, flattened bill is well shaped for scooping up aquatic snails, shrimps, worms, mussels, and insect larvae. The horny pads lining the platypus jaws grind up and make short work of those shelled and hard-cuticled meals. Back at the burrow, the platypus uses the strong claws that project from its hind feet for digging underground chambers.

Like a submarine, a platypus closes the hatches, so to speak, when it dives into the water. Its nostrils as well as a fleshy groove around the eyes and ears snap shut. No need for eyes or ears to zero in on prey; many of the 800,000 sensory receptors in its bill can detect tiny oscillations in water pressure as prey swim past. Other receptors detect changes in electric fields as weak as those initiated by the flicks of a shrimp tail.

The platypus is one of 4,500 kinds of mammals that now occupy the vertebrate branch of the animal family tree (Figure 24.2). Its vertebrate relatives include fishes, amphibians, reptiles, and birds, which are described in this chapter. This chapter surveys all vertebrate groups. It concludes with a look at the evolutionary history of the human species, starting with its mammalian and primate stocks. As you consider each group, keep this key concept in mind: *Each animal is a mosaic of traits, many conserved from remote ancestors and others unique to its branch on the animal family tree.*

KEY CONCEPTS

1. The chordate branch of the family tree for animals includes invertebrate and vertebrate species. All are bilateral animals. While they are embryos or larvae, each typically develops a supporting rod for the body (notochord), a dorsal nerve cord, a pharynx, gill slits in the pharynx wall, and a tail that extends past the anus. Some or all of these features persist in adults.

2. Existing invertebrate chordates include the tunicates and the lancelets.

3. Of eight classes of vertebrates, seven have living representatives. These are jawless fishes, cartilaginous fishes, bony fishes, amphibians, reptiles, birds, and mammals. The other class, the placoderms, became extinct early in vertebrate history.

4. Four major trends occurred when certain vertebrate lineages evolved. Support and movement of the body came to depend less on the notochord and more on a backbone. After jaws evolved, the nerve cord evolved into a spinal cord and brain. When some lineages moved onto land, gills became less important than lungs for gas exchange, a trend enhanced by the evolution of more efficient circulatory systems. Among pioneers on land, fleshy fins with skeletal supports evolved into limbs, which evolved in diverse ways among the amphibians, reptiles, birds, and mammals.

5. In the primate branch of the mammalian lineage are prosimians, tarsioids, and anthropoids (monkeys, apes, and humans). Apes and humans are hominoids. Humans and some extinct species with a mosaic of apelike and humanlike traits are further classified as hominids.

6. Starting in the Miocene, a long-term cooling trend led to seasonal changes in habitats and food sources. As food sources became scarcer, hominids spread out through Africa and, later, entered Europe.

7. Unlike earlier hominids, *Homo erectus* and *H. sapiens* displayed remarkable behavioral flexibility and creative experimentation with their environment, as when they started using fire. This characteristic allowed them to survive the challenges of dispersing into novel and often harsh environments around the world.

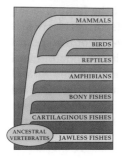

Few of us make a career of studying life beneath the surface of the seas and other bodies of water, so fishes are not widely recognized as being the world's dominant vertebrates. Their numbers exceed those of all other vertebrate groups combined. They also show far more diversity; we have identified more than 21,000 species of bony fishes alone.

The form and behavior of a fish tell us something about the challenges it faces in water. For example, being about 800 times more dense than air, water resists fast motions. As an adaptation to this constraint, sharks and other predatory marine fishes are streamlined for pursuit (Figure 24.8*a*). Their long, trim body reduces friction; their tail muscles are organized for propulsive force and forward motion. By contrast, some bottom-dwelling fishes, such as the rays, have a flattened body (Figure 24.8*b*). Theirs is not a high-speed body plan; it is good for hiding out from predators or prey.

A motionless trout, suspended in shallow water, is another example of adaptation to water's density. Like many fishes, it maintains neutral buoyancy with a **swim bladder**, an adjustable flotation device that exchanges gases with blood. When a fish gulps air at the water's surface, it is adjusting the volume of its swim bladder.

Cartilaginous Fishes

Cartilaginous fishes (Chondrichthyes) include about 850 species of skates, sharks, and chimaeras, as shown in Figure 24.8. These marine predators are equipped with pronounced fins, a skeleton of cartilage, and five to seven gill slits on both sides of the pharynx. Most species have a few scales or many rows of them. **Scales** are small, bony plates at the body surface. Fish scales often protect the body without weighing it down.

Skates and rays are mainly bottom dwellers with flattened teeth suitable for crushing hard-shelled prey. Both have enlarged fins that extend onto the side of the head. The largest species, the manta ray, measures up to six meters from fin tip to fin tip. A venom gland in the tail of stingrays probably helps to deter predators. Other rays have electric organs in the tail or fins that can stun prey with as much as 200 volts of electricity.

At fifteen meters from head to tail, some sharks are among the largest living vertebrates. As Figure 24.6*d* shows, sharks have formidable jaws. They continually shed and replace sharp-edged teeth (modified scales), which they use to grab prey and rip off chunks of flesh. The relatively few shark attacks on humans have given the group a bad reputation. Surfboards with human legs dangling over the side are a recent temptation for some

Figure 24.8 Cartilaginous fishes: (**a**) shark, (**b**) blue-spotted reef ray, and (**c**) chimaera, sometimes called a ratfish.

sharks, which have been hunting invertebrates, fishes, and marine mammals for many millions of years.

The thirty or so species of chimaeras feed mostly on mollusks. With their bulky body and slender tail, they look a bit like a rat; hence the common name, ratfishes. They have a venom gland in front of the dorsal fin.

Bony Fishes

Bony fishes (Osteichthyes) are the most numerous and diverse vertebrates. They make up all but 4 percent of modern fish species. Their ancestors arose in Silurian times and gave rise to three lineages: ray-finned fishes, lobe-finned fishes, and lungfishes. Descendants of the early forms radiated into nearly all aquatic habitats.

Body plans vary greatly. Marine predators typically have a torpedo shape, a flexible body, and strong tail fins that aid in swift pursuit. Many reef dwellers are small finned and box shaped; they move easily through narrow spaces. Those with an elongated, flexible body, including the moray eel, wriggle through mud and slip into concealing crevices. Sea horses and many bottom-dwelling species have a cryptic body shape that helps conceal them from prey or from predators. Figure 24.9 shows examples of these varied body plans.

In ray-finned fishes, outward-projecting rays that are derived from the dermis (one of the skin's layers)

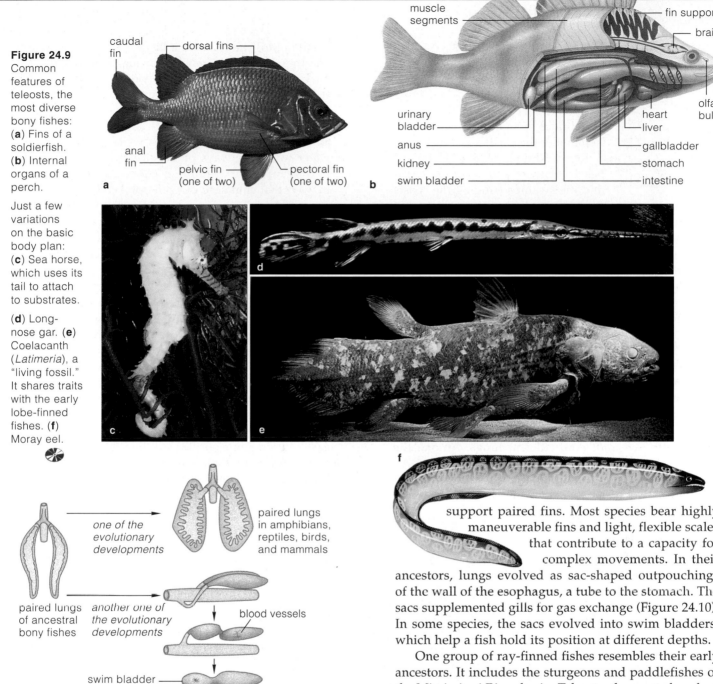

Figure 24.9 Common features of teleosts, the most diverse bony fishes: (**a**) Fins of a soldierfish. (**b**) Internal organs of a perch.

Just a few variations on the basic body plan: (**c**) Sea horse, which uses its tail to attach to substrates.

(**d**) Long-nose gar. (**e**) Coelacanth (*Latimeria*), a "living fossil." It shares traits with the early lobe-finned fishes. (**f**) Moray eel.

a
caudal fin
dorsal fins
anal fin
pelvic fin (one of two)
pectoral fin (one of two)

b
muscle segments
fin supports
brain
olfactory bulb
heart
liver
gallbladder
stomach
intestine
urinary bladder
anus
kidney
swim bladder

one of the evolutionary developments
paired lungs in amphibians, reptiles, birds, and mammals

paired lungs of ancestral bony fishes
another one of the evolutionary developments
blood vessels
swim bladder

Figure 24.10 Evolution of swim bladders and lungs. Respiratory surfaces are coded *pink* and the esophagus (a tube leading to the stomach), *gold*. Lungs evolved as outpouchings of the esophagus. Being in close contact with blood vessels, they increased the surface area for gas exchange in oxygen-poor habitats. The lung sacs became swim bladders (buoyancy devices) in some species.

Trout and other less specialized fishes surface and gulp air, which enters the swim bladder through a duct from the esophagus. Most bony fishes have no such duct. Gases dissolved in blood diffuse into a swim bladder that has a dense mesh of arteries and veins. Blood flow through the vessels increases the concentrations of gases in the swim bladder. Another area of the swim bladder enhances reabsorption of gases by cells in body tissues.

support paired fins. Most species bear highly maneuverable fins and light, flexible scales that contribute to a capacity for complex movements. In their ancestors, lungs evolved as sac-shaped outpouchings of the wall of the esophagus, a tube to the stomach. The sacs supplemented gills for gas exchange (Figure 24.10). In some species, the sacs evolved into swim bladders, which help a fish hold its position at different depths.

One group of ray-finned fishes resembles their early ancestors. It includes the sturgeons and paddlefishes of the Mississippi River basin. Teleosts, the most abundant fishes, include salmon, tuna, rockfish, catfish, perch, minnows, moray eels, flying fish, sculpins, blennies, scorpionfish, and pikes. All are thin scaled or scaleless.

By contrast, only one species of lobe-finned fishes and three genera of lungfishes made it to the present. As their name suggests, a **lobe-finned fish** is unique in having paired fins that incorporate fleshy extensions from the body. As you will see next, neither it nor the lungfishes have changed much from their ancient stock.

Of all existing vertebrates, the bony fishes are the most spectacularly diverse and the most abundant.

www

AMPHIBIANS

Origin of Amphibians

The coelacanth, the lobe-finned fish shown in Figure 24.9e, is a "living fossil," a relic of an early time when vertebrates first moved onto land. Remember the earlier description of living conditions in the Devonian? Sea levels rose and fell repeatedly, and the swamps fringing the coasts flooded and drained numerous times. The ancestors of lobe-finned fishes evolved in those trying times. They probably used their lobed fins to pull themselves from dried-up ponds to ones that still had water and were habitable. What else made their pond-to-pond lurchings possible? They gulped air, and they had lungs.

Existing lungfishes provide more clues to how the ancestral forms might have made it through stressful times. They live in stagnant water but surface to gulp air. In dry seasons, when streams dwindle to mud, they encase themselves in a slathering of mud and slime that keeps them from drying out until the next rainy season.

Devonian lobe-finned fishes lurched over land only as a way of reaching more hospitable ponds. Yet the very act of traveling out of water favored the evolution of stronger fins and more efficient lungs. Among the evolving forms were the ancestors of amphibians. An **amphibian** (Amphibia) is a vertebrate with a body plan and reproductive mode somewhere between fishes and reptiles. Most kinds have a largely bony endoskeleton, and they have four legs (or four-legged ancestors).

Early amphibians were spending time on land by the close of the Devonian (Figure 24.11a). For them, life in those drier habitats was dangerous—and promising. Temperatures shifted more on land than in the water, air didn't support the body as well as water did, and water was not always plentiful. However, air is richer in oxygen. Amphibian lungs continued to be modified in ways that enhanced oxygen uptake. Also, circulatory systems became better at rapidly distributing oxygen to living cells throughout the body. Both modifications increased the energy base for more active life-styles.

New sensory information also challenged the early amphibians. Swamp forests supported vast numbers of edible insects and other invertebrate prey. Animals with good vision, hearing, and balance—the senses that are most advantageous on land—were favored. And brain regions concerned with those senses expanded.

All of the salamanders, frogs, toads, and caecilians alive today are descended from those first amphibians. None has escaped the water entirely. Even when gills or lungs are present, amphibians can use their thin skin as a respiratory surface (that is, for gas exchange). But respiratory surfaces must be kept moist, and skin dries easily in air. Some species still live their entire lives in water; others lay their eggs in water or produce aquatic larvae. Even species that have adapted to habitats on land must lay their eggs in moist places.

a

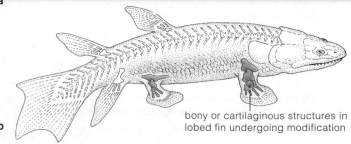

b

bony or cartilaginous structures in lobed fin undergoing modification

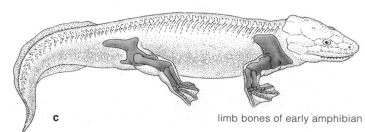

c

limb bones of early amphibian

Figure 24.11 (**a**) *Ichthyostega*, one of the first of the Devonian amphibians. Fossils of this species have been recovered in Greenland. The skull, deep tail, and fins were decidedly fishlike. Unlike fishes, this species had four limbs adapted for moving on land, and a short neck intervened between its head and the rest of the body. Its vertebral column and rib cage were adapted to support the body's weight out of water. (**b,c**) Proposed evolution of skeletal elements inside the lobed fins of certain fishes into the limb bones of early amphibians.

Salamanders

Diverse salamanders and their kin, the newts, live in the world's north temperate regions and tropical parts of Central and South America. Most are less than fifteen centimeters long. And most have legs at right angles to the body, with forelimbs and hindlimbs about the same size (Figure 24.12a). Like fishes and early amphibians, salamanders bend from side to side when they walk:

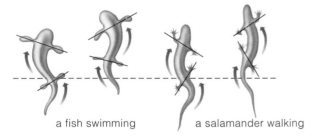

a fish swimming   a salamander walking

Probably the first four-legged vertebrates also walked this way. Both the larvae and adults are carnivores. Adults of some species retain several larval features. For example, the Mexican axolotl retains the external gills of the larval form, and the development of its teeth and bones is arrested at an early stage. As in some other species, its larvae are sexually precocious; they breed.

Frogs and Toads

With more than 3,000 species, frogs and toads are the most successful amphibians (Figure 24.12b,c). Their long

Figure 24.12 Amphibians. (**a**) Terrestrial stage in the life cycle of a red-spotted salamander. (**b**) A frog, splendidly jumping. (**c**) An American toad. (**d**) A caecilian.

hindlimbs and powerful muscles allow them to catapult into the air or barrel through the water. Most frogs flip a sticky-tipped, prey-capturing tongue from their mouth. An adult eats just about any animal it can catch; only its head size limits the size of prey. Skin glands of some species produce toxins, and poisonous types often have bright warning coloration. The skin of frogs, including the South African clawed frog (*Xenopus laevis*), harbors antibiotics against many pathogens living in the water. In Unit VI, you will have opportunities to take a look at how frogs are put together and how they function.

Caecilians

The ancestors of caecilians lost their limbs and most of their scales. They gave rise to decidedly worm-shaped amphibians (Figure 24.12d). Nearly all of the 160 or so existing species burrow through soft, moist soil as they pursue insects and earthworms. A few live in shallow freshwater habitats. Adults of most species are blind.

Amphibians are halfway between fishes and reptiles in their body form and behavior. Regardless of how far they venture onto land, they have not fully escaped dependency on aquatic or moist habitats to complete the life cycle.

WWW

THE RISE OF REPTILES

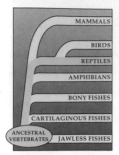

In the late Carboniferous, amphibians gave rise to **Reptiles** (Reptilia). The reptiles were the first vertebrates to escape from aquatic habitats. They did so through four adaptations that clearly distinguish them from fishes and amphibians. *First*, reptiles have tough, dry, scaly skin that restricts water loss from their body (Figure 24.13). *Second*, their eggs are fertilized internally. A copulatory organ deposits sperm into a female's body; sperm do not need free water to reach eggs. Third, reptilian kidneys are good at conserving water. Fourth, most species have **amniote eggs**, within which embryos develop to an advanced stage before being hatched or born into dry habitats (Figure 24.14). The eggs have membranes that retain water and protect or metabolically support embryos. Most have a leathery or calcified shell. We will take a look at the structure and development of amniote eggs in Chapter 38.

Compared to amphibians, the early reptiles chased prey with far greater cunning and speed. With their well-muscled jawbones and formidable teeth, reptiles seized and applied sustained, crushing force on prey. Their prey included insects and other vertebrates. Also, their limbs generally were better at supporting the body's trunk on land. For many species, the nervous system became more complex. A reptile's brain is small compared to the rest of the body mass, but it governs forms of behavior unknown among amphibians. The cerebral cortex, where information from sensory organs is integrated and stored, first evolved among reptiles.

Reptiles called crocodilians were the first animals with a muscular, four-chambered heart fully separated into two halves. (Blood enters the first chamber of each half, and the second pumps it out.) Such a separation allows oxygen-rich blood to travel from lungs to the rest of the body, and oxygen-poor blood from the body to the lungs, in two separate circuits. Also, gas exchange

Figure 24.14
(**a**) Reconstruction of the nest of a duck-billed, plant-eating dinosaur (*Maiasaura*), discovered in Montana. This species, which lived 80 million years ago, showed social behavior. It traveled in herds and cared for the young. (**b**) Eastern hognose snakes, shown emerging from leathery shelled amniote eggs.

across skin, so vital for amphibians, was abandoned by reptiles—nearly all of which have well-developed lungs.

Early reptiles underwent adaptive radiations that led to fabulous diversity. The group called dinosaurs evolved in the Triassic and for the next 125 million years were the dominant land vertebrates. Fossilized nests of one kind (*Maiasaura*) contain not only eggs but also juveniles a few months old, at most (Figure 24.14). Such evidence implies that at least some dinosaurs showed parental behavior; they took care of their young during a period of dependency. When the Cretaceous ended abruptly, so did the last dinosaurs (Sections 19.6 and 19.7). The reptiles that made it to the present are the crocodilians, turtles, tuataras, snakes, and lizards.

The name Reptilia is from the Latin *repto* (to creep). Maybe the first reptiles did creep about, but a capacity to move fast and with agility evolved among some of the descendants, which race, lumber, and slither about.

Figure 24.13 Body plan of a representative reptile, a crocodile.

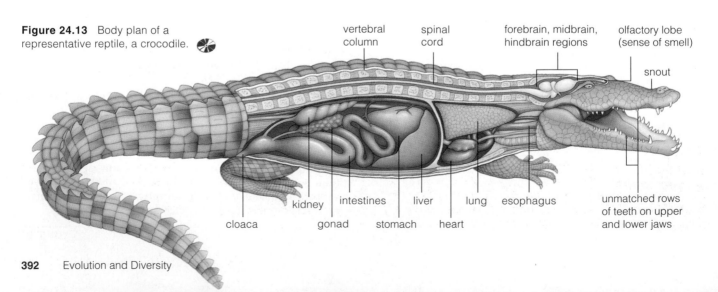

vertebral column · spinal cord · forebrain, midbrain, hindbrain regions · olfactory lobe (sense of smell) · snout · unmatched rows of teeth on upper and lower jaws · esophagus · lung · heart · liver · stomach · gonad · intestines · kidney · cloaca

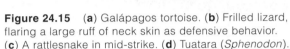

Figure 24.15 (**a**) Galápagos tortoise. (**b**) Frilled lizard, flaring a large ruff of neck skin as defensive behavior. (**c**) A rattlesnake in mid-strike. (**d**) Tuatara (*Sphenodon*).

Among the modern crocodiles and alligators are the largest living reptiles and the closest living relatives of birds. Crocodiles and alligators have powerful jaws, a slender snout, and sharp teeth (Figure 24.13). All live near or in water. The "man-eaters" of southern Asia and Nile crocodiles drag prey into water, spin violently to tear it apart, then gulp the torn chunks. Crocodilians control body temperature by adjustments in behavior and physiology, as other reptiles and birds do. Like the birds, they show complex social behavior, as when they guard nests and assist hatchlings into water.

The 250 existing species of turtles live inside a shell attached to the skeleton (Figure 24.15a). Most pull their head and limbs inside if threatened. The body plan works well; it has been around since Triassic times. Only sea turtles and other highly mobile types have a reduced shell. Turtles have tough, horny plates for gripping and chewing food. They have strong jaws and often a fierce disposition. All turtles lay eggs on land, then leave them to develop on their own.

About 95 percent of modern-day reptiles, the lizards and snakes, are distantly related to dinosaurs. Most are small. But the Komodo monitor lizard is large enough to prey on young water buffalo, and the longest snake would stretch across ten yards of a football field.

Most lizards, including iguanas and geckos, live in deserts and tropical forests. Most grab insect prey with small, peglike teeth; chameleons capture them with flicks of a tongue that is longer than the body (Section 28.4). Some lizards startle predators and intimidate rivals by flaring a throat fan (Figure 24.15b). Many give up their tail when grabbed. The detached tail wriggles and may be distracting enough to permit a getaway.

In the late Cretaceous, elongated, limbless snakes evolved from some of the short-legged, long-bodied lizards. Some pythons and other snakes have bony remnants of ancestral hindlimbs (Section 16.1). Most of the 2,300 or so existing species slither in S-shaped waves, much like salamanders. The movements of "sidewinders" across loose sand or sediments leave a pattern of J-shaped tracks that is easy to recognize.

Snakes have amazingly movable jaws; some swallow animals wider than they are. The pythons and boas coil around prey and suffocate it into submission. Fanged types, including rattlesnakes and coral snakes, bite and subdue prey with venom (Figure 24.15c). Snakes are not usually aggressive toward humans, but each year in the United States alone, rattlesnakes (one of the pit vipers) and other poisonous types bite 8,000 or so people and kill about 12 of them. In India, king cobras and other types bite 200,000 or so and kill about 9,000 of them.

Even the most feared snakes are vulnerable during their life cycle; birds and other predators relish snake eggs. The female snakes store sperm and lay several clutches of fertilized eggs at intervals after they mate, which improves the odds that at least some will hatch.

Two species of tuataras live on small islands near New Zealand (Figure 24.15d). Their body plan has not changed much since the Mesozoic. Tuataras look like lizards but are of a more ancient lineage. At the top of their head is a third "eye," with retina and lens. Being covered with skin, it only detects changes in daylength and light intensity. It might function in the hormonal control of reproduction (compare Section 31.8). Like turtles, the tuataras may live longer than sixty years.

Reptiles, with their tough, scaly skin, reliance on internal fertilization, water-conserving kidneys, and amniote eggs, were the first vertebrates to escape dependency on free-standing water in their habitats. The move onto land also involved major modifications in the nervous, circulatory, and respiratory systems.

WWW

24.8 BIRDS

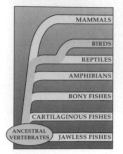

Of all organisms, only **birds** (Aves) grow feathers. **Feathers**, lightweight structures derived from skin, are used for flight, body insulation, or both. Judging from the fossil record, birds are descended from tiny reptiles that ran around on two legs 160 million years ago. *Archaeopteryx* was on or near the lineage leading to modern birds. It had many reptilian traits as well as feathers and other avian traits (Section 16.3).

In many respects, birds still resemble reptiles. For example, they have scales on their legs and a number of the same internal structures. They, too, lay eggs and commonly engage in parenting behavior, as when they guard the nest (Figure 24.16). Thomas Huxley, one of Darwin's supporters, was the first to argue that birds are glorified reptiles. Today, most biologists do indeed classify birds as a branching from the reptilian lineage.

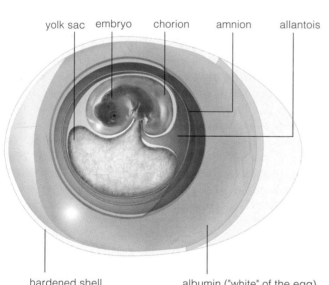

yolk sac   embryo   chorion   amnion   allantois

hardened shell   albumin ("white" of the egg)

Figure 24.16 Characteristics of birds. (**a**) Generalized sketch of an amniote egg. All birds lay these eggs. (**b**) Flight. Of all living vertebrates, only birds and bats fly by flapping their wings. (**c**) Feathers, the key defining characteristic of birds. The flamboyant plumage of this male pheasant is an outcome of sexual selection. This bird is a native of the Himalaya Mountains of India, and it is an endangered species. As is the case for many other kinds of birds, its bright, jewel-colored feathers end up adorning humans—in this case, on the caps of native tribespeople.

(**d**) Many birds, including these Canada geese, display migratory behavior. *Animal migration* is a recurring pattern of movement between two or more places in response to environmental rhythms. For instance, seasonal change in daylength is a cue that acts on internal timing mechanisms (biological clocks), which trigger physiological and behavioral changes in individuals. Such changes induce migratory birds to make round trips between distant regions that differ in climate. Canada geese spend the summer nesting in marshes and lakes in the northern United States and Canada. Their wintering grounds are in New Mexico and other parts of the southern United States.

(**e**) Speckled, hard-shelled eggs of a magpie, unhatched in the nest.

There are almost 9,000 named species of birds. They show stunning variation in their body size, proportions, coloration, and capacity for flight. One of the smallest hummingbirds barely tips the scales at 2.25 grams (0.08 ounce). The largest existing bird, the ostrich, weighs about 150 kilograms (330 pounds). Ostriches cannot fly, but they are impressively long-legged sprinters (Figure 16.2). Many birds, such as warblers and other perching types, differ markedly in feather coloration and in their territorial songs. Bird songs and other complex forms of social behavior are topics of later chapters.

Flight demands high metabolic rates, which require a good flow of oxygen through the body. Just as you do,

birds have a large, durable, four-chambered heart. The heart pumps oxygen-enriched blood to lungs by one route and to the rest of the body by another, as it probably did in the reptilian ancestors. Also, a bird's respiratory system incorporates a unique ventilating apparatus that greatly enhances the uptake of oxygen. Bird respiration is described in Section 35.3.

Flight also demands an airstream, low weight, and a powerful downstroke that can provide lift (a force at right angles to the airstream). The bird wing, a modified forelimb, consists of feathers and of lightweight bones attached to powerful muscles. Bird bones weigh very little because of profuse air cavities in the bone tissue.

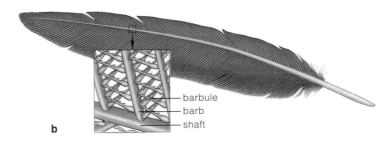

Figure 24.17 (a) Body plan of a typical bird. Flight muscles attach to the large, keeled breastbone (sternum). (b) A bird wing is a complex system of lightweight bones and feathers. Feathers gain strength from a hollow central shaft and from tiny barbules that are interlocked in a latticelike array.

For example, the skeleton of a frigate bird, which has a seven-foot wingspan, weighs only four ounces. That is less than the feathers weigh! A bird's flight muscles attach to an enlarged breastbone, or sternum, and to upper limb bones next to it (Figure 24.17). Contraction of the muscles creates the powerful downstroke required for flight. The wings, with their long flight feathers, serve as the airfoils. Usually, a bird spreads out long feathers on a downstroke and thus increases the size of the surface pushing against air (Figure 24.16b). On the upstroke, it folds the feathers somewhat, so that each wing presents the least possible resistance to the air.

Of all animals, birds alone have feathers, which they use in flight, in heat conservation, and in socially significant communication displays. They all lay hard-shelled eggs.

WWW

THE RISE OF MAMMALS

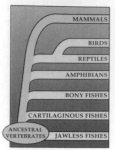

Mammals are the only vertebrates with hair and mammary glands. The name of class Mammalia refers to the Latin *mamma* (breast). Females feed the young with milk, a nutrient-rich fluid of mammary glands that moves through ducts to openings at the body surface (Figure 24.18*a*).

Mammals also care for the young for extended periods, and adults are models for their behavior. Mammals are born with a capacity to learn and to repeat behavior with survival value. They show stunning behavioral flexibility—a capacity to embroider basic reflex responses with novel behavior—although this is more pronounced in some species than in others. Behavioral flexibility coevolved with expansion of brain regions, and it became most notable among the mammals called primates.

Unlike most reptiles, which swallow prey whole, most mammals secure, cut, and sometimes chew food first. They also differ from reptiles in **dentition** (in the type, number, and size of teeth). Mammals have four kinds of upper and lower teeth that match up and work together to cut, crush, or grind food (Figure 24.18*b*). Incisors (flat chisels or cones) nip or cut food and are large in grazing mammals. Canines, with piercing points, are longest in meat eaters. Premolars and molars (cheek teeth) are a platform with surface bumps (cusps) used to crush, grind, and shear food. If a mammal has large, flat-surfaced cheek teeth, then its ancestors evolved in places where fibrous plants were abundant foods.

During the Triassic, a genetic divergence from small, hairless reptiles (synapsids) gave rise to therapsids, the ancestors of mammals. By Jurassic times, diverse plant-eating and meat-eating mammals called therians had evolved. Most were as small as a mouse, with hair and with modifications in jaws, teeth, and body form. For instance, they held their four limbs upright, under the body. That arrangement made it easier to walk erect but was less stable than a trunk held closer to the ground. During this time the cerebellum, a brain region dealing with balance, started expanding in volume.

Therians coexisted with dinosaurs in the Cretaceous. When the last dinosaurs vanished, diverse adaptive zones opened up for three mammalian lineages. In one lineage, young hatched from eggs laid by the females. In another, young were born live and completed early development in a permanent pouch on the mother's body surface. In the third lineage, a placenta evolved. A **placenta** is a spongy, blood-filled tissue of maternal and fetal membranes. It forms in a female's uterus, a chamber in which embryos complete development in relative freedom from harsh conditions on the outside (Figure 24.19). A placenta delivers nutrients and oxygen

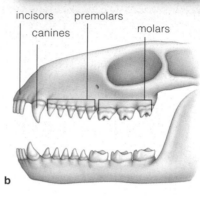

incisors premolars
canines molars

b

Figure 24.18 Three distinctly mammalian traits. (**a**) A human baby busily demonstrating a key defining feature: it derives nourishment from mammary glands. (**b**) Unlike the teeth of reptilian ancestors, the upper and lower rows of mammalian teeth match up. (**c**) Furry juvenile raccoons.

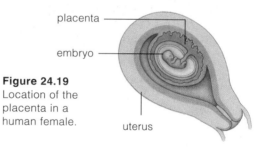
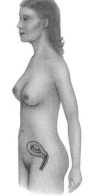

placenta

embryo

Figure 24.19 Location of the placenta in a human female.

uterus

to the embryo and removes its wastes. Compared to embryos of other lineages, those of placental mammals grow faster. Many are fully formed at birth.

Such differences in traits, in combination with new opportunities on the changing geologic stage, put the early mammalian lineages on different paths. By the late Jurassic, ancestors of egg-laying and pouched mammals were living in the southern part of Pangea. After that supercontinent split up, species on the land mass that became Australia evolved in isolation from ancestors of placental mammals. On the land mass that became South America, egg layers were replaced by the other two lineages. In North America, placental mammals became highly evolved. In Pliocene times they migrated across a land bridge into South America and rapidly drove many of the previously isolated mammals to extinction. Only the opossums and a few other species successfully invaded North America from the south.

The only living egg-laying mammals are the duck-billed platypus, described in the chapter introduction,

Figure 24.20 Representative mammals. (**a**) From Australia, a marsupial: a young kangaroo (joey) inside its mother's pouch. (**b**) A spiny anteater (*Tachyglossus*), an egg-laying mammal. A few placental mammals: (**c**) One of the bats, the only truly flying mammals, which dominate the night sky vacated by birds. (**d**) A herd of camels traversing hot desert sand with ease. (**e**) A manatee. It lives underwater and feeds on submerged vegetation. (**f**) An arctic fox, with thick, insulative fur that camouflages it from prey. In summer, its light-brown fur blends with golden-brown grasses. In winter, the fur turns white and blends with snow-covered ground.

and two species of spiny anteaters (in Australia and in New Guinea.) Spiny anteaters are small, burrowing mammals. Like porcupines, they bristle with protective spines that are modified hairs. Most of the 260 existing species of pouched mammals, or marsupials, are native to Australia and neighboring islands; a few live in the Americas. Newborns are tiny, blind, and hairless. All other modern descendants of the therians are placental mammals, a few of which are shown in Figure 24.20. Appendix I lists the major groups. Later chapters touch on their body form, function, behavior, and ecology.

Having been geographically isolated for so long, the three mammalian lineages are evolutionarily distant. Yet many of their ancestors evolved in similar ways in similar habitats, and they now resemble one another in form and function. Thus they are a prime example of **convergent evolution** (Section 18.2). Australia's spiny anteater, South America's giant anteater, and Africa's aardvark are alike in body form and life-styles. Several families of tree dwelling, nut- and seed-eating flying squirrels occur in Southeast Asia, Africa, and Australia. North America's wolf and Australia's Tasmanian wolf resemble each other in form and function. So do South America's howler monkey, Africa's colobus monkey, Madagascar's woolly lemur, and Australia's koala—all tree-dwelling, leaf-eating mammals.

In terms of diversity, mammals do not approach, say, the mollusks or arthropods. Yet in size, body form, functioning, and life-styles, the 4,500 known species are breathtaking. They range from the 1.5-gram Kitti's hognosed bat to 100-ton whales. They outlasted the dinosaurs only to confront a new challenge—a human population bent on hunting them, encroaching on their habitats, and introducing novel species or favoring a few domesticated ones at the expense of wild stock. Think of it. In the 1800s, bison were nearly driven to extinction as part of a program to starve the Plains Indians into submission by killing off their major food source. Modern whaling industries of Japan, Norway, and elsewhere are exterminating the very mammals they hunt. About 300 mammalian species, including most whales, wild cats, otters, and primates other than humans, currently have the bad luck to be members of the endangered species club.

Mammals alone have hair and mammary glands. They have distinctive dentition, a highly developed nervous system, and a notable capacity for behavioral flexibility.

Much of mammalian history was a matter of luck, of species with particular traits being in the right or wrong places on a changing geologic stage at particular times.

WWW

Having traversed the mammalian branch of the animal family tree, we are now ready to follow it along the branchings leading to primates, then to humans. As we explore the branchings, keep this key point in mind: *"Uniquely" human traits emerged through modification of traits that evolved earlier, in ancestral forms.*

In the order **Primates** are prosimians, tarsioids, and anthropoids (Appendix I). Figure 24.21 shows a few. The first prosimians were arboreal (tree-dwelling). They dominated northern forests for millions of years before monkeys and apes evolved and almost displaced them. Small tarsiers of Southeast Asia, the only living tarsioids, are somewhere between prosimians and anthropoids in traits. Monkeys, apes, and humans are all anthropoids. In biochemistry and body form, the apes are closer to humans than to monkeys. Apes, humans, and extinct species of their lineages are classified as hominoids. The **hominids** are all humanlike and human species of a line of descent that started with its divergence from the last shared ancestor of apes and humans.

Most primates live in tropical or subtropical forests, woodlands, or savannas, which are open grasslands with a few stands of trees. Like their ancestors, the vast majority are tree dwellers. Yet no one feature sets them apart from other mammals, and each lineage has its own defining traits. Five trends define the lineage that led to humans. They were set in motion when primates started adapting to life in the trees. *First*, there was less reliance on the sense of smell and more on daytime vision. *Second*, skeletal changes led to upright walking, which freed the hands for novel tasks. *Third*, changes in bones and muscles led to refined hand movements. *Fourth*, teeth became less specialized. *Fifth*, changes in the brain and behavior became interlocked with each other and with cultural evolution. **Culture** is the sum of behavior patterns of a social group, handed down among the generations through learning and symbolic behavior, especially language.

Origins and Early Divergences

Primates evolved from mammals more than 60 million years ago, in tropical forests of the Paleocene. Like the small rodents and tree shrews they resembled (Figure 24.22), they had huge appetites and foraged at night for insects, seeds, buds, and eggs beneath the trees. They had a long snout and a good sense of smell, suitable for snuffling food and predators. They clawed their way up stems and branches, although not with much speed or grace. During the Eocene (between 54 and 38 million years ago), some primates were staying in the trees. They had a shorter snout, enhanced daytime vision, a larger brain, and better ways to grasp objects. *How did these traits evolve?*

Figure 24.21 A few primates. (**a**) A gibbon's body and limbs are adapted for swinging arm over arm in trees. The existing apes are gibbons, orangutans, gorillas, and chimpanzees. (**b**) Spider monkey, a four-legged climber, leaper, and runner. (**c**) Tarsiers, which are vertical climbers and leapers.

Figure 24.22 Nocturnal tree shrew of Indonesia.

Consider the trees. They offered abundant food and safety from ground-dwelling predators. They also were habitats of uncompromising selection. Picture an Eocene morning with dappled sunlit boughs swaying in the breeze, colorful fruit, and predatory birds. A long, odor-sensitive snout would not have been of much use in the trees, where air currents disperse odors. But a brain able to evaluate motion, depth, shape, and color—and one that worked fast when its owner ran and leaped about—would have been a plus. (The brain simultaneously had to estimate body weight, distance, wind speed, and the suitability of a destination—and adjustments had to be quick.) Also, skeletal changes that increased fitness would have been favored. (For instance, forward-directed eye sockets would have contributed to enhanced depth perception.)

By 36 million years ago, tree-dwelling anthropoids had

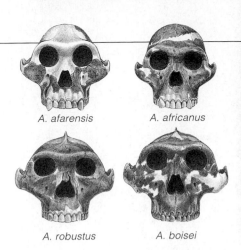

A. afarensis   *A. africanus*

A. robustus   *A. boisei*

Figure 24.23 Australopith skulls (part fossil, part reconstruction), which were apelike in size and proportion. Males of *A. afarensis*, *A. robustus*, and *A. boisei* had a crested skull, like a gorilla's, which afforded attachment sites for large jaw muscles. Also like apes, *A. afarensis* had a large face and projecting jaws. But its canines and molars were apelike in some ways and humanlike in others.

Figure 24.24 (**a**) Remains of "Lucy" (*A. afarensis*), who lived 3.2 million years ago. (**b**) At Laetoli, Tanzania, Mary Leakey found footprints made 3.7 million years ago. (**c,d**) The arch, big toe, and heel marks of the footprints are signs of bipedal hominids. Unlike apes, the early hominids did not have a splayed big toe, as the chimpanzee in (**c**) obligingly demonstrates.

evolved that were on or close to the lineage that gave rise to monkeys and apes. One had forward-directed eyes, a snoutless, flattened face, and an upper jaw with shovel-shaped front teeth. It must have used its hands to grab food. Some early anthropoids lived in swamps infested with predatory reptiles and rarely ventured to the ground. Maybe that is one reason why it became imperative to think fast and grip strongly. Slip-ups were always possible; primates still fall out of the trees.

Between 25 and 5 million years ago, in the Miocene, apelike forms—*the first hominoids*—evolved and began an adaptive radiation into Africa, southern Asia, and Europe. Most forms became extinct. But genetic analyses suggest that, between 10 million and 5 million years ago, divergences from some of the Miocene apes led to gorillas and chimps, and to the first *hominids*.

The First Hominids

Most of the early hominids we know about lived in the East African Rift Valley. By the dawn of the Pliocene, 5 million years ago, a long-term cooling trend was well under way, a result of drifting continents and changes in ocean circulation patterns. The climate became more seasonal, and the tropical forests—with their bounty of edible soft fruits, leaves, and insects—were giving way to dry woodlands and grasslands. The African savanna was emerging. Foods were getting harder (and harder to find). Hominids that had evolved in the forests had two options: move into new adaptive zones or die out.

Between 4.5 and 4 million years ago, a confounding variety of hominids appeared; it was a "bushy" period of evolution (Figures 24.23 and 24.24). At present it is

impossible to figure out their family tree, so they are simply grouped as the **australopiths** (southern apes). *Australopithecus anamensis* is the oldest of these. Like *A. afarensis* and *A. africanus*, it was gracile (slightly built). The ones called *A. boisei* and *A. robustus* were robust (muscular and heavily built).

Like apes, austalopiths had a large face, protruding jaws, and small skull (and brain) size. Yet they differed from earlier hominoids. For example, their molars had thicker enamel and could grind harder foods. And they were good at walking upright. We know this from their fossilized hip bones and limb bones. More telling, some *A. afarensis* left footprints. About 3.7 million years ago, they walked on newly fallen volcanic ash, which a light rain turned into quick-drying cement (Figure 24.24*b*).

Bipedal hominids had started to evolve earlier, in the late Miocene, when they were forced down to the ground. In the trees, their hands had become adapted to grip objects strongly and with precision. Now, rather than becoming specialized in running fast on all fours, they used their manipulative skills to advantage. They kept their hands free to carry offspring and probably to carry precious food during their foraging expeditions.

Primates evolved from small, rodentlike mammals that moved into arboreal habitats 60 million years ago. During the Miocene, the first hominoids (apelike forms) evolved and radiated through Africa, Europe, and southern Asia.

Between 10 million and 5 million years ago, divergences from Miocene apes led to australopiths, the first hominids. They were apelike in many skeletal details, but they were humanlike in a crucial respect: They walked upright.

WWW

EMERGENCE OF EARLY HUMANS

What can fossilized fragments of the early hominids tell us about our own origins? The record is still too sketchy for us to know how the diverse australopiths were related to one another, let alone which ones may have been ancestral to humans. Besides, what *are* the traits that characterize **humans**—members of the genus *Homo*? Well, what about the brain? The brain of modern humans is the basis of great analytical and verbal skills, complex social behavior, and technological innovation. It easily sets us apart from the apes, which have a much smaller skull volume and brain size. Yet this feature alone cannot tell us when certain hominids made the leap to being human, because their brain size probably fell within the range for apes. They made simple tools, but so do chimps and certain parrots. It goes without saying that behavior did not lend itself to fossilization.

And so we are left to speculate on a continuum of physical traits among a number of fossils—a skeleton adapted for bipedalism, manual dexterity, and larger brain volume; a smaller face; and smaller, more thickly enameled teeth. Those traits, which originated in the late Miocene, were evident in possibly the earliest humans —*Homo habilis*, a name that means "handy man."

Between 2.5 and 1.6 million years ago, one or two forms of *H. habilis* lived in dry woodlands interspersed in the savannas of eastern and southern Africa. Based on their dentition, the diet of early humans included hard-shelled nuts and seeds as well as soft fruits, leaves, and insects. Supplies of such foods changed with the seasons. Most likely, *H. habilis* had to think ahead, to plan when to venture about to gather and possibly to store foods that would help it survive the cold dry seasons.

H. habilis shared its habitat with predators such as saber-tooth cats. The cats had teeth that could impale prey and shear off flesh but could not crush open the bones for marrow. Carcasses of the kills, with shreds of meat clinging to the marrow bones, were concentrated stores of nutrients in places that were nutrient-stingy. *H. habilis* was a forager, not a full-time carnivore, but it opportunistically supplemented its diet by scavenging the carcasses (Figure 24.25a).

Fossil hunters have found a great many stone tools that date to the time of *H. habilis*. But they cannot say with certainty that *H. habilis* was the only species that made them. Possibly australopiths as well as *H. habilis* used sticks and other perishable tools before then, as modern apes do, but we have no way of knowing.

Maybe individuals on the road to modern humans started down a toolmaking road by picking up rocks to crack marrow bones. Maybe they started to scrape flesh from bones with small, sharp flakes that had fractured naturally from rocks. Eventually, early humans started *shaping* stone implements. Paleoanthropologist Mary Leakey was the first to find evidence of toolmaking at

Figure 24.25 (**a**) Artist's view of *Homo habilis* males alert to australopiths in an East African woodland. Two of the 37,000+ stone tools from Olduvai Gorge: (**b**) chopper and (**c**) cleaver.

Africa's Olduvai Gorge, which cuts through a great sequence of sedimentary rock layers. The most ancient tools at the site are crudely chipped pebbles buried in the deepest layers (Figure 24.25b,c). Possibly they were used to smash marrow bones, dig for roots, and poke insects from tree bark. More recent layers contain more complex tools. Large numbers of bones and tools also have been discovered along the shores of ancient lakes that would have beckoned plenty of thirsty animals.

At such sites we find fossils of one form of *H. habilis* that was twice as brainy as australopiths and obviously ate well. There apparently was no selection pressure for more creativity in securing food resources; stone tools did not change much for the next 500,000 years.

The ancestors of modern humans apparently stayed put in Africa until about 2 million years ago. At that time, a genetic divergence from early members of *Homo* led to *Homo erectus*, a species on the evolutionary road to fully modern humans (Figure 24.26). Its name means "upright man." Its forerunners also were upright, two-legged walkers, but *H. erectus* populations did the name justice; they walked on out of Africa, turned left into Europe, and right into Asia. It took some of them a long time, but they traversed the 14,000 kilometers to China; fossils from Southeast Asia and the former Soviet republic of Georgia are 1.8 million and 1.6 million years old. And their treks were strenuous. More than once, *H. erectus* endured ice ages as glaciers advanced into northern Europe, southern Asia, and North America.

Whatever selection pressures triggered the far-flung travelings, this was a time of physical changes, as in skull size and leg length. It also was a time of cultural lift-off for the human lineage. *H. erectus* had a larger brain. It was a more creative toolmaker. And its social

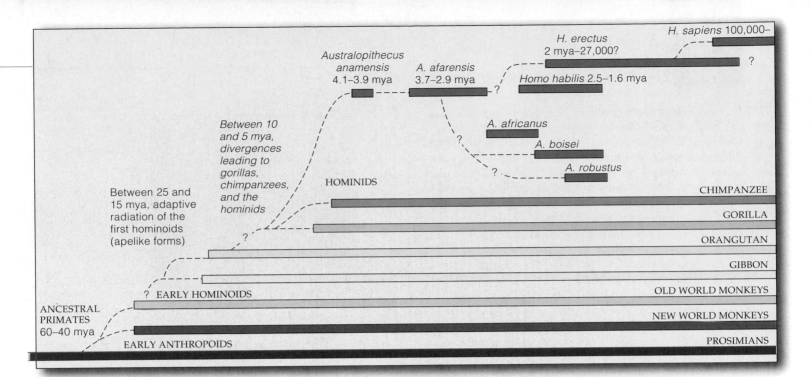

Figure 24.26 Presumed evolutionary branchings leading to modern humans in the primate family tree.

organization and communication skills must have been well developed. How else can we explain that hominid's successful geographic dispersal? From southern Africa to England, members of different populations used the same variety of hand axes and other tools to pound, scrape, shred, chop, cut, and whittle material. *H. erectus* could withstand environmental challenges by building fires and using furs for clothing. Clear evidence of fire use dates from an ice age in the early Pleistocene.

As fossils from the Middle East show, *Homo sapiens* had evolved by 100,000 years ago (Figure 24.26). The origin and geographic dispersal of early *H. sapiens* are hotly debated subjects (Section 24.12). Early *H. sapiens* had smaller teeth and jaws than *H. erectus*, and often a chin was present. Facial bones were smaller, the skull was higher and rounder; the brain was larger. Analysis of fossils indicates that the early forms might have had the capacity for complex language.

One group of early humans, the Neandertals, lived in Europe and in the Near East from 200,000 to 35,000 years ago. Massively built and large brained, some of them became the first to adapt to the coldest regions (Figure 24.27). Their disappearance coincided with the appearance of anatomically modern humans in the same regions about 40,000 to 30,000 years ago. There is no evidence that the Neandertals warred or interbred with the later arrivals. We don't know yet what happened to them. Gene sequencing of Neandertal DNA revealed unique sequences, so they may not have contributed to the gene pools of modern European populations.

From 40,000 years ago to today, human evolution has been almost entirely cultural, not biological. And so we leave the story with these conclusions: Humans

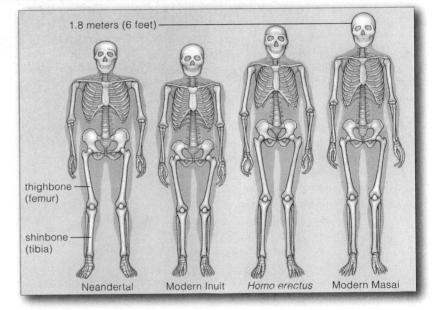

Figure 24.27 Body build correlated with climate. Humans adapted to cold climates have a heat-conserving body (stockier, shorter legs, compared with humans adapted to hot climates).

spread rapidly through the world by devising *cultural* means to deal with a broad range of environments. Compared to their predecessors, they developed rich, varied cultures. Although hunters and gatherers persist in parts of the world, others moved from "stone-age" technology to the age of "high tech," attesting to the great plasticity and depth of human adaptations.

Cultural evolution has outpaced the biological evolution of the only remaining human species, *H. sapiens*. Today, humans everywhere rely on cultural innovation to adapt rapidly to a broad range of environmental challenges.

WWW

24.12 OUT OF AFRICA—ONCE, TWICE, OR . . .

If researchers are interpreting the fossil record of human evolution correctly, then it would seem that Africa was the cradle for us all. At this writing, at least, no one has found any fossils of humans older than 1.8 million years *except* in Africa. *H. erectus* coexisted for a time with earlier humans (*H. habilis*) before dispersing from Africa to the cooler grasslands, forests, and mountains of Europe and Asia. They apparently left Africa in waves between about 2 million and 500,000 years ago. Judging from *H. erectus* fossils from Java, some isolated populations may have endured until 27,000–53,000 years ago.

Where, on the larger geologic stage, do we place the origin of *H. sapiens*? *Here we find a good example of how the same body of evidence can be interpreted in different ways.* These interpretations are called the multiregional model and African emergence model for modern human origins. Both attempt to explain the world distribution of fossils of particular ages and the measured genetic distances among existing human populations. For example, biochemical and immunological studies imply that the greatest genetic distance separates *H. sapiens* populations native to Africa from populations everywhere else; the next greatest distance separates Southeast Asia (and Australia) from everywhere else (Figure 24.28). As another example, Figure 24.29 correlates *H. sapiens* fossils to specific times.

MULTIREGIONAL MODEL By this model, *H. erectus* populations had spread through much of the world by about 1 million years ago. Those geographically separated groups were subject to different selection pressures, so their traits evolved in regionally distinctive ways. Then the subpopulations ("races") of *H. sapiens* evolved from them in different places. They differed phenotypically, but they did not evolve into separate species because gene flow continued among them, even to the present day. (Thus, for example, while the armies of Alexander the Great were sweeping eastward, they also contributed "blue-eye genes" from the Greeks to the allele pool of generally brown-eyed subpopulations in Africa, the Near East, and Asia.)

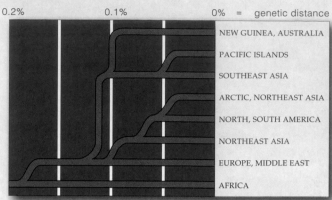

Figure 24.28 One proposed family tree for populations of modern humans (*Homo sapiens*) that are native to different regions of the world. Branch points show presumed genetic divergences. The tree is based on nucleic acid hybridization studies of many genes (including those for mitochondrial DNA and the ABO blood group) and immunological comparisons.

AFRICAN EMERGENCE MODEL This model does not dispute fossil evidence that populations of *H. erectus* evolved in distinctive ways in different regions. But it holds that *H. sapiens*—modern humans—originated in sub-Saharan Africa somewhere between 200,000 and 100,000 years ago. Only later did *H. sapiens* populations move out of Africa, then into other regions along the routes indicated by the Figure 24.29 map. In each region that *H. sapiens* populations settled, they replaced the archaic *H. erectus* populations that had preceded them. Only then did regional phenotypic differences become superimposed on the original *H. sapiens* body plan.

In support of this model, the oldest known *H. sapiens* fossils are from Africa. Also, in Zaire, a finely wrought barbed-bone harpoon and other tools suggest that the African populations were as skilled at making tools as *Homo* populations known earlier from Europe. Also, in 1998, researchers at the University of Texas and in China announced findings from the Chinese Human Genome Diversity Project. Detailed analysis of gene patterns from forty-three ethnic groups in Asia suggest that modern humans moved from central Asia, along the coast of India, then on into Southeast Asia and southern China. Populations later moved north and northwest into China, then into Siberia, and then on down into the Americas. WWW

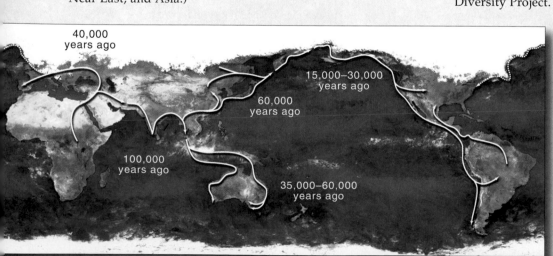

Figure 24.29 Estimated times when populations of early *H. sapiens* were colonizing different parts of the world, based on radiometric dating of fossils. The presumed dispersal routes (white lines) seem to support the African emergence model.

1. Almost all chordate embryos have a notochord, a dorsal hollow nerve cord, a pharynx with gill slits (or hints of these), and a tail that extends past the anus. Some or all of these traits persist in adults. Chordates with a backbone are vertebrates. Tunicates (including sea squirts) and lancelets are invertebrate chordates.

2. The eight classes of vertebrates are jawless fishes, jawed armored fishes (now extinct), cartilaginous fishes, bony fishes, amphibians, reptiles, birds, and mammals.

3. The earliest vertebrates, the jawless fishes that arose in the Cambrian, included ostracoderms. Their modern descendants are lampreys and hagfishes. Jawed fishes also arose in the Cambrian. A once-dominant lineage, the placoderms, became extinct. Other lineages gave rise to the cartilaginous and bony fishes.

4. Four trends occurred during the evolution of certain vertebrate lineages:

   a. A vertebral column supplanted the notochord as a structural element against which muscles act. This led to swift predatory animals.

   b. Jaws evolved from gill-supporting elements. This led to increased predator-prey competition; it favored far more efficient nervous systems and sensory organs.

   c. In one lineage of bony fishes, paired fins evolved into fleshy lobes having internal structural elements. The lobes were the forerunners of paired limbs.

   d. In some fish lineages, respiration by gills came to be supplemented by paired lungs, which proved to be adaptive during the invasion of the land. In a related development, the circulatory system became far more efficient at distributing oxygen through the body.

5. Amphibians were the first vertebrates to invade the land, but they never fully escaped the water. Their skin dries out, and aquatic stages persist in most life cycles.

6. Unlike amphibians, reptiles escaped dependence on standing free water. The key adaptations that allowed them to do so include toughened, scaly skin that can restrict loss of water from the body; a reliance upon internal fertilization; more efficient, water-conserving kidneys; and amniote eggs, often leathery or shelled, which protect and metabolically support the embryos.

7. Reptiles, and birds and mammals (which descended from certain reptilian lineages), have highly efficient circulatory and respiratory systems. They also have well-developed nervous system and sensory organs.

8. Of all vertebrates, birds alone have feathers, which they use in flight, heat conservation, and social displays.

9. Of all animals, mammals alone have milk-producing mammary glands, and they have hair or thick skin that functions in insulation. They have distinctive dentition and a highly developed cerebral cortex, the brain's most complex region. Adults nurture their young through an extended period of dependency and learning.

10. Primates include the prosimians (lemurs and related forms), the tarsioids, and the anthropoids (which include monkeys, apes, and humans). Apes and humans alone are hominoids. Only the anatomically modern humans (*H. sapiens*) and others of their lineage (from *A. afarensis* to *H. erectus*) are further classified as hominids.

11. The first primates were small, rodentlike mammals that evolved by 60 million years ago in tropical forests.

12. Apelike forms (the first hominoids) evolved between 25 and 13 million years ago, in the Miocene. In Africa, some had given rise to australopiths, the earliest known hominids, before 4 million years ago. Their evolution correlated with strong selection pressures that emerged during a long-term trend from tropical climates to cooler, drier climates. The African savanna emerged, and the hominids that survived were the ones with physical modifications that permitted upright walking (bipedalism), a more varied diet (changes in dentition), and more brain complexity to figure out how to gather scarcer and seasonally available food.

13. *H. habilis*, the earliest known species of the genus *Homo*, evolved by 2.5 million years ago. It might have been the first stone toolmaker. *Homo erectus*, presumed ancestor of modern humans, evolved by 2 million years ago. *H. erectus* populations radiated out of Africa, into Asia and Europe. The oldest known fossils of early modern humans (*H. sapiens*) date from 100,000 years ago. About 40,000 years ago, cultural evolution outstripped biological evolution of the human form.

14. Modern humans are adapted to a wide range of environments. This capacity resulted from evolutionary modifications in certain primate lineages. Starting with arboreal primate ancestors, there was less reliance on the sense of smell and more on enhanced daytime vision. Also among the arboreal primates, manipulative skills increased as hands began to be freed from load-bearing functions. Starting with Miocene apes, there was a shift from four-legged climbing to bipedalism, a shift from specialized to omnivorous eating habits, and increases in brain complexity and behavior.

Review Questions

1. Which features distinguish chordates from other groups of animals? *24.1*

2. List four major trends that occurred during the evolution of at least some vertebrate lineages. *24.3*

3. Describe some features of jawless fishes, cartilaginous fishes, and bony fishes. Which are the most abundant? *24.3*

4. Which traits distinguish reptiles from amphibians? *24.6, 24.7*

5. Which traits distinguish birds from reptiles? *24.8*

6. List some characteristics that distinguish each of the three mammalian lineages from the reptiles. *24.9, 24.10*

7. What is the difference between a "hominoid" and a "hominid"? *24.10*

8. Briefly describe some of the conserved physical traits that connect anatomically modern humans with their mammalian ancestors, then their primate ancestors. *24.10, 24.11*

Self-Quiz (Answers in Appendix III)

1. Only _____ have a notochord, a tubular dorsal nerve cord, a pharynx with slits in the wall, and a tail extending past the anus.
   a. echinoderms
   b. tunicates and lampreys
   c. vertebrates
   d. both b and c
   e. all are correct

2. Gills function in _____ .
   a. respiration
   b. circulation
   c. food trapping
   d. water regulation
   e. both a and c

3. A shift from a reliance on _____ to reliance on _____ was pivotal in the evolution of all vertebrates.
   a. the notochord; a backbone
   b. filter feeding; jaws
   c. gills; lungs
   d. all are correct

4. The first vertebrates were _____ .
   a. bony fishes
   b. jawless fishes
   c. jawed fishes
   d. both a and b

5. Of all existing vertebrates, _____ are the most diverse.
   a. cartilaginous fishes
   b. bony fishes
   c. amphibians
   d. reptiles
   e. birds
   f. mammals

6. The only amphibians to entirely escape dependence on aquatic habitats are _____ .
   a. salamanders
   b. desert toads
   c. caecilians
   d. none is correct

7. Reptiles moved fully onto land owing to _____ .
   a. tough skin
   b. internal fertilization
   c. good kidneys
   d. amniote eggs
   e. both b and d
   f. all are correct

8. A four-chambered heart evolved first in _____ .
   a. bony fishes
   b. amphibians
   c. birds
   d. mammals
   e. crocodilians
   f. both c and d

9. _____ have highly efficient circulatory and respiratory systems, and a complex nervous system and sensory organs.
   a. Reptiles
   b. Birds
   c. Mammals
   d. all are correct

10. Birds use feathers in _____ .
    a. flight
    b. heat conservation
    c. social functions
    d. all are correct

11. Various mammals _____ .
    a. hatch
    b. complete their embryonic development in pouches
    c. complete their embryonic development in the uterus
    d. both b and c
    e. all are correct

12. Early humans _____ .
    a. adapted to a wide range of environments
    b. adapted to a narrow range of environments
    c. had flexible bones that cracked easily
    d. were limber enough to swing through the trees

13. Match the organisms with the appropriate features.
    ____ jawless fishes
    ____ cartilaginous fishes
    ____ bony fishes
    ____ amphibians
    ____ reptiles
    ____ birds
    ____ mammals

    a. complex cerebral cortex, thick skin or hair
    b. respiration by skin and lungs
    c. include coelacanths
    d. include hagfishes
    e. include sharks and rays
    f. complex social behavior, feathers
    g. first with amniote eggs

Critical Thinking

1. Describe the factors that might contribute to the collapse of native fish populations in a lake following the introduction of a novel predator, such as the lamprey.

2. Think about the flight muscles of birds and their demands for oxygen and ATP energy. What type of organelle would you expect to be profuse in these muscles? Explain your reasoning.

3. Kathie and Gary, two amateur fossil hunters, have unearthed the complete fossilized remains of a mammal. How can they detmine whether it is a herbivore, a carnivore, or an omnivore?

4. In Australia, many marsupials are competing with recently introduced placental mammals (such as rabbits) for resources but are not winning. Explain how placental mammals that did not even evolve in Australian habitats show greater fitness than the native mammals.

5. When it comes to human origins, different researchers "read" the fossil record in different parts of the world in different ways. Is this indication that the primate fossils they are investigating might have nothing whatsoever to do with human origins? Why or why not?

Selected Key Terms

amniote egg *24.7*	fin *24.3*	nerve cord *24.1*
amphibian *24.6*	gill *24.3*	notochord *24.1*
australopith *24.10*	gill slit *24.2*	ostracoderm *24.3*
bird *24.8*	hagfish *24.4*	pharynx *24.1*
bony fish *24.5*	hominid *24.10*	placenta *24.9*
cartilaginous fish *24.5*	human *24.11*	placoderm *24.3*
chordate *24.1*	jaw *24.3*	Primates *24.10*
convergent	lamprey *24.4*	reptile *24.7*
evolution *24.9*	lancelet *24.2*	scale (fish) *24.5*
culture *24.10*	lobe-finned	swim bladder *24.5*
dentition *24.9*	fish *24.5*	tunicate *24.2*
feather *24.8*	lung *24.3*	vertebra *24.3*
filter feeder *24.2*	mammal *24.9*	vertebrate *24.1*

Readings See also www.infotrac-college.com

Beardsley, T. April 1996. "Out of Food?" *Scientific American*, 20–22. Update on *H. erectus* migrations.

Gould, S. J. (general editor). 1993. *The Book of Life*. New York: Norton. Splendid, easy-to-read essays, gorgeous illustrations.

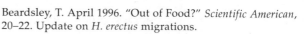

WWW *http://www.brookscole.com/biology*

Practice quiz questions, hypercontents, BioUpdates, and critical thinking. The Brooks/Cole Biology Resource Center provides a wealth of information fully organized and integrated by chapter.

FACING PAGE: *A flowering plant (Prunus) busily doing what it does best: producing flowers for the fine art of reproduction.*

V

PLANT TISSUES

Plants Versus the Volcano

On a clear spring day in 1980, in a richly forested part of the Cascade Range of southwestern Washington, Mount Saint Helens exploded and 500 million metric tons of ash were blown skyward. Within minutes, the shock waves from the blast flattened or incinerated hundreds of thousands of mature trees growing near the northern flanks of the mountain. Hot volcanic ash surged down the slopes at rates exceeding forty-four meters per second. It quickly turned into rivers of cementlike mud when the intense heat melted and released more than 75 billion liters of water from the mountain's snowfields and glacial ice.

In one mind-numbing moment, 40,500 hectares—100,000 acres—of magnificent forests dominated by hemlock and Douglas fir were transformed into barren sweeps of land (Figure 25.1*a,b*). In the aftermath of the violent eruption, more than a few observers gained stunning insight into what the world must have looked

Figure 25.1 (**a**,**b**) A grim reminder of what our world would be like without plants: the devastation following the violent eruption of Mount Saint Helens in 1980. Nothing remained of the extensive forest that had surrounded this Cascade volcano. (**c**) In less than a decade, however, seed-bearing vascular plants were making a comeback. (**d**) Merely twelve years after the volcanic eruption, seedlings of a dominant species (Douglas fir) were starting to reclaim the land.

like long ago, before the first plants started to colonize habitats on land.

Yet it did not take long for existing plants to move back into habitats that their ancestors had claimed. In less than a year's time, seeds of a variety of flowering plants, including fireweed and blackberry, sprouted near the grayed trunks of fallen trees around Mount Saint Helens. Even before ten years had passed, young willows and alders took hold near riverbanks, and low shrubs cloaked the land (Figure 25.1c). Their presence afforded pockets of shade, which favored germination of the seeds of slower growing but ultimately dominant species, the hemlocks and Douglas fir (Figure 25.1d). In less than a century, the forest will be as it once was.

With this example, we open a unit dedicated to seed-bearing vascular plants, with emphasis on the flowering types. In terms of distribution and diversity, they are the most successful plants on Earth.

This first chapter provides an overview of plant tissues and body plans. Chapter 26 explains how the seed-bearing plants absorb water and mineral ions, conserve water, and distribute organic substances among roots, stems, and leaves. Chapter 27 takes a closer look at their patterns of growth, development, and reproduction. As you will see, their structure and physiology (that is, the functioning of the plant body) help them survive sometimes hostile conditions on land—even momentary takeovers by volcanoes.

KEY CONCEPTS

1. Angiosperms (flowering plants) and, to a lesser extent, gymnosperms are groups that now dominate the plant kingdom. These seed-bearing vascular plants all have complex aboveground shoot systems, which consist of stems, leaves, flowers, and some other structures. They also have complex root systems that typically grow downward and outward through soil.

2. There are three major categories of tissue systems in these plants. A ground tissue system makes up the bulk of the plant body. A vascular tissue system distributes water, dissolved minerals, and photosynthetic products through roots and shoots. A dermal tissue system covers and protects plant surfaces exposed to the surroundings.

3. The simple plant tissues—parenchyma, collenchyma, and sclerenchyma—are each composed of no more than one type of cell.

4. Complex plant tissues incorporate two or more types of cells. Xylem and phloem, which are vascular tissues, are like this. So are the dermal tissues called epidermis and periderm.

5. Plants grow by way of mitotic cell divisions and cell enlargements at meristems, which are localized regions of self-perpetuating, embryonic cells.

6. Each growing season, shoots and roots lengthen. The lengthening, called primary growth, originates only at apical meristems, in the tips of shoots and roots.

7. Each growing season, the shoots and roots of many plants also thicken. Typically, lateral meristems inside shoots and roots give rise to an increase in diameter, which is called secondary growth. Wood is one outcome of secondary growth.

PLANT NUTRITION AND TRANSPORT

Flies for Dinner

How often do we think that plants actually do anything impressive? Being mobile, intelligent, and emotional, we tend to be fascinated more with ourselves than with immobile, expressionless plants. Yet plants don't just stand around soaking up sunlight. Consider the Venus flytrap (*Dionaea muscipula*), a native flowering plant of bogs in North and South Carolina. Its two-lobed, spine-fringed leaves open and close much like a steel trap (Figure 26.1a–d). Like all other plants, it cannot grow properly without nitrogen and other nutrients, which happen to be scarce in the soil of bogs. But plenty of insects fly in from regions around the bogs.

Sticky sugars ooze from epidermal glands onto the surface of the flytrap's leaf. The sugars entice insects to land. As they do, they brush against hairlike structures that project from the leaf surface. These are triggers for the trap. When an insect touches two hairs at the same time or the same hair twice in rapid succession, the two lobes of the leaf snap shut. Now digestive juices pour out from cells of the leaf. They pool around the insect,

dissolve it, and so release nutrients from it. In other words, the Venus flytrap makes its own nutrient-rich water, which it proceeds to absorb!

The Venus flytrap is only one of several species of **carnivorous plants**. We call them this even though it takes a leap of the imagination to put their peculiar mode of nutrient acquisition (a form of extracellular digestion and absorption) in the same category as the chompings of lions, dogs, and similar meat eaters. Not all carnivorous plants have active traps. Some species lure in prey and then simply let them drown (Figure 26.2). But they all evolved in habitats where nitrogen and other nutrients are hard to come by.

Plants with bizarre nutrient-acquiring habits also live in the shallow waters of many freshwater lakes and streams, which contain only dilute concentrations of dissolved minerals.

Given the variety and numbers of insects and other animals that attack plants, you can just imagine how endearing the carnivorous plants are to botanists. With their plucky modes of nutrition, these plants also are a fine way to start thinking about **plant physiology**—the study of adaptations by which plants function in their environment. As you already know, nearly all plants are photoautotrophs that use energy from sunlight to drive the synthesis of organic compounds from water, carbon dioxide, and some minerals. Like people, they do not have unlimited supplies of the resources required to nourish themselves. Of every 1 million molecules of air, only 350 are carbon dioxide. Unlike the soggy habitats

VENUS FLYTRAP, OPEN FOR DINNER

Figure 26.1 Do plants take nutrition seriously? You bet. (**a**) A Venus flytrap (*Dionaea muscipula*). This carnivorous plant makes up for scarce nutrients by turning the dinner table on animals that alight upon its leaves. (**b**) A fly stuck in the sugary goo on a lobed leaf. (**c**) It brushes against hairlike triggers on the leaf surface; the base of one is shown here. (**d**) Activated, the leaf snaps shut.

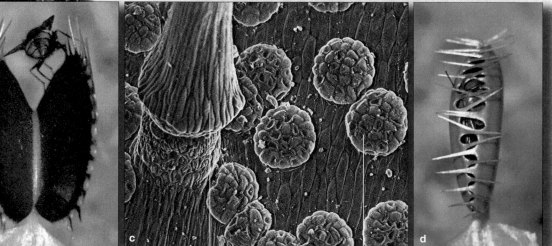

base of epidermal hair   epidermal gland

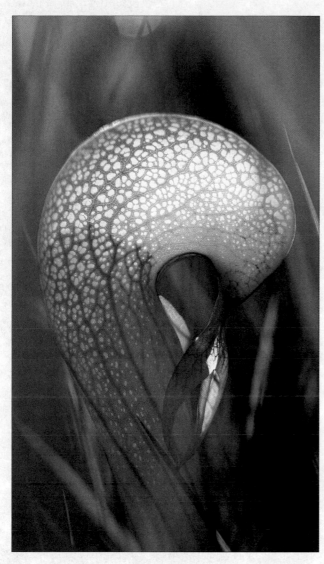

Figure 26.2 Cobra lily (*Darlingtonia californica*). Its leaves form a "pitcher" that is partly filled with digestive juices. Insects lured in by irresistible odors often cannot find the way back out; light shining through the pitcher's patterned dome confuses them. They just wander around and down, adhering to downward-pointing leaf hairs—which become slickened with wax above the potent vat.

of Venus flytraps, most soils are frequently dry. And nowhere except in overfertilized gardens does soil water hold lavish amounts of dissolved minerals. As you continue with these chapters on vascular plants, keep this point in mind: *Many aspects of plant structure and function are responses to low concentrations of vital resources in the environment.*

KEY CONCEPTS

1. Many aspects of the structure and function of plants are adaptive responses to low concentrations of water, minerals, and other environmental resources.

2. A plant's root system takes up water and mines the soil for nutrients. For many land plants, mycorrhizae and bacterial symbionts assist in nutrient uptake. The properties of soil in a given habitat influence the availability of nutrients and water.

3. A plant's cuticle and its many stomata function in the conservation of water, a scarce resource in most habitats on land. Stomata are passageways across the epidermis of leaves and, to a lesser extent, stems. They help control water loss, yet they permit gas exchange.

4. During the day, stomata remain open so that carbon dioxide, required for photosynthesis, is able to diffuse into leaves. Water loss is rapid when stomata are open. Most plants conserve water by closing stomata at night.

5. In flowering plants and other vascular plants, the flow of water and solutes through xylem and phloem functionally connects all living cells of the roots, stems, and leaves. Xylem serves in the uptake and distribution of water and dissolved nutrients. Phloem serves in the distribution of photosynthetically produced sugars and other organic compounds through the plant body.

6. Water absorbed from soil moves on up through xylem and into leaves. By the process of transpiration, the dry air around leaves promotes evaporation through stomata. The force of evaporation is enough to pull continuous columns of water molecules, which are hydrogen-bonded to one another, from roots to aboveground parts.

7. By the energy-requiring process of translocation, sucrose and other organic compounds are distributed throughout the plant. Organic compounds produced by photosynthetic cells in leaves are loaded into conducting cells of phloem. They are unloaded at actively growing regions or storage regions of the plant body.

Properties of Soil

At the surface of most habitats on land is a thin cloak of soil. **Soil** consists of particles of minerals mixed with variable amounts of decomposing organic material, or **humus**. Weathering of hard rocks produces the soil's minerals. Dead organisms and litter from them (fallen leaves, feces, and so on) are the source of humus. Water and oxygen occupy the spaces in between the particles and organic bits.

In different regions, even in different parts of the same habitat, soils vary in the proportions of mineral particles and in the extent to which these have become compacted together. Generally, those particles come in three sizes, known as sand, silt, and clay. If you have ever dribbled beach sand through your fingers, you already have an idea that sand particles are large (0.05 to 2 millimeters across). Rub some silt from pond mud between your fingers. You won't be able to distinguish among individual particles; they are only 0.002 to 0.05 millimeter across. Clay particles are the finest of all.

How suitable is a given soil for plant growth? Is it gummy when wet because it does not have enough air spaces? Does it form hard clods when dry? The answer depends partly on the relative proportions of sand, silt, and clay. The more clay, the finer the soil's texture.

Each clay particle consists of thin, stacked layers of aluminosilicates with negatively charged ions at their surfaces. Clay attracts and holds (adsorbs) positively charged mineral ions dissolved in the water trickling through soil, as well as water molecules themselves. Ions and water cling reversibly to the clay, and this is crucial for plants. With its high adsorption capacity, clay holds on to many nutrients for plants even as the water percolates on past and drains away.

Too much clay, however, is bad for plants. Clay particles pack together so tightly, they do not leave enough space for the oxygen that root cells require for aerobic respiration. The tight packing also retards penetration of water into the soil. In heavy clay soils, runoff is pronounced, so water and nutrients dissolved in it are not available for plant growth. Plants do best in **loams**, which are soils having roughly the same proportions of sand, silt, and clay.

The amount of humus in a given soil also affects plant growth. Generally, humus has an abundance of negatively charged organic acids, so that it can weakly bind with and retain dissolved mineral ions of opposite charge. Humus also has a high capacity to absorb and swell with water, then shrink as the water is gradually released. Its alternating swelling and shrinking aerate the soil. As decomposers work it over, humus slowly releases nutrients, which become available for plants.

In general, soils that are 10 to 20 percent humus are most favorable for plant growth. The worst have less than 10 percent humus or more than 90 percent humus, which is characteristic of swamps and bogs.

Soils can be classified by *profile* properties. The word refers to the layered characteristics of soils, which are in different stages of development in different places. Figure 26.3 is an example. **Topsoil**, the uppermost part of soil, is called the A horizon. This is the most essential layer for plant growth, and its depth is highly variable.

Nutrients Essential for Plant Growth

We have mentioned nutrients in passing, but what does the word mean? **Nutrients** are elements that are essential for a given organism because, directly or indirectly, they have roles in metabolism (hence growth and survival) that no other element can fulfill. For plants, the essential elements include oxygen, hydrogen, and carbon, which the plants get from water and carbon dioxide during

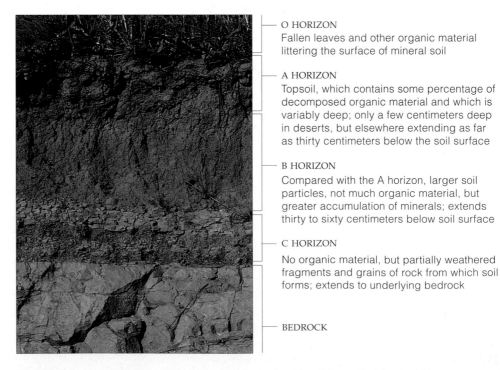

O HORIZON
Fallen leaves and other organic material littering the surface of mineral soil

A HORIZON
Topsoil, which contains some percentage of decomposed organic material and which is variably deep; only a few centimeters deep in deserts, but elsewhere extending as far as thirty centimeters below the soil surface

B HORIZON
Compared with the A horizon, larger soil particles, not much organic material, but greater accumulation of minerals; extends thirty to sixty centimeters below soil surface

C HORIZON
No organic material, but partially weathered fragments and grains of rock from which soil forms; extends to underlying bedrock

BEDROCK

Figure 26.3 Some of the soil horizons that have developed in one habitat in Africa.

Table 26.1 Essential Elements and Plant Function

MACRONUTRIENT	Some Functions	Some Deficiency Symptoms
Carbon Hydrogen Oxygen	Basic ingredients for photosynthesis	Available in abundance from water and from carbon dioxide in the air
Nitrogen	Component of proteins, nucleic acids, coenzymes, chlorophylls	Stunted growth; light-green older leaves; older leaves yellow and die (these symptoms define a condition called chlorosis)
Potassium	Activation of enzymes; key role in maintaining water-solute balance and so influences osmosis*	Reduced growth; curled, mottled, or spotted older leaves; burned leaf edges; weakened plant
Calcium	Regulation of many cell functions; cementing of cell walls	Leaves deformed; terminal buds die; poor root growth
Magnesium	Component of chlorophyll; activation of enzymes	Chlorosis; drooped leaves
Phosphorus	Component of nucleic acids, phospholipids, ATP	Purplish veins; stunted growth; fewer seeds, fruits
Sulfur	Component of most proteins, two vitamins	Light-green or yellowed leaves; reduced growth

MICRONUTRIENT	Some Functions	Some Deficiency Symptoms
Chlorine	Role in root and shoot growth; role in photolysis	Wilting; chlorosis; some leaves die
Iron	Roles in chlorophyll synthesis and in electron transport	Chlorosis; yellow and green striping in grasses
Boron	Roles in germination, flowering, fruiting, cell division, nitrogen metabolism	Terminal buds, lateral branches die; leaves thicken, curl, and become brittle
Manganese	Chlorophyll synthesis; coenzyme action	Dark veins, but leaves whiten and fall off
Zinc	Role in formation of auxin, chloroplasts, and starch; enzyme component	Chlorosis; mottled or bronzed leaves; abnormal roots
Copper	Component of several enzymes	Chlorosis; dead spots in leaves; stunted growth
Molybdenum	Part of enzyme used in nitrogen metabolism	Pale green, rolled or cupped leaves

* All mineral elements contribute to the water-solute balance, but potassium is notable because there is so much of it.

Figure 26.4 In agricultural fields, the erosive force of water can form gullies that channel runoff from the land. When gullies grow deeper and wider, erosion becomes more and more rapid. When topsoil is depleted, productivity declines and fertilizers usually must be trucked in to replace the lost nutrients.

plant's dry weight (weighed after all the water has been removed from it). The other elements are *micro*nutrients; they make up traces (a few parts per million, usually) of the dry weight. As Table 26.1 indicates, even those trace amounts are essential for normal growth.

Leaching and Erosion

Leaching refers to the removal of some of the nutrients in soil as water percolates through it. Leaching is most pronounced in sandy soils, which are not as good as clay at binding nutrients. It is not the same as **erosion**, which is the movement of land under the force of wind, running water, and ice (Figure 26.4). For example, each year, erosion from farmlands in the Mississippi River watershed puts about 25 billion metric tons of topsoil into the Gulf of Mexico. Whether by leaching or erosion, the loss of nutrients from soil is bad for plants and for all the organisms that depend on plants for survival.

The mineral component of soil includes particles ranging from large-grained sand to silt and fine-grained clay. These particles, clay especially, reversibly bind water molecules and dissolved mineral ions and thereby make them more accessible for uptake by plant roots.

Soil also contains humus—a reservoir of organic material, rich in organic acids, in different stages of decay.

Most plants grow best in soils having equal proportions of sand, silt, and clay and 10 to 20 percent humus.

Nutrients are essential elements; no other element can substitute for their direct or indirect roles in the metabolic activities that sustain growth and keep organisms alive.

photosynthesis. Besides these elements, plants depend on the uptake of at least thirteen others, listed in Table 26.1. Typically these elements are dissolved in soil water in ionic forms that can reversibly bind with clay. Ions of calcium (Ca^{++}) and potassium (K^+) are examples. Plants give up hydrogen ions to the clay in exchange for those weakly bound elements.

Nine essential elements are *macro*nutrients. Normally they are required in amounts above 0.5 percent of the

WWW

HOW DO ROOTS ABSORB WATER AND MINERAL IONS?

Energetically speaking, mining the soil for mineral ions and water molecules that are clinging to clay particles is an expensive task. Plants spend considerable energy on building extensive root systems. Wherever the soil's texture and composition change, new roots must form to replace old ones and snake out in different regions. It isn't that roots "explore" soil for resources. Rather, the patches of soil where concentrations of water and mineral ions are greater stimulate the outward growth.

Absorption Routes

Think back on the preceding chapter's discussion of a typical root's structure (Section 25.5). Water molecules in soil are only weakly bound to clay particles, so they readily move across the root epidermis and continue to the **vascular cylinder**, a column of vascular tissue at the root's center. There a cylindrical layer of cells, an **endodermis**, wraps around the column. A waxy band, a **Casparian strip**, is deposited in abutting parts of endodermal cell walls (Figure 26.5). Water molecules are not able to penetrate the wax, so they are directed to wall regions that are not waxed, then into the cells proper, and out the other side. This is the only way water and solutes can get into the vascular cylinder. Like all cells, endodermal cells have many transport proteins embedded in the plasma membrane. The proteins let some solutes but not others cross it (Section 5.6). *Transport proteins of endodermal cells are control points where the plant adjusts the quantity and types of solutes that it absorbs from soil water.*

Many roots have an **exodermis**, a cell layer beneath their surface (Figure 26.5a). Walls of exodermal cells may have a Casparian strip that functions just like the one next to the root vascular cylinder.

Specialized Absorptive Structures

ROOT HAIRS Vascular plants require huge amounts of water. Roots of a mature corn plant absorb more than three liters of water daily. They could not do so without **root hairs**. Recall, from the preceding chapter, that root hairs are slender extensions of specialized epidermal cells. They greatly increase the surface area available for absorption (Section 25.5 and Figure 26.6). When a plant is putting on primary growth, its system of roots may develop millions or billions of root hairs.

ROOT NODULES Bacteria and fungi help many plants absorb nutrients and receive something in return. A two-way flow of benefits between species, remember, is a symbiotic interaction known as **mutualism** (Section 21.5).

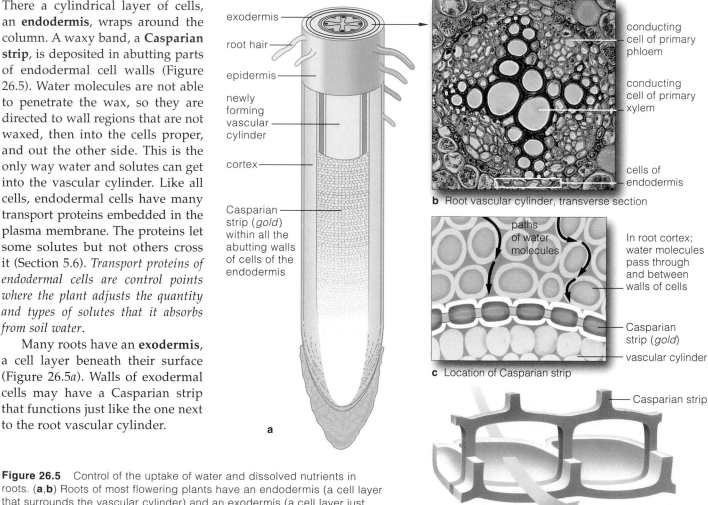

exodermis
root hair
epidermis
newly forming vascular cylinder
cortex
Casparian strip (*gold*) within all the abutting walls of cells of the endodermis

a

conducting cell of primary phloem
conducting cell of primary xylem
cells of endodermis

b Root vascular cylinder, transverse section

paths of water molecules
In root cortex; water molecules pass through and between walls of cells
Casparian strip (*gold*)
vascular cylinder

c Location of Casparian strip

Casparian strip

d Two endodermal cells with the cytoplasm removed. Water and solutes can only move into the vascular cylinder by passing through the cells, not in between them or through their walls.

Figure 26.5 Control of the uptake of water and dissolved nutrients in roots. (**a**,**b**) Roots of most flowering plants have an endodermis (a cell layer that surrounds the vascular cylinder) and an exodermis (a cell layer just beneath the epidermis). (**c**) Abutting walls of cells of both layers contain a waxy Casparian strip. The strip keeps water from moving indiscriminately *around* the cells and into the vascular column. It makes water move *through* the cells (**d**). In this way, transport proteins that span the plasma membrane of those cells can selectively control the uptake of water and nutrients.

Figure 26.7 (**a**) Nutrient uptake at root nodules of legumes that are mutualists with nitrogen-fixing bacteria (*Rhizobium* and *Bradyrhizobium*). When infected by such bacteria, root hair cells are induced to form an "infection thread" of cellulose deposits. The bacteria use the thread like a highway to invade plant cells in the root cortex. (**b,c**) Infected plant cells and the bacterial cells within them divide repeatedly, forming a swollen mass that becomes a root nodule. The bacteria start fixing nitrogen when plant cell membranes surround them. The plant takes up some of the nitrogen; the bacteria take up some photosynthetic compounds. (**d**) To the *left* in this photograph, rows of soybean plants growing in nitrogen-poor soil. The plants in the rows to the *right* were inoculated with *Rhizobium* cells and developed root nodules.

For example, think of how nitrogen deficiency limits plant growth. There is an abundance of gaseous nitrogen (N≡N) in the air. But plants do not have the metabolic means to engage in **nitrogen fixation**. By this process, certain enzymes break all three of the covalent bonds in gaseous nitrogen, and then they attach the nitrogen atoms to organic compounds. To get high crop yields, farmers apply nitrogen-rich fertilizers, or they encourage growth of nitrogen-fixing bacteria that naturally inhabit soil. Such bacteria convert gaseous nitrogen to forms they—and plants—can use. String beans, peas, alfalfa, clover, and other legumes have an advantage in this respect. Nitrogen-fixing bacteria live in their roots, inside local swellings called **root nodules**. Figure 26.7*b* shows an example. The symbiotic bacteria withdraw some photosynthetically produced organic compounds that were distributed from leaves to the roots. But in return they provide the plants with some of the nitrogen that they secured.

MYCORRHIZAE Also think back on the **mycorrhizae** (singular, mycorrhiza). As described in Section 21.5, a mycorrhiza is a symbiotic interaction between a young root and a fungus. Hence the name, meaning "fungus-root." Fungal filaments (hyphae) form a velvety cover around roots or they penetrate root cells. Collectively, hyphae have a large surface area that absorbs mineral ions from a larger volume of soil than the roots can do. The fungus absorbs sugars and nitrogen-containing compounds from root cells. The root cells obtain some scarce minerals that the fungus is better able to absorb.

Gymnosperms and flowering plant roots control the uptake of water and dissolved nutrients at the vascular cylinder's endodermis and at a similar layer near the root surface.

Root hairs, root nodules, and mycorrhizae enhance the uptake of water and scarce nutrients at the root.

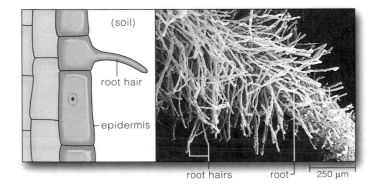

Figure 26.6 Sketch of a root hair (a slender extension of a specialized root epidermal cell), and a scanning electron micrograph of the root hairs of a portion of a young root.

LIFE CYCLES END, AND TURN AGAIN

Senescence

While leaves and fruits are growing, cells inside them produce auxin (IAA), which moves into stems. There it interacts with cytokinins and gibberellins to maintain growth. When autumn approaches and the length of daylight decreases, plants start to withdraw nutrients from leaves, stems, and roots, then distribute them to flowers, fruits, and seeds. Deciduous plants, which shed leaves when a growing season ends, channel nutrients to storage sites in twigs, stems, and roots before the leaves die and drop. The dropping of flowers, leaves, fruits, and other plant parts is called **abscission**. The abscission zone consists of thin-walled parenchyma cells at the base of a petiole or some other part about to drop from the plant. Figure 27.26 shows one example.

Senescence is the sum total of processes that lead to death of a plant or some of its parts. The recurring cue for senescence is a decrease in daylength, but other environmental factors, such as drought, wounds, and nutrient deficiencies, also bring it about. Either way, the cue triggers a decline in IAA production in leaves and fruits. A different signal, possibly produced by the plant itself, stimulates abscission zone cells to produce ethylene. The cells enlarge, deposit suberin in their walls, and produce enzymes that digest cellulose and pectin in middle lamellae. A middle lamella is the cementing layer between plant cell walls (Section 4.10). As cells continue to enlarge and their walls are digested, they separate from one another. The leaf or other plant part above the abscission zone inevitably drops away.

Interrupt the diversion of nutrients into flowers, seeds, or fruits, and you can prevent senescence in a plant's leaves, stems, and roots. For example, if you remove each new flower or seed pod from a plant, its leaves and stems will remain vigorous and green much longer. Gardeners routinely remove flower buds from many kinds of plants to maintain vegetative growth.

Entering and Breaking Dormancy

As autumn approaches and days grow shorter, many perennials and biennials start to shut down growth. They do so even when temperatures are still mild, the sky is bright, and water is plentiful. When a plant stops growing under conditions that seem (to us) suitable for growth, it has entered a state of **dormancy** in which its metabolic activities idle. Ordinarily, the plant's buds will not resume growth until there is a convergence of precise environmental cues in early spring.

Short days and long, cold nights are strong cues for dormancy. (So is dry, nitrogen-deficient soil.) Test this for yourself by interrupting the long dark period of, say, Douglas firs with a short period of red light. The

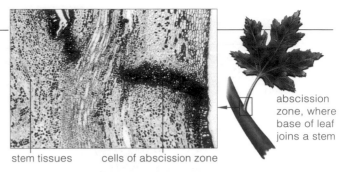

Figure 27.26 Abscission zone in maple (*Acer*). The section is through the base of a leaf petiole.

stem tissues   cells of abscission zone

abscission zone, where base of leaf joins a stem

Three Douglas fir plants exposed to different day-night cycles during three year-long experiments

Figure 27.27 Effect of the relative length of day and night on Douglas firs. The plant at left was exposed to 12-hour light/12-hour dark cycles for a year. Its buds became dormant; daylength was too short. The plant at right was exposed to 20-hour light/4-hour dark cycles; growth continued. The middle plant was exposed to 12-hour light/11-hour dark cycles with 1 hour of light in the middle of the dark period. The light interruption prevented bud dormancy; it caused Pfr formation at a sensitive time in the normal day-night cycle.

plants will respond as if nights are shorter and days are longer. They will continue to grow taller (Figure 27.27). In this experiment, conversion of Pr to Pfr by red light during the dark period prevented dormancy. In nature, buds might enter dormancy because less Pfr can form when daylength shortens in late summer.

A dormancy-breaking process is at work between fall and spring. The temperature may become milder, and rains and nutrients become available again. With the return of favorable conditions, the cycling of life begins anew. Seeds germinate; buds resume growth and give rise to new leaves, then to flowers. Depending on the species of plant, breaking dormancy probably involves gibberellins and abscisic acid, and it requires exposure to low winter temperatures at specific times.

Multiple cues from the environment influence hormonal secretions that stimulate or inhibit processes of growth and development during the life cycle of plants. The cues include changes in daylength, temperature, moisture, and nutrient availability.

SUMMARY

1. Sexual reproduction is the main reproductive mode of flowering plant life cycles. A diploid sporophyte (spore-producing plant body) alternates with haploid gametophytes (gamete-producing bodies).

a. The sporophyte is a multicelled vegetative body with roots, stems, leaves, and, at some point, flowers. Air currents, water currents, and animals coevolved with flowers and serve as their pollinating agents. The gametophytes form in male and female floral parts.

b. Many flowering plants also reproduce asexually. They do this naturally (as by runners, rhizomes, and bulbs) and artificially (as by cuttings and grafting).

2. Flowers typically have sepals, petals, one or more stamens (the male reproductive structures), and carpels (female reproductive structures), most or all attached to a receptacle, the modified end of a floral shoot.

a. Anthers of stamens contain pollen sacs in which cells divide by meiosis. A wall develops around each resulting haploid cell (microspore), which develops into a sperm-bearing pollen grain (male gametophyte).

b. A carpel (or two or more fused carpels) has an ovary where eggs develop, fertilization takes place, and seeds mature. A stigma (sticky or hairy surface tissue) above the ovary captures pollen grains and promotes their germination. Ovules form on the ovary's inner wall. Each consists of a female gametophyte with an egg cell, endosperm mother cell, a surrounding tissue, and one or two protective layers called integuments.

3. At double fertilization, one sperm nucleus fuses with an egg nucleus and forms a diploid zygote. The other sperm nucleus and both nuclei of another cell inside the female gametophyte fuse to form a cell that will give rise to endosperm, a nutritive tissue.

4. After fertilization, the endosperm forms, the ovule expands, the embryo sporophyte develops, and the integuments harden and often thicken. A fully matured ovule is a seed (its integuments have become the seed coat). While the seeds form, ovaries develop into fruits, which will help protect and disperse the seeds.

5. After dispersal, seeds germinate—the embryo inside absorbs water, resumes growth, and breaks through the seed coat. The seedling increases in volume and mass. Its tissues and organs develop. Later, fruits and new seeds form; older leaves drop. Plant hormones govern these and other patterns of growth and development. They also help adjust the patterns in response to local conditions and to environmental rhythms, including seasonal changes in daylength and temperature.

6. We know of five classes of plant hormones. Auxins and gibberellins promote stem elongation. Cytokinins promote cell division and leaf expansion, and retard leaf aging. Abscisic acid promotes bud and seed dormancy, and limits water loss by triggering stomatal closure. Ethylene promotes fruit ripening and abscission.

7. Plant parts make tropic responses to light, gravity, and other environmental conditions. Hormones induce a difference in the rate and direction of growth on two sides of the part, which causes it to turn or move.

8. A biological clock is an internal, time-measuring mechanism that has a biochemical basis.

a. In photoperiodism, plants respond to a change in the relative length of daylight and darkness, as occurs seasonally. The blue-green pigment phytochrome is the switching mechanism of the clock that promotes or inhibits germination, stem elongation, leaf expansion, stem branching, and flower, fruit, and seed formation.

b. Long-day plants flower during spring or summer, when there are more hours of daylight than darkness. Short-day plants flower when daylength is less. Day-neutral plants flower regardless of daylength.

9. Senescence is the sum of processes leading to the death of a plant or plant structure. Dormancy is a state in which a biennial or perennial plant stops growing even when conditions appear suitable for continued growth. A decrease in Pfr levels may trigger dormancy.

Review Questions

1. Label the floral parts. Explain floral function by relating the parts to events in the life cycle of flowering plants. *27.1*

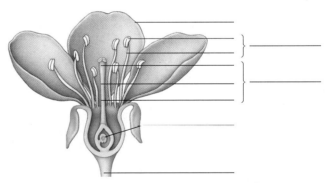

2. Consider the peach (*right*). Is a peach classified as a berry, drupe, or pome? *27.4*

3. Distinguish between:
   a. Sporophyte, gametophyte *27.1*
   b. Megaspore, microspore *27.3*
   c. Pollination, fertilization *27.3*
   d. Pollen grain, pollen tube *27.1, 27.3*
   e. Ovule, female gametophyte *27.3*
   f. Seed, fruit *27.4*
   g. Senescence, dormancy *27.11*

fleshy fruit

seed

4. Describe the steps by which a seven-cell, eight-nucleate embryo sac (a female gametophyte) forms. *27.3*

5. Define and describe one mode of vegetative growth. *27.5*

6. Name the three key factors that interact to dictate patterns of plant growth and development. *27.6*

7. List five types of plant hormones and briefly describe the known functions of each. *27.7*

7. Define plant tropism and give a specific example. *27.9*

8. What is phytochrome, and what is its role in flowering or some other process? *27.10*

Self-Quiz *(Answers in Appendix III)*

1. The _____ , which bears flowers, roots, stems, and leaves, dominates the life cycle of flowering plants.
   a. sporophyte   b. gametophyte

2. The flowers of many species coevolved with insects, birds, and other agents that function as _____ .

3. Seeds are mature _____ ; fruits are mature _____ .
   a. ovaries; ovules   c. ovules; ovaries
   b. ovules; stamens   d. stamens; ovaries

4. A _____ is a vessel, the lower portion of which is an ovary in which eggs develop, fertilization occurs, and seeds mature.
   a. pollen sac   b. carpel   c. receptacle   d. sepal

5. After meiosis within pollen sacs, haploid _____ form.
   a. megaspores   c. stamens
   b. microspores   d. sporophytes

6. Following meiosis in ovules, _____ megaspores form.
   a. two   b. four   c. six   d. eight

7. Cotyledons develop as part of all flowering plant _____ .
   a. seeds   b. fruits   c. embryos   d. ovaries

8. Which of the following statements is false?
   a. Auxins and gibberellins promote stem elongation.
   b. Cytokinins promote cell division but retard leaf aging.
   c. Abscisic acid promotes water loss and dormancy.
   d. Ethylene promotes fruit ripening and abscission.

9. Plant hormones _____ .
   a. interact with one another
   b. are influenced by environmental cues
   c. are active in plant embryos within seeds
   d. are active in adult plants
   e. all of the above

10. Light of _____ is the strongest stimulus for phototropism.
    a. red wavelengths   c. green wavelengths
    b. far-red wavelengths   d. blue wavelengths

11. The flowering process is a _____ response.
    a. phototropic   c. photoperiodic
    b. gravitropic   d. thigmotropic

12. Match the terms with the most suitable description.
    ____ double fertilization
    ____ ovule
    ____ mature female gametophyte
    ____ asexual reproduction
    ____ coevolution
        a. formation of zygote and first cell of endosperm
        b. outcome of two species interacting in close ecological fashion over geologic time
        c. contains a female gametophyte and has potential to be a seed
        d. an embryo sac, commonly with seven cells (one has two nuclei)
        e. mitotic cell division at bud or node produces a new plant

Critical Thinking

1. Observe several kinds of flowers growing in the area where you live. Given the coevolutionary links between flowering plants and their pollinators, describe what sorts of pollination agents your floral neighbors might depend on.

2. Before cherries, apples, peaches, and many other fruits ripen and the seeds inside them mature, their flesh is bitter or sour. Only later does it become tasty to animals that assist in seed dispersal. Develop a hypothesis of how this feature improves the odds for the plant's reproductive success.

3. Given what you know about the primary growth of plants (Section 25.1), propose an explanation of why plant hormones need not travel very far from the cells that secrete them.

4. Plant growth depends on photosynthesis, which depends on inputs of energy from the sun. How, then, can seedlings that were germinated in a dark room grow taller than seedlings that germinated in the sun?

5. *Solar tracking* refers to the observation that many plants are able to maintain the flat blades of their leaves at right angles to the sun throughout the day. Sunflowers provide us with a good example. This tropic response maximizes light harvest by leaves. Suggest the name of one type of molecule that might be involved in the response.

6. Belgian scientists isolated a mutant of wall cress (*Arabidopsis thaliana*) that produces excess amounts of auxin. Predict what some of this mutant plant's phenotypic traits might be.

7. Cattle receive somatotropin, an animal hormone that makes them grow bigger (added weight means more profits). There is growing concern that such hormones may have unforeseen effects on beef-eating humans. Would you think plant hormones applied to crop plants can affect humans also? Why or why not?

Selected Key Terms

abscisic acid *27.7*	endosperm *27.3*	phytochrome *27.10*
abscission *27.11*	ethylene *27.7*	plant tropism *27.9*
auxin *27.7*	flower *CI*	pollen grain *27.1*
biological clock *27.10*	fruit *27.4*	pollination *27.3*
carpel *27.1*	gametophyte *27.1*	pollinator *CI*
circadian rhythm *27.10*	germination *27.6*	seed *27.4*
coevolution *CI*	gibberellin *27.7*	senescence *27.11*
cotyledon *27.4*	gravitropism *27.9*	sporophyte *27.1*
cytokinin *27.7*	growth *27.7*	stamen *27.1*
development *27.7*	hormone *27.7*	statolith *27.9*
dormancy *422*	megaspore *27.3*	thigmotropism *27.9*
double fertilization *27.3*	microspore *27.3*	tissue culture propagation *27.5*
	ovary *27.1*	vegetative growth *27.5*
	ovule *27.3*	
	photoperiodism *27.10*	
	phototropism *27.9*	

Readings   *See also www.infotrac-college.com*

Grant, M. October 1993. "The Trembling Giant." *Discover* 14(10): 82–89.

Proctor, M., and P. Yeo. 1973. *The Pollination of Flowers.* New York: Taplinger.

Rost, T., M. Barbour, C. R. Stocking, and T. Murphy. 1998. *Plant Biology.* Belmont, California: Wadsworth.

WWW   *http://www.brookscole.com/biology*

Practice quiz questions, hypercontents, BioUpdates, and critical thinking. The Brooks/Cole Biology Resource Center provides a wealth of information fully organized and integrated by chapter.

FACING PAGE: *How many and what kinds of body parts does it take to function as a lizard in a tropical forest? Make a list of what comes to mind as you start reading Unit VI, then see how resplendent the list can become at the unit's end.*

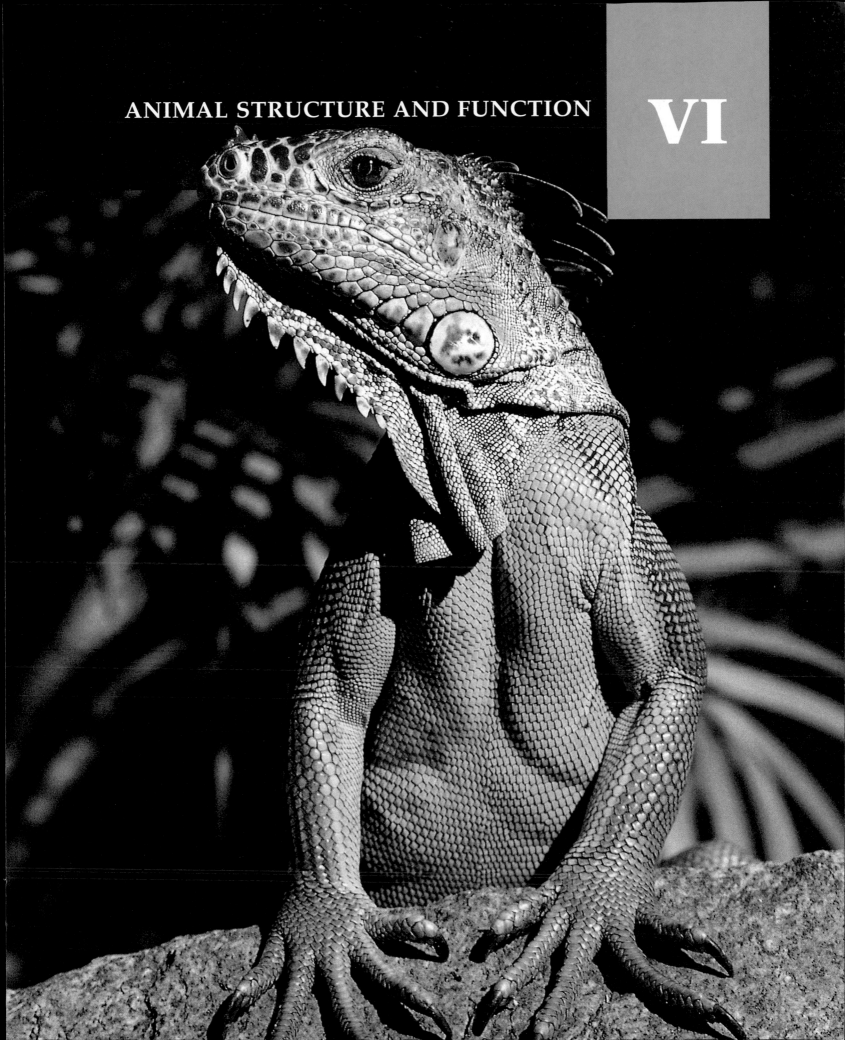

28

TISSUES, ORGAN SYSTEMS, AND HOMEOSTASIS

Meerkats, Humans, It's All the Same

After a cold night in Africa's Kalahari Desert, animals small enough to fit inside a coat pocket emerge stiffly from their burrows. These "meerkats" are a type of mongoose. They stand on their hind legs and face east, exposing their chilled bodies to the warm rays of the morning sun (Figure 28.1). Meerkats don't know it, but sunning behavior helps their enzymes. If their body's internal temperature were to fall below a tolerable range, the action of countless enzyme molecules in their cells would falter—and metabolism would suffer.

Once meerkats warm up, they fan out from their burrows and look for food. Into the meerkat gut go insects and an occasional lizard. These are pummeled, dissolved, and digested into glucose and other nutritious bits small enough to move across the gut wall, into the blood, and on to the body's cells. Aerobic machinery in the cells cracks apart molecules of glucose and other organic compounds and releases energy. A respiratory system contributes to supplying the machinery with oxygen and taking away its carbon dioxide leftovers.

All of this activity changes the composition and volume of the **internal environment**, which consists of blood and interstitial fluid (tissue fluid) that bathes the living cells of any complex animal. Drastic changes in those fluids would kill the cells, but a urinary system works to keep this from happening. Governing that system and all others are the nervous and endocrine systems. The two interact as a central command post to mobilize the body as a whole for everything from simple housekeeping tasks to heart-thumping flights from predators.

And so meerkats start us thinking about this unit's topics: how the animal body is structurally put together (its *anatomy*) and how it functions (its *physiology*). This chapter is an overview of the animal tissues and organ systems we will be considering. It also introduces the key concept of **homeostasis**. With respect to animals, the word refers to stable operating conditions in the internal environment, as brought about by coordinated activities of cells, tissues, organs, and organ systems.

Amazingly, all complex animals consist of only four basic types of tissues. These are epithelial, connective, muscle, and nervous tissues. A **tissue** is a group of cells and intercellular substances, all interacting in one or more tasks. As one example, muscle tissue takes part in contraction. An **organ** consists of different tissues that are organized in specific proportions and patterns. Thus every vertebrate heart has predictable proportions and arrangements of epithelial, connective, muscle, and nervous tissues. An **organ system** consists of two or

more organs that are interacting physically, chemically, or both in a common task, as when interconnected arteries and other vessels transport blood through the body under the driving force of a beating heart.

Cells, tissues, organs, and organ systems split up the work, so to speak, in ways that contribute to survival of the body as a whole. This is sometimes known as a **division of labor**. By the end of this unit, you may have an abiding appreciation of the sheer magnitude of the division of labor among the separate parts and the extent to which their activities are integrated.

As you will see, whether you look at a flatworm or salmon, a meerkat or human being, the animal body is structurally and physiologically adapted to perform four overriding tasks:

1. *Maintain conditions in the internal environment within ranges that are most favorable for cell activities.*

2. *Acquire nutrients and other raw materials, distribute them through the body, and dispose of wastes.*

3. *Afford protection against injury or attack from viruses, bacteria, and other agents of disease.*

4. *Reproduce, then often help nourish and protect the new individuals during their early growth and development.*

Figure 28.1 In the Kalahari Desert, gray meerkats (*Suricata suricatta*) face the warming rays of the morning sun, just as they do every morning. This simple behavior helps maintain internal body temperature. How animals function in their environment is the subject of this unit.

KEY CONCEPTS

1. The cells of most animals interact at three levels of organization—in tissues, many of which are combined in organs, which are components of organ systems.

2. Most animals are constructed of four types of tissues, which are called epithelial, connective, muscle, and nervous tissues.

3. Each animal cell engages in basic metabolic activities that ensure its own survival. At the same time, animal cells of a given tissue perform one or more activities that contribute to the survival of the animal as a whole.

4. The internal environment consists of all fluids that are not inside the body's cells—that is, blood and interstitial fluid.

5. The combined contributions of cells, tissues, organs, and organ systems help maintain stability in the internal environment, which is required for the survival of each individual cell. This concept helps us understand the functions of any organ or organ system.

6. Homeostasis is the formal name for stable operating conditions in the internal environment.

General Characteristics

We commonly refer to an epithelial tissue as **epithelium** (plural, epithelia). This tissue has a free surface, which faces either a body fluid or the outside environment. *Simple* epithelium, with a single layer of cells, functions as a lining for body cavities, ducts, and tubes. *Stratified* epithelium, which has two or more cell layers, typically functions in protection, as it does in your skin. Figure 28.2 shows examples of this type of animal tissue.

All cells in epithelium are close together, with little intervening material. As is the case for nearly all animal tissues, specialized junctions afford both structural and functional links between its individual cells, which absorb, make, and secrete a variety of substances.

Cell-to-Cell Contacts

In epithelium and other tissues, we observe three classes of cell junctions. **Tight junctions** help stop substances from leaking across a tissue. **Adhering junctions** are spot welds; they cement neighbor cells together. **Gap junctions** are channels connecting the cytoplasm of abutting cells. By promoting rapid transfers of ions and small molecules, they help cells communicate with each other (Figure 28.3).

Think of gastric fluid inside the stomach to get an idea of how important such junctions are. If this highly acidic fluid were to leak past the stomach's epithelial lining, it would digest proteins of your own body instead of those brought in with meals. Actually, that happens in patients with peptic ulcers (Section 36.4). Many tight junctions in such linings form a leakproof barrier between all the cells near their free surface.

free surface of epithelium

epithelial cell

basement membrane

underlying connective tissue

a Some characteristics of epithelium

Figure 28.2 (**a**) Some basic characteristics of epithelium. Epithelia have a free surface exposed to a body fluid or to the environment. Between the epithelium's other surface and the connective tissue on which it rests is a basement membrane. The diagram below the athlete shows this arrangement for a simple epithelium, which has a single layer of cells. The light micrograph shows the upper part of stratified epithelium. This type of epithelium has more than one layer of cells, which are flattened out near the surface.

(**b**) Light micrographs and sketches of three simple epithelia, showing the three shapes of cells that are common in this type of tissue.

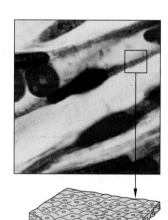

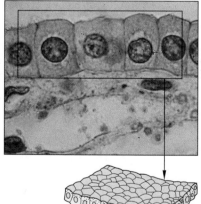

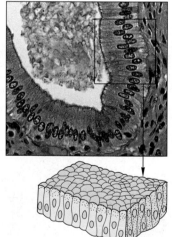

TYPE: Simple squamous
DESCRIPTION: Single layer of flattened cells
COMMON LOCATIONS: Blood vessel walls; air sacs of lungs
FUNCTION: Diffusion

TYPE: Simple cuboidal
DESCRIPTION: Single layer of cubelike cells; free surface may have microvilli (absorptive structures)
COMMON LOCATIONS: Glands and tubular parts of nephrons in kidneys
FUNCTION: Secretion, absorption

TYPE: Simple columnar
DESCRIPTION: Single layer of tall, slender cells; free surface may have microvilli
COMMON LOCATIONS: Part of lining of gut and respiratory tract
FUNCTION: Secretion, absorption

b Examples of cell shape in simple epithelium

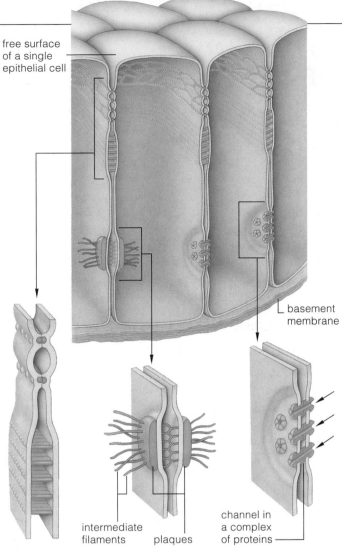

free surface of a single epithelial cell

basement membrane

intermediate filaments

plaques

channel in a complex of proteins

a
TIGHT JUNCTION

Strands (rows of proteins) running parallel with the free surface of the tissue; they stop leaks between adjoining cells.

b
ADHERING JUNCTION

Adjoining cells adhere at a mass of proteins (a plaque) anchored beneath their plasma membranes by many intermediate filaments of the cytoskeleton.

c
GAP JUNCTION

Cylindrical arrays of proteins that span the plasma membrane of adjoining cells pair up, forming open channels between the cytoplasm of both cells.

Figure 28.3 *Left:* Examples of cell junctions.

(**a**) In some epithelia, protein strands ring each cell. They form tight seals and fuse the cell with its neighbors. Tight junctions stop most substances from leaking across the epithelium's free surface. Substances reach the tissues below only by passing *through* epithelial cells. Built-in controls make the plasma membrane of these cells selectively permeable. They allow some substances but not others to move across, through the interior of transport proteins (Sections 5.5 and 5.6).

(**b**) Adhesion junctions (spot welds or collars) hold cells of epithelium and all other tissues together, so that they function as a unit. They are profuse in the skin's surface layer and in other tissues subjected to abrasion.

(**c**) Gap junctions are channels across the plasma membrane that allow the cytoplasm of adjoining cells to interconnect. Gap junctions promote chemical communication (the diffusion of ions and small molecules from cell to cell). They are especially abundant in the heart, stomach, and other organs in which cell activities must be rapidly coordinated.

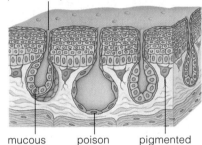

pore that opens at skin surface

mucous gland

poison gland

pigmented cell

Figure 28.4 Section through glandular epithelium of a frog. The photograph shows a frog of the genus *Dendrobates*, which produces one of the most lethal glandular secretions known. (Some natives of a tribe in Colombia use the exocrine gland secretion to poison dart tips, which they shoot through blowguns.) Pigment-rich cells that branch through the epithelium impart color to the skin. The striking coloration of all poisonous frogs serves as a strong warning signal to potential predators.

Glandular Epithelium and Glands

Epithelia of structurally simple invertebrates contain **gland cells**, which secrete (release) products, unrelated to their own metabolism, that are to be used elsewhere. In more complex animals, such cells occur in glandular epithelium and in glands, which are secretory organs derived from epithelia. ("Secretion" is not the same as "excretion," a concentration and removal of metabolic wastes or excess substances of no use to the body.)

Exocrine glands secrete mucus, saliva, earwax, oil, milk, digestive enzymes, and other cell products. Most products are released onto the free epithelial surface through ducts or tubes, as in Figure 28.4. By contrast, **endocrine glands** do not have ducts. Their products are hormones, which are secreted directly into the fluid bathing the gland. From there, the hormone molecules typically enter the bloodstream, which distributes them to target cells elsewhere in the body.

Epithelia are sheetlike tissues lining the body's surface and its cavities, ducts, and tubes. Epithelia have one free surface facing a body fluid or the outside environment. Their cells are structurally and functionally connected at junctions.

Glands are secretory organs derived from epithelium.

WWW

CONNECTIVE TISSUE

Of all tissues in complex animals, connective tissues are the most abundant and widely distributed. They range from soft connective tissues to specialized types, which include cartilage, bone, adipose tissue, and blood (Table 28.1). In all connective tissues except blood, cells secrete fibers of structural proteins—collagen or elastin. (This is the collagen, suctioned from patients, that plastic surgeons use to plump wrinkled skin and lips, as well as sunken acne scars.) The cells also secrete modified polysaccharides, which accumulate between cells and fibers as the connective tissue's "ground substance."

Soft Connective Tissues

These tissues all have mostly the same components but in different proportions. **Loose connective tissue** has fibers and cells loosely arranged in a semifluid ground substance (Figure 28.5*a*). Often it serves as a support framework for epithelium. It contains fibroblasts (cells that produce and secrete fibers) and white blood cells. When bacteria enter skin through small cuts or breach the lining of the digestive, respiratory, or urinary tracts, those white blood cells mount an early counterattack.

Dense, irregular connective tissue has fibroblasts and many fibers (mostly collagen-containing ones) that are oriented every which way. This tissue is present in skin and forms protective capsules around organs that

Table 28.1 Categories of Connective Tissues

SOFT	SPECIALIZED
Loose connective tissue	Cartilage
Dense, irregular connective tissue	Bone tissue
Dense, regular connective tissue (ligaments, tendons)	Adipose tissue
	Blood

do not stretch much (Figure 28.5*b*). In **dense, regular connective tissue**, the fibroblasts occur in rows between many parallel bundles of fibers. Tendons, which attach skeletal muscle to bone, have this tissue. The bundles of collagen fibers help tendons resist being torn (Figure 28.5*c*). Dense, regular connective tissue also is present in elastic ligaments, which attach one bone to another. The elastic fibers allow movement around joints.

Specialized Connective Tissues

Like rubber, the intercellular material called **cartilage** is solid and pliable, and resists compression. Cells that made and released the material became imprisoned in small cavities in their own secretions (Figure 28.5*d*). Cartilage deposited in vertebrate embryos serves as the structural models for bones. Bone replaces most of the models. In adults, some cartilage maintains the shape

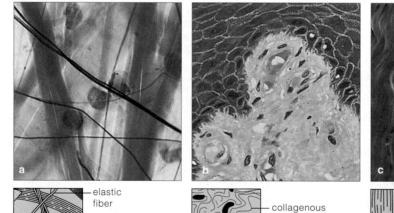
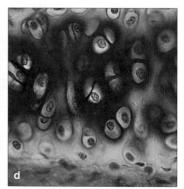
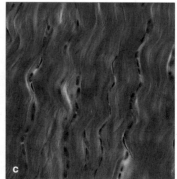
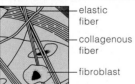

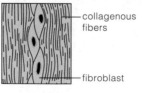

TYPE: Loose connective tissue

DESCRIPTION: Fibroblasts, other cells, plus fibers loosely arranged in semifluid ground substance

COMMON LOCATIONS: Under the skin and most epithelia

FUNCTION: Elasticity, diffusion

TYPE: Dense, irregular connective tissue

DESCRIPTION: Collagenous fibers, fibroblasts, less ground substance

COMMON LOCATIONS: In skin and capsules around some organs

FUNCTION: Support

TYPE: Dense, regular connective tissue

DESCRIPTION: Collagen fibers in parallel bundles, long rows of fibroblasts, little ground substance

COMMON LOCATIONS: Tendons, ligaments

FUNCTION: Strength, elasticity

TYPE: Cartilage

DESCRIPTION: Cells embedded in pliable, solid ground substance

COMMON LOCATIONS: Ends of long bones, nose, parts of airways, skeleton of vertebrate embryos

FUNCTION: Support, flexibility, low-friction surface for joint movement

Figure 28.5 Examples of soft connective tissues and specialized connective tissues.

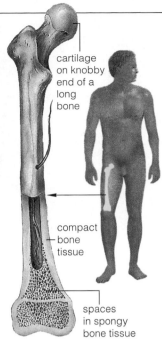

cartilage on knobby end of a long bone

compact bone tissue

spaces in spongy bone tissue

Figure 28.6 Cartilage and bone tissue. Spongy bone tissue has tiny, needlelike hard parts with spaces in between. Compact bone tissue is more dense. Bone is a load-bearing tissue that resists compression. Over time, it served as a basis for increases in the body size of many land vertebrates, including giraffes. It gives large animals selective advantages. Among other things, they can ignore most predators with impunity, roam farther for food and water, and heat up and cool off more slowly than small animals (because of a lower surface-to-volume ratio and greater heat production).

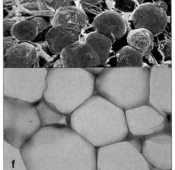

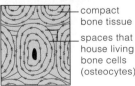

compact bone tissue

spaces that house living bone cells (osteocytes)

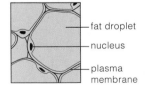

fat droplet

nucleus

plasma membrane

TYPE: Bone tissue

DESCRIPTION: Collagen fibers, ground substance hardened with calcium

COMMON LOCATIONS: Bones of vertebrate skeleton

FUNCTION: Movement, support, protection

TYPE: Adipose tissue

DESCRIPTION: Large, tightly packed fat cells occupying most of ground tissue

COMMON LOCATIONS: Under skin, around heart, kidneys

FUNCTION: Energy reserves, insulation, padding

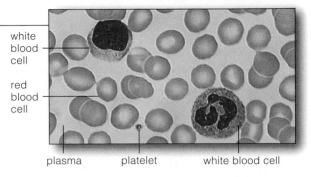

white blood cell

red blood cell

plasma   platelet   white blood cell

Figure 28.7 Some components of human blood. This tissue's straw-colored, liquid matrix (plasma) is mostly water in which many nutrients, diverse proteins, oxygen, carbon dioxide, ions, and other substances are dissolved.

of the nose, outer ear, and other body parts. It cushions joints between adjacent bones of the vertebral column, limbs, hands, and elsewhere.

Bone tissue is mineral-hardened; its collagen fibers and ground substance are rich in calcium salts (Figures 28.5*e* and 28.6). It is the main tissue of bones. In vertebrate skeletons, different bones support and protect softer tissues and organs. Limb bones, such as the long bones of your own legs, serve weight-bearing functions. They also interact with skeletal muscles attached to them to bring about movements. Parts of some bones also are production sites for blood cells.

Excess carbohydrates and proteins that cells do not use at once are converted to storage forms, especially fats. A few fat droplets collect in the cytoplasm of various cells. But cells in **adipose tissue**, located mainly beneath skin, specialize in fat storage; fat droplets practically fill their cytoplasm (Figure 28.5*f*). A rich blood supply in the tissue is an accessible medium through which fats move.

Blood is derived primarily from connective tissue, and that is why many authors consider it a connective tissue. Blood has transport functions. Circulating within *plasma*, its fluid medium, are a great many red blood cells, white blood cells, and platelets (Figure 28.7). Red blood cells efficiently deliver oxygen to metabolically active tissues, and then carry carbon dioxide and other wastes away from them. Plasma is largely water, but it has a great number of different kinds of proteins, ions, and other substances dissolved in it. We take a closer look at this complex tissue in Section 33.2.

Diverse types of connective tissues bind together, support, strengthen, protect, and insulate other tissues in the body.

Soft connective tissues consist of protein fibers as well as a variety of cells arranged in a ground substance.

Cartilage, bone, blood, and adipose tissue are specialized connective tissues. Cartilage and bone are both structural materials. Blood is a fluid tissue with transport functions. Adipose tissue is a reservoir of stored energy.

WWW

MUSCLE TISSUE

In all muscle tissues, cells *contract* (that is, shorten) in response to stimulation from the outside, then lengthen and so return to their uncontracted state. These tissues have many long, cylindrical cells arranged in parallel arrays. Muscle layers and muscular organs contract and relax in coordinated fashion. Their action moves

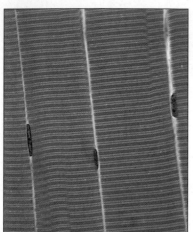

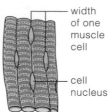

width of one muscle cell

cell nucleus

TYPE: Skeletal muscle

DESCRIPTION: Bundles of long, cylindrical, striated, contractile cells

LOCATION: Associated with skeleton

FUNCTION: Locomotion, movement of body parts

a

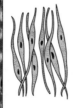

cells, teased apart for clarity

TYPE: Smooth muscle

DESCRIPTION: Contractile cells with tapered ends

LOCATION: Wall of internal organs, such as stomach

FUNCTION: Movement of internal organs

b

junction between adjacent cells

TYPE: Cardiac muscle

DESCRIPTION: Cylindrical, striated cells that have specialized end junctions

LOCATION: Wall of heart

FUNCTION: Pump blood within circulatory system

c

Figure 28.8 Characteristics and examples of skeletal muscle, smooth muscle, and cardiac muscle tissues.

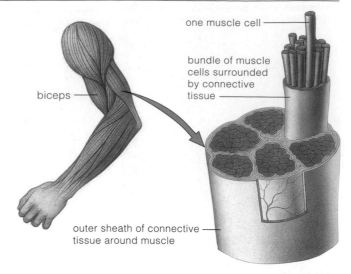

one muscle cell

bundle of muscle cells surrounded by connective tissue

biceps

outer sheath of connective tissue around muscle

Figure 28.9 Location and general arrangement of muscle cells in a typical skeletal muscle.

the body through the environment, and it maintains and changes the positions of the body's various parts. The three types of muscle tissue are called skeletal, smooth, and cardiac muscle tissues.

Skeletal muscle tissue is located in muscles that are securely fastened to skeletal bones (Figure 28.8*a*). In a typical muscle, such as the biceps, striated skeletal muscle cells are bundled closely together, in parallel. (*Striated* means striped.) A sheath of tough connective tissue encloses several bundles of muscle cells, as you can see from Figure 28.9. The structure and function of skeletal muscle tissue are topics of Chapter 32.

The contractile cells of **smooth muscle tissue** taper at both ends (Figure 28.8*b*). Cell junctions hold them together, and they are bundled together in a connective tissue sheath. The wall of internal organs—including blood vessels, the stomach, and the intestines—contains this type of muscle tissue. Contraction of smooth muscle action is sometimes said to be "involuntary," because we usually are not able to make it contract merely by thinking about it (as we can do with skeletal muscle).

Cardiac muscle tissue is a contractile tissue that is present only in the heart (Figure 28.8*c*). Cell junctions fuse the plasma membranes of cardiac muscle cells and make them stick together. Communication junctions at some fusion points allow the cells to contract as a unit. That is, when one cell receives a signal to contract, its neighbors are stimulated to contract, also.

Muscle tissue, which can contract (shorten) in response to stimulation, helps move the body and specific body parts.

Skeletal muscle is the only muscle tissue attached to bones. Smooth muscle is a component of internal organs. Cardiac muscle alone makes up the contractile walls of the heart. Connective tissue sheathes all three types of tissues.

𝒲𝒲𝒲

NERVOUS TISSUE

Of all tissues, **nervous tissue** exerts the greatest control over the body's responsiveness to changing conditions. In your own body, **neuroglia** makes up more than one-half the volume of nervous tissue. That word refers to a variety of cells that protect, structurally support, and metabolically support neurons, which make up the rest of your nervous tissue. **Neurons** are excitable cells, the communication units of most nervous systems.

When a neuron is suitably stimulated, an electrical disturbance swiftly travels along its plasma membrane. Arrival of the disturbance at the neuron's endings, or output zone, triggers events that may cause stimulation or inhibition of adjacent neurons and other cells.

cell body of one neuron

a

b

Figure 28.10 (**a**) A few motor neurons from the human nervous system. This type relays signals from the brain or spinal cord to muscles and glands. Collectively, different neurons detect and process a great number and great variety of signals about environmental change, and initiate suitable responses. Without neurons, for instance, a chameleon (**b**) could not detect an edible insect, calculate its distance, and command a long, sticky, prey-capturing tongue to uncoil with stunning speed.

For example, more than a hundred billion neurons are organized as communication lines throughout your body. Some detect specific changes in environmental conditions. Others coordinate immediate and long-term responses to change. The type shown in Figure 28.10*a* relays signals from the brain to muscles and glands. How these cells function is a topic of later chapters.

Neurons are the basic units of communication in nervous tissue. Different kinds detect specific stimuli, integrate information, and issue or relay commands for response.

WWW

FRONTIERS IN TISSUE RESEARCH

As you probably will gather after thinking about the many tissues that are used as examples throughout this unit, a tissue is more than a sum of its cells. As each new animal grows and develops, its cells interact and become organized in particular ways to give rise to the body's diverse tissues. And as each new tissue develops, its cells synthesize specific gene products that are vital for normal body functioning.

For many decades, medical researchers have attempted to find a way to construct artificial tissues in quantity in the laboratory. Currently, they can grow extensive sheets of epidermis from a patient and use it to regenerate skin that was destroyed through third-degree burns and other types of severe damage. In some laboratories, researchers are taking small sections of epidermis from the skin of patients and exposing them to a culture medium that contains nutrients and growth factors. The cells proliferate and form *laboratory-grown epidermis*. When surgeons place an epidermal sheet over a wound, the cells in the sheet interact biochemically and structurally with underlying cells. As an outcome of the interactions, the damaged or missing tissue is regenerated.

On the horizon are *designer organs*. These encapsulated, selected groups of living cells might synthesize specific hormones, enzymes, and other substances. The idea is to surgically snip a bit of epithelium from a patient and then use it   to enclose the cells, like a capsule. Because   the capsule is derived from epithelial cells of the patient's own body, it will not be chemically recognized as "foreign" and attacked by the patient's immune system. As you will see in Chapter 34, rejection of tissue and organ implants can have serious medical consequences.

Currently, biotechnologists are close to understanding how to synthesize molecular cues that will allow designer organs to stick to appropriate sites in the body. Once the synthetic organs stick, they may become integral parts of normal body functioning.

Ultimately, the goal of this research is to put together packages of cells capable of producing specific life-saving substances that are absent in patients who suffer from severe genetic disorders or chronic diseases. For example, imagine the potential for people who have type I *diabetes mellitus*. This metabolic disorder results in an elevated concentration of glucose in the blood. Affected people produce little if any insulin, which is the hormone that signals cells to take up glucose from the blood. Unless they receive insulin injections on a regular basis, they will die. However, if a customized, insulin-secreting organ can be successfully installed inside the body of such individuals, their daily injections of insulin might be a thing of the past.

WWW

Overview of the Major Organ Systems

Figure 28.11 gives an overview of the organ systems of a typical vertebrate, the adult human. Figure 28.12 lists some terms that are used when describing positions of various organs. It also shows the major body cavities in which a number of important organs are located.

Each organ system contributes to the survival of all living cells in the animal body. You may think this is stretching things a bit. For example, how could muscles and bones be helping each microscopically small cell to stay alive? And yet interactions between the skeletal and muscular systems allow us to move about—toward sources of nutrients and water, for example. Some parts of the two organ systems help keep blood circulating to cells, as when leg muscle contractions help move blood in veins back to the heart. Blood inside the circulatory system rapidly transports oxygen, nutrients, and other substances to cells, and transports products and wastes away from them. The respiratory system swiftly delivers oxygen from the outside air to the circulatory system and takes up carbon dioxide wastes from it, skeletal muscles assist the respiratory system—and so it goes, throughout the entire body.

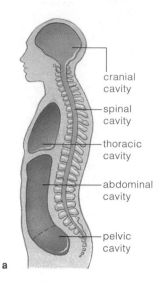

Figure 28.12 (a) The major cavities in the human body. (b,c) Directional terms and planes of symmetry for the vertebrate body. Notice how the *midsagittal* plane divides the body into right and left halves. Most vertebrates, such as fishes and rabbits, move with the main body axis parallel with the Earth's surface. For them, *dorsal* pertains to their back or upper surface, and *ventral* pertains to the opposite, lower surface.

cranial cavity

spinal cavity

thoracic cavity

abdominal cavity

pelvic cavity

a

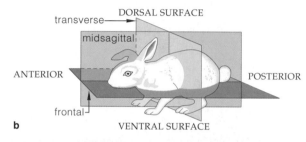

DORSAL SURFACE

transverse

midsagittal

ANTERIOR

POSTERIOR

frontal

VENTRAL SURFACE

b

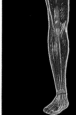
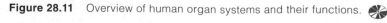

INTEGUMENTARY SYSTEM	MUSCULAR SYSTEM	SKELETAL SYSTEM	NERVOUS SYSTEM	ENDOCRINE SYSTEM	CIRCULATORY SYSTEM
Protect body from injury, dehydration, and some pathogens; control its temperature; excrete some wastes; receive some external stimuli.	Move body and its internal parts; maintain posture; generate heat (by increases in metabolic activity).	Support and protect body parts; provide muscle attachment sites; produce red blood cells; store calcium, phosphorus.	Detect both external and internal stimuli; control and coordinate responses to stimuli; integrate all organ system activities.	Hormonally control body function; work with nervous system to integrate short-term and long-term activities.	Rapidly transport many materials to and from cells; help stabilize internal pH and temperature.

Figure 28.11 Overview of human organ systems and their functions.

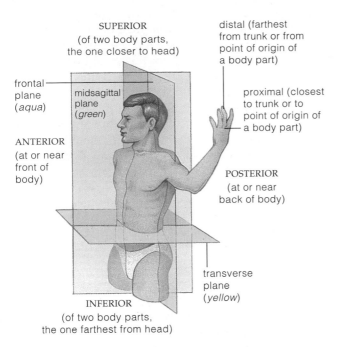

SUPERIOR
(of two body parts, the one closer to head)

distal (farthest from trunk or from point of origin of a body part)

frontal plane (*aqua*)

midsagittal plane (*green*)

ANTERIOR (at or near front of body)

proximal (closest to trunk or to point of origin of a body part)

POSTERIOR (at or near back of body)

transverse plane (*yellow*)

INFERIOR
(of two body parts, the one farthest from head)

c Unlike quadrupedal animals, humans walk upright, with their main body axis perpendicular to the ground. *Anterior* refers to the front of the body; it corresponds to ventral, as shown in (**b**). *Posterior* refers to the back; it corresponds to dorsal.

Tissue and Organ Formation

Where do the tissues of organ systems come from? To get a sense of how they originate, start with a sperm and egg. Recall that sperm and eggs develop from germ cells, which are immature reproductive cells. (All other cells in the body are "somatic," after the Greek word for body.) After a zygote forms at fertilization, mitotic cell divisions form an early embryo. In vertebrates, the cells become arranged as three primary tissues—ectoderm, mesoderm, and endoderm. These three are embryonic forerunners of all tissues in the adult. **Ectoderm** gives rise to the skin's outer layer and to tissues of a nervous system. **Mesoderm** gives rise to tissues of the muscles, bones, and most of the circulatory, reproductive, and urinary systems. **Endoderm** gives rise to the lining of the digestive tract and to organs derived from it.

In general, vertebrates have the same kinds of organ systems. Each organ system serves specialized functions, such as gas exchange, blood circulation, and locomotion.

Vertebrate tissues, organs, and organ systems arise from three primary tissues in the developing embryo. These primary tissues are ectoderm, mesoderm, and endoderm.

WWW

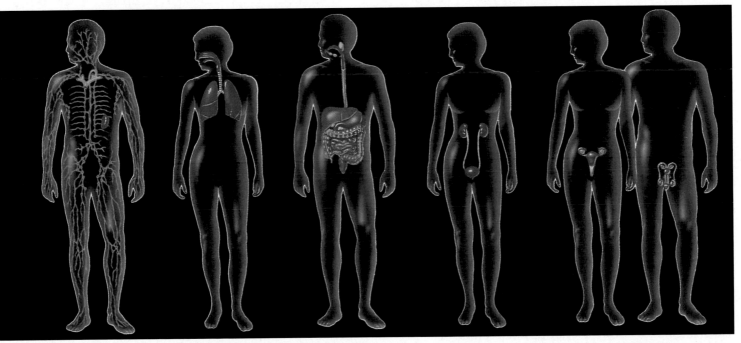

LYMPHATIC SYSTEM	RESPIRATORY SYSTEM	DIGESTIVE SYSTEM	URINARY SYSTEM	REPRODUCTIVE SYSTEM
Collect and return some tissue fluid to the bloodstream; defend the body against infection and tissue damage.	Rapidly deliver oxygen to the tissue fluid that bathes all living cells; remove carbon dioxide wastes of cells; help regulate pH.	Ingest food and water; mechanically, chemically break down food, and absorb small molecules into internal environment; eliminate food residues.	Maintain the volume and composition of internal environment; excrete excess fluid and blood-borne wastes.	*Female:* Produce eggs; after fertilization, afford a protected, nutritive environment for the development of new individual. *Male:* Produce and transfer sperm to the female. Hormones of both systems also influence other organ systems.

Concerning the Internal Environment

To stay alive, your cells must remain bathed in a fluid that offers nutrients and carries away metabolic wastes. In this they are no different from an amoeba or some other free-living, single-celled organism. The difference is, trillions of cells coexist in your body. They must draw nutrients from and dump wastes into the same fifteen liters of fluid, which is less than sixteen quarts.

The fluid *not* inside cells is **extracellular fluid**. Much of it is *interstitial*, meaning it occupies spaces between cells and tissues. The remainder is *plasma*, which is the fluid portion of blood. The interstitial fluid exchanges substances with the cells it bathes and with blood.

In functional terms, changes in extracellular fluid cause changes in cells. That is why drastic changes in the composition and volume of extracellular fluid have drastic effects on cell activities. The type and number of ions are especially crucial, for they must be maintained at concentrations that are compatible with metabolism. Otherwise, the animal itself cannot survive.

It makes no difference whether an animal is simple or complex. *The component parts of any animal work together to maintain the stable fluid environment required by all of its living cells.* This concept is absolutely central to understanding the structure and function of animals, and its key points may be summarized as follows. First, each cell of the animal body engages in basic metabolic activities that ensure its own survival. Second, the cells of a given tissue perform one or more activities that contribute to the survival of the whole organism. Third, the combined contributions of individual cells, tissues, organs, and organ systems engaged in a division of labor help maintain a stable internal environment—extracellular fluid—required for individual cell survival.

Mechanisms of Homeostasis

Homeostasis, recall, means stable operating conditions in the internal environment. Three components interact to maintain this state. They are called sensory receptors, integrators, and effectors. **Sensory receptors** are cells or cell parts that can detect a **stimulus**, which is a specific change in the environment. When someone kisses you, for example, there is a change in pressure on your lips. Receptors in the skin of your lips translate the stimulus into a signal that can be sent to the brain. Your brain is an **integrator**, a central command post where different bits of information are pulled together in the selection of a response. The brain sends signals to your muscles or glands (or both). Muscles and glands are **effectors**—they carry out the response. In this particular case, the response may include flushing with pleasure and kissing the person back. Of course, you cannot engage in a kiss indefinitely, for this would prevent you from eating and carrying out other activities necessary to maintain operating conditions inside your body.

So how does your brain reverse the physiological changes induced by the kiss? Receptors only provide it with information about how things *are* operating. The brain also receives information about how things *should be* operating—that is, information from "set points." When physical or chemical conditions deviate sharply from a set point, the brain functions to bring them back to an effective operating range. It does this by way of signals that cause specific muscles and specific glands to increase or decrease their activity.

NEGATIVE FEEDBACK Feedback mechanisms are among the controls that operate to keep physical and chemical aspects of the body within tolerable ranges. To give one example, with a **negative feedback mechanism**, some activity alters a condition in the internal environment, and this triggers a response that reverses the altered condition (Figure 28.13).

Think of a furnace with a thermostat. A thermostat senses the air temperature and "compares" it against a preset point on a thermometer built into the furnace's control system. When the temperature falls below the preset point, the thermostat sends signals to a switching mechanism that can turn on the furnace. When the air becomes heated enough to match the prescribed level, the thermostat again signals the switching mechanism, which shuts off the furnace.

Similarly, feedback mechanisms help keep the body temperature of meerkats, humans, huskies, and many other animals near 37°C (98.6°F), even during hot or cold weather. Visualize a young husky running around on a hot summer day. Soon its body gets hot, and receptors trigger events that slow down the whole dog *and* its cells. The husky searches for shade and rests under a tree. Moisture from its respiratory system evaporates from the tongue and carries away some body heat with

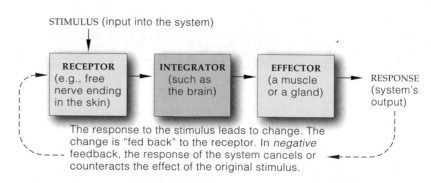

Figure 28.13 Components necessary for negative feedback at the organ level.

STIMULUS

The husky is overactive on a hot, dry day and its body surface temperature rises.

```
RECEPTORS          An INTEGRATOR        Some EFFECTORS
in skin and        (the hypothalamus,   (pituitary gland
elsewhere          a brain region)      and thyroid
detect the         compares input       gland) trigger
temperature        from the receptors   widespread
change.            against a set point. adjustments.
```

RESPONSE

Temperature of circulating blood starts decreasing.

Many EFFECTORS carry out specific responses:

SKELETAL MUSCLES	SMOOTH MUSCLE IN BLOOD VESSELS	SALIVARY GLANDS	ADRENAL GLANDS
Husky rests, starts to pant (behavioral changes).	Blood carrying metabolically generated heat shunted to skin, some heat lost to surroundings.	Secretions from glands increase; evaporation from tongue. Both have a cooling effect, especially on the brain.	Output drops, husky is less stimulated.

Activity of the body in general slows down (behavioral change).

The overall slowdown in activities results in less metabolically generated heat.

Figure 28.14 Homeostatic controls over the internal temperature of a husky's body. *Blue* arrows indicate the main control pathways. The *dashed line* shows how a feedback loop is completed.

it, as Figure 28.14 shows. These control mechanisms and others counter overheating by curbing the activities that naturally generate metabolic heat and by giving up the body's excess heat to the surrounding air.

POSITIVE FEEDBACK   In some cases, **positive feedback mechanisms** operate. These controls set in motion a chain of events that *intensify* a change from an original condition—and after a limited time, the intensification reverses the change. Positive feedback is associated with instability in a system. For example, during sexual intercourse, chemical signals from a female's nervous system can induce her to make intense physiological responses to her sexual partner. Her responses stimulate changes in her partner that stimulate the female even more—and so on until she reaches an explosive, climax level of excitation. Normal conditions now return, and homeostasis prevails.

As another example, at childbirth, a fetus exerts pressure on the wall of its mother's uterus. Pressure stimulates the production and secretion of oxytocin, a hormone. Oxytocin causes wall muscles to contract and exert pressure on the fetus, which exerts more pressure on the wall, and so on until the fetus is expelled.

What we have been describing is a general pattern of detecting, evaluating, and responding to a continual flow of information about the animal's internal and external environments. During all of this activity, organ systems operate together in astoundingly coordinated fashion. Throughout this unit of the book, you will be asking the following questions about their operation:

1. *What physical or chemical aspects of the internal environment are organ systems working to maintain as conditions change?*

2. *By what means are organ systems kept informed of the various changes?*

3. *By what means do they process incoming information?*

4. *What mechanisms are set in motion in response?*

As you will see in later chapters, operation of all organ systems is under neural and endocrine control.

Each living cell of an animal body engages in basic metabolic activities that ensure its own survival. Concurrently, the cells of any given tissue are performing one or more activities that contribute to the survival of the whole animal.

The combined contributions of cells, tissues, organs, and organ systems maintain the stable internal environment (extracellular fluid) required for individual cell survival.

Homeostatic control mechanisms help maintain physical and chemical aspects of the body's internal environment within ranges that are most favorable for cell activities.

SUMMARY

1. A tissue is an aggregation of cells and intercellular substances that perform a common task. An organ is a structural unit of different tissues combined in definite proportions and patterns that allow them to perform a common task. An organ system has two or more organs interacting chemically, physically, or both in ways that contribute to the survival of the body as a whole.

2. Epithelial tissues cover external body surfaces and line internal cavities and tubes. These tissues have one free surface exposed to body fluids or the environment.

3. Different connective tissues bind together, support, strengthen, protect, and insulate other tissues. Most have fibers of structural proteins (especially collagen), fibroblasts, and other cells within a ground substance.

   a. Loose connective tissue, with a semifluid ground substance, is present under skin and most epithelia.

   b. Dense, irregular connective tissue contains mostly collagen fibers and fibroblasts. It is present in skin and forms protective capsules around a number of organs.

   c. Dense, regular connective tissue, such as that in tendons, contains parallel bundles of collagen fibers. It protects and structurally supports organs.

   d. Cartilage, with its solid yet pliable intercellular material, has structural and cushioning roles. Bone is the weight-bearing tissue of vertebrate skeletons, which interacts with skeletal muscle to bring about movement.

   e. Blood, a specialized connective tissue, consists of plasma, cellular components, and dissolved substances. Adipose tissue, another specialized connective tissue, is a reservoir of energy; it consists mainly of fat cells.

4. Muscle tissues contract (shorten), then return to the resting position. They help move the body or parts of it. The three types of muscle tissue are skeletal muscle, smooth muscle, and cardiac muscle tissue.

5. Nervous tissue intercepts and integrates information about internal and external conditions, and governs the body's responses to change. Neurons of this tissue are the basic units of communication in nervous systems.

6. Tissues, organs, and organ systems work together to maintain the stable internal environment (that is, the extracellular fluid) required for individual cell survival. At homeostasis, conditions in the internal environment are balanced at levels most favorable for cell activities.

7. Feedback controls help maintain internal operating conditions for the body's cells. With negative feedback, for example, a change in a particular condition triggers a response that results in a reversal of the change.

8. Homeostasis depends on receptors, integrators, and effectors. Receptors detect stimuli, which are specific changes in the environment. Integrating centers, such as a brain, process information and direct muscles and glands (the body's effectors) to carry out responses.

Review Questions

1. Describe the characteristics of epithelial tissue in general. Then describe the various types of epithelial tissues in terms of specific characteristics and functions. *28.1*

2. List the major types of connective tissues; add the names and characteristics of their specific types. *28.2*

3. Identify and describe the following tissues: *28.1–28.3*

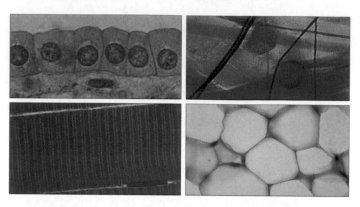

4. Identify this category of tissue and its characteristics: *28.4*

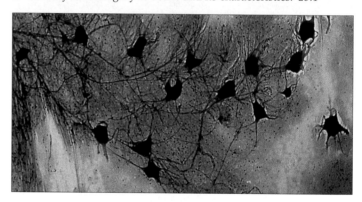

5. What type of cell serves as the basic unit of communication in nervous systems? *28.4*

6. Define animal tissue, organ, and organ system. List and define the functions of the eleven major organ systems of the human body. *CI, 28.6*

7. Define extracellular fluid, interstitial fluid, and plasma. *28.7*

8. Define homeostasis. *CI, 28.7*

9. Briefly describe two major categories of the homeostatic mechanisms operating in the human body. *28.7*

Self-Quiz (*Answers in Appendix III*)

1. _____ tissues have closely linked cells and one free surface.
   a. Epithelial   c. Nervous
   b. Connective   d. Muscle

2. In most _____ , cells secrete fibers of collagen and elastin.
   a. epithelial tissue   c. muscle tissue
   b. connective tissue   d. nervous tissue

3. _____ has a semifluid ground substance and occurs under most epithelia.
   a. Dense, irregular   c. Dense, regular
      connective tissue   connective tissue
   b. Loose connective tissue   d. Cartilage

4. _____ , a specialized connective tissue, is mostly plasma with cellular components and various dissolved substances.
   a. Irregular connective tissue   c. Cartilage
   b. Blood   d. Bone

5. After you eat too many carbohydrates and proteins, your body converts the excess to storage fats, which accumulate in _____ .
   a. connective tissue proper   c. adipose tissue
   b. dense connective tissue   d. both b and c

6. Components of _____ detect and coordinate information about changes and control responses to those changes.
   a. epithelial tissue   c. muscle tissue
   b. connective tissue   d. nervous tissue

7. In your own body, _____ can shorten (contract).
   a. epithelial tissue   c. muscle tissue
   b. connective tissue   d. nervous tissue

8. Cells of complex animals _____ .
   a. survive by their own metabolic activities
   b. contribute to the survival of the whole animal
   c. help maintain extracellular fluid
   d. all of the above

9. At _____ , physical and chemical conditions in the internal environment are within tolerable ranges.
   a. positive feedback   c. homeostasis
   b. negative feedback   d. metastasis

10. With negative feedback mechanisms, _____ .
   a. a stimulus brings about a response that tends to return internal operating conditions to the original state
   b. a stimulus suppresses internal operating conditions to levels below a set point for the body
   c. a stimulus raises internal operating conditions to levels above a set point for the body
   d. fewer solutes are fed back to the affected cells

11. Of the components that exert feedback control of organ activity, _____ detect specific changes in the environment, an _____ pulls together different bits of information and selects a suitable response, and _____ carry out the response.

12. Match the terms with the suitable description.
   ____ epithelium   a. strong, pliable; like rubber
   ____ cartilage   b. covers or lines body surfaces
   ____ homeostasis   c. stable internal environment
   ____ muscles and glands   d. integrating center
   ____ positive feedback   e. most common homeostatic
   ____ negative feedback   control mechanism
   ____ brain   f. effectors
         g. chain of events intensifies original condition in body

Critical Thinking

1. *Anhidrotic ectodermal dysplasia*, a genetic disorder described in Section 14.3, has been associated with a recessive allele on the mammalian X chromosome. As one symptom of this disorder, affected males and females have no sweat glands in the tissues where the recessive allele is expressed. What type of tissue are we talking about?

2. Adipose tissue and blood are often said to be "atypical" connective tissues. Compared with other connective tissues, which of their features are *not* typical?

3. *Porphyria* is a genetic disorder that shows up in about 1 in 25,000 individuals. Affected individuals lack certain enzymes that are part of a metabolic pathway leading to formation of heme, which is the iron-containing group of hemoglobin. An accumulation of porphyrins, which are intermediates of the pathway, causes awful symptoms, especially after exposure to sunlight. Lesions and scars form on the skin. Hair grows thickly on the face and hands. As gums retreat from teeth, the canines take on a fanglike appearance. Symptoms worsen upon exposure to a variety of substances, including garlic and alcohol. Affected individuals avoid sunlight and aggravating substances. They also may receive injections of heme from normal red blood cells.

If you are familiar with vampire stories, which date from the Middle Ages or earlier, speculate on how they may have evolved among superstitious folk who did not have medical knowledge of porphyria.

4. After graduating from high school, Jeff and Ryan set out on a trip through the desert roads of California and Arizona. One hot, dry morning in Joshua Tree National Monument, they saw an unusual rock formation that didn't appear to be too far from the road. They left the car and started hiking toward it. Their destination turned out to be farther away than they thought, and the sun's rays were more relentless than they had anticipated. They reached the shade of the rocks in the early afternoon. Their canteen was nearly empty, and the physiological meaning of "thirst" made itself known in a scary way. They knew they had to locate and drink water (or some other fluid), which is what the brain usually tells us to do when dehydration sets in.

From what you read in this chapter, would you suspect that Ryan and Jeff's *thirst behavior* is part of a positive or negative feedback control mechanism?

Selected Key Terms

adhering junction 28.1	gland cell 28.1
adipose tissue 28.2	homeostasis CI
blood 28.2	integrator 28.7
bone tissue 28.2	internal environment CI
cardiac muscle tissue 28.3	loose connective tissue 28.2
cartilage 28.2	mesoderm 28.6
dense, irregular	negative feedback mechanism 28.7
connective tissue 28.2	nervous tissue 28.4
dense, regular	neuroglia 28.4
connective tissue 28.2	neuron 28.4
division of labor CI	organ CI
ectoderm 28.6	organ system CI
effector 28.7	positive feedback mechanism 28.7
endocrine gland 28.1	sensory receptor 28.7
endoderm 28.6	skeletal muscle tissue 28.3
epithelium 28.1	smooth muscle tissue 28.3
exocrine gland 28.1	stimulus 28.7
extracellular fluid 28.7	tight junction 28.1
gap junction 28.1	tissue CI

Readings *See also www.infotrac-college.com*

Bloom, W., and D. W. Fawcett. 1995. *A Textbook of Histology*. Twelfth edition. Philadelphia: Saunders.

Leeson, C. R., T. Leeson, and A. Paparo. 1988. *Textbook of Histology*. Philadelphia: Saunders.

Telford, I., and C. Bridgman. 1995. *Introduction to Functional Histology*. Second edition. New York: HarperCollins.

WWW *http://www.brookscole.com/biology*

Practice quiz questions, hypercontents, BioUpdates, and critical thinking. The Brooks/Cole Biology Resource Center provides a wealth of information fully organized and integrated by chapter.

Why Crack the System?

Suppose your biology instructor asks you to volunteer for an experiment. You will get a microchip implanted in your brain. It will make you feel really good. But it may mess up your health, lop ten years off your life, and destroy a good part of your brain. Your behavior will change for the worse, so you might have trouble completing school, getting or keeping a job, or even having a normal family life.

The longer the chip is implanted, the less you will want to give it up. You won't get paid. You will pay the experimenter—first at bargain rates, then a little more each week. The chip is illegal. If you get caught using it, you and the experimenter will go to jail.

Sometimes Jim Kalat, a professor at North Carolina State University, proposes this experiment, which of course is hypothetical. Hardly any students volunteer. Then he substitutes *drug* for microchip and *dealer* for experimenter, and an amazing number of students come forward! Like 30 million other Americans, the "volunteers" seem ready to engage in self-destructive uses of drugs that alter emotional and behavioral states.

The destruction shows up in unexpected places. Each year, for instance, about 300,000 newborns are already addicted to crack, thanks to their addicted mothers. *Crack* is a cheap form of cocaine. It causes relentless stimulation of brain regions that govern the sense of pleasure. It dampens normal urges to eat and to sleep, and blood pressure rises. Elation and sexual desire intensify. In time, however, the brain cells that produce the stimulatory chemicals cannot keep up with the abnormal demand. The chemical vacuum makes crack users frantic and then profoundly depressed. Only crack makes them feel good again.

Addicted babies cannot know all of this. They can only quiver with "the shakes" and respond to the world with chronic irritation. And they are abnormally small. While they were developing inside their mother, their body tissues simply were not provided with enough oxygen and nutrients. As one of its side effects, crack causes blood vessels to constrict, and maternal blood vessels are the only supply lines that the developing individual has.

Figure 29.1
Owners of an evolutionary treasure—a complex brain that is the foundation for our memory and reasoning, and our future.

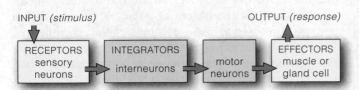

INPUT (stimulus)　　　　　　　　　　OUTPUT (response)

RECEPTORS
sensory
neurons

INTEGRATORS
interneurons

motor
neurons

EFFECTORS
muscle or
gland cell

Figure 29.2 Functional organization of nervous systems.

Paradoxically, crack babies are exceptionally fussy, yet they cannot respond to rocking and other forms of stimulation that normally have soothing effects. It may be a year or more before they recognize even their own mother. Without treatment, they are likely to grow up as emotionally unstable children, prone to aggressive outbursts and stony silences. Why? Their mother's drug habit crippled their nervous system.

Think about it. The **nervous system** evolved as a way to sense and respond, with exquisite precision, to changing conditions inside and outside the body. Awareness of sounds and sights, of odors, of hunger and passion, fear and rage—all such things begin with the flow of information along the communication lines of the nervous system. Do those lines remain silent until they receive outside signals, much as telephone lines wait to carry calls from all over the country? Absolutely not. Even before you were born, excitable cells called neurons became organized into extensive gridworks in your newly forming tissues and started chattering among themselves. All through your life, in moments of danger or reflection, supreme excitement or sleep, their chattering has never ceased and will not cease until the time you die.

Three classes of neurons interact in vertebrate nervous systems. **Sensory neurons** are adapted to respond to specific stimuli and relay information about them to the spinal cord and brain. A **stimulus** is a specific form of energy, such as light and pressure. In the spinal cord and brain you find **interneurons**. These receive sensory input, integrate it with other incoming information and with stored information, then influence the activity of other neurons. **Motor neurons** relay information from the brain and spinal cord to effectors—muscles or glands—which carry out the specified responses (Figure 29.2).

Neurons are not the only cells in vertebrate nervous systems. A variety of cells, collectively called neuroglia, metabolically assist, protect, and structurally support the neurons. In your own body, they make up more than one-half the volume of the nervous system.

KEY CONCEPTS

1. Nervous systems are composed of neurons and a variety of cells called neuroglia, which structurally and functionally support the neurons. Collectively, neurons detect and integrate information about external and internal conditions, and command both muscles and glands to carry out suitable responses.

2. Neurons are a type of excitable cell, which means a suitable stimulus can disturb the distribution of electric charge across their plasma membrane. Such disturbances are the basis of messages, called action potentials, that travel from the input zone of each neuron to its output zone near a neighboring cell.

3. Information flow through nervous systems depends on moment-by-moment integration of excitatory and inhibitory signals that act upon the neurons of given pathways.

4. The simplest nervous systems are the nerve nets of radial animals, such as hydras and sea anemones. The nervous systems of most animals show pronounced cephalization and bilateral symmetry.

5. Vertebrate nervous systems are functionally divided into central and peripheral regions. The brain and spinal cord make up the central nervous system. Paired nerves that thread through the rest of the body are the key components of the peripheral nervous system.

6. The vertebrate brain has three functional divisions, called the hindbrain, midbrain, and forebrain. Its most ancient parts deal with reflex control of breathing, blood circulation, and other basic functions that are essential for staying alive. Later, complex centers for receiving, integrating, storing, and retrieving information evolved in the forebrain, especially among birds and mammals.

The preceding section gave an overview of a neuron's structure and function. Let's now take a closer look at how signals arise and are propagated at its surface.

Approaching Threshold

Weakly stimulate a neuron at its input zone and you disturb the ion balance across the membrane, but not much. Suppose your toes gently tap a cat snoozing at your feet and put a bit of pressure on its skin. Tissues beneath the skin surface have receptor endings—input zones of sensory neurons. Patches of plasma membrane at the endings deform under the pressure. Some ions now flow across, so the voltage difference across the membrane changes slightly. The pressure has produced a graded, local signal.

STIMULUS

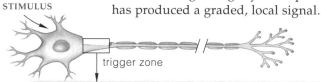

trigger zone

29.6). With the influx, the cytoplasmic side of the membrane becomes less negative. This causes more gates to open and more sodium to enter. The ever increasing inward flow of sodium is a case of **positive feedback**; the event intensifies as a result of its own occurrence:

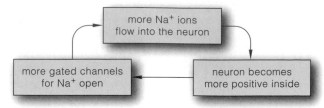

At threshold, the opening of more sodium gates no longer depends on stimulus strength. Now the positive-feedback cycle is under way, so the inward-rushing sodium itself is enough to open more sodium gates.

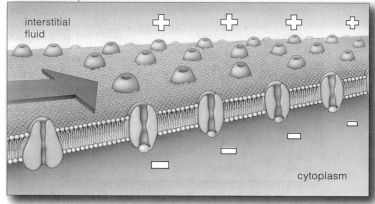

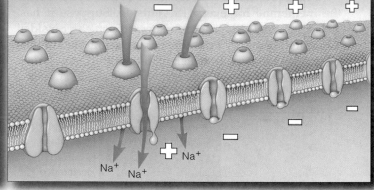

a Membrane at rest (inside negative with respect to the outside). An electrical disturbance (*red* arrow) spreads from an input zone to an adjacent trigger region of the membrane, which has a great number of gated sodium channels.

b A strong disturbance initiates an action potential. Sodium gates open. The sodium inflow decreases the negativity inside the neuron. The change causes more gates to open, and so on, until threshold is reached and the voltage difference across the membrane reverses.

Figure 29.6 Propagation of an action potential along the axon of a motor neuron.

Graded means signals arising at an input zone can vary in magnitude. Such signals may be large or small, depending on how intense the stimulus is. *Local* means the signals don't spread far from the site of stimulation. It takes specialized types of ion channels to propagate a signal, and input zones simply don't have them.

When a stimulus is intense or long lasting, graded signals spread from the input zone into an adjacent trigger zone. There, a certain amount of change in the voltage difference across the plasma membrane triggers an action potential. That amount is the threshold level. *Threshold can be reached at any membrane patch that has voltage-sensitive, gated channels for sodium ions.*

The stimulus causes positively charged sodium ions to flow across the membrane, into the neuron (Figure

An All-or-Nothing Spike

Figure 29.7 shows a recording of the voltage difference across the plasma membrane before, during, and after an action potential. Notice how the membrane potential spikes once threshold is reached. Every single action potential in the neuron spikes to the same level above threshold as an *all-or-nothing* event. That is, once the positive-feedback cycle starts, nothing will stop the full spiking. If threshold is not reached, the disturbance to the plasma membrane will subside just as soon as the stimulus is removed.

Each spike lasts only for a millisecond or so. Why? At the membrane site of the charge reversal, the gated sodium channels closed and shut off the sodium inflow.

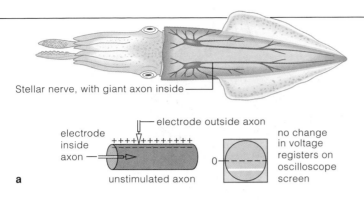

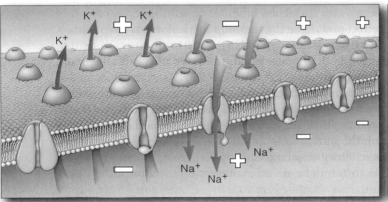

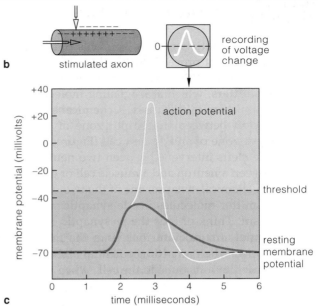

Figure 29.7 Action potentials. (**a**) When early researchers studied neural function, the squid *Loligo* provided them with evidence of action potential spiking. The squid's "giant" axons were large enough to slip electrodes inside. (**b**) When such an axon was stimulated, electrodes positioned inside and outside detected voltage changes, which showed up as deflections in a beam of light across the screen of an oscilloscope. (**c**) This is a typical waveform (*yellow* line) for an action potential on an oscilloscope screen. The *red* line represents a recording of a local signal that did not reach the threshold of an action potential; spiking did not occur.

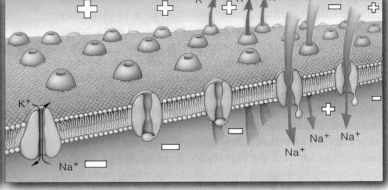

c With the reversal, sodium gates shut and potassium gates open (*pink* arrows). Potassium follows its gradient out of the neuron. Voltage is restored. The disturbance triggers an action potential at the adjacent site, and so on, away from the point of stimulation.

d Following each action potential, the inside of the plasma membrane becomes negative once again. However, the sodium and potassium concentration gradients are not yet fully restored. Active transport at sodium-potassium pumps restores them.

Also, about halfway through the reversal, potassium channels opened, so many more potassium ions flowed out and restored the original voltage difference across the membrane. And sodium-potassium pumps restored the ion gradients. Later on, after the resting membrane potential has been restored, most potassium gates close and sodium gates are in their initial state, ready to be opened with the arrival of a suitable disturbance.

Propagation of Action Potentials

The membrane disturbances leading up to an action potential are self-propagating, and they do not diminish in magnitude. As they spread to an adjacent membrane patch, an equivalent number of gated channels open. With this new disturbance, gated channels open in the *next* adjacent patch, then the next, and so on. For a brief period after each membrane patch has been excited, it is insensitive to stimulation. Sodium gates in the patch are inactivated and ions cannot move through them. This is the reason why action potentials do not spread back to the trigger zone (the site where they were initiated) but rather are self-propagating away from it.

The cytoplasm next to the plasma membrane of a neuron at rest is more negative than the interstitial fluid just outside the membrane.

During an action potential, the inside of a disturbed patch of membrane becomes more positive than the outside.

After an action potential, resting conditions are restored at the membrane patch.

PATHS OF INFORMATION FLOW

Blocks and Cables of Neurons

Through synaptic integration, messages arriving at any neuron in the body might be reinforced and sent on to neighboring neurons. What determines the direction in which a given message will travel? That depends on the organization of neurons in different body regions.

For example, your brain deals with its staggering numbers of neurons in a manner analogous to block parties. Regional blocks of hundreds or thousands of neurons receive excitatory and inhibitory signals. They integrate signals entering the block, then send out new ones in response. Some regions have neurons organized as *divergent* circuits, with processes fanning out from one block to form connections with others. Some have neurons arranged in *convergent* circuits, with signals from many sent on to just a few. In still other regions, neurons synapse back on themselves and repeat signals like gossip that just won't go away. Such *reverberating* circuits include the ones that make your eye muscles rhythmically twitch while you sleep.

In the cablelike **nerves**, long axons of many sensory neurons, motor neurons, or both permit long-distance communication between the brain or spinal cord and the rest of the body. Connective tissue bundles most of the axons in parallel array (Figure 29.11). Each axon has a **myelin sheath** that enhances the rate of action potential propagation. The sheath is a series of Schwann cells, a

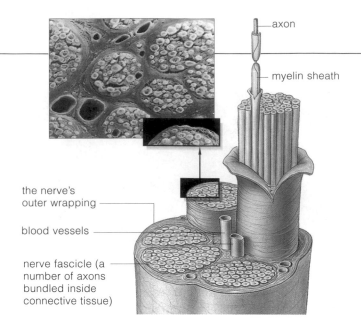

Figure 29.11 Structure of a nerve. Axons in the nerve are bundled together inside wrappings of connective tissue.

type of neuroglial cell, wrapped like jellyrolls around the long axon. An exposed node, or gap, separates each cell from adjacent ones. There, voltage-sensitive, gated sodium channels pepper the plasma membrane (Figure 29.12). The sheathed regions in between nodes hamper ion movements across the membrane. Ion disturbances tend to flow along the membrane until the next node in line. At each node, ion flow can produce a new action potential. In large sheathed axons, action potentials are propagated at a remarkable 120 meters per second.

In *multiple sclerosis*, myelin sheaths around axons in the spinal cord degenerate slowly. A mutant gene may predispose a person to the disease, but viral infection may trigger it. Symptoms include serious progressive weakening of muscles, fatigue, and numbness.

Reflex Arcs

Figure 29.13 is a specific example of the direction of information flow through nervous systems. It shows how sensory and motor neurons of certain nerves take part in a path known as the stretch reflex. **Reflexes** are simple movements made in response to stimuli. In the simplest reflex arcs, sensory neurons synapse directly on motor neurons.

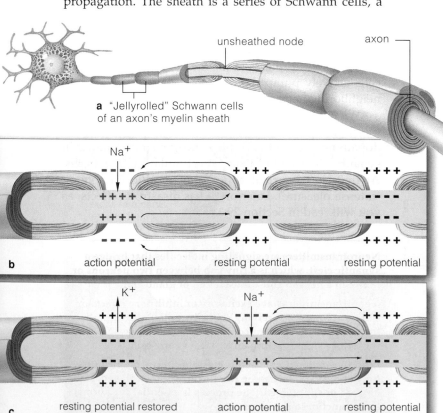

a "Jellyrolled" Schwann cells of an axon's myelin sheath

Figure 29.12 Action potential propagation along sheathed neurons. (**a**) A myelin sheath is a series of Schwann cells, each wrapped like a jellyroll around an axon. It blocks ion movements across the membrane, but ions cross at unsheathed nodes in between the jellyrolls. The nodes have dense arrays of gated sodium channels. (**b**) Disturbance caused by an action potential spreads along the axon. When it reaches a node, sodium gates open, sodium ions rush inward, and another action potential results. (**c**) The new disturbance spreads swiftly to the next node and triggers a new action potential, and so on down the line.

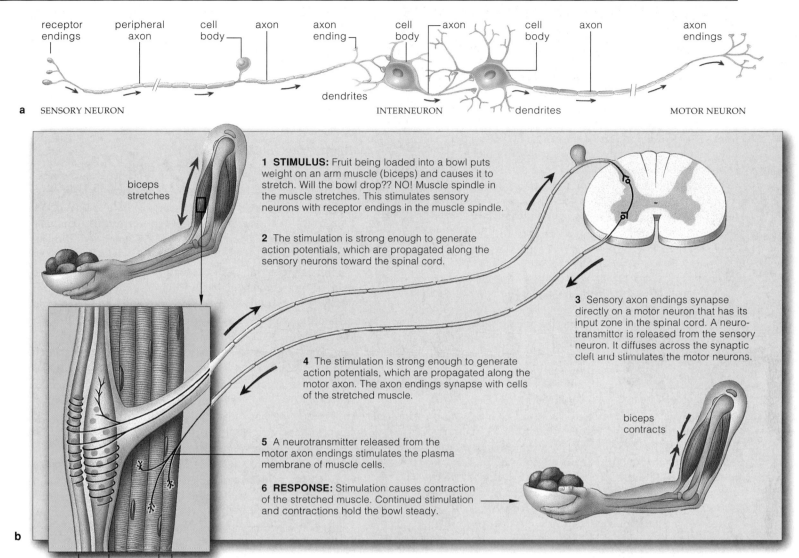

a SENSORY NEURON

receptor endings · peripheral axon · cell body · axon · axon ending · dendrites · INTERNEURON · cell body · axon · cell body · axon · dendrites · MOTOR NEURON · axon endings

b

1 **STIMULUS:** Fruit being loaded into a bowl puts weight on an arm muscle (biceps) and causes it to stretch. Will the bowl drop?? NO! Muscle spindle in the muscle stretches. This stimulates sensory neurons with receptor endings in the muscle spindle.

biceps stretches

2 The stimulation is strong enough to generate action potentials, which are propagated along the sensory neurons toward the spinal cord.

3 Sensory axon endings synapse directly on a motor neuron that has its input zone in the spinal cord. A neurotransmitter is released from the sensory neuron. It diffuses across the synaptic cleft and stimulates the motor neurons.

4 The stimulation is strong enough to generate action potentials, which are propagated along the motor axon. The axon endings synapse with cells of the stretched muscle.

biceps contracts

5 A neurotransmitter released from the motor axon endings stimulates the plasma membrane of muscle cells.

6 **RESPONSE:** Stimulation causes contraction of the stretched muscle. Continued stimulation and contractions hold the bowl steady.

muscle spindle · muscle cell

Figure 29.13 (**a**) General direction of information flow in nervous systems. Sensory neurons relay information *into* the spinal cord and brain, where they synapse with interneurons. Interneurons *within* the spinal cord and brain integrate signals. Many synapse with motor neurons, which carry signals *away* from the spinal cord and brain. (**b**) Organization of nerves in a reflex arc that deals with muscle stretching. In a skeletal muscle, stretch-sensitive receptors of a sensory neuron are located in muscle spindles. The stretching generates action potentials, which reach axon endings in the spinal cord. These synapse with a motor neuron that carries signals to contract, from the spinal cord back to the stretched muscle.

The stretch reflex works to contract a muscle after gravity or some other load has caused the muscle to stretch. Suppose you hold out a large bowl and keep it stationary as someone puts several peaches into it. The peaches add weight to the bowl, and when your hand starts to drop, a muscle in your arm (the biceps) is stretched.

In the muscle, stretching activates receptor endings that are a part of muscle spindles—sensory organs in which specialized cells are enclosed in a sheath that runs parallel with the muscle. The receptor endings are the input zones of sensory neurons, the axons of which synapse with motor neurons within the spinal cord (Figure 29.13). Axons of the motor neurons lead back to the stretched muscle. Action potentials that reach the axon endings trigger the release of ACh, which initiates contraction. As long as receptor activity continues, the motor neurons are excited even more, and this allows them to maintain your hand's position.

In the vast majority of reflex pathways, the sensory neurons also interact with a number of interneurons, which then activate or suppress all the motor neurons necessary for a coordinated response.

Vertebrates have interneurons organized in information-processing blocks. Cablelike nerves that have long axons of sensory neurons, motor neurons, or both connect the brain and spinal cord with the rest of the body.

Reflex arcs, in which sensory neurons synapse directly on motor neurons, are the simplest paths of information flow.

WWW

What happens if something disrupts information flow in a nervous system? Ask *Clostridium botulinum*, an anaerobic, endospore-forming bacterium that normally lives in soil. It can cause the disease *botulism*. Improperly preserved, canned, or stored food may contain its endospores. These can germinate in someone who ingests the food. One metabolic product of this bacterium is toxic. It enters neurons that synapse with muscle cells and blocks the release of acetylcholine (ACh). Without ACh, muscles cannot contract. They become more and more flaccid and paralyzed. Unless antitoxins are administered within ten days, breathing becomes impossible and the heart stops beating.

A related bacterium, *C. tetani*, lives in the gut of horses, cattle, and other grazing animals, even many people. Its endospores survive in soil, especially manure-rich ones. They persist for years if sunlight and oxygen do not reach them. They resist disinfectants, heat, and boiling water. When they enter the body through a deep puncture or a deep cut, they germinate in anaerobic, dead tissues. These bacteria do not spread from the dead tissue. They produce a toxin that the blood or nerves deliver to the spinal cord and brain. In the spinal cord, the toxin affects interneurons that help control motor neurons. It blocks the release of inhibitory neurotransmitters (GABA and glycine) and frees the motor neurons from normal inhibitory control. Symptoms of the disease *tetanus* are about to begin.

After four to ten days, the overstimulated muscles stiffen and go into spasm. They cannot be released from contraction, and prolonged, spastic paralysis follows. Fists and jaws may remain clenched (the disease is also called lockjaw). The back may arch severely and permanently. Once respiratory and cardiac muscles become paralyzed, death nearly always follows.

Vaccines were not available for soldiers of early wars, when dead cavalry horses and manure littered battlefields (Figure 29.14). Today, vaccines have all but eradicated tetanus in the United States.

Figure 29.14 Painting of a young victim of a contaminated battle wound, as he lay dying in a military hospital.

So far, our focus has been on the nature of the messages that travel through any nervous system. Let's turn now to the organization of the systems themselves. They differ in major ways among animals. As you will see, the differences correspond to differences in life-styles.

Regarding the Nerve Net

Animals first evolved in the seas, and it is in the seas that we still find the animals with the simplest nervous systems. They are sea anemones, jellyfishes, and other cnidarians (Section 23.4). These invertebrates display *radial* symmetry. Their body parts are arranged about a central axis, much like spokes of a bike wheel.

A cnidarian has a **nerve net**, a loose mesh of nerve cells intimately associated with epithelial tissue (Figure 29.15). These interact with sensory cells and contractile cells along reflex pathways in the same epithelium. In reflex pathways, remember, sensory stimulation directly triggers simple, stereotyped movements. In cnidarians, one pathway concerned with feeding behavior extends from sensory receptors in tentacles, along nerve cells, to contractile cells around the mouth. In jellyfishes, other reflexes permit slow swimming movements and keep the body right-side up.

A nerve net extends throughout the body, The flow of information through it is not highly focused. Signals only travel diffusely, in all directions, from a point of stimulation. You might not find this type of nervous system impressive. However, if you did little more than slowly contract and expand a body wall or move tentacles through water, it would be quite sufficient. Besides this, a nerve net is equally responsive to food or potential danger coming from any direction.

On the Importance of Having a Head

Flatworms are the simplest animals having a bilateral nervous system (Figure 29.16). *Bilateral* symmetry means having equivalent body parts on the left and right sides of the body's midsagittal plane. (Imagine sliding down a staircase banister that turns into a razor and you may never forget the location of the midsagittal plane.) Both sides have the same array of muscles that function in moving the body forward. Both have the same array of nerves to control the muscles, and so on.

A flatworm has a ladderlike nervous system. Its two cordlike nerves run longitudinally through the body and have many side branches. (A nerve, recall, is like a cable in which sensory axons, motor axons, or both are bundled in a sheath of connective tissue.) Also, in the head end of some flatworms are two ganglia (singular, ganglion), local clusters of nerve cell bodies that act as local integrating centers. Flatworm ganglia coordinate

Figure 29.15 Nerve net of a sea anemone, one of the cnidarians. Its nerve cells interact with sensory and contractile cells. The cells are all embedded in an epithelial tissue, between the outer epidermis and a jellylike middle layer (mesoglea) of the body wall. ✦

nerve cells of nerve net

epithelial cell

sensory cell

part of sheet of contractile extensions of epithelial cells

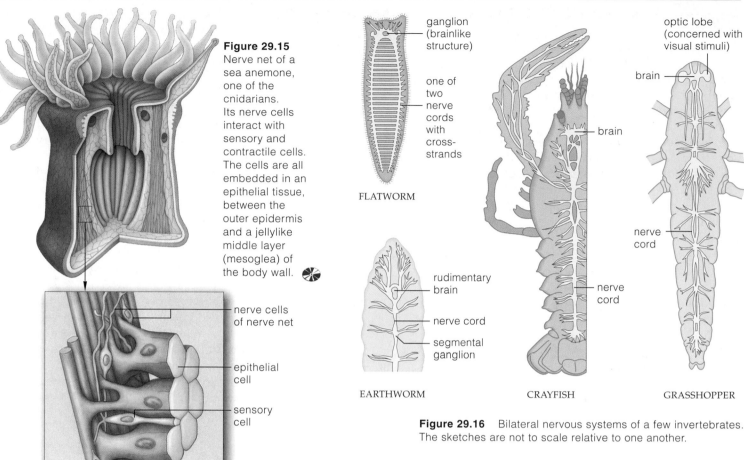

FLATWORM

ganglion (brainlike structure)

one of two nerve cords with cross-strands

EARTHWORM

rudimentary brain

nerve cord

segmental ganglion

CRAYFISH

brain

nerve cord

GRASSHOPPER

optic lobe (concerned with visual stimuli)

brain

nerve cord

Figure 29.16 Bilateral nervous systems of a few invertebrates. The sketches are not to scale relative to one another.

signals from paired sensory organs (two eyespots, for example), and provide a bit of control over the nerves.

Did bilateral nervous systems evolve from nerve nets? Maybe. In nearly all animals more complex than flatworms, we find local nerve nets (plexuses), such as the one in your intestinal wall. More intriguing, a self-feeding larval stage called a planula develops in some cnidarian life cycles. Like flatworms, a planula has a flattened body and uses cilia to swim or crawl about:

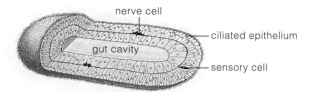

nerve cell

ciliated epithelium

gut cavity

sensory cell

Imagine a few ancient planulas crawling on the seafloor in Cambrian times. By chance, mutations of regulatory genes caused the scheduled transformation into adults to be blocked but had no effect on the maturing of the reproductive organs. (This happens to larvae of some salamanders and other animals, as mentioned in Section

24.6.) Such planulas kept on crawling. If their offspring inherited the mutant genes that gave the capacity for forward mobility, they could explore new places that were food-rich, food-poor, or potentially dangerous. In such environments, the mutation had survival value. Thus a concentration of sensory cells in the leading end, not the trailing end, was favored, because it allowed for more rapid, effective responses to varied stimuli.

Natural selection must have favored a concentration of sensory cells at the body's leading end. Cephalization (formation of a head) and bilateral symmetry may have started this way. As Chapter 23 describes, both features occur in most lineages of invertebrates. As you will see next, patterns of bilateral symmetry and cephalization also are evident in the paired nerves, paired muscles, paired sensory structures, and paired brain centers of yourself and all other vertebrates.

Radial animals have a nerve net, a diffuse mesh of nerve cells that take part in simple reflex pathways involving sensory cells and contractile cells of an epithelial tissue.

Nervous systems of cephalized, bilateral animals include a brain (or ganglia at the head end) as well as paired nerves and paired sensory structures.

Nerves are cablelike communication lines of sensory axons, motor axons, or both bundled in a connective tissue sheath.

Evolutionary High Points

Hundreds of millions of years ago, the earliest fishlike vertebrates were evolving (Section 24.3). A column of bony segments was taking over the functions of the notochord, a long rod of stiffened tissue that worked with muscles to bring about movements. Above the notochord, a hollow, tubular nerve cord was evolving. It was the forerunner of a spinal cord and brain. These developments were the foundation for new life-styles. Not long afterward, genetic divergences from lineages of filter-feeding, scavenging fishes gave rise to swift, jawed predators of the seas.

At first, simple reflex pathways prevailed. Sensory neurons synapsed directly with motor neurons, which directly signaled muscles to contract. No other neurons altered the flow of signals from reception of a stimulus to the response. In the changing world of fast-moving vertebrates, however, the predators and prey that were better equipped to sense the presence of food or danger had the competitive edge.

The senses of smell, hearing, and balance became keener among the vertebrates that invaded land. Bones and muscles evolved in ways that allowed specialized motor activities. The brain became variably thickened with nervous tissue that could integrate rich sensory information and issue orders for complex responses.

The oldest parts of the vertebrate brain still deal with reflex coordination of breathing and other vital functions. But many interneurons now synapse on the sensory and motor neurons of ancient pathways, and with one another in the newer brain regions. In the most complex vertebrates, interneurons receive, store, retrieve, and compare information about experiences. They weigh possible responses. And they give our own species the capacity to reason, remember, and learn.

The nerve cord persists in all vertebrate embryos. We call it the **neural tube**. While the embryo grows and develops, the tube expands and becomes regionally modified into the brain and spinal cord (Figure 29.17a). A vertebrate column encloses the spinal cord. Adjacent tissues give rise to nerves that thread through all body regions and connect with the spinal cord and brain.

Functional Divisions of the Vertebrate Nervous System

Figure 29.18 gives you a general sense of the expansions of nervous tissue in the human nervous system. This diagram shows the major paired nerves of this bilateral system. What it cannot possibly show are the 100 billion interneurons in the brain alone. To be sure, humans do have the most intricately wired nervous system in the animal world. Even so, you find similar patterns among other vertebrates.

FOREBRAIN. Receives, integrates sensory information from nose, eyes, and ears; in land-dwelling vertebrates, contains the highest integrating centers

MIDBRAIN. Coordinates reflex responses to sight, sounds

HINDBRAIN. Reflex control of respiration, blood circulation, other basic tasks; in complex vertebrates, coordination of sensory input, motor dexterity, and possibly mental dexterity

(start of spinal cord)

a Diagram of how the anterior end of the dorsal, hollow nerve cord of vertebrates expanded into functionally distinct regions, which also increased in complexity in certain lineages

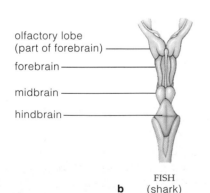
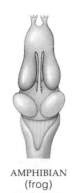
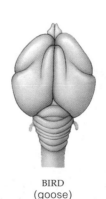

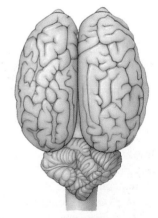

olfactory lobe (part of forebrain)

forebrain

midbrain

hindbrain

b

| FISH (shark) | AMPHIBIAN (frog) | REPTILE (alligator) | BIRD (goose) | MAMMAL (horse) |

Figure 29.17 Evolutionary trend toward an expanded, more complex brain, as suggested by comparing the brains of some existing vertebrates. These dorsal views are not to the same scale.

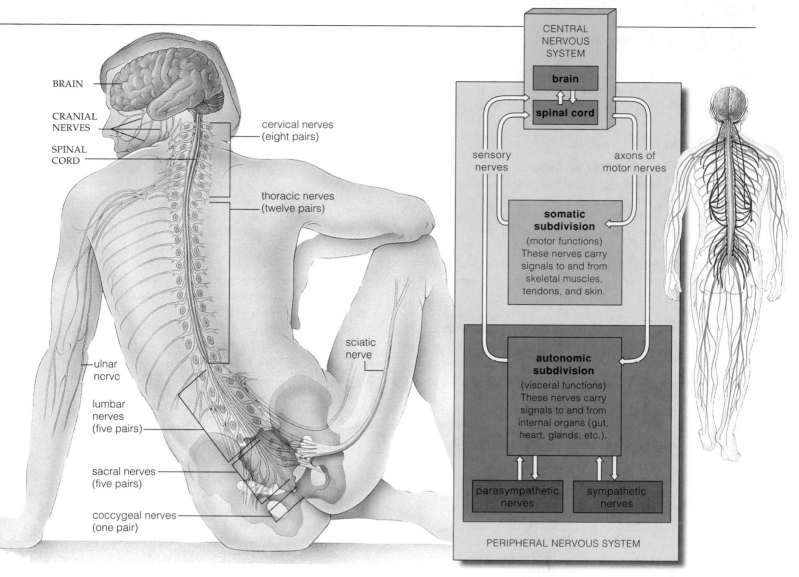

BRAIN

CRANIAL NERVES

SPINAL CORD

cervical nerves (eight pairs)

thoracic nerves (twelve pairs)

sciatic nerve

ulnar nerve

lumbar nerves (five pairs)

sacral nerves (five pairs)

coccygeal nerves (one pair)

CENTRAL NERVOUS SYSTEM

brain

spinal cord

sensory nerves

axons of motor nerves

somatic subdivision

(motor functions) These nerves carry signals to and from skeletal muscles, tendons, and skin.

autonomic subdivision

(visceral functions) These nerves carry signals to and from internal organs (gut, heart, glands, etc.).

parasympathetic nerves

sympathetic nerves

PERIPHERAL NERVOUS SYSTEM

Figure 29.18 Sketch of the human nervous system, showing the brain, spinal cord, and some of the major peripheral nerves. This nervous system also includes twelve pairs of cranial nerves that originate from the brain. Other vertebrates have a similar system.

Figure 29.19 Functional divisions of the nervous system. The central nervous system is color-coded *blue*, the somatic nerves *green*, and the autonomic nerves *red*. Sometimes the nerves carrying sensory input to the central nervous system are said to be *afferent* (a word meaning "to bring to"). The ones carrying motor output away from the central nervous system to muscles and glands are *efferent* ("to carry outward").

Investigators typically approach the complexity of the vertebrate nervous system by functionally dividing it into central and peripheral regions (Figure 29.19). All of its interneurons are confined to the **central nervous system**, or the spinal cord and brain. The **peripheral nervous system** consists mainly of nerves that thread through the rest of the body and carry signals into and out of the central nervous system.

Inside the brain and spinal cord, communication lines are called tracts, not nerves. The tracts making up white matter have axons with glistening white myelin sheaths and specialize in rapid signal transmission. By contrast, gray matter consists of unmyelinated axons, dendrites, nerve cell bodies, and neuroglial cells. The neuroglial cells, remember, protect or structurally and functionally support neurons. They make up more than half the volume of vertebrate nervous systems.

The coevolution of nervous, sensory, and motor systems made possible more complex life-styles among vertebrates.

The vertebrate nervous system has become so intricately wired that the structure and functions of its central and peripheral regions are described separately.

The central nervous system consists of the brain and spinal cord. The peripheral nervous system consists of nerves that thread through the rest of the body and carry signals into and out of the central region.

Myelinated axons make up the white matter of tracts inside the spinal cord and the brain. Neuroglia, unmyelinated axons, dendrites, and cell bodies of neurons make up the gray matter of those tracts.

WWW

THE MAJOR EXPRESSWAYS

Let's now take a look at the peripheral nervous system and the spinal cord. The two interconnect as the major expressways for information flow through the body.

Peripheral Nervous System

SOMATIC AND AUTONOMIC SUBDIVISIONS In humans, the peripheral nervous system includes thirty-one pairs of *spinal* nerves, which connect with the spinal cord. The system also includes twelve pairs of *cranial* nerves, which connect directly with the brain.

Cranial and spinal nerves can be further classified according to function. The ones that carry signals about moving your head, trunk, and limbs are the **somatic nerves**. The sensory axons inside these nerves deliver information from receptors in skin, skeletal muscles, and tendons to the central nervous system. Their motor axons deliver commands from the brain and spinal cord to the body's skeletal muscles.

By contrast, spinal and cranial nerves dealing with smooth muscle, cardiac (heart) muscle, and glands are the **autonomic nerves**. They deal with visceral parts of the body—in other words, with its internal organs and structures. Autonomic nerves carry signals to and from these organs and structures.

SYMPATHETIC AND PARASYMPATHETIC NERVES Figure 29.20 shows the two categories of autonomic nerves. We call them parasympathetic and sympathetic. Normally they work antagonistically, with the signals from one opposing signals from the other. However, both carry excitatory and inhibitory signals to internal organs. Often their signals arrive at the same time at muscle or gland cells and compete for control over them. In such instances, synaptic integration at the cellular level leads to minor adjustments in the organ's level of activity.

Parasympathetic nerves dominate when the body is not receiving much outside stimulation. They tend to

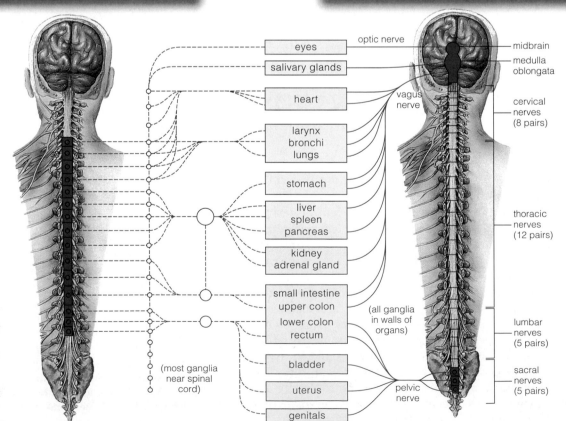

SYMPATHETIC OUTFLOW FROM THE SPINAL CORD

Examples of Responses
Heart rate increases
Pupil of eyes dilate (widen, let in more light)
Glandular secretions in airways to lungs decrease
Salivary gland secretions thicken
Stomach and intestinal movements slow down
Sphincters (rings of muscle) contract

PARASYMPATHETIC OUTFLOW FROM THE SPINAL CORD AND BRAIN

Examples of Responses
Heart rate decreases
Pupil of eyes constrict (keep more light out)
Glandular secretions in airways to lungs increase
Salivary gland secretions become dilute
Stomach and intestinal movements increase
Sphincters (rings of muscle) relax

Figure 29.20 Autonomic nervous system. This diagram shows the major sympathetic nerves and parasympathetic nerves leading out from the central nervous system to some major organs. Remember, there are *pairs* of both kinds of nerves, servicing the right and left halves of the body. The ganglia are clusters of the cell bodies of neurons that have their axons bundled together in nerves.

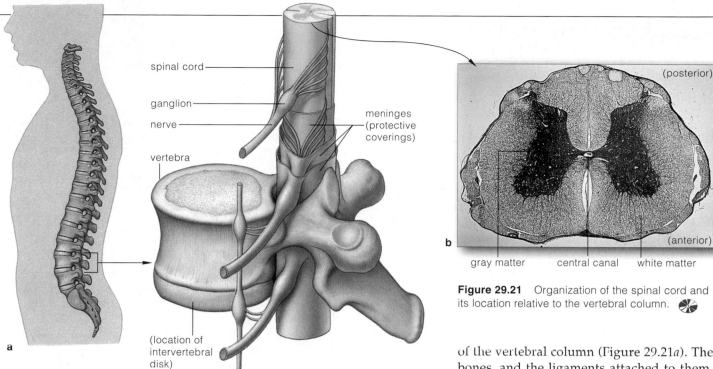

spinal cord

ganglion

nerve

vertebra

meninges
(protective
coverings)

(location of
intervertebral
disk)

a

b

gray matter   central canal   white matter

(posterior)

(anterior)

Figure 29.21 Organization of the spinal cord and its location relative to the vertebral column.

slow down the body overall and divert energy to basic "housekeeping" tasks, such as digestion.

Sympathetic nerves dominate in times of sharpened awareness, excitement, or danger. They tend to shelve housekeeping tasks. At the same time, they prepare the animal to fight or escape (when threatened) or to frolic (as in play or sexual behavior). Right now, sympathetic nerves are commanding your heart to beat a bit faster, and parasympathetic nerves are commanding it to beat a bit slower. Integration of the opposing signals adjusts the heart rate. If something scares or excites you, the parasympathetic input to the heart drops. Sympathetic signals cause the release of epinephrine, which makes your heart beat faster. It also makes you breathe faster and start to sweat. In this state of intense arousal, you are primed to fight (or play) hard or to get away fast. Hence the name *fight-flight response.*

Suppose the stimulus for the fight-flight response ends. Sympathetic activity may decrease abruptly, and parasympathetic activity may rise suddenly. You might observe this "rebound effect" after someone has been instantly mobilized to rush onto a highway to save a child from an oncoming car. The person may well faint as soon as the child has been swept out of danger.

The Spinal Cord

By definition, a **spinal cord** is a vital expressway for signals between the peripheral nervous system and the brain. Here also, sensory and motor neurons make direct reflex connections. For instance, the stretch reflex described earlier arcs through the spinal cord this way. The spinal cord threads through a canal made of bones

of the vertebral column (Figure 29.21*a*). The bones, and the ligaments attached to them, protect the cord. So do the **meninges**, three tough, tubelike coverings around the spinal cord and brain. The coverings are not impervious to all attacks. *Meningitis*, an often-fatal disease, results from certain viral or bacterial infections. Disease symptoms include severe headaches, fever, a stiff neck, and nausea.

Signals swiftly travel up and down the spinal cord in bundles of axons having glistening myelin sheaths. The sheaths distinguish the cord's white matter from its gray matter (Figure 29.21*b*). Gray matter consists of dendrites and cell bodies of neurons, and neuroglial cells. It plays an important role in controlling reflexes for limb movements (as when you walk or wave your arms) and organ activity (such as bladder emptying).

Experiments with frogs provide evidence for these pathways. Between the frog's spinal cord and brain are neural circuits that deal with straightening legs after they have been bent. If these circuits are severed at the base of the brain, the legs become paralyzed—but only for about a minute. So-called extensor reflex pathways in the spinal cord recover quickly and have the frog hopping about in no time. Recovery is minimal after similar damage in humans and other primates—the vertebrates with the greatest cephalization.

The nerves of the peripheral nervous system connect the brain and spinal cord with the rest of the body.

Its somatic nerves deal with skeletal muscle movements. Its autonomic nerves deal with the functions of internal organs, such as the heart and glands.

The spinal cord is a vital expressway for signals between the brain and peripheral nerves. Some of its interneurons also exert direct control over certain reflex pathways.

WWW

THE VERTEBRATE BRAIN

Functional Divisions

The spinal cord merges with the **brain**, a master control center that receives, integrates, stores, and retrieves sensory information. The brain coordinates responses to information by adjusting activities throughout the body. Bones, fluid, and membranes protect it.

The vertebrate forebrain, midbrain, and hindbrain, recall, develop from a neural tube. The most ancient nervous tissue persists in all three regions and still has basic reflex centers. We call it the **brain stem**. Expanded layers of gray matter developed from the brain stem when the senses of smell, hearing, and vision started to become increasingly important in early vertebrates.

HINDBRAIN The medulla oblongata, cerebellum, and pons are hindbrain components. The first has reflex centers for respiration, circulation, and other basic tasks. It coordinates motor responses with complex reflexes, such as coughing, and influences brain centers for sleep and arousal. The cerebellum integrates sensory input from eyes, ears, and muscle spindles with motor signals from the forebrain. It helps control motor dexterity and motor memories, such as dance routines. Many bands of axons extend from both sides of the cerebellum to the pons (meaning bridge). The pons is a major traffic center for information passing between the cerebellum and higher integrating centers of the forebrain.

MIDBRAIN The midbrain coordinates reflex responses to sights and sounds. In fishes and amphibians, its roof of gray matter—the tectum—is a center for coordinating nearly all sensory input and initiating motor responses. (Surgically remove a frog's highest brain center and its activities will not change much if its tectum is intact.) In most vertebrates (not mammals), the midbrain has a pair of optic lobes—brain centers for input from eyes. In mammals, the eyes formed functional connections with the forebrain and the tectum became a reflex center that only relays sensory signals to the forebrain.

FOREBRAIN For much of vertebrate history, chemical odors in aquatic environments were the major clues to survival. An olfactory lobe dealing mainly with odors from predators, prey, and mates was a key forebrain structure (Figure 29.17). So were paired outgrowths of the brain stem where olfactory input and responses to it became integrated. Especially after the invasion of land, the outgrowths expanded greatly to become the two hemispheres of the cerebrum. A thin layer of gray matter above the brain's axons became the **cerebral cortex**.

Another forebrain region, the thalamus, evolved as a coordinating center for sensory input and as a relay station to the cerebrum. Below this, the hypothalamus

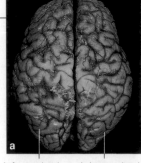

Figure 29.22 (**a**) Human brain. This dorsal view shows how a longitudinal fissure separates the two cerebral hemispheres. (**b**) Sagittal view of the right cerebral hemisphere. Not visible is the reticular formation, which extends between the upper spinal cord and the cerebrum.

left cerebral hemisphere   right cerebral hemisphere

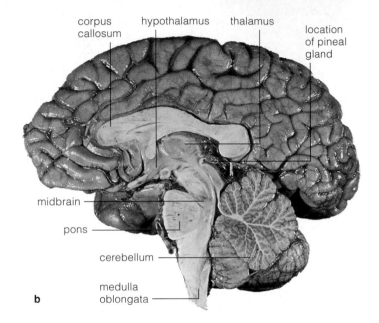

corpus callosum   hypothalamus   thalamus   location of pineal gland

midbrain

pons

cerebellum

medulla oblongata

b

evolved as the main center for homeostatic control over the internal environment. The hypothalamus is central to control of behaviors related to internal organs, such as thirst responses, hunger, and sex; and to the physical expression of emotions, such as sweating with fear.

You may be thinking the brain is tidily subdivided into three main regions. However, an ancient mesh of interneurons extends from the upper spinal cord and on through the brain stem, to higher integrative centers of the cerebral cortex. This network of interneurons, the **reticular formation**, persists as a low-level pathway to motor centers of the medulla oblongata and spinal cord. It also activates integrating centers in the cerebrum and thus influences activities of the whole nervous system.

THE HUMAN BRAIN The human brain weighs 1,300 grams (3 pounds) in a person of average size, and more than half of it consists of neuroglial cells. But it also has at least 100 billion interneurons that contribute most to your humanness. A fissure divides the cerebrum into left and right cerebral hemispheres (Figure 29.22). Most centers for analytical skills, mathematics, and speech reside in the left hemisphere. Most centers for spatial relations, music, and other nonverbal (abstract) skills are in the right hemisphere. Each hemisphere receives,

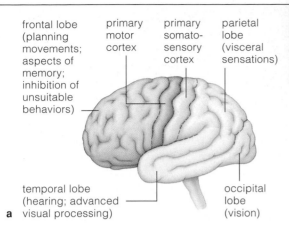

frontal lobe
(planning
movements;
aspects of
memory;
inhibition of
unsuitable
behaviors)

primary
motor
cortex

primary
somato-
sensory
cortex

parietal
lobe
(visceral
sensations)

temporal lobe
(hearing; advanced
visual processing)

occipital
lobe
(vision)

a

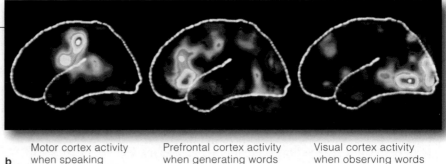

b

Motor cortex activity
when speaking

Prefrontal cortex activity
when generating words

Visual cortex activity
when observing words

Figure 29.23 (**a**) Primary receiving and integrating centers for the human cerebral cortex. Primary cortical areas receive signals from receptors on the body's periphery. Association areas coordinate and process sensory input from different receptors. (**b**) These three PET scans identified which brain regions were active when a person performed three specific tasks: speaking, generating words, and observing words.

processes, and coordinates responses to sensory input mostly from the opposite side of the body. A transverse band of nerve tracts, the corpus callosum, carries signals both ways to coordinate activities of both hemispheres.

A **limbic system** in the cerebrum governs emotions and influences memory. Distantly related to the olfactory lobes, it still deals with the sense of smell. That is one reason why you might feel warm and fuzzy when you recall the cologne of a special person. Connections from the cerebral cortex and other brain centers pass through the system, which includes the amygdala, hippocampus, hypothalamus, and parts of the thalamus. That wiring correlates organ activities with self-gratifying behavior, such as eating and sex. Rage, hatred, and other "gut" reactions arise here, but signals based on reasoning in the cerebral cortex often can override or dampen them.

Different regions of the cerebral cortex receive and process different signals. EEGs and PET scans pinpoint activity in four tissue lobes of each hemisphere—the occipital, temporal, parietal, and frontal lobes. (An *EEG*, for electroencephalogram, is a recording of summed electrical activity in a brain region. Section 2.2 describes PET scans.) In the occipital lobe are centers for vision (Figure 29.23). Severe blows to it may destroy part of the visual field, the outside world that the individual sees. The temporal lobe near each temple has centers for hearing, visual associations, and emotional behavior. These are highly developed among humans and other primates. A blow to this lobe can impair the ability to recognize complex visual patterns, such as faces.

Brain Cavities and Canals

The hollow neural tube that first develops in vertebrate embryos persists in adults, as a continuous system of fluid-filled cavities and canals. It contains cerebrospinal fluid, a clear extracellular fluid that cushions the brain and spinal cord against jarring movements (Figure 29.24).

A mechanism called the **blood-brain barrier** protects the brain and spinal cord by exerting some control over which solutes enter the cerebrospinal fluid. No other

Figure 29.24 Cerebrospinal fluid (*blue*) in a human brain. This is an extracellular fluid that fills four interconnected cavities (cerebral ventricles) in the brain and spinal cord's central canal.

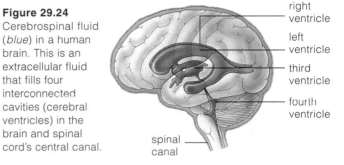

right ventricle

left ventricle

third ventricle

fourth ventricle

spinal canal

portion of extracellular fluid has solute concentrations maintained within such narrow limits. Even normal changes in composition that accompany, say, eating or exercising are opposed here. Why? Some blood-borne hormones and amino acids can affect the functioning of neurons. Also, shifting levels of certain ions (such as K^+) can skew the threshold for action potentials.

The barrier operates at the plasma membrane of cells. The cells make up the wall of blood capillaries that service the brain. In most brain regions, continuous tight junctions fuse abutting walls of cells, so water-soluble substances must move *through* cells to reach the brain. Membrane transport proteins let vital nutrients (such as glucose) and some ions move into and out of the cells but bar some metabolic wastes (such as urea) and toxins. The barrier doesn't keep out fat-soluble substances, such as oxygen, carbon dioxide, anesthetics, alcohol, caffeine, and nicotine. It is nonexistent around the hypothalamus and the brain stem's vomiting center (guess why).

The vertebrate brain develops from a hollow neural tube, which persists in adults as a system of cavities and canals filled with cerebrospinal fluid. The fluid cushions nervous tissue from sudden, jarring movements. The nervous tissue is subdivided into the hindbrain, forebrain, and midbrain.

The brain stem, the nervous tissue first to evolve in all three regions, affords reflex control over basic functions necessary for survival. The highest integrative centers are in the forebrain, especially in the cerebral hemispheres.

WWW

Memory refers to a brain's capacity to store and retrieve information about past sensory experience. Without it, learning and adaptive modifications of behavior would be impossible. Information gets stored in stages. In *short-term* storage, neural excitation lasts a few seconds to a few hours. This stage is limited to a few bits of sensory information—numbers, words of a sentence, and so on. In *long-term* storage, seemingly unlimited amounts of information get tucked away more or less permanently.

Only part of the sensory input reaching the cerebral cortex is selected for transfer to short-term memory bins, where information is processed for relevance. If irrelevant, it is forgotten. Otherwise it is consolidated with information in long-term storage structures. Also, the human brain processes facts separately from skills.

Facts, soon forgotten or filed in long-term storage, include dates, names, faces, words, odors, and other bits of explicit information along with the circumstance in which they were learned. That's why you may associate, say, the smell of warm watermelon with a special picnic held long ago. Fact memory involves the amygdala, the limbic system's gatekeeper between the sensory cortex and the thalamus, hypothalamus, and basal ganglia. It involves the hippocampus, which is part of the limbic system that mediates learning and spatial relations. Information also flows to the prefrontal cortex, where multiple banks of fact memories can be retrieved and used to stimulate or inhibit other parts of the brain.

We gain *skills* by practicing specific motor actions. Slam-dunking a basketball or playing a concerto is best recalled by performing it, rather than by recalling the circumstances in which the skill was first learned. Skill memory involves brain structures that promote motor responses. Motor skills involve muscle conditioning. The circuit extends to the cerebellum, the brain region that coordinates motor activity.

Amnesia is a loss of memory. Its severity depends on whether the hippocampus, amygdala, or both are damaged, as by a severe head blow. Amnesia does not affect the capacity to learn new skills. Basal ganglia are destroyed and learning ability is lost during *Parkinson's disease*, yet skill memory is retained. *Alzheimer's disease* usually starts late in life and involves structural changes in the cerebral cortex and hippocampus. Often affected people are able to remember long-standing facts, such as their Social Security number, but they have difficulty remembering what has just happened to them. In time they become confused, depressed, and incoherent.

Memory, the storage and retrieval of sensory information, results from circuits between the cerebral cortex and parts of the limbic system, thalamus, and hypothalamus. Sensory input is processed through short-term and long-term storage.

Broadly speaking, a drug is a substance introduced into the body to provoke a specific physiological response. Some drugs help a person cope with illness or emotional stress. Others artificially fan pleasure associated with sex and other self-gratifying behaviors.

Many drugs are habit-forming. That is, even if the body is able to function well without them, a person continues to use drugs for the real or imagined relief they afford. Often the body develops tolerance of such drugs; it takes larger or more frequent doses to produce the same effect. Habituation and tolerance are signs of **drug addiction**, a chemical dependence on a drug (Table 29.1). *With an addiction, the drug has assumed an "essential" biochemical role in the body.* Abruptly deprive addicts of the drug, and they go through physical pain and mental anguish. The entire body goes through a period of biochemical upheavals. Stimulants, depressants, hypnotics, narcotic analgesics, hallucinogens, and psychedelics have such effects.

STIMULANTS   Stimulants increase alertness and body activity—then cause depression. *Caffeine* in coffee, tea, chocolate, and many soft drinks is a common stimulant. Low doses arouse the cerebral cortex first and increase alertness. Higher doses act at the medulla oblongata to disrupt motor coordination and mental coherence. Another stimulant, the *nicotine* in tobacco, has potent effects on the nervous system. It mimics acetylcholine and directly stimulates a variety of sensory receptors. In the short term, nicotine results in water retention, irritability, high blood pressure, and gastric upsets.

Possibly 2 million Americans are *cocaine* abusers. This stimulant gives a rush of pleasure by blocking reabsorption of norepinephrine, dopamine, and other neurotransmitters. Because the molecules accumulate in synaptic clefts, they incessantly stimulate postsynaptic cells for an extended period. Heart rate, blood pressure, and sexual appetite increase. In time the molecules diffuse away. However, neurons cannot synthesize replacements fast enough to

Table 29.1   Warning Signs of Drug Addiction*
1. Tolerance—it takes increasing amounts of the drug to produce the same effect.
2. Habituation—it takes continued drug use over time to maintain self-perception of functioning normally.
3. Inability to stop or curtail use of the drug, even if there is persistent desire to do so.
4. Concealment—not wanting others to know of the drug use.
5. Extreme or dangerous behavior to get and use the drug—as by stealing, asking more than one doctor to write drug prescriptions, or jeopardizing employment by drug use at work.
6. Deterioration of professional and personal relationships.
7. Anger and defensive behavior when someone suggests there may be a problem.
8. Preference of drug use over previously customary activities.

* Three or more of these signs may be cause for concern.

WWW

counter the loss, and the sense of pleasure evaporates as hypersensitized postsynaptic cells demand stimulation. After prolonged, heavy use of cocaine, "pleasure" is no longer possible. Addicts become anxious and depressed. They lose weight and cannot sleep properly. The immune system weakens, and heart abnormalities set in.

Granular cocaine is inhaled (snorted). Abusers burn crack cocaine and inhale the smoke. As suggested at the start of this chapter, crack is extremely addictive. Its highs are higher, its crashes are more devastating, and the social and economic tolls are extreme.

Amphetamines induce massive release of dopamine and norepinephrine. Addicts smoke, snort, inject, or swallow the form called *crank*. Effects last two to twelve hours and range from euphoria and sexual arousal to a pounding heart, dry mouth, tremors, agitation, inability to sleep, and paranoia. Crank also wipes out appetite; initially it attracted women who want to lose weight. In time, the production of dopamine and norepinephrine dwindles as the brain depends more on artificial stimulation. Long-term abuse leads to malnutrition, psychosis, depression, memory loss, and damage to the brain, heart, lungs, and liver. Crank is made cheaply in home kitchens (dubbed Beavis and Butthead labs) from such ingredients as drain cleaners. Abuse of this "poor man's cocaine" is epidemic, especially in the American West and Midwest.

DEPRESSANTS, HYPNOTICS These drugs lower activity in nerves and parts of the brain. Some act at synapses in the reticular formation and thalamus. Depending on the dosage, physiological state, and emotional state, responses range from emotional relief through drowsiness, sleep, anesthesia, and coma to death. Low doses have the most effect on inhibitory synapses, so a person feels excited or euphoric at first. Increased doses suppress excitatory synapses as well, and they lead to depression. Depressants and hypnotics are addictive. One amplifies another, as when alcohol plus barbiturates heightens the depression.

Alcohol, or ethyl alcohol, acts directly on the plasma membrane to alter cell function. Like nicotine and cocaine, it is lipid soluble and easily crosses the blood-brain barrier to exert rapid effects. Some people mistakenly think of it as a harmless stimulant because of the initial "high" that it produces. However, alcohol is one of the most powerful psychoactive drugs and a major factor in many deaths. In the short term, small amounts cause disorientation, uncoordinated motor functions, and diminished judgment. Long-term addiction can lead to *cirrhosis*. Then, connective tissue permanently replaces damaged liver cells, so the self-regenerating capacity of liver tissue slowly diminishes. In time, the outcome is chronic liver failure and death.

ANALGESICS When severe stress leads to physical or emotional pain, the brain produces natural *analgesics*, or pain relievers. Endorphins and enkephalins are two examples. They act on many parts of the nervous system,

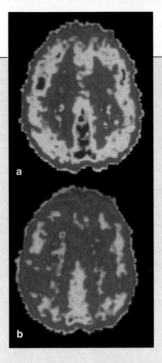
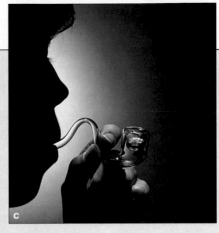

Figure 29.25 (**a**) Normal brain activity revealed by a PET scan. (**b**) PET scan of a comparable section showing cocaine's effect. *Red* indicates greatest activity; and *yellow*, *green*, and *blue* indicate increasingly inhibited activity. (**c**) In less than eight seconds, smoking crack puts cocaine in the brain.

including brain centers for pain and emotions. *Codeine, heroin,* and other narcotic analgesics sedate the body and relieve pain. They also are among the most addictive substances known. Deprivation after massive doses of heroin results in hyperactivity and anxiety, fever, chills, violent vomiting, cramping, and diarrhea.

PSYCHEDELICS, HALLUCINOGENS These drugs skew sensory perception by interfering with the action of acetylcholine, norepinephrine, or serotonin. *LSD* (lysergic acid diethylamide) alters serotonin's roles in inducing sleep, controlling body temperature, and mediating sensory perception. Even in small doses, LSD warps perceptions. For example, some users "perceived" they could fly and "flew" off buildings.

Marijuana, another hallucinogen, is made from crushed leaves, flowers, and stems of the plant *Cannabis*. In low doses it is like a depressant. It slows down but does not impair motor activity; it relaxes the body and elicits mild euphoria. It also causes disorientation, anxiety bordering on panic, delusions, and hallucinations. Like alcohol, it affects the performance of complex tasks, such as driving a car. In one study, pilots showed a marked deterioration in instrument-flying ability for more than two hours after smoking marijuana. In time, marijuana smoking impairs the immune system and mental functions.

Each of us possesses a body of great complexity. Its architecture, its functioning are legacies of millions of years of evolution. Its nervous system is unparalleled in the living world. One of its most astonishing products is language—the encoding of shared experiences of groups of individuals in time and space. Through the evolution of our nervous system, the sense of history was born, and the sense of destiny. Through this system we can ask how we have come to be what we are, and where we are headed from here. Perhaps the sorriest consequence of drug abuse is its implicit denial of this legacy—the denial of self when we cease to ask, and cease to care.

WWW

SUMMARY

1. Nervous systems sense and interpret specific aspects of the environment and issue commands for responses to them. In nearly all animals, communication lines are made of sensory neurons, interneurons, and motor neurons, which activate muscle and gland cells.

2. A neuron's dendrites and cell body are input zones. If arriving signals spread to a trigger zone (such as the start of an axon), they may trigger an action potential that can be propagated to output zones (axon endings).

3. For a neuron at rest, the ion distribution between the cytoplasm and extracellular fluid results in a difference in electric charge (voltage difference) across the plasma membrane. The cytoplasmic fluid is more negatively charged by an amount known as the resting membrane potential. An action potential is an abrupt, short-lived reversal in the voltage difference across the membrane in response to adequate stimulation.

    a. Stimuli give rise to local, graded potentials, which vary in their magnitude and do not spread far from the point of stimulation. A disturbance from a number of graded potentials may spread to a trigger zone and drive the membrane to the threshold of an action potential.

    b. A strong disturbance opens many gated channels for sodium ions in an ever accelerating, self-propagating way. Thus the voltage difference across the membrane reverses abruptly. Accelerated openings proceed in patch after patch of membrane, to the neuron's output zone.

    c. In between action potentials, sodium-potassium pumps counter small ion leaks across the membrane. By doing so, they help maintain the gradients. After an action potential, they restore the gradients.

4. Action potentials arriving at an output zone trigger the release of neurotransmitters. These diffuse across chemical synapses, where the neuron forms a junction with another neuron, a muscle cell, or a gland cell. They may excite the cell's plasma membrane (drive it closer to threshold) or inhibit it (drive it away from threshold).

5. Integration is the moment-by-moment summation of excitatory and inhibitory signals reaching all of the synapses on a neuron. It is a means of playing down, suppressing, reinforcing, or sending on information to other neurons of the nervous system.

6. A nervous system's most basic operations are reflexes: simple, stereotyped movements made in direct response to stimuli. The simplest nervous systems are nerve nets, such as those of cnidarians and other radial animals. The meshworks of nerve cells form reflex connections with contractile and sensory cells of epithelium. Most animals have a bilateral, cephalized nervous system, with a brain or ganglia at the anterior end. They have cordlike nerves (a number of axons of sensory neurons, motor neurons, or both bundled inside a sheath).

Table 29.2 Features of the Vertebrate Brain and Spinal Cord*

FOREBRAIN	Cerebrum	Localizes, processes sensory inputs; initiates, controls skeletal muscle activity. In the most complex vertebrates, has roles in memory, mediating emotions, and abstract thought
	Olfactory lobes	Relaying of sensory input from the nose to olfactory centers of the cerebrum
	Thalamus	Relay stations for conducting sensory signals to and from cerebral cortex; role in memory
	Hypothalamus	With pituitary gland, homeostatic control center over volume, composition, and temperature of internal environment. Governs behavior affecting organ functions (e.g., hunger, thirst), physical expressions of emotions (e.g., sweating)
	Limbic system	A complex of structures. Governs emotions, has roles in memory
	Pituitary gland (Chapter 31)	With hypothalamus, endocrine control of metabolism, growth, and development
	Pineal gland (Chapter 31)	Control of some circadian rhythms; role in mammalian reproductive physiology
MIDBRAIN	Tectum	In fishes, amphibians, coordinating centers for sensory input (including optic lobes) and motor responses. In mammals, mainly reflex centers that rapidly relay sensory input to forebrain
HINDBRAIN	Pons	A "bridge" of tracts between cerebrum and cerebellum; its other tracts connect forebrain and spinal cord. Works with medulla oblongata to control rate, depth of respiration
	Cerebellum	Coordinates motor activity for limb movements, maintaining posture, and spatial orientation
	Medulla	Contains tracts between pons and spinal oblongata cord; reflex centers involved in control of heart rate, blood vessel diameter, respiratory rate, vomiting, coughing, other vital functions
SPINAL CORD		Reflex connections for limb movements. Many of its tracts carry signals between the brain and the peripheral nervous system

* The reticular formation extends from the spinal cord to the cerebral cortex.

7. The vertebrate central nervous system consists of a brain and spinal cord (Table 29.2). Its peripheral nervous system consists mostly of nerves, which carry signals between all body regions and the spinal cord and brain.

    a. Somatic nerves of the peripheral nervous system deal with the skeletal muscles. Autonomic nerves deal with the heart, lungs, glands, and other internal organs.

    b. Autonomic nerves are either parasympathetic or sympathetic. Parasympathetic nerves dominate when outside stimulation is low and tend to divert energy to basic housekeeping tasks. Overall, sympathetic nerves step up body activities during heightened awareness or danger. They govern the fight-flight response.

Review Questions

1. Describe sensory neurons, interneurons, and motor neurons in terms of their structure and functions. *CI, 29.1, 29.4*

2. Define resting membrane potential, graded potential, and action potential. *29.1, 29.2*

3. Label the functional zones of the neuron shown above. *29.1*

4. With respect to action potentials, explain what is meant by threshold level, all-or-nothing spikes, and self-propagation of an action potential. *29.2*

5. Define chemical synapse and neurotransmitter. Choose an example of a neurotransmitter and state where it acts. *29.3*

6. Define synaptic integration. Include definitions of EPSPs and IPSPs as part of your answer. *29.3*

7. What is a myelin sheath? Do all neurons have one? *29.4*

8. Define reflex, then give an example of a reflex arc. *29.4*

9. What is a nerve net? Is it found only in radial animals? *29.6*

10. Describe the components of the vertebrate central nervous system and then of the peripheral nervous system. *29.7*

11. Distinguish between (a) somatic and autonomic nerves, and (b) sympathetic and parasympathetic nerves. *29.8*

12. Name the components of the vertebrate hindbrain, midbrain, and forebrain and list their main functions. *29.7, 29.9, Table 29.2*

13. Define cerebrospinal fluid and the blood-brain barrier. *29.9*

14. Briefly describe one of the brain centers of the cerebral cortex. Mention the tissue lobe in which it is located. *29.9*

Self-Quiz (*Answers in Appendix III*)

1. Action potentials occur when _____ .
   a. a neuron receives adequate stimulation
   b. sodium gates open in an ever accelerating way
   c. sodium-potassium pumps kick into action
   d. both a and b

2. Compared with interstitial fluid near the plasma membrane of a neuron at rest, the cytoplasm near the membrane's inner surface carries a slight _____ charge.
   a. positive   c. graded and local
   b. negative   d. both b and c

3. A nerve may consist of bundled-together axons of _____ .
   a. sensory neurons   c. sensory and motor neurons
   b. motor neurons   d. all are correct

4. The resting membrane potential is maintained by _____ .
   a. ion leaks   c. neurotransmitters
   b. ion pumps   d. both a and b

5. In sea anemones and jellyfishes, the _____ is the loose mesh of nerve cells that interact with sensory and contractile cells of an epithelium to bring about reflex responses to stimuli.

6. Is this statement true or false: White matter and gray matter are components of the spinal cord alone.

7. Overall, _____ nerves slow down the body and divert energy to digestion and other basic housekeeping tasks; and _____ nerves slow down housekeeping tasks and increase overall activity in times of heightened awareness or excitement.
   a. autonomic; somatic   c. sympathetic; parasympathetic
   b. peripheral; central   d. parasympathetic; sympathetic

8. Match the terms with their most suitable description.
   ___ synaptic integration   a. occurs at neuron's input zone
   ___ muscle spindle   b. moment-by-moment summing
   ___ graded, local   of all signals arriving at
         potential   neuron
   ___ action potential   c. arises at trigger zone
                            d. stretch-sensitive receptor

Critical Thinking

1. With *multiple sclerosis*, recall, myelin sheaths around axons in the spinal cord slowly degenerate. Affected individuals progressively lose the ability to control movements of their skeletal muscles. Reflect on the role of myelin sheaths in the nervous system, then formulate a hypothesis of how their degeneration results in a loss of control over skeletal muscles.

2. *Epilepsy* is a neurological disorder characterized by short, recurring episodes of sensory and motor malfunctioning. The episodes, or epileptic seizures, might arise when reverberating circuits involving millions of interneurons in the brain become abnormally activated. Affected individuals typically experience involuntary muscle contractions and often sense lights, sounds, and odors even if receptors in the eyes, ears, and nose have not been stimulated. The drug valproic acid can eliminate or lessen the severity of the seizures by stimulating the body's synthesis of GABA. Why would stepping up GABA production help?

3. In human newborns and premature babies, the blood-brain barrier is not fully developed. Explain why this might be reason enough to pay careful attention to their diet.

4. When Jennifer was only six years old, a man lost control of his car and hit a tree in front of her house. She ran over to the car and screamed when she saw blood from a deep head wound dripping onto a huge bouquet of red roses. Thirty-five years later, someone gave Jennifer a bottle of *Tea Rose* perfume for a birthday present. When she sniffed the perfume, she became frightened and extremely anxious. A few minutes later she also had a vivid recollection of the accident. Explain this incident in terms of what you have learned about memory.

Selected Key Terms

acetylcholine (ACh) *29.3*	myelin sheath *29.4*
action potential *29.1*	nerve *29.4*
autonomic nerve *29.8*	nerve net *29.6*
axon *29.1*	nervous system *CI*
blood-brain barrier *29.9*	neural tube *29.7*
brain *29.9*	neurotransmitter *29.3*
brain stem *29.9*	parasympathetic nerve *29.8*
central nervous system *29.7*	peripheral nervous system *29.7*
chemical synapse *29.3*	positive feedback *29.2*
dendrite *29.1*	reflex *29.4*
drug addiction *29.11*	resting membrane potential *29.1*
forebrain *29.9*	reticular formation *29.9*
hindbrain *29.9*	sensory neuron *CI*
interneuron *CI*	sodium-potassium pump *29.1*
limbic system *29.9*	somatic nerve *29.8*
memory *29.10*	spinal cord *29.8*
meninges *29.8*	stimulus *CI*
midbrain *29.9*	sympathetic nerve *29.8*
motor neuron *CI*	synaptic integration *29.3*

Readings   *See also www.infotrac-college.com*

Delcomyn, F. 1998. *Foundations of Neurobiology.* New York: Freeman. Notably accessible writing and clear illustrations.

Shepherd, G. 1994. *Neurobiology.* Third edition. New York: Oxford. Paperback.

WWW   *http://www.brookscole.com/biology*

Practice quiz questions, hypercontents, BioUpdates, and critical thinking. The Brooks/Cole Biology Resource Center provides a wealth of information fully organized and integrated by chapter.

30

SENSORY RECEPTION

Different Strokes for Different Folks

Even though you might be reluctant to pet a snake or scratch a bat behind the ears, you have to give them credit for being two of your vertebrate relatives with some special sensory traits.

Think of a python. Figure 30.1*a* shows one of these reptiles, which have thermoreceptors in rows of pits above and below their mouth. Thermoreceptors detect infrared energy—in this case, the body heat of small, night-foraging mammals that are the snake's prey of choice. When stimulated, the receptors send messages to the brain, where several centers process the signals and issue commands to muscle cells. In no time at all, a strike is aimed and executed with stunning accuracy.

A motionless, edible frog would not have the same stimulatory effect on the receptors, so the snake would slither past it. The skin of a frog is not warm, and it blends with the colors of the frog's habitat. A python does not have the types of receptors that can detect frogs or a neural program for responding to them.

Or think of the mammals called bats. Nearly all bats sleep during the day and spread webbed wings at dusk. Different kinds take to the air in search of nectar, fruit, frogs, or insects. Many sensory receptors located in their eyes, nose, ears, mouth, and skin are not all that different from yours. However, other receptors provide bats with a sense of hearing that you cannot begin to match. Even tiny-eyed, nearly blind species navigate and capture flying insects with precision in the dark!

Bats are masters of **echolocation**. They emit calls, as the bat in Figure 30.1*b* is doing. Sound waves of the calls bounce off insects, trees, and other objects. Inside the bat ears, acoustical receptors detect the echoes and send signals about them to the brain.

As an echolocating bat flies, it emits a steady stream of about ten clicking sounds per second—sounds we cannot hear. They are intense "ultrasounds" above the range of sound waves that receptors in human ears can detect.

When a bat hears a pattern of distant echoes from, say, an airborne mosquito or

moth, it canincrease the rate of ultrasonic clicks to as many as 200 per second, faster than a machine gun fires bullets. In the few milliseconds of silence between clicks, sensory receptors detect the echoes and deliver messages about them to the bat brain. The brain rapidly constructs a "map" of the sounds that the bat uses in its maneuvers through the night world.

With this chapter, we turn to **sensory systems**, the means by which animals receive signals from their external and internal environments. Such systems, the front doors of the nervous system, receive information about specific changes inside and outside the body, then notify the spinal cord and brain of what is going on. They consist of sensory receptors, nerve pathways from receptors to the brain, and brain regions where sensory information is translated into sensations. A **sensation** is conscious awareness of a stimulus. It is not the same as **perception**: understanding of what a sensation means. Information about different stimuli is even integrated at the same time to give rise to compound sensations. For instance, "wetness" is not a single stimulus, for our perception of it arises from simultaneous inputs that concern pressure, touch, and temperature.

Continuing with our analogy, sensory receptors are the receptionists at the front door of nervous systems.

Figure 30.1 Examples of sensory receptors. (**a**) In pits above and below this python's mouth, thermoreceptors detect body heat, or infrared energy, of nearby prey. (**b**) Some bats listen to echoes of their own high-frequency sounds. The bat brain constructs a sound map, based on echoes bouncing back from objects in the surroundings. Such maps help this night-flying bat capture insects in midair, without the help of eyes.

We group most of them into six categories, based on the type of stimulus energy that each type detects:

Mechanoreceptors detect forms of mechanical energy (changes in pressure, position, or acceleration).

Thermoreceptors detect infrared energy (heat).

Pain receptors (nociceptors) detect tissue damage.

Chemoreceptors detect chemical energy of specific substances dissolved in the fluid surrounding them.

Osmoreceptors detect changes in water volume (solute concentration) in the surrounding fluid.

Photoreceptors detect visible and ultraviolet light.

Depending upon the kinds and numbers of their sensory receptors, animals sample the environment in different ways and differ in their awareness of it. Unlike bees, you have no receptors for ultraviolet light and do not see many flowers the way they do. Unlike many bats, you and bees have no receptors for ultrasound. Unlike pythons, you, bees, and bats have no receptors for detecting warm-blooded prey in the dark. In this chapter, you will encounter processes and molecular structures responsible for these fascinating differences.

b

KEY CONCEPTS

1. Sensory systems are portions of the nervous system. Each consists of specific types of sensory receptors, nerve pathways from receptors to the brain, and brain regions that receive and process the sensory information.

2. A stimulus is a form of energy that activates a specific type of sensory receptor, which is either a sensory neuron or a specialized cell adjacent to it. Photoreceptors detect light energy, thermoreceptors detect infrared energy, and so on.

3. A sensation is conscious awareness of change in some aspect of the external or internal environment. It begins when sensory receptors detect a specific stimulus. The stimulus energy becomes converted to a graded, local signal that may contribute to initiating an action potential.

4. Information concerning the stimulus becomes encoded in the number and frequency of action potentials sent to the brain along particular nerve pathways. Specific brain regions translate the information into a sensation.

5. The somatic sensations include touch, pressure, pain, temperature, and muscle sense.

6. Taste, smell, hearing, and vision are special senses.

OVERVIEW OF SENSORY PATHWAYS

Recall, from Chapter 29, that sensory axons carry signals from receptors to the brain. Before they can do so, the stimulus energy must be converted to action potentials, the basis of neural messages. Briefly, when a stimulus disturbs the plasma membrane of a receptor's ending, certain ions flow across a local patch of the membrane. In other words, the disturbance triggers a local, graded potential. This type of signal does not spread far from the point of stimulation, and it can vary in magnitude.

When a stimulus is intense or repeated fast enough for a summation of local signals, action potentials may result. Action potentials propagate themselves from a receptor to axon endings of a sensory neuron (Figure 30.2). The neuron releases neurotransmitter that affects the electrical activity of another interneuron or a motor neuron. The disturbance may trigger action potentials in the neighboring cell, which is part of an information pathway leading to the brain.

Action potentials traveling along a sensory neuron are not like a wailing ambulance siren. *They do not vary in amplitude.* How, then, does the brain assess the nature of a given stimulus? The answer lies with (1) *which* nerve pathways happen to be carrying action potentials, (2) the *frequency* of action potentials traveling along each axon that is part of the pathway, and (3) the *number* of axons that the stimulus recruited.

First, the wiring in each animal's brain is genetically programmed to interpret action potentials only in certain ways. That's why you "see stars" when one of your eyes gets poked, even in a darkened room. Photoreceptors in that eye were mechanically disturbed enough to trigger messages that traveled along one of two optic nerves to the brain. And your brain always interprets any signals that are arriving from an optic nerve as "light."

Second, when a stimulus is strong, receptors fire action potentials more frequently than they do with a weak stimulus. The same receptor can detect the sounds of a throaty whisper and a wild screech. And the brain senses the difference through frequency variations in signals that the receptor sends to it.

Third, a stronger stimulus can recruit more sensory receptors than a weaker stimulus can do. Gently tap a spot of skin on one of your arms and you activate a few receptors. Press hard on the spot and you activate more receptors in a larger area. The increased

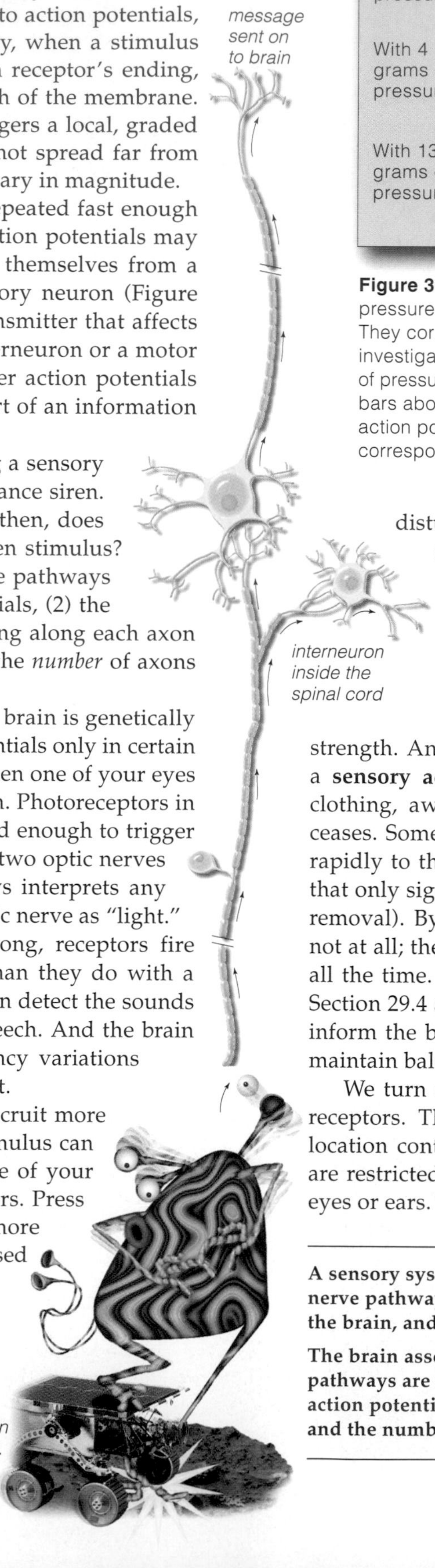

message sent on to brain

interneuron inside the spinal cord

Figure 30.2 Example of a sensory nerve pathway that leads away from a sensory receptor to the brain. The sensory neuron is coded *red* and the interneurons *yellow*.

NASA rover runs over Martian foot; the receptor endings of a sensory neuron are stimulated.

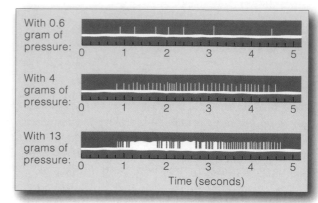

Figure 30.3 Recordings of action potentials from a single pressure receptor with endings in the skin of a human hand. They correspond to variations in stimulus strength. Here, an investigator pressed a thin rod against skin with the amount of pressure indicated on the left side of each diagram. Vertical bars above each thick horizontal line represent individual action potentials. In each case, the increases in frequency correspond to increases in the strength of the stimulus. 🌀

disturbance translates into action potentials in many sensory axons at the same time. The brain interprets the combined activity as an increase in the stimulus intensity. Figure 30.3 is an example of this effect.

In some cases, the frequency of action potentials decreases or stops even when a stimulus is being maintained at constant strength. Any decrease in the response to a stimulus is a **sensory adaptation**. For example, after you put on clothing, awareness of its pressure against your skin ceases. Some mechanoreceptors inside your skin adapt rapidly to the sustained stimulation; they are of a type that only signal a change in a stimulus (its onset and its removal). By contrast, other receptors adapt slowly or not at all; they help the brain monitor particular stimuli all the time. Those stretch receptors you read about in Section 29.4 are like this. A number of them continually inform the brain about the length of muscles that help maintain balance and posture.

We turn next to some specific examples of sensory receptors. The types present at more than one body location contribute to **somatic sensations**. Other types are restricted to particular locations, such as inside the eyes or ears. They contribute to the **special senses**.

A sensory system has sensory receptors for specific stimuli, nerve pathways that conduct information from receptors to the brain, and brain regions that receive the information.

The brain assesses a given stimulus based on which nerve pathways are carrying action potentials, the frequency of action potentials traveling along each axon of that pathway, and the number of axons that have been recruited to action.

SOMATIC SENSATIONS

Somatic sensations begin with receptors in the body's surface tissues, skeletal muscles, and walls of internal organs. The receptors are most highly developed in birds and mammals; amphibians have few, and apparently fishes have none. Receptor inputs travel into the spinal cord and on into the **somatosensory cortex**, part of the gray matter of the cerebral hemispheres (Section 29.9). The interneurons in this brain region are organized like maps corresponding to particular parts of the body's surface. Map regions with the largest areas correspond to body parts that have the greatest sensory acuity and that require the most intricate control. Such body parts include the fingers, thumbs, and lips (Figure 30.4).

You and other mammals discern sensations of touch, pressure, cold, warmth, and pain. **Pain** is perception of injury to some body region. Sensations of *somatic* pain start with signals from pain receptors in skin, skeletal muscles, joints, and tendons. The sensations of *visceral* pain, associated with the internal organs, are related to excessive chemical stimulation, muscle spasms, muscle fatigue, inadequate blood flow to organs, distension of digestive tract regions, and other abnormal conditions.

Free nerve endings are the simplest receptors. These thinly myelinated or unmyelinated (naked) branched endings of sensory neurons are distributed in skin and internal tissues. Different types are mechanoreceptors, thermoreceptors, and pain receptors. All adapt slowly to stimulation. One subpopulation gives rise to a sense of prickling pain, as when you jab a finger with a pin. Another contributes to itching or warming sensations that are provoked by chemicals, including histamine. Two thermoreceptive types have peak sensitivities that are higher and lower than normal body temperature, respectively. One mechanoreceptive type coils around hair follicles and detects hair movements (Figure 30.5).

Encapsulated receptors are common near the body surface. The Meissner's corpuscle adapts very slowly to low-frequency vibrations. It is abundant in fingertips, lips, eyelids, nipples, and genitals. The bulb of Krause, an encapsulated thermoreceptor, is activated at 20°C (68°F) or lower. Below 10°C, it contributes to painful freezing sensations. Ruffini endings are slowly adapting encapsulated types. They respond to steady touching and pressure, and to temperatures above 45°C (113°F).

The Pacinian corpuscle, an encapsulated receptor, is widely distributed deep in the skin, where it responds to fine textures. It also is located near freely movable joints and in some internal organs. Onionlike layers of membrane alternating with fluid-filled spaces enclose this sensory ending (see Figure 30.5). The construction enhances the receptor's ability to detect rapid pressure changes associated with touch and vibrations.

Sensing limb movements and the body's position in space requires mechanoreceptors in skeletal muscle,

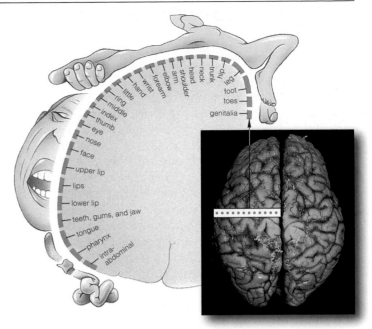

Figure 30.4 Differences in the representation of different body parts in the primary somatosensory cortex. This region is a strip of cerebral cortex, a little wider than 2.5 centimeters (about an inch), from the top of the head to above the ear.

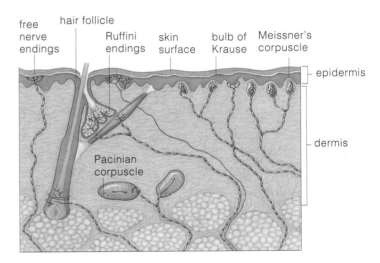

Figure 30.5 A sampling of receptors in human skin.

joints, tendons, ligaments, and skin. Examples include the stretch receptors of muscle spindles. As described in Section 29.4, these sensory organs run parallel with skeletal muscle cells. The degree of their responses to stimulation depends on how much and how fast the muscle is being stretched.

Diverse sensory receptors near the body surface and in internal organs detect touch, pressure, temperature, pain, motion, and positional changes of body parts. Their input travels through the spinal cord to the somatosensory cortex and other brain regions where somatic sensations arise.

WWW

SENSES OF HEARING AND BALANCE

Inner Ear Functions

Most animals assess and respond to displacements from their natural, *equilibrium* position, in which the body is balanced in relation to gravity, velocity, acceleration, and other forces that influence its positions and movement. For example, animals right themselves after tilting or turning upside-down. Most vertebrates have a pair of **inner ears** on both sides of the brain. These evolved first as organs of equilibrium and had little, if anything, to do with hearing. The inner ears have fluid-filled sacs and canals that detect rotational, linear, and accelerated head motions. In amphibians, birds, and mammals, the brain integrates sensory input from the inner ears, eyes, skin, and joints to arrive at sensations of balance.

In each of your inner ears is a **vestibular apparatus** (Figure 30.6). When your head rotates, fluid in a canal corresponding to the direction of rotation is displaced the opposite way. The resulting fluid pressure activates arrays of hair cells, mechanoreceptors that bend under pressure. Deforming their plasma membrane may trigger action potentials, and signals may reach nearby sensory neurons. A different part of the vestibular apparatus is activated when you start and stop moving in a straight line. *Motion sickness* results when monotonous linear, angular, or vertical motion overstimulates the hair cells.

Properties of Sound

Many arthropods and most vertebrates have a sense of **hearing**—sound perception. Sounds are vibrations, or wavelike forms of mechanical energy. Clap your hands to produce waves of compressed air. Each time they clap together, molecules are forced outward and a low-pressure state is created in the vacated region. Such pressure variations are depicted as wave forms in which the *amplitude* corresponds to loudness (Figure 30.7). We typically measure amplitude in decibels. Every ten of

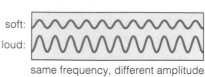

Sounds differ in *amplitude* (they vary in pressure, depicted here as wave forms) and in *frequency* (number of wave cycles/unit time).

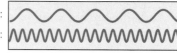

A sound's *intensity* (loudness) depends on its amplitude.

same frequency, different amplitude

A sound's *pitch* (tone) depends on its frequency.

same loudness, different pitch

Figure 30.7 Properties and examples of sound waves. Compared to the pure tone of a tuning fork, most sounds are combinations of sound waves of different frequencies. Combinations of overtones give sounds their timbre, or quality, which is one of the properties that help you recognize, say, voices of people you know over the phone.

these units represents a tenfold increase in intensity above the faintest sound that humans can hear. The *frequency* of a sound is the number of wave cycles per second. Each cycle extends from the start of one wave peak to the start of the next peak. The more cycles per second, the higher are the frequency and perceived pitch.

Evolution of the Vertebrate Ear

After vertebrates invaded the land, hearing became far more crucial than it had been in aquatic habitats, where sounds are muffled. Although parts of the internal ear dealing with equilibrium did not change much, other parts evolved into structures that could receive and process even faint sounds traveling through air. Also, a

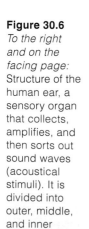

Figure 30.6
To the right and on the facing page: Structure of the human ear, a sensory organ that collects, amplifies, and then sorts out sound waves (acoustical stimuli). It is divided into outer, middle, and inner regions. **a**

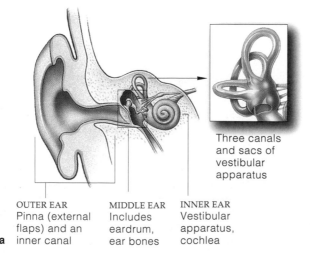

OUTER EAR
Pinna (external flaps) and an inner canal

MIDDLE EAR
Includes eardrum, ear bones

INNER EAR
Vestibular apparatus, cochlea

Three canals and sacs of vestibular apparatus

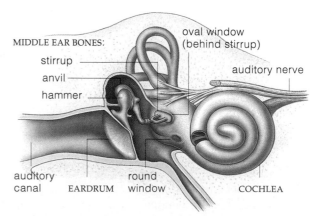

MIDDLE EAR BONES:
stirrup
anvil
hammer

oval window (behind stirrup)

auditory nerve

auditory canal   EARDRUM   round window   COCHLEA

b Closer look at components of the human ear.

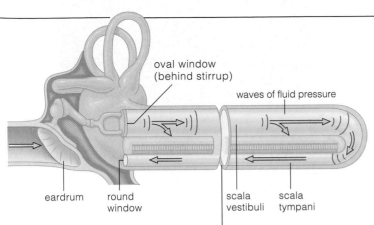

oval window (behind stirrup)

waves of fluid pressure

eardrum   round window

scala vestibuli   scala tympani

c Sound reception in the inner ear. an eardrum make it vibrate. Bones the vibrations, and they amplify the the force of the pressure waves to a window. The "window" is an elastic to the part of the inner ear called in this diagram for clarity).

Sound waves that arrive at of the middle ear pick up stimulus by transmitting smaller surface, the oval membrane over the entrance the cochlea (shown uncoiled

When the oval window bows in and fluid pressure in two ducts that are and scala tympani. The waves reach round window. When this membrane pressure, fluid moves back and forth

out, it produces waves of called the scala vestibuli another membrane, the bulges under incoming in the inner ear.

The pressure waves are sorted out third duct inside the coiled inner ear. (the basilar membrane) starts out broader and flexible deeper in the the basilar membrane vibrates more different frequencies.

at the cochlear duct, the One of its membranes narrow and stiff. It is coil. At different points, strongly to sounds of

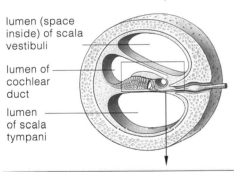

lumen (space inside) of scala vestibuli

lumen of cochlear duct

lumen of scala tympani

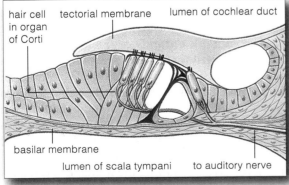

hair cell in organ of Corti   tectorial membrane   lumen of cochlear duct

basilar membrane

lumen of scala tympani   to auditory nerve

ORGAN OF CORTI
(organ of hearing that rests on top of basilar membrane)

d At the organ of Corti, 16,000 hair cells project into a jellylike flap (tectorial membrane) over the basilar membrane. When the flap vibrates suitably, hair cells bend, action potentials arise, and messages travel along an auditory nerve to the brain.

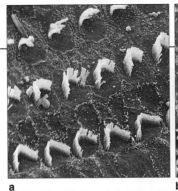
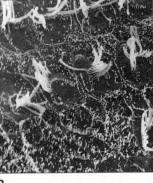

a   b

Figure 30.8 Results of an experiment on the effect of intense sound on the inner ear. (**a**) From a guinea pig's ear, three rows of hair cells that project into the tectorial membrane in the organ of Corti. (**b**) Hair cells in the same organ after twenty-four hours of exposure to noise levels comparable to extremely loud music.

To give you a sense of what "loud" is, a ticking watch measures 10 decibels (100 times louder than the threshold of hearing for humans). A normal conversation is about 60 decibels (about a million times louder), a food blender operating at high speed is about 90 decibels (a billion times louder), and a raging rock concert, about 120 decibels (a trillion times louder).

middle ear evolved that could amplify and transmit air waves to the inner ear. In reptiles, a depression on each side of the head became an eardrum (a thin membrane that rapidly vibrates in response to air waves). Behind the eardrum of existing crocodiles, birds, and mammals is an air-filled cavity and small bones, which transmit vibrations to the inner ear. Among ancient fishes, the bones structurally supported gill pouches, then became part of the jaw joint. Thus, among reptiles, birds, and mammals, bones that once functioned in gas exchange became specialized for feeding—then for hearing.

Among mammals, an **external ear** that functions in collecting sound waves became well developed. In nearly all species it has a pinna (sound-collecting, skin-covered flaps of cartilage, with elaborate folds, that project from both sides of the head) and a channel to the middle ear.

Figure 30.7b–d shows the external, middle, and inner ear of humans. Like other mammals (and birds), it has a cochlea, a part of the inner ear that receives and sorts out sound waves. Its acoustical receptors, **hair cells**, bend in response to pressure waves. This starts a flow of information along an auditory nerve that leads from the receptors to brain centers. Prolonged exposure to intense sounds can permanently damage the hair cells. The cells are not adapted to amplified music, jet planes, and other recent developments (Figure 30.8).

Organs of equilibrium help keep the body balanced in relation to gravity, velocity, acceleration, and other forces that influence its position and movement. In humans, such organs occur in the vestibular apparatus of the inner ear.

Hearing among land vertebrates involves structures that collect, amplify, and sort out sound waves traveling through air. In the inner ear, sound waves produce fluid pressure variations, which hair cells transduce into action potentials.

WWW

VISUAL PERCEPTION

The sensory pathway from a retina to the brain receives, transmits, and combines raw visual information. And it leads to awareness of light and shadows, of colors, of near and distant objects in the outside world. The flow of information starts when light strikes dense arrays of **rod cells** and **cone cells**, two classes of photoreceptors (Figure 30.14). Rod cells detect very dim light. At night or in dark places, they contribute to coarse perception of movement by detecting changes in the intensity of light across the visual field. Cone cells detect bright light; they contribute to sharp vision and color perception during the day. Layers of neurons occur above the rods and cones. First bipolar sensory neurons, then ganglion cells, accept the signals from rods and cones. Axons of the ganglion cells converge to form two optic nerves to the brain.

Before leaving the retina, signals from 125 million photoreceptors converge dramatically, on a mere 1 million ganglion cells. Also, signals flow laterally among other sensory cells that act in concert to dampen or enhance signals before the ganglion cells receive them. As you can see, *a great deal of synaptic integration and processing proceeds even before visual information is sent on to the brain.*

A rod cell's outer segment contains several hundred membrane disks, each peppered with 10^8 molecules of a visual pigment called rhodopsin. Membrane stacking and the high density of pigments greatly increase the chances of intercepting packets of light energy (photons of particular wavelengths). Action potentials that result from the absorption of even a few photons can lead to conscious awareness of objects in dim surroundings. Rhodopsin effectively absorbs photons corresponding to blue-to-green wavelengths (Section 6.2). Absorption changes the pigment's shape, which triggers a cascade of reactions that alter activity at ion channels and ion pumps across the rod cell's plasma membrane. Gated sodium channels close and the voltage difference across the membrane changes. The result: a graded potential that dampens the ongoing release of a neurotransmitter having inhibitory effects on adjacent sensory neurons. Released from inhibition, these neurons start sending signals about the visual stimulus toward the brain.

The sense of color and of daytime vision starts with photon absorption by red, green, and blue cone cells, each with a different visual pigment. Photon absorption reduces the release of a neurotransmitter that otherwise inhibits neurons next to the photoreceptors. In back of the eyeball is the fovea, a funnel-shaped depression near the center of the retina. This retinal area has the greatest density of photoreceptors and affords the greatest visual acuity. Its cone cells are the ones that can discriminate most precisely between adjacent points in space.

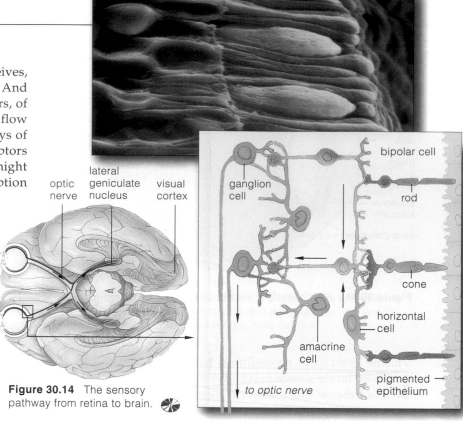

Figure 30.14 The sensory pathway from retina to brain.

The retina's surface is organized into receptive fields, restricted areas that influence the activity of individual sensory neurons. For instance, each ganglion cell's field is a tiny circle. Some cells respond best to a spot of light ringed by dark, in the field's center. Others respond to a rapid change in light intensity, to a spot of one color, or to motion. Some cells respond to diagonally oriented objects, not to diffuse, uniform illumination.

Remember, an individual's visual field is the portion of the outside world that it actually sees. The right side of both retinas intercepts light from the *left* half of the visual field; the left side intercepts light from the *right* half. The optic nerve leading out of each eye delivers the signals about a stimulus from the *left* visual field to the right cerebral hemisphere, and signals from the *right* visual field to the left hemisphere (Figure 30.14).

Axons of the optic nerves end in a layered brain region, the lateral geniculate nucleus. Each layer has a map corresponding to the receptive fields, and it deals with one aspect of a stimulus—form, movement, depth, color, texture, and so on. After initial processing, signals rapidly and simultaneously reach different portions of the visual cortex. There, final integration organizes the electrical activity and produces the sensation of sight.

Eyes collect and integrate information about distance, shape, brightness, position, and movement of visual stimuli. Their sensory pathways start at receptive fields in the retina and proceed along optic nerves to the brain, where signals are further processed and finally integrated in the visual cortex.

$\mathcal{WWW}$

30.6 SENSES OF TASTE AND SMELL

We leave this chapter with two final examples of the special senses, taste and smell. Both are *chemical* senses; their sensory pathways start at chemoreceptors, which are activated when they bind a chemical substance that is dissolved in the fluid bathing them. These receptors wear out and new ones replace them on an ongoing basis. In both pathways, sensory input travels from the receptors through the thalamus and on to the cerebral cortex, where perceptions of the stimulus take shape and undergo fine-tuning. The input also travels to the limbic system, which integrates it with emotional states and stored memories (Sections 29.9 and 29.10).

Different animals taste substances with their mouth, antennae, legs, tentacles, or fins. It all depends on where the chemoreceptors called **taste receptors** are located. In your throat and mouth, especially the tongue's upper surface, they are located in about 10,000 sensory organs called taste buds (Figure 30.15). A taste bud has a pore through which fluids in the mouth contact the surface of receptor cells. Of thousands of perceived tastes, all are some combination of four primary sensations: sweet (elicited by sucrose, glucose, and other simple sugars), sour (acids), salty (NaCl and other salts), and bitter (alkaloids and other toxic plant substances).

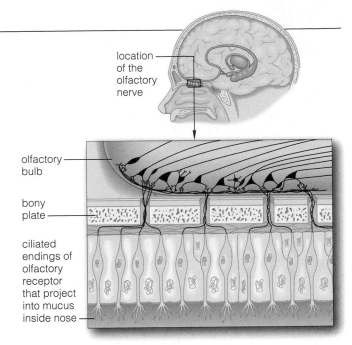

Figure 30.16 Sensory pathway from the sensory endings of olfactory receptors in the nose to the cerebral cortex and limbic system. Receptor axons pass through holes in a bony plate between the nasal lining and the brain. In the earliest vertebrates, an olfactory bulb and olfactory lobe dominated the forebrain. In certain lineages, the olfactory bulb became reduced in importance, and the cerebrum expanded greatly in size and functions, as described in Sections 29.7 and 29.9.

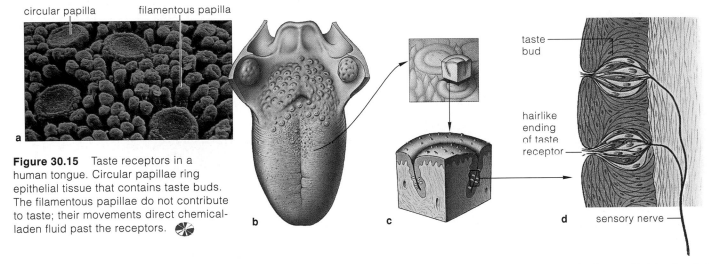

Figure 30.15 Taste receptors in a human tongue. Circular papillae ring epithelial tissue that contains taste buds. The filamentous papillae do not contribute to taste; their movements direct chemical-laden fluid past the receptors.

Olfactory receptors detect water-soluble or volatile (easily vaporized) substances. A bloodhound nose has more than 200 million; a human nose has about 5 million. The receptor axons lead into one of two small brain structures called olfactory bulbs (Figure 30.16). There they synapse with groups of cells that sort out the components of a given scent and relay the information onward, along an olfactory tract, for further processing.

A vomeronasal organ, or "sexual nose," is common among animals, including humans. Its receptors detect **pheromones**, a type of signaling molecule with roles in the social aspects of reproduction. These exocrine gland secretions affect the behavior of other individuals of the same species, especially potential mates. Think about the effect of bombykol on olfactory receptors of a male silk moth. Contact with merely one bombykol molecule per second sends action potentials to the moth brain. They help a male locate a female in the dark, even if she is more than a kilometer upwind from him.

The senses of taste and smell both start at chemoreceptors. Both involve sensory pathways that lead to processing regions in the cerebral cortex and in the limbic system.

SUMMARY

1. A stimulus is a specific form of energy that the body detects by means of sensory receptors. A sensation is an awareness that stimulation has occurred. Perception is understanding what the sensation means.

2. Sensory receptors are endings of sensory neurons or specialized cells next to them. They respond to stimuli, which are specific forms of energy, such as mechanical pressure and light. Animals only respond to aspects of the internal or external environment when they have receptors that are sensitive to the energy of stimuli.

   a. Mechanoreceptors, such as free nerve endings and hair cells, detect mechanical energy related to touch, pressure, and motion and changes in position.

   b. Thermoreceptors detect radiant energy (heat).

   c. Pain receptors (nociceptors) detect tissue damage.

   d. Chemoreceptors, such as olfactory receptors and taste receptors, detect chemical substances dissolved in the fluids bathing them.

   e. Osmoreceptors detect changes in water volume (hence solute concentrations) in the surrounding fluid.

   f. Photoreceptors detect light. Rods and cones of the retina in the human eye are examples.

3. A sensory system has sensory receptors for specific stimuli and nerve pathways from those receptors to receiving and processing centers in the brain. The brain assesses a particular stimulus based upon which nerve pathway is delivering signals, the frequency of signals traveling along individual axons of that pathway, and the number of axons that were recruited into action.

4. Somatic sensations include touch, pressure, pain, temperature, and muscle sense. The receptors associated with these sensations are not localized in a single organ or tissue. Special senses include taste, smell, hearing, balance, and vision. Receptors associated with these senses typically reside in sensory organs, such as eyes, or some other particular body region.

5. Organs of equilibrium, as in the vertebrate inner ear, detect gravity, velocity, acceleration, and other forces that affect the body's positions and movements. The vertebrate sense of hearing (sound perception) requires components of the outer, middle, and inner ear that respectively collect, amplify, and sort out sound waves.

6. Vision requires eyes: sensory organs having a dense array of photoreceptors, as in a retina (Table 30.1). It requires a capacity for image formation in the brain, based upon incoming patterns of visual stimulation. The sense of vision and discrimination among objects evolved first among fast-moving, predatory animals.

7. The senses of taste and smell both involve sensory pathways from chemoreceptors to processing regions in the cerebral cortex and limbic system.

Table 30.1	Summary of Vertebrate Eye Components
THREE LAYERS FORMING WALL OF EYEBALL	
Outer layer	*Sclera.* Protects eyeball
	Cornea. Focuses light
Middle layer	*Choroid.* Blood vessels nutritionally support wall cells; pigments prevent light scattering
	Ciliary body. Its muscles control lens shape; its fine fibers hold lens in upright position
	Iris. Adjustments here control incoming light
	Pupil. Serves as entrance for light
Inner layer	*Retina.* Absorbs and transduces light energy
	Fovea. Increases visual acuity
	Start of optic nerve. Carries signals to brain
INTERIOR OF EYEBALL	
Lens	Focuses light onto photoreceptors
Aqueous humor	Transmits light, maintains pressure
Vitreous body	Transmits light, supports lens and eyeball

Review Questions

1. Define stimulus. *Chapter 29 CI, Chapter 30 CI*

2. Name six categories of sensory receptors and the type of stimulus energy that each kind detects. *CI*

3. What are the basic components of a sensory system? How does the brain assess the nature of a given stimulus? *30.1*

4. How do somatic sensations differ from special senses? *30.1*

5. What is pain? Describe one type of pain receptor. *30.2*

6. Which evolved first, a sense of balance or sense of hearing? Do both involve the outer, middle, and inner ear? *30.3*

7. How does vision differ from light sensitivity? What sensory organs and structures does vision require? *30.4, 30.5*

8. Label the component parts of the human eye: *30.4*

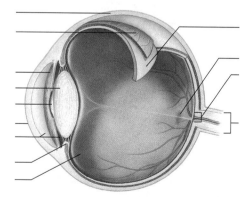

9. Which substances elicit four primary taste sensations? *30.6*

10. Define pheromone and give an example. *30.6*

Self-Quiz *(Answers in Appendix III)*

1. A _____ is a specific form of energy that is detected by a sensory receptor.

2. Conscious awareness of a stimulus is called a _____ .

3. _____ is understanding what particular sensations mean.

Figure 30.17 Flamingos at a large lake in Tanzania, East Africa.

what Kim sees

what Francesca sees

4. Each sensory system consists of _____ .
   a. nerve pathways from specific receptors to the brain
   b. sensory receptors
   c. brain regions that deal with sensory information
   d. all of the above

5. _____ is (are) a decrease in the response to an ongoing stimulus.
   a. Perception
   b. Sensory adaptation
   c. Visual accommodation
   d. both b and c

6. _____ detect mechanical energy associated with changes in pressure, position, or acceleration.
   a. Chemoreceptors
   b. Mechanoreceptors
   c. Photoreceptors
   d. Thermoreceptors

7. Detecting light energy is the function of _____ .
   a. chemoreceptors
   b. mechanoreceptors
   c. photoreceptors
   d. thermoreceptors

8. Vision requires _____ .
   a. a tissue with dense arrays of photoreceptors
   b. eyes
   c. image-forming centers in the brain
   d. all of the above

9. The outer layer of the human eyeball includes the _____ .
   a. lens and choroid
   b. sclera and cornea
   c. retina
   d. both a and c

10. The inner layer of the human eyeball includes the _____ .
    a. lens and choroid
    b. sclera and cornea
    c. retina
    d. both a and c

11. Match each term with the most suitable description.
   _____ somatic senses
   _____ stimulus
   _____ special senses
   _____ variations in stimulus intensity
   _____ sensory receptor

   a. sensory neuron endings or specialized cells next to them
   b. taste, smell, hearing, balance, and vision
   c. form of energy that a specific sensory receptor can detect
   d. encoded in the frequency and number of action potentials
   e. touch, pressure, temperature, pain, and muscle sense

Critical Thinking

1. Wayne, on standby for the last flight from San Francisco to New York, was assigned the last available seat on the plane. It was in the last row, where vibrations and noise from the engines are most pronounced. When Wayne got off the plane in New York, he was speaking very loudly and having trouble hearing what other people were saying. The next day, things were back to normal. Speculate on what happened to his sense of hearing during the flight.

2. Juanita made an appointment with her doctor because she was experiencing recurring episodes of dizziness. Her doctor immediately asked whether "dizziness" meant she had a sense of lightheadedness, as if she were going to faint, or whether it meant she had sensations of *vertigo*—that is, a feeling that she herself or objects near her were spinning around. Why do you suppose her doctor considered this clarification important early in his evaluation of her condition?

3. To Tim, a graduate student studying the social behavior of flamingos in Tanzania, the birds closest to him along the shore of a lake are in sharp focus, but the ones flying some distance away look fuzzy (Figure 30.17*a*). Francesca, another student, says the birds closest to her look fuzzy but those in the distance are in focus (Figure 30.17*b*). Is Tim nearsighted or farsighted, and is the focal point of light rays from the flamingos falling in front or behind his retinas? What about Francesca?

Selected Key Terms

camera eye *30.4*	olfactory	sensory system *CI*
chemoreceptor *CI*	receptor *30.6*	somatic sensation *30.1*
cone cell *30.5*	osmoreceptor *CI*	somatosensory
cornea *30.4*	pain *30.2*	cortex *30.2*
echolocation *CI*	pain receptor *CI*	special sense *30.1*
external ear *30.3*	perception *CI*	taste receptor *30.6*
eye *30.4*	pheromone *30.6*	thermoreceptor *CI*
hair cell *30.3*	photoreceptor *CI*	vestibular
hearing *30.3*	retina *30.4*	apparatus *30.3*
inner ear *30.3*	rod cell *30.5*	vision *30.4*
lens *30.4*	sensation *CI*	visual
mechanoreceptor *CI*	sensory	accommodation *30.4*
middle ear *30.3*	adaptation *30.1*	visual field *30.4*

Readings See also www.infotrac-college.com

Delcomyn, F. 1998. *Foundations of Neurobiology*. New York: Freeman.

Romer, A., and T. Parsons. 1986. *The Vertebrate Body*. Sixth edition. New York: Saunders.

Sherwood, L. 1997. *Human Physiology*. Third edition. Belmont, California: Wadsworth.

Wright, Karen. April 1994. "The Sniff of Legend." *Discover* 15(4): 60–67. Speculation on the existence of a sensory pathway activated by human pheromones.

Zeki, S. September 1992. "The Visual Image in Mind and Brain." *Scientific American* 267(3): 68–76.

WWW *http://www.brookscole.com/biology*

Practice quiz questions, hypercontents, BioUpdates, and critical thinking. The Brooks/Cole Biology Resource Center provides a wealth of information fully organized and integrated by chapter.

31

ENDOCRINE CONTROL

Hormone Jamboree

In the 1960s, at her camp in a forest by the shore of Lake Tanganyika in Tanzania, the primatologist Jane Goodall let it be known that bananas were available. One of the first chimpanzees attracted to the delicious food was a female—Flo, as she came to be called (Figure 31.1). Flo brought along her two offspring, an infant female and a juvenile male, and exhibited commendable parental behavior toward them.

Three years passed, and Goodall observed that Flo's preoccupation with motherhood gave way to a preoccupation with sex. She also observed that male chimpanzees followed Flo to the camp by the lake and stayed for more than the bananas.

Sex, as Goodall discovered, is *the* premier binding force in the social life of chimpanzees.

These primates do not mate for life as eagles do, or wolves. Before the rainy season begins, mature females that are entering their fertile cycle become sexually active. As one of the more dramatic responses to changing concentrations of hormones in their bloodstream, the external sexual

Figure 31.1 (a) Primatologist Jane Goodall in Gombe National Park, near the shores of Lake Tanganyika, scouting for chimpanzees. (b) Flo and three of her offspring, all subjects of long-term field observations. Flo helped Goodall clarify the central role of sex—and of the hormones that orchestrate it—in the social life of these primate relatives of humans.

organs of the females become swollen and vivid pink. The swellings are strong visual signals to males. They are the flags of sexual jamborees, of great gatherings of highly stimulated chimps in which any males present may copulate, in sequence, with the same females.

The gathering of many flag-waving females draws together individuals that forage alone or in small family groups for most of the year. It reestablishes the bonds that unite a rather fluid community.

As infants and juveniles play with one another and with the adults, their aggressive and submissive jostlings help map out future dominance hierarchies. Flo happened to be a high-ranking individual in the social hierarchy. Through her sexual attractiveness and direct solicitations, she built alliances with many male chimps. Through her status and aggressive behavior, she helped her male offspring win confrontations with other young male chimps.

The hormone-induced swelling during the fertile period of Flo and other female chimps lasts somewhere between ten and sixteen days. Yet they are fertile for only one to five days. Sex hormones induce swelling even after a female gets pregnant. We can hypothesize that sexual selection has favored prolonged swelling. Males groom a sexually attractive female more often, protect her, give her more food, and let her tag along to new foraging sites. The more that males accept a female, the higher she rises in the social hierarchy—and the more her individual offspring benefit.

Through their effects, hormones help orchestrate the growth, development, and reproductive cycles of nearly all animals, from the invertebrate worms to chimpanzees and humans. They influence minute-by-minute and day-to-day metabolic functions. Through interplays with one another and with the nervous system, hormones have profound influence over the physical appearance, the well-being, and the behavior of individuals. Even the behavior of individuals affects whether they will survive, either on their own or as part of social groups.

This chapter focuses mainly on hormones—on their sources, targets, and interactions as well as the mechanisms involved in their secretion. If the details start to seem remote, remember that this is the stuff of life. Hormones underwrote Flo's appearance, behavior, and rise through the chimpanzee social hierarchy. Just imagine what they have been doing for you.

KEY CONCEPTS

1. For nearly all animals, hormones and other signaling molecules have central roles in integrating the activities of individual cells in ways that benefit the whole body.

2. Only the cells with molecular receptors for a specific hormone are its targets. Hormones operate by serving as signals for change in the activities of their targets.

3. Many types of hormones influence gene transcription and protein synthesis in target cells. Other types call for alterations in existing molecules and structures in cells. Some exert their effects by binding with and altering membrane characteristics, such as the permeability of the plasma membrane to a particular solute.

4. Some hormones help the body adjust to short-term shifts in diet and in levels of activity. Others help induce long-term adjustments in cell activities that bring about the body's growth, development, and reproduction.

5. Among vertebrates, the hypothalamus and pituitary gland interact in ways that coordinate the activities of a number of endocrine glands. Together, they exert control over many aspects of the body's functioning.

6. Besides hormonal signals, neural signals, changes in local chemical conditions, and environmental cues serve as the triggers for hormonal secretion.

Hormones and Other Signaling Molecules

Throughout their lives, cells must respond to changing conditions by taking up and releasing various chemical substances. In vertebrates, the responses of millions to many billions of cells must be integrated in ways that are essential to the functions of the whole body.

Integrating activities within and between cells, and between tissues, depends on signaling molecules. These include hormones, neurotransmitters, local signaling molecules, and pheromones. Each type acts on target cells (any cells that have receptors for the molecule and that may alter activities in response to it). A target may or may not be adjacent to the cell that sends the signal.

By definition, **hormones** are secretory products of endocrine glands, endocrine cells, and certain neurons, and they travel the bloodstream to nonadjacent target cells. They are this chapter's focus.

By contrast, **neurotransmitters** are released from axon endings of neurons, and then act swiftly on target cells by diffusing across the tiny gap that separates them. Section 29.3 describes their sources and their actions. Also, **local signaling molecules**, released by many types of body cells, alter conditions in localized regions of tissues. Prostaglandins, for instance, target smooth muscle cells in bronchiole walls, which then constrict or dilate and so alter air flow in lungs.

Pheromones, nearly odorless secretions of certain exocrine glands, diffuse through water or air to targets outside the animal body. These hormone-like secretions act on cells of other individuals of the same species. They help integrate social behavior, such as behaviors related to sexual reproduction. Pheromones are common signals among animals. As examples, female silk moths secrete bombykol as a sex attractant (Section 44.4), and soldier termites secrete alarm signals when ants attack their colony (Section 44.9). The vomeronasal organ, a common pheromone detector in animals, was recently discovered in humans. Its presence invites speculation: Do human pheromones act at the subconscious level to trigger those inexplicable impressions, such as instant good or bad "feelings" about somebody you just met? We do not know the answer, but studies are under way.

Discovery of Hormones and Their Sources

The word "hormone" dates to the early 1900s. W. Bayliss and E. Starling were trying to find out what triggers the secretion of pancreatic juices when food is traveling through the canine gut. As they knew, acids are mixed with food in the stomach, and the pancreas secretes an alkaline solution after the acidic mixture is propelled forward into the small intestine. Was the nervous system or something else stimulating the pancreatic response?

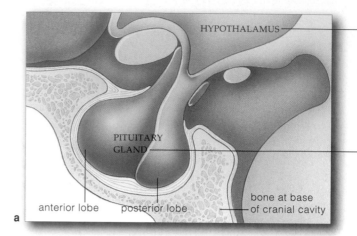

a

Figure 31.2 (**a**) A major neural-endocrine control center. The pituitary gland interacts intimately with the hypothalamus, a brain region that secretes some hormones. (**b**) *Facing page:* Overview of the key components of the human endocrine system and of the primary effects of their main hormonal secretions. The system also includes endocrine cells of many organs, including the liver, kidneys, heart, small intestine, and skin.

To find an answer, Bayliss and Starling blocked nerves—but not blood vessels—leading to a laboratory animal's upper small intestine. Later, when acidic food entered the small intestine, the pancreas still secreted the alkaline solution. Even more telling, extracts of cells from the intestinal lining—a glandular epithelium— also induced the response. Glandular cells had to be the source of the pancreas-stimulating substance.

The substance came to be called secretin. Proof of its existence and mode of action confirmed a centuries-old idea: *The bloodstream picks up internal secretions that can influence the activities of organs inside the body.* Starling coined the word "hormone" for such internal glandular secretions (after *hormon,* meaning to set in motion).

Later on, researchers identified many other kinds of hormones and their sources. Figure 31.2 is a simplified picture of the locations of the following major sources of hormones in the human body. Bear in mind, these sources are typical of most vertebrates:

Pituitary gland
Adrenal glands (*two*)
Pancreatic islets (*numerous cell clusters*)
Thyroid gland
Parathyroid glands (*in humans, four*)
Pineal gland
Thymus gland
Gonads (*two*)
Endocrine cells of the hypothalamus, stomach, small intestine, liver, kidneys, heart, placenta, skin, adipose tissue, and other organs

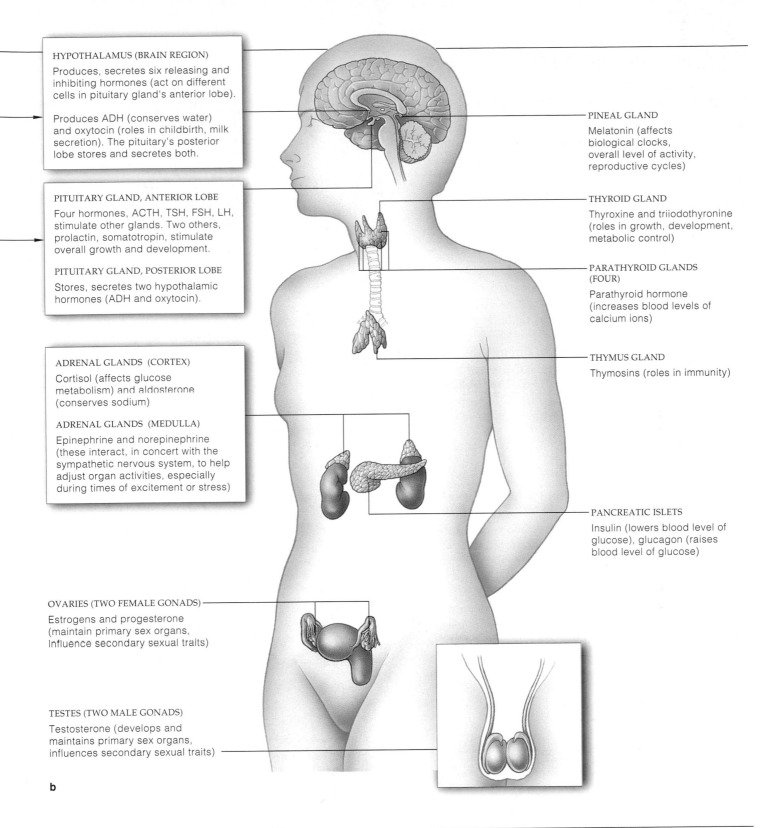

HYPOTHALAMUS (BRAIN REGION)

Produces, secretes six releasing and inhibiting hormones (act on different cells in pituitary gland's anterior lobe).

Produces ADH (conserves water) and oxytocin (roles in childbirth, milk secretion). The pituitary's posterior lobe stores and secretes both.

PITUITARY GLAND, ANTERIOR LOBE

Four hormones, ACTH, TSH, FSH, LH, stimulate other glands. Two others, prolactin, somatotropin, stimulate overall growth and development.

PITUITARY GLAND, POSTERIOR LOBE

Stores, secretes two hypothalamic hormones (ADH and oxytocin).

ADRENAL GLANDS (CORTEX)

Cortisol (affects glucose metabolism) and aldosterone (conserves sodium)

ADRENAL GLANDS (MEDULLA)

Epinephrine and norepinephrine (these interact, in concert with the sympathetic nervous system, to help adjust organ activities, especially during times of excitement or stress)

OVARIES (TWO FEMALE GONADS)

Estrogens and progesterone (maintain primary sex organs, Influence secondary sexual traits)

TESTES (TWO MALE GONADS)

Testosterone (develops and maintains primary sex organs, influences secondary sexual traits)

PINEAL GLAND

Melatonin (affects biological clocks, overall level of activity, reproductive cycles)

THYROID GLAND

Thyroxine and triiodothyronine (roles in growth, development, metabolic control)

PARATHYROID GLANDS (FOUR)

Parathyroid hormone (increases blood levels of calcium ions)

THYMUS GLAND

Thymosins (roles in immunity)

PANCREATIC ISLETS

Insulin (lowers blood level of glucose), glucagon (raises blood level of glucose)

b

Collectively, the body's sources of hormones came to be called the **endocrine system**. The name implies that there is a separate control system for the body, apart from the nervous system. (*Endon* means within; *krinein* is taken to mean secrete.) Later, biochemical research and electron microscopy studies revealed that endocrine sources and the nervous system function in intricately connected ways, as you will see shortly.

Integration of cell activities depends on hormones and other signaling molecules. Each type of signaling molecule acts on target cells, which are any cells that have receptors for it and that may alter their activities in response to it.

Components of the endocrine system and certain neurons produce and secrete hormones, which the bloodstream takes up and distributes to nonadjacent target cells.

WWW

The Nature of Hormonal Action

Hormones and other signaling molecules interact with protein receptors of target cells, and their interactions have diverse effects on physiological processes. Some kinds of hormones induce a target cell to increase its uptake of glucose, calcium, or some other substance from the surroundings. Others stimulate or inhibit the target in ways that alter its rates of protein synthesis, modify the structure of proteins or other elements in the cytoplasm, or change cell shape.

Two factors greatly influence responses to hormonal signals. *First,* different hormones activate different kinds of mechanisms in target cells. *Second,* not all types of cells are equipped to respond to a particular signal. For example, many cell types have receptors for cortisol, so this hormone has widespread effects through the body. By contrast, only a few cell types have the receptors for hormones that stimulate a highly directed response.

Let us now briefly consider the effects of two main categories of these signaling molecules: the steroid and peptide hormones (Table 31.1).

Characteristics of Steroid Hormones

Steroid hormones, recall, are lipid-soluble molecules derived from cholesterol (Section 3.3). The cholesterol remodeling proceeds at multiple-enzyme systems in the endoplasmic reticulum and in mitochondria of steroid-secreting cells. We find such cells in adrenal glands and primary reproductive organs. Testosterone, one of the sex hormones, is an example of one such product.

Think of testosterone's effects on the development of secondary sexual traits associated with maleness. The developmental steps will proceed normally only if target cells have working receptors for testosterone. In a genetic disorder called *testicular feminization syndrome,* these receptors are defective. Genetically, the affected person is male (XY); he has functional testes that are

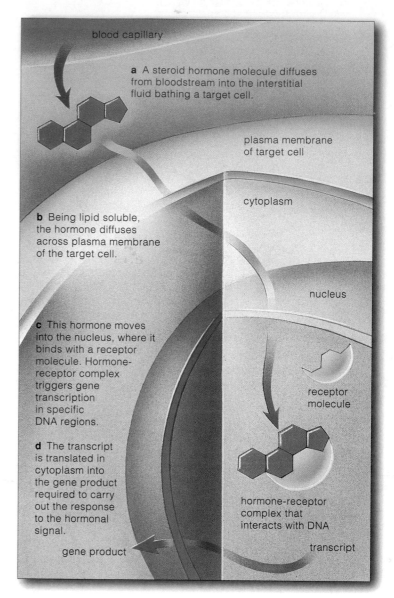

a A steroid hormone molecule diffuses from bloodstream into the interstitial fluid bathing a target cell.

blood capillary

plasma membrane of target cell

cytoplasm

b Being lipid soluble, the hormone diffuses across plasma membrane of the target cell.

nucleus

c This hormone moves into the nucleus, where it binds with a receptor molecule. Hormone-receptor complex triggers gene transcription in specific DNA regions.

receptor molecule

d The transcript is translated in cytoplasm into the gene product required to carry out the response to the hormonal signal.

hormone-receptor complex that interacts with DNA

gene product

transcript

Figure 31.3 Example of a mechanism by which a steroid hormone initiates changes in a target cell's activities.

able to secrete testosterone. Target cells cannot respond to the hormone, however, so secondary sexual traits that do develop are like those of females.

How does a steroid hormone such as testosterone exert effects on cells? Being lipid soluble, it may diffuse directly across the lipid bilayer of a target cell's plasma membrane (Figure 31.3). Once it is inside the cytoplasm, the hormone molecule usually moves into the nucleus and binds to some type of protein receptor. Or it binds to a receptor molecule in the cytoplasm in some cases, then the hormone-receptor complex enters the nucleus. The shape of the complex allows it to interact with a specific region of the cell's DNA. Different complexes inhibit or stimulate transcription of target gene regions

Table 31.1	Two Main Categories of Hormones
Type	Examples
Steroid and steroid-like hormones	Estrogens (feminizing effects), progestins (related to pregnancy), androgens (such as testosterone; masculinizing effects), cortisol, aldosterone. In addition, thyroid hormones and vitamin D act like steroids.
Peptide hormones:	
Peptides	Glucagon, ADH, oxytocin, TRH
Proteins	Insulin, somatotropin, prolactin
Glycoproteins	FSH, LH, TSH

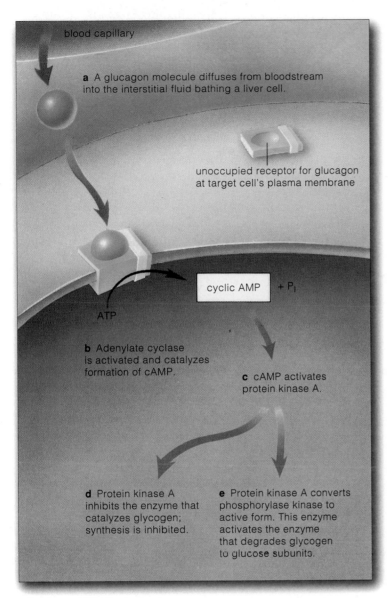

blood capillary

a A glucagon molecule diffuses from bloodstream into the interstitial fluid bathing a liver cell.

unoccupied receptor for glucagon at target cell's plasma membrane

cyclic AMP + P$_i$

ATP

b Adenylate cyclase is activated and catalyzes formation of cAMP.

c cAMP activates protein kinase A.

d Protein kinase A inhibits the enzyme that catalyzes glycogen; synthesis is inhibited.

e Protein kinase A converts phosphorylase kinase to active form. This enzyme activates the enzyme that degrades glycogen to glucose subunits.

Figure 31.4 Example of a mechanism by which a peptide hormone initiates changes in a target cell's activities. When the hormone glucagon binds at a receptor, it initiates reactions inside the cell. In this case cyclic AMP, which is one type of second messenger, relays the signal into the cell interior.

Characteristics of Peptide Hormones

Traditionally, **peptide hormones** have been defined as water-soluble signaling molecules having anywhere from 3 to 180 amino acids. Many peptides, polypeptides, and glycoproteins fall in this category. Each peptide hormone binds to a receptor at the plasma membrane of a cell. Then the receptor activates specific membrane-bound enzyme systems, which in turn initiate reactions leading to the cellular response.

Let's focus on a liver cell with receptors for glucagon, a peptide hormone. This type of receptor spans the plasma membrane. Part of it extends into the cytoplasm. When a receptor binds glucagon, a **second messenger** is produced. Such messengers are small molecules in the cytoplasm that relay signals from hormone-receptor complexes at the plasma membrane into a cell. In this case, cAMP (cyclic adenosine monophosphate) is the second messenger. Glucagon binding triggers activity at a membrane-bound enzyme system (Figure 31.4).

An activated form of adenylate cyclase, an enzyme, starts a cascade of reactions by converting ATP to cyclic AMP. Many molecules of cyclic AMP form. They act as signals to convert many molecules of a protein kinase, another type of enzyme, to active form. These act on other enzymes, and so on until a final reaction converts stored glycogen in the cell to glucose. In short order, the number of molecules involved in the final response to the glucagon-receptor complex is enormous.

Or consider a muscle cell with receptors for insulin, a protein hormone. Among other things, when insulin binds to the receptor, the complex induces molecules of proteins called glucose transporters to move through the cytoplasm and insert themselves into the plasma membrane. They help cells take up glucose faster. The signal also activates enzymes of glucose storage.

Bear in mind, there are other hormone categories, including the catecholamines such as epinephrine. Like glucagon, epinephrine combines with specific surface receptors and triggers the release of cyclic AMP as a second messenger that assists in the cellular response.

Hormones interact with receptors at the plasma membrane or inside the cytoplasm of target cells. Hormone binding directly influences protein synthesis in target cells.

Steroid hormones interact with a cell's DNA after entering the nucleus directly or after binding with a receptor in the cytoplasm. Some may bind to plasma membrane receptors.

Peptide hormones do not exert their effects by entering a cell. When they bind to a membrane receptor, the binding itself is a signal for enzyme-mediated, intracellular events. Often a second messenger inside the cytoplasm relays the signal into the cell interior.

---

into mRNA. The translation of such mRNA transcripts results in enzymes and other proteins that can carry out a response to the hormonal signal.

Steroid hormones also may exert effects in another way. It now appears some may have receptors on cell membranes and may alter the membrane properties in ways that modify the functions of the target cell.

One other point: Thyroid hormones and vitamin D are not steroid hormones, but they behave like them. Also, the genes coding for their receptors are part of the same group that codes for steroid hormones.

THE HYPOTHALAMUS AND PITUITARY GLAND

Deep in the forebrain is the **hypothalamus**. This brain region monitors internal organs and activities related to their functioning, such as eating. It influences certain forms of behavior, such as sexual behavior. It also secretes some hormones. Suspended from its base by a slender stalk of tissue is a lobed, pea-size gland. This **pituitary gland** and the hypothalamus interact closely as a major neural-endocrine control center.

The pituitary's *posterior* lobe secretes two hormones that are synthesized by the hypothalamus. Its *anterior* lobe produces and secretes its own hormones, most of which help control the release of hormones from other endocrine glands (Table 31.2). The pituitary of many vertebrates—*not* humans—has an intermediate lobe as well. In many cases, the third lobe secretes a hormone that governs reversible changes in skin or fur color.

Posterior Lobe Secretions

Figure 31.5*a* shows the cell bodies of some neurons in the hypothalamus. Their axons extend down into the posterior lobe, then terminate next to a capillary bed. The neurons produce antidiuretic hormone (ADH) and oxytocin, then store them in the axon endings. After either hormone is released into the interstitial fluid, it diffuses into capillaries, then travels the bloodstream to targets. ADH acts on cells of nephrons and collecting ducts in the kidneys, which filter blood and rid the body of excess water and salts (in urine). ADH promotes water reabsorption when the body must conserve water. Oxytocin has certain roles in reproduction. It triggers contractions of the uterus during labor, and it causes milk release when offspring are being nursed.

Anterior Lobe Secretions

ANTERIOR PITUITARY HORMONES Inside the pituitary stalk, a capillary bed picks up hormones secreted by the hypothalamus and delivers them into a *second* capillary bed in the anterior lobe. There, the hormones leave the bloodstream and then act on target cells. As Figure 31.6 shows, different cells of the anterior pituitary secrete six hormones that they themselves produce:

Corticotropin	ACTH
Thyrotropin	TSH
Follicle-stimulating hormone	FSH
Luteinizing hormone	LH
Prolactin	PRL
Somatotropin (or growth hormone)	STH (or GH)

All of these hormones have widespread effects through the body. ACTH and TSH orchestrate secretions from the adrenal glands and thyroid gland, respectively, as described shortly. FSH and LH act through the gonads

Table 31.2 Hormones Released From the Mammalian Pituitary Gland

Pituitary Lobe	Secretions	Designation	Main Targets	Primary Actions
POSTERIOR Nervous tissue (extension of hypothalamus)	Antidiuretic hormone	ADH	Kidneys	Induces water conservation as required during control of extracellular fluid volume and solute concentrations
	Oxytocin	OCT	Mammary glands	Induces milk movement into secretory ducts
			Uterus	Induces uterine contractions
ANTERIOR Glandular tissue, mostly	Corticotropin	ACTH	Adrenal cortex	Stimulates release of cortisol, an adrenal steroid hormone
	Thyrotropin	TSH	Thyroid gland	Stimulates release of thyroid hormones
	Follicle-stimulating hormone	FSH	Ovaries, testes	In females, stimulates estrogen secretion, egg maturation; in males, helps stimulate sperm formation
	Luteinizing hormone	LH	Ovaries, testes	In females, stimulates progesterone secretion, ovulation, corpus luteum formation; in males, stimulates testosterone secretion; sperm release
	Prolactin	PRL	Mammary glands	Stimulates and sustains milk production
	Somatotropin (or growth hormone)	STH (GH)	Most cells	Promotes growth in young; induces protein synthesis, cell division; roles in glucose, protein metabolism in adults
INTERMEDIATE* Glandular tissue, mostly	Melanocyte-stimulating hormone	MSH	Pigmented cells in skin and other integuments	Induces color changes in response to external stimuli; affects some behaviors

* Present in most vertebrates (not humans). MSH is associated with the anterior lobe in humans.

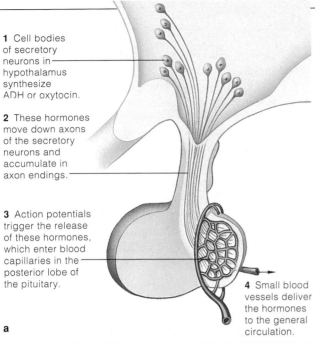

1 Cell bodies of secretory neurons in hypothalamus synthesize ADH or oxytocin.

2 These hormones move down axons of the secretory neurons and accumulate in axon endings.

3 Action potentials trigger the release of these hormones, which enter blood capillaries in the posterior lobe of the pituitary.

4 Small blood vessels deliver the hormones to the general circulation.

a

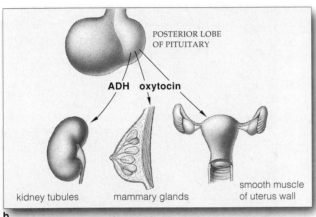

POSTERIOR LOBE OF PITUITARY

ADH   oxytocin

kidney tubules   mammary glands   smooth muscle of uterus wall

b

Figure 31.5 (**a**) Functional links between the hypothalamus and the posterior lobe of the pituitary gland. (**b**) Main targets of the posterior lobe secretions.

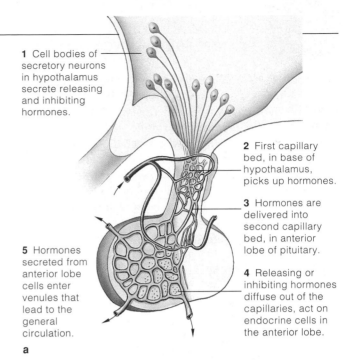

1 Cell bodies of secretory neurons in hypothalamus secrete releasing and inhibiting hormones.

2 First capillary bed, in base of hypothalamus, picks up hormones.

3 Hormones are delivered into second capillary bed, in anterior lobe of pituitary.

4 Releasing or inhibiting hormones diffuse out of the capillaries, act on endocrine cells in the anterior lobe.

5 Hormones secreted from anterior lobe cells enter venules that lead to the general circulation.

a

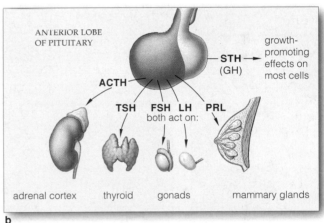

ANTERIOR LOBE OF PITUITARY

STH (GH) growth-promoting effects on most cells

ACTH

TSH   FSH LH   PRL
both act on:

adrenal cortex   thyroid   gonads   mammary glands

b

Figure 31.6 (**a**) Functional links between the hypothalamus and the anterior lobe of the pituitary gland. (**b**) Main targets of the anterior lobe secretions.

to influence gamete formation and secretion of the sex hormones required in sexual reproduction, the central topic of Chapter 38. Somatotropin affects metabolism in many tissues and triggers liver secretions that affect growth of bone and soft tissues (Table 31.2 and Figure 31.6*b*). Prolactin acts on different cell types. However, it is best known for stimulating and then sustaining milk production in mammary glands, after other hormones prime the tissues. In some species, prolactin also affects hormone production in the ovaries.

REGARDING THE HYPOTHALAMIC TRIGGERS   Most of the hypothalamic hormones acting in the anterior lobe of the pituitary are **releasers**, meaning they stimulate secretion of hormones from target cells. For example, GnRH (gonadotropin-releasing hormone) brings about secretion of FSH and LH, which are gonadotropins. As

another example, TRH (thyrotropin-releasing hormone) stimulates the secretion of thyrotropin.

Other hypothalamic hormones are **inhibitors** of the secretions from their targets in the anterior pituitary. For instance, the one called somatostatin brings about a decrease in somatotropin and thyrotropin secretion.

The hypothalamus and pituitary gland produce eight kinds of hormones and interact to control their secretion.

The posterior lobe of the pituitary stores and secretes two hypothalamic hormones, ADH and oxytocin, both of which target specific cell types.

The anterior lobe of the pituitary produces and secretes six hormones, ACTH, TSH, FSH, LH, PRL, and STH. These trigger release of other hormones from other glands, with a variety of effects throughout the body.

WWW

EXAMPLES OF ABNORMAL PITUITARY OUTPUT

The body does not churn out enormous quantities of hormone molecules. Two researchers, Roger Guilleman and Andrew Schally, realized this when they isolated the first known releasing hormone. In their four-year attempt to secure TRH, they dissected 500 metric tons of brains and 7 metric tons of hypothalamic tissue from sheep and ended up with only a single milligram of it.

Yet normal body function depends on the tiny but significant amounts of endocrine gland secretions, which commonly are released in short bursts. Elegant controls over the frequency of secretion prevent the underproduction as well as the overproduction of a hormone. If something interferes with the controls, the body's form and functioning may be altered in abnormal ways.

For instance, *gigantism* results from overproduction of somatotropin. Affected adults are proportionally like an average-size person but larger (Figure 31.7a,b). *Pituitary dwarfism* results from underproduction of somatotropin. Affected adults are proportionally similar to an average person but much smaller (Figure 31.7b).

What if somatotropin output becomes excessive in adulthood, when long bones no longer are lengthening? Bone, cartilage, and other connective tissues in hands, feet, and jaws thicken abnormally. So do epithelia of the skin, nose, eyelids, lips, and tongue. This outcome is called *acromegaly* (Figure 31.7c).

Figure 31.7 Examples of the outcome of abnormalities in the secretion of a hormone.

(**a**) Manute Bol, an NBA center, is 7 feet 6-3/4 inches tall owing to excessive secretion of somatotropin (STH) during childhood.

(**b**) More examples of the effect of STH on body growth. The male at the center of this photograph is affected by gigantism, which resulted from excessive STH production in childhood. The person at the right displays pituitary dwarfism, which resulted from underproduction of STH in childhood. The person at the left is of average height.

(**c**) Acromegaly, which resulted from excessive production of STH during adulthood. Before this female reached maturity, she was symptom-free.

As another example, ADH secretion may dwindle or stop if the pituitary's posterior lobe is damaged, as by a blow to the head. This is one cause of *diabetes insipidus*. Symptoms include copious excretion of dilute urine and life-threatening dehydration. Patients respond to hormone replacement therapy based on either injections or nasal spray applications of synthetic ADH.

Generally, endocrine glands release very small amounts of hormones. The frequency of those releases depends on control mechanisms. When controls fail, the resulting oversecretion or undersecretion may cause disorders.

---

www

SOURCES AND EFFECTS OF OTHER HORMONES

Table 31.3 lists hormones from endocrine sources other than the pituitary. The remainder of this chapter will provide you with a few examples of their effects and of the controls over their output. The examples will make more sense if you keep the following points in mind.

First, hormones often interact with one another. In other words, one or more hormones may oppose, add to, or prime target cells for another hormone's effects. *Second*, negative feedback mechanisms often control the secretions. When a hormone's concentration increases or decreases in some body region, the change triggers events that respectively dampen or stimulate further secretion. *Third*, a target cell may react differently to a hormone at different times. Its response depends on the hormone's concentration as well as on the functional state of the cell's receptors. *Fourth*, environmental cues may be important mediators of hormonal secretion.

The secretion of a hormone and its effects are influenced by hormone interactions, feedback mechanisms, variations in the state of target cells, and sometimes environmental cues.

Table 31.3   Hormone Sources Other Than the Mammalian Hypothalamus and Pituitary

Source	Secretion(s)	Main Targets	Primary Actions
ADRENAL CORTEX	Glucocorticoids (including cortisol)	Most cells	Promote protein breakdown and conversion to glucose
	Mineralocorticoids (including aldosterone)	Kidney	Promote sodium reabsorption (sodium conservation); help control the body's salt–water balance
ADRENAL MEDULLA	Epinephrine (adrenaline)	Liver, muscle, adipose tissue	Raises blood level of sugar, fatty acids; increases heart rate and force of contraction
	Norepinephrine	Smooth muscle of blood vessels	Promotes constriction or dilation of blood vessel diameter
THYROID	Triiodothyronine, thyroxine	Most cells	Regulate metabolism; have roles in growth, development
	Calcitonin	Bone	Lowers calcium level in blood
PARATHYROIDS	Parathyroid hormone	Bone, kidney	Elevates calcium level in blood
GONADS:			
Testes (in males)	Androgens (including testosterone)	General	Required in sperm formation, development of genitals, maintenance of sexual traits; growth, development
Ovaries (in females)	Estrogens	General	Required for egg maturation and release; preparation of uterine lining for pregnancy and its maintenance in pregnancy; genital development; maintenance of sexual traits; growth, development
	Progesterone	Uterus, breasts	Prepares, maintains uterine lining for pregnancy; stimulates development of breast tissues
PANCREATIC ISLETS	Insulin	Liver, muscle, adipose tissue	Lowers sugar level in blood
	Glucagon	Liver	Raises sugar level in blood
	Somatostatin	Insulin-secreting cells	Inhibits digestion of nutrients, hence their absorption from gut
THYMUS	Thymosins, etc.	Lymphocytes	Have roles in immune responses
PINEAL	Melatonin	Gonads (indirectly)	Influences daily biorhythms, seasonal sexual activity
STOMACH, SMALL INTESTINE	Gastrin, secretin, etc.	Stomach, pancreas, gallbladder	Stimulate activities of stomach, pancreas, liver, gallbladder required for food digestion, absorption
LIVER	Somatomedins	Most cells	Stimulate cell growth and development
KIDNEYS	Erythropoietin	Bone marrow	Stimulates red blood cell production
	Angiotensin*	Adrenal cortex, arterioles	Helps control secretion of aldosterone (hence sodium reabsorption, and blood pressure)
	1,25-hydroxyvitamin D_6* (calcitriol)	Bone, gut	Enhances calcium reabsorption from bone and calcium absorption from gut
HEART	Atrial natriuretic hormone	Kidney, blood vessels	Increases sodium excretion; lowers blood pressure

* Kidneys produce *enzymes* that modify precursors of this substance, which then enters the general circulation as an activated hormone.

By considering just a few of the endocrine glands listed in Table 31.3, you can sense how feedback mechanisms control hormonal secretions. Briefly, the hypothalamus, pituitary, or both signal these glands to alter secretory activity. The outcome is a change in the concentration of a secreted hormone in blood or elsewhere. With the shift in chemical signals, a feedback mechanism trips into action and blocks or promotes further change.

In cases of **negative feedback**, an increase in the concentration of a secreted hormone triggers activities that *inhibit* further secretion. With **positive feedback**, an increase in the concentration of a secreted hormone triggers events that *stimulate* further secretion.

Negative Feedback From the Adrenal Cortex

Humans have a pair of adrenal glands, one above each kidney (Figure 31.2b). Some cells of the **adrenal cortex**, the outer part of an adrenal gland, secrete hormones such as glucocorticoids. Glucocorticoids help increase the level of glucose in blood. Cortisol is one of those hormones. It comes into play when the body is under stress, as when the blood glucose level declines below a set point. Glucose itself provides the information for a negative feedback mechanism that works to counter the decline.

Take a look at Figure 31.8. When the hypothalamus detects the decline, it secretes CRH in response. This releasing hormone stimulates the anterior pituitary to secrete corticotropin (ACTH). In turn, ACTH stimulates cells of the adrenal cortex to secrete cortisol. In response to cortisol, liver cells break down glycogen to glucose. Muscle cells especially break down proteins. Adipose cells break down fats. Now glucose, amino acids, and fatty acids enter the blood and are mobilized as energy sources or building blocks to build or repair damaged cell structures. The hypothalamus and pituitary detect the increase in the blood glucose level and respond by inhibiting cortisol secretion.

During chronic stress, injury, or illness, the nervous system initiates a stress response in which cortisol also helps to suppress inflammation. Unchecked, prolonged inflammation can damage tissues. That is why doctors may prescribe cortisol-like drugs for asthma and other chronic inflammatory disorders.

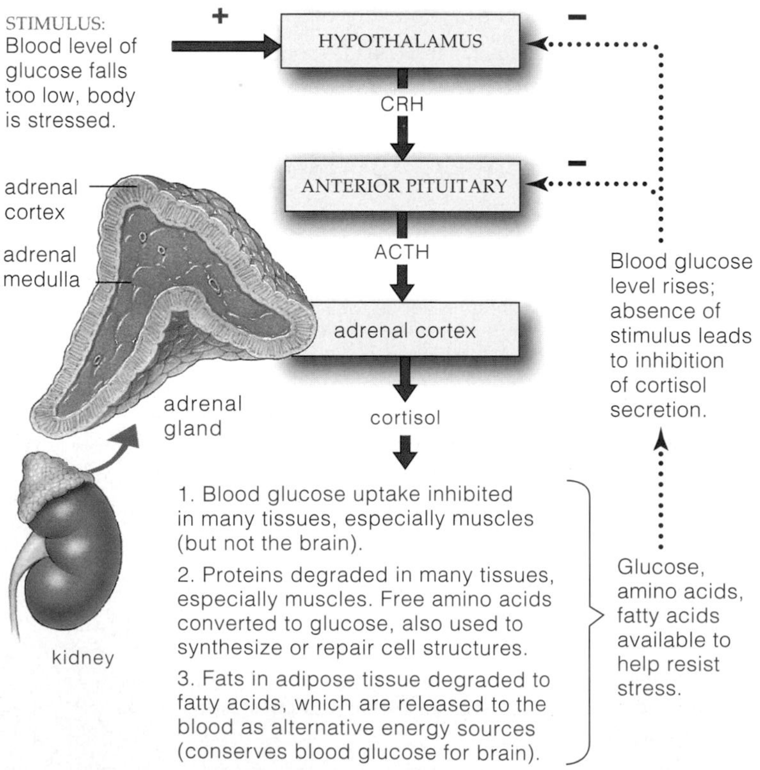

Local Feedback in the Adrenal Medulla

The **adrenal medulla** is the inner part of the adrenal gland (Figure 31.8). It has hormone-secreting neurons that release epinephrine and norepinephrine. (These substances are neurotransmitters in some contexts and hormones in others.) Suppose axons of sympathetic nerves carry hypothalamic signals that call for secretion of norepinephrine. Molecules of norepinephrine collect in the synaptic cleft between the axon endings and the target cells. In this case, a localized negative feedback mechanism operates at receptors on the axon endings. The excess norepinephrine binds to the receptors and causes a shutdown of its further release.

In times of excitement or stress, epinephrine and norepinephrine help adjust blood circulation and fat and carbohydrate metabolism. They increase heart rate, trigger vasoconstriction and vasodilation of arterioles in different regions, and dilate airways to the lungs. The controlled activity directs more of the total volume of blood to heart and muscle cells, and more oxygen flows to energy-demanding cells through the body. These are features of the *fight-flight response* (Section 29.8).

Cases of Skewed Feedback From the Thyroid

The human **thyroid gland** is located at the base of the neck in front of the trachea, or windpipe (Figures 31.2b and 31.9a,b). Thyroxine and triiodothyronine, its main hormones, have widespread effects. In their absence, many tissues cannot develop normally. Also, the overall metabolic rates of warm-blooded animals, including

STIMULUS:
Blood level of glucose falls too low, body is stressed.

+ → HYPOTHALAMUS ←····· −

CRH

ANTERIOR PITUITARY ←····· −

ACTH

adrenal cortex

cortisol

Blood glucose level rises; absence of stimulus leads to inhibition of cortisol secretion.

adrenal cortex

adrenal medulla

adrenal gland

kidney

1. Blood glucose uptake inhibited in many tissues, especially muscles (but not the brain).

2. Proteins degraded in many tissues, especially muscles. Free amino acids converted to glucose, also used to synthesize or repair cell structures.

3. Fats in adipose tissue degraded to fatty acids, which are released to the blood as alternative energy sources (conserves blood glucose for brain).

Glucose, amino acids, fatty acids available to help resist stress.

Figure 31.8 Location of the adrenal glands. One rests on top of each kidney. The diagram shows a negative feedback loop that governs secretion of cortisol from the adrenal cortex.

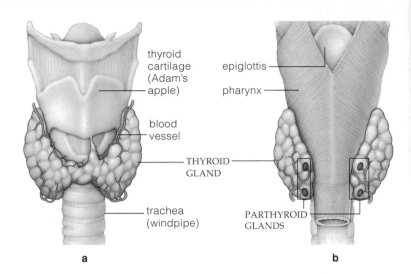

Figure 31.9 Human thyroid gland. (**a**) Anterior and (**b**) posterior views showing the location of four parathyroid glands. (**c**) A mild case of goiter, displayed by Maria de Medici in the year 1625. A rounded neck was considered to be a sign of great beauty during the late Renaissance. It occurred regularly in parts of the world where iodine supplies were insufficient for normal thyroid function.

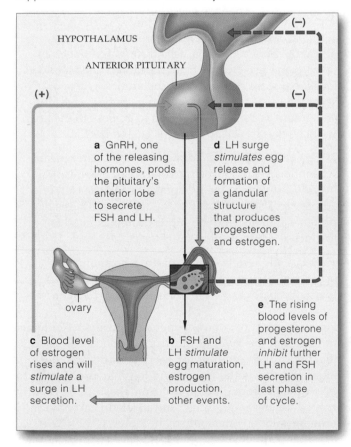

Figure 31.10 Feedback loops to the hypothalamus and pituitary gland from the ovaries during the menstrual cycle, a recurring reproductive event. Positive feedback triggers egg release from an ovary. Negative feedback after its release prevents release of another egg until the cycle is completed.

humans, depend upon them. The importance of feedback control of the secretions of these hormones comes into sharp focus with cases of abnormal output of thyroid hormones.

As a case in point, the synthesis of thyroid hormones requires iodine, which we obtain from food. Iodine is converted to an iodized form, iodide, when absorbed from the gut. Without iodide, the blood levels of thyroid hormones decrease. The anterior pituitary responds by secreting its thyroid-stimulating hormone (TSH). But without iodine, thyroid hormones cannot be made. The feedback signal continues, and so does TSH secretion. Over time, a sustained response is made to the low blood level of TSH level and the thyroid gland enlarges. The enlargement is one form of *goiter*. Goiter resulting from iodine deficiency is no longer common in countries where people use iodized salt (Figure 31.9c).

When blood levels of thyroid hormones are too low, *hypothyroidism* results. Hypothyroid adults commonly are overweight, sluggish, dry-skinned, intolerant of cold, confused, and depressed. Affected women commonly show menstrual disturbances.

When blood levels of the thyroid hormones are too high, *hyperthyroidism* results. Affected adults show an increased heart rate, heat intolerance, elevated blood pressure, profuse sweating, and weight loss even when caloric intake increases. Affected individuals typically are nervous and agitated, and have trouble sleeping.

Feedback Control of the Gonads

Gonads are *primary* reproductive organs, which make gametes and sex hormones, which they secrete. Testes (singular, testis) in males and ovaries in females are examples. Testes secrete testosterone; ovaries secrete estrogens and progesterone. These hormones influence secondary sexual traits (as they did for those chimps described earlier), and feedback controls govern their secretion. Figure 31.10 is a preview of the feedback loops from ovaries to the hypothalamus and pituitary during the menstrual cycle, a key topic of Chapter 38.

Feedback mechanisms control secretions from endocrine glands. In many cases, feedback loops to the hypothalamus, pituitary, or both govern the secretory activity.

With negative feedback, further secretion of a hormone slows down. With positive feedback, further secretion is enhanced.

WWW

Some endocrine glands or cells don't respond primarily to signals from other hormones or nerves. They respond homeostatically to chemical changes in the immediate surroundings, as the following examples illustrate.

Secretions From Parathyroid Glands

Humans have four **parathyroid glands**, located on the posterior surface of the thyroid gland (Figure 31.9b). The glands secrete parathyroid hormone, or PTH, the main regulator of the calcium level in blood. Calcium ions, recall, have roles in muscle contraction, enzyme action, blood clot formation, and other tasks. The parathyroids secrete PTH in response to a low calcium level in blood. Their secretory activity slows when the calcium level rises. PTH acts on cells of the skeleton and kidneys.

PTH induces living bone cells to secrete enzymes that digest bone tissue and thereby release calcium and other minerals to interstitial fluid, then to blood. It enhances calcium reabsorption from the filtrate flowing through the nephrons of kidneys. PTH also prods some kidney cells to secrete enzymes that act on blood-borne precursors of the active form of vitamin D_3, a hormone (Table 31.3). The vitamin D_3 stimulates intestinal cells to increase the absorption of calcium from the lumen of the gut. In children who have vitamin D deficiency, not enough calcium and phosphorus is absorbed so the rapidly growing bones develop improperly. The resulting bone disorder, *rickets*, is characterized by bowed legs, a malformed pelvis, and in many cases a malformed skull and rib cage (Figure 31.11).

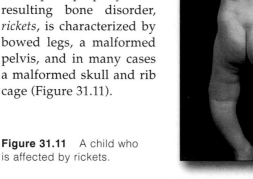

Figure 31.11 A child who is affected by rickets.

Effects of Local Signaling Molecules

Many cells detect changes in the surrounding chemical environment and alter their activity, often in ways that counteract or amplify those changes. The cells secrete various local signaling molecules, the effects of which are confined to the immediate vicinity of change. Target cells take up most signaling molecules so rapidly that few enter the general circulation.

Local signaling molecules include prostaglandins. Cells in many tissues continually produce and release a variety of prostaglandins. The rate of synthesis often increases in response to local chemical changes. Sections 38.6 and 38.9 include fine examples of this response.

Other examples are growth factors, which influence cell division rates in tissues. An epidermal growth factor (EGF) discovered by Stanley Cohen acts on many cell types. A nerve growth factor (NGF) discovered by Rita Levi-Montalcini helps neurons survive and guides their direction of growth in embryos. Experimenters have demonstrated that certain immature neurons survive indefinitely in tissue culture when NGF is present but die within a few days when it is absent.

Secretions From Pancreatic Islets

The pancreas is one gland with exocrine and endocrine functions. Its *exocrine* cells secrete digestive enzymes into the small intestine. It also contains about 2 million clusters of *endocrine* cells. Each tiny cluster, a **pancreatic islet**, contains three types of hormone-secreting cells:

1. *Alpha* cells in the pancreas secrete the hormone glucagon. Between meals, cells throughout the body take up and use glucose from the blood. The blood level of glucose decreases. At such times, glucagon secretion causes glycogen (a storage polysaccharide) and amino acids to be converted to glucose in the liver. In such ways, *glucagon raises the glucose level*.

2. *Beta* cells secrete the hormone insulin. After meals, when the level of glucose circulating in blood is high, insulin stimulates glucose uptake by muscle cells and adipose cells especially. It promotes the synthesis of proteins and fats, and it inhibits protein conversion to glucose. Thus, *insulin lowers the glucose level*.

3. *Delta* cells secrete somatostatin, a hormone that helps control digestion and absorption of nutrients. It also can block secretion of insulin and glucagon.

Figure 31.12 shows how pancreatic hormones interact to maintain the level of glucose in blood even though the times and amounts of food intake vary. Bear in mind, insulin is the only hormone that prods cells to take up and store glucose in forms that can be rapidly tapped when required. Its central role in carbohydrate, protein, and fat metabolism becomes clear when we consider people who cannot produce enough insulin or who lack body cells that can respond to it.

For example, insulin deficiency may lead to *diabetes mellitus*, a disorder in which excess glucose accumulates in blood, then in urine. Urination becomes excessive, so the body's water-solute balance is disrupted. Affected

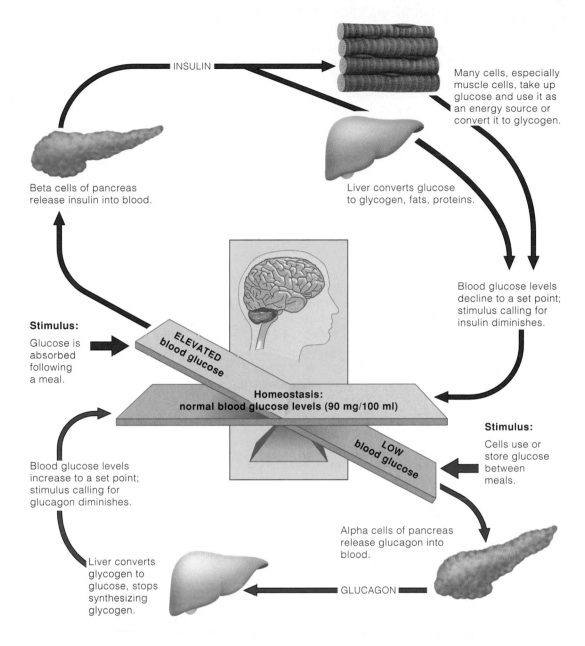

Figure 31.12 Some of the homeostatic controls over glucose metabolism.

Following a meal, glucose enters the bloodstream faster than cells can use it. The blood glucose level rises, and pancreatic beta cells are stimulated to secrete insulin. Insulin's main targets—liver and muscle cells—not only use glucose but store excess amounts of it in the form of glycogen.

Between meals, the blood glucose level decreases. Pancreatic alpha cells are stimulated to secrete glucagon. The target cells with receptors for this hormone convert glycogen back to glucose, which enters the blood.

Glucose metabolism also is subject to indirect controls. For example, the hypothalamus commands the adrenal medulla to secrete certain hormones. The hormones speed the conversion of glycogen to glucose in the liver and slow the reverse process, especially in cells of the liver and muscle tissue.

INSULIN

Many cells, especially muscle cells, take up glucose and use it as an energy source or convert it to glycogen.

Beta cells of pancreas release insulin into blood.

Liver converts glucose to glycogen, fats, proteins.

Stimulus:
Glucose is absorbed following a meal.

ELEVATED blood glucose

Blood glucose levels decline to a set point; stimulus calling for insulin diminishes.

**Homeostasis:
normal blood glucose levels (90 mg/100 ml)**

LOW blood glucose

Stimulus:
Cells use or store glucose between meals.

Blood glucose levels increase to a set point; stimulus calling for glucagon diminishes.

Alpha cells of pancreas release glucagon into blood.

Liver converts glycogen to glucose, stops synthesizing glycogen.

GLUCAGON

people become dehydrated and thirsty—abnormally so. Without a steady supply of glucose, their body cells start depleting their own fats and proteins as sources of energy. Weight loss is one outcome. Another is that ketones accumulate in the blood and urine. Ketones are normal acidic products of fat breakdown. When they accumulate, they contribute to water losses and alter the body's acid-base balance. Such imbalances disrupt brain function. In extreme cases, death may follow.

In "type 1 diabetes," the body mistakenly mounts an autoimmune response against its insulin-secreting beta cells. Certain lymphocytes identify the beta cells as foreign and destroy them. A combination of genetic susceptibility and environmental triggers produces the disorder, which is less common but more immediately dangerous than other forms of diabetes. Usually, the

symptoms first appear in childhood and adolescence (the disorder is also known as juvenile-onset diabetes). Type 1 diabetic patients survive with insulin injections.

In "type 2 diabetes," insulin levels are close to or above normal, but the target cells are not equipped to respond to insulin. As affected persons grow older, their beta cells produce less and less insulin. Symptoms of type 2 diabetes usually emerge during middle age. Affected persons lead normal lives by controlling their diet and weight, and sometimes by taking prescription drugs to enhance insulin action or secretion.

The secretions from some endocrine glands and cells are direct homeostatic responses to a change in the localized chemical environment.

WWW

This last section of the chapter invites you to reflect on a key point. An individual's growth, development, and reproduction begin with genes and hormones, and so does behavior. *But certain environmental factors commonly influence gene expression and hormonal secretion, and they do so in predictable ways.* Chapter 44 invites analysis of the environmental influence on animal behavior. For now, it is enough to consider the following examples.

Daylength and the Pineal Gland

Embedded in the brain is a photosensitive organ, the **pineal gland** (Section 29.9). In the absence of light, the gland secretes the hormone melatonin. Thus the level of melatonin in the blood varies from day to night, and over the seasons. The variations influence the growth and development of gonads, the primary reproductive organs. In a variety of species, they have important roles in reproductive cycles and reproductive behavior.

Think about a hamster. In winter, when nighttime darkness is longest, the blood level of melatonin is high and sexual activity is suppressed. In the summer, when daylength is longest, the blood level of melatonin is low and hamster sex reaches its peak. Or think about a male white-throated sparrow (Figure 31.13a). In the fall and winter, melatonin indirectly suppresses growth of its gonads by inhibiting gonadotropin secretion. It does so until days start to lengthen in spring. Now, stepped-up gonadal activity leads to production of hormones that influence singing behavior, as described in Section 44.1. With his distinctive song, the male sparrow defines his territory and may hold the interest of a mate.

Does melatonin also influence human behavior? Perhaps. Clinical observations and studies suggest that decreased melatonin secretion might trigger **puberty**, the age during which human reproductive organs and structures start to mature. For example, in cases where disease caused the destruction of an individual's pineal gland, puberty began prematurely.

Melatonin is known to act on certain neurons that lower your body's core temperature and that make you drowsy after sunset, when light is waning. At sunrise, when melatonin secretion slows, your core temperature increases and you wake up and become active.

An internal, biological clock governs the cycle of sleep and arousal. It seems to tick in synchrony with daylength. Think of night workers who try to sleep in the morning but end up staring groggily at sunbeams on the ceiling. Think of travelers from the United States to Paris who go through four days of "jet lag." Two or three hours past midnight they are sitting up in bed, wondering where the coffee and croissants are. Two hours past noon they are ready for bed. They will shift to a new routine when the melatonin signals arrive at their target neurons on Paris time.

In winter, some individuals experience *winter blues.* They get abnormally depressed, go on carbohydrate binges, and have an almost overwhelming desire to sleep (Figure 31.13b). Winter blues might arise when a biological clock is out of sync with the seasonally shorter daylengths. Intriguingly, clinically administered doses of melatonin make the seasonal symptoms worse. And exposure to intense light, which can shut down pineal activity, may lead to dramatic improvement.

Figure 31.13 (a) A male white-throated sparrow, belting out a song that began, indirectly, with an environmentally induced decline in melatonin secretion. (b) Annie blanketing her winter blues.

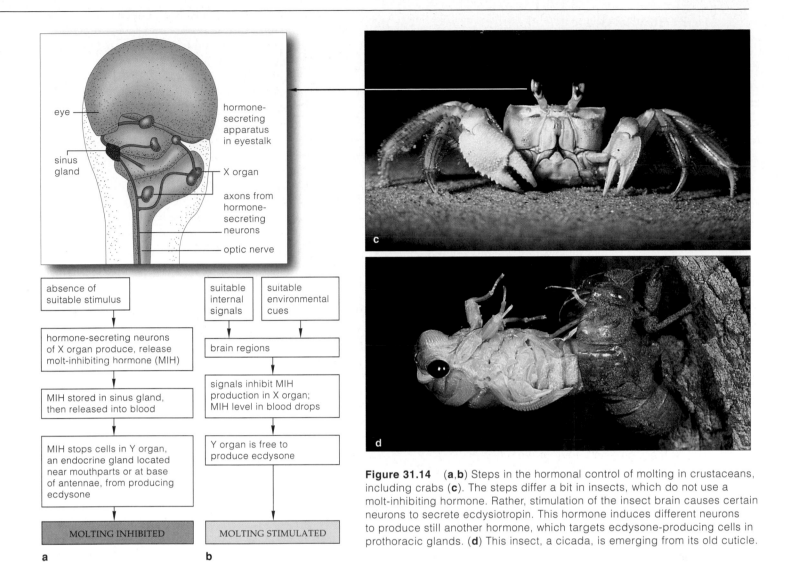

Figure 31.14 (a,b) Steps in the hormonal control of molting in crustaceans, including crabs (c). The steps differ a bit in insects, which do not use a molt-inhibiting hormone. Rather, stimulation of the insect brain causes certain neurons to secrete ecdysiotropin. This hormone induces different neurons to produce still another hormone, which targets ecdysone-producing cells in prothoracic glands. (d) This insect, a cicada, is emerging from its old cuticle.

Flow chart labels (panel a, b, and eyestalk diagram):

eye
hormone-secreting apparatus in eyestalk
sinus gland
X organ
axons from hormone-secreting neurons
optic nerve

absence of suitable stimulus

hormone-secreting neurons of X organ produce, release molt-inhibiting hormone (MIH)

MIH stored in sinus gland, then released into blood

MIH stops cells in Y organ, an endocrine gland located near mouthparts or at base of antennae, from producing ecdysone

MOLTING INHIBITED

a

suitable internal signals   suitable environmental cues

brain regions

signals inhibit MIH production in X organ; MIH level in blood drops

Y organ is free to produce ecdysone

MOLTING STIMULATED

b

Comparative Look at a Few Invertebrates

Although this chapter's focus has been on vertebrates, do not lose sight of the fact that all organisms produce signaling molecules of one sort or another. Let's look at hormonal control of **molting**, a periodic discarding and replacement of a hardened cuticle that otherwise would limit increases in body mass. As described in Section 23.11, molting occurs during the life cycle of all insects, crustaceans, and other invertebrates with thick cuticles.

Although details vary from group to group, molting is largely under the control of ecdysone. This steroid hormone is derived from cholesterol and is chemically related to many important vertebrate hormones. In the insects and crustaceans, molting glands produce and store ecdysone, then release it for distribution through the body at molting time. Hormone-secreting neurons in the brain seem to regulate its release. The hormone-secreting neurons apparently respond to a combination of environmental cues, including light and temperature, as well as internal signals.

Figure 31.14 gives examples of the control steps, which differ in crustaceans and insects. During premolt and molting periods, coordinated interactions among ecdysone and other hormones bring about structural and physiological changes. The interactions make the old cuticle separate from the epidermis and muscles. They induce events that dissolve the inner layers of the cuticle and recycle the remnants. The interactions also trigger shifts in metabolism and in the composition and volume of the internal environment. They promote cell divisions, secretions, and pigment formation, all of which go into producing a new cuticle. Simultaneously, hormonal interactions control heart rate, muscle action, color changes, and other physiological processes.

Environmental cues, such as changes in light intensity from day to night and seasonal changes in daylength, influence certain hormonal secretions.

SUMMARY

1. The cells of complex animals continually exchange substances with the body's internal environment. Their myriad withdrawals and secretions are integrated in ways that ensure cell survival through the whole body.

2. Integration of cell activities requires the stimulatory or inhibitory effects of signaling molecules.

   a. Signaling molecules are chemical secretions from a cell that adjust the behavior of other, target cells.

   b. Any cell with molecular receptors for a signaling molecule is a target. The target cells may or may not be next to the cell that sends the signal.

   c. There are different kinds of signaling molecules. Hormones as well as neurotransmitters, local signaling molecules, and pheromones are the main kinds.

   d. Certain steroids, steroidlike molecules, amines, peptides, proteins, and glycoproteins are hormones.

3. In target cells, hormones influence gene activation, protein synthesis, and alterations in existing enzymes, membranes, and other cellular components. Hormones exert their physiological effects through interactions with specific protein receptors at the plasma membrane or in the cytoplasm of target cells.

   a. Steroid hormones enter a target cell's nucleus directly or after binding with an intracellular receptor; some may bind with plasma membrane receptors.

   b. Being water soluble, the protein hormones cannot enter target cells. They bind to membrane receptors at the cell surface. Their effect is exerted with the help of transport proteins and of second messengers inside the cytoplasm, some of which trigger the actual response.

4. The posterior lobe of the pituitary stores and secretes two hypothalamic hormones, ADH and oxytocin. ADH targets cells in kidneys and affects extracellular fluid volume. Oxytocin acts on cells in mammary glands and the uterus to influence reproductive events.

5. The hypothalamic hormones called releasing and inhibiting hormones control secretions from different cells of the anterior lobe of the pituitary gland.

6. The anterior lobe makes and secretes six hormones, ACTH, TSH, FSH, LH, PRL, and STH. These trigger secretions from the adrenal cortex, thyroid, gonads, and mammary glands. They exert a variety of responses throughout the body.

7. The vertebrate body has other sources of hormones, including the adrenal medulla, parathyroid, thymus, and pineal glands; pancreatic islets; and endocrine cells in the stomach, small intestine, liver, and heart.

8. Hormone interactions, feedback mechanisms, the number and kind of receptors on target cells, variations in the state of target cells, and often environmental cues influence secretion of a hormone and its effects.

9. Many cellular responses to hormones help the body adjust to short-term shifts in diet and levels of activity. Other kinds help bring about long-term adjustments for growth, development, and reproduction.

   a. In general, secretion of hormones such as insulin and parathyroid hormone can change rapidly when the extracellular concentration of some substance must be controlled homeostatically.

   b. Hormones such as somatotropin have prolonged, gradual, often irreversible effects, as on development.

Review Questions

1. Name the endocrine glands that occur in most vertebrates and state where each is located in the human body. *31.1*

2. Distinguish among hormones, neurotransmitters, local signaling molecules, and pheromones. *31.1*

3. A hormone molecule binds to a receptor on a cell membrane. It does not enter the cell. Binding activates a second messenger in the cell, which triggers an amplified response to the hormonal signal. State whether the molecule is a steroid hormone or a peptide hormone. *31.2*

4. Which secretions of the posterior lobe of the pituitary gland have the targets indicated? (Fill in the blanks.) *31.3*

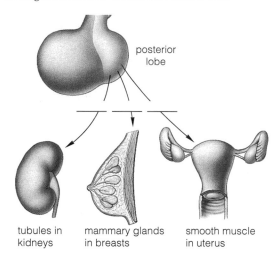

posterior lobe

tubules in kidneys   mammary glands in breasts   smooth muscle in uterus

5. Which secretions of the anterior lobe of the pituitary gland have the targets indicated? (Fill in the blanks.) *31.3*

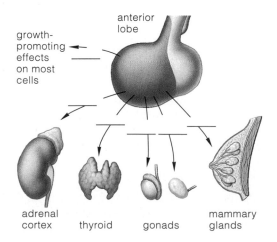

anterior lobe

growth-promoting effects on most cells

adrenal cortex   thyroid   gonads   mammary glands

Self-Quiz (Answers in Appendix III)

1. _____ are molecules released from a signaling cell that have effects on target cells.
   a. Hormones
   b. Neurotransmitters
   c. Pheromones
   d. Local signaling molecules
   e. both a and b
   f. a through d

2. Hormones are products of _____ .
   a. endocrine glands
   b. some neurons
   c. exocrine cells
   d. a and b
   e. a and c
   f. a, b, and c

3. ADH and oxytocin are hypothalamic hormones secreted from the _____ lobe of the pituitary gland.
   a. anterior
   b. posterior
   c. intermediate
   d. secondary

4. _____ has effects on body tissues in general.
   a. ADH
   b. Oxytocin
   c. Bombykol
   d. Somatotropin

5. Which do *not* stimulate hormone secretions?
   a. neural signals
   b. local chemical changes
   c. hormonal signals
   d. environmental cues
   e. All of the above can stimulate secretion

6. _____ lowers blood sugar levels; _____ raises it.
   a. Glucagon; insulin
   b. Insulin; glucagon
   c. Gastrin; insulin
   d. Gastrin; glucagon

7. The pituitary detects a rising hormone concentration in blood and inhibits the gland secreting the hormone. This is a _____ feedback loop.
   a. positive
   b. negative
   c. long-term
   d. b and c

8. Second messengers include _____ .
   a. steroid hormones
   b. protein hormones
   c. cyclic AMP
   d. both a and b

9. Match the hormone source with the closest description.
   _____ adrenal gland
   _____ thyroid gland
   _____ parathyroids
   _____ pancreatic islets
   _____ pineal gland
   _____ thymus gland
   a. affected by daylength
   b. key roles in immunity
   c. raise blood calcium level
   d. epinephrine source
   e. insulin, glucagon
   f. hormones require iodine

Critical Thinking

1. The zebra offspring being nursed in Figure 31.15 is too young to nourish itself by eating grasses. Its source of nutrients is its mother's milk. Explain how secretions from the hypothalamus and both lobes of the pituitary gland influence the production and secretion of milk.

2. In winter, with its far fewer daylight hours compared to the summer, Maxine became very depressed, craved carbohydrate-rich foods, and stopped exercising regularly. And she put on a great deal of weight. Her doctor diagnosed her condition as *seasonal affective disorder* (SAD), or the winter blues. Maxine was advised to purchase a cluster of intense, broad-spectrum lights and to sit near them for at least an hour every day. The cloud of depression started to lift quickly. Use your understanding of the secretory activity of the pineal gland to explain why Maxine's symptoms appeared and why the prescribed therapy worked.

3. Marianne is affected by *type 1 insulin-dependent diabetes*. One day, after injecting herself with too much insulin, she starts to shake and feels confused. Her doctor recommends a glucagon injection. What caused her symptoms? How would an injection of glucagon help?

Figure 31.15 Female zebra nursing her offspring.

4. By application of recombinant DNA technology, somatotropin (growth hormone) is now commercially available for treating pituitary dwarfism. Although it is illegal to do so, some athletes use somatotropin instead of anabolic steroids. (Refer to *Critical Thinking Question 4* in Section 32.9.) Why? Somatotropin cannot be detected by the drug test procedures employed in sports medicine. Explain how the athletes might believe this hormone can improve their performance.

5. *Osteoporosis* is a condition in which loss of calcium results in thin, brittle bones. Combined with other treatments, vitamin D_3 injections are sometimes recommended. Explain why.

Selected Key Terms

adrenal cortex *31.6*
adrenal medulla *31.6*
endocrine system *31.1*
gonad *31.6*
hormone *31.1*
hypothalamus *31.3*
inhibitor (hypothalamic) *31.3*
local signaling molecule *31.1*
molting *31.8*
negative feedback *31.6*
neurotransmitter *31.1*
pancreatic islet *31.7*
parathyroid gland *31.7*
peptide hormone *31.2*
pheromone *31.1*
pineal gland *31.8*
pituitary gland *31.3*
positive feedback *31.6*
puberty *31.8*
releaser (hypothalamic) *31.3*
second messenger *31.2*
steroid hormone *31.2*
thyroid gland *31.6*

Readings *See also www.infotrac-college.com*

Goodall, J. 1986. *The Chimpanzees of Gombe*. Cambridge, Massachusetts: Belknap Press of Harvard University Press.

Goodman, H. 1994. *Basic Medical Endocrinology*. Second edition. New York: Raven Press.

Hadley, M. 1995. *Endocrinology*. Fourth edition. Englewood Cliffs, New Jersey: Prentice-Hall.

Sherwood, L. 1997. *Human Physiology*. Third edition. Belmont, California: Wadsworth.

WWW   *http://www.brookscole.com/biology*

Practice quiz questions, hypercontents, BioUpdates, and critical thinking. The Brooks/Cole Biology Resource Center provides a wealth of information fully organized and integrated by chapter.

32

PROTECTION, SUPPORT, AND MOVEMENT

Of Men, Women, and Polar Huskies

In 1989 Will Steger and his dog-sled team walked on ice for seven months, endured temperatures of −113°F, and lived through a blizzard that lasted for more than seven weeks. They crossed Antarctica—all 6,023 kilometers (3,741 miles) of it. In 1995 that legendary polar explorer set out with four men, two women, and thirty-three sled dogs to cross 3,220 kilometers of the Arctic Ocean in one season. Ice blankets this northernmost ocean in winter, but the ice becomes treacherously thin during the spring thaw. The sled dogs were with the team for two-thirds of the journey. They were flown out only when the team encountered too much melting ice and had to switch to using canoes.

To Steger's mind, the polar huskies were the heroes of the polar crossings, the members of the team that worked hardest and pulled all the weight (Figure 32.1). They are a mixed breed, the traits of which have been modified through years of artificial selection among Canadian and Greenland huskies (bred for size and strength), Siberian huskies (bred for intelligence), and Alaskan racing dogs (bred for spirit and endurance). Steger's polar huskies show a combination of these traits as well as the loyalty of cared-for pets.

A husky's leg bones are sturdy yet lightweight. Its forelegs move freely, thanks to a rib cage that is deep but not too broad. Its hind legs have massive muscles. These are not the muscles of sprinting greyhounds or cheetahs. They are the muscles of a load-pulling, long-distance runner. The husky also has tough, calloused foot pads—cushions against sharp ice and frozen rock. Like many other mammals, it has a fur coat. The coat's underhair, a dense, soft, insulative layer, traps heat. Its coarser, longer, and slightly oily guard hairs protect the insulative layer from wear and tear. On winter nights, the husky settles into a comfortable position and covers its nose with its furry tail, oblivious of drifting snow.

Steger and his teammates, Victor Boyarsky, Julie Hanson, Martin Hignell, Paul Pregont, and Takako Takano, could not even approach the polar husky's stamina and built-in protection against the elements. Long before the polar crossings, they were adhering to a regimen of diet and exercise to put their arm and leg muscles in peak condition for the extraordinary effort that lay ahead. Human legs are not adapted for load-pulling motion, but rather for long-distance walking. Also, human skin cannot withstand bitter

Figure 32.1 In Ely, Minnesota, Will Steger and his polar huskies warming up for an Arctic crossing.

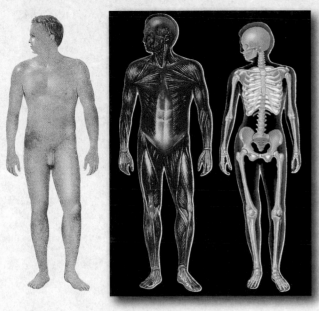

Figure 32.2 From left to right, overview of the human integumentary, muscular, and skeletal systems.

cold. Lacking the fur coat of mammals that evolved in polar climates, the team had to depend on special clothing that could insulate and protect them from cold without restricting body movements. From this perspective, it was human ingenuity that allowed humans to keep company with the huskies, which are supremely adapted for the challenges of life on ice.

With this chapter, we turn to three systems that together are responsible for the superficial features, shape, and movements of most animals. Figure 32.2 serves as the starting point for our consideration of the structural organization and functions of those systems. Traveling from the outside in, it depicts the integumentary system, muscle system, and skeletal system of one of the more familiar vertebrates.

Animals ranging from invertebrate worms to humans have an outer covering, or **integument** (after the Latin *integere*, meaning to cover). Most coverings are tough, pliable barriers against many environmental insults. Integuments of roundworms and insects, crabs, and other arthropods are protective cuticles hardened with chitin. Chitin, remember, is a polysaccharide that incorporates nitrogen atoms (Sections 3.2 and 23.11). For vertebrates, the integument is a covering called **skin**, which includes various structures derived from epidermal cells of the skin's outer tissue layers.

Movement of the animal body or parts of it require contractile cells and some medium or structure against which contractile force can be applied. As you will see, hydrostatic skeletons, exoskeletons, and endoskeletons receive the applied force among different animal groups.

KEY CONCEPTS

1. Nearly all animals have an integument, some type of skeleton, and muscles. An integument is the outer covering of the animal body, and a prime example is vertebrate skin.

2. Skin protects the body from abrasion, ultraviolet radiation, bacterial attack, and other environmental insults. It also contributes to overall body functioning, as when it helps control moisture loss.

3. Three categories of skeletal systems are common in the animal kingdom. We call them hydrostatic skeletons, exoskeletons, and endoskeletons. Each has body fluids or structural elements, such as bones, against which a contractile force can be applied.

4. Bones are collagen-rich, mineralized organs. They function in movement, protection and support of soft organs, and mineral storage. Blood cells form in some. Ligaments or cartilage bridges joints between bones, and tendons attach them to skeletal muscles.

5. Many responses to changes in external and internal conditions involve muscles that move the animal body or parts of it. In response to suitable stimulation, the cells of muscle tissue contract, or shorten.

6. Smooth muscle and cardiac muscle are responsible for the motions of internal organs. Skeletal muscle helps move the body's limbs and other structural elements and maintain their spatial positions.

7. In each muscle cell, many threadlike structures called myofibrils are divided into sarcomeres. The sarcomere is the basic unit of contraction. It has parallel arrays of actin and myosin filaments. ATP-driven interactions between the actin and myosin shorten the sarcomeres of a muscle and collectively account for its contraction.

VERTEBRATE SKIN

No garment even comes close to having the qualities of skin. What besides skin holds its shape after repeated stretchings and washings, blocks harmful rays from the sun, kills many bacteria on contact, holds in moisture, fixes small cuts or burns, *and* lasts as long as its owner? Skin makes vitamin D, required for calcium metabolism. It has a passive role in adjusting internal temperature, for the nervous system can rapidly adjust the flow of blood (which transports metabolic heat) to and from the skin's great numbers of tiny blood vessels. And signals from sensory receptor endings in skin help the brain assess what is going on in the outside world.

Vertebrate skin has two regions: an outer **epidermis** and an underlying **dermis**. The dermis is mainly dense connective tissue with many elastin fibers (which resist stretching) and collagen fibers (which impart strength). Blood vessels, lymph vessels, and receptor endings of sensory nerves thread through it. Below the dermis, a tissue region (hypodermis) anchors skin to underlying structures yet still allows it to move a bit (Figure 32.3). Fats stored in the hypodermis help insulate the body and cushion some of its parts. Beyond the typical array of epithelial tissues and glands, there is great variation within and between vertebrates. For instance, birds have epidermal derivatives called feathers, bills, and claws, many plant-eating mammals have hooves and horns, porcupines have quills, you have nails, and so on.

Like puff pastry, epidermis has sheetlike layers; it is a stratified epithelium. Many junctions structurally and functionally join its cells (Section 28.1). Ongoing, rapid mitotic cell divisions push epidermal cells from deeper layers toward the skin's free surface. Owing to pressure from the continually growing cell mass and from normal wear and tear at the surface, older cells are dead and flattened by the time they reach the outer layers. There, they are abraded off or flake away on an ongoing basis.

Keratinocytes, the most abundant epidermal cells, make keratin, a water-insoluble protein. Its **melanocytes** produce and donate melanin, a brown-black pigment, to the keratinocytes. Melanin blocks harmful ultraviolet radiation. Humans generally have the same number of melanocytes, but skin color varies owing to differences in the distribution and metabolic activity of those cells. For example, melanocytes in *albinos* cannot produce all of the enzymes necessary for melanin production. Pale skin contains little melanin, so the pigment hemoglobin inside red blood cells is not masked. The skin appears pink because hemoglobin's red color shows through thin-walled blood vessels and the epidermis itself, both of which are transparent. Carotene, which is a yellow-orange pigment, also contributes to skin color.

Human skin contains sweat glands, oil glands, and hair follicles (husklike cavities). Fluid secreted by sweat glands is 99 percent water, with dissolved salts, traces

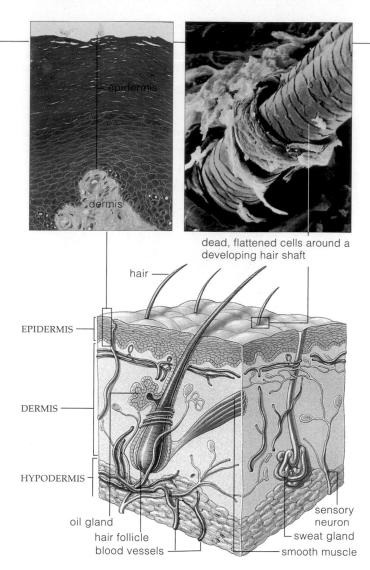

Figure 32.3 Structure of human skin. The uppermost portion is the epidermis; the lower portion is the dermis.

of ammonia, vitamin C, and other substances. Except on the palms of hands and the soles of feet, human skin contains oil glands (sebaceous glands) that lubricate and soften hair and skin. Their secretions also kill many harmful bacteria. *Acne* is a type of skin inflammation that develops after bacteria infect oil gland ducts.

A **hair** is a flexible structure of mostly keratinized cells, rooted in skin with a shaft above its surface (Figure 32.3). Cells divide near the base, are pushed upward, then flatten and die. Flattened cells of the shaft's outer layer overlap like roof shingles. If mechanically abused, these frizz out as "split ends." An average human scalp has about 100,000 hairs. Genes, nutrition, hormones, and stress influence hair growth and density, however.

The body of most animals has an integument, a protective covering that usually is tough yet pliable. A number of invertebrate species have a cuticle. Vertebrates have skin, which consists of epidermis and dermis.

WWW

SUNLIGHT AND SKIN

THE VITAMIN D CONNECTION Even when you do little more than sit outside in the sun, you are giving some of the epidermal cells in your skin the opportunity to make **vitamin D**, or cholecalciferol. This steroid-like compound helps the body absorb calcium from food. When exposed to sunlight, some type of cell in the skin produces it from a precursor molecule that is related to cholesterol. The cells then release vitamin D to the bloodstream, which transports it to absorptive cells in the intestinal lining. This is a hormone-like action—which means skin acts like an endocrine gland when exposed to sunlight.

Humans, remember, evolved beneath the intense sun of the African savanna, so skin alone would have supplied our early ancestors with enough vitamin D. Only after some human populations moved out of the tropics and into caves, animal skins, and clothing did they start to depend more on dietary sources of the essential vitamin D.

SUNTANS AND SHOE-LEATHER SKIN Do you like to tan your body by rotating under the sun's rays or a tanning lamp, like a chicken in an oven broiler? If so, think about what a broiler does to the chicken. The sun's ultraviolet wavelengths stimulate melanin production in skin cells. Continued exposure increases melanin concentrations in light skin and visibly darkens it, thereby producing the "tan" that so many people covet (Figure 32.4). Tanning does protect the body against ultraviolet radiation. Even in naturally dark skin, however, prolonged exposure to sunlight causes the elastin fibers in connective tissue of the dermis to clump together, so skin loses its resiliency. In time it starts to look like old shoe leather.

In this respect, tanning accelerates the *aging* of skin. As any person grows older, epidermal cells divide less often. Skin gets thinner and more vulnerable to injury. Glandular

secretions that once kept it soft and moistened dwindle. Collagen and elastin fibers in the dermis break down and become sparser, so skin loses elasticity and its wrinkles deepen. Either by itself or in combination with excessive tanning, prolonged exposure to dry wind and tobacco smoke also accelerates the aging process.

SUNLIGHT AND THE FRONT LINE OF DEFENSE Besides having cells that produce melanin and keratin, skin also contains two other cell types that help protect the body against invasion by pathogenic cells and against cancer. The defenders of the epidermis are called Langerhans cells and Granstein cells.

Langerhans cells are phagocytes that develop in bone marrow, then take up stations in skin. After they engulf virus particles or bacterial cells, they display molecular alarm signals at the surface of their plasma membrane. Those signals can mobilize the body's immune system.

Ultraviolet radiation can damage Langerhans cells. This might be why sunburns can trigger *cold sores*, the small, painful blisters that announce the recurrence of a *Herpes simplex* infection. Nearly everyone harbors the *H. simplex* virus. It remains hidden within the face, inside a ganglion. (A ganglion, remember, is a cluster of neuron cell bodies.) Sunburns and other stress factors can activate the virus. Virus particles move down the neurons to the axonal endings in skin. There they infect epithelial cells and cause skin eruptions.

When ultraviolet radiation damages Langerhans cells, it weakens one of the body's first lines of defense. It also might activate proto-oncogenes and trigger cancerous transformation of skin cells. As described in Chapter 14's introduction, skin cancers grow rapidly and can spread to adjacent lymph nodes unless surgically removed.

In some as-yet-undetermined way, **Granstein cells** apparently interact with the white blood cells that can put the brakes on immune responses in the skin. By issuing suppressor signals, they help keep the responses from spiraling out of control. Even though the functions of Granstein cells are not completely understood, researchers have learned that these cells are less vulnerable than the Langerhans cells to the damaging effects of ultraviolet radiation.

WWW

Figure 32.4 Demonstration of how shoe-leather skin forms.

SKELETAL SYSTEMS

Many of the responses an animal makes to external and internal conditions involve the movement of the whole body or portions of it. The activation, contraction, and relaxation of muscle cells bring about the movements. But muscle cells alone cannot produce them. *All muscles require the presence of some medium or structural element against which the force of contraction may be applied.* A skeletal system fulfills this requirement.

Three types of skeletons predominate in the animal world. With a **hydrostatic skeleton**, the muscles work against an internal body fluid and redistribute it within a confined space. Like a filled waterbed, the confined fluid resists compression. With an **exoskeleton**, rigid, *external* body parts, such as cuticles or shells, take the applied force of muscle contraction. An **endoskeleton** has rigid, *internal* body parts that receive the applied force of muscle contraction.

We find hydrostatic skeletons among the softbodied invertebrates, such as a sea anemone with a soft, vase-shaped body and saclike gut (Figure 32.5). Its body wall incorporates longitudinal and radial muscles. Between meals, longitudinal muscles are contracted (shortened) and radial ones are relaxed (lengthened), so the body is short and squat. When the body lengthens upright, radial muscles are contracting (and forcing some fluid from the gut cavity), and longitudinal ones are relaxing. And earthworms, remember, use coelomic chambers of their segmented body as a hydrostatic skeleton (Section 23.10). They move forward by contracting and relaxing the fluid-filled segments in sequence. By coordinating contractions on one side or the other of the segments, they also thrash sideways and move forward and back.

Recall how an arthropod has a hinged exoskeleton? It moves some hard parts like levers by sets of muscles attached to them. Small contractions bring about large movements. This is especially true of a winged insect's cuticle. It extends over certain body segments and gaps between segments. The cuticle is pliable at the gaps. It functions like a hinge when certain muscles raise either the wing or body parts to which the wings are attached. Figure 32.6 shows how this works.

Most vertebrates have an endoskeleton of bone (in sharks and rays, it is of cartilage). The human skeleton, shown in Figure 32.7, has 206 bones. Its pectoral girdle (at the shoulders), pelvic girdle (at the hips), and paired arms, hands, legs, and feet are the *appendicular* portion. Its pectoral girdles have slender collarbones and flat shoulder blades. Fall on an outstretched arm and you might dislocate a shoulder blade or fracture a flimsily arranged collarbone, the bone most frequently broken. Skull bones, twelve pairs of ribs, a breastbone, and twenty-six vertebrae are the *axial* portion. Its **vertebrae** (singular, vertebra) are bony segments of the curved backbone, or vertebral column. The backbone extends

a Resting position   **b** Feeding position

Figure 32.5 Outcomes of contractile force applied against the sea anemone's hydrostatic skeleton. This invertebrate's body wall has contractile cells running longitudinally (parallel with the body axis) and radially around the gut cavity. (**a**) In this case, radial cells are relaxed and longitudinal ones are contracted. Anemones typically are in this resting position at low tide, when currents cannot bring food morsels to them. (**b**) Radial cells are contracted, longitudinal ones are relaxed, and the body is extended to its upright feeding position.

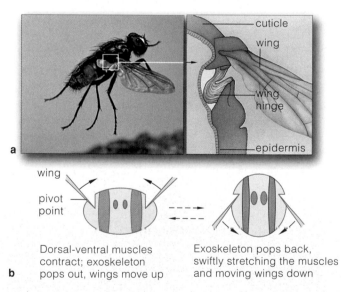

Figure 32.6 Housefly wing movement, an outcome of the contraction of sets of muscles that extend from dorsal to ventral regions of the exoskeleton. The contractile force works against the exoskeleton, near hinge points where wings are attached. When the muscles contract, the exoskeleton pops inward and wings move up. When the exoskeleton pops outward by elastic force, the wings move down. The quick-stretched muscles invite a fast repeat of the events.

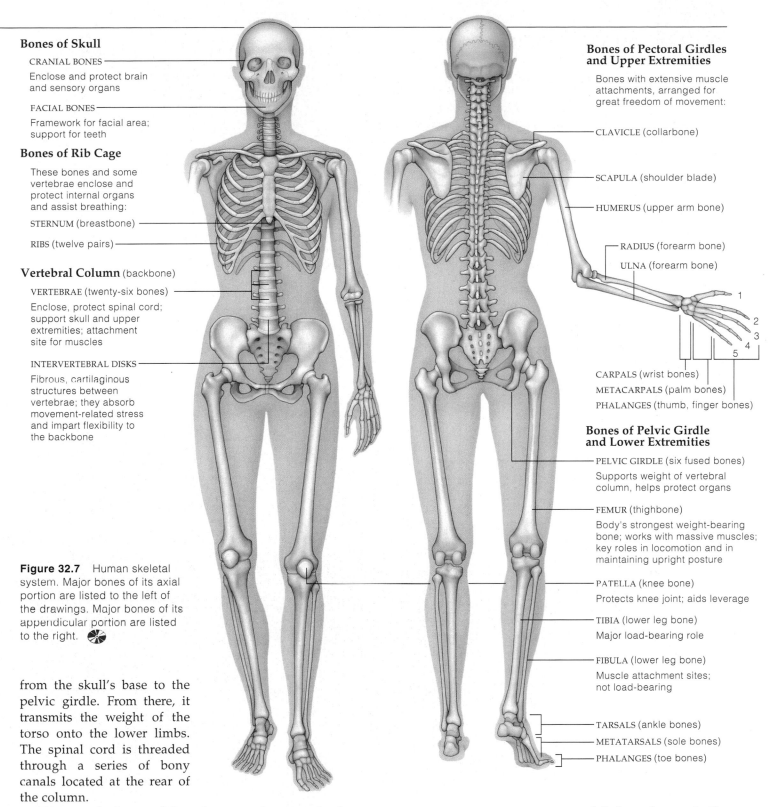

Bones of Skull

CRANIAL BONES

Enclose and protect brain and sensory organs

FACIAL BONES

Framework for facial area; support for teeth

Bones of Rib Cage

These bones and some vertebrae enclose and protect internal organs and assist breathing:

STERNUM (breastbone)

RIBS (twelve pairs)

Vertebral Column (backbone)

VERTEBRAE (twenty-six bones)

Enclose, protect spinal cord; support skull and upper extremities; attachment site for muscles

INTERVERTEBRAL DISKS

Fibrous, cartilaginous structures between vertebrae; they absorb movement-related stress and impart flexibility to the backbone

Figure 32.7 Human skeletal system. Major bones of its axial portion are listed to the left of the drawings. Major bones of its appendicular portion are listed to the right.

Bones of Pectoral Girdles and Upper Extremities

Bones with extensive muscle attachments, arranged for great freedom of movement:

CLAVICLE (collarbone)

SCAPULA (shoulder blade)

HUMERUS (upper arm bone)

RADIUS (forearm bone)

ULNA (forearm bone)

CARPALS (wrist bones)

METACARPALS (palm bones)

PHALANGES (thumb, finger bones)

Bones of Pelvic Girdle and Lower Extremities

PELVIC GIRDLE (six fused bones)

Supports weight of vertebral column, helps protect organs

FEMUR (thighbone)

Body's strongest weight-bearing bone; works with massive muscles; key roles in locomotion and in maintaining upright posture

PATELLA (knee bone)

Protects knee joint; aids leverage

TIBIA (lower leg bone)

Major load-bearing role

FIBULA (lower leg bone)

Muscle attachment sites; not load-bearing

TARSALS (ankle bones)

METATARSALS (sole bones)

PHALANGES (toe bones)

from the skull's base to the pelvic girdle. From there, it transmits the weight of the torso onto the lower limbs. The spinal cord is threaded through a series of bony canals located at the rear of the column.

Between the bones of the column are **intervertebral disks**—cartilaginous shock absorbers and flex points that permit some movement. Severe or rapid shock sometimes forces a disk to slip out of place or rupture. The painful, *herniated disks* are a less-than-advantageous outcome of bipedalism. Remember, our first primate ancestors were quadrupedal. The first hominids started to walk upright about 4 million years ago, and a long-

term outcome was a pronounced S-shaped curve in the backbone. Today, the older we get, the longer we have been fighting gravity in a compromised way, and the more back pain we suffer.

Animal skeletons have structural elements or body fluids against which the force of contraction can be applied.

32.4 CHARACTERISTICS OF BONE

Bone Structure and Function

By definition, **bones** are complex organs that function in movement, protection, support, mineral storage, and formation of blood cells (Table 32.1). Bones that support and anchor skeletal muscles help maintain or change positions of body parts. Some form hard compartments that enclose and protect the brain, the lungs, and other internal organs. Bones are reservoirs for ions of calcium and phosphorus. Deposits and withdrawals of those ions from bone help maintain blood levels of calcium and phosphrus, and thus support metabolic activities. Some bones (not all) are sites of blood cell formation.

Human bones range in size from middle earbones smaller than lentils to clublike femurs, or thighbones. Different bones are long, short (or cubelike), flat, and irregular. All contain connective tissues—bone tissue especially—and epithelia. The bone tissues have living cells and collagen fibers in a calcium-hardened ground substance. Cartilage models for many bones form in vertebrate embryos. Bone-forming cells secrete material inside the model's shaft and on its surface. Cartilage inside breaks down to open up a marrow cavity. In time, bone cells become imprisoned by their secretions.

For the thighbone in Figure 32.8, *compact* bone tissue in the shaft and at both ends resists mechanical shock. This bone tissue is deposited as many thin, cylindrical, dense layers around interconnected canals for blood vessels and nerves that service living bone cells. Each array is a Haversian system. *Spongy* bone tissue in the bone ends and shaft imparts strength but not much weight. Abundant spaces make the tissue appear spongy, but its flattened parts are firm. **Red marrow**, a major site of blood cell formation, fills spaces in some bones, such as a breastbone. Cavities in most mature bones contain **yellow marrow**. Yellow marrow consists largely of fat. It converts to red marrow and produces new red blood cells when blood loss from the body is severe.

Even though a bone doesn't seem to change much, its component mineral deposits are being turned over

Table 32.1 Functions of Bone

1. *Movement.* Bones interact with skeletal muscle to maintain or change the position of body parts.

2. *Support.* Bones support and anchor muscles.

3. *Protection.* Many bones form hard compartments that enclose and protect soft internal organs.

4. *Mineral storage.* Bones are a reservoir for calcium and phosphorus, the deposits and withdrawals of which help maintain ion concentrations in body fluids.

5. *Blood cell formation.* Some bones contain regions where blood cells are produced.

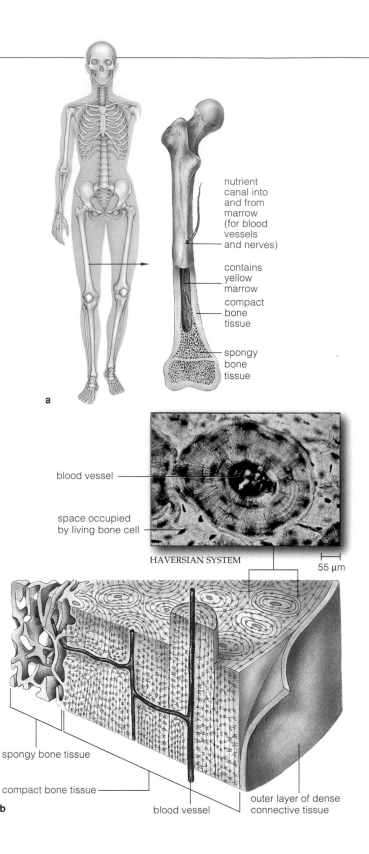

nutrient canal into and from marrow (for blood vessels and nerves)

contains yellow marrow

compact bone tissue

spongy bone tissue

a

blood vessel

space occupied by living bone cell

HAVERSIAN SYSTEM

55 μm

spongy bone tissue

compact bone tissue

b

blood vessel

outer layer of dense connective tissue

Figure 32.8 (**a**) Structure of a thighbone. (**b**) Appearance of its spongy and compact bone tissue. Thin, dense layers of compact bone tissue form cylindrical, interconnected arrays around canals that contain blood vessels and nerves. Each array is a Haversian system. The blood vessel threading through its center services osteocytes, living bone cells in small spaces in the tissue. Tunnels connect adjacent spaces.

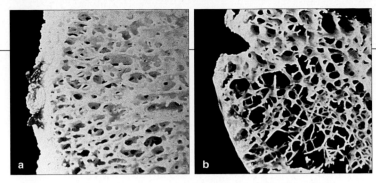

Figure 32.9 Bone affected by osteoporosis. (**a**) Section through normal bone tissue; mineral deposits continually replace withdrawals. (**b**) After the onset of osteoporosis, mineral replacements lag behind withdrawals. The tissue erodes; bones become hollow and brittle.

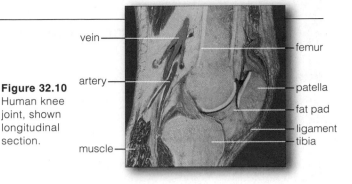

Figure 32.10 Human knee joint, shown longitudinal section.

vein — artery — muscle — femur — patella — fat pad — ligament — tibia

on an ongoing basis. Because mineral deposition and removal (resorption) usually proceed at the same time, the bone maintains its structure even while it is being remodeled. This bone tissue turnover serves two key functions: it adjusts bone strength and helps maintain the concentration of calcium ions (Ca^{++}) in blood.

Mineral deposition exceeds withdrawals when bone is physically stressed or subjected to compression. Thus the bones of people who rigorously exercise are more dense and stronger than those of couch potatoes. The balance tips to withdrawals when bone is injured. Also, as a person ages, the backbone, and other bones decline in mass, a condition called *osteoporosis* (Figure 32.9). Decreased physical activity, declining bone-forming cell activity, calcium loss, excessive protein intake, and sex hormone deficiencies contribute to the disorder.

The blood level of calcium is one of the most tightly controlled aspects of metabolism, owing to calcium's roles in neural function, muscle contraction, and other vital activities. Bones and teeth store all but 1 percent or so of the body's calcium. Enzymes from bone-dissolving cells (osteoclasts) release calcium and other ions from bone. The rates of calcium release and uptake depend on negative feedback loops between two glands and the blood itself. When the blood level of calcium rises, the thyroid gland releases calcitonin. This hormone slows calcium movement into blood and suppresses activity of osteoclasts. When the blood level of calcium falls, the parathyroid glands release parathyroid hormone (PTH), which promotes calcium movement from bone, and from filtrate in the kidneys, back to the blood. PTH also enhances the activity of osteoclasts, and it helps activate vitamin D. That vitamin stimulates calcium absorption from the intestine (compare Section 31.7).

Skeletal Joints

Each **joint**, an area of contact or near-contact between bones, has a distinctive bridge of connective tissue. Short connecting fibers join bones at *fibrous* joints. Straps of cartilage join them at *cartilaginous* joints. **Ligaments**, long straps of dense connective tissue, bridge the gap between bones at the *synovial* joints.

Fibrous joints hold teeth in sockets and connect the flat skull bones of a fetus. At birth, the loose connections let the bones slide over each other a bit and prevent skull fractures. The skull of a newborn still has fibrous joints as well as membranous areas called "soft spots," or fontanels. In childhood, the fibrous tissue hardens and skull bones are fused into a single unit.

Cartilaginous joints bridging vertebrae, ribs, and the breastbone permit slight movements. Synovial joints, such as knee joints, move freely. Ligaments stabilize the knee joint (Figure 32.10). Where one bone touches another, cartilage cushions them. A flexible capsule of dense connective tissue surrounds the area of contact. Cells of a membrane that lines the inside of the capsule secrete a fluid that lubricates the joint.

Like many other joints, the knee joint is vulnerable to stress. This is the joint that lets you swing, bend, and turn the long bones below it. When you run, it absorbs the force of your weight each time the foot below it hits the ground. Stretch or twist the joint suddenly and too far, and you may *strain* it. Tear its ligaments or tendons and you *sprain* it. Move the wrong way and you may well dislocate the attached bones. During football and other "collision sports," blows to the knees frequently sever a ligament. The severed part must be reattached surgically before ten days pass. Phagocytes in the joint normally clean up after everyday wear and tear. When ligaments tear, they indiscriminately turn tissue to mush.

Joint inflammation and degeneration are collectively known as "arthritis." In *osteoarthritis*, cartilage at freely movable joints wears off as a person ages. Joints in the fingers, knees, hips, and vertebral column are affected most often. In *rheumatoid arthritis*, synovial membranes in the joints become inflamed and thickened, cartilage degenerates, and bone deposits build up. This disorder may be triggered by a bacterial or viral infection, but it also has a genetic component. It can begin at any age, but symptoms usually emerge before age fifty.

Bones are collagen-rich, mineralized organs that function in movement, protection, support, storage of calcium and other minerals, and blood cell formation.

Joints are areas of contact or near-contact between bones, and each has a distinctive bridge of connective tissue.

How Muscles and Bones Interact

Skeletal muscles are the functional partners of bones. Each skeletal muscle contains bundles of hundreds to many thousands of muscle cells, which look like long, striped fibers. In muscle tissue, remember, muscle cells contract (shorten) in response to adequate stimulation. They lengthen in response to gravity and other loads. Whenever you dance, breathe, scribble notes, or tilt your head, contracting muscle cells are helping to move your body or change the positions of some of its parts.

Connective tissue bundles the muscle cells together and extends past them to form **tendons**. Each tendon, a cord or strap of dense connective tissue, attaches some muscle to bone (Figure 32.11). Most of the attachment sites are like a car's gearshift. *They form a lever system, in which a rigid rod is attached to a fixed point but is able to move about at it.* Muscles connect to bones (rigid rods) near a joint (fixed point). As the muscles contract, they transmit force to bones and make them move.

Tendons commonly rub against bones. However, they slide inside fluid-filled, membranous sheaths that help reduce the friction. Your knees, wrists, and finger joints have such sheaths (Figure 32.11).

Skeletal muscles interact with one another as well as with bones. Some are arranged in pairs or groups that work together to promote the same movement. Others work in opposition, with the action of one opposing or reversing the action of another. Figure 32.12 shows how opposing muscle groups work to move frog legs.

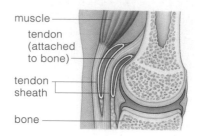

Figure 32.11 Tendon sheath. This is not the same as a bursa, a small sac also filled with synovial fluid. Bursae are cushions interposed *between* bone and skin or bone and tendons.

muscle
tendon (attached to bone)
tendon sheath
bone

Also look at Figure 32.13. Extend your right arm forward, then place your left hand over the biceps in the upper right arm and slowly "bend the elbow." Feel the biceps contract? Even when your biceps contracts only a bit, it causes a large motion of the bone connected to it. This is true of most leverlike arrangements.

Bear in mind, only *skeletal* muscle is the functional partner of bone. As mentioned earlier, smooth muscle is mainly a component of the walls of internal organs, such as the stomach (Section 28.3). Cardiac muscle is present only in the wall of the heart. We will consider the structure and functioning of smooth muscle and cardiac muscle in later chapters in this unit.

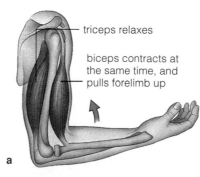

c Now the first muscle group in the frog's upper hindlimb contracts again and draws the limb back toward the body.

b Then an opposing muscle group that is attached to the same limb forcefully contracts and pulls it back. The contractile force, applied against the ground, propels the frog forward.

a A frog muscle attached to each upper hindlimb contracts and pulls the leg slightly forward.

Figure 32.12 A frog demonstrating how a small decrease in the length of contracting muscles can produce a large movement. Its leap depends on opposing muscle groups attached to the upper limb bone of each hind leg. (**a**) One muscle group pulls the limb forward a bit, toward the body's midline. (**b**) Another pulls it back and away from the body.

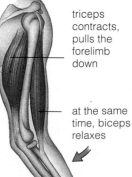

triceps relaxes

biceps contracts at the same time, and pulls forelimb up

triceps contracts, pulls the forelimb down

at the same time, biceps relaxes

a

b

Figure 32.13 Two opposing muscle groups in a human arm. When the triceps relaxes and its opposing partner (biceps) contracts, the elbow joint flexes and the forearm bends upward. When the triceps contracts, the forearm straightens out.

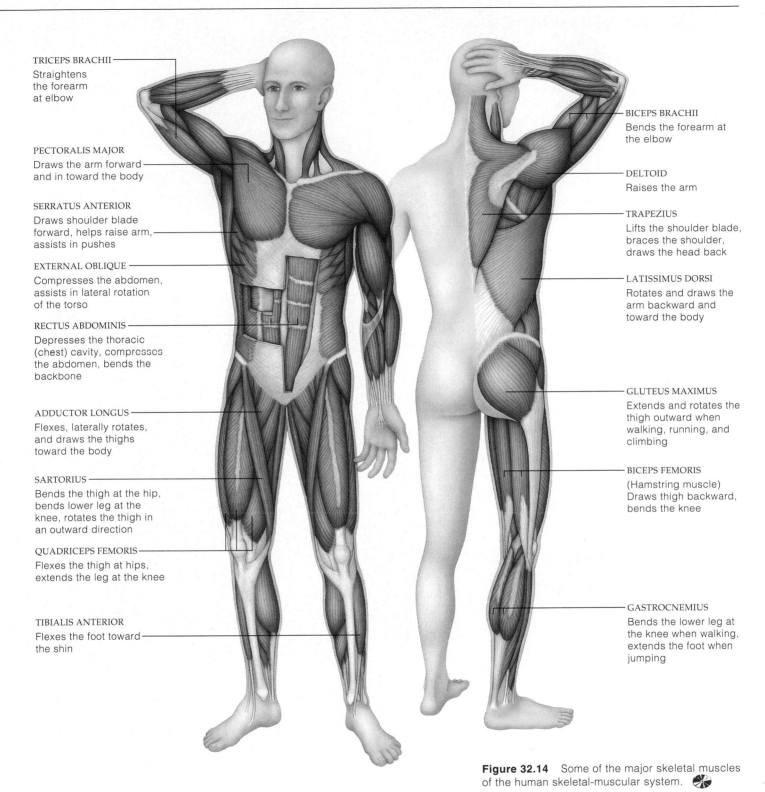

TRICEPS BRACHII
Straightens the forearm at elbow

PECTORALIS MAJOR
Draws the arm forward and in toward the body

SERRATUS ANTERIOR
Draws shoulder blade forward, helps raise arm, assists in pushes

EXTERNAL OBLIQUE
Compresses the abdomen, assists in lateral rotation of the torso

RECTUS ABDOMINIS
Depresses the thoracic (chest) cavity, compresses the abdomen, bends the backbone

ADDUCTOR LONGUS
Flexes, laterally rotates, and draws the thighs toward the body

SARTORIUS
Bends the thigh at the hip, bends lower leg at the knee, rotates the thigh in an outward direction

QUADRICEPS FEMORIS
Flexes the thigh at hips, extends the leg at the knee

TIBIALIS ANTERIOR
Flexes the foot toward the shin

BICEPS BRACHII
Bends the forearm at the elbow

DELTOID
Raises the arm

TRAPEZIUS
Lifts the shoulder blade, braces the shoulder, draws the head back

LATISSIMUS DORSI
Rotates and draws the arm backward and toward the body

GLUTEUS MAXIMUS
Extends and rotates the thigh outward when walking, running, and climbing

BICEPS FEMORIS
(Hamstring muscle) Draws thigh backward, bends the knee

GASTROCNEMIUS
Bends the lower leg at the knee when walking, extends the foot when jumping

Figure 32.14 Some of the major skeletal muscles of the human skeletal-muscular system.

Human Skeletal-Muscular System

The human body has more than 600 skeletal muscles, some superficial, others deep in the body wall. Some, such as facial muscles, attach to the skin. The trunk has muscles of the thorax, backbone, abdominal wall, and pelvic cavity. Other groups of muscles attach to upper and lower limb bones. Figure 32.14 shows a few of the main skeletal muscles and lists their functions. We turn next to the mechanisms underlying their contraction.

Skeletal muscles transmit contractile force to bones and make them move. Tendons strap skeletal muscles to bones.

WWW

Skeletal Muscle Structure and Function

Bones move—they are pulled in some direction—when the skeletal muscles attached to them shorten. When a skeletal muscle shortens, its component muscle cells are shortening. And when a muscle cell shortens, many units of contraction within that cell are shortening. The basic units of contraction are called **sarcomeres**.

Figure 32.15 shows how bundles of cells in a skeletal muscle run parallel with the muscle. In each muscle cell are myofibrils, threadlike structures bundled together in parallel. Each myofibril is functionally divided into many sarcomeres, arranged one after another along its length. Dark bands called Z lines define the two ends of each sarcomere. A sarcomere contains many filaments, oriented parallel with its long axis. Certain differences in length and positioning give skeletal (and cardiac) muscle a striped appearance. Some filaments are thin; others are thick. Each *thin* filament is like two strands of pearls, twisted together. The "pearls" are molecules of **actin**, a globular protein with contractile functions:

one actin molecule   portion of one thin filament

Other proteins (coded *green*) are near actin's surface grooves. Each *thick* filament is made of molecules of **myosin**, another protein with contractile functions, in parallel array. A myosin molecule has a long tail and a double head that projects from the filament's surface:

one myosin molecule   part of one thick filament

Thus myofibrils, muscle cells, and muscle bundles all run in the same direction. What is the function of this consistent, parallel orientation? It focuses the force of muscle contraction onto a bone in a particular direction.

How do sarcomeres shorten to contract a muscle? The answer starts in each sarcomere, with sliding and pulling interactions between Z lines. Actin filaments extend from each Z line to a sarcomere's center. A set of myosin filaments partly overlaps them but does not extend all the way to the Z lines (Figure 32.15). When a muscle contracts, actin filaments from opposite sides of the sarcomere slide over myosin filaments, which remain stationary. And so the sarcomere shortens.

By a **sliding-filament model** of contraction, myosin and actin interact through formation of cross-bridges: attachments between myosin heads and binding sites on actin. When activated, the myosin heads attach to an adjacent actin filament and tilt in a short power stroke,

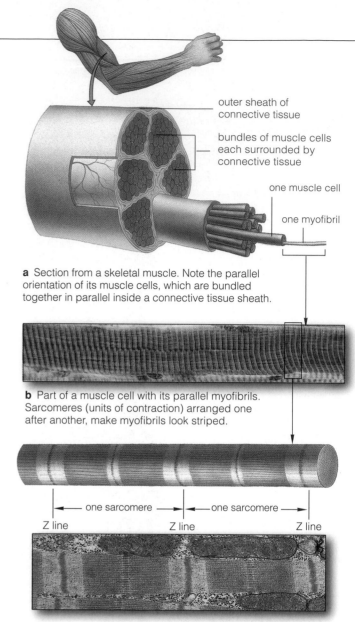

a Section from a skeletal muscle. Note the parallel orientation of its muscle cells, which are bundled together in parallel inside a connective tissue sheath.

b Part of a muscle cell with its parallel myofibrils. Sarcomeres (units of contraction) arranged one after another, make myofibrils look striped.

c Diagram and electron micrograph of two sarcomeres from one myofibril. This view shows how two portions called Z lines define the two ends of each sarcomere. Mitochondria (the oval organelles next to the sarcomeres) provide ATP energy for muscle action.

Figure 32.15 Components of a skeletal muscle.

driven by ATP energy, toward the sarcomere's center (Figure 32.16). The heads pull the actin filament along with them. Another energy input makes the heads let go, attach to another part of the filament, tilt in another power stroke, and so on down the filament's length. A single contraction of each sarcomere requires a whole series of power strokes.

Sources of Energy for Contraction

All cells require ATP, but only in muscle cells does the demand skyrocket in so short a time. When a muscle cell at rest is called upon to contract, phosphate donations from ATP must proceed twenty to a hundred times faster.

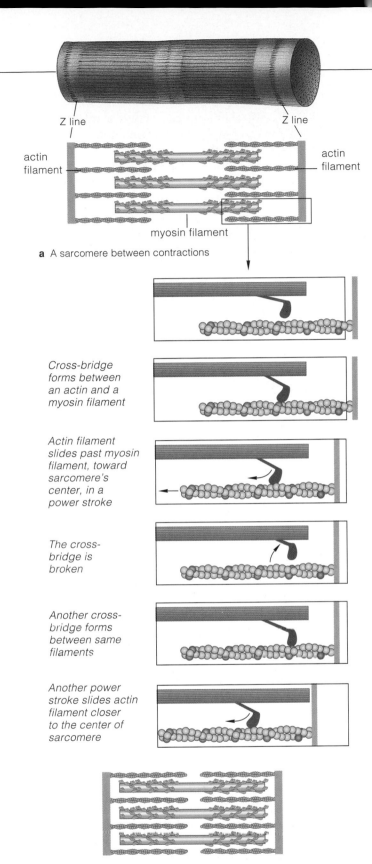

a A sarcomere between contractions

Cross-bridge forms between an actin and a myosin filament

Actin filament slides past myosin filament, toward sarcomere's center, in a power stroke

The cross-bridge is broken

Another cross-bridge forms between same filaments

Another power stroke slides actin filament closer to the center of sarcomere

b Same sarcomere, contracted

Figure 32.16 (**a**) Simplified picture of how actin filaments and myosin filaments are arranged in a sarcomere. Interactions between the two kinds of filaments shorten the width of each sarcomere. (**b**) Diagram of the sliding-filament model of contraction as it proceeds in the sarcomeres of muscle cells. For simplicity, the action of only one myosin head is shown. ✺

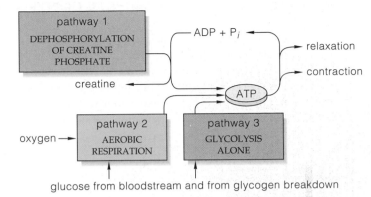

Figure 32.17 Three metabolic routes by which ATP forms in muscle cells in response to the demands of physical exercise.

But a cell has only a small supply of ATP at the start of contractile activity. At such times, it forms ATP by a fast reaction. An enzyme simply transfers phosphate from **creatine phosphate**, an organic compound, to ADP. A cell has about five times as much creatine phosphate as ATP, so this reaction is good for a few contractions. And it is enough to buy time for a relatively slower ATP-forming pathway to kick in (Figure 32.17).

During prolonged, moderate exercise, the oxygen-requiring reactions of aerobic respiration typically can supply most of the ATP needed for contraction. Suppose a muscle cell taps its store of glycogen for glucose (the starting substrate) in the first five to ten minutes. For the next half hour or so of sustained activity, that muscle cell depends on glucose and fatty acid deliveries from the blood. For contractile activity longer than this, fatty acids are the main fuel source (Section 7.6).

What happens when exercise is so intensive that it exceeds the capacity of the respiratory and circulatory systems to deliver oxygen for the aerobic pathway? At such times, glycolysis alone will contribute more of the total ATP that is being produced. Remember, by this set of anaerobic reactions, a glucose molecule is not fully broken down, so the net ATP yield is small. But muscle cells use this metabolic route as long as glycogen stores continue to provide glucose before fatigue occurs.

After intense exercise, deep, rapid breathing helps repay the body's **oxygen debt**, incurred when ATP use by muscles exceeded the aerobic pathway's deliveries.

A skeletal muscle shortens by combined decreases in the length of its sarcomeres, the basic units of contraction.

The parallel orientation of a muscle's components directs the force of contraction toward a bone that must be pulled in some direction. By energy-driven interactions between the myosin and actin filaments, sarcomeres shorten and collectively account for its contraction.

During exercise, the availability of ATP inside muscle cells affects whether contraction will proceed, and for how long.

WWW

CONTROL OF MUSCLE CONTRACTION

When skeletal muscles contract, they move the body and its parts at certain times, in certain ways. They do so in response to commands from the nervous system. The commands stimulate or inhibit the release of ACh, a neurotransmitter, from motor neurons. Muscle cells are a major target of this neurotransmitter (Section 29.3).

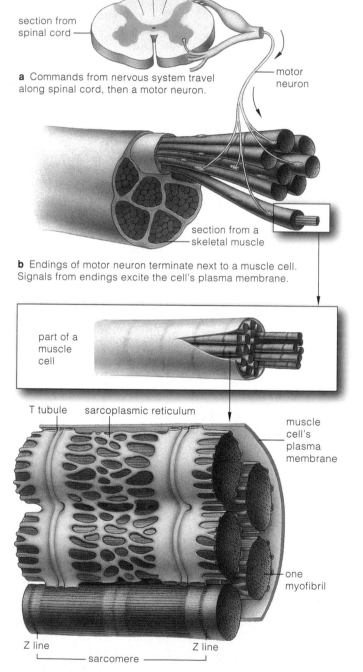

a Commands from nervous system travel along spinal cord, then a motor neuron.

section from spinal cord

motor neuron

section from a skeletal muscle

b Endings of motor neuron terminate next to a muscle cell. Signals from endings excite the cell's plasma membrane.

part of a muscle cell

T tubule   sarcoplasmic reticulum

muscle cell's plasma membrane

one myofibril

Z line   Z line

sarcomere

c Action potentials arise, spread to T tubules (the extensions of plasma membrane connected to the calcium-storing sarcoplasmic reticulum). Arrival of action potentials induces calcium release. Calcium diffuses into myofibrils, where it binds to sites in thin filaments that allow cross-bridge formation and contraction.

Like all cells, a muscle cell shows a difference in electric charge across its plasma membrane. That is, the cytoplasm just beneath the membrane is a bit more negative than fluid outside it. But only in muscle cells, neurons, and other *excitable* cells does this difference in charge reverse suddenly though briefly in response to sufficient stimulation.

Figure 32.18 (*Left*) Pathway for signals from the nervous system that stimulate contraction of skeletal muscle. The plasma membrane of muscle cells connects with their myofibrils and with the inward-threading T tubules. These membranous tubes are close to the sarcoplasmic reticulum, a calcium-storing system that functions in the control of contraction.

That abrupt reversal in charge, an **action potential**, occurs as charged ions flow across the membrane in an ever accelerating way. The excitation swiftly spreads out from the stimulation site and travels along the membrane without diminishing.

Suppose action potentials arise in a muscle cell. They spread rapidly away from the stimulation point along small, tubelike extensions of the plasma membrane. The small tubes connect with a system of membrane-bound chambers that thread lacily around the myofibrils in a muscle cell. That system, the **sarcoplasmic reticulum**, takes up, stores, and releases calcium ions in controlled ways (Figure 32.18).

The arrival of action potentials causes an outward flow of calcium ions from the sarcoplasmic reticulum. The released ions diffuse into the myofibrils and reach actin filaments. Before this happened, the muscle was at rest (it was not contracting). Myosin binding sites on the actin filaments were blocked; hence cross-bridges could not form. However, the arrival of calcium allows cross-bridge attachments, so contraction can proceed. Afterward, calcium ions are actively transported back into the membrane storage system and muscles relax.

What blocks cross-bridge binding sites in a muscle at rest? Tropomyosin and troponin are two accessory proteins located in or near the surface grooves of actin filaments (here you may wish to compare Sections 4.9 and 32.6). When the calcium level is low, the proteins are joined so tightly together, the tropomyosin is forced slightly outside the groove—and its position blocks the cross-bridge binding site. When the calcium level rises, calcium binds to troponin and alters its shape. When troponin's shape is altered, it has a different molecular grip on tropomyosin—which is now free to move into the groove and expose the binding site.

Commands from the nervous system initiate action potentials in muscle cells by way of ACh. Those action potentials are signals for cross-bridge formation—hence for contraction.

PROPERTIES OF WHOLE MUSCLES

Muscle Tension and Muscle Fatigue

Whether a muscle actually shortens during cross-bridge formation depends on the external forces acting on it. Collectively, the cross-bridges exert **muscle tension**. By definition, this is a mechanical force that a contracting muscle exerts on an object, such as a bone. Opposing it is a load, either the weight of an object or gravity's pull on the muscle. Only when muscle tension exceeds the load does a stimulated muscle shorten.

An *isometrically* contracting muscle develops tension but does not shorten. It supports a load in a constant position, as when you hold a glass of lemonade in front of you. An *isotonically* contracting muscle shortens and moves a load. With *lengthening* contraction, though, an external load is greater than the muscle tension, so the muscle lengthens during the period of contraction. This happens to leg muscles when you walk down stairs.

A muscle's tension relates to the formation of cross-bridges in its cells and the number of cells recruited into action. Consider a **motor unit**: a motor neuron and all muscle cells that form junctions with its endings. By stimulating the motor unit with an electrical impulse, we can induce an action potential and make a recording of an isometric contraction. It takes a few milliseconds for tension to increase, then it peaks and declines. This response is a **muscle twitch** (Figure 32.19a). Its duration depends on the load and cell type. For example, fast-acting muscle cells rely on glycolysis (not efficient but fast) and use up ATP faster than slow-acting cells do.

If another stimulus is applied before the response is over, the muscle twitches again. **Tetanus** is a large contraction resulting from the repeated stimulation of a motor unit, so that twitches mechanically run together. (In a disease by the same name, toxins interfere with muscle relaxation.) Figure 32.19d shows a recording of tetanic contraction.

Continuous, high-frequency stimulation that keeps a muscle in a state of tetanic contraction leads to *muscle fatigue*, or a decline in tension. After a few minutes of rest, a fatigued muscle will contract again in response to stimulation. The extent of recovery depends largely on how long and how frequently it was stimulated before. Muscles associated with brief, intense exercise (such as weightlifting) fatigue fast but recover fast. The muscles associated with prolonged, moderate exercise fatigue slowly but take longer to recover, often up to twenty-four hours. The molecular mechanisms causing muscle fatigue are unknown, but glycogen depletion is a factor.

Effects of Exercise and Aging

A muscle's properties depend on how often, how long, and how intensely it is put to use. With regular **exercise**

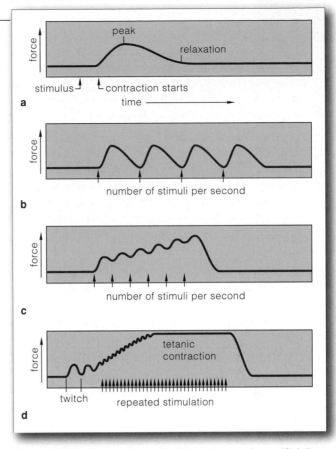

Figure 32.19 Recordings of twitches in muscles artificially stimulated in different ways. (**a**) A single twitch. (**b**) Four stimulations per second cause a series of twitches. (**c**) Six per second cause a summation of twitches, and (**d**) about twenty per second cause tetanic contraction.

(that is, increased levels of contractile activity), muscle cells do not increase in number. But they increase in size and metabolic activity, and become more resistant to fatigue. Think of *aerobic exercise*, which is not intense but is long in duration. Aerobic exercise increases the number of mitochondria in both fast and slow muscle cells, and it increases the blood capillaries that service them. Such physiological changes improve endurance. By contrast, *strength training* (intense, short-duration exercise such as weightlifting) affects fast-acting muscle cells. These form more myofibrils and more enzymes of glycolysis. Strong, bulging muscles may result, although they don't have much endurance. They fatigue rapidly.

Muscle tension decreases in adult humans after age thirty or forty. These people may exercise just as long and intensely as younger ones, but their muscles cannot adapt (change) in response to the same extent. Even so, some adaptation can be beneficial. Aerobic exercise can improve blood circulation. And, as it turns out, modest strength training slows the loss of muscle tissue that is an inevitable part of the aging process.

Properties of muscles vary with age and levels of activity.

32.9

SUMMARY

1. Most animals have an integumentary system, which covers the body's surface. Examples include the cuticle of roundworms and arthropods, as well as vertebrate skin and the structures derived from it.

2. Skin protects the body against abrasion, ultraviolet radiation, dehydration, and many harmful bacteria. It helps control internal temperature (blood flow to skin can dissipate heat). Receptors in skin detect external stimuli. Skin exposed to sunlight has an endocrine function; it makes vitamin D, a hormone-like substance that is required for absorption of calcium from food.

3. Skin has two regions: an outermost epidermis and underlying dermis, where rapid cell divisions produce replacements for cells shed on an ongoing basis. Most abundant are the keratinocytes (keratin-forming cells). Also present are melanocytes (melanin producers), and Langerhans cells and Granstein cells (these help defend the body against pathogens and cancer cells).

4. Movement of the animal body or parts of it requires contractile cells and some medium or structure against which contractile force can be applied.

   a. In hydrostatic skeletons, as in sea anemones, some body fluid accepts contractile force and is redistributed inside a confined space.

   b. In exoskeletons, as in insects and other arthropods, rigid external body parts accept the contractile force.

   c. In endoskeletons, rigid internal body parts (bones, mostly) receive the applied force of contraction.

5. Bones are complex organs with living bone cells in a mineralized, collagen-rich ground substance. Bones function in movement, protection, and support of body parts; in mineral storage. Also, the bones with red and yellow marrow serve in blood cell formation.

6. Skeletal joints are areas of contact or near-contact where connective tissue bridges gaps between bones. Joints are fibrous (short fibers), cartilaginous (made of cartilage), or synovial (with straplike ligaments).

7. Tendons attach skeletal muscles to bones. Skeletal muscles and bones interact as a system of levers, with rigid rods (bones) moving at fixed points (joints). Many muscles work together or in opposition to bring about movement or positional changes in body parts.

8. Cells of smooth, cardiac, and skeletal muscle contract (shorten) in response to suitable stimulation. A skeletal muscle cell has many myofibrils arranged parallel with its long axis. These structures contain actin and myosin filaments, also organized in parallel. Each myofibril is functionally divided into sarcomeres, the basic units of contraction. The parallel orientation of all the muscle's components directs the force of contraction at a bone to be pulled in some direction.

9. In response to stimulation from the nervous system, skeletal muscles shorten by decreases in the length of all of its sarcomeres. Here are the key points of the sliding-filament model of muscle contraction:

   a. Action potentials cause the release of calcium ions from a membrane system (sarcoplasmic reticulum) that threads around the cell's myofibrils. Calcium diffuses into sarcomeres, and binds to and pulls aside accessory proteins on actin filaments to expose binding sites for myosin. Cross-bridges can now form.

   b. Each cross-bridge is a brief attachment between a myosin head and an actin binding site. Cross-bridges form during repeated, ATP-driven power strokes. The repeated strokes make actin filaments slide past myosin filaments, and collectively they shorten the sarcomere.

10. Muscle cells obtain the ATP required for contraction by three metabolic pathways:

    a. Dephosphorylation of creatine phosphate. This is direct, fast, and good for a few seconds of contraction.

    b. Aerobic respiration. This pathway predominates during prolonged, moderate exercise.

    c. Glycolysis. This pathway takes over when intense exercise exceeds the body's capacity to deliver oxygen to muscle cells.

11. Muscle tension refers to a mechanical force created by cross-bridge formation. The load (gravity or weight of objects) is an opposing force. A stimulated muscle shortens when tension exceeds the load and lengthens when tension is less than the load. Levels of exercise and aging affect the properties of muscles.

    a. A motor unit is a motor neuron together with all muscle cells that form junctions with its endings.

    b. A muscle twitch is a brief, weak contraction in response to a single action potential at a motor unit. Tetanus is a large contraction resulting from repeated stimulation at a motor unit.

Review Questions

1. List the functions of skin, then describe the distinguishing features of its regions. Is the hypodermis part of skin? *32.1*

2. Identify four cell types in vertebrate skin and briefly describe their functions. *32.1, 32.2*

3. Distinguish between:
   a. hydrostatic skeleton, exoskeleton, and endoskeleton *32.3*
   b. vertebra and intervertebral disk *32.3*
   c. ligament and tendon *32.4, 32.5*

4. What are the functions of bones? What is a joint? *32.4*

5. Which hormones help control calcium ion concentrations in blood? Name their general effects on bone tissue turnover. *32.3*

6. In what respect are tendons like a car's gearshift? *32.5*

7. Review Figure 32.15. On your own, sketch and label the fine structure of a muscle, down to one of its individual myofibrils. Identify the basic unit of contraction in the myofibrils. *32.6*

8. What role does calcium play in control of contraction? What role does ATP play, and by what routes does it form? *32.6, 32.7*

Figure 32.20
Pumped-up muscles of a human male.

Self-Quiz (Answers in Appendix III)

1. Which is *not* a function of skin?
   a. resist abrasion   c. initiate movement
   b. restrict dehydration   d. help control temperature

2. _____ are shock pads and flex points.
   a. Vertebrae   c. Marrow cavities
   b. Femurs   d. Intervertebral disks

3. Blood cells form in _____ .
   a. red marrow   c. certain bones only
   b. all bones   d. a and c

4. The _____ is the basic unit of contraction.
   a. myofibril   c. muscle fiber
   b. sarcomere   d. myosin filament

5. Muscle contraction requires _____ .
   a. calcium ions   c. action potential arrival
   b. ATP   d. all of the above

6. ATP for muscle contraction can be formed by _____ .
   a. aerobic respiration   c. creatine phosphate breakdown
   b. glycolysis   d. all of the above

7. Match the M words with their defining feature.
   ____ muscle   a. actin's partner
   ____ muscle twitch   b. all in the hands
   ____ muscle tension   c. blood cell production
   ____ melanin   d. decline in tension
   ____ myosin   e. brownish-black pigment
   ____ marrow   f. motor unit response
   ____ metacarpals   g. force exerted by cross-bridges
   ____ myofibrils   h. muscle cells bundled in
   ____ muscle fatigue   connective tissue
        i. threadlike parts in muscle cell

Critical Thinking

1. The nose, lips, tongue, navel, and private parts are common targets for *body piercing*, cutting holes into the body so jewelry and other objects can be threaded through them. *Tattooing*, using permanent dyes to make patterns in skin, is another common fad. Besides being painful, both fads invite bacterial infections, chronic viral hepatitis, AIDS, and other nasty diseases if piercers and tattooers reuse unsterilized needles, razors, dye, gloves, swabs, or trays. Months may pass before disease symptoms arise, so the cause-and-effect connection is not always obvious. If you dismiss such risks and decide unregulated tissue invasions are okay, how could you make sure the equipment used is sterile?

2. For young women, the recommended daily allowance (RDA) of calcium is 800 milligrams/day. During Hilde's pregnancy, the RDA is 1,200 milligrams/day. Why, and what might happen to a pregnant woman's bones without the larger amount?

3. Compared to most people, Joe and other long-distance runners have a greater number of muscle fibers with more mitochondria. Sprinters have muscle fibers that contain more of the enzymes necessary for glycolysis but fewer mitochondria. Think about how these two forms of exercise differ, then explain why the muscle fibers differ between the two kinds of runners.

4. Extreme sexual dimorphism in body size involves differences in muscle mass. This might be a measure of how much male mammals invest in fighting capacity. Maybe the cost of securing and using resources for growth and maintenance of a massive body is offset by great reproductive rewards. What are the "rewards" for overly muscled human athletes (Figure 32.20)? Each year, millions of athletes and even adolescent boys use anabolic steroids to increase "brute power." *Anabolic steroids* are synthetic hormones that mimic testosterone, a sex hormone that governs secondary sexual traits and stimulates heightened aggressive behavior, which is often associated with maleness.

By stimulating protein synthesis, these anabolic steroids induce very rapid gains in muscle mass and strength during weight-training and exercise programs. Results of most studies are based on too few subjects. Also, high blood levels of these steroids trigger a sharp decline in testosterone production, resulting in acne, baldness, shrinking testes, infertility, and maybe early heart disease in men. Even brief or occasional use may damage kidneys or set the stage for cancers of the liver, testes, and prostate gland. Among women, anabolic steroids deepen the voice and produce pronounced facial hair. Their menstrual cycles become irregular and breasts may shrink.

Not every steroid user develops severe side effects. More common are mental difficulties, called *'roid rage* or *body-builder's psychosis*. Users become irritable and increasingly aggressive. Some men have become uncontrollably aggressive, manic, and delusional. One steroid user accelerated his car to high speed and deliberately drove it into a tree.

Using natural selection theory as the basis of your argument, would you say the advantages of anabolic steroid use outweigh the disadvantages? This is not a trick question.

Selected Key Terms

actin 32.6	integument CI	sarcomere 32.6
action potential 32.7	intervertebral	sarcoplasmic
bone 32.4	disk 32.3	reticulum 32.7
creatine	joint 32.4	skeletal
phosphate 32.6	keratinocyte 32.1	muscle 32.5
dermis 32.1	Langerhans cell 32.2	skin CI
endoskeleton 32.3	ligament 32.4	sliding-filament
epidermis 32.1	melanocyte 32.1	model 32.6
exercise 32.8	motor unit 32.8	tendon 32.5
exoskeleton 32.3	muscle tension 32.8	tetanus 32.8
Granstein cell 32.2	muscle twitch 32.8	vertebra
hair 32.1	myosin 32.6	(vertebrae) 32.3
hydrostatic	oxygen debt 32.6	vitamin D 32.2
skeleton 32.3	red marrow 32.4	yellow marrow 32.4

Readings   See also www.infotrac-college.com

Sherwood, L. 1997. *Human Physiology*. Third edition. Belmont, California: Wadsworth.

Weeks, O. December 1989. "Vertebrate Skeletal Muscle: Power Source for Locomotion." *BioScience* 39(11): 791–798.

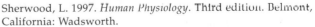

WWW   *http://www.brookscole.com/biology*

Practice quiz questions, hypercontents, BioUpdates, and critical thinking. The Brooks/Cole Biology Resource Center provides a wealth of information fully organized and integrated by chapter.

33 CIRCULATION

Heartworks

For Dr. Augustus Waller, Jimmie the bulldog was no ordinary pooch. Connected to wires and soaked to his ankles in buckets of salty water, Jimmie was a four-footed explorer of the workings of the heart (Figure 33.1a). Press your fingers to your chest a few inches left of the center, between the fifth and sixth ribs, and feel the repetitive thumpings of your heart. The same rhythms intrigued Waller and other nineteenth-century physiologists. They wondered: Does each beat of the heart produce a pattern of electrical currents? Could they find out by devising a painless way to record such currents at the body surface?

That's where Jimmie and the buckets of salty water came in. Saltwater happens to be an efficient conductor of electricity. In Waller's experiment, it picked up faint signals from Jimmie's beating heart through the skin of his legs and conducted them to a crude monitoring device. With that device, Waller made one of the first recordings of heart activity (Figure 33.1c). Today we call such recordings ECGs, an abbreviation for electrocardiograms.

A graph of the normal electrical activity of your own heart would look much the same. The pattern emerged a few weeks after you started growing, by way of mitotic cell divisions, from a single fertilized egg inside your mother. Early on, some of the embryonic cells differentiated into cardiac muscle cells, and these started to contract spontaneously. One small patch of the cells took the lead, and it has functioned as your heart's pacemaker ever since.

If all goes well, that patch of cardiac muscle cells will continue to contract as it should until the day you die. It is a natural pacemaker; it sets the baseline rate at which blood is pumped out of the heart, through the blood vessels, then back to the heart. The rate is moderate, about seventy beats every minute. But commands from the nervous

a

BUCKET

b

JIMMIE   DR. WALLER

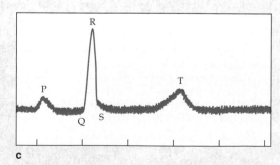

c

Figure 33.1 A bit of history in the making.
(**a**) Jimmie the bulldog, taking part in a painless experiment. (**b**) Augustus Waller and his beloved pet bulldog sharing a quiet moment in Waller's study after the experiment, which yielded one of the world's first electrocardiograms (**c**).

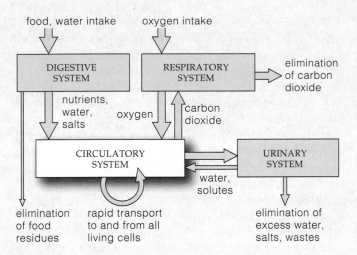

food, water intake   oxygen intake

DIGESTIVE SYSTEM   RESPIRATORY SYSTEM   elimination of carbon dioxide

nutrients, water, salts   oxygen   carbon dioxide

CIRCULATORY SYSTEM   URINARY SYSTEM

water, solutes

elimination of food residues   rapid transport to and from all living cells   elimination of excess water, salts, wastes

Figure 33.2 Diagram of the functional connections between the circulatory, respiratory, and digestive systems, which interact in transporting substances to and from all living cells in the animal body. Their integrated activities help maintain favorable operating conditions in the internal environment.

system and endocrine system continually adjust it. When you jog, for example, your skeletal muscle cells demand much more blood-borne oxygen and glucose than they do when you sleep. At such times, your heart starts pounding more than twice as fast, and this helps deliver sufficient blood to the cells.

We have come a long way from Waller and Jimmie in our monitorings of the heart. Sensors can now detect the faint signals characteristic of an impending heart attack. Internists now use computers to analyze a patient's beating heart, and they use ultrasound probes to build images of it on a video screen. Cardiologists routinely substitute battery-powered pacemakers for malfunctioning natural ones.

With this chapter, we turn to the circulatory system, the means by which substances move rapidly to and from the interstitial fluid that bathes living cells in nearly all animals. The circulatory system continually accepts the oxygen, nutrients, and other substances that the animal secures, by way of respiratory and digestive systems, from the outside environment (Figure 33.2). Simultaneously it picks up carbon dioxide and other wastes from cells and delivers them to the respiratory and urinary systems for disposal. Its smooth operation is absolutely central to maintaining operating conditions in the internal environment within a tolerable range— a state we call homeostasis.

KEY CONCEPTS

1. All cells survive by exchanging substances with their surroundings. In most animals, substances rapidly move to and from cells by way of a closed circulatory system.

2. Blood, a fluid connective tissue, is the transport medium of circulatory systems. It transports oxygen, carbon dioxide, plasma proteins, vitamins, hormones, lipids, and other solutes. It also transports metabolically generated heat.

3. In birds and mammals, a four-chambered, muscular heart pumps blood through two separate circuits of blood vessels, both of which lead back to the heart. One flows through the lungs and picks up oxygen. The other delivers oxygenated blood to all body regions and picks up carbon dioxide wastes.

4. Arteries are large-diameter, low-resistance vessels that rapidly transport oxygenated blood from the heart. Veins are large-diameter, low-resistance vessels that transport oxygen-poor blood to the heart and serve as blood volume reservoirs.

5. Arterioles are sites where the flow volume through each organ is controlled. In response to signals, their diameters widen in some regions and narrow in others. Their coordinated responses divert more of the flow volume to organs that are most active at a given time.

6. Numerous small-diameter, thin-walled capillaries interconnect in capillary beds, which are diffusion zones. As a volume of blood spreads out through a bed, it slows owing to the total cross-sectional area of the capillaries, which is greater than that of arterioles. The slowdown allows time for exchanges with interstitial fluid.

7. Pressure forces some fluid out of capillaries, into interstitial fluid. Drainage vessels of the lymphatic system return it to the blood. Organs of that system work to cleanse blood of infectious agents and other threats to the body.

CIRCULATORY SYSTEMS—AN OVERVIEW

General Characteristics

Imagine that an earthquake has closed off the highways around your neighborhood. Grocery trucks can't enter and waste-disposal trucks can't leave, so food supplies dwindle and garbage piles up. Cells would face similar predicaments if your body's highways were disrupted. The highways are part of a **circulatory system**, which functions in the rapid internal transport of substances to and from cells. The system helps maintain favorable neighborhood conditions, so to speak, and this is vital work. All of your differentiated cells perform specialized tasks and cannot fend for themselves. Different types interact in coordinated ways to maintain the volume, composition, and temperature of **interstitial fluid**, the tissue fluid that bathes them. A circulating connective tissue—**blood**—interacts with interstitial fluid. Blood makes continual deliveries and pickups that help keep conditions tolerable for enzymes and other molecules that carry out cell activities. Together, interstitial fluid and blood are the body's "internal environment."

Blood flows inside blood vessels—tubes that differ in wall thickness and in diameter. A muscular pump, the **heart**, generates pressure that keeps blood flowing. Like many animals, you have a *closed* circulatory system, in which blood is confined within continuously connected walls of the heart and blood vessels. This is not true of *open* circulatory systems of arthropods, such as insects, and most mollusks. In open systems, blood flows out through the vessels and into sinuses, which are small spaces in body tissues. There, blood mingles with tissue fluids, then moves back to the heart through openings in the blood vessels or the heart wall (Figure 33.3a).

Think about the overall "design" of a closed system. Because the heart pumps incessantly, the *volume* of blood flowing through blood vessels has to equal the heart's output at any given time. The flow's *velocity* (speed) is highest in large-diameter transport vessels. It decreases in specific body regions, where there must be enough time for blood to exchange substances with cells. The required slowdown proceeds at **capillary beds**, where blood spreads out through many small-diameter blood vessels called **capillaries**. During any interval, the same volume of blood is moving forward through the beds as elsewhere in the system. But it is doing so at a more leisurely pace, owing to the great total cross-sectional area of the blood capillaries. The simple analogy given in Figure 33.3e may help you grasp this concept.

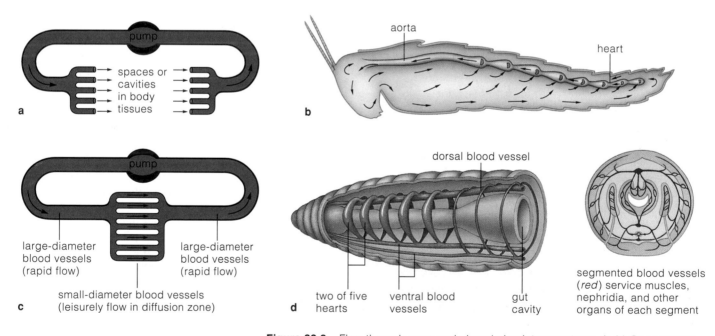

Figure 33.3 Flow through open and closed circulatory systems. (**a,b**) Open system of a grasshopper. A "heart" pumps blood through a vessel (aorta). Blood moves into tissue spaces and mingles with fluid bathing cells, then reenters the heart at openings in the heart wall. (**c,d**) Closed system of an earthworm. Blood is confined within several pairs of muscular "hearts" near the head end and within blood vessels.

(**e**) Relation between flow velocity and total cross-sectional area in a closed circulatory system. Visualize two fast rivers flowing into and out of a lake. The flow *rate* is the same in all three places; an identical volume of water moves from point *1* to point *3* during the same interval. Flow *velocity* decreases in the lake, for the volume spreads through a larger cross-sectional area and moves forward a shorter distance during the same time.

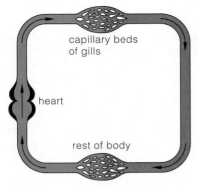
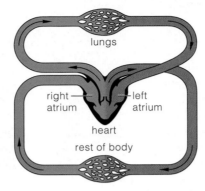

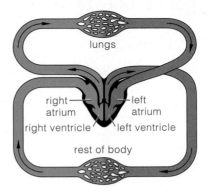

a In fishes, a two-chambered heart (atrium, ventricle) pumps blood in one circuit. Blood picks up oxygen in gills, delivers it to rest of body. Oxygen-poor blood flows back to heart.

b In amphibians, a heart pumps blood through two partially separate circuits. Blood flows to lungs, picks up oxygen, returns to heart. It mixes with oxygen-poor blood still in heart, flows to rest of body, returns to heart.

c In birds and mammals, the heart is fully partitioned into two halves. Blood circulates in two circuits: from the heart's right half to lungs and back, then from the heart's left half to oxygen-requiring tissues and back.

Figure 33.4 Comparison of the closed circulatory systems of vertebrate groups.

Evolution of Vertebrate Circulatory Systems

Humans and other existing vertebrates have a closed circulatory system, although the pump and plumbing for fishes, amphibians, birds, and mammals differ in their details. The differences evolved over hundreds of millions of years and corresponded to the move of some vertebrate lineages onto land. Recall, from Section 24.3, that fishes, the first vertebrates, had gills. Gills, like all respiratory structures, have a thin, moist surface that oxygen and carbon dioxide can diffuse across. Later, in the ancestors of all land vertebrates, lungs evolved that supplemented gas exchange. Being *internally moistened* sacs, lungs had advantages for the move onto dry land. Also advantageous were concurrent modifications in circulatory systems, which pick up oxygen from lungs and deliver carbon dioxide wastes to them.

Consider this: In fishes, blood flows in *one* circuit (Figure 33.4a). Pressure generated by a two-chambered heart forces it through capillary beds of gills, into the largest artery, capillary beds of organs, then back to the heart. Extensive gill capillaries offer so much resistance to flow that pressure drops considerably before blood enters the main artery. Blood delivery is fine for the activity level of most fishes. It would not be sufficient for the more active life-styles of most land vertebrates.

When amphibians were evolving, their heart became partitioned into right and left halves—only partly so, but enough to pump blood through *two* partially separated circuits (Figure 33.4b). The separation of flow continued in crocodilians, which are reptiles. It became complete in birds and mammals. Their heart acts like two side-by-side pumps (Figure 33.4c). The *right* half of their heart pumps oxygen-poor blood to the lungs, where the blood picks up oxygen and gives up carbon dioxide. Then the freshly oxygenated blood flows to the heart's left half. This route is the **pulmonary circuit.**

In the **systemic circuit**, the heart's *left* half pumps the freshly oxygenated blood to every tissue and organ where oxygen is used and carbon dioxide forms. Then the oxygen-poor blood flows to the heart's right half.

The double circuit is a rapid and efficient mode of blood delivery. It supports the high levels of activity typical of vertebrates whose ancestors evolved on land.

Links With the Lymphatic System

The heart's pumping action puts pressure on blood flowing through the circulatory system. Partly because of the pressure, small amounts of water and a few of the proteins dissolved in blood are forced out of capillaries and become part of interstitial fluid. However, a rather elaborate network of drainage vessels picks up excess interstitial fluid and reclaimable solutes, then returns them to the circulatory system. This network is part of the **lymphatic system**. Later, you will see how other parts of the lymphatic system help cleanse bacteria and other threats from fluid being returned to the blood.

Closed circulatory systems confine blood within one or more hearts and a network of blood vessels. In the open systems, blood also intermingles with tissue fluids.

The closed system of vertebrates transports substances to and from interstitial fluid bathing the body's cells. It is functionally connected with the lymphatic system.

Blood flows rapidly in large-diameter vessels between the heart and capillary beds. There, the flow velocity slows and exchanges are made between blood and interstitial fluid.

In fishes, blood flows in one circuit from and back to the heart. In birds and mammals, blood flows in two circuits, through a heart partitioned as two side-by-side pumps. The double circuit supports the high levels of activity typical of vertebrates that evolved on land.

CHARACTERISTICS OF BLOOD

Functions of Blood

Blood is a connective tissue with multiple functions. It transports oxygen, nutrients, and other solutes to cells. It carries away their metabolic wastes and secretions, including hormones. Blood helps stabilize internal pH. Blood also serves as a highway for phagocytic cells that fight infections and scavenge debris in tissues. In birds and mammals, blood helps equalize body temperature. It does so by carrying excess heat from skeletal muscles and other regions of high metabolic activity to the skin, where heat can be dissipated.

Blood Volume and Composition

The volume of blood depends on body size and on the concentrations of water and solutes. Blood volume for average-size adult humans is about 6 to 8 percent of the total body weight. That amounts to about four or five quarts. As is the case for all vertebrates, human blood is a sticky fluid, thicker than water and slower flowing. Its components are the **plasma**, **red blood cells**, **white blood cells**, and **platelets.** Plasma normally accounts for 50 to 60 percent of the total blood volume.

PLASMA Prevent a blood sample in a test tube from clotting, and it separates into a red, cellular portion and the plasma, a straw-colored liquid, which floats on the cellular portion (Figure 33.5). Plasma is mostly water, and it serves as a transport medium for blood cells and platelets. Plasma also functions as a solvent for ions

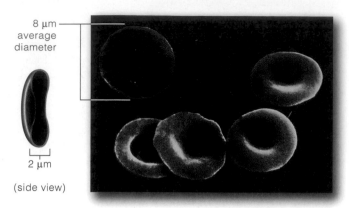

8 μm average diameter

2 μm

(side view)

Figure 33.6 Size and shape of red blood cells.

and molecules, including hundreds of different plasma proteins. Some of the plasma proteins transport lipids and fat-soluble vitamins through the body. Others have roles in blood clotting or in defense against pathogens. Collectively, the concentration of the plasma proteins affects the blood's fluid volume, for it influences the movement of water between the blood and interstitial fluid. Glucose and other simple sugars, as well as lipids, amino acids, vitamins, and hormones, are dissolved in plasma. So are oxygen, carbon dioxide, and nitrogen.

RED BLOOD CELLS Erythrocytes, or red blood cells, are biconcave disks, like doughnuts with a squashed-in center instead of a hole (Figure 33.6). They transport the oxygen used in aerobic respiration, and they carry away some carbon dioxide wastes. When oxygen diffuses into

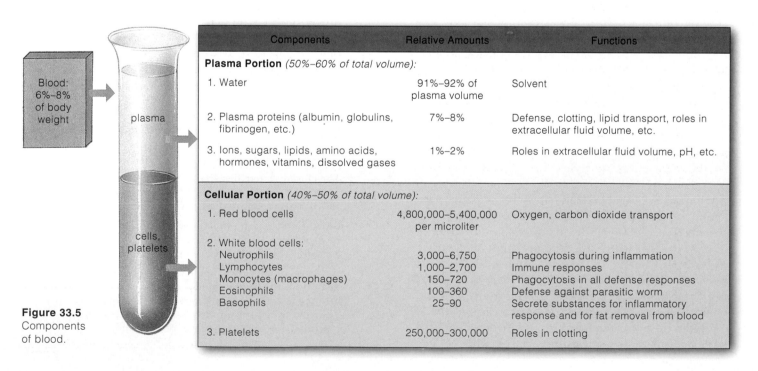

Figure 33.5 Components of blood.

Blood: 6%–8% of body weight

plasma

cells, platelets

Components	Relative Amounts	Functions
Plasma Portion *(50%–60% of total volume):*		
1. Water	91%–92% of plasma volume	Solvent
2. Plasma proteins (albumin, globulins, fibrinogen, etc.)	7%–8%	Defense, clotting, lipid transport, roles in extracellular fluid volume, etc.
3. Ions, sugars, lipids, amino acids, hormones, vitamins, dissolved gases	1%–2%	Roles in extracellular fluid volume, pH, etc.
Cellular Portion *(40%–50% of total volume):*		
1. Red blood cells	4,800,000–5,400,000 per microliter	Oxygen, carbon dioxide transport
2. White blood cells:		
Neutrophils	3,000–6,750	Phagocytosis during inflammation
Lymphocytes	1,000–2,700	Immune responses
Monocytes (macrophages)	150–720	Phagocytosis in all defense responses
Eosinophils	100–360	Defense against parasitic worm
Basophils	25–90	Secrete substances for inflammatory response and for fat removal from blood
3. Platelets	250,000–300,000	Roles in clotting

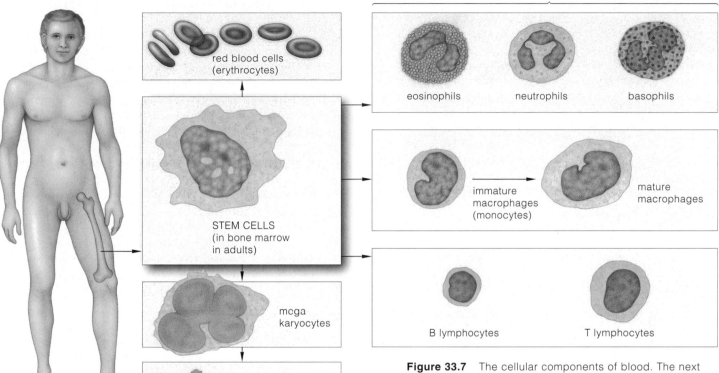

white blood cells (leukocytes)

red blood cells (erythrocytes)

eosinophils   neutrophils   basophils

STEM CELLS (in bone marrow in adults)

immature macrophages (monocytes)   mature macrophages

mega karyocytes

B lymphocytes   T lymphocytes

platelets

Figure 33.7 The cellular components of blood. The next chapter gives a closer look at white blood cell functions. Section 33.10 describes the role of platelets in blood clotting.

blood, it binds with hemoglobin—the iron-containing pigment that gives red blood cells their color. (Here you may wish to review hemoglobin's molecular structure, as shown in Section 3.5.) Oxygenated blood is bright red. Poorly oxygenated blood is darker red but appears blue inside blood vessel walls near the body surface.

Red blood cells are derived from stem cells in bone marrow (Figure 33.7). Generally speaking, **stem cells** remain unspecialized and retain the capacity for mitotic cell division. Their daughter cells also divide, but only a portion go on to differentiate into specialized types.

Mature red blood cells no longer have their nucleus, nor do they require it. They have enough hemoglobin, enzymes, and other proteins to function for about 120 days. Phagocytes continually engulf the oldest red blood cells or those already dead. But ongoing replacements keep the cell count fairly stable. A **cell count** is a measure of the number of cells of a given type in a microliter of blood. For example, the average number of red blood cells is 5.4 million in males and 4.8 million in females.

WHITE BLOOD CELLS Leukocytes, or white blood cells, arise from stem cells in bone marrow. They function in daily housekeeping and defense. Many patrol tissues, where they target or engulf damaged or dead cells and anything chemically recognized as foreign to the body. Many others are massed together in the lymph nodes

and spleen, which are organs of the lymphatic system. There they divide to produce armies of cells that battle specific viruses, bacteria, and other invaders.

White blood cells differ in size, nuclear shape, and staining traits. They fall in five categories: neutrophils, eosinophils, basophils, monocytes, and lymphocytes (Figure 33.7). Their numbers can change, depending on whether an individual is active, healthy, or under siege, as described in the next chapter. The neutrophils and monocytes are search-and-destroy cells. The monocytes follow chemical trails to inflamed tissues. There they develop into macrophages ("big eaters") that can engulf invaders and debris. Two classes of lymphocytes, B cells and T cells, make highly specific defense responses.

PLATELETS Some stem cells in bone marrow give rise to megakaryocytes. These "giant" cells shed cytoplasmic fragments wrapped in a bit of plasma membrane. The membrane-bound fragments are what we call platelets. Each platelet only lasts five to nine days, but hundreds of thousands are always circulating in blood. They can release substances that initiate blood clotting.

Vertebrate blood has roles in transport, defense, clotting, and maintaining the volume, composition, and temperature of the internal environment.

WWW

BLOOD DISORDERS

The body continually replaces blood cells for good reason. Besides aging and dying off regularly, blood cells typically encounter a variety of pathogens that use them as places to complete their life cycle. Besides this, sometimes blood cells malfunction as a result of gene mutations.

RED BLOOD CELL DISORDERS   Consider the **anemias**, disorders that result from too few red blood cells or deformed ones. Oxygen levels in blood cannot be kept high enough to support normal metabolism. Shortness of breath, fatigue, and chills follow. *Hemorrhagic* anemias result from sudden blood loss, as from a severe wound; *chronic* anemias result from ongoing but slight blood loss, as from an undiagnosed bleeding ulcer or hemorrhoids, or during menstruation.

Certain infectious bacteria and parasites replicate inside blood cells, then escape by lysis; they cause some *hemolytic* anemias. Insufficient iron in the diet results in *iron deficiency* anemia, for red blood cells cannot produce enough normal hemoglobin without iron. B_{12} *deficiency* anemia is a potential hazard for strict vegetarians and alcoholics. Red blood cells form but they cannot divide without vitamin B_{12}. Normally, meats, poultry, and fish in the diet provide enough of the vitamin.

As described elsewhere in the book, a gene mutation that gives rise to an abnormal form of hemoglobin can result in *sickle-cell* anemia. Another mutation blocks or lowers the synthesis of the globin chains that make up hemoglobin, and causes *thalassemias*. Too few red blood cells can form; those that do are thin and fragile.

Polycythemias—symptoms of far too many red blood cells—make blood flow sluggish. Some bone marrow cancers can result in this condition. So can "blood doping" by some athletes who compete in strenuous events. They withdraw and store their own red blood cells, then reinject them a few days prior to competition. The withdrawal triggers red blood cell formation as the body attempts to replace "lost" cells, so the cell count is bumped up when withdrawn cells are put back in the body. The idea is to increase oxygen-carrying capacity and endurance. Blood doping leads to temporary high blood pressure and lowers blood's viscosity. Some believe the practice works, but others call it unethical. It is banned from the Olympics.

WHITE BLOOD CELL DISORDERS   You may have heard of *infectious mononucleosis*. An Epstein-Barr virus causes this highly contagious disease, which results from too many monocytes and lymphocytes. Following a few weeks of fatigue, aches, low fever, and a chronic sore throat, the patient usually recovers.

Recovery is chancy for *leukemias*. As you read in Chapter 11, these cancers suppress or impair the formation of white blood cells in bone marrow. The cancer cells can be killed by radiation therapy and chemotherapy, but side effects are severe. Remissions may last for months or years. A remission is a symptom-free period of a chronic illness.

www

Concerning Agglutination

Whenever blood volume or blood cell counts decline, countermeasures kick in automatically. However, if the volume were to decrease by more than 30 percent, then circulatory shock would follow and could lead to death.

Blood from donors can be transfused into patients affected by a blood disorder or substantial blood loss. Such **blood transfusions** cannot be hit-or-miss events. Why not? *A potential donor and the recipient may not have the same kinds of recognition proteins at the surface of their red blood cells.* Some of the proteins are "self" markers; they identify the cells as belonging to one's own body. During a blood transfusion, if cells from a donor have the "wrong" marker, the recipient's immune system will recognize them as foreign, with serious consequences.

Figure 33.8 shows what happens when blood from incompatible donors and recipients intermingles. In a defense response called **agglutination**, proteins called

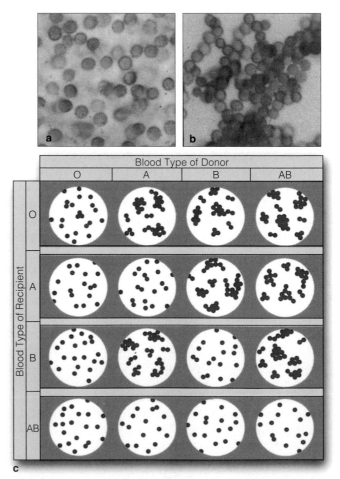

Figure 33.8   Light micrographs showing (**a**) the absence of agglutination in a mixture of two different yet compatible blood types and (**b**) agglutination in a mixture of incompatible types. (**c**) Agglutination responses in blood types A, B, AB, and O when mixed with samples of the same and different types.

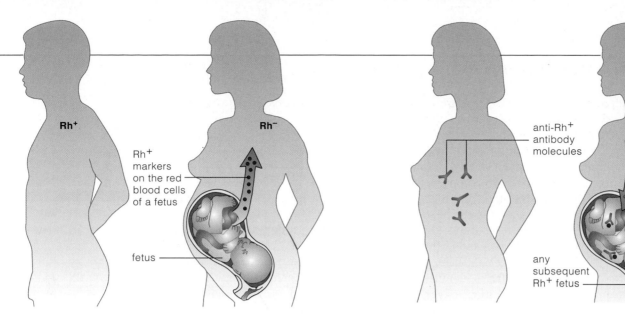

a A forthcoming child of an Rh⁻ woman and Rh⁺ man inherits the gene for the Rh⁺ marker. During childbirth, some of its cells bearing the marker may leak into the maternal bloodstream.

b The foreign marker stimulates antibody formation. If she becomes pregnant again and if this second fetus (or any other) inherits the gene for the marker, her circulating anti-Rh⁺ antibodies will act against it.

Figure 33.9 Antibody production in response to Rh⁺ markers on red blood cells of a fetus.

antibodies that are circulating in the plasma act against the foreign cells and cause them to clump together. (As described in the next chapter, antibodies bind to specific foreign markers and target the cell or particle bearing them for destruction by the immune system.) The sheer number of "foreign" cells transfused into the recipient translates into numerous clumps, which can clog small blood vessels and damage tissues. Without treatment, death may follow. The same thing can happen during certain pregnancies if antibodies diffuse from a mother's circulatory system into that of her unborn child.

Based on an understanding of cell surface markers and antibodies, scientists have devised ways to analyze the forms of self markers on a person's red blood cells. Blood typing is based on these analyses.

ABO Blood Typing

Molecular variations in one kind of self-marker on red blood cells are analyzed with **ABO blood typing**. The genetic basis of such variation is described in Section 10.4. People with one form of the marker are said to have type A blood; those with another form have type B blood. Those with both forms of the marker on their red blood cells have type AB blood. Others do not have either form of the marker; they have type O blood.

If you are type A, your antibodies ignore A markers but will act against B markers. If you are type B, your antibodies will ignore B markers but will act against A markers. If you are type AB, your antibodies ignore both forms of the marker, so you can tolerate donations of type A, B, AB, or O blood. If you are type O, you have antibodies against both forms of the marker, so your options are limited to type O donations.

Rh Blood Typing

Rh blood typing is based on the presence or absence of an Rh marker (so named because it was first identified in blood samples of *Rh*esus monkeys). If you are type Rh⁺, your blood cells bear this marker at their surface. If you are type Rh⁻, they don't. Ordinarily, people do not have antibodies against Rh markers. But a recipient of transfused Rh⁺ blood produces antibodies against them, and the antibodies remain in the blood.

If an Rh⁻ woman becomes impregnated by an Rh⁺ man, there is a chance that the fetus will be Rh⁺. During childbirth, some fetal red blood cells may leak into the woman's bloodstream. If they do, then her body will produce antibodies against Rh (Figure 33.9). And if she becomes pregnant *again*, Rh antibodies will enter the bloodstream of this new fetus. If its blood is type Rh⁺, then her antibodies will cause its red blood cells to swell, rupture, and release hemoglobin.

Erythroblastosis fetalis is a severe outcome of mixing Rh⁺ and Rh⁻ types. The fetus dies; too many cells are destroyed. If the condition is diagnosed before birth, the newborn can survive if its blood is slowly replaced with transfusions that are free of Rh antibodies. Today, a known Rh⁻ woman can be treated right after her first pregnancy with an anti-Rh gamma globulin (RhoGam) that will protect her next fetus. The drug will inactivate any Rh⁺ fetal blood cells that are circulating through her bloodstream before she can become sensitized and begin producing potentially dangerous antibodies.

To avoid the symptoms of blood incompatibilities, red blood cells should be typed before transfusions or pregnancies.

WWW

"Cardiovascular" comes from the Greek *kardia* (heart) and Latin *vasculum* (vessel). In a human cardiovascular system, blood is pumped by a muscular heart into large-diameter **arteries**. Then it flows into small, muscular **arterioles**, which branch into the even smaller diameter capillaries introduced earlier. Blood flows continuously from capillaries into small **venules**, and from there into large-diameter **veins** that return blood to the heart.

As in most vertebrates, a partition separates a human heart into a double pump, which drives blood through two cardiovascular circuits (Figure 33.10). Each circuit has its own arteries, arterioles, capillaries, venules, and veins. The *pulmonary* circuit, a short loop, oxygenates blood. It leads from the heart's right half to capillary beds in both lungs, then returns to the heart's left half. The *systemic* circuit is a longer loop starting at the left half of the heart. Its main artery, the **aorta**, accepts the oxygenated blood, which flows on through arterioles, capillary beds in all body regions, and then veins that deliver the oxygen-poor blood to the heart's right half. Figure 33.11 identifies the location of the major blood vessels of both circuits and defines their functions.

For most of the systemic branchings, a given volume of blood flows through one capillary bed. The branch to the intestines is one of the exceptions. First blood picks up glucose and other absorbed substances from one bed, then moves through another capillary bed, in the liver —an organ with a key role in nutrition. The second bed gives the liver time to process absorbed substances.

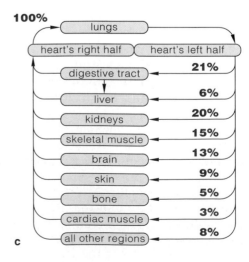

b Systemic circuit for blood flow

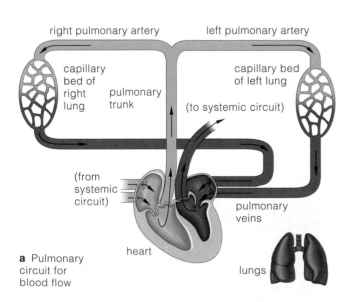

a Pulmonary circuit for blood flow

Figure 33.10 (**a**,**b**) Diagrams of the pulmonary and systemic circuits for blood flow through the human cardiovascular system. The blood vessels that are carrying oxygenated blood are color-coded *red*. Those carrying oxygen-poor blood are color-coded *blue*. (**c**) Distribution of the heart's output in a person at rest. The lungs receive all blood pumped out of the heart's right half. The organs serviced by the systemic circuit receive the portions indicated from the heart's left half. Control mechanisms adjust the flow distribution when necessary.

100%

lungs	
heart's right half	heart's left half
digestive tract	21%
liver	6%
kidneys	20%
skeletal muscle	15%
brain	13%
skin	9%
bone	5%
cardiac muscle	3%
all other regions	8%

c

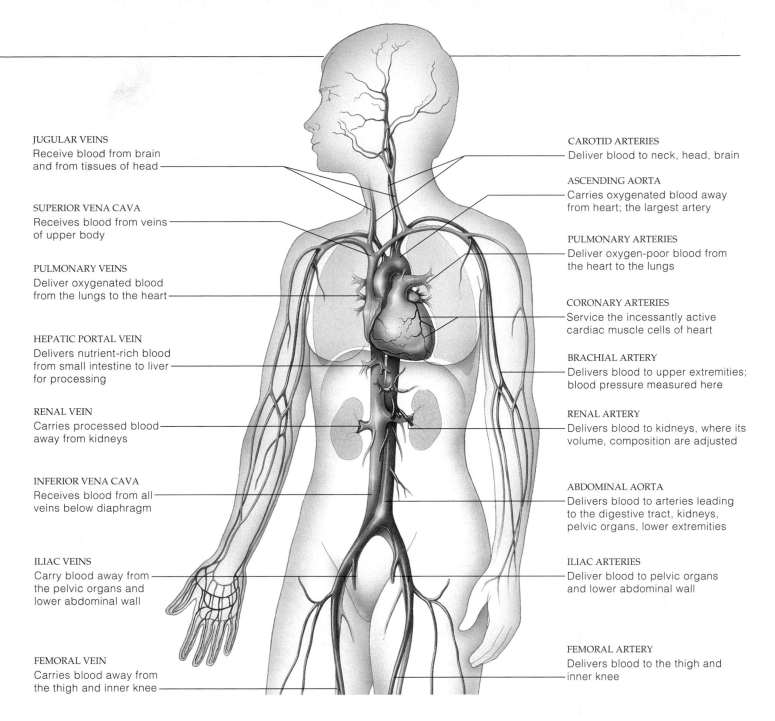

JUGULAR VEINS
Receive blood from brain and from tissues of head

SUPERIOR VENA CAVA
Receives blood from veins of upper body

PULMONARY VEINS
Deliver oxygenated blood from the lungs to the heart

HEPATIC PORTAL VEIN
Delivers nutrient-rich blood from small intestine to liver for processing

RENAL VEIN
Carries processed blood away from kidneys

INFERIOR VENA CAVA
Receives blood from all veins below diaphragm

ILIAC VEINS
Carry blood away from the pelvic organs and lower abdominal wall

FEMORAL VEIN
Carries blood away from the thigh and inner knee

CAROTID ARTERIES
Deliver blood to neck, head, brain

ASCENDING AORTA
Carries oxygenated blood away from heart; the largest artery

PULMONARY ARTERIES
Deliver oxygen-poor blood from the heart to the lungs

CORONARY ARTERIES
Service the incessantly active cardiac muscle cells of heart

BRACHIAL ARTERY
Delivers blood to upper extremities; blood pressure measured here

RENAL ARTERY
Delivers blood to kidneys, where its volume, composition are adjusted

ABDOMINAL AORTA
Delivers blood to arteries leading to the digestive tract, kidneys, pelvic organs, lower extremities

ILIAC ARTERIES
Deliver blood to pelvic organs and lower abdominal wall

FEMORAL ARTERY
Delivers blood to the thigh and inner knee

Figure 33.11 Location and functions of major blood vessels of the human cardiovascular system.

Figure 33.10c shows how the pulmonary and systemic circuits distribute the heart's output to different organs. The percentages listed are for a person at rest. As you will see, the flow distribution along the systemic route is adjusted in response to changes in levels of physical activity and to shifts in external and internal conditions. For instance, the flow to and from a skeletal muscle varies greatly, depending on what it is being called upon to do at a given time. The same is true of skin. When the body gets too cold, the flow of blood (and metabolic heat) is diverted away from skin's vast capillary beds.

When the body is hot, flow to the skin's beds increases and heat radiates away from the skin's surface. Only the brain is exempt; it cannot tolerate flow variations.

The human cardiovascular system consists of two separate circuits, pulmonary and systemic, for blood flow.

The pulmonary circuit is a short loop from the heart's right half, through both lungs, to the heart's left half. Oxygen-poor blood flowing into the circuit rapidly picks up oxygen and gives up carbon dioxide at capillary beds in the lungs.

The longer systemic circuit starts at the heart's left half and aorta. It delivers oxygen and accepts carbon dioxide at capillary beds of all metabolically active regions; then its veins deliver oxygen-poor blood to the heart's right half.

WWW

Heart Structure

Think about the 2.5 billion times a human heart beats during a seventy-year life span, and you know it must be a durable pump. Its structure speaks of its durability (Figure 33.12). The pericardium, a double sac of tough connective tissue, protects the heart and anchors it to nearby structures (*peri*, around). A fluid in between its layers lubricates the heart during its perpetual twisting motions. The inner layer is actually the outer portion of the heart wall. The bulk of that wall, the myocardium, consists of cardiac muscle cells tethered to elastin and collagen fibers. The fibers are so densely crisscrossed, they act as the "skeleton" against which contractile force is applied. The oxygen-demanding cardiac muscle cells have their own, *coronary* circulation. Coronary arteries branch off the aorta and lead into a capillary bed that services only them. The heart wall's glistening, inner layer consists of connective tissue and endothelium, a one-cell-thick epithelial sheet. Endothelium is a special tissue, present only in the heart and blood vessels.

Each half of the heart has two chambers: an atrium (plural, atria) and a ventricle. Blood flows into the atria, down into the ventricles, then out through great arteries (the aorta or pulmonary trunk). Between each atrium and ventricle is an AV valve (short for atrioventricular). Between each ventricle and the artery leading out from it is a semilunar valve. Both are *one-way* valves that are alternately forced open and shut to help keep the blood moving in the forward direction only.

Cardiac Cycle

Each time the heart beats, its four chambers go through phases of contraction (systole) and relaxation (diastole). The sequence of contraction and relaxation is a **cardiac cycle**. When relaxed, the atria fill with blood. Increasing fluid pressure forces the AV valves open. Blood flows into the ventricles, which completely fill when the atria contract (Figure 33.13*a*). As the filled ventricles start to contract, the rising fluid pressure forces the AV valves shut. It rises so sharply above the pressure in the great

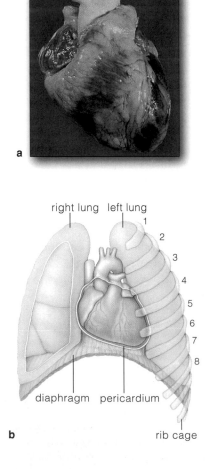

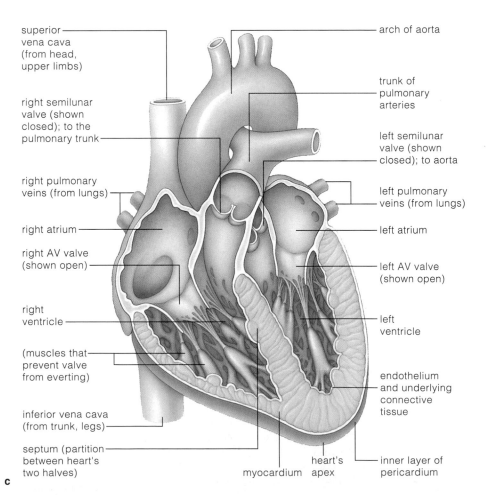

Figure 33.12 (**a**) Photograph of the human heart and (**b**) its location in the thoracic cavity. (**c**) Cutaway view of the human heart, showing its wall and internal organization.

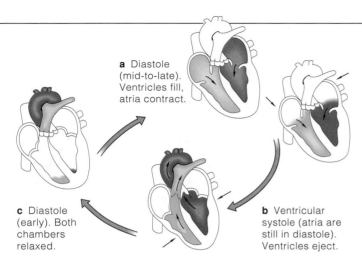

a Diastole (mid-to-late). Ventricles fill, atria contract.

c Diastole (early). Both chambers relaxed.

b Ventricular systole (atria are still in diastole). Ventricles eject.

Figure 33.13 Blood flow during part of a cardiac cycle. Blood and heart movements generate a "lub-dup" sound at the chest wall. At each "lub," AV valves are closing as ventricles contract. At each "dup," semilunar valves are closing as ventricles relax.

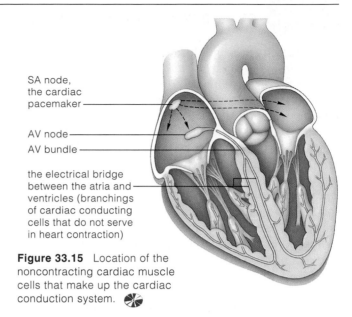

SA node, the cardiac pacemaker

AV node

AV bundle

the electrical bridge between the atria and ventricles (branchings of cardiac conducting cells that do not serve in heart contraction)

Figure 33.15 Location of the noncontracting cardiac muscle cells that make up the cardiac conduction system.

arteries, it forces the semilunar valves open—and blood leaves the heart (Figure 33.13b). The ventricles relax, the semilunar valves close, and the already filling atria are ready to repeat the cycle. Thus, in a cardiac cycle, atrial contraction simply helps fill the ventricles. *Contraction of the ventricles is the driving force for blood circulation.*

Mechanisms of Contraction

Like skeletal muscle, cardiac muscle is striated (striped). In response to action potentials, the sarcomeres of its cells contract in the manner predicted by the sliding-filament model (Section 32.6). This tissue also requires energy for contraction; large numbers of mitochondria in the myocardium provide the ATP. But cardiac muscle cells are structurally unique. They are branching, short, and joined at the end regions. Communication junctions span the plasma membranes of abutting regions and let

action potentials spread swiftly among cells, in waves of excitation that wash over the heart (Figure 33.14).

In addition, about 1 percent of the cardiac muscle cells *do not* contract. They function instead as a **cardiac conduction system**. About seventy times each minute, these specialized cells initiate and propagate waves of excitation that travel in a rhythmic, orderly sequence from the atria to ventricles. The synchronized excitation underlies the heart's efficient pumping. Each wave starts at the SA node, a cluster of cell bodies in the wall of the right atrium (Figure 33.15). It passes through the wall to another cell body cluster, the AV node. This is the only electrical bridge between the atria and ventricles, which connective tissue insulates everywhere else. After the AV node, conducting cells are arranged as a bundle in the partition between the heart's two halves. These cells branch, and the branches deliver the excitatory wave up the ventricle walls. The ventricles contract in response with a twisting movement, upward from the heart's apex, that ejects blood into the great arteries.

The SA node fires action potentials faster than the rest of the system and serves as the **cardiac pacemaker**. Its spontaneous, rhythmic signals are the foundation for the normal rate of heartbeat. The nervous system can only *adjust* the rate and strength of contractions dictated by the pacemaker. Even if all nerves leading to a heart are severed, the heart will keep on beating!

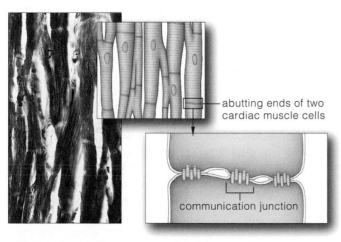

abutting ends of two cardiac muscle cells

communication junction

Figure 33.14 Light micrograph of cardiac muscle cells and sketches showing communication junctions between them.

The heart's construction reflects its role as a durable pump. Under the spontaneous, rhythmic signals from the cardiac pacemaker, its branching, abutting cardiac muscle cells contract in synchrony, almost as if they were a single unit.

Although the heart has four chambers (each half has one atrium and one ventricle), contraction of the ventricles is the driving force for blood circulation away from the heart.

WWW

BLOOD PRESSURE IN THE CARDIOVASCULAR SYSTEM

As you have seen, blood flows to and from the human heart through arteries, arterioles, capillaries, venules, and finally veins. Figure 33.16 shows the structure of these vessels. The arteries are the major transporters of oxygenated blood throughout the body. Arterioles in each region are sites where the volume of blood flow through each organ can be controlled. The capillaries and, to a lesser extent, venules are diffusion zones, and veins are the major transporters of oxygen-poor blood to the heart.

Two key factors influence the rate of flow through each type of blood vessel. First, the flow rate is directly proportional to the pressure gradient between the start and end of the vessel. Second, the flow rate is inversely proportional to the vessel's resistance to flow.

Blood pressure is fluid pressure imparted to blood by heart contractions. The beating heart establishes the higher blood pressure at the beginning of a vessel. As flowing blood rubs against the vessel's inner wall, the friction causes some energy (in the form of pressure) to be lost. The lower pressure at the end of the vessel is due to the frictional losses. Even small changes in blood vessel diameter have major influence over the flow rate. In the pulmonary and systemic circuits, the diameters decrease and present more resistance to flow. And so, because of heart contractions and subsequent events (mainly frictional losses), blood pressure is highest in the contracting ventricles, still high at the beginning of arteries, then continues to drop until the relaxed atria, where it reaches its lowest value (Figure 33.17).

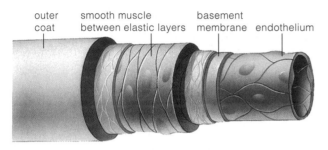

a ARTERY

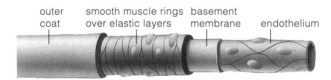

b ARTERIOLE

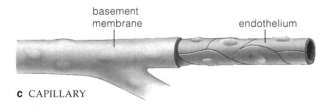

c CAPILLARY

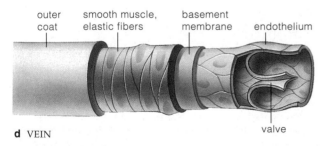

d VEIN

Figure 33.16 Structure of blood vessels. The basement membrane around the endothelium of each vessel is a noncellular layer, rich with proteins and polysaccharides.

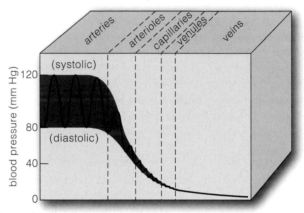

Figure 33.17 Blood pressure. This is a plot of measurements made of the drop in fluid pressure for a given volume of blood moving through the systemic circuit.

Arterial Blood Pressure

From Figure 33.16*a* and the preceding discussion, you see that arteries have a large diameter and present low resistance to flow. Thus they serve as rapid transporters of oxygenated blood. They also are pressure reservoirs that smooth out pulsations in pressure caused by each cardiac cycle. Their thick, muscular, elastic wall bulges as ventricular contraction forces a large volume of blood into them. Then the wall recoils and so forces the blood forward through the circuit while the heart is relaxing.

Suppose you decide to measure your blood pressure daily when you are resting. Typically, you would take a reading from the brachial artery of an upper arm, as in Figure 33.18. *Systolic* pressure is the peak pressure that the contracting ventricles exert against the artery's wall during a cardiac cycle. *Diastolic* pressure is the lowest pressure when blood is draining into the vessels after it. You can estimate the mean arterial pressure as *diastolic pressure + 1/3 pulse pressure.* (The difference between the highest and lowest readings is the pulse pressure).

Figure 33.18 Measuring blood pressure.

A hollow cuff attached to a pressure gauge is wrapped around a person's upper arm. The cuff is inflated with air to a pressure above the highest pressure of the cardiac cycle (at systole, when the ventricles contract). Above this pressure, no sounds can be heard through a stethoscope positioned below the cuff and above the artery (because no blood is flowing through the vessel). Air in the cuff is slowly released, so that some blood flows into the artery. The turbulent flow causes soft tapping sounds, and when this first occurs, the value on the gauge is the systolic pressure, or about 120 mm mercury (Hg) in young adults at rest. This particular value means that the measured pressure would force mercury to move upward 120 millimeters in a narrow glass column.

More air is released from the cuff. Just after the sounds become dull and muffled, blood flows continuously. So the turbulence and tapping sounds stop. The silence corresponds to the diastolic pressure (at the end of a cardiac cycle, just before the heart pumps out blood). The reading is usually about 80 mm Hg. In this example, pulse pressure (the difference between the highest and the lowest pressure readings) is 120 – 80, or 40 mm Hg. Assuming you are an adult in good health, this average resting value stays fairly constant over a few weeks or even months.

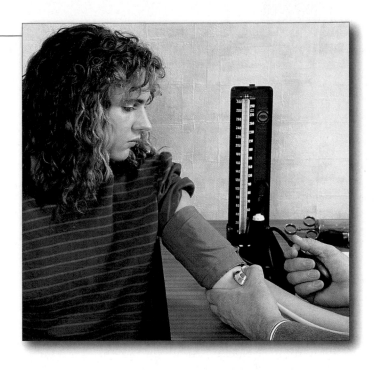

How Can Arterioles Resist Flow?

Track a volume of blood through the systemic circuit and you find the greatest pressure drop at arterioles (Figure 33.17), for these offer the greatest resistance to blood flow. The slowdown at arterioles allows time for control mechanisms to divert greater or lesser portions of the total flow volume to different regions. Control is exerted at arteriole walls, which incorporate rings of smooth muscle. Some control signals cause the smooth muscle cells to relax, which results in **vasodilation**. The word means an enlargement (dilation) of the blood vessel diameter. Other signals cause contraction of the smooth muscle cells, which results in a decrease in the blood vessel diameter, or **vasoconstriction**.

The nervous and endocrine systems exert control over arteriole diameter. For example, sympathetic nerve endings terminate on the smooth muscle cells of many arterioles. Increased activity along the nerves triggers action potentials in the cells to bring about contraction, hence widespread vasoconstriction. Epinephrine and angiotensin are two of the signaling molecules that can trigger changes in arteriole diameter.

Arteriole diameter also is adjusted when changes in metabolic activity shift the localized concentrations of substances in a tissue. Such local chemical changes are "selfish," in that they invite or divert blood flow to meet the tissue's own metabolic needs. Think of what happens as you run. The oxygen level drops in skeletal muscle tissue, and levels of carbon dioxide, hydrogen and potassium ions, and other substances increase. The changed conditions trigger vasodilation of arterioles in the vicinity. Now more blood can flow through active muscles to deliver more raw materials and carry away cell products and metabolic wastes. When muscles relax and demand fewer blood deliveries, the oxygen level increases and triggers vasoconstriction of the arterioles.

Controlling Mean Arterial Blood Pressure

Mean arterial pressure depends on cardiac output and on total resistance through the vascular system. Cardiac output is influenced by controls over the rate and strength of heartbeats, and total resistance mainly by vasoconstriction at the arterioles. Given that organs and tissues vary in their demands for blood, maintaining blood pressure is not easy. The balance between cardiac output and total resistance must be continually juggled.

A **baroreceptor reflex** is the main short-term control over arterial pressure. It starts at baroreceptors, which detect fluctuations in arterial mean pressure and pulse pressure. The most critical of these sensory receptors are located in the carotid arteries (which supply blood to the brain) and aortic arch (which supplies blood to the rest of the body). They continually generate action potentials, and their signals reach a control center in a hindbrain region, the medulla oblongata (Section 29.9). In response, the center adjusts its commands flowing along sympathetic and parasympathetic nerves to the heart and blood vessels. For example, when the mean arterial pressure rises, the center commands the heart to beat more slowly and contract less forcefully.

Long-term control of blood pressure is exerted at kidneys, which adjust the volume and composition of the blood. And that is a topic of another chapter.

Mean arterial pressure is an outcome of controls over the heart's output and over the total resistance to blood flow through the vascular system, as exerted mainly at arterioles.

WWW

Capillary Function

THE NATURE OF THE EXCHANGES Capillary beds, again, are the diffusion zones for exchanges between blood and interstitial fluid. When the living cells in any part of the body are deprived of those exchanges, they die quickly; brain cells are dead within four minutes.

By some estimates, the human body has between 10 billion and 40 billion capillaries, which collectively represent an astounding surface area for the exchanges. These components of the circulatory system extend into nearly every tissue, and at least one is as close as 0.001 centimeter to every living cell. Their proximity to cells is crucial, for shorter distances mean faster diffusion rates. (Think about this: It would take years for oxygen to diffuse on its own from your lungs all the way down your legs. By that time your toes, and the rest of you, would long be dead.) Although capillaries are three to seven micrometers across, they deform amazingly; red blood cells (eight micrometers across) squeeze through them single file. Therefore, these oxygen-transporting cells, as well as the substances dissolved in plasma, are in direct contact with the exchange area—the capillary wall—or only a short distance away from it.

And remember, taken as a whole, capillaries present a greater cross-sectional area than the arterioles leading into them, so the flow velocity cannot be as great in the capillary beds. The slowdown means there is plenty of time for interstitial fluid to exchange substances with the 5 percent or so of the total blood volume that is moving forward through the beds at a given time. Here you may wish to reflect on Figure 33.3e.

Each capillary is a tube of endothelial cells, a single layer thick (Figure 33.16c). The cells abut one another, but the "fit" is different in different parts of the body. In most cases, narrow, water-filled clefts occur between the endothelial cells. In capillaries that service the brain, tight junctions join the cells, and clefts are nonexistent; substances cannot "leak" between cells but must move through them. Thus the junctions are a functional part of the blood-brain barrier, as described in Section 29.9.

Oxygen, carbon dioxide, and most other small lipid-soluble substances cross the capillary wall by diffusing through the lipid portion of an endothelial cell's plasma membrane and through its cytoplasm. Certain proteins enter and leave the cells by endocytosis or exocytosis. Small, water-soluble substances such as ions enter and leave at the clefts between endothelial cells. So do white blood cells, as you will see in the chapter to follow.

MECHANISMS OF EXCHANGE Diffusion is not the only means by which substances are exchanged across the wall of capillaries. At any time, fluid pressures acting on the wall may not be balanced, and there may be bulk flow one way or the other across it. Bulk flow, recall, is a movement of water and solutes in the same direction in response to fluid pressure.

As Figure 33.19 shows, the concentrations of water and solutes in blood and interstitial fluid influence the direction of flow. At the beginning of a capillary bed, the outward-directed force of blood pressure is greater than the inward-directed osmotic force because of the plasma proteins. A small amount of protein-free plasma is pushed out in bulk through clefts in the capillary wall. This process is called **ultrafiltration**. Farther on in the bed, the balance shifts. Whereas blood pressure has been continually declining, the osmotic force has stayed the same. When the inward-directed osmotic force exceeds the outward force of blood pressure, tissue fluid moves through clefts between endothelial cells and into the capillary. This process is called **reabsorption**.

Normally, the outcome is a very small *net* outward movement of fluid from a capillary bed. The lymphatic system returns the fluid to the blood. Such bulk flow helps maintain the fluid balance between blood and interstitial fluid. This is important, for blood pressure is maintained only when there is adequate blood volume. When the blood volume plummets, as happens during hemorrhage, interstitial fluid may be tapped by way of reabsorption to help counter the loss.

Sometimes blood pressure increases so much that it triggers excessive ultrafiltration. When excess fluid accumulates in interstitial spaces, we call this condition *edema*. Edema occurs to some extent during physical exercise, when arterioles dilate in many tissue regions. It also can result from an obstructed vein or from heart failure. Edema becomes extreme during *elephantiasis*. As described in Section 23.7, elephantiasis is a disease brought on by roundworm infection and a subsequent obstruction of lymphatic vessels.

Venous Pressure

What happens to the flow velocity after it continues on past the vast cross-sectional area of the capillary beds? Remember, the capillaries merge into venules, or "little veins." These in turn merge into large-diameter veins. The total cross-sectional area is once again reduced, and flow velocity increases as blood returns to the heart.

In functional terms, venules are a bit like capillaries. Some solutes diffuse across their wall, which is only a bit thicker than the (thin) capillary wall. Some control over capillary pressure also is exerted at these vessels.

Veins are large-diameter, low-resistance transport tubes to the heart (Figure 33.16d). They have valves that prevent backflow. When gravity beckons, venous flow reverses direction and pushes the valves shut. The vein wall can bulge considerably under pressure, more so

ARTERIOLE END OF CAPILLARY BED

Outward-Directed Pressure:

Hydrostatic pressure of blood in capillary:	35 mm Hg
Osmotic force due to plasma proteins:	28 mm Hg

Inward-Directed Pressure:

Hydrostatic pressure of interstitial fluid:	0
Osmotic force due to interstitial proteins:	3 mm Hg

Net ultrafiltration pressure:

$$(35 - 0) - (28 - 3) = 10 \text{ mm Hg}$$

ULTRAFILTRATION FAVORED

VENULE END OF CAPILLARY BED

Outward-Directed Pressure:

Hydrostatic pressure of blood in capillary:	15 mm Hg
Osmotic force due to plasma proteins:	28 mm Hg

Inward-Directed Pressure:

Hydrostatic pressure of interstitial fluid:	0
Osmotic force due to interstitial proteins:	3 mm Hg

Net reabsorption pressure:

$$(15 - 0) - (28 - 3) = -10 \text{ mm Hg}$$

REABSORPTION FAVORED

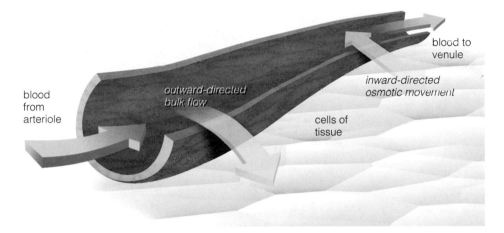

Figure 33.19 Bulk flow in an idealized capillary bed. The fluid movement plays no significant role in diffusion of solutes. But it is important in maintaining the distribution of extracellular fluid between the blood and interstitial fluid.

The fluid movements across a capillary wall result from the two opposing forces of ultrafiltration and reabsorption.

At the arteriole end of a capillary, the difference between blood pressure and interstitial fluid pressure forces some plasma (but very few plasma proteins) to leave the capillary. They do so by bulk flow through clefts between endothelial cells of the capillary wall. Ultrafiltration is the bulk flow of fluid *out* of the capillary.

Reabsorption is an osmotic movement of some interstitial fluid *into* the capillary. It results from a difference in water concentration between the plasma and interstitial fluid. With its dissolved protein components, plasma has a greater solute concentration, and therefore a lower water concentration.

Reabsorption near the end of a capillary bed tends to balance ultrafiltration at the beginning. Normally, there is only a small *net* filtration of fluid, which the lymphatic system returns to the blood.

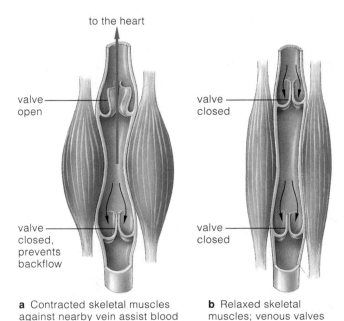

a Contracted skeletal muscles against nearby vein assist blood flow to heart.

b Relaxed skeletal muscles; venous valves shut, no backflow.

Figure 33.20 How the bulging of contracting skeletal muscles helps increase fluid pressure inside a vein. Notice the structure of each valve (membrane flaps) in the vein's lumen.

than an arterial wall. Thus veins can serve as reservoirs for variable volumes of blood. Collectively, the human body's veins can hold up to 50 to 60 percent of the total blood volume.

A vein wall contains some smooth muscle. When blood must circulate faster, as during physical exercise, the smooth muscle contracts. The wall stiffens and the vein bulges less, so that venous pressure rises and drives more blood to the heart. Also, when limbs are moving, skeletal muscles bulge against veins in their vicinity. They help raise the venous pressure and drive blood back to the heart (Figure 33.20). Rapid breathing also contributes to increased venous pressure. Inhaled air pushes down on internal organs and thereby alters the pressure gradient between the heart and veins.

Capillary beds are diffusion zones for exchanges between blood and interstitial fluid. Here also, a slight amount of bulk flow helps maintain the fluid balance between blood and interstitial fluid.

Venules overlap somewhat with capillaries in function.

Veins are highly distensible blood volume reservoirs and help adjust flow volume back to the heart.

33.9 CARDIOVASCULAR DISORDERS

Cardiovascular disorders are the leading cause of death in the United States. They affect at least 40 million people and kill about 1 million each year. The most common are *hypertension*, which is sustained high blood pressure; and *atherosclerosis*, a progressive thickening of the arterial wall and narrowing of the arterial lumen. Both affect blood circulation and so cause most *heart attacks* (damaged or destroyed heart muscle) and *strokes* (brain damage).

In most heart attacks, a "crushing" pain behind the breastbone lasts a half hour or more. Mild to severe pain may radiate into the left arm, shoulder, or neck. Sweating, shortness of breath, erratic heartbeat, nausea, vomiting, and dizziness or fainting may accompany an attack.

RISK FACTORS   Researchers correlate the following risk factors with cardiovascular disorders:

1. Smoking (Section 35.7)
2. Genetic predisposition to heart failure
3. High level of cholesterol in the blood
4. High blood pressure
5. Obesity (Section 36.10)
6. Lack of regular exercise
7. Diabetes mellitus (Section 31.7)
8. Age (the older you get, the greater the risk)
9. Gender (until age fifty, males are at much greater risk than females)

The risk factors can be controlled by exercising regularly, eating properly, and not smoking. With too little exercise and too much food, adipose tissue masses increase in the body, more blood capillaries develop to service them, and the heart has to work harder to pump blood through the increasingly divided vascular circuit. In people who smoke, the nicotine in tobacco makes the adrenal glands secrete epinephrine. This hormone causes blood vessel diameters to constrict, and thereby boosts heart rate and blood pressure. Also, carbon monoxide in cigarette smoke competes with oxygen for binding sites on hemoglobin, so the heart has to pump harder to get enough oxygen to cells. Smoking also predisposes people to atherosclerosis even when their blood level of cholesterol is normal.

The next paragraphs describe some of the damage that results from cardiovascular disorders.

HYPERTENSION   Hypertension results from gradual increases in the flow resistance through small arteries. In time, blood pressure remains above 140/90, even when a person is resting. Heredity may be a factor; the disorder tends to run in families. Diet is also a factor. For instance, high salt intake can raise blood pressure in susceptible people and increase the heart's workload. The heart may eventually enlarge and fail to pump blood effectively. High blood pressure also may contribute to a "hardening" of arterial walls that hampers delivery of oxygen to the brain, heart, and other vital organs.

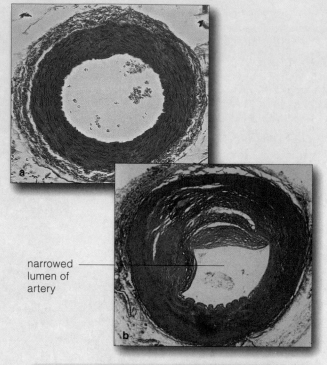

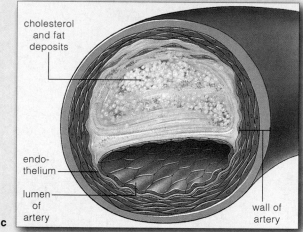

narrowed lumen of artery

cholesterol and fat deposits

endo-thelium

lumen of artery

wall of artery

Figure 33.21   Sections from (**a**) a normal artery and (**b**) one with a narrowed lumen. (**c**) Sketch of an atherosclerotic plaque.

Hypertension is called a "silent killer" because affected individuals may show no outward symptoms. Even when they know their blood pressure is high, some people tend to resist helpful medication, changes in diet, and regular exercise. Of 23 million hypertensive Americans, most do not undergo treatment. About 180,000 die each year.

ATHEROSCLEROSIS   With *arteriosclerosis*, arteries thicken and lose their elasticity. With atherosclerosis, the condition worsens as cholesterol and other lipids build up in the arterial wall and cause the lumen to narrow (Figure 33.21).

Recall, from the introduction to Chapter 15, that the liver produces enough cholesterol to satisfy the body's needs. Habitually eating cholesterol-rich foods typically increases the blood cholesterol level. When circulating in blood, cholesterol is bound to proteins as *low-density* lipoproteins, or **LDLs**. These are able to bind to receptors on cells throughout the body. Cells take up the LDLs—with the cholesterol cargo—for use in cell activities. They don't take up all of it. Excess cholesterol becomes attached to proteins as *high-density* lipoproteins, or **HDLs**, and is transported back to the liver, where it is metabolized.

In some people, for a variety of reasons, not enough LDL is removed from the blood. As the blood level of LDL increases, so does the risk of atherosclerosis. LDLs, with their bound cholesterol, infiltrate the walls of arteries. In those walls, abnormal smooth muscle cells multiply, connective tissue components increase in mass, and cholesterol accumulates in endothelial cells and the clefts between them. On top of the lipids, calcium deposits actually form microscopic slivers of bone. A fibrous net forms over the entire mass. This *atherosclerotic plaque* sticks out into the arterial lumen (Figure 33.21c).

The bony slivers of the plaque shred the endothelium. Platelets gather at the damaged site, and, in the manner described in Section 33.10, they secrete chemicals that initiate clot formation. The condition worsens as fatty globules in the plaque become oxidized. Many globules take on a form that resembles the surface components of common bacteria—including a type that instigates the formation of bonelike calcium deposits in the lungs. As an awful consequence, the call goes out to bacteria-fighting monocytes. Soon an inflammatory response is under way, and certain chemicals released during the fray activate the genes for bone formation. Normally, this is a good thing; it helps wall off invaders and prevents infection from spreading. It is bad news for arteries.

As plaques and clots grow, they can narrow or block an artery. Blood flow to the tissues serviced by the artery may dwindle to a trickle or stop. A clot that stays in place is a *thrombus*. If it becomes dislodged and then travels the bloodstream, it is an *embolus*. Think of coronary arteries, which have narrow diameters. They are highly vulnerable to clogging by a plaque or clot. When they narrow to one-quarter of their former diameter, the outcome ranges from *angina pectoris*, or mild chest pains, to a heart attack.

Physicians can diagnose atherosclerosis in coronary arteries by *stress electrocardiograms*. These are recordings of the electrical activity of the cardiac cycle while a person is exercising on a treadmill. They also can diagnose the condition by *angiography*. In this procedure, a dye that is opaque to x-rays is injected into the bloodstream.

Severe blockage may require surgery. With *coronary bypass surgery*, a section of an artery from the chest is stitched to the aorta and to the coronary artery below the narrowed or blocked region, as shown in Figure 33.22. With *laser angioplasty*, laser beams directed at the plaques vaporize them. With *balloon angioplasty*, a small balloon inflated within a blocked artery breaks up the plaques.

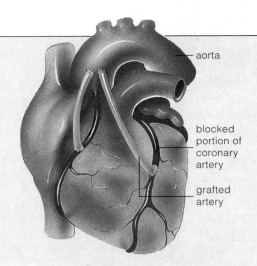

Figure 33.22 Two coronary bypasses (color-coded *green*), which extend from the aorta around two clogged portions of coronary arteries.

aorta

blocked portion of coronary artery

grafted artery

ARRHYTHMIAS ECGs can be used to detect *arrhythmias*, or irregular heart rhythms (Figure 33.23). Arrhythmias are not always a sign of abnormal conditions. Endurance athletes, for example, may have a below-average resting cardiac rate, or *bradycardia*. As an adaptation to ongoing strenuous exercise, their nervous system has adjusted the cardiac pacemaker's rate of contraction downward. Exercise or stress often causes 100+ heartbeats a minute, or *tachycardia*.

Atrial fibrillation, an irregular heartbeat, affects more than 10 percent of the elderly and young people with various heart diseases. Coronary occlusions or some other disorder may cause irregular rhythms that rapidly lead to a dangerous condition called ventricular fibrillation. In portions of the ventricles, cardiac muscle haphazardly contracts and blood pumping falters. Within seconds, the individual loses consciousness, which might signify impending death. A strong electric shock to the chest may restore normal cardiac function. WWW

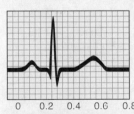

0   0.2   0.4   0.6   0.8
a   time (seconds)

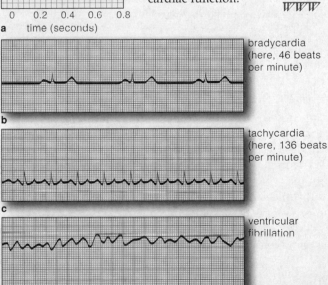

b   bradycardia (here, 46 beats per minute)

c   tachycardia (here, 136 beats per minute)

d   ventricular fibrillation

Figure 33.23 (**a**) ECG of a single, normal human heartbeat. (**b–d**) Three examples of recordings of arrhythmias.

Small blood vessels are vulnerable to ruptures, cuts, and similar injuries. **Hemostasis**, a process involving blood vessel spasm, platelet plug formation, and blood coagulation, may repair the damage and stop blood loss.

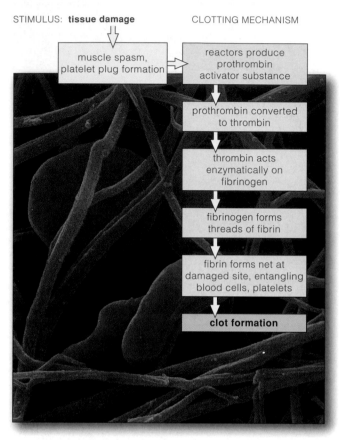

STIMULUS: **tissue damage**   CLOTTING MECHANISM

muscle spasm, platelet plug formation

reactors produce prothrombin activator substance

prothrombin converted to thrombin

thrombin acts enzymatically on fibrinogen

fibrinogen forms threads of fibrin

fibrin forms net at damaged site, entangling blood cells, platelets

clot formation

Figure 33.24 Plasma proteins involved in clot formation. Blood coagulates when damage exposes the collagen fibers in blood vessel walls. Exposure initiates reactions that cause rod-shaped plasma proteins (fibrinogens) to stick together as long, insoluble threads. These adhere to the exposed collagen, forming a net that traps blood cells and platelets. The entire mass is a clot.

As shown in Figure 33.24, smooth muscle located in the damaged wall contracts in an automatic response called a spasm. The vasoconstriction is a temporary fix; it staunches blood flow. Platelets clump together as a temporary plug in the damaged wall. They also release substances that can help prolong the spasm and attract more platelets. Then blood coagulates, or converts to a gel, and forms a clot. Finally, the clot retracts, forming a compact mass, and the breach in the wall is sealed.

The body routinely repairs damage to small blood vessels and prevents blood loss. The repair process includes blood vessel spasm, platelet plug formation, and coagulation.

We conclude this chapter with a brief section on the manner in which the lymphatic system supplements blood circulation. Think of this section as a bridge to the next chapter, on immunity, for the lymphatic system also helps defend the body against injury and attack. The system is composed of drainage vessels, lymphoid organs, and lymphoid tissues. The tissue fluid that has moved into the vessels is called lymph.

Lymph Vascular System

A portion of the lymphatic system called the lymph vascular system consists of many tubes that collect and transport water and solutes from interstitial fluid to ducts of the circulatory system. Its main components are lymph capillaries and lymph vessels (Figure 33.25).

The lymph vascular system serves three functions. First, its vessels are drainage channels for water and plasma proteins that have leaked away from blood at capillary beds and that must be delivered back to the blood circulation. Second, the system also takes up fats that the body has absorbed from the small intestine and delivers them to the blood circulation, in the manner described in Section 33.5. Third, it delivers pathogens, foreign cells and material, and cellular debris from the

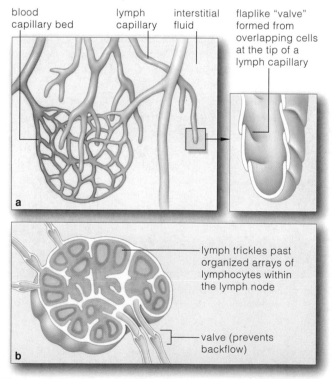

blood capillary bed   lymph capillary   interstitial fluid   flaplike "valve" formed from overlapping cells at the tip of a lymph capillary

a

lymph trickles past organized arrays of lymphocytes within the lymph node

valve (prevents backflow)

b

Figure 33.25 (**a**) Diagram of lymph capillaries at the start of a drainage network, the lymph vascular system. (**b**) Cutaway diagram of a lymph node. Its inner compartments are packed with organized arrays of infection-fighting white blood cells.

TONSILS
Defense against bacteria and other foreign agents

RIGHT LYMPHATIC DUCT
Drains right upper portion of the body

THYMUS GLAND
Site where certain white blood cells acquire means to chemically recognize specific foreign invaders

THORACIC DUCT
Drains most of the body

SPLEEN
Major site of antibody production; disposal site for old red blood cells and foreign debris; site of red blood cell formation in the embryo

SOME OF THE LYMPH VESSELS
Return excess interstitial fluid and reclaimable solutes to the blood

SOME OF THE LYMPH NODES
Filter bacteria and many other agents of disease from lymph

BONE MARROW
Marrow in some bones is production site for infection-fighting blood cells (as well as red blood cells and platelets)

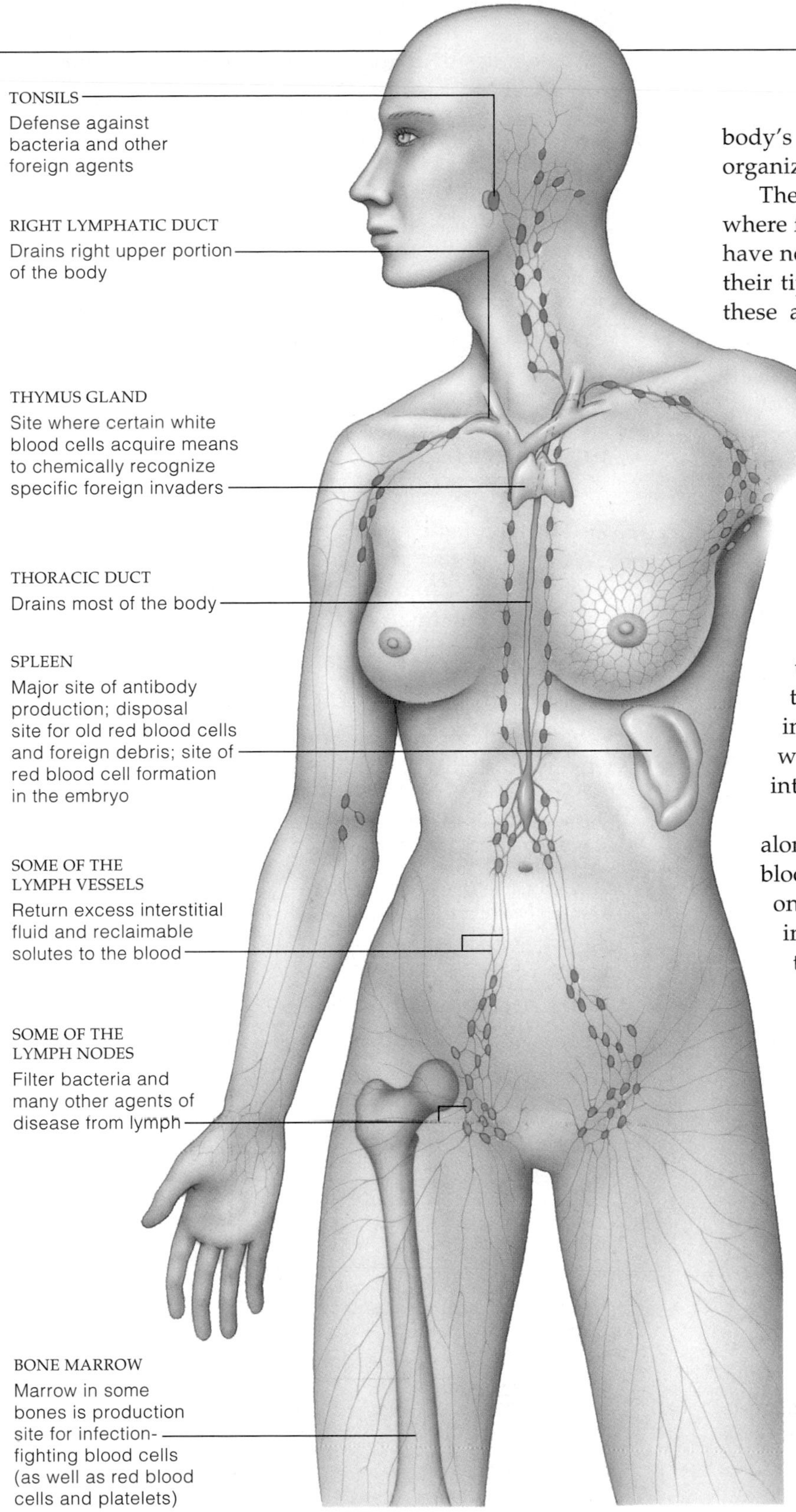

Figure 33.26 Components of the human lymphatic system and their functions. The small *green* ovals represent some of the major lymph nodes. Not shown are patches of lymphoid tissue in the small intestine and in the appendix, which also are part of the lymphatic system.

body's tissues to the lymph vascular system's efficiently organized disposal centers, the lymph nodes.

The lymph vascular system starts at capillary beds, where fluid enters the lymph capillaries. The capillaries have no obvious entrance; water and solutes move into their tips at flaplike "valves." As Figure 33.25a shows, these are areas where endothelial cells overlap. The lymph capillaries merge into lymph vessels, which have a larger diameter. The lymph vessels contain some smooth muscle in their wall as well as valves in their lumen that prevent backflow. They converge into collecting ducts that drain into veins in the lower neck (Figure 33.26).

Lymphoid Organs and Tissues

Other portions of the lymphatic system, called the *lymphoid* organs and tissues, are players in the body's defenses against injury or attack. They include the lymph nodes, spleen, and thymus, as well as tonsils and patches of tissue in the small intestine and appendix.

Lymph nodes are strategically located at intervals along lymph vessels (Figure 33.26). Before entering blood, lymph is filtered as it trickles through at least one node. Masses of lymphocytes take up stations in the nodes after forming in bone marrow. When they recognize an invader, they multiply rapidly and form large armies to destroy it.

The **spleen** is the largest lymphoid organ. It filters pathogens and used-up blood cells from the blood itself. One of its compartments, the red pulp, is a huge reservoir of red blood cells. In human embryos, the red pulp also produces these cells. The other compartment, the white pulp, has masses of lymphocytes associated with blood vessels. If and when a specific invader reaches the spleen during a severe infection, the lymphocytes become mobilized to destroy it, just as they are in lymph nodes.

It is in the **thymus gland** that immature T lymphocytes differentiate in ways that allow them to recognize and respond to specific pathogens. The thymus produces hormones that influence these events. It is central to immunity, the focus of the chapter to follow.

The lymph vascular portion of the lymphatic system returns water and solutes from tissue fluid to blood, and delivers fats and foreign material to lymph nodes. The system's lymph nodes and other lymphoid organs help defend the body against tissue damage and infectious diseases.

WWW

SUMMARY

1. The closed circulatory system of humans and other vertebrates consists of a heart (muscular pump), many blood vessels (arteries, arterioles, capillaries, venules, and veins), and blood. The system functions in the rapid internal transport of substances to and from cells.

2. Blood helps maintain favorable conditions for cells. This fluid connective tissue consists of plasma, red and white blood cells, and platelets. It delivers oxygen and other substances to the interstitial fluid around cells. It also picks up cell products and wastes from that fluid.

a. Plasma, the liquid portion of blood, is a transport medium for blood cells and platelets. Plasma is a solvent for plasma proteins, simple sugars, lipids, amino acids, mineral ions, vitamins, hormones, and several gases.

b. Red blood cells transport oxygen from the lungs to all body regions. They are packed with hemoglobin, an iron-containing pigment that binds reversibly with oxygen. Red blood cells transport some carbon dioxide from interstitial fluid to the lungs.

c. Some phagocytic white blood cells cleanse tissues of dead cells, cellular debris, and anything else detected as not belonging to the body. Other white blood cells (lymphocytes) form great armies that destroy specific bacteria, viruses, and other disease agents.

3. The human heart is an incessantly beating double pump. Each half of the heart has two chambers, called an atrium and a ventricle. Blood flows into atria, then ventricles, then into the great arteries. One-way valves enforce the forward-directed flow.

4. The heart's partition separates blood flow into two circuits, one pulmonary and the other systemic.

a. The pulmonary circuit loops between the heart and lungs. Oxygen-poor blood from the systemic veins enters the *right* atrium, is pumped through pulmonary arteries to the two lungs, picks up oxygen, then flows through pulmonary veins to the heart's left atrium.

b. The systemic circuit loops between the heart and all body tissues. Oxygenated blood in the *left* atrium flows into the left ventricle, is pumped into the aorta, then is distributed to capillary beds. There, the blood gives up oxygen and picks up carbon dioxide. Systemic veins return it to the heart's right atrium.

5. Ventricular contraction drives blood through both circuits. Blood pressure is high in contracting ventricles. It drops successively in arteries, arterioles, capillaries, venules, and veins. It is lowest in relaxed atria.

a. Arteries are rapid-transport vessels and a pressure reservoir (they smooth out pressure changes resulting from heartbeats and thereby smooth out blood flow).

b. Arterioles are vessels at which the volume of flow through organs can be controlled.

c. Capillary beds are diffusion zones where blood and interstitial fluid exchange substances.

d. Venules overlap capillaries in function. Veins are rapid-transport vessels and a blood volume reservoir that is tapped to adjust flow volume back to the heart.

6. The cardiac conduction system serves as the basis for the heart's rhythmic, spontaneous contractions.

a. One percent of the cardiac muscle cells are not contractile. They are specialized to initiate and distribute action potentials, independently of the nervous system. The nervous system only adjusts the rate and strength of the basic contractions; it does not initiate them.

b. The system's SA node fires action potentials the fastest and is the cardiac pacemaker. Waves of excitation starting here wash over the atria, then down the heart's partition, then up the ventricles. The ventricles contract in a wringing motion that ejects blood from the heart.

7. The lymphatic system has these functions:

a. Its vascular portion takes up water and plasma proteins that seep out of blood capillaries, then returns them to circulating blood. It transports absorbed fats and delivers pathogens and foreign material to disposal centers. Lymph capillaries and vessels are components.

b. Its lymphoid organs and tissues have production centers for lymphocytes. Some are battlegrounds where organized arrays of lymphocytes fight disease agents.

Review Questions

1. Define the functions of the circulatory system and lymphatic system. Distinguish between blood and interstitial fluid. *33.1*

2. Describe the cellular components of blood. Describe the plasma portion of blood. *33.2*

3. Distinguish between systemic and pulmonary circuits. *33.1*

4. State the functions of arteries, arterioles, capillaries, veins, venules, and lymph vessels. *33.1, 33.5, 33.7, 33.8*

5. Distinguish between the functions of the human heart's atria and ventricles. Then label the heart's components: *33.6*

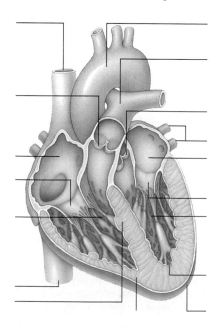

Self-Quiz (Answers in Appendix III)

1. Cells directly exchange substances with _____ .
   a. blood vessels   c. interstitial fluid
   b. lymph vessels   d. both a and b

2. Which are *not* components of blood?
   a. plasma
   b. blood cells and platelets
   c. gases and other dissolved substances
   d. all of the above are components of blood

3. The _____ produces red blood cells, which transport _____ and some _____ .
   a. liver; oxygen; mineral ions
   b. liver; oxygen; carbon dioxide
   c. bone marrow; oxygen; hormones
   d. bone marrow; oxygen; carbon dioxide

4. The _____ produces white blood cells, which function in _____ and _____ .
   a. liver; oxygen transport; defense
   b. lymph nodes; oxygen transport; pH stabilization
   c. bone marrow; housekeeping; defense
   d. bone marrow; pH stabilization; defense

5. In the pulmonary circuit, the heart's _____ half pumps blood to lungs, then _____ blood flows to the heart.
   a. right; oxygen-poor   c. right; oxygen-rich
   b. left; oxygen-poor   d. left; oxygen-rich

6. In the systemic circuit, the heart's _____ half pumps _____ blood to all body regions.
   a. right; oxygen-poor   c. right; oxygen-rich
   b. left; oxygen-poor   d. left; oxygen-rich

7. Blood pressure is high in _____ and lowest in _____ .
   a. arteries; veins   c. arteries; ventricles
   b. arteries; relaxed atria   d. arterioles; veins

8. _____ contraction drives blood through the pulmonary circuit and the systemic circuit; blood pressure is highest in contracting _____ .
   a. Atrial; ventricles   c. Ventricular; arteries
   b. Atrial; atria   d. Ventricular; ventricles

9. Which is *not* a function of the lymphatic system?
   a. delivers disease agents to disposal centers
   b. produces lymphocytes
   c. delivers oxygen to cells
   d. returns water and plasma proteins to blood

10. Match the type of blood vessel with its major function.
    _____ arteries   a. diffusion
    _____ arterioles   b. control of blood volume distribution
    _____ capillaries   c. transport, blood volume reservoirs
    _____ venules   d. overlap capillary function
    _____ veins   e. transport, pressure reservoirs

11. Match the components with their most suitable description.
    _____ capillary beds   a. two atria, two ventricles
    _____ lymph vascular system   b. bathes body's living cells
    _____ heart chambers   c. driving force for blood
    _____ blood   d. zones of diffusion
    _____ heart contractions   e. starts at capillary beds
    _____ interstitial fluid   f. fluid connective tissue

Critical Thinking

1. Shirelle, who is using a light microscope to examine a human tissue specimen, sees red blood cells moving single file through thin-walled tubes. She makes a photomicrograph (Figure 33.27). What type of blood vessel has she captured on film?

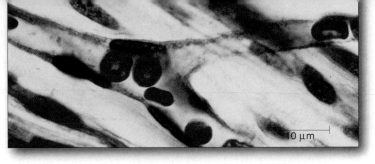

Figure 33.27 Light micrograph of a branching blood vessel.

2. In individuals who have weak or leaky valves in their veins, fluid pressure associated with the backflow of blood causes venous walls below the valves to bulge outward. In time, the walls become stretched and flabbily distorted, a condition called *varicose veins*. Some people are genetically predisposed to develop the condition, but the cumulative mechanical stress associated with prolonged standing, pregnancy, and aging can contribute to it. With chronic varicosity, the legs themselves become swollen. Explain why this swelling might happen. Also explain why veins close to the leg surface are more susceptible to varicosity than those deeper in the leg tissues.

3. Infection by a hemolytic bacterium (*Streptococcus pyogenes*) may trigger an inflammation that ultimately damages valves in the heart. The disease symptoms of *rheumatic fever* follow. Explain how this disease must affect the heart's functioning and what kinds of symptoms would arise as a consequence.

Selected Key Terms

ABO blood typing *33.4*	LDL *33.9*
agglutination *33.4*	lymph node *33.11*
anemia *33.3*	lymphatic system *33.1*
aorta *33.5*	plasma *33.2*
arteriole *33.5*	platelet *33.2*
artery *33.5*	pulmonary circuit *33.1*
baroreceptor reflex *33.7*	reabsorption *33.8*
blood *33.1*	red blood cell *33.2*
blood pressure *33.7*	Rh blood typing *33.4*
capillary (blood) *33.1*	spleen *33.11*
capillary bed *33.1*	stem cell *33.2*
cardiac conduction system *33.6*	systemic circuit *33.1*
cardiac cycle *33.6*	thymus gland *33.11*
cardiac pacemaker *33.6*	transfusion (blood) *33.4*
cell count *33.2*	ultrafiltration *33.8*
circulatory system *33.1*	vasoconstriction *33.7*
heart *33.1*	vasodilation *33.7*
hemostasis *33.10*	vein *33.5*
HDL *33.9*	venule *33.5*
interstitial fluid *33.1*	white blood cell *33.2*

Readings *See also www.infotrac-college.com*

Baraga, M. 3 May 1996. "Finding New Drugs to Treat Stroke." *Science*, 274.

Little, R., and W. Little. 1989. *Physiology of the Heart and Circulation*. Fourth edition. Chicago: Year Book Medical.

Zamir, M. September–October 1996. "Secrets of the Heart." *The Sciences*. Effects of exercise on heart function.

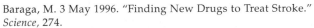

WWW *http://www.brookscole.com/biology*

Practice quiz questions, hypercontents, BioUpdates, and critical thinking. The Brooks/Cole Biology Resource Center provides a wealth of information fully organized and integrated by chapter.

IMMUNITY

Russian Roulette, Immunological Style

Until about a century ago, smallpox epidemics swept repeatedly through the world's cities. Some outbreaks were so severe that only half of those who were stricken survived. The survivors were left with permanent scars on the face, neck, shoulders, and arms. But they seldom contracted the same disease again; they were said to be "immune" to smallpox.

No one knew what caused smallpox, but the idea of acquiring immunity was dreadfully appealing. In twelfth-century China, people in good health gambled with deliberate infections. They sought out people with mild cases of smallpox (who displayed only mild scarring), then removed bits of crusts from the scars, ground them up, and inhaled the powder.

By the seventeenth century, Mary Montagu, wife of the English ambassador to Turkey, was championing inoculation. She went so far as to poke bits of smallpox scabs into the skin of her own children. Other people soaked threads in fluid from smallpox sores, then poked the threads into their own skin.

Some individuals who survived the chancy practices acquired immunity to smallpox, but many developed raging infections. As if the odds were not dangerous enough, the crude inoculation procedures also invited the acquisition of several other infectious diseases.

While this immunological version of Russian roulette was going on, Edward Jenner was growing up in the countryside of England. At the time, it was common knowledge that people who contracted cowpox never got smallpox. (Cowpox is a mild disease that can be transmitted from cattle to humans.) No one thought much about this until 1796, when Jenner, by now a physician, injected material from a cowpox sore into the arm of a healthy boy. Six weeks later, after the reaction subsided, Jenner injected some material from *smallpox* sores into the boy (Figure 34.1*a*). He had hypothesized that the earlier injection might provoke immunity to

Figure 34.1 (**a**) Statue honoring Edward Jenner's work to develop an immunization procedure against smallpox, one of the most dreaded diseases in human history. (**b**) False-color scanning electron micrograph of a white blood cell being attacked by HIV (*blue particles*), the virus that causes AIDS. Immunologists are working to develop effective weapons against this modern-day scourge.

smallpox. Fortunately he was right; the boy remained free of smallpox. Jenner had developed an effective immunization procedure against a specific pathogen.

The French mocked Jenner's procedure by calling it **vaccination**. The word literally means "encowment." Much later Louis Pasteur, an influential French chemist, developed similar immunization procedures for other diseases. In acknowledgment of Jenner's pioneering work, Pasteur called his procedures vaccinations, also, and only then did the word become respectable.

By Pasteur's time, improved light microscopes were revealing diverse bacteria, fungal spores, and other previously invisible forms of life. As Pasteur himself discovered, microorganisms abound in ordinary air. Did some cause diseases? Probably. Could they settle into food or drink and make it spoil? He demonstrated that they could and did.

Pasteur also found a way to kill most of the suspect disease agents in food or beverages. As he and others knew, boiling killed the suspects. He also knew people cannot boil wine (or beer or milk, for that matter) and end up with the same beverage. He devised a way to heat the beverages at a temperature low enough not to ruin them but high enough to kill most of the resident microorganisms. We still depend on this antimicrobial method, which was named pasteurization in his honor.

In the late 1870s Robert Koch, a German physician, linked a specific microorganism to a specific disease; namely, anthrax. In one experiment, Koch injected blood from animals with anthrax into healthy ones. Recipients of the injection ended up with blood that teemed with cells of the bacterium *Bacillus anthracis*—and they developed anthrax. Even more convincing, injections of bacterial cells that were cultured outside the animal body also caused the disease!

Thus, by the beginning of the twentieth century, the promise of understanding the basis of infectious disease and immunity loomed large—and the battles against those diseases were about to begin in earnest. Since that time, spectacular advances in microscopy, biochemistry, and molecular biology have increased our understanding of the body's defenses. We now have greater insight into responses to tissue damage in general and immune responses to specific pathogens or tumor cells. The responses are the focus of this chapter.

KEY CONCEPTS

1. The vertebrate body has physical, chemical, and cellular defenses against pathogenic microorganisms, malignant tumor cells, and other agents that can destroy tissues and even the individual itself.

2. In the early stages of tissue invasion and damage, white blood cells and certain proteins in the plasma portion of blood escape from capillaries. They execute a rapid counterattack in response to a general alarm, not to the presence of a particular pathogen—so we call this a nonspecific inflammatory response. Phagocytic white blood cells ingest invading agents and clean up tissue debris. Plasma proteins promote phagocytosis, and some also destroy invaders directly.

3. If the invasion persists, certain white blood cells make immune responses. Those cells can chemically recognize distinct configurations on molecules that are abnormal or foreign to the body, such as those on bacterial cells and viruses. If the foreign or abnormal molecule triggers an immune response, it is called an antigen.

4. In one type of immune response, some of the white blood cells produce enormous quantities of antibodies. Antibodies are molecules that bind to a specific antigen and tag it for destruction.

5. In another type of immune response, executioner cells directly destroy body cells that have become abnormal, as by infection or by a tumor-producing process.

You continually cross paths with astoundingly diverse **pathogens**—the viruses, bacteria, fungi, protozoans, parasitic worms, and other agents that cause diseases. Having coevolved with most of them, you and other vertebrates have three lines of defense, so you need not lose sleep over this. Most pathogens cannot breach the body surface. If they do, they face chemical weapons and white blood cells, or leukocytes, that attack anything perceived as foreign. Other white blood cells zero in on specific targets. Table 34.1 lists the three lines of defense.

Surface Barriers to Invasion

Most often, pathogens cannot get past skin or the other linings of body surfaces. Think of skin as a habitat of low moisture, low pH, and thick layers of dead cells. Normally harmless bacteria tolerate these conditions. Few pathogens can compete with dense populations of established types unless conditions change. Repeatedly subject your toes to warm, damp shoes, for example, and you may be inviting certain fungi to penetrate the sodden, weakened tissues and cause *athlete's foot*.

Similarly, resident bacteria on the mucous lining of the gut and vagina help protect you—as when lactate, a fermentation product of *Lactobacillus* populations in the vagina, helps maintain a low pH that most bacteria and fungi cannot tolerate. Barriers in branching, tubular

airways leading to your lungs stop airborne pathogens. Air rushing down the tubes flings bacteria against their sticky, mucus-coated walls. In the mucus are protective substances such as **lysozyme**. This enzyme digests cell walls of bacteria and so invites their death. As a final touch, broomlike cilia lining the airways sweep out the trapped and enzymatically whapped pathogens.

The body has additional defenses. Lysozyme and other substances in tears, saliva, and gastric fluid offer protection, as when an outpouring of tears gives the eyes a sterile washing. Urine's low pH and flushing action help bar pathogens from moving into the urinary tract. Also, diarrhea swiftly flushes irritating pathogens from the gut. Diarrhea must be controlled in children because it causes dehydration, but blocking its action in adults may prolong infection.

Nonspecific and Specific Responses

All animals react to tissue damage. Even small aquatic invertebrates have phagocytic cells and antimicrobial substances, including lysozymes. But the more complex the animal, the more complex are the systems to defend it. Think back on the evolution of the circulatory and lymphatic systems of vertebrates that invaded the land (Section 33.1). As the circulation of body fluids became more efficient, so did the means to defend the body. Phagocytic cells as well as plasma proteins could move swiftly to tissues under attack, and they could intercept pathogens trickling along the vascular highways.

Vertebrates came to be equipped with sets of plasma proteins. Some proteins promote rapid clot formation after tissue damage; others destroy invading pathogens or target them for phagocytosis. In addition, exquisitely focused responses to specific dangers evolved.

In short, *internal defenses against a tremendous variety of pathogens are in place even before damage occurs*. Ready and waiting are specialized white blood cells as well as plasma proteins. They take part in *nonspecific* responses to tissue damage in general, not to one pathogen or another. Other white blood cells may recognize unique molecular configurations of a *specific* pathogen. If they do, the resulting "immune" response will run its course whether tissues are damaged or not.

Table 34.1   The Vertebrate Body's Three Lines of Defense Against Pathogens

BARRIERS AT BODY SURFACES (*nonspecific* targets)

Intact skin; mucous membranes at other body surfaces

Infection-fighting chemicals in tears, saliva, etc.

Normally harmless bacterial inhabitants of skin and other body surfaces that can outcompete pathogenic visitors

Flushing effect of tears, saliva, urination, and diarrhea

NONSPECIFIC RESPONSES (*nonspecific* targets)

Inflammation:
1. Fast-acting white blood cells (neutrophils, eosinophils, and basophils)
2. Macrophages (also take part in immune responses)
3. Complement proteins, blood-clotting proteins, and other infection-fighting substances

Organs with pathogen-killing functions (such as lymph nodes)

Some cytotoxic cells (e.g., NK cells) with a range of targets

---

IMMUNE RESPONSES (*specific* targets only)

T cells and B cells; macrophages interact with them

Communication signals (e.g., interleukins) and chemical weapons (e.g., antibodies, perforins)

Intact skin, mucous membranes, antimicrobial secretions, and other barriers at the body surface constitute the first line of defense against invasion and tissue damage.

Inflammation and other internal, nonspecific responses to invasion are the second line of defense.

Immune responses against specific invaders, as executed by armies of white blood cells and their chemical weapons, are the third line of defense.

COMPLEMENT PROTEINS

A set of plasma proteins has roles in nonspecific *and* specific defenses. Collectively, they are the **complement system**. About twenty kinds circulate in the blood in inactive form. If even a few molecules of one kind are activated, they trigger cascading reactions that activate many molecules of another complement protein. Each of these activates many molecules of another kind of protein at the next reaction step, and so on. Deployment of so many molecules has the following effects.

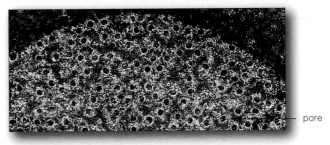

Figure 34.2 Micrograph of a cell surface, showing pores formed by membrane attack complexes.

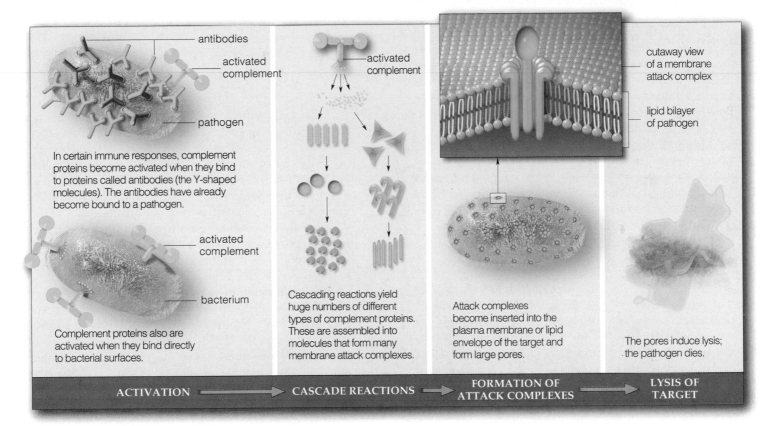

antibodies

activated complement

pathogen

In certain immune responses, complement proteins become activated when they bind to proteins called antibodies (the Y-shaped molecules). The antibodies have already become bound to a pathogen.

activated complement

bacterium

Complement proteins also are activated when they bind directly to bacterial surfaces.

activated complement

Cascading reactions yield huge numbers of different types of complement proteins. These are assembled into molecules that form many membrane attack complexes.

cutaway view of a membrane attack complex

lipid bilayer of pathogen

Attack complexes become inserted into the plasma membrane or lipid envelope of the target and form large pores.

The pores induce lysis; the pathogen dies.

ACTIVATION ⟹ **CASCADE REACTIONS** ⟹ **FORMATION OF ATTACK COMPLEXES** ⟹ **LYSIS OF TARGET**

Some complement proteins join together and form pore complexes. These molecular structures have an inner channel (Figures 34.2 and 34.3). When inserted in the plasma membrane of many pathogens, they form pores that induce lysis. **Lysis**, remember, is the gross structural disruption of a plasma membrane or cell wall that leads to cell death. Pore complexes also become inserted into the wall of Gram-negative bacteria. Such cell walls consist of a lipid-rich, outer surface above a peptidoglycan layer. Molecules of lysozyme can diffuse through the pores and digest the peptidoglycan.

Some activated proteins promote inflammation, a nonspecific defense response described next. Through their cascades of reactions, they create concentration gradients that attract phagocytic white blood cells to an irritated or damaged tissue. The complement proteins also encourage phagocytes to dine. They can bind to the surface of many types of invaders. When they do so, an

Figure 34.3 Formation of membrane attack complexes. One reaction pathway starts as complement proteins bind to bacterial surfaces. Another operates in immune responses to specific invaders. Both produce membrane pore complexes that induce lysis in pathogens.

invader ends up with a complement "coat." The coat adheres to phagocytes, and this is rather like putting a basted turkey on a dinner table.

In such ways, the complement proteins target many bacteria, parasitic protistans, and enveloped viruses.

The complement system, a set of twenty or so kinds of plasma proteins circulating in blood, takes part in cascades of reactions that help defend against many bacteria, some parasitic protistans, and enveloped viruses.

The complement system takes part in both nonspecific and specific defenses.

The Roles of Phagocytes and Their Kin

Certain types of white blood cells take part in an initial response to tissue damage. White blood cells, recall, arise from stem cells in bone marrow. Many of the cells circulate in blood and lymph. A great many take up stations in lymph nodes as well as in the spleen, liver, kidneys, lungs, and brain. Here you may wish to scan Sections 33.2 and 33.11 once more; they introduce the components of blood and the lymphatic system.

Like SWAT teams, three kinds of white blood cells react swiftly to danger in general but are not adapted for sustained battles. **Neutrophils**, the most abundant kind, phagocytize bacteria. They ingest, kill, and digest bacterial cells to simple molecular bits. **Eosinophils** secrete enzymes that make holes in parasitic worms. **Basophils** secrete histamine and other substances that may help keep inflammation going after it starts.

Although slower to act, the white blood cells called **macrophages** are the "big eaters." Figure 34.4 shows one of these phagocytic cells. A macrophage engulfs and digests just about any foreign agent. It also helps clean up damaged tissues. The immature macrophages circulating in blood are classified as monocytes.

The Inflammatory Response

An inflammatory response develops when something damages or kills cells of any given tissue. Infections, punctures, burns, and other insults are the triggers. By a mechanism known as **acute inflammation**, the fast-acting phagocytes and complement proteins, as well as other plasma proteins, escape from the bloodstream at capillary beds in the damaged tissue. Localized signs that they have entered interstitial fluid and that acute inflammation is under way include redness, swelling, heat, and pain, as listed in Table 34.2.

Table 34.2	Localized Signs of Inflammation and Their Causes
Redness	Arteriolar vasodilation; increase in blood flow to the affected site
Warmth	Arteriolar vasodilation; more blood, carrying more metabolic heat, arrives at site
Swelling	Chemical signals increase capillary permeability; plasma proteins leak out, disrupt fluid balance across wall of capillaries; localized edema
Pain	Nociceptors (pain receptors) stimulated by increased fluid pressure, local chemical signals

Mast cells, which reside in connective tissues and function like basophils, take part in an inflammatory response. They synthesize and release **histamine** and other substances into interstitial fluid. Their secretions are local chemical signals, which trigger vasodilation of arterioles that snake through the damaged tissue. As you read earlier, vasodilation is an increase in a blood vessel's diameter after smooth muscle in its wall has relaxed. When arterioles become engorged with blood, an affected tissue reddens and becomes warmer owing to blood-borne metabolic heat.

Released histamine also increases the permeability of the thin-walled capillaries in the tissue. It induces endothelial cells making up the capillary wall to pull apart farther at the narrow clefts between them. Thus the capillaries become "leaky" to plasma proteins that normally do not leave the blood. When some proteins leak out, osmotic pressure increases in the surrounding interstitial fluid. In combination with the higher blood pressure brought about by the increased blood flow to the tissue, ultrafiltration increases and reabsorption

Figure 34.4
(**a**) White blood cell squeezing out of a blood capillary, at a cleft between endothelial cells. (**b**) Macrophage about to engulf a yeast cell.

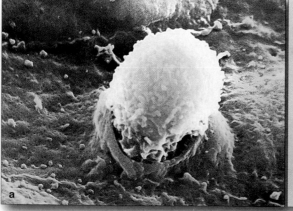
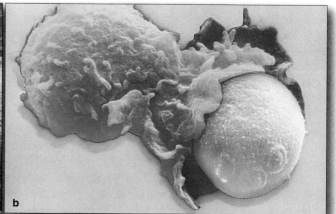

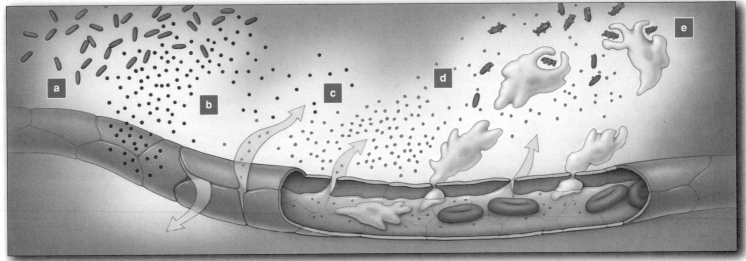

a Bacteria invade a tissue and directly kill cells or release metabolic products that damage tissue.

b Mast cells in tissue release histamine, which then triggers arteriolar vasodilation (hence redness and warmth) as well as increased capillary permeability.

c Fluid and plasma proteins leak out of capillaries; localized edema (tissue swelling) and pain result.

d Complement proteins attack bacteria. Clotting factors wall off inflamed area.

e Neutrophils, macrophages, and other phagocytes engulf invaders and debris. Macrophage secretions attract even more phagocytes, directly kill invaders, and call for fever and for T and B cell proliferation.

Figure 34.5 Acute inflammation in response to a bacterial invasion. The response involves delivering phagocytes and plasma proteins to the tissue. Together, these components of blood inactivate, destroy, or isolate the invaders, remove chemicals and cellular debris, and prepare the tissue for subsequent repair. These are their functions in all inflammatory responses.

decreases across the capillary wall. Localized edema is the outcome of the shift in the fluid balance across the capillary wall. (Here you may wish to review Section 33.8.) The tissue swells with fluid, and its nociceptors give rise to sensations of pain. Typically an individual avoids voluntary movements that might aggravate the pain. This behavior promotes tissue repair.

Within hours of the first physiological responses to the damage, neutrophils are squeezing across capillary walls. They swiftly go to work. Monocytes arrive later, differentiate into macrophages, and engage in sustained action (Figure 34.5). While macrophages are engulfing invaders and debris, they secrete chemical mediators. Mediators called *chemotaxins* attract more phagocytes. One of the *interleukins* stimulates formation of B and T cell armies, as described shortly. *Lactoferrin* directly kills bacteria. *Endogenous pyrogen* might trigger the release of prostaglandins, which in turn trigger an increase in the "set point" on the hypothalamic thermostat that controls body temperature. What we call a **fever** is a core temperature that has reached the higher set point.

A fever of about 39°C (100°F) is not a bad thing. It increases body temperature to a level that is "too hot" for the functioning of most pathogens. It also promotes an increase in a host's defense activities. *Interleukin-1* induces drowsiness, which reduces the body's demands for energy, so more energy can be diverted to the tasks

of defense and tissue repair. Macrophages take part in the cleanup and repair operations.

Among the plasma proteins that leak into the tissue are complement proteins and clotting factors of the sort described in Section 33.10. Upon exposure to chemicals secreted by phagocytes and to tissue thromboplastin, fibrin forms and clots develop in the spaces around the inflamed tissue. The clots wall off the inflamed area and typically prevent or delay the spread of invaders and toxic chemicals into the surrounding tissues. After the inflammation subsides, anticlotting factors that had also escaped from the capillaries dissolve the clots.

An inflammatory response develops in a local tissue when cells are damaged or killed, as by infection. It proceeds during both nonspecific and specific defenses of tissues.

Mast cells in damaged or invaded tissues secrete histamine, which causes arterioles to vasodilate and increases capillary permeability to fluid and plasma proteins. The localized vasodilation reddens and warms the tissue. Edema results from the fluid imbalance across the capillary wall. The tissue swelling causes pain.

The response involves phagocytes such as macrophages, which engulf invaders and debris and secrete chemical mediators. It requires plasma proteins, such as complement proteins that target invaders for destruction, as well as clotting factors that wall off the inflamed tissue.

WWW

Defining Features

Sometimes physical barriers and inflammation are not enough to overwhelm an invader, so an infection may become well established. Then, white blood cells called **B** and **T lymphocytes** form armies that engage in battle.

B and T cells are central to the body's third line of defense—the immune system. We define the **immune system** by two key features. The first is *immunological specificity*, whereby certain kinds of lymphocytes zero in on specific pathogens and eliminate them. The second feature is *immunological memory*, whereby a portion of T and B cells formed during a first-time confrontation is set aside for a future battle with the same pathogen.

The operating principle for the immune system is this: *Each kind of cell, virus, or substance bears unique molecular configurations that give it a unique identity.* The unique configurations on an individual's own cells function as markers of *self*. Lymphocytes recognize self markers and will normally ignore them. They also can recognize *nonself* molecular configurations, which are unique to specific foreign agents. When that happens, the lymphocytes are stimulated to divide repeatedly, by way of mitosis, and form huge populations.

As divisions proceed, subpopulations of the new cells become specialized to respond to the foreign agent in different ways. Some consist of *effector* cells—fully differentiated cells that engage and destroy the enemy. Other subpopulations are *memory* cells. These enter a resting phase. Instead of engaging in the attack on the agent that triggered the initial response, memory cells "remember" it. And if the same kind of agent shows up again, they will join a larger, more rapid response to it.

Any molecular configuration that triggers formation of lymphocyte armies and is their target is an **antigen**. The most important antigens are certain proteins at the surface of pathogens or tumor cells. As you will see, lymphocytes synthesize receptor molecules that are able to bind to the configurations. That is how lymphocytes are able to recognize nonself.

In short, immunological specificity and memory involve three events: *recognition* of antigen, *repeated cell divisions* that form huge populations of lymphocytes, and *differentiation* into subpopulations of effector and memory cells with receptors for one kind of antigen.

Antigen-Presenting Cells—The Triggers for Immune Responses

The plasma membrane of every human nucleated cell incorporates diverse proteins. Among the proteins are **MHC markers**, named after the genes that encode the instructions for making them. Certain MHC markers are common to each nucleated cell in the body. Others are unique to its macrophages and lymphocytes.

Think of what happens after a cut allowed bacteria to enter a tissue inside your finger. Lymph vessels pick up some of the inflamed tissue's interstitial fluid and deliver it, along with some invading cells, to the lymph nodes in the vicinity. There, macrophages join the fray. Foreign cells are engulfed and enclosed in vesicles with digestive enzymes that cleave antigen molecules into fragments. The fragments bind to MHC molecules and form **antigen-MHC complexes**. Later, vesicles fuse with the plasma membrane, and this means the complexes are automatically displayed at the macrophage surface.

Any cell displaying processed antigen that is bound with a suitable MHC molecule is an **antigen-presenting cell**. When lymphocytes do encounter such a cell, they take notice (Figure 34.6). *This is the antigen recognition that promotes the divisions that form lymphocyte armies.*

Key Players in Immune Responses

The same categories of white blood cells are called into action during each immune response. Figure 34.7 is an overview of how the cells interact. In brief, recognition of antigen-MHC complexes activates **helper T cells**. These produce and secrete substances that induce any responsive T or B lymphocyte to divide and give rise to large populations of effector cells and memory cells. Recognition also activates **cytotoxic T cells**, which can eliminate infected body cells or tumor cells by "touch killing." When they contact a target, they deliver

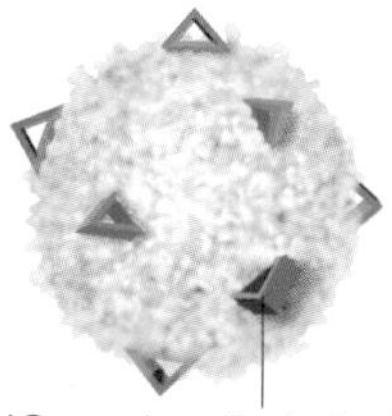

MHC marker that designates "self" (it occurs only at the surface of body's own cells)

T cells and B cells ignore this

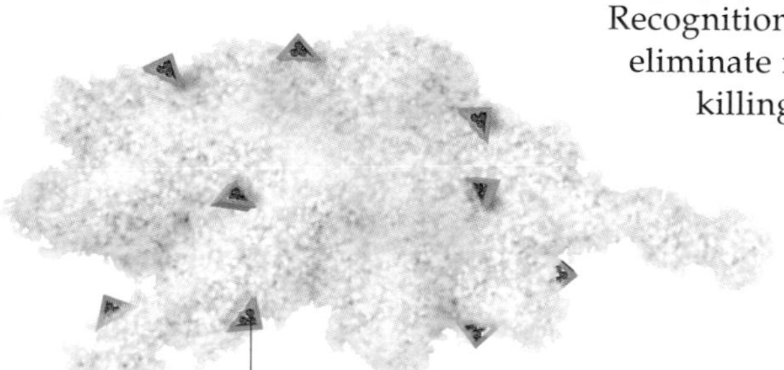

Figure 34.6 Molecular cues that T and B cells either ignore or recognize as a signal to initiate immune responses.

processed antigen, bound to MHC marker, at surface of an antigen-presenting cell

T cells initiate an immune response

antigen (any unprocessed foreign or abnormal molecular configuration that lymphocytes recognize as nonself)

B cells initiate an immune response

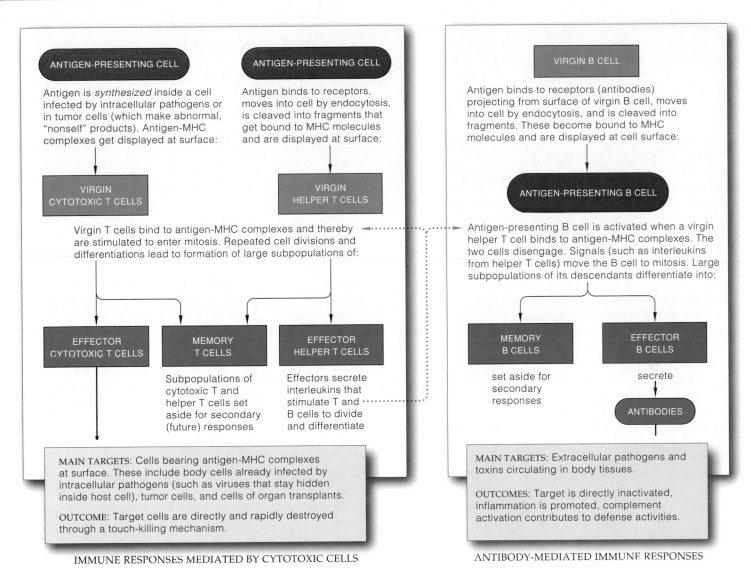

IMMUNE RESPONSES MEDIATED BY CYTOTOXIC CELLS   ANTIBODY-MEDIATED IMMUNE RESPONSES

Figure 34.7 Overview of key interactions among B and T lymphocytes during an immune response. Most often, both types of white blood cells are activated when antigen has been detected. An antigen is any large molecule that lymphocytes recognize as not being "self" (normal body molecules). A first-time encounter with antigen elicits a *primary* response. A subsequent encounter with the same type of antigen elicits a *secondary* immune response. A secondary response is larger and more rapid. Memory cells that formed but that were not used during the first battle immediately engage in the second one.

cell-killing chemicals into it. By contrast, B cells make antigen-binding receptor molecules called **antibodies**. When a response is under way, effector B cells secrete staggering numbers of antibody molecules. Only B cells are the basis of *antibody-mediated* responses.

Control of Immune Responses

Antigen provokes an immune response—and removal of antigen stops it. For example, by the time the tide of battle turns, effector cells and their chemical secretions have already killed most of the antigen-bearing agents inside the body. With fewer antigen molecules around

to stimulate the cells, the response declines, then stops. As a final example, inhibitory signals from cells with suppressor roles help shut down immune responses.

Antigens are any nonself molecular configurations which, when recognized by certain lymphocytes, trigger immune responses. Helper T cells, cytotoxic T cells, B cells, and their secretions execute these responses.

Following antigen recognition, large T and B cell armies form by repeated mitotic cell divisions. These differentiate into subpopulations of effector cells and memory cells, all of which are sensitized to that one kind of antigen.

The antigen-presenting cells and lymphocytes we have just introduced interact within lymphoid organs that promote immune responses (Figures 33.26 and 34.8).

Think about the tonsils and other lymph nodules located beneath mucous membranes of the respiratory, digestive, and reproductive systems. Just after invaders penetrate surface barriers, antigen-presenting cells and lymphocytes housed here intercept them.

Or think about antigen in interstitial fluid that is entering the lymph vascular system. Because lymph vessels eventually drain into the expressways for blood transport, antigen could become distributed to every body region. However, before antigen can reach the blood, it must trickle through lymph nodes—which are packed with defending cells. Even in those few cases where antigen does manage to enter blood, defending cells in the spleen intercept it.

Inside the lymph nodes, the defending cells are organized for utmost effectiveness. The antigen-presenting cells make up the front line and engulf invaders. They process and display antigen, thus calling their lymphocyte comrades into action.

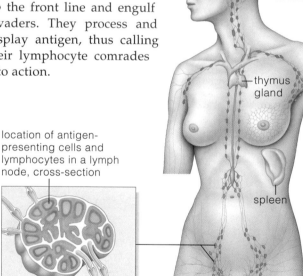

location of antigen-presenting cells and lymphocytes in a lymph node, cross-section

tonsils

thymus gland

spleen

Figure 34.8 Organized arrays of antigen-presenting cells and lymphocytes in lymph nodes.

The cell divisions that produce subpopulations of effector and memory cells proceed in lymph nodes. As lymph drains through, it moves effector activities to the back of the organ and beyond. And all the while, virgin and memory cells circulate through the lymph node, reconnoitering at the front line.

Antigen-presenting cells and lymphocytes intercept and battle pathogens in organized ways within lymphoid organs and tissues, especially the lymph nodes.

T Cell Formation and Activation

Let's first consider the functions of T lymphocytes in an immune response. As you read in Section 33.2, T cells arise from stem cells in bone marrow. However, they do not fully develop in bone marrow. Rather, they travel to an organ called the thymus, where they become fully differentiated into helper T cells and cytotoxic T cells. Specifically, these immature cells acquire their **TCRs** (short for *T-Cell Receptors*) in the thymus. Bristling with receptors, the cells now leave the thymus. They circulate in blood or take up stations in lymph nodes and the spleen, as virgin T cells. In this context, "virgin" means the cells are as yet undisturbed (by antigen).

The TCRs of virgin T cells ignore unadorned MHC markers on the body's cells. They ignore free antigen. *But they recognize and bind with antigen-MHC complexes at the surface of antigen-presenting cells.* As Figure 34.9 shows, binding induces the T cells to divide repeatedly and give rise to large clones. (A clone is a population of genetically identical cells.) Then the clonal descendants differentiate into subpopulations of effector cells and memory cells. *And every one of those descendants has the same TCR for one kind of antigen-MHC complex.*

Functions of Effector T Cells

What actions do subpopulations of effector T cells take? Effector helper T cells secrete interleukins, the chemical mediators that fan repeated mitotic cell divisions and then differentiation of any responsive T and B cells, as described shortly. Effector cytotoxic T cells are killers; they respond to antigen-MHC complexes at the surface of tumor cells and of body cells already attacked by intracellular pathogens or viruses. This complex is a "double signal" that tells the killers to attack any cell bearing it (Figure 34.9*e*).

Effector cytotoxic T cells destroy infected cells with a touch-kill mechanism. They secrete *perforins*, protein molecules that form doughnut-shaped pores in a target cell's plasma membrane. (The pores look similar to the ones shown in Figure 34.3.) These effectors also secrete chemicals that induce cell death by way of **apoptosis**. As Section 14.5 describes, the target commits suicide. Its cytoplasm dribbles out, its organelles are disrupted, and its DNA becomes fragmented. The cytotoxic T cell, having made its lethal hit, quickly disengages from the doomed target and resumes its reconnoitering.

Cytotoxic T cells also contribute to the rejection of tissue and organ transplants. Parts of MHC markers on donor cells are different enough from the recipient's to be recognized as antigens. But other parts are similar enough to complete the double signal. MHC typing and matching donors to recipients minimize the risk.

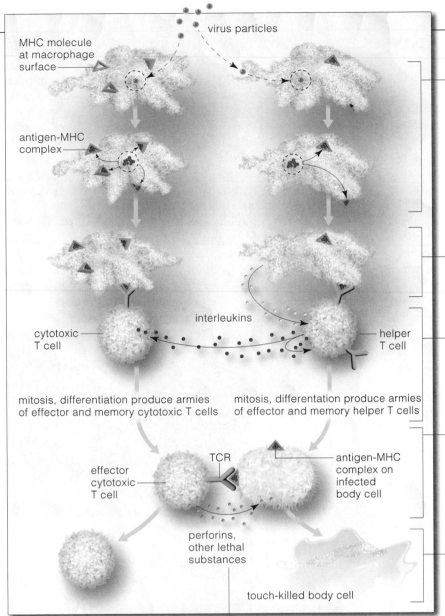

MHC molecule at macrophage surface

antigen-MHC complex

cytotoxic T cell

interleukins

helper T cell

mitosis, differentiation produce armies of effector and memory cytotoxic T cells

mitosis, differentation produce armies of effector and memory helper T cells

TCR

effector cytotoxic T cell

antigen-MHC complex on infected body cell

perforins, other lethal substances

touch-killed body cell

a A virus inserts its genetic material into a macrophage. The host cell's pirated metabolic machinery synthesizes viral proteins, which are antigenic. These are processed into fragments, bound to MHC molecules, and displayed at the cell surface. TCRs of cytotoxic T cells recognize antigens that have been synthesized inside a cell.

Another macrophage engulfs a particle of the same virus and encloses it in an endocytic vesicle. Digestive enzymes cleave the particle's antigen into fragments. These bind to MHC molecules and are presented at the macrophage surface. *TCRs of helper T cells recognize such engulfed and processed antigens.*

b A responsive T cell binds with the antigen-MHC complexes. Binding stimulates the macrophage to secrete interleukins (*black* dots).

c These communication signals stimulate the helper T cell to secrete different kinds of interleukins (*blue* dots). These new signals stimulate cell divisions and differentiations by which large populations of effector T cells and memory T cells form.

d The same virus penetrated a cell in the lining of the respiratory tract, and antigen that was produced in the cytoplasm has been processed. Antigen is bound to MHC molecules and displayed at the cell's surface.

An effector cytotoxic T cell encounters the target. This type of effector specializes in touch-killing. It releases perforins and toxic substances (*green* dots) onto its target and so programs it for death.

e The effector disengages from the doomed cell and reconnoiters for more targets. Meanwhile, perforins make holes in the cell's plasma membrane. Toxins can move into the cell, disrupt its organelles, and make the DNA disassemble. The infected cell dies.

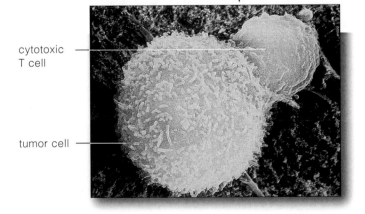

cytotoxic T cell

tumor cell

Figure 34.9 *Above:* Diagram of a T cell-mediated immune response. In this example, the response involves an antigen-MHC complex that activates T cells. *Left:* Scanning electron micrograph of a cytotoxic T cell, caught in the act of touch-killing a tumor cell.

Their arousal does not depend on the double signal (an antigen-MHC complex). NK cells reconnoiter for tumor cells and virus-infected cells, then proceed to touch-kill their finds. Possibly these killer cells can recognize odd molecular configurations at the surface of their targets.

T cells arise in bone marrow. Later, in the thymus, they acquire TCRs (receptors for self markers and bound antigen).

Effector helper T cells secrete interleukins that trigger the cell divisions and differentiation into huge armies against specific antigens. Effector cytotoxic cells touch-kill infected cells or tumor cells, even foreign cells of transplants.

Regarding the Natural Killer Cells

Other cytotoxic cells, including **natural killer cells** (NK cells), also arise from stem cells in bone marrow. They appear to be a type of lymphocyte, but not a T or B cell.

ANTIBODY-MEDIATED RESPONSES

B Cells and the Targets of Antibodies

Like T cells, B cells also arise from stem cells in bone marrow and start down a pathway that will culminate in full differentiation. Along *their* pathway, however, B cells start synthesizing many, many copies of a single kind of antibody molecule.

All antibodies are proteins, but the antigen binding site of each kind matches only one kind of antigen. Antibody molecules are more or less Y-shaped, with a tail and two arms that bear identical antigen receptors. Section 34.9 provides a closer look at these molecules. For now, simply think of them as Y-shaped structures of the sort shown in Figure 34.10.

When a B cell is maturing, each freshly synthesized antibody molecule moves to the plasma membrane. Its tail becomes embedded in the membrane's lipid bilayer and its two arms project above it. Soon the cell bristles with antigen receptors (the bound antibodies), and it is ready to join the body's defenses as a virgin B cell.

When its antigen receptors lock on to a target, the B cell does something you might not expect. *It becomes an antigen-presenting cell.* First, an endocytic vesicle forms around the bound antigen and moves it into the cell for digestion into fragments. The fragments bind to MHC molecules and are presented at the B cell surface. Now suppose the TCRs of a responsive helper T cell bind to the antigen-MHC complex, and signals are transferred

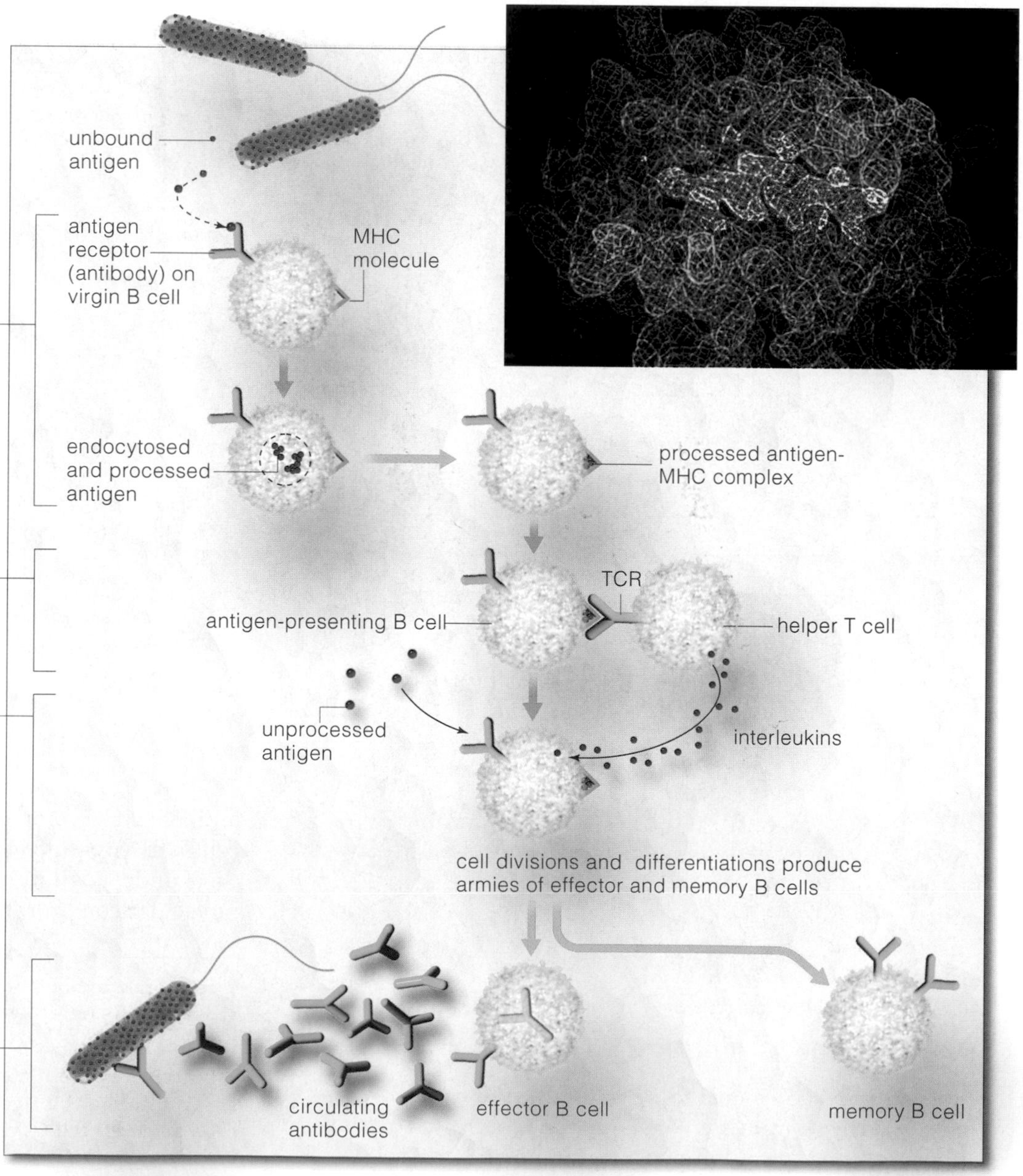

Figure 34.10 Antibody-mediated immune response. This example is a response to a bacterial invasion. The inset is a computer model of an antigen fragment (*pink*) bound to the cleft of an MHC protein (*blue*).

a A virgin B cell encounters unbound antigen in tissue fluid. Antigen receptors (in this case, membrane-bound antibodies) bind the antigen. An endocytic vesicle moves bound antigen into the cell for processing. Antigen-MHC complexes are displayed at the cell surface; the B cell has become an antigen-presenting cell.

b TCRs of a helper T cell bind to antigen-MHC complexes on the B cell. Binding activates the T cell and stimulates the B cell to prepare for mitosis. Then the cells disengage.

c Unprocessed antigen binds to the B cell. Meanwhile, the helper T cell secretes interleukins (*black* dots). Both events trigger repeated cell divisions and differentiations that yield large armies of antibody-secreting effector B and memory B cells.

d Antibody molecules released from the effector B cells enter extracellular fluid. When they contact a bacterial cell that is the target, they bind to antigen on its surface. Binding tags the cell for destruction. (Compare Figures 34.3 and 34.7).

between the B cell and a T cell. The cells soon disengage. Then the B cell encounters unprocessed antigen, and its surface antibodies bind to it. The binding, along with interleukins secreted from nearby helper T cells, drives the B cell to mitosis. Its clonal descendants differentiate into effector B and memory B cells. The effectors (also called plasma cells) make and secrete huge numbers of antibody molecules. When freely circulating antibody binds antigen, it tags an invader for destruction, as by phagocytes and complement activation.

The main targets of antibody-mediated responses are extracellular pathogens and toxins, which are freely circulating in tissues or body fluids. *Antibodies cannot bind to pathogens or toxins that are hidden in a host cell.*

The Immunoglobulins

During immune responses, B cells produce four classes of antibodies in abundance, and lesser quantities of another class. Collectively, the five classes of antibodies are known as the **immunoglobulins**, or **Igs**. They are the protein products of gene shufflings that proceed while B cells mature and while an immune response is under way. The molecules in each class have antigen-binding sites *and* other sites with specialized functions.

IgM antibodies are the first to be secreted during immune responses. They trigger the cascade reactions that produce complement and bind targets together in clumps that are more handily destroyed by phagocytes. (Remember the agglutination responses described in Section 33.4?) The *IgD* antibodies associate with IgM on virgin B cells, but their function is not yet understood.

IgG antibodies activate complement proteins and neutralize many toxins. These long-lasting antibodies are the only ones that can cross the placenta. They help protect both the fetus and newborn with the mother's acquired immunities. IgGs also are secreted into the early milk produced by mammary glands and then are absorbed into the suckling newborn's bloodstream.

IgA antibodies enter mucus-coated surfaces of the respiratory, digestive, and reproductive tracts, where they may neutralize infectious agents. Mother's milk delivers them to the mucous lining of a newborn's gut.

IgE triggers inflammation after attacks by parasitic worms and other pathogens. As described later, it also figures in allergies. The tails of IgE antibodies bind to basophils and mast cells, and the antigen receptors face outward. Antigen binding induces basophils and mast cells to release substances that promote inflammation.

Antibodies that are secreted by B cells bind to antigens of extracellular pathogens or toxins and tag them for disposal, as by phagocytes and complement activation.

Carcinomas, sarcomas, leukemia—these chilling words refer to malignant tumors in skin, bone, and other tissues. Such tumors arise when viral attack, irradiation, or chemicals alter genes and cells turn cancerous (Section 14.5). The transformed cells divide repeatedly. Unless they can be destroyed or surgically removed, they kill the individual.

Often, transformed cells bear abnormal proteins and protein fragments bound to MHC. Immune responses to them may make a tumor regress but may not be enough to kill it. Also, some tumors "hide" by releasing copies of the abnormal proteins, which saturate antigen receptors. Hidden tumors that reach critical mass may overwhelm the immune system. The idea behind *immunotherapy* is to enlist white blood cells to destroy this threat and others.

MONOCLONAL ANTIBODIES   Two decades ago, Cesar Milstein and Georges Kohler made limited amounts of antibodies that home in on tumor-specific antigens. They injected an antigen into mice, which made antibodies against it. Then they fused antibody-producing B cells from the mice with cells extracted from the B cell tumors. Clonal descendants of such hybrid cells made *monoclonal antibodies* (identical copies of antibodies). However, mice can't produce useful amounts. Later, genetic engineers designed cows that secrete antibodies into milk, but it may be difficult to isolate these from any bacteria or other pathogens present in the milk.

Today, genetically engineered plants can manufacture monoclonal antibodies. Cornfields might mass-produce them. Besides being cost-effective, "plantibodies" are safer than antibodies from cattle (few plant pathogens infect people). The first plantibody used on human volunteers prevented infection by a bacterial agent of tooth decay.

MULTIPLYING THE TUMOR KILLERS   Lymphocytes often infiltrate tumors. Researchers have removed them from a tumor and exposed them to an interleukin (lymphokine). Large populations of tumor-infiltrating lymphocytes with enhanced killing abilities have formed. These *LAK* cells (short for *Lymphokine-Activated Killers*) appear to be somewhat effective when injected back into a patient.

THERAPEUTIC VACCINES   On the horizon are *therapeutic vaccines*—wake-up calls against specific tumor cells. One idea is to genetically engineer antigen so that it becomes obvious to killer lymphocytes. For example, remember Langerhans cells (Section 32.2)? Like other *dendritic* cells, these phagocytes prowl on many pseudopods. They also are antigen-presenting cells that swiftly migrate to lymph nodes, where they sound the alarm. Dirk Schadendorf harvested dendritic cells from melanoma patients. He cultured them with ground-up tumor cells, then injected them at intervals into his patients' lymph nodes or skin. A year later, two patients had no trace of melanoma. In three others, tumors shrank more than half their size. Such vaccines may be available to the public within five years.

How Do Antigen-Specific Receptors Form?

The variety of antigens in your surroundings is mind-boggling. Collectively, however, antigen receptors of all T and B cell populations in your body show staggering diversity—enough to recognize about a billion different antigens. How does antibody diversity arise?

For the answer, start with the knowledge that all antigen receptors of a single T or B cell are identical—and all are proteins. For example, a Y-shaped antibody molecule consists of four polypeptide chains bonded together (Figure 34.11). Certain parts of each chain, the *variable* regions, fold in ways that produce grooves and bumps with a certain charge distribution. Only antigen that has complementary grooves, bumps, and charge distribution will be able to bind with them.

Receptor diversity begins with rearrangements in DNA sequences that code for such variable regions. Many of the sequences are V segments and, some distance away, J segments (Figure 34.12). As a B cell matures, one of its V segments randomly contacts and binds with one of its J segments. The stretch of DNA intervening between the two joined segments loops outward, and enzymes snip it off. The B cell ends up with a unique sequence, randomly put together from a choice of segments.

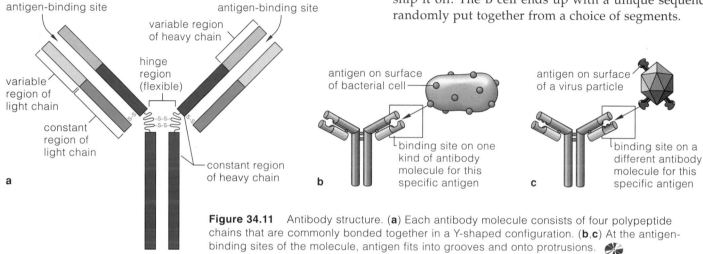

Figure 34.11 Antibody structure. (**a**) Each antibody molecule consists of four polypeptide chains that are commonly bonded together in a Y-shaped configuration. (**b**,**c**) At the antigen-binding sites of the molecule, antigen fits into grooves and onto protrusions.

Figure 34.12 Generation of antibody diversity. Antibodies are proteins, and instructions for building proteins are encoded in genes. In the chromosomes having genes for antibodies, extensive DNA regions contain different versions of segments that code for the variable regions of an antibody molecule.

The different V and J segments shown here are examples. They code for the variable region of a light chain (compare Figure 34.11*a*). As each B cell matures, a recombination event occurs in this region. In this example, any one of the V segments may be joined to any one of the J segments. Afterward, the DNA intervening between them is excised. The new sequence is attached to a C (*C*onstant) segment, and this completes a rearranged antibody gene—which also will be present in all of the descendants of the cell.

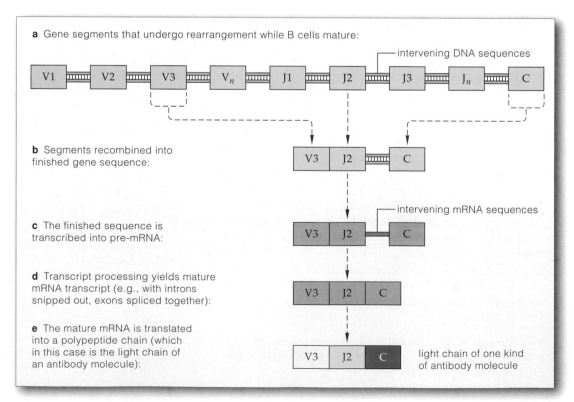

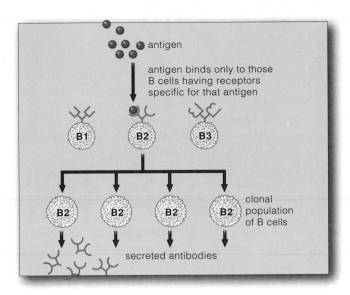

Figure 34.13 Clonal selection of a B cell that produced the specific antibody that can combine with a specific antigen. Only antigen-selected B cells (and T cells) are activated and give rise to a clonal population of immunologically identical cells.

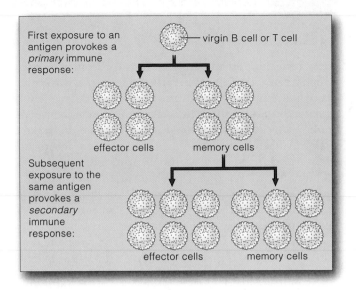

Figure 34.14 Immunological memory. Not all B and T cells are used in a primary immune response to an antigen. A large number continue to circulate as memory cells, which become activated during a secondary immune response.

The same kinds of DNA rearrangements have roles in producing the variable regions of the TCR molecules of maturing T cells.

Some time ago, Macfarlane Burnet devised a *clonal selection* hypothesis that helped point the way to our current view of receptor diversity. He proposed that antigen "chooses" (binds to) one lymphocyte from all the various types in the body, because that lymphocyte has the receptor specific for it. Repeated mitotic cell divisions then give rise to a clone of cells that carry out the response (Figure 34.13).

Immunological Memory

The clonal selection theory explains how an individual can have "immunological memory" of a first encounter with antigen. The term refers to the body's capacity to make a *secondary* immune response to any subsequent encounter with the same type of antigen that provoked the primary response (Figure 34.14).

Memory cells that form during a primary immune response do not engage in that battle. They circulate for years or for decades. Compared to the virgin cells that initiate a primary response, the patrolling battalions consist of far more cells and intercept antigen far sooner. Effector cells form sooner, in greater numbers, so the infection is terminated before the host gets sick. Even greater numbers of memory T and B cells form during a secondary response. Figure 34.15 has an example of this. In evolutionary terms, the advance preparations against subsequent encounters with a pathogen bestow a survival advantage on the individual.

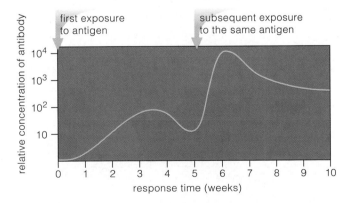

Figure 34.15 One example of the differences in magnitude and duration between a primary and a secondary immune response to the same antigen. The primary response peaked twenty-four days after it started. The secondary response peaked after only seven days (the span between weeks five and six). Antibody concentration during the secondary response was 100 times greater (10^2 to 10^4).

By recombination of segments, drawn at random from the receptor-encoding regions of DNA, each T cell or B cell receives a gene sequence for one of a billion possible kinds of antigen receptors.

Immunological specificity means the clonal descendants of an antigen-selected cell will react only with the selecting antigen.

Immunological memory refers to the capacity to make a secondary (faster, greater) immune response to a pathogen that caused a primary response in an individual.

Immunization

After reflecting on the chapter introduction, you may already sense that **immunization** refers to a variety of processes which promote increased immunity against specific diseases. With *active* immunization, an antigen-containing preparation called a **vaccine** is either taken orally or injected into the body, as in Figure 34.16. An initial injection triggers a primary immune response to an antigen. A subsequent injection (a booster) elicits a secondary response, with the rapid formation of more effector cells and memory cells that can provide long-lasting protection against the disease.

Many vaccines are manufactured from weakened or killed pathogens. Others use inactivated natural toxins, such as a bacterial toxin that causes tetanus. Others are made of harmless viruses that are genetically engineered so genes from three or more pathogens are inserted into their DNA or RNA. After vaccination, an individual's cells use the new genes to produce the antigens, and immunity is established.

Passive immunization is used for individuals who are already infected with pathogens that cause diphtheria, tetanus, measles, hepatitis B, and some other diseases. A person at risk receives injections of purified antibody, the best source of which is some other individual who already has made a large amount of the required antibody. The effects do not last long, for the patient's B cells haven't made any antibodies. But antibody injections may counter the initial attack.

Vaccines may fail or have adverse effects. In a few cases, they result in chronic immunological or neurological problems. It is important to assess the risks and benefits before agreeing to a procedure.

Figure 34.16 From the Centers for Disease Control and Prevention, the 1998 immunization guidelines for children in the United States. Doctors routinely immunize infants and children. Low-cost or free vaccinations are available at many community clinics and health departments.

Allergies

In many people, normally harmless substances provoke inflammation, excess mucus secretion, and sometimes immune responses. Such substances are **allergens**, and hypersensitivity to them is called an **allergy**. Common allergens are pollen, many drugs and foods, dust mites, fungal spores, insect venom, and cosmetics.

Some people are genetically inclined to develop allergies. Infections, emotional stress, or changes in air temperature also trigger reactions that otherwise might not occur. Upon exposure to an antigen, IgE antibodies are secreted and bind to mast cells. When the IgE binds antigen, mast cells secrete prostaglandins, histamine, and other substances that fan inflammation. Copious amounts of mucus are secreted, and airways constrict. Stuffed sinuses, labored breathing, a drippy nose, and sneezing are key symptoms of the allergic response in *asthma* and *hay fever* (Figure 34.17).

In a few cases, the inflammatory reactions wash through the body in a life-threatening event called *anaphylactic shock*. For example, someone allergic to wasp or bee venom can die within minutes of one sting. Airways constrict massively, and fluid escapes rapidly from grossly permeable capillaries. Blood pressure plummets, and circulation may collapse.

Antihistamines (anti-inflammatory drugs) often relieve the mild, short-term symptoms of allergies. Over time, a patient can follow a desensitization program. First, skin tests may identify offending allergens. Inflammatory responses to some can be blocked if the patient can be stimulated to make

RECOMMENDED VACCINES	RECOMMENDED AGES
Hepatitis B	Birth–2 months
Hepatitis B booster	1–4 months
Hepatitis B booster	6–18 months
Hepatitis B assessment	11–12 years
DTP (Diphtheria; Tetanus; and Pertussin, or whooping cough)	2, 4, and 6 months
DTP booster	15–18 months
DTP booster	4–6 years
DT	11–16 years
HiB (*Hemophilus influenzae*)	4 and 6 months
HiB booster	12–15 months
Polio	2 and 4 months
Polio booster	6–18 months
Polio booster	4–6 years
MMR (Measles, Mumps, Rubella)	12–15 months
MMR booster	4–6 years
MMR assessment	11–12 years
Varicella	12–18 months
Varicella assessment	11–12 years

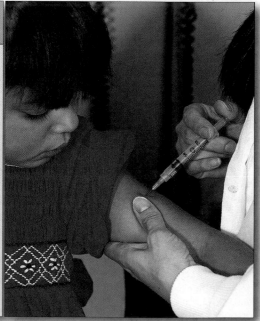

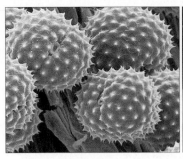

RAGWEED POLLEN . . .   . . . AND SOMETHING IT CAN PROVOKE

Figure 34.17 One of the effects of pollen and other allergens in sensitive people. Allergy sufferers who moved to deserts to escape pollen brought it with them. Half the human population in Tucson, Arizona, is now sensitized to pollen from olive and mulberry trees, planted far and wide in cities and the suburbs.

IgG instead of IgE. Larger doses of specific allergens are administered slowly. Each time, the body makes more circulating IgG and memory cells. IgG binds with the allergen to block its attachment to IgE, and thereby blocks inflammation.

Autoimmune Disorders

In an **autoimmune response**, the immune system acts against self antigens. For instance, in *Grave's disorder*, the body makes too many thyroid hormone molecules, which control metabolic rates and the growth of many tissues. Affected individuals produce antibodies that bind to receptors on cells that make the hormones. The antibodies do not respond to the feedback mechanisms that normally control production of thyroid hormones, so too many molecules are produced. Typical symptoms are elevated rates of metabolism, heart fibrillations, excessive sweating, nervousness, and weight loss.

As other examples, *myasthenia gravis*, a progressive weakening of muscles, arises when antibodies bind to receptors on skeletal muscle cells and interfere with normal acetylcholine action. *Rheumatoid arthritis* is a chronic inflammation of skeletal joints. Patients are genetically predisposed to this disorder. Macrophages, T cells, and B cells are activated when antigens act at skeletal joints. Immune responses are made against the body's own collagen molecules and against antibody that has become bound to as-yet-unidentified antigen. Complement activation and inflammation cause more damage in tissues of skeletal joints. So do skewed repair mechanisms. Eventually, joints fill with synovial membrane cells and become immobilized.

Deficient Immune Responses

When the body has inadequate numbers of functioning lymphocytes, immune responses are not effective. Such *severe combined immunodeficiencies* (SCIDs) result from

Figure 34.18 A case of severe combined immunodeficiency, or SCID. Ashanthi DeSilva was born without an immune system. She has a mutated gene for ADA (adenosine deaminase), an enzyme. Without ADA, her cells cannot break down adenosine. As a result, a reaction product accumulates that is toxic to lymphocytes. The disorder's symptoms are caused by infections that cannot be controlled. They include high fever, severe ear and lung infections, diarrhea, and an inability to gain weight.

Ashanthi's parents consented to the first federally approved gene therapy for a human patient. Genetic engineers spliced the ADA gene into the genetic material of a harmless virus. Then they used the modified virus rather like a hypodermic needle: they allowed it to deliver copies of the "good" gene into Ashanthi's bone marrow cells. Some of those cells incorporated the gene in their DNA and started to synthesize the missing enzyme. At this writing, Ashanthi is in her teens. Like other ADA-deficient patients who have undergone this gene therapy, she is doing well.

In other cases, researchers take bone marrow stem cells from blood in the umbilical cord of affected newborns. (The cord, which connects the fetus and placenta during pregnancy, is discarded after birth.) They expose the cells to viruses that deliver copies of the ADA gene into them and to factors that stimulate mitotic cell division and growth. The cells are reinserted into newborns.

heritable disorders as well as from various assaults on the body by outside agents. Deficient or nonexistent immune responses make the person highly vulnerable to infections that are not life threatening to the general population. Figure 34.18 describes one of the heritable disorders. Another is the acquired immunodeficiency syndrome (AIDS). The next section describes how HIV, the virus that causes AIDS, replicates inside certain lymphocytes and destroys the body's capacity to fight infections. We also return to this topic in Section 38.19.

Immunization programs boost immunity to specific diseases.

Certain heritable disorders or attacks by certain pathogens and other agents can result in misdirected, compromised, or nonexistent immunity.

WWW

34.11 AIDS—THE IMMUNE SYSTEM COMPROMISED

CHARACTERISTICS OF AIDS   AIDS is a constellation of disorders that follow infection by HIV (short for human immunodeficiency virus). The virus cripples the immune system, so that the body becomes highly susceptible to usually harmless infections and some otherwise rare forms of cancer. Worldwide, more than 30 million are now infected (Table 34.3). Of the 16,000 new infections each day, most are among people between ages 15 and 24. Researchers have not yet developed a vaccine to work against the known forms of the virus (HIV-1 and HIV-2). *At this writing, there is no cure for those already infected.*

At first an infected person might appear to be in good health, suffering no more than a bout of "the flu." Then symptoms that foreshadow AIDS emerge. Typically, they include chronic weight loss, fever, many enlarged lymph nodes, fatigue, and bed-drenching night sweats. In time, diseases result from certain opportunistic infections. Rare in the population at large, the diseases are signs of AIDS. Among these are yeast infections of the mouth, vagina, esophagus, and elsewhere, and a form of pneumonia caused by *Pneumocystis carinii*. Bruises often develop, on legs and feet especially (Figure 34.19). They are signs of Kaposi's sarcoma, a cancer of blood vessel endothelium. The infections or cancers end up killing the person.

HOW HIV REPLICATES   HIV infects antigen-presenting macrophages and helper T cells (which immunologists call CD4 lymphocytes). HIV is a retrovirus. Each virus particle has an outermost lipid envelope—a bit of plasma membrane that surrounded the particle when it departed from an infected cell. Spiking outward from the envelope are HIV proteins that became inserted into it. Under the lipid envelope, two protein coats—the core proteins—surround two RNA strands and several copies of reverse transcriptase, a viral enzyme (Sections 15.1 and 20.5).

Once inside a host cell, the viral enzyme uses the RNA as a template for making DNA, which then is inserted into a host chromosome (Figure 34.20*a*). In some host cells, the inserted viral genes remain silent, only to become activated during a later round of infection. Whenever it occurs, transcription yields copies of the viral RNA. Some transcripts are translated into viral proteins. Others become enclosed by viral coat proteins when new particles are put together; they will function as the viral hereditary material. As Figure 34.20*b* shows, the particles are released by budding from the host cell's plasma membrane, after which they start a new round of infection in other cells. With each round, more of the body's macrophages, antigen-presenting cells, and helper T cells are impaired or killed.

A TITANIC STRUGGLE BEGINS   Infection marks the onset of a titanic battle between the enemy and the immune system. B cells synthesize antibodies in response to HIV antigenic proteins. These are the antibodies that are the basis of diagnostic tests to identify HIV infection. Armies of helper T cells and cytotoxic T cells also form. However, the virus infects an estimated 2 billion helper T cells and produces 100 million to 1 billion virus particles per day during certain phases of the infection. Every two days, the immune system destroys about half the virus particles and replaces half the helper T cells lost in the battle. Immense reservoirs of HIV and masses of infected T cells accumulate in lymph nodes. As the battle proceeds, the number of virus particles in the general circulation rises. Gradually, the numbers tilt. The body produces fewer and fewer helper T cells to replace the ones it lost. It may take a decade or more, but erosion of the helper T cell count inevitably causes the body to lose its capacity to mount effective immune responses.

Some viruses, including the measles virus, produce far more virus particles in a given day, but the immune system usually wins out. Other viruses, including herpes viruses, can lurk in the body for a lifetime, but the immune system keeps them in check. With HIV, the immune system loses, and infections and tumors always kill the person.

HOW HIV IS TRANSMITTED   Like any other virus that infects humans, HIV requires a medium that allows it to leave one host, survive in the environment into which it is released, then enter another host. HIV is transmitted when some body fluid of an infected person enters tissues of another person. In the United States, transmission initially occurred most often between males who engaged in homosexual activities, especially anal intercourse. It has spread among intravenous drug abusers who share blood-contaminated syringes and needles. It also has spread in the heterosexual population, increasingly by vaginal intercourse.

HIV travels from infected mothers to offspring during pregnancy, birth, and breast-feeding. Before 1985, contaminated blood supplies accounted for some AIDS cases. Health care providers have since implemented careful screening. Tissue transplants caused a few infections. In several developing countries, health care providers have accidentally spread HIV by way of contaminated transfusions and reuse of unsterile needles and syringes.

The molecular structure of HIV does not remain stable outside the human body, which is why it must

Table 34.3	1997 Reported AIDS Cases
Sub-Saharan Africa:	20,800,000
South/Southeast Asia:	6,000,000
Latin America:	1,300,000
North America:	860,000
Western Europe:	530,000
East Asia and Pacific:	440,000
Caribbean Islands:	310,000
Middle East/North Africa:	210,000
Central and Eastern Europe/ Central Asia:	150,000
Australia/New Zealand:	12,000

Figure 34.19
Lesions that are signs of Kaposi's sarcoma.

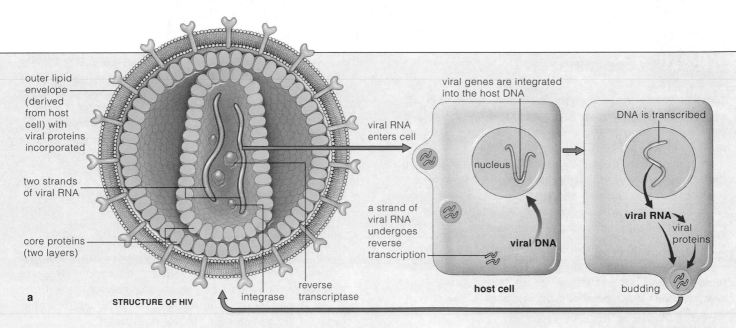

outer lipid
envelope
(derived
from host
cell) with
viral proteins
incorporated

two strands
of viral RNA

core proteins
(two layers)

integrase

reverse
transcriptase

a STRUCTURE OF HIV

viral RNA
enters cell

a strand of
viral RNA
undergoes
reverse
transcription

viral genes are integrated
into the host DNA

nucleus

viral DNA

host cell

DNA is transcribed

viral RNA

viral
proteins

budding

Figure 34.20 Life cycle of HIV, one of the retroviruses.

b Electron
micrographs of
an HIV particle
budding from
a host cell

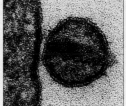

25 µm

be directly transmitted from one host to another. At this writing, there is no evidence that HIV can be effectively transmitted by way of food, air, water, casual contact, or insect bites. The virus *has* been isolated from human blood, semen, vaginal secretions, saliva, tears, urine, breast milk, amniotic fluid, urine, and cerebrospinal fluid. Probably it is present in other fluids, secretions, and excretions. Even so, only infected blood, semen, vaginal secretions, and breast milk contain HIV particles in concentrations that appear to be high enough for successful transmission.

REGARDING PREVENTION AND TREATMENT Developing effective drugs or vaccines against HIV is a formidable challenge. High mutation rates characterize the viral genome, owing to HIV's replication mechanisms and the staggering number of replications in an infected person. Inevitably, natural selection operates in patients who are undergoing drug therapies; it favors drug resistance and

Figure 34.21
Computer model of
gp120, the HIV protein
that can bind to two
receptors of host cells
(CD4 receptor and
chemokine). The most
stable parts of gp120
are hidden by a thick
"forest " of sugar
molecules or at the
bottom of a crevice,
where they elude the
rather bulky antibody
molecules. One part
becomes exposed,
but only briefly, after
it hooks on to a CD4
receptor and before it
hooks on to a nearby
chemokine receptor.

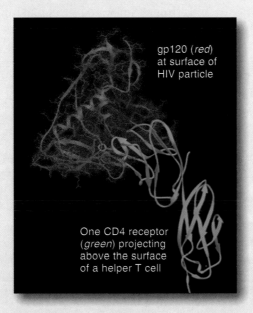

gp120 (*red*)
at surface of
HIV particle

One CD4 receptor
(*green*) projecting
above the surface
of a helper T cell

fans the evolution of drug-resistant HIV populations. Chemical "cocktails" help slow replication. Among the current drugs of choice are protease inhibitors, AZT (azidothymidine), and ddI (dideoxyinosine). However, they cannot *cure* infected people; drugs cannot eliminate HIV genes that are already incorporated in their DNA.

High mutation rates have another worrisome outcome: they give rise to variations in HIV antigens. The variation makes it hard for researchers to select effective antigens for vaccines. Another problem is the scarcity of HIV strains with poor cell-killing abilities. Such strains are central to producing antigen that can stimulate the formation of cytotoxic T cell armies—the best protection against HIV.

Over a decade ago, many people in Australia received blood transfusions from a donor whose infection had not been diagnosed. At this writing, neither they nor the donor show any immunodeficiency. And they all carry HIV with a similar defect in the same gene (the nef gene).

Recently, x-ray diffraction methods revealed the three-dimensional structure of gp120, the viral protein that binds to helper T cell receptors (Figure 34.21). Portions of gp120 must remain stable; if they were to change, HIV would lose its binding capacity. They hide behind an oligosaccharide "forest" and in crevices that are too small for antibody molecules to reach. Genetic engineers might be able to design molecules small enough to infiltrate the crevices and inactivate the binding site. Such an approach may or may not work, but research is under way.

In short, *until researchers develop effective vaccines and treatments, checking the spread of HIV depends absolutely on persuading people to avoid or modify social behaviors that put them at risk.* We return to this topic in Section 38.19.

WWW

1. Vertebrates fend off many pathogens with physical and chemical barriers at body surfaces. They also are protected by nonspecific and specific defense responses of the white blood cells listed in Table 34.4.

   a. Nonspecific responses to irritation or damage of tissues include inflammation and involve organs with phagocytic functions, such as the spleen and liver.

   b. Immune responses are mounted against specific pathogens, foreign cells, or abnormal body cells.

2. Skin and mucous membranes lining body surfaces are physical barriers to infection. The chemical barriers include glandular secretions (as in tears, saliva, and gastric fluid) and metabolic products of bacteria that normally reside on body surfaces.

3. An inflammatory response develops in body tissues that have become damaged, as by infection.

   a. An inflammatory response starts with arteriole vasodilation that increases the blood flow to the tissue, which reddens and becomes warmer as a result. Blood capillary permeability increases and the resulting local edema causes swelling and pain.

   b. Pathogens and dead or damaged body cells release the substances that trigger increased permeability of capillaries. White blood cells leave the blood and enter the tissue, where they release a number of chemical mediators and engulf invaders. Plasma proteins also enter the tissue. Complement proteins bind pathogens and induce their lysis. They also attract phagocytes. Blood-clotting proteins wall off the damaged tissue.

4. An immune response has these characteristics:

   a. It shows specificity, meaning it is directed against antigen. An antigen is a specific molecular configuration that lymphocytes recognize as foreign (nonself).

   b. Each response shows memory, which means that a subsequent encounter with the same antigen triggers a more rapid, secondary response, of greater magnitude.

   c. Normally, an immune response is not mounted against the body's own self-marker proteins.

5. Antigen-presenting cells process and bind fragments of antigen to their own MHC markers. Lymphocyte receptors can bind to displayed antigen-MHC complexes. Binding is a start signal for immune responses.

6. After recognition of antigen, an immune response proceeds by repeated cell divisions that form clones of B and T cells. These differentiate into subpopulations of effector and memory cells. Chemical mediators such as interleukins secreted by white blood cells drive the responses. The effector helper T cells, cytotoxic T cells, effector B cells, and antibodies act at once. The memory cells are set aside for secondary responses.

7. T cells arise in bone marrow but continue to develop in the thymus, where they acquire TCRs. These T-cell receptors recognize and bind antigen-MHC complexes on antigen-presenting cells. B cells arise in bone marrow. As they mature, they synthesize antigen receptors (that is, antibodies) that become positioned at their surface.

8. Effector cytotoxic T cells directly kill virus-infected cells, tumor cells, and cells making up tissue or organ transplants. Effector B cells (plasma cells) produce and secrete great numbers of antibodies that freely circulate.

9. Antibodies are protein molecules, often Y-shaped, each with binding sites for one kind of antigen. Only B cells make them. When antibody binds antigen, toxins are neutralized, pathogens are tagged for destruction, or attachment of pathogens to body cells is prevented.

10. In active immunization, vaccines provoke immune responses, with production of effector and memory cells. In passive immunization, injections of purified antibodies help the individual through an infection.

11. Allergic reactions are immune responses to some generally harmless substance. Autoimmune responses are misguided attacks triggered by configurations on the body's own cells. Immunodeficiency is a weakened or nonexistent capacity to mount an immune response.

Table 34.4	Summary of White Blood Cells and Their Roles in Defense
Cell Type	Main Characteristics
MACROPHAGE	Phagocyte; acts in nonspecific and specific responses; presents antigen to T cells; cleans up and helps repair tissue damage
NEUTROPHIL	Fast-acting phagocyte; takes part in inflammation, not in sustained responses; most effective against bacteria
EOSINOPHIL	Secretes enzymes that attack certain parasitic worms
BASOPHIL AND MAST CELL	Secrete histamine, other substances that act on small blood vessels to produce inflammation; contribute to allergies
LYMPHOCYTES:	*All take part in most immune responses; following antigen recognition, form clonal populations of effector and memory cells.*
B cell	Effectors secrete four types of antibodies (IgA, IgE, IgG, and IgM) that protect the host in specialized ways
Helper T cell	Effectors secrete interleukins that stimulate rapid divisions and differentiation of both B cells and T cells
Cytotoxic T cell	Effectors kill infected cells, tumor cells, and foreign cells by a touch-kill mechanism
NATURAL KILLER (NK) CELLS	Cytotoxic cell of undetermined affiliation; kills virus-infected cells and tumor cells by a touch-kill mechanism

Review Questions

1. While jogging barefoot along a seashore, some of your toes accidentally land on a jellyfish. Soon the toes are swollen, red, and warm to the touch. Describe the events that result in these signs of inflammation. *34.3*

2. Distinguish between:
   a. neutrophil and macrophage *34.3*
   b. cytotoxic T cell and natural killer cell *34.4, 34.6*
   c. effector cell and memory cell *34.4*
   d. antigen and antibody *34.4*

3. Describe the events by which a macrophage turns into an antigen-presenting cell. *34.4*

4. Why is a vaccine to control AIDS so elusive? *34.11*

Self-Quiz *(Answers in Appendix III)*

1. _____ are barriers to pathogens at body surfaces.
   a. Intact skin, mucous membranes   d. Urine flow
   b. Tears, saliva, gastric fluid   e. All of the above
   c. Resident bacteria

2. Macrophages are derived from _____ .
   a. basophils   b. neutrophils   c. monocytes   d. eosinophils

3. Complement functions in defense by _____ .
   a. neutralizing toxins   d. forming pore complexes that
   b. enhancing resident   cause lysis of pathogens
      bacteria   e. both a and b are correct
   c. promoting inflammation   f. both c and d are correct

4. _____ are certain molecules that lymphocytes recognize as foreign and that elicit an immune response.
   a. Interleukins   d. Antigens
   b. Antibodies   e. Histamines
   c. Immunoglobulins

5. Immunoglobulins _____ increase antimicrobial activity in mucus-coated surfaces of some organ systems.
   a. IgA   b. IgE   c. IgG   d. IgM   e. IgD

6. Antibody-mediated responses work best against _____ .
   a. intracellular pathogens   d. both b and c
   b. extracellular pathogens   e. all of the above
   c. extracellular toxins

7. The most important antigens are _____ .
   a. nucleotides   c. steroids
   b. triglycerides   d. proteins

8. _____ would be a target of an effector cytotoxic T cell.
   a. Extracellular virus particles in blood
   b. A virus-infected body cell or tumor cell
   c. Parasitic flukes in the liver
   d. Bacterial cells in pus
   e. Pollen grains in nasal mucus

9. Development of a secondary immune response is based on populations of _____ .
   a. memory cells   d. effector cytotoxic T cells
   b. circulating antibodies   e. mast cells
   c. effector B cells

10. Match the immunity concepts.
    ____ inflammation   a. neutrophil
    ____ antibody secretion b. effector B cell
    ____ a phagocyte   c. nonspecific response
    ____ immunological   d. deliberately provoking
         memory   an immune response
    ____ vaccination   e. basis of secondary response
    ____ allergy   f. nonprotective immune response

Critical Thinking

1. As described in the chapter introduction, Edward Jenner lucked out. He performed a potentially harmful experiment on a boy who managed to survive it. What would happen if a would-be Jenner tried to do the same thing today?

2. Rob's bumper sticker reads, "Have you thanked your resident bacteria today?" Explain why he appreciates the bacteria that normally reside on the body's skin and mucous membranes.

3. Researchers have been attempting to develop a way to get the immune system to accept foreign tissue as "self." Speculate on some of the clinical applications of such a development.

4. Before each flu season, you get an influenza vaccination. This year you come down with "the flu" anyway. What do you suppose happened? (There are at least three explanations.)

5. Infection by *Ebola* virus results in a hemorrhagic fever with a 90 percent mortality rate (Section 20.11). A patient received a blood serum transfusion from another who survived the disease. Explain why the transfusion might increase chances of survival.

6. Ellen developed *chicken pox* when she was in kindergarten. Later in life, when her children developed chicken pox, she remained healthy even though she was exposed to countless virus particles daily. Explain why.

7. Quickly review Section 28.7 on homeostasis. Then write a short essay on how the immune response contributes to stability in the internal environment.

Selected Key Terms

allergen *34.10*
allergy *34.10*
antibody *34.4*
antigen *34.4*
antigen-MHC complex *34.4*
antigen-presenting cell *34.4*
apoptosis *34.6*
autoimmune response *34.10*
B lymphocyte (B cell) *34.4*
basophil *34.3*
complement system *34.2*
cytotoxic T cell *34.4*
eosinophil *34.3*
fever *34.3*
helper T cell (CD4 lymphocyte) *34.4*
histamine *34.3*
immune system *34.4*
immunization *34.10*
immunoglobulin (Ig) *34.7*
inflammation, acute *34.3*
lysis *34.2*
lysozyme *34.1*
macrophage *34.3*
mast cell *34.3*
MHC marker *34.4*
natural killer (NK) cell *34.6*
neutrophil *34.3*
pathogen *34.1*
T lymphocyte (T cell) *34.4*
TCR *34.6*
vaccination *CI*
vaccine *34.10*

Readings   *See also www.infotrac-college.com*

Golub, E., and D. Green. 1991. *Immunology: A Synthesis.* Second edition. Sunderland, Massachusetts: Sinauer.

Janeway, C., Jr. September 1993. "How the Immune System Recognizes Invaders." *Scientific American,* 72–79.

Nowak, M., and A. McMichael. August 1995. "How HIV Defeats the Immune System." *Scientific American* 273(2): 58–65.

Tizard, I. 1995. *Immunology: An Introduction.* Fourth edition. Philadelphia: Saunders.

WWW   http://www.brookscole.com/biology

Practice quiz questions, hypercontents, BioUpdates, and critical thinking. The Brooks/Cole Biology Resource Center provides a wealth of information fully organized and integrated by chapter.

RESPIRATION

Conquering Chomolungma

To experienced climbers, Chomolungma may be the ultimate challenge (Figure 35.1). The summit of this Himalayan mountain, also known as Everest, is 9,700 meters (29,128 feet) above sea level. It is the highest place on Earth. Iced-over vertical rock, driving winds, blinding blizzards, and heart-stopping avalanches await the challengers. So does the extreme danger that oxygen-poor air poses to the brain.

Most of us live at low elevations. Of the air that we breathe, one molecule in five is oxygen. When we travel 2,400 meters (about 8,000 feet) or more above sea level, the Earth's gravitational pull is weaker, gas molecules spread out more, and the breathing game changes. We face *hypoxia*, or cellular oxygen deficiency. Sensing the

deficiency, the brain makes us breathe faster and deeper than usual, or hyperventilate. Past 3,300 meters (10,000 feet), hyperventilating causes significant ion imbalances in cerebrospinal fluid. It may lead to heart palpitations, shortness of breath, headaches, nausea, and vomiting. These are strong clues that the body's cells are, in a manner of speaking, screaming for oxygen.

Living for months at high elevations helps climbers adapt physiologically to the thinner air. For example, mechanisms kick in that raise the red blood cell count, hence the body's oxygen-carrying capacity. Professional climbers know that, above 7,000 meters, oxygen scarcity and low air pressure combine to make blood capillaries leaky. More plasma escapes through the expanded gaps

Figure 35.1 A climber inching toward the summit of Chomolungma, where oxygen is brutally scarce.

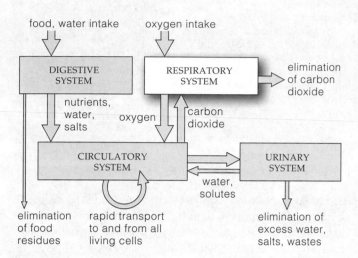

food, water intake   oxygen intake

DIGESTIVE SYSTEM

RESPIRATORY SYSTEM

elimination of carbon dioxide

nutrients, water, salts

oxygen

carbon dioxide

CIRCULATORY SYSTEM

URINARY SYSTEM

water, solutes

elimination of food residues

rapid transport to and from all living cells

elimination of excess water, salts, wastes

Figure 35.2 Interactions between the respiratory system and other organ systems in complex animals.

between endothelial cells of the capillary walls. In the brain and lungs, tissues swell with excess fluid. When the edema is not reversed, climbers become comatose and die. Rescuers zip stricken climbers inside airtight bags, then use a device that pumps in bottled oxygen and removes carbon dioxide until the "air" inside the bag more closely approximates the air at 2,400 meters.

Few of us will ever find ourselves near the peak of Chomolungma, pushing our reliance on oxygen to the limits. Here in the lowlands, disease, smoking, and other environmental insults push it in more ordinary ways, although the risks can be just as great.

The point is this: *Each animal has a body plan that is adapted to the oxygen levels of a particular habitat.* One way or another, by a physiological process known as **respiration**, the body plan allows oxygen to move into the internal environment and carbon dioxide to move out. Why is this important? Animals, remember, have great energy demands. Their cells require a great deal of oxygen—especially for aerobic respiration. This metabolic pathway, remember, produces carbon dioxide wastes as it releases energy from organic compounds.

This chapter samples a few **respiratory systems**, which function in the exchange of gases between the body and the environment. Together with other organ systems, they also contribute to homeostasis—that is, to maintaining internal operating conditions for all of the body's living cells (Figure 35.2).

KEY CONCEPTS

1. Of all organisms, multicelled animals require the most energy to drive their metabolic activities. At the cellular level, the energy comes mainly from aerobic respiration, an ATP-producing metabolic pathway that requires oxygen and produces carbon dioxide wastes.

2. By a physiological process called respiration, animals move oxygen into their internal environment and give up carbon dioxide to the external environment.

3. Oxygen diffuses into the animal body as a result of a pressure gradient. The pressure of this gas is higher in air than it is in metabolically active tissues, where cells rapidly use oxygen. Carbon dioxide follows its own gradient, in the opposite direction. Its pressure is higher in tissues, where it is a by-product of metabolism, than it is in the air.

4. In most respiratory systems, oxygen and carbon dioxide diffuse across a respiratory surface, such as the thin, moist epithelium of the human lungs. The blood flowing through the body's circulatory system picks up oxygen and gives up carbon dioxide at this respiratory surface.

5. Gas exchange is most efficient when the rate of air flow matches the rate of blood flow. The nervous system brings the rates into balance by controlling the rhythmic pattern and magnitude of breathing.

THE NATURE OF RESPIRATION

The Basis of Gas Exchange

A concentration gradient, recall, is a difference in the number of molecules of a substance between two regions. Like all substances, oxygen or carbon dioxide tends to diffuse down its concentration gradient, or show a net outward movement from the region where its molecules are colliding more frequently (Section 5.6). The process of respiration is based on the tendency of both gases to follow their respective concentration gradients—or, as we say for gases, down **pressure gradients**—that form between the animal body and its surroundings.

The gases do not exert the *same* pressure. Pump air into a flat tire near a beach in San Diego or Miami, and you fill it with about 78 percent nitrogen, 21 percent oxygen, 0.04 percent carbon dioxide, and 0.96 percent other gases. This is true of dry air anywhere at sea level. At sea level, atmospheric pressure is about 760 mm Hg, as measured by a mercury barometer of the sort shown in Figure 35.3. Oxygen exerts only part of that total pressure on the tire wall, and its "partial" pressure is greater than that of carbon dioxide. Said another way, the **partial pressure** of oxygen—its contribution to the total atmospheric pressure—is 760 × 21/100, or about 160 mm Hg. When similarly measured, carbon dioxide's partial pressure is about 0.3 mm Hg.

Gases enter and leave the animal body by crossing a **respiratory surface**. A respiratory surface is a thin layer of epithelium or some other tissue. It must be kept moist at all times, for gaseous molecules cannot diffuse across it without being dissolved in fluid. What dictates the *number* of gas molecules moving across a respiratory surface in a given interval? According to **Fick's law**, the more extensive the surface area and the larger the partial pressure gradient, the faster will be the diffusion rate.

Which Factors Influence Gas Exchange?

SURFACE-TO-VOLUME RATIO All animal body plans promote favorable rates of inward diffusion of oxygen and outward diffusion of carbon dioxide. For example, animals with no respiratory organs are tiny, tubelike, or flattened, and gases diffuse directly across the body surface. These body plans meet a constraint imposed by the surface-to-volume ratio, as Section 4.2 describes. To get a sense of the ratio's effects, imagine a flatworm growing in all directions, like an inflating balloon. The worm's surface area does not increase at the same rate as its volume. Once its girth exceeds a single millimeter, the diffusion distance between the body's surface and internal cells will be so great that the flatworm will die.

VENTILATION Large-bodied, highly active animals have great demands for gas exchange—more than diffusion

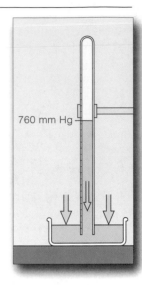

Figure 35.3 Atmospheric pressure as measured with a device called a mercury barometer. Part of the device is a glass tube in which the height of a column of mercury (Hg) can increase or decrease, depending upon air pressure outside the device. At sea level, the mercury rises to about 760 millimeters (29.91 inches) from the base of the tube. At this level, the pressure that the column of mercury exerts inside the tube is equal to the atmospheric pressure on the outside.

760 mm Hg

alone could satisfy. A variety of adaptations make the exchange rate more efficient. For example, above their gills (respiratory organs), many fishes have tissue flaps that stir the surrounding water by moving back and forth. Stirred water puts more dissolved oxygen closer to the gills and carries more carbon dioxide away from them. Among vertebrates, a circulatory system rapidly transports oxygen to cells and carbon dioxide to gills or lungs for disposal. As another example, you breathe in and out to ventilate your lungs.

TRANSPORT PIGMENTS Rates of gas exchange get a boost with transport pigments—mainly **hemoglobin**—that help maintain the steep pressure gradients across a respiratory surface. For example, at the respiratory surfaces in human lungs, the oxygen concentration is high, and each hemoglobin molecule in blood weakly binds as many as four oxygen molecules. (Here you may wish to refer to Section 3.5.) The circulatory system swiftly transports hemoglobin away from the respiratory surface. In oxygen-poor tissues, the oxygen follows its gradient and diffuses out of hemoglobin. By its oxygen-transporting activity, then, hemoglobin helps maintain a pressure gradient that entices oxygen into the blood.

As the chapter introduction states, respiration is a process by which the animal body takes in oxygen for aerobic respiration (an ATP-producing pathway), then removes the pathway's carbon dioxide wastes.

Oxygen and carbon dioxide enter and leave the body by diffusing across a moist respiratory surface. Like other atmospheric gases, they tend to move down their respective pressure gradients. Each gas exerts only part of the total pressure across a respiratory surface.

Gas exchange depends on steep partial pressure gradients between the outside and inside of the animal body. The greater the area of the respiratory surface and the larger the partial pressure gradient, the faster diffusion will proceed.

Flatworms, earthworms, and many other invertebrates are not massive, and their life-styles do not depend on high metabolic rates (Figure 35.4*a*). Demands for gas exchange are simply met by **integumentary exchange**, in which gases diffuse directly across the body's surface covering (integument). This mode of respiration only works when the surface is moist, and invertebrates that rely exclusively on it are restricted to aquatic or damp habitats. (By contrast, amphibians and some other large animals also use integumentary exchange, but only as a supplement to other modes of respiration.)

Many invertebrates of aquatic habitats have moist, thin-walled respiratory organs called **gills**. Extensively folded gill walls have an increased respiratory surface area that enhances the rates of exchange between blood or some other body fluid and the surroundings. Figure 35.4*b* shows the much-folded gill of a sea hare (*Aplysia*). By supplementing integumentary exchange, the gill helps provide adequate oxygen for this large mollusk. Some sea hares are 40 centimeters (nearly 16 inches) long.

Spiders and other invertebrates of dry habitats have small, thick, or hardened integuments that are not well

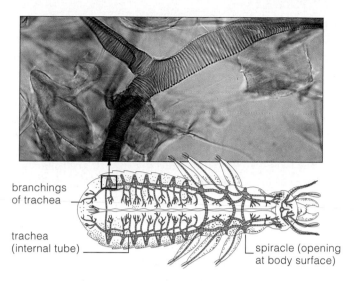

Figure 35.5 General plan of insect tracheal systems. Chitin rings reinforce many of the branching tubes of the system.

branchings of trachea

trachea (internal tube)

spiracle (opening at body surface)

endowed with blood vessels. Their integuments hold in precious water but are not good respiratory surfaces. Such animals have *internal* respiratory surfaces. Most spiders, for instance, have book lungs: respiratory organs with thin, folded walls that resemble book pages (Section 23.12). Some species rely on a system of internal tubes for **tracheal respiration**. Most insects, millipedes, and centipedes also have such a system.

Figure 35.5 shows the tracheal system of an insect. Small openings perforate the insect integument. Each opening (a spiracle) is the start of a tube that branches within the body. Each of the last branchings dead-ends at its fluid-filled tip, where gases diffuse directly into tissues. The tube tips are especially profuse in muscle and other tissues with high oxygen demands.

We find hemoglobin or other respiratory pigments in many invertebrates, although they are rare in insects. Respiratory pigments increase the capacity of body fluids to transport oxygen, as they do in vertebrates. In species having a well-developed head, oxygenated blood tends to circulate first through the head end and then through the rest of the body.

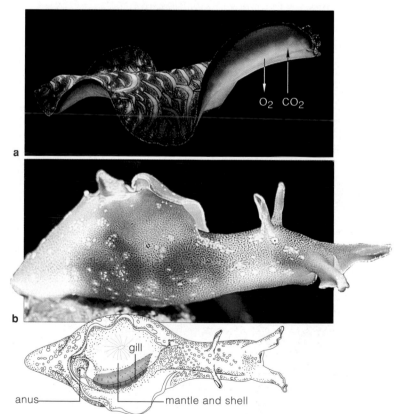

O₂ CO₂

gill

anus

mantle and shell

Figure 35.4 Invertebrates of aquatic habitats. (**a**) A flatworm, small enough to get along well without an oxygen-transporting circulatory system. Dissolved oxygen in such habitats reaches individual cells by diffusing across the body surface. (**b**) Gill of a sea hare (*Aplysia*), one of the gastropods.

Flatworms and some other invertebrates of aquatic or moist habitats do not have a massive body and use integumentary exchange. By this mode of respiration, oxygen and carbon dioxide diffuse directly across the body surface.

Most marine invertebrates and many freshwater types have gills of one sort or another. These respiratory organs have moist, thin, and often highly folded walls.

Most insects, millipedes, centipedes, and some spiders use tracheal respiration. Gases flow through open-ended tubes that start at the body surface and end directly in tissues.

Gills of Fishes and Amphibians

Gills are respiratory organs of many vertebrates. A few kinds of fish larvae and a few amphibians have *external* gills projecting into the water. Adult fishes have a pair of *internal* gills. These are rows of slits or pockets at the back of the mouth that extend to the body surface, as in Figure 35.6*a*. Whatever its form, each gill has a wall of moist, thin, vascularized epithelium.

In fishes, water flows into the mouth and pharynx, then over arrays of filaments in the gills (Figure 35.6*b*). Blood vessels thread through respiratory surfaces in each filament. First the water flows past a vessel that leads to the rest of the body. The blood inside has less oxygen than the water does, so oxygen diffuses into the blood. The same volume of water flows over a vessel leading into the gills. The water already gave up some oxygen, but it still has more than the blood inside the vessel does, so more oxygen diffuses into the filament. The movement of two fluids in opposing directions is called **countercurrent flow**. By this mechanism, a fish extracts about 80 to 90 percent of the dissolved oxygen flowing past. That is more than the fish would get from a one-way flow mechanism, at far less energy cost.

Lungs

Some fishes and all amphibians, birds, and mammals have a pair of **lungs**, internal respiratory surfaces in the shape of a cavity or sac. Lungs originated in some lineages of fishes more than 450 million years ago, as pouches off the anterior part of the gut wall (Section 24.5). They evolved rapidly by way of natural selection, probably because they afforded survival advantages. Paired lungs increased the surface area available for gas exchange in oxygen-poor habitats. They worked better than gills could when some vertebrates moved onto dry land. Gills stick together and cannot function unless water flows through them and keeps them moist.

Lungfishes of oxygen-poor habitats still have gills. They also use tiny lungs as backups.

Amphibians never completed the transition to land. Their skin is a respiratory surface for integumentary exchange, in salamanders especially. Frogs and toads rely more on small lungs for oxygen uptake, but most of their carbon dioxide wastes diffuse outward across the skin. Frogs also are heavy-duty breathers; they *force* air into their lungs, which they empty by contracting muscles in their body wall (Figure 35.7).

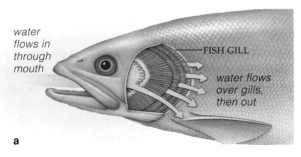

water flows in through mouth

FISH GILL

water flows over gills, then out

a

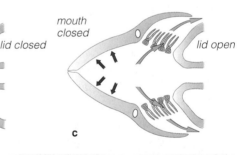

mouth open

lid closed

mouth closed

lid open

b   **c**

Figure 35.6 Example of fish gills. (**a**) One of a pair of gills. Each gill is located under a bony lid, removed for this sketch.

(**b**) Ventilation of gills. Water is pulled across gills when the mouth opens and the lid closes. (**c**) Water is forced back out when the mouth closes and the lid is opens.

(**d**) Gas exchange at gills. Filaments in the gill have vascularized respiratory surfaces. Tips of adjacent filaments touch, so water passing over them is directed past the gas exchange surfaces before leaving the body. (**e**) One blood vessel from tissues deep in the body delivers oxygen-poor blood into a filament. Another carries oxygenated blood away from it. Blood flowing from one vessel into the other runs counter to the direction of the water flowing over the gas exchange surfaces. Countercurrent flow favors the movement of oxygen (down its partial pressure gradient) from water to the blood.

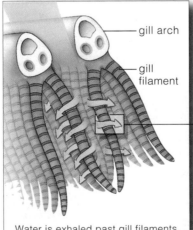

gill arch

gill filament

Water is exhaled past gill filaments. The tips of these are in contact, so that the water is directed past the gas exchange surface.

d

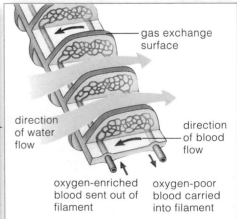

gas exchange surface

direction of water flow

direction of blood flow

oxygen-enriched blood sent out of filament

oxygen-poor blood carried into filament

A blood vessel carries oxygen-poor blood into each filament. Another carries oxygenated blood out. Blood flowing from one vessel to the other runs counter to the direction of water flowing over gas exchange surfaces.

e

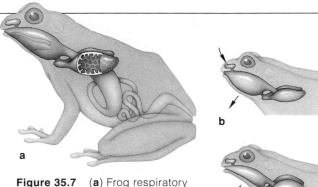

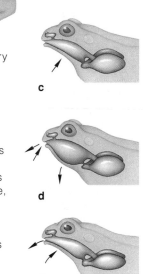

Figure 35.7 (**a**) Frog respiratory system. Two lungs fill when air is *forced* into them. (**b**) The frog lowers the floor of its mouth and inhales air through two nostrils. (**c**) Then it closes the nostrils, opens the glottis, and elevates the floor of its mouth. The air has nowhere to go except down into the lungs. (**d**) The frog ventilates its mouth rhythmically for a while, which helps move more oxygen into the mouth and more carbon dioxide out from it. (**e**) Muscles in the body wall above the lungs contract, the lungs elastically recoil, and air is forced out.

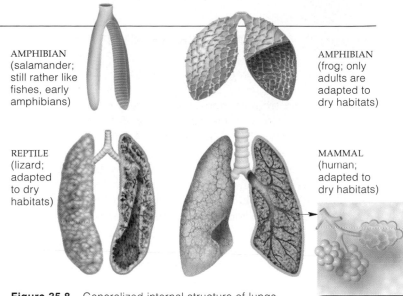

Figure 35.8 Generalized internal structure of lungs from four vertebrates, suggestive of an evolutionary trend from simple gas exchange sacs to complex lungs with a greater respiratory surface area.

AMPHIBIAN (salamander; still rather like fishes, early amphibians)

AMPHIBIAN (frog; only adults are adapted to dry habitats)

REPTILE (lizard; adapted to dry habitats)

MAMMAL (human; adapted to dry habitats)

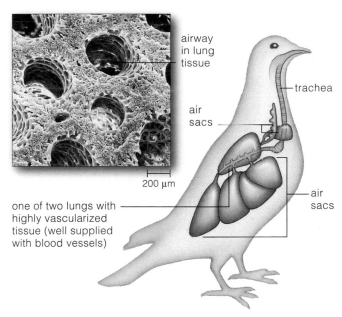

airway in lung tissue

trachea

air sacs

200 μm

one of two lungs with highly vascularized tissue (well supplied with blood vessels)

air sacs

Figure 35.9 Bird respiratory system. A number of air sacs are attached to each of two small, inelastic lungs. When a bird inhales, air is drawn into air sacs through tubes, open at both ends, that thread through the vascularized lung tissue. This tissue is the respiratory surface, where gases are exchanged.

When the bird exhales, air is forced out of the sacs, through the small tubes, and out of the trachea. Thus, air is not merely drawn into the bird lungs. Air is continuously drawn through them and across respiratory surfaces. This unique ventilating system supports the high metabolic rates that birds require for flight and other energy-intensive activities.

Frogs also use their lungs for sound production. So do all mammals except whales. Sound originates near the entrance to a larynx, an airway to the lungs. Here, part of a mucous membrane folds perpendicularly to the airway. The folds are **vocal cords**, and the gap between them is a **glottis**. Frogs produce sounds by forcing air through the glottis, back and forth between the lungs and paired pouches on the floor of the mouth. Air flow causes the cords to vibrate. Like other vertebrates, frogs produce different sounds by controlling the vibrations.

Paired lungs are the dominant respiratory organs in reptiles, birds, and mammals (Figure 35.8). Breathing moves air by bulk flow into and out of the lungs, where blood capillaries wrap lacily around the respiratory surface. In the lungs, oxygen and carbon dioxide have steep concentration gradients and diffuse rapidly across the respiratory surface. Oxygen enters the capillaries and is circulated quickly through the body. In regions with low oxygen levels, oxygen diffuses into interstitial fluid, then into cells. Carbon dioxide moves rapidly in the opposite direction and is expelled from the lungs.

This mode of gas exchange is embellished a bit only in birds. As described in Figure 35.9, birds are unique in that air not only flows into and out from their lungs; it also flows *through* their lungs. With that exception in mind, we turn next to the human respiratory system, for its operating principles apply to most vertebrates.

A countercurrent flow mechanism in fish gills compensates for low oxygen levels in aquatic habitats. Internal air sacs—lungs—are more efficient in dry land habitats. Amphibians use integumentary exchange, and they also force air into and out of small lungs. Ventilation of paired lungs is the major mode of respiration in reptiles, birds, and mammals.

Functions of the Respiratory System

Getting oxygen from air and expelling carbon dioxide are the key functions of the human respiratory system. Breathing, a form of ventilation, alternately moves air into and out of a pair of lungs, each of which has about 300 million outpouchings. Each outpouching is a tiny air sac: an **alveolus** (plural, alveoli). Controls adjust the rate of breathing so that the inflow and outflow of air match the metabolic demands for gas exchange.

The respiratory system's role in respiration ends at alveoli, where the circulatory system takes over the task of moving gases. Oxygen and carbon dioxide move by diffusion between alveoli and pulmonary capillaries adjacent to them. (The Latin *pulmo* means lung.)

The respiratory system has other functions. Breathing is used for vocalizations, such as speech. It enhances the return of venous blood to the heart, and it helps rid the body of excess body heat and water. Controls over breathing adjust the body's acid-base balance. Carbon dioxide, remember, is used in the formation of carbonic acid (H_2CO_3), which works with bicarbonate (HCO_3^-) as a buffer system. HCO_3^- accepts or releases hydrogen ions (H^+), depending on the blood's pH (Section 2.6). Rapid, deep breathing expels more carbon dioxide, so less carbonic acid forms—and so the body loses acid.

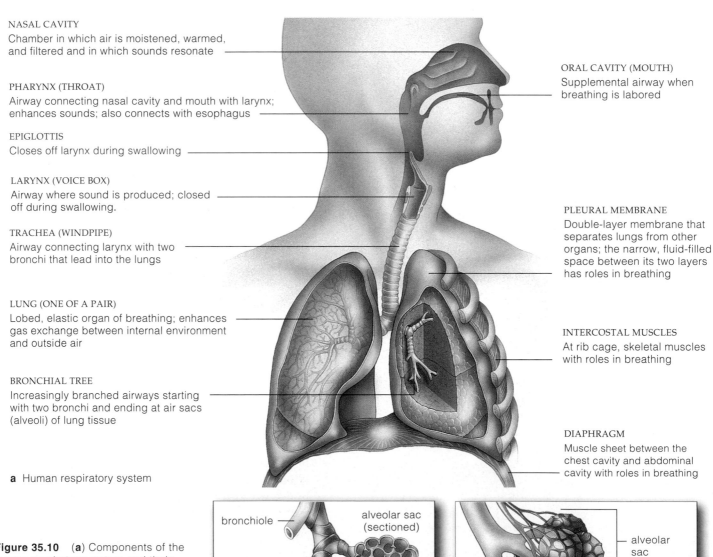

NASAL CAVITY
Chamber in which air is moistened, warmed, and filtered and in which sounds resonate

PHARYNX (THROAT)
Airway connecting nasal cavity and mouth with larynx; enhances sounds; also connects with esophagus

EPIGLOTTIS
Closes off larynx during swallowing

LARYNX (VOICE BOX)
Airway where sound is produced; closed off during swallowing.

TRACHEA (WINDPIPE)
Airway connecting larynx with two bronchi that lead into the lungs

LUNG (ONE OF A PAIR)
Lobed, elastic organ of breathing; enhances gas exchange between internal environment and outside air

BRONCHIAL TREE
Increasingly branched airways starting with two bronchi and ending at air sacs (alveoli) of lung tissue

ORAL CAVITY (MOUTH)
Supplemental airway when breathing is labored

PLEURAL MEMBRANE
Double-layer membrane that separates lungs from other organs; the narrow, fluid-filled space between its two layers has roles in breathing

INTERCOSTAL MUSCLES
At rib cage, skeletal muscles with roles in breathing

DIAPHRAGM
Muscle sheet between the chest cavity and abdominal cavity with roles in breathing

a Human respiratory system

bronchiole

alveolar sac (sectioned)

alveolar duc

alveoli

alveolar sac

pulmonary capillary

Figure 35.10 (**a**) Components of the human respiratory system and their functions. Muscles, including the diaphragm, and parts of the axial skeleton have secondary roles in respiration. (**b,c**) Location of the alveoli relative to bronchioles and to the pulmonary (lung) capillaries.

b

c

Slow, shallow breathing has the opposite effect; carbon dioxide builds up, more H_2CO_3 forms—and the body gains acid.

The respiratory system also has built-in mechanisms for dealing with airborne foreign cells and substances inhaled with air. Finally, it removes and inactivates or otherwise modifies a number of blood-borne substances before they can circulate through the rest of the body.

From Airways Into the Lungs

It will take at least 300 million breaths to get you to age seventy-five. You might find yourself going without food for a few hours or days. But stop breathing even for five minutes and normal brain function is over.

Take a deep breath. Now look at Figure 35.10 to get an idea of where the air will travel in your respiratory system. Unless you are out of breath and panting, air just entered two nasal cavities, not the mouth. There, mucus secretions warm and moisten it. In the cavities, ciliated epithelium and hairs filter dust and particles from air. Their olfactory receptors function in the sense of smell (Section 30.6). Now air is poised at the **pharynx**, or throat. The pharynx is the entrance to the **larynx**, an airway with two paired folds of mucous membrane. The lower pair are vocal cords (Figure 35.11). When you breathe, air is forced in and out through the glottis, the gap between them. Air flow makes the cords vibrate in ways that can be controlled to produce different sounds.

Within the folds are thick bands of elastic ligaments connected to various cartilage tissues. When muscles of the larynx contract and relax, the ligaments tighten or slacken, which changes the extent to which the folds are stretched. Under commands from the nervous system, the coordinated action of muscles narrows or widens the glottis. For example, by increasing muscle tension in the vocal cords, you can decrease the gap between them and make high-pitched sounds or squeaks. The lips, teeth, tongue, and the soft roof over the tongue are enlisted to modify the different sounds into patterns of vocalization, such as speech and song.

When vocal cords are inflamed as an outcome of an infection or irritation, swelling of their mucous lining interferes with their capacity to vibrate. If the swelling causes hoarseness, this condition is called *laryngitis*.

When the **epiglottis** (a tissue flap at the entrance to the larynx) is pointing upward, the air moves into the **trachea**, or windpipe. When you swallow, the epiglottis points downward and so closes off the entrance to the trachea. At such times, food or fluid being swallowed enters the esophagus, a tube connecting the pharynx with the stomach.

The trachea branches into two airways, one leading into the tissue of each lung. Each airway is a **bronchus**

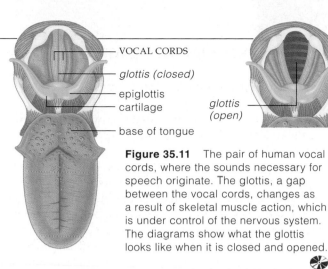

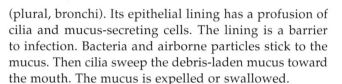

Figure 35.11 The pair of human vocal cords, where the sounds necessary for speech originate. The glottis, a gap between the vocal cords, changes as a result of skeletal muscle action, which is under control of the nervous system. The diagrams show what the glottis looks like when it is closed and opened.

(plural, bronchi). Its epithelial lining has a profusion of cilia and mucus-secreting cells. The lining is a barrier to infection. Bacteria and airborne particles stick to the mucus. Then cilia sweep the debris-laden mucus toward the mouth. The mucus is expelled or swallowed.

Human lungs are elastic, cone-shaped organs of gas exchange. They are located inside the rib cage, to the left and right of the heart and above the diaphragm. The **diaphragm** is a muscular partition between the thoracic and abdominal cavities. A thin, saclike pleural membrane lines the outer surface of the lungs and inner surface of the thoracic cavity wall. The thoracic wall and the lungs press its opposing surfaces together. By analogy, think of the lungs as two baseballs pushed into a partially inflated balloon. The baseballs take up so much space, they press the balloon's opposing sides together.

Only a thin film of lubricating fluid separates the pleural membrane surfaces and decreases the friction between them. In a respiratory ailment called *pleurisy*, the membrane becomes inflamed and swollen, friction follows, and breathing can be painful.

Inside each lung, air moves through finer and finer branchings of a "bronchial tree" (Figure 35.10*a*). These airways are **bronchioles**. Their endings, the *respiratory* bronchioles, bear the cup-shaped alveoli. Most often, alveoli are clustered as larger pouches called alveolar sacs (Figure 35.10*b,c*). Collectively, alveolar sacs offer a tremendous surface area for gas exchanges with blood. If all the alveolar sacs were stretched out in one layer, they would cover the floor of a racquetball court!

Oxygen uptake and carbon dioxide removal are the major functions of the human respiratory system. In its paired lungs, the circulatory system takes over the remaining tasks of respiration.

The respiratory system also has roles in moving venous blood to the heart, in vocalization, in adjusting the body's acid-base balance, in defense against harmful airborne cells or particles, in removing or modifying a number of blood-borne substances, and in the sense of smell.

WWW

The Respiratory Cycle

There is a cyclic pattern to breathing, which ventilates the lungs. Each **respiratory cycle** consists of two actions: *inhalation* (a single breath of air drawn into the airways) and *exhalation* (a single breath out). Inhalation always is an active, energy-requiring action. When someone is breathing quietly, it is brought about by the contraction of the diaphragm and, to a lesser extent, the external intercostal muscles. The outcome is an increase in the thoracic cavity volume. Breathe hard, and the volume increases further because neck muscles contract and so elevate the sternum and first two ribs attached to them.

During each respiratory cycle, the thoracic cavity's volume increases, then decreases. *And pressure gradients between air inside and outside the respiratory tract change.* Let's think about the different pressures exerted during a respiratory cycle. The *atmospheric* pressure, 760 mm Hg at sea level, is exerted by the combined weight of all atmospheric gases on all the airways. Before inhalation, *intrapulmonary* pressure (the pressure inside all alveoli) is also 760 mm Hg (Figure 35.12*c*).

Another pressure gradient helps keep the lungs close to the thoracic cavity wall during a respiratory cycle, even during exhalation, when lungs have a far smaller volume than the thoracic cavity (Figure 35.12*b*). When that cavity expands, so do the lungs, as a result of a pressure gradient across the lung wall.

In a person at rest, the *intrapleural* pressure (inside the pleural sac) averages 756 mm Hg, which is less than atmospheric pressure. This pressure is exerted outside the lungs—that is, within the thoracic cavity. While intrapleural pressure is pushing inward on a lung wall, the intrapulmonary pressure is pushing outward. The difference in pressure between them (4 mm Hg) is great enough to make the lungs stretch and fill the thoracic cavity.

The cohesiveness of water molecules in the intrapleural fluid (the fluid inside the pleural sac) also helps keep lungs close to the thoracic wall. By analogy, wet two small panes of glass and press them together. The panes easily slide back and forth but strongly resist being pulled apart. Similarly, intrapleural fluid "glues" the lungs to the wall. What is the outcome? *When the thoracic cavity expands during inhalation, the lungs must expand, also.*

Figure 35.12*a* shows what happens as you start to inhale. The dome-shaped diaphragm flattens down, and the rib cage is lifted upward and outward. As the thoracic cavity expands, the lungs expand along with it. At this time, air pressure in all alveolar sacs combined is lower than the atmospheric pressure. Fresh air follows the gradient and flows down the airways, then into the alveoli.

When someone breathes quietly, the second action of the respiratory cycle is passive. The muscles that brought about inhalation relax and the lungs passively recoil, without any further energy outlays. The resultant decrease in lung volume compresses the air in the alveolar sacs. At this time, pressure in the sacs is greater than atmospheric pressure. Air follows the gradient, out from the lungs (Figure 35.12*b*). Exhalation only becomes active and and energy-requiring

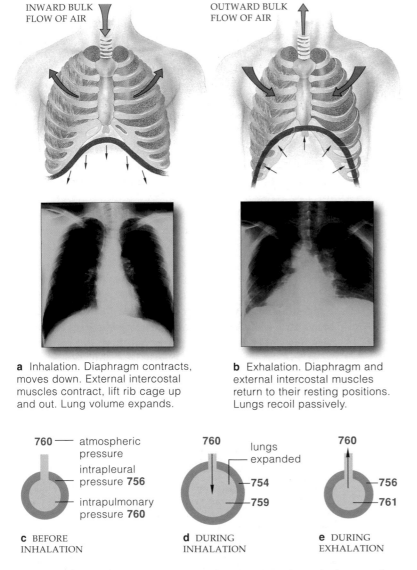

a Inhalation. Diaphragm contracts, moves down. External intercostal muscles contract, lift rib cage up and out. Lung volume expands.

b Exhalation. Diaphragm and external intercostal muscles return to their resting positions. Lungs recoil passively.

c BEFORE INHALATION — 760 atmospheric pressure, intrapleural pressure 756, intrapulmonary pressure 760

d DURING INHALATION — 760, lungs expanded, 754, 759

e DURING EXHALATION — 760, 756, 761

Figure 35.12 (**a**,**b**) Changes in the size of the thoracic cavity in a respiratory cycle. The *blue* line represents the diaphragm. The x-ray images show how the maximum inhalation possible changes the thoracic cavity volume. (**c–e**) Changes in lung volume and intrapulmonary pressure during a respiratory cycle.

when you exercise vigorously and more air must be expelled.

With active exhalation, the muscles in the abdominal wall contract, so pressure inside the abdomen increases and exerts upward-directed force on the diaphragm. As the diaphragm is pushed upward, the thoracic cavity volume decreases. How? The internal intercostal muscles contract and pull the thoracic wall downward and inward. The chest wall flattens, and so the dimensions of the thoracic cavity decrease even more. The lung volume decreases as well when the elastic tissue of the lungs recoils.

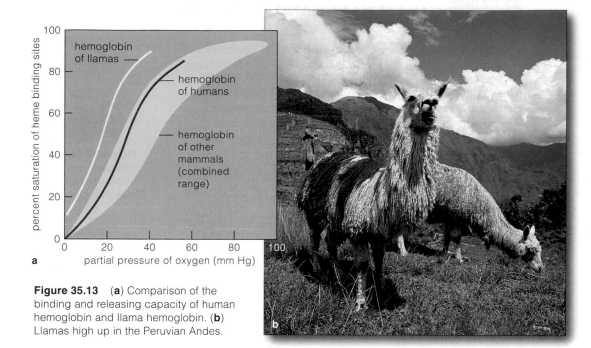

Figure 35.13 (**a**) Comparison of the binding and releasing capacity of human hemoglobin and llama hemoglobin. (**b**) Llamas high up in the Peruvian Andes.

Breathing at High Altitudes

From the chapter introduction, you already know that breathing becomes an agonizing challenge for climbers of Chomolungma and other cloud-piercing peaks. Even so, like llamas and other species that evolved at high elevations, millions of people live quite comfortably 4,800 meters (16,000 feet) or more above sea level.

Compared with people living at lower elevations, the permanent residents of high mountains have lungs with far more alveoli and blood vessels, which formed as they were growing up. Also, their heart developed larger ventricles, so it pumps larger volumes of blood. And more mitochondria formed in their muscle tissue.

Llamas have another advantage. Compared to human hemoglobin, llama hemoglobin has greater affinity for oxygen. It picks up oxygen more efficiently at the lower pressures characteristic of high altitudes (Figure 35.13).

Does this mean a healthy person who grew up by the seashore cannot ever pick up roots and move to the mountains? No. By mechanisms of **acclimatization**, an individual often can make long-lasting physiological and behavioral adaptations to a new environment that is markedly different from the one left behind. At high altitudes, gradual changes in the pattern and magnitude of breathing and in the heart's output can supplant the initially acute compensations that a person makes to cellular oxygen deficiency (hypoxia).

Within a few days, the reduction in oxygen delivery stimulates kidney cells to secrete more **erythropoietin**, a hormone that induces stem cells in bone marrow to divide repeatedly and give rise to red blood cells. Each second in adult humans, 2 million to 3 million red blood cells are churned out as replacements for the ones that continually die off. Stepped-up erythropoietin secretion can increase that astounding pace by six times during extreme stress. Increased numbers of circulating red blood cells increase blood's oxygen-delivery capacity. When the oxygen concentration increases sufficiently, the kidneys slow down erythropoietin secretion.

Erythropoietin is the major hormonal mediator of red blood cell production. However, human males have a larger muscle mass and greater demands for oxygen. They depend as well upon testosterone. Among other things, this male sex hormone stimulates increases in the basic rate of red blood cell production. That is why a sample of blood from males normally has a greater percentage of red blood cells than an equivalent sample from females. The compensatory increase in red blood cells comes at a cost. Having many more cells in the bloodstream increases resistance to blood flow, because the blood is more viscous, or "thicker." The heart must work harder to pump blood.

Breathing, which ventilates the lungs, has a cyclic pattern. The respiratory cycle consists of inhalation (one breath of air in) and exhalation (one breath of air out).

Inhalation is always an active, energy-requiring process involving contractions mainly of the diaphragm and the external intercostal muscles.

During quiet breathing, exhalation is a passive process. Muscles relax, the thoracic cavity volume decreases, and the lungs recoil elastically. Forceful exhalation is an active process that requires abdominal muscle contraction.

Breathing reverses pressure gradients between the lungs and the air outside the body.

WWW

35.7   WHEN THE LUNGS BREAK DOWN

In large cities, in certain workplaces, even in the cloud around a cigarette smoker, airborne particles and certain gases are present in abnormally high concentrations. And they put extra workloads on the respiratory system.

BRONCHITIS   Ciliated, mucous epithelium lines the inner wall of your bronchioles (Figure 35.16). It is one of the built-in defenses that protect you from respiratory infections. Toxins in cigarette smoke and other airborne pollutants irritate the lining and may lead to *bronchitis*. With this respiratory ailment, epithelial cells along the airways become irritated and secrete excessive mucus. As mucus accumulates, so do the bacteria and other particles stuck in it. Coughing brings up some of the gunk. If the source of irritation persists, so does coughing.

Initial attacks of bronchitis are treatable. But when the aggravation continues, bronchioles become chronically inflamed. Bacteria, chemical agents, or both attack the bronchiole walls. Ciliated cells in the walls are destroyed, and the mucus-secreting cells multiply. Fibrous scar tissue forms and in time may narrow or obstruct the airways.

EMPHYSEMA   Thick mucus clogs the airways during persistent bronchitis. Inside the lungs, tissue-destroying bacterial enzymes attack the stretchable, thin walls of alveoli. The walls break down, and inelastic fibrous tissue forms around them. Gas exchange proceeds at fewer alveoli, which become enlarged. In time, the lungs remain distended and inelastic, so the fine balance between air flow and blood flow is permanently compromised.

Compare Figure 35.17*a* with *b* to get a sense of what happens. It becomes hard to run, walk, and exhale. These are symptoms of *emphysema*, a respiratory ailment that affects about 1.3 million people in the United States alone.

A few people are genetically predisposed to developing emphysema. They do not have a workable gene for the enzyme antitrypsin, which can inhibit bacterial attack on alveoli. Poor diet and persistent or recurring colds and

Figure 35.16 False-color scanning electron micrograph of ciliated and mucus-secreting cells that line a bronchus.

— *free surface of a mucus-secreting cell*

— *free surface of a cluster of ciliated cells*

other respiratory infections also invite emphysema later in life. However, *smoking is the major cause of the disease*. Emphysema may develop slowly, over twenty or thirty years. When severe damage is not detected in time, the lung tissues cannot be repaired.

EFFECTS OF SMOKING   Worldwide, 3 million people die each year as a result of complications arising from tobacco smoking. The World Health Organization, the Harvard School of Health, and the World Bank estimate that deaths from smoking may surpass even the AIDS epidemic by the year 2020. In the United States, every thirteen seconds one of 50 million smokers dies from emphysema, chronic bronchitis, or heart disease, and one nonsmoker dies of ailments brought on by *secondhand smoke*—of prolonged exposure to tobacco smoke in the surrounding air. And how much thought is given to children who breathe secondhand smoke? Parents and others who otherwise care about children indirectly heighten their vulnerability to allergies and lung ailments. Smoking is the major cause of lung cancer. Yet every day, 3,000 to 5,000 Americans light a cigarette for the first time. Even children spend a billion dollars a year on cigarettes. Each year, direct medical costs of treating smoke-induced respiratory disorders drain 22 billion dollars from the economy.

How does cigarette smoke damage lungs? Noxious particles in the smoke from just one cigarette immobilize cilia in the bronchioles for several hours. The particles also trigger mucus secretions, which in time clog the airways. They can kill infection-fighting

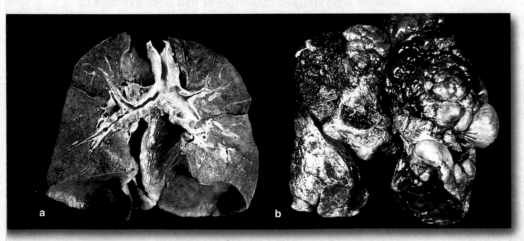

Figure 35.17   (**a**) Normal appearance of tissues of human lungs. (**b**) Lungs from someone affected by emphysema.

RISKS ASSOCIATED WITH SMOKING:	REDUCTION IN RISKS BY QUITTING:
SHORTENED LIFE EXPECTANCY: Nonsmokers live 8.3 years longer on average than those who smoke two packs daily from the midtwenties on.	Cumulative risk reduction; after 10 to 15 years, life expectancy of ex-smokers approaches that of nonsmokers.
CHRONIC BRONCHITIS, EMPHYSEMA: Smokers have 4–25 times more risk of dying from these diseases than do nonsmokers.	Greater chance of improving lung function and slowing down rate of deterioration.
LUNG CANCER: Cigarette smoking is the major cause of lung cancer.	After 10 to 15 years, risk approaches that of nonsmokers.
CANCER OF MOUTH: 3–10 times greater risk among smokers.	After 10 to 15 years, risk is reduced to that of nonsmokers.
CANCER OF LARYNX: 2.9–17.7 times more frequent among smokers.	After 10 years, risk is reduced to that of nonsmokers.
CANCER OF ESOPHAGUS: 2–9 times greater risk of dying from this.	Risk proportional to amount smoked; quitting should reduce it.
CANCER OF PANCREAS: 2–5 times greater risk of dying from this.	Risk proportional to amount smoked; quitting should reduce it.
CANCER OF BLADDER: 7–10 times greater risk for smokers.	Risk decreases gradually over 7 years to that of nonsmokers.
CORONARY HEART DISEASE: Cigarette smoking is a major contributing factor.	Risk drops sharply after a year; after 10 years, risk reduced to that of nonsmokers.
EFFECTS ON OFFSPRING: Women who smoke during pregnancy have more stillbirths, and weight of liveborns averages less (hence, babies are more vulnerable to disease, death).	When smoking stops before fourth month of pregnancy, risk of stillbirth and lower birthweight eliminated.
IMPAIRED IMMUNE SYSTEM FUNCTION: Increase in allergic responses, destruction of defensive cells (macrophages) in respiratory tract.	Avoidable by not smoking.
BONE HEALING: Evidence suggests that surgically cut or broken bones require up to 30 percent longer to heal in smokers, possibly because smoking depletes the body of vitamin C and reduces the amount of oxygen reaching body tissues. Reduced vitamin C and reduced oxygen interfere with production of collagen fibers, a key component of bone. Research in this area is continuing.	Avoidable by not smoking.

Figure 35.18 From the American Cancer Society, a list of the risks incurred by smoking and the benefits of quitting. The photograph shows a few swirls of cigarette smoke poised at the entrance to the two bronchi that lead into the lungs.

macrophages in the respiratory tract. What starts out as "smoker's cough" can end in bronchitis and emphysema.

Or consider how cigarette smoke contributes to lung cancer. Inside the body, certain compounds in coal tar and in cigarette smoke become converted to highly reactive intermediates. These are the carcinogens; they provoke uncontrolled cell divisions in lung tissues. On the average, 90 of every 100 smokers who develop lung cancer will die from it. If you now smoke, or are thinking about starting or quitting, you may wish to give serious thought to the information in Figure 35.18.

EFFECTS OF MARIJUANA SMOKE In 1994, 17 million Americans smoked *pot*—marijuana (*Cannabis*)—at least once to induce light-headed euphoria. The number is rising, especially in junior high and high schools. About 1.5 million are chronic users. They become enamored of the euphoria. However, the sense of well being does not last long; it fades to apathy, depression, and fatigue. Users keep smoking to avoid the negative effects.

Besides psychological dependency, chronic use can result in throat irritation, persistent coughing, bronchitis, and emphysema.

WWW

36

DIGESTION AND HUMAN NUTRITION

Lose It—And It Finds Its Way Back

America's fixation on Beautiful People who have nary an ounce of extra fat on their utterly perfect selves is bad enough. After all, we can always rationalize our own extra ounce with the truism that beauty is only skin deep. But the American Medical Association's dire warnings of the Fat Connection to atherosclerosis, heart attacks, strokes, colon cancer, and other deadly ailments is beyond rationalization.

By current standards, the proportion of body fat relative to total tissue mass should be 18 to 24 percent for human females who are less than thirty years old. For males, it should be no more than 12 to 18 percent. An estimated 34 million Americans do not even come close to the standards.

No matter how much we diet, the lost weight seems to find its way back. This physiological dilemma is an outcome of our evolutionary heritage. Like most other mammals, we have an abundance of fat-storing cells, in adipose tissue. Collectively, the cells are an adaptation for survival, an energy warehouse that opens up when food is not available. Once those fat-storing cells have formed, they are in the body to stay. Variations in food intake only influence how empty or full each one gets.

When we diet and the fat warehouse opens, the brain interprets this as "starvation," and it issues commands for a metabolic slowdown. The body uses energy far more efficiently even for the basics, such as breathing and digesting food. It now takes less food to do the same things!

As you have probably heard, dieting does no good without a long-term commitment to physical exercise. Why? Your skeletal muscles also adapt to "starvation," and they burn less energy than before. Jog for four hours or play tennis for eight hours straight, and you may lose a pound of fat. Meanwhile,

your appetite surges. If and when you do stop dieting, "starved" fat cells quickly refill. Unless you eat less *and* exercise moderately throughout your life, you can't keep off extra weight (Figure 36.1).

What about gaining weight? A weight gain triggers an *increase* in metabolism. Now the body uses 15 to 20 percent more energy than before—until weight drops back to what it was.

As if this were not enough, your emotions influence weight gains and losses. As an extreme case, *anorexia nervosa* is a potentially fatal eating disorder based on a seriously flawed assessment of body weight. Typically, anorexics have an overwhelming dread of being fat *and* hungry. They starve themselves and often overexercise. They commonly have fears about growing up and of being sexually mature. They also have irrationally high expectations about their personal performances.

We find another extreme case in *bulimia*, an out-of-control, "oxlike appetite." During an hour-long eating binge, a bulimic might take in about 50,000 kilocalories' worth of food, then vomit or use laxatives to purge the body. Some follow a binge-purge routine as an "easy" way to lose weight. Others suffer severe emotional

Figure 36.1 Organisms positively engaged in countering imbalances between caloric intake and energy output.

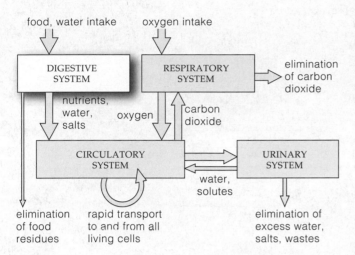

food, water intake   oxygen intake

DIGESTIVE SYSTEM

RESPIRATORY SYSTEM

elimination of carbon dioxide

nutrients, water, salts

oxygen

carbon dioxide

CIRCULATORY SYSTEM

URINARY SYSTEM

water, solutes

elimination of food residues

rapid transport to and from all living cells

elimination of excess water, salts, wastes

Figure 36.2 Functional links between the digestive, respiratory, circulatory, and urinary systems. These organ systems and others work together to supply cells with raw materials and eliminate wastes.

stress. They may not even like to eat, but at some level the purging relieves them of anger and frustration.

Some bulimics binge-purge once a month. Others do so several times a day. Do they know that repeated purgings can damage the gut? That chronic vomiting brings gastric fluid into the mouth and erodes teeth to stubs? That, at its extreme, bulimia can rupture the stomach, and cause heart or kidney failure?

Are severe eating disorders rare? No. In the United States, an estimated 7 million females and 1 million males are anorexic or bulimic. Most are in their teens and early twenties, but the number also is increasing among preadolescents. Each year, 5 to 6 percent of the weight-obsessed individuals die from complications arising from their disorders.

And with these sobering thoughts in mind, we start a tour of **nutrition**. The word encompasses all those processes by which an animal ingests and digests food, then absorbs the released nutrients for later conversion to the body's own carbohydrates, lipids, proteins, and nucleic acids.

The nutritional processes proceed at the **digestive system**. This body tube or cavity mechanically and chemically reduces food to particles, then to molecules that are small enough to be absorbed into the internal environment. It also eliminates unabsorbed residues. Other organ systems, especially those shown in Figure 36.2, contribute to the nutritional processes.

KEY CONCEPTS

1. In complex animals, interactions among digestive, circulatory, respiratory, and urinary systems supply the body's cells with raw materials, dispose of wastes, and maintain the volume and composition of extracellular fluid.

2. Most digestive systems include specialized regions for food transport, processing, and storage. Different regions mechanically break apart and chemically break down food, absorb breakdown products, and eliminate the unabsorbed residues.

3. To maintain an acceptable body weight and overall health, energy intake must balance energy output by way of metabolic activity, physical exertion, and so on. Complex carbohydrates are the main source of dietary glucose, which typically is the body's main source of immediately usable energy.

4. Nutrition also involves the intake of foods that are good sources of vitamins, minerals, and a number of amino acids and fatty acids that the body itself cannot produce.

Incomplete and Complete Systems

Recall, from Chapter 23, that we don't see many organ systems until we get to the flatworms. These are among the invertebrates with an **incomplete digestive system**, a saclike, branching gut cavity with a single opening at the start of a pharynx (Figure 36.3*a*). Food enters the sac, then is partly digested and circulated to cells even as residues are being sent back out.

Flatworms, cnidarians, and a few rotifers and other small invertebrates require little more than two-way traffic through a saclike gut that has a single opening. But most animals have a **complete digestive system**— basically, a tube with a mouth (an opening at one end for food intake) and an anus (an opening at the other end for eliminating unabsorbed residues). Between the two openings, the tube is subdivided into specialized regions that function in one-way transport, processing, and storage of raw materials.

Think about the complete digestive system of a frog (Figure 36.3*b*). Between the frog's mouth and the anus are a pharynx, stomach, and small and large intestines. Functionally connected with the small intestine are the liver, gallbladder, and pancreas, which are organs with accessory roles in digestion. Birds, too, have a complete digestive system that also has a few unique regional specializations (Figure 36.3*c*).

Regardless of its complexity, a complete digestive system carries out five overall tasks:

1. **Mechanical processing and motility**. Movements that break up, mix, and propel food material.

2. **Secretion**. Release of digestive enzymes and other substances into the space inside the tube.

3. **Digestion**. Breakdown of food into particles, then into nutrient molecules small enough to be absorbed.

4. **Absorption**. Passage of digested nutrients and fluid across the tube wall and into body fluids.

5. **Elimination**. Expulsion of the undigested and unabsorbed residues from the end of the gut.

Correlations With Feeding Behavior

You can correlate the specializations of any digestive system with feeding behavior. Look at the pigeon in Figure 36.3*c*. Its *food-gathering* region is a bill adapted to pecking seeds from the ground. Seeds travel into a mouth, a long tube (esophagus), then a *food-processing* region that is compactly centered in the body mass. The compact arrangement helps a pigeon balance its body during flight. As in other seed-eating birds, part of the esophagus balloons out as a *crop* (a stretchable storage organ). A crop lets the bird gulp down a lot of food and

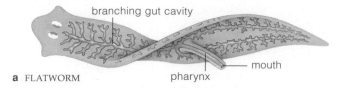

a FLATWORM

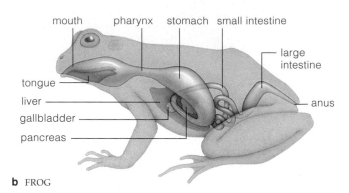

b FROG

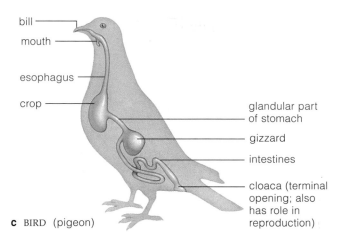

c BIRD (pigeon)

Figure 36.3 (**a**) Incomplete digestive system of a flatworm, with two-way traffic of food and undigested material through one opening. (**b,c**) Examples of a complete digestive system, a tube with specialized regions and an opening at each end.

make quick getaways from predators. Like all birds, this one eats during the day. Its crop fills before the sun goes down, then it releases food in the first few hours after dark and so lessens the time of overnight fasting. The first part of the stomach, with its glandular lining, secretes enzymes and other substances that function in digestion. The second part, the gizzard, is a muscular organ that grinds food much as teeth and jaws do. The intestines are proportionally shorter than they are in ducks, ostriches, and all other birds that feed on plant parts rich in tough, fibrous cellulose—which requires a much longer processing time.

You also can correlate feeding behavior with the digestive system of pronghorn antelope (*Antilocapra americana*). Fall through winter, on the mountain ridges from central Canada down into northern Mexico, these

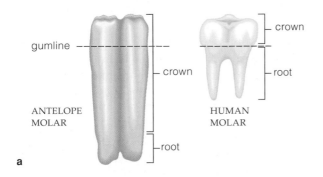

gumline

crown

ANTELOPE
MOLAR

root

crown

root

HUMAN
MOLAR

a

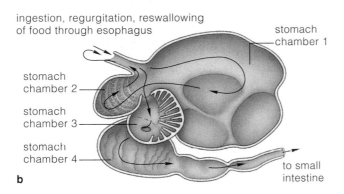

ingestion, regurgitation, reswallowing
of food through esophagus

stomach
chamber 1

stomach
chamber 2

stomach
chamber 3

stomach
chamber 4

to small
intestine

b

Figure 36.4 Some regional specializations of the complete digestive system of pronghorn antelope (*Antilocapra americana*).

(**a**) Comparison of antelope and human molars. (**b**) Multiple-chambered antelope stomach. The first chamber is a large pouch. The second is smaller, with a honeycombed inner surface (what chefs call "tripe"). In both chambers, food is mixed with fluid, kneaded, and exposed to fermentation activities of bacterial and protozoan symbionts. Some of the symbionts degrade cellulose; others synthesize organic compounds, fatty acids, and vitamins. The host uses a portion of these substances. Kneaded food is regurgitated into the mouth, rechewed, then swallowed again. It enters the third chamber, where it is pummeled once more before entering the last stomach chamber.

Figure 36.5 Exceptionally Big Mac of the snake world.

animals browse on wild sage. Come spring, they move to open grasslands and deserts, there to browse on new growth (Figure 36.4).

Now think of your own cheek teeth, or molars, each with a flattened crown that acts as a grinding platform. The crown of an antelope's molars dwarfs yours (Figure 36.4*a*). Why the difference? You probably do not brush your mouth against dirt while eating. An antelope does. Abrasive bits of soil enter its mouth along with tough plant material, so its teeth wear down rapidly. Natural selection favored more crown to wear down.

Antelopes are **ruminants**, a type of hoofed mammal having multiple stomach chambers in which cellulose is slowly broken down. The breakdown gets under way inside the first two of four stomach sacs (Figure 36.4*b*). There, bacterial symbionts produce cellulose-digesting enzymes, to the antelope's benefit as well as their own. As enzymes act, the antelope regurgitates and rechews the contents of the first two sacs, then swallows again. (That is what "chewing cud" means.) Pummeling the plant material more than once exposes more surface area to the enzymes and gives them more time to act. Thus the stomach system of ruminants accepts a steady flow of plant material during lengthy feeding times, then slowly liberates nutrients when the animal rests.

Predatory or scavenging mammals have different specializations. They gorge on food when they can get it, then may not eat again for some time (Figure 36.5). Part of their digestive system stores the food, which is digested and absorbed at a more leisurely pace. As you

will see, other organs with accessory roles in digestion help ensure that there will be an adequate distribution of nutrients between meals.

An incomplete digestive system is a saclike body cavity. Both food and residues enter and leave through the same opening. A complete digestive system is a tube with two openings (mouth and anus) and regional specializations.

Complete digestive systems carry out the following tasks in controlled ways: movements that break up, mix, and propel food material; secretion of digestive enzymes and other substances; breakdown of food; absorption of nutrients; and expulsion of undigested and unabsorbed residues.

OVERVIEW OF THE HUMAN DIGESTIVE SYSTEM

Let's look at your own body as an example of the kinds of organs in a complete digestive system. The human digestive system is a tube with two openings and many specialized organs (Figure 36.6). If the tube of an adult were fully stretched out, it would extend 6.5 to 9 meters (21 to 30 feet). From beginning to end, mucus-coated epithelium lines all surfaces facing the lumen. (A *lumen* is the space within a tube.) The thick, moist mucus protects the wall of the tube and enhances diffusion across its inner lining. Substances advance in one direction, from the mouth through the pharynx, esophagus, and gastrointestinal tract (gut). The **gut** starts at the stomach and extends through the small intestine, the large intestine (colon) and rectum, to the anus. Accessory organs called the salivary glands, gallbladder, liver, and pancreas secrete a variety of substances into different regions of the tube.

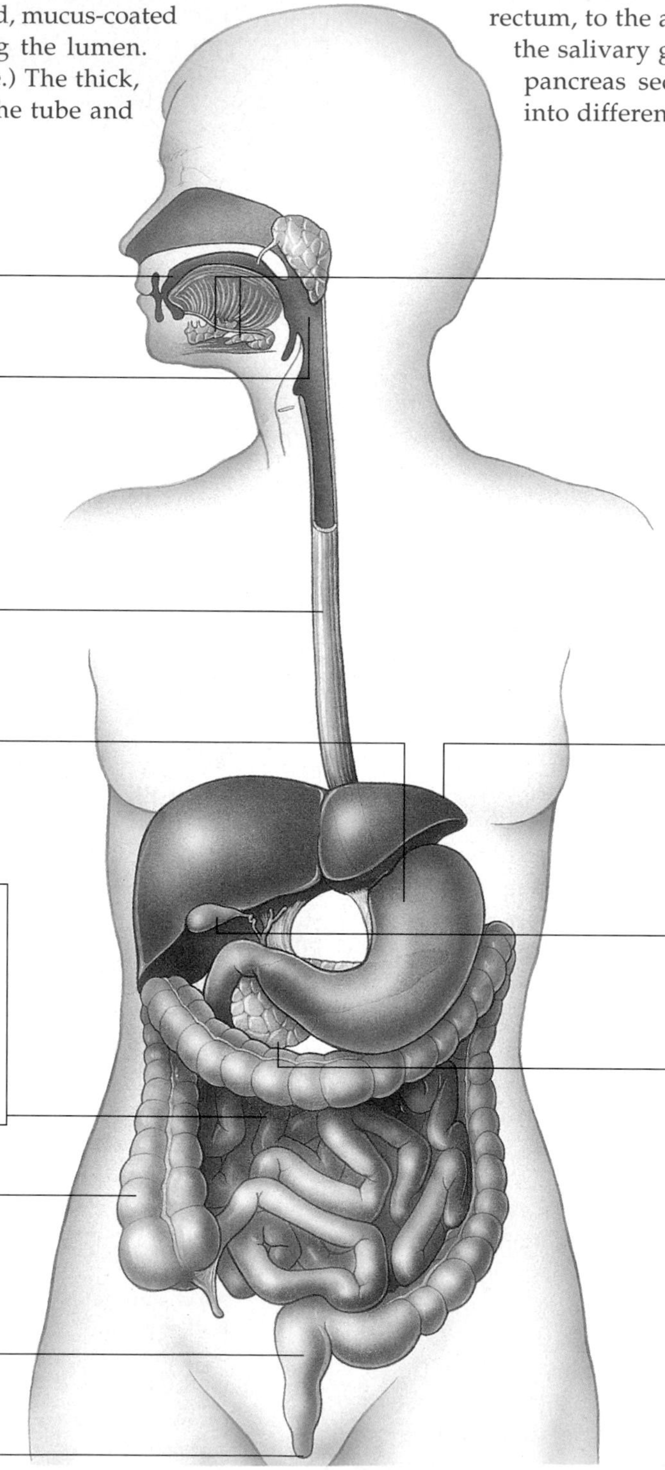

Major Components:

MOUTH (ORAL CAVITY)

Entrance to system; food is moistened and chewed; polysaccharide digestion starts.

PHARYNX

Entrance to tubular part of system (and to respiratory system); moves food forward by contracting sequentially.

ESOPHAGUS

Muscular, saliva-moistened tube that moves food from pharynx to stomach.

STOMACH

Muscular sac; stretches to store food taken in faster than can be processed; gastric fluid mixes with food and kills many pathogens; protein digestion starts.

SMALL INTESTINE

First part (duodenum, C-shaped, about 10 inches long) receives secretions from liver, gallbladder, and pancreas.

In second part (jejunum, about 3 feet long), most nutrients are digested and absorbed.

Third part (ileum, 6–7 feet long) absorbs some nutrients; delivers unabsorbed material to large intestine.

LARGE INTESTINE (COLON)

Concentrates and stores undigested matter by absorbing mineral ions, water; about 5 feet long; divided into ascending, transverse, and descending portions.

RECTUM

Distension stimulates expulsion of feces.

ANUS

End of system; terminal opening through which feces are expelled.

Accessory Organs:

SALIVARY GLANDS

Glands (three main pairs, many minor ones) that secrete saliva, a fluid with polysaccharide-digesting enzymes, buffers, and mucus (which moistens and lubricates food). *1 liter a day.*

- *Teeth*
- *Tongue*

LIVER

Secretes bile (for emulsifying fat); roles in carbohydrate, fat, and protein metabolism.

GALLBLADDER

Stores and concentrates bile that the liver secretes. *(fat)*

PANCREAS

Secretes enzymes that break down all major food molecules; secretes buffers against HCl from the stomach.

Small intestine common bile duct.

Figure 36.6 Overview of the component parts of the human digestive system and their specialized functions. Organs with accessory roles in digestion are also listed. 🔅

WWW

INTO THE MOUTH, DOWN THE TUBE

Food is chewed and polysaccharide breakdown begins in the oral cavity, or mouth. Thirty-two teeth typically project into an adult's mouth (Figure 36.7). Each **tooth** has an enamel coat (hardened calcium deposits), dentin (a thick, bonelike layer), and an inner pulp (with nerves and blood vessels). It is an engineering marvel, able to withstand years of chemical insults and mechanical stress. Recall, from Section 24.9, that the chisel-shaped incisors shear off chunks of food. Cone-shaped canines tear it. Premolars and molars, with broad crowns and rounded cusps, are good at grinding and crushing food.

Also in the mouth is a **tongue**, an organ consisting of membrane-covered skeletal muscles that function in positioning food in the mouth, swallowing, and speech. On the tongue's surface are many circular structures with taste buds embedded in their tissues (Figure 36.8). Each bud contains sensory receptors that respond to chemical differences in dissolved substances. The brain uses this information to give rise to our sense of taste.

Chewing mixes food with **saliva**. This fluid contains an enzyme (salivary amylase), a buffer (bicarbonate, or HCO_3^-), mucins, and water. Salivary glands, beneath and in back of the tongue, produce and secrete saliva through ducts to the free surface of the mouth's lining. Salivary amylase breaks down starch. The HCO_3^- helps maintain the mouth's pH when you eat acidic foods. Modified proteins called mucins help form the mucus that binds food into a softened, lubricated ball (bolus).

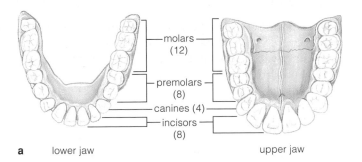

a lower jaw   upper jaw

molars (12)
premolars (8)
canines (4)
incisors (8)

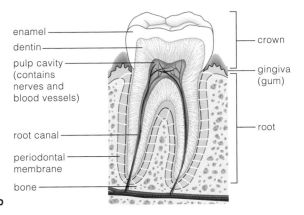

enamel
dentin
pulp cavity (contains nerves and blood vessels)
root canal
periodontal membrane
bone

crown
gingiva (gum)
root

b

Figure 36.7 (**a**) Number and arrangement of human teeth. (**b**) Closer look at a molar's two main regions, the crown and root. Enamel caps the crown. It consists of calcium deposits and is the hardest substance in the body.

Normally harmless bacteria live on and between teeth. Daily flossing, gentle brushing, and avoidance of too many sweets help keep their populations in check. In the absence of such preventive measures, conditions favor bacterial infections. These may result in *caries* (tooth decay), *gingivitis* (inflamed gums), or both. In addition, infections can spread to the periodontal membrane, which anchors teeth to the jawbone. With *periodontal disease*, infection slowly destroys the bone tissue around a tooth.

When you swallow, contractions of tongue muscles force boluses into the **pharynx**, the tubular entrance to the esophagus *and* the trachea, an airway to the lungs. As food leaves the pharynx, the epiglottis (a flaplike valve) and the vocal cords close off the trachea and so prevent breathing. The controlled closure is why you normally do not choke on food (Sections 35.4 and 35.8). The **esophagus** connects the pharynx with the stomach. Contractions of its muscular wall propel food past a sphincter and into the stomach. A **sphincter** is a ring of smooth muscles, the contractions of which can close off a passageway or an opening to the body surface.

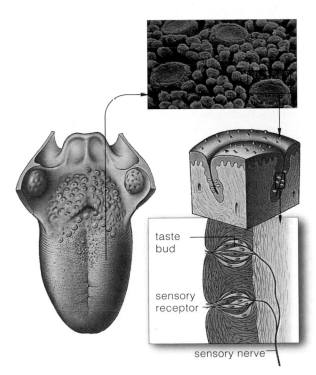

taste bud

sensory receptor

sensory nerve

Figure 36.8 Human tongue. The circular structures at its surface house many taste buds. Filamentous structures adjacent to them help move food in the mouth.

By the action of the mouth's teeth and tongue, food gets chewed, mixed with saliva, and bound into soft, lubricated balls that will be propelled through tubes to the stomach. Enzymes in saliva start the digestion of polysaccharides.

WWW

DIGESTION IN THE STOMACH AND SMALL INTESTINE

We arrive now at the premier food-processing organs, the stomach and small intestine. Both have layers of smooth muscles, the contractions of which break apart, mix, and directionally move food. The lumen of both organs receives digestive enzymes and other secretions that help break up nutrients into fragments, then into molecules small enough to be absorbed. Table 36.1 lists the names and sources of the major digestive enzymes. Carbohydrate breakdown *starts* in the mouth. Protein breakdown *starts* in the stomach. However, digestion of nearly all of the carbohydrates, lipids, proteins, and nucleic acids in food is *completed* in the small intestine.

The Stomach

The **stomach**, a muscular, stretchable sac (Figure 36.9a), serves three major functions. First, it mixes and stores ingested food. Second, its secretions help dissolve and degrade the food, particularly proteins. Third, it helps control passage of food into the small intestine.

The stomach wall surface exposed to the lumen is lined with glandular epithelium. Each day, the lining's glandular cells secrete two liters or so of hydrochloric acid (HCl), mucus, pepsinogens, and other substances that make up **gastric fluid** (stomach fluid). The potent stomach acidity, in combination with strong stomach contractions, converts food into a thick liquid mixture called **chyme**. The high acidity kills many pathogens that are ingested with food. It also can cause *heartburn* when gastric fluid backs up into the esophagus.

Protein digestion starts when the high acidity alters the structure of proteins and exposes the peptide bonds. It also converts pepsinogens to active enzymes (pepsins)

that cleave the bonds. Protein fragments accumulate in the stomach's lumen. Meanwhile, glandular cells of the stomach lining secrete gastrin. This hormone stimulates the cells that secrete HCl and pepsinogen.

A *peptic ulcer*, an eroded wall region in the stomach or small intestine, results from insufficient secretion of mucus and buffers (or excess pepsin). Heredity, chronic emotional stress, smoking, and excessive use of alcohol or aspirin are contributing factors. In addition, nearly everyone who has intestinal ulcers and 70 percent of those having stomach ulcers are infected by *Helicobacter pylori*. (Section 20.1 has a micrograph of this bacterium.) A full course of antibiotic therapy cures peptic ulcers in patients who test positive for *H. pylori*.

The stomach empties by waves of contraction and relaxation, of a type called peristalsis. The waves mix chyme and gain force as they approach a sphincter at the base of the stomach (Figure 36.9b). With the arrival of a strong contraction, the sphincter closes, so most of the chyme is squeezed backward. Only a small amount moves into the small intestine at a given time. In due course, however, it all moves out of the stomach.

The Small Intestine

Figure 36.6 shows three regions of the small intestine, called the duodenum, jejunum, and ileum. The stomach and the ducts from three organs, the **pancreas**, **liver**, and **gallbladder**, deliver about 9 liters of fluid to the duodenum every day. At least 95 percent of the fluid is absorbed across the small intestine's epithelial lining.

Cells of the intestinal lining and the pancreas secrete enzymes that digest food to monosaccharides (such as

Table 36.1 Major Digestive Enzymes and Their Breakdown Products

Enzyme	Source	Where Active	Substrate	Main Breakdown Products
CARBOHYDRATE DIGESTION				
Salivary amylase	Salivary glands	Mouth, stomach	Polysaccharides	Disaccharides
Pancreatic amylase	Pancreas	Small intestine	Polysaccharides	Disaccharides
Disaccharidases	Intestinal lining	Small intestine	Disaccharides	MONOSACCHARIDES* (e.g., glucose)
PROTEIN DIGESTION				
Pepsins	Stomach lining	Stomach	Proteins	Protein fragments
Trypsin and chymotrypsin	Pancreas	Small intestine	Proteins	Protein fragments
Carboxypeptidase	Pancreas	Small intestine	Protein fragments	AMINO ACIDS*
Aminopeptidase	Intestinal lining	Small intestine	Protein fragments	AMINO ACIDS*
FAT DIGESTION				
Lipase	Pancreas	Small intestine	Triglycerides	FREE FATTY ACIDS, MONOGLYCERIDES*
NUCLEIC ACID DIGESTION				
Pancreatic nucleases	Pancreas	Small intestine	DNA, RNA	NUCLEOTIDES*
Intestinal nucleases	Intestinal lining	Small intestine	Nucleotides	NUCLEOTIDE BASES, MONOSACCHARIDES*

* Breakdown products small enough to be absorbed into the internal environment.

glucose), monoglycerides (each consists of glycerol with one fatty acid tail), free fatty acids, amino acids, and nucleotide bases and nucleotides. For instance, like pepsin secreted from cells in the stomach lining, two pancreatic enzymes (trypsin and chymotrypsin) break down proteins to peptides. Another cleaves peptides to free amino acids. The pancreas also secretes bicarbonate, a buffer that helps neutralize HCl from the stomach.

The Role of Bile in Fat Digestion

Fat digestion requires enzyme action. It also requires **bile**, a fluid that the liver continually secretes. The fluid contains bile salts and pigments, plus cholesterol and lecithin, a phospholipid. When the stomach is empty, a sphincter closes off the main bile duct from the liver. Bile backs up into the saclike gallbladder, where it is stored and concentrated.

By a process called **emulsification**, bile salts speed up fat digestion. Most of the fats in human diets are triglycerides. Triglycerides are insoluble in water, and they tend to cluster into large fat globules in the chyme. When muscles in the intestinal wall (Figure 36.9c) move and agitate chyme, fat globules break apart into small droplets, which become coated with bile salts. Because the bile salts carry negative charges, the coated droplets repel each other and remain separated. This suspension of fat droplets, formed by mechanical and chemical action, is the "emulsion."

Compared to fat globules, emulsion droplets give fat-digesting enzymes a much greater surface area to act upon. Thus triglycerides can be broken down much more rapidly to fatty acids and monoglycerides.

Controls Over Digestion

Homeostatic controls, recall, work to counter changes in the internal environment. By contrast, controls over digestion act *before* food is absorbed into the internal environment. The nervous and endocrine systems, plus a mesh of nerves in the gut wall, exert the control.

For example, incoming food distends the stomach and stimulates mechanoreceptors in the stomach wall. The resulting signals travel along short reflex pathways to smooth muscles and glands; they also travel longer reflex pathways to the brain. Either way, they stimulate muscles in the gut wall to contract or glandular cells to

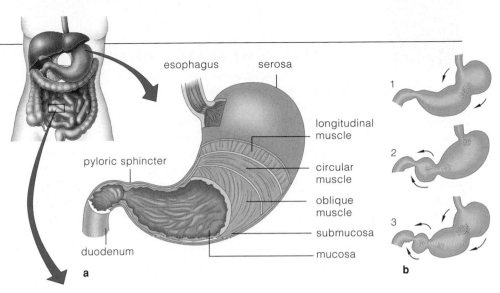

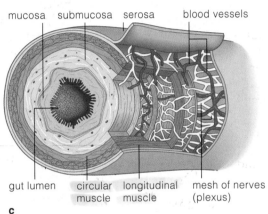

Figure 36.9 (a) Structure of the stomach. (b) Peristaltic wave down the stomach. (c) Structure of the small intestine of humans. Generally, the gut wall starts with the mucosa. This epithelium and underlying layer of connective tissue face the lumen. The submucosa, a connective tissue layer with blood and lymph vessels, has a mesh of nerves that offer local control over digestion. The intestinal wall has smooth muscle layers that differ in orientation (hence direction of contraction). The serosa is an outer layer of connective tissue.

secrete enzyme-rich fluids into the lumen or hormones into blood. Stomach emptying depends on the chyme's volume and composition. For example, a large meal activates more receptors in the stomach wall, and so contractions get more forceful and emptying proceeds faster. High acidity or a high fat content in the small intestine calls for secretion of hormones that trigger a slowdown in stomach emptying, so food is not moved faster than it can be processed. Fear, depression, and other emotional upsets also trigger such slowdowns.

Gastrointestinal hormones are part of the controls. If chyme contains amino acids and peptides, cells of the stomach lining will secrete the gastrin that stimulates secretion of acid into the stomach. Secretin prods the pancreas to secrete bicarbonate. CCK (cholecystokinin) promotes pancreatic enzyme secretion and gallbladder contraction. When the small intestine contains glucose and fat, GIP (glucose insulinotropic peptide) calls for insulin secretion, which prods cells to absorb glucose.

Carbohydrate breakdown starts in the mouth, and protein breakdown starts in the stomach.

In the small intestine, most large organic compounds are digested to molecules small enough to be absorbed into the internal environment.

ABSORPTION IN THE SMALL INTESTINE

Structure Speaks Volumes About Function

Unlike the stomach, which can only absorb alcohol and a few other substances, the small intestine is the main site where the vast majority of nutrients are absorbed.

Figure 36.10 shows the structure of the intestinal wall. Focus first on the profuse folds of the mucosa, which project into the lumen. Look closer and you see amazing numbers of even tinier projections from each one of the large folds. Look closer still and you see that epithelial cells at the surface of these tiny projections have a brushlike crown of even tinier projections, all exposed to the intestinal lumen.

What is the significance of so much convolution? It has a highly favorable surface-to-volume ratio (Section 4.2). Collectively, all the projections from the intestinal mucosa ENORMOUSLY increase the surface area available for interacting with the chyme and absorbing nutrients

from it. Without that immense surface area, absorption would proceed hundreds of times more slowly, which of course would not be enough to sustain human life.

Figure 36.10c,d shows some **villi** (singular, villus), the absorptive structures on each fold of the intestinal mucosa. Each villus is only about a millimeter long, but millions of them project outward from the mucosa. Their density gives the mucosa a velvety appearance. Inside each villus is an arteriole, a venule, and a lymph vessel that function in moving substances to and from the general circulation (Figure 36.10e).

Most cells in the epithelial lining of each villus bear **microvilli** (singular, microvillus), which are ultrafine, threadlike projections from their free surface. Each of these cells has about 1,700 microvilli. Hence its name, "brush border" cell. Glandular cells in the lining make and secrete digestive enzymes (Table 36.1), Also, some phagocytic cells patrol and help protect the lining.

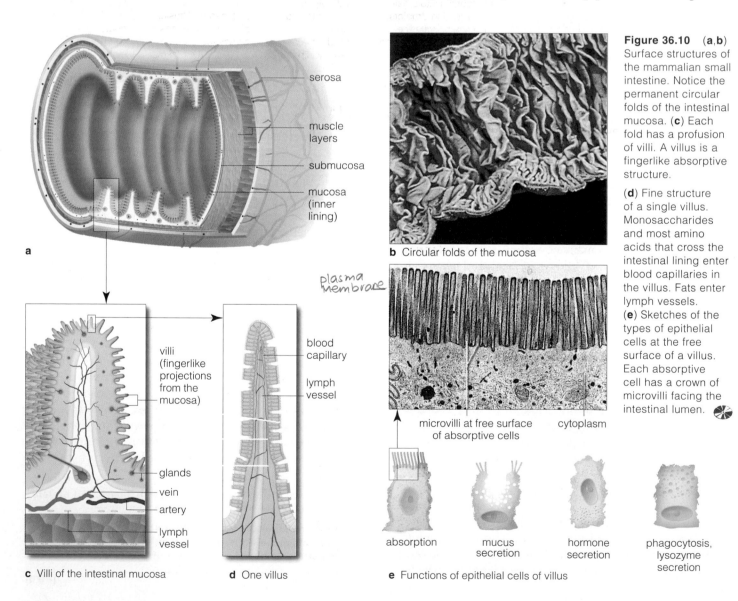

a

serosa

muscle layers

submucosa

mucosa (inner lining)

b Circular folds of the mucosa

plasma membrane

microvilli at free surface of absorptive cells

cytoplasm

c Villi of the intestinal mucosa

villi (fingerlike projections from the mucosa)

glands

vein

artery

lymph vessel

d One villus

blood capillary

lymph vessel

absorption

mucus secretion

hormone secretion

phagocytosis, lysozyme secretion

e Functions of epithelial cells of villus

Figure 36.10 (**a**,**b**) Surface structures of the mammalian small intestine. Notice the permanent circular folds of the intestinal mucosa. (**c**) Each fold has a profusion of villi. A villus is a fingerlike absorptive structure.

(**d**) Fine structure of a single villus. Monosaccharides and most amino acids that cross the intestinal lining enter blood capillaries in the villus. Fats enter lymph vessels. (**e**) Sketches of the types of epithelial cells at the free surface of a villus. Each absorptive cell has a crown of microvilli facing the intestinal lumen.

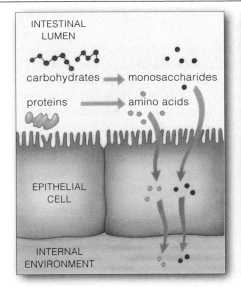

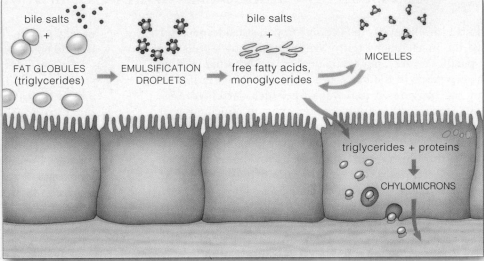

a Digestion of carbohydrates to monosaccharides, and proteins to amino acids, is completed with enzymes secreted by the pancreas and by cells of the epithelial lining.

b Monosaccharides and amino acids are actively transported across the plasma membrane of the epithelial cells, then out of the same cells and into the internal environment.

c Emulsification: The constant movement of the intestinal wall breaks up fat globules into small, emulsification droplets. Bile salts prevent the globules from re-forming. Pancreatic enzymes digest the droplets to fatty acids and monoglycerides.

d Micelles form as bile salts combine with digestion products and phospholipids. Products can readily slip into and out of the micelles.

e Concentration of monoglycerides and fatty acids in micelles enhances gradients that lead to diffusion of both substances across the lipid bilayer of the plasma membrane of cells making up the lining.

f Products of fat digestion reassemble into triglycerides in cells of the intestinal lining. These become coated with proteins, then are expelled (by exocytosis from the cells) into the internal environment.

Figure 36.11 Digestion and absorption in the small intestine

What Are the Absorption Mechanisms?

Absorption, recall, is the passage of nutrients, water, salts, and vitamins into the internal environment. The vast absorptive surface of the intestinal wall facilitates the process, but so does the action of smooth muscle in the intestinal wall. In **segmentation**, rings of circular muscle contract and relax repeatedly. This creates an oscillating (back and forth) movement that constantly mixes and forces the contents of the lumen against the wall's absorptive surface:

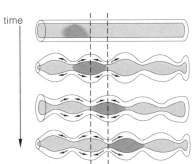

time

By the time a meal is halfway through the small intestine, most of it has been broken apart and digested. Water molecules move across the intestinal lining by osmosis. Mineral ions are selectively absorbed. Some breakdown products, including monosaccharides and amino acids, are absorbed in straightforward fashion, as transport proteins in the plasma membrane of brush border cells actively shunt them across the lining. By contrast, fatty acids and monoglycerides must diffuse across the lipid bilayer of the plasma membrane. And bile salts must help them get across (Figure 36.11).

By a process called **micelle formation**, the bile salts combine with the products of fat digestion and with phospholipids to form tiny droplets (micelles). Product molecules in the micelles continuously exchange places with those dissolved in the chyme. But the micelles concentrate them next to the intestinal lining, and when they are concentrated enough, the gradients favor their diffusion out of micelles and into epithelial cells. Fatty acids and monoglycerides recombine in the cells to form triglycerides. Then triglycerides combine with proteins into particles (chylomicrons) that leave the cells, by way of exocytosis, and enter the internal environment.

Once absorbed, the glucose and amino acids enter blood vessels directly. The triglycerides enter lymph vessels, which eventually drain into blood vessels.

With its richly folded intestinal mucosa, millions of villi, and hundreds of millions of microvilli, the small intestine offers a vast surface area for absorbing nutrients.

Substances pass through the brush border cells that line the free surface of each villus by active transport, osmosis, and diffusion across the lipid bilayer of plasma membranes.

DISPOSITION OF ABSORBED ORGANIC COMPOUNDS

Earlier in the book, in Section 7.6, you considered some of the mechanisms that govern organic metabolism—specifically, the disposition of glucose and other organic compounds in the body as a whole. You saw examples of the conversion pathways by which carbohydrates, fats, and proteins can be broken apart to molecules that serve as intermediates in the ATP-producing pathway of aerobic respiration. Here, Figure 36.12 rounds out the picture by showing all of the major routes by which organic compounds obtained from the diet are shuffled and reshuffled in the body as a whole.

energy source. There is no net breakdown of protein in muscle tissue or any other tissue during this period.

Between meals, the body taps fat and glycogen stores. In adipose tissue, cells dismantle fats to glycerol and fatty acids and release these to the blood. Cells in the liver break apart glycogen and release glucose, which also enters blood. Body cells take up the fatty acids and glucose and use them for ATP production.

Bear in mind, the liver does more than store and interconvert organic compounds. For example, it helps maintain their concentrations in blood. It inactivates

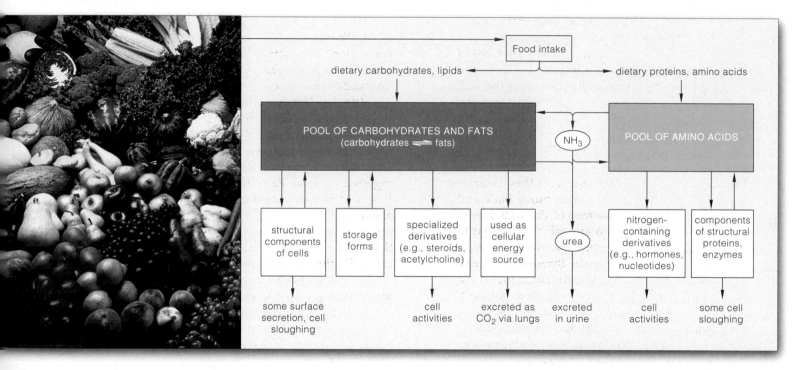

Figure 36.12 Summary of the major pathways of organic metabolism. Cells continually synthesize and tear down carbohydrates, fats, and proteins. Urea forms mainly in the liver. See also Sections 7.6 and 31.7.

All living cells in the body continually break apart most of their carbohydrates, lipids, and proteins, then use the breakdown products as energy sources or as building blocks. The nervous and endocrine systems interact to integrate this massive molecular turnover, but a treatment of how they do this would be beyond the scope of this book.

For now, simply reflect on a few key points about organic metabolism. When you eat, your body adds to its pools of materials. Excess amounts of carbohydrates and other organic compounds absorbed from the gut are transformed mostly into fats, which become stored in adipose tissue. Some are converted to glycogen in the liver and muscle tissue. *While* compounds are being absorbed and stored, most cells use glucose as the main

most hormone molecules, which are sent to the kidneys for excretion, in urine. The liver also removes worn-out blood cells from the circulation and inactivates many potentially toxic compounds. For example, ammonia (NH_3), produced during amino acid breakdown, can be toxic in high concentrations. The liver converts NH_3 to urea, a less toxic waste product. Urea is excreted from the kidneys, as a component of urine.

Just after a meal, the body's cells take up the glucose being absorbed from the gut and use it as a quick energy source. Excess amounts of glucose and other organic compounds become converted mainly to fats that are stored in adipose tissue. Some of the excess is converted to glycogen, which gets stored mainly in the liver and in muscle tissue.

Between meals, the body taps the fat reservoirs as the main energy source. Fats get converted to glycerol and fatty acids, both of which can enter ATP-producing pathways.

THE LARGE INTESTINE

What happens to the material *not* absorbed in the small intestine? It moves into the large intestine, or **colon**. The colon concentrates and stores feces: a mixture of water, undigested and unabsorbed matter, and bacteria. The colon starts at a cup-shaped pouch, the cecum. It ascends on the right side of the abdominal cavity, continues across to the other side, then descends and connects with a short tube, the rectum (Figures 36.6 and 36.13).

Colon Functioning

Material becomes concentrated as water moves across the colon's lining. Cells of the lining actively transport sodium ions out of the lumen. As ion concentrations in the lumen fall, the water concentration increases. Thus water also moves out of the lumen, by osmosis. Cells of the lining secrete mucus, which lubricates the feces and helps keep them from mechanically damaging the wall of the colon. Additionally, the cells secrete bicarbonate, which helps buffer the acidic fermentation products of ingested bacteria that managed to colonize the colon.

Short, longitudinal bands of smooth muscle in the colon wall are gathered at their ends, like a series of full skirts nipped in at elastic waistbands. As they contract and relax, the contents of the lumen move back and forth against the absorptive surface of the wall. Nerve plexuses largely control the motion, which is similar to segmentation in the small intestine but much slower.

Whereas the quick transit time through the small intestine does not favor the rapid population growth of ingested bacteria, the colon's slower motion favors it. Generally speaking, the colonizers do no harm unless they breach the wall and enter the abdominal cavity.

With each new meal, hormonal signals (gastrin) and commands from autonomic nerves induce large portions of the ascending and transverse colon to contract at the same time. Within a few seconds, the lumen's existing contents move as much as three-fourths of the colon's length and make way for incoming food.

The contents are stored in the last part of the colon until they trigger defecation—they distend the rectal wall enough to trigger a reflex action that leads to their expulsion. The nervous system controls the expulsion by stimulating or inhibiting contraction of a muscle sphincter at the anus, the terminal opening of the gut.

The volume of cellulose fiber and other undigested material that cannot be decreased by absorption in the colon is called **bulk**. Bulk contributes to the volume and normal transit time of material through the colon.

Colon Malfunctioning

The normal frequency of defecation ranges from three times a day to once a week. Aging, emotional stress, a

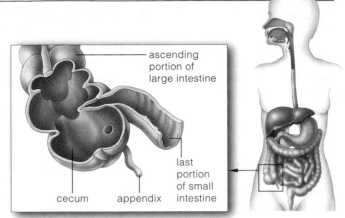

Figure 36.13 Cecum and appendix of the large intestine (colon).

ascending portion of large intestine

last portion of small intestine

cecum   appendix

low-bulk diet, injury, or disease can result in delayed defecation, or *constipation*. The longer the delay, the more water is absorbed, so feces become hard and dry. The abdominal discomfort is accompanied by loss of appetite, headaches, and often nausea and depression.

Hard feces may become lodged inside the **appendix**, a narrow projection from the cecum (Figure 36.13). The appendix has no known digestive functions. But it is colonized by lymphocytes that defend the body against ingested bacteria. Feces that obstruct the normal blood flow and mucus secretion to this projection can cause *appendicitis*, or an inflamed appendix. Unless surgically removed, an inflamed appendix may rupture. Bacteria that enter the abdominal cavity through the breach can cause life-threatening infections.

The colon also is vulnerable to cancer, which occurs most often among the world's wealthiest and best-fed populations. Too many of these people skip meals, eat too much and too fast when they do sit down at the table, and generally give their gut erratic workouts. Their diet tends to be rich in sugar, cholesterol, and salt, and low in bulk. Too little bulk extends the transit time of feces through the colon. The longer that irritating, potentially carcinogenic material is in contact with the colon wall, the more damage it may cause. Symptoms of *colon cancer* include a change in bowel functioning, rectal bleeding, and blood in feces. Some people appear to be genetically predisposed to develop colon cancer, but a low-fiber diet might also be a factor.

Colon cancer is almost nonexistent in rural India and Africa, where people can't afford to eat much more than fiber-rich whole grains. When those same people move to cities in the affluent nations and change their eating habits, the incidence of colon cancer increases.

The large intestine, or colon, functions in the absorption of water and mineral ions from the gut lumen. It also functions in the compaction of undigested residues into feces, for expulsion at the terminal opening of the gut.

36.10 WEIGHTY QUESTIONS, TANTALIZING ANSWERS

Have you ever asked yourself *why* you want to weigh a given amount? Are you merely fearful of "being fat"— that is, obese? By definition, **obesity** is an excess of fat in adipose tissues, most often caused by imbalances between caloric intake and energy output. To be sure, one standard of what constitutes obesity is simply cultural, and it varies from one culture to the next. Think of the female student who despaired over her plumpness until she took part in a graduate studies program in Africa. There, great numbers of males considered her to be one of the most desirable females on the planet. However, a lot depends on how healthy you want to be and how long you want to live. Thinner people generally live longer.

In the United States, obesity is now the second leading cause of death; 300,000 or so people die each year from preventable, weight-related conditions. Each year, the weight-related cases of type 2 diabetes, heart disease, hypertension, breast cancer, colon cancer, gout, gallstones, and osteoarthritis add up to a 100-billion-dollar drain on the economy. The future doesn't look any rosier: children today are 42 percent fatter than they were in 1980.

IN PURSUIT OF THE "IDEAL" WEIGHT Figure 36.16 shows charts for estimating the "ideal" weight for adults. Another indicator of obesity-related health risk is the *body mass index* (BMI), as determined by the formula

$$BMI = \frac{weight\ (pounds) \times 700}{height\ (inches)^2}$$

Shape Up America, a national initiative for promoting health, will quickly do the calculation for you. Their Web site is http://www.shapeup.org. If your BMI value is 27

or higher, the health risk increases dramatically. Other factors that influence the risk are smoking habits, family history of heart disorders, intake of sex hormones after menopause, and fat distribution (fat stored above the belt, as with "beer bellies," definitely is not good).

Dieting alone cannot lower a BMI value. Eat less, and the body slows its metabolic rate to conserve energy. So how do you keep functioning normally over the long term while maintaining acceptable weight? *You must balance caloric intake with energy output.* For most of us, this means eating controlled portions of low-calorie, nutritious foods *and* exercising regularly.

To figure out how many kilocalories you should take in daily to maintain a desired weight, multiply that weight (in pounds) by 10 if you are not active physically, by 15 if moderately active, and by 20 if highly active. From the value you get this way, subtract the following amount:

Age:	25–34	Subtract:	0
	35–44		100
	45–54		200
	55–64		300
	Over 65		400

For instance, if you wish to weigh 120 pounds and are very active, 120 × 20 = 2,400 kilocalories. If you are thirty-five years old, then (2,400 − 100) or 2,300 kilocalories. Such calculations provide a rough estimate of caloric intake. Other factors, including height, must also be taken into consideration. An active person 5 feet, 2 inches tall does not require as much energy as an active person who weighs the same but is 6 feet tall.

Figure 36.16 How to estimate the "ideal" weight for an adult. The values shown are consistent with a long-term Harvard study into the link between excessive weight and increased risk of cardiovascular disorders. Depending on certain factors (such as having a small, medium, or large skeletal frame), the "ideal" might vary by plus or minus 10 percent.

Weight Guidelines for Women:

Starting with an ideal weight of 100 pounds for a woman who is 5 feet tall, add five additional pounds for each additional inch of height. Examples:

Height (feet)	Weight (pounds)
5' 2"	110
5' 3"	115
5' 4"	120
5' 5"	125
5' 6"	130
5' 7"	135
5' 8"	140
5' 9"	145
5' 10"	150
5' 11"	155
6'	160

Weight Guidelines for Men:

Starting with an ideal weight of 106 pounds for a man who is 5 feet tall, add six additional pounds for each additional inch of height. Examples:

Height (feet)	Weight (pounds)
5' 2"	118
5' 3"	124
5' 4"	130
5' 5"	136
5' 6"	142
5' 7"	148
5' 8"	154
5' 9"	160
5' 10"	166
5' 11"	172
6'	178

MY GENES MADE ME DO IT Some people have far more trouble keeping off excess weight than others do. On average, one of every three Americans is like this, and *genes* have a lot to do with it. Researchers suspected as much for a long time. Studies of identical twins provided clues. (*Identical twins* are born with identical genes.) As an outcome of family problems, certain twins who were studied had been separated at birth and had been raised apart, in different households. Yet, at adulthood, body weights of those separated twins were similar! People, it seemed, were born with a "set point" for body fat. Could it be that whatever set point an individual inherits is the one he or she will be stuck with for life?

Many experiments yielded evidence that confirmed this suspicion. The first experiments started in 1950 with a seriously obese mouse (Figure 36.17). In 1995, molecular geneticists identified one of its genes that influences the set point for body fat. They named it the *ob* gene (guess why). Which clues led to its discovery? The nucleotide sequence in the gene region of obese mice differs from that in normal mice. The gene is active in adipose tissue, nowhere else. Its biochemical message translates into a type of protein released from cells and picked up by the bloodstream. The gene, apparently, specifies a hormone.

Jeffrey Friedman discovered the hormone in 1994 and named it **leptin**. A nearly identical hormone also was isolated in humans. Leptin is only one of a number of factors that influence the brain's commands to suppress or whip up appetite, but it's a crucial one. By assessing the blood concentrations of incoming hormonal signals, an appetite control center in the hypothalamus "decides" whether the body has taken in enough food to provide enough fat for the day. If so, hypothalamic commands go out to increase metabolic rates—and to stop eating.

If the *ob* gene is mutated, then its expression alone might be enough to disrupt the control center so that a mouse's appetite skyrockets and metabolic furnaces burn less. By extension, the same disruptions might occur in a human with a similar gene mutation even if he or she fully understands the inherent health risks. During one experiment that supports the hypothesis, researchers injected obese mice with leptin. The mice quickly shed their excess weight.

Will genetic researchers go on to develop therapies for obesity in humans? Possibly, but probably not for five to ten years. As you saw in Chapter 31, hormonal control in humans is tricky business.

WWW

Figure 36.17 Chronology of research developments that revealed the identity of a key factor in the genetic basis of body weight.

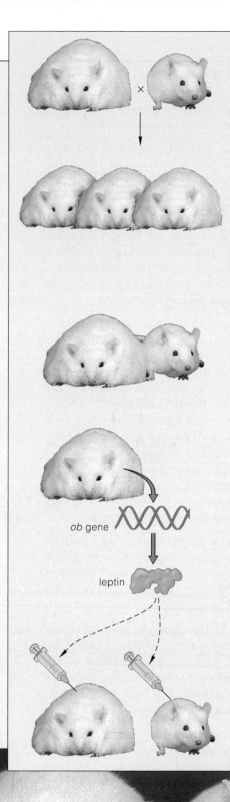

ob gene

leptin

a *1950.* Researchers at the Jackson Laboratories in Maine notice that one of their laboratory mice is extremely obese, with an uncontrollable appetite. Through crossbreeding of this apparent mutant individual with a normal mouse, they produce a strain of obese mice.

b *Late 1960s.* Douglas Coleman of the Jackson Laboratories surgically joins the bloodstreams of an obese mouse and a normal one. The obese mouse now loses weight. Coleman hypothesizes that a factor circulating in blood may be influencing its appetite, but he is not able to isolate it.

c *1994.* Late in the year, Jeffrey Friedman of Rockefeller University discovers a mutated form of what is now called the *ob* gene in obese mice. Through DNA cloning and gene sequencing, he defines the protein that the mutated gene encodes. The protein, now called leptin, is a hormone that influences the brain's commands to suppress appetite and increase metabolic rates.

d *1995.* Three different research teams develop and use genetically engineered bacteria to produce leptin, which, when injected in obese and normal mice, triggers significant weight loss, apparently without harmful side effects.

SUMMARY

1. For animals, *nutrition* refers to processes by which the body takes in, digests, absorbs, and uses food.

2. Most animals have a complete digestive system, a tube that has two openings (a mouth and an anus), and regional specializations between them. Mucus-coated epithelium lines and protects all exposed surfaces of the tube and facilitates diffusion across the tube wall.

3. As summarized in Table 36.4, the human digestive system includes a mouth, pharynx, esophagus, stomach, small intestine, large intestine (colon), rectum, and anus. Salivary glands and the liver, gallbladder, and pancreas have accessory roles in the system's functions.

4. These activities proceed in the digestive system:
   a. Mechanical processing and motility (movements that break up, mix, and propel food material).
   b. Secretion (release of digestive enzymes and other substances from the pancreas and liver, as well as from glandular epithelium, into the gut lumen).
   c. Digestion (breakdown of food into particles, then into nutrient molecules small enough to be absorbed).
   d. Absorption (the diffusion or transport of digested organic compounds, fluid, and ions from the gut lumen into the internal environment).
   e. Elimination (the expulsion of undigested as well as unabsorbed residues at the end of the system).

5. Starch digestion starts in the mouth, and protein digestion starts in the stomach. Digestion is completed and most nutrients are absorbed in the small intestine. The pancreas secretes the main digestive enzymes. Bile from the liver assists in fat digestion.

6. In absorption, cells of the intestinal lining actively transport glucose and most amino acids out of the gut lumen. Fatty acids and monoglycerides diffuse across the lipid bilayer of these cells, then are recombined in the cytoplasm as triglycerides. The cells release these triglycerides, by exocytosis, into interstitial fluid.

7. The nervous and endocrine systems, and meshes of neurons (plexuses) in the gut wall, interact to govern the digestive system. Many controls operate in response to the volume and composition of food in the stomach and intestines. They cause changes in muscle activity and in the rate at which hormones and enzymes are secreted.

8. Nutritionists advise a daily food intake in certain proportions. For example, for an adult male of average body weight: 55–60 percent complex carbohydrates, 15–20 percent proteins (less for females), 20–25 percent fats and other lipids. A well-balanced diet of whole foods normally provides all required vitamins and minerals.

9. To maintain a particular body weight and overall health, caloric intake must balance energy output.

Table 36.4 Summary of the Digestive System

MOUTH (oral cavity)	Start of digestive system, where food is chewed and moistened; polysaccharide digestion begins here
PHARYNX	Entrance to tubular part of digestive and respiratory systems
ESOPHAGUS	Muscular tube, moistened by saliva, that moves food from pharynx to stomach
STOMACH	Sac where food mixes with gastric fluid and protein digestion begins; stretches to store food taken in faster than can be processed; gastric fluid destroys many microbes
SMALL INTESTINE	The first part (duodenum) receives secretions from the liver, gallbladder, and pancreas
	Most nutrients are digested and absorbed in the second part (jejunum)
	Some nutrients are absorbed in the last part (ileum), which delivers unabsorbed material to the colon
COLON (large intestine)	Concentrates and stores undigested matter (by absorbing mineral ions and water)
RECTUM	Distension triggers expulsion of feces
ANUS	Terminal opening of digestive system

Accessory Organs:

SALIVARY GLANDS	Glands (three main pairs, many minor ones) that secrete saliva, a fluid with polysaccharide-digesting enzymes, buffers, and mucus (which moistens and lubricates ingested food)
PANCREAS	Secretes enzymes that digest all major food molecules; secretes buffers against HCl from the stomach
LIVER	Secretes bile (used in fat emulsification); roles in carbohydrate, fat, and protein metabolism
GALLBLADDER	Stores and concentrates bile from the liver

Review Questions

1. Define the five key tasks carried out by a complete digestive system. Then correlate some organs of such a system with the feeding behavior of a particular kind of animal. *36.1*

2. Using the diagram on the next page, list the organs and the accessory organs of the human digestive system. On a separate sheet of paper, list the main functions of each. *36.2; Table 36.4*

3. Name the breakdown products small enough to be absorbed across the intestinal lining, into the internal environment. *36.5*

4. Define segmentation. Does it proceed in the stomach? Does it proceed in the small intestine, colon, or both? *36.5, 36.7*

Self-Quiz (Answers in Appendix III)

1. The _____ maintains the internal environment, supplies cells with raw materials, and disposes of metabolic wastes.
   a. digestive system
   b. circulatory system
   c. respiratory system
   d. urinary system
   e. interaction of all of the systems listed

2. Most digestive systems have regions for _____ food.
   a. transporting
   b. processing
   c. storing
   d. all of the above

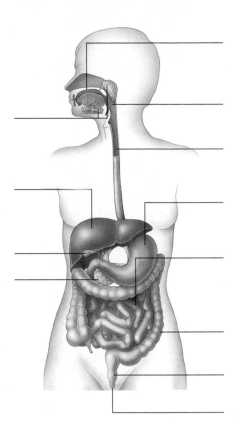

3. Maintaining good health and normal body weight requires that _____ intake be balanced by _____ output.

4. Most of our caloric intake should come from _____ .
   a. complex carbohydrates   c. proteins
   b. simple carbohydrates   d. lipids

5. On its own, the human body cannot produce all of the _____ it requires.
   a. vitamins and minerals   d. a through c
   b. fatty acids   e. a and c
   c. amino acids

6. _____ secretions do not assist in digestion and absorption.
   a. Salivary gland   c. Liver
   b. Thymus gland   d. Pancreas

7. Digestion is completed and most nutrients are absorbed in the _____ .
   a. mouth   c. small intestine
   b. stomach   d. colon

8. Glucose and most amino acids are absorbed across the gut lining _____ .
   a. by active transport   c. at lymph vessels
   b. by diffusion   d. as fat droplets

9. Bile has roles in _____ digestion and absorption.
   a. carbohydrate   c. protein
   b. fat   d. amino acid

10. Match the organ with its key digestive function(s).
    _____ gallbladder   a. secrete bile and bicarbonate
    _____ stomach   b. digest, absorb most nutrients
    _____ colon   c. store, mix, dissolve food; start
    _____ pancreas   protein breakdown
    _____ salivary   d. store, concentrate bile
           gland   e. concentrate undigested matter
    _____ small   f. secrete substances that moisten food,
           intestine   start polysaccharide breakdown
    _____ liver   g. secrete digestive enzymes and
                            bicarbonate

Critical Thinking

1. A glassful of whole milk contains lactose, proteins, butterfat (mostly triglycerides), vitamins, and minerals. Explain what will happen to each component in your digestive tract.

2. When people age, the number of their body cells steadily decreases and energy needs decline. If you were planning an older person's diet, which foods would you emphasize, and why? Which ones would you deemphasize?

3. Using Section 36.10 as a reference, determine your ideal weight and design a well-balanced program of diet and exercise that will help you achieve or maintain that weight.

4. Often, holiday meals are larger than everyday ones and have a high fat content. After stuffing themselves at an early dinner on Thanksgiving Day, Richard and other members of his family feel uncomfortably full for the rest of the afternoon. Based on what you have learned about controls over digestion, propose a biochemical explanation for their discomfort.

5. Your digestive system affords some effective protection against many pathogenic bacteria that may contaminate the kinds of food you eat. Explain some of the ways it can destroy or wash away these microorganisms. (You may wish to refer to Section 34.1, also.)

Selected Key Terms

appendix 36.7	liver 36.4
bile 36.4	micelle formation 36.5
bulk 36.7	microvillus (microvilli) 36.5
chyme 36.4	mineral 36.9
colon 36.7	nutrition CI
complete digestive system 36.1	ob gene 36.10
digestive system CI	obesity 36.10
emulsification 36.4	pancreas 36.4
esophagus 36.3	pharynx 36.3
essential amino acid 36.8	ruminant 36.1
essential fatty acid 36.8	saliva 36.3
food pyramid 36.8	segmentation 36.5
gallbladder 36.4	sphincter 36.3
gastric fluid 36.4	stomach 36.4
gut 36.2	tongue 36.3
incomplete digestive system 36.1	tooth 36.3
kilocalorie 36.8	villus (villi) 36.5
leptin 36.10	vitamin 36.9

Readings   See also www.infotrac-college.com

Blaser, M. J. February 1996. "The Bacteria Behind Ulcers." *Scientific American* 274.

Kassirer, J. 17 September 1998. *New England Journal of Medicine.* Editorial introduces reports on unproven and harmful uses of dietary supplements, some of which were contaminated with heavy metals and with a stimulant that can cause cardiac arrest.

Sherwood, L. 1997. *Human Physiology.* Third edition. Monterey, California: Brooks-Cole.

Withers, P. 1992. *Comparative Animal Physiology.* Chapter 18. New York: Saunders/Harcourt Brace Javonovich.

WWW *http://www.brookscole.com/biology*

Practice quiz questions, hypercontents, BioUpdates, and critical thinking. The Brooks/Cole Biology Resource Center provides a wealth of information fully organized and integrated by chapter.

37

THE INTERNAL ENVIRONMENT

Tale of the Desert Rat

Look closely at a fish or some other marine animal, and you will find that the cells of its body are exquisitely adapted to life in a salty fluid. And yet, some lineages of animals that evolved in the seas moved onto dry land about 375 million years ago. They were able to do so partly because they brought some salty fluid along with them, as an *internal* environment for their cells. Even so, it was not a simple transition. On land, those pioneers and their descendants encountered intense sunlight, dry winds, more pronounced swings in temperature, water of dubious salt content, and sometimes no water at all.

How did the pioneers conserve or replace the water and the salts that they lost as a result of their everyday activities? How did they manage to stay comfortably warm when their surroundings became too cold or too hot? They must have done those things, otherwise the volume, composition, and temperature of their internal environment would have spun out of control. In other words, *how did the land-dwelling descendants of marine animals maintain operating conditions inside their body and thereby prevent cellular anarchy?*

Observe any of their existing descendants and you get a sense of what some answers might be. Suppose you focus on a tiny mammal, a kangaroo rat living in an isolated desert of New Mexico (Figure 37.1). After a brief rainy season, the sun bakes the desert sand for months. The only obvious water is imported, sloshing about in the canteens of an occasional researcher or tourist. Yet with nary a sip of free water, a kangaroo rat routinely counters threats to its internal environment.

It waits out the daytime heat inside a burrow, then forages in the cool of night for dry seeds and maybe a succulent. It is not sluggish about this. It hops rapidly and far, searching for seeds and fleeing from coyotes and snakes. All that hopping requires ATP energy and water. Seeds, chockful of energy-rich carbohydrates, provide both. Metabolic reactions that release energy from carbohydrates and other organic compounds also yield water. Each day, such "metabolic water" represents a whopping 90 percent of a kangaroo rat's total water intake. By comparison, it is only 12 percent or so of the total intake for your body.

	KANGAROO RAT	HUMAN
WATER GAIN (milliliters):		
By ingesting solids	6.0	850
By ingesting liquids	0	1,400
By way of metabolism	54.0	350
	60.0	2,600
WATER LOSS (milliliters):		
In urine	13.5	1,500
In feces	2.6	200
By evaporation	43.9	900
	60.0	2,600

Figure 37.1 Kangaroo rat, a master of water conservation in a New Mexico desert. The chart gives you an idea of how kangaroo rats and humans gain and lose water. They do so in different ways, Yet, in both cases, *losses balance out the gains*— just as they do in all animals. How this happens, and why it absolutely must happen, is the first focus of this chapter.

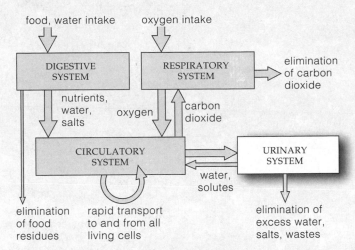

food, water intake   oxygen intake

DIGESTIVE SYSTEM   RESPIRATORY SYSTEM   elimination of carbon dioxide

nutrients, water, salts   oxygen   carbon dioxide

CIRCULATORY SYSTEM   URINARY SYSTEM

water, solutes

elimination of food residues   rapid transport to and from all living cells   elimination of excess water, salts, wastes

Figure 37.2 Links between the urinary system and other organ systems that contribute to homeostasis, or stability in favorable operating conditions in the animal body.

Inside its cool burrow, the kangaroo rat conserves and recycles water. As it inhales cool air into its warm lungs, water vapor condenses on the epithelial lining inside its nose—and some of that water diffuses back into the body. Also, after a busy night of foraging, it quickly empties its cheek pouches of seeds—which soak up the small amount of water dripping from its nose. When the kangaroo rat eats the dripped-upon seeds, it reclaims the water.

The kangaroo rat cannot lose water by perspiring; it has no sweat glands. It loses water by urinating, but its specialized kidneys do not let it piddle away much. Kidneys filter the blood's water and solutes, including dissolved salts. They adjust *how much water* and *which solutes* return to the blood or leave the body as urine.

Overall, kangaroo rats and all other animals take in enough water and solutes to replace their daily losses (Figure 37.1). How they accomplish their balancing acts will be our initial focus in the chapter. Later, we will consider some of the means by which mammals withstand hot, cold, and sometimes unpredictable changes in environmental temperatures on land.

As a starting point, remind yourself of the kinds of fluids inside most animals. **Interstitial fluid** fills the spaces between living cells and other components of tissues. Another fluid, **blood**, transports substances to and from all tissue regions by way of a circulatory system. Taken together, the interstitial fluid and blood are the **extracellular fluid**. In many animals, a well-developed urinary system helps keep the volume and composition of extracellular fluid within tolerable ranges. As you will see, other organ systems, especially those indicated in Figure 37.2, interact with the urinary system in the performance of this homeostatic task.

KEY CONCEPTS

1. Animals are continually gaining and losing water and dissolved substances (solutes). They continually produce metabolic wastes. Even with all the inputs and outputs, the overall volume and composition of the extracellular fluid in the body remain relatively constant.

2. In humans as in other vertebrates, a urinary system is crucial to balancing the intake and output of water and solutes. This system continually filters water and solutes from the blood. It reclaims some amount of both and eliminates the rest. Different amounts are reclaimed at different times, depending on what is necessary to maintain extracellular fluid.

3. Kidneys are blood-filtering organs, and the urinary system of vertebrates has a pair of them. Packed inside each kidney are a great number of nephrons.

4. At its beginning, each nephron cups around a set of blood capillaries and receives water and solutes from them. The nephron returns most of the filtrate to the blood by giving it up to a second set of capillaries, which intertwine around the nephron's tubular parts.

5. Water and solutes not returned to the blood leave the body as a fluid called urine. During any interval, control mechanisms influence whether the urine is concentrated or dilute. Two hormones, ADH and aldosterone, have key roles in the adjustments.

6. The internal body temperature of animals depends on the balance between heat produced through metabolism, heat absorbed from the environment, and heat lost to the environment.

7. The internal body temperature is maintained within a favorable range through controls over metabolic activity and adaptations in body form and behavior.

URINARY SYSTEM OF MAMMALS

The Challenge—Shifts in Extracellular Fluid

Different solid foods and fluids intermittently enter a mammal's gut. Afterward, variable amounts of absorbed water, nutrients, and other substances move into the blood, then into interstitial fluid and on into cells. Such events could easily shift the volume and composition of extracellular fluid beyond tolerable limits. But the body makes compensatory adjustments that balance out the gains and losses. Within a given time frame, it takes in as much water and solutes as it gives up.

WATER GAINS AND LOSSES To start, think about humans and other mammals. They *gain* water mainly by two processes:

> *Absorption from the gut*
> *Metabolism*

Considerable water is absorbed from solids and liquids inside the gut. As you know, water also forms as a normal by-product of many metabolic reactions. In land mammals, how much water enters the gut in the first place depends on a thirst mechanism. When the body loses too much water, land mammals seek out streams, waterholes, and so on. We will be looking at the thirst mechanism later in the chapter.

Normally, the mammalian body *loses* water mainly by way of four physiological processes, as listed here:

> *Urinary excretion*
> *Evaporation from lungs and the skin*
> *Sweating, by mammals that sweat*
> *Elimination, in feces*

Urinary excretion affords the most control over water loss. The process eliminates excess water and solutes as urine, the fluid that forms in a urinary system such as that shown in Figure 37.3. Some water also evaporates from respiratory surfaces. Some animal species lose it in sweat. A mammal in good health loses very little water from the gut (most is absorbed, not eliminated in feces).

SOLUTE GAINS AND LOSSES There are several ways in which mammals *gain* solutes. But they do so mainly by four processes:

> *Absorption from the gut*
> *Secretion from cells*
> *Respiration*
> *Metabolism*

KIDNEY (one of a pair)
Constantly filters water and all solutes except proteins from blood; reclaims water and solutes as the body requires and excretes the remainder, as urine

URETER (one of a pair)
Channel for urine flow from a kidney to the urinary bladder

URINARY BLADDER
Stretchable container for temporarily storing urine

URETHRA
Channel for urine flow between the urinary bladder and body surface

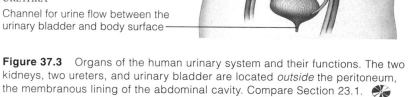

Figure 37.3 Organs of the human urinary system and their functions. The two kidneys, two ureters, and urinary bladder are located *outside* the peritoneum, the membranous lining of the abdominal cavity. Compare Section 23.1.

For example, nutrients and mineral ions are absorbed from the gut. So are drugs and food additives. Cellular secretions and wastes, including carbon dioxide, enter interstitial fluid, then the blood. The respiratory system puts oxygen into blood, and aerobically respiring cells put carbon dioxide into it.

Mammals typically *lose* solutes by these processes:

> *Urinary excretion*
> *Respiration*
> *Sweating, by some species*

The urine of mammals includes various wastes formed by the breakdown of organic compounds. For example, ammonia forms as amino groups are split from amino acids. Then **urea**, a major waste, forms in the liver when two ammonia molecules join with carbon dioxide. Also in urine are uric acid (from nucleic acid breakdown), hemoglobin breakdown products (which give the urine much of its color), drugs, and food additives. Besides these solute losses, some mammals also lose mineral ions in sweat. And all mammals lose carbon dioxide, the most abundant waste, by way of respiration.

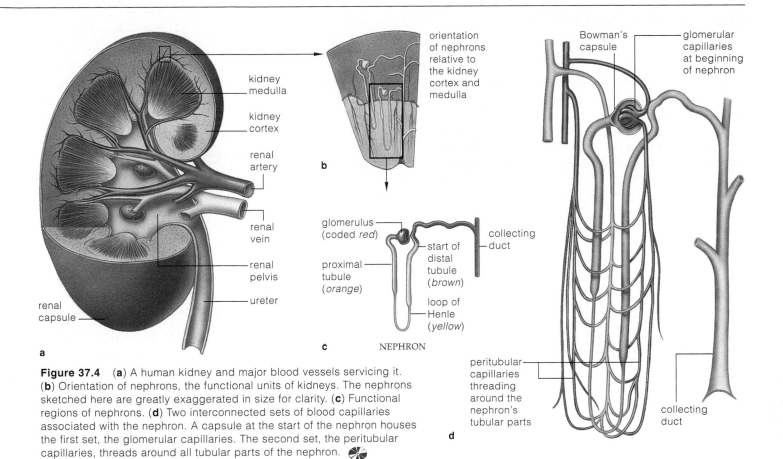

Figure 37.4 (**a**) A human kidney and major blood vessels servicing it. (**b**) Orientation of nephrons, the functional units of kidneys. The nephrons sketched here are greatly exaggerated in size for clarity. (**c**) Functional regions of nephrons. (**d**) Two interconnected sets of blood capillaries associated with the nephron. A capsule at the start of the nephron houses the first set, the glomerular capillaries. The second set, the peritubular capillaries, threads around all tubular parts of the nephron.

Components of the Urinary System

Mammals routinely counter shifts in the composition and volume of extracellular fluid, mainly with a **urinary system**. This consists of two kidneys, two ureters, one urinary bladder, and one urethra. **Kidneys** are a pair of bean-shaped organs, about as big as a fist in an adult human (Figures 37.3 and 37.4). Each has an outer capsule of connective tissue. Blood capillaries thread through its two inner regions, the cortex and medulla.

Kidneys filter water, mineral ions, organic wastes, and other substances from the blood, then adjust the filtrate's composition and return all but about 1 percent to the blood. The small portion of unreclaimed water and solutes is urine. By definition, **urine** is a fluid that rids the body of water and solutes that are in excess of the amounts required to maintain extracellular fluid.

Urine flows from each kidney into a **ureter**, a tubular channel between it and the **urinary bladder**. Urine is briefly stored in that muscular sac before flowing into the **urethra**. The end of this muscular tube opens at the body surface. Flow from the bladder (urination) is a reflex action. When the bladder is filled, a sphincter around its neck opens as smooth muscle in its balloonlike wall contracts, and urine is forced out through the urethra. Skeletal muscle surrounds the urethra. Its contraction, which is under voluntary control, prevents urination.

Nephrons—Functional Units of Kidneys

Each human kidney has more than a million **nephrons**, slender tubules packed inside lobes that extend from the kidney cortex down through the medulla. *It is at the nephrons that water and solutes are filtered from blood and the amounts to be reclaimed are adjusted.*

Each nephron starts as a **Bowman's capsule**. Here its wall cups around *glomerular* capillaries. The cupped wall region and blood vessels are a blood-filtering unit called a **glomerulus** (Figure 37.4c). Next, the nephron has a **proximal tubule** (closest to the capsule), then a hairpin-shaped **loop of Henle** and **distal tubule** (most distant from the capsule). It ends as a **collecting duct**, which is part of a duct system leading into the kidney's central cavity (renal pelvis) and then into a ureter.

Blood does not give up all of its water and solutes. The unfiltered part flows into a second set of capillaries around the nephron's tubular parts. In these *peritubular* capillaries, the blood reclaims water and solutes, then it flows into veins and back to the general circulation.

A urinary system counters unwanted shifts in the volume and composition of extracellular fluid. In its paired kidneys, water and solutes are filtered from blood. The body reclaims most of this, but the excess leaves the kidneys as urine.

Urine forms in nephrons by three processes: filtration, tubular reabsorption, and tubular secretion. All three depend on properties of cells of the nephron wall. The cells differ in membrane transport mechanisms and in permeability from one part of the nephron to the next.

Blood pressure generated by the heart's contractions drives **filtration**, which takes place at the glomerulus (Figure 37.5). Pressure "filters" blood by forcing water and all solutes except proteins out from the glomerular capillaries. The protein-free filtrate moves out from the cupped part of the nephron, into the proximal tubule.

Tubular reabsorption proceeds along the nephron's tubular regions. Most of the filtrate's water and solutes move out from the nephron's lumen (the space inside) and into peritubular capillaries (Figure 37.6).

Tubular secretion occurs at the tubule wall but in the opposite direction of reabsorption. Cells making up the wall accept solutes from the peritubular capillaries, then secrete them into the nephron's lumen. The main solutes are hydrogen and potassium ions, H+ and K+. The process also prevents certain metabolites (such as uric acid) and foreign substances (such as drugs) from accumulating in blood.

Factors Influencing Filtration

Each minute, about 1.5 liters (1-1/2 quarts) of blood flow through an adult's kidneys, and 120 milliliters of water and small solutes are filtered into the nephrons. That's 180 liters of filtrate per day! The high filtration rate is possible mainly because glomerular capillaries are 10 to 100 times more permeable to water and small solutes than other capillaries are. Also, blood pressure is high in glomerular capillaries. Why? Arterioles delivering blood to the glomerulus have a wider diameter and less resistance to flow than those carrying blood away. As a result, blood dams up inside the glomerulus, which increases the glomerular capillary blood pressure.

Filtration rates also depend on flow volume to the kidneys. Neural, endocrine, and local controls maintain the flow even when blood pressure changes. When you run in a race or dance until dawn, the nervous system diverts an above-normal volume of blood away from kidneys, toward the heart and skeletal muscles. It also directs coordinated vasoconstriction and vasodilation in different regions of the body, so less of the blood flow reaches the kidneys.

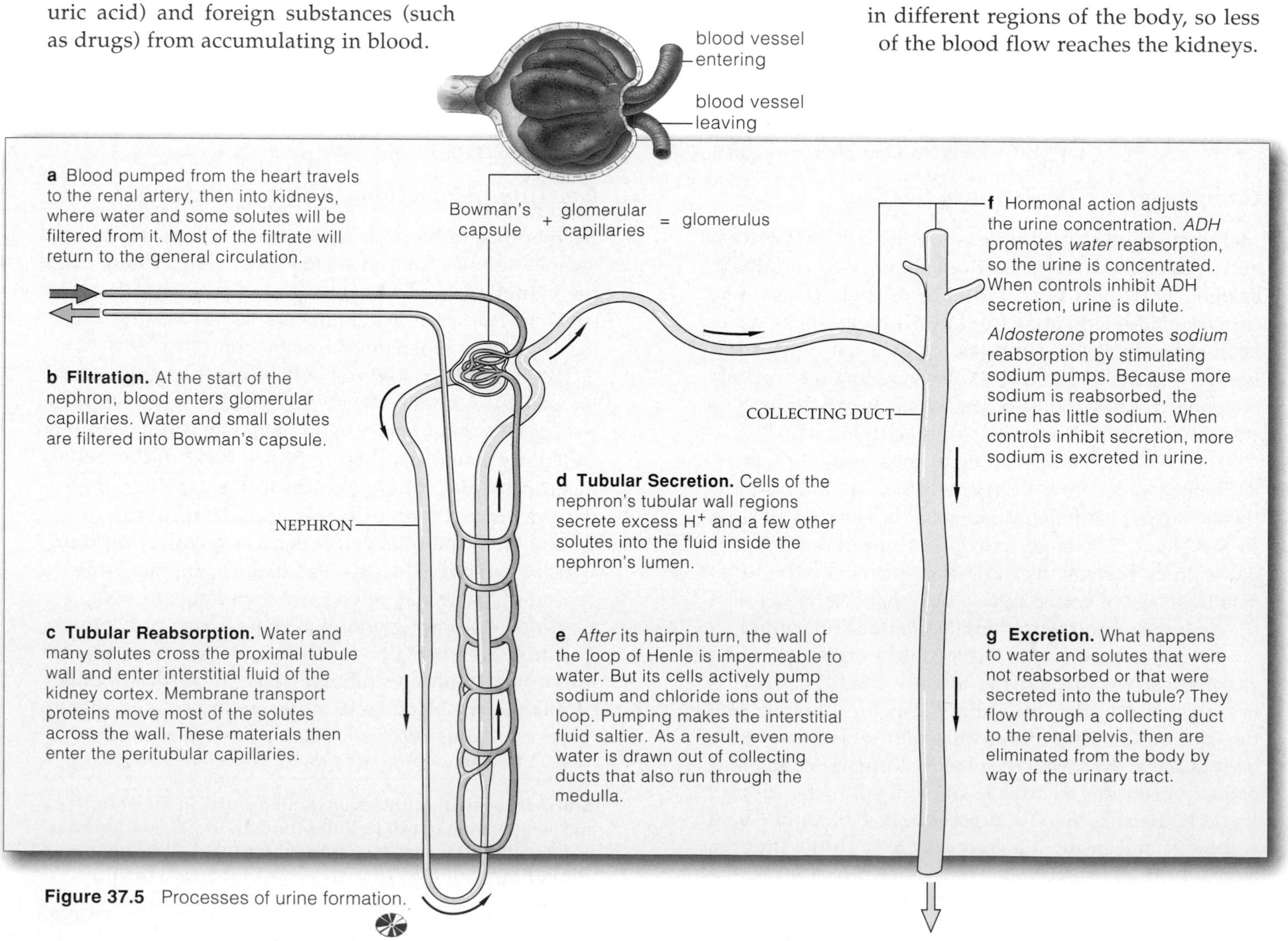

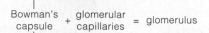

blood vessel entering

blood vessel leaving

a Blood pumped from the heart travels to the renal artery, then into kidneys, where water and some solutes will be filtered from it. Most of the filtrate will return to the general circulation.

Bowman's capsule + glomerular capillaries = glomerulus

f Hormonal action adjusts the urine concentration. *ADH* promotes *water* reabsorption, so the urine is concentrated. When controls inhibit ADH secretion, urine is dilute.

Aldosterone promotes *sodium* reabsorption by stimulating sodium pumps. Because more sodium is reabsorbed, the urine has little sodium. When controls inhibit secretion, more sodium is excreted in urine.

b Filtration. At the start of the nephron, blood enters glomerular capillaries. Water and small solutes are filtered into Bowman's capsule.

COLLECTING DUCT

d Tubular Secretion. Cells of the nephron's tubular wall regions secrete excess H+ and a few other solutes into the fluid inside the nephron's lumen.

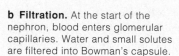

NEPHRON

c Tubular Reabsorption. Water and many solutes cross the proximal tubule wall and enter interstitial fluid of the kidney cortex. Membrane transport proteins move most of the solutes across the wall. These materials then enter the peritubular capillaries.

e *After* its hairpin turn, the wall of the loop of Henle is impermeable to water. But its cells actively pump sodium and chloride ions out of the loop. Pumping makes the interstitial fluid saltier. As a result, even more water is drawn out of collecting ducts that also run through the medulla.

g Excretion. What happens to water and solutes that were not reabsorbed or that were secreted into the tubule? They flow through a collecting duct to the renal pelvis, then are eliminated from the body by way of the urinary tract.

Figure 37.5 Processes of urine formation.

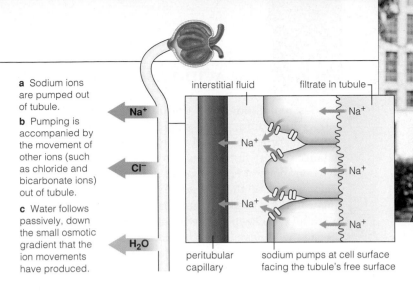

a Sodium ions are pumped out of tubule.

b Pumping is accompanied by the movement of other ions (such as chloride and bicarbonate ions) out of tubule.

c Water follows passively, down the small osmotic gradient that the ion movements have produced.

interstitial fluid   filtrate in tubule

Na^+

Na^+

Cl^-

H_2O

peritubular capillary   sodium pumps at cell surface facing the tubule's free surface

Figure 37.6 Membrane pumps in the nephron wall. Sodium and water can reabsorbed here, depending on the responses required to deal with, say, gulping too much water or ingesting too much salt.

Also, cells in the walls of arterioles leading to the glomeruli respond to pressure changes. When blood pressure decreases, arterioles vasoconstrict, so kidneys receive less of the total blood volume. But when blood pressure rises, they vasodilate, so more blood flows in.

Reabsorption of Water and Sodium

REABSORPTION MECHANISM Kidneys precisely adjust how much water and sodium ions the body excretes or conserves. Suppose you drink too much or too little water or wolf down salty potato chips or lose too much sodium in sweat. Responses start promptly as filtrate enters proximal tubules of nephrons. Cells in the tubule wall actively transport some sodium out of the filtrate. Other ions follow sodium into interstitial fluid. Water also leaves the filtrate, by osmosis. The nephron wall is highly permeable in this region, so about two-thirds of the filtrate's water is reabsorbed here (Figure 37.5c,d).

Interstitial fluid is saltiest around the hairpin turn of the loop of Henle. Water moves out of the filtrate, by osmosis, *before* the turn. The fluid left behind becomes saltier until it matches the interstitial fluid. The loop wall *after* the turn is impermeable to water. But sodium is pumped out by active transport mechanisms (Figure 37.5e). The interstitial fluid becomes saltier and attracts more water out of the filtrate just entering the loop.

Fluid arriving at the distal tubule is dilute. The stage is set for adjustments, which can lead to urine that is highly dilute, concentrated, or anywhere in between.

HORMONE-INDUCED ADJUSTMENTS The cells of distal tubules and collecting ducts have receptors for ADH and aldosterone. If the hypothalamus detects a drop in extracellular fluid volume, it calls for secretion of **ADH** (antidiuretic hormone). ADH makes the tubule walls more permeable to water, so more water is reabsorbed and urine gets more concentrated (Figure 37.5f). When

the body holds too much water, ADH secretion decreases. The walls become less permeable, less water is reabsorbed, and urine remains dilute.

Aldosterone promotes reabsorption of sodium. The extracellular fluid volume falls when too much sodium is lost. Sensory receptors in the heart and blood vessels detect the decreases and signal gland cells in the wall of the arteriole at the glomerulus. The cells secrete renin, an enzyme that splits away part of a plasma protein. Then the protein fragment is converted to **angiotensin II**. This hormone acts on aldosterone-secreting cells of the adrenal cortex, the outer part of a gland perched on each kidney (Figure 37.3). In response, the cells step up their aldosterone secretion, which stimulates cells of the distal tubules and collecting ducts to reabsorb sodium faster and excrete less sodium. Conversely, when there is too much sodium, aldosterone secretion is inhibited. Less sodium is reabsorbed, so more is excreted.

THIRST BEHAVIOR A **thirst center** in the hypothalamus induces water-seeking behavior. Osmoreceptors located nearby warn of decreases in blood volume and rising blood solute levels. Its signals activate the thirst center *and* neighboring ADH-secreting cells. Thus, while water intake is being encouraged, urinary output is reduced.

In addition to stimulating secretion of aldosterone, angiotensin II also acts on the brain to promote thirst and secretion of ADH. Thirst behavior also is initiated when free nerve endings detect dryness in the mouth, which is an early sign of dehydration.

Concentrated or dilute urine forms in kidneys by filtration, tubular reabsorption, and tubular secretion.

Filtration rates depend mainly on heart contractions, which generate high hydrostatic pressure at the glomerulus of each nephron. They also depend on neural, endocrine, and local controls over blood flow being directed to the kidneys.

Reabsorption, which can be adjusted by hormonal controls, helps maintain extracellular fluid. The adjustments rid the body of suitable amounts of water and solutes, in urine.

ADH promotes water conservation and concentrated urine. When ADH secretion is inhibited, urine is dilute.

Aldosterone promotes sodium conservation. When its secretion is inhibited, more sodium is excreted in urine.

WWW

By this point in the chapter, you probably have sensed that the functioning of nephrons is central to good health. Whether by illness or accident, when the nephrons of both kidneys become damaged and no longer perform their regulatory and excretory functions, we call this **renal failure**. Chronic renal failure is irreversible.

Infectious agents that reach the kidneys by way of the bloodstream or through the urethra can cause renal failure. So can ingestion of lead, arsenic, pesticides, and other toxins. Continued high doses of aspirin and some other drugs can do the same thing. Abnormal retention of metabolic wastes, such as the by-products of protein breakdown, results in *uremic toxicity*. Atherosclerosis, heart failure, hemorrhage, and shock diminish blood flow and skew filtration pressure in the kidneys.

In a rare type of disease called *glomerulonephritis*, the kidneys become inflamed, but not as a result of being infected. Antibody-antigen complexes become trapped in the glomeruli. Unless phagocytes remove them, they go on activating complement and other agents that bring about widespread inflammation and tissue damage.

Kidney stones form when uric acid, calcium salts, and other wastes settle out of urine and collect in the renal pelvis. These hard deposits are usually passed in urine but can become lodged in the ureter or urethra. If they disrupt urine flow, they must be medically or surgically removed to prevent renal failure.

About 13 million people in the United States alone suffer from renal failure. A *kidney dialysis machine* often can restore the proper solute balances. Like the kidney, it helps maintain extracellular fluid by selectively adjusting solutes in blood. "Dialysis" is an exchange of substances across an artificial membrane that is interposed between two solutions of different compositions.

In *hemodialysis*, the machine is connected to an artery or to a vein. Then the patient's blood is pumped on through tubes made of a material that is similar to sausage casing or cellophane. The tubes are submerged in a warm saline bath. The mix of salts, glucose, and other substances of the bath sets up the correct concentration gradients with blood. The blood then returns to the body. For kidney dialysis to have optimum effect, it must be performed three times a week. Each time, the procedure takes about four hours, because the patient's blood must circulate repeatedly to improve the solute concentrations in her or his body. In *peritoneal dialysis*, fluid of an appropriate composition is introduced into the patient's abdominal cavity, left in place for a specific length of time, then drained out. In this case, the cavity's lining itself, the peritoneum, serves as the membrane for dialysis.

Bear in mind, kidney dialysis is used as a temporary measure in reversible kidney disorders. In chronic cases, the procedure must be used for the rest of the patient's life or until a transplant operation provides her or him with a functional kidney. With treatment and controlled diets, many patients can resume fairly normal activity.

Besides maintaining the volume and composition of extracellular fluid, kidneys help keep it from becoming too acidic or too basic (alkaline). The overall **acid–base balance** of that fluid is an outcome of controls over its concentrations of H^+ and other dissolved ions. *Metabolic acidosis* hints at the importance of the balancing acts. This condition results when the kidneys cannot secrete enough H^+ to keep pace with all the H^+ that is forming during metabolism. It is life-threatening.

Buffer systems, respiration, and urinary excretion all work in concert to provide control over the acid–base balance. A buffer system, remember, consists of weak acids or bases that can reversibly latch onto and release ions, and so help minimize shifts in pH (Section 2.6).

Normally, the extracellular pH of the human body should be maintained between 7.37 and 7.43. As you know, acids lower the pH and bases raise it. A variety of acidic and basic substances enter blood after they are absorbed from the gut and as an outcome of normal metabolism. Typically, cell activities produce an excess of acids. These dissociate into H^+ and other fragments, and pH decreases. The effect is minimized when excess hydrogen ions react with buffer molecules. An example is the *bicarbonate–carbon dioxide* buffer system:

$$H^+ \; + \; HCO_3^- \; \underset{\longleftarrow}{\overset{\longrightarrow}{}} \; \underset{\text{CARBONIC ACID}}{H_2CO_3} \; \underset{\longleftarrow}{\overset{\longrightarrow}{}} \; CO_2 \; + \; H_2O$$

$$\underset{\text{BICARBONATE}}{}$$

In this case, the buffer system neutralizes excess H^+, and the carbon dioxide that forms during the reactions is exhaled from the lungs. Like other buffer systems in the body, however, this one has only temporary effects; it does not *eliminate* excess H^+. Only the urinary system can do so and thereby restore the buffers.

The same reactions proceed in reverse in cells of the nephron's tubular walls. HCO_3^- formed by the reverse reactions moves into interstitial fluid, then peritubular capillaries. Afterward, it enters the general circulation and buffers excess acid. The H^+ formed in the cells is secreted into the nephron and may join with HCO_3^-. The CO_2 that formed may be returned to blood, then exhaled. H^+ also may combine with phosphate ions or ammonia (NH_3), then leave the body in urine.

The kidneys work in concert with buffering systems, which neutralize acids, and with the respiratory system to help keep the extracellular fluid from becoming too acidic or too basic (alkaline).

A bicarbonate–carbon dioxide buffer system temporarily neutralizes excess hydrogen ions. The urinary system alone eliminates the excess ions and so restores these buffers.

The bicarbonate–carbon dioxide buffer system is one of the key mechanisms that help maintain the acid–base balance.

WWW

ON FISH, FROGS, AND KANGAROO RATS

Now that you have a general sense of how your body maintains water and solute levels, consider what goes on in some other vertebrates, including that kangaroo rat hopping about at the start of the chapter.

Bony fishes and amphibians of freshwater habitats gain water and lose solutes (Figure 37.7a). Water moves into their internal environment by osmosis; they don't gain water by drinking it. The water diffuses across thin gill membranes in fishes and across the skin in adult amphibians. Excess water leaves as dilute urine, formed inside a pair of kidneys. In both groups of vertebrates, solute losses are balanced when food provides more solutes and when gill or skin cells pump sodium in.

Body fluids of herring, snapper, and other marine fishes are about three times less salty than seawater. These fishes lose water by osmosis and replace it by drinking more. They excrete ingested solutes against concentration gradients (Figure 37.7b). Fish kidneys do not have loops of Henle, so urine cannot ever become saltier than body fluids. Cells in the fish gills actively pump out most of the excess solutes in blood.

Figure 37.7c describes the water-solute balancing act in a salmon. This type of fish spends part of its life cycle in freshwater and another part in seawater.

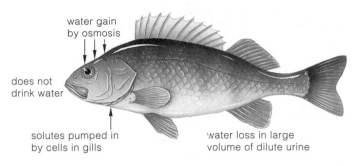

a Freshwater bony fish (body fluids far saltier than surroundings)

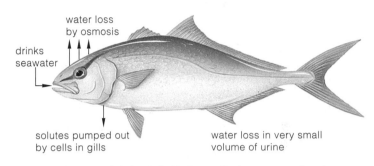

b Marine bony fish (body fluids less salty than surroundings)

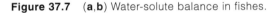

Figure 37.7 (**a**,**b**) Water-solute balance in fishes.

(**c**) Water-solute balance by salmon, a type of fish that lives in saltwater and in freshwater. Salmon hatch in streams and later move downstream to the open sea, where they feed and mature. They return to home streams to spawn.

For most salmon, salt tolerance is one outcome of changing concentrations of certain hormones. The changes appear to be triggered by increasing daylength in spring. Prolactin, a pituitary hormone, plays a key role in sodium retention in freshwater habitats. We know this because a freshwater fish that has its pituitary gland removed will die from sodium loss—but that fish will live if prolactin is administered to it.

Cortisol, a steroid hormone secreted by the adrenal cortex, is crucial to the development of salt tolerance in salmon. Cortisol secretions correlate with an increase in sodium excretion, in sodium-potassium pumping by cells in the salmon's gills, and in absorption of ions and water in the gut. In young salmon, cortisol secretion increases prior to the seaward movement—and so does salt tolerance.

salmon avoiding grizzly while maintaining solute-water balance

c

And about that kangaroo rat! Proportionally, the loops of Henle of its nephrons are astonishingly long, compared to yours. This means a great deal of sodium is pumped out from the nephron. Therefore, the solute concentration in the interstitial fluid around the loops becomes very high. The osmotic gradient between the fluid in the loops of Henle and the urine is *so* steep that nearly all of the water that does reach the equally long collecting ducts gets reabsorbed. In fact, kangaroo rats give up only a tiny volume of urine. And that urine is three to five times more concentrated than the most concentrated urine of humans.

The urinary systems of vertebrates differ in their details, such as the length of the nephron's loop of Henle. They are adapted to balance the body's gains in water and solutes with its losses of water and solutes in particular habitats.

MAINTAINING BODY TEMPERATURE

We turn now to another major aspect of the internal environment, its temperature. Many physiological and behavioral responses to change help maintain the **core temperature** within the range of tolerance of the body's enzymes. "Core" means the body's internal temperature, as opposed to temperatures of tissues near its surface.

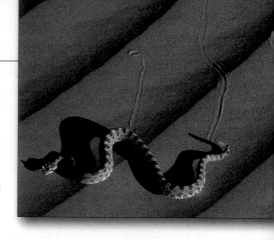

Figure 37.8 In hot desert sand at dusk, the J-shaped track of a sidewinder. As near as you can tell, how might this rattlesnake be gaining and losing heat?

Heat Gains and Heat Losses

Each metabolic reaction generates heat. If heat were to accumulate internally, the core temperature would rise. But a warm body loses heat to cooler surroundings. Its core temperature stabilizes when the rate of heat loss balances the rate of metabolic heat production. The heat content of any complex animal depends on a balance between heat gains and losses, as summarized here:

$$\begin{matrix}\text{CHANGE IN} \\ \text{BODY HEAT}\end{matrix} = \begin{matrix}\text{HEAT} \\ \text{PRODUCED}\end{matrix} + \begin{matrix}\text{HEAT} \\ \text{GAINED}\end{matrix} - \begin{matrix}\text{HEAT} \\ \text{LOST}\end{matrix}$$

Heat is gained and lost by way of exchanges at skin and other surfaces. Four processes—radiation, conduction, convection, and evaporation—drive the exchanges.

With **radiation**, an animal gains heat after exposure to radiant energy (as from sunlight) or to any surface that is warmer than the surface temperature of its body.

With **conduction**, an animal contacting a solid object directly exchanges heat with it in response to a thermal gradient between them. Animals lose heat when resting on objects cooler than they are. Animals gain heat when they are in direct contact with objects warmer than they are, such as hot sand (Figure 37.8).

With **convection**, moving air or water transfers heat. The process involves conduction; heat follows a thermal gradient between the body and air or water next to it. It also involves mass transfer; air currents carry heat to or away from the body. When heated, air becomes less dense and moves away. The body also loses heat when its movements create convection.

With **evaporation**, a liquid converts to gaseous form and heat is lost in the process. As described earlier, in Section 2.5, the heat content of a liquid provides energy for the conversion. Evaporation from a body surface has a cooling effect, because the escaping water molecules carry away some energy with them.

Evaporation rates depend on humidity and on the rate of air movement. If air next to the body is already saturated with water (that is, when the local relative humidity is 100 percent), water will not evaporate. If air next to the body is hot and dry, evaporation may be the only means of countering the metabolic production of heat and the heat gains from radiation and convection.

Animals adjust the amount of heat lost or gained through changes in behavior and physiology, although some do so better than others. Like most animals, reptiles

have low metabolic rates and poor body insulation. They rapidly absorb and gain heat, especially the small species. Reptiles maintain a core temperature mainly by gaining environmental heat. Hence they are among the **ectotherms**, which means "heat from outside."

When outside temperatures change, ectotherms use *behavioral* temperature regulation. Thus a lizard basks on warm rocks and gains heat by conduction. It keeps reorienting its body to expose the most surface area to the sun's infrared radiation. It loses heat after sunset. Before its metabolic rate drops, it hides under rocks or in crevices to conserve heat and avoid predators.

Most birds and mammals are **endotherms** ("heat from within"). With their high metabolic rates, they can stay active under a wide temperature range. (Compared to a foraging lizard of the same weight, a foraging mouse uses up to thirty times *more* energy.) Endotherms balance metabolism, controlled heat loss and conservation, and complex behavior. Also, their body forms are adapted to conserve or dissipate metabolic heat. As examples, fur, fat layers, feathers, and clothing reduce heat loss (Figure 37.9); and some mammals in cold habitats have more massive bodies than close relatives in warmer ones.

Certain birds and mammals are **heterotherms**. Their core temperature shifts some of the time; at other times they control heat exchange. For example, hummingbirds have remarkably high metabolic rates, given their tiny size. They locate and sip nectar only during the day. At night, their metabolic rates plummet and they become almost as cool as their surroundings. That way, they conserve precious energy.

Warm, humid regions favor ecotherms, which need not spend much energy to maintain core temperatures. They can devote more energy to reproduction and other tasks. In numbers and diversity, reptiles exceed mammals in the tropics. Endotherms have the edge in moderate to cold regions. With their high metabolic rates, snowshoe hares, arctic foxes, and some other endotherms occupy polar habitats, where you would never find a lizard.

Responses to Stressful Temperatures

Control centers that maintain the core temperature of a mammal reside in the hypothalamus (Figure 29.22). They constantly receive input from peripheral thermoreceptors in skin and from central thermoreceptors. When the core temperature deviates from a set point, control centers integrate responses involving smooth muscle in skin arterioles, skeletal muscle, and often sweat glands. (Here you may wish to review Section 28.7.) Negative feedback loops to the hypothalamus stop the responses when a suitable temperature returns.

Figure 37.9 How might the intrepid tourist relaxing on a deck chair of a ship off the coast of Antarctica be gaining and losing heat?

RESPONSES TO COLD STRESS In mammals, cold stress brings on **peripheral vasoconstriction.** The diameter of arterioles constricts, so the blood's convective delivery of heat to body surfaces lessens. When your fingers are chilled, all but 1 percent of blood that would otherwise flow to skin is diverted to other body regions.

Cold stress also makes hairs and feathers "stand up." This **pilomotor response** creates a layer of still air next to skin and reduces convective and radiative heat loss. Behavioral changes also minimize exposed surface areas and reduce heat loss, as when a cat curls up or when you hold both arms tightly against the body.

A **shivering response**, made to prolonged cold, is not useful for long. Rhythmic tremors start as muscles contract, ten to twenty times a second. Heat production increases by several times at high energy cost.

Prolonged or severe cold exposure also leads to a hormonal response that causes a rise in metabolic rates. This **nonshivering heat production** is greatest in *brown* adipose tissue. Human infants, hibernating animals, and cold-acclimated animals have this connective tissue. So do career divers, such as the ama of Japan and Korea who dive for shellfish six hours a day in frigid water. Adults who are not similarly adapted have little of it.

In hypothermia, core temperature is below normal. In humans, a drop of a few degrees alters brain function; more cooling leads to coma and death. Many mammals can recover from hypothermia. Frozen human cells die unless tissues thaw under medical supervision. Tissue destruction by localized freezing is called *frostbite.*

RESPONSES TO HEAT STRESS When a mammal gets too hot, **peripheral vasodilation** kicks in. The hypothalamus signals blood vessels in the skin to dilate. More blood flows to the skin, which dissipates the excess heat.

Evaporative heat loss is another response to heat stress. It occurs at moist respiratory surfaces and across skin. Animals that sweat lose more water this way. For instance, humans and some other mammals have sweat glands that release water and specific solutes through pores at the skin surface. In an adult human of average size, 2-1/2 million or more sweat glands can produce 2 liters of sweat per hour. For each liter that evaporates, 600 kilocalories of heat energy are lost.

Bear in mind, sweat that is dripping from skin does not dissipate heat. Outside temperatures must be high enough to cause water from sweat to *evaporate.* On hot, humid days, evaporation rates cannot match the rate of sweat secretion; air's high water content slows it down.

During strenuous exercise, sweating may balance high rates of heat production in skeletal muscle. With extreme sweating, as in a marathon race, the body loses a vital salt—sodium chloride—as well as water. Such losses disrupt the extracellular fluid's composition and volume. Runners collapse and faint with large losses.

What about mammals that sweat little or not at all? Some make behavioral responses, such as licking fur, panting, or resting in the shade. "Panting" is shallow, very rapid breathing that increases evaporative water loss from the respiratory tract (compare Section 28.7). Cooling results when water evaporates from the nasal cavity, mouth, and tongue.

Sometimes peripheral blood flow and evaporative heat loss cannot counter heat stress, and *hyperthermia* results. The core temperature increases above normal. For humans and other endotherms, increases of only a few degrees above normal can be dangerous.

A *fever*, recall, is part of the inflammatory response to tissue damage. The hypothalamus resets the body's thermostat, which dictates what the core temperature is supposed to be (Sections 28.7 and 34.3). Mechanisms that increase metabolic heat production and decrease heat loss operate, but their operation maintains a higher temperature. A person feels chilled at a fever's onset. When the fever "breaks," peripheral vasodilation and sweating increase as the body attempts to restore the normal core temperature. Then, the person feels warm.

By bringing down a fever, anti-inflammatory drugs such as aspirin may actually prolong healing time. Only when fevers approach dangerous levels should drugs be prescribed, under medical supervision.

The internal, core temperature of an animal's body is being maintained when heat gains and heat losses are in balance.

Metabolic reactions generate heat inside the body. Radiation, conduction, and convection can move heat down thermal gradients that exist between the body and its surroundings. Evaporative heat loss carries heat away from the body.

Besides being morphologically adapted to their habitats, animals can make behavioral and physiological adjustments to environmental temperatures.

SUMMARY

Control of Extracellular Fluid

1. For animal cells, the environment consists of certain types and amounts of substances that are dissolved in water. Extracellular fluid fills tissue spaces and blood vessels. Its volume and composition are maintained when the animal's daily intake and output of water as well as solutes are in balance. The following processes maintain the balance in mammals:

a. Water is gained by absorption from the gut and by metabolism. It is lost by urinary excretion, evaporation from lungs and skin, sweating, and elimination of feces.

b. Solutes are gained by absorption from the gut, secretion, respiration, and metabolism. They are lost by excretion, respiration, and sweating.

c. Losses of water and solutes are controlled mainly by adjusting the volume and composition of urine.

2. The vertebrate urinary system has a pair of kidneys, a pair of ureters, a urinary bladder, and a urethra.

3. Kidneys have many nephrons that filter blood and form urine. Each nephron interacts intimately with two sets of blood capillaries: glomerular and peritubular.

a. The start of a nephron is cup-shaped (Bowman's capsule). It continues as three tubular regions (proximal tubule, loop of Henle, and distal tubule, which empties into a collecting duct).

b. Together, Bowman's capsule and the set of highly permeable glomerular capillaries within it are a blood-filtering unit (glomerulus). Blood pressure forces water and small solutes out of the capillaries, into the cup. Most of the filtrate is reabsorbed at tubular regions and is returned to blood. A portion is excreted as urine.

4. Urine forms in the nephron by three processes:

a. Filtration of blood at the glomerulus, which puts water and small solutes into the nephron.

b. Reabsorption. Water and solutes to be retained leave the nephron's tubular parts and enter capillaries that thread around them. A small volume of water and solutes remains in the nephron.

c. Secretion. A few substances can leave peritubular capillaries and enter the nephron, for disposal in urine.

5. Urine is made more concentrated or less so by two hormones that act on cells of the wall of distal tubules and collecting ducts. ADH conserves water by enhancing reabsorption across the wall. In its absence, more water is excreted. Aldosterone enhances sodium reabsorption. In its absence, sodium is excreted. Angiotensin II induces aldosterone secretion (sodium conservation) and thirst; it also promotes ADH secretion (water conservation).

6. The urinary system acts in concert with buffers and with the respiratory system to maintain the acid-base balance of extracellular fluid.

Control of Body Temperature

1. Maintaining an animal's core (internal) temperature depends on balancing metabolically produced heat and the heat absorbed from and lost to the environment.

2. Animals exchange heat with their environment by four processes:

a. Radiation. Emission from the body of infrared and other wavelengths. Radiant energy can be absorbed at the body surface, then converted to heat energy.

b. Conduction. Direct transfer of heat energy from one object to another object in contact with it.

c. Convection. Heat transfer by air or water currents; involves conduction and mass transfer of heat-bearing currents away from or toward the animal body.

d. Evaporation. Conversion of liquid to a gas, driven by energy inherent in the heat content of the liquid. Some animals lose heat by evaporative water loss.

3. Core temperatures depend on metabolic rates and anatomical, behavioral, and physiological adaptations.

a. For ectotherms, core temperature depends more on heat exchange with the environment than on heat generated by metabolism.

b. For endotherms, core temperature depends more on high metabolic rates and precise controls over heat produced and heat lost.

c. For heterotherms, the core temperature fluctuates some of the time, and controls over heat balance come into play at other times.

Review Questions

1. State the function of the urinary system in terms of gains and losses for the internal environment. Name the components of the mammalian urinary system and state their functions. *37.1*

2. Label the component parts of this kidney and nephron. *37.1*

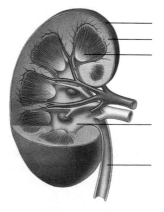 

3. Define filtration, tubular reabsorption, and secretion. Explain how urine formation helps maintain the internal environment. *37.2*

4. Which hormone or hormones promote (a) water conservation, (b) sodium conservation, and (c) thirst behavior? *37.2*

5. Name and define the physical processes by which animals gain and lose heat. What are the main physiological responses to cold stress and to heat stress in mammals? *37.6*

Self-Quiz (Answers in Appendix III)

1. In mammals, water intake depends on _____ .
   a. absorption from gut c. a thirst mechanism
   b. metabolism d. all of the above

2. In mammals, water is lost by way of the _____ .
   a. skin d. urinary system
   b. respiratory system e. c and d
   c. digestive system f. a through d

3. Water and small solutes enter nephrons during _____ .
   a. filtration c. tubular secretion
   b. tubular reabsorption d. both a and c

4. Kidneys return water and small solutes to blood by _____ .
   a. filtration c. tubular secretion
   b. tubular reabsorption d. both a and b

5. A few substances move out of the peritubular capillaries that thread around tubular parts of the nephron. The substances are moved into the nephron during _____ .
   a. filtration c. tubular secretion
   b. tubular reabsorption d. both a and c

6. A nephron's reabsorption mechanism depends on _____ .
   a. osmosis across nephron wall
   b. active transport of sodium across nephron wall
   c. a steep solute concentration gradient
   d. all of the above

7. _____ promotes water conservation.
   a. ADH c. Low extracellular fluid volume
   b. Aldosterone d. Both a and c

8. _____ enhances sodium reabsorption.
   a. ADH c. Low extracellular fluid volume
   b. Aldosterone d. Both b and c

9. Match the term with the most suitable description.
   ___ glomerulus a. surrounded by saltiest fluid
   ___ distal tubule b. extra-long loops of Henle
   ___ loop of Henle c. involves buffer systems
   ___ acid-base balance d. blood-filtering unit
   ___ kangaroo rat e. ADH, aldosterone act here

10. Match the term with the most suitable description.
    ___ ectotherm a. heat transfer by air or water currents
    ___ endotherm b. body temperature fluctuates some-
    ___ evaporation times, is controlled other times
    ___ heterotherm c. emission of wavelength energy
    ___ radiation d. direct heat transfer between an object
    ___ conduction and another object in contact with it
    ___ convection e. metabolism dictates core temperature
    f. environment dictates core temperature
    g. conversion of liquid to gas

Critical Thinking

1. Fatty tissue holds kidneys in place. Rarely, extremely rapid weight loss may cause the tissue to shrink and kidneys to slip from their normal position. If slippage puts a kink in one or both ureters and blocks urine flow, what may happen to the kidneys?

2. Drink one quart of water in an hour. What changes can you expect in your kidney function and in urine composition?

3. In 1912, the ocean liner *Titanic* left Europe on her maiden voyage across the Atlantic to America. In that same year, a chunk of the leading edge of a Greenland glacier broke off and floated out to sea. Late at night, off the Newfoundland coast, the iceberg and the *Titanic* made an ill-fated rendezvous (Figure 37.10). The *Titanic* was said to be unsinkable. Survival drills had been neglected. There were not enough lifeboats to hold even

Figure 37.10 Sinking of the *Titanic*, based on eyewitness accounts.

half the 2,200 passengers. The *Titanic* sank in about 2-1/2 hours. In less than two hours, rescue ships were on the scene, but 1,517 bodies were recovered from a calm sea. All the dead had on life jackets. None had drowned. Probably they died from _____ . If so, how did their blood flow, metabolism, and skeletal muscle action change prior to death?

4. When iguanas have an infection, they rest for a prolonged period in the sun. Propose a hypothesis to explain why.

5. Out on a first date, Jon takes Geraldine's hand in a darkened theater. "*Aha!*" he thinks. "*Cold hands, warm heart!*" What does this tell us about the regulation of core temperature, let alone Jon?

Selected Key Terms

Salt-Water Balance:
acid–base balance 37.4
ADH 37.2
aldosterone 37.2
angiotensin II 37.2
blood CI
Bowman's capsule 37.1
collecting duct 37.1
distal tubule 37.1
extracellular fluid CI
filtration 37.2
glomerulus 37.1
interstitial fluid CI
kidney 37.1
loop of Henle 37.1
nephron 37.1
proximal tubule 37.1
renal failure 37.3

thirst center 37.2
tubular
 reabsorption 37.2
tubular
 secretion 37.2
urea 37.1
ureter 37.1
urethra 37.1
urinary bladder 37.1
urinary
 excretion 37.1
urinary system 37.1
urine 37.1

Body Temperature:
(*All in Section 37.6*)
conduction
convection

core temperature
ectotherm
endotherm
evaporation
evaporative
 heat loss
heterotherm
nonshivering
 heat production
peripheral
 vasoconstriction
peripheral
 vasodilation
pilomotor
 response
radiation
shivering
 response

Readings See also www.infotrac-college.com

Flieger, K. March 1990. "Kidney Disease: When Those Fabulous Filters Are Foiled." *FDA Consumer* 24: 26–29.

Sherwood, L. 1997. *Human Physiology*. Second edition. Monterey, California: Brooks-Cole.

Smith, H. 1961. *From Fish to Philosopher*. New York: Doubleday.

WWW *http://www.brookscole.com/biology*

Practice quiz questions, hypercontents, BioUpdates, and critical thinking. The Brooks/Cole Biology Resource Center provides a wealth of information fully organized and integrated by chapter.

REPRODUCTION AND DEVELOPMENT

From Frog to Frog and Other Mysteries

With a quavering, low-pitched call that only a female of its kind could find seductive, a male frog proclaims the onset of warm spring rains, of ponds, of sex in the night. By August the summer sun will have parched the earth, and his pond dominion will be gone. But tonight is the hour of the frog!

Through the dark, a hormone-primed female moves toward the vocal male. They meet; they dally in the behaviorally prescribed ways of their species. He clamps his forelegs above her swollen abdomen and gives her a prolonged squeeze (Figure 38.1a). Out into the water streams a ribbon of hundreds of eggs, which the male blankets with a milky cloud of sperm. Soon afterward, fertilized eggs—**zygotes**—are suspended in the water.

For the leopard frog, *Rana pipiens*, a drama begins to unfold that has been reenacted each spring, with only minor variations, for many millions of years. Within a few hours after fertilization, each zygote divides into two cells, the two divide into four, then the four into eight. In less than twenty hours after fertilization, the mitotic cell divisions have produced a ball of cells no larger than the zygote. It is an early embryonic stage, of a type known as a blastocyst.

The cells continue to divide, but now they start to interact by way of their surface structures and chemical secretions. At prescribed times, many change shape and migrate to prescribed positions. They all inherited the same genetic instructions from the zygote, yet now they start to differ in appearance and function!

Through their associations, the cells form layers of embryonic tissues and then embryonic organs. A pair of tissue regions at the embryo's surface interact with the tissues beneath them. Together they give rise to a pair of eyes. Within the embryo a heart is forming, and soon it starts an incessant, rhythmic beating. In less than a week, events have transformed the frog embryo into a swimming, algae-eating larva called a tadpole.

Several months pass. Legs form; the tail shortens and disappears. The mouth develops jaws that snap shut on insects and worms. Eventually the transformations lead to an adult frog. With luck the frog will avoid predators, disease, and other threats in the months ahead. In time

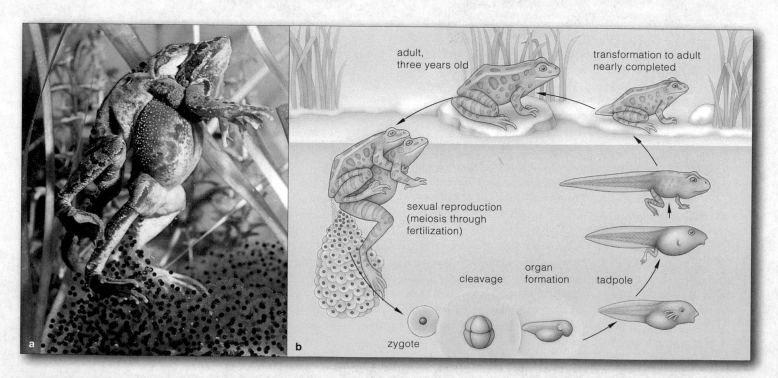

Figure 38.1 Reproduction and development of *Rana pipiens*, the leopard frog. (**a**,**b**) We zoom in on the life cycle as a male clasps a female in a reproductive behavior called amplexus. The female releases eggs into the water. The male releases sperm over the eggs. A zygote forms when an egg nucleus and a sperm nucleus fuse at fertilization. (**c**) Frog embryos suspended in the water. (**d**) A tadpole. (**e**) A transitional form between a tadpole and the young adult frog (**f**).

it may even call out quaveringly across a moonlit pond, and the life cycle will turn again.

Many years ago you, too, started a developmental journey when a zygote carved itself up. Three weeks into the journey, your embryonic body had the stamp of "vertebrate" on it. A mere five weeks after that, it was a recognizable human in the making!

With this chapter we turn to one of life's greatest dramas—the development of offspring in the image of sexually reproducing parents. The guiding question is this: *How does a single-celled zygote of a frog, human, or any other complex animal become transformed into all the specialized cells and structures of the adult form?* Some answers will become apparent as we move through a survey of basic principles, then through a case study of human reproduction and development.

DEVELOPING EMBRYO

KEY CONCEPTS

1. Sexual reproduction dominates the life cycles of nearly all animals. The separation into sexes requires specialized reproductive structures, hormonal control mechanisms, and forms of behavior. Having two distinct sexes affords a great selective advantage—variation in traits among the offspring. This advantage offsets the biological cost of the separation.

2. The life cycles of humans and many other animals proceed through six stages of embryonic development—gamete formation, fertilization, cleavage, gastrulation, organ formation, and growth and tissue specialization.

3. Each stage of embryonic development builds on the tissues and structures that formed during the stage that preceded it.

4. In a developing embryo, the fate of each type of cell depends partly on cleavage, which distributes different regions of the fertilized egg's cytoplasm to different daughter cells. It also depends on interactions among cells of the embryo. These activities are the foundation for cell differentiation and morphogenesis.

5. With cell differentiation, each cell type selectively uses certain genes and synthesizes proteins not found in other types, and so becomes unique in structure and function. With morphogenesis, tissues and organs change in size, shape, and proportion. They become organized relative to one another in prescribed patterns.

6. Human males continually produce sperm from puberty onward. The hormones testosterone, LH, and FSH control male reproductive functions.

7. Human females are fertile on a cyclic basis. Each month during their reproductive years, an egg is released from one of a pair of ovaries, and the lining of the uterus is primed for a possible pregnancy. The hormones estrogen, progesterone, FSH, and LH regulate the cyclic activity.

THE BEGINNING: REPRODUCTIVE MODES

Sexual Versus Asexual Reproduction

In earlier chapters, you learned about the cellular basis of **sexual reproduction**. Briefly, by this reproductive mode, meiosis and gamete formation typically proceed in two prospective parents. At fertilization, a gamete from one parent fuses with a gamete from the other to form the zygote, the first cell of the new individual. You also read about **asexual reproduction**, by which a single parent organism produces offspring by various means (but not by gamete formation). We now turn to a few structural, behavioral, and ecological aspects of the two reproductive modes.

Picture a scuba diver accidentally kicking a sponge. A tissue fragment breaks away from the sponge body, then grows and develops by mitotic cell divisions and cell differentiation into a new sponge. Or picture one of the flatworms undergoing transverse fission while it glides along through the water. First its body constricts below the midsection. The part behind the constriction grips a substrate and starts a tug-of-war with the part in front. After a few hours, it splits off. Both parts go their separate ways, regenerate the missing part, and become a whole worm. Only some species do this.

In such cases of *asexual* reproduction, all offspring are genetically the same as their individual parent, or nearly so. Phenotypically they are much the same, also. We can speculate that phenotypic uniformity is useful when each individual's gene-encoded traits are highly adapted to a limited and more or less consistent set of environmental conditions. Most variations introduced into the finely tuned gene package would not do much good, and often they would do harm.

However, most animals live where opportunities, resources, and danger are highly variable. For the most part, such animals reproduce sexually, with female and male parents bestowing different mixes of alleles on their offspring (Section 9.1). The resulting variation in traits improves the odds that some of those offspring, at least, will survive and reproduce even if conditions change in the environment.

Costs and Benefits of Sexual Reproduction

Among animals, separation into sexes is not without cost. Some cells that can serve as gametes must be set aside and nurtured. Housing and delivering gametes near or within a prospective mate require specialized reproductive structures. Often, mating require special forms of behavior, such as courtship, that can promote fertilization. Mating also requires built-in controls that can synchronize the timing of gamete formation, sexual readiness, even parental behavior in two individuals.

Figure 38.2 Examples of where invertebrate and vertebrate embryos develop, how they are nourished, and how (if at all) parents protect them.

(**a**) Snails are *oviparous*, meaning they are egg producers (*ovi–*, egg; *parous*, produce). Hermaphroditic parents release fertilized eggs, which develop on their own, unprotected.

(**b**) Birds also are oviparous. Their fertilized eggs have large yolk reserves, and they develop and hatch outside the mother's body. Unlike snails, one or both parents expend considerable energy feeding and caring for the young.

Facing page: Some fishes, all lizards, and many snakes are *ovoviviparous*. Their fertilized eggs develop within the mother; then offspring are born live. Yolk reserves, not the mother's own tissues, sustain the eggs. The copperhead in (**c**) is an example. Her offspring are born live inside the relics of egg sacs.

Most mammals are *viviparous*; their young are born live (*vivi–*, alive). (**d**,**e**) In kangaroos and some other species, embryos are born in unfinished form. The young of this marsupial complete development in a pouch on the mother's ventral surface. Juvenile stages (joeys) continue to draw nourishment from mammary glands located in the mother's pouch. (**f**) By contrast, a human female retains the fertilized egg inside her body. Maternal tissues nourish the developing individual until the time of birth, in the manner described later in the chapter.

Just look at the question of *reproductive timing*. How do mature sperm in one individual become available exactly when the eggs mature in a different individual? Timing depends upon energy outlays for constructing, maintaining, and operating neural as well as hormonal control mechanisms in each parent. Also, parents must produce mature gametes in response to the same cues, such as a seasonal change in daylength, that mark the onset of the most suitable time of reproduction for their species. For example, male and female moose become sexually active only in late summer and early fall. The coordinated timing ensures that their offspring will be born next spring—when the weather will improve and food will be plentiful for many months.

liveborn snake inside egg sac

c

d

e

f

Finding and actually recognizing a potential mate of the same species is another challenge. Different species invest energy when they synthesize pheromones. They build visual mating signals such as feathers of certain colors and patterns, and complex sensory receptors to detect the specific signals being sent. In addition, males often expend astonishing amounts of energy to execute courtship routines, as you will read in Chapter 44.

Assuring the survival of offspring is also costly. For example, many invertebrates, bony fishes, and frogs simply release eggs and motile sperm into the watery surroundings, as in Figure 38.1*a*. If each adult produced only *one* sperm or *one* egg each season, the odds for fertilization would not be good. Such species invest energy in producing numerous gametes, often thousands of them. As another example, nearly all land animals rely on internal fertilization, the union of sperm and egg *inside* the female's body. They invest energy to construct elaborate reproductive organs, such as a penis (by which a male deposits its sperm inside a female) and a uterus (a chamber inside the female where the embryo grows and develops).

Finally, animals set aside energy in forms that can *nourish the developing individual* until it is developed enough to feed itself. For instance, nearly all animal eggs contain **yolk**, which is a protein-rich, lipid-rich substance that nourishes embryonic stages. The eggs of some species have much more yolk than others. Sea urchins make tiny eggs with very little yolk, release large numbers of them, and thus limit the biochemical investment in each one. Within twenty-four hours, each fertilized egg has developed into a self-feeding, free-moving larva. Sea stars and other predators consume most of the eggs. So for sea urchins, bestowing as little as possible on as many gametes as possible does pay off in terms of reproductive success.

By contrast, mother birds lay truly yolky eggs. Yolk nourishes the bird embryo through an extended period of development, inside an eggshell that forms after the egg is fertilized. Your mother put tremendous demands on her body to protect and nourish you through nine months of development inside her, starting from the time you were an egg with almost no yolk. After you

implanted yourself into her uterus, physical exchanges with her tissues supported you through the extended pregnancy (Figure 38.2*f*).

As these few examples suggest, animals show great diversity in reproduction and development. However, as you will see in the sections to follow, some patterns are widespread throughout the animal kingdom, and they can serve as a framework for our reading.

Separation into male and female sexes requires special reproductive cells and structures, neural and hormonal control mechanisms, and forms of behavior. A selective advantage—variation in traits among offspring—offsets biological costs associated with the separation into sexes.

STAGES OF DEVELOPMENT—AN OVERVIEW

Embryos are a class of transitional forms on the road from a fertilized egg to an adult. Although they all start out as a single cell, embryos of different species often look different as they grow and develop. For example, you do not look like a frog now, and you did not look like one when you and the frog were early embryos, either. However, despite the differences in appearance, it is possible to identify certain patterns in the way the embryos of nearly all animal species develop.

Figure 38.3 is an overview of the stages of animal development. During **gamete formation**, the first stage, eggs or sperm develop inside the reproductive organs of one parent's body. **Fertilization**, the second stage, starts when the plasma membrane of a sperm fuses with the plasma membrane of an egg. It is over when the egg nucleus and sperm nucleus fuse and form a zygote.

The third stage, **cleavage**, is a program of mitotic cell divisions that divide the volume of egg cytoplasm into a number of **blastomeres**—smaller cells, each with its own nucleus. There is no growth during this stage. Cleavage only increases the number of cells; it does not change the original volume of the egg cytoplasm.

As cleavage draws to a close, the pace of mitotic cell division slackens. The embryo enters **gastrulation**. This fourth stage of animal development is a time of major cellular reorganization. The newly formed cells become arranged into two or three primary tissues, often called germ layers. The cellular descendants of the primary tissues will form all the tissues and organs of the adult:

1. **Ectoderm**. This is the *outermost* primary tissue layer, the one that forms first in the embryos of every animal. Ectoderm is the embryonic forerunner of the cell lineages that give rise to tissues of the nervous system and to the integument's outer layer.

2. **Endoderm**. Endoderm is the *innermost* primary tissue layer. It is the embryonic forerunner of the gut's inner lining and organs derived from the gut.

3. **Mesoderm**. This *intermediate* primary tissue layer is the forerunner of muscle and most of the skeleton; of circulatory, reproductive, and excretory organs; and of connective tissue layers of the gut and integument. Mesoderm originated hundreds of millions of years ago, and it was a pivotal step in the evolution of nearly all large, complex animals.

After they have formed, the primary tissue layers give rise to subpopulations of cells. This marks the onset of **organ formation**. The subpopulations become unique in structure and function, and their descendants give rise to different kinds of tissues and organs.

During the final stage of animal development, **growth and tissue specialization**, organs increase in size, and they gradually assume specialized functions. This stage continues into adulthood.

Now take a look at Figure 38.4. Its photographs and sketches show several stages in the embryonic development of a typical animal, a frog. Take a moment to study this

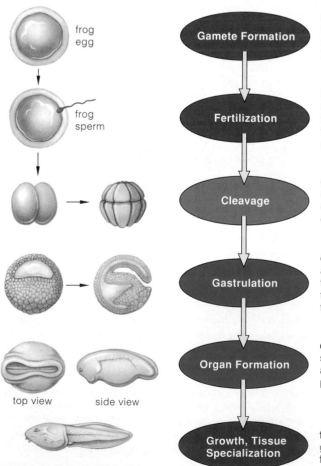

frog egg

frog sperm

top view side view

Gamete Formation

a Eggs form and mature in female reproductive organs, and sperm form and mature in male reproductive organs.

Fertilization

b A sperm and an egg fuse at their plasma membrane, then the nucleus of one fuses with the nucleus of the other to form the zygote.

Cleavage

c By a series of mitotic cell divisions, different daughter cells receive different regions of the egg cytoplasm.

Gastrulation

d Cell divisions, migrations, and rearrangements produce two or three primary tissues, the forerunners of specialized tissues and organs.

Organ Formation

e Subpopulations of cells are sculpted into specialized organs and tissues in prescribed spatial patterns at prescribed times.

Growth, Tissue Specialization

f Organs increase in size and gradually assume specialized functions.

Figure 38.3 Overview of the stages of animal development. We use a few forms that appear during the frog life cycle as examples.

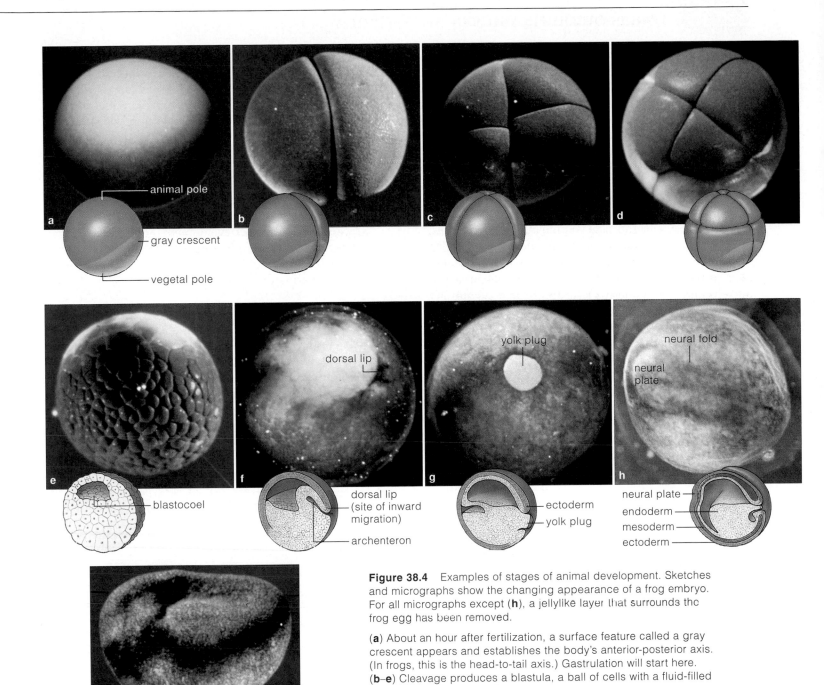

Figure 38.4 Examples of stages of animal development. Sketches and micrographs show the changing appearance of a frog embryo. For all micrographs except (**h**), a jellylike layer that surrounds the frog egg has been removed.

(**a**) About an hour after fertilization, a surface feature called a gray crescent appears and establishes the body's anterior-posterior axis. (In frogs, this is the head-to-tail axis.) Gastrulation will start here. (**b–e**) Cleavage produces a blastula, a ball of cells with a fluid-filled cavity called a blastocoel.

(**f,g**) Cells move about and become rearranged during gastrulation. The primary tissue layers form, then a primitive gut cavity (in this case, an archenteron) develops. (**h,i**) Early structures form, and so does a fluid-filled body cavity in which vital organs will be suspended. Cell differentiation proceeds, moving the embryo on its way to becoming the functional tadpole shown in Figure 38.1.

illustration, for it reinforces an important concept. The structures that form during one stage of development serve as the foundation for the stage that comes after it. Successful development depends on the formation of all of those structures according to normal patterns, in prescribed sequence.

Animal development proceeds from gamete formation and fertilization through cleavage, gastrulation, then organ formation, and finally growth and tissue specialization.

Development cannot proceed properly unless each stage is successfully completed before the next begins.

Information in the Egg Cytoplasm

Why don't you have an arm attached to your nose or toes growing from your navel? The patterning of body parts for any complex animal, including yourself, is partly mapped out in the cytoplasm of an immature egg—an **oocyte**—even before a sperm enters the picture. A **sperm**, remember, consists of paternal DNA and a bit of equipment (such as a tail) that helps the sperm reach and penetrate an egg. Compared to a sperm, an oocyte is much larger and more complex (Section 9.5).

As an oocyte matures, its volume increases. Enzymes, mRNA transcripts, and other factors become stockpiled in different parts of the cytoplasm and will be activated after fertilization. Typically they take part in early rounds of DNA replication and cell division. Also present are tubulin molecules, plus factors that will govern the angle and timing of tubulin assembly into the microtubules of a mitotic spindle. Such factors influence the pattern of cleavage. Also, the cytoplasm has yolk. The amount and distribution of yolk dictate where cleavage cuts will proceed and how large the blastomeres will be.

Such regionally localized aspects of the oocyte are "maternal messages." We find evidence of their effects as early as fertilization. For example, a frog egg has pigment granules concentrated near one pole and yolk near the other. The egg's cortex (the plasma membrane and the cytoplasm just under it) undergoes structural reorganization when a sperm fertilizes the egg. Part of the cortex shifts toward the point of sperm entry, and it exposes a crescent-shaped area of yolky cytoplasm. This area of intermediate pigmentation, a gray crescent, forms near the frog egg's midsection. And it establishes the anterior-posterior body axis. (Section 28.6 shows the directional planes for this body axis and others.)

In itself, a gray crescent is not evidence of regional differences in maternal messages. Such evidence comes from experiments of the sort shown in Figure 38.5. It also comes from observing embryos in which localized cytoplasmic differences are pronounced enough to be tracked easily during development. For example, if you were to continue tracking the development of fertilized frog eggs, you would see that a gray crescent is the site where gastrulation normally begins.

Cleavage—The Start of Multicellularity

Once fertilization is finished, the zygote enters cleavage. Beneath the plasma membrane, its midsection bears a ring of microfilaments, made of the contractile protein actin. The ring tightens as microfilaments slide past one another and pinch in the cytoplasm. The cell surface above the tightening ring is drawn inward as a cleavage furrow (Section 8.4). And it is this force of contraction that splits the cytoplasm into blastomeres. A new ring forms from actin subunits in each daughter cell.

Simply by virtue of where they form, blastomeres end up with different maternal messages. This outcome

Figure 38.5 Examples of experiments that illustrate how the cytoplasm of a fertilized egg has localized differences that help determine the fate of cells in a developing embryo. The cortex of frog eggs contains granules of dark pigment, concentrated near one pole. At fertilization, a portion of the granule-containing cortex shifts toward the point of sperm entry. The shift exposes lighter colored, yolky cytoplasm in a crescent-shaped gray area:

Normally, the first cleavage puts part of the gray crescent in both of the resulting blastomeres.

(**a**) During one experiment, the first two blastomeres that formed were physically separated from each other. Each blastomere still gave rise to a complete tadpole.

(**b**) For another experiment, a fertilized egg was manipulated so the cut through the first cleavage plane missed the gray crescent entirely. Only one of the two blastomeres received the gray crescent. It alone developed into a normal tadpole. The blastomere deprived of maternal messages in the gray crescent gave rise to a ball of undifferentiated cells. ✿

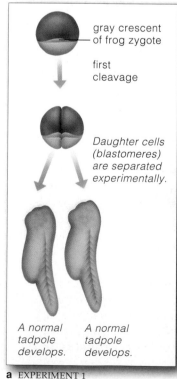

gray crescent of frog zygote

first cleavage

Daughter cells (blastomeres) are separated experimentally.

A normal tadpole develops.

A normal tadpole develops.

a EXPERIMENT 1

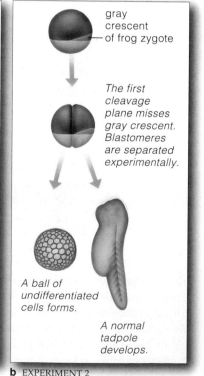

gray crescent of frog zygote

The first cleavage plane misses gray crescent. Blastomeres are separated experimentally.

A ball of undifferentiated cells forms.

A normal tadpole develops.

b EXPERIMENT 2

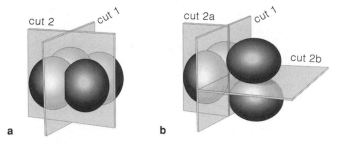

a

b

Figure 38.6 Comparison of the early cleavage planes for the eggs of (**a**) sea urchins and (**b**) mammals. These early cuts are radial in sea urchins and rotational in mammals.

Figure 38.7
Identical twins, who started out from the same zygote. The first two blastomeres that formed during cleavage, the inner cell mass, or another early embryonic stage split. The split was the start of two genetically identical, look-alike individuals.

of cleavage, called **cytoplasmic localization**, helps seal the developmental fate of each cell's descendants. Its cytoplasm alone may have molecules of a protein that can activate, say, a gene coding for a certain hormone. And its descendants alone will make that hormone.

Cleavage Patterns

In the simplest cleavage pattern, complete cuts occur after each nuclear division, and each cleavage plane is perpendicular to the mitotic spindle. In effect, the cuts parcel out a copy of the nucleus to every blastomere, which makes all the blastomeres genetically equivalent.

Blastomeres do not grow in size before they divide again. Cleavage divides the volume of cytoplasm into increasingly smaller cells, often quite rapidly.

In most animals, the zygote's genes are silent during early cleavage. How fast the cuts proceed and how the blastomeres become arranged are under the control of the proteins and mRNAs stockpiled in the cytoplasm. In mammals, however, certain genes must be activated first. The proteins specified by those genes take part in cleavage, which cannot be completed without them.

We can correlate cleavage patterns of major animal groups with the amount and distribution of yolk, which typically inhibits cleavage. If an egg has little yolk, the cuts may go right through it. But if yolk is concentrated at one end, early cuts will only go partway through the cytoplasm. Such eggs have polarity. Their yolk-rich end is the *vegetal* pole; the *animal* pole is the end closest to the nucleus. A frog egg is like this (Figure 38.4).

Cleavage differs among animals with little yolk in their eggs, so heritable factors must influence the cuts. For instance, frog eggs and sea urchin eggs undergo *radial* cleavage; the cleavage furrows run horizontally and vertically with the animal-vegetal axis (Figure 38.6a). The successive cuts produce a **blastula**, a stage with blastomeres around a fluid-filled cavity (a blastocoel). A sea urchin egg is nearly yolkless, and the blastula has horizontal rows of blastomeres. The yolk of a frog egg impedes cuts near the vegetal pole. Cleavage is faster and yields more, smaller blastomeres near the animal

pole and an offset blastocoel, as in Figure 38.4e. Between the time that 16 to 64 blastomeres form, any amphibian blastula resembles a mulberry and is called a morula, which is Latin for mulberry.

What about eggs of reptiles, birds, and most fishes? They undergo *incomplete* cleavage. The large volume of yolk restricts those early cuts to a small, caplike region near the animal pole. The result is two flattened layers of cells with a narrow cavity in between.

What about a mammal's egg? It undergoes *rotational* cleavage. The first cut passes through both poles. One of the resulting blastomeres is cut the same way but the other is cut horizontally (Figure 38.6b). Blastomeres divide slowly, at different times. After a third cleavage, they abruptly huddle into a compact ball, and ongoing divisions form a 16- to 32-cell morula. Descendants of the outer cells (the trophoblasts) secrete fluid into the morula. As a cavity forms, they become organized as a thin, single layer, and in time will give rise to part of the placenta. The inner cells mass together to one side of the fluid-filled cavity. This particular kind of blastula is called a **blastocyst**. In humans and other mammals, the inner cell mass gives rise to the embryo proper. You will see an example of this in Section 38.12.

Sometimes the first two blastomeres, or the inner cell mass, or even a later stage split. Any such split may result in *identical twins*, which have the same genetic makeup and share a common placenta (Figure 38.7). By contrast, *fraternal twins* arise from two oocytes that matured and were fertilized during the same menstrual cycle. Each is serviced by its own placenta.

The egg cytoplasm contains maternal instructions in the form of regionally distributed enzymes and other proteins, mRNAs, cytoskeletal elements, yolk, and other factors.

Cleavage divides a zygote into blastomeres, each with a localized part of maternal messages inherent in the egg cytoplasm. This outcome is called cytoplasmic localization.

Differences in the amount and distribution of yolk and other, heritable factors give rise to different patterns of cleavage that affect body plans of different animal groups.

HOW DO SPECIALIZED TISSUES AND ORGANS FORM?

Nearly all animals have a gut, with tissues and organs that function in digestion and absorption of nutrients. They have surface parts that protect internal parts and detect what is going on outside. In between, most have organs, such as those dealing with structural support, movement, and blood circulation.

This type of three-layer body plan starts to emerge as cleavage ends and gastrulation begins. The embryo's size increases little, if any. But cells start migrating to new positions to form primary tissue layers—ectoderm, endoderm, and mesoderm. Figure 38.8 shows some outer cells of a sea urchin embryo migrating inward to form a lining for a cavity that will become a gut. Figure 38.9 shows the formation of a bird embryo's anterior-posterior axis, the outcome of cell migrations and other rearrangements. In all vertebrates, the anterior-posterior axis defines where the neural tube (the forerunner of a brain and spinal cord) will form. All such organs start forming by cell differentiation and morphogenesis.

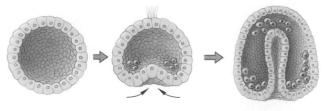

a Blastula   **b** Cell migrations in early gastrula

Figure 38.8 (**a**,**b**) A few early stages of development of a sea urchin (*Lytechinus*), cross-section. The blastula that resulted from cleavage becomes transformed into a gastrula. (**c**) Scanning electron micrograph of the gastrula's surface cells and region of inward migration.

endoderm cells **c** migrating inward

Cell Differentiation

All cells of an embryo descend from the same zygote, so they have the same number and kinds of genes. They all activate genes that specify histones, enzymes of glucose metabolism, and other proteins that are basic to cell survival. However, from gastrulation onward, certain groups of genes are activated in some cells but not others. When a cell selectively activates genes and synthesizes proteins not found in other cell types, we call this process **cell differentiation**. You read about the molecular basis of selective gene expression in Section 14.3. Basically, it results in proteins that are required for distinctive cell structures, products, and functions.

For instance, when your eye lenses developed, some cells started synthesizing crystallin, a family of proteins

that become incorporated in transparent fibers of each lens. Only those cells could activate the required genes. Long crystallin fibers formed and forced the cells to lengthen and flatten. Collectively, those differentiated cells impart unique optical properties to each lens. And those crystallin-producing cells are only 1 of 150 or so differentiated cell types now present in your body.

As many experiments tell us, nearly all cells become fully differentiated without loss of genetic information. For example, John Gurdon removed the nucleus from unfertilized eggs of the African clawed frog (*Xenopus laevis*). Then he isolated intestinal cells from tadpoles of the same species. He ruptured their plasma membrane, left the nucleus and most of the cytoplasm intact, then

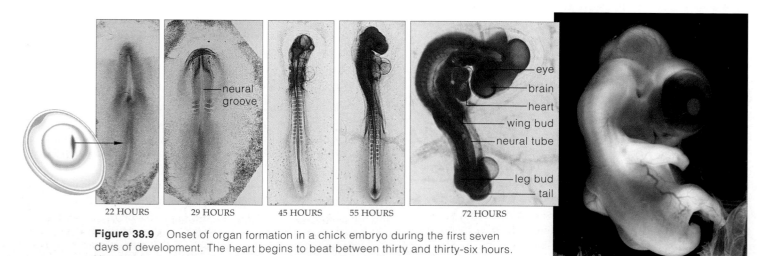

neural groove

22 HOURS   29 HOURS   45 HOURS   55 HOURS   72 HOURS

eye
brain
heart
wing bud
neural tube
leg bud
tail

168 HOURS (SEVEN DAYS OLD)

Figure 38.9 Onset of organ formation in a chick embryo during the first seven days of development. The heart begins to beat between thirty and thirty-six hours. You may have observed such embryos at the yolk surface of raw, fertilized eggs.

inserted the nucleus into the enucleated egg. In some experiments, the nucleus directed the development of a complete frog! The intestinal cell had the same number and kinds of genes as the zygote, so its nucleus had all of the genes necessary to make all of the cell types that make up a frog. Its genes had not been somehow lost when it differentiated into an intestinal cell.

Human cells that have become part of an embryonic stage still retain the capacity to give rise to an entire individual. This is what happens after the spontaneous separation of human blastomeres at the two-cell stage. The split does not result in two half-embryos, but rather in identical twins.

Morphogenesis

Morphogenesis refers to a program of orderly changes in an embryo's size, shape, and proportions, the result being specialized tissues and early organs. As part of the program, cells divide, grow, migrate, and change in size. Tissues expand and fold, and the cells in some of them die in controlled ways at prescribed locations.

Consider active cell migration. *Cells send out and use pseudopods that move them along prescribed routes.* When they reach their destination, they establish contact with cells already there. For example, forerunners of neurons interconnect this way as a nervous system is forming.

How do cells "know" where to move? They respond to adhesive cues, as when migrating Schwann cells stick to adhesion proteins on the surface of axons but not blood vessels. And they respond to chemical gradients. Their migrations may be coordinated by the synthesis, release, deposition, and removal of specific chemicals in the extracellular matrix. Adhesive cues also tell cells when to stop. Cells will migrate to regions of strongest adhesion, but once there, further migration is impeded. Section 20.8 describes the chemical-induced migrations among cells of a slime mold, *Dictyostelium discoideum*.

Also, *whole sheets of cells expand and fold inward and outward.* Microtubules lengthen and microfilament rings constrict within the cells. The assembly and disassembly of these components of the cytoskeleton are part of a controlled program of changes in cell shape.

Through such controlled, localized events, the size, shape, and proportion of body parts emerge. As one example, take a look at Figure 38.10, which shows what happens after three primary tissues form in the embryos of amphibians, reptiles, birds, and mammals. At an embryo's midline, ectodermal cells elongate to form a neural plate, the first sign that a region of ectoderm is on its way to becoming nervous tissue. Microtubules lengthen in some cells, and other cells become wedge shaped as a microfilament ring contracts and constricts each cell at one end. Collectively, the changes in shape

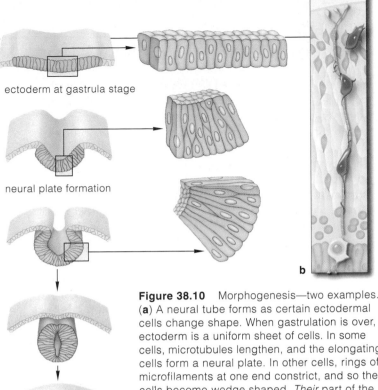

ectoderm at gastrula stage

neural plate formation

a

neural tube

b

Figure 38.10 Morphogenesis—two examples. (**a**) A neural tube forms as certain ectodermal cells change shape. When gastrulation is over, ectoderm is a uniform sheet of cells. In some cells, microtubules lengthen, and the elongating cells form a neural plate. In other cells, rings of microfilaments at one end constrict, and so the cells become wedge shaped. *Their* part of the ectodermal sheet folds over the neural plate to form the tube. (**b**) Cell migration. A nerve cell "climbs" through the developing brain to its final position, using a glial cell as its highway.

cause tissue flaps to fold over and meet at the midline, thus forming the neural tube.

As a last example, *programmed cell death helps sculpt body parts*. Cells that function for only limited periods in embryonic tissues execute themselves. By this form of cell death, called apoptosis, molecular signals from some cells activate weapons of self-destruction already stockpiled in other, target cells. Section 14.5 describes what happens to the targets at the molecular level. For now, simply think about a hand of a human embryo, which first looks like a paddle (Section 8.4). Cartilage models of the digits are forming inside it. Cells in the tissue zones between digits have receptors for certain proteins, which are produced when certain genes are expressed in the embryo. When the receptors bind those proteins, the cells commit mass suicide and the digits separate from one another. In some humans, a gene mutation blocks apoptosis and digits stay webbed.

In cell differentiation, a cell selectively uses certain genes and makes certain proteins, not found in other cell types, for distinctive cell structures, products, and functions.

Morphogenesis is a program of orderly changes in body size, shape, and proportions. It results in specialized tissues and organs at prescribed locations, at prescribed times.

Morphogenesis involves cell division, active cell migration, tissue growth and foldings, changes in cell size and shape, and programmed cell death by way of apoptosis.

PATTERN FORMATION

In some animals, such as *Drosophila*, the fate of each cell is sealed in advance by localized gene products in the cytoplasm. A different style of cell determination predominates in the embryos of vertebrates. Although each cell retains a capacity to alter its fate (it still has all the instructions for making a whole animal), once it is locked in place in an embryo, it becomes committed to starting cell lineages that will form body parts in places where we expect those parts to form. Either way, *the sculpting of nondescript clumps of cells into specialized embryonic tissues and organs normally follows an ordered, spatial pattern that completes the body plan.*

Based upon studies of gene mutations in *Drosophila* and other organisms, researchers put together a **theory of pattern formation**. Here are its key points:

1. During development, classes of master genes are activated in orderly sequence, at prescribed times.

2. Interactions among the master genes are guided by regulatory proteins. They result in the appearance of different gene products that are spatially organized relative to one another in the embryo.

3. Different genes are activated and suppressed in cells along the embryo's anterior-posterior axis and dorsal-ventral axis. Certain protein products of this selective gene expression diffuse through the embryo and create chemical gradients that help define each cell's identity.

4. Homeotic genes are a class of master genes that specify the development of specific body parts.

Genes as Master Organizers

The developmental biologist Edward Lewis discovered that the products of certain mutated genes introduce bizarre blips into pattern formation in *Drosophila*. For example, one gene mutation makes a leg grow out of the head, where an antenna should grow (Figure 38.11).

Look at the *Drosophila* zygote in Figure 38.12. A **fate map** of its surface is a map showing where each kind of differentiated cell in the adult originates. The embryo develops with a head end and a tail end. In between, a series of body segments form. Each has a gene-specified identity. Only the first segment will grow legs; the next will grow legs and wings, the next will grow legs and balancing devices, and so on.

Researchers came up with a model to explain how polarity in the egg gives rise to such segmented polarity in the adult. First, *maternal effect* genes for regulatory proteins are transcribed, translated, or both. The mRNAs and proteins that form become localized in the egg cytoplasm. When gene products appear in the zygote, they activate or inhibit *gap* genes, which map out broad body regions. Later, as one gap gene product diffuses

Figure 38.11 Experimental evidence that certain genes control the specification of body parts. In a *Drosophila* larva, a cluster of cells gives rise to antennae on the head. If the larva carries a mutated form of an *antennapedia* gene, the adult fly will have legs on its head instead.

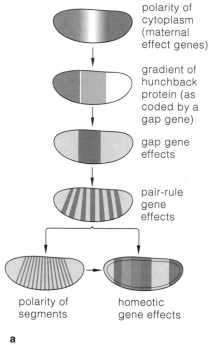

polarity of cytoplasm (maternal effect genes)

gradient of hunchback protein (as coded by a gap gene)

gap gene effects

pair-rule gene effects

polarity of segments   homeotic gene effects

a

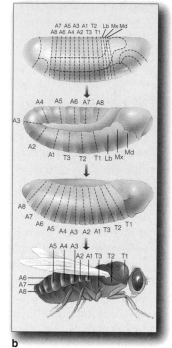

b

Figure 38.12 (a) One model of pattern formation in *Drosophila*, as described in the text. (b) Fate map of a *Drosophila* zygote. Dashed lines show regions that normally develop into body segments with specialized structures and appendages. For this organism, each segment's fate is sealed at the time of cytoplasmic localization.

through the blastocyst, it creates a chemical gradient that affects expression of other gap genes. Differences in concentrations of gap gene products activate *pair-rule* genes, the products of which accumulate in bands. Then these products activate *segment polarity* genes, which divide the embryo into segment-size units.

Interactions among products of gap, pair-rule, and segment polarity genes control expression of another class of genes—*homeotic* genes that collectively govern the developmental fate of each body segment. Mutated homeotic genes cause one body segment to look like another. For the fruit fly in Figure 38.11, two antennae should have formed on its head. But an antennapedia gene mutation led to abnormal responses to regulatory proteins. As a result, the wrong set of homeotic genes was activated, and they gave rise to a pair of legs.

Embryonic Induction

Exposure to a gene product released from one tissue can change the developmental fate of an adjacent embryonic tissue. We call this effect **embryonic induction**. Modify a developing chick wing experimentally, as in Figure 38.13, and you see examples. Mitotic cell divisions in mesoderm and ectoderm produce a bud where a chick wing will grow. The mesoderm induces the overlying ectodermal cells to lengthen and to form a ridge at the bud's apex. This apical ectodermal ridge (AER) forms on the newly developing limbs of mammals as well as birds. Surgically remove the AER before the bud grows fully, and development of that body part stops.

Rapid cell divisions in mesoderm beneath the AER generate the tissues that elongate a developing limb. As the newest cells push out others from this zone, they give rise to tissues of the limb's outermost parts. Cells pushed out early on give rise to the bones closest to the body's trunk. Cells that are pushed out much later are using genes differently. They are biochemically different and they respond to different signals. These cells give rise to the limb's outermost bones and other structures.

Slowly degradable proteins known as morphogens create key concentration gradients as they diffuse from an inducing tissue into adjoining tissues. Their signals are strongest at the start of the gradients and weaken with distance. This means cells at different positions along the gradients are exposed to different chemical information—which guides the selective expression of different genes in different parts of the embryo.

Morphogens and other inducers switch on blocks of genes in sequence. Their targets include homeotic genes, which belong to animals as evolutionarily distant as *Caenorhabditis elegans* (a roundworm) and vertebrates. As organs are first forming, the gene products interact with regulatory elements. Together, they activate and inhibit blocks of genes in similar ways among major animal groups. They direct the inductions that map out the overall body plan, including its major axes. If they fail to do so when organs are forming, major disasters follow—as when a heart gets constructed at the wrong location. Once the basic plan is set, however, the many remaining inductions will have only localized effects. If a lens fails to form, for instance, only the eye is affected.

Why do we find no more than a few dozen body plans among all animals? Apparently, the first stages of development are open to evolutionary change but early organ formation is not. We've known about the *physical* constraints (such as the surface-to-volume ratio) and *architectural* constraints (as imposed by body axes). The kinds of events you just read about indicate there also are *phyletic* constraints on change. These are imposed on each lineage by master organizer genes that operate when organs first form and that govern induction of

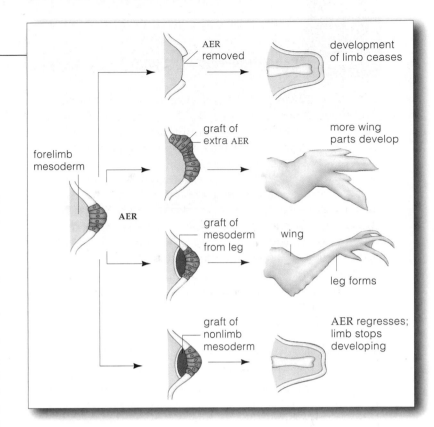

Figure 38.13 Experiments that revealed an interaction between mesoderm and ectoderm when a chick wing forms. This mesoderm induces overlying ectodermal cells to elongate and form a narrow ridge at its apex. An apical ectodermal ridge (AER) is a population of self-sustaining cells that do not mingle with the cells around them. Remove AER when a wing bud is developing, and the bud develops no more. Graft extra AER onto the bud, and the wing that forms has extra parts. Graft a bit of leg mesoderm beneath the AER, and part of a leg (with toes) develops. Graft a bit of nonlimb mesoderm below the AER, and the AER regresses and wing development stops.

the basic body plan. So once a body part has formed owing to their interactions, it's hard to start over again.

Biologists have now decoded all 19,900 genes of *C. elegans*, and about 70 percent of its genes are identical or similar to many of yours. They know how all 957 of its cells develop in sequence and interact. This worm's organ systems and yours operate in the same basic ways. In short, *C. elegans* is the first fully described organism that can serve as a model for understanding the development of organisms as complex as humans.

Pattern formation is the orderly, sequential sculpting of embryonic cells into specialized tissues and organs. Cells differentiate partly through cytoplasmic localization and, later, according to their position in the developing embryo.

Inductive interactions among classes of master genes map out the basic body plan and specify where and how body parts develop. Products of their selective expression create chemical gradients in the embryo that help seal each cell's developmental fate.

There are physical, architectural, and phyletic constraints on the evolution of animal body plans.

WWW

REPRODUCTIVE SYSTEM OF HUMAN MALES

So far, we have looked at the basic principles of animal reproduction and development. The remainder of the chapter shows how you can apply these principles to humans, starting with the male reproductive system.

As Figure 38.14 and Table 38.1 show, an adult male has a pair of gonads, of a type called **testes** (singular, testis). These are his primary reproductive organs, the equivalent of ovaries in females. Testes produce sperm and sex hormones. The hormones are central to male reproductive function. They also are necessary for the development of *secondary* sexual traits. Such traits do not play a direct role in reproduction but are distinctly associated with maleness (or femaleness). Examples are the amount of body fat, hair, and skeletal muscle.

Where Sperm Form

In an embryo that is destined to become male, a pair of testes form on the abdominal cavity wall. Before birth, the testes descend from the abdominal cavity into the scrotum, an outpouching of skin that hangs below the pelvic region. At the time of birth, testes are fully formed miniatures of the adult organs. They start producing sperm at puberty, the time of sexual maturation. Boys typically enter puberty between ages twelve and sixteen.

Figure 38.14*a* shows the scrotum's position in an adult male. If sperm cells are to develop properly, the temperature inside the scrotum must remain a few degrees cooler than the rest of the body's normal core temperature. A control mechanism operates by stimulating or inhibiting the contraction of smooth muscles in the scrotum's wall. It helps assure that the scrotum's internal temperature does not stray far from 95°F. When the air just outside the body becomes too cold, contractions draw the pouch closer to the body mass, which is warmer. When it is warmer outside, muscles relax and thereby lower the pouch.

Packed inside each testis are a great number of small, highly coiled tubes called the **seminiferous tubules**. Sperm formation begins in these tubules, in the manner described in Section 38.7.

Where Semen Forms

The sperm of every mammal travel from each testis through a series of ducts that lead to the urethra. When sperm enter the first duct, they are not quite mature. The first duct, called the epididymis, is lengthy and coiled. Secretions from glandular cells of the duct wall trigger events that put the finishing touches on sperm maturation. Fully mature sperm are stored in the last stretch of each epididymis until the time when they are ejaculated from the body.

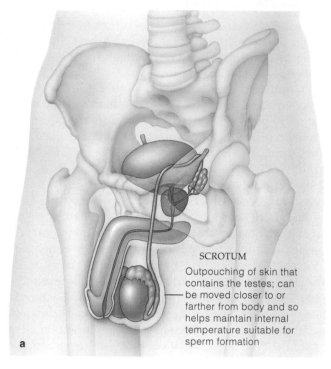

SCROTUM
Outpouching of skin that contains the testes; can be moved closer to or farther from body and so helps maintain internal temperature suitable for sperm formation

a

Figure 38.14 *Above and facing page*: (**a**) Position of the human male reproductive system relative to the pelvic girdle and urinary bladder. The midsagittal section in (**b**) shows the components of the system and lists their functions.

Table 38.1	Organs and Accessory Glands of the Male Reproductive Tract
REPRODUCTIVE ORGANS	
Testis (2)	Production of sperm, sex hormones
Epididymis (2)	Sperm maturation site and sperm storage
Vas deferens (2)	Rapid transport of sperm
Ejaculatory duct (2)	Conduction of sperm to penis
Penis	Organ of sexual intercourse
ACCESSORY GLANDS	
Seminal vesicle (2)	Secretion of large part of semen
Prostate gland	Secretion of part of semen
Bulbourethral gland (2)	Production of lubricating mucus

When a male is sexually aroused, muscle contractions in the walls of the reproductive organs propel mature sperm into and through a pair of thick-walled tubes, the vasa deferentia (singular, vas deferens). From there, contractions propel sperm through a pair of ejaculatory ducts, then on through the urethra. That last tube in the series threads through the interior of the penis, the male sex organ, and opens at its tip. The urethra is a duct that also functions in urinary excretion.

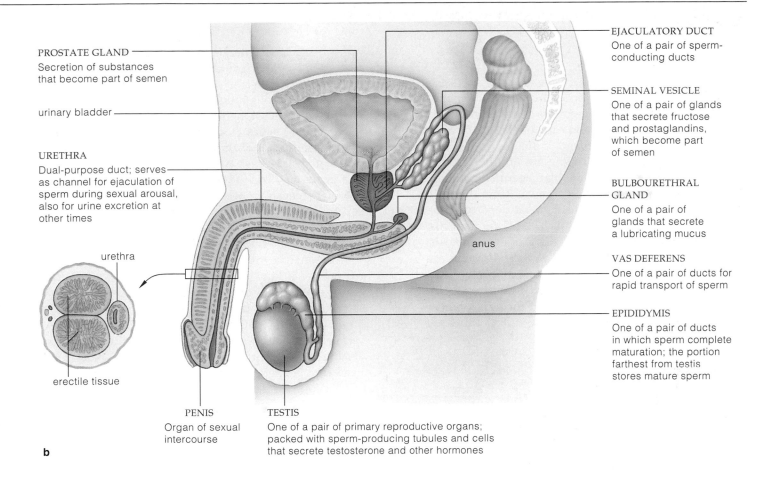

PROSTATE GLAND
Secretion of substances
that become part of semen

urinary bladder

URETHRA
Dual-purpose duct; serves
as channel for ejaculation of
sperm during sexual arousal,
also for urine excretion at
other times

urethra

erectile tissue

EJACULATORY DUCT
One of a pair of sperm-
conducting ducts

SEMINAL VESICLE
One of a pair of glands
that secrete fructose
and prostaglandins,
which become part
of semen

BULBOURETHRAL
GLAND
One of a pair of
glands that secrete
a lubricating mucus

anus

VAS DEFERENS
One of a pair of ducts for
rapid transport of sperm

EPIDIDYMIS
One of a pair of ducts
in which sperm complete
maturation; the portion
farthest from testis
stores mature sperm

PENIS
Organ of sexual
intercourse

TESTIS
One of a pair of primary reproductive organs;
packed with sperm-producing tubules and cells
that secrete testosterone and other hormones

b

Glandular secretions become mixed with sperm as they travel to the urethra. The result is semen, a thick fluid that eventually is expelled from the penis during sexual activity. Early in the formation of semen, a pair of seminal vesicles secrete fructose. The sperm use this sugar as an energy source. Seminal vesicles also secrete certain kinds of prostaglandins that can induce muscle contractions. Possibly these signaling molecules take effect during sexual activity. At that time, they might induce contractions in the female's reproductive tract and thereby assist sperm movement through it.

Secretions from a prostate gland may help buffer the acidic conditions in the female reproductive tract. (The pH of vaginal fluid is about 3.5–4.0, but sperm are able to swim more efficiently at pH 6.) Two bulbourethral glands secrete some mucus-rich fluid into the urethra when a male is sexually aroused.

Cancers of the Prostate and Testis

Until recently, cancers of the male reproductive tract didn't get much media coverage in the United States. Yet in 1996 there were an estimated 317,000 new cases of *prostate cancer,* which kills more than 41,000 older men each year. The death rate for breast cancer, which is highly publicized, is not much higher. Also, *testicular*

cancer is a frequent cause of death among young men. About 5,000 cases are diagnosed each year.

Both cancers are painless in their early stages. When they are not detected in time, they spread silently into lymph nodes of the abdomen, chest, neck, then lungs. Once the cancer metastasizes, prospects are not good. Testicular cancer, for example, kills as many as half of those stricken.

Doctors can detect prostate cancer through physical examinations and blood tests for an increase in prostate-specific antigen (PSA). Once a month from high school on, men also should examine each testis after a warm bath or shower, when scrotal muscles are relaxed. The testis should be rolled gently between the thumb and the forefinger to check for any enlargement, hardening, or lump. Such changes may or may not cause noticeable discomfort. But they must be reported so a physician can order a full examination. The treatment of testicular cancer has one of the highest success rates, provided that the cancer is caught before it starts to spread.

Human males have a pair of testes—primary reproductive organs that produce sperm and sex hormones—as well as accessory glands and ducts. The hormones influence sperm formation and the development of secondary sexual traits.

WWW

Sperm Formation

Each testis is only about 5 centimeters long, which is smaller than a golfball. Yet packed inside are 125 meters of seminiferous tubules. As many as 300 wedge-shaped lobes partition the interior, and each holds two or three coiled tubules (Figures 38.15 and 38.16).

Inside the wall of each tubule are undifferentiated cells called spermatogonia (singular, spermatogonium). Ongoing cell divisions force them away from the wall, toward the interior. During their forced departure, the cells undergo mitotic cell division. Their daughter cells, primary spermatocytes, are the ones that enter meiosis.

As Figure 38.16a indicates, the nuclear divisions are accompanied by *incomplete* cytoplasmic divisions. Thus the developing cells are interconnected by cytoplasmic bridges. Ions and molecules required for development move freely across the bridges, so successive divisions from each spermatogonium produce clones of cells that mature together. Sertoli cells, the only other type of cell in the seminiferous tubules, provide the forerunners of sperm with nourishment and molecular signals.

Secondary spermatocytes form by way of meiosis I. The chromosomes of these haploid cells are still in the duplicated state; each consists of two sister chromatids. (Here you may wish to review the general overview of spermatogenesis in Section 9.5.) The sister chromatids separate from each other at meiosis II, then daughter cells form. These are haploid spermatids that gradually develop into sperm, the male gametes.

Each mature sperm is a flagellated cell with a head region and a long tail having a core of microtubules. An enzyme-containing cap (acrosome) covers most of the head, which contains a DNA-packed nucleus (Figure 38.16b). During fertilization, enzymes released from the cap help the sperm penetrate the extracellular material around an egg. Mitochondria in a midpiece behind the head supply energy for the tail's whiplike movements.

The formation of each sperm takes nine to ten weeks. From puberty onward, human males produce sperm on a continual basis, so that many millions of cells are in different stages of development on any given day.

Hormonal Controls

Coordinated secretions of LH, FSH, and testosterone govern reproductive function in males. Take a look at Figure 38.15b. It shows the Leydig cells located between the lobes in the testes. Leydig cells secrete **testosterone**, a steroid hormone that helps govern the growth, form, and functions of the male reproductive tract. Besides having the main role in sperm formation, it stimulates sexual behavior and aggressive behavior. It promotes the development of secondary sexual traits in males,

seminal vesicle

prostate gland

bulbo-urethral gland

urethra

vas deferens

epididymis

penis

testis

SEMINIFEROUS TUBULE

a

wall of seminiferous tubule   Leydig cells between tubules

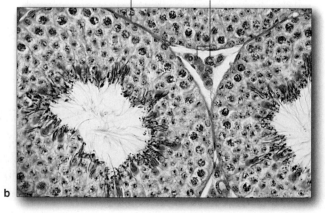

b

Figure 38.15   (**a**) Male reproductive tract, posterior view. The arrows show the route that sperm take before ejaculation from a sexually aroused male. (**b**) Light micrograph of cells inside three adjacent seminiferous tubules, cross-section. Leydig cells occupy tissue spaces between the tubules.

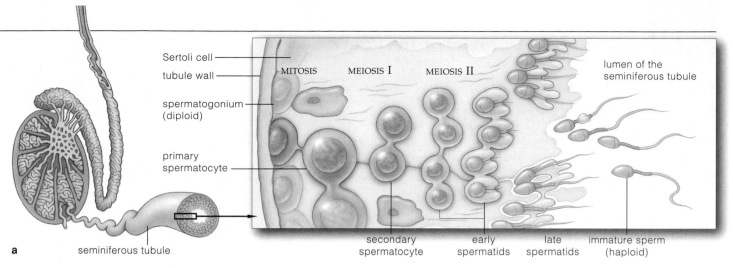

Sertoli cell
tubule wall
spermatogonium (diploid)
primary spermatocyte

MITOSIS MEIOSIS I MEIOSIS II

lumen of the seminiferous tubule

secondary spermatocyte
early spermatids
late spermatids
immature sperm (haploid)

a seminiferous tubule

Figure 38.16 (**a**) Sperm formation. The process starts with a spermatogonium (a diploid germ cell). Mitosis, meiosis, and incomplete divisions of the cytoplasm result in a clone of immature haploid cells that differentiate and develop into mature sperm. (**b**) Structure of a mature sperm from a human male.

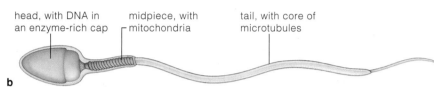

head, with DNA in an enzyme-rich cap
midpiece, with mitochondria
tail, with core of microtubules

b

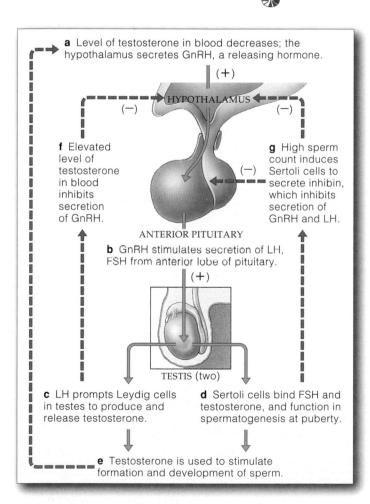

a Level of testosterone in blood decreases; the hypothalamus secretes GnRH, a releasing hormone.

(+)

HYPOTHALAMUS

(−)   (−)

f Elevated level of testosterone in blood inhibits secretion of GnRH.

g High sperm count induces Sertoli cells to secrete inhibin, which inhibits secretion of GnRH and LH.

(−)

ANTERIOR PITUITARY

b GnRH stimulates secretion of LH, FSH from anterior lobe of pituitary.

(+)

TESTIS (two)

c LH prompts Leydig cells in testes to produce and release testosterone.

d Sertoli cells bind FSH and testosterone, and function in spermatogenesis at puberty.

e Testosterone is used to stimulate formation and development of sperm.

Figure 38.17 Negative feedback loops to the hypothalamus and to the anterior lobe of the pituitary gland from the testes. Through these loops, excess testosterone production shuts off the mechanisms leading to its production. This helps maintain the testosterone level in amounts required for sperm formation.

including an obvious increase in the growth of facial hair and a deepening of the voice at puberty.

LH and **FSH**, remember, are secreted by the anterior lobe of the pituitary gland (Section 31.3). Initially, both hormones were named for their effects in females. (One abbreviation stands for *Luteinizing Hormone*, the other for *Follicle-Stimulating Hormone*.) Later on, researchers discovered that the molecular structure of LH and FSH is identical in both males and females.

The hypothalamus, a part of the forebrain, controls sperm formation by controlling secretion of LH, FSH, and testosterone. As Figure 38.17 shows, in response to low blood levels of testosterone and other factors, the hypothalamus secretes GnRH. This releasing hormone stimulates the anterior lobe to step up the release of LH and FSH, which have targets in the testes.

LH prods Leydig cells to secrete testosterone, which helps stimulate sperm formation and development. Sertoli cells have receptors for FSH, which is necessary to start up spermatogenesis at puberty. We don't know if FSH is also necessary for the normal functioning of mature human testes.

Figure 38.17 also shows how feedback loops to the hypothalamus lead to decreased testosterone secretion and sperm formation. An elevated testosterone level in blood slows down the release of GnRH. Also, when the sperm count is high, Sertoli cells release inhibin. This protein hormone acts on the hypothalamus and pituitary to cut back on the release of GnRH and FSH.

Sperm formation depends on the hormones LH, FSH, and testosterone. Negative feedback loops from the testes to the hypothalamus and pituitary gland control their secretion.

REPRODUCTIVE SYSTEM OF HUMAN FEMALES

The Reproductive Organs

We turn now to the reproductive system of a human female. Figure 38.18 shows its components, and Table 38.2 summarizes their functions. The female's *primary* reproductive organs, her pair of ovaries, produce eggs and secrete sex hormones. Her immature eggs are called oocytes. After each oocyte is released from an ovary, it passes through the entrance of an adjacent oviduct. A female has a pair of oviducts. Each serves as a channel to her **uterus**, a hollow, pear-shaped organ in which the embryo can grow and develop.

A thick layer of smooth muscle (the myometrium) makes up most of the uterine wall. As you will see, the wall's inner lining, called the **endometrium**, is central to embryonic development. It consists of connective tissues, glands, and blood vessels. The narrowed-down portion of the uterus is the cervix. A muscular tube, the vagina, extends from the cervix to the surface of the body. The vagina receives sperm and functions as part of the birth canal.

At the body surface are external genitals (vulva), which include organs for sexual stimulation. Outermost are a pair of fat-padded skin folds (the labia majora). They enclose a smaller pair of skin folds (labia minora) that are highly vascularized but have no fatty tissue. The smaller folds partly enclose the clitoris, a sex organ that is sensitive to stimulation. At the body's surface, the opening of the urethra is positioned about midway between the clitoris and the vaginal opening.

Overview of the Menstrual Cycle

Most mammalian females follow an *estrous* cycle. They are fertile and in heat (sexually receptive to males) only at certain times of year. By contrast, female primates, including humans, follow a **menstrual cycle**. They are fertile intermittently, on a cyclic basis, and heat is not synchronized with their fertile periods. Said another way, female primates of reproductive age can become pregnant only at certain times of year, but they may be receptive to sex at any time.

Briefly, during a menstrual cycle, an oocyte matures *(immature eggs)* and escapes from an ovary. Also during each cycle, the endometrium becomes primed to receive and nourish a forthcoming embryo *if* a sperm penetrates the oocyte and fertilization occurs. However, if the oocyte does not become fertilized, then four to six tablespoons of blood-rich fluid from the uterus will start flowing out through the vaginal canal. Such a recurring blood flow is called menstruation. It means "there is no embryo at this time," and it marks the first day of a new cycle. The uterine nest is being sloughed off and is about to be constructed once again.

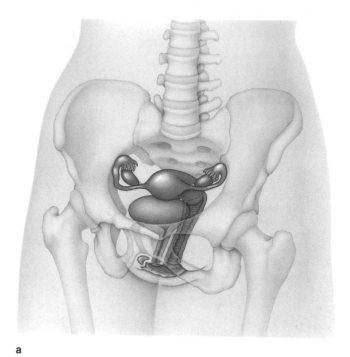

a

Figure 38.18 *Above and facing page*: (**a**) Position of the human female reproductive system relative to the pelvic girdle and urinary bladder. The midsagittal section in (**b**) shows the components of the system and lists their functions.

Table 38.2	Female Reproductive Organs
Ovaries	Oocyte production and maturation, sex hormone production
Oviducts	Ducts for conducting oocyte from ovary to uterus; fertilization normally occurs here
Uterus	Chamber in which new individual develops
Cervix	Secretion of mucus that enhances sperm movement into uterus and (after fertilization) reduces embryo's risk of bacterial infection
Vagina	Organ of sexual intercourse; birth canal

The events just sketched out proceed through three phases. The cycle starts with a *follicular* phase. This is the time of menstruation, endometrial breakdown and rebuilding, and oocyte maturation. The next phase is restricted to the release of an oocyte from the ovary. We call it **ovulation**. During the *luteal* phase of the cycle, an endocrine structure (corpus luteum) forms, and the endometrium is primed for pregnancy (Table 38.3).

All three phases are governed by feedback loops to the hypothalamus and pituitary gland from the ovaries. FSH and LH promote cyclic changes in the ovaries. As

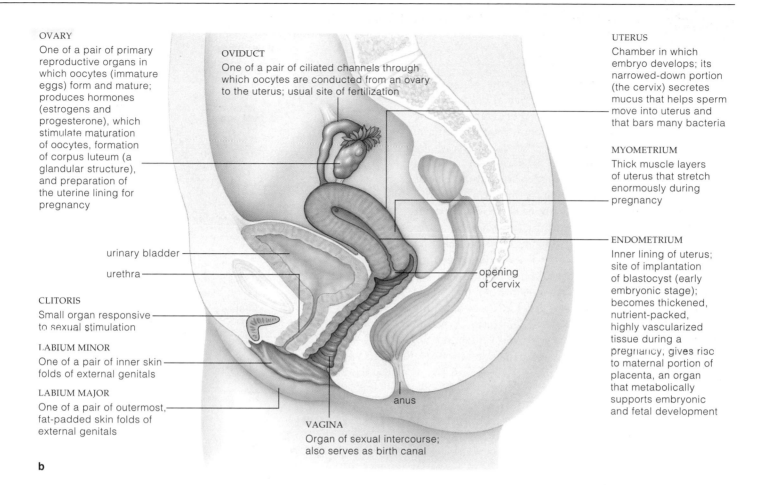

OVARY
One of a pair of primary reproductive organs in which oocytes (immature eggs) form and mature; produces hormones (estrogens and progesterone), which stimulate maturation of oocytes, formation of corpus luteum (a glandular structure), and preparation of the uterine lining for pregnancy

OVIDUCT
One of a pair of ciliated channels through which oocytes are conducted from an ovary to the uterus; usual site of fertilization

UTERUS
Chamber in which embryo develops; its narrowed-down portion (the cervix) secretes mucus that helps sperm move into uterus and that bars many bacteria

MYOMETRIUM
Thick muscle layers of uterus that stretch enormously during pregnancy

urinary bladder

urethra

opening of cervix

ENDOMETRIUM
Inner lining of uterus; site of implantation of blastocyst (early embryonic stage); becomes thickened, nutrient-packed, highly vascularized tissue during a pregnancy; gives rise to maternal portion of placenta, an organ that metabolically supports embryonic and fetal development

CLITORIS
Small organ responsive to sexual stimulation

LABIUM MINOR
One of a pair of inner skin folds of external genitals

LABIUM MAJOR
One of a pair of outermost, fat-padded skin folds of external genitals

anus

VAGINA
Organ of sexual intercourse; also serves as birth canal

b

Table 38.3	Events of the Menstrual Cycle	
Phase	Events	Days of Cycle*
Follicular phase	Menstruation; endometrium breaks down	1–5
	Follicle matures in ovary; endometrium rebuilds	6–13
Ovulation	Oocyte released from ovary	14
Luteal phase	Corpus luteum forms, secretes progesterone; the endometrium thickens and develops	15–28

* Assuming a 28-day cycle.

you will see, FSH and LH also stimulate the ovaries to secrete sex hormones—**estrogens** and **progesterone**—to promote the cyclic changes in the endometrium.

A human female's menstrual cycles start between ages ten and sixteen. Each cycle lasts for about twenty-eight days, but this is merely the average. It runs longer for some women and shorter for others. The menstrual cycles continue until a woman is in her late forties or early fifties, when her supply of eggs is dwindling and hormonal secretions slow down. This is the onset of *menopause*, the twilight of reproductive capacity.

You may have heard about *endometriosis*, a condition that arises when endometrial tissue abnormally spreads and grows outside the uterus. Endometrial scar tissue may form on ovaries or oviducts and lead to infertility. In the United States, 10 million women may be affected annually. Possibly the condition arises when menstrual flow backs up through the oviducts and spills into the pelvic cavity. Or perhaps some embryonic cells became positioned in the wrong place before birth and were stimulated to grow during puberty, when sex hormones became active. Whatever the case, estrogen still acts on cells in the mislocated tissue. The resulting symptoms include pain during menstruation, sex, or urination.

Ovaries, the female primary reproductive organs, produce oocytes (immature eggs) and sex hormones. Endometrium lines the uterus, a chamber in which embryos develop.

Secretion of the sex hormones estrogen and progesterone is coordinated on a cyclic basis through the reproductive years.

A menstrual cycle starts with menstruation, breakdown and rebuilding of the endometrium, and maturation of an oocyte. After an oocyte escapes from an ovary (ovulation), the cycle ends with the formation of a corpus luteum and an endometrium that has become primed for pregnancy.

WWW

Cyclic Changes in the Ovary

Take a moment to review Section 9.5, the generalized picture of meiosis in an oocyte. A normal baby girl has about 2 million primary oocytes in her ovaries. By the time she is seven years old, only about 300,000 remain; her body resorbed the rest. Her primary oocytes have already entered meiosis I, but this nuclear division process was arrested in a genetically programmed way. Meiosis resumes in one oocyte at a time, starting with the first menstrual cycle. Only about 400 to 500 oocytes will be released during the reproductive years.

In Figure 38.19, sketch *a* is a primary oocyte near an ovary's surface. A cell layer (granulosa cells) surrounds and nourishes it. We call the primary oocyte and the cell layer around it a follicle. At the start of a menstrual cycle, the hypothalamus is secreting GnRH in amounts that cause the anterior pituitary to step up *its* secretion of FSH and LH. The blood concentration of these two hormones increases. That increase causes the follicle to grow (Figure 38.20).

The oocyte starts to increase in size, and more layers of cells form around it. Glycoprotein deposits accumulate between the oocyte and the layers. As they do, they widen the space between them. In time, all the deposits form the zona pellucida, a noncellular coating around the oocyte.

FSH and LH stimulate cells outside the zona pellucida to secrete estrogens. An estrogen-containing fluid accumulates in the follicle, and estrogen levels in blood begin to increase. About eight to ten hours before being released from the ovary, the oocyte completes meiosis I. And then its cytoplasm divides, forming two cells.

One cell, the secondary oocyte, ends up with nearly all of the cytoplasm. The other cell is the first of three polar bodies. The meiotic parceling of the chromosomes among all four cells gives the secondary oocyte a haploid chromosome number—which is the exact number required for gametes and for sexual reproduction.

About halfway through the menstrual cycle, the pituitary gland detects the rise in the blood level of estrogens. It responds with a brief outpouring of LH. The LH surge causes rapid vascular changes that make the follicle swell quickly. The surge also induces enzymes to digest the bulging follicle wall. The weakened wall ruptures. Fluid escapes and carries the secondary oocyte with it (Figure 38.19).

Thus, the midcycle surge of LH triggers ovulation—the release of a secondary oocyte from the ovary.

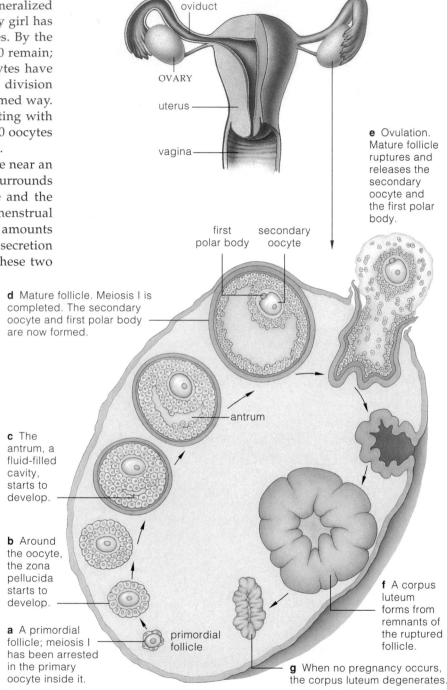

e Ovulation. Mature follicle ruptures and releases the secondary oocyte and the first polar body.

d Mature follicle. Meiosis I is completed. The secondary oocyte and first polar body are now formed.

first polar body

secondary oocyte

antrum

c The antrum, a fluid-filled cavity, starts to develop.

b Around the oocyte, the zona pellucida starts to develop.

a A primordial follicle; meiosis I has been arrested in the primary oocyte inside it.

primordial follicle

f A corpus luteum forms from remnants of the ruptured follicle.

g When no pregnancy occurs, the corpus luteum degenerates.

oviduct

OVARY

uterus

vagina

Figure 38.19 Cyclic events in a human ovary, cross-section.

A follicle stays in the same place in an ovary all through the menstrual cycle. It does not move around as in this diagram, which only shows the *sequence* in which events occur. In the cycle's first phase, the follicle grows and matures. At ovulation, the second phase, the mature follicle ruptures and releases a secondary oocyte. During the third phase, a corpus luteum forms from the follicle's remnants. If the woman does not get pregnant during the cycle, the corpus luteum self-destructs.

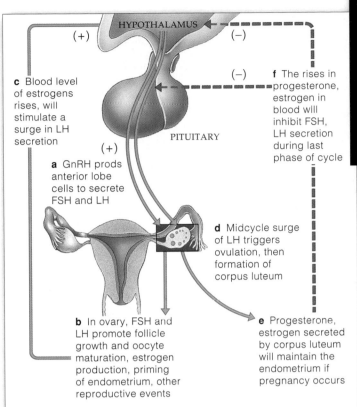

c Blood level of estrogens rises, will stimulate a surge in LH secretion

a GnRH prods anterior lobe cells to secrete FSH and LH

HYPOTHALAMUS

PITUITARY

(+) (+) (−) (−)

f The rises in progesterone, estrogen in blood will inhibit FSH, LH secretion during last phase of cycle

d Midcycle surge of LH triggers ovulation, then formation of corpus luteum

b In ovary, FSH and LH promote follicle growth and oocyte maturation, estrogen production, priming of endometrium, other reproductive events

e Progesterone, estrogen secreted by corpus luteum will maintain the endometrium if pregnancy occurs

secondary oocyte surrounded by follicle cells

surface of ovary

Figure 38.20 Feedback control of hormonal secretion during a menstrual cycle. A positive feedback loop from an ovary to the hypothalamus causes a surge in LH secretion. The surge triggers ovulation. The light micrograph shows a secondary oocyte being released from an ovary at this time. Afterward, negative feedback loops to the hypothalamus and pituitary inhibit FSH secretion. They prevent another follicle from maturing until the cycle is over.

Cyclic Changes in the Uterus

The estrogens released early in the menstrual cycle also help pave the way for a possible pregnancy. Estrogens stimulate growth of the endometrium and its glands. Just before the midcycle surge of LH, follicle cells start to secrete some progesterone as well as estrogens. Blood vessels grow rapidly in the thickened endometrium. At ovulation, the estrogens act on tissue around the cervical canal, the narrowed portion of the uterus that leads to the vagina. The cervix starts to secrete large quantities of a thin, clear mucus that is an ideal medium for sperm to swim through.

The midcycle surge of LH that triggers ovulation also induces the formation of a **corpus luteum** in the ovary. This yellowish glandular structure forms by cell differentiation among granulosa cells left behind in the follicle. (Its name means yellow body.) Secretions from the corpus luteum influence the rest of the cycle.

The corpus luteum secretes progesterone and some estrogen. Progesterone prepares the reproductive tract for the arrival of a blastocyst, the type of blastula that forms from a fertilized mammalian egg (Section 38.3). For example, this hormone makes cervical mucus turn thick and sticky. The mucus may keep normal bacterial inhabitants of the vagina out of the uterus. Progesterone also will maintain the endometrium during pregnancy.

A corpus luteum persists for about twelve days. All the while, the hypothalamus is calling for minimal FSH secretion, which stops other follicles from developing. If a blastocyst does not burrow into the endometrium, the corpus luteum will self-destruct during the last days of the cycle. It will secrete certain prostaglandins that apparently can disrupt its own functioning.

After this, progesterone and estrogen levels in blood decline rapidly, and so the endometrium starts to break down. Deprived of oxygen and nutrients, the lining's blood vessels constrict and tissues die. Blood escapes as the walls of weakened capillaries start rupturing. Blood as well as the sloughed endometrial tissues make up a menstrual flow, which continues for three to six days. Then the cycle begins anew, with rising estrogen levels stimulating the repair and growth of the endometrium.

By menopause, the supply of oocytes is dwindling, hormone secretions slow down, and in time menstrual cycles—and fertility—will be over. Eggs that are still to be released late in a woman's life are at some risk of alterations in chromosome number or structure when meiosis resumes. A newborn with Down syndrome, as described in Section 11.10, is one possible outcome.

During a menstrual cycle, FSH and LH stimulate growth of an ovarian follicle (a primary oocyte and the surrounding layer of cells). The first meiotic cell division results in a secondary oocyte and the first polar body.

A midcycle surge of LH triggers ovulation, the release of the secondary oocyte and the polar body from the ovary.

Early on, estrogens call for endometrial repair and growth. Then estrogens and progesterone prepare the endometrium and other parts of the reproductive tract for pregnancy.

WWW

VISUAL SUMMARY OF THE MENSTRUAL CYCLE

By now, you probably have come to the conclusion that the menstrual cycle is not a simple tune on a biological banjo. It is a full-blown hormonal symphony!

Before continuing with your reading, take a moment to review Figure 38.21. It correlates the cyclic events in the ovary and uterus with all the coordinated changes in hormone levels that bring about those events. The illustration may leave you with a better understanding of what goes on.

Figure 38.21 Changes in the ovary and uterus, as correlated with changing hormone levels during each turn of the menstrual cycle. *Green* arrows indicate which hormones dominate the cycle's first phase (the time when the follicle matures), then the second phase (when the corpus luteum forms).

(**a–c**) FSH and LH secretions bring about the changes in ovarian structure and function. (**d,e**) Estrogen and progesterone secretions from the ovary stimulate the changes in the endometrium.

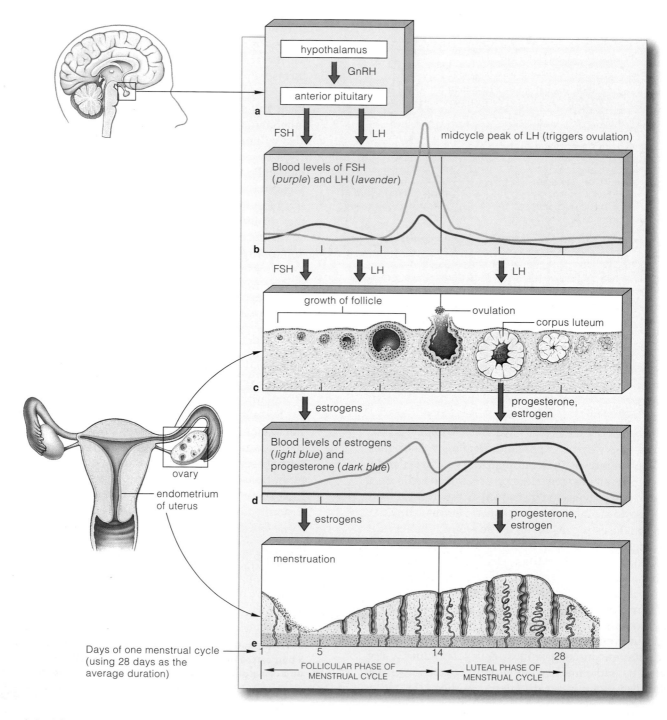

Sexual Intercourse

Suppose the secondary oocyte is on its way down an oviduct when a female and male are engaged in sexual intercourse (coitus). The male sex act requires erection, whereby a normally limp penis stiffens and lengthens; and ejaculation, a forceful expulsion of semen into the urethra and out from the penis. As Figure 38.14 shows, the penis incorporates cylinders of spongy tissue. Many friction-activated sensory receptors occur on the glans penis, the mushroom-shaped tip of this organ. In sexually unaroused males, large blood vessels leading into the cylinders are vasoconstricted. In aroused males, they vasodilate. Blood flows into the penis faster than it flows out, so blood collects in the spongy tissue. The resulting engorgement stiffens and lengthens the organ, this being helpful for penetration into the vaginal canal.

During coitus, pelvic thrusts stimulate the penis, the female's clitoris, and the vaginal wall. The mechanical stimulation triggers involuntary contractions in the male's reproductive tract. These rapidly force sperm out of each epididymis. They force the contents of the seminal vesicles and the prostate gland into the urethra. The substances mix together and form semen. When semen is ejaculated, a sphincter closes the neck of the bladder and prevents urination. During coitus, the semen is released into the vagina.

Emotional intensity, hard breathing, heart pounding, as well as generalized skeletal muscle contractions accompany the rhythmic throbbing of the pelvic muscles. At *orgasm,* the end of the sex act, strong sensations of physical release, warmth, and relaxation dominate. Similar sensations typify female orgasm. It is a common misconception that a female cannot become pregnant if she doesn't reach orgasm. Don't believe it.

Fertilization

Now sperm are in the vagina. A single ejaculation can put 150 million to 350 million there. If they arrive a few days before or after ovulation or anytime in between, fertilization may be the outcome. Less than thirty minutes after ejaculation, muscle contractions move the sperm deeper into the female's reproductive tract. Only a few hundred sperm actually reach the upper portion of the oviduct, where fertilization usually takes place.

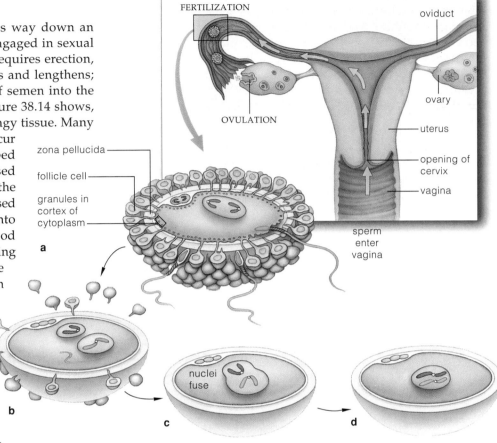

Figure 38.22 Fertilization. (**a**) Many sperm surround a secondary oocyte. Acrosomal enzymes clear a path through the zona pellucida. (**b**) When a sperm does penetrate the secondary oocyte, granules in the egg cortex release substances that make the zona pellucida impenetrable to other sperm. Penetration also stimulates meiosis II of the oocyte's nucleus. (**c**) The sperm's tail degenerates and its nucleus enlarges and fuses with the oocyte nucleus. (**d**) With fusion, fertilization is over. The zygote has formed.

The stunning micrograph that opens Unit II shows living sperm around a secondary oocyte. When sperm contact an oocyte, they release enzymes that clear a path through the zona pellucida (Figure 38.22). Although many sperm might get this far, usually only one fuses with the oocyte. Only its nucleus and centrioles do not degenerate in the oocyte's cytoplasm. Penetration induces the secondary oocyte and the first polar body to finish meiosis II. There are now three polar bodies and a mature egg, or **ovum** (plural, ova). As the sperm and egg nuclei fuse, their chromosomes restore the diploid number for a brand new zygote.

The intense physiological events that accompany coitus have one function: to put sperm on a collision course with an egg. Fertilization is over with the fusion of a sperm nucleus and egg nucleus, which results in a diploid zygote.

WWW

FORMATION OF THE EARLY EMBRYO

Pregnancy lasts an average of thirty-eight weeks from the time of fertilization. It takes about two weeks for a blastocyst to form. The time span from the third to the end of the eighth week is the *embryonic* period, when the major organ systems form. When it ends, the new individual has distinctly human features and is called a **fetus**. In the *fetal* period, from the start of the ninth week until birth, organs enlarge and become specialized.

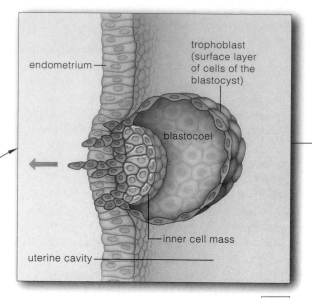

endometrium

trophoblast (surface layer of cells of the blastocyst)

blastocoel

inner cell mass

uterine cavity

d DAY 5. A blastocoel (a fluid-filled cavity) forms in the morula as a result of secretions from the surface cells. By the thirty-two-cell stage, cells of an inner cell mass are already differentiating. They will give rise to the embryo proper. This embryonic stage is called a blastocyst.

c DAY 4. By 96 hours, there is a ball of sixteen to thirty-two cells that is shaped like a mulberry. It is a morula (after *morum*, Latin for mulberry). Cells of the surface layer will function in implantation and will give rise to a membrane, the chorion.

b DAY 3. After the third cleavage, the cells suddenly huddle together into a compacted ball, which becomes stabilized by numerous tight junctions among the outer cells. Gap junctions form among the interior cells and enhance intercellular communication.

a DAYS 1–2. Cleavage begins within 24 hours after fertilization. The first cleavage furrow extends between the two polar bodies. Subsequent cuts are rotational, so the resulting cells are not symmetrically arranged (compare Section 38.3). Until the eight-cell stage forms, the cells are loosely arranged, with considerable space between them.

inner cell mass

FERTILIZATION

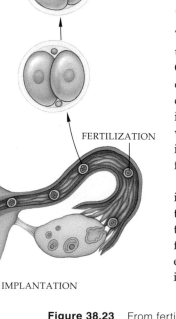

oviduct   uterus

ovary

endometrium

IMPLANTATION

e DAYS 6–7. Surface cells of the blastocyst attach to the endometrium and start to burrow into it. Implantation is under way.

actual size

We typically call the first three months of pregnancy the *first* trimester. The *second* trimester extends from the start of the fourth month to the end of the sixth. The *third* trimester extends from the seventh month until birth. Beginning with Figure 38.23, the next series of illustrations shows the characteristic features of the new individual at progressive stages of development.

Cleavage and Implantation

Three to four days after fertilization, the zygote is already undergoing cleavage as it tumbles through the oviduct. Genes are already being expressed; the early divisions depend on their products. At the eight-cell stage, the cells huddle into a compact ball. By the fifth day, there is a surface layer of cells (a trophoblast), a cavity filled with their secretions (a blastocoel), and a tiny cluster of interior cells (an inner cell mass). These are the defining features of a human blastocyst (Figure 38.23d).

Six or seven days after fertilization, **implantation** is under way. By this process, the blastocyst adheres to the uterine lining, some of its cells send out projections that invade the mother's tissues, and connections start forming that will metabolically support the developing embryo through the months ahead. While the invasion is proceeding, the inner cell mass develops into two

Figure 38.23 From fertilization through implantation. Cleavage produces a blastocyst. Within the blastocyst, the inner cell mass gives rise to the disk-shaped early embryo. Three of the extraembryonic membranes (amnion, chorion, and yolk sac) start forming. The fourth extraembryonic membrane (allantois) forms after the blastocyst is implanted.

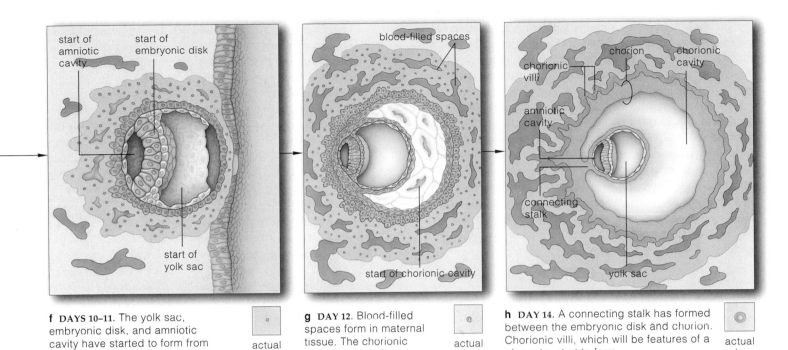

f DAYS 10–11. The yolk sac, embryonic disk, and amniotic cavity have started to form from parts of the blastocyst.

actual size

g DAY 12. Blood-filled spaces form in maternal tissue. The chorionic cavity starts to form.

actual size

h DAY 14. A connecting stalk has formed between the embryonic disk and chorion. Chorionic villi, which will be features of a placenta, start to form.

actual size

cell layers of a flattened and somewhat circular shape. The two layers make up the embryonic disk—and in short order they will give rise to the embryo proper.

Extraembryonic Membranes

As implantation progresses, membranes start to form outside the embryo. First a fluid-filled *amniotic* cavity opens up between the embryonic disk and part of the blastocyst's surface (Figure 38.23f). Then cells migrate around the wall of the cavity and form the **amnion**, a membrane that will enclose the embryo. Fluid inside the cavity will function as a buoyant cradle where the embryo can grow, move freely, and be protected from abrupt temperature changes and mechanical impacts.

While the amnion forms, other cells migrate around the inner wall of the blastocyst's first cavity. They form a lining that becomes the **yolk sac**. This extraembryonic membrane speaks of the evolutionary heritage of land vertebrates (Sections 24.7 and 24.8). For most animals that produce shelled eggs, the sac holds nutritive yolk. In humans, part of the yolk sac becomes a site of blood cell formation, and part will give rise to germ cells, the forerunners of gametes.

Before the blastocyst is fully implanted, spaces open in maternal tissues and fill with blood seeping in from ruptured capillaries. Inside the blastocyst, another cavity opens around the amnion and yolk sac. Now fingerlike projections start to form on the cavity's lining, which is the **chorion**. This new membrane will become part of a spongy, blood-engorged tissue called the placenta.

After the blastocyst is finally implanted, another extraembryonic membrane will form as an outpouching of the yolk sac. This third membrane will become the **allantois**. An allantois functions differently in different animal groups. Among reptiles, birds, and some of the mammals, it has roles in respiration and in the storage of metabolic wastes. In humans, the urinary bladder as well as blood vessels for the placenta form from it.

One more point should be made here. Cells of the blastocyst secrete the hormone HCG (*Human Chorionic Gonadotropin*), which stimulates the corpus luteum to keep on secreting progesterone and estrogen. Thus the blastocyst itself prevents menstrual flow and works to avoid being sloughed off until the placenta takes over the task, some eleven weeks later. By the start of the third week, HCG can be detected in the mother's blood or urine. At-home *pregnancy tests* use a treated "dip-stick" that changes color when HCG is present in urine.

A human blastocyst is composed of a surface layer of cells around a fluid-filled cavity (blastocoel) and an inner cell mass, which will give rise to the embryo proper.

Six or seven days after fertilization, the blastocyst implants itself in the endometrium. Now projections from its surface invade maternal tissues, and connections start to form that in time will metabolically support the developing embryo.

Some parts of the blastocyst give rise to an amnion, yolk sac, chorion, and allantois. These extraembryonic membranes serve different functions. Together they are vital for the structural and functional development of the embryo.

EMERGENCE OF THE VERTEBRATE BODY PLAN

By the time a woman has missed her first menstrual period, cleavage is completed and gastrulation is under way. Gastrulation, recall, is a stage when extensive cell divisions, migrations, and rearrangements result in the primary tissue layers (Section 38.4).

By now, the embryonic disk is surrounded by the amnion and chorion, except at the point where a stalk connects it to the chorion's inner wall. Before then, the inner cell mass had behaved *as if* it were still perched on the large ball of yolk that evolved in the reptilian ancestors of mammals. (Hence the flattened, two-layer embryonic disk, which also forms in reptiles and birds.) Cell divisions and migrations of one layer produced the lining of the yolk sac. The other layer now starts to generate the embryo proper.

For example, by the eighteenth day, two neural folds have appeared on the embryonic disk (Figure 38.24*b*). They will merge to form a neural tube. Early in the third week, a sausage-shaped outpouching forms on the yolk sac. It is the allantois (Greek *allas*, meaning sausage). In humans, remember, the allantois only takes part in the formation of a urinary bladder and blood vessels for the placenta. Some mesoderm folds into a tube that becomes the notochord. A human notochord is only a structural framework; bony segments of the vertebral column soon form around it. Toward the end of the third week, some mesoderm gives rise to the **somites**. These paired segments are the embryonic source of most bones, of skeletal muscles of the head and trunk, as well as of the dermis overlying these regions.

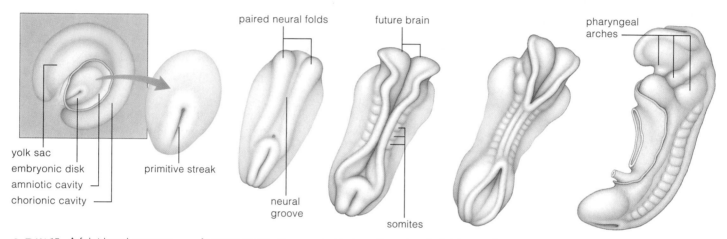

a DAY 15. A faint band appears around a depression along the axis of the embryonic disk. This is the primitive streak, and it marks the onset of gastrulation in vertebrate embryos.

b DAYS 18–23. Organs start to form through cell divisions, cell migrations, tissue folding, and other events of morphogenesis. Neural folds will merge to form the neural tube. Somites (bumps of mesoderm) appear near the embryo's dorsal surface. They will give rise to most of the skeleton's axial portion, skeletal muscles, and much of the dermis.

c DAYS 24–25. By now, some embryonic cells have given rise to pharyngeal arches. These will contribute to the formation of the face, neck, mouth, nasal cavities, larynx, and pharynx.

Figure 38.24 Hallmarks of the embryonic period of humans and other vertebrates. A primitive streak, then a notochord form. Neural folds, somites, and pharyngeal arches form later. (**a**,**b**) Dorsal views of the embryo's back. (**c**) Side view. Compare the diagrams with the photographs in Figure 38.26.

Through cell divisions and migrations, the midline of that layer thickens faintly around a depression at its surface. This "primitive streak" lengthens and thickens further the next day. It marks the onset of gastrulation (Figure 38.24*a*). It also defines the anterior-posterior axis of the embryo and, in time, its bilateral symmetry. Here, cells migrating inward give rise to endoderm and mesoderm. Now pattern formation begins, as described in Section 38.5. Interactions among classes of master genes map out the basic body plan characteristic of all vertebrates. Through embryonic inductions, specialized tissues and organs start forming in orderly sequence in prescribed parts of the embryo.

Pharyngeal arches start to form. They will contribute to the face, neck, mouth, nose, larynx, and pharynx. Small spaces open up in parts of the mesoderm. In time, all the spaces will interconnect as the coelomic cavity.

During the third week after fertilization, a time when the woman has missed her first menstrual period, the basic vertebrate body plan emerges in the new individual.

A primitive streak, neural tube, somites, and pharyngeal arches form during the embryonic period of all vertebrates. Formation of the primitive streak establishes the body's anterior-posterior axis and its bilateral symmetry.

WWW

WHY IS THE PLACENTA SO IMPORTANT?

Even before the onset of the embryonic period, the extraembryonic membranes have been collaborating with the uterus to sustain the embryo's rapid growth. By the third week, tiny fingerlike projections from the chorion have grown profusely into the maternal blood that has pooled in endometrial spaces. The projections, the chorionic villi, enhance the exchange of substances between the mother and the new individual. They are functional components of the **placenta**.

The placenta, a blood-engorged organ, consists of endometrial tissue and extraembryonic membranes. At full term, it will make up a fourth of the inner surface of the uterus (Figure 38.25).

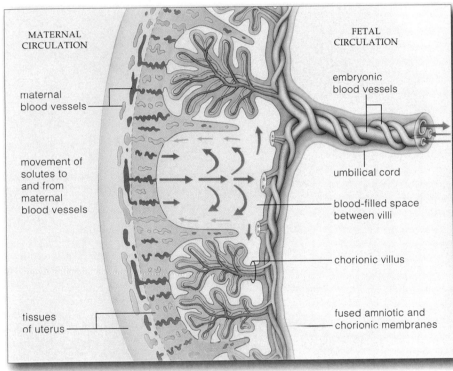

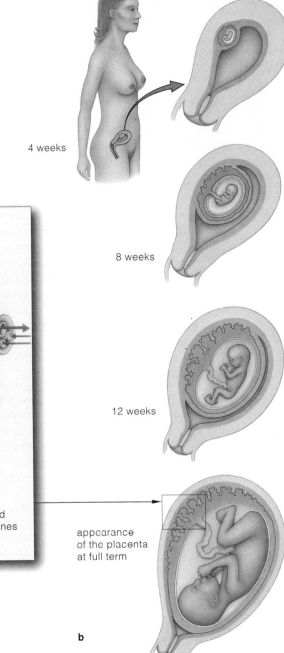

4 weeks

8 weeks

12 weeks

appearance of the placenta at full term

b

Figure 38.25 (**a**) Relationship between fetal and maternal blood circulation in a full-term placenta (**b**). Blood vessels extend from the fetus, through the umbilical cord, and into chorionic villi. Maternal blood spurts into spaces between the villi, but the two bloodstreams do not intermingle. Oxygen, carbon dioxide, and other small solutes diffuse across the placental membrane surface.

A placenta is the body's way of sustaining the new individual while allowing its blood vessels to develop apart from the mother's. Oxygen and nutrients diffuse out of the maternal blood vessels, across the placenta's blood-filled spaces, then into embryonic blood vessels. (The vessels converge in the umbilical cord, the lifeline between the placenta and the new individual.) Carbon dioxide and other wastes diffuse in the other direction. The mother's lungs and kidneys quickly dispose of them.

After the third month, the placenta secretes progesterone and estrogens—and so maintains the uterine lining.

The placenta is a blood-engorged organ of endometrial and extraembryonic membranes. It permits exchanges between the mother and the new individual without intermingling their bloodstreams. Thus it sustains the individual and allows its blood vessels to develop apart from the mother's.

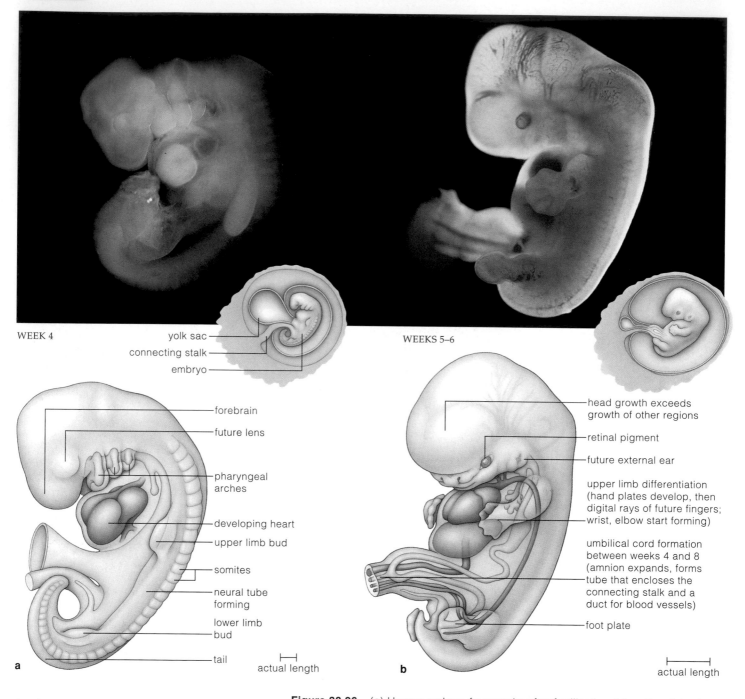

WEEK 4

yolk sac
connecting stalk
embryo

WEEKS 5–6

forebrain

future lens

pharyngeal arches

developing heart

upper limb bud

somites

neural tube forming

lower limb bud

tail

actual length

a

head growth exceeds growth of other regions

retinal pigment

future external ear

upper limb differentiation (hand plates develop, then digital rays of future fingers; wrist, elbow start forming)

umbilical cord formation between weeks 4 and 8 (amnion expands, forms tube that encloses the connecting stalk and a duct for blood vessels)

foot plate

actual length

b

As the fourth week of the embryonic period draws to a close, the embryo has grown to 500 times its original size. The placenta has been sustaining the growth spurt, but now the pace slows as details of organs fill in. Limbs form. Fingers and toes develop from embryonic paddles. The umbilical cord also develops, and the circulatory system becomes intricate. Growth of the all-important head now surpasses that of any other body region (Figure 38.26). The embryonic period ends after the eighth week is over. No longer is

Figure 38.26 (**a**) Human embryo four weeks after fertilization. Like all vertebrates, it has a tail and pharyngeal arches. (**b**) Embryo at five to six weeks. (**c**) Embryo at the boundary between the embryonic and fetal periods. It now has human features. It floats in amniotic fluid. The chorion covers the amniotic sac but has been pulled aside here. (**d**) Fetus at sixteen weeks. Movements begin as nerves make functional connections with forming muscles. Legs kick, arms wave, fingers grasp, the mouth puckers. These reflex actions will be vital skills in the world outside the uterus.

the embryo merely "a vertebrate." By now its features clearly define it as a human fetus.

During the second trimester, the fetus is moving its facial muscles. It frowns; it squints. It busily practices

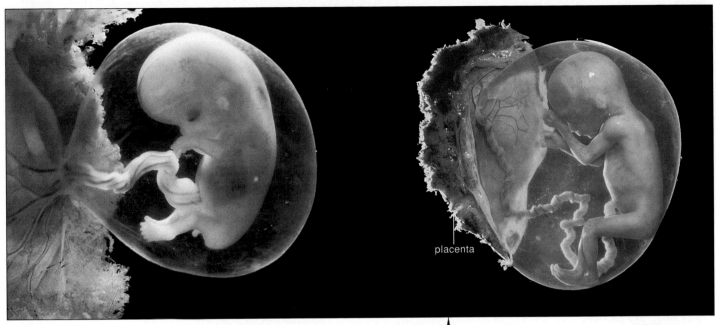

WEEK 8

final week of embryonic
period; embryo looks
distinctly human
compared to other
vertebrate embryos

upper and lower limbs well
formed; fingers and then
toes have separated

primordial tissues of
all internal, external
structures now developed

tail has become stubby

c   actual length

WEEK 16 ⟶
Length: 16 centimeters
          (6.4 inches)
Weight: 200 grams
          (7 ounces)

WEEK 29
Length: 27.5 centimeters
          (11 inches)
Weight: 1,300 grams
          (46 ounces)

WEEK 38 (full term) ⟶
Length: 50 centimeters
          (20 inches)
Weight: 3,400 grams
          (7.5 pounds)

During fetal period, length
measurement extends
from crown to heel (for
embryos, it is the longest
measurable dimension, as
from crown to rump).

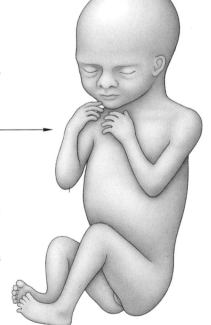

d

the sucking reflex, as shown in the next section. Now
the mother can easily sense movements of the fetal
arms and legs. When the fetus is five months old, she
can hear its heart through a stethoscope positioned on
her abdomen. Soft, fuzzy hair (the lanugo) covers the
fetal body. The skin is wrinkled, reddish, and protected
from abrasion by a thick, cheesy coating. In the sixth
month, delicate eyelids and eyelashes form. During the
seventh month, the eyes open.

A fetus born prematurely (before 22 weeks) cannot
survive. The situation also is grave for births before 28

weeks, mainly because the lungs have not developed
sufficiently. The risks start to drop after this. But a fetus
born before the optimal birthing time (about 38 weeks
after the estimated time of fertilization) still has trouble
breathing and maintaining a normal core temperature
even with the best medical care. By 36 weeks, however,
the survival rate is 95 percent.

**In the fetal period, primary tissues that formed in the early
embryo become sculpted into distinctly human features.**

MOTHER AS PROVIDER, PROTECTOR, POTENTIAL THREAT

A woman who decides to become pregnant is committing a large part of her body's resources and functions to the development of a new individual. From fertilization until birth, her future child is absolutely at the mercy of her diet, health habits, and life-style (Figure 38.27).

SOME NUTRITIONAL CONSIDERATIONS How does a pregnant woman best provide nutrients for the embryo, then the fetus? The same balanced diet that is good for her should provide her future child with all of the required carbohydrates, lipids, and proteins. (Chapter 36 is one starting point for an understanding of human nutritional requirements.) The woman's own demands for vitamins and minerals increases during pregnancy. The placenta absorbs enough for her embryo from the bloodstream, except for folate (folic acid). She can reduce the risk that her embryo will develop severe neural tube defects by taking more B-complex vitamins (under supervision by her doctor) before conception and during early pregnancy. Besides this, one study showed that women who smoked while pregnant had depressed blood levels of vitamin C even when their vitamin C intake was identical to that of a control group. *And so did their fetuses.* Smoking may affect utilization of other nutrients as well.

A pregnant woman also must eat enough so that her body weight increases by between 20 and 25 pounds, on the average. If her weight gain is a great deal lower than that, she is stacking the deck against the fetus. Compared to newborns of normal weight, significantly underweight newborns have more postdelivery complications. They are also at greater risk of having impaired mental functions later in life.

As birth approaches, a fetus makes greater nutritional demands of the mother. Clearly her diet will profoundly influence the remaining developmental events. The brain, like most of the other fetal organs, is especially vulnerable in the weeks just before and after birth, when it undergoes its greatest expansion. By now, all of its neurons have formed. Poor nutrition now will have repercussions on intelligence and other neural functions later in life.

RISK OF INFECTIONS The antibodies circulating in a pregnant woman's blood continually move across the placenta. They protect the new individual from all but

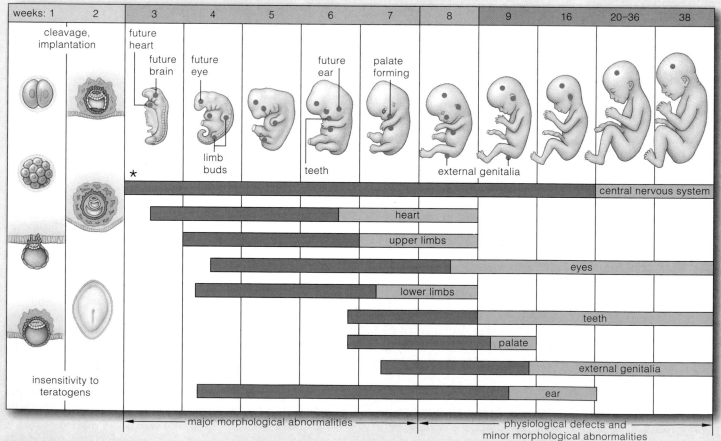

Figure 38.27 Sensitivity to teratogens during pregnancy. *Teratogens* are drugs and any other environmental factors that may induce deformities in the embryo or fetus. They usually have no effect before the onset of organ formation. After that, they can block or abnormally stimulate growth, tissue remodeling, and tissue resorption.

★ *dark blue* denotes highly sensitive period; *light blue* denotes less severe sensitivity to teratogens

the most serious bacterial infections. Some virus-induced diseases can be dangerous during the first six weeks after fertilization, which is a critical time of organ formation. Suppose a pregnant woman contracts *rubella* (German measles) during this critical period. There is a 50 percent chance that some organs of her embryo will not form properly. For example, if she becomes infected when embryonic ears are forming, her newborn may be deaf. If she becomes infected during or after the fourth month of pregnancy, the disease will have no notable effect. However, a woman can avoid this risk entirely, because vaccination before pregnancy can prevent rubella.

EFFECTS OF PRESCRIPTION DRUGS   A pregnant woman absolutely should not take any drugs unless she is under close medical supervision. Think about what happened when the tranquilizer *thalidomide* was being routinely prescribed in Europe. Women who had used thalidomide during the first trimester gave birth to infants who were either missing arms and legs or had grossly deformed ones. As soon as its connection with the deformities was apparent, thalidomide was withdrawn from the market.

However, other tranquilizers, sedatives, and barbiturates are still being prescribed. There is a risk that they may cause similar, although less severe, damage. Even certain *anti-acne drugs* increase the risk of facial and cranial deformities. Or consider two overprescribed antibiotics. One of these, tetracycline, yellows teeth. Streptomycin causes hearing problems and may adversely affect the nervous system.

EFFECTS OF ALCOHOL   As a fetus grows, its physiology becomes increasingly like its mother's. Because the fetus is so small and grows so rapidly, alcohol is more harmful to it—and alcohol passes freely across the placenta.

Drinking alcohol while pregnant invites *fetal alcohol syndrome*, or FAS. Symptoms include reduced brain and head size, mental retardation, facial deformities, poor growth and coordination, and often heart defects (Figure 38.28b). One in every 750 newborns shows the abnormal features. The symptoms are irreversible; children affected by FAS never "catch up," physically or mentally.

Many researchers suspect there is no "safe" drinking level; any alcohol may harm the fetus. More doctors are now urging near- or total abstinence during pregnancy.

EFFECTS OF COCAINE   A pregnant woman who uses cocaine, crack especially, disrupts the nervous system of her future child as well as her own. Here you may wish to

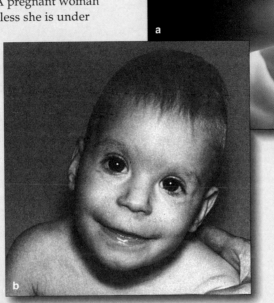

Figure 38.28   (**a**) The fetus at eighteen weeks. (**b**) An infant with fetal alcohol syndrome, or FAS. Obvious symptoms are low and prominent ears, improperly formed cheekbones, and an abnormally wide, smooth upper lip. Growth problems and abnormalities of the nervous system can be expected. This disorder currently affects about 1 in 750 newborns in the United States.

read again the start of Chapter 29, which describes the kind of future that offspring of a crack addict face.

EFFECTS OF CIGARETTE SMOKE   Cigarette smoke impairs fetal growth and development. Also, a long-term study at Toronto's Hospital for Sick Children showed that toxic elements in tobacco accumulate even in the fetuses of pregnant nonsmokers who are exposed to *secondhand smoke* at home or work.

Daily smoking during pregnancy leads to underweight newborns even if the woman's weight, nutritional status, and all other relevant variables match those of pregnant nonsmokers. Smoking has other effects. In Great Britain, all infants born the same week were tracked for seven years. Those of smokers were smaller, died of more post-delivery complications, and had twice as many heart abnormalities. At age seven, they were nearly half a year behind children of nonsmokers in average "reading age."

The mechanisms by which smoking affects a fetus are not known. Its demonstrated effects are evidence that the placenta, marvelous structure that it is, cannot prevent all assaults on the fetus that the human mind can dream up.

Giving Birth

On average, pregnancy ends thirty-eight weeks after fertilization, give or take a few weeks. We call the birth process *labor* (or delivery). Labor involves dilation of the cervical canal, so the fetus can move out from the uterus, through the vagina, and out into the world. It also requires strong uterine contractions as the driving force behind the expulsion.

In the last trimester, mild uterine contractions begin. The cervix softens as its connective tissues loosen. Relaxin, a peptide hormone secreted from the corpus luteum and placenta, may bring about the softening. Relaxin also induces the connections between pelvic bones to loosen up. Meanwhile the fetus "drops," or is shifted downward, usually with its head in contact with the cervix (Figure 38.29). A *breech birth* is the likely outcome when any part of the fetus other than the head becomes positioned near the birth canal first.

Rhythmic, usually painless contractions herald the onset of labor. They increase in frequency and intensity over the next two to eighteen hours.

Just before birth, the amnion typically ruptures, and "water" (amniotic fluid) rushes from the vagina. Usually contractions expel the fetus less than an hour after the cervix has fully dilated. Fifteen to thirty minutes later, contractions make the placenta detach from the uterus; it is expelled as the "afterbirth." The contractions also constrict blood vessels at the placental attachment site and prevent hemorrhaging. Someone ties and cuts the umbilical cord. A few days after it shrivels, the stump is the newborn's navel. Once the lifeline to the mother is severed, the newborn embarks on its course of post-embryonic development, starting with an extended time of dependency and learning that is typical of primates.

Many new mothers commonly sink into postpartum depression, the *afterbaby blues*. Corticotropin releasing hormone (CRH) produced by the placenta may have a role in this. Its level in blood rises by as much as three times during pregnancy and may influence the timing of labor. Possibly cortisol, a stress hormone, is secreted in response to elevated CRH levels to help the mother cope with extraordinary stresses that pregnancy and labor place on her body. Once the placenta is expelled, her CRH levels crash to levels typical of some clinical depressions. Afterbaby blues continue until secretion of CRH from the hypothalamus returns to normal.

Nourishing the Newborn

Survival of the newborn requires an ongoing supply of milk or its nutritional equivalent. Milk production, or *lactation,* occurs in mammary glands in the mother's breasts. Before pregnancy, the breasts consist mainly of

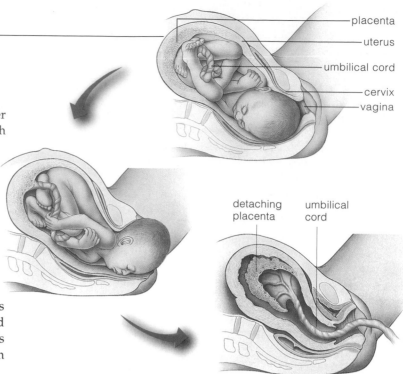

Figure 38.29   Expulsion of a human fetus during the process of birth. Contractions result in the expulsion of the afterbirth (the placenta, tissue fluid, and blood).

adipose tissue and an undeveloped duct system (Figure 38.30a). Their size depends on how much fat they hold, not on milk-producing ability. During pregnancy, they respond to estrogen and progesterone, and a glandular system of milk production develops (Figure 38.30b).

For the first few days after birth, mammary glands make a fluid rich in proteins and lactose. The anterior pituitary secretes prolactin, a hormone that induces the synthesis of the enzymes required for milk production (Section 31.3). As a newborn suckles, the pituitary also releases oxytocin. This hormone triggers contractions that force milk into breast tissue ducts and that shrink the uterus back to its previous size.

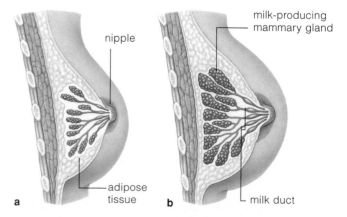

Figure 38.30   (**a**) Breast of a woman who is not pregnant. (**b**) Breast of a lactating woman.

Regarding Breast Cancer

On average, each year 100,000 women develop *breast cancer*. Obesity, high cholesterol, and high estrogen levels contribute to the cancerous transformation. The chances of cure are excellent with early detection and treatment. Once a month, about a week after she has menstruated, a woman should examine her breasts. For recommended examination procedures, she can contact her physician or the American Cancer Society (check a local telephone directory or Web site http://www.cancer.org).

Postnatal Development and Aging

After birth, a new individual grows and develops into an adult, the mature form of the species. Its lifespan is divided into *prenatal* (before birth) and *postnatal* (after birth) stages. Figure 38.31 is an example of proportional changes that occur in the body as the stages unfold. Postnatal growth is most rapid between ages thirteen and nineteen. Then, sex hormone secretions step up and bring about secondary sexual traits and sexual maturity. Not until adulthood are bones fully mature. Normally, body tissues are maintained in peak condition during early adulthood. As the years pass, it is more difficult to maintain and repair existing tissues, and gradually the body deteriorates. Through processes collectively called **aging**, cell structure and function start to break down, and this leads to structural changes and gradual loss of body functions. All animals that show extensive cell differentiation undergo aging and eventual death.

No one knows what causes aging, but research gives us interesting things to think about. For example, years ago, Paul Moorhead and Leonard Hayflick cultured normal human embryonic cells. All of the cell lines divided about fifty times, then the entire populations died off. Hayflick took some cultured cells partway through the divisions and froze them for several years. Afterward, he thawed the cells and placed them in a culture medium. The cells proceeded to complete the cycle of fifty doublings—then all died on schedule.

Such experiments suggest that normal types of cells have a limited division potential. That is, mitosis may be genetically programmed to decline at a certain time in the life cycle. This is certainly true of neurons. They generally stop undergoing mitotic cell division after birth, then deteriorate over time. But does shutdown of mitosis *cause* aging or is it a *result* of aging?

DNA's capacity for self-repair is gradually lost as mutations accumulate. Free radicals also contribute to aging. (These rogue molecular fragments, which have an unpaired electron, can avidly attack DNA and thereby alter its structure and functions.) Whether by chemical attacks or mutations, *changes in DNA compromise the*

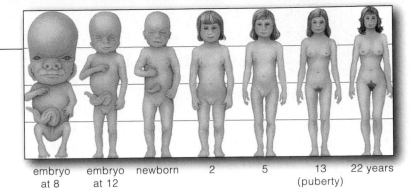

embryo at 8 weeks   embryo at 12 weeks   newborn   2   5   13 (puberty)   22 years

Figure 38.31 Observable, proportional changes in the human body during prenatal and postnatal growth. Changes in overall physical appearance are gradual but quite noticeable until the teenage years. For example, the head becomes proportionally smaller, compared to what it was during the embryonic period. The legs become longer, and the trunk becomes shorter.

production of functional enzymes and other gene products required for normal life processes.

For example, researchers have correlated *Werners' syndrome*, an aging disorder, with mutation of the gene coding for helicase. DNA replication and repair cannot occur without this enzyme. As another example, three mutated genes have been correlated with *Alzheimer's disease*, which affects people late in life. Apparently the normal gene products are part of a toxic waste removal system in the brain; they clear away protein fragments that can damage brain cells. In the United States, more than 4 million elderly people have been diagnosed with this aging disorder.

Or think about collagen, a structural component of bone and other connective tissues throughout the body. Collagen influences exchanges of materials between the cytoplasm and the extracellular fluid. If something shuts down or mutates the collagen-encoding genes, the missing or altered gene product will affect the flow of oxygen, nutrients, hormones, and other important substances to and from living cells throughout the body. Skin, bones, and many other tissues will deteriorate.

Similarly, the genes for membrane proteins that are self markers might also undergo mutation. If the self markers change, then T lymphocytes might perceive the body's own cells as foreign and attack them. If such autoimmune responses increase as a person ages, they certainly might invite greater vulnerability to disease and stress—a common problem among the elderly.

What can counteract some of the common effects of aging? Adhering to a low-fat diet and a low-impact aerobic exercise program throughout adulthood helps. Consistent exercise also improves cardiovascular and respiratory functioning. It minimizes bone loss in older adults by helping to maintain bone density.

The human life cycle flows naturally from the time of birth, growth, and development, to production of the individual's own offspring, and on through aging to the time of death.

Some Ethical Considerations

The transformation of a zygote into an adult of intricate detail raises profound questions. *When does development begin?* As you have seen, major developmental events unfold even before fertilization. *When does life begin?* During her lifetime, a woman can produce as many as 500 eggs, all of which are alive. During one ejaculation, a man can release a quarter of a billion sperm, which are alive. Even before sperm and egg merge by chance and establish the genetic makeup of a new individual, they are as much alive as any other form of life. It is scarcely tenable, then, to say life begins at fertilization. *Life began billions of years ago—and every gamete, every zygote, every sexually mature individual is but a fleeting stage in the continuation of that beginning.*

This greater perspective on life cannot diminish the meaning of conception. It is no small thing to entrust a new individual with the gift of life, wrapped in the unique evolutionary threads of our species and handed down through an immense sweep of time.

Yet how can we reconcile the marvel of individual birth with growing awareness of the astounding birth rate for our whole species? In 1998, more than 15,000 newborns entered the world every hour. By the time you go to bed tonight, there will be about 364,409 more newborns on Earth than there were last night at that hour. In three months there will be 33,000,000 more—about as many people as there are now in the entire state of California. *Within three months.*

Worldwide, human population growth is rapidly outstripping resources. Each year, many millions face the horrors of starvation. Living as we do on one of the most productive continents on Earth, few of us know what it means to give birth to a child, to give it the gift of life, and have no food to keep it alive.

And how can we reconcile the marvel of birth with the confusion that surrounds unwanted pregnancies? Even highly developed countries do not have adequate educational programs concerning fertility. And many people are not inclined to exercise control. Each year, there are 800,000 unintended teenage pregnancies in the United States. (Many parents actually encourage early boy-girl relationships without thinking through the risks of premarital intercourse and unplanned pregnancy. Advice is often condensed to a terse "Don't do it. But if you do it, be careful!") And each year, there are more than 1 million abortions among all age groups.

The motivation to engage in sex has been evolving for more than 500 million years. And a few centuries of moral and ecological arguments for its suppression have not stopped all that many unwanted pregnancies. Besides this, complex social factors have contributed to a population growth rate that is out of control.

How will we reconcile our biological past and the need for a stabilized cultural present? Whether and how human fertility is to be controlled is one of the most volatile issues of our time. We will return to this issue in the next chapter, in the context of principles that govern the growth and stability of populations. Here, we briefly consider some control options.

Birth Control Options

The most effective method of birth control is complete *abstinence*, no sexual intercourse whatsoever. Data show that it is unrealistic to expect many people to practice it.

A less reliable variation of abstinence is the *rhythm method.* The idea is to avoid intercourse in the woman's fertile period, starting a few days before ovulation and ending a few days after. A woman identifies and tracks her fertile period. She also keeps records of the length of her menstrual cycles, takes her temperature each morning when she wakes up, or both. (The body's core temperature rises by one-half to one degree just before a fertile period.) But ovulation may not be regular, and miscalculations are frequent. Also, sperm deposited in the vagina a few days before ovulation may survive until ovulation. The rhythm method *is* inexpensive; it costs nothing after you buy a thermometer. It does not require fittings and periodic medical checkups. But its practitioners run a risk of pregnancy (Figure 38.32).

Withdrawal, or removing the penis from the vagina before ejaculation, dates back at least to biblical times. But withdrawal requires very strong willpower, and the practice may fail anyway. Fluid released from the penis just before ejaculation may contain some sperm.

Douching (rinsing the vagina with a chemical right after intercourse) is next to useless. Sperm move past the cervix and out of reach of the douche within ninety seconds after ejaculation.

Controlling fertility by surgical intervention is less chancy. In *vasectomy*, a physician makes a tiny incision in a man's scrotum, then severs and ties off each vas deferens. The operation takes only twenty minutes and requires only a local anesthetic. After the operation, sperm cannot leave the testes and cannot be present in semen. So far, there is no firm evidence that vasectomy disrupts hormonal functions or adversely affects sexual activity. Vasectomies can be reversed. But half of those who submit to surgery later develop antibodies against sperm and may not be able to regain fertility.

Females may have a *tubal ligation.* By this surgical intervention, oviducts are cauterized or cut and tied off, usually in a hospital. When the operation is performed correctly, tubal ligation is the most effective means of birth control. A few women have recurring pain in the pelvic area. The operation sometimes can be reversed.

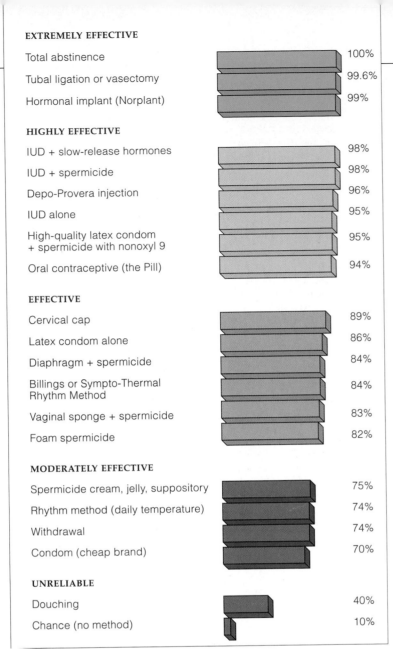

EXTREMELY EFFECTIVE

Total abstinence	100%
Tubal ligation or vasectomy	99.6%
Hormonal implant (Norplant)	99%

HIGHLY EFFECTIVE

IUD + slow-release hormones	98%
IUD + spermicide	98%
Depo-Provera injection	96%
IUD alone	95%
High-quality latex condom + spermicide with nonoxyl 9	95%
Oral contraceptive (the Pill)	94%

EFFECTIVE

Cervical cap	89%
Latex condom alone	86%
Diaphragm + spermicide	84%
Billings or Sympto-Thermal Rhythm Method	84%
Vaginal sponge + spermicide	83%
Foam spermicide	82%

MODERATELY EFFECTIVE

Spermicide cream, jelly, suppository	75%
Rhythm method (daily temperature)	74%
Withdrawal	74%
Condom (cheap brand)	70%

UNRELIABLE

Douching	40%
Chance (no method)	10%

Figure 38.32 Comparison of the effectiveness of methods of contraception in the United States. Percentages reflect the number of unplanned pregnancies per 100 couples who used only that method of birth control for a year. For example, "94% effectiveness" for oral contraceptives (the Pill) means that 6 of every 100 women will still become pregnant, on the average.

Less drastic methods of controlling fertility involve physical or chemical barriers that prevent sperm from entering the uterus and oviducts. *Spermicidal foam* and *spermicidal jelly* are toxic to sperm. They are transferred from an applicator into the vagina before intercourse. These products aren't always reliable unless used with another device, such as a diaphragm or condom. *IUDs* are coils inserted into the uterus by a physician. They sometimes invite pelvic inflammatory disease.

A *diaphragm* is a flexible, dome-shaped device. It is inserted into the vagina and positioned over the cervix before intercourse. It is relatively effective when fitted initially by a doctor, used with foam or jelly before

each sexual contact, inserted correctly each time, and left in place for a prescribed length of time.

Good brands of *condoms*—thin, tight-fitting sheaths worn over the penis during intercourse—are up to 95 percent effective if used with a spermicide. Only latex condoms offer protection against sexually transmitted diseases (Section 38.19). However, condoms often tear and leak, at which time they become absolutely useless.

The *birth control pill* is an oral contraceptive made of synthetic estrogens and progestins (progesterone-like hormones). Taken daily except for the last five days of a menstrual cycle, it suppresses oocyte maturation and ovulation. Often it corrects erratic menstrual cycles and reduces cramps, but in some women it causes nausea, weight gain, tissue swelling, and headaches. With more than 50 million women taking it, "the Pill" is the most-used fertility control method. It is 94 percent effective. Earlier formulations were linked to blood clots, high blood pressure, and maybe breast cancer. Lower doses of newer formulations might lessen the risk of breast and endometrial cancer. The possibility of a correlation between the Pill and breast cancer is still under study.

Progestin injections or implants inhibit ovulation. A *Depo-Provera* injection works for three months and is 96 percent effective. *Norplant* (six rods implanted under the skin) works for five years and is 99 percent effective. Both may cause sporadic, heavy bleeding, and doctors have some trouble surgically removing Norplant rods.

A pregnancy test doesn't register positive until after implantation. According to one view, a woman is not pregnant until that time, so *morning-after pills* intercept pregnancy. Actually, such pills work up to seventy-two hours after unprotected intercourse by interfering with the hormones that control events between ovulation and implantation. One is Preven, a kit with high doses of birth control pills that suppress ovulation and block corpus luteum function. RU-486, more commonly used as an abortion pill, also can be used as a morning-after pill. It can block fertilization and progesterone-dependent implantation. It also may induce abortion when taken within seven weeks of implantation (progesterone also is essential to maintain pregnancy). As is the case with oral contraceptives, RU-486 may trigger elevated blood pressure, blood clots, or breast cancer in some women. At this writing, RU-486 is available in Europe, but its use in the United States is controversial. By contrast, Preven had been used previously as a contraceptive, and its repackaging as a morning-after pill has not been a subject of controversy. It gained FDA approval in 1998.

Whether and how human fertility is to be controlled is a volatile issue. It centers on reconciling our biological past with seriously divergent views about our cultural present.

WWW

At some point in their life, at least one of every four people in the United States who engage in sexual intercourse will probably be infected by the pathogens that cause **sexually transmitted diseases**, or **STDs**. Tens of millions are already infected; STDs have reached epidemic proportions. Two-thirds are under age twenty-five; one-fourth are teenagers. Women and children are hardest hit. Worse yet, antibiotic-resistant strains of bacteria are on the rise, and some viral diseases simply cannot be cured.

Urban poverty, prostitution, intravenous drug abuse, and sex-for-drugs are fanning the epidemic. Yet STDs also are rampant in high schools and colleges, where students too often think, "It can't happen to me." They reject the idea that no sex—abstinence—is the only safe sex. In one poll of high school students, two-thirds of the respondents said they don't use condoms. More than 40 percent of them have two or more sex partners.

The economics of this health problem are staggering. In 1993, the Centers for Disease Control tallied the annual cost of treating the most prevalent STDs: *Herpes*, $759 million; gonorrhea, $1 billion; chlamydial infection, $2.4 billion; and pelvic inflammatory disease, $4.2 billion. This does not include the accelerating cost of treating patients with AIDS. In many developing countries, AIDS alone threatens to overwhelm health-care delivery systems and to unravel decades of economic progress.

The social consequences are sobering. Mothers bestow a chlamydial infection on 1 of every 20 newborns in the United States alone. They bestow type II *Herpes* virus on 1 in 10,000 newborns; one-half of the infected babies die, and one-fourth have severe neural defects. Every year, 1 million women develop pelvic inflammatory disease, and 100,000 to 150,000 become infertile.

AIDS   Someone can become infected by HIV, the human immunodeficiency virus, and not even know it. However, the infection marks the start of a titanic battle that the immune system almost certainly will lose (Section 34.11). At first there may be no outward symptoms. Five to ten years later, a set of chronic disorders develops. They are evidence of *AIDS* (acquired immune deficiency syndrome). When the immune system finally does give up, the stage is set for opportunistic infections. Normally harmless, resident bacteria are the first to take advantage of the lowered resistance. Dangerous pathogens also take their toll. In time, the immune-compromised person simply is overwhelmed.

Most commonly, HIV spreads through vaginal, anal, and oral intercourse and by IV drug users. Most infections occur by the transfer of blood, semen, urine, or vaginal

secretions between people. Cuts or abrasions on the penis, vagina, rectum, and possibly oral membranes serve as HIV's entrances to the internal environment of a new host.

Today there is no vaccine against HIV and no effective treatment for AIDS. If you get it, you die. *There is no cure.*

HIV was not identified until 1981, but it was present in some parts of Central Africa for at least several decades before that. In the 1970s and early 1980s, it spread to the United States and other developed countries where most of those initially infected were male homosexuals. Today, many in the heterosexual population are infected or at risk.

Free or low-cost, confidential testing for HIV exposure is available at public health facilities and at many doctor's offices. People who are worried should know there may be a time lag between their first exposure and the first test to come out positive. It takes a few weeks to six months or more before detectable amounts of antibodies will form in response to infection. (The presence of antibodies in the blood indicates exposure to the virus.) Anyone who tests positive is capable of spreading the virus.

Public education programs attempt to stop the spread of HIV. Most health-care workers advocate safe sex, yet there is great confusion about what "safe" means. Many advocate the use of high-quality latex condoms *and* a spermicide that contains nonoxynol-9 to help block the transmission of the virus. Even then, there is a slight risk. As an unfortunate couple learned in 1997, no one should participate in open-mouthed kissing with someone who tests positive for HIV. Caressing is not risky *if* there are no lesions or cuts where body fluids that harbor the virus can enter the body. Pronounced lesions caused by other sexually transmitted diseases may increase susceptibility to HIV infection.

In sum, AIDS reached epidemic proportions mainly for three reasons. First, it took a while to discover that the virus can travel in semen, blood, and vaginal fluid and that *behavioral* controls can limit its spread. Second, it took time to develop ways to test symptom-free carriers, who infect others. Third, many still don't realize the medical, economic, and social consequences affect everyone.

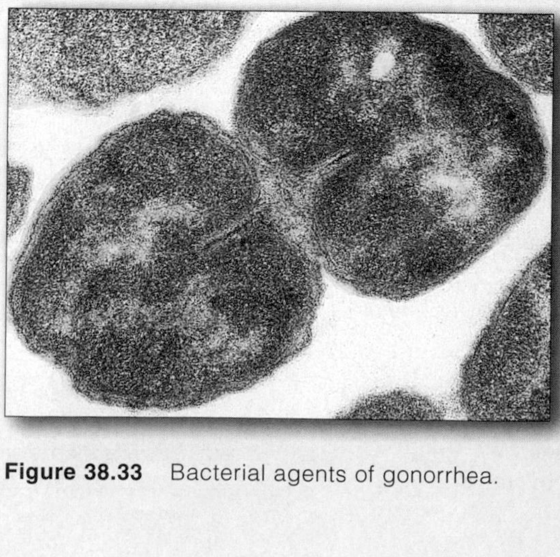

Figure 38.33   Bacterial agents of gonorrhea.

GONORRHEA   During intercourse, *Neisseria gonorrhoeae* (Figure 38.33) can enter the body at mucous membranes of the urethra, cervix, and anal canal. This bacterium causes *gonorrhea*. An infected female may notice a slight vaginal discharge or burning sensation while she is urinating. If the infection spreads to her oviducts, it may induce severe cramps, fever, vomiting, and scarring, which may cause sterility. Males have more obvious symptoms. Within a week of infection, the penis discharges yellow pus. Urinating is more frequent and may be painful.

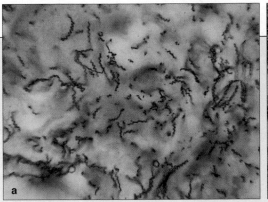
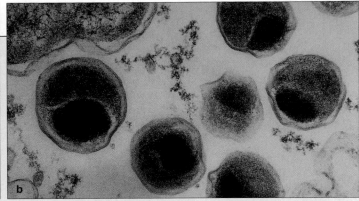

Figure 38.34 The bacterial agents of (**a**) syphilis and (**b**) chlamydia.

This STD is rampant even though prompt treatment quickly cures it. Why? Women experience no troubling symptoms during early stages. Also, infection does not confer immunity to the bacterium, perhaps because there are sixteen or more different strains of it. Contrary to common belief, someone can get gonorrhea over and over again. Also, oral contraceptives, used so widely, promote infection by altering vaginal pH. Populations of resident bacteria decline—and *N. gonorrhoeae* is free to move in.

SYPHILIS *Syphilis*, a dangerous STD, is caused by a spirochete, *Treponema pallidum* (Figure 38.34*a*). Sex with an infected partner puts the motile, spiral bacterium on the surface of genitals or into the cervix, vagina, or oral cavity. It can enter the body through tiny epidermal cuts. One to eight weeks later, new treponemes are twisting about in a flattened, painless chancre (local ulcer). This first chancre is a symptom of the primary stage of syphilis. By then, treponemes are in the blood. The treponemes can cross the placenta of an infected, pregnant woman. The result will be a miscarriage, stillbirth, or syphilitic newborn.

Usually a chancre heals, but in mucous membranes, joints, bones, eyes, spinal cord, and brain, treponemes are multiplying. More chancres and a skin rash develop in this infectious, secondary stage of syphilis. Symptoms subside; immune responses counter the disease in about 25 percent of the cases. Another 25 percent remain infected but symptom-free. In the rest, minor to significant lesions and scars appear in the skin and liver, bones, and other internal organs. Few treponemes occur in this tertiary stage. But the host's immune system is hypersensitive to them. Chronic immune reactions may severely damage the brain and spinal cord and lead to general paralysis.

Probably because the symptoms are so alarming, more people seek early treatment for syphilis than for gonorrhea. Later stages require prolonged treatment.

PELVIC INFLAMMATORY DISEASE *Pelvic inflammatory disease* (PID) is one of the most serious complications of gonorrhea, chlamydial infections, and other STDs. It also can arise when normal bacterial residents of the vagina ascend to the pelvic region. Most commonly, the uterus, oviducts, and ovaries are affected. There is bleeding and vaginal discharge. Pain in the lower abdomen may be as severe as an acute appendicitis attack. The oviducts may become scarred and invite abnormal pregnancies as well as sterility.

GENITAL HERPES About 25 million Americans have *genital herpes*, caused by the type II *Herpes simplex* virus. Infection requires direct contact with active *Herpes* viruses or sores that contain them. Mucous membranes of the mouth or the genitals are highly susceptible to invasion. Symptoms often are mild or absent. Among infected women, small, painful blisters may appear on the vulva, cervix, urethra, or anal tissues. Among men, blisters form on the penis and anal tissues. Within three weeks, the virus enters latency, and the sores crust over and heal.

This virus is reactivated sporadically. Each time, it causes new, painful sores at or near the original site of infection. Sexual intercourse, menstruation, emotional stress, or other infections can trigger *Herpes* infections. Acyclovir, an antiviral drug, decreases the healing time and often decreases the pain.

GENITAL WARTS More than sixty types of the human papillomaviruses (HPV) are known. A few cause *genital warts*, or benign, bumplike growths. HPV infection of the genitals and anus has become the most prevalent STD in the United States. Type 16 HPV does not usually cause obvious warts but may be linked to precancerous sores and cancers of the cervix, vagina, vulva, penis, and anus. In one Seattle study, 22 percent of female college students who were examined tested positive for the virus.

CHLAMYDIAL INFECTION *Chlamydia trachomatis* (Figure 38.34*b*), a parasitic bacterium, spends part of its life cycle in cells of the genital and urinary tracts. It causes several diseases, including *NGU* (chlamydial nongonococcal urethritis). NGU is far more common than syphilis or gonorrhea. *N. gonorrhoeae* and *C. trachomatis* often are transmitted at the same time. Prompt doses of penicillin cure the gonorrhea—but not NGU, which requires both tetracycline and sulfonamide.

NGU leads to inflammation of the cervix and, in both sexes, the urethra. Infected people may notice a burning sensation while urinating. But when they are symptom-free, they don't seek treatment and complications develop. In males, the prostate becomes swollen and inflamed. In females, the infection may spread into the uterus and the oviducts to cause pelvic inflammatory disease.

Symptom-free infected individuals are unwittingly spreading destructive chlamydial infections through all ethnic groups, among poor and affluent alike. *WWW*

38.20 TO SEEK OR END PREGNANCY

Because of sterility or infertility, about 15 percent of all couples in the United States cannot conceive. For example, hormonal imbalances in the female body may prevent ovulation, or the male's sperm count may be so low that fertilization would be next to impossible without medical intervention.

In vitro fertilization means conception outside the body (literally "in glass" petri dishes or test tubes). It may occur if the couple can produce normal sperm and oocytes. First the woman receives injections of a hormone that prepares her ovaries for ovulation. Afterward, the preovulatory oocyte is removed from her with a suction device. In the meantime, sperm from the male are placed in a solution that simulates fluid in the oviducts. A few hours after the sperm and suctioned oocyte make contact in a petri dish, fertilization may result. Twelve hours later, the zygote is transferred to a solution that can sustain it through the initial cleavages. Two to four days later, the resulting ball of cells is transferred to the female's uterus.

Each attempt at in vitro fertilization costs about 8,000 dollars, and most attempts aren't successful. In 1994, each "test-tube" baby cost the nation's health-care system about 60,000 to 100,000 dollars, on average. The childless couple may believe no cost is too great. But in an era of increased population growth and shrinking medical coverage, is the cost too great for society to bear? Court battles are being waged over this issue.

At the other extreme is *abortion*, the dislodging and removal of the blastocyst, embryo, or fetus from the uterus. At one time in the United States, abortions were forbidden by law unless the pregnancy endangered the mother's life. Later the Supreme Court ruled that the government does not have the right to forbid abortions during the early stages of pregnancy (typically up to five months). Before this ruling, there were dangerous, traumatic, and often fatal attempts to abort embryos, either by pregnant women themselves or by quacks.

From a clinical standpoint, vacuum suctioning and other methods make abortion painless for the woman, rapid, and free of complications when performed during the first trimester. Abortions performed in the second and third trimesters are extremely controversial unless the mother's life is threatened. Even so, for both medical and humanitarian reasons, the majority of people in this country generally agree that the preferred route to birth control is not through abortion. Rather, it is through sexually responsible behavior that prevents unwanted pregnancy from happening in the first place.

This biology textbook cannot offer you the "right" answer to a moral question because of the reasons given in Section 1.7. It *can* provide you with a serious, detailed description of how a new human individual develops. Your choice of the "right" answer to the question of the morality of abortion will be just that—your choice—and one that can be based on objective insights into the nature of life.

WWW

38.21 SUMMARY

Principles of Reproduction and Development:

1. For animals, sexual reproduction is the dominant reproductive mode. It requires specialized reproductive structures, control mechanisms, and forms of behavior that assist fertilization and support the offspring.

2. Among animals, embryonic development commonly proceeds through six stages:

   a. Gamete formation, when oocytes (immature eggs) and sperm form and develop in reproductive organs. Molecular and structural components are stockpiled and become localized in different parts of the oocyte.

   b. Fertilization, from the time a sperm penetrates an egg to the fusion of sperm and egg nuclei that results in a zygote (fertilized egg).

   c. Cleavage, when mitotic cell divisions transform a zygote into smaller cells (blastomeres). Cleavage does not increase the original volume of egg cytoplasm; it only increases the number of cells. It regionally divides the yolk, mRNAs, proteins, cytoskeletal elements, and other "maternal messages" among new blastomeres, an outcome called cytoplasmic localization.

   d. Gastrulation, when primary tissue layers (germ layers) form. Cell divisions, cell migrations, and other events lead to the formation of endoderm, ectoderm, and (in most species) mesoderm. All tissues of the adult body develop from primary tissue layers.

   e. The onset of organ formation. Different organs start developing by a tightly orchestrated program of cell differentiation and morphogenesis.

   f. Growth and tissue specialization, when organs enlarge overall and acquire their specialized chemical and physical properties. The maturation of tissues and organs continues into post-embryonic stages.

3. An embryo cannot develop properly unless each stage of development is successfully completed before the next stage begins.

4. In cell differentiation, a cell selectively uses certain genes and synthesizes proteins not found in other cell types. The outcome is subpopulations of specialized lineages of cells that differ from one another in their structure, biochemistry, and functioning.

5. Morphogenesis starts at gastrulation. It brings about changes in the embryo's size, shape, and proportions.

6. Pattern formation means the emergence of the basic body plan and body parts in specific regions of the embryo, in orderly sequence. Cytoplasmic localization helps seal the fate of cells. Cell determination depends also on interactions among classes of master genes that specify certain products. These products are spatially organized and create chemical gradients. The gradients influence differentiation, hence the identity of each cell lineage in the embryo.

Human Reproduction and Development:

1. Humans have a pair of primary reproductive organs (sperm-producing testes in males, and egg-producing ovaries in females), accessory ducts, and glands. Testes and ovaries also produce sex hormones that influence reproductive functions and secondary traits.

  a. The hormones LH, FSH, and testosterone control sperm formation. They are part of feedback loops from the testes to the hypothalamus and anterior lobe of the pituitary gland.

  b. The hormones estrogen, progesterone, FSH, and LH control the maturation and release of oocytes from the ovary, as well as cyclic changes in the endometrium (the inner lining of the uterus). Their secretion is part of feedback loops from ovaries to the hypothalamus and to the anterior lobe of the pituitary gland.

2. A menstrual cycle is a recurring cycle of intermittent fertility in the reproductive years of female humans and other primates. These events occur during a cycle:

  a. Follicular phase: One of many follicles matures inside an ovary. Each follicle is an oocyte and the cell layer surrounding it. Meanwhile, the endometrium is prepared for a possible pregnancy. It breaks down each cycle if pregnancy does not occur.

  b. Ovulation: A midcycle surge of the blood level of LH triggers ovulation, the release of a secondary oocyte from the ovary.

  c. Luteal phase: After ovulation the corpus luteum, a glandular structure, develops from the remnants of the follicle. It secretes progesterone and some estrogen that prime the endometrium for fertilization. If fertilization occurs, the corpus luteum is maintained.

3. After fertilization, a blastocyst forms by cleavage and becomes implanted in the endometrium. The three primary tissue layers form. They give rise to all organs. They are ectoderm, endoderm, and mesoderm.

4. Four extraembryonic membranes form as embryos of humans (and other vertebrates) develop:

  a. The amnion becomes a fluid-filled sac around the embryo, which it protects from drying out, mechanical shock, and abrupt temperature changes.

  b. The yolk sac stores nutritive yolk in most shelled eggs. In humans, part of the sac becomes a major site of blood formation and some of its cells give rise to germ cells. (Germ cells later give rise to sperm and eggs.)

  c. The chorion is protective membrane surrounding the embryo and the other membranes. It also becomes a major component of the placenta.

  d. In humans, the blood vessels for the placenta arise from the allantois, as does the urinary bladder.

5. The placenta, a blood-engorged organ, is composed of endometrium and extraembryonic membranes. The placenta allows embryonic blood vessels to develop independently of the mother's, but it allows oxygen, nutrients, and wastes to diffuse between them.

Review Questions

1. What is the main benefit of sexual reproduction? What are some of its biological costs? *38.1*

2. Define the key events during gamete formation, fertilization, cleavage, gastrulation, and organ formation. At which stage is the frog embryo in the photograph at right? *38.2*

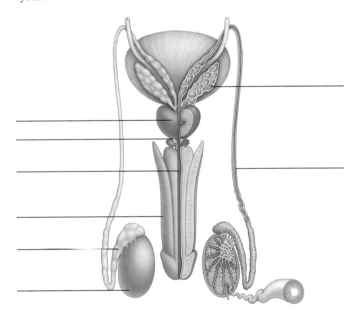

3. Does cleavage increase the volume of cytoplasm, the number of cells, or both, compared to the zygote? *38.2*

4. Give an example of cytoplasmic localization. *38.3*

5. Define cell differentiation and morphogenesis. *38.4*

6. Define pattern formation. Then explain the key points of the theory of pattern formation. *38.5*

7. Label the components of the human male reproductive system and state their functions: *38.6, 38.7*

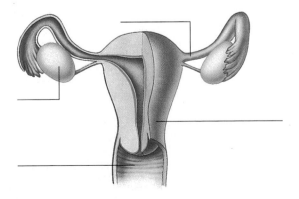

8. Label the components of the human female reproductive system and state their functions: *38.8, 38.9*

9. Does sperm formation require mitosis, meiosis, or both? *38.7*

10. Study Figure 38.17. Then, on your own, sketch the feedback loops to the hypothalamus and anterior pituitary from the testes that govern sperm formation. Include the names of the releasing hormone, hormones, and cells in the testes involved in these loops. *38.7*

11. Study Figure 38.20. Then, on your own, sketch the feedback loops to the hypothalamus and the anterior pituitary from the ovaries. These are the loops that govern the menstrual cycle. Include the names of the releasing hormone, hormones, and ovarian structures involved in these loops. *38.9*

12. State the embryonic source of the amnion, yolk sac, chorion, and allantois. State the role each extraembryonic membrane plays in the structure or functioning of an embryo *38.12*

Self-Quiz *(Answers in Appendix III)*

1. During cleavage, the new blastomeres are allocated different regions of the fertilized egg's cytoplasm. This is called _____ .
   a. cytoplasmic localization   c. cell differentiation
   b. embryonic induction   d. morphogenesis

2. Primary tissue layers first appear _____ .
   a. in the egg cortex   c. in the gastrula
   b. during cleavage   d. in primary organs

3. In organisms ranging from fruit flies to humans, the body's organization relative to its long axis depends on _____ .
   a. morphogens   c. embryonic induction
   b. homeotic genes   d. all of the above

4. The _____ , a fluid-filled sac, surrounds and protects the embryo from mechanical shocks and keeps it from drying out.
   a. yolk sac   c. amnion
   b. allantois   d. chorion

5. At full term, a placenta _____ .
   a. is composed of extraembryonic membranes
   b. directly connects maternal and fetal blood vessels
   c. keeps maternal and fetal blood vessels separated

6. Match the development stage with its description.
   _____ cleavage   a. egg and sperm mature in parents
   _____ gamete   b. sperm nucleus, egg nucleus fuse
         formation   c. formation of primary tissue layers
   _____ organ   d. cytoplasmic localization, not
         formation   gene interactions, dominate in
   _____ growth, tissue   most animals (not in mammals)
         specialization   e. organs, tissues increase in size,
   _____ gastrulation   acquire specialized properties
   _____ fertilization   f. starts when primary tissue layers
                          split into subpopulations of cells

Critical Thinking

1. Experimentally, it is possible to divide an amphibian egg so that the gray crescent is wholly within one of the two cells formed. If the two cells are separated from each other, only the cell with the gray crescent will form an embryo with a long axis, notochord, nerve cord, and back musculature. The other cell will form a shapeless mass of immature gut and blood cells. Reflect on Section 38.3. Does cytoplasmic localization or embryonic induction play a greater role in these outcomes?

2. Infection by the rubella virus apparently has an inhibitory effect on mitosis. Serious birth defects result when a woman is infected during the first trimester of pregnancy, but not later.

Review the developmental events that unfold during pregnancy and explain why this might be so.

3. Imagine you are an obstetrician advising a woman who has just learned she is pregnant. What instructions would you provide concerning her diet and behavior during pregnancy?

4. In the United States, teenage pregnancies and STD infections are rampant. Suppose the office of the Surgeon General asks you to join a task force that will recommend practices to reduce the incidence of both. What practices might be most successful? What kind of enthusiasm or resistance might they provoke among the teenagers in your community? Among adults? Explain why.

5. Despite obvious differences in early stages of development, chick and human embryos resemble one another once early organ formation is under way. Using the theory of pattern formation, speculate on the role of embryonic induction in bringing about the similarities between these vertebrate lineages.

Selected Key Terms

Principles:
asexual reproduction *38.1*
blastocyst *38.3*
blastomere *38.2*
blastula *38.3*
cell differentiation *38.4*
cleavage *38.2*
cytoplasmic localization *38.3*
ectoderm *38.2*
embryo *38.2*
embryonic induction *38.5*
endoderm *38.2*
fate map *38.5*
fertilization *38.2*
gamete formation *38.2*
gastrulation *38.2*
growth, tissue specialization *38.2*
mesoderm *38.2*
morphogenesis *38.4*
oocyte *38.3*
organ formation *38.2*
pattern formation, theory of *38.5*
sexual reproduction *38.1*
sperm *38.3*
yolk *38.1*
zygote *CI*

Human Case Study:
aging *38.17*
allantois *38.12*
amnion *38.12*
chorion *38.12*
corpus luteum *38.9*
endometrium *38.8*
estrogen *38.8*
fetus *38.12*
FSH *38.7*
implantation *38.12*
LH *38.7*
menstrual cycle *38.8*
ovulation *38.8*
ovum *38.11*
placenta *38.14*
progesterone *38.8*
seminiferous tubule *38.6*
sexually transmitted
  disease (STD) *38.19*
somite *38.13*
testis (testes) *38.6*
testosterone *38.7*
uterus *38.8*
yolk sac *38.12*

Readings   See also www.infotrac-college.com

Gilbert, S. 1994. *Developmental Biology.* Fourth edition. Sunderland, Massachusetts: Sinauer.

McGinnis, W., and M. Kuziora. February 1994. "The Molecular Architects of Body Design." *Scientific American* 270(2): 58–66.

Nusslein-Volhard, C. August 1996. "Gradients That Organize Embryonic Development. *Scientific American*, 54–61.

WWW   *http://www.brookscole.com/biology*

Practice quiz questions, hypercontents, BioUpdates, and critical thinking. The Brooks/Cole Biology Resource Center provides a wealth of information fully organized and integrated by chapter.

FACING PAGE: *Two organisms—a fox in the shadows cast by a snow-dusted spruce tree. What are the nature and consequences of their interactions with each other, with other organisms, and with their environment? By the end of this last unit, you possibly will see worlds within worlds in such photographs.*

VII

POPULATION ECOLOGY

Tales of Nightmare Numbers

Across from Sausalito, California, the steep flanks of Angel Island rise from the waters of San Francisco Bay (Figure 39.1*a*). The island, set aside as a game reserve, escaped urban development. It did not escape from the descendants of a few deer that well-meaning nature lovers shipped over in the early 1900s. With no natural predators to keep them in check, the few deer became many—far too many for the limited food supply of their isolated habitat. Yet the island attracted a steady stream of picnickers from the mainland. They felt sorry for the malnourished animals and made sure to load the picnic baskets with extra food for them.

The visitors imported so much food that scrawny deer kept on living and reproducing. In time, the herd nibbled away the native grasses, the roots of which had helped slow soil erosion on the steep hillsides. Hungry deer chewed off all the new leaves of seedlings; they killed small trees by stripping the bark and its phloem. The herd was destroying the environment.

In desperation, game managers proposed using a few skilled hunters to thin the herd. They were strongly denounced as being cruel. They proposed importing

a few coyotes to the island to thin the herd naturally. Animal rights advocates opposed that solution, also.

As a compromise, about 200 of the 300+ deer were captured, loaded onto a boat, and shipped to suitable mainland habitats. A number of them received collars with radio transmitters so that game managers could track them after the release. In less than sixty days, dogs, coyotes, bobcats, hunters, and speeding cars and trucks had killed off most of them. In the end, relocating each surviving deer had cost taxpayers almost 3,000 dollars. The State of California refused to do it again. And no one else, anywhere, volunteered to pick up future tabs.

It is not difficult to define the boundaries of Angel Island or track its inhabitants, so it is easy to draw a lesson from this tale: *A population's growth depends on the resources of its environment. And attempts to "beat nature" by altering the sometimes cruel outcome of limited resources only postpone the inevitable.* Does the same lesson apply to other populations, in other places? Yes, it does, as the next tale makes clear.

When 1998 drew to a close, there were nearly 6 billion people on Earth. About 2 billion already live in poverty. Each year 40 million more join the ranks of the starving. Next to China, India is the most populous country, with 984 million inhabitants (Figure 39.1*b*). By 2010 it may

Figure 39.1 (**a**) Angel Island, which turned out to be a laboratory for studying population growth. (**b**) Bathers crowded along a bank of the Ganges River in India—just a small sampling of a human population that is approaching 6 billion. In this chapter we turn to principles that govern the growth and sustainability of all populations.

reach 1.82 billion. Forty percent of those people live in rat-infested shantytowns, without enough food or water. They are forced to wash their clothes and dishes in open sewers. Land available to raise their food shrinks by 365 acres a day, on average. Why? Irrigated soil becomes too salty when it drains poorly and there is not enough water to flush away the salts.

Can wealthier, less densely populated nations help? After all, they use most of the world's resources. Maybe they should learn to get by more efficiently, on less. For example, people might limit their meals to cereal grains and water; give up their private cars, living quarters, air conditioners, televisions, and dishwashers; stop taking vacations and stop laundering so much; close all the malls, restaurants, and theaters at night; and so on.

Maybe wealthier nations also should donate more surplus food than they already donate to less fortunate ones. Then again, would huge donations help, or would they encourage dependency and spur more increases in population size? And what if surpluses run out?

It is a monumental dilemma. At one extreme, the redistribution of resources on a global scale would allow the greatest number of people to survive, but at the lowest comfort level. At the other extreme, foreign aid rationed only to nations that restrict population growth would allow fewer individuals to be born, but the quality of life would be greater.

Currently, the foreign aid program of the United States is based on two premises: (1) that individuals of every nation have an irrevocable right to bear children, even if unrestricted reproduction ruins the environment that must sustain them; and (2) that because human life is precious above all else, the wealthiest nations have an absolute moral obligation to save lives everywhere.

Regardless of the positions that nations take on this issue, ultimately they must come to terms with this fact: *Certain principles govern the growth and sustainability of populations over time.* These principles are the bedrock of **ecology**—the systematic study of how organisms interact with one another and with their physical and chemical environment. Ecological interactions start within and between populations, and they extend on through communities, ecosystems, and the biosphere. They are the focus of this last unit of the book.

In this chapter we look first at the relationships that influence the size, structure, and distribution of populations. Later, we will apply the basic principles of population growth to the past, present, and future of the human species.

KEY CONCEPTS

1. Certain ecological principles govern the growth and sustainability of all populations, including our own.

2. In addition to genetic factors, a population's size, density, distribution, and the number of individuals in various age categories influence its patterns of growth.

3. When the birth rate exceeds the death rate, and when immigration and emigration are in balance, populations may show a pattern of exponential growth.

4. With exponential growth, a population expands by ever increasing increments during successive intervals, because the number of individuals in its reproductive age category (or about to enter it) becomes ever larger.

5. A shortage of any resource that individuals require is a limiting factor on population growth.

6. For all populations, carrying capacity is the maximum number of individuals that can be sustained indefinitely by the resources of a given environment. That number may rise or fall with changes in resource availability.

7. A population may show a pattern of logistic growth. By this pattern, population density—that is, the number of individuals in a specified area at a given point in time—is initially low. Then the population size rapidly increases. It finally levels off as resource scarcity limits its further increase or triggers a decline in numbers.

8. Some populations show fluctuations in numbers that cannot be explained by a single growth model.

9. All populations face limits to growth, because no environment can indefinitely sustain a successively increasing number of individuals. Competition, disease, predation, and other factors control population growth. The controls vary in their relative effects on populations of different species, and they vary over time.

CHARACTERISTICS OF POPULATIONS

By this point in the book you know that a population is a group of individuals of the same species occupying a given area. As described in Section 16.4, its individuals share a gene pool, which is the basis for characteristic ranges of morphological, physiological, and behavioral traits. When studying a population, ecologists consider the genetic make-up as well as the reproductive modes and reproductive behavior of individuals. In addition, they consider the **demographics**, or vital statistics of a population, and that will be our focus here. Among the vital statistics are population size, density, distribution, and age structure.

Population size is the number of individuals that contribute to a population's gene pool. A population's **age structure** is the number of individuals in each of several to many age categories. For instance, all members may be grouped into *pre-reproductive, reproductive,* and *post-reproductive* ages. Individuals in the first category will have the capacity to produce offspring at maturity. Together with the actually and potentially reproducing individuals in the second category, they help make up the population's **reproductive base**.

Population density is the number of individuals in some specified area or volume of a habitat, such as the number of guppies per liter of water in a stream. (A *habitat*, remember, is the type of place where a species normally lives. We characterize a habitat by its physical features, chemical features, and the presence of other species.) **Population distribution** is the general pattern in which individuals of the population are dispersed through a specified area.

Crude density is a measured number of individuals in a specified area. Section 39.2 outlines how ecologists make their counts, which help them track changes in population density over time. The counts do not reveal how much of a habitat is used as living space. Even areas that appear to be uniform, such as a long, sandy beach, are more like fine tapestries of light, moisture, temperature, mineral composition, and other variables. Also, the tapestry of environmental conditions often changes with the seasons. So a population might find one part of a habitat more suitable for occupancy than other parts, and it may do so some or all of the time.

Also, different species typically share the same area, and they compete for energy, nutrients, living space, and other resources. Most interact as predators, prey, or parasites. Later chapters describe these interactions. For now, simply note that species interactions influence population density and dispersion through a habitat.

Theoretically, populations show three distribution patterns. By these patterns, individuals are dispersed in clumps, nearly uniformly, or randomly (Figure 39.2). The individuals of most populations form aggregations at specific sites in a habitat. Why is clumping the most

Figure 39.2 Three generalized patterns of population distribution.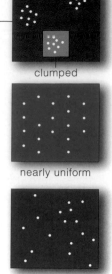

clumped

nearly uniform

random

common dispersion pattern? Three possible reasons come to mind.

First, each species is adapted to a limited set of ecological conditions, which usually are patchy through a habitat. For example, some parts of a habitat offer more or less shade, moisture, hiding spots, and hunting spots. Second, many animal species gather in social groups that afford advantages in terms of survival and reproduction. The groups offer more opportunities for mating and mutual defense against predators. Third, adults of many species cannot disperse their seeds, larvae, or other immature forms of the new generation over large distances. For example, sponge larvae can swim, but not very far from a parent sponge. They simply settle down on substrates near parents.

Where the individuals are more evenly spaced than they would be by chance alone, we call this a nearly uniform disperal pattern. Regularly spaced fruit trees in orchards are often cited as examples. Few populations live this way in nature. You might see the pattern where habitat conditions are fairly uniform and where there is strong competition for resources or territorial behavior.

Creosote bushes (*Larrea*) that live in arid habitats of the American Southwest may show this pattern. Large, mature plants deplete soil water all around them. Seed-eating ants and rodents forage near the plants, which offer cover from predators and from the hot sun. Most seeds are devoured, which puts seeds and seedlings at a competitive disadvantage. So the seeds typically take hold only at patches where established, mature plants are weakened or have died. As a result, creasote bushes are not clumped together (Figure 39.3*a*). Neither are they dispersed at random; seeds that rodents and ants do overlook and that await opportunity for life in the sun cannot travel far from the mature parent plants.

We observe random dispersion only when habitat conditions are nearly uniform, resource availability is fairly steady, and individuals of the population neither attract nor avoid one another. For instance, wolf spiders are solitary hunters on forest floors. Each generation might well be randomly spaced. However, a dispersion pattern such as this is extremely rare in nature.

Each population has its own gene pool and range of traits. It also has a characteristic size, density, distribution pattern, and age structure. Environmental conditions and species interactions influence these characteristics.

WWW

39.2 ELUSIVE HEADS TO COUNT

Does it seem like Bambis are just about everywhere, munching through gardens and smacking into more and more cars on the roads? The estimated number of deer in the United States has risen from fewer than 1 million at the turn of the century to more than 18 million today. Suppose you live in a northern California county and wonder how many deer live there with you. How would you go about determining their population density?

To get some idea of what your study will entail, you start with a literature search. Ecologists already have baselines of approximate population densities for many organisms in their natural habitats. For example, in some habitats, baselines indicate that you can expect to find as many as 5 million diatoms per cubic meter, 500 trees per hectare (a hectare is 2.47 acres), 250 field mice per hectare, and so on. Your search turns up one estimate of 4 deer per square kilometer. And this should tell you that deer will be easier to count than, say, diatoms.

Total counts are the most straightforward way to measure the absolute density of a population. Census takers do this every ten years for the human population in the United States. Ecologists do the same for large animals in small areas, such as songbirds in a forest, northern fur seals at their breeding grounds, sea stars in a tidepool, and creosote bushes in a specified part of a desert (Figure 39.3a). More often, ecologists must content themselves with sampling a small part of the population and then estimating the total density.

You could get a map of your county and divide it into a number of small plots, or quadrats. Quadrats are sampling areas of the same size and shape (such as rectangles, squares, and hexagons). Then you could count all the deer in several plots and extrapolate the average for the entire county. Ecologists conduct such counts for migrating herds and flocks. For example, each year, the North American Breeding Bird Survey makes counts along more than 2,000 flyways for migratory waterfowl (Figure 39.3b).

Deer are among the animals that do not stay put in their habitat. How can you be absolutely sure that the individuals you count in one plot aren't the same individuals you counted earlier in a different plot?

For mobile animals, ecologists sample population density by using a **capture-recapture method**. The idea is to capture individual animals and mark them in some way. (Squirrels get tattoos, salmon get tags, migratory birds get leg rings, deer get bright collars, and so on.) Marked animals are released at time 1, then animals are captured and checked for marks at time 2. In later samples taken from the population, the proportion of marked individuals should represent the proportion marked in the whole population:

$$\frac{\text{Marked individuals in sampling 2}}{\text{Total capture in sampling 2}} = \frac{\text{Marked individuals in sampling 1}}{\text{Total population size}}$$

Ideally, marked and unmarked individuals of the population are captured at random, no marked animal dies during the study interval, and none of the marked animals leaves the population or is overlooked. In the real world, however, recapturing of marked individuals might *not* be random. Squirrels that were marked after being attracted to yummy bait in boxes might now be trap-happy or trap-shy, so they might overrepresent or underrepresent the population. Instead of mailing the tags of marked fish back to ecologists, some fishermen keep them as good-luck charms. Birds may lose their leg rings. Undocumented immigrants may not answer the door when the census taker comes knocking.

Your estimate also depends on the time of year when you do a sampling. Population distribution varies over time, as in migratory responses to environmental rhythms. Few places yield abundant resources all year long, and many animals often move from one habitat to another with the seasons (Figure 39.3b). Deer are like this, so you would probably use mark-recapture methods more than once a year, for several years. *WWW*

Figure 39.3 (**a**) Example of nearly uniform spacing, represented by creosote bushes near Death Valley, California. (**b**) Snow geese, one of many species of waterfowl that migrate through 3.4 million square kilometers of air space above Canada and the United States.

Gains and Losses in Population Size

Populations are dynamic units of nature. Depending on the species, they might add or lose individuals every half hour or every day, season, or year. Sometimes they glut portions of the habitat with individuals. At other times individuals may be scarce. Populations even can drive themselves or be driven to extinction. During a specified interval, such changes in population size can be measured in terms of birth rates, death rates, and the number of individuals entering and leaving.

Population size increases as a result of (1) births and (2) **immigration**—the arrival of new residents from other populations of the species. It decreases as a result of (1) deaths and (2) **emigration**, whereby individuals permanently move out of the population.

For many species, population size also changes on a predictable basis as a result of daily or seasonal events called **migrations**. However, migration is a recurring round trip between two areas, so we need not consider its transient effects during this initial consideration of population size.

From Ground Zero to Exponential Growth

For our purposes, assume that immigration is balancing emigration over time, so that we can ignore the effects of both on population size. Doing so allows us to define

zero population growth as an interval during which the number of births is balanced out by the number of deaths. The population size is stabilized during such an interval, with no overall increase or decrease.

Births, deaths, and other variables that might affect population size can be measured in terms of **per capita** rates, or rates per individual. (*Capita* means heads, as in head counts.) Visualize 2,000 mice living in a cornfield. Twenty or so days after their eggs are fertilized, the females produce a litter, then nurse their offspring for a month or so, then get pregnant again. If the female mice collectively give birth to 1,000 mice per month, the birth rate would be 1,000/2,000 = 0.5 per mouse per month. If 200 of the 2,000 die during that interval, the death rate would be 200/2,000 = 0.1 per mouse per month.

If we assume the birth rate and death rate remain constant, we can combine both into a single variable—the **net reproduction per individual per unit time**, or r for short. For that mouse population in the cornfield, r is $0.5 - 0.1 = 0.4$ per mouse per month. This example gives us a way to represent population growth:

$$\begin{pmatrix} \text{population} \\ \text{growth per} \\ \text{unit time} \end{pmatrix} = \begin{pmatrix} \text{net population} \\ \text{growth rate} \\ \text{per individual} \\ \text{per unit time} \end{pmatrix} \times \begin{pmatrix} \text{number of} \\ \text{individuals} \end{pmatrix}$$

or, more simply, $G = rN$.

	Net Monthly Increase:	New Population Size:
$G = r \times$ 3,920 =	1,568 =	5,488
$r \times$ 5,488 =	2,195 =	7,683
$r \times$ 7,683 =	3,073 =	10,756
$r \times$ 10,756 =	4,302 =	15,058
$r \times$ 15,058 =	6,023 =	21,081
$r \times$ 21,081 =	8,432 =	29,513
$r \times$ 29,513 =	11,805 =	41,318
$r \times$ 41,318 =	16,527 =	57,845
$r \times$ 57,845 =	23,138 =	80,983
$r \times$ 80,983 =	32,393 =	113,376
$r \times$ 113,376 =	45,350 =	158,726
$r \times$ 158,726 =	63,490 =	222,216
$r \times$ 222,216 =	88,887 =	311,103
$r \times$ 311,103 =	124,441 =	435,544
$r \times$ 435,544 =	174,218 =	609,762
$r \times$ 609,762 =	243,905 =	853,667
$r \times$ 853,677 =	341,467 =	1,195,134

a

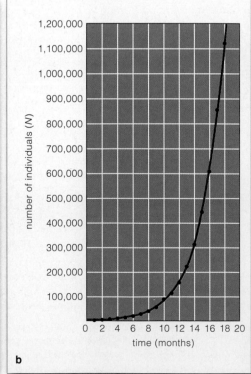

b

c

Figure 39.4 (**a**) Data showing the net monthly increases in a population of field mice living in a cornfield. From start to finish, the numbers listed show a pattern that is typical of exponential growth. (**b**) Graphing the data yields a J-shaped growth curve.

(**c**) Do such growth curves seem far removed from your own experience? Think about the growth of populations of, say, the resident bacteria in your mouth after you present them with a buffet of nutrients in candy or some other sugar-laden food.

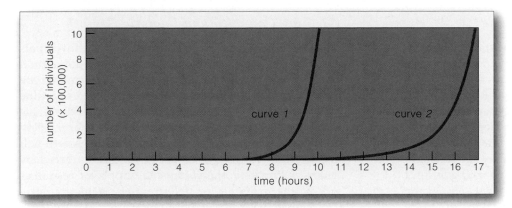

Figure 39.5 Effect of deaths on the rate of increase in two bacterial populations. Growth curve *1* represents a population of bacterial cells that reproduced every half hour. Growth curve *2* represents a different population in which cells divided every half hour, but 25 percent died between divisions.

The graph shows that deaths can slow the rate of increase but cannot in themselves stop exponential growth.

As the next month begins, 2,800 mice are scurrying about. With the net increase of 800 fertile critters, the reproductive base has expanded. If *r* does not change, the population size will expand this month also, by a net increase of $0.4 \times 2,800 = 1,120$ mice. The population is now 3,920. In the months ahead, it just so happens that *r* remains constant, and a growth pattern emerges. As Figure 39.4*a* shows, *less than two years from the time we started counting, the number of mice running around in the cornfield increased from 2,000 to more than a million!*

Plot the monthly increases against time, as in Figure 39.4*b*, and you end up with a graph line in the shape of a "J." When growth of a population over increments of time plots out as a J-shaped curve, you know that you are tracking **exponential growth**.

"Exponential" refers to a relationship in which one variable increases much faster than another in a specific mathematical way. In this case, a population's size is a variable that depends on how many individuals make up the reproductive base in successive increments of time. *The larger the reproductive base, the greater will be the expansion in population size during each specified interval.*

Now look at other aspects of exponential growth by supplying a single bacterium in a culture flask with all the nutrients required for growth. Thirty minutes later, the one cell divides in two. Thirty minutes pass, the two cells divide, and so on every thirty minutes. Assuming no cells die between divisions, the population size will double in each interval—from 1 to 2, then 4, 8, 16, 32, and so on. The length of time it takes for a population to double in size is its **doubling time**.

The larger the population gets, the more cells there are to divide. After 9–1/2 hours (nineteen doublings), it has more than 500,000 bacterial cells. After 10 hours (twenty doublings), it has more than 1 million. Plot the doublings in size against time, and you end up with the J-shaped curve typical of unrestricted, exponential growth. Figure 39.5, curve *1*, shows this.

To examine the effect of deaths on the growth rates, start over with one bacterium. Assume that 25 percent

of the cells die in each thirty-minute interval. Because of deaths, it takes about seventeen hours (not ten) for population size to reach a million. *But deaths changed only the time scale for population growth.* You still end up with a J-shaped curve—which is curve 2 in Figure 39.5.

What Is the Biotic Potential?

Finally, visualize a population occupying a place where conditions are ideal. Every one of its individuals has adequate shelter, food, and other vital resources. No predators, pathogens, or pollutants lurk anywhere in the habitat. The population might well display its **biotic potential**, which is the maximum rate of increase per individual under ideal conditions.

Each species has a characteristic maximum rate of increase. For many bacteria, it is 100 percent every half hour or so. For humans and other large mammals, it is between 2 and 5 percent per year. But the *actual* rate depends on the age at which each generation starts to reproduce, how often each individual reproduces, and how many offspring are produced. Now think about this. A human female is biologically capable of bearing twenty or more children, yet in each generation, many females do not reproduce at all. The human population has not been displaying its biotic potential. Even so, since the mid-eighteenth century, its growth has been exponential, for reasons that will soon be apparent.

During a specified interval, population size is generally an outcome of births, deaths, immigration, and emigration.

With exponential growth, population size expands by ever increasing increments during successive time intervals, because the reproductive base becomes ever larger.

Population size against time plots out as a characteristic J-shaped curve if the population is growing exponentially.

As long as the per capita birth rate remains even slightly above the per capita death rate, a population can grow exponentially.

WWW

What Are the Limiting Factors?

Most of the time, environmental circumstances prevent any population from fulfilling its biotic potential. That is why sea stars, the females of which could produce 2,500,000 eggs each year, do not fill up the oceans with sea stars. That also is why humans will never fill up the planet, even with our capacity for exponential growth.

In natural environments, complex interactions exist within and between populations of different species, so it is not easy to identify all the factors working to limit population growth. To get a sense of what *some* of the factors might be, start again with a lone bacterium in a culture flask, where you can control the variables. First you enrich the culture medium with glucose and other nutrients required for bacterial growth. Then you allow bacterial cells to reproduce for many generations. At first the growth pattern appears to be exponential. Then growth slows, and, after that, population size remains rather stable. But after the stable period, the population size declines rapidly and all of the bacteria die.

What happened? When the population expanded by ever increasing amounts, cells tapped more and more nutrients. The dwindling nutrients signaled the cells to stop dividing (Section 8.2). And when the existing cells exhausted the nutrient supply, they starved to death.

Any essential resource that is in short supply is a **limiting factor** on population growth. Food, minerals of certain types, refuge from predators, living quarters, and a pollution-free environment are examples. The number of such factors can be huge, and their effects can vary. Even so, one factor alone is often enough to put the brakes on population growth at any given time.

Suppose you keep on freshening the supply of all required nutrients for the growing bacterial culture. After an episode of exponential growth, the population still crashes. Like all other organisms, bacteria produced metabolic wastes. The wastes of that huge population of bacterial cells were so great, they drastically altered living conditions in the culture. Therefore, by its own metabolic activities, the population polluted the experimentally designed habitat and put a stop to exponential growth.

Figure 39.6 Idealized S-shaped curve characteristic of logistic growth. After a rapid growth phase (time B to C), growth slows and the curve flattens out as the carrying capacity is reached (time C to D). S-shaped growth curves can show variations, as when changes in the environment bring about a decreased carrying capacity (time D to F). This happened to the human population of Ireland before 1900, when late blight, a disease caused by a water mold, destroyed potato crops that were the mainstay of the diet (Section 20.8).

Carrying Capacity and Logistic Growth

Now visualize a small population in which individuals are dispersed through the habitat. As the population increases in size, more and more individuals must share nutrients, living quarters, and other resources. As the share available to each diminishes, fewer individuals may be born and more may die by starvation or nutrient deficiencies. Then, the population's rate of growth will decline until births are balanced or outnumbered by deaths. Ultimately, the *sustainable* supply of resources will be the major factor determining population size. The name **carrying capacity** refers to the maximum number of individuals of a population (or species) that a given environment can sustain indefinitely.

The pattern of **logistic growth** is a fine example of how carrying capacity can affect a population. By this pattern, a population at low density starts growing slowly in size, then grows rapidly, and finally levels off in size once the carrying capacity is reached. How can you represent the pattern? Starting with the exponential growth equation given earlier, in Section 39.3, add the term $(K - N)/K$, where K designates carrying capacity:

$$G = r_{max} N \frac{K - N}{K}$$

The term indicates how many individuals can be added to the population, based on the proportion of resources still available. When population size is small, $(K - N)$ is close to 1. It approaches 0 when the size is close to the carrying capacity. Representing this another way,

$$\begin{pmatrix} \text{population} \\ \text{growth per} \\ \text{unit time} \end{pmatrix} = \begin{pmatrix} \text{maximum} \\ \text{net population} \\ \text{growth rate} \\ \text{per individual} \\ \text{per unit time} \end{pmatrix} \times \begin{pmatrix} \text{number} \\ \text{of} \\ \text{individuals} \end{pmatrix} \times \begin{pmatrix} \text{proportion} \\ \text{of resources} \\ \text{not yet used} \end{pmatrix}$$

A plot of logistic growth gives an S-shaped curve (Figure 39.6). Such curves are only an approximation of

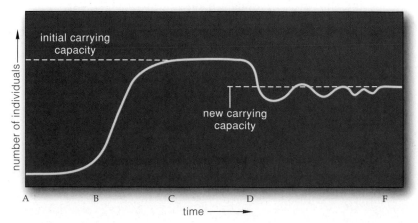

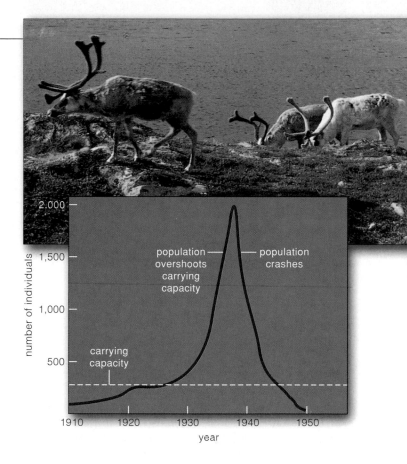

Figure 39.7 Carrying capacity and a reindeer herd. In 1910, four male and twenty-two female reindeer were introduced on St. Matthew Island in the Bering Sea. In less than thirty years, the size of the herd increased to two thousand. Individuals had to compete for dwindling vegetation, and overgrazing destroyed most of it. In 1950, the herd plummeted to eight members. The growth pattern reflects how the population size overshot the carrying capacity, then crashed.

what goes on in nature. For example, a population that grows too rapidly can overshoot the carrying capacity. The death rate skyrockets and the birth rate plummets. These two responses drive the number of individuals down to the carrying capacity—or lower (Figure 39.7).

Density-Dependent Controls

The logistic growth equation just described deals with **density-dependent control** of populations. That is, a small population may grow rapidly, but when it finally bumps up against its carrying capacity, so to speak, its size will stabilize or even decline.

In addition to carrying capacity, other natural factors help control population density. When they come into play, they, too, put the number of individuals below the maximum sustainable level. For example, individuals are more likely to die under crowded conditions. Then, predators, parasites, and pathogens are able to interact more intensely with their host population and trigger a decline in density. Once population size declines, the density-dependent interactions relax, and population size may increase once again.

Bubonic plague and *pneumonic plague*, two dangerous diseases, are examples of density-dependent controls. Both are caused by the bacterium *Yersinia pestis*. A large reservoir of *Y. pestis* persists in rabbits, rats, and certain other small mammals. Hungry, biting fleas transmit it to new hosts. The population of bacteria increases so much in the flea gut that it interferes with digestion. The host flea must feed more often on more hosts, and so the disease spreads.

A devastating episode of bubonic plague occurred in the fourteenth century. The plague swept through European cities where humans were crowded together, sanitary conditions were poor, and rats were abundant. By the time the epidemic subsided, urban populations in Europe had declined by 25 million.

Both diseases are still threats. For example, in 1994 in India, they raced through rat-infested cities where garbage and animal carcasses had piled up for months in the streets. Terrified residents fled by the thousands. Some carried the pathogen with them as far away as London. Global panic ensued before concerted efforts to burn the garbage, poison the rats and the fleas, and rapidly dispense antibiotics averted a pandemic.

Density-Independent Factors

Sometimes events result in more deaths or fewer births regardless of population density. For example, every year, monarch butterflies migrate from Canada to spend the winter in forested mountains of Mexico. But logging opened up stands of the forest trees—which normally buffer temperatures. In 1995 a sudden freeze combined with the deforestation to kill millions of butterflies. Both were **density-independent factors**, meaning their effects came into play independently of population density.

Similarly, heavy applications of pesticides in your backyard may also kill most insects, mice, cats, birds, and other animals. They will do so regardless of how uncrowded or dense the populations are.

Resources in short supply put limits on population growth. Together, all of the limiting factors acting on a population dictate how many individuals can be sustained.

Carrying capacity is the maximum number of individuals of a population that can be sustained indefinitely by the resources in a given environment. The number may rise or fall with changes in resource availability.

The size of a low-density population may increase slowly, go through a rapid growth phase, then level off once the carrying capacity for the population is reached. This is a logistic growth pattern.

Density-dependent controls and density-independent factors can bring about decreases in population size.

WWW

LIFE HISTORY PATTERNS

So far, we have looked at populations as if all of their members are identical throughout a given interval. For most species, however, individuals are at different stages of development and are interacting in different ways with other organisms and with the environment. At different times in the life cycle, they may be adapted to exploiting different resources, as when caterpillars eat leaves and butterflies prefer nectar. Also, individuals at different stages might be more or less vulnerable to danger. In short, each species has a **life history pattern**. Let's look at a few of the environmental variables that influence age-specific life history patterns.

Life Tables

Although each species has a characteristic life span, few of its individuals reach the maximum age possible. Death looms larger at some ages than at others. Also, individuals of a species tend to reproduce or emigrate during a characteristic age interval.

Age-specific patterns in populations first intrigued life insurance and health insurance companies, then ecologists. Typically, such investigators track a cohort (a group of individuals from the time of birth until the last one dies). They also track the number of offspring born to individuals during each age interval. Life tables list data gathered on a population's age-specific death schedule. Often they are converted to a more cheery "survivorship" schedule, or the number of individuals that reach some specified age (x). The life table in Table 39.1 is a typical example. This one was constructed for the 1989 human population of the United States.

Dividing a population into age classes and assigning birth rates and mortality risks to each one has practical applications. Unlike a crude census (head count), the data might be a basis for informed policy decisions in pest management, endangered species protection, social planning for human populations, and other areas. For instance, birth/death schedules for the northern spotted owl figured in federal court rulings to halt mechanized logging in old-growth forests, the owl's habitat.

Patterns of Survival and Reproduction

Evolutionarily, we measure an individual's reproductive success in terms of the number of surviving offspring. Yet the number differs among species, which differ in how much energy and time is allocated to producing gametes, securing mates, and parenting, and the size of offspring. Many tradeoffs have been made in response to selection pressures, such as species interactions and conditions characteristic of their habitat.

For each species, a **survivorship curve** is a graph line that emerges when ecologists plot a cohort's age-specific

Table 39.1 Life Table, U.S. Human Population, 1989*

Age Interval	Survivorship*	Mortality**	Life Expectancy***	Reported Live Births for Total Population
0–1	100,000	896	75.3	
1–5	99,104	192	75.0	
5–10	98,912	117	71.1	
10–15	98,795	132	66.2	11,486
15–20	98,663	429	61.3	506,503
20–25	98,234	551	56.6	1,077,598
25–30	97,683	606	51.9	1,263,098
30–35	97,077	737	47.2	842,395
35–40	96,340	936	42.5	293,878
40–45	95,404	1,220	37.9	44,401
45–50	94,184	1,766	33.4	1,599
50–55	92,418	2,727	28.9	
55–60	89,691	4,334	24.7	
60–65	85,357	6,211	20.8	
65–70	79,146	8,477	17.2	
70–75	70,669	11,470	13.9	
75–80	59,199	14,598	10.9	
80–85	44,601	17,448	8.3	
85+	27,153	27,153	6.2	
				TOTAL: 4,040,958

* Number alive at start of age interval per 100,000 individuals.
** Number dying during the age interval.
*** Average lifetime remaining at start of age interval.

survival in a habitat. Three types of survivorship curves are common in nature.

Type I curves reflect high survivorship until fairly late in life, then a large increase in deaths. Such curves are typical of elephants and other large mammals that bear only one or a few large offspring at a time, then engage in extended parental care (Figure 39.8a). For example, a female elephant gives birth to four or five calves and devotes several years to parenting each one.

Type I curves also are typical of human populations in which individuals have access to good health care services. However, today as in the past, in places where health care is poor, infant deaths cause a sharp drop at the start of the curve. Following the drop, it levels off from childhood to early adulthood.

Type II curves reflect a fairly constant death rate at all ages. They are typical of organisms just as likely to be killed or to die of disease at any age, such as lizards, small mammals, and some songbirds (Figure 39.8b).

Type III curves signify a death rate that is highest early in life. They are typical of species that produce many small offspring, then show little, if any, parental behavior. Figure 39.8c shows how the curve plummets for sea stars. Sea stars produce mind-boggling numbers of tiny larvae, which must rapidly eat, grow, and finish developing on their own without support, protection, or guidance from the parents. Corals and other animals swiftly eat most of them. The plummeting survivorship curve is typical of many other marine invertebrates, most insects, and many fishes, plants, and fungi.

At one time, ecologists thought selection processes favored *either* early, rapid production of many small

Figure 39.8 Three generalized types of survivorship curves. Type I populations show high survivorship until some age, then show high mortality. Type II populations show a fairly constant death rate. Type III populations show low survivorship early in life.

offspring *or* the late production of a few large ones. They now know that the two patterns are extremes, at opposite ends of a range of possible life histories. Also, both life-history patterns—as well as intermediate ones—are sometimes evident in different populations of the same species, as the next section makes clear.

Tracking a cohort (a group of individuals from birth until the last one dies) reveals patterns of reproduction, death, and migration that typify the populations of a species.

Survivorship curves can reveal differences in age-specific survival among species. In some cases, such differences exist even between populations of the same species.

WWW

Several years ago, evolutionary biologists David Reznick and John Endler were netting guppies in the mountains of Trinidad, a Caribbean island. Different populations of these live-bearing fishes occupy different streams and sometimes different parts of the same stream. There they encounter different predators. A type of killifish shares some streams with guppies. It is not an especially large fish. It successfully preys on smaller, immature guppies but not on larger adults. A bigger fish, a pike-cichlid, shares different streams with guppies. It preys on larger, sexually mature guppies and tends to ignore small ones.

Using natural selection theory, Reznick and Endler hypothesized that predation influences guppy life history patterns. In streams where pike-cichlids reign supreme, individual guppies mature faster and their body size is smaller at maturity, compared with guppies in killifish-dominated streams. In addition, guppies targeted by pike-cichlids reproduce earlier in life, they produce far more offspring, and they do so more frequently.

Did other variables influence the life history patterns in killifish streams versus pike-cichlid streams? To check this possibility, the researchers shipped two groups of guppies from the two kinds of streams to a laboratory in the United States. There, the guppies reproduced for two generations, free from predation. Aquarium conditions were identical for the two experimental populations. In body size, generation time, and other features, offspring of both experimental populations were like the guppies in the wild, so the differences must have a genetic basis. After many generations, body size of the guppy lineage that was subjected to predation killifish became larger at maturity. The lineage raised with pike-cichlids showed a trend toward earlier maturity.

GUPPY FROM A KILLIFISH STREAM

GUPPY FROM A PIKE-CICHLID STREAM

When Reznick and Endler first visited Trinidad, they did a field experiment. They captured guppies that had evolved with pike-cichlids and introduced them to a site upstream from a small waterfall that had been a barrier to dispersal (it had prevented guppies and all predators except killifish from moving upstream). They designated that part of the stream as the control site.

Eleven years later, researchers revisited the stream and discovered the guppies had evolved. They made comparisons between guppies from the experimental site and the control site. Just as Reznick and Endler had predicted, the body size, frequency of reproduction, and other aspects of guppy life history patterns correlated with the preferences of the neighborhood predator. Later, a laboratory experiment involving two generations of guppies confirmed that the differences were indeed genetic. The researchers concluded that the differences are the product of natural selection.

WWW

In 1998, the human population approached 6 billion. The year before, rates of increase for different nations ranged from below 1 percent to 3+ percent, for an annual average of 1.47 percent. Visualize enough people to fill another New York City every month and you get the picture. Now think about this: Annual additions to the base population will result in a *larger* absolute increase every year into the future.

Our staggering population growth continues even though as many as 2 billion are already malnourished or starving, without clean drinking water and adequate shelter. It continues when 1.5 billion are already going without the benefits of health care delivery and sewage treatment facilities. And it continues mainly in regions that are already overcrowded. Nearly 6 billion live on 10 percent of the land; 4 billion of us crowd together within 480 kilometers (about 300 miles) of the seas.

Suppose it were possible, by monumental efforts, to double the food supply to keep pace with growth. We would do little more than maintain marginal living conditions for most. The annual deaths from starvation could still be 20 million to 40 million. Even this would come at great cost, for we are drastically modifying the environment that must sustain us. Salted-out cropland, desertification, deforestation, pollution—these are some of the consequences you will read about in Chapter 43, and they do not bode well for our future.

For a while, it will be like the Red Queen's garden in Lewis Carroll's *Through the Looking Glass*, where one is forced to run as fast as one can to remain in the same place. But what happens when our population doubles again? Can you brush the doubling aside as being too far in the future to warrant concern? *It is no further removed from you than the sons and daughters of the next generation.*

How We Began Sidestepping Controls

How did we get into this predicament? For most of its history, the human population grew slowly. During the past two centuries, increases in growth rates became astounding. There are three possible reasons for this:

1. Humans steadily developed the capacity to expand into new habitats and new climate zones.

2. Humans increased the carrying capacity in their existing habitats.

3. Human populations sidestepped limiting factors.

The human population did all three of these things.

Reflect on the first point. The first humans evolved in woodlands and savannas. They were vegetarians, for the most part, but they also scavenged bits of meat. Starting about 2 million years ago, small bands of hunter-gatherers moved out of Africa. And by 40,000 years ago,

different populations of their descendants had become established in much of the world (Section 24.12).

Most other species could not have expanded into such a broad range of habitats. Humans, with their highly complex brains, could use learning and memory to figure out how to build fires, assemble shelters, make clothes and tools, and plan community hunts. Learned experiences did not die with individuals. They spread quickly from one band to another through language— the capacity for extraordinary cultural communication. *Therefore, the human population expanded into diverse new environments in an exceptionally short time, compared to the long-term geographic dispersal of other kinds of organisms.*

What about the second possibility? Starting about 11,000 years ago, many of the hunter-gatherers shifted to agriculture. Instead of following migratory game herds, they settled in fertile valleys and other places where the seasonal harvesting of fruits and grains was favored. By doing so, they developed a more dependable basis for life. A pivotal factor was the domestication of wild grasses, including species ancestral to modern wheats and rice. People harvested, stored, *and planted* seeds in one place. They domesticated animals and kept them close to home for food and for pulling plows. They dug ditches and diverted water to irrigate their croplands.

Their agricultural practices increased productivity. And the larger, more dependable food supplies favored increases in population growth rates. Towns and cities developed. A social hierarchy emerged that provided a labor base for more intensive agriculture. Much later, food supplies increased again by the use of fertilizers, herbicides, and pesticides. As transportation improved, so did food distribution. *Therefore, even at its simplest, management of food supplies through agriculture increased the carrying capacity for the human population.*

What about the third possibility—the sidestepping of limiting factors? Think about what happened when medical practices and sanitary conditions improved. Until about 300 years ago, poor hygiene, malnutrition, and contagious disease kept death rates high enough to more or less balance birth rates. Contagious diseases, which are density-dependent controls, swept unchecked through overcrowded settlements and cities that were infested with fleas and rodents. Then came plumbing and methods of sewage treatment. Over time, vaccines, antitoxins, and antibiotics were developed as weapons against many of the pathogens. Death rates dropped sharply. Births began to exceed deaths—and extremely rapid population growth was under way.

In the mid-eighteenth century, people discovered how to harness the energy stored in fossil fuels, starting with coal. Within a few decades, large industrialized societies started forming in western Europe and North America. Then efficient technologies developed after

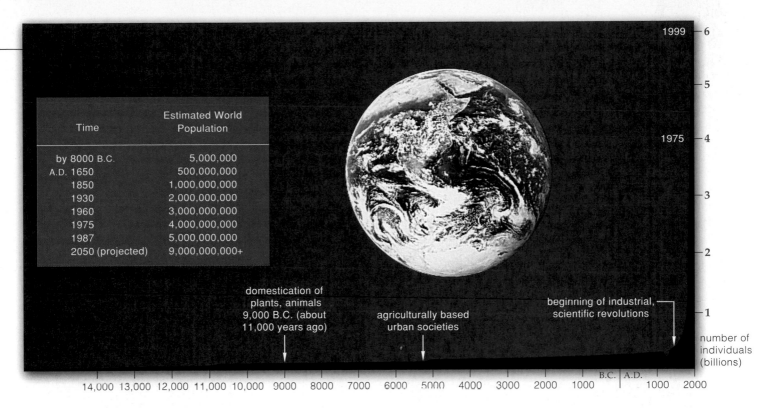

Time	Estimated World Population
by 8000 B.C.	5,000,000
A.D. 1650	500,000,000
1850	1,000,000,000
1930	2,000,000,000
1960	3,000,000,000
1975	4,000,000,000
1987	5,000,000,000
2050 (projected)	9,000,000,000+

domestication of plants, animals 9,000 B.C. (about 11,000 years ago)

agriculturally based urban societies

beginning of industrial, scientific revolutions

number of individuals (billions)

14,000 13,000 12,000 11,000 10,000 9000 8000 7000 6000 5000 4000 3000 2000 1000 | B.C. | A.D. | 1000 2000

Figure 39.9 Growth curve (*red*) for the human population. The graph's vertical axis represents world population, in billions. (The dip between the years 1347 and 1351 is the time when 25 million people died in Europe as a result of bubonic plague.) Agricultural revolutions, industrialization, and improvements in health care have been sustaining the accelerated growth pattern for the past two centuries. The data in the *blue* box show how long it took for the human population to increase from 5 million to 5 billion.

World War I. Large factories mass-produced tractors, cars, and other affordable goods. Machines replaced many of the farmers that had produced the food, and fewer farmers were able to support a larger population.

Therefore, by controlling many disease agents and by tapping into concentrated, existing forms of energy (fossil fuels), humans managed to sidestep major factors that had previously limited their population growth.

Present and Future Growth

Where have all the far-flung dispersals and spectacular advances in agriculture, industrialization, and health care taken us? Starting with *Homo habilis* (Section 24.11), it took about 2.5 million years for human population size to reach 1 billion. As Figure 39.9 shows, it took only 80 years to reach 2 billion, 30 to reach 3 billion, 15 to reach 4 billion, and 12 to reach 5 billion. And it took merely 11 more years to reach 6 billion!

From what we know about the principles governing population growth—and unless more breakthroughs in technology can increase the carrying capacity—we can expect a dramatic increase in death rates. *Although the stupendously accelerated growth of the human population continues, it cannot be sustained indefinitely.*

Besides having adverse effects on resource supplies, these skyrocketing numbers invite density-dependent controls. For example, infection by *Vibrio cholerae* results in a disease called cholera. The bacterium enters hosts who drink water or eat food that is contaminated with raw sewage. It multiplies in the gut, where it produces a toxin that triggers severe diarrhea and massive fluid loss. Within two to seven days, infected people who are not treated can die from extreme dehydration.

For centuries, *V. cholerae* has thrived and mutated into different strains in India's sewage-enriched rivers (Figure 39.1). In the slums of Calcutta alone, millions are forced to bathe in polluted ponds and waterways. In 1992, cholera struck tens of thousands in that city. Since then, the largest cholera epidemic of this century has been sweeping through southern Asia. Like the six previous epidemics, it began in India and then spread through Africa, the Middle East, and the Mediterranean countries. It has reached Central America and the United States. And this latest round of cholera resurgence may claim 5 million people. Existing vaccines do not work against the mutated bacterial strain causing it.

At this writing, a new era of human migration has begun. By some estimates, economic hardship and civil strife have put 50 million people on the move within and between many countries. Will the relocations prove peaceable? Where will they find sustainable supplies of food, clean water, and other basic resources?

Through expansion into new habitats, cultural intervention, and technological innovation, the human population has temporarily skirted environmental resistance to growth. Its accelerated growth cannot be sustained indefinitely.

WWW

CONTROL THROUGH FAMILY PLANNING

Figure 39.10 shows the 1997 annual rates of increase for populations in different parts of the world. The average rate of 1.47 percent is expected to decline. Even so, the world population is still projected to reach more than 9 billion by 2050.

It is mind numbing to think about all the natural resources required to sustain that many people. We will have to increase food production, supplies of drinkable water, energy reserves, and supplies of wood products and other materials to meet everyone's basic needs—something we are not even doing now. Most likely, the gross manipulation of resources will intensify pollution, which will adversely affect water supplies, air quality, and food productivity on land and in the seas.

Population growth, resource depletion, pollution, and the quality of life are all interconnected. Currently, most governments are attempting to lower birth rates, as through family planning programs. The programs educate individuals about choosing how many children they will have, and when. Their details vary from country to country, but all provide information on the available methods of fertility control, as outlined in Section 38.18. Carefully developed and administered programs might bring about a long-term decline in birth rates.

To arrive at zero population growth, the average "replacement rate" must be slightly higher than two children per couple, because some female children die before reaching reproductive age. The replacement rate is about 2.5 children per woman in the less developed countries, and 2.1 in the more developed countries. Yet even if each couple on the planet decided to have only two children, the population would keep on growing for another sixty years! Why? A huge number of existing children will soon be reproducing.

A more useful measure of global trends is the **total fertility rate**: the average number of children born to women during their reproductive years, as estimated on the basis of current age-specific birth rates. In 1997, the average rate was 3 children per woman. This is an impressive decline from 1950, when the total fertility rate was 6.5. It is still too far above the replacement level.

Figure 39.11 shows some age structure diagrams for populations growing at different rates. The central part of each diagram includes individuals of reproductive age. The lower part includes children who will move into the reproductive age category over the next fifteen

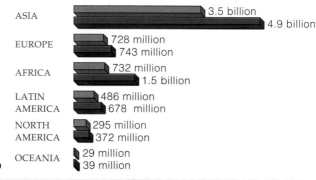

ASIA	3.5 billion / 4.9 billion
EUROPE	728 million / 743 million
AFRICA	732 million / 1.5 billion
LATIN AMERICA	486 million / 678 million
NORTH AMERICA	295 million / 372 million
OCEANIA	29 million / 39 million

b

Figure 39.10 (a) The 1997 average annual population growth rate in different regions of the world. (b) Population sizes in 1997 (*orange* bars) and projected for the year 2025 (*blue* bars).

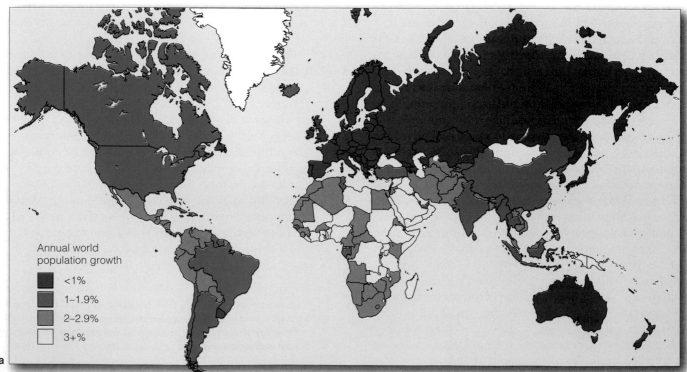

Annual world population growth

- <1%
- 1–1.9%
- 2–2.9%
- 3+%

a

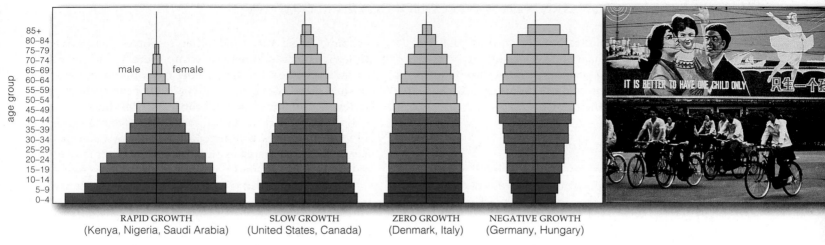

Figure 39.11 Age structure diagrams for countries with rapid, slow, zero, and negative population growth rates. *Dark green* signifies pre-reproductive years; *purple*, reproductive years; and *light blue*, post-reproductive years. The portion of the population to the left of the vertical axis in each diagram represents males, and that to the right, females. Bar widths correspond to the proportion of individuals in each age group. The photograph shows a family planning billboard in China.

years, with 15–44 the average range for childbearing years. The population of the United States has a relatively narrow base and is an example of slow growth. Figure 39.12 tracks its 78 million *baby-boomers*, a cohort that started to form in 1946 when soldiers returned home after World War II and started families. By contrast, age structure diagrams for rapidly growing populations have a broad base.

At present, *more than one-third of the world population falls in the broad pre-reproductive base.* The numbers give an idea of the magnitude of the effort it will take to control the world population growth.

The birth rate slows when women bear children during their early thirties, rather than in the mid-teens or early twenties. Delayed reproduction slows the rate of growth and lowers the average number of children in families. In China, for example, the government has established the world's most extensive family planning program. It strongly discourages premarital sex. It calls for pledges to postpone marriage and limit family size to one child only. Contraceptives, induced abortion, and sterilization are free to married couples; paramedics and mobile units ensure access to these measures in remote rural areas.

Couples who pledge to have only one child receive more food, free medical care, better housing, and salary bonuses. Their child will be granted free tuition and preferential treatment when he or she enters the job market. Those who break the pledge forgo benefits and pay higher taxes. Are the measures inhumane? Between

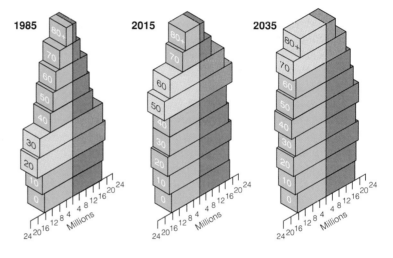

Figure 39.12 Age structure diagrams for the United States population. *Gold* bars track the baby-boom generation.

1958 and 1962 alone, possibly 30 million Chinese died of starvation as a consequence of widespread famines.

After 1972, China's total fertility rate declined from 5.7 to 1.8. Even so, the population time bomb has not stopped ticking. Currently its population is more than 1.23 billion—and 340 million of its young women are moving into the reproductive age category. By 2010, China's population is projected to surpass 1.3 billion.

Family planning programs on a global scale are designed to help stabilize the size of the human population.

Even if it reaches a level of zero population growth, the human population will continue to grow for sixty years, for its reproductive base already consists of a staggering number of individuals.

WWW

Today, more and more people are aware of a connection between economic development and population growth rates. When individuals are economically secure, they apparently are under less pressure to produce a large number of children to help them survive.

Demographic Transition Model

We can correlate changes in population growth with changes that typically unfold in four stages of economic development. The four stages are at the heart of the **demographic transition model** (Figure 39.13).

According to this model, living conditions are harsh during a *preindustrial* stage, before widespread use of technology and medical advances. Birth and death rates are both high, so the rate of population growth is low.

Next, during the *transitional* stage, industrialization begins, food production as well as health care improves. Death rates drop, but birth rates stay fairly high. Thus the population grows rapidly. And its growth continues at high rates for an extended length of time. Annual growth rates tend to be 2.5 to 3 percent, on the average. When living conditions improve and birth rates begin to decline, the growth starts to level off.

During the *industrial* stage, when industrialization is in full swing, the population's growth slows dramatically. A slowdown emerges mostly because people move from the country to cities —and urban couples tend to control the size of their families. Many couples get caught up in the accumulation of goods, and often decide that the time and cost of raising more than a few children conflict with that goal. As you will see, this is a sticky issue with respect to population size.

In the *postindustrial* stage, zero population growth is reached. The birth rate falls below the death rate and population size slowly decreases.

The United States, Canada, most of western Europe, Australia, Japan, and the nations of the former Soviet Union are in the industrial stage. Their growth rate is slowly decreasing. In Germany, Bulgaria, Hungary, and some other countries, the death rates exceed the birth rates, and the populations are getting smaller.

Mexico and other less developed countries are in the transitional stage. But they do not have enough skilled workers to complete the transition to a fully industrial economy. Fossil fuels and other resources that drive industrialization are being used up there as well as in the industrialized countries. Fuel costs might become prohibitive for countries at the bottom of the economic ladder even before they enter the industrial stage.

If population growth continues to outpace economic growth, death rates will increase. Thus many countries may be stuck in the transitional stage. And some may return to the harsh conditions of the preceding stage.

Enormous disparities in economic development are driving forces for immigration and emigration. In 1996 close to 916,000 legal immigrants and 75,000 political refugees crossed the borders of the United States. So did 275,000 undocumented immigrants, mainly from Latin America and Asia. They brought the estimated total of undocumented immigrants residing in this developed country to 5 million.

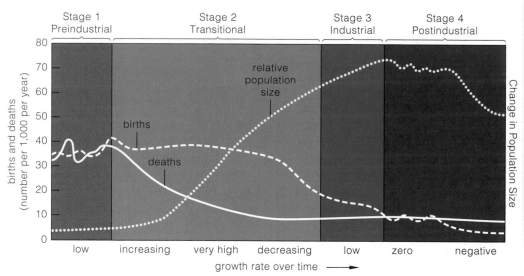

Figure 39.13 Diagram of the demographic transition model of changes in the growth characteristics and size of populations, correlated with changing economic development. The model explains past changes in western Europe and other industrialized regions.

Or think of how California's population increased by 12 *million* between 1970 and 1996. With its 2.5 percent growth rate, mainly among legal and illegal newcomers, it may jump by another 16 million by the year 2020. For some time, California has been a symbol of the good life, drawing people from the world over. However, its schools are overcrowded, funds for welfare and other social services are shrinking, sewage treatment plants are nearing capacity, and water shortages are severe and frequent. Also, both salinization of croplands and air pollution are eroding California's agricultural base, which is a major factor in the state's economic health.

Claiming that population growth affects economic health, many governments restrict immigration. Only the United States, Canada, Australia, and a few other

countries allow large annual increases in size. Elsewhere, appalling living conditions, harsh governmental policies, and civil strife combine to promote emigration of far more people than the better-off nations can handle.

A Question of Resource Consumption

This chapter opened with a brief look at the conditions that confront most people in India, with its whopping 16 percent of the world human population. By comparison, the United States has merely 4.7 percent.

Yet which country is the most "overpopulated"—not in terms of actual numbers, but rather in terms of resource consumption and instigation of environmental damage?

The highly industrialized United States produces 21 percent of the world's goods and services. And its people consume fifty times more goods and services than the average person in India. Its people use 25 percent of the world's processed minerals and available, nonrenewable sources of energy. Also, they generate at least 25 percent of the world's pollution and trash. By contrast, India produces about 1 percent of all goods and services. It uses 3 percent of the available minerals and nonrenewable energy resources. And it generates about 3 percent of the pollution and trash.

Extrapolating from these numbers, Tyler Miller, Jr., estimated that it would take 12.9 billion impoverished individuals living in India to have as much impact on the environment as 258 million Americans.

Differences in population growth among countries correlate with levels of economic development, hence with economic security (or lack of it) of individuals.

Growth rates are low in preindustrial, industrial, and post-industrial stages. They are greatest in the transitional stage.

For us, as for all species, the biological implications of extremely rapid growth are staggering. Yet so are the social implications of what will happen when (and if) the human population decreases to the point of zero population growth—and stays there.

For instance, in a growing population, most people fall in younger age brackets. If living conditions ensure constant growth over time, the age distribution should guarantee availability of a future work force. This has social implications. Why? *It takes a large work force to support the older age brackets.* In the United States, older, nonproductive people expect the Federal government to provide them with subsidized medical care, low-cost housing, and many other social programs. However, as a result of improved medicine and hygiene, people in those brackets are living far longer than the elderly did when the nation's social security program was set up. The current cash benefits exceed the contributions that they made to the program when *they* were younger.

If the population ever does reach and maintain zero growth, a larger proportion of individuals will end up in the older age brackets. Even slower growth will pose problems. What happens when baby-boomers retire? Will all those nonproductive members continue to get goods and services if productive ones must carry more and more of the economic burden? This is not an abstract question. Put it to yourself. Exactly how much economic hardship are you willing to bear for the sake of your parents? For your grandparents? And how much will your children be willing or able to bear for you?

We have arrived at a major turning point, not only in our biological evolution but in our cultural evolution as well. The decisions awaiting us are among the most difficult we will ever have to make—yet it is clear that they must be made, and soon.

All species face limits to growth. We might think we are different from the rest, for our special ability to undergo rapid cultural evolution has enabled us to postpone the action of most of the factors that limit growth. But the crucial word is *postpone*. No amount of cultural intervention can hold back the ultimate check of limited resources and a damaged environment.

We have sidestepped a number of the smaller laws of nature. In doing so, we have become more vulnerable to those laws which cannot be repealed. Today, there may be only two options available. Either we make a global effort to limit population growth in accordance with environmental carrying capacity, or we wait until the environment does it for us.

In the final analysis, no amount of cultural intervention can repeal the ultimate laws governing population growth, as imposed by the carrying capacity of the environment.

WWW

SUMMARY

1. A population is a group of individuals of the same species occupying a given area. It has a characteristic size, density, distribution, and age structure as well as characteristic ranges of heritable traits.

2. The growth rate for a population during a specified interval can be determined by calculating the rates of birth, death, immigration, and emigration. To simplify calculations, we may put aside effects of immigration and emigration, and combine the birth and death rates into a variable r (net reproduction per individual per unit time). Then we may represent population growth (G) as $G = rN$, where N is the number of individuals during the interval specified.

    a. In cases of exponential growth, the reproductive base of a population increases and its size expands by ever increasing increments during successive intervals. This trend plots out as a J-shaped growth curve.

    b. Any population will show exponential growth as long as its per capita birth rate stays even slightly above its per capita death rate.

    c. In logistic growth, a low-density population slowly increases in size, goes through a rapid growth phase, then levels off in size once carrying capacity is reached.

3. Carrying capacity is the name ecologists give to the maximum number of individuals in a population that can be sustained indefinitely by the resources available in their environment.

4. The availability of sustainable resources as well as other factors that limit growth dictates population size during a specified interval. The limiting factors vary in their relative effects and vary over time, so population size also changes over time.

5. Limiting factors such as competition for resources, disease, and predation are density-dependent. Density-independent factors, such as weather on the rampage, tend to increase the death rate or decrease the birth rate more or less independently of population density.

6. Patterns of reproduction, death, and migration vary over the life span of a species. Environmental variables also help shape the life history (age-specific) patterns.

7. The human population approached 6 billion in 1998. Its annual growth rate now varies from below zero in a few developed countries to above 3 percent in some less developed countries. In 1997 the annual growth rate for the entire human population was 1.47 percent.

8. The human population's rapid growth in the past two centuries occurred through its capacity to expand into new habitats, and because of agricultural, medical, and technological developments that increased the carrying capacity. Ultimately, we must confront the reality of the carrying capacity and limits to our population growth.

Review Questions

1. Define population size, population density, and population distribution. Describe a typical population in terms of several categories for its age structure. *39.1*

2. Define exponential growth. Be sure to state what goes on in the age category that is a foundation for its occurrence. *39.3*

3. Define carrying capacity and describe its effect as evidenced by a logistic growth pattern. *39.4*

4. Give examples of the limiting factors that come into play when a population of mammals (for example, rabbits or humans) reaches very high density. *39.4, 39.5*

5. Define doubling time. In what year is the human population expected to surpass 9 billion? *39.3, 39.7*

6. How did earlier human populations expand steadily into new environments? How did they increase the carrying capacity in their habitats? Have they avoided some limiting factors on population growth? Or is the avoidance an illusion? *39.7*

Self-Quiz (*Answers in Appendix III*)

1. _____ is the study of how organisms interact with one another and with their physical and chemical environment.

2. A _____ is a group of individuals of the same species that occupy a certain area.

3. The rate at which a population grows or declines depends upon the rate of _____ .
    a. births   c. immigration   e. all of the above
    b. deaths   d. emigration

4. Populations grow exponentially when _____ .
    a. birth rate exceeds death rate, immigration and emigration rates stay the same
    b. birth rate and death rate stabilize
    c. emigration rates exceed immigration rates
    d. both b and c

5. For a given species, the maximum rate of increase per individual under ideal conditions is the _____ .
    a. biotic potential   c. environmental resistance
    b. carrying capacity   d. density control

6. Resource competition, disease, and predation are _____ controls on population growth rates.
    a. density-independent   c. age-specific
    b. population-sustaining   d. density-dependent

7. Which of the following factors does *not* affect sustainable population size?
    a. predation   c. resources   e. all of the above can
    b. competition   d. pollution   affect population size

8. In 1998, the average annual growth rate for the human population was _____ percent.
    a. 0   b. 1.05   c. 1.47   d. 1.55   e. 2.7   f. 4.0

9. Match each term with its most suitable description.
    _____ carrying capacity   a. depends on rates of birth,
    _____ exponential growth   death, emigration, immigration
    _____ population   b. slow growth then rapid growth
       growth rate   phase, then leveling off
    _____ limiting factor   c. the maximum number of
    _____ logistic growth   individuals sustainable by
       an environment's resources
       d. population growth plots out
       as a J-shaped curve
       e. short supply of any resource
       essential to population growth

Critical Thinking

1. If house cats that have not been neutered or spayed live up to their biotic potential, two can be the start of many kittens—12 the first year, 72 the second year, 429 the third, 2,574 the fourth, 15,416 the fifth, 92,332 the sixth, 553,019 the seventh, 3,312,280 the eighth, and 19,838,741 kittens in the ninth year. Is this a case of logistic growth? Exponential growth? Irresponsible cat owners?

2. A third of the world population is below age fifteen. Describe the effect of this age distribution on the future growth rate of the human population. If you conclude that it will have severe impact, what sorts of humane recommendations would you make to encourage individuals of this age group to limit family size? What are some social, economic, and environmental factors that might keep them from following the recommendations?

3. Write a short essay about a population having one of the age structures shown below. Describe what may happen to younger and older groups when individuals move into new categories.

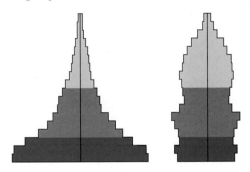

4. Figure 39.14 charts the legal immigration to the United States between 1820 and 1995. (The Immigration Reform and Control Act of 1986 accounted for the most recent dramatic increase; it granted legal status to illegal immigrants who could prove they had lived in the country for years.) During the 1980s and 1990s, an economic downturn fanned resentment against newcomers. Many people now say legal immigration should be restricted to 300,000–450,000 annually and we should crack down on illegal immigrants. Others argue that such a policy would diminish our reputation as a land of opportunity. They also say it would discriminate against legal immigrants during crackdowns on others of the same ethnic background. Do some research, then write an essay on the pros and cons of both positions. (You may wish to start with Web site http://www.census.gov.)

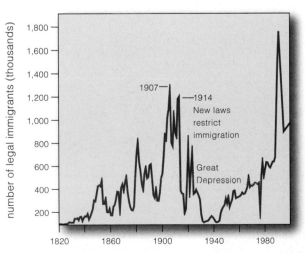

Figure 39.14 Chart of legal immigration to the United States between 1820 and 1997.

5. In his book *Environmental Science*, Miller points out that the unprecedented projected increase in the human population to 9+ billion by the year 2050 raises serious questions. Will there be enough food, energy, water, and other resources to sustain twice as many people? Will governments be able to provide adequate education, housing, medical care, and other social services for all of them? Computer models suggest that the answers are no (Figure 39.15). Yet some people claim we can adapt socially and politically to a far more crowded world, assuming harvests improve through technological innovation, every inch of arable land is cultivated, and everyone eats only grain.

There are no easy answers to these questions. If you have not yet been doing so, start following the arguments in your local newspapers, in magazines, on television, and on the Web. This will allow you to become an informed participant in a global debate that surely will have impact on your future.

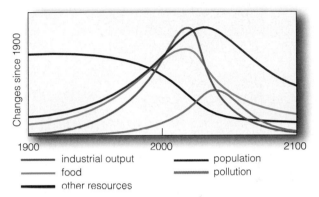

Figure 39.15 Computer-based projection of what may happen if the size of the human population continues to skyrocket without dramatic policy changes and technological innovation. The assumptions are that the population has already overshot the carrying capacity and current trends continue unchanged.

Selected Key Terms

age structure *39.1*	life history pattern *39.5*
biotic potential *39.3*	limiting factor *39.4*
capture-recapture method *39.2*	logistic growth *39.4*
carrying capacity *39.4*	migration *39.3*
demographic transition model *39.9*	per capita *39.3*
	population density *39.1*
demographics *39.1*	population distribution *39.1*
density-dependent control *39.4*	population size *39.1*
density-independent factor *39.4*	r (net reproduction per individual per unit time) *39.3*
doubling time *39.3*	reproductive base *39.1*
ecology *CI*	survivorship curve *39.5*
emigration *39.3*	total fertility rate *39.8*
exponential growth *39.3*	zero population growth *39.3*
immigration *39.3*	

Readings *See also www.infotrac-college.com*

Cohen, J. E. 1995. *How Many People Can the Earth Support?* New York: Norton. No pat answer to title's question.

Miller, G. T. 1999. *Environmental Science*. Seventh edition. Belmont, California: Wadsworth.

WWW *http://www.brookscole.com/biology*

Practice quiz questions, hypercontents, BioUpdates, and critical thinking. The Brooks/Cole Biology Resource Center provides a wealth of information fully organized and integrated by chapter.

40

COMMUNITY INTERACTIONS

No Pigeon Is An Island

Flying through the rain forests of New Guinea is an extraordinary pigeon with cobalt blue feathers and lacy plumes on its head (Figure 40.1). It is about as big as a turkey, and it flaps so slowly and noisily that its flight sounds like an idling truck. As is true of eight species of smaller pigeons living in the same forest, it perches on branches to eat fruit. How is it possible that *nine* species of large and small fruit-eating pigeons live in the space of the same forest? Wouldn't you think that competition for food would leave one the winner? In fact, in that rain forest, every species lives, grows, and reproduces in a characteristic way, as defined

Figure 40.1 A cobalt blue, turkey-size Victoria crowned pigeon, one of nine species of pigeons living in the same tropical rain forest of New Guinea. Within this habitat, each species has its own niche.

by its relationships with other organisms and with the surroundings.

Big pigeons perch on the sturdiest branches when they feed, and they eat big fruit. Smaller pigeons, with their smaller bills, cannot open big fruit. They eat small fruit hanging from slender branches that are not sturdy enough to support the weight of a turkey-size pigeon. The species of trees in the forest differ with respect to the diameter of their fruit-bearing branches and the size of their fruit. So they attract different pigeons with different characteristics. In such ways, the nine species of pigeons *partition* the fruit supply.

And how do individual trees benefit from enticing the pigeons to dine? The seeds inside their fruits have tough coats, which can resist the action of digestive enzymes inside the pigeon gut. During the time it takes for ingested seeds to travel through the gut, the pigeons fly about, so they dispense seed-containing droppings in more than one place. In this way, the pigeons tend to disperse seeds some distance away from the parent plants. Later, when seedlings grow, the odds are better that at least some will not have to compete with their parents for sunlight, water, and nutrients. Seeds that drop close to home cannot compete in any significant way with the resource-gathering capacity of mature trees, which already have extensive, well-developed roots and leafy crowns.

Within the same forest, leaf-eating, fruit-munching, and bud-nipping insects interact with other organisms and the surroundings in certain ways. So do the nectar-drinking, flower-pollinating bats, birds, and insects. And so do great numbers of beetles, worms, and other invertebrates that busily extract energy from remains and wastes of other organisms on the forest floor. By their activities, they cycle nutrients back to the trees.

Like humans, then, no pigeon is an island, isolated from the rest of the living world. The nine species of New Guinea pigeons eat fruit of different sizes. They disperse seeds from different sorts of trees. Dispersal influences where new trees will grow and where the decomposers will flourish. Ultimately, tree distribution and decomposition activities influence how the entire forest community is organized.

Directly or indirectly, interactions among coexisting populations organize the community to which they belong. With this chapter, we turn to community interactions that influence all populations over time and in the space of their environment.

KEY CONCEPTS

1. A habitat is the type of place where individuals of a species normally live. A community is an association of all the populations of species that occupy the same habitat.

2. Every species in the community has its own niche, defined as the sum of all activities and relationships in which its individuals engage as they secure and use the resources required for their survival and reproduction.

3. The structure of a community starts with adaptive traits that give individuals of each species the capacity to respond to the physical and chemical features of their habitat, and to levels and patterns of resource availability over time.

4. Interactions among species influence the structure of a community. They include mutually beneficial interactions, competition, predation, and parasitism.

5. Community structure also depends on the geographic location and size of the habitat, the rates at which the member species arrive and disappear, and the history of physical disturbances to the habitat.

6. The first species to occupy a particular type of habitat are replaced by others, which are replaced by still others, and so on in sequence. This process, known as primary succession, produces a climax community. A climax community is a stable, self-perpetuating array of species in balance with one another and with the environment.

7. Different stages of succession often exist in the same habitat, owing to local differences in the soil and other environmental factors, recurring disturbances such as seasonal fires, and chance events.

Think of a clownfish darting above a coral reef, a maple tree on a Vermont hillside, or a mole burrowing in your lawn. The type of place where you will normally find a clownfish, maple, or mole is its **habitat**. The habitat of an organism is characterized by physical and chemical features, such as temperature and salinity, and by the array of other species living in it. Directly or indirectly, the populations of all species in a habitat associate with one another as a **community**.

Five factors shape the structure of a community. *First*, interactions between climate and topography help dictate the habitat's temperatures, rainfall, soil types, and other conditions. *Second*, the kinds and amounts of food and other resources that become available through the year influence which species can live there. *Third*, individuals of each species have adaptive traits that allow them to survive and exploit specific resources in the habitat. *Fourth*, species in the habitat interact, as by competition, predation, and mutually helpful activities. *Fifth*, community structure is influenced by the overall pattern of population size (and the actual history of changes in its size), by the arrival and disappearances of species, and by physical disturbances to the habitat.

Together, the five factors help dictate the number of species at different "feeding levels," starting with the producers and continuing through levels of consumers. They influence population sizes. They also help dictate the overall number of species. For example, high solar radiation, warm temperatures, and high humidity in tropical habitats favor growth of many kinds of plants, which support many kinds of animals. Conditions in arctic habitats do not favor great numbers of species.

The chapters to follow deal with energy flow through feeding levels and with geographic factors influencing community structure. Here we begin with interactions among species, using the niche concept as our guide.

The Niche

If the organisms within a community all share the same habitat—that is, if they all have the same "address"—in what respects do they differ? Each kind is distinct in terms of its "profession" within the community—that is, in the sum of activities and relationships in which it engages to secure and use the resources necessary for its survival and reproduction. This is its **niche**.

For each species, the *fundamental* niche is the one that might prevail in the absence of competition and other factors that could constrain its acquisition and use of resources. However, as you will see, such constraining factors do come into play in all communities. They tend to bring about a more constrained, *realized* niche that shifts in large and small ways over time, as individuals of the species respond to a mosaic of changes.

Categories of Species Interactions

Dozens to hundreds of species interact in diverse ways, even in simple communities. In spite of this diversity, we can identify six categories of interactions that have different effects on population growth (Table 40.1).

Table 40.1　　Categories of Two-Species Interactions*		
Type of Interaction	Direct Effect on Species 1	Direct Effect on Species 2
Neutral relationship	0	0
Commensalism	+	0
Mutualism	+	+
Interspecific competition	−	−
Predation	+	−
Parasitism	+	−

* 0 means no direct effect on population growth;
  + means positive effect; − means negative effect.

Each species has a *neutral* relationship with most species in its habitat. For example, Canadian lynx and grasses do not affect each other directly. Interactions with other species link them only indirectly. The lynx preys on and decreases the number of snowshoe hares that eat plants; and plants fatten the hares, the prey of lynx.

Commensalism directly helps one species but does not affect the other much, if at all. For instance, some birds use tree branches for roosting sites. The trees get nothing but are not harmed. With **mutualism**, benefits flow both ways between the interacting species. (Don't think of this as cozy cooperation. Benefits flow from a two-way exploitation.) With **interspecific competition**, disadvantages flow both ways between species. Finally, **predation** and **parasitism** are interactions that directly benefit one species (either the predator or the parasite) and directly hurt the other (the prey or host).

Commensalism, mutualism, and parasitism are all examples of **symbiosis**, which means "living together." For at least part of their life cycle, individuals of one species live near, in, or on individuals of another species.

A habitat is the type of place where individuals of a species normally live. A community consists of all populations that live in the habitat.

Community structure arises from the habitat's physical and chemical features, resource availability over time, adaptive traits of its members, how the members interact, and the history of the habitat and its occupants. Species may have neutral, positive, or negative effects on one another.

A niche is the sum of all activities and relationships in which individuals of a species engage as they secure and use the resources necessary to survive and reproduce.

WWW

Mutualistic interactions, in which positive benefits flow both ways, abound in nature. When trees provide New Guinea pigeons with food and when the pigeons help disperse seeds from the trees to new germination sites, both function as mutualists. Many other kinds of plants and animals enter into such interactions. For example, most flowering plants and the insects and birds, bats, and other animals that pollinate them are mutualists. The introduction to Chapter 27 gives vivid examples.

of absorptive structures interact in two ways. Either the fungal hyphae penetrate root cells or they grow as a dense, velvety mat around them. The fungus is good at absorbing mineral ions from soil, and the plant comes to depend on tapping into some of these. In turn, the fungus withdraws some sugar molecules that the plant makes by photosynthesis. It also depends on the plant for its own reproductive success. When photosynthesis ceases, the fungus stops producing spores.

Figure 40.2 One mutualistic interaction on a rocky slope of Colorado's high desert.

(**a**) Different kinds of flowering plants of the genus *Yucca* are each pollinated exclusively by one species of yucca moth (**b**). This insect cannot complete its life cycle with any other plant.

The adult stage of the moth life cycle coincides with blossoming of yucca flowers. By using her specialized mouthparts, a female moth gathers sticky pollen and rolls it into a ball. Then she wings her way to another flower. She pierces the wall of the flower's ovary, where seeds form and develop, and lays her eggs inside. As she crawls out of the flower, she pushes a ball of pollen onto a pollen-receiving surface.

Pollen grains germinate and grow down through tissues of the ovary. They deliver sperm to the flower's eggs. The seeds develop after fertilization. Meanwhile, the moth eggs develop to the larval stage. (**c**) When larvae emerge, they eat a few seeds, then gnaw their way out of the ovary. The seeds that moth larvae do not eat give rise to new yucca plants.

Some forms of mutualism are *obligatory*. That is, the individuals of one species cannot grow and reproduce in the absence of intimate dependency with individuals of another species during the life cycle. This is the case for an interaction between yucca plants and yucca moths. Each plant of this genus (*Yucca*) can be pollinated only by one species of the yucca moth genus. In addition, larval stages of the moth grow only in the yucca plant; they eat only yucca seeds (Figure 40.2).

We also find cases of obligatory mutualism between many fungi and plants. Mycorrhizae, remember, are intimate ecological interactions between fungal hyphae and young roots (Chapters 21 and 26). The two kinds

And what about the apparent endosymbiotic origins of eukaryotes? Long ago, phagocytic bacteria may have engulfed other, aerobic bacteria—which resisted digestion, used nutrients from its host, and reproduced independently of it. The hosts came to depend on the ATP produced by its guests, which evolved into mitochondria and chloroplasts. If those prokaryotic cells had not become so mutually interdependent, you and all other eukaryotic organisms would not even be around today (Section 19.4).

In cases of mutualism, each of the participating species reaps benefits from the interaction.

WWW

COMPETITIVE INTERACTIONS

Organisms do not come into the world with guarantees that they will secure enough energy, nutrients, living space, and other necessities to survive and reproduce. They typically compete for a share of limited resources. *Intraspecific* competition means that individuals of the same population or species compete with one another. As you probably deduced from the preceding chapter, their interactions can be fierce. *Interspecific* competition, which occurs between populations of different species, usually is not as intense. Why? *The requirements of two species might be similar, but they never can be as close as they are for individuals of the same species.*

Think about two forms of competitive interactions. Sometimes individuals have equal access to a required resource, but some are better than others at exploiting it. In such cases, competition tends to reduce the supply of a shared, limited resource. (When you and a friend both use straws to share a small milkshake, you might not get nearly as much if your friend uses a jumbo straw.) In other cases, some individuals control access to a resource and partially or completely prevent others from using it regardless of its scarcity or abundance. (For instance, even if you shared a ten-gallon milkshake, you still would get less if your friend pinched your straw.)

Competition abounds in nature. Chipmunks exclude other species of chipmunks from their habitats (Figure 40.3). A strangler fig tree wraps around other trees as a framework for its own growth and finally kills them. From early spring until late summer, a male *broadtailed* hummingbird busily chases other males and females of its species from his blossoming territory in the Rockies. In August, however, *rufous* hummingbirds of the Pacific Northwest migrate through the Rockies on their way to wintering grounds in Mexico. The rufous males prove to be more aggressive and stronger competitors for the available food. And they evict the male broadtails from broadtail territories all along their migratory route.

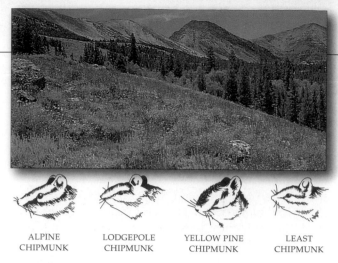

ALPINE CHIPMUNK   LODGEPOLE CHIPMUNK   YELLOW PINE CHIPMUNK   LEAST CHIPMUNK

Figure 40.3 Example of competition in nature. On the eastern slopes of the Sierra Nevada, different chipmunk species occupy different habitats. The alpine habitat is at the highest elevation. Below it are the lodgepole pine, piñon pine, then the sagebrush habitats. The least chipmunk lives at the base of the mountains in sagebrush. Its adaptations would allow it to move up into the piñon pine habitat, but the aggressively competitive behavior of yellow pine chipmunks in that habitat won't let it. Food preferences keep the yellow pine chipmunk out of the sagebrush habitat.

Competitive Exclusion

To a greater or lesser extent, any two species differ in their adaptations for getting food and avoiding enemies. Thus one usually competes more effectively for scarce resources. The two are less likely to coexist in the same habitat when they use resources in very similar ways. G. Gause demonstrated this by growing two species of *Paramecium* separately, then together (Figure 40.4*a–c*). Both exploited the same food (bacteria) and competed intensely for it. As the experimental results suggested, two species requiring identical resources cannot coexist indefinitely. Many subsequent studies also support this concept, which is now called **competitive exclusion**.

Gause also studied two other species of *Paramecium*, which did not overlap as much in their requirements. When grown together, one tended to feed on bacteria suspended in liquid inside a culture tube. The other fed on yeast cells at the tube bottom. The population growth rate slowed down for both species, but the overlap in resource use was not enough for one species to fully exclude the other. In other words, they continued to coexist.

Similarly, field experiments with keystone species reveal the effects of competition. A **keystone species** is a dominant species that dictates community structure. Robert Paine identified one type through removal experiments in the intertidal zones along North America's west coast. Pounding surf, tides, and storms

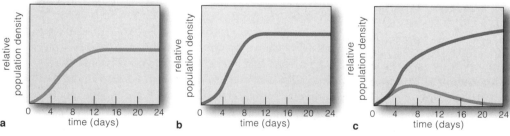

a — relative population density — time (days) — 0 4 8 12 16 20 24
b — relative population density — time (days) — 0 4 8 12 16 20 24
c — relative population density — time (days) — 0 4 8 12 16 20 24

Figure 40.4 Results of competitive exclusion between two protistan species that compete for the same food. (**a**) *Paramecium caudatum* and (**b**) *P. aurelia* were grown in separate culture tubes and established stable populations. The S-shaped growth curves in these graphs indicate stability. (**c**) Then the populations were grown together. *P. aurelia* (*red* curve) drove the other species toward extinction (*blue* curve in **c**). This experiment and others suggest that two species cannot coexist indefinitely in the same habitat *when they require identical resources*. If their requirements do not overlap much, one might influence the population growth rate of the other, but they may still coexist.

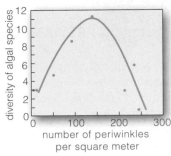

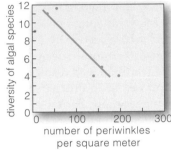

d Algal diversity in tidepools

e Algal diversity on rocks that are alternately exposed and submerged

Figure 40.5 Effect of competition and predation on community structure. (**a**) Periwinkles (*Littorina littorea*) influence the number of algal species in different ways in different marine habitats. (**b**) *Chondrus* and (**c**) *Enteromorpha*, two algae in their natural habitat. (**d**) In tidepools, periwinkles graze on a dominant alga (*Enteromorpha*). Thus less competitive algae that might otherwise be overwhelmed have an advantage. (**e**) Algal diversity is lower on rocks exposed at low tide; periwinkles do not eat the dominant species (*Chondrus* and other red algae).

repeatedly disturb the zones, and living space is scarce. Paine put a sea star (*Pisaster*) and its prey in his control plots. *Pisaster*, a keystone species, controls abundances of mussels (*Mytilus*), limpets, chitons, and barnacles.

After Paine removed all sea stars from experimental plots, mussels took over and crowded out seven other invertebrate species. Mussels, the main prey of sea stars, become the strongest competitors when sea stars are absent. Thus sea star predation normally maintains the diversity of prey species because it blocks competitive exclusion (by mussels). Remove the sea stars, and the community shrinks from fifteen species to eight.

Similarly, Jane Lubchenco found that the periwinkle (*Littorina littorea*), an alga-eating mollusk, promotes *or* lowers diversity in different habitats. In tidepools, it eats dominant algal species and thus helps many other, less competitive algae survive. But on rocks exposed only at high tide, *L. littorea* ignores tough, unpalatable species and prefers the competitively weak algal species that it ignores in tidepools. Thus *L. littorea* promotes algal diversity in tidepools yet lowers it on rocks exposed at high tide (Figure 40.5).

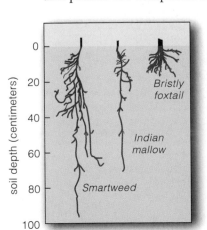

Figure 40.6 Resource partitioning among three annual plants in a plowed, abandoned field. All require water and mineral ions but differ in their adaptations for securing them.

Resource Partitioning

Think back on the nine species of fruit-eating pigeons in the same New Guinea forest. They require the same resource: fruit. Yet they overlap only slightly in their use of the resource, because each specializes in fruits of a particular size. Together, they are a good example of **resource partitioning**—the *subdividing* of some category of similar resources that lets competing species coexist.

A similar competitive situation arises among three species of annual plants in certain plowed, abandoned fields. Like other plants, all three require sunlight, water, and dissolved mineral ions. Yet each species is adapted to exploiting a different portion of the habitat (Figure 40.6). Drought-tolerant foxtail grasses have a shallow, fibrous root system that quickly absorbs rainwater. They grow where moisture in the soil varies from day to day. Mallow plants, with a taproot system, grow in deeper soil that is moist early in the growing season but drier later. Smartweed's taproot system branches in topsoil and soil below the roots of the other species. It grows where soil is continuously moist. In sum, species may coexist in the same habitat even if their niches overlap.

In some competitive situations, all individuals have equal access to a resource that they all require, but some are better than others at exploiting it. In other competitive situations, some individuals control access to a resource.

The more that two species in the same habitat differ in their use of resources, the more likely they can coexist.

Two competing species also may coexist by sharing the same resource in different ways or at different times.

PREDATION AND PARASITISM

To keep things simple, let's use two broad definitions for interactions between various consumers and their victims. **Predators** are animals that feed on other living organisms, which are called their **prey**, but do *not* take up residence on or in them. Their prey may or may not die from the interaction. **Parasites** live in or on other living organisms—their **hosts**—and feed upon specific host tissues for part of their life cycle. Their hosts may or may not die as a result of the interaction.

Many adaptations of predators (or parasites) and their victims arose by **coevolution**, the joint evolution of two (or more) species that exert selection pressure on each other as a result of close ecological interaction. Suppose, as an outcome of mutation, a prey organism displays a new, heritable means of defense, which later spreads through the prey population. Some individual predators are more effective than others at countering the defense. They tend to eat more and have a better chance to survive and reproduce. In time the forms of traits that are most effective at overcoming the defense increase in frequency in the population. Now the more effective predators exert selection pressure that favors better defenses among prey—and so on through time.

Dynamics of Predator–Prey Interactions

In any specified interval, the outcome of predator–prey interactions depends partly on the carrying capacity of the prey population. *Carrying capacity*, recall, is defined as the maximum number of individuals that resources in a given environment can maintain indefinitely. The reproductive rates of predator and prey populations also influence the outcome. So do predator responses to increases in prey density.

During the times when predation is keeping a prey population from exceeding the carrying capacity, both populations tend to coexist at fairly steady levels. More

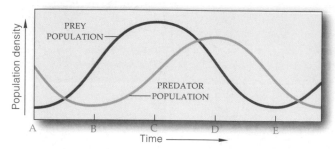

Figure 40.7 Idealized cycling of abundances of predator and prey. (For clarity, the diagram exaggerates predator density; predators usually are less common than their prey during the cycle.) This pattern arises owing to time lags in predator responses to changes in prey abundance. At time A, prey density is low; predators have a harder time getting food; their population is declining. In response to the predator decline, prey increase. But predators do not start increasing until they start reproducing at time B. Both populations grow until the predator increase causes the prey population to decline (time C to E). Predators continue to increase and take out more prey. But lower prey density leads to starvation among predators, and their growth rate slows (starting at time D). At time E, a new cycle starts.

prey are around; predators reproduce promptly and eat more prey. Population densities fluctuate if predators don't reproduce as fast as the prey, if they can eat only so many prey organisms in a given interval, and when the carrying capacity for the prey population is high.

Figure 40.7 shows an idealized pattern of the cyclic changes corresponding to time lags in the predator's response to changes in prey abundance. In nature, we sometimes observe such a correspondence, but other factors also influence the pattern. Consider Figure 40.8.

Figure 40.8 Predator–prey interactions between the Canadian lynx and snowshoe hare. The correspondence between the abundances of both populations is based on counts of pelts sold by trappers to Hudson's Bay Company over ninety years. The *dashed* line tracks lynx abundances, and the *solid* line, the hare abundances. The chart is a good test of whether you readily accept someone else's conclusions without questioning their scientific basis. (Remember the discussion of scientific methods in Chapter 1?) What other factors might have influenced the cycle? Did weather vary, with more severe winters imposing greater demand for food (required for animals to keep warmer) and higher death rates? Did the lynx have to compete with other predators for prey? Did predators turn to alternative kinds of prey during low points of the hare cycle? Did trapping increase with rising fur prices in Europe, and did it decrease as the pelt supply outstripped the demand?

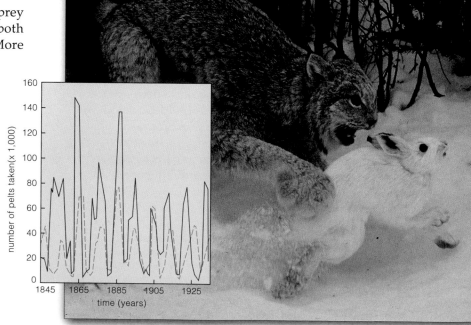

Figure 40.9
Adult roundworms (*Ascaris*), a type of endoparasite, packed inside a small intestine from a host pig.

It reveals a ten-year cycling in populations of the Canadian lynx and showshoe hare. Charles Krebs led a long-term study in Alaska's Yukon to identify the basis of this cycling. For eight years, Krebs tracked hare densities in 1-square-kilometer control plots and experimental plots. In some plots, electrified fences kept out mammalian predators but not hares. In others, hares were supplied with extra food. In still other plots, fertilizers enhanced plant growth for the hares. Cyclic peaks and declines corresponded to increased hare densities in plots that were predator-free and had extra food. In these plots, densities were thirty-six times greater than they were in control plots. By itself, increased vegetation (in fertilized plots) had negligible effect on hare densities. A simple predator-prey model or plant-herbivore model cannot explain the results. This cycle seems to turn on plants, herbivores, and carnivores, a *three-level* interaction.

Dynamics of Parasite–Host Interactions

Parasites have pervasive influences in populations. By draining hosts of nutrients, they alter how much energy and how many nutrients the population is withdrawing from the habitat. Weakened hosts are more vulnerable to predation and not as appealing to potential mates. Some infections result in sterility. Some may alter the ratio of host males to females. In such ways, parasitic infections lower the birth rate, raise the death rate, and influence intraspecific and interspecific competition.

Sometimes the slow drain of nutrients in a parasitic infection indirectly causes death, for a host may become so weakened that it dies from secondary infections. In evolutionary terms, however, killing a host is not good for the parasite's reproductive success. An infection of longer duration gives a parasite more time to produce more offspring. Thus natural selection tends to favor parasite and host adaptations that promote some level of mutual tolerance and less-than-fatal effects. Usually, death results when a parasite attacks a novel host (which has no coevolved defenses against it) or when too many individual parasites are active at the same time.

*Ecto*parasites live on a host's surface; *endo*parasites live inside a host's body. Some species live on or in one or more hosts for their entire life cycle. Others are free-living some of the time. Many use arthropods, such as insects, like taxicabs between hosts. The **microparasites** among them include many bacteria, viruses, protozoans, and sporozoans. Most are microscopically small and reproduce rapidly. The larger, **macroparasites** include many of the flatworms, roundworms (nematodes), and arthropods such as fleas and ticks (Figure 40.9).

Then there are parasitic plants. Nonphotosynthetic types withdraw nutrients and water from young roots of host plants. Members of the broomrape family do this to oak and beech tree roots. Other plants retain the capacity for photosynthesis but still withdraw nutrients and water from host plants. Mistletoe is an example; its extensions invade the sapwood of host trees.

Not all parasites feed on host tissues. **Social parasites** complete their life cycle by drawing on social behaviors of another species. For instance, a cowbird never builds a nest, incubates eggs, or cares for offspring. It removes an egg from another bird's nest and proceeds to lay a "replacement." Some birds hatch the alien egg and raise the hatchling. Usually, the young cowbird aggressively pushes the smaller, remaining rightful occupants out of the nest or demands and gets most of the food.

Many parasites are being commercially raised and selectively released as *biological controls*. They are touted as an alternative to chemical pesticides. But less than 20 percent of existing selections qualify as effective defenses against pests. As outlined by C. Huffaker and C. Kennett, effective biological control agents display five attributes. They are well adapted to a host species and to their habitat. They are exceptionally good at searching for a host. Their population growth rate is high relative to that of the host species. Their offspring are mobile enough to ensure adequate dispersal. And the lag time between their responses to changes in the numbers of the host population is minimal.

Releasing more than one kind of biological control agent in an area may trigger competition among them and eventually lessen their overall effectiveness. Also, a shotgun approach to biological control is risky. There is a possibility that the parasites may attack nontargeted species. In 1983, for example, F. Howarth reported that the populations of butterflies and moths native to the Hawaiian Islands are declining. Introductions of wasps that were supposed to serve as biological controls over something else are partly to blame. Later in the chapter, we will return to the effects of introducing new species into a habitat on native populations.

Predators and their prey, as well as parasites and their hosts, are locked in long-term, coevolutionary contests.

Predator and prey populations may coexist at fairly steady levels. Others may undergo recurring cycles of abundance and crashes, erratic cycles, or prey extinction.

Natural selection favors parasitic species that temper their demands in ways that ensure an adequate supply of hosts.

WWW

40.5

THE COEVOLUTIONARY ARMS RACE

Populations of other species are part of any organism's environment, and the ones that interact as predators and prey exert continual selection pressure on each other. One must defend itself and the other must overcome the defenses. This is the basis of a coevolutionary arms race that has resulted in some truly amazing adaptations.

CAMOUFLAGE Consider prey species that **camouflage** themselves; they can hide in the open. Such organisms have adaptations in form, patterning, color, and behavior that help them blend with their surroundings and escape detection. Figure 40.10 shows classic examples, including a desert plant (*Lithops*) that resembles a small rock. Only during a brief rainy season does *Lithops* flower. That is when other plants grow profusely and divert herbivores from *Lithops*—and when free water instead of juicy plant tissues is available to quench an animal's thirst.

WARNING COLORATION Many prey species taste bad, are highly toxic, or inflict pain on attackers. Often the toxic types have **warning coloration**, or conspicuous patterns and colors that predators learn to recognize as "avoid me" signals. For example, maybe a young, inexperienced bird will spear a yellow-banded wasp or an orange-patterned monarch butterfly—once. It will quickly learn to associate the distinctive colors and patterning with a painful sting or with vomiting foul-tasting butterfly toxins.

Truly dangerous or repugnant species make little or no attempt to conceal themselves. Skunks are like this. So are frogs of the genus *Dendrobates*; they are among the most vivid and most poisonous organisms (Section 28.1).

MIMICRY Many prey organisms bear close resemblance to dangerous, unpalatable, or hard-to-catch species. Any close resemblance in form, behavior, or both between one species that serves as a *model* for deception and another that is its *mimic* is called **mimicry**. Figure 40.11 shows how some tasty but weaponless mimics physically resemble species that predators have learned to ignore. In "speed" mimicry, sluggish prey species look like swift-moving species that predators have given up trying to catch.

MOMENT-OF-TRUTH DEFENSES When luck runs out, survival of prey organisms that are cornered or under attack may turn on a last-ditch trick. Suppose a leopard runs down a tasty baboon. By turning abruptly and displaying formidable canines, the baboon may startle and confuse this predator long enough for a getaway (see the last page of Section 44.11). Other cornered animals spew chemicals as disgusting repellents or toxins. Skunks, earwigs, and stink beetles produce awful odors. Several beetle species take aim and let loose with noxious sprays.

Similarly, many plants synthesize predator repellents. Tannins in the foliage and seeds of certain plants taste

Figure 40.10 A few prey organisms demonstrating the fine art of camouflage. (**a**) What bird??? When a predator approaches its nest, the least bittern stretches its neck (which is colored like the surrounding withered reeds), thrusts its beak upward, and sways gently like reeds in the wind. (**b**) An unappetizing bird dropping? No. The body coloration of this caterpillar and the stiff positions it takes help it hide in the open from birds that prey on it. (**c**) Find the plants (*Lithops*) hiding in the open from herbivores, owing to their stonelike form, pattern, and coloring.

Figure 40.11 A few examples of mimicry among the insects.

Many predators avoid prey that taste awful, secrete toxins, or inflict painful bites or stings. Commonly, such prey display warning coloration (bright colors, bold markings, or both), and many do not even bother to hide. Many species of prey unrelated to the dangerous or unpalatable ones have evolved striking behavioral and morphological resemblances to them. (**a**) The yellowjacket shown here stings aggressively and probably is the model for nonstinging, edible wasps (**b**) and beetles (**c**) that have a similar appearance. The inedible butterfly in (**d**) is a model for the edible mimic *Dismorphia* (**e**).

a The dangerous model . . .   **b** . . . one of its edible mimics . . .   **c** . . . and another edible mimic

bitter and make the plant tissues hard to digest. Make the mistake of nibbling on the seemingly luscious yellow petals of a buttercup (*Ranunculus*), and you will inflict a chemical burn upon the lining of your mouth.

PREDATOR RESPONSES TO PREY As part of the coevolutionary arms race, predators counter prey defenses with their own marvelous adaptations. Among the countering measures are stealth, camouflage, and clever ways of avoiding repellents.

Consider the edible beetles that direct sprays of noxious chemicals at their attackers. Grasshopper mice grab such beetles and plunge the "sprayer" end into the ground, then feast on the unprotected head (Figure 40.12*a,b*). Chameleons typically hold themselves motionless for extended intervals. Prey might not even "see" them until the amazingly swift chameleon tongue zaps them (Section 28.4). And it is no accident that stealthy predators blend with backgrounds. Think of snow-white polar bears camouflaged against snow, golden tigers crouched in tall-stalked, golden grasses, and pastel predatory insects lurking in pastel flowers. And hope you never step barefoot on a scorpionfish concealed on the seafloor.

Figure 40.12 Examples of adaptive responses of predators to prey defenses. (**a**) Certain beetles spray noxious chemicals at attackers, which works as a deterrent some of the time. (**b**) At other times, however, grasshopper mice plunge the chemical-spraying tail end of their beetle prey into the ground and feast on the head end. (**c**) Find the scorpionfish, a venomous predator with camouflaging fleshy flaps, multiple colors, and profuse spines. (**d**) Where do pink parts of the flower end and pink parts of the praying mantis begin?

A Successional Model

By now you might be asking: How does a community come into being? By the classical model of **ecological succession**, a community develops in sequence, from pioneer species to an end array of species that remain in equilibrium over some region. **Pioneer species** are opportunistic colonizers of vacant or vacated habitats. They enjoy high dispersal rates and rapid growth. In time, more competitive species replace the pioneers, then are themselves replaced until the array of species stabilizes under the prevailing habitat conditions. This persistent array of species is the **climax community**.

Primary succession is a process that begins as pioneer species colonize a barren habitat such as a new volcanic island or land exposed by a retreating glacier (Figure 40.13). Typical pioneer species are small plants having brief life cycles and adaptations for growing in exposed sites with intense sunlight, large temperature changes, and nutrient-deficient soil. Each year they produce great numbers of small seeds, which are quickly dispersed.

Once established, the pioneers improve conditions for other species and often set the stage for their own replacement. Many types are mutualists with nitrogen-fixing bacteria and initially outcompete other plants in nitrogen-poor habitats. Their accumulated remains and wastes add volume and nutrients to the soil that help other species to take hold. The pioneers also form low-growing mats that shelter the seeds of later species yet cannot shade out seedlings. In time, later successional species crowd out the pioneers, whose seeds travel as fugitives on the wind or water—destined, perhaps, for a new but temporary habitat.

In **secondary succession**, a disturbed area within a community recovers and moves again toward a climax state. The pattern is typical of abandoned fields, burned forests, and storm-battered intertidal zones. It emerges after falling trees open part of an established forest's canopy. Sunlight reaches seeds and seedlings that are already on the forest floor and spurs their growth.

By one hypothesis, the colonizers *facilitate* their own replacement. By another, the earliest colonizers compete against the species that could replace them, so that the sequence of succession depends on who gets there first.

The Climax-Pattern Model

At one time, some ecologists thought the same general type of community always develops in a given region because of constraints imposed by climate. However, stable communities other than "the climax community" commonly persist in a region, as when tallgrass prairie extends from the west into Indiana's deciduous forests. By the **climax-pattern model**, a community is adapted

Figure 40.13 (**a**) Primary succession in Alaska's Glacier Bay area. This glacier has been retreating since 1794. (**b**) As a glacier retreats, meltwater leaches newly exposed soil of minerals, including nitrogen. Less than ten years ago, ice buried this nutrient-poor soil. (**c**) *Facing page:* Seeds of mountain avens (*Dryas*), a pioneer species that benefits from nitrogen-fixing activities of mutualistic microorganisms, are the first invaders. *Dryas* grows and spreads rapidly over the glacial till.

(**d**) Twenty years later, shrubby deciduous alders, cottonwood, and willows take hold. They are symbionts with nitrogen-fixing microbes. In time, the alders form dense thickets. As the alder thickets mature, the cottonwood as well as hemlock trees and a few evergreen spruce trees grow rapidly. (**e**) After eighty years, spruce crowd out the mature alders. (**f**) In areas deglaciated for more than a century, dense forests of Sitka spruce and western hemlock dominate.

to many environmental factors—topography, climate, soil, wind, species interactions, recurring disturbances, chance events, and so on—that vary in their influence over a region. As a result, even in the same region you may see one climax stage extending into another along gradients of influential environmental conditions.

Cyclic, Nondirectional Changes

Many small-scale changes occur over and over again in patches of a habitat. Such recurring changes contribute to the internal dynamics of the community as a whole. Thus, observe a disturbed patch of habitat, and you might conclude that great shifts in species composition are afoot. Observe the community on a larger scale, and you may find that the overall composition includes all of the pioneer species that are colonizing the patch as well as the dominant climax species.

c   d   e

Think of a tropical forest. It slowly develops through phases of colonization by pioneer species, increases in species diversity, and then reaches maturity. Whipping sporadically through the successional pattern, however, are heavy winds that cause treefalls. Where trees fall, gaps open in the forest canopy and more light reaches the forest floor. In that local patch, conditions favor the growth of previously suppressed small trees as well as germination of pioneers or shade-intolerant species.

Or think of the groves of sequoia trees in the Sierra Nevada in California. Some trees in this type of climax community are giants, more than 4,000 years old. Their persistence depends partly on recurring brush fires that sweep through parts of the forests. Sequoia seeds only germinate in the absence of smaller, shade-tolerant plant species. Too much litter on the forest floor inhibits germination. Modest fires eliminate trees and shrubs that compete with young sequoias but do not damage the sequoias themselves. Mature sequoias have thick bark that burns poorly and insulates the living phloem cells of the giant trees against modest heat damage.

At one time, fires were prevented in many sequoia groves in national and state parks—not just accidental fires from campsites and discarded cigarettes, but also natural fires sparked by lightning. Litter builds up if rangers prevent small fires. Fire-susceptible species can take hold, and underbrush gradually thickens. Dense underbrush prevents germination of sequoia seeds and fuels hotter fires that damage the giants. Park rangers now set controlled fires. By removing underbrush early on, fires promote conditions that are the most favorable for the community's cyclic replacements.

Restoration Ecology

Think back on the regrowth after Mount Saint Helens' violent eruption in 1980 (Chapter 25), and you know that secondary succession can heal large-scale disturbances. In terms of a human lifetime, such *natural* restoration of the climax community often takes a long time. Today we also have *active* restoration: attempts to reestablish biodiversity in large areas disturbed by farming, mining, and other human activities. For example, in 1972, "lost" natural prairie species were discovered in old cemeteries and other neglected bits of land in Illinois. Volunteers transplanted them to a site measuring 180 hectares (445 acres) and now maintain the restored patch of prairie by controlled burns and hand weeding. They hope to find and transplant all of the 150–200 original prairie species.

Similarly, volunteers build artificial reefs near coasts and also repair wetlands. **Wetlands** are transitional zones between aquatic and terrestrial ecosystems where plants live in periodically or permanently saturated soil. They include *marshes* submerged in 15 to 300 centimeters of water, *riparian zones* of most rivers (including shrubby bottomlands), and *mangrove swamps* (Figure 40.14). Many species must complete at least part of their life cycle in wetlands. Migratory birds cannot journey without resting at them. Wetlands also replenish groundwater supplies, improve water quality, and control erosion and flooding.

For years, wetlands have been drained for farming, industry, home building, and other activities. Only 50 percent of the spectacular Florida Everglades remains. Only 9 percent of California's original wetlands are left; 10 percent remain in Indiana and Missouri. Government agencies, private agencies, Native American groups, and individuals are working to reverse the damage.

Figure 40.14
Mangrove swamp. Mangrove trees take root in tidal zones of tropical regions. The salty water would kill most plants. Living cells of mangrove trees accumulate so many organic solutes that they can take up fresh water by osmosis.

Community structure is an outcome of a balance of forces, including predation and competition, operating over time.

A climax community is a stable, self-perpetuating array of species in equilibrium with one another and their habitat.

Similar climax stages can persist along gradients dictated by environmental factors and by species interactions. Also, recurring, small-scale changes help shape many communities.

Natural or deliberate ecological restoration often can repair a damaged climax community, provided that suitable species are available to reinstate the original biodiversity.

WWW

COMMUNITY INSTABILITY

The preceding sections might lead you to believe that all communities become stabilized in predictable ways. But this is not always the case. *Community stability is an outcome of forces that have come into uneasy balance.* Resources are sustained, as long as populations do not flirt dangerously with the carrying capacity. Predators and prey coexist, as long as neither wins. Competitors have no sense of fair play. Mutualists are stingy, as when plants produce as little nectar as necessary to attract pollinators, and pollinators take as much nectar as they can for the least effort.

In the short term, disturbances can hurt the growth of some populations. You saw an example of this in Section 40.3. Also, long-term changes in climate or some other environmental variable often have destabilizing effects. If the instability is great enough, a community might change in ways that will persist even when the disturbance ends or is reversed. If some of its member species happen to be rare or don't compete well with others, they might become extinct.

For example, sometimes the residents of established communities move out from their home range and then successfully take up residence elsewhere. **Geographic dispersal** is the name for such a directional movement, and it proceeds in three ways. First, over a number of generations, a population expands its home range by slowly invading outlying regions that prove hospitable. Second, some individuals are rapidly transported across great distances. This is called *jump* dispersal. It often takes an individual across a region where it could not survive on its own, as when an insect travels from the mainland to Maui in a ship's cargo. Third, a population moves away from its home range with imperceptible slowness, as brought about by continental drift.

The dispersal and colonization of vacant places can be amazingly rapid as well as successful. Consider one of Amy Schoener's experiments in the Bahamas. She set out small plastic sponges on the barren sandy floor of Bimini Lagoon. How fast did aquatic species take up residence in chambers of the artificial hotels? Schoener recorded occupancy by 220 species in thirty days.

Or think about the 4,500 or so exotic species (the non-natives) that successfully established themselves in the United States following jump dispersal. These are just the ones we know about. Some, including rice, soybeans, wheat, corn, and potatoes, have been put to good use as food resources. As the next section makes clear, however, most destabilize natural communities or adversely affect farmlands.

Long-term shifts in climate, the rapid introduction and successful establishment of a new species, and many other disturbances can permanently alter community structure.

WWW

EXOTIC AND ENDANGERED SPECIES

When you hear someone bubbling enthusiastically about an **exotic species,** you can safely bet the speaker is not an ecologist. This is a name for a resident of an established community that has moved from its home range and successfully taken up residence elsewhere. It makes no difference whether the importation was deliberate or accidental. Unlike most imports, which cannot take hold outside their home range, an exotic species insinuates itself into the new community.

Occasionally the addition is harmless and even has beneficial effects. More often, they make native species **endangered species**, which by definition are extremely vulnerable to extinction. Of all species that are now on rare or endangered lists or have already become extinct, *close to 70 percent owe their precarious existence or demise to displacement by exotic species.*

A KALEIDOSCOPE OF IMPORTS If you reside in the northeastern United States, you know of imported gypsy moths. They escaped from a research facility near Boston in 1869. Today their descendants defoliate whole forests. You may know of zebra mussels, which probably entered the Great Lakes on a cargo ship's hull. They attach to and block water intake pipes for cities. The estimated damage exceeds 5 billion dollars. And *Cryphonectria parasitica*, a fungus introduced to North America almost a century ago, killed nearly all of the great American chestnut trees.

If you live in the southeastern United States, you know of the hot sting of fire ants (*Solenopsis invicta*). Nearly half a century ago, fire ants were accidentally imported from South America, quite possibly in the cargo of a ship that docked in Mobile, Alabama. Each year, their painful stings make 70,000 or so people seek medical attention.

And remember those African bees released in South America? As you read in Chapter 7, their Africanized, "killer bee" descendants traveled northward, through Mexico and on into the United States. At this writing, swarms of these aggressive bees have already killed more than a thousand people in Latin America. They also have attacked 140 Americans, one of whom died.

THE RABBITS THAT ATE AUSTRALIA During the 1800s, British settlers in Australia just couldn't bond with koalas and kangaroos, so they started to import familiar animals from their homeland. In 1859, in what would be the start of a wholesale disaster, a landowner in northern Australia imported and released two dozen wild European rabbits (*Oryctolagus cuniculus*). Good food and good sport hunting, that was the idea. An ideal rabbit habitat with no natural predators—that was the reality.

Six years later, the landowner had killed 20,000 rabbits and was besieged by 20,000 more. The rabbits displaced livestock, even kangaroos. Now Australia has 200 to 300 million hippityhopping through the southern half of the country. They overgraze perennial grasses in good times and strip bark from shrubs and trees during droughts. You

Figure 40.15 Part of the fence against 200–300 million rabbits that are destroying Australia's vegetation.

Figure 40.16 Kudzu (*Pueraria lobata*) taking over part of Lyman, South Carolina. You can now find kudzu vines from East Texas to Florida and as far north as Pennsylvania.

know where they've been; they transform grasslands and shrublands into eroded deserts (Figure 40.15). They have been shot and poisoned. Their warrens have been plowed under, fumigated, and dynamited. Even when all-out assaults reduced their population size by 70 percent, the rapidly reproducing imports made a comeback in less than a year. Did the construction of a 2,000-mile-long fence protect western Australia? No. Rabbits made it to the other side before workers completed the fence.

In 1951, government researchers introduced myxoma virus by way of mildly infected South American rabbits, its normal hosts. This virus causes *myxomatosis*. The disease has mild effects on the South American rabbits that coevolved with the virus but nearly always had lethal effects on *O. cuniculus*. Biting insects, mainly mosquitoes and fleas, quickly transmit the virus from host to host. Having no coevolved defenses against the virus, the European rabbits died in droves. As you might expect, natural selection has since favored the rapid growth of populations of *O. cuniculus* that are resistant to the virus.

In 1991, on an uninhabited island in Spencer Gulf, Australian researchers released a population of rabbits that they had injected with a calicivirus. The rabbits died quickly and relatively painlessly from blood clots in their lungs, heart, and kidneys. In 1995, the test virus escaped from the island, possibly on insect vectors. It has been killing 80 to 95 percent of the adult rabbits in Australian regions. At this writing, researchers are questioning whether the calicivirus should be used on a widespread scale, whether it can jump boundaries and infect animals other than rabbits (such as humans), and what the long-term consequences will be.

THE PLANT THAT ATE GEORGIA A vine called kudzu (*Pueraria lobata*) was deliberately brought over from Japan to the United States, where it faces no serious threats from herbivores, pathogens, or competitor plants. In temperate parts of Asia, kudzu is a well-behaved legume with a well-developed root system. It *seemed* like a good idea to import it for erosion control on hills and near highways in the southeastern United States. but with nothing to stop it, kudzu's shoots can grow one-third of a meter per day. Vines blanket streambanks, trees, telephone poles, houses, hills, and almost everything else in their path (Figure 40.16). Attempts to dig them up or burn them are futile. Grazing goats and herbicides help, but goats are indiscriminate eaters and herbicides contaminate water supplies. If the global temperature continues to rise, kudzu could reach the Great Lakes by the year 2040.

On the bright side, a Japanese firm is constructing a kudzu farm and processing plant in Alabama. Asians use a starch extract from kudzu in beverages, candy, and herbal medicines. The idea is to export the starch to Asia, where the demand currently exceeds the supply. (And perhaps this gives new meaning to the expression, What comes around, goes around.) Also, kudzu might eventually help reduce the extent of logging operations. At the Georgia Institute of Technology, researchers have reported that kudzu may be useful as an alternative source of paper.

WWW

PATTERNS OF BIODIVERSITY

As you might well conclude from your consideration of the preceding discussions, biodiversity differs greatly from one habitat to another. In addition, as many field investigations by ecologists have revealed, biodiversity differs greatly between geographic regions.

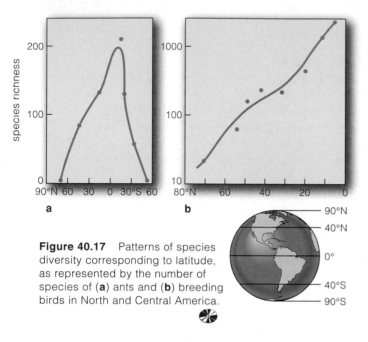

Figure 40.17 Patterns of species diversity corresponding to latitude, as represented by the number of species of (**a**) ants and (**b**) breeding birds in North and Central America.

Figure 40.18 Aerial photograph of Surtsey, a volcanic island, at the time of its formation in the sea. Newly formed, isolated islands are natural laboratories for ecologists. The chart gives the number of colonizing species of mosses and vascular plants recorded on Surtsey between 1965 and 1973.

What Causes Mainland and Marine Patterns?

The most striking pattern of biodiversity corresponds to distance from the equator. For most groups of plants and animals, *the number of coexisting species on land and in the seas is greatest in the tropics, and it systematically declines from the equator to the poles.* Figure 40.17 shows two examples of this biodiversity pattern. What factors underlie it? Three are paramount.

First, as described in Chapter 42, tropical latitudes intercept more sunlight of consistently greater intensity, the rainfall is heavier, and the growing season is longer. As one result, *resource availability tends to be higher and more reliable in the tropics than elsewhere.* All year long, different tree species in humid tropical forests put out new leaves, flowers, and fruit. Year in and year out, they support diverse herbivores, nectar foragers, and fruit eaters. Such diverse specializations cannot evolve to a comparable degree in temperate and arctic regions.

Second, *species diversity might be self-reinforcing.* The diversity of tree species in tropical forests is far greater than in comparable forests at higher latitudes. When more plant species compete and coexist, more herbivore species also evolve and coexist, partly because no single herbivore can overcome the chemical defenses of all of the different plants. Typically, then, more predators and parasites evolve in response to the diversity of prey and hosts. The same applies to diversity on tropical reefs.

Third, evolutionary history tells us this: *The rates of speciation in the tropics have exceeded those of background extinction.* Decreases in biodiversity have occurred, but mainly during mass extinctions. As you read in Section 17.4, global temperatures drop at such times. Species adapted to cool climates survive; those adapted to the temperatures of the tropics do not. Also, millions of species in tropical forests may disappear within the next decade, for reasons described in chapters to follow.

What Causes Island Patterns?

Islands often serve as terrific laboratories for studying biodiversity. For example, in 1965, a volcanic eruption formed a new island southwest of Iceland. Within six months, bacteria, fungi, seeds, flies, and some seabirds were established on it. One vascular plant appeared after two years, and a moss two years after that (Figure 40.18). As soils improved, the number of plant species increased. All the colonists originated in Iceland. None

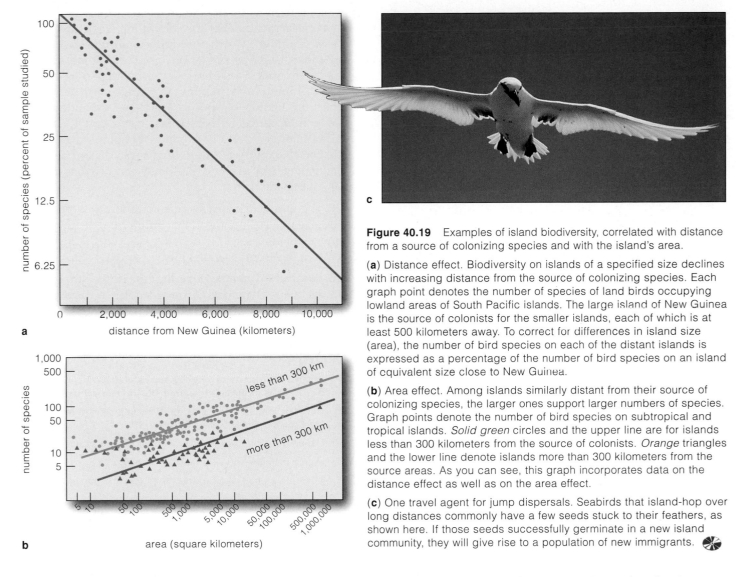

Figure 40.19 Examples of island biodiversity, correlated with distance from a source of colonizing species and with the island's area.

(**a**) Distance effect. Biodiversity on islands of a specified size declines with increasing distance from the source of colonizing species. Each graph point denotes the number of species of land birds occupying lowland areas of South Pacific islands. The large island of New Guinea is the source of colonists for the smaller islands, each of which is at least 500 kilometers away. To correct for differences in island size (area), the number of bird species on each of the distant islands is expressed as a percentage of the number of bird species on an island of equivalent size close to New Guinea.

(**b**) Area effect. Among islands similarly distant from their source of colonizing species, the larger ones support larger numbers of species. Graph points denote the number of bird species on subtropical and tropical islands. *Solid green* circles and the upper line are for islands less than 300 kilometers from the source of colonists. *Orange* triangles and the lower line denote islands more than 300 kilometers from the source areas. As you can see, this graph incorporates data on the distance effect as well as on the area effect.

(**c**) One travel agent for jump dispersals. Seabirds that island-hop over long distances commonly have a few seeds stuck to their feathers, as shown here. If those seeds successfully germinate in a new island community, they will give rise to a population of new immigrants.

originated on the island, which was named Surtsey. Yet the number of new species won't increase indefinitely. Why not? Many studies of the community patterns on islands around the world provide us with two models.

First, islands far from a source of potential colonists receive few colonizing species, and the few that arrive are adapted for long-distance dispersal (Figure 40.19). This is the **distance effect**. Second, larger islands tend to support more species than smaller islands at equivalent distances from source areas. This is the **area effect**. The larger islands tend to have more and varied habitats; most have more complex topography and extend higher above sea level. Thus they favor species diversity. Also, being bigger targets, they may intercept more colonists.

Most importantly, extinctions suppress biodiversity on small islands. There, the small populations are far more vulnerable to storms, volcanic eruptions, diseases, and random shifts in birth and death rates. As for any island, the number of species reflects a balance between

immigration rates for new species and extinction rates for established ones. Small islands that are distant from a source of colonists have low immigration rates and high extinction rates, so they support few species once the balance has been struck for their populations.

For most groups of organisms, the number of coexisting species is highest in the tropics and systematically declines from the equator to the poles.

A region's biodiversity depends on many factors, such as its climate, topographical variation, possibilities for dispersal, and evolutionary history—including extinctions and the time available for speciation.

Habitat disturbances tend to work against competitive exclusion and therefore favor increases in biodiversity.

Biodiversity on islands represents a balance between the immigration rates for new species and extinction rates for established species.

SUMMARY

1. A habitat is the type of place where individuals of a given species normally live—that is, their "address." A community consists of all populations of all species that occupy a habitat and that directly or indirectly associate with one another.

2. Each species has its own niche, or "profession," in the community. This is defined as the sum of activities and relationships in which its members engage as they secure and use the resources they require to survive and reproduce.

3. Mutualism, commensalism, competition, predation, and parasitism are species interactions that directly or indirectly link the populations in a community.

4. Two species that require the same limited resource tend to compete, as by using the resource as rapidly or efficiently as possible or by interfering with use of it.

   a. According to the competitive exclusion concept, when two (or more) species require identical resources, they cannot coexist indefinitely.

   b. Species are more likely to coexist if they differ in their use of resources. Also, they may coexist by using a shared resource in different ways or at different times.

5. Some predator and prey populations may coexist at stable levels if predators keep prey from overshooting the carrying capacity. Others show recurring or erratic cycles of abundance and crashes. Delays in predator responses to changes in prey density, and changes in the prey's food supplies, influence the cycles.

6. Predators and their prey coevolve. After a novel, heritable trait appears and gives prey an edge in their coevolutionary contest, selection pressure works on the predator populations, which may evolve in response. The same occurs when a novel, heritable trait arises in a predator population.

   a. Evolved defenses by prey include threat displays, chemical weapons, mimicry, and camouflaging.

   b. Predators overcome prey defenses by adaptive behavior (including stealth), camouflaging, and so on.

7. Parasites and their hosts coevolve in ways that favor resistant hosts and only moderately harmful parasites.

8. By the classical model of ecological succession, a community develops in predictable sequence, from its pioneer species to an end array of species that persists over an entire region.

9. A stable, self-perpetuating array of species that are in equilibrium with one another and with a particular environment is called a climax community.

10. Similar climax stages may persist within the same region, yet show variation as a result of environmental gradients and species interactions.

11. Recurring, small-scale changes are a part of the internal dynamics of communities. Other changes, such as long-term shifts in climate or species introductions, may permanently alter the community structure.

   a. Community structure reflects an uneasy balance of forces, such as predation (as by keystone species) and competition, that have been operating over time.

   b. Community structure can change permanently by the introduction of species that have expanded their geographic range. Dispersals may occur slowly, as when a population gradually expands its home range. Jump dispersals rapidly put individuals into distant habitats. Extremely slow dispersals also have occurred as a result of continental drift.

12. The number of species in a community depends on the size of a region, the colonization rate, disturbances, and extinction rates. It depends also on the level and pattern of resource availability.

13. Biodiversity tends to be highest in the tropics and to decline systematically toward polar regions, except during times of mass extinction.

Review Questions

1. What is the difference between the habitat and the niche of a species? Why is it difficult to define "the human habitat"? *40.1*

2. Describe competitive exclusion. How might two species that compete for the same resource coexist? *40.3*

3. Define the difference between a predator and a parasite. *40.4*

4. Define primary and secondary succession. *40.6*

5. What is a climax community? How does the climax-pattern model help explain its structure? *40.6*

Self-Quiz *(Answers in Appendix III)*

1. A habitat _____ .
   a. has distinguishing physical and chemical features
   b. is where individuals of a species normally live
   c. is occupied by various species
   d. both a and b
   e. a through c

2. A two-way flow of benefits in mutualistic interactions between species is an outcome of _____ .
   a. close cooperativeness   c. resource partitioning
   b. two-way exploitation   d. competitive coexistence

3. A niche _____ .
   a. is the sum of activities and relationships in a community by which individuals of a species secure and use resources
   b. is unvarying for a given species
   c. shifts in large and small ways
   d. both a and b
   e. both a and c

4. Two species in the same habitat can coexist when they _____ .
   a. differ in their use of resources
   b. share the same resource in different ways
   c. use the same resource at different times
   d. all of the above

Figure 40.20 (**a**) A flesh fly and (**b**) a weevil (*Zygops*).

Figure 40.21 Water hyacinths choking a waterway in Florida.

5. A predator population and prey population _____ .
   a. always coexist at relatively stable levels
   b. may undergo cyclic or irregular changes in density
   c. cannot coexist indefinitely in the same habitat
   d. both b and c

6. All parasites _____ .
   a. tend to kill their hosts   c. consume host tissues
   b. can kill novel hosts   d. both b and c

7. In _____ , a disturbed site in a community recovers and moves again toward the climax state.
   a. the area effect   c. primary succession
   b. the distance effect   d. secondary succession

8. The number of coexisting species is _____ in the tropics and _____ toward the poles.
   a. highest; declines systematically
   b. lowest; increases systematically
   c. highest; declines sporadically
   d. lowest; increases sporadically

9. Match the terms with the most suitable descriptions.
   _____ jump dispersal
   _____ area effect
   _____ pioneer species
   _____ climax community
   _____ keystone species

   a. opportunistic colonizer of barren or disturbed places
   b. dominates community structure
   c. rapid transport over great distance
   d. more biodiversity on large islands than small at same distance from source
   e. stable, self-perpetuating array of species

Critical Thinking

1. Think of possible examples of competitive exclusion besides the ones used in the chapter, as by considering some of the animals and plants living in your own neighborhood.

2. Flesh flies (Figure 40.20*a*) have a gray and black body, red eyes, and red tail ends. They are fast fliers, and bird predators soon give up trying to catch them. Many sluggish insects, such as the weevil *Zygops rufitorquis*, resemble flesh flies (Figure 40.20*b*). This appears to be a case of _____ .

3. The water hyacinth (*Eichhornia crassipes*) is an aquatic plant native to South America. In the 1880s, someone fancied the plant's blue flowers and displayed it at an exposition in New Orleans. Other flower fanciers took home clippings and put them in ponds and streams. Unchecked by natural predators, the fast-growing hyacinths spread through nutrient-rich waters, displaced many native species, and choked rivers and canals (Figure 40.21). They have spread as far west as San Francisco. Do some research to see whether the United States now limits imports and, if so, how effective the restrictions have been.

4. Somewhere between predators and parasites are *parasitoids*, insect larvae that always kill what they eat. Peter Price studied the evolutionary effects of a parasitoid wasp that lays eggs on sawfly cocoons. Sawflies lay eggs in trees. In time, fertilized eggs give rise to fly larvae, which drop to the forest floor. The larvae spin cocoons after burrowing into leaf litter. Some burrow deeper than others. The first adult sawflies emerge from the cocoons that were the least deeply buried. Later in the season, more sawflies emerge from the more deeply buried ones.

These parasitoid wasps tend to lay eggs on the cocoons closest to the surface of the leaf litter. They exert strong selection pressure on the sawfly population, for the deep-burrowing sawfly individuals are more likely to escape detection. There are only so many flies near the surface, so the wasp able to locate cocoons deeper in the litter will be competitive in securing food for her larvae. Explain how the host species stays ahead in this coevolutionary contest.

Selected Key Terms

area effect (on island biodiversity) *40.9*
camouflage *40.5*
climax community *40.6*
climax-pattern model *40.6*
coevolution *40.4*
commensalism *40.1*
community *40.1*
competitive exclusion *40.3*
distance effect (on island biodiversity) *40.9*
ecological succession *40.6*
endangered species *40.8*
exotic species *40.8*
geographic dispersal *40.7*
habitat *40.1*
host *40.4*
interspecific competition *40.1*
keystone species *40.3*
macroparasite *40.4*
microparasite *40.4*
mimicry *40.5*
mutualism *40.1*
niche *40.1*
parasite *40.4*
parasitism *40.1*
pioneer species *40.6*
predation *40.1*
predator *40.4*
prey *40.4*
primary succession *40.6*
resource partitioning *40.3*
secondary succession *40.6*
social parasite *40.4*
symbiosis *40.1*
warning coloration *40.5*
wetland *40.6*

Readings   See also www.infotrac-college.com

Begon, M., et al. 1990. *Ecology: Individuals, Populations, and Communities.* Second edition. Sunderland, Massachusetts: Sinauer.

Krebs, C., et al. 25 August 1995. "Impact of Food and Predation on the Snowshoe Hare Cycle." *Science* 269: 1112–1115.

Moore, P., 1987. "What Makes a Forest Rich?" *Nature* 329: 292.

Smith, R. 1996. *Elements of Ecology.* Fifth edition. New York: HarperCollins.

Worthington, E. B. 1994. "African Lakes Reviewed: Creation and Destruction of Biodiversity." *Environmental Conservation*, 201–204.

WWW   *http://www.brookscole.com/biology*

Practice quiz questions, hypercontents, BioUpdates, and critical thinking. The Brooks/Cole Biology Resource Center provides a wealth of information fully organized and integrated by chapter.

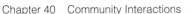

41 ECOSYSTEMS

Crêpes for Breakfast, Pancake Ice for Dessert

Think of Antarctica, and you think of ice. Mile-thick slabs of the stuff hide all but a small fraction of a vast continent whipped by fierce winds and kept frozen by murderously low temperatures, on the order of –100°F. Yet, on patches of exposed rocky soil and on nearby islands, mosses and lichens grow. There, during the breeding season, a variety of penguins and seals form great noisy congregations, then reproduce and raise offspring. They cruise offshore or venture out in the open ocean in pursuit of krill, fishes, and squids.

The first explorers of Antarctica called it "The Last Place on Earth." For all of its harshness, however, the beauty of this ecosystem fired the imagination of those first travelers and others who followed them.

In 1961, thirty-eight nations signed a treaty to set aside Antarctica as a reserve for scientific research. This was the start of scientific outposts—and of the trashing of Antarctica. At first, the researchers discarded a few oil drums and old tires. Then prefabricated villages went up. Onto the ice and into the water went garbage and used equipment. Just offshore, marine life became acquainted with sewage and chemical wastes.

Antarctica even became a destination of cruise ships (Figure 41.1a). Every summer thousands of tourists, fortified by three sumptuous meals a day plus snacks, are ferried from ship to land across channels glistening with pancake ice. They trample the sparse vegetation and bob around the penguins, cameras clicking.

We do not know what tourism's long-term impact will be, but it pales by comparison to other assaults. Nations looking for new sources of food started licking their chops over Antarctica's krill. Word got out about

Figure 41.1 (**a**) A boatload of tourists crossing a channel of "pancake ice" off the coast of Antarctica.
(**b**) On the shore of an island near Antarctica, tourists get close to nature—maybe too close.

potentially rich deposits of uranium, oil, and gold. By 1988, treaty nations were poised to authorize digs and drillings. But oil spills or commercial harvesting of krill could easily destroy the fragile ecosystem. For example, tiny, shrimplike krill are the only food source for Adélie penguins and a key food for baleen whales and other marine animals which are, in turn, food for still others.

In 1991, the treaty nations thought about all of this and imposed a fifty-year ban on mineral exploration. Research stations started to bury or incinerate wastes, treat raw sewage, and take other steps to curb pollution. Tour operators promised to supervise tourists. Krill are not being harvested on a massive scale. Yet.

Because Antarctica seems remote from the rest of the world, it is easier to see how species interconnect there. We can ask whether harm to one species or one habitat will lead to collapse of the whole and be fairly sure of the answer. What about places not as sharply defined? Are they as vulnerable to disturbance or more resilient? We won't know for sure until researchers gain deeper insights into the evolutionary and ecological histories of their species, including the suspected great numbers of species we haven't even discovered yet.

In the meantime, biodiversity is rapidly declining in ecosystems on land and in the seas. Competition from exotic species is one threat; the human propensity to overexploit species and destroy habitats is another. To counter the threats, individuals around the world are engaged in *conservation biology*. They are working to lower current extinction rates and to design management programs that will help sustain the most vulnerable ecosystems over time. To appreciate the magnitude of the task before them, start with a few principles of how ecosystems work—which is the topic of this chapter.

KEY CONCEPTS

1. An ecosystem is an association of organisms and their physical environment, interconnected by an ongoing flow of energy and a cycling of materials through it.

2. Every ecosystem is an open system, in that it has inputs and outputs of both energy and nutrients.

3. Over time, energy flows in only one direction through an ecosystem. Most commonly, energy flow begins when photosynthetic autotrophs harness sunlight energy and convert it to forms that they and other organisms of the ecosystem can use. Autotrophs are primary producer organisms for the ecosystem.

4. Energy-rich organic compounds that the primary producers synthesize become incorporated in their body tissues. They are stored forms of energy, and they serve as the foundation for the ecosystem's food webs. Such webs consist of a number of interconnected food chains.

5. Each chain in a food web extends in a straight-line sequence from the producers through all the consumers, decomposers, and detritivores. And it cross-connects with other chains at different feeding levels.

6. The water, carbon, nitrogen, phosphorus, and other substances required for primary productivity move through biogeochemical cycles that are global in scale. Ions or molecules of the substances move slowly from environmental reservoirs, then rapidly among organisms of food webs, then back to their reservoirs.

7. Depending on which part of the environment serves as its largest reservoir, each substance moves through a hydrologic, atmospheric, or sedimentary cycle.

8. Human activities are disrupting the natural cycles of materials and are thereby endangering ecosystems.

Overview of the Participants

Diverse natural systems abound on the Earth's surface. In climate, landforms, soil, vegetation, animal life, and other features, deserts differ from hardwood forests, which differ from tundra and prairies. In biodiversity and physical properties, seas differ from reefs, which differ from lakes. *Yet despite the differences, such systems are alike in many aspects of their structure and function.*

With few exceptions, each system runs on energy that plants and other photosynthetic organisms capture from the sun. The photosynthesizers, remember, are the most common autotrophs (self-feeders). They convert captured sunlight energy to chemical energy and use it to construct organic compounds from inorganic raw materials. Because they can obtain energy directly from their environment, autotrophs are **primary producers** for the entire system (Figure 41.2).

All other organisms in the system are heterotrophs, not self-feeders. They extract energy from compounds that the primary producers put together. **Consumers** feed on the tissues of other organisms. The consumers called *herbivores* eat only plants, *carnivores* eat animals, *omnivores* eat both, and *parasites* extract energy from living hosts that they live in or on. Other heterotrophs, the **decomposers**, include fungi and many bacteria that get energy by breaking down the remains or products

of organisms. Others, the **detritivores**, obtain nutrients from the decomposing particles of organic matter. Crabs and earthworms are examples of detritivores.

Autotrophs secure nutrients as well as energy for the entire system. As they grow, they take up water and carbon dioxide from the environment (as sources of oxygen, carbon, and hydrogen) and dissolved minerals (such as nitrogen and phosphorus). Such materials are building blocks for their carbohydrates, lipids, proteins, and nucleic acids. When decomposers and detritivores degrade organic matter to small inorganic compounds, they also release nutrients. Unless something removes the nutrients, as when a creek takes dissolved minerals away from a meadow, autotrophs typically reuse them.

What we have just described in broad outline is an ecosystem. An **ecosystem** is an array of organisms and their physical environment, interacting by a one-way flow of energy and a cycling of materials. It is an open system, unable to sustain itself. Each ecosystem runs on *energy inputs* (as from the sun) and often *nutrient inputs* (as from a creek that delivers dissolved minerals into a lake). It has *energy outputs* and *nutrient outputs*. Energy cannot be recycled. Over time, most of the energy that the autotrophs fix is lost to the environment, mainly as metabolically generated heat. Nutrients are cycled, but some still slip away. Most of this chapter deals with the inputs, internal transfers, and outputs of ecosystems.

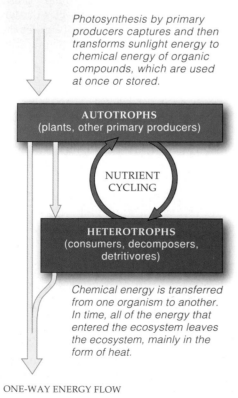

Figure 41.2
Model of ecosystems. Energy flows in only one direction: into the ecosystem, through all of its living organisms, and out from it. Nutrients typically are cycled among the ecosystem's autotrophs and heterotrophs. In this case, energy flow starts with autotrophs that capture sunlight energy.

Photosynthesis by primary producers captures and then transforms sunlight energy to chemical energy of organic compounds, which are used at once or stored.

AUTOTROPHS
(plants, other primary producers)

NUTRIENT CYCLING

HETEROTROPHS
(consumers, decomposers, detritivores)

Chemical energy is transferred from one organism to another. In time, all of the energy that entered the ecosystem leaves the ecosystem, mainly in the form of heat.

ONE-WAY ENERGY FLOW

Table 41.1 Trophic Levels in a Deciduous Forest

Type of Organism at Trophic Level	Energy Source	Examples of Organisms
FOURTH TROPHIC LEVEL		
Tertiary consumers (heterotrophs):		
Secondary carnivores, parasites of carnivores	Secondary consumers	Owls, weasels, parasitic flies
THIRD TROPHIC LEVEL		
Secondary consumers (heterotrophs):		
Primary carnivores	Primary consumers	Spiders, mice, beetles
SECOND TROPHIC LEVEL		
Primary consumers (heterotrophs):		
Herbivores, decomposers, detritivores	Primary producers	Moth larvae, earthworms
FIRST TROPHIC LEVEL		
Primary producers (autotrophs):		
Photoautotrophs	Sunlight	Algae, trees
Chemoautotrophs	Inorganic substances	Nitrifying bacteria

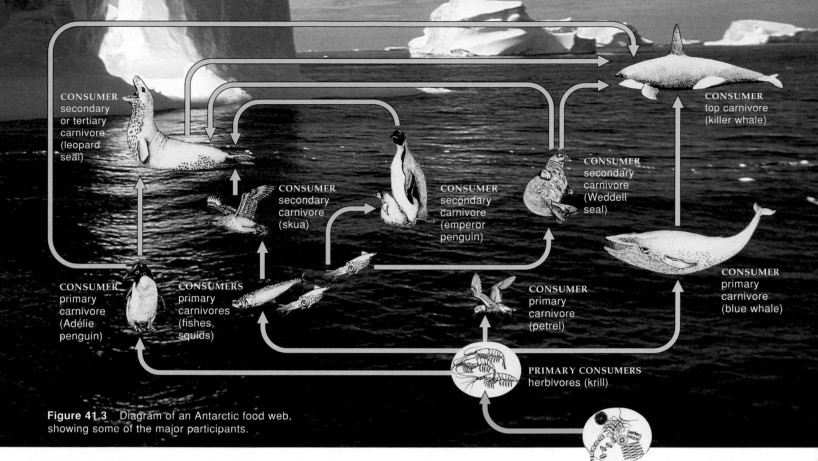

Figure 41.3 Diagram of an Antarctic food web, showing some of the major participants.

CONSUMER
secondary or tertiary carnivore (leopard seal)

CONSUMER
top carnivore (killer whale)

CONSUMER
secondary carnivore (skua)

CONSUMER
secondary carnivore (emperor penguin)

CONSUMER
secondary carnivore (Weddell seal)

CONSUMER
primary carnivore (Adélie penguin)

CONSUMERS
primary carnivores (fishes, squids)

CONSUMER
primary carnivore (petrel)

CONSUMER
primary carnivore (blue whale)

PRIMARY CONSUMERS
herbivores (krill)

Structure of Ecosystems

TROPHIC LEVELS We can classify all of the organisms of an ecosystem in terms of their functional roles in a hierarchy of feeding relationships, called **trophic levels** (after *troph*, meaning nourishment). "Who eats whom?" we can ask. As organism **B** eats organism **A**, energy is transferred to **B** from **A**. All the organisms at a given trophic level are the same number of transfer steps away from some energy input into the ecosystem.

Table 41.1, for example, lists organisms of a forest ecosystem. Being closest to the energy input (sunlight), trees and other primary producers are at the first trophic level. Primary consumers, which feed directly upon the primary producers, include various herbivores, such as leaf-eating larvae. They also include earthworms and other detritivores, which eat organic debris. Primary producers are at the second trophic level. At the third trophic level we find primary carnivores (an assortment of beetles, spiders, and birds), which prey on primary consumers. At the fourth trophic level are secondary carnivores, including owls, which feed on the species of the third trophic level.

By this classification scheme, all of the organisms at a given trophic level are interacting with the same sets of predators, prey, or both. But remember this: Many of the omnivores, such as decomposers and people, feed at several trophic levels. So they must be partitioned among different levels or assigned one of their own.

FOOD WEBS Commonly a straight-line sequence of who eats whom in an ecosystem is called a **food chain**.

And yet we have a hard time   finding such simple, isolated cases in nature. Why? Most often, the same species belongs to more than one food chain, especially when it is at a low trophic level. It is more accurate to visualize food chains as *cross-connecting* with one another—as **food webs**. Figure 41.3 shows the major players in a food web in the seas around Antarctica.

A bad-luck story about a fisherman can clarify the difference between a food chain and food web. Suppose a fisherman nets fish that were feeding on algae near the ocean surface. At lunchtime, he cooks some fish. He later loses his footing and falls into the water, where sharks lurk. You may think this is a simple food chain: algae⟶ fish⟶ fisherman⟶ shark. But alternate feeding relationships cut into it. Crustaceans also were grazing on the algae. Small squids and midsize fishes were feeding on the crustaceans. And larger fishes were feeding on small ones. Sharks may have been about to feed on those larger and midsize fishes. The fisherman ate cooked fish, garlic, and onions and therefore shifted between trophic levels (as carnivore and herbivore). He was even more omnivorous than this, for he sipped wine, a beverage derived from the fermenting activities of the decomposer organisms called yeasts.

An ecosystem consists of producers, consumers, decomposers, and detritivores and the physical environment, connected by a one-way energy flow and a cycling of materials.

A food web is a network of crossing, interlinked food chains involving primary producers, consumers, and decomposers.

𝑊𝑊𝑊

Primary Productivity

To get an idea of how energy flow is studied, focus on an ecosystem on land. Such ecosystems typically have multicelled plants as the primary producers. The rate at which the primary producers capture and store a given amount of energy in their tissues during some specified interval is the **primary productivity** of the ecosystem. How much energy actually gets stored in those tissues depends on how many individual plants live there. It also depends on the balance between photosynthesis (energy trapped) and aerobic respiration (energy used). Whereas *gross* primary productivity is the total rate of photosynthesis for the ecosystem during the specified interval, the *net* amount is the rate of energy storage in plant tissues in excess of the rate of aerobic respiration by the plants.

Other factors influence the amount of net primary production, its seasonal patterns, and its distribution through a given habitat. They do so in ecosystems in the seas as well as on land (Figure 41.4). For example, the body size and form of primary producers affect how much energy they can trap and store. So do availability of minerals, the temperature range, and the amount of sunlight and rainfall during each growing season. The more harsh the environment, the less new growth on plants—and the lower the productivity.

Major Pathways of Energy Flow

In what direction does energy flow through ecosystems on land? Plants fix only a small part of the energy from the sun. They store half of that in new tissues but lose the rest as metabolic heat. (Remember, no metabolic reaction is 100 percent efficient.) Other organisms tap into energy stored in plant tissues, remains, or wastes; and they, too, lose heat. *All of these heat losses represent a one-way flow of energy out of the ecosystem.*

Energy from a primary source flows in one direction through two categories of food webs. In **grazing food webs**, the energy flows from photosynthetic organisms to herbivores, then through an array of carnivores. By contrast, in **detrital food webs**, energy flows primarily from photosynthetic species through detritivores and decomposers. In most ecosystems, the two kinds of food webs cross-connect. For instance, this happens when a herring gull of a grazing food web opportunistically gulps down a crab of a detrital food web (Figure 41.5).

The amount of energy moving through food webs differs from one ecosystem to the next and often varies with the seasons. In most cases, however, most of the net primary production passes through the detrital food webs. You may doubt this. After all, when cattle graze heavily on pasture plants, about half the net primary production enters a grazing food web. But cattle do not use all the stored energy. Quantities of undigested plant parts and feces become available for decomposers and detritivores. Marshes along coasts are a similar case in point. In these aquatic ecosystems, most of the stored energy is not used until parts of marsh grasses die and become available for detrital food webs.

Ecological Pyramids

Often ecologists will represent the trophic structure of an ecosystem in the form of an **ecological pyramid**. In such pyramids, the primary producers form a base for successive tiers of consumers above them.

Figure 41.4 A summary of three years of satellite data on the Earth's primary productivity. *Dark green* denotes the most highly productive regions, such as rain forests. *Yellow* denotes the deserts (regions of low productivity). Productivity in the oceans, ranging from high to low, is color-coded *red* down through *orange*, *yellow*, *green*, and *blue*. Also compare the satellite maps in Section 6.8.

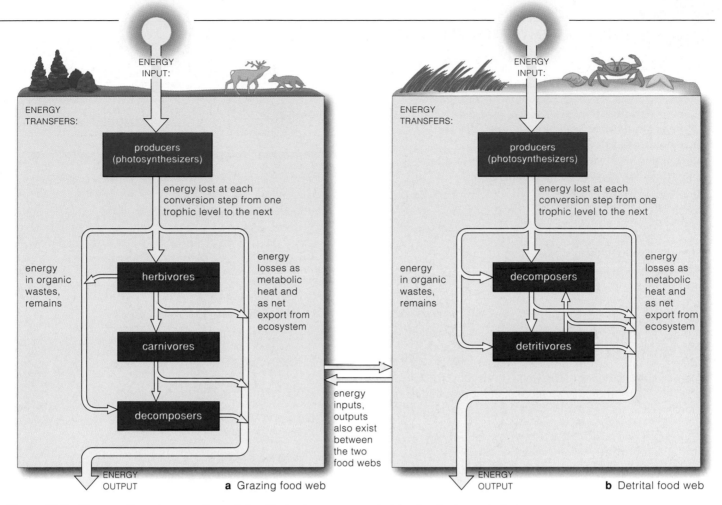

Figure 41.5 One-way flow of energy through two kinds of cross-connected food webs in ecosystems.

Some pyramids are based on biomass, or the weight of all the members at each trophic level. For example, for Silver Springs, Florida (a small aquatic ecosystem), a pyramid of biomass (measured as grams/square meter during one specified interval) would look like this:

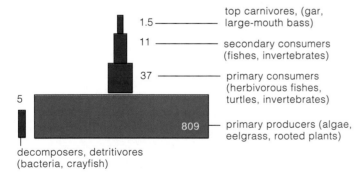

This is a typical pyramid of biomass. Some others are "upside-down," with the smallest tier on the bottom. A small pond is like this. The pond's biomass of primary producers consists of rapidly growing and reproducing phytoplankton. It supports a much greater biomass of zooplankton, in which the individuals are bigger, grow slower, and consume less energy per unit of weight.

An **energy pyramid** is a more useful way to depict an ecosystem's trophic structure. Such pyramids show the energy losses at each transfer to a different trophic level in the ecosystem. They have a large energy base at the bottom and are always "right-side up." As you will see in the next section, they provide a better picture of how energy flows in ever diminishing amounts through successive trophic levels of an ecosystem.

Energy flows into food webs of ecosystems from an outside source, usually the sun. Energy leaves ecosystems mainly by losses of metabolic heat, which each organism generates.

Gross primary productivity is an ecosystem's total rate of photosynthesis during a specified interval. The *net* amount is the rate at which primary producers store energy in tissues in excess of their rate of aerobic respiration. Heterotrophic consumption affects the rate of energy storage.

Tissues of living photosynthesizers are the basis of grazing food webs. Remains and wastes of photosynthesizers and consumers are the basis of detrital food webs.

The loss of metabolic heat and the shunting of food energy into organic wastes mean that usable energy flowing through consumer trophic levels declines at each energy transfer.

WWW

41.3 ENERGY FLOW AT SILVER SPRINGS

Imagine you are with ecologists who are bent on gathering data to construct an energy pyramid for a small freshwater spring over the course of one year. You observe them as they measure the energy that each type of individual in the spring takes in, loses as metabolic heat, stores in its body tissues, and loses in waste products. You see that they multiply the energy per individual by population size, then they calculate energy inputs and outputs. In such ways, the ecologists are able to express the flow per unit of water (or land) per unit of time.

The energy pyramid shown in Figure 41.6 summarizes the data from a long-term study of a grazing food web in an aquatic ecosystem—Silver Springs, Florida. The larger diagram in Figure 41.7 shows some of the calculations that ecologists used to construct the pyramid.

Given the metabolic demands of organisms and the amount of energy lost in their organic wastes, only about 6 to 16 percent of the energy entering one trophic level becomes available for organisms at the next level.

Because the efficiency of the energy transfers is so low, ecosystems in general have no more than four consumer trophic levels.

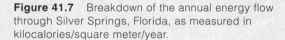

Figure 41.6 Pyramid of energy flow through Silver Springs, Florida, as measured in kilocalories/square meter/year.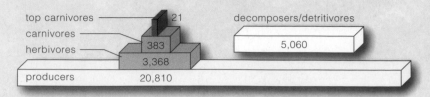

Figure 41.7 Breakdown of the annual energy flow through Silver Springs, Florida, as measured in kilocalories/square meter/year.

The primary producers in this small spring are mostly aquatic plants. The carnivores are insects and small fishes; top carnivores are larger fishes. The energy source (sunlight) is available throughout the year. The spring's detritivores and decomposers cycle organic compounds from the other trophic levels.

The producers trapped 1.2 percent of the incoming solar energy, and only a little more than a third of this amount became fixed in new plant biomass (4,245 + 3,368). The producers used more than 63 percent of the fixed energy for their own metabolism.

About 16 percent of the fixed energy was transferred to herbivores, and most of it was used for metabolism or transferred to detritivores and decomposers.

Of the energy that did get transferred to herbivores, only 11.4 percent reached the next trophic level (carnivores). These carnivores used all but about 5.5 percent, which was transferred to top carnivores.

By the end of the specified time interval, all of the 5,060 kilocalories of energy that had been transferred through the system appeared as metabolically generated heat.

Bear in mind, this energy flow diagram is oversimplified, because no community is isolated from others. New individual organisms and substances continually drop from overhead leaves and branches into the springs. Also, organisms and substances are gradually lost by way of a stream that leaves the spring.

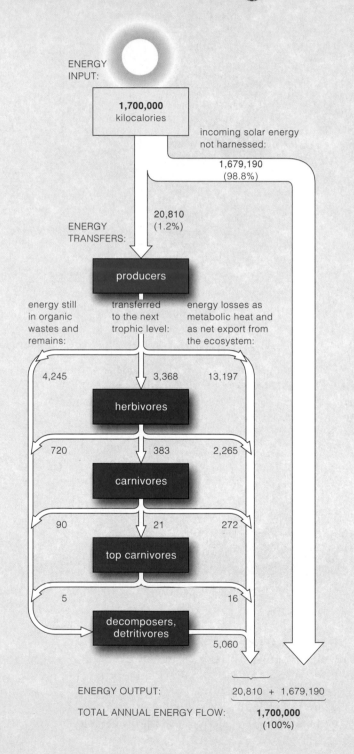

BIOGEOCHEMICAL CYCLES—AN OVERVIEW

Availability of nutrients as well as energy profoundly influences the structure of ecosystems. Photosynthetic producers require carbon, hydrogen, and oxygen, which they obtain from water and air. They require nitrogen, phosphorus, and other mineral ions. A scarcity of even one of those minerals has widespread adverse effects, for it lowers the ecosystem's primary productivity.

In a **biogeochemical cycle**, ions or molecules of a nutrient are transferred from the environment into organisms, then back to the environment, part of which functions as a vast reservoir for them. Transfer rates of nutrients into and out of a reservoir are usually lower than the exchange rates between and among organisms.

Figure 41.8 is a simple model of the relationship between most ecosystems and the geochemical part of the cycles. This model is based on four factors.

First, mineral elements that producer organisms utilize as nutrients usually are available in the form of mineral ions, such as ammonium (NH_4^+). *Second*, inputs from the physical environment, together with nutrient cycling activities of all of the decomposers and detritivores, maintain the ecosystem's nutrient reserves. *Third*, the actual amount of a nutrient that is being cycled on through most major ecosystems is greater than the amount entering and departing per year. *Fourth*, rainfall, snowfall, and the slow weathering of rocks are common sources of environmental inputs into an ecosystem's nutrient reserves. So are the combined effects of metabolic activities, such as nitrogen fixation.

Ecosystems on land also have typical outputs from the nutrient reserves. The loss of mineral ions by way of runoff from irrigated cropland is an example.

There are three categories of biogeochemical cycles, based on the part of the environment that contains the greatest portion of the specified ion or molecule. As you will see, in the *hydrologic* cycle, oxygen and hydrogen move in the form of water molecules. This movement is also known as the global water cycle. In *atmospheric*

cycles, a large percentage of the nutrient is in the form of an atmospheric gas. For example, this is true of gaseous forms of nitrogen and carbon (mainly carbon dioxide). *Sedimentary* cycles involve phosphorus and other solid nutrients that do not have gaseous forms. Solid nutrients move from land to the seafloor and return to dry land only by way of geological uplifting,

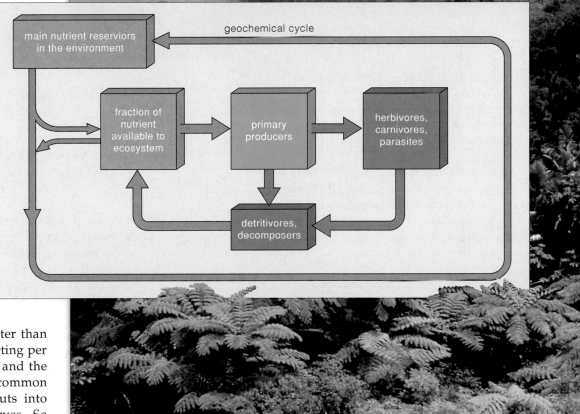

Figure 41.8 Generalized model of nutrient flow through a land ecosystem. The overall movement of nutrients from the physical environment, through organisms, and back to the environment constitutes a biogeochemical cycle.

which may take millions of years. The Earth's crust is the largest storehouse for sedimentary cycles.

The primary productivity on which ecosystems depend is profoundly influenced by the availability of nutrients.

In a biogeochemical cycle, ions or molecules of a nutrient move slowly through the environment, then rapidly among organisms, then back to the environmental reservoir for them.

Driven by ongoing inputs of solar energy, the Earth's waters move slowly, on a vast scale, from the oceans into the atmosphere, to land, and back to the oceans—the main reservoir. Water evaporating into the lower atmosphere initially stays aloft as vapor, clouds, and ice crystals. It returns to Earth as precipitation—mostly rain and snow. Ocean currents and prevailing wind patterns influence the global **hydrologic cycle**, as shown in Figure 41.9.

Water is vital for all organisms. It also is a transport medium that moves nutrients into and out of ecosystems. Its role in moving nutrients became clear in long-term studies in watersheds. A **watershed** is any region in which precipitation becomes funneled into a single stream or river. It may be any size of interest to an investigator. For example, the Mississippi River watershed extends across about a third of the continental United States. Watersheds in New Hampshire's Hubbard Brook Valley average 14.6 hectares (36 acres).

Most water entering a watershed seeps into soil or becomes part of surface runoff, which enters streams. Plants withdraw water and its dissolved minerals from the soil, then lose it by transpiration.

Measurements of watershed inputs and outputs have many practical applications. For instance, cities that draw on surface supplies of watersheds adjust usage according to seasonal shifts in water volume. Measurements also reveal the influence

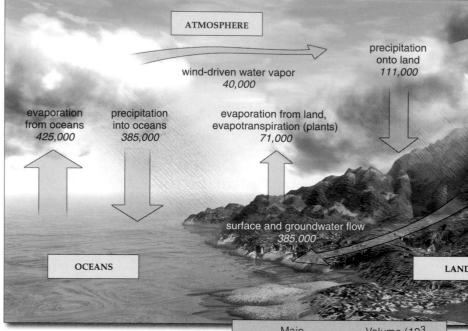

a

Figure 41.9 (a) Hydrologic cycle. Water moves from oceans to the atmosphere, the land, and back to the oceans. *Gold* boxes signify the main reservoirs. Arrow labels identify the processes involved in water movement between reservoirs, measured in cubic kilometers per year. (**b**) The global water budget.

Main Reservoirs	Volume (10^3 cubic kilometers)
Oceans	1,370,000
Polar ice, glaciers	29,000
Groundwater	4,000
Lakes, rivers	230
Soil moisture	67
Atmosphere (water vapor)	14

b

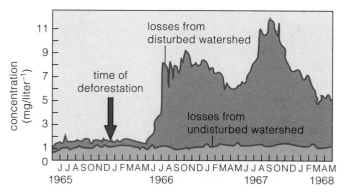

Figure 41.10 Results from experiments involving disturbances to forests of experimental watersheds in Hubbard Brook Valley. Researchers stripped the vegetation cover but did not disturb the soil. They applied herbicides to the soil for three years to prevent regrowth. All water that drained from each watershed was measured as it flowed over V-notched concrete catchments. The arrow marks the time of deforestation. Concentrations of calcium and other minerals were compared against those in water passing over a control catchment in an undisturbed area. Calcium losses were *six times* greater from the deforested area.

of vegetation cover over the movement of nutrients through the ecosystem phase of biogeochemical cycles.

For example, you might think that water draining a watershed would rapidly leach calcium ions and other minerals. But in young, undisturbed forests in Hubbard Brook watersheds, each hectare lost only 8 kilograms or so of calcium. Also, rainfall and weathering of rocks were bringing in calcium replacements. And tree roots were "mining" the soil, so calcium was being stored in a growing biomass of tree tissues.

In experimental watersheds, deforestation caused a shift in nutrient outputs (Figure 41.10). Calcium and other nutrients cycle so slowly that deforestation may disrupt nutrient availability for entire ecosystems. This is the case for forests that cannot regenerate themselves over the short term. Coniferous forests are like this.

In the hydrologic cycle, water slowly moves on a global scale from the oceans (the main reservoir), through the atmosphere, onto land, then back to the oceans.

In ecosystems on land, plants stabilize the soil and absorb dissolved minerals. By doing so, they minimize the loss of soil nutrients in runoff from land.

A Look at the Phosphorus Cycle

We continue our survey of biogeochemical cycling with one of the key sedimentary cycles. In the **phosphorus cycle**, phosphorus passes through food webs as it moves from land, to ocean sediments, then back to land (Figure 41.11). The Earth's crust is its largest reservoir, just as it is for other minerals.

Rock formations incorporate phosphorus mainly as phosphates. By natural processes of weathering and soil erosion, phosphates enter streams and rivers that transport them to ocean sediments. Mainly on the submerged "shelves" of continents, phosphorus slowly accumulates, along with other minerals, and forms insoluble deposits. Millions of years go by. Where great movements of crustal plates uplift part of the seafloor, phosphates become exposed on drained land surfaces. Over time, weathering releases phosphates from the exposed rocks, and so the cycle's geochemical phase begins again.

The ecosystem phase of the cycle is more rapid than the long-term geochemical phase. All organisms require phosphorus for synthesizing phospholipids, NADPH, ATP, nucleic acids, and other compounds. Plants take up dissolved, ionized forms of phosphorus so rapidly and efficiently that they often reduce its concentrations in soil to extremely low levels. Herbivorous animals get phosphates by dining on plants; carnivores obtain it by dining on the herbivores. Both types of animals excrete phosphates as a waste in urine and feces.

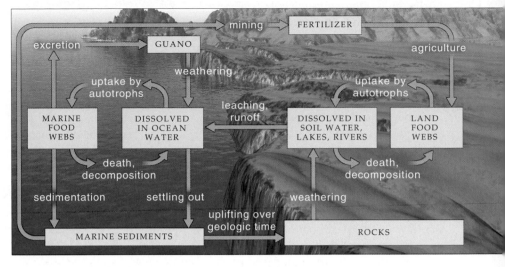

Figure 41.11 The phosphorus cycle. Most of the phosphorus moves in the form of phosphate ions (PO_4^{3-}).

Decomposition activities of soil organisms release phosphates. Plants take up the mineral and help ensure its rapid cycling within the ecosystem.

Eutrophication

Fertilizers use phosphates as key ingredients. Where the fertilizers are heavily applied, phosphorus that becomes concentrated in runoff from agricultural fields will alter living conditions in lakes and other aquatic ecosystems. Three key nutrients that algae and aquatic plants use during growth are nitrogen, potassium, and phosphorus. Bacteria fix enough nitrogen. Freshwater ecosystems usually have excess potassium. Most of the phosphorus is locked in sediments, so it tends to be a limiting factor on plant growth. So water enriched with phosphorus-containing fertilizers promotes dense algal blooms.

Most mineral elements enter sedimentary cycles. In their dissolved forms, they serve as micronutrients and macronutrients for the growth of primary producers in ecosystems on land as well as in the water provinces. Activities that increase the concentrations of dissolved nutrients can lead to **eutrophication**. The term refers to nutrient enrichment of any aquatic ecosystem. Thus an outpouring of raw sewage into a river or lake, and even logging over the land surrounding it, can increase the nutrient levels and trigger eutrophication. Many field experiments, such as the one illustrated in Figure 41.12, nicely demonstrate this outcome.

Figure 41.12 Experiment demonstrating lake eutrophication in Ontario, Canada. Researchers stretched a plastic curtain across a channel between two basins of the same lake. They added phosphorus, carbon, and nitrogen to one basin (the basin in the *background*) and carbon and nitrogen to the other (the basin in the *foreground*). Within two months, the phosphorus-enriched basin showed signs of accelerated eutrophication; a dense algal bloom turned the water green.

The Earth's crust is the largest reservoir for phosphorus and for other minerals that move through ecosystems as part of sedimentary cycles.

CARBON CYCLE

In a vital atmospheric cycle, carbon moves through the atmosphere and food webs on its way to and from the ocean, sediments, and rocks (Figure 41.13). Its global movement is called the **carbon cycle**. Sediments and rocks hold most of the carbon, followed by the ocean, soil, atmosphere, and land biomass. It enters the atmosphere as cells engage in aerobic respiration, as fossil fuels burn, and when volcanoes erupt and release it from rocks deep in the Earth's crust. Most carbon in the atmosphere and dissolved in the ocean is in the form of carbon dioxide (CO_2).

If you watch bubbles escaping from a glass of carbonated soda left in the sun, you might wonder: Why doesn't all the CO_2 dissolved in warm surface waters of oceans escape to the atmosphere? Driven by winds and regional differences in water density, water makes a gigantic loop from the surface of the Pacific and Atlantic oceans to the Atlantic and Antarctic seafloors. There its CO_2 moves into deep storage reservoirs before the water loops up again (Figure 41.14).

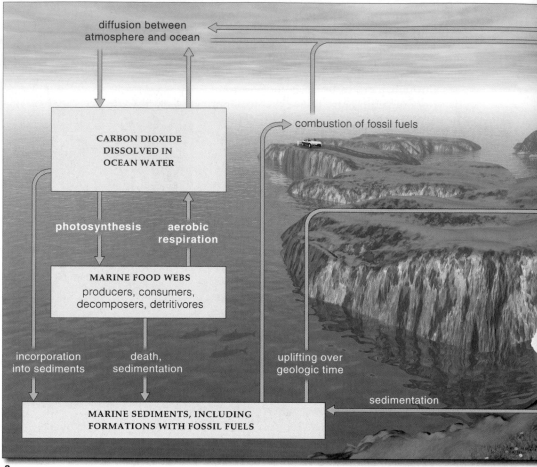

a

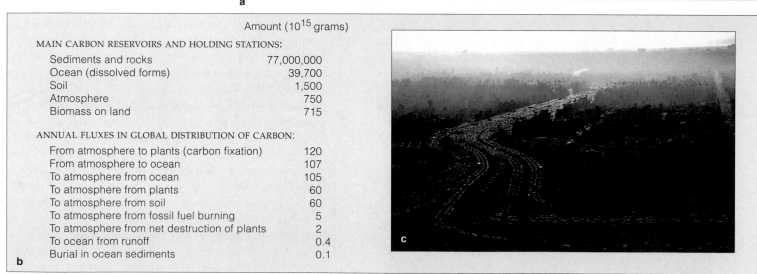

Amount (10^{15} grams)	
MAIN CARBON RESERVOIRS AND HOLDING STATIONS:	
Sediments and rocks	77,000,000
Ocean (dissolved forms)	39,700
Soil	1,500
Atmosphere	750
Biomass on land	715
ANNUAL FLUXES IN GLOBAL DISTRIBUTION OF CARBON:	
From atmosphere to plants (carbon fixation)	120
From atmosphere to ocean	107
To atmosphere from ocean	105
To atmosphere from plants	60
To atmosphere from soil	60
To atmosphere from fossil fuel burning	5
To atmosphere from net destruction of plants	2
To ocean from runoff	0.4
Burial in ocean sediments	0.1

b

c

Figure 41.13 (**a**) Global carbon cycle. The part of the diagram on this page shows carbon's movement through typical marine ecosystems. The part on the facing page shows its movement through ecosystems on land. *Gold* boxes indicate key reservoirs. (**b**) Present-day global carbon budget. (**c**) Part of a Los Angeles freeway under its self-generated blanket of smog at twilight. Each day, fossil fuel burning (by vehicles, industries, and homes) puts carbon and other substances into the atmosphere. ✳

When photosynthetic autotrophs engage in carbon dioxide fixation, they lock up billions of metric tons of carbon atoms in organic compounds each year (Section 6.8). However, the average time that an ecosystem holds any carbon atom varies greatly. For example, organic wastes and remains decompose rapidly in tropical rain forests, so not much carbon accumulates at the surface

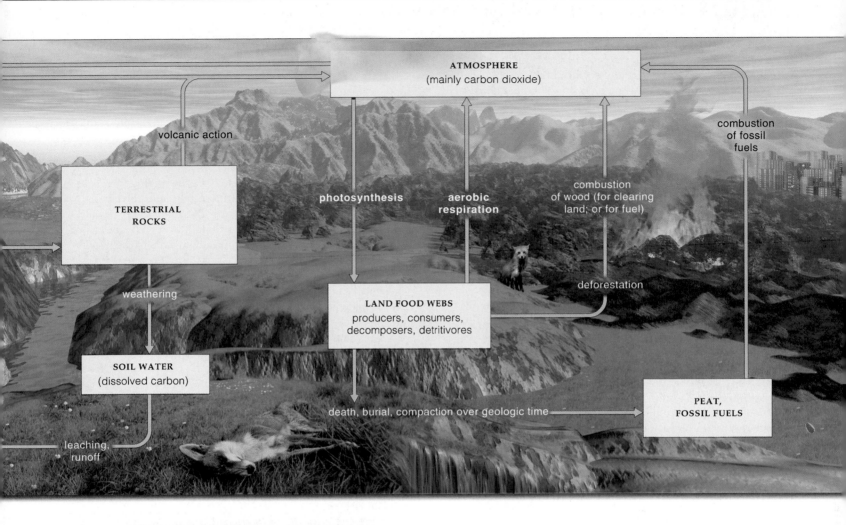

ATMOSPHERE
(mainly carbon dioxide)

volcanic action

combustion
of fossil
fuels

TERRESTRIAL
ROCKS

photosynthesis

aerobic
respiration

combustion
of wood (for clearing
land; or for fuel)

weathering

deforestation

LAND FOOD WEBS
producers, consumers,
decomposers, detritivores

SOIL WATER
(dissolved carbon)

PEAT,
FOSSIL FUELS

death, burial, compaction over geologic time

leaching,
runoff

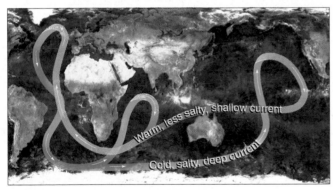

Warm, less salty, shallow current

Cold, salty, deep current

Figure 41.14 Loop of ocean water that delivers carbon dioxide to its deep ocean reservoir. It sinks in the cold, salty North Atlantic and rises in the warmer Pacific.

of soils. In bogs, marshes, and other anaerobic habitats, decomposers cannot degrade organic compounds to smaller bits, so carbon gradually accumulates in peat and other forms of compressed organic matter. Also, in food webs of ancient aquatic ecosystems, carbon was incorporated in shells and other hard parts. The shelled organisms died and sank through water, then became

buried in sediments. Carbon remained buried for many millions of years in deep sediments until part of the seafloor was uplifted above the ocean surface through geologic forces (Section 18.6).

As a final example, plants of the vast swamp forests of the Carboniferous also incorporated carbon atoms in their organic compounds (Section 22.3). In time, those compounds became converted to petroleum, coal, and gas reserves—which we now tap as fossil fuels. Fossil fuel burning and other human activities are currently putting more carbon into the atmosphere than can be cycled naturally to the ocean reservoirs. (As you might deduce from Figure 41.13b, the oceans can remove only about 2 percent of the excess carbon being pumped into the air). The excess is amplifying the greenhouse effect. For that reason, it may be contributing to global warming. The section to follow describes this effect and some possible outcomes of its modification.

The ocean and atmosphere interact in the global cycling of carbon. Fossil fuel burning and other human activities may be contributing to imbalances in the global carbon budget.

www

41.8 FROM GREENHOUSE GASES TO A WARMER PLANET?

GREENHOUSE EFFECT   Atmospheric concentrations of gaseous molecules play a profound role in shaping the average temperature near the surface of the Earth. That temperature has enormous effects on the global climate.

Countless molecules of carbon dioxide, water, ozone, methane, nitrous oxide, and chlorofluorocarbons are key players in interactions that dictate the global temperature. Collectively, the gases act somewhat like a pane of glass in a greenhouse—hence their name, the "greenhouse gases." Wavelengths of visible light can pass around them and reach the Earth's surface. However, greenhouse gases impede the escape of longer, infrared wavelengths— that is, heat—from the Earth into space. How? Gaseous molecules can absorb those wavelengths and reradiate much of the absorbed energy back toward the Earth, as shown in Figure 41.15.

The constant reradiation of heat energy from the greenhouse gases proceeds lockstep with the constant bombardment and absorption of wavelengths from the sun, and so heat builds up in the lower atmosphere. The **greenhouse effect** is the name for this warming action.

GLOBAL WARMING DEFINED   Without the action of greenhouse gases, the Earth's surface would be cold and lifeless. However, there can be too much of a good thing. Largely as an outcome of human activities, greenhouse gases are building to atmospheric levels that are higher than they were in the past. Figure 41.16 documents the increase. As many researchers suspect, greenhouse gases may be contributing to long-term higher temperatures at the Earth's surface, an effect called **global warming**.

What is so alarming about a warmer planet? Suppose the temperature of the lower atmosphere were to rise by only 4°C (7°F). The increase might cause sea levels to rise by about 0.6 meter (2 feet). Why? Temperatures near the ocean surface would increase—and water expands when heated. Also, global warming could make glaciers and the polar ice sheets melt faster. The volume of water released this way alone would flood low coastal regions.

Imagine a long-term rise in sea level, combined with high tides and storm waves. The waterfronts of Vancouver, Seattle, San Diego, New York, Boston, Galveston, Hilo, and all other cities perched along the rim of the world's ocean would be submerged. So would the agricultural lowlands and deltas in India, China, and Bangladesh— where much of the world's rice is grown. Huge tracts of Florida and Louisiana would face saltwater intrusions.

And what if global warming disturbed the regional patterns of precipitation and temperature? We might predict deserts to expand and the interiors of the great continents to become much drier than they already are. Will the nations that control the world's great grain belts be affected? Will other nations be better or worse off? Imagine the economic and political consequences.

Some predicted effects of climate change are emerging. For example, in 1996, we found that the water 200 meters below the Arctic icecap is 1°C warmer than it was just five years ago. If the rapid warming continues, the icecap may disappear in the next century. The rise in temperature may already be disrupting the loop of ocean water that extends like a gigantic conveyor belt from the surface to the bottom of the great oceans (Section 41.7). In addition, during the 1990s, record-breaking hurricanes, flooding, and prolonged droughts have adversely affected crop production around the world. Sandy beaches have been diminishing by 0.6 meter or more per year.

EVIDENCE OF AN INTENSIFIED GREENHOUSE EFFECT
During the late 1950s, researchers on the highest peak in the Hawaiian Islands started measuring the atmospheric concentrations of different greenhouse gases. They chose this remote site because it is almost free of local airborne

a Rays of sunlight penetrate the lower atmosphere and warm the Earth's surface.

b The Earth's surface radiates heat (infrared wavelengths) to the lower atmosphere. Some heat escapes into space. But greenhouse gases and water vapor absorb some infrared wavelengths and reradiate a portion back toward the Earth.

c As concentrations of greenhouse gases increase in the atmosphere, more heat is trapped near the Earth's surface. The surface temperature of the world ocean rises, more water evaporates into the atmosphere, and the Earth's surface temperature rises.

Figure 41.15   The greenhouse effect, as executed mainly by carbon dioxide, water, ozone, methane, nitrous oxide, and chlorofluorocarbons in the atmosphere.

Figure 41.16 (**a**) Changing carbon dioxide levels in the atmosphere, as correlated with the glaciations and interglacial periods over the past 160,000 years. (**b**) Changes in global temperatures for the past 160,000 years relative to the current mean temperature.

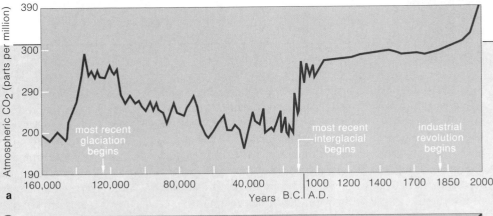

a

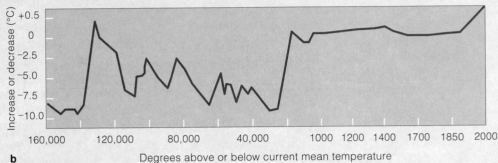

b

Degrees above or below current mean temperature

contamination and is representative of the typical conditions for the Northern Hemisphere. Monitoring activities are still proceeding.

The introduction to Chapter 3 outlined what researchers found out about carbon dioxide. Briefly, in the Northern Hemisphere, atmospheric carbon dioxide levels follow annual cycles of plant growth. They decline in summer, when photosynthesis rates are highest. They rise in winter, when photosynthesis declines and aerobic respiration still proceeds.

The troughs and peaks around the graph line in Figure 41.17*a* represent annual lows and highs. For the first time, scientists glimpsed the integrated effects of carbon balances for land and water ecosystems of an entire hemisphere. Notice the midline of the troughs and peaks in the cycle.

It is steadily increasing. Many take this as evidence of a buildup in carbon dioxide levels. They predict that it will intensify the greenhouse effect over the next century.

Probably global burning of fossil fuels is contributing most to the rise in carbon dioxide levels. Deforestation is also adding to it; when wood burns, its carbon is released. Especially during the past four decades, vast tracts of the world's great forests have been decimated, by deliberate burns and other practices, at astounding rates (Section 43.4). Also, as plant biomass plummets, global absorption of carbon dioxide in photosynthesis may decline.

Will atmospheric levels of greenhouse gases continue to increase until the middle of the twenty-first century? Will global temperature rise by several degrees? If the warming trend is already in motion, we will not be able to reverse it simply by waiting until the last minute to stop fossil fuel burning and deforestation.

There is widespread agreement among scientists that nations must begin preparing for the consequences. For example, we might increase funding levels for genetic engineering studies to develop drought-resistant and salt-resistant plants. Such plants may prove crucial in regions of saltwater intrusions and climatic change.

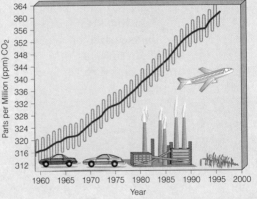

Figure 41.17 Recently documented increases in atmospheric concentrations of four greenhouse gases.

a CARBON DIOXIDE (CO$_2$). Of all human activities, fossil fuel burning and deforestation are contributing most to increases in atmospheric levels of this gas.

WWW

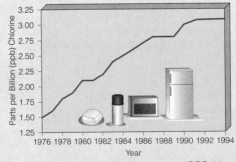

b CHLOROFLUOROCARBONS (CFCs). Until recent restrictions, CFCs were prevalent in plastic foams, refrigerators, air conditioners, and industrial solvents.

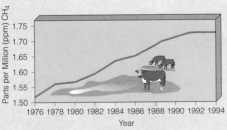

c METHANE (CH$_4$). Termite activity and anaerobic bacteria in swamps, landfills, and the stomachs of cattle and other ruminants produce large amounts of methane as a by-product of metabolism.

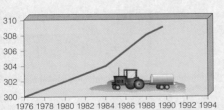

d NITROUS OXIDE (N$_2$O). Denitrifying bacteria produce N$_2$) as a metabolic by-product. The gas also is released in great amounts from fertilizers and animal wastes, as in livestock feedlots.

Since the time of life's origin, the atmosphere and ocean have contained nitrogen. This component of all proteins and nucleic acids moves in an atmospheric cycle called the **nitrogen cycle**. Gaseous nitrogen (N_2) makes up about 80 percent of the atmosphere, the largest nitrogen reservoir. Triple covalent bonds hold its atoms together ($N \equiv N$). Few organisms have the metabolic means to break them. Only certain bacteria, volcanic action, and lightning convert N_2 into forms that enter food webs.

Of all nutrients required for plant growth, nitrogen often is the scarcest. Nearly all of the nitrogen in soils was put there by nitrogen-fixing organisms. Ecosystems lose it when other bacteria "unfix" the fixed nitrogen. Land ecosystems lose more by leaching, although this is the basis of nitrogen inputs to aquatic ecosystems such as streams, lakes, and the oceans (Figure 41.18).

Cycling Processes

Let's follow nitrogen atoms through the portion of the nitrogen cycle that proceeds in an ecosystem on land. They are objects of nitrogen fixation, assimilation and biosynthesis, decomposition, and ammonification.

In **nitrogen fixation**, a few kinds of bacteria convert N_2 to ammonia (NH_3), which quickly dissolves in the cytoplasm, thus forming ammonium (NH_4^+). In aquatic ecosystems, *Anabaena, Nostoc*, and other cyanobacteria are nitrogen fixers. In many land ecosystems, *Rhizobium* and *Azotobacter* fix nitrogen. Collectively, these bacteria fix about 200 million metric tons of nitrogen each year!

Decomposition and **ammonification** are processes by which bacteria and fungi degrade nitrogenous wastes and remains of organisms. Decomposers use part of the released proteins and amino acids during metabolism. But most of the nitrogen remains in the decay products, in the form of ammonia or ammonium, which plants take up. Nitrifying bacteria also act on the ammonia or ammonium. In **nitrification**, they strip the compounds of electrons, and nitrite (NO_2^-) is the result. Other kinds of bacteria use the nitrite during their metabolism and produce nitrate (NO_3^-), which plants then take up.

Most species of plants growing in nitrogen-poor soil are mutualists with fungi. Their roots form mycorrhizae with fungal hyphae, which have an enormous surface area for absorbing mineral ions from soil. Clover, beans, peas, and other legumes are mutualists with nitrogen-fixing bacteria that form root nodules. (The Chapter 21 introduction, as well as Sections 21.5 and 26.2, describes these interactions.)

What do the plants do with nitrogen provided by the bacterial activities? They assimilate and use much of the nitrogen to synthesize amino acids, proteins, and nucleic acids. Plant tissues are the exclusive nitrogen source for animals, which directly or indirectly feed on plants.

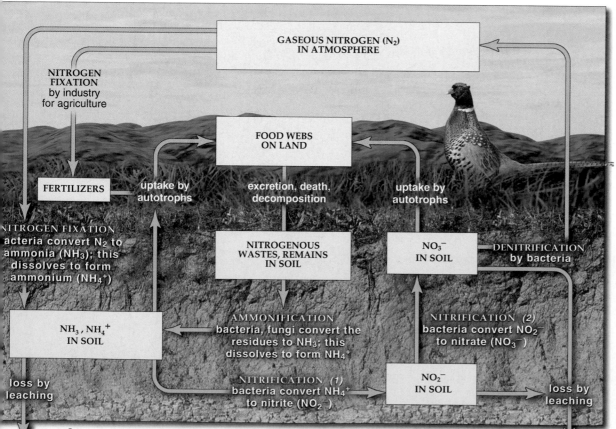

	Amount (10^{12} grams)
Atmospheric N_2	4×10^9
Sediments	4×10^8
Ocean	2.3×10^7
Soil	9.5×10^4
Land biomass	3.5×10^4
Atmospheric N_2O	1.4×10^3
Marine biomass	4.7×10^2

b

Figure 41.18 (a) The nitrogen cycle in an ecosystem on land. The action of nitrogen-fixing bacterial species makes some nitrogen available to plants. Other bacteria cycle nitrogen atoms within the ecosystem by converting organic wastes to ammonium and nitrates. (b) The atmosphere is the largest reservoir of nitrogen.

Labels in figure (a):

GASEOUS NITROGEN (N_2) IN ATMOSPHERE

NITROGEN FIXATION by industry for agriculture

FOOD WEBS ON LAND

FERTILIZERS

NITROGEN FIXATION bacteria convert N_2 to ammonia (NH_3); this dissolves to form ammonium (NH_4^+)

uptake by autotrophs

excretion, death, decomposition

uptake by autotrophs

NITROGENOUS WASTES, REMAINS IN SOIL

NO_3^- IN SOIL

DENITRIFICATION by bacteria

NH_3, NH_4^+ IN SOIL

AMMONIFICATION bacteria, fungi convert the residues to NH_3; this dissolves to form NH_4^+

NITRIFICATION (2) bacteria convert NO_2^- to nitrate (NO_3^-)

loss by leaching

NITRIFICATION (1) bacteria convert NH_4^+ to nitrite (NO_2^-)

NO_2^- IN SOIL

loss by leaching

a

Nitrogen Scarcity

Given the nitrogen cycling processes, you might think that plants have no trouble locating enough nitrogen. But the ammonium, nitrite, and nitrate that form during the cycle are highly vulnerable to leaching and runoff. Leaching, recall, is the removal of some of the nutrients in soil as water percolates through it (Section 26.1).

Besides losses from leaching, some nitrogen also is lost to the air through **denitrification**. By this process, certain bacteria convert nitrate or nitrite to N_2 and a bit of nitrous oxide, or N_2O. Ordinarily, most denitrifying bacteria rely on aerobic respiration. In waterlogged or poorly aerated soil, they switch to anaerobic pathways and use nitrate, nitrite, or nitrous oxide (not oxygen) as the final electron acceptor. (Section 7.5 describes this metabolic strategy.) During the reactions, fixed nitrogen is converted to N_2, much of which escapes to the air.

Also, nitrogen fixation comes at high metabolic cost to the plants that are mutualists with nitrogen fixers. In exchange for nitrogen, plants give up sugars and other photosynthetic products that require big investments of ATP and NADPH. Such plants have the competitive edge in nitrogen-poor soil. However, in nitrogen-rich soil, species that do not have to pay the metabolic price often displace them.

Nitrogen losses from soil are great in agricultural fields. With each harvest, some nitrogen departs from the fields (in tissues of harvested plants). Soil erosion and leaching remove even more. In Europe and North America, farmers traditionally have rotated crops, as when they alternate wheat with legumes. With other conservation practices, crop rotation has helped keep soils stable and productive, sometimes for thousands of years. Now, however, intensive agriculture relies on the application of nitrogen-rich fertilizers.

We can't get something for nothing. Production of fertilizers requires huge amounts of energy from fossil fuels—not from free and unending sunlight. Once, few believed that fossil fuel supplies might run out, so there was little concern about fertilizer costs. It still is quite common to pour more energy into soil (as in fertilizers) than we get out of it (in the form of food). As long as the human population continues to grow exponentially, farmers will be engaged in a race to grow as much food as they can, as fast as they can, for as many people as possible. Enrichment of soil with nitrogen-containing fertilizers is part of the race, as it is now being run.

The cycling of nitrogen in natural ecosystems depends on the activity of nitrogen-fixing bacteria and on mycorrhizae. Human disruptions to the nitrogen cycle may damage the ecosystems.

$W\!W\!W$

41.10 ECOSYSTEM MODELING

We now come full circle to a premise that opened this chapter—that disturbances to one part of an ecosystem often can have unexpected effects on other, seemingly unrelated parts. One approach to predicting unforeseen effects of disturbances is through **ecosystem modeling**. By this method, researchers identify crucial bits of data on different ecosystem components, then use computer programs and models to combine the bits. The resulting data help them predict outcomes of the next disturbance. The danger is that investigators may not have identified all of the key relationships in the ecosystem under study and incorporated them accurately into the computer model. The most crucial fact may be one that we do not yet know, as the following case study makes clear.

DDT, a relatively stable hydrocarbon, is a synthetic organic pesticide. It is nearly insoluble in water, so you might think that it would act only where applied. But winds carry DDT in vapor form; water transports fine particles of it. DDT also is highly soluble in fats, so it can accumulate in the tissues of organisms. Thus, DDT can show **biological magnification**. By this occurrence, a nondegradable or slowly degradable substance becomes more and more concentrated in the tissues of organisms at the higher trophic levels of a food web. Most of the DDT from all organisms that a consumer feeds on during its lifetime ends up in its own tissues. DDT and modified forms of it disrupt metabolic activities and are often toxic to many aquatic and terrestrial animals.

Several decades ago, DDT started to spread around the globe, infiltrate food webs, and affect organisms in ways that no one had predicted. In cities where DDT was sprayed to control Dutch elm disease, songbirds started dying. In streams flowing through forests where DDT was sprayed to kill larvae of budworms, fish started dying. In fields sprayed to control one kind of pest, new pests moved in. *DDT was indiscriminately killing natural predators that had been keeping pest populations in check!*

Then side effects of biological magnification started showing up in places far removed from the areas of DDT application—*and much later in time.* Most vulnerable were species at the top of food webs, such as bald eagles and brown pelicans. A product of DDT breakdown interferes with vital physiological processes. As one outcome, birds produced eggs with brittle shells. As a result, many of their chick embryos simply did not make it to hatching time. Some species were at the brink of extinction.

Since the 1970s, DDT has been banned in the United States, except for restricted cases where public health is endangered. Today, many species that were hit hardest have partially recovered. Yet even now, some birds are still laying thin-shelled eggs; they are picking up DDT at winter ranges in Latin America. Even as late as 1990, the California State Department of Health recommended closure of a fishery off the coast near Los Angeles. DDT from industrial waste discharges that ended twenty years before is still contaminating that ecosystem.

$W\!W\!W$

1. An ecosystem is an array of producers, consumers, detritivores, and decomposers and their environment. It is an open system, with inputs and outputs of energy and nutrients. There is a one-way flow of energy into and out from an ecosystem and a cycling of materials among its organisms (Figure 41.19).

2. Sunlight is the initial energy source for nearly all ecosystems. Photoautotrophs, the primary producers, convert that energy to other forms, such as chemical bond energy of ATP. They also assimilate many of the nutrients required by the ecosystem's heterotrophs.

    a. The heterotrophs include consumers. Herbivores feed upon algae and plants. Carnivores ingest animals. Parasites withdraw nutrients from the tissues of living hosts. Omnivores use a variety of food sources.

    b. Decomposers and detritivores are heterotrophs, too. The major decomposers (certain fungi and bacteria) get energy and nutrients from organic remains and wastes. Detritivores, such as crabs and earthworms, ingest bits of dead or decomposing material.

3. Feeding relationships are structured as trophic levels (a hierarchy of energy transfers in an ecosystem, which is sometimes referred to as "Who eats whom").

    a. Primary producers are at the first trophic level. Herbivores are at the second level, and carnivores are at successively higher levels.

    b. Decomposers, humans, and many other kinds of organisms obtain energy from more than one source and cannot be assigned to a single trophic level.

4. Isolated food chains (straight-line sequences of who eats whom in an ecosystem) are rare in nature. They cross-connect with one another, as food webs.

5. The rate at which primary producers capture and store a given amount of energy in a given time interval is the primary productivity. The rate of energy storage in primary producers in excess of the rate of aerobic metabolism is the net primary productivity.

6. Energy fixed by photosynthesizers passes through grazing food webs and detrital food webs, which often interconnect within the same ecosystem. The amount of useful energy that flows through successive levels of consumers declines at each energy transfer in a food web (because of metabolic heat loss and shunting of food energy into organic wastes).

7. Primary productivity, hence ecosystem structure, depends on water and nutrients. These substances move gradually through the physical environment, rapidly through organisms, and back to the environment, in biogeochemical cycles.

    a. Water moves through a hydrologic cycle. In land ecosystems, plants stabilize soil and minimize nutrient loss during the cycle, as by runoff.

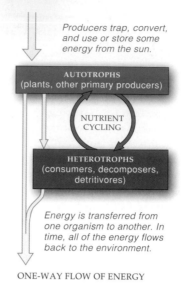

Producers trap, convert, and use or store some energy from the sun.

AUTOTROPHS (plants, other primary producers)

NUTRIENT CYCLING

HETEROTROPHS (consumers, decomposers, detritivores)

Energy is transferred from one organism to another. In time, all of the energy flows back to the environment.

ONE-WAY FLOW OF ENERGY

Figure 41.19 Summary of the one-way flow of energy and the nutrient cycling in ecosystems.

    b. In atmospheric cycles, a nutrient is mainly in gaseous form (such as carbon dioxide).

    c. In a sedimentary cycle, some nutrient is in solid form, such as phosphorus. The main reservoir is Earth's crust.

8. Carbon dioxide is the main atmospheric gas in the carbon cycle. The ocean is its greatest reservoir (in dissolved form).

9. Burning fossil fuels and converting natural ecosystems to farming or grazing add to imbalances in a global carbon budget and may be factors in long-term global warming.

10. Supplies of nitrogen are a limiting factor in the total net primary productivity of ecosystems on land. Gaseous nitrogen is abundant in the atmosphere. Nitrogen-fixing bacteria convert the N_2 to ammonia and nitrates that producers take up. Mycorrhizae and root nodules (two symbiotic interactions) enhance the nitrogen uptake.

11. Most mineral elements enter sedimentary cycles and become available, in dissolved form, for primary producers. High concentrations of phosphorus especially trigger eutrophication (nutrient enrichment of aquatic ecosystems) and promote harmful algal blooms.

12. Disturbance of one aspect of an ecosystem often has unexpected effects on other, seemingly unrelated parts.

13. To predict the consequences of a disruption to an ecosystem, as by computer modeling, researchers must identify all of the ecosystem's key relationships and then incorporate them into the model.

Review Questions

1. Define an ecosystem in terms of inputs and outputs. *41.1*

2. Define primary producer, consumer, decomposer, and detritivore. Give an example of each. *41.1*

3. Define and describe trophic levels. Which class of organisms is farthest from the energy input to an ecosystem? *41.1*

4. How does a food web differ from a food chain? *41.1*

5. Characterize grazing and detrital food webs. *41.2*

6. Distinguish between energy pyramid and ecological pyramid. Which kind was constructed for Silver Springs? *41.2*

7. Define three types of biogeochemical cycles and give an example of each. *41.4*

8. Define and describe the connections among nitrogen fixation, nitrification, ammonification, and denitrification. *41.9*

Self-Quiz (Answers in Appendix III)

1. Ecosystems have _____ .
   a. energy inputs and outputs
   b. one trophic level
   c. nutrient cycling but not outputs
   d. a and b

2. Trophic levels are _____ .
   a. structured feeding relationships
   b. who eats whom in an ecosystem
   c. a hierarchy of energy transfers
   d. all of the above

3. Primary productivity on land is affected by _____ .
   a. photosynthesis and respiration by plants
   b. how many plants are neither eaten nor decomposed
   c. rainfall and temperature
   d. all of the above

4. Eutrophication is _____ .
   a. nutrient enrichment of soil in terrestrial ecosystem
   b. nutrient enrichment of an aquatic ecosystem
   c. nutrient loss from soil, as by leaching
   d. nutrient loss from an aquatic ecosystem, as by runoff

5. Match the ecosystem terms with the suitable description.
   _____ producers   a. herbivores, carnivores, omnivores
   _____ consumers   b. feed on partly decomposed matter
   _____ decomposers   c. degrade organic remains, wastes
   _____ detritivores   d. photoautotrophs

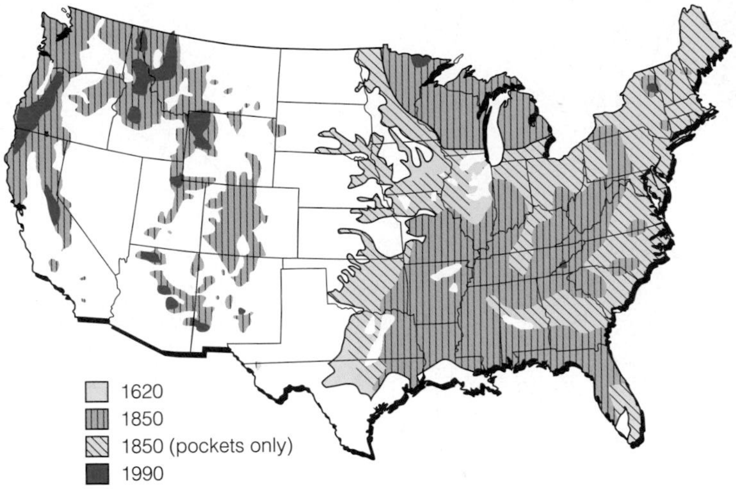

Critical Thinking

1. Imagine and describe an extreme situation in which you would be a participant in a food chain rather than a food web.

2. Marguerite is growing a vegetable garden in Maine. What are the variables that can affect its net primary production?

3. List as many of the agricultural products and manufactured goods as you can identify that you depend upon. Are any implicated in the amplified greenhouse effect?

4. Of all crops grown in the United States, less than 10 percent are free of pesticide or fungicide applications. Such *organically grown produce* costs more, spoils faster, and is nonuniform in size and appearance. Do you buy such produce? Why or why not?

5. In 1995, the biologists Reed Noss, J. Michael Scott, and Edward LaRoe issued a report on *endangered ecosystems* in the United States. They categorize thirty natural ecosystems as being in danger of disappearing; the once-vast domains have diminished in size by more than 98 percent (Figure 41.20). Agricultural conversion, urban development, and other human activities account for the losses.

In the next chapter, you will read about the type and extent of the ecosystems in question. The once-largest among them include tallgrass prairie, oak-studded savannas, and eastern deciduous forests, as well as pine forests that once covered much of the coastal plains of the southeastern states. As Figure 41.20 shows, the most imperiled regions are largely in the eastern part of the country. As the map indicates, more biodiversity was lost at the ecosystem level than is generally recognized. The finding may impact how (and whether) the government amends its conservation laws, such as the Endangered Species Act. Do we as a nation protect entire ecosystems, not just an endangered species? That is a goal of many conservationists. But property-rights advocates criticize the goal; they view it as a threat to the long-standing tradition of private ownership of land. They also worry that merely identifying ecosystems in need of protection will endanger property values.

Do some research on a natural habitat that is part of an imperiled ecosystem. Get a sense of its decline in biodiversity, of the kinds of organisms that once flourished there. Then ask yourself: Would you participate in efforts to set aside land for ecological restoration? Or would you consider such efforts too intrusive on individual rights of property ownership?

6. *Polar ice shelves* are vast, enormously thickened sheets of ice floating on seawater. Measurements taken between 1978 and 1994 suggest the Arctic ice shelves are retreating. Figure 41.21 shows how a gargantuan chunk of the Larsen Ice Shelf broke away from Antarctica in January 1995. Do you suppose these events are evidence of a long-term trend in global warming? If not, what factors might have triggered them?

Figure 41.20 Map showing the extent of deforestation in the United States, starting from the year 1620 through 1990.

- 1620
- 1850
- 1850 (pockets only)
- 1990

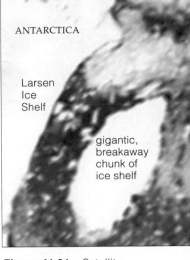

ANTARCTICA

Larsen Ice Shelf

gigantic, breakaway chunk of ice shelf

Figure 41.21 Satellite image of an iceberg as big as Rhode Island. This one broke away from Antarctica in 1995.

Selected Key Terms

ammonification *41.9*
biogeochemical cycle *41.4*
biological magnification *41.10*
carbon cycle *41.7*
consumer *41.1*
decomposer *41.1*
decomposition *41.9*
denitrification *41.9*
detrital food web *41.2*
detritivore *41.1*
ecological pyramid *41.2*
ecosystem *41.1*
ecosystem modeling *41.10*
energy pyramid *41.2*
eutrophication *41.6*
food chain *41.1*
food web *41.1*
global warming *41.8*
grazing food web *41.2*
greenhouse effect *41.8*
hydrologic cycle *41.5*
nitrification *41.9*
nitrogen cycle *41.9*
nitrogen fixation *41.9*
phosphorus cycle *41.6*
primary producer *41.1*
primary productivity *41.2*
trophic level *41.1*
watershed *41.5*

Readings *See also www.infotrac-college.com*

Krebs, C. 1994. *Ecology.* Fourth edition. New York: HarperCollins.

WWW *http://www.brookscole.com/biology*

Practice quiz questions, hypercontents, BioUpdates, and critical thinking. The Brooks/Cole Biology Resource Center provides a wealth of information fully organized and integrated by chapter.

THE BIOSPHERE

Does a Cactus Grow in Brooklyn?

Suppose you live in the American Southwest but find yourself touring one of the deserts of Africa. There you come across a flowering plant with spines, tiny leaves, and columnlike, fleshy stems—just like some cactus plants back home (Figure 42.1). Or suppose you live in the coastal hills of California and decide to tour the Mediterranean coast, the southern tip of Africa, or even central Chile. There you come across many-branched, tough-leafed woody plants—very much like the many-branched, tough-leafed chaparral plants back home.

In both cases, the plants are separated by enormous geographic and evolutionary distances. Why, then, are they so much alike? The question intrigues you, so you decide to compare their locations on a global map. As you quickly discover, the American and African desert plants grow approximately the same distance from the equator. Chaparral plants and their distant look-alikes grow along the western or southern coasts of continents between latitudes 30° and 40°. As Charles Darwin and other naturalists did long ago, you have just stumbled onto one of many predictable patterns in nature.

What causes such patterns? With this question, we turn to **biogeography**–the study of the distribution of organisms, past and present, and of diverse processes that underlie the distribution patterns. For example, "accidents of history" put many species in particular places. Think back on the colossal breakup of Pangea, more than 100 million years ago. When chunks of that supercontinent began to drift apart, many species had no choice but to go along for the ride. Over evolutionary time, the travelers were dispersed to different isolated locations. And there they proceeded to undergo genetic

Figure 42.1 Life in the environment—a case of morphological convergence. (**a**) *Echinocerus*, of the cactus family, grows in deserts of the American Southwest. (**b**) In deserts of southwestern Africa, we find *Euphorbia*, of the spurge family (Euphorbiaceae). The lineages appear to be closely related, but they are geographically and evolutionarily distant. Long ago, plants in both lineages put similar structures (including leaves) to similar uses in similar habitats—and their descendants ended up resembling each other.

divergence from the parent populations. Among those changing populations on the fragment that became Australia were the ancestors of eucalyptus trees and platypuses, of wombats and kangaroos.

Species also owe their distribution to topography, climate, and species interactions. With diligence, you might grow a cactus under artificial lights in a heated room in Brooklyn or some other New York City borough. Plant that cactus outside, and it won't last one winter.

This last example reminds us that we humans tinker with the distribution of species. Not all of our tinkering is as harmless as growing a cactus in Brooklyn. Think of our predatory effects on the world's fisheries or the effects of our pesticide battles with insect competitors for food. Earlier chapters provided you with a general picture of predation, competition, and other species interactions. This chapter considers the physical forces shaping the biosphere itself. It serves as a foundation for addressing the impact of the human species on the biosphere—the topic of the chapter to follow.

Start with the definition of the **biosphere**—the sum total of all places in which organisms live. They live in the waters of the Earth, including the ocean, polar ice caps, and other forms of liquid and frozen water (the hydrosphere). They live in the soils and sediments of the Earth's outer, rocky layer (the lithosphere). And they live in the lower **atmosphere**, which is made up of gases and airborne particles that envelop the Earth. The air in the upper atmosphere, about 17 kilometers above the Earth's surface, is too thin to support life.

In this vast biosphere you will find ecosystems that range from continent-straddling forests to tiny pools in cup-shaped leaves of plants. As you will see, climate profoundly influences all ecosystems except for a few at hydrothermal vents on the ocean floor.

Climate refers to the average weather conditions, such as temperature, humidity, wind speed, cloud cover, and rainfall, over time. Many factors contribute to climate. The main ones are variations in the amount of incoming solar radiation, the Earth's daily rotation and its path around the sun, the world distribution of continents and oceans, and landmass elevations. All of these factors interact to produce prevailing winds and ocean currents that influence the global patterns of climate. Climate affects the physical and chemical development of sediments and soils.

Climate, together with the composition of sediments and soils, influences the growth of primary producers— and, through them, the distribution of ecosystems.

KEY CONCEPTS

1. Energy from the sun is the initial energy source for nearly all ecosystems on Earth. In addition, solar energy influences the global distribution of ecosystems. It does so by continually providing heat energy that warms the atmosphere and drives the Earth's weather systems.

2. Interactions among global air circulation patterns, ocean currents, and diverse topographic features result in regional variations in patterns of temperature and rainfall. The variations influence the composition of soils and sediments, the growth and distribution of primary producers, and, through them, the distribution of ecosystems.

3. A biome is a large, regional unit of land that can be characterized by the climax vegetation of the ecosystems within its boundaries. Deserts and broadleaf forests are examples. Their distribution corresponds roughly with regional variations in climate, topography, and soil type.

4. The water provinces cover more than 71 percent of the Earth's surface. The oceans hold all but 3 percent of the free-flowing water. The rest is in inland seas, estuaries, and bodies of fresh water, including lakes, streams, and rivers.

5. Each freshwater and marine ecosystem has gradients in light availability, temperature, and dissolved gases. The gradients vary daily and seasonally. They affect primary productivity and the composition of species.

AIR CIRCULATION PATTERNS AND REGIONAL CLIMATES

Each winter Pacific Grove, California, is host to great gatherings of tourists and monarch butterflies (Figure 42.2). Similarly, caribou, Canada geese, sea turtles, and whales undertake vast migrations, moving to and from overwintering grounds. Other kinds of animals stay put. They have adaptations in form, physiology, and behavior that allow them to endure seasonal change.

Figure 42.2 Monarch butterflies, migratory insects that each winter gather in the trees of California coastal regions and of central Mexico. Monarchs travel hundreds of kilometers south to those places, which are cool and humid in winter. If they were to remain in their northern breeding grounds, they would risk being killed by far more severe conditions.

Throughout Canada and the United States, flowering plants form leaves, flower, bear fruit, and drop leaves. In the ocean, uncountable numbers of photosynthetic microorganisms undergo seasonal bursts of primary productivity. Clearly, organisms are exquisitely attuned to climatic conditions, which differ from one region to the next and as one season gives way to the next.

Climate begins with incoming rays from the sun. Of the total amount of solar radiation arriving at the outer atmosphere, only about half gets through to the Earth's surface. Molecules of ozone (O_3) and of oxygen (O_2) in the atmosphere absorb most wavelengths of ultraviolet radiation. Such wavelengths, recall, are lethal for most organisms. Absorption is greatest in a region between 17 and 27 kilometers above sea level, where molecules of ozone are the most concentrated (Figure 42.3). Hence the name, "ozone layer." Clouds, bits of dust, and water vapor suspended in the atmosphere absorb some other wavelengths or reflect them back into space.

Radiation that penetrates the atmosphere warms the surface of the Earth, which gives up heat by way of radiation or evaporation. The molecules making up the lower atmosphere absorb some heat and reradiate part of it toward the Earth. The effect is somewhat like heat retention in a greenhouse, which lets in the sun's rays while retaining heat that is being lost from the plants and soil inside (Section 41.8). Why is the greenhouse effect important? *Heat energy derived from the sun warms the atmosphere, and it ultimately drives the Earth's great weather systems.*

The sun's effect differs from one latitude to the next. Its rays are more concentrated at the equator than at the poles, so air is heated more at the equator (Figure 42.4). The global pattern of air circulation starts as warm air at the equator rises and spreads north and south. Under the moving air, the Earth rotates faster at the equator than it does at the poles, and this creates worldwide belts of prevailing east and west winds. The differences in solar heating at different latitudes and the modified air circulation patterns help define the world's major **temperature zones** (Figure 42.5*a*).

Global air circulation patterns give rise to differences in rainfall at different latitudes as well. Think about this: Warm air can hold more moisture than cool air. At the equator, air picks up moisture from the seas and rises to cooler altitudes. There it gives up moisture as rain, which supports the luxuriant growth of tropical

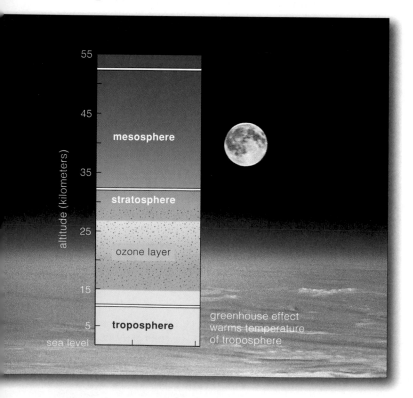

Figure 42.3 Earth's atmosphere. Most of the global air circulation proceeds in the troposphere, where temperature decreases rapidly with altitude. Most of the ultraviolet wavelengths in the sun's rays are absorbed in the upper atmosphere, mainly at the ozone layer. The ozone layer ranges between 17 and 27 kilometers above sea level.

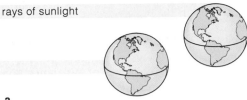

a

Figure 42.4 Global air circulation patterns, as brought about by three interrelated factors. *First*, the sun's rays are less spread out in equatorial regions than in polar regions (**a**). Warm equatorial air rises and spreads northward and southward to produce the initial pattern of air circulation (**b**).

Second, the nonuniform distribution of landmasses creates variations in air pressure. Land absorbs and gives up heat faster than the ocean does, so some parcels of air rise (or sink) faster than others. Air pressure decreases where air rises (and increases where air sinks). The differences create winds that disrupt the overall movement from the equator to the poles.

Third, the rotation and overall shape of the Earth introduce easterly and westerly deflections in wind directions. Each time the ball-shaped Earth makes a full rotation, its surface turns faster beneath air masses at the equator and slower beneath those at the poles. Therefore, a rising air mass cannot really move "straight north" or "straight south," but rather is deflected to the east or west. This is the source of prevailing east and west winds (**c**).

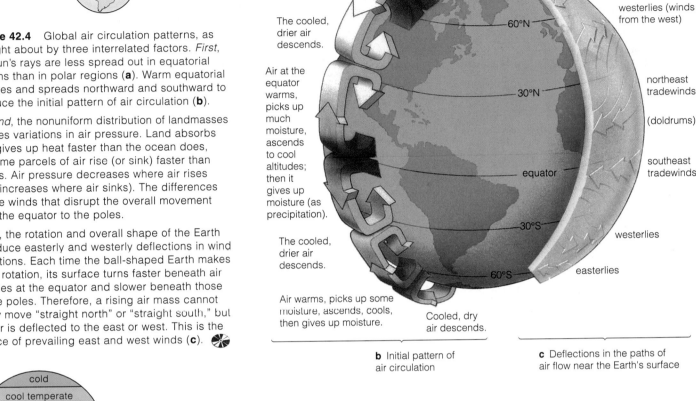

b Initial pattern of air circulation **c** Deflections in the paths of air flow near the Earth's surface

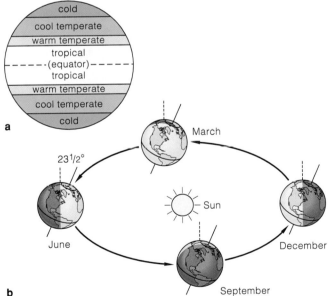

a

b

Figure 42.5 (**a**) World temperature zones. (**b**) Annual variation in incoming solar radiation. The northern end of the Earth's fixed axis tilts toward the sun in June and away from it in December. As a result, the equator's position relative to the boundary of illumination between day and night varies throughout the year. Such variations in sunlight intensity and in daylength give rise to seasonal variations in temperature in the two hemispheres. The seasonal change becomes more pronounced with distance from the equator. Change is greatest in the central regions of continents, where the moderating effects of oceans are minimal.

forests. The air, which is drier now, moves away from the equator. Later, it descends at latitudes of about 30°, and it becomes warmer and drier during this descent. Deserts tend to form at those latitudes. Farther from the equator, air again picks up moisture, ascends to high altitudes, and creates another moist belt at latitudes of about 60°. Then it descends in polar regions, where the low temperatures and almost nonexistent precipitation give rise to the cold, dry polar deserts.

Finally, the amount of solar radiation reaching the surface varies seasonally because of the Earth's rotation around the sun (Figure 42.5*b*). This leads to seasonal changes in daylength, prevailing wind directions, and temperature. And so primary productivity alternately rises and falls on land and in the seas—and migrations shift many kinds of animals to new locations.

Latitudinal differences in the amount of solar radiation reaching the Earth produce global air circulation patterns. The Earth's rotation and its overall shape affect the patterns to produce latitudinal belts of temperature and rainfall.

Also, the Earth's annual rotation around the sun introduces seasonal changes in winds, temperatures, and rainfall.

These factors influence the locations of different ecosystems.

OCEAN, LANDFORMS, AND REGIONAL CLIMATES

Ocean Currents and Their Effects

The **oceans** are continuous bodies of water that cover almost three-fourths of the Earth. The uppermost 10 percent circulates in currents, it distributes nutrients in marine ecosystems and affects regional climates. Solar heating and wind friction drive the surface currents.

Latitudinal and seasonal variations in solar heating warm and cool ocean water on a vast scale. The volume of water increases if heated and decreases if cooled. So the sea level is about 8 centimeters (3 inches) higher at the (hot) equator than it is at the (cold) poles. The sheer volume of water in this "slope" is enough to get the surface waters moving in response to gravity, generally from the equator to the poles. Along the way, surface waters warm the air parcels above them. Every second at midlatitudes, about *10 million billion* calories of heat energy get transferred from warm water to air!

The "tug" of mainly the trade winds and westerlies causes currents—rapid mass flows of water. The Earth's rotation, positions of landmasses, and shapes of ocean basins affect the direction and properties of the currents. Water circulates clockwise in the Northern Hemisphere and counterclockwise in the Southern Hemisphere, as shown in Figure 42.6.

Swift, deep, and narrow currents of nutrient-poor waters move parallel with the east coast of continents. For instance, 55 million cubic meters per second of warm water move northward in the Gulf Stream, along the east coast of North America. The slower, shallow, and broad currents paralleling the west coast of continents move cold water toward the equator. As you will see, these currents move deep, nutrient-rich waters to the surface.

Why do cool, mild, foggy summers prevail along the Pacific Northwest coastline? The California Current moves cold water near the coast as it flows toward the equator, and winds approaching the coast cool off as they give up heat to the cold water. Why are Baltimore and Boston so muggy in summer? Air above the Gulf Stream gains heat and moisture, which winds from the south and east carry to the city. Compared to Ontario in central Canada, why are winters milder in London and Edinburgh, which are at the same latitude? The North Atlantic Current picks up warm water from the Gulf Stream, and as it flows past northwestern Europe, it gives up heat energy to the prevailing winds.

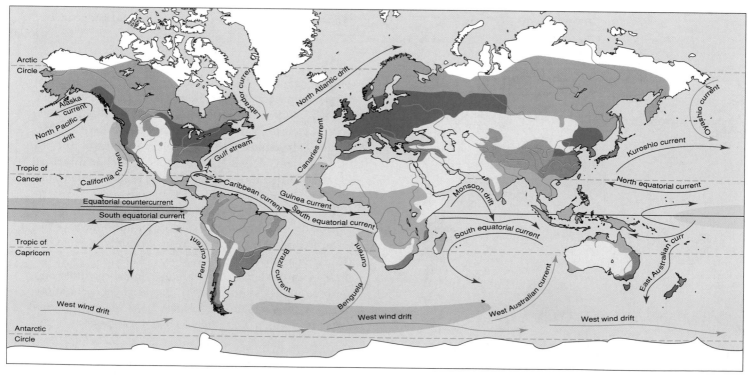

Figure 42.6 Generalized map of major climate zones, correlated with surface currents and drifts of the oceans. Warm surface currents move from the equator toward the poles. Water temperatures differ with latitude and depth and contribute to regional differences in temperature and rainfall. In which direction does a given current flow? It depends on prevailing winds, the Earth's rotation, gravity, the shapes of ocean basins, and the positions of landmasses.

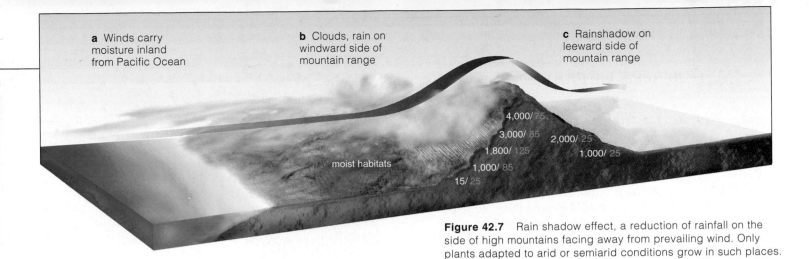

a Winds carry moisture inland from Pacific Ocean

b Clouds, rain on windward side of mountain range

c Rainshadow on leeward side of mountain range

4,000/ 75
3,000/ 85   2,000/ 25
1,800/ 125   1,000/ 25
1,000/ 85
moist habitats
15/ 25

Figure 42.7 Rain shadow effect, a reduction of rainfall on the side of high mountains facing away from prevailing wind. Only plants adapted to arid or semiarid conditions grow in such places. *Blue* numbers show average yearly precipitation (in centimeters), measured at different locations on both sides of the mountain range. *White* numbers signify elevation (in meters).

Regarding Rain Shadows and Monsoons

Mountains, valleys, and other aspects of topography influence regional climates. "Topography" means the physical features of a region, such as its elevation. For example, imagine a warm air mass picking up moisture off California's western coast. Moving inland, it reaches the Sierra Nevada, a mountain range that parallels the coast. The air cools as it ascends to higher altitudes, then loses moisture as rain (Figure 42.7). The vegetation belts at different elevations correspond to changes in air temperature and moisture. Grasslands with species adapted to semiarid conditions prevail at the western base of the range. At higher elevations, more moisture and cooler air support deciduous and evergreen forests. Higher still, a subalpine belt supports a few species of evergreen trees that withstand the rigors of a colder habitat. Above the subalpine belt, only low, nonwoody plants and dwarfed shrubs can grow.

After air flows over mountain crests and starts its descent, it warms up. Warm air can hold more water, so now it retains moisture and draws more out of plants and soil. The outcome is a **rain shadow**, a semiarid or arid region of sparse rainfall on the leeward side of high mountains. *Leeward* is the direction not facing a wind; *windward* is the direction from which the wind blows. Europe's mountain ranges, Asia's Himalayas, and South America's Andes create rain shadows. Tropical forests thrive on the windward side of Hawaii's high volcanic peaks, and arid conditions prevail on the leeward side.

Monsoons are air circulation patterns that influence the continents north or south of warm oceans. The land heats so intensely, low-pressure air parcels form above it. The low pressure causes moisture-laden air above the neighboring ocean to move inland, as from the Bay of Bengal into India and Bangladesh or from the Gulf of Mexico into the hot deserts of the American Southwest. Fantastic summer thunderstorms are one outcome. In a zone called the doldrums, trade winds converge and the equatorial sun causes intense heating and heavy rain. The air circulates north and south seasonally. It creates patterns of wet and dry seasons, as in East Africa.

The recurring sea breezes along coastlines might be viewed as "mini-monsoons." Here again, the water and the adjacent land have different heat capacities. During the morning, the water's temperature does not rise as rapidly as the land's temperature. When the warmed air above the land rises, the cooler marine air moves in. After sunset, the land loses heat faster than the water; land breezes flow in the reverse direction (Figure 42.8).

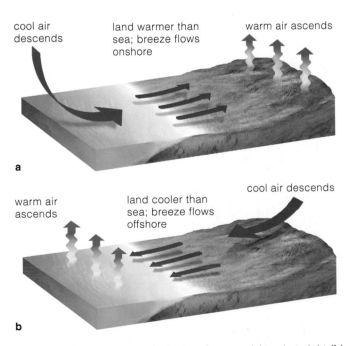

cool air descends

land warmer than sea; breeze flows onshore

warm air ascends

a

warm air ascends

land cooler than sea; breeze flows offshore

cool air descends

b

Figure 42.8 Coastal breeze in the afternoon (**a**) and at night (**b**).

Surface ocean currents, in combination with global air circulation patterns, influence regional climates and help distribute nutrients in marine ecosystems.

Air circulation patterns, ocean currents, and landforms interact in ways that influence regional temperatures and moisture levels. Thus they also influence the distribution and dominant features of ecosystems.

WWW

Biogeographic Distribution

The circulation patterns in the atmosphere and at the ocean surface, in concert with topography, give rise to regional differences in temperature and moisture. This data helps explain why tundra, forests, grasslands, and deserts form in some places but not others. It helps us understand the distribution of ecosystems and their species, each adapted to regional conditions.

For example, the information provides clues to why many evolutionarily distant species look alike. Often such species are results of convergent evolution, which Section 18.2 describes. Briefly, their ancestors were not closely related but lived in similar environments and faced similar selection pressures. By natural selection, their body plan underwent similar modifications and ended up looking alike. Ancestors of those cactus and

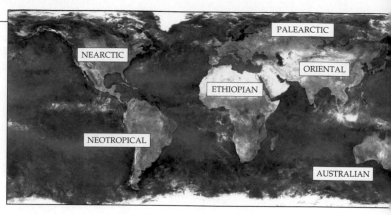

Figure 42.9 Biogeographic realms.

spurge plants in Figure 42.1 both evolved in hot, dry deserts, where water is scarce. Fleshy plant stems that have thick cuticles conserve precious water. And rows of sharp spines help deter hungry herbivores, which might otherwise chew on juicy plant parts.

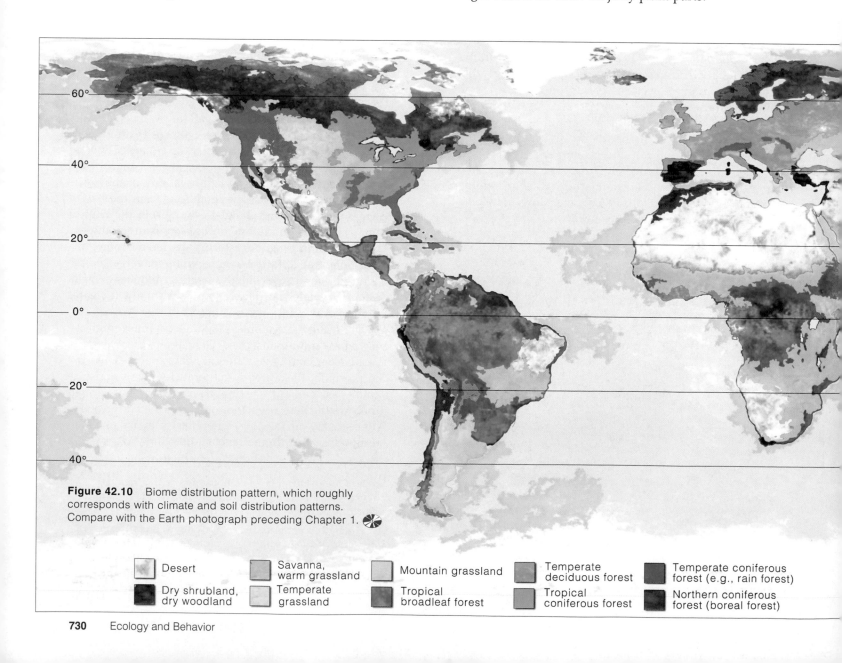

Figure 42.10 Biome distribution pattern, which roughly corresponds with climate and soil distribution patterns. Compare with the Earth photograph preceding Chapter 1.

- Desert
- Dry shrubland, dry woodland
- Savanna, warm grassland
- Temperate grassland
- Mountain grassland
- Tropical broadleaf forest
- Temperate deciduous forest
- Tropical coniferous forest
- Temperate coniferous forest (e.g., rain forest)
- Northern coniferous forest (boreal forest)

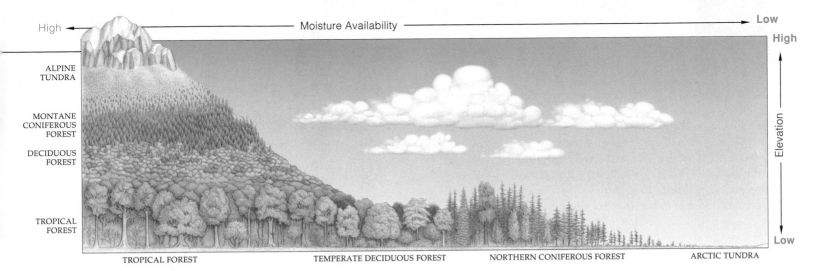

General changes in plant form along environmental gradients for North America.

ALPINE TUNDRA

MONTANE CONIFEROUS FOREST

DECIDUOUS FOREST

TROPICAL FOREST

High ← Moisture Availability → Low

High / Elevation / Low

TROPICAL FOREST   TEMPERATE DECIDUOUS FOREST   NORTHERN CONIFEROUS FOREST   ARCTIC TUNDRA

Figure 42.11 General changes in plant form along environmental gradients for North America. This diagram indicates gradients in elevation and in the availability of water, both of which have great influence on a biome's primary productivity. Other factors, such as the mean annual air temperature and soil drainage characteristics, also have effects on plant form.

Long ago W. Sclater, then Alfred Wallace, attached the name **biogeographic realms** to six vast land areas, each with distinctive kinds and numbers of plants and animals (Figure 42.9). Biogeographic realms maintain their identity partly because of climate. They also tend to maintain a distinct identity when mountain ranges, oceans, and other physical barriers restrict gene flow and thus keep their component species isolated.

Biomes and Ecoregions

Each biogeographic realm may be further divided into biomes. A **biome** is a large region of land characterized by habitat conditions and by its community structure, including endemic species, which evolved nowhere else. Take a look at Figure 42.10, then see how it correlates with the environmental factors shown in Figure 42.11.

As you might deduce, distinctive biomes prevail at certain latitudes and elevations. For example, biomes that are dominated by species of short plants prevail in dry regions, at high elevations, and at high latitudes. Biomes dominated by tall, leafy plant species prevail at tropical and temperate latitudes, and at low elevations with warm temperatures and high rainfall. As you will see shortly, soils also affect biome distribution.

Conservationists are working to locate, inventory, and protect "hot spots," These are portions of certain biomes that are the richest in biodiversity and the most vulnerable to species loss. Twenty-four of the hot spots hold more than half of all land species.

In its Global 200 program, the World Wildlife Fund is targeting ecoregions as well as hot spots. **Ecoregions** are large areas representative of globally important biomes *and* water provinces that are vulnerable to extinction. Figure 42.10 identifies the marine ecoregions.

The land and the seas contain realms of biodiversity. The unique identity of each biogeographic realm, biome, and ecoregion is an outcome of habitat conditions, the kinds and numbers of species, and their evolutionary history.

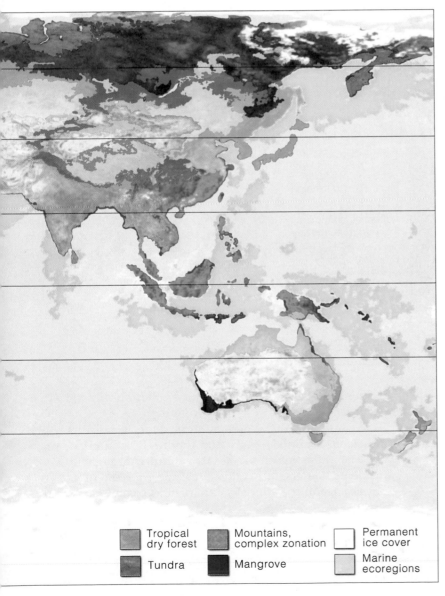

Tropical dry forest

Tundra

Mountains, complex zonation

Mangrove

Permanent ice cover

Marine ecoregions

WWW

SOILS OF MAJOR BIOMES

Most regions of land have **soils**, which are mixtures of mineral particles and variable amounts of decomposing organic material (humus). As described in Section 26.1, the weathering of hard rocks produces coarse-grained gravel, then sand, silt, and finely grained clay. Water and air infiltrate spaces between particles. How much the particles compact together and the proportions of each kind vary within and between regions.

Soils have a layered structure, or *profile*, that reflects their stage of development. Figure 42.12 shows a few examples. Uppermost is topsoil. It has the most humus and is the most vulnerable to erosion. Topsoil may be less than a centimeter thick on steep slopes and more than a meter thick in grasslands. The lowest layer of soil consists of rocks in varying degrees of weathering.

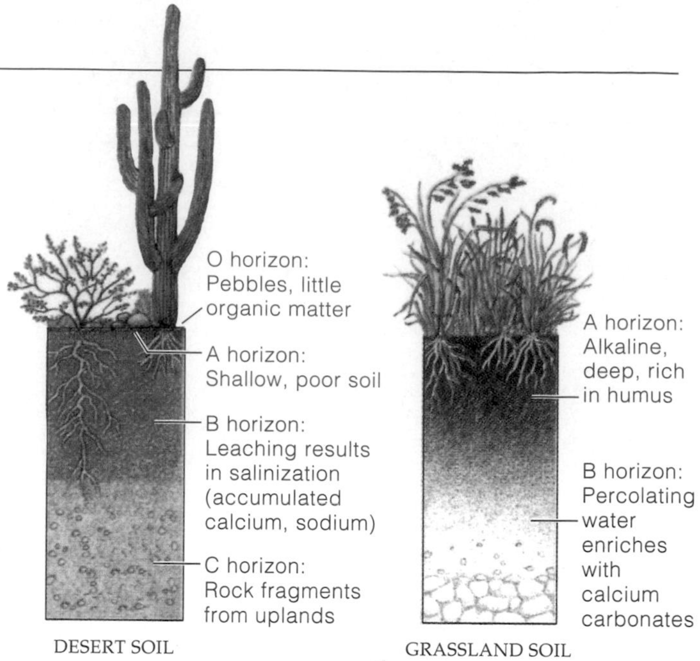

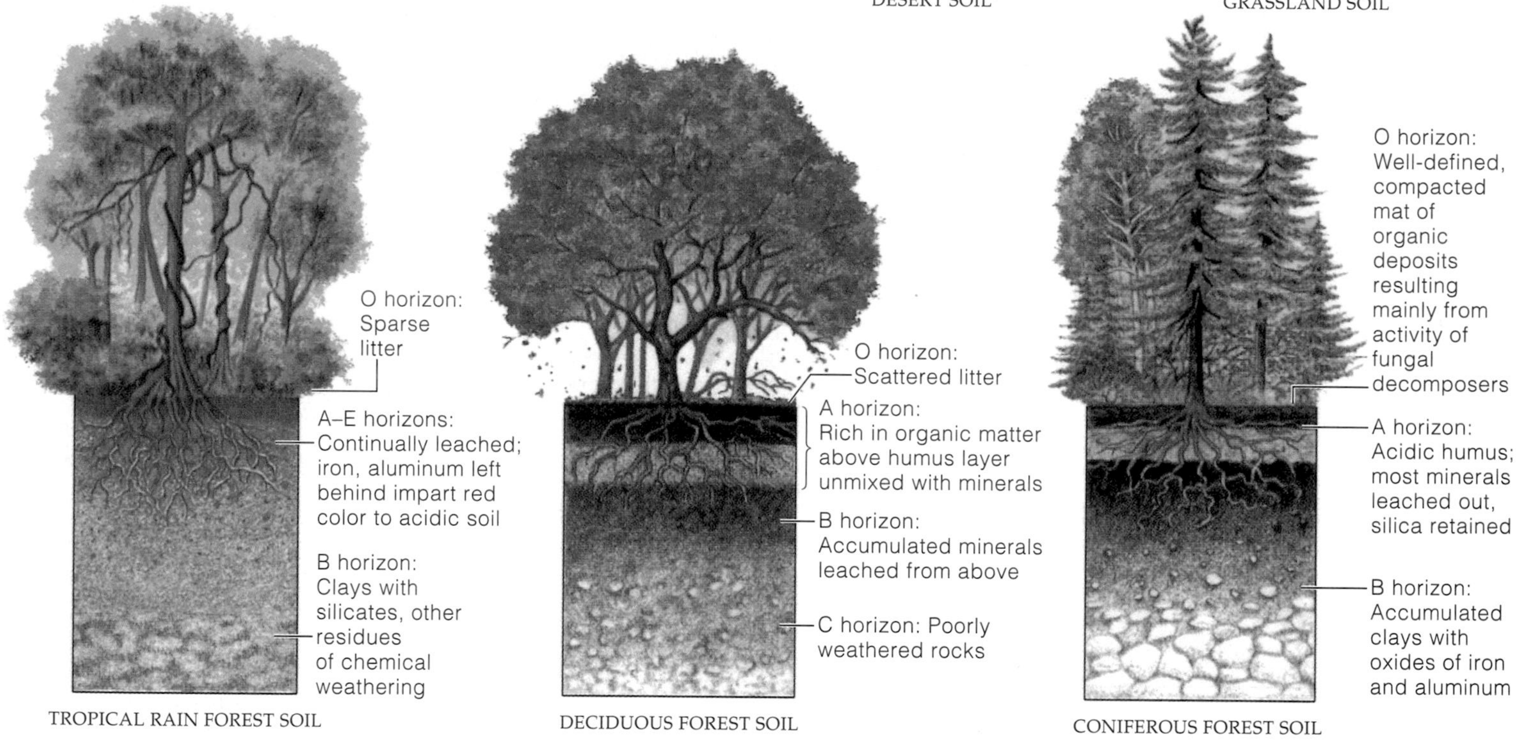

DESERT SOIL

O horizon: Pebbles, little organic matter

A horizon: Shallow, poor soil

B horizon: Leaching results in salinization (accumulated calcium, sodium)

C horizon: Rock fragments from uplands

GRASSLAND SOIL

A horizon: Alkaline, deep, rich in humus

B horizon: Percolating water enriches with calcium carbonates

TROPICAL RAIN FOREST SOIL

O horizon: Sparse litter

A–E horizons: Continually leached; iron, aluminum left behind impart red color to acidic soil

B horizon: Clays with silicates, other residues of chemical weathering

DECIDUOUS FOREST SOIL

O horizon: Scattered litter

A horizon: Rich in organic matter above humus layer unmixed with minerals

B horizon: Accumulated minerals leached from above

C horizon: Poorly weathered rocks

CONIFEROUS FOREST SOIL

O horizon: Well-defined, compacted mat of organic deposits resulting mainly from activity of fungal decomposers

A horizon: Acidic humus; most minerals leached out, silica retained

B horizon: Accumulated clays with oxides of iron and aluminum

Figure 42.12 Soil profiles from a few representative biomes. Plants are not drawn to the same scale. Compare Section 26.1.

The growth of most plants suffers in poorly aerated, poorly draining soils. Remember, loam topsoils have the best mix of sand, silt, and clay for agriculture. They have enough coarse particles to promote drainage and enough fine particles to retain water-soluble mineral ions that serve as nutrients for plant growth. Gravelly or sandy soils promote rapid leaching, which depletes them of water and vital minerals. Clay soils with fine, closely packed particles are poorly aerated and do not drain well. Few plants grow in waterlogged clay soils.

As farmers know, soil affects primary productivity. They grow most crops in cleared, former grasslands. Many burn or clear-cut tropical forests for agriculture. But these biomes have very little topsoil above poorly draining sublayers. Clearing exposes the topsoil, and heavy rains leach most of the nutrients.

We turn now to the deserts, shrublands, woodlands, grasslands, various forests, and tundra. Bear in mind, none of these major biomes is uniform throughout. Inside its borders, local climates, landforms, and other physical features favor patches of distinct communities.

Primary productivity depends on the soil profile and its proportions of sand, silt, clay, gravel, and humus.

DESERTS

Deserts form in lands with less than ten centimeters or so of annual rainfall and high potential for evaporation. Such conditions prevail at latitudes of about 30° north and south. There we find great deserts of the American Southwest, northern Chile, Australia, northern and southern Africa, and Arabia. Farther north are the high deserts of eastern Oregon, and Asia's vast Gobi and the Kyzyl-Kum east of the Caspian Sea. Rain shadows are the main reason why these northern deserts are so arid.

the rains. Deep-rooted plants, including mesquite and cottonwood, commonly grow near the few streambeds that have a permanent underground water supply.

Of all land surfaces, more than one-third is arid or semiarid, without enough rainfall to support crops. Crops growing in California's Imperial Valley and some other deserts require unflagging soil management and extensive irrigation. Without drainage programs, such croplands become unproductive from waterlogging and

Figure 42.13 Warm desert near Tucson, Arizona. Primary producers include creosote bush, multistemmed ocotillo, tall saguaro cacti, and prickly pear cacti with rounded pads.

Deserts do not have lush vegetation. Rain falls in heavy, brief, infrequent pulses that swiftly erode the exposed topsoil. Humidity is so low that the sun's rays easily penetrate the air. They quickly heat the ground surface, which radiates heat and cools quickly at night.

Although arid or semiarid conditions do not favor large, leafy plants, deserts show plenty of biodiversity. In one patch of Arizona's desert, you may come across deep-rooted, evergreen, woody shrubs, such as creosote bush and fleshy-stemmed, shallow-rooted cacti (Figure 42.13). You may see tall saguaro, short prickly pear, and ocotillo, which drops leaves more than once a year and grows new ones after a rain. Desert perennials and annuals flower profusely, spectacularly, but briefly after

salt buildup. Alarmingly, many parts of the world are now undergoing **desertification**. The term refers to the conversion of vast grasslands and similarly productive biomes to desertlike wastelands with low biodiversity. We take a closer look at this trend in Section 43.6.

Where the potential for evaporation greatly exceeds sparse rainfall, deserts form. This condition prevails at latitudes 30° north and south and in rain shadows.

WWW

DRY SHRUBLANDS, DRY WOODLANDS, AND GRASSLANDS

Dry shrublands and **dry woodlands** prevail in western or southern coastal regions of the continents between latitudes 30° and 40°. These semiarid regions get more rain than deserts, but not much more. Most falls in mild winters; the summers are long, hot, and dry. Dominant plants often have hardened, tough, evergreen leaves.

We find examples of dry shrublands, which get less than 25 to 60 centimeters of rain per year, in California, South Africa, and Mediterranean regions. Their exotic local names include fynbos and chaparral. California has 2.4 million hectares (6 million acres) of chaparral. In summer, lightning-sparked, wind-driven firestorms can sweep through these biomes (Figure 42.14). Shrubs that have highly flammable leaves burn swiftly to the ground. Yet they are highly adapted to episodes of fire and quickly resprout from their root crowns. Trees do not fare as well during firestorms. The shrubs, which "feed" the fires, have the competitive edge.

Dry woodlands dominate where annual rainfall is about 40 to 100 centimeters. The dominant trees can be tall, but they do not form a dense, continuous canopy. Eucalyptus woodlands of southwestern Australia and oak woodlands of California and Oregon are like this.

Grasslands cloak much of the interior of continents, in zones between deserts and temperate forests. Warm temperatures prevail in the summer, and winters are extremely cold. The 25 to 100 centimeters of annual rainfall keeps deserts from forming but isn't enough to support forests. Drought-tolerant primary producers survive strong winds, light and infrequent rainfall, and rapid evaporation. Dominant animals are grazing and burrowing species. Their grazing activities, combined with periodic fires, keep shrublands and forests from encroaching on the fringes of many grasslands.

The main grasslands of North America are *shortgrass* prairie and *tallgrass* prairie. Usually they form where the land is flat or rolling, as in Figure 42.15a. Roots of perennial plants extend profusely through the topsoil. In the 1930s, shortgrass prairie of the American Great Plains was overgrazed and also plowed under to grow wheat, which requires more water than this region sometimes receives. Strong winds, prolonged droughts, and unsuitable farming practices turned much of the prairie into a Dust Bowl (Section 43.6). John Steinbeck's *The Grapes of Wrath* and James Michener's *Centennial*, which are two historical novels, speak eloquently of the consequences.

Also in the interior of North America, tallgrass prairie once extended west from the biomes of temperate deciduous forests. Diverse legumes and composites, including daisies, thrived in the interior, with its richer topsoil and slightly more frequent rainfall. Farmers converted nearly all of the original tallgrass prairie for agriculture, but some areas are now being restored (Section 40.6).

Between the tropical forests and deserts of Africa, South America, and Australia are broad belts of grasslands with a smattering of shrubs or trees. These are the **savannas** (Figure 42.15c). Rainfall averages 90 to 150 centimeters a year, and prolonged seasonal droughts are common. Where rainfall is low, fast-growing grasses dominate. Acacia and other shrubs grow in regions that get a bit more moisture. Where the rainfall is higher,

Figure 42.14 California chaparral. The dominant plants are multibranched, woody, and typically only a few meters tall. Without periodic fires, they form a nearly impenetrable vegetation cover. The boxed inset shows a firestorm racing through a chaparral-choked canyon above Malibu.

Figure 42.15 (**a**) Rolling shortgrass prairie to the east of the Rocky Mountains. Once, bison were the dominant large herbivore of North American grasslands. At one time, the astoundingly abundant grasses supported 60 million of these hefty ungulates. (Ungulates are hooved, plant-eating mammals.)

(**b**) A rare patch of natural tallgrass prairie in eastern Kansas.

(**c**) The African savanna—warm grasslands with scattered stands of shrubs and trees. More varieties and greater numbers of large ungulates live here than anywhere else. They include migratory wildebeests (shown in this photograph), giraffes, Cape buffalo, zebras, and impalas.

savannas grade into tropical woodlands with tall, coarse grasses, hardy shrubs, and low trees. *Monsoon* grasslands form in southern Asia where heavy rains alternate with a dry season. Dense stands of tall, coarse grasses form, then die back and often burn in the dry season.

Plants of dry shrublands, dry woodlands, and grasslands are supremely adapted to surviving strong winds, grazing animals, and recurring episodes of drought and fire.

WWW

TROPICAL RAIN FORESTS AND OTHER BROADLEAF FORESTS

In forest biomes, tall trees grow close together and form a fairly continuous canopy over a broad stretch of land. In general, there are three types of trees. Which type prevails in a region depends partly on distance from the equator. Evergreen broadleafs dominate between latitudes 20° north and south. Deciduous broadleafs are common at moist, temperate latitudes where winters are not severe. Evergreen conifers are common at high, colder latitudes and in mountains of temperate zones.

Evergreen broadleaf forests sweep across tropical zones of Africa, the East Indies and Malay Archipelago, southeast Asia, South America, and Central America. Annual rainfall can exceed 200 centimeters and is never less than 130 centimeters. One forest biome, the **tropical rain forest**, depends on regular and heavy rainfall, an annual mean temperature of 25°C, and humidity of 80 percent or more (Figure 42.16). In this highly productive forest, evergreen trees produce leaves and shed old ones throughout the year. Even so, litter doesn't accumulate; decomposition and mineral cycling are extremely rapid in the hot, humid climate. Soils are weathered, humus-deficient, and poor nutrient reservoirs. We will take a closer look at these forests in the next chapter.

Leaving tropical rain forests, we enter regions where temperatures stay mild but rainfall dwindles during part of the year. This is the start of **deciduous broadleaf**

Figure 42.16 Tropical forests, biomes of spectacular biodiversity. Among millions of known species are jaguars and *Rafflesia*, a leafless, foul-smelling plant with a fly-pollinated flower that grows three meters across. Bromeliads, orchids, and other epiphytes grow on trees, obtaining minerals from organic remains of leaves, insects, and other litter that have dissolved in tiny pockets of water.

SPRING

SUMMER

WINTER

AUTUMN

forests. Trees of *tropical* deciduous forests drop some or all of their leaves during a pronounced dry season. The *monsoon* forests of India and southeastern Asia also have such trees. Farther north, in the temperate zone, rainfall is even lower. The winters are cold, and water is locked away as snow and ice. *Temperate* deciduous forests, such as those of the southeastern United States, prevail here (Figure 42.17). Decomposition is not as rapid as in the humid tropics, and many nutrients are conserved in accumulated litter on the forest floor.

Complex forests of ash, beech, birch, chestnut, elm, and deciduous oaks once stretched across northeastern North America, Europe, and east Asia. They declined

Figure 42.17 Part of a temperate deciduous forest south of Nashville, Tennessee, in spring, summer, autumn, and winter.

drastically when farmers cleared the land. Pathogenic species introduced to North America destroyed nearly all of the chestnuts and many elms (Section 40.8). Now, maple and beech predominate in the Northeast. Farther west, oak-hickory forests prevail, then oak woodlands that eventually grade into tallgrass prairie.

In forest biomes, conditions favor dense stands of tall trees that form a continuous canopy over a broad region.

WWW

Conifers (cone-bearing trees) are primary producers of **coniferous forests**. Most have needle-shaped leaves with a thick cuticle and recessed stomata—adaptations that help the trees conserve water through dry times. They dominate the boreal forests, montane coniferous forests, temperate rain forests, and pine barrens.

Boreal forests stretch across northern Europe, Asia, and North America. (Such a forest also is called *taiga*, meaning "swamp forest.") Most are in glaciated regions with cold lakes and streams (Figure 42.18*a*). It rains mostly in summer, and evaporation is low in the cool summer air. The cold, dry winters are more severe in

biome. One bright note: In 1994, a timber corporation surrendered its rights to log one of the largest intact temperate rain forests outside of the tropics. The trees, 800 years old, cloak British Columbia's Kitlope Valley.

Southern pine forests dominate the coastal plains of the south Atlantic and the Gulf states. Pine species are adapted to the dry, sandy, nutrient-poor soil and to natural fires or controlled burns. These fires do not get hot enough to damage the trees, and they open up the understory. Forests of pine, scrub oak, and wiregrass grow in New Jersey. Palmettos grow beneath pines and loblolly in the Deep South. During a heat wave in 1998,

eastern parts of these biomes than in the west, where ocean winds moderate the climate. Spruce and balsam fir dominate the boreal forests of North America. Pines, birches, and aspens take hold in the burned or logged areas. Where soil is poorly drained, highly acidic bogs dominated by peat mosses, shrubs, and stunted trees prevail. Boreal forests become much less dense to the north, where they grade into arctic tundra.

In the Northern Hemisphere, montane coniferous forests extend southward through the great mountain ranges. Spruces and firs dominate in the north and at higher elevations. They give way to firs and pines in the south and at lower elevations (Figure 42.18*b*).

Some temperate lowlands support coniferous forests. One temperate rain forest paralleling the coast from Alaska into northern California contains some of the world's tallest trees—Sitka spruce to the far north and redwoods to the south. Logging destroyed much of this

Figure 42.18 (**a**) Spruce-dominated boreal forest. (**b**) Montane coniferous forest of Yosemite Valley in the Sierra Nevada, California. (**c**) A fighter retreats from a wall of fire in a pine-oak forest near Daytona Beach, Florida.

fires swept through more than 320,000 acres in Florida. Smaller burns had been suppressed in counties where housing and other developments encroached on the forests. Dense, tinder-dry understory accumulated and was the basis for the destruction (Figure 42.18*c*).

Coniferous forests prevail in regions in which a cold, dry season alternates with a cool, rainy season.

WWW

TUNDRA

Tundra is derived from *tuntura*, a Finnish word for the great treeless plain between the polar ice cap and belts of boreal forests in Europe, Asia, and North America. Much of this *arctic* tundra is flat, windswept, and wet (Figure 42.19a). Temperatures are cool in summer and below freezing in winter, so not much water evaporates. Little rain or snow falls. Sunlight is nearly continuous during the summer months. Then, short plants grow and flower profusely, and seeds ripen fast.

Snow does not cloak arctic tundra all year long, but summers are too short to thaw much more than surface soil. Just beneath the surface is a perpetually frozen

Figure 42.19 (**a**) Extensive ponding in the arctic tundra. The rain and snowmelt cannot percolate downward because of the permafrost. (**b**) Short, hardy plants typical of alpine tundra.

layer, the **permafrost**. It is more than 500 meters thick in some places. Because permafrost prevents drainage, the soil above it remains waterlogged even in summer. The anaerobic conditions and low temperatures limit nutrient cycling. Organic matter decomposes slowly, so it accumulates in soggy masses of organic matter. All but about 5 percent of the carbon in the arctic tundra is locked up in peat bogs.

A similar type of biome prevails at high elevations in mountains throughout the world, although there is no permafrost beneath the soil. Figure 42.19b shows an example of this *alpine* tundra. Dominant plants often form low cushions and mats that are able to withstand the buffeting of strong winds. Winter temperatures fall below freezing. Even in the summer, shaded patches of snow persist. The thin, fast-draining soil of an alpine tundra is nutrient-poor, so primary productivity is low.

At high latitudes with short, cool summers and long, very cold winters, we find arctic tundra. In high mountains where seasonal changes vary with latitude but where the climate is too cold to support forests, we find alpine tundra.

WWW

FRESHWATER PROVINCES

Freshwater and saltwater provinces cover more of the Earth's surface than all biomes combined. They include lakes, ponds, estuaries, wetlands, coral reefs, oceans, and hydrothermal vents. "Typical" examples are not that easy to find. Some ponds can be waded across; Lake Baikal in Siberia is more than 1.7 kilometers deep. All aquatic ecosystems have gradients in light penetration, temperature, and dissolved gases, but the values differ greatly. All we can do here is sample the diversity.

Lake Ecosystems

A **lake** is a body of standing freshwater produced by geologic processes, as when an advancing glacier carves a basin in the Earth. When the glacier retreats, water collects in the exposed basin (Figure 42.20). Over time, erosion and sedimentation alter the dimensions of the lake, which usually ends up filled or drained completely.

A lake has littoral, limnetic, and profundal zones (Figure 42.21). The littoral extends all around the shore, to a depth at which rooted aquatic plants stop growing. The water is shallow and usually well lit. Diversity is greatest here. The limnetic is open, sunlit water past the littoral; it extends to a depth where photosynthesis is insignificant. Communities of tiny organisms (plankton) abound here. Cyanobacteria, diatoms, and green algae dominate the *phyto*plankton. The *zoo*plankton include rotifers and copepods. The lake's profundal includes all open water below the depth at which wavelengths suitable for photosynthesis can penetrate. Detritus sinks through this zone to bottom sediments. Communities of diverse bacterial decomposers live in and on bottom sediments, and they enrich the water with nutrients.

SEASONAL CHANGES IN LAKES In temperate regions, where summers are warm and winters cold, lakes show seasonal changes in density and temperature from the surface to the bottom. A layer of ice forms over many of them in midwinter. Water near the freezing point is the least dense and accumulates beneath the ice. Water at 4°C is the most dense. It collects in deeper layers that are a bit warmer than the surface layer in midwinter.

Figure 42.20 In the Canadian Rockies, the basin of Moraine Lake, formed by glacial action during the last ice age.

In spring, daylength increases and the air is not as cold. Lake ice melts, the surface water slowly warms to 4°C, and temperatures turn uniform throughout. Winds blowing over the surface now cause a **spring overturn**: strong vertical movements carry dissolved oxygen from the surface layer to its depths, and nutrients released by decomposition move from sediments to the surface.

By midsummer, the surface layer is above 4°C. Now the lake has a *thermocline*—a midlayer of water where the temperature changes abruptly and blocks vertical mixing (Figure 42.22). Being warmer and less dense, the surface water floats on the thermocline. In cooler water beneath it, decomposers typically deplete the dissolved oxygen. Come autumn, the upper layer cools, becomes denser, then sinks, so the thermocline vanishes. During this **fall overturn**, water mixes vertically, so dissolved oxygen moves down and nutrients move up.

Primary productivity corresponds with the seasons. After a spring overturn, the longer daylengths and the cycled nutrients support higher rates of photosynthesis. Phytoplankton and rooted aquatic plants quickly take up phosphorus, nitrogen, and other nutrients. During a growing season, the thermocline cuts off vertical mixing. Nutrients locked in the remains of organisms sink to deeper water, so photosynthesis slows. By late summer,

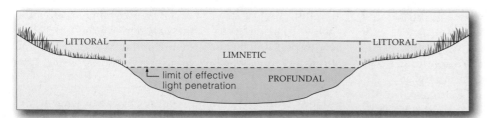

Figure 42.21 Lake zonation. The littoral extends around the lake's edge, from the shore to the depth where aquatic plants stop growing. The profundal is all water below the depth of light penetration. Above the profundal are open, sunlit waters of the limnetic zone.

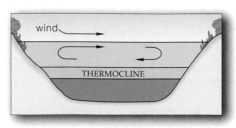

Figure 42.22 Thermal layering in a temperate lake in summer.

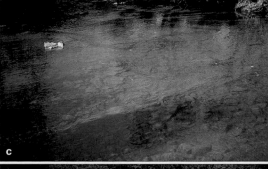

shortages are limiting photosynthesis. A fall overturn cycles nutrients to the surface, and this drives a burst of primary productivity. But the shorter daylight hours do not sustain a long-lasting burst. Not until spring will primary productivity increase again.

TROPHIC NATURE OF LAKES The topography, climate, and geologic history of a lake dictate the numbers and kinds of residents, their dispersal, and the cycling of nutrients among them. Soils of the lake basin and the surrounding regions contribute to the type and amount of nutrients available to support organisms.

Interplays among the climate, soil, basin shape, and metabolic activities of a lake's residents contribute to conditions ranging from oligotrophy to eutrophy. Water in *oligotrophic* lakes often is deep, clear, and nutrient poor. Its primary productivity is not great. The water in *eutrophic* lakes typically is shallow, nutrient rich, and rather high in primary productivity. These conditions develop naturally, as sediments gradually accumulate in the lake basin. The water becomes less deep and less transparent, compared with water in oligotrophic lakes, and phytoplankton dominate the communities. More sediments usually accumulate and the final successional stage is a filled-in basin—hence no more lake.

Eutrophication is a name for processes that enrich a body of water with nutrients. As Section 41.6 described, human activities can bring it about. Here, think about Val Smith's report on the way people brought about eutrophication of Lake Washington in Seattle—and then reversed it. From 1941 to 1963, the lake received too much phosphorus-rich sewage, which promoted huge blooms of cyanobacteria. These formed thick, slimy mats over the lake each summer and made it useless for recreation. With phosphorus enrichment, nitrogen became the lake's limiting resource. So cyanobacteria—which are superior competitors for nitrogen by their ability to fix N_2—became dominant. From 1963 to 1968, sewage discharges slowed, then stopped. By 1975, the lake neared full recovery. Lower density populations of diatoms and green algae became dominant.

Stream Ecosystems

The flowing-water ecosystems called **streams** start out as freshwater springs or seeps. They grow and merge as they flow downslope, then often combine to form a river. Between the headwaters and the river's end, we

Figure 42.23 Stream habitats in North Carolina and Virginia. (**a**) Pool. (**b**) Pool leading into a riffle. (**c**) A run, Sinking Creek. (**d**) Leaf detritus in a riffle. (**e**) Closer look at a riffle.

find three kinds of habitats—riffles, pools, and runs. The riffles are shallow, turbulent stretches where water flows swiftly over a rough bottom of sand and rock. Figure 42.23 shows two examples. The pools have deep water flowing slowly over a smooth, sandy, or muddy bottom. The runs are smooth-surfaced but fast-flowing stretches over bedrock or rock and sand.

A stream's average flow volume and temperature depend on rainfall, snowmelt, geography, altitude, and even the shade cast by plants. Its solute concentrations are influenced by the streambed's composition as well as by agricultural, industrial, and urban wastes.

Especially in forests, streams import most of the organic matter that supports food webs. Where trees cast shade, photosynthesis is diminished. Most of the organic matter is litter that enters detrital food webs. The aquatic organisms continually take up and release nutrients as water flows downstream. (Nutrients move upstream only as components of migrating fishes and other animals.) Think of nutrients as spiraling between water and aquatic organisms as the stream flows on its one-way course, usually to a river that flows to the sea.

Ever since cities formed, streams have been sewers for industrial and municipal wastes. The wastes, as well as other pollutants from poorly managed farmlands, choked many streams with sediments and chemically poisoned them. However, streams are resilient. They can recover impressively when pollution is controlled.

Freshwater and saltwater provinces are far more extensive than the Earth's biomes. All of the aquatic ecosystems in these water provinces have characteristic gradients in light penetration, temperature, and dissolved gases.

WWW

Beyond land's end are two vast provinces of the open oceans, which cover almost three-fourths of the Earth's surface. The *benthic* province includes all sediments and rocks of the ocean bottom (Figure 42.24). It starts at continental shelves and extends to deep-sea trenches. The *pelagic* province is the full volume of ocean water. Its neritic zone is all water above continental shelves. Its oceanic zone is the water of the ocean basins.

Walk along an ocean and you might be humbled by its vastness (Figure 42.25). An immense volume of water extends to the distant horizon, with no mountains or plains or valleys to break it up visually into something less overwhelming. What you do not see are *submerged* mountains and valleys and plains—which are as varied as the ones configuring the dry continents.

Primary Productivity in the Oceans

Within an ocean's upper surface waters, photosynthesis proceeds on a stupendous scale, just as it does on land. Primary productivity there varies seasonally, just as it does on land (Section 6.8 and Figure 42.26). Its huge "pastures" of phytoplankton are the start of food webs that include copepods, shrimplike krill, whales, squids, and diverse fishes. Organic remains and wastes from all of these marine communities slowly sink through the water. They become the basis of detrital food webs for most communities in the benthic province.

About 70 percent of the primary productivity near the ocean surface might be the work of **ultraplankton**. These photosynthetic bacterial cells are only about two micrometers wide (40 millionths of an inch), at most. In tropical and subtropical seas once thought to be nearly devoid of producers, 0.035 gram (one ounce) of water holds as many as 3 million of the cells. In deeper ocean water too dark to support photosynthesis, food webs start with **marine snow**. These bits of organic matter collectively drift down as the food base for staggering biodiversity in the mid-ocean waters—possibly as many as 10 million species. In what may be the greatest of all circadian migrations, some species rise thousands of feet to feed at night in shallower water, then move back down the following morning. Siphonophores, a type of soft-bodied cnidarian, are among the top carnivores. At thirty-nine meters, *Praya dubia*—a newly discovered siphonophore—is one of the world's longest animals!

Hydrothermal Vents

In 1977, researchers at the ocean floor near the Galápagos Islands discovered

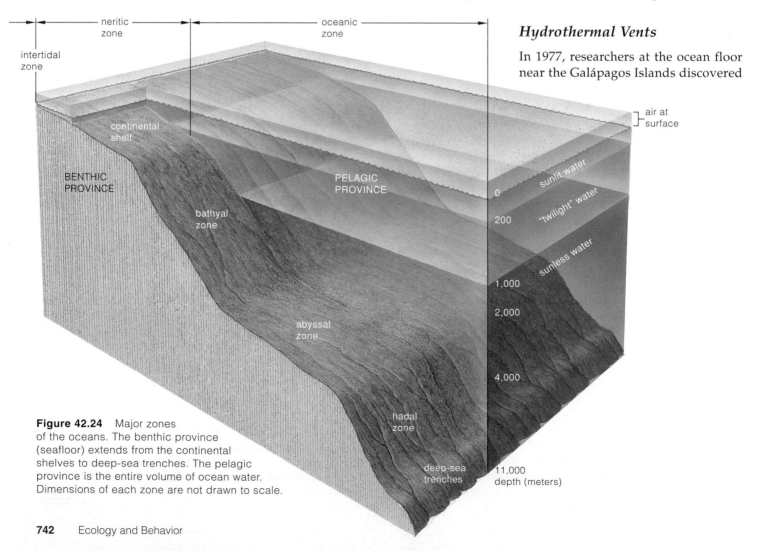

Figure 42.24 Major zones of the oceans. The benthic province (seafloor) extends from the continental shelves to deep-sea trenches. The pelagic province is the entire volume of ocean water. Dimensions of each zone are not drawn to scale.

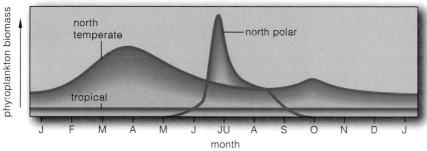

Figure 42.25 (**a**) From a rocky shore, a view of the open ocean. (**b**) Near a British Columbia coastline, a whale breaching (leaping out of the water). Crustaceans (**c**) and tube worms (**d**) near a hydrothermal vent ecosystem on the ocean floor.

Figure 42.26 Seasonal variations in primary production in the oceans corresponding to latitude. The strongest peak in the *dark-green* graph lines for the north polar and temperate seas corresponds to phytoplankton blooms, brought about by a seasonal increase in daylength. The lower peak corresponds to an increase in availability of nutrients, something like the fall overturn in lakes. In most tropical seas, daylength and nutrient availability do not vary much; neither does primary productivity.

a new ecosystem. In the Galápagos Rift, a volcanically active boundary between two crustal plates, they found communities thriving near **hydrothermal vents**. At this boundary, near-freezing water seeps into fissures and becomes heated to extremely high temperatures. As the pressure forces up heated water, minerals are leached from rocks before the water spews out through vents in the seafloor. Iron, zinc, copper sulfides, and sulfates of magnesium and calcium are dissolved in hydrothermal outpourings. These settle out and form rich mineral deposits. Sulfides in the deposits are energy sources for chemoautotrophs, the starting point for hydrothermal vent communities. Chemoautotrophic bacteria are the primary producers for food webs that include red tube worms, crustaceans, clams, and fishes (Figure 42.25*c,d*).

Still more hydrothermal vent ecosystems have been found in the South Pacific near Easter Island; the Gulf of California, about 150 miles south of the tip of Baja California, Mexico; and the Atlantic. In 1990, a team of United States and Russian scientists found one in Lake Baikal, the world's deepest lake. This lake basin seems

to be splitting apart (hence the vents) and may mark the beginning of a new ocean.

Did life originate in such nutrient-rich places on the seafloor? Conditions at the surface of the early Earth were about as inhospitable as you might imagine. At the least, cells at the seafloor would have had protection from the destructive radiation bombarding the Earth before the oxygen-rich atmosphere formed. Were some descendants of those cells carted closer to the surface as the seafloor was uplifted during episodes of crustal crunchings? These are some unanswered questions that evolutionary detectives are now asking. Some of their interesting hypotheses are described in Section 6.7.

Although we are most familiar with features of the land, the great oceans dominate the Earth's surface. It now seems their primary productivity and biodiversity are staggering.

Communities of organisms thrive near hydrothermal vents on the ocean floor, which suggests to some that life itself may not have originated in the shallow waters of the Earth.

WWW

CORAL REEFS AND CORAL BANKS

The Master Builders

Each wave-resistant formation called a **coral reef** began with accumulated remains of countless organisms. The hard parts of corals became its structural foundation. Coralline algae added to it; deposits of calcium and magnesium carbonates hardened the cell walls of these red algae, including *Corallina* (Figure 42.27). Secretions from other organisms helped cement things together.

The massive, pocketed spine of an existing reef is home to hundreds of species of corals, and a staggering variety of red algae and other organisms. Section 23.4 and Figure 42.27 merely hint at the wealth of warning colors, spines, tentacles, and stealthy behavior—clues to dangers and fierce competition for resources among species that are packed together in limited space.

Fringing reefs, barrier reefs, and atolls are the main reef formations (Figure 42.27). Most of the substantial reefs form in clear, warm waters between latitudes 25° north and south. Farther north or south, solitary corals or small colonies of them have formed **coral banks** in temperate waters, even in cold waters of continental shelves. Among these are smooth, vertical banks in the cold, deep waters near Japan, California, England, and New Zealand. A colonial branched coral, *Lophelia*, has built great banks in the cold waters of Norway's fjords.

The Once and Future Reefs

Colorful dinoflagellates often live as symbionts in the tissues of reef-building corals. In return for protection, they provide a coral polyp with oxygen and recycle its mineral wastes. When stressed, the polyps expel their protistan symbionts. When stressed for more than a few months, they die, and only bleached hard parts remain. Abnormal, widespread bleaching in the Caribbean and tropical Pacific began in the 1980s. So did an increase in sea surface temperature, and this may be a key stress factor. If the damage is one outcome of global warming, as marine biologists Lucy Bunkley-Williams and Ernest Williams suggest, the future looks grim for the reefs.

Human activities are destroying reefs in more direct ways than this. Where raw sewage and other pollutants are being discharged into the nearshore waters around populated islands, including some Hawaiian islands, reefs are dead or dying. Where commercial fishermen from Japan, Indonesia, and Kenya move in, reef life is being decimated. No simple nets for these fellows. They drop dynamite in the water, so fish hiding among the coral float dead to the surface. They squirt sodium cyanide into the water, and stunned but not dead fish float to the surface. These are the tropical fish in pet stores as well as the rare fish transported live to Asian restaurants, where they are dispatched and served up

CORALLINE ALGA

coral reef   island   lagoon   open ocean

Figure 42.27 A sampling of the stunning biodiversity of tropical reefs. The diagrams show three types of coral reefs. *Atolls* are ring-shaped coral reefs that fully or partially enclose nothing but a shallow lagoon. *Fringing reefs* form at the edge of the land in regions of limited rainfall, as on the leeward side of the big island of Hawaii. *Barrier reefs* form around islands or parallel with the shore of a continent, and a calm lagoon forms behind them. Section 23.4 shows one.

LIONFISH

CRAB

MORAY EEL

CHAMBERED NAUTILUS

SEA ANEMONE

DAISY CORAL

PILLAR CORAL

CROWN-OF-THORNS SEA STAR

as exorbitantly expensive status symbols. On small, native-owned islands, fishing rights are traded for small sums. Then the reefs are destroyed, and they can no longer sustain the small human populations that once depended on them for their own sustenance.

Coral reefs generally form in clear, warm waters between latitudes 25° north and south. Vertical coral banks form in the temperate or cold waters of continental shelves.

All show spectacular biodiversity, and vulnerability.

𝍢𝍢𝍢

Remarkable ecosystems exist in estuaries and intertidal zones along coasts. Like freshwater ecosystems, they differ in physical and chemical properties, such as light penetration and water temperature, depth, and salinity.

Estuaries

An **estuary** is a partially enclosed coastal region where seawater swirls in with nutrient-rich freshwater from rivers, streams, and runoff from the surrounding land. The confined conditions, slow mixing of water, and tidal action combine to trap the dissolved nutrients. The continually freshened water replenishes nutrients and allows estuaries to support productive ecosystems.

Primary producers are phytoplankton, salt-tolerant plants that withstand submergence at high tide, and algae living in mud and on plants. Much of the primary production enters detrital food webs in which bacterial and fungal decomposers are the first to feed. Detritus (and bacteria on and in it) feeds nematodes, snails, crabs, and fish. Filter feeders such as clams eat food particles suspended in the slowly moving water. So many larval and juvenile stages of invertebrates and some fishes are present that estuaries are often called marine nurseries. Also, many migratory birds use estuaries as rest stops.

Chesapeake Bay, Mobile Bay, and San Francisco Bay are broad, shallow estuaries. Estuaries in Alaska and British Columbia are narrow and deep; so are Norway's fjords. In Texas and Florida, estuaries lie behind long spits of sand and mud. The New England coast has many estuarine salt marshes (Figure 42.28). In tropical regions, we find mangrove swamps that function much like salt marshes in estuarine ecology.

Today, many estuarine ecosystems are under serious assault from raw sewage; agricultural runoff; industrial, urban, and suburban wastes; and upstream diversion of freshwater for human use. Normal conditions, such as suitable salinity levels, cannot be maintained without inflows of unpolluted freshwater.

The Intertidal Zone

Along rocky and sandy coastlines we find ecosystems of the **intertidal zone**, which is not renowned for its creature comforts. Waves batter its resident organisms. Tides alternately submerge and expose them. The higher they are, the more they dry out, freeze in winter, or bake in summer, and the less food comes their way. The lower they are, the more they must compete in limited spaces. At low tides, birds, rats, and raccoons move in to feed on them. High tides bring the predatory fishes.

Generalizing about coastlines isn't easy, for waves and tides constantly resculpt them. But one feature that rocky and sandy shores do share is vertical zonation.

Rocky shores often have three zones (Figure 42.29*a*). The highest zone, the upper littoral, is submerged only during the highest tide of the lunar cycle; it is sparsely populated. The midlittoral (middle zone) is submerged during the highest regular tide and exposed during the lowest. In its tidepools we typically find red, brown, and green algae, hermit crabs, nudibranchs, sea stars, and small fishes (Figure 42.29*b*). Diversity is greatest in the lowest zone, the lower littoral, which is exposed only during the lowest tide of the lunar cycle. In all three shore zones, rapid erosion prevents detritus from accumulating, so grazing food webs prevail.

Waves and currents continually rearrange stretches of loose sediments, called *sandy* and *muddy* shores. Few large plants grow in these unstable places, so you will not find many grazing food webs. Detrital food webs start with organic debris imported from offshore or nearby landforms. Vertical zonation is less obvious than along rocky shores. Below the low tide mark in temperate areas are sea cucumbers and blue crabs. Marine worms, crabs, and other invertebrates live between high and low tide marks. At night, at the high tide mark, beach hoppers and ghost crabs pop out of their burrows, seeking food.

Figure 42.28 Salt marsh of a New England estuary where *Spartina*, a marsh grass, is the main producer. Its microbe-enriched litter feeds consumers in the creeks and sound.

SALT MARSH (estuary)

open ocean   sound   shallow bay   creek   tidal river

UPPER LITTORAL

MID-LITTORAL

LOWER LITTORAL

a

b

Figure 42.29 (**a**) Vertical zonation in the intertidal zone of a rocky shore in the Pacific Northwest. The difference between high and low tides is about three meters. Elsewhere, it varies from a few centimeters (in the Mediterranean Sea) to more than fifteen meters (in the Bay of Fundy, next to Nova Scotia). (**b**) Tidepool region typical of the Pacific Northwest.

Upwelling Along the Coasts

In the Northern Hemisphere, prevailing winds from the north parallel the west coasts of continents and tug on the ocean surface. Wind friction causes the surface waters to begin moving. Under the force of the Earth's rotation, the slow-moving water is deflected westward, away from a coast. And cold, deep, often nutrient-rich water moves in vertically to replace it (Figure 42.30). Whenever cold, deep water moves up this way, we call it **upwelling**. It happens in equatorial currents as well as along the coasts of continents in both hemispheres, and it cools air masses above it. For example, fogbanks along the California coast are an outcome of warm air encountering upwellings of cold water.

a Wind from the north starts surface ocean water moving.   **b** Force of Earth's rotation deflects the moving water westward.

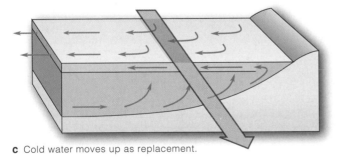

c Cold water moves up as replacement.

Figure 42.30 Upwelling along the west coasts in the Northern Hemisphere. Prevailing winds from the north start surface water moving. The force of Earth's rotation deflects the moving water westward. Cold, deeper water moves up to replace it.

In the Southern Hemisphere, commercial fishing industries of Peru and Chile depend on wind-induced upwelling. When prevailing coastal winds blow in from the south and southeast, they tug surface water away from shore; and cold, deeper water that the Humboldt Current delivers to the continental shelf moves to the surface. Tremendous amounts of nitrate and phosphate are pulled up, then carried northward by the cold Peru Current. Phytoplankton that depend on these nutrients are the basis of one of the world's richest fisheries.

Every three to seven years, warm surface waters of the western equatorial Pacific Ocean move eastward. This massive displacement of warm water affects the prevailing wind direction. The eastward flow speeds up so much that it influences the movement of water along the coastlines of Central and South America.

Waters driven toward any coastline will be forced downward and will flow seaward along the continental shelf. Near Peru's coast, a prolonged "downwelling" displaces the cooler waters of the Humboldt Current—and prevents upwelling. The phenomenon, which local fishermen named **El Niño**, has catastrophic effects on productivity, on seabirds that feed on anchovetas and other fishes, and on fishing industries. The next section provides a closer look at the El Niño phenomenon.

Major shifts in the circulation patterns of the ocean and atmosphere can have repercussions on the functioning of ecosystems that are global in scope.

www

42.14 RITA IN THE TIME OF CHOLERA

We turn now to an application that reinforces a unifying concept of this chapter—that events in the atmosphere, in the ocean, and on land interconnect in ways that can profoundly influence the world of life.

EL NIÑO SOUTHERN OSCILLATION   Let's set the stage for our story with a glimpse at a climatic event called the *El Niño Southern Oscillation* (ENSO). Massive dislocations in global rainfall patterns characterize this event, which corresponds to changes in sea surface temperatures and air circulation patterns. "Southern Oscillation" refers to a recurring seesaw in atmospheric pressure in the western equatorial Pacific—the world's largest reservoir of warm water. More warm air rises here than anywhere else. It is the source of heavy rainfall, which releases heat energy that drives the world's air circulation system.

Between ENSOs, the warm reservoir and heavy rainfall associated with it move to the west (Figure 42.31). In an ENSO, surface winds prevailing in the western equatorial Pacific pick up speed and "drag" the surface waters east (Figure 42.32). As eastward water transport increases, westward transport slows. Even more warm water in the vast reservoir moves eastward—and so on in a feedback loop between the ocean and the atmosphere. Sea surface temperatures rise, evaporation from the ocean accelerates, and air pressure falls. The humid updrafts trigger violent storms and flooding along coasts and often heavier rain inland. Elsewhere, extended droughts prevail.

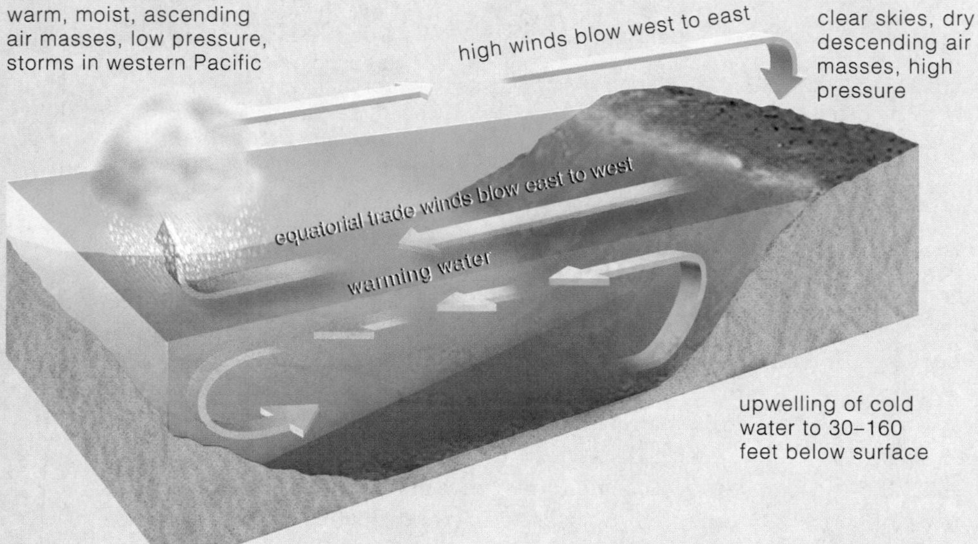

As you read earlier, reversal in a westward flow of air and water displaces the cold, deep Humboldt Current and stops upwelling along the western coast of South America. Usually a warm, nutrient-poor current from the east reaches the coast around Christmas. Peru's fishermen named the current "El Niño" ("the little one," a reference to the baby Jesus).

Today, satellites and ocean-bobbing buoys gather data on winds, currents, and the sea surface temperatures, which are loaded into supercomputers that simulate global weather patterns. The simulations are getting good: In 1997, scientists were able to predict an ENSO a season in advance. It turned out to be the most powerful ENSO of the century. Average sea surface temperatures rose nine degrees in the eastern Pacific. Warm water extended 9,660 kilometers west from Peru's coastline and 320 kilometers north (Figure 42.33).

Storms slamming into California, Mexico, Ecuador, and Peru caused heavy flooding and mudslides. An ice storm hit New England and knocked out power for 4 million people in one of Canada's worst natural disasters. Tornadoes devastated parts of the southeastern United States. A numbing heat wave caused hundreds of deaths and triggered wildfires in Mexico and the United States. Australia reeled under a prolonged drought. Heavy rains and floods struck Kenya and Somalia. Monsoons were delayed in Southeast Asia—and fires set to clear tracts of tropical rain forests swept out of control through 400,000 hectares (1 million acres). The strongest hurricane ever hit the eastern Pacific; eight cyclones hit the central Pacific, compared to two the year before.

A CHOLERA CONNECTION   During the 1997–1998 El Niño episode, 30,000 cases of cholera were reported in Peru, compared to only 60 from January to August 1997. For

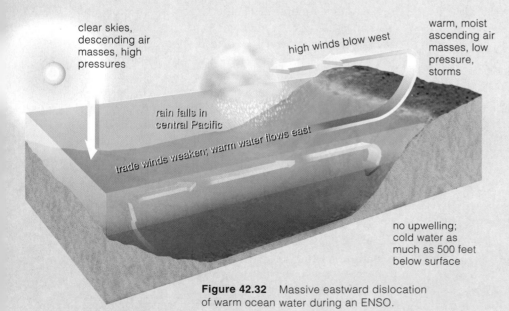

warm, moist, ascending air masses, low pressure, storms in western Pacific

high winds blow west to east

clear skies, dry descending air masses, high pressure

equatorial trade winds blow east to west

warming water

upwelling of cold water to 30–160 feet below surface

Figure 42.31   Westward flow of cold, equatorial surface water between ENSOs.

clear skies, descending air masses, high pressures

high winds blow west

warm, moist ascending air masses, low pressure, storms

rain falls in central Pacific

trade winds weaken; warm water flows east

no upwelling; cold water as much as 500 feet below surface

Figure 42.32   Massive eastward dislocation of warm ocean water during an ENSO.

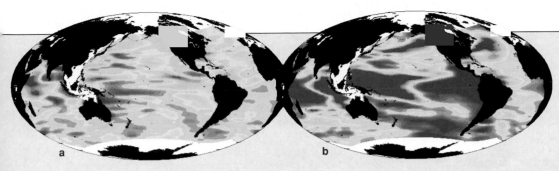

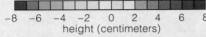

Figure 42.33 (**a**) Sea levels for 1993–1994, a winter between ENSOs. (**b**) Sea levels for 1997–1998, an El Niño winter. Values are based on measurements of increases in sea surface temperatures.

-8 -6 -4 -2 0 2 4 6 8
height (centimeters)

some time, people had known that cholera was linked to water contaminated by *Vibrio cholerae* (Figure 42.34*a*). They assumed epidemics follow after this bacterium hitches rides in the gut of human travelers, then enters water and food supplies by way of feces expelled during the severe diarrhea that characterizes the disease.

What people did *not* know was where the bacterium lurked between cholera outbreaks. Year after year, no one could find it in humans or in water supplies. Then cholera would emerge simultaneously in places some distance apart—most often in coastal cities where the poor drew water from rivers that entered the sea.

Rita Colwell, a microbiologist who later became director of the National Science Foundation, suspected that humans did not harbor the pathogen between the outbreaks. Was there an environmental reservoir? Maybe, but nobody had detected any cells of *V. cholerae* in water samples subjected to standard culturing. Then Colwell had a flash of insight: What if no one could find the pathogen because it changes form and enters a dormant, sporelike stage between outbreaks?

During a cholera outbreak in Louisiana, she decided to employ labeled antibodies that would bind to a certain protein at the bacterium's surface. Later, antibody tests in Bangladesh pinpointed the bacterium in fifty-one of fifty-two water samples—but culture methods missed it in all but seven of the samples.

V. cholerae thrives in brackish water, rivers, estuaries, and the seas. Colwell knew that plankton (communities of mostly microscopic aquatic species) also thrive in those environments. She focused her search on the waters near Bangladesh, where outbreaks of cholera are endemic and seasonal. In time she discovered the

dormant stage of *V. cholerae* in copepods, which graze on algae and other phytoplankton (Figure 42.34*b*). And the abundances of copepods—hence of *V. cholerae*—rise and fall with changes in the abundance of phytoplankton, the "pastures of the seas."

As it happened, Colwell already knew about seasonal variations in sea surface temperatures. Remember the saying, Chance favors the prepared mind? She chanced upon a correlation between seasonal temperature peaks in the Bay of Bengal with cholera admissions to hospitals in the area. Her correlation held for the 1990–1991 and 1997–1998 El Niño episodes (Figure 42.34*c,d*). Four to six weeks after sea surface temperatures go up, so do the cases of cholera!

Together with her colleague Anwarul Huq, Colwell is working to identify other factors, such as shifts in salinity and nutrient content, that trigger algal blooms, hence rises in copepod population size. Their goal is to integrate the data into a model that can be used to predict precisely where cholera outbreaks will occur and to give advance warning to filter drinking water. Meanwhile, Colwell is advising women in Bangladesh to use four layers of old sari cloth as filters, which remove 99 percent of *V. cholerae* cells from the water (Figure 42.34*e*). They can be rinsed in clean water, sun-dried, and used again and again.

WWW

Figure 42.34 (**a**) *Vibrio cholerae*. (**b**) Copepod. (**c, d**) Two satellilte images correlating algal blooms with sea surface temperatures near South America during an El Niño episode. (**e**) In Bangladesh, Rita Colwell with unfiltered and filtered drinking water.

SUMMARY

1. The biosphere encompasses the Earth's waters, the lower atmosphere, and the uppermost portions of its crust in which organisms live. Energy flows one way through the biosphere, and materials move through it on a grand scale to influence ecosystems everywhere.

2. The distribution of species through the biosphere is an outcome of the Earth's history, topography, climate, and interactions among species.

   a. "Climate" (average weather conditions, including temperature, humidity, wind velocity, cloud cover, and rainfall over time) results from differences in the amount of solar radiation reaching equatorial and polar regions, the Earth's daily rotation and annual path around the sun, the distribution of continents and oceans, and the elevation of landmasses.

   b. Interacting climatic factors produce the prevailing winds and ocean currents, which shape global weather patterns. The weather affects soil composition and water availability, which affects the growth and distribution of primary producers in ecosystems.

3. The world's landmasses are classified as six major biogeographic realms. Each is more or less isolated by oceans, mountain ranges, or desert barriers, which tend to restrict gene flow between other realms. As a result, each tends to maintain a characteristic array of species.

4. A biome (a category of major ecosystems on land) is shaped by regional variations in climate, landforms, and soils. Dominant plant species are adapted to the set of conditions prevailing in the major biomes: deserts, dry shrublands, dry woodlands, grasslands, broadleaf forests (such as tropical forests), coniferous forests, and tundra.

5. Water provinces cover more than 71 percent of the Earth's surface. They include standing freshwater (such as lakes), running freshwater (such as streams), as well as the world oceans and seas. All aquatic ecosystems show gradients in light penetration, water temperature, salinity, and dissolved gases. These factors vary daily and seasonally. They influence primary productivity.

6. Estuaries, intertidal zones, rocky and sandy shores, tropical reefs, and regions of the open ocean are major marine ecosystems. Photosynthetic activity is greatest in shallow coastal waters and in regions of upwelling. Upwelling is an upward movement of deep, cool ocean water that often carries nutrients to the surface.

Review Questions

1. Define biosphere. As part of your answer, include definitions of the atmosphere, lithosphere, and hydrosphere. *CI*

2. List the major interacting factors that influence climate. *CI*

3. List some of the ways in which air currents, ocean currents, or both may influence the region where you live. *42.1, 42.2*

4. Indicate where these biomes tend to be located in the world, and describe some of their defining features. *42.3, 42.5–42.9*
   - a. deserts
   - b. dry shrublands
   - c. dry woodlands
   - d. grasslands
   - e. evergreen broadleaf forests
   - f. deciduous forests
   - g. coniferous forests
   - h. arctic tundra
   - i. alpine tundra

5. Define soils, then explain how the composition of regional soils affects ecosystem distribution. *42.4*

6. Describe the littoral, limnetic, and profundal zones of a large temperate lake in terms of seasonal primary productivity. *42.10*

7. Define the two major provinces of the world ocean. Is the open ocean devoid of life? *42.11*

8. Define and characterize the following ecosystems. *42.12, 42.13*
   - a. estuary
   - b. rocky shore
   - c. coral bank
   - d. coral reef

Self-Quiz (*Answers in Appendix III*)

1. Solar radiation drives the distribution of weather systems and so influences the distribution of _____ .
   - a. every ecosystem
   - b. land ecosystems
   - c. all ecosystems except those at hydrothermal vents

2. The _____ is a shield against ultraviolet wavelengths.
   - a. upper atmosphere
   - b. lower atmosphere
   - c. ozone layer
   - d. greenhouse effect

3. Regional variations in the global patterns of rainfall and temperature depend on _____ .
   - a. global air circulation
   - b. ocean currents
   - c. topography
   - d. all of the above

4. A rain shadow is a reduction in rainfall on the _____ of a mountain range.
   - a. windward side
   - b. leeward side
   - c. highest elevation
   - d. lowest elevation

5. Biogeographic realms are _____ .
   - a. land and water provinces
   - b. six major land provinces
   - c. divided into biomes
   - d. b and c

6. Biome distribution corresponds roughly with regional variations in _____ .
   - a. climate b. soils c. topography d. all of the above

7. Dominant plants of _____ are highly adapted to recurring episodes of lightning-sparked fires.
   - a. dry shrublands
   - b. grasslands
   - c. southern pine forests
   - d. all of the above

8. During _____ , deeper, often nutrient-rich water moves to the surface of a body of water.
   - a. spring overturns
   - b. fall overturns
   - c. upwellings
   - d. all of the above

9. Match the terms with the most suitable description.
   - ____ boreal forest
   - ____ permafrost
   - ____ chaparral
   - ____ tropical rain forest
   - a. high productivity; rapid cycling of nutrients (poor reservoirs)
   - b. "swamp forest"
   - c. a type of dry shrubland
   - d. feature of arctic tundra

10. Match the terms with the most suitable description.
   - ____ plankton
   - ____ upwelling
   - ____ eutrophication
   - ____ estuary
   - ____ benthic province
   - a. deep, cool, often nutrient-rich ocean water moves upward
   - b. sediments, rocks of ocean bottom
   - c. partially enclosed mix of seawater, and freshwater
   - d. nutrient enrichment of body of water; reduced transparency, phytoplankton blooms
   - e. aquatic community, tiny species

OLIGOTROPHIC LAKE	EUTROPHIC LAKE
Deep, steeply banked	Shallow with broad littoral
Large deep-water volume relative to surface-water volume	Small deep-water volume relative to surface-water volume
Highly transparent	Limited transparency
Water blue or green	Water green to yellow or brownish-green
Low nutrient content	High nutrient content
Oxygen abundant through all levels throughout year	Oxygen depleted in deep water during summer
Not much phytoplankton; green algae and diatoms dominant	Abundant, thick masses of phytoplankton; cyanobacteria dominant
Abundant aerobic decomposers favored in profundal zone	Anaerobic decomposers
Low biomass in profundal	High biomass in profundal

b

Figure 42.35 (**a**) Crater Lake, Oregon, a collapsed volcanic cone filled with water from rains and melted snow. Like other volcanoes of the Cascade range, it started forming as a result of tectonic forces that prevailed at the dawn of the Cenozoic . (**b**) Major characteristics of oligotrophic and eutrophic lakes.

Critical Thinking

1. Kangaroos are furry, pouched mammals. Raccoons are furry placental mammals. Speculate on why Australia but not North America is the original home of kangaroos and why the reverse is true of raccoons. Use your knowledge of geologic history when devising an explanation. (Compare Section 24.9.)

2. Reflect on the world distribution of landmasses and ocean water, then develop an explanation of why grassland biomes tend to form in the interior of continents, not along their coasts.

3. *Wetlands*, remember, are transitional zones between aquatic and terrestrial ecosystems in which plants are adapted to grow in periodically or permanently saturated soil. As mentioned in Section 40.6, they include riparian zones and mangrove swamps. Wetlands everywhere are being converted for agriculture, home building, and other human activities. Does the great ecological value of wetlands for the nation as a whole outweigh rights of private citizens—who own to many of the remaining wetlands in the United States? Should the private owners be forced to transfer title to the government? If so, who should decide the value of a parcel of land *and* pay for it? Or would such seizure of private property violate the United States Constitution?

4. Observe conditions in a lake near your home or a place where you vacation. If lakes aren't part of your life, think about Oregon's Crater Lake instead (Figure 42.35*a*). Using the data in Figure 42.35*b* as a guide, would you conclude Crater Lake is oligotrophic or eutrophic? Will it remain so over time?

5. The United States Forest Service and Geophysical Dynamics Laboratory ran supercomputer programs to predict possible outcomes of a *global warming* trend. (Here you may wish to review Section 41.8.) According to the results, by the year 2030, the United States will experience recurring, devastating forest fires. Severe storms and more frequent rainfall in the western states will accelerate erosion, especially along the coasts and in steep foothills and mountain ranges.

Increases in the deposition of sediments will have adverse effects on rivers, streams, and estuaries. Dry shrublands and woodlands may replace hardwood forests in Minnesota, Iowa, and Wisconsin, and in parts of Missouri, Illinois, Indiana, and Michigan. The breadbasket of the American Midwest will shift north into Canada. Think about where you live now and where you plan to live in the future. How would such changes affect you? Should nations get serious about finding ways to counter global warming? Or do you believe that the scientists who are concerned about this are merely alarmist and that nothing bad will happen? Explain why you have reached your conclusion.

6. Think about the ocean, which covers the majority of the Earth's surface. Its deepest regions are remote from human populations, to say the least. Now think about a modern-day problem—where to dispose of nuclear wastes and other truly hazardous materials. You will be reading about this serious problem in the next chapter; for now, simply be aware that the United States government has stockpiles of very dangerous wastes and has nowhere to put them. Would it be feasible, let alone ethical, to use the deep ocean as a "burying ground" for them? Why or why not?

Selected Key Terms

atmosphere *CI*
biogeographic realm *42.3*
biogeography *CI*
biome *42.3*
biosphere *CI*
boreal forest *42.8*
climate *CI*
coniferous forest *42.8*
coral bank *42.12*
coral reef *42.12*
deciduous broadleaf forest *42.7*
desert *42.5*
desertification *42.5*
dry shrubland *42.6*

dry woodland *42.6*
ecoregion *42.3*
El Niño *42.13*
estuary *42.13*
eutrophication *42.10*
evergreen broadleaf forest *42.7*
fall overturn *42.10*
grassland *42.6*
hydrothermal vent *42.11*
intertidal zone *42.13*
lake *42.10*
marine snow *42.11*
monsoon *42.2*
ocean *42.2*

permafrost *42.9*
rain shadow *42.2*
savanna *42.6*
soil *42.4*
southern pine forest *42.8*
spring overturn *42.10*
stream *42.10*
temperature zone *42.1*
tropical rain forest *42.7*
tundra *42.9*
ultraplankton *42.11*
upwelling *42.13*

Readings See also www.infotrac-college.com

Brown, J., and A. Gibson. 1983. *Biogeography*. St. Louis: Mosby.

Dold, C., February 1999. "The Cholera Lesson." Discover, 71–76.

Garrison, T. 1996. *Oceanography*. Second edition. Belmont, California: Wadsworth.

Olson, D. M., and Dinerstein, E. 1998. "The Global 200: A Representation Approach to Conserving the Earth's Most Biologically Valuable Ecoregions. *Conservation Biology* 12(3).

Smith, R. 1996. *Ecology and Field Biology*. Fifth edition. New York: HarperCollins.

WWW   *http://www.brookscole.com/biology*

Practice quiz questions, hypercontents, BioUpdates, and critical thinking. The Brooks/Cole Biology Resource Center provides a wealth of information fully organized and integrated by chapter.

43

HUMAN IMPACT ON THE BIOSPHERE

An Indifference of Mythic Proportions

Of all the concepts introduced in the preceding chapter, the one that should be foremost in your mind is this: *The atmosphere, ocean, and land interact in ways that help dictate living conditions throughout the biosphere.* Driven by energy streaming in continually from the sun, these stupendous interactions give rise to globe-spanning temperatures and circulation patterns upon which life ultimately depends. With this chapter, we turn to a related concept of equal importance. Simply put, *we have become major players in these interactions even before we fully comprehend how they work.*

To gain perspective on what is happening, think about something we take for granted—the air around us. The composition of the present-day atmosphere is a result of geologic and metabolic events—most notably, photosynthesis—that began billions of years ago. The first humans, recall, evolved about 2.5 million years ago. Like us, they breathed oxygen from an atmosphere of ancient origins. Like us, they were protected from the sun's ultraviolet radiation by an ozone shield in the stratosphere. The size of their population was not much

to speak of, and their effect on the biosphere was trivial. About 11,000 years ago, however, agriculture began in earnest, and it laid the foundation for huge increases in population size. A few centuries ago, medical and industrial revolutions expanded that foundation—and human population growth skyrocketed in a mere blip of evolutionary time.

Today we are extracting huge amounts of energy and resources from the environment and giving back monumental amounts of wastes. As we do so, we are destabilizing ecosystems everywhere, even though the magnitude of change might not even be recognized when measured on the scale of a lifetime (Figure 43.1).

In a few developed countries, population growth has more or less stabilized, and the resource use per individual has dropped a bit. But resource utilization levels in those places are already high. In developing countries in Central America, Africa, and elsewhere, population sizes and resource consumption are rapidly growing, even though hundreds of millions of people are already malnourished or starving to death.

1900   1940   1954   1962

Many of the problems sketched out in this chapter are not going to disappear tomorrow. It will take many decades, even centuries, to reverse some trends that are already in motion—and not everyone is ready to make the effort. A few enlightened individuals in Michigan or Alberta or New South Wales can commit themselves to resource conservation and to minimizing pollution. But scattered attempts will not be enough. Individuals of all nations will unite to reverse global trends only when they perceive that the dangers of *not* doing so outweigh the personal benefits of ignoring them.

Does this seem pessimistic? Think of the exhaust fumes released into the air each time you drive a car or truck. Think of oil refineries, food-processing plants, and paper mills that supply you with goods and also release chemical wastes into the nation's waterways. Think of Mexico and other developing countries that produce cheap food by using an unskilled labor force and toxic pesticides. Unregulated pesticide applications poison the people who work the land, the land itself, and sometimes people who buy the exported produce.

Who changes behavior first? We have no answer to the question. We can suggest, however, that a strained biosphere can rapidly impose an answer upon us. As an individual, you might choose to cherish or brood about or ignore any aspect of the world of life. Whatever your choice may be, the bottom line is that you and all other organisms are in this together. Our lives interconnect, to degrees that we are only now starting to comprehend.

Figure 43.1 Tracking human population growth in one small part of the world, based on historical data and satellite imaging. *Red* denotes areas of dense human settlement in and around the San Francisco Bay Area and Sacramento since 1900.

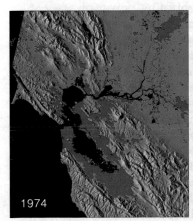
1974

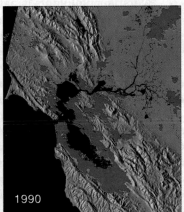
1990

KEY CONCEPTS

1. Human population growth has been skyrocketing ever since the mid-eighteenth century. At present, humans have the population size, the technology, and the cultural inclination to use energy and alter the environment at astonishing rates.

2. Pollutants are substances with which ecosystems have had no prior evolutionary experience, in terms of kinds or amounts, so adaptive mechanisms that can deal with them are not in place. In a more restrictive sense, pollutants are substances that accumulate in amounts that have adverse effects on human health, activities, or survival.

3. Many pollutants, including the kinds that contribute to the formation of smog or acid rain, exert regionally harmful effects. The effects of other pollutants, including chlorofluorocarbons that attack the ozone layer in the stratosphere, are global in scale.

4. Conversion of marginally fertile lands for agriculture, rampant deforestation, and other practices required to meet the demands of the growing human population are leading to loss of soil fertility and desertification. They are also resulting in a decline in the quality and quantity of one of the most crucial of all resources—fresh water.

5. Ultimately, the world of life depends on energy inputs from the sun. That energy drives the complex interactions among the atmosphere, ocean, and land. Our activities are disrupting the interactions in ways that may have severe consequences in the near future.

6. We as a species must come to terms with principles of energy flow and principles of resource utilization that govern all systems of life on Earth.

DEFORESTATION—AN ASSAULT ON FINITE RESOURCES

At one time, tropical forests cloaked regions that were, collectively, twice the size of Europe. For ten thousand years or more, those forests endured in rich complexity, as the homes of an estimated 50 to 90 percent of all land-dwelling species. In less than four decades, human populations destroyed more than half of those ancient forests, and most of their spectacular arrays of species may be lost forever. With every passing year, another 38 million acres is logged over. That is the equivalent of leveling thirty-four city blocks every minute.

The destruction extends beyond the tropics. Today, highly mechanized logging operations are proceeding in the once-vast temperate forests of the United States, Canada, Europe, Siberia, and elsewhere (Section 22.6).

We have a name for the removal of all trees from large tracts of land for logging, agriculture, and grazing operations. It is **deforestation**. Why are we doing this? Paralleling the rapid increases in the size of the human population are rapidly increasing demands for lumber, fuel, and other forest products, as well as for cropland and grazing land. More and more people compete for diminishing resources. They do so for economic profit, but also because alternative ways of life simply are not available to most individuals and families.

The world's great forests have profound influences on ecosystems. Like enormous sponges, the watersheds of forested regions absorb, hold, and then release water gradually. By intervening in the downstream flow of water, they help control soil erosion, flooding, and the accumulation of sediments that can clog rivers, lakes, and reservoirs. When the vegetation cover gets stripped away, the exposed soil becomes vulnerable to leaching of nutrients and to erosion, especially on steep slopes.

Today, deforestation is greatest in Brazil, Indonesia, Colombia, and Mexico. If the clearing and destruction continue at present rates, only Brazil and Zaire will still have large tropical forests in the year 2010. By 2035, most of their forested regions will be cleared, also.

Figures 43.7 and 43.8 show close-up and panoramic views of what is now happening in South America's Amazon basin. Section 22.6 includes examples of the effects of rampant deforestation in North America.

In tropical regions, the clearing of forests for agriculture sets the stage for long-term losses in productivity. The irony is that tropical forests are one of the worst places to grow crops or to raise pasture animals. In intact forests, litter cannot accumulate, for the high temperatures and the heavy, frequent rainfall promote the rapid decomposition of organic remains and wastes. As fast as the decomposers release nutrients, the trees and other plants take them up. Deep, nutrient-rich topsoils simply cannot form.

Long before the advent of highly mechanized logging practices, people were practicing **shifting cultivation** (once referred to as "slash-and-burn

Figure 43.7 Countries that are allowing the greatest destruction of tropical forests. *Red* denotes the regions where 2,000 to 14,800 square kilometers are deforested every year. *Orange* denotes "moderate" deforestation, which encompasses areas of 100 to nearly 2,000 square kilometers.

ANDES

SMOKE FROM FIRES

Figure 43.8 The vast Amazon River basin of South America in September 1988. Smoke from fires that had been deliberately set during the dry season to clear tropical forests, pasturelands, and croplands completely obscured its features. The smoke extended to the Andes Mountains near the western horizon, about 650 miles (1,046 kilometers) away. That smoke cover was the largest that astronauts had ever observed. It extended almost 175 million square kilometers (1,044,000 square miles).

The smoke plume near the center of the photograph covered an area comparable to the huge forest fire in Yellowstone National Park in that year. During the El Niño–induced drought of 1997, set fires burned out of control and a smoke cover formed again.

Massive deforestation is not confined to the equatorial regions. For instance, during the past century, 2 million acres of redwood forests along the coast of California were logged over. Most of the destruction of such temperate forests has been proceeding since 1950, owing to the widespread use of chainsaws and tractors and the practice of exporting many logs to lumbermills overseas, where wages are low.

agriculture"). They cut and burn trees, then till ashes into the soil. The nutrient-rich ashes can sustain crops for one to several seasons. Afterward, cleared plots are abandoned, for heavy leaching leaves the soil infertile. When shifting cultivation is practiced on small, widely scattered plots, a forest ecosystem does not necessarily suffer extensive damage. But soil fertility plummets with increases in population size. Then, larger areas are cleared, and plots are cleared again at shorter intervals.

Figure 43.9 Wangari Maathai, a Kenyan who organized the internationally acclaimed Green Belt Movement in 1977. The 50,000 members of this women's group are committed to establishing nurseries, raising seedlings, and planting a tree for each of the 27 million Kenyans. Together with half a million schoolchildren, they had planted more than 10 million trees by 1990. Their success inspired similar programs in more than a dozen countries in Africa. Dr. Maathai's efforts are not appreciated by her own government. Kenyan police have jailed her twice, and in 1992 they severely beat her because of her efforts on behalf of the environment.

In the larger picture, deforestation alters rates of evaporation, transpiration, runoff, and perhaps regional patterns of rainfall. For example, trees release between 50 and 80 percent of the water vapor above tropical forests. In the logged-over regions, annual precipitation declines, and rain swiftly drains away from the exposed, nutrient-poor soil. The regions are now hotter and drier, and soil fertility and moisture have declined. In time, sparse grassland or desertlike conditions might prevail instead of the formerly rich, forested biomes.

Also, tropical forests absorb much of the sunlight reaching the equatorial regions of the Earth's surface. Deforested land is shinier, so to speak, and it reflects more incoming energy back into space. And because of the combined photosynthetic activity of so many trees, the forests help sustain the global cycling of carbon and oxygen. With extensive tree harvesting and burning, the carbon stored in the tree biomass is released to the atmosphere, as carbon dioxide. Thus deforestation may be a factor in the amplification of the greenhouse effect.

Conservation biologists are attempting to reverse the trend. As three examples, a coalition of 500 groups is dedicated to preserving Brazil's remaining tropical forests. In India, women have already built and installed 300,000 inexpensive, smokeless wood stoves. Over the past decade, the stoves saved more than 182,000 metric tons' worth of trees by reducing demands for fuelwood. In Kenya, women have planted millions of trees to hold soil in place and to provide fuelwood (Figure 43.9).

Once-vast forests helped sustain rapid increases in human population growth. Recent and highly mechanized modes of deforestation are rapidly depleting these finite resources.

WWW

43.5 YOU AND THE TROPICAL RAIN FOREST

Developing countries in Latin America, Southeast Asia, and Africa have the fastest-growing populations but limited food, fuel, and lumber. Thanks to the relatively recent inventions of chainsaws, tractors, and logging trucks, most of their forests will probably disappear within your lifetime.

Why does it matter? For purely ethical reasons, many condemn the destruction of so much biodiversity. Tropical rain forests have the greatest variety and numbers of insects, and the world's largest ones. They are home to the most bird species and to plants with the largest flowers. Within the forest canopy and understory are monkeys, tapirs, and jaguars in South America and apes, okapi, and leopards in Africa. Massive vines twist around trees. Orchids, mosses, lichens, and other organisms grow on branches, absorbing minerals that rains deliver to them. Entire communities of microbes, insects, spiders, and amphibians live, breed, and die in small pools of water that collect in furled leaves.

For practical reasons, the destruction affects your life. Only a few strains of crop plants and livestock, vulnerable to evolving pathogens, sustain most human populations. Tissue-culture specialists and genetic engineers use genes of forest species to develop new or hybrid strains that can make our food base less vulnerable. Geneticists use them to develop more effective antibiotics and vaccines. Aspirin, the most widely used painkiller, is based on a chemical blueprint of an extract from tropical willow leaves. Many ornamental plants and spices and foods, including cocoa, cinnamon, and coffee, originated in the tropics. So did latex, gums, resins, dyes, waxes, and oils for tires, shoes, toothpaste, ice cream, shampoo, compact discs, condoms, and perfumes. And think about this: Rampant burning of forests is releasing enough air pollutants to change the air you breathe and help heat the planet during your lifetime.

Thus conservation biologists rightly decry the mass extinction, the assaults on species diversity, the depletion of much of the world's genetic reservoir. Yet something else is going on here. Too many of us become uneasy when we hike or drive through destroyed forests in our own country. Is it because we are losing the comfort of our heritage—a connection with our evolutionary past? Many millions of years ago, our earliest primate ancestors moved into the trees of tropical forests. Through countless generations, their nervous and sensory systems evolved and became highly responsive to information-rich, arboreal worlds.

Does our neural wiring still resonate with rustling leaves, with shafts of light and mosaic shadows? Are we innately attuned to the forests of Eden—or have time and change buried recognition of home?

WWW

Figure 43.10 Tropical rain forest in Southeast Asia.

WHO TRADES GRASSLANDS FOR DESERTS?

Long-term shifts in climate can convert grasslands to deserts. So can human populations. **Desertification** is the name for the conversion of large tracts of natural grasslands to a more desertlike condition. It applies also when conversions of rain-fed or irrigated croplands result in a 10 percent or greater decline in agricultural productivity. Over the past fifty years, 9 million square kilometers worldwide have become desertified. At least

water from plants. They also are better at conserving water; they lose little in feces, compared to cattle.

In 1978 a biologist, David Holpcraft, began ranching antelopes, zebras, giraffes, ostriches, and other native herbivores. He also raised cattle as control groups to compare costs and meat yields on the same land. The native herds increased and yielded tasty meat. Range conditions did not deteriorate; they improved. Vexing

Figure 43.11 An awe-inspiring dust storm approaching Prowers County, Colorado, in 1934.

The Great Plains of the American Midwest are dry, windy grasslands that are subjected to pronounced, recurring droughts. Extensive conversion of these grasslands to agriculture began in the 1870s. Overgrazing left the ground bare across vast tracts. In May 1934, a cloud of topsoil that blew off the land blanketed the entire eastern portion of the United States, giving the Great Plains a dubious new name—the Dust Bowl. About 3.6 million hectares (9 million acres) of cropland were destroyed. Today, without large-scale irrigation and intensive conservation farming, desertlike conditions could prevail.

Figure 43.12 Desertification in the Sahel, a region of West Africa that forms a belt between the hot, dry Sahara Desert and tropical forests. This savanna country is undergoing rapid desertification as a result of overgrazing, overfarming, and prolonged drought.

200,000 square kilometers are still being converted annually. Prolonged droughts accelerate the process, as one did in the Great Plains years ago (Figure 43.11). At present, overgrazing of livestock on marginal lands is the main cause of large-scale desertification.

In Africa, for example, there are too many cattle in the wrong places. Cattle require more water than the region's native wild herbivores, so they move back and forth between grazing areas and watering holes more often. As they do, they trample grasses and compact the soil surface (Figure 43.12). By contrast, gazelles and other endemic herbivores get most (if not all) of their

problems remained. Certain tribes in Africa have their own idea of what constitutes "good" meat, and some view cattle as the symbol of wealth in their society.

Without irrigation and conservation practices, grasslands that were converted for agriculture often end up as deserts.

WWW

A GLOBAL WATER CRISIS

The Earth has a tremendous supply of water, but most is too salty for human consumption or for agriculture. Imagine all that water in a bathtub. Withdraw all of the fresh, renewable portion (from lakes, rivers, reservoirs, groundwater, and other sources of surface water) and it would barely fill a teaspoon.

Why not consider **desalinization**—removal of salt from seawater? Basically, we have unlimited supplies of seawater. Desalinization processes exist. They either distill seawater or force it through membranes (a method called reverse osmosis). Costly fuel energy drives these processes, so they may be feasible only in Saudi Arabia and a few other countries with limited population sizes, large reserves of energy, and lots of cash. Actually, in some situations they might be the only alternative to running out of drinkable water, as Santa Barbara and some other California cities nearly did during a recent prolonged drought. Yet desalinization can't solve the core problem. It may never be cost-effective for large-scale agriculture, and it makes mountains of salts.

Figure 43.13 Irrigated crops in the Sahara Desert, in Algeria.

Consequences of Heavy Irrigation

Large-scale agriculture accounts for nearly two-thirds of the human population's use of freshwater. In many cases, irrigation water from surface sources is piped into vast fields where water-demanding crops will not grow on their own. At its most extreme, irrigation has turned some hot deserts into lush gardens, although believing these can be maintained over the long term is a bit delusional (Figure 43.13).

Irrigation itself can change the land's suitability for agriculture. Concentrations of mineral salts commonly are high in piped-in water. In regions where soil drains poorly, evaporation may cause **salinization**: a buildup of salt in soil. Salinization can stunt the growth of crop plants, eventually kill them, and decrease yields.

Land that drains poorly also becomes waterlogged. When water accumulates underground, it slowly raises the **water table**, the upper limit at which the ground is fully saturated. When the water table rises close to the ground's surface, soil gets saturated with saline water, which can damage plant roots. Properly managing the water-soil system can correct salinization and water-logging. The economic cost of doing so is high.

Worse, groundwater overdrafts (the amount nature doesn't replenish) are high (Figure 43.14). For example, overdrafts have depleted half of the Ogallala aquifer, which supplies irrigation water for 20 percent of the croplands in the United States. So much groundwater has been withdrawn for irrigation in some parts of the San Joaquin Valley in California that the surface of the water table has subsided by as much as six meters.

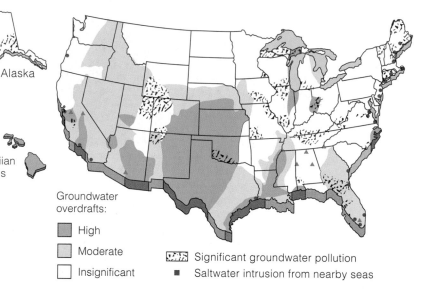

Groundwater overdrafts:

- ■ High
- ■ Moderate
- □ Insignificant

▨ Significant groundwater pollution
■ Saltwater intrusion from nearby seas

Figure 43.14 Areas of greatest aquifer depletion, saltwater intrusion, and groundwater contamination in the United States.

Water Pollution

Water pollution amplifies the problem of water scarcity. Inputs of sewage, animal wastes, and toxic chemicals make water unfit to drink, even to swim in. Pollutants encourage contamination by pathogens. Agricultural runoff pollutes water with sediments, pesticides, and plant nutrients. Power-generating plants and factories pollute water with chemicals, radioactive materials, and excess heat (thermal pollution).

Pollutants collect in lakes, rivers, and bays before reaching the oceans. Many cities throughout the world dump untreated sewage into coastal waters. The cities

Figure 43.15 An experimental wastewater treatment facility in Rhode Island. Treatment begins when sewage flows into rows of large water tanks in which water hyacinths, cattails, and other aquatic plants are growing. Decomposers in the tank degrade wastes—which contain nutrients that promote plant growth. Heat from incoming sunlight speeds the decomposition. From these tanks, water flows through an artificial marsh of sand, gravel, and bulrushes that filter out algae and organic wastes. Then it flows into aquarium tanks, where zooplankton and snails consume microorganisms suspended in the water—and where zooplankton become food for crayfishes, tilapia, and other fishes that can be sold as bait. After ten days, the now-clear water flows into a second artificial marsh for final filtering and cleansing.

along rivers and harbors maintain shipping channels by dredging the polluted muck and barging it out to sea. They also barge sewage sludge—coarse, settled solids that contain bacteria, viruses, and toxic metals.

In the United States, about 15,000 facilities partially treat liquid wastes from 70 percent of the population and 87,000 industries. The remaining wastes are mostly from suburban and rural populations. These are treated in lagoons or septic tanks or are directly discharged—untreated—into waterways.

There are three levels of **wastewater treatment**. In *primary* treatment, screens and settling tanks remove sludge, which is dried, burned, dumped in landfills, or treated further. Chlorine often is used to kill pathogens. It does not kill them all, and it produces carcinogens whenever it reacts with certain industrial chemicals.

In *secondary* treatment, microbial populations break down organic matter after primary treatment but before chlorination. Wastewater trickles past microorganisms in gravel beds or is aerated in tanks and seeded with microorganisms. Toxic solutes can poison the microbes. At such times, the facilities shut down until populations of microorganisms become reestablished. Primary and secondary treatments remove most of the suspended solids and oxygen-demanding wastes, but not all of the nitrogen, phosphorus, and toxic substances, including heavy metals and pesticides. Then the water is usually chlorinated before being released into the waterways.

Tertiary treatment adequately reduces pollution but is largely experimental and expensive. It is applied to only 5 percent of our nation's wastewater.

In short, most wastewater is not treated adequately. A pattern gets repeated thousands of times along our waterways. Water for drinking is drawn upstream from a city, and wastes from industry and sewage treatment are discharged downstream. It takes no great leap of the imagination to see that water pollution intensifies as rivers flow to the oceans. In Louisiana, waters drained from the central states flow toward the Gulf of Mexico. Its high pollution levels threaten public health as well as ecosystems. Water destined for drinking gets treated to remove pathogens, but treatment does not remove toxic wastes dumped by numerous factories upstream.

This rather bleak picture might be numbing to most of us, but not to biologist John Todd. He constructed experimental wastewater treatment facilities in several greenhouses and artificial lagoons (Figure 43.15). When it works properly, the solar-aquatic treatment system produces water fit to drink. Such natural alternatives cannot work for very large urban areas. But they are an attractive alternative for small towns and rural areas.

The Coming Water Wars

If the current rates of population growth and water depletion hold, the amount of freshwater available for everyone on the planet will soon be 55 to 66 percent less than what it was in 1976. Already in the past decade, thirty-three nations have been engaged in conflicts over reductions in water flow, pollution, and silt buildup in major aquifers, rivers, and lakes. The United States and Mexico, Pakistan, India, and Israel and the occupied territories are among the squabblers.

Remember the Persian Gulf War, mainly about oil? Unless we pull off a blue revolution equivalent to the green one, we may be in for upheavals and wars over water rights. Does this sound farfetched? By building dams and irrigation systems at the headwaters of the Tigris and Euphrates rivers, Turkey can, in the view of one of its dam-site managers, stop the water flow into Syria and Iraq for as long as eight months "to regulate their political behavior." Regional, national, and global planning for the future is long overdue.

Water, not oil, may become the most important fluid of the twenty-first century. National, regional, and global policies for water usage and water rights have yet to be developed.

WWW

A QUESTION OF ENERGY INPUTS

Paralleling the J-shaped curve of human population growth is a dramatic rise in total and per capita energy consumption. It is due to increased numbers of energy users and to extravagant consumption and waste. For example, in one of the most pleasant of all climates, a major university constructed seven- and eight-story buildings with narrow, sealed windows. The windows can't be opened to catch prevailing ocean breezes. The buildings and windows were not designed or aligned to use sunlight for passive solar heating and breezes for passive cooling. Massive energy-demanding cooling and heating systems were installed.

When you hear talk of abundant energy supplies, bear in mind that there is a large difference between the *total* and net amounts available. *Net* refers to the amount left over after subtracting the energy required to locate, extract, transport, store, and deliver energy to consumers. Some sources, such as direct solar energy, are renewable. And others, such as coal, are not (Figure 43.16).

Fossil Fuels

Fossil fuels are the legacy of forests that disappeared many hundreds of millions of years ago, so they are nonrenewable resources (Section 22.3). Over time, the carbon-containing remains of the plants were buried in sediments, compacted, and chemically transformed into coal, petroleum (oil), and natural gas.

Even if conservation becomes strict, we may use up known petroleum and natural gas reserves in the next century. As known reserves run out in accessible areas, we explore wilderness areas in Alaska and other fragile environments, such as continental shelves. Net energy declines when the cost of extraction and transportation to and from remote areas increases. The environmental costs of extraction and transportation escalate. The widespread damage and clean-up costs following the 11-million-gallon oil spill from the supertanker *Valdez*, off the coast of Alaska, are a classic example.

What about coal? In theory, known world reserves may meet the energy needs of the human population for at least several centuries. However, coal burning has been the primary source of air pollution. Most of the known coal reserves contain low-quality material with a high sulfur content. Unless sulfur is removed before or after fuel is burned, sulfur dioxides enter the air and contribute to global acid deposition. Fossil fuel burning also releases carbon dioxide and adds to the greenhouse effect.

Extensive strip mining of coal reserves close to the surface carries its own problems. It reduces the land available for agriculture, grazing, and wildlife. Most strip mines are located in arid and semiarid regions where the absence of sufficient water supplies and poor soils make restoration efforts difficult.

Nuclear Energy

NUCLEAR REACTORS In 1945, as Hiroshima burned, people recoiled in horror from the destructive force of nuclear energy. During the 1950s, however, many were championing nuclear energy as an instrument of progress. A number of energy-poor industrialized countries, including France, now depend heavily upon nuclear power. Yet, construction of more nuclear plants has been delayed or even cancelled in most countries. After 1970 in the United States alone, the plans to build 117 nuclear power plants were shelved. Plants already being constructed were abandoned before completion. A few are being converted, at great cost, to fossil fuel burning. What happened? We started to question the operating cost, efficiency, safety record, and environmental impact of reliance on nuclear power.

By 1990, nuclear power was generating electricity at a cost only slightly above that of coal-burning plants. Putting aside other factors, its electricity-generating costs are now lower. However, by the year 2000, solar energy with natural gas backup also should cost less.

What about safety? Compared to coal-burning plants of the same capacity, a nuclear plant emits less radioactivity and carbon dioxide, and no sulfur dioxide. However, there is greater danger in their potential for **meltdown**. As nuclear fuel undergoes radioactive decay, it releases considerable heat. Water

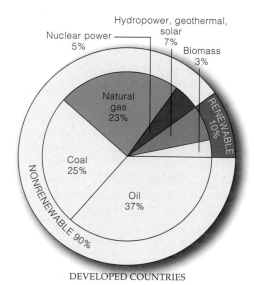

Figure 43.16 Energy consumption in the developed and developing countries, which differ greatly in sources of energy and average per capita energy use. The values indicated do not take into account energy from the sun, which is the foundation for agriculture.

Figure 43.17 Incident at Chernobyl. (**a**) On April 26, 1986, errors in judgment during a routine test procedure resulted in runaway reactions, explosions, and a full core meltdown at the Chernobyl power plant in Ukraine. Helicopter pilots who were supposed to drop 5,000 tons of lead, sand, clay, and other materials on the blazing core to suppress further release of radioactivity missed the target. Unimpeded and uncovered, nuclear fuel burned for nearly ten days, right on through a six-foot-thick steel and gravel barrier beneath it.

Between 185 and 250 million curies of radioactive matter may have escaped in those ten days alone. Inhaling as little as ten-millionths of a curie of plutonium can cause cancer. Thirty-one people died at once; others died of radiation sickness in the following weeks. Inhabitants of entire villages were relocated; their former homes were bulldozed under. In time, concrete entombed the 180 tons of partially burned nuclear fuel.

(**b**) Afterward, the number of people opposed to nuclear plants rose sharply, even in France. We get a sense of why opposition increased dramatically by studying maps of the global distribution of radioactive fallout within two weeks of the meltdown. The fallout put 300 to 400 million people at risk for leukemia and other radiation-induced disorders. Throughout Europe, hundreds of millions of dollars were lost when the fallout made crops and livestock unfit for consumption. In 1994, rainwater and air were still moving through 11,000 square meters of holes in the power plant's concrete. By 1998, the rate of thyroid abnormalities in children living immediately downwind from Chernobyl was nearly seven times as high as for those living upwind; their thyroid gland had collected iodine radioisotopes from the fallout.

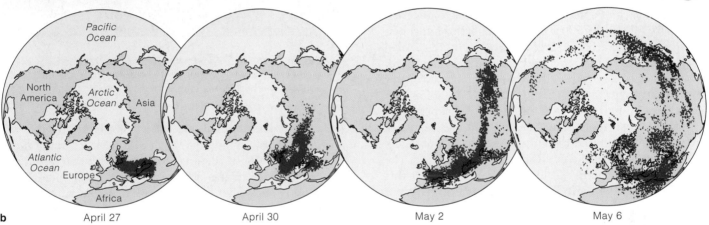

b   April 27   April 30   May 2   May 6

typically circulates over the nuclear fuel, absorbs heat, and so produces steam to drive electricity-generating turbines. If the circulating water system develops a leak, water levels may plummet around the fuel, which may then heat past its melting point. Melting fuel on a generator floor would instantly convert the remaining water to steam. Combined with other reactions, steam formation could blow the system apart and release radioactive material. An overheated core could melt through its greatly thickened concrete containment slab. Figure 43.17 describes one incident that yielded compelling evidence of the consequences.

NUCLEAR WASTE DISPOSAL Unlike coal, nuclear fuel cannot be burned to harmless ashes. After three years or so, the fuel elements are spent, but they still contain uranium fuel as well as hundreds of new radioisotopes that formed in the reactions. The wastes are extremely radioactive and dangerous. They get extremely hot as they undergo radioactive decay, so they are plunged at once into water-filled pools. The water cools them and

keeps radioactive material from escaping. Even after being stored for several months, the isotopes remaining are lethal. Some must be kept isolated for at least 10,000 years. If a certain isotope of plutonium (^{239}Pu) is not removed, the wastes must be kept isolated for a quarter of a million years! After nearly fifty years of research, scientists still cannot agree on what is the best way to store high-level radioactive wastes. Even if they could, there is no politically acceptable solution. No one wants radioactive wastes anywhere near where they live.

Finally, as if we don't have enough to worry about, following the Soviet Union's breakup, some underpaid workers of a Russian nuclear power plant have been selling fuel elements on the black market. The buyers? Some developing nations that want to produce nuclear weapons—and possibly deliver them into the hands of terrorist organizations.

The nuclear genie is out of the bottle, exploitable by the best and worst elements of the human population.

WWW

ALTERNATIVE ENERGY SOURCES

Less than thirty years from now, the projected size of the human population will be such that the demands for fuel will increase by 30 percent—and for electricity by 265 percent. More efficient use and conservation of our existing energy sources alone will not do the trick. We must make the transition to reliance on alternative sources of energy, of the sort described next.

Solar-Hydrogen Energy

Each year, incoming sunlight contains about ten times more energy than that in all of the known fossil fuel reserves. That is 15,000 times as much energy as the human population uses now. Isn't it about time we start to collect it in earnest, in something besides crops?

For example, when exposed to sunlight, electrodes in "photovoltaic cells" produce an electric current that splits water molecules into oxygen and hydrogen gas (H_2)—which can be used directly as fuel or to generate electricity. The technology to tap such **solar–hydrogen energy** has been around since the 1940s (Figure 43.18*a*). Such energy is stored efficiently for as long as required. It costs less to distribute H_2 than electricity, and water is the only by-product of using it. Space satellites run on it. Here on Earth, fossil fuels are still "cheaper."

Unlike fossil fuels, however, sunlight and seawater are virtually unlimited resources. And the technology's potential to protect the environment is staggering. The environmental scientist G. Tyler Miller, Jr., puts it this way: "If we make the transition to an energy-efficient solar-hydrogen age, we can say good-bye to smog, oil spills, acid rain, and nuclear energy, and perhaps to global warming. The reason is simple. As hydrogen burns in air, it reacts with oxygen gas to produce water vapor—not a bad thing to have coming out of tailpipes, chimneys, and smokestacks." Also, if this technology becomes cost-effective for the developing countries, the great forests now being destroyed for timber and fuel might still be around for future generations.

In the United States, the largest supplier of natural gas and a manufacturer of photovoltaic cells combined forces to build a solar facility in the Nevada desert. It will generate enough energy to supply a city of 100,000 at less cost than electricity from fossil fuels. If the plan succeeds, they may revolutionize the energy industry.

Wind Energy

As you know, solar energy also is converted into the mechanical energy of winds. Where prevailing winds travel faster than 7.5 meters per second, we find **wind farms**. These arrays of cost-effective turbines exploit wind patterns that arise from latitudinal variations in the intensity of incoming sunlight (Figure 43.18*b*). One

Figure 43.18 Harnessing solar energy. (**a**) A large array of electricity-producing photovoltaic cells in panels that collect sunlight energy. (**b**) A field of turbines harvesting wind energy.

percent of California's electricity comes from wind farms. Possibly the winds of North and South Dakota alone can meet all but 20 percent of the current energy needs of the United States. Wind energy also has potential for islands and other remote areas far from utility grids. One drawback: winds don't blow on a regular schedule, so they can't be an exclusive or major energy source.

What About Fusion Power?

The sun's gravitational force is enough to compress atomic nuclei to high densities, and its temperatures are high enough to force atomic nuclei to fuse. We call this **fusion power**. Similar conditions do not exist on Earth, but maybe we can mimic them. Researchers confine a certain fuel (a heated gas of two isotopes of hydrogen) in magnetic fields, then bombard it with lasers. The fuel implodes, it is compressed to extremely high densities—and energy is released. The more energetic the lasers, the greater the compression, and the more the fuel will burn. The bad news is, although the amount of energy released has been steadily increasing, it will be at least fifty years before fusion reactors might be operating, and the costs will probably be high. The good news is, that's about the time fossil fuels will start running out.

Sunlight may end up sustaining the energy needs of the human population in more ways than one.

$\mathcal{WWW}$

43.10 BIOLOGICAL PRINCIPLES AND THE HUMAN IMPERATIVE

Molecules, single cells, tissues, organs, organ systems, multicelled organisms, populations, communities, then ecosystems and the biosphere. These are architectural systems of life, assembled in increasingly complex ways during the past 3.8 billion years. We are latecomers to this immense biological building program. Yet within the relatively short span of 10,000 years, our activities have been changing the very character of the land, ocean, and atmosphere, even the genetic character of species.

It would be presumptuous to think we alone have had profound effects on the world of life. As long ago as the Proterozoic era, photosynthetic organisms irrevocably changed the course of biological evolution by gradually enriching the atmosphere with oxygen. During the past as well as the present, competitive adaptations ensured the rise of some groups, whose dominance assured the decline of others. Change is nothing new to this biological building program. What *is* new is the capacity of a species —our own—to comprehend what might be going on.

We now have the population size, the technology, and the cultural inclination to use up energy and modify the environment at frightening rates. Where will rampant, accelerated change lead us? Will feedback controls begin to operate as they do, for example, when population growth exceeds the carrying capacity of the environment? In other words, will negative feedback controls come into play and keep things from getting too far out of hand?

Feedback control will not be enough, for it operates only when deviation already exists. Patterns of resource consumption and growth rates for the human population are founded on an illusion of unlimited resources and a forgiving environment. A prolonged, global shortage of food or the passing of a critical threshold for the global climate can come too fast to be corrected. At some point, deviations may have too great an impact to be reversed.

What about feedforward mechanisms that might serve as early warning systems? For example, when sensory receptors near the surface of skin sense a drop in outside air temperature, each sends messages to the nervous system. That system responds by triggering mechanisms that raise the body's core temperature before the body itself becomes dangerously chilled. If we develop feedforward control mechanisms, maybe we can begin corrective measures before we have altered the environment too significantly.

By themselves, feedforward controls won't work, for they start operating when change is under way. Think of the DEW line—the Distant Early Warning system. It is like a vast sensory receptor, one that can detect the launching of intercontinental ballistic missiles against North America. By the time the system actually detects what it is designed to detect, it may be too late to stop widespread destruction.

It would be naive to assume we can ever reverse who we are at this point in evolutionary time, to de-evolve ourselves culturally and biologically into becoming less complex in the hope of averting disaster. However, there is reason to believe we can avert disaster by using a third kind of control mechanism—a capacity to anticipate events even before they happen. We are not locked into responding only after irreversible change has begun. We have the capacity to anticipate the future—it is the essence of our visions of utopia or of nightmarish hell. *We all have the capacity to adapt to a future that we can partly shape.*

We can, for example, stop trying to "beat nature" and learn to work with it. Individually and collectively, we can work to develop long-term policies at the local, regional, and global levels—policies that take into account biotic and abiotic limits on population growth. Far from being a surrender, this would be one of the most complex and intelligent behaviors of which we are capable.

Having a capacity to adapt and actually using it are not the same thing. We have already put the world of life on dangerous ground because we have not yet mobilized ourselves as a species to work toward self-control.

Our survival depends on predicting possible futures. It depends on preserving, restoring, and even designing and constructing ecosystems that are in harmony with our definition of basic human values and with the biological models available to us. Human values can change; our expectations can and must be adapted to biological reality. *For the principles of energy flow and resource utilization, which govern the survival of all systems of life, do not change.* It is our biological and cultural imperative that we come to terms with these principles, and ask ourselves what our long-term contribution will be to the world of life.

3.11 SUMMARY

1. Accompanying the extraordinarily rapid growth of the human population are increases in energy demands and in environmental pollution.

2. Pollutants are substances with which ecosystems have had no prior evolutionary experience (in terms of kinds and amounts) and therefore have no mechanisms for absorbing or cycling them. Many pollutants result from certain human activities, and they adversely affect the health, activities, or survival of all organisms.

3. Smog, a form of air pollution, arises in industrialized and urban regions that rely on fossil fuels. It becomes most concentrated in land basins that promote thermal inversion (a layer of cool, dense air trapped below a warm air layer). Industrial smog forms in industrial coal-burning regions with cold, wet winters. In warm climates, large cities with many fuel-burning vehicles have photochemical smog. (Mainly, sunlight makes nitric oxide emitted from vehicles react with hydrocarbons to form photochemical oxidants such as PANs.)

4. During dry weather, acidic air pollutants, especially oxides of nitrogen and sulfur, fall to Earth as dry acid deposition. They also dissolve in atmospheric water, then fall to Earth as wet acid deposition, or acid rain.

5. Seasonal thinning of the ozone layer at high latitudes has become pronounced as CFCs (chlorofluorocarbons) and other air pollutants rise to the stratosphere, where they deplete ozone and allow more harmful ultraviolet radiation from the sun to reach the Earth's surface.

6. Human population growth presently depends upon the expansion of agriculture, made possible by large-scale irrigation and extensive applications of fertilizers and pesticides. Global freshwater supplies are limited, and yet they are being polluted by agricultural runoff (which includes sediments, pesticides, and fertilizers), industrial wastes, and human sewage.

7. Human populations are damaging land surfaces by:
    a. Passively accepting the dumping, burning, and burial of solid wastes rather than making a concerted effort to recycle or reuse materials and to reduce waste.
    b. Engaging in rampant deforestation (destruction of vast tracts of tropical and temperate forest biomes).
    c. Contributing to desertification (the conversion of natural grasslands, croplands, and grazing lands, on a large scale, to desertlike conditions).

8. Energy in the form of fossil fuels is nonrenewable, dwindling, and environmentally costly to extract and use. Nuclear energy in itself is less polluting, but the costs and risks associated with fuel containment and storing radioactive wastes are enormous. The challenge is to develop affordable alternatives that are based on renewable energy resources, such as solar energy.

Review Questions

1. Define pollution and list some specific examples of water pollutants. 43.1, 43.7

2. Distinguish among the following conditions: 43.1
    a. industrial smog   c. dry acid deposition
    b. photochemical smog d. wet acid deposition

3. Define CFCs and describe how they apparently contribute to seasonal thinning of the ozone layer in the stratosphere. 43.2

4. What percent of the Earth's land masses is under cultivation? What percent is available for new cultivation? 43.3

5. Define and describe possible consequences of deforestation and of desertification. 43.4, 43.5, 43.6

6. Which human activity uses the most freshwater? 43.7

Self-Quiz (Answers in Appendix III)

1. Since the mid-eighteenth century, human population growth has been _____ .
    a. leveling off   c. accelerating
    b. growing slowly   d. not much to speak of

2. Pollutants disrupt ecosystems because _____ .
    a. their components differ from those of natural substances
    b. only humans have uses for them
    c. there are no evolved mechanisms to deal with them
    d. their only effect is on ecosystems, not humans

3. During a thermal inversion, weather conditions trap a layer of _____ air under a layer of _____ air.
    a. warm; cool   c. warm; sooty
    b. cool; warm   d. cool; sooty

4. _____ is (are) a case of regional air pollution.
    a. Smog   c. Ozone layer thinning
    b. Acid rain   d. a and b

5. _____ is (are) a case of air pollution with global effects.
    a. Smog   c. Ozone layer thinning
    b. Acid rain   d. b and c

6. Worldwide, two-thirds of the freshwater used annually goes to _____ .
    a. urban centers   c. treatment facilities
    b. agriculture   d. a and c

7. The upper limit at which the ground is fully saturated with water is called _____ .
    a. groundwater   c. the water table
    b. an aquifer   d. the salinization limit

8. Energy from fossil fuels is _____ ; their extraction and use come at _____ cost to the environment.
    a. renewable; low   c. renewable; high
    b. nonrenewable; low   d. nonrenewable; high

9. Nuclear energy normally pollutes _____ than fossil fuels; it poses _____ dangers than fossil fuels.
    a. less; lesser   b. more; greater   c. more; lesser   d. less; greater

10. Match each term with the most suitable description.
    _____ desertification   a. probably one of our best options
    _____ deforestation   b. soil loss, watershed damage,
    _____ green   altered rainfall patterns follow
         revolution   c. attempt to improve crop
    _____ solar-hydrogen   production on existing land
         power   d. converting large tracts of natural
    _____ CFCs   grasslands to desertlike state
                               e. invisible, odorless compounds
                                   that contribute to ozone thinning

Critical Thinking

1. Investigate where the water for your own city comes from and where it has been. You may find the answer illuminating.

2. Make a list of advantages you personally enjoy as a member of an affluent, industrialized society. List some drawbacks. Do the benefits outweigh the costs? This is not a trick question.

3. List activities you pursue each day that may be contributing to regional or global pollution. Also list ways in which you might help reduce that contribution.

4. Kristen, a recent college graduate, is finding her idealism on a collision course with reality. She strongly believes people who live in the United States are obliged to make the world a level playing field for all human beings, with equality in resources, health, education, economic security, and a pristine environment for all. Yet she also understands that the sheer size of the human population makes this impossible. Kristen recently said she cannot be party to hard choices and actions that go against her ideals and just wants nature to solve the problem for us. Comment on this true story.

5. It has been said that economic wars, more than military wars, will determine the winners and losers among nations in the near future. The economic growth of certain nations, including some in the former Soviet Union and the Far East, has had devastating impact on the environment. Elsewhere, efforts of conservation biologists to maintain standards of environmental protection adds to the cost of goods produced and puts practicing nations at serious disadvantage in this global competition. Should the United States loosen some of its existing environmental laws to help ensure its economic survival? Why or why not? Can you think of some pressures that might be imposed on indifferent nations to encourage them to change their harmful practices?

6. Populations of every species utilize resources and produce wastes. Use Figure 43.19 as a starting point for a brief essay on the accumulation and uses of energy and materials, including wastes, in a human population that is concentrated in a large city. Contrast your description with the flow of energy and the materials cycling that proceed in a natural population of some other organism.

Selected Key Terms

acid rain *43.1*
chlorofluorocarbon (CFC) *43.2*
deforestation *43.4*
desalinization *43.7*
desertification *43.6*
dry acid deposition *43.1*
fossil fuel *43.8*
fusion power *43.9*
green revolution *43.3*
industrial smog *43.1*
meltdown *43.8*
ozone thinning *43.2*
PAN (peroxyacyl nitrate) *43.1*
photochemical smog *43.1*
pollutant *43.1*
salinization *43.7*
shifting cultivation *43.4*
solar-hydrogen energy *43.9*
thermal inversion *43.1*
wastewater treatment *43.7*
water table *43.7*
wind farm *43.9*

Readings *See also www.infotrac-college.com*

Collins, M. 1990. *The Last Rain Forests*. New York: Oxford University Press.

Frosh, R. September 1995. "The Industrial Ecology of the Twenty-First Century." *Scientific American* 273(3): 178–181.

Gruber, D. 1989. "Biological Monitoring and Water Resources." *Endeavour* 13(3): 135–140.

Miller, G. T., Jr. 1999. *Environmental Science*. Seventh edition. Belmont, California: Wadsworth. This author consistently puts information from numerous sources into a current, accessible survey of the present and future state of the environment.

Plucknett, D., and D. Winkelmann. September 1995. "Technology for a Sustainable Agriculture." *Scientific American* 273(3): 182–186.

Western, D., and M. Pearl. 1989. *Conservation for the Twenty-First Century*. New York: Oxford University Press.

Wilson, E. 1988. *Biodiversity*. Washington, D.C.: National Academy of Sciences.

WWW *http://www.brookscole.com/biology*

Practice quiz questions, hypercontents, BioUpdates, and critical thinking. The Brooks/Cole Biology Resource Center provides a wealth of information fully organized and integrated by chapter.

Figure 43.19 City as ecosystem.

AN EVOLUTIONARY VIEW OF BEHAVIOR

Deck the Nest With Sprigs of Green Stuff

In 1890, bird fanciers imported more than a hundred starlings (*Sturnus vulgaris*) from Europe and released them in New York City's Central Park. In less than a century, the descendants of the introduced species had expanded their geographic range from coast to coast. They are so good at gleaning food from the agricultural fields of North America that they have outmultiplied native birds. They also have evicted great numbers of them from scarce nesting sites in the cavities of trees.

Once a male and female commandeer a tree cavity, they gather dry grass and twigs and use them to build or rebuild a nest (Figure 44.1). Theirs is no ordinary nest. They *decorate* the nest bowl with sprigs, freshly plucked.

Why do starlings do this? Do the sprigs of greenery camouflage the nest from predators? Not likely. The nests are already concealed, inside the cavities of trees. Do the sprigs function as insulative material that might help keep the forthcoming eggs warm? Actually, the still-moist, green plant parts would promote heat *loss*, not heat conservation. Well, then, do the green sprigs combat parasites? Mites, for example, parasitize birds and infest their nest cavities. In short order, even a few tiny mites can produce thousands of descendants. When present in large numbers, mites can suck enough blood from a nestling to weaken it. They interfere with the nestling's growth, development, and survival.

Larry Clark and Russell Mason decided to test the third hypothesis. As both biologists knew, starlings do not weave just any green plant material into the nests. They choosily favor the leaves of certain plants, such as wild carrot (Figure 44.1). Thus Clark and Mason built a set of experimental nests, some with freshly cut wild carrot leaves and some without. They removed the natural nests that pairs of starlings had constructed and had already started using. Fifty percent of the nesting pairs received replacement nests decorated with sprigs of wild carrot. No sprigs decorated the replacement nests for the other pairs of starlings.

Figure 44.2 shows the results. The number of mites in greenery-free nests was consistently greater than the number in nests decorated with greenery. At the end of one experiment, for example, sprig-free nests teemed with an average of 750,000 mites. The ones with sprigs contained a mere 8,000.

Shoots of wild carrot happen to contain a highly aromatic steroid compound. Almost certainly, the compound repels herbivores and thus helps the plant survive. By coincidence, the compound also prevents mites from becoming sexually mature—and thereby

Figure 44.1 A most excellent fumigator in nature—a European starling (*Sturnus vulgaris*). It combats infestations of mites by decorating previously owned nests with fresh sprigs of wild carrot (*Daucus carota*), shown at the left.

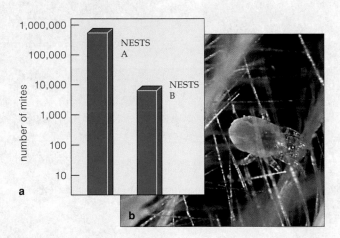

number of mites

1,000,000
100,000
10,000
1,000
100
10

NESTS A

NESTS B

a

b

Figure 44.2 (**a**) Experimental results that point to the adaptive value of starling nest-decorating behavior. Nests designated *A* were kept free of fresh sprigs of wild carrot and other plants that contain aromatic compounds. The compounds prevent baby mites from developing into (**b**) adult mites. Other nests, designated *B*, had fresh sprigs added every seven days. Head counts of mites (*Ornithonyssus sylviarum*) infesting the nests were made at the time the starling chicks left the nest, twenty-one days after this experiment was under way.

prevents mite population explosions inside the nests festooned with wild carrot.

Far from being a trivial behavior, then, "decorating" a nest with aromatic greenery has adaptive value. It fumigates the nest and thereby increases an individual's chance of producing healthy, surviving offspring.

And so starlings lead us into the fascinating world of research into behavior. As you will see, some studies focus on the adaptive value of some behavioral trait to an individual's reproductive success. Others focus on internal mechanisms that enable individuals to behave as they do.

As a starting point, recall that information encoded in genes directs the formation of tissues and organs that make up the animal body, including those of its nervous system. That system detects, processes, and integrates information about stimuli in the internal and external environments, then commands muscles and glands to make suitable responses. Other gene products called hormones contribute to the responses. Because genes specify all of the substances required for the structure and function of nervous and endocrine systems, *they are the heritable foundation for responses that animals make to stimuli.* We call the observable, coordinated responses to stimuli **animal behavior**.

KEY CONCEPTS

1. "Behavior" refers to the coordinated responses that an animal makes to stimuli. A stimulus is a form of energy that activates a specific type of sensory receptor, either a sensory neuron or a specialized cell adjacent to it.

2. Genes can affect the behavioral ability of individuals, as when instructions encoded in certain genes govern the development of the nervous and endocrine systems. These organ systems allow an animal to detect, process, and issue commands for behavioral responses to stimuli.

3. Most newly born or hatched animals have the neural wiring and motor systems necessary to perform some behaviors without having to learn them through actual experience. Such instinctive behaviors are triggered by sign stimuli—simple, well-defined environmental cues.

4. For nearly all animal species, the nervous system has a capacity to process and retain information about specific experiences in the environment, then to use the information to vary or change behavioral responses.

5. Like other traits having a genetic basis, behavior has evolved by way of natural selection, which occurs when genetically different individuals in a population have different numbers of surviving offspring. Behavioral mechanisms that enhance the ability of individuals to pass on their genes to offspring have been favored.

6. Evolved modes of communication underlie social behavior. Communication signals, which are encoded in stimuli such as species-specific odors, body coloration and patterning, postures, and movements, hold clear meaning for both the sender and the receiver of signals.

7. Living in a social group has costs and benefits, as measured by an individual's ability to pass on its genes to offspring. Not every environment favors the evolution of such groups. In some cases, a solitary life-style allows an individual to leave more descendants than belonging to a large group would allow.

8. In altruism, an individual helps others in ways that require it to sacrifice its own reproductive success. The evolution of altruism requires special circumstances in which individuals can propagate their genes indirectly, by helping their relatives reproduce successfully.

BEHAVIOR'S HERITABLE BASIS

Genes and Behavior

An animal's nervous system, recall, is wired to detect, interpret, and issue commands for response to stimuli—that is, to specific aspects of the external and internal environments. One way or another, each of the steps required to assemble and operate the system's receptors, nerve pathways, and brain involves the animal's genes. If we define animal behavior as coordinated responses to stimuli, then behavior starts with genes.

Stevan Arnold found evidence of the genetic basis of behavior by studying the feeding preferences of coastal and inland populations of a snake species in California. Garter snakes near the coast prefer banana slugs (Figure 44.3a). Snakes living inland prefer tadpoles and small fishes. Offer them a banana slug and they ignore it.

In one set of experiments, Arnold offered captive newborn garter snakes a chunk of slug as the first meal. Offspring of coastal snakes usually ate the chunk. They even flicked their tongue at cotton swabs drenched in essence of slug. (A snake "smells" by tongue-flicking, which draws chemical odors into the mouth.) But the offspring of snakes living inland ignored those swabs, and only rarely did they eat slug meat. Here was a clear difference between captive baby snakes that had no prior, direct experience with slugs. Arnold concluded that the snakes had been programmed before birth to accept or reject slugs; they certainly didn't learn their feeding preferences through taste trials. Maybe the coastal and inland populations had allelic differences for the genes that affect the formation of odor-detecting mechanisms when a garter snake embryo is developing.

To test his hypothesis, Arnold crossbred coastal and inland snakes. If the different food preferences between snake populations has a genetic basis, then

Figure 44.3 (**a**) Banana slug, food for (**b**) an adult garter snake of coastal California. (**c**) A newborn garter snake from a coastal population, tongue-flicking at a cotton swab drenched with tissue fluids from a banana slug.

hybrid offspring might make an *intermediate* response to slug chunks and odors. Results of the crosses matched the prediction. Compared with typical newborn inland snakes, many baby snakes of mixed parentage tongue-flicked more often at the slug-scented cotton swabs—but less often than newborn coastal snakes did.

Hormones and Behavior

The gene products called hormones also contribute to bird song and other behaviors. Look at the male zebra finch in Figure 44.4. Each spring he repeats a song with spendid consistency and clarity. In part, his singing is caused by seasonal differences in how much melatonin, a hormone, is secreted from his pineal gland (Section 31.8). A high level of melatonin in blood suppresses the growth and functions of songbird gonads. Its effect is reversed when photoreceptors in the pineal gland absorb sunlight energy and issue signals that lead to a decrease in melatonin secretion.

Winter has fewer hours of daylight than spring; there is not enough light to inhibit secretion of melatonin. In

Figure 44.4 Hormones and the territorial song of zebra finches. (**a**) Sound spectrogram of the male's song. The spectrogram is a visual record of each note's pitch (that is, its frequency, measured in kilohertz). (**b**) Sound spectrogram of a female zebra finch that was experimentally converted to a singer. When she was a nestling, she received extra estrogen. At adulthood, she received an implant of testosterone—and started singing. Usually, estrogen organizes the song system (certain parts of the brain) in developing songbird embryos. Testosterone activates the system in adult males.

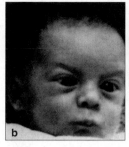

Figure 44.5 Examples of instinctive responses that human offspring make to sign stimuli. (**a**) A close-up face of an adult triggers smiling behavior in very young infants. (**b**) Somewhat older infants instinctively attempt to imitate facial expressions of adults.

spring, when daylength increases, gonads are released from hormonal suppression. They increase in size and step up their secretions of estrogen and testosterone. These sex hormones trigger gender-related differences in bird singing behavior. How? While embryos of most songbirds are developing, estrogen controls formation of a **song system**. The song system consists of several brain structures that will govern the activity of a vocal organ's muscles. Male and female birds differ in the size and structural components of their sound system.

Unlike humans, females of songbird species are XY and males are XX. Certain genes on the Y chromosome specify products that block estrogen production. Before a *male* bird hatches, his gonads secrete estrogen which, at high levels, stimulates development of a masculinized brain. (Estrogen gets converted to testosterone in his brain.) Later on, his gonads enlarge. At the start of the breeding season, his gonads secrete more testosterone. When the hormone binds to receptors on cells in the song system, it induces metabolic changes that will prepare the male to sing when he is suitably stimulated.

Instinctive Behavior Defined

With their tongue-flicking, body orientation, and strikes at prey, newborn garter snakes offer a fine example of **instinctive behavior**. This means a particular behavior is performed without having been learned by actual experience in the environment. Instead, the nervous system of a newly born or hatched animal is already wired to recognize one or two simple, well-defined cues in the environment, or **sign stimuli**, that can trigger a suitable response. The snake's response is a stereotyped motor program. When a snake recognizes certain sign stimuli, a **fixed action pattern** follows. It is a program of coordinated muscle activity that runs to completion independently of feedback from the environment.

As another example, when human infants are two or three weeks old, they tend to smile instinctively when an adult's face comes close to their own (Figure 44.5a). Infants make the same response to an overly simplified

Figure 44.6 Example of a social parasite. European cuckoos lay eggs in the nests of other species. In response to an environmental cue, this hatchling is executing a behavior that is innate. It has inherited the knowledge of what to do without having to learn it. Even before its eyes open, it responds to the spherical shape of the host's eggs and shoves them out of the nest. The foster parents keep on feeding the usurper.

stimulus—a flat, face-size mask with two dark spots where eyes would be on a human face. A mask with one "eye" won't do the trick. As Figure 44.5b suggests, older infants continue to show instinctive behavior.

And remember the cuckoo, a social parasite? Adult females lay their eggs in nests of other species. If newly hatched cuckoos eliminate the natural-born offspring, they will receive the undivided attention of their foster parents (Section 40.4). They are blind at birth. But if they contact an egg or any other round object, they will carry out a fixed action pattern. They maneuver the egg onto their back, then push it from the nest (Figure 44.6).

Clearly, animals have a genetically based capacity to respond automatically to certain environmental cues. But they also have the means to process information about specific experiences, then use the information to vary or change their responses. As you will see next, the outcome is what we call learned behavior.

Genes underlie animal behavior—coordinated responses to stimuli. Certain gene products are essential in constructing and operating the nervous system, which governs behavior. Other gene products called hormones also influence the mechanisms required for particular forms of behavior.

Animals start out life neurally wired to recognize simple but important cues in the environment. These sign stimuli trigger suitable responses, such as fixed action patterns.

WWW

Animals process and integrate information gained from experiences, then use it to vary or change responses to stimuli. We call this **learned behavior**. For example, a young toad's nervous system commands it to flip its sticky tongue instinctively at any dark object crossing its field of vision. In the toad world, such objects usually are edible insects. What if a dark object is a bumblebee that stings the tongue? After this experience, the toad will avoid "bumblebee-size black-and-yellow-banded objects that sting." **Imprinting**, another example, is a time-dependent form of learning triggered by exposure to sign stimuli, most often at a sensitive period when

Figure 44.7   No one can tell these imprinted baby geese that Konrad Lorenz is not Mother Goose. (Refer to Table 44.1.)

Table 44.1   A Few Categories of Learned Behavior

IMPRINTING   This time-dependent form of learning involves exposure to sign stimuli, most often early in development. For instance, in response to a moving object and probably to certain sounds, baby geese imprint on their mother and follow her during a short, sensitive period after hatching. They are neurally wired to learn crucial information—the identity of the individual that will protect them in the months ahead. Normally that individual is the mother or father. Imprinting occurs among many animals.

Konrad Lorenz, an early ethologist, must have presented the baby geese in Figure 44.7 with sign stimuli that made them form an attachment to him. As another outcome of imprinting, birds also direct sexual attention to members of the species they had been sexually imprinted on when young.

CLASSICAL CONDITIONING   Ivan Pavlov's classic experiments with dogs are an example of classical conditioning. Dogs will salivate just before eating. Pavlov's dogs were conditioned to salivate, even in the absence of food. They did so in response to the sound of a bell or a flash of light that was initially associated with the presentation of food. In this case, the animals learned to associate an automatic, unconditioned response with a novel stimulus that does not normally trigger the response.

OPERANT CONDITIONING   An animal learns to associate a voluntary activity with its consequences, as when a toad learns to avoid stinging insects after its first attempt to eat them.

HABITUATION   An animal learns by experience *not* to respond to a situation if the response has neither positive nor negative consequences. Thus pigeons and other birds living in cities learn not to flee from people who pose no threat to them.

SPATIAL OR LATENT LEARNING   By inspecting its environment, an animal acquires a mental map of a particular region, often by learning the position of local landmarks. For example, blue jays can store information about the position of dozens, if not hundreds, of places where they have stashed food.

INSIGHT LEARNING   An animal abruptly solves some problem without trial-and-error attempts at the solution. Chimpanzees often exhibit insight learning in captivity when they suddenly solve a novel problem that their captors devise for them. Some chimps abruptly stacked and stood on several boxes *and* used a stick to reach bananas suspended out of their reach.

the animal is young. Imprinted baby geese, shown in Figure 44.7, are a classic example. Table 44.1 lists others.

Don't fall into the trap of thinking that instinctive behavior is governed only by genes and learned behavior only by the environment. Behavior arises through gene expression *and* experiences. A bird does have a nervous system prewired to recognize sounds (acoustical cues), but what it *actually hears* affects its response to them. Male white-crown sparrows have an internal capacity to sing—but birds in different habitats use variations, or dialects, of the species song. As Peter Marler found out, male songbirds acquire the full song ten to fifty days after hatching by listening to other birds sing it. During the sensitive period, their primed learning mechanism responds to select information from the environment.

Also, male birds learn parts of a song by picking up cues when other males sing. Marler raised nestlings to maturity in soundproof chambers so they wouldn't hear adult males singing. At maturity, the song of captive males had none of the detailed structure of a typical adult's song. Marler also exposed isolated captives to recordings of white-crown sparrows *and* song sparrows. As adults, the captives sang only the white-crown song. They even mimicked the dialects. Evidently, birdsong partly requires *a genetically based capacity to learn* from specific sounds, or acoustical cues.

This isn't the whole story. In another experiment, Marler let young, hand-reared white-crowns interact with a "social tutor" of a different species, as opposed to listening to taped songs. The males tended to learn the tutor's song. Social experience as well as acoustical cues must influence their primed learning mechanisms.

In instinctive behavior, animals make complex, stereotyped responses to specific, often simple environmental cues. In learned behavior, responses may vary or change as a result of individual experiences in the environment.

Whether instinctive or learned, behavior develops through interactions between genes and environmental experiences.

WWW

THE ADAPTIVE VALUE OF BEHAVIOR

If you accept that genes directly or indirectly govern forms of behavior, then it follows that forms of behavior are subject to evolution by natural selection. **Natural selection** is a result of differences in reproductive success among individuals of a population that differ from one another in their heritable traits. Some versions of a trait may be better than others at helping individuals leave offspring. Thus alleles for those versions increase in a population, while other alleles do not. (Alleles, recall, are different molecular forms of the same gene.) In time, genetic changes that yield greater reproductive success for individuals spread through the population.

By using the theory of evolution by natural selection as our point of departure, we should be able to develop and test possible explanations of why a behavior has endured. We should be able to discern how some actions bestow reproductive benefits that offset reproductive costs (disadvantages) associated with them. If a behavior is adaptive, it must promote the *individual's* production of offspring. Keep this thought in mind while you read through the following list of terms, which you will use repeatedly in the remainder of this chapter:

1. **Reproductive success**: The number of surviving offspring that the individual produces.

2. **Adaptive behavior**: Any behavior that promotes the propagation of an individual's genes and tends to occur at increased frequency in successive generations.

3. **Social behavior**: Cooperative, interdependent relationships among individuals of the species.

4. **Selfish behavior**: Within a population, any form of behavior that increases an individual's chance to produce or protect offspring of its own, regardless of the consequences for the group as a whole.

5. **Altruistic behavior**: Within a population, a self-sacrificing behavior. The individual behaves in a way that helps others but diminishes or precludes its own chance of producing offspring.

When biologists speak of selfish or altruistic behavior, they don't mean an individual is consciously aware of what it is doing or of the behavior's reproductive goal. A lion doesn't have to know that eating zebras is good for its reproductive success. Its nervous system simply calls for hunting behavior when the lion is hungry and sees a zebra. The behavior persists in the population because the genes responsible for it are persisting also.

As a case in point, Norwegian lemmings disperse from a population when density skyrockets and food is scarce. Many die during the exodus. Are they helping the species by committing suicide to get rid of "excess" individuals? Or do some just happen to die as a result of starvation, predation, or accidental drowning while scurrying to new locations where they may reproduce?

An instructive cartoon by Gary Larson depicts a population of lemmings plunging over a cliff above water, presumably in the act of suicide. But one has an inflated inner tube about its waist! If many lemmings were suicidally altruistic, then only "selfish" lemmings would reproduce. Over time, then, genes underlying altruism would disappear from the population. More likely, individual lemmings disperse to less crowded places where each will have a better chance to survive and reproduce.

Another case: In northern forests, ravens scavenge carcasses of deer, elk, or moose, which are few and far between. When one of these large birds comes across a carcass—even in winter, when food is scarce—it often calls loudly and attracts a crowd of similarly hungry ravens. The calling behavior puzzled Bernd Heinrich, for it might seem to go against the caller's interests. Wouldn't a quiet raven eat more, increase its chances of surviving, and also leave more descendants than an "unselfish" vocalizer? If the behavior is an outcome of natural selection, then the cost in terms of lost calories and nutrients must be offset by a reproductive benefit for the individual caller. But what could be the benefit?

Maybe a lone bird picking at a carcass is vulnerable to predators that lie in wait for it. If that were so, then other ravens attracted by the calls could help keep an eye out for danger. But ravens are large, agile birds that have very few known enemies. Heinrich stealthily watched ravens feeding alone and feeding in pairs at a carcass. He never saw a predator attack any of them.

Then he realized that territory may have something to do with it. A **territory** is an area that one or more individuals defend against competitors. After hauling a cow carcass into a Maine forest, Heinrich observed that solitary or paired ravens do not always vocally advertise food. Maybe those silent ravens were adults that had already staked out a large territory, which happened to include the spot where he put the carcass.

A pair of ravens would gain nothing by attracting others to their territory. But what if that Maine forest had been *subdivided* into territories and was defended by powerful adults? A wandering young bird would be lucky to eat at all in an aggressive pair's territory. But recruiting a gang of other, nonterritorial ravens might overwhelm the resident pair's defensive behavior.

As it turned out, only wandering ravens advertise carcasses. Basically, their calling behavior is selfish and adaptive. It gives them a shot at otherwise off-limits food and therefore promotes their reproductive success.

Behavioral biologists generally find it profitable to look for evidence of natural selection of the individual's traits rather than something that benefits the species as a whole.

44.4 COMMUNICATION SIGNALS

Competing for food, defending territory, alerting others to danger, advertising sexual readiness, forming bonds with a potential mate, caring for offspring—these are the kinds of intraspecific interactions that depend on communication among animals, as in the case of those vocalizing ravens.

The Nature of Communication Signals

Intraspecific interactions depend on intricate mixes of instinctive and learned behaviors by which individuals send and respond to **communication signals**. These are information-laden cues, encoded in stimuli that hold unambiguous meaning for members of the species. The cues include specific odors, colors, patterning, sounds, postures, and movements.

Communication signals sent by one individual, the **signaler**, induce behavioral changes in others of the species. Responding individuals are **signal receivers**. A signal will evolve or persist when it tends to increase the reproductive success of both the sender and the receiver. If a signal proves disadvantageous to either party, then natural selection will favor individuals that either do not send the signal or do not respond to it.

Remember **pheromones**? They are chemical signals between individuals of the same species. As you know, chemical odors from food and danger were the most important stimuli that early animals had to deal with. As animals evolved, most started to rely on *signaling* pheromones to bring about rapid responses from a receiver. Some, such as chemical alarm signals, call for aggressive or defensive behaviors. Others, such as the bombykol molecules released by female silk moths, serve as sex attractants (Section 31.1). The

Figure 44.8 A dog soliciting play behavior with a play bow.

priming pheromones call for generalized physiological responses. One, a volatile odor in the urine of certain male mice, triggers and enhances estrus in female mice.

Acoustical signals also abound in nature, as when male songbirds sing to secure territory and attract a female. Similarly, tungara frogs issue nighttime calls to females and rival males. This frog call is a distinctive "whine," followed by an equally distinctive "chuck."

Some signals never vary. For instance, ears laid back against the head of a zebra convey hostility, but ears pointing up convey its absence. Other signals convey the intensity of the signaler's message. A zebra with laid-back ears isn't too riled up when its mouth is just a bit open. But when its mouth is gaping, watch out. The combination is a **composite signal**: a communication signal with information encoded in two cues (or more).

Signals can take on different meaning in different contexts. A lion may emit a spine-tingling roar to keep in touch with others of its pride or to threaten a rival. Also, one signal often conveys information about other signals to follow. Dogs and wolves solicit play behavior by means of a play bow, as in Figure 44.8. Because of the bow, subsequent behavioral patterns that a signal receiver construes as aggressive, sexual, or exploratory in other contexts will be construed as playful only.

Examples of Communication Displays

The play bow is a **communication display**—a pattern of behavior that is a social signal. Another common pattern, the **threat display**, is an unambiguous message that a signaler is prepared to attack a signal receiver. When a rival for a receptive female confronts him, a dominant male baboon will role his eyes upward and "yawn" to expose his formidable canines (Figure 44.9*a*).

Figure 44.9 (**a**) Exposed canines, part of a male baboon's threat display. (**b**) Part of a courtship display, with visual, acoustical, and tactile signals, that commonly precedes copulation. Here a male albatross spreads his wings, an information-laden cue for the female. See also Section 17.1.

Figure 44.10 Dances of the honeybees, examples of tactile displays. (**a**) Honeybees that visit food sources close to their hive perform a *round* dance on the honeycomb. Worker bees that maintain contact with the forager through the dance will then search for food close to the hive.

(**b**) Bees trained to visit feeding stations more than 100 meters from the hive perform a *waggle* dance. During the dance, a bee makes a straight run and waggles its abdomen. The slower the waggles during this part of the dance, the more distant the food. (**c**) As Karl von Frisch discovered, the orientation of the straight run varies, depending on the direction in which food is located. When he put a dish of honey on a direct line between the hive and the sun, foragers that located it returned to the hive and oriented their straight runs right up the honeycomb. When he put the honey at right angles to a line between the hive and the sun, the foragers made their straight runs at 90 degrees to vertical. Thus, a honeybee "recruited" into foraging for food can orient its flight *with respect to the sun and the hive*. By doing so, it will waste less time and energy during its food-gathering expedition.

Waggle dancers also vary the speed of their dance to convey even more information about the distance of a food source. When a site is 150 meters away, the dance is executed much faster, with more waggles per straight run, compared with a dance concerning a food source that is, say, 500 meters away.

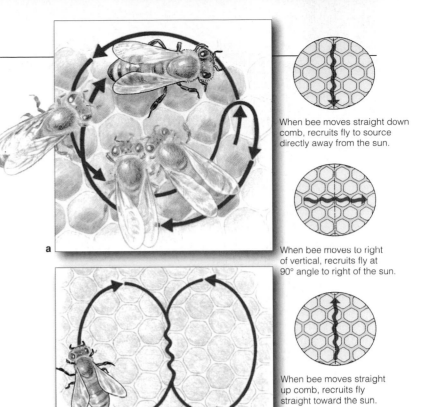

When bee moves straight down comb, recruits fly to source directly away from the sun.

When bee moves to right of vertical, recruits fly at 90° angle to right of the sun.

When bee moves straight up comb, recruits fly straight toward the sun.

Often the rival backs down. If so, the signaler benefits, for he retains access to the female without putting up a fight. The signal receiver benefits, because he avoids a serious beating, infected wounds, and possibly death.

Such displays are ritualized, with intended changes in the function of common behavior patterns. Normal movements may be exaggerated yet simplified; postures may be frozen. Body parts such as feathers, manes, and claws are often conspicuously enlarged, patterned, and colored. Ritualization is often developed to an amazing degree in **courtship displays** between potential mates. For example, food-enticing behavior is common among many birds. A male bird might emit calls while bowing low, as if to peck the ground for food. When a female comes running, he may spread his wings, as if to focus attention on the ground in front of him (Figure 44.9*b*). Here, one behavior (searching for food) has changed to attract a female, the intent being copulation. Another example: Courtship displays of fireflies and some other nocturnal animals incorporate bioluminescent flashes (Chapter 5). A male firefly uses a light-generating organ to emit a bright, flashing signal. A few seconds later, a receptive female of his species might answer with a flash. Both might flash back and forth until they meet.

Similarly, in **tactile displays**, a signaler touches the receiver in ritualized ways. After discovering a source of pollen or nectar, a foraging honeybee returns to its colony, a hive, and performs a complex dance. It moves about in a circle, jostling in the dark with a crowd of workers. Other bees may follow and maintain physical contact with the dancer. Honeybees that do this acquire information about the general location, distance, and direction of pollen or nectar (Figure 44.10).

Illegitimate Signalers and Receivers

Sometimes the wrong parties intercept communication signals. Termites will respond with defensive behavior if they catch a whiff of a scent from an invading ant. That scent is meant to identify the ant as a member of an ant colony and elicits cooperative behavior from other ants. Thus the scent has evolved functions, but announcing an invasion of a termite colony is not one of them. A termite also might detect the scent and kill the ant. In that case, it is an **illegitimate receiver** of a signal *meant for individuals of a different species*.

We even see **illegitimate signalers**. Certain assassin bugs hook dead and drained bodies of termite prey on their dorsal surface to acquire the odor of their victims. By deceptively signaling that they "belong" to a termite colony, they hunt termite victims more easily. Another illegitimate signaler is the female of certain predatory fireflies. If one observes a flash from a male firefly, she flashes back. If she can lure him into attack range, he becomes her meal—an evolutionary cost of having an otherwise useful response to a come-hither signal.

A communication signal between individuals of the same species is an action that has a net beneficial effect on both the signaler and the receiver. Natural selection tends to favor communication signals that promote reproductive success.

WWW

For reasons that we need not explore here, most people find the mating and parenting behaviors of different animals fascinating. How useful is selection theory in helping us interpret such behavior? Let's take a look.

Sexual Selection Theory and Mating Behavior

Competition among members of one sex for access to mates is common. So is choosiness in selecting a mate. Recall, from Section 16.9, that such activities are forms of a microevolutionary process called **sexual selection**. Sexual selection favors traits that give the individual a competitive edge in attracting and holding on to mates.

Typically, sexual selection occurs because the male animals produce great numbers of tiny sperm and the females produce considerably fewer but larger eggs. Reproductive success for a male generally depends on how many eggs he fertilizes. For the female, success depends largely on how many eggs she produces or how many offspring she can care for. In most cases, *the key factor influencing her sexual preference is the quality of a mate, not the quantity of partners.* Hangingflies, sage grouse, and bison (Figures 44.11 through 44.13) provide wonderful examples of how such females dictate the rules of male competition. They also illustrate how the males employ tactics that might help them fertilize as many eggs as possible.

Female hangingflies (*Harpobittacus apicalis*) select males that offer them superior material goods. And so you might observe males capturing and killing a moth or some other insect. After males do this, they release a sex pheromone that might attract females to the "nuptial

Figure 44.11 A male hangingfly dangling a moth as a nuptial present for a future mate. The females of certain hangingfly species choose sexual partners on the basis of the size of prey that males offer to them.

Figure 44.12 Ornamental feathers of a male sage grouse. These visual signals function in a courtship display—a dance performed at a lek. Males vigorously defend a small patch of this communal mating area. Females (the smaller brown birds) observe the prancing males before choosing the one they will mate with.

gift" (Figure 44.11). A female chooses males that offer the larger, calorie-rich offerings. She permits a male to mate, but only *after* she's been eating a gift for five minutes or so. Then she will accept sperm, hold them in a special storage sac, but only as long as the food holds out. Before twenty minutes are up, she can break off the mating at any point. If she does so, she might well mate with another male and accept his sperm. And doing so will dilute the reproductive success of her first partner.

In western regions of North and South Dakota, female sage grouse (*Centrocercus urophasianus*) are dispersed among stands of sagebrush. You never find them far from the protective cover of sage during spring courtship and summer nesting. Male sage grouse make no attempt to maintain a large territory. In the breeding season they congregate in a **lek**, a type of communal display ground. There, every male stakes out a few square meters as his territory. Females are attracted to the lek—not to feed or nest, but rather to observe males. With tail feathers erect and big neck pouches puffed, each male emits booming calls and stamps about, like a wind-up toy, on his display ground (Figure 44.12). Female sage grouse tend to select and mate with one male only. Then they go off to nest alone in sagebrush, unassisted by a sexual partner. Many females often choose the same male, so most of the males never mate.

As another example, females of some species cluster in defendable groups when they are sexually receptive. Where you find such a group, most likely you will observe male competition for access to the clusters. Competition for ready-made harems favors extremely combative male lions, sheep, elk, elephant seals, and bison, to name a few animals (Figures 1.7g and 44.13).

Costs and Benefits of Parenting

What happens after mating, when the offspring arrive? Until the offspring have developed enough to survive on their own, the parents of some species will care for them. For example, adult Caspian terns (Figure 44.14) incubate the eggs, shelter the nestlings and feed them, then accompany and protect them after they start to fly. Such parental behavior comes at a reproductive cost. It drains time and energy that might otherwise be given to improving their own chances of living to reproduce

Figure 44.13 Sexual competition between male bison, which are fighting for access to a cluster of females.

Figure 44.14 Male and female Caspian terns, protecting their chicks. Parental care has costs as well as benefits.

another time. Yet for many species, parenting improves the likelihood that the current generation of offspring will survive. The benefit of devoting time and energy to immediate reproductive success outweighs the cost of reduced reproductive success at some later time.

Selection theory helps explain some aspects of mating behavior, for sexual selection favors behavioral traits that give the individual a competitive edge in reproductive success. Selection theory applies also to parental behavior that contributes to reproductive success.

WWW

BENEFITS OF LIVING IN SOCIAL GROUPS

Survey the animal kingdom and you observe a range of social groupings. For some species, individuals spend most of their lives alone or in small family groups. For some other species, individuals live in huge groups of thousands of related individuals. Termite and honeybee societies are like this. Individuals of still other species live in social units composed primarily of nonrelatives. Populations of the human species are like this.

Given such differences, how might we gain insight into the basis for any social group? To find answers, evolutionary biologists commonly take a cost-benefit approach. That is, they attempt to identify the costs and benefits of sociality in terms of *reproductive success of the individual*, as measured by its contribution to the gene pool of the next generation.

Cooperative Predator Avoidance

First, look at how a group of animals that are acting cooperatively against a predator can reduce the net risk to any one individual. In a flock or herd vulnerable to predation, simply having more pairs of eyes to scan the surrounding area helps individuals detect predators sooner. Individuals of the group also may join forces to make a counterattack or perform a defensive behavior, as the musk-oxen in Figure 44.15 are doing.

The biologist Birgitta Sillén-Tullberg found tangible evidence of such benefits of sociality. She was studying Australian sawfly caterpillars, which live together in clumps on tree branches. Figure 44.16 shows one of the clumps. When something disturbs the caterpillars, they collectively rear up from the branch and writhe about, all the while regurgitating partially digested food. The food of choice is eucalyptus leaves. These leaves are heavily impregnated with chemical compounds that are toxic to most kinds of predatory animals—including the songbirds that prey on the caterpillars.

Sillén-Tullberg hypothesized that individual sawfly caterpillars benefit from the coordinated act of repelling bird predators. She used her hypothesis to predict that the birds are more likely to eat a solitary individual, not a cluster of individuals.

To test her prediction, Sillén-Tullberg offered young, hand-reared Great Tits (*Parus major*) a chance to feed on sawfly caterpillars, which she offered either one by one or in a group of twenty per offering. She did this for a standard number of presentations. Ten birds that were offered one individual at a time consumed an average of 5.6 caterpillars. But ten birds that were each offered a clump of caterpillars only ate an average of 4.1. As she had predicted for this experiment, the individuals were somewhat safer in a group than on their own.

The Selfish Herd

Simply by their physical position within a group, some individuals form a living shield against predation on others in the group. They all belong to a **selfish herd**, a

Figure 44.16 Social defensive behavior of Australian sawfly caterpillars, which clump together in tree branches. These larvae collectively regurgitate and hold fluid in their mouth (the yellow blobs), where it serves to repel predators that try to grab them. After danger passes, they will swallow the fluid, which is toxic to most animals.

Figure 44.15 Social defensive behavior of musk-oxen (*Ovibos moschatus*). In the presence of a perceived threat (usually wolf predators), adults form a circle around their young. They face outward, and the "ring of horns" successfully deters the wolves.

By contrast, some individuals may *help* others survive and reproduce at their own expense. Their helpful behavior may be a cost of belonging to the social group. And it may be that they do not entirely give up their chance at reproductive success. We find some evidence of both possibilities in **dominance hierarchies**, a type of social group in which some of the individuals have adopted subordinate status to others.

In baboon troops, for example, individuals help one another, but reproductive opportunity is unequal. Upon receiving a threat signal from a dominant member of the troop, subordinate members give up safe sleeping places, choice bits of food, and receptive females. Each member recognizes its social status with respect to all the others (Figure 44.17). Jane Goodall has studied the dominance hierarchies that also exist among the chimpanzees of Gombe. You read about their social behaviors in Chapter 31.

Figure 44.17 Appeasement behavior between baboons. Notice the assured position of the dominant animal (*left*) and the abject stare and groveling posture of the subordinate one, who is making little conciliatory smacking noises with its lips.

relatively simple society held together by reproductive self-interest, although not consciously so. Investigators have tested the selfish-herd hypothesis for male bluegill sunfishes, which build adjacent nests on the bottom of lakes. Males use their fins to hollow out a depression in lake mud, where females deposit eggs.

If a colony of bluegill males is a selfish herd, then we can predict competition for the "safe" sites—at the center of the colony. Compared to the periphery, eggs laid in nests at the center are less likely to be attacked by snails and largemouth bass. The competition does indeed exist. The largest, most powerful males tend to claim central locations. Other, smaller males assemble around them and bear the brunt of predatory attacks. Even so, they are better off in the group than on their own, fending off a bass singlehandedly, so to speak.

Dominance Hierarchies

As you have seen, individuals of a selfish herd make no personal sacrifice for the others; the personal benefits of living with others simply seem to outweigh the costs.

Why do subordinate adults remain in a group where they have such low social status? As is the case for the bluegill sunfish, they probably reap long-term benefits that offset their low status. Consider that it simply may not be possible to survive alone, outside the protection afforded by the group. A solitary baboon out in the open surely quickens the pulse of the first leopard that spies it (examine Figure 44.25 at the chapter's end). Besides, challenging a stronger member may result in injuries that could shorten a life. And self-sacrificing behavior may give a subordinate individual a chance to reproduce, if it lives long enough and if predation or weakness in old age removes dominant peers. Some subordinate wolves and baboons do eventually move up the social ladder if dominant members slip down a rung or fall off. In short, accepting subordinate status can pay off in the long term for the patient individual.

We may evaluate the costs and benefits of social life in terms of reproductive success, as measured by the genes that each individual contributes to the next generation.

WWW

COSTS OF LIVING IN SOCIAL GROUPS

So far, we have used individual selection theory to help illuminate the advantages of communication signals, territoriality and dominance hierarchies, mating tactics, parenting, and other aspects of living in social groups. Now the question becomes this: If social behavior is so advantageous, *then why are there so few social species among most groups of animals?* One plausible answer is that, in certain environments, costs to the individual outweigh the benefits of life in a social group.

Most obviously, when more individuals of a species live together in their habitat, the competition for food increases. For example, royal penguins, herring gulls, cliff swallows, and prairie dogs are among the animals

Another cost is the risk of being killed or exploited by others in the group. For example, when presented with the opportunity to do so, breeding pairs of herring gulls will cannibalize the eggs or young chicks of their neighbors in an instant. Or consider lions. These long-lived cats compete for permanent hunting territories, even though they can live for an extended time between kills. When three or more lions hunt cooperatively, they are better at capturing large prey. Yet George Schaller's data revealed that individuals in prides of three or more actually eat *less* well than one or a pair of lions. Also, males intent on taking over a pride show infanticidal behavior; they kill the cubs. For lionesses, then, group

Figure 44.18 Colony of royal penguins on Macquarie Island, between New Zealand and Antarctica.

that live in huge colonies (Figure 44.18). Great numbers of individuals must compete for a fair share of the same pie, so to speak.

In addition, living in social groups can encourage the spread of contagious diseases and parasites. Under crowded living conditions, the individual as well as its offspring is more likely to be weakened by pathogens and parasites that are readily transmitted from host to host in crowded groups. Similarly, plagues spread like wildfire through densely crowded human populations. As described in Section 39.7, this is especially the case for settlements and cities with recurring infestations of rats and fleas, and with very poor or nonexistent sewage treatment and medical care.

living costs, in terms of food intake and reproductive success. They still stick together. Maybe doing so helps them defend their territories against smaller groups of rivals. Besides, aggressive males almost always kill the cubs of a single lioness, but occasionally a group of two or more lionesses can save some of the cubs.

Individuals that belong to a social group pay costs in terms of increased competition for food, living quarters, mates, and other limited resources. And they are more vulnerable to contagious diseases and parasitic infections.

Individuals also may risk being exploited or having their offspring killed by others in the social group.

WWW

EVOLUTION OF ALTRUISM

A subordinate animal that gives way to a dominant one acts in its own interest. What about truly altruistic animals? We find them among many vertebrate groups. Consider the wolf pack. Although the females of most mammalian groups make the largest investment in raising offspring, male wolves also make contributions by providing their pups with prey, defending a feeding territory, and driving off infanticidal intruders. Unlike herbivores, which forage as individuals, wolves are carnivores that share captured prey with others of the social group. Therefore, a female's reproductive success increases when she monopolizes benefits that males offer. Usually, a wolf pack contains only one dominant breeding female and male (Figure 44.19). The others are nonbreeding sisters, aunts, brothers, and uncles. They altruistically hunt and bring back food to members that stay inside the den and guard the pups. Nonbreeding females may ovulate and court males, but they fail to reproduce when a dominant female is present.

Altruistic behavior is extreme in some insect societies, as demonstrated when a worker bee plunges its stinger into an invader of the hive. With this act of defending the colony, the bee is commiting suicide. Section 44.9 focuses on termite and honeybee altruists.

Yet if the altruistic individuals of a social group do not contribute their genes to the next generation, then how are genes for their altruistic behavior perpetuated over evolutionary time? According to William Hamilton's **theory of indirect selection**, genes associated with caring for one's *relatives*, not one's direct descendants, may be favored in certain situations. Remember, when a sexually reproducing parent cares for offspring, it is not helping exact genetic copies of itself. Commonly, such parents are diploid; they have pairs of genes. Each gamete they produce, hence each offspring, has one-half of its genes. If other individuals of the social group have the same ancestors, they share the same genes with parents. Two siblings (brothers, sisters) are as genetically similar as a parent is to one of its offspring. Nephews and nieces have inherited about one-fourth of their uncle's genes.

Therefore, we can think of selective altruism as an extension of parenting. For example, suppose an uncle helps his niece survive long enough to reproduce. He has made an *indirect* genetic contribution to the next generation, as measured in terms of the genes that he and his niece share. Altruism costs him; he might lose his own opportunities to reproduce. But if the cost is less than the benefit, altruistic actions will propagate the uncle's genes and favor the spread of his kind of altruism in the species. If an uncle saves two nieces, this is equivalent to saving his own daughter.

Similarly, nonbreeding workers in insect societies indirectly promote their "self-sacrifice" genes through altruistic behavior directed toward relatives. Colonies of honeybees, ants, and termites actually are extended families. The family's worker force labors on behalf of their siblings, some of which are the future kings and queens. Thus, when a guard bee drives her stinger into a raccoon, she inevitably dies—but her siblings in the hive might perpetuate some of her genes. Sterility and extreme self-sacrifice are rare among social groups of

Figure 44.19 Dominant male of a wolf pack, surrounded by subordinate members of the pack engaged in a ritual display involving lowered heads and drooped tails.

vertebrates. The known exceptions include naked mole-rats. Section 44.9 describes one study of the genetic relationships in a clan of these nearly hairless mammals.

Altruistic behavior occurs among many vertebrate groups such as wolf packs. It reaches its most extreme among some insect societies, such as honeybee and termite colonies.

Altruistic behavior may persist when individuals pass on genes indirectly, by helping relatives survive and reproduce.

By the theory of indirect selection, genes associated with altruistic behavior that is directed toward relatives may spread through a population in certain situations.

44.9 WHY SACRIFICE YOURSELF?

CONSIDER THE TERMITE Picture yourself in a eucalyptus forest in Queensland, Australia. You see a narrow, brittle tube running along the trunk of a dead tree and decide to chip a few fragments from it. Sunlight pours in through the breached tube wall. Some small, nearly white insects bang their heads against the wall, then scurry away.

Head banging, an acoustical signal, makes vibrations that alert golden-brown, soldier termites. Soldiers run to the breach and make a defensive stand (Figure 44.20). Each one has a swollen, eyeless, tapering head with a long, pointed "nose." Disturb a soldier termite and it shoots thin jets of silvery goo out of its nose! The silvery strands release volatile compounds that attract still more soldier termites. Together the soldiers battle the danger, which more typically is an invasion by ants.

Insects in that ruptured tunnel belong to a complex termite colony. Whereas the soldier termites protect the colony, the pale ones are workers, which build tunnels that lead to places where they can safely gather fibers of wood. They carry fibers to an underground nest, where termites live and cultivate an edible fungus. Fungal hyphae grow into the wood fibers and absorb nutrients from them. The termites eat some portions of the fungus. They chew the wood into small particles that they can swallow. Unlike most termite species, they don't produce enzymes that can digest cellulose in wood. Symbiotic microorganisms in their gut do this, and the symbionts and termites both absorb the released nutrients.

Soldier termites and worker termites are sterile; neither can reproduce. They engage in self-sacrificing behavior that contributes to the survival of others. A single queen and one or more kings serve as "parents" for the entire society.

COOPERATIVE HONEYBEES The only fertile female in a honeybee colony, or hive, is the queen bee. Figure 44.21*a* shows one, with her court of sterile worker daughters. The daughters feed the queen and relay her pheromones throughout the hive, and these influence the activities of all members. The queen is much larger than the workers, partly because of the relatively enormous egg-producing ovaries in her abdomen. Unlike the ovaries of her sterile daughters, hers are fully developed.

Only at certain times of year do the stingless drones develop and mature (Figure 44.21*b*). Drones do not work for the colony. Instead, they leave the colony and attempt to mate with queens of other hives. If successful, they may perpetuate part of their family's genetic lineage.

Figure 44.20 Soldier termites defending their social colony by guarding a break in a foraging tunnel, which worker termites have constructed on a dead tree trunk. The members of the soldier caste shoot out thin strands of glue from their pointed "nose." The sticky glue can entangle intruders, such as invading ants.

egg   nearly mature pupa

Figure 44.21 Life in a honeybee colony. (**a**) A court of sterile worker daughters surrounds a queen bee, the only female of the colony that reproduces. (**b**) A stingless drone. (**c**) One of 30,000 to 50,000 worker bees in the hive. (**d**) Worker bees store honey or pollen in honeycomb, which also houses offspring. Young workers feed the larvae. (**e**) Worker bees transferring food to one another. (**f**) Worker females at the hive entrance serve as guards.

Figure 44.22 Naked mole-rats (*Heterocephalus glaber*).

The hive has 30,000 and 50,000 workers (Figure 44.21c). They feed bee larvae, clean and maintain the hive, and construct honeycomb from wax secretions. Workers store honey or pollen in honeycomb, which houses bee generations from eggs, through a series of larval stages, to pupae, to the adults (Figure 44.21d).

Adult workers live for about six weeks in spring and summer. Workers also engage in scent-fanning. Fanned air passes over a bee's exposed scent gland. As pheromones waft away from the gland, they help bees orient to the hive's entrance on their way to or from foraging expeditions. When the foragers return to the hive after locating a rich source of nectar or pollen, they start a dance—a tactile display that recruits more workers into taking off for the source (Figure 44.10).

In other cooperative behaviors, workers transfer food to one another, and worker females at the hive entrance act as guards (Figure 44.21 e,f). Guard bees readily sacrifice themselves to repel intruders.

After these glimpses into the lives of self-sacrificing insects, see if you can answer a few basic questions. First, *by what means do the members of a termite or honeybee colony cooperate and benefit the group and themselves?* For instance, identify some of their communication signals and the manner in which individuals respond to them. Second, *what is the adaptive value of some of their social behaviors?* How, for instance, might sterility increase an individual's chances for genetic success?

ABOUT THOSE NAKED MOLE-RATS The highly social mammals called naked mole-rats (*Heterocephalus glaber*) look rather like bucktoothed sausages with wrinkled, pink skin and a few sprouts of hairs (Figure 44.22). They live in arid regions of eastern Africa, in cooperatively excavated burrows. They always live in clans (small social units) of anywhere from 25 to 300 individuals.

A single reproducing female dominates each clan, and she mates with one to three males. As is the case for honeybees and termites—and not for any other known vertebrate—other members of the clan are nonbreeding. They live out their lives protecting and caring for the "queen" and "king" (or kings) and their offspring.

Nonreproducing "diggers" busily excavate extensive subterranean tunnels and special chambers, which serve as living rooms or waste-disposal centers. They locate large tubers growing underground. These they chop up into edible bits, which they then deliver to the queen, her retinue of males, and her offspring.

Also, digger mole-rats deliver food to certain other helpers that loaf about, shoulder to shoulder and belly to back, with the reproductive royals. Usually the "loafers" are larger than the diggers. And they aren't really loafers. They spring to action when a snake or some other enemy threatens the clan. Collectively, at great personal risk, they chase away or attack and kill the predator.

Using indirect selection theory as a guide, we can formulate a hypothesis to account for the altruism of the mole-rats that never do breed.

If their helpful behavior is genetically advantageous to some individuals in a naked mole-rat clan (*the hypothesis*), then it follows that helpers will be related to reproductive members of the clan that benefit from their altruistic behavior (*the prediction*).

To *test* the prediction, we require information about genetic relationships among the members of the naked mole-rat colony. One way to obtain it would be to know which animals were offspring of which others. However, to construct a family tree of an entire colony would be most laborious and difficult, because there are several breeding males and queens that die and are replaced.

H. Kern Reeve and his colleagues decided on a more practical approach. They relied upon *DNA fingerprinting*, a method of establishing degrees of genetic relatedness among individuals. As described in Section 15.3, this laboratory method starts with the formation of restriction fragments of DNA molecules. Then technicians construct a visual record of different sets of fragments of DNA from different individuals. Essentially, a set of identical twins would have identical DNA fingerprints (their DNA would have the same base sequence). Individuals having the same father and mother would have similar DNA; and so they would have similar although not identical DNA fingerprints. On average, DNA fingerprints of genetically unrelated individuals should differ much more than the DNA fingerprints of siblings or other relatives.

When Reeve constructed DNA fingerprints for naked mole-rats, he discovered that all individuals from a clan are *very* close relatives, and are very different genetically from members of other clans. The findings suggest that each naked mole-rat clan is highly inbred, a result of generations of brother-sister, mother-son, and father-daughter matings. As you know from Section 16.11, inbreeding among individuals of a population results in extremely reduced genetic variability.

Therefore, a self-sacrificing naked mole-rat is helping to perpetuate a high proportion of the alleles it carries. As it turns out, the genotypes of helpers and the helped might be as much as 90 percent identical!

WWW

If we can analyze the evolutionary basis of the behavior of termites, naked mole-rats, and other animals, would it not be rewarding to analyze such a basis of human behavior also? Many people resist the idea. It seems they believe that attempts to identify the adaptive value of a particular human trait are attempts to define its moral or social advantage. Clearly, however, there is a difference between trying to explain something in terms of its evolutionary history and attempting to justify it. "Adaptive" does not mean "morally right." It means valuable with respect to the transmission of an individual's genes.

Consider the case of altruistic behavior that we call **adoption**—the acceptance of offspring of other individuals as one's own. If we wish to gain understanding of adoptive behavior, we might use evolutionary theory to formulate a hypothesis about it. But this does not mean we are passing judgment on whether adoption is moral or even socially desirable. These are two separate issues—about which biologists have no more to say than anybody else.

Figure 44.23 Two adult emperor penguins competing to adopt an orphan, which penguins accept as substitute offspring.

Start with a premise that all adaptations have costs and benefits. One cost is that certain adaptive behaviors may be *redirected* under rare or unusual circumstances. In many species, adults that lost offspring will adopt a substitute. Some cardinals have fed goldfish. A whale tried to lift a log out of water as if it were a distressed infant that needed help in reaching the water's surface. Emperor penguins fight to adopt orphans (Figure 44.23).

As John Alcock suggests, such examples constitute a test of a hypothesis about adoption by humans: Human adoptive behavior occurs "by mistake" when otherwise adoptive parental behavior is directed toward unrelated offspring. Suppose it arises through a frustrated desire to bear children (the hypothesis). If so, then we can predict that couples who have lost an only child or who are sterile should be especially prone to adopt strangers.

Now consider this. For most of their evolutionary history, humans have lived in small groups and, later, small villages. They probably had few opportunities or showed little inclination to adopt strangers. They were likely to accept the child of a deceased relative. But how do adoptive parents gain genetic representation in the next generation? According to theories of natural selection and indirect selection, individuals act in ways that promote genetic self-interest. So we might predict that parents with dependent, care-requiring children of their own are much less likely to adopt substitutes, compared with adults who have no children or who already raised children and are living by themselves. We could test the prediction by piecing together a large enough sampling of information, ideally from diverse existing human cultures, on possible connections between adoption and a childless condition.

Indirect selection also favors adults who focus their parenting behavior on relatives and thereby perpetuate their shared genes in an indirect way. We might even predict that such adults are more likely to be related to the adopted child than we would expect by chance alone. A researcher, Joan Silk, tested the prediction in traditional societies. Her results showed that people will adopt related children far more often than nonrelated children. Particularly in the large, industrialized societies with welfare agencies and other means of adoption assistance, individuals become parents of nonrelated children. Such societies create evolutionary environments. Parenting mechanisms evolved in the past, and it may be that their redirection toward nonrelatives says more about our evolutionary history than it does about the transmission of one's genes.

The point is this: Evolutionary hypotheses about the adaptive value of behavior lend themselves to testing. And through such testing, we can gain understanding about the evolution of human behavior.

It is possible to test evolutionary hypotheses regarding the adaptive value of human behaviors.

There may be greater acceptance of such testing when more people come to understand that adaptive behavior and socially desirable behavior are separate issues.

In biology, "adaptive" means only that some specified trait has proved beneficial in the transmission of the genes of an individual that are responsible for that trait.

SUMMARY

1. Animal behavior (coordinated responses to stimuli) originates with genes that directly or indirectly specify products required for the development and operation of the nervous, endocrine, and skeletal-muscular systems.

2. A behavior performed without having been learned by actual experience is instinctive, a prewired response to one or two simple, well-defined environmental cues (sign stimuli) that trigger a suitable response, such as a fixed action pattern. Individual experiences can lead to variations or changes in responses (learned behavior). Both are outcomes of genetic and environmental inputs.

3. Behavior with a genetic basis is subject to evolution by natural selection. It evolved as a result of individual differences in reproductive success in past generations, when the reproductive benefits of a particular behavior exceeded its reproductive costs (or disadvantages).

4. Members of the same species often create obstacles to one another's reproductive success, as by selective mate choice and by competition for access to mates.

5. Social groups require cooperative interdependency among individuals of a species. Sociality is promoted by communication signals, cues sent by one individual (a signaler) that can change the behavior of another of the same species (a signal receiver). A signal benefits both the signaler and the receiver.

6. Chemical, visual, acoustical, and tactile signals are components of communication displays.

　a. Pheromones serve in chemical communication. Signaling pheromones, such as sex attractants and alarm signals, can induce immediate change in the behavior of a receiver. Priming pheromones elicit a generalized physiological response by the receiver.

　b. Visual signals (observable actions or cues) are key components of courtship displays and threat displays.

　c. Acoustical signals (sounds with precise, species-specific information) include the mating calls of frogs.

　d. Tactile signals are specific forms of physical contact between a signaler and a receiver.

7. Costs and benefits of social life are reflected in the individual's reproductive success, as measured by its genetic contribution to the next generation.

8. Social groups have costs, such as competition for limited resources and greater exposure to contagious diseases and parasites. Benefits outweigh the costs in some cases, as when predation pressure is severe.

9. In self-sacrificing (altruistic) behavior, individuals give up chances to reproduce while helping others of their social group. In most social groups, individuals do not sacrifice reproductive chances to help others.

　a. Dominance hierarchies are social rankings in which some of the individuals give way to others of

their group. Often, the dominant individuals force their subordinates to relinquish food or some other resource.

　b. When subordinates give in to dominant members of their group, they may receive compensatory benefits of group living, such as safety from predators. In some species, subordinates may reproduce if they live long enough or if dominant ones slip in the hierarchy or die.

10. By one theory of indirect selection, genes associated with caring for relatives (not one's direct descendants but individuals that bear a fraction of the same genes) are favored in some cases. Indirect selection can lead to extreme altruism, as among workers of some species of social insects and naked mole-rats. The workers usually do not reproduce. Altruism helps reproducing relatives survive, so altruistic individuals pass on "by proxy" the genes underlying the development of this behavior.

Review Questions

1. Explain how genes and their products, including hormones, influence the mechanisms required for forms of behavior. 44.1

2. Define these terms: instinctive behavior, sign stimulus, fixed action pattern, and learned behavior. 44.1, 44.2

3. Contrast altruistic behavior with selfish behavior. Why does either form of behavior persist in a population? 44.3

4. Describe some characteristics of communication signals. Then give an example of a communication display. 44.4

5. Describe some feeding behavior or mating behavior in terms of natural (individual) selection. 44.3, 44.5

6. List some of the benefits and costs of sociality. 44.6, 44.7

Self-Quiz (Answers in Appendix III)

1. "Starlings festoon their nests with sprigs of wild carrot and so minimize nest mites." This statement is _____ .
　a. an untested hypothesis　　c. a test of a hypothesis
　b. a prediction　　　　　　　d. a proximate conclusion

2. Genes affect the behavior of individuals by _____ .
　a. influencing the development of nervous systems
　b. affecting the kinds of hormones in individuals
　c. governing the development of muscles and skeletons
　d all of the above

3. Many kinds of female mammals live in herds. Which of these mating systems might you see in such herds?
　a. lek mating system
　b. males competing for clusters of receptive females
　c. territorial defense of feeding sites by males
　d. males capturing food to lure females

4. A statement that lemmings dispersing from an overcrowded population bring it in line with available resources _____ .
　a. is consistent with Darwinian evolutionary theory
　b. is based on a theory of evolution by group selection
　c. is supported by the finding that most animals behave altruistically during their lives

5. Altruism is helpful behavior that _____ .
　a. cannot evolve
　b. can spread through a species only if it raises the altruist's reproductive success
　c. reduces an individual's reproductive success
　d. always helps spread the altruist's genes

Figure 44.24 Caterpillar under siege.

Figure 44.25 Moment of truth for a baboon.

6. The genetic similarity between an uncle and his nephew is _____ .
   a. the same as between a parent and his offspring
   b. greater than between two full siblings
   c. dependent on how many other nephews the uncle has
   d. less than that between a mother and her daughter

7. Match the terms with their most suitable description.
   _____ fixed action pattern
   _____ altruism
   _____ basis of instinctive *and* learned behavior
   _____ imprinting

   a. time-dependent form of learning requiring exposure to key stimulus
   b. genes and environmental experience
   c. stereotyped motor program that runs to completion independently of feedback from environment
   d. assisting another individual at one's own expense

Critical Thinking

1. A large caterpillar is crawling along a tree branch in a tropical forest. You poke it, and it partly responds by letting go of the branch and puffing up the anterior end of its body (Figure 44.24). Propose a mechanism that may underlie this defensive behavior. Also propose how the behavior has adaptive value. How would you test your two hypotheses?

2. A leopard chases a baboon into a clearing (Figure 44.25). Having no place to hide, the baboon swiftly turns around and faces the formidable predator. It displays its large canines, rolls its eyes upward, and makes its body appear more massive by way of the pilomotor response (its hairs "stand up"). As you know from Section 44.4, this is a ritualized communication display, laden with social meaning for other baboons. However, is the display also an evolved communication signal used to convey information to leopards and other large predators? What would you need to know to determine whether this is the case?

Figure 44.26 Behaviorally confused rooster.

3. You observe mallard ducks in a small pond, then you notice a rooster wading out to the ducks with amorous intent (Figure 44.26). What probably happened to the rooster early in life?

4. Develop an adaptive value hypothesis for the observation that male lions kill offspring of females that they acquire after chasing away previous pride holders that had mated with those females. How would you test your infanticide hypothesis?

5. What are the likely evolutionary costs and benefits to a male hangingfly of offering a nuptial gift to its potential mate? How might a female's choosiness depend on those costs and benefits?

Selected Key Terms

adaptive behavior *44.3*	lek *44.5*
adoption *44.10*	natural selection *44.3*
altruistic behavior *44.3*	pheromone *44.4*
animal behavior *CI*	reproductive success *44.3*
communication display *44.4*	selfish behavior *44.3*
communication signal *44.4*	selfish herd *44.6*
composite signal *44.4*	sexual selection *44.5*
courtship display *44.4*	sign stimulus *44.1*
dominance hierarchy *44.6*	signal receiver *44.4*
fixed action pattern *44.1*	signaler *44.4*
illegitimate receiver *44.4*	social behavior *44.3*
illegitimate signaler *44.4*	song system *44.1*
imprinting *44.2*	tactile display *44.4*
indirect selection *44.8*	territory *44.3*
instinctive behavior *44.1*	threat display *44.4*
learned behavior *44.2*	

Readings See also www.infotrac-college.com

Alcock, J. 1998. *Animal Behavior: An Evolutionary Approach*. Sixth edition. Sunderland, Massachusetts: Sinauer.

Dawkins, R. 1989. *The Selfish Gene*. New York: Oxford University Press.

Frisch, K. von. 1961. *The Dancing Bees*. New York: Harcourt Brace Jovanovich. A classic on the behavior of honeybees.

Wilson, E. O. 1975. *Sociobiology: The New Synthesis*. Cambridge, Massachusetts: Harvard University Press.

WWW   *http://www.brookscole.com/biology*

Practice quiz questions, hypercontents, BioUpdates, and critical thinking. The Brooks/Cole Biology Resource Center provides a wealth of information fully organized and integrated by chapter.

APPENDIX I. CLASSIFICATION SCHEME

The classification scheme that follows is a composite of several that microbiologists, botanists, and zoologists use. The major groupings are agreed upon, more or less. There is not always agreement, however, on what to name a particular grouping or where it may fit within the overall hierarchy. There are several reasons for the lack of total consensus.

First, the fossil record varies in its quality and in its completeness. Therefore, the phylogenetic relationship of one group to others is sometimes open to interpretation. Comparative studies at the molecular level are firming up the picture, but this work is still under way.

Second, ever since the time of Linnaeus, classification schemes have been based on the perceived morphological similarities and differences among organisms. Although some original interpretations are now open to question, we are so used to thinking about organisms in certain ways that reclassification often proceeds slowly.

Traditionally, for example, birds and reptiles have been considered to be separate classes (Reptilia and Aves). And yet there are now compelling arguments for grouping the lizards and snakes together in one class, and the crocodilians, dinosaurs, and birds in a different class.

Third, researchers in microbiology, mycology, botany, zoology, and the other fields of biological inquiry have inherited a wealth of literature, based on classification schemes that were developed over time in each of those fields. Many see no good reason to give up the established terminology and thereby disrupt access to the past.

For instance, many microbiologists and botanists use *division*, and zoologists *phylum*, for taxa that are equivalent in the hierarchy of classification. Also, opinions are still polarized with respect to the kingdom Protista, certain members of which could just as easily be grouped in the kingdoms of plants, fungi, or animals. Indeed, the term protozoan is a holdover from an earlier scheme in which amoebas and certain other single-celled organisms were ranked as simple animals.

Given the problems, why do we even bother imposing artificial frameworks on the history of life? We do this for the same reason that a writer might decide to break up the history of civilization into several volumes, a number of chapters, and many paragraphs. Both efforts are attempts to impart obvious structure to what might otherwise be an overwhelming body of knowledge and to enhance the retrieval of information from it.

Finally, bear in mind that we include this classification scheme primarily for your reference purposes. Besides being open to revision, it also is by no means complete. It does not include the most recently discovered species, as from the mid-ocean. Many existing and extinct organisms of the so-called lesser phyla are not represented here. Our strategy is to focus mainly on organisms mentioned in the text. A few examples of organisms also are listed.

SUPERKINGDOM PROKARYOTA. Prokaryotes. Almost all microscopic species. DNA organized at nucleoid (a cytoplasmic region), not inside a membrane-bound nucleus. All are bacteria, either single cells or simple associations of cells. Autotrophs and heterotrophs. Table A on the following page lists representative types. Reproduce by prokaryotic fission, sometimes by budding and by bacterial conjugation.

Bergey's Manual of Systematic Bacteriology, the authoritative reference in bacteriology, calls this "a time of taxonomic transition." It groups bacteria mostly by numerical taxonomy (Section 22.6), not on phylogeny. The scheme presented here reflects strong evidence of evolutionary relationships for at least some bacterial groupings.

KINGDOM EUBACTERIA. Gram-negative and gram-positive forms. Peptidoglycan in cell wall. Photosynthetic autotrophs, chemosynthetic autotrophs, and heterotrophs.

PHYLUM GRACILICUTES. Typical Gram-negative, thin wall. Autotrophs (photosynthetic and chemosynthetic) and heterotrophs. *Anabaena* and other cyanobacteria. *Escherichia, Pseudomonas, Neisseria, Myxococcus.*

PHYLUM FIRMICUTES. Typical Gram-positive, thick wall. Heterotrophs. *Bacillus, Staphylococcus, Streptococcus, Clostridium, Actinomycetes.*

PHYLUM TENERICUTES. Gram-negative, wall absent. Heterotrophs (saprobes, pathogens). *Mycoplasma.*

KINGDOM ARCHAEBACTERIA. Methanogens, halophiles, thermophiles. Evolutionarily closer to eukaryotic cells than to eubacteria. Strict anaerobes. Distinctive in cell wall, membrane lipids, ribosomes, and RNA sequences. *Methanobacterium, Halobacterium, Sulfolobus.*

SUPERKINGDOM EUKARYOTA. Eukaryotes. Both single-celled and multicelled species. Cells start out life with a nuclcus (encloses the DNA) and usually other membrane-bound organelles. Chromosomes have many histones and other proteins attached.

KINGDOM PROTISTA. Diverse single-celled, colonial, and multicelled eukaryotic species. Existing species are unlike bacteria in characteristics and most like the earliest, structurally simple eukaryotes. Autotrophs, heterotrophs, or both (Table 20.4). Reproduce sexually and asexually (by meiosis, mitosis, or both). Many related evolutionarily to plants, fungi, and possibly animals.

PHYLUM CHYTRIDIOMYCOTA. Chytrids. Heterotrophs; saprobic decomposers or parasites. *Chytridium.*

PHYLUM OOMYCOTA. Water molds. Heterotrophs. Decomposers, some parasites. *Saprolegnia, Phytophthora, Plasmopara.*

PHYLUM ACRASIOMYCOTA. Cellular slime molds. Heterotrophs with free-living, phagocytic amoeboid cells and spore-bearing stages. *Dictyostelium.*

PHYLUM MYXOMYCOTA. Plasmodial slime molds. Heterotrophs with free-living, phagocytic amoeboid cells and spore-bearing stages. Aggregate into streaming mass of cells that discard plasma membranes. *Physarum.*

Table A Representative Eubacteria and Archaebacteria Grouped on the Basis of Numerical Taxonomy

Some Major Groups	Main Habitats	Characteristics	Representatives
EUBACTERIA			
Photoautotrophs:			
Cyanobacteria, green sulfur bacteria, and purple sulfur bacteria	Mostly lakes, ponds; some marine, terrestrial habitats	Photosynthetic; use sunlight energy, carbon dioxide; cyanobacteria use oxygen-producing noncyclic pathway; some also use cyclic route	*Anabaena, Nostoc, Rhodopseudomonas, Chloroflexus*
Photoheterotrophs:			
Purple nonsulfur and green nonsulfur bacteria	Anaerobic, organically rich muddy soils, and sediments of aquatic habitats	Use sunlight energy; organic compounds as electron donors; some purple nonsulfur may also grow chemotrophically	*Rhodospirillum, Chlorobium*
Chemoautotrophs:			
Nitrifying, sulfur-oxidizing, and iron-oxidizing bacteria	Soil; freshwater, marine habitats	Use carbon dioxide, inorganic compounds as electron donors; influence crop yields, cycling of nutrients in ecosystems	*Nitrosomonas, Nitrobacter, Thiobacillus*
Chemoheterotrophs:			
Spirochetes	Aquatic habitats; parasites of animals	Helically coiled, motile; free-living and parasitic species; some major pathogens	*Spirochaeta, Treponema*
Gram-negative aerobic rods and cocci	Soil, aquatic habitats; parasites of animals, plants	Some major pathogens; some fix nitrogen (e.g., *Rhizobium*)	*Pseudomonas, Neisseria, Rhizobium, Agrobacterium*
Gram-negative facultative anaerobic rods	Soil, plants, animal gut	Many major pathogens; one bioluminescent (*Photobacterium*)	*Salmonella, Escherichia, Proteus, Photobacterium*
Rickettsias and chlamydias	Host cells of animals	Intracellular parasites; many pathogens	*Rickettsia, Chlamydia*
Myxobacteria	Decaying organic material; bark of living trees	Gliding, rod-shaped; aggregation and collective migration of cells	*Myxococcus*
Gram-positive cocci	Soil; skin and mucous membranes of animals	Some major pathogens	*Staphylococcus, Streptococcus*
Endospore-forming rods and cocci	Soil; animal gut	Some major pathogens	*Bacillus, Clostridium*
Gram-positive nonsporulating rods	Fermenting plant, animal material; gut, vaginal tract	Some important in dairy industry, others major contaminators of milk, cheese	*Lactobacillus, Listeria*
Actinomycetes	Soil; some aquatic habitats	Include anaerobes and strict aerobes; major producers of antibiotics	*Actinomyces, Streptomyces*
ARCHAEBACTERIA			
Methanogens	Anaerobic sediments of lakes, swamps; animal gut	Chemosynthetic; methane producers; used in sewage treatment facilities	*Methanobacterium*
Halophiles	Brines (extremely salty water)	Heterotrophic; also, unique photosynthetic pigments (bacteriorhodopsin) form in some	*Halobacterium*
Extreme thermophiles	Acidic soil, hot springs, hydrothermal vents	Heterotrophic or chemosynthetic; use inorganic substances as electron donors	*Sulfolobus, Thermoplasma*

PHYLUM SARCODINA.   Amoeboid protozoans. Heterotrophs, free-living or endosymbiotic, some pathogens. Soft- or shelled bodies, locomotion by pseudopods. The rhizopods (naked amoebas, foraminiferans), *Amoeba proteus, Entomoeba.* Also the actinopods (radiolarians, heliozoans).

PHYLUM MASTIGOPHORA.   Animal-like flagellated protozoans. Heterotrophs, free-living, many internal parasites. All with one to several flagella. *Trypanosoma, Trichomonas, Giardia.*

PHYLUM CILIOPHORA.   Ciliated protozoans. Heterotrophs, predators or symbionts, some parasitic. All have cilia. Free-living, sessile, or motile. *Paramecium,* hypotrichs.

APICOMPLEXA.   Heterotrophs, many parasitic. Complex of rings, tubules, other structures at head end. Most familiar members called sporozoans. *Plasmodium, Toxoplasma.*

PHYLUM EUGLENOPHYTA.   Euglenoids. Mostly heterotrophs, some autotrophs (photosynthetic), some switch depending on the environmental conditions. Flagellated. *Euglena.*

PHYLUM PYRRHOPHYTA.   Dinoflagellates. Photosynthetic, mostly, but some heterotrophs. *Gymnodinium breve.*

PHYLUM CHRYSOPHYTA.   Golden algae, yellow-green algae, diatoms. Photosynthetic. Some flagellated, others not. *Mischococcus, Synura, Vaucheria.*

PHYLUM RHODOPHYTA.   Red algae. Mostly photosynthetic, some parasitic. Nearly all marine, some in freshwater habitats. *Porphyra. Bonnemaisonia, Euchema.*

PHYLUM PHAEOPHYTA.   Brown algae. Photosynthetic, nearly all in temperate or marine waters. *Macrocystis, Fucus, Sargassum, Ectocarpus, Postelsia.*

PHYLUM CHLOROPHYTA.   Green algae. Mostly photosynthetic, some parasitic. Most freshwater, some marine or terrestrial. *Chlamydomonas, Spirogyra, Ulva, Volvox, Codium, Halimeda.*

KINGDOM FUNGI.   Nearly all multicelled eukaryotic species. Heterotrophs, mostly saprobic decomposers, some parasites. Nutrition based on extracellular digestion of organic matter and absorption of nutrients by individual cells. Multicelled species form absorptive mycelia within substrates and structures that produce asexual spores (and sometimes sexual spores).

PHYLUM ZYGOMYCOTA.   Zygomycetes. Zygosporangia (zygote inside thick wall) formed by sexual reproduction. Bread molds, related forms. *Rhizopus, Philobolus.*

PHYLUM ASCOMYCOTA.   Ascomycetes. Sac fungi. Sac-shaped cells form sexual spores (ascospores) Most yeasts and molds, morels, truffles. *Saccharomycetes, Morchella Neurospora, Sarcoscypha, Claviceps, Ophiostoma.*

PHYLUM BASIDIOMYCOTA.   Basidiomycetes. Club fungi. Most diverse group. Produce basidiospores inside club-shaped structures. Mushrooms, shelf fungi, stinkhorns. *Agaricus, Amanita, Puccinia, Ustilago.*

IMPERFECT FUNGI.   Sexual spores absent or undetected. The group has no formal taxonomic status. If better understood, a given species might be grouped with sac fungi or club fungi. *Verticillium, Candida, Microsporum, Histoplasma.*

LICHENS.   Mutualistic interactions between fungal species and a cyanobacterium, green alga, or both. *Usnea, Cladonia.*

KINGDOM PLANTAE.   Multicelled eukaryotes. Nearly all photosynthetic autotrophs with chlorophylls *a* and *b*. Some parasitic. Nonvascular and vascular species, generally with well-developed root and shoot systems. Nearly all adapted in form and function to survive dry conditions in land habitats; a few in aquatic habitats. Sexual reproduction predominant; also asexual reproduction by vegetative propagation and other mechanisms.

PHYLUM RHYNIOPHYTA.   Earliest known vascular plants; muddy habitats. Extinct. *Cooksonia, Rhynia.*

PHYLUM PROGYMNOSPERMOPHYTA.   Progymnosperms. Ancestral to early seed-bearing plants; extinct. *Archaeopteris.*

PHYLUM PTERIDOSPERMOPHYTA.   Seed ferns. Fernlike gymnosperms; extinct. *Medullosa*

PHYLUM CHAROPHYTA.   Stoneworts.

PHYLUM BRYOPHYTA.   Bryophytes: mosses, liverworts, hornworts. Seedless, nonvascular, haploid dominance. *Marchantia, Polytrichum, Sphagnum.*

PHYLUM PSILOPHYTA.   Whisk ferns. Seedless, vascular. No obvious roots, leaves on sporophyte. *Psilotum.*

PHYLUM LYCOPHYTA.   Lycophytes, club mosses. Seedless, vascular. Leaves, branching rhizomes, vascularized roots and stems. *Lycopodium, Selaginella.*

PHYLUM SPHENOPHYTA.   Horsetails. Seedless, vascular. Some sporophyte stems photosynthetic, others nonphotosynthetic, spore-producing. *Equisetum.*

PHYLUM PTEROPHYTA.   Ferns. Largest group of seedless vascular plants (12,000 species), mainly tropical, temperate habitats.

PHYLUM CYCADOPHYTA.   Cycads. Type of gymnosperm (vascular, bears "naked" seeds). Tropical, subtropical. Palm-shaped leaves, simple cones on male and female plants *Zamia.*

PHYLUM GINKGOPHYTA.   Ginkgo (maidenhair tree). Type of gymnosperm. Seeds with fleshy outer layer. *Ginkgo.*

PHYLUM GNETOPHYTA.   Gnetophytes. Only gymnosperms with vessels in xylem and double fertilization (but endosperm does not form). *Ephedra, Welwitchia.*

PHYLUM CONIFEROPHYTA.   Conifers. Most common and familiar gymnosperms. Generally cone-bearing with needle-like or scale-like leaves.
  Family Pinaceae. Pines, firs, spruces, hemlock, larches, Douglas firs, true cedars. *Pinus.*
  Family Cupressaceae. Junipers, cypresses. *Juniperus.*
  Family Taxodiaceae. Bald cypress, redwoods, Sierra bigtree, dawn redwood. *Sequoia.*
  Family Taxaceae. Yews.

PHYLUM ANTHOPHYTA.   Angiosperms (flowering plants). Largest group of vascular seed-bearing plants. Only organisms that produce flowers, fruits.
  Class Dicotyledonae. Dicotyledons (dicots). Some families of several different orders are listed:
    Family Nymphaeaceae. Water lilies.
    Family Papaveraceae. Poppies.
    Family Brassicaceae. Mustards, cabbages, radishes.
    Family Malvaceae. Mallows, cotton, okra, hibiscus.
    Family Solanaceae. Potatoes, eggplant, petunias.
    Family Salicaceae. Willows, poplars.
    Family Rosaceae. Roses, apples, almonds, strawberries.
    Family Fabaceae. Peas, beans, lupines, mesquite.
    Family Cactaceae. Cacti.
    Family Euphorbiaceae. Spurges, poinsettia.
    Family Cucurbitaceae. Gourds, melons, cucumbers, squashes.
    Family Apiaceae. Parsleys, carrots, poison hemlock.
    Family Aceraceae. Maples.
    Family Asteraceae. Composites. Chrysanthemums, sunflowers, lettuces, dandelions.
  Class Monocotyledonae. Monocotyledons (monocots). Some families of several different orders are listed:
    Family Liliaceae. Lilies, hyacinths, tulips, onions, garlic.
    Family Iridaceae. Irises, gladioli, crocuses.
    Family Orchidaceae. Orchids.
    Family Arecaceae. Date palms, coconut palms.
    Family Cyperaceae. Sedges.
    Family Poaceae. Grasses, bamboos, corn, wheat, sugarcane.
    Family Bromeliaceae. Bromeliads, pineapples, Spanish moss.

KINGDOM ANIMALIA.   Multicelled eukaryotes, nearly all with tissues, organs, and organ systems and with motility during at least part of the life cycle. Heterotrophs, predators (herbivores, carnivores, omnivores), parasites, detritivores. Reproduce sexually and, in many species, asexually by a variety of mechanisms. Embryonic development proceeds through a series of continuous stages.

PHYLUM PLACOZOA.   Marine. Simplest known animal. Two cell layers, no mouth, no organs. *Trichoplax.*

PHYLUM MESOZOA.   Ciliated, wormlike parasites, about the same level of complexity as *Trichoplax.*

PHYLUM PORIFERA.   Sponges. No symmetry, tissues, or organs.

PHYLUM CNIDARIA.   Radial symmetry, tissues, nematocysts.
  Class Hydrozoa. Hydrozoans. *Hydra, Obelia, Physalia, Prya.*
  Class Scyphozoa. Jellyfishes. *Aurelia.*
  Class Anthozoa. Sea anemones, corals. *Telesto.*

PHYLUM CTENOPHORA.   Comb jellies. Modified radial symmetry,

PHYLUM PLATYHELMINTHES.   Flatworms. Bilateral, cephalized; simplest animals with organ systems. Saclike gut.
  Class Turbellaria. Triclads (planarians), polyclads. *Dugesia.*
  Class Trematoda. Flukes. *Schistosoma.*
  Class Cestoda. Tapeworms. *Taenia.*

PHYLUM NEMERTEA.   Ribbon worms.

PHYLUM NEMATODA.   Roundworms. *Ascaris, Trichinella.*

PHYLUM ROTIFERA.   Rotifers.

PHYLUM MOLLUSCA.   Mollusks.
  Class Polyplacophora. Chitons.

Class Gastropoda. Snails (periwinkles, whelks, limpets, abalones, cowries, conches, nudibranchs, tree snails, garden snails), sea slugs, land slugs.

Class Bivalvia. Clams, mussels, scallops, cockles, oysters, shipworms.

Class Cephalopoda. Squids, octopuses, cuttlefish, nautiluses. *Loligo.*

PHYLUM BRYOZOA.   Bryozoans (moss animals).

PHYLUM BRACHIOPODA.   Lampshells.

PHYLUM ANNELIDA.   Segmented worms.

Class Polychaeta. Mostly marine worms.

Class Oligochaeta. Mostly freshwater and terrestrial worms, but many marine. *Lumbricus* (earthworms).

Class Hirudinea. Leeches.

PHYLUM TARDIGRADA.   Water bears.

PHYLUM ONYCHOPHORA.   Onychophorans. *Peripatus.*

PHYLUM ARTHROPODA.

Subphylum Trilobita. Trilobites; extinct.

Subphylum Chelicerata. Chelicerates. Horseshoe crabs, spiders, scorpions, ticks, mites.

Subphylum Crustacea. Shrimps, crayfishes, lobsters, crabs, barnacles, copepods, isopods (sowbugs).

Subphylum Uniramia.
  Superclass Myriapoda. Centipedes, millipedes.
  Superclass Insecta.
    Order Ephemeroptera. Mayflies.
    Order Odonata. Dragonflies, damselflies.
    Order Orthoptera. Grasshoppers, crickets, katydids.
    Order Dermaptera. Earwigs.
    Order Blattodea. Cockroaches.
    Order Mantodea. Mantids.
    Order Isoptera. Termites.
    Order Mallophaga. Biting lice.
    Order Anoplura. Sucking lice.
    Order Homoptera. Cicadas, aphids, leafhoppers, spittlebugs.
    Order Hemiptera. Bugs.
    Order Coleoptera. Beetles.
    Order Diptera. Flies.
    Order Mecoptera. Scorpion flies. *Harpobittacus.*
    Order Siphonaptera. Fleas.
    Order Lepidoptera. Butterflies, moths.
    Order Hymenoptera. Wasps, bees, ants.

PHYLUM ECHINODERMATA.   Echinoderms.

Class Asteroidea. Sea stars.
Class Ophiuroidea. Brittle stars.
Class Echinoidea. Sea urchins, heart urchins, sand dollars.
Class Holothuroidea. Sea cucumbers.
Class Crinoidea. Feather stars, sea lilies.
Class Concentricycloidea. Sea daisies.

PHYLUM HEMICHORDATA.   Acorn worms.

PHYLUM CHORDATA.   Chordates.

Subphylum Urochordata. Tunicates, related forms.

Subphylum Cephalochordata. Lancelets.

Subphylum Vertebrata. Vertebrates.
  Class Agnatha. Jawless vertebrates (lampreys, hagfishes).
  Class Placodermi. Jawed, heavily armored fishes; extinct.
  Class Chondrichthyes. Cartilaginous fishes (sharks, rays, skates, chimaeras).
  Class Osteichthyes. Bony fishes.
  Subclass Dipnoi. Lungfishes.
  Subclass Crossopterygii. Coelacanths, related forms.
  Subclass Actinopterygii. Ray-finned fishes.
    Order Acipenseriformes. Sturgeons, paddlefishes.
    Order Salmoniformes. Salmon, trout.
    Order Atheriniformes. Killifishes, guppies.
    Order Gasterosteiformes. Seahorses.
    Order Perciformes. Perches, wrasses, barracudas, tunas, freshwater bass, mackerels.

Order Lophiiformes. Angler fishes.
Class Amphibia. Mostly tetrapods; embryo enclosed in amnion.
  Order Caudata. Salamanders.
  Order Anura. Frogs, toads.
  Order Apoda. Apodans (caecilians).
Class Reptilia. Skin with scales, embryo enclosed in amnion.
  Subclass Anapsida. Turtles, tortoises.
  Subclass Lepidosaura. *Sphenodon*, lizards, snakes.
  Subclass Archosaura. Dinosaurs (extinct), crocodiles, alligators.
Class Aves. Birds. (In some of the more recent schemes, dinosaurs, crocodilians, and birds are grouped in the same category.)
  Order Struthioniformes. Ostriches.
  Order Sphenisciformes. Penguins.
  Order Procellariiformes. Albatrosses, petrels.
  Order Ciconiiformes. Herons, bitterns, storks, flamingoes.
  Order Anseriformes. Swans, geese, ducks.
  Order Falconiformes. Eagles, hawks, vultures, falcons.
  Order Galliformes. Ptarmigan, turkeys, domestic fowl.
  Order Columbiformes. Pigeons, doves.
  Order Strigiformes. Owls.
  Order Apodiformes. Swifts, hummingbirds.
  Order Passeriformes. Sparrows, jays, finches, crows, robins, starlings, wrens.
Class Mammalia. Skin with hair; young nourished by milk-secreting glands of adult.
  Subclass Prototheria. Egg-laying mammals (duckbilled platypus, spiny anteaters).
  Subclass Metatheria. Pouched mammals or marsupials (opossums, kangaroos, wombats).
  Subclass Eutheria. Placental mammals.
    Order Insectivora. Tree shrews, moles, hedgehogs.
    Order Scandentia. Insectivorous tree shrews.
    Order Chiroptera. Bats.
    Order Primates.
      Suborder Strepsirhini (prosimians). Lemurs, lorises.
      Suborder Haplorhini (tarsioids and anthropoids).
        Infraorder Tarsiiformes. Tarsiers.
        Infraorder Platyrrhini (New World monkeys).
          Family Cebidae. Spider monkeys, howler monkeys, capuchin.
        Infraorder Catarrhini (Old World monkeys and hominoids).
          Superfamily Cercopithecoidea. Baboons, macaques, langurs.
          Superfamily Hominoidea. Apes and humans.
            Family Hylobatidae. Gibbon.
            Family Pongidae. Chimpanzees, gorillas, orangutans.
            Family Hominidae. Existing and extinct human species (*Homo*) and australopiths.
    Order Carnivora. Carnivores.
      Suborder Feloidea. Cats, civets, mongooses, hyenas.
      Suborder Canoidea. Dogs, weasels, skunks, otters, raccoons, pandas, bears.
    Order Proboscidea. Elephants; mammoths (extinct).
    Order Sirenia. Sea cows (manatees, dugongs).
    Order Perissodactyla. Odd-toed ungulates (horses, tapirs, rhinos).
    Order Artiodactyla. Even-toed ungulates (camels, deer, bison, sheep, goats, antelopes, giraffes).
    Order Edentata. Anteaters, tree sloths, armadillos.
    Order Tubulidentata. African aardvarks.
    Order Cetacea. Whales, porpoises.
    Order Rodentia. Most gnawing animals (squirrels, rats, mice, guinea pigs, porcupines).

Metric-English Conversions

Length

English		Metric
inch	=	2.54 centimeters
foot	=	0.30 meter
yard	=	0.91 meter
mile (5,280 feet)	=	1.61 kilometer

To convert	multiply by	to obtain
inches	2.54	centimeters
feet	30.00	centimeters
centimeters	0.39	inches
millimeters	0.039	inches

Weight

English		Metric
grain	=	64.80 milligrams
ounce	=	28.35 grams
pound	=	453.60 grams
ton (short) (2,000 pounds)	=	0.91 metric ton

To convert	multiply by	to obtain
ounces	28.3	grams
pounds	453.6	grams
pounds	0.45	kilograms
grams	0.035	ounces
kilograms	2.2	pounds

Volume

English		Metric
cubic inch	=	16.39 cubic centimeters
cubic foot	=	0.03 cubic meter
cubic yard	=	0.765 cubic meters
ounce	=	0.03 liter
pint	=	0.47 liter
quart	=	0.95 liter
gallon	=	3.79 liters

To convert	multiply by	to obtain
fluid ounces	30.00	milliliters
quart	0.95	liters
milliliters	0.03	fluid ounces
liters	1.06	quarts

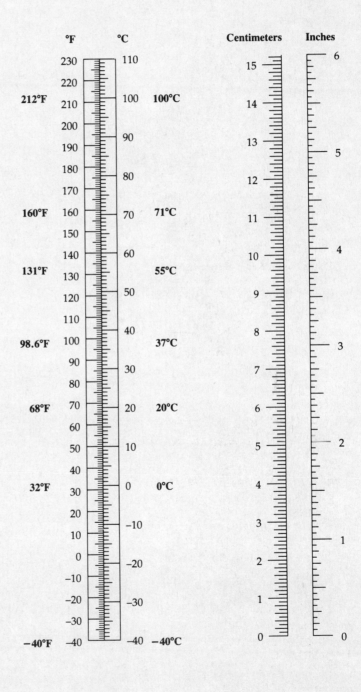

CHAPTER 1
1. Metabolism *5*
2. Homeostasis *5*
3. Cell *6*
4. Adaptive *10*
5. Mutations *10*
6. d *4*
7. d *4, 10*
8. d *10*
9. d *13*
10. c *13*
11. a *16*
12. c *10*
   e *12*
   d *13*
   b *11*
   a *12*

CHAPTER 2
1. b *22*
2. f *26–27*
3. a *27–28*
4. e *29*
5. f *30–31*
6. acid, base *30*
7. c *20*
   a *31*
   b *24*
   d *29*

CHAPTER 3
1. d *36*
2. e *37*
3. f *38*
4. b *40*
5. d *42,47*
6. d *47*
7. d *45*
8. c *42–43*
   e *47*
   b *41*
   d *47*
   a *38*

CHAPTER 4
1. c *52–53*
2. d *56*
3. d *69*
4. False *66*
5. False; many cells have a wall *68, 69*
6. c *52, 70*
7. e *64*
   d *65*
   a *56t, 62*
   b *62*
   c *62*

CHAPTER 5
1. d *77*
2. d *78–79*
3. d *78–79*
4. e *80–82*
5. a *87*
6. d *83*
7. b *85*

8. f *78–79*
   g *83, 84–85*
   a *79*
   d *79*
   e *79*
   b *79, 80–81*
   c *79*

CHAPTER 6
1. Carbon dioxide; sunlight *92*
2. d *94, 95i*
3. b *94*
4. e *94, 100*
5. d *98*
6. b *94, 98*
7. c *101*
8. c *101*
9. c *101*
   d *94–95, 101*
   e *98–99*
   b *94, 98–99*
   f *98*
   a *96–97, 98*

CHAPTER 7
1. d *110*
2. c *111i, 112*
3. b *110, 114*
4. c *110, 116*
5. d *118*
6. c *118*
7. b *118*
8. d *120–121*
9. b *112*
   c *118*
   a *114–115*
   d *111, 116*

CHAPTER 8
1. a *128*
2. b *128*
3. b *128*
4. c *129*
5. a *128, 131*
6. d *128, 136t*
7. b *129*
8. d *130–131*
   b *130, 131*
   c *131*
   a *130, 131*

CHAPTER 9
1. d *139, 140*
2. d *139, 141*
3. a *139, 140*
4. b *140–141*
5. d *140*
6. c *141*
7. c *142*
8. d *143*
9. d *140*
   a *140*
   c *142, 145*
   b *141, 143*

CHAPTER 10
1. a *155*
2. b *155*
3. a *155*
4. c *156*
5. b *156*
6. a *156, 157*
7. d *158*
8. b *158*
   d *156*
   a *155*
   c *155*

CHAPTER 11
1. c *172*
2. c *177*
3. e *182–183*
4. c *184*
5. d *184*
6. d *179, 182–184*
7. d *172, 179*
8. c *177*
   e *183*
   b *182*
   a *170*

CHAPTER 12
1. c *194*
2. d *195*
3. d *195*
4. c *195*
5. a *196–197*
6. d *197*
7. d *197*
   b *195*
   c *196*
   a *195*

CHAPTER 13
1. c *202*
2. b *202*
3. c *202*
4. c *204*
5. a (UAG = STOP) *204*
6. a *204–205*
7. e *209*
   c *206*
   a *203*
   f *204*
   d *204*
   g *203*
   b *204*

CHAPTER 14
1. d *214*
2. d *214*
3. d *214*
4. a *214*
5. a *214*
6. d *214–215*
7. e *216–217*
8. b *216*
9. d *217*
10. b *219*
11. b *218*
12. d *217*
   e *216*
   c *218*
   a *214*
   b *218*

CHAPTER 15
1. d *222*
2. c *224*
3. Plasmids *224–225*
4. a *225*
5. b *225*
6. a *226*
7. b *227*
8. d *228*
9. d *227*
   c *231*
   f *223*
   e *229*
   b *232*
   a *233*

CHAPTER 16
1. populations *246*
2. d *246*
3. c *255*
4. c *249*
5. b *250–251*
6. c *253*
7. c *255*
   d *249*
   a *247*
   b *256*

CHAPTER 17
1. d *262–263*
2. d *262*
3. c *263*
4. c *268*
5. b *268*
6. c *268*
   e *268*
   a *268*
   d *269*
   b *269*

CHAPTER 18
1. a *278–279*
2. a *277*
3. b *282*
4. c *282*
5. d *282*
6. e *282*
   b *274*
   f *275*
   c *276*
   a *280–281*
   d *277*

CHAPTER 19
1. c *291, 296*
2. e *291, 307*
3. b *293*
4. b *292*
5. d *296–297; also compare 292*
6. d *297*
7. b *296–297*
   d *297, 307i*
   e *300–301, 307i*
   a *302, 307i*
   c *305, 307i*

CHAPTER 20
1. c *314*
2. b *314*
3. c *316*
4. c *316*
5. d *317*
6. a *318*
7. a *320–321*
8. c *322; 331t*
9. f *323*
10. a *326*
11. d *326*
12. d *314*
   e *314*
   b *317*
   c *324*
   a *316*

CHAPTER 21
1. c *334*
2. c *335*
3. b *332, 338*
4. a *334*
5. a *334*
6. e *334, 336*
   c *337i*
   d *334*
   a *334*
   b *336*
   f *334, 336*

CHAPTER 22
1. d *342, 342i*
2. b *344, 344i*
3. c *342, 346*
4. c *343, 348*
5. b *343, 349*
6. d *350*
   e *342*
   g *346*
   h *351*
   f *344*
   a *342*
   b *342*
   d *351*

CHAPTER 23
1. b *358*
2. c *362*
3. a *372*
4. a *372*
5. b *364–366*
6. c *358, 364*
7. d *360*
   f *362*
   c *364–365, 366*
   g *367*
   e *381*
   i *368*
   j *370*
   h *372*
   a *378*
   b *358t*

CHAPTER 10

1. a. *AB*

   b. *AB, aB*

   c. *Ab, ab*

   d. *AB, Ab, aB, ab*

2. a. All of the offspring will be *AaBB*.

   b. 1/4 *AABB* (25% each genotype)
   1/4 *AABb*
   1/4 *AaBB*
   1/4 *AaBb*

   c. 1/4 *AaBb* (25% each genotype)
   1/4 *Aabb*
   1/4 *aaBb*
   1/4 *aabb*

   d. 1/16 *AABB*   (6.25%)
   1/8 *AaBB* (12.5%)
   1/16 *aaBB* (6.25%)
   1/8 *AABb* (12.5%)
   1/4 *AaBb* (25%)
   1/8 *aaBb* (12.5%)
   1/16 *AAbb* (6.25%)
   1/8 *Aabb* (12.5%)
   1/16 *aabb* (6.25%)

3. Yellow is recessive. Because F_1 plants have a green phenotype and must be heterozygous, green must be dominant over the recessive yellow.

4. a. *ABC*

   b. *ABc, aBc*

   c. *ABC, aBc, ABc, aBc*

   d. *ABC, aBC, AbC, abC,*
   ABc, aBc, Abc, abc

5. Because all F_1 plants of this dihybrid cross had to be heterozygous for both genes, then 1/4 (25%) of the F_2 plants will be heterozygous for both genes.

6. a. The mother must be heterozygous for both genes. Both the father and their first child are homozygous recessive for both genes.

   b. The probability that their second child will not be a tongue roller and will have detached earlobes is 1/4 (25%).

7. a. Bill *AaEeSs*, and Marie *AAEESS*

   b. There is a 100% probability that all their children will have Bill's phenotype (flat feet, long eyelashes, and tendency to sneeze).

   c. Zero probability; no child of theirs will have high arches or short eyelashes, and none will be sneezy.

8. a. The mother must be heterozygous $I^A i$. The male with type B blood could have fathered the child if he were heterozygous $I^B i$.

   b. Genotype alone cannot prove the accused male is the father. Even if he happens to be heterozygous, *any* male who carries the *i* allele could be the father, including those heterozygous for type A blood ($I^A i$) or type B blood ($I^B i$) and those with type O blood (*ii*).

9. A mating between a mouse from a true-breeding, white-furred strain and a mouse from a true-breeding. brown-furred strain would provide you with the most direct evidence. Because true-breeding strains typically are homozygous for a trait being studied, all F_1 offspring from this mating should be heterozygous. Record the phenotype of each F_1 mouse, then let them to mate with one another. Assuming only one gene locus is involved, these are possible outcomes for the F_2 offspring:

   a. All F_1 mice are brown, and their F_2 offspring segregate 3 brown : 1 white. *Conclusion*: Brown is dominant to white.

   b. All F_1 mice are white, and their F_2 offspring segregate 3 white : 1 brown. *Conclusion*: White is dominant to brown.

   c. All F_1 mice are tan, and the F_2 offspring segregate 1 brown : 2 tan : 1 white. *Conclusion*: The alleles at this locus show incomplete dominance.

10. You cannot guarantee that the puppies won't develop the disorder without more information about Dandelion's genotype. You could do so only if she is a heterozygous carrier, if the male is free of the alleles, and if the alleles are recessive.

11. Fred could use a testcross to find out if his pet's genotype is *WW* or *Ww*. He can let his black guinea pig mate with a white guinea pig having the genotype *ww*.

   If any F_1 offspring are white, then his pet's genotype is *Ww*. If the two parents are allowed to mate repeatedly and all the offspring of the matings are black, there is a high probability that his pet is *WW*. (If, say, ten offspring all are black, then the probability that the male is *WW* is about 99.9 percent. The greater the number of offspring, the more confident Fred can be of his conclusion.)

12. a. 1/2 red, 1/2 pink

    b. All pink

    c. 1/4 red, 1/2 pink, 1/4 white

    d. 1/2 pink, 1/2 white

13.   9/16 walnut comb
    3/16 rose comb
    3/16 pea comb
    1/16 single comb

14. Because both parents are heterozygotes (Hb^AHb^S), the following are the probabilities for each child:

    a. 1/4 Hb^SHb^S

    b. 1/4 Hb^AHb^A

    c. 1/2 Hb^AHb^S

15. A mating of two M^LM cats yields:

    1/4 homozygous dominant (MM)

    1/2 heterozygous (M^LM)

    1/4 homozygous recessive (M^LM^L)

Because M^LM^L is lethal, the probability that any one kitten among the survivors will be heterozygous is 2/3.

16. a. Both parents must be heterozygotes (Aa). Their children may be albino (aa) or unaffected (AA or Aa).

    b. All are aa.

    c. The albino father must be aa. They have an albino child, so the mother must be Aa. (If she were AA, they could not have an albino child.) The albino child is aa. The three unaffected children are Aa. There is a 50% chance that any child of theirs will be albino. The observed 3:1 ratio is not surprising, given the small number of offspring.

17. a. All of the offspring will have medium-red color corresponding to the genotype $A^1A^2B^1B^2$.

    b. All possible genotypes could appear in the following proportions:

1/16	$A^1A^2B^1B^1$	dark red
1/8	$A^1A^1B^1B^2$	medium-dark red
1/16	$A^1A^1B^2B^2$	medium red
1/8	$A^1A^2B^1B^1$	medium-dark red
1/4	$A^1A^2B^1B^2$	medium red
1/8	$A^1A^2B^2B^2$	light red
1/16	$A^2A^2B^1B^1$	medium red
1/8	$A^2A^2B^1B^2$	light red
1/16	$A^2A^2B^2B^2$	white

CHAPTER 11

1. a. Human males (XY) inherit their X chromosome only from their mother.

    b. In males, an X-linked allele will only be found on his one X chromosome. Males can produce two kinds of gametes: one kind with a Y chromosome free of the gene, and the other kind with an X chromosome bearing the X-linked allele.

    c. One. Each gamete of a woman who is homozygous for an X-linked allele will have an X chromosome that carries the allele.

    d. Two. If a female is heterozygous for an X linked allele, half of the gametes that she produces will contain one of the alleles and the other half will contain the other allele.

2. All of the offspring should be heterozygous for the gene and all should have long wings. However, because some have vestigial wings, the dominant allele might have mutated because of radiation.

3. Because the phenotype appeared in every generation shown in the diagram, this must be a pattern of autosomal dominant inheritance.

4. Because this is a case of autosomal dominant inheritance and because one parent bears the allele, the probability of any child inheriting the allele is 50%.

5. a. Females normally do not have a Y chromosome, hence they would not have a Y-linked gene.

    b. All sons inherit their Y chromosome from their father, but a daughter can only inherit his X chromosome. So the allele for hairy pinnae can be transmitted only to his sons, not to his daughters.

6. A daughter could develop this disorder only if she were to inherit two X-linked recessive alleles, one from her father and one from her mother. But if a male bears the allele on his X chromosome, it will be expressed, he will develop the disorder, and most likely he won't father children because of early death.

7. If no crossover occurs between the two genes, then half the chromosomes will carry alleles AB and half will carry alleles $a\,b$.

8. If the alleles are close together, it is highly unlikely that a crossover will separate them from each other during meiosis. The greater the distance between the two loci, the greater the probability that a crossover will separate them from each other.

9. a. Nondisjunction could occur in anaphase I or anaphase II of meiosis.

    b. As a result of a translocation, chromosome 21 (which is small) may become attached to the end of chromosome 14. Even though the chromosome number of the new individual would be 46, its somatic cells would contain the translocated chromosome 21, in addition to two normal chromosomes 21.

10. In the mother, a crossover between the two genes at meiosis generates an X chromosome that carries neither mutant allele.

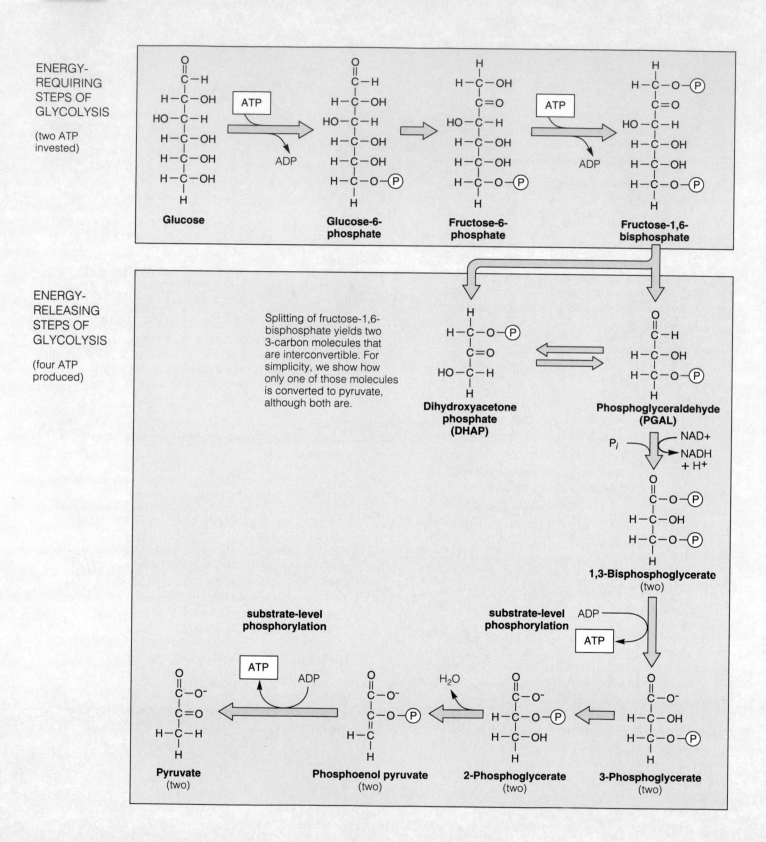

Figure A Glycolysis, ending with two 3-carbon pyruvate molecules for each 6-carbon glucose entering the reactions. The *net* energy yield is two ATP molecules (two invested, four produced).

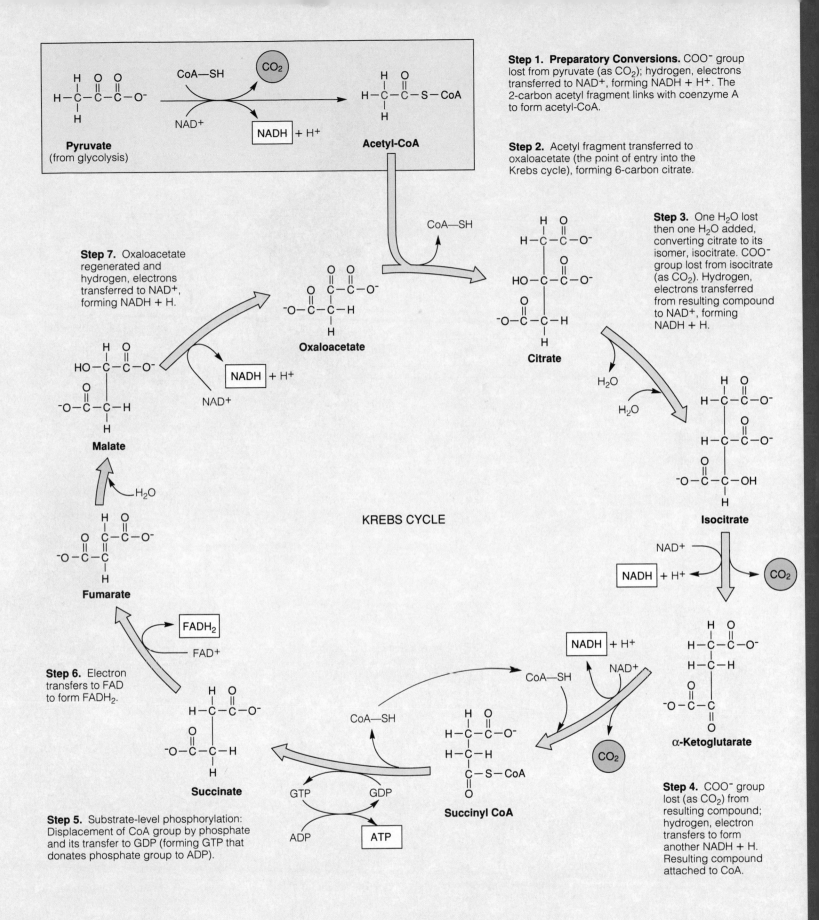

Step 1. Preparatory Conversions. COO⁻ group lost from pyruvate (as CO_2); hydrogen, electrons transferred to NAD⁺, forming NADH + H⁺. The 2-carbon acetyl fragment links with coenzyme A to form acetyl-CoA.

Step 2. Acetyl fragment transferred to oxaloacetate (the point of entry into the Krebs cycle), forming 6-carbon citrate.

Step 3. One H_2O lost then one H_2O added, converting citrate to its isomer, isocitrate. COO⁻ group lost from isocitrate (as CO_2). Hydrogen, electrons transferred from resulting compound to NAD⁺, forming NADH + H.

Step 7. Oxaloacetate regenerated and hydrogen, electrons transferred to NAD⁺, forming NADH + H.

Step 6. Electron transfers to FAD to form $FADH_2$.

Step 5. Substrate-level phosphorylation: Displacement of CoA group by phosphate and its transfer to GDP (forming GTP that donates phosphate group to ADP).

Step 4. COO⁻ group lost (as CO_2) from resulting compound; hydrogen, electron transfers to form another NADH + H. Resulting compound attached to CoA.

KREBS CYCLE

Figure B Krebs cycle (also known as the citric acid cycle). *Red* identifies the carbon atoms entering the cyclic pathway by way of acetyl-CoA.

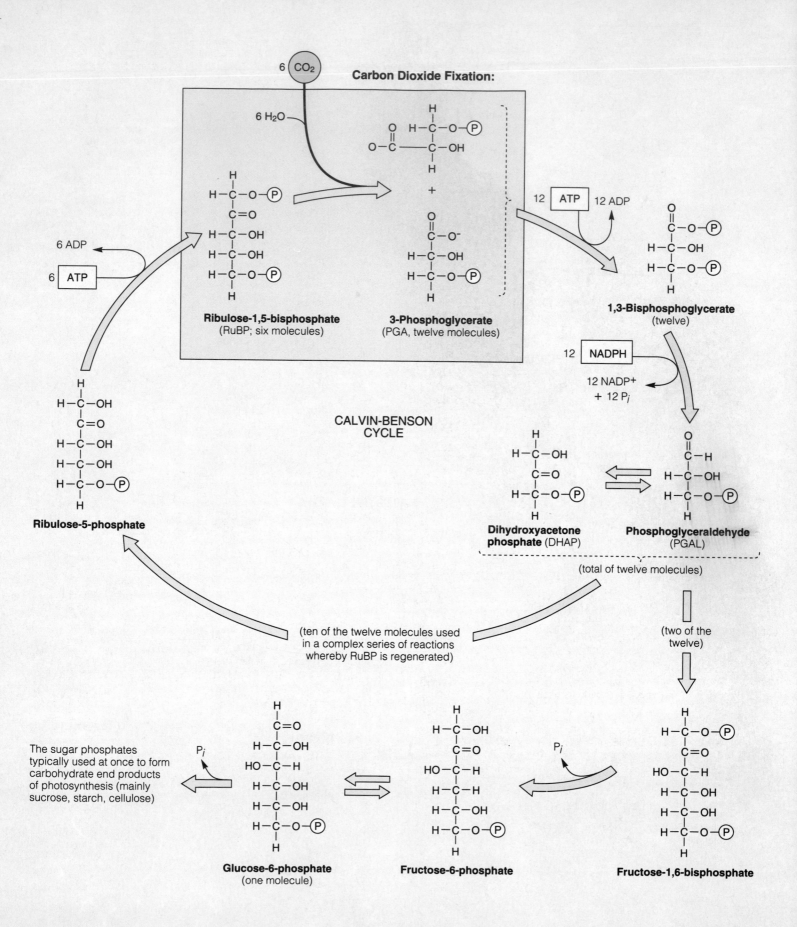

Figure C Calvin-Benson cycle of the light-independent reactions of photosynthesis.

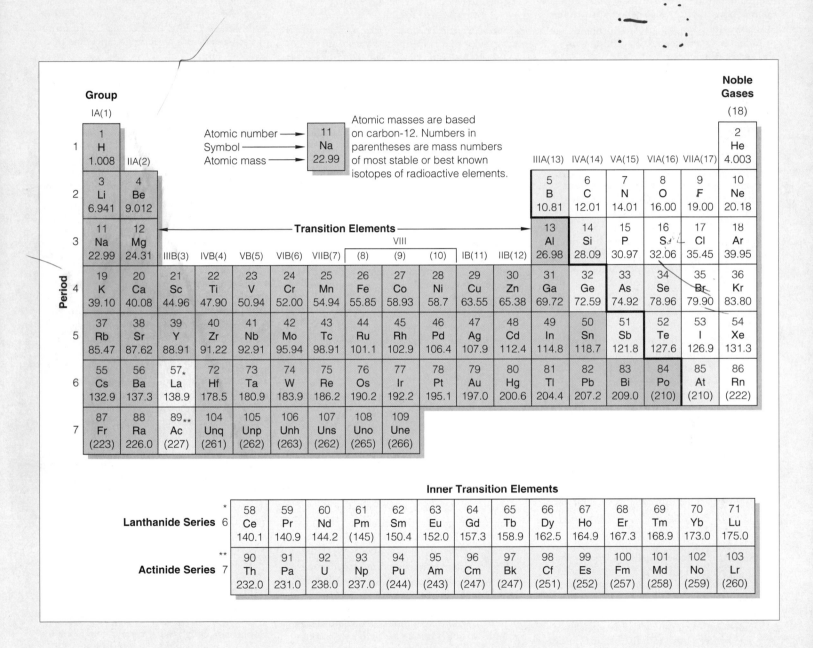

APPENDIX VI. PERIODIC TABLE OF THE ELEMENTS

Group

Noble Gases

IA(1)

(18)

Atomic number → 11
Symbol → Na
Atomic mass → 22.99

Atomic masses are based on carbon-12. Numbers in parentheses are mass numbers of most stable or best known isotopes of radioactive elements.

Period

Transition Elements

VIII

Inner Transition Elements

Lanthanide Series 6	58 Ce 140.1	59 Pr 140.9	60 Nd 144.2	61 Pm (145)	62 Sm 150.4	63 Eu 152.0	64 Gd 157.3	65 Tb 158.9	66 Dy 162.5	67 Ho 164.9	68 Er 167.3	69 Tm 168.9	70 Yb 173.0	71 Lu 175.0
Actinide Series 7	90 Th 232.0	91 Pa 231.0	92 U 238.0	93 Np 237.0	94 Pu (244)	95 Am (243)	96 Cm (247)	97 Bk (247)	98 Cf (251)	99 Es (252)	100 Fm (257)	101 Md (258)	102 No (259)	103 Lr (260)

ABO blood typing Method of using the presence of proteins A, B, or both at surface of red blood cells to characterize an individual's blood. O signifies absence of both proteins.

abortion Premature, spontaneous or induced expulsion of the embryo or fetus from uterus. Spontaneous abortion also called miscarriage.

abscisic acid (ab-SISS-ik) Plant hormone that promotes stomatal closure, bud dormancy, and seed dormancy.

abscission (ab-SIH-zhun) [L. abscindere, to cut off] Hormone-induced dropping of leaves, flowers, fruits, or other parts from a plant.

absorption Uptake of water and solutes from the environment by cell or multicelled organism; e.g., movement of nutrients, fluid, and ions across gut lining and into internal environment.

accessory pigment Any of a variety of light-trapping pigments that extend the range of wavelengths for photosynthesis beyond those absorbed by chlorophylls.

acetyl-CoA (uh-SEED-ul) Coenzyme A with a two-carbon fragment from pyruvate attached. In the second stage of aerobic respiration, it transfers the fragment to oxaloacetate for the Krebs cycle.

acidity Of a solution, an excess of hydrogen ions relative to hydroxyl ions.

acid [L. acidus, sour] Any dissolved substance that donates hydrogen ions to other solutes or to water molecules.

acid rain Wet acid deposition; falling of rain (or snow) rich in sulfur and nitrogen oxides.

acoelomate (ay-SEE-luh-mate) Absence of a fluid-filled cavity between gut and body wall.

acoustical signal Sounds used as a form of intraspecific communication.

actin (AK-tin) Cytoskeletal protein; subunit of microfilaments.

action potential Abrupt, brief reversal in the resting membrane potential of a neuron and other excitable cells.

activation energy For each type of reaction, the minimum amount of collision energy that will drive reactant molecules to an activated state, from which the reaction will proceed spontaneously.

active site Crevice in the surface of an enzyme molecule where a specific reaction is catalyzed (made to proceed more rapidly than it would spontaneously).

active transport Pumping of a specific solute across a cell membrane, through the interior of a transport protein, against its concentration gradient. Requires an energy boost, as from ATP.

adaptation [L. adaptare, to fit] Of evolution, being adapted (or becoming more adapted) to a set of environmental conditions. Of a sensory neuron, a decrease or cessation in the frequency of action potentials when a stimulus is maintained at constant strength.

adaptive radiation Macroevolutionary pattern; a burst of genetic divergences from a lineage that gives rise to many new species, each adapted to using a novel resource or a new (or recently vacated) habitat.

adaptive trait Any aspect of form, function, or behavior that helps the individual survive and reproduce under prevailing conditions.

adaptive zone A way of life available for organisms that are physically, ecologically, and evolutionarily equipped to live it, such as "catching insects in the air at night."

adenine (AH-de-neen) A purine; a nitrogen-containing base in certain nucleotides.

ADH Antidiuretic hormone. Hypothalamic hormone that induces water conservation; helps control the solute concentrations and volume of extracellular fluid.

adipose tissue A connective tissue having an abundance of fat-storing cells and blood vessels for transporting fats.

ADP Adenosine diphosphate (ah-DEN-uh-seen die-FOSS-fate). Nucleotide coenzyme; typically accepts inorganic (unbound) phosphate or a phosphate group, thus becoming ATP.

aerobic respiration (air-OH-bik) [Gk. aer, air, + bios, life] Main ATP-forming pathway; proceeds from glycolysis through Krebs cycle and then electron transport phosphorylation. Final electron acceptor is oxygen. Typical net energy yield: 36 ATP per glucose molecule.

age structure Number of individuals in each age category for a population.

agglutination (ah-glue-tin-AY-shun) Forced clumping together of nonself markers when antibodies circulating in blood chemically recognize them. Clumping makes bearers of those markers more easily destroyed by phagocytes. Potential problem in recipients of transfused blood of a different type.

aging Of any multicelled organism showing extensive cell differentiation, a gradual and expected deterioration of the body over time.

AIDS Acquired immunodeficiency syndrome. A set of chronic disorders following infection by the human immunodeficiency virus (HIV), which destroys cells of the immune system.

alcohol Organic compound that includes one or more hydroxyl groups (—OH); it dissolves readily in water. Sugars are examples.

alcoholic fermentation One of the anaerobic ATP-forming pathways. The pyruvate from glycolysis is degraded to acetaldehyde, which accepts electrons from NADH to form ethanol; NAD^+ needed for the reactions is regenerated. Net yield: two ATP.

aldosterone (al-DOSS-tuh-rohn) Hormone of the adrenal cortex; helps control the body's reabsorption of sodium from nephrons.

allantois (ah-LAN-twahz) [Gk. allas, sausage] An extraembryonic membrane. Functions in respiration and in storing metabolic wastes of embryos of reptiles, birds, and certain mammals. In humans, the urinary bladder and placental blood vessels form from it.

allele (uh-LEEL) For a given gene locus, one of two or more slightly different molecular forms of a gene that arise through mutation and that code for different versions of the same trait.

allele frequency For a given locus, the relative abundance of each kind of allele among all the individuals of a population.

allergen Any normally harmless substance that provokes inflammation, excessive mucus secretion, and often immune responses.

allergy Hypersensitivity to an allergen.

allopatric speciation [Gk. allos, different, + L. patria, native land]. Model of what may be the most common speciation route. A physical barrier arises and separates populations or subpopulations of a species, stops gene flow, and favors divergences that end in speciation.

altruism (AL-true-IZ-um) Behavior that helps other members of a species but diminishes an individual's own chance of reproductive success.

alveolus (ahl-VEE-uh-lus), plural **alveoli** [L. alveus, small cavity] Cupped, thin-walled outpouching of respiratory bronchiole where oxygen diffuses from lungs into blood, and carbon dioxide diffuses from blood to lungs.

amino acid (uh-MEE-no) An organic molecule with a hydrogen atom, an amino group, an acid group, and an R group, all covalently bonded to a carbon atom. Twenty kinds are the subunits of polypeptide chains.

ammonification (uh-moan-ih-fih-KAY-shun) Process by which some soil bacteria and fungi break down nitrogenous wastes and organic remains; part of nitrogen cycle.

amnion (AM-nee-on) An extraembryonic membrane; the boundary layer of a fluid-filled sac (amniotic cavity) in which the embryos of some vertebrate embryos grow and develop, move freely, and are protected from sudden impacts and temperature shifts.

amniote egg Egg that has extraembryonic membranes and often a shell. Contributed to successful invasion of land by vertebrates.

amphibian Only type of vertebrate making the transition from water to land (evolutionarily and in their embryonic development). Existing gropus are salamanders, frogs, toads, and caecilians.

anaerobic electron transport ATP-forming pathway in which electrons stripped from an organic compound move through transport systems in the bacterial plasma membrane; an inorganic compound in the environment often serves as the final electron acceptor. Variable but always small net energy yield.

anaerobic pathway (an-uh-ROW-bik) [Gk. *an*, without, + *aer*, air] Metabolic pathway in which a substance other than oxygen is final acceptor of electrons stripped from substrates.

analogous structures (ann-AL-uh-gus) [Gk. *analogos*, similar to one another] Body parts that once differed in evolutionarily distant lineages but converged in their structure and function in response to similar environmental pressures.

anaphase (AN-uh-faze) Of mitosis, stage when sister chromatids of each chromosome move apart to opposite spindle poles. In anaphase I of meiosis, each duplicated chromosome and its homologue move to opposite spindle poles. In anaphase II of meiosis, sister chromatids of each chromosome move to opposite poles.

aneuploidy (AN-yoo-ploy-dee) Having one extra or one less chromosome relative to the parental chromosome number.

angiosperm (AN-gee-oh-spurm) [Gk. *angeion*, vessel, and *spermia*, seed] Flowering plant.

animal Multicelled heterotroph that feeds on other organisms, is motile for at least part of life cycle, develops through embryonic stages, has tissues (except for *Trichoplax* and sponges), and most often organs and organ systems.

Animalia Kingdom of animals.

annelid Type of invertebrate; a segmented worm (e.g., oligochaete, leech, or polychaete).

annual Flowering plant that completes its life cycle in one growing season.

anther [Gk. *anthos*, flower] Part of a stamen; pollen forms in it and is dispersed from it.

antibiotic Metabolic product of soil microbes that kills bacterial competitors for nutrients.

antibody [Gk. *anti*, against] Antigen-binding receptor. Only B cells make antibodies, then position them at their surface or secrete them.

anticodon Sequence of three nucleotide bases in a tRNA molecule that can base-pair with a codon in an mRNA molecule.

antigen (AN-tih-jen) [Gk. *anti*, against, + *genos*, race, kind] Any molecular configuration that certain lymphocytes recognize as nonself and that triggers an immune response.

antigen-presenting cell Cell that processes and bears antigen fragments, bound to MHC molecules, at its surface. The antigen–MHC complexes promote immune responses.

aorta (ay-OR-tah) Of vertebrates, main artery of systemic circulation; transports the volume of blood that the heart pumps into it.

apical dominance Growth-inhibiting effect of a terminal bud on growth of lateral buds.

apical meristem (AY-pih-kul MARE-ih-stem) [L. *apex*, top, + Gk. *meristos*, divisible] Mass of self-perpetuating cells underlying primary growth at root tip and shoot tip.

apoptosis (APP-oh-TOE-sis) Programmed cell death. Molecular signals activate weapons of self-destruction in body cells that finished their prescribed functions or became altered, as by infection or cancerous transformation.

appendicular skeleton (ap-en-DIK-yoo-lahr) Bones of limbs, hips, and shoulders.

Archaebacteria Kingdom of prokaryotes more like eukaryotic cells than eubacteria; includes methanogens, halophiles, and thermophiles.

archipelago An island chain some distance away from a continent.

area effect Idea that larger islands support more species than smaller ones at equivalent distances from sources of colonizing species.

arteriole (ar-TEER-ee-ole) Type of blood vessel between arteries and capillaries. Controls over arteriolar dilation and constriction selectively distribute blood volume throughout body.

artery Large-diameter rapid-transport vessel with thick, muscular wall; smooths out blood pressure pulses caused by heart contractions.

arthropod Invertebrate having a hardened exoskeleton, specialized body segments, and jointed appendages. Insects are examples.

artificial selection Selection of traits among individuals of a population in an artificial environment, under contrived conditions.

asexual reproduction Any of a number of modes of reproduction by which offspring arise from a single parent and inherit the genes of that parent only.

atmosphere The volume of gases, airborne particles, and water vapor enveloping Earth.

atmospheric cycle Any biogeochemical cycle in which the element occurs in gaseous phase (e.g., carbon dioxide of the carbon cycle).

atom Smallest particle unique to an element; has one or more positively charged protons, electrons, and (except for hydrogen), neutrons.

atomic number Number of protons in the nucleus of each atom of an element.

ATP Adenosine triphosphate (ah-DEN-uh-seen try-FOSS-fate). Energy-carrying nucleotide with adenine, ribose, and three phosphate groups. Phosphate-group transfers from ATP drive most energy-requiring metabolic reactions.

australopith (OHSS-trah-low-pith) [L. *australis*, southern, + Gk. *pithekos*, ape] One of earliest known hominids; a primate that may be on or near evolutionary road to modern humans.

autoimmune response Misdirected immune response in which lymphocytes mount an attack against normal body cells.

autonomic nervous system (auto-NOM-ik) All nerves from central nervous system to the smooth muscle, cardiac muscle, and glands of viscera (soft internal organs and structures).

autosome Type of chromosome that is the same in males and in females of the species.

autotroph (AH-toe-trofe) [Gk. *autos*, self, + *trophos*, feeder] Organism that makes its own organic compounds using an environmental energy source (e.g., sunlight) along with carbon dioxide as its carbon source.

auxin (AWK-sin) Plant hormone; promotes stem lengthening and responses to gravity and light.

axial skeleton (AX-ee-uhl) Skull, backbone, ribs, and breastbone (sternum).

axon Cylindrical extension of neuron cell body, specialized for the rapid propagation of action potentials.

B cell B lymphocyte; only cell that produces antibodies. Key player in immune responses.

bacterial conjugation The transfer of plasmid DNA from one bacterial cell to another.

bacteriophage (bak-TEER-ee-oh-fahj) [Gk. *baktērion*, small staff, rod, + *phagein*, to eat] Category of viruses that infect bacterial cells.

Barr body In body cells of female mammals, one of either of the two X chromosomes that was condensed to inactivate its genes.

basal body A centriole which, after giving rise to microtubules of a flagellum or cilium, remains attached to its base in the cytoplasm.

base Any substance that accepts hydrogen ions when dissolved in water.

base sequence Particular order in which one nucleotide base follows the next in a strand of DNA or RNA. The order is unique in at least some regions for each species.

basophil Fast-acting white blood cell; its secretions (e.g., histamine) cause vasodilation during an inflammatory response.

behavior Response to external and internal stimuli based on sensory, neural, endocrine, and effector components. Has a genetic basis, can evolve, and can be modified by learning.

benthic province All sediments and rocky formations of the ocean bottom.

biennial (bi-EN-yul) Flowering plant that completes life cycle in two growing seasons.

bilateral symmetry Body plan in which left and right halves generally are mirror images.

binary fission Asexual reproductive mode of protozoans and some other animals. The body divides into two parts of the same or different sizes. *Compare* prokaryotic fission.

binomial system Of taxonomy, assigning a generic and a specific name to each species.

biogeochemical cycle Slow movement of an element from the environment, through food webs, then back to the environment.

biogeographic realm [Gk. *bios*, life, + *geographein*, to describe the Earth's surface] One of six vast land areas, each with distinctive kinds and numbers of plants and animals.

biological clock Internal time-measuring mechanism that helps adjust an organism's daily activities, seasonal activities, or both in response to environmental cues.

biological magnification The increasing concentration of a nondegradable or slowly degradable substance in body tissues as it is passed along food chains.

biological species concept Defines a species as one or more populations of individuals that are interbreeding under natural conditions, that are producing fertile offspring, and that are isolated reproductively from other such populations. Applies to sexually reproducing species only.

bioluminescence Any organism flashing with fluorescent light by way of an ATP-driven reaction involving enzymes (luciferases).

biomass Combined weight of all organisms at a given trophic level in an ecosystem.

biome Large subdivision of a biogeographic realm, distinctive in its habitat conditions, community structure, and endemic species.

biosphere [Gk. *bios*, life, + *sphaira*, globe] All regions of the Earth's waters, crust, and atmosphere in which organisms live.

biosynthetic pathway Metabolic pathway by which organic compounds are synthesized.

biotic potential Of population growth for a given species, the maximum rate of increase per individual under ideal conditions.

bipedalism Habitually walking upright on two feet, as by ostriches and hominids.

bird Only vertebrate that produces feathers; evolutionarily linked with dinosaurs.

blastocyst (BLASS-tuh-sist) [Gk. *blastos*, sprout, + *kystis*, pouch] Type of blastula. Blastomeres form a surface layer, a cavity filled with their own secretions, and an inner cell mass.

blastomere One of the small, nucleated cells that form during cleavage of animal zygote.

blastula (BLASS-chew-lah) An early outcome of cleavage; a number of blastomeres enclosing a fluid-filled cavity.

blood Fluid connective tissue of water, solutes, and formed elements (blood cells and platelets). Blood transports substances to and from cells, and helps maintain internal environment.

blood pressure Fluid pressure, generated by heart contractions, that circulates blood.

blood-brain barrier Mechanism that controls which solutes enter cerebrospinal fluid; helps protect the brain and spinal cord.

bone Vertebrate organ with mineral-hardened connective tissue (bone tissue); helps move body, protects other organs, stores minerals. Some (e.g., breastbone) produce blood cells.

bottleneck Severe reduction in population size brought about by intense selection pressure or by a natural calamity.

Bowman's capsule Cup-shaped portion of a nephron that receives water and solutes being filtered from blood.

brain Of most nervous systems, integrating center that receives and processes sensory input and issues coordinated commands for responses by muscles and glands.

brain stem Of vertebrates, nervous tissue that evolved first and still persists in the hindbrain, midbrain, and forebrain.

bronchiole Finely branched airway; part of the bronchial tree inside the lung.

bronchus, plural **bronchi** (BRONG-cuss, BRONG-kee) [Gk. *bronchos*, windpipe] Tubular airway that branches from trachea and leads into lungs.

brown alga Photoautotrophic protistan with chlorophylls *a*, c_1, and c_2, and carotenoids (e.g., fucoxanthin). Mostly marine. Range in size from microscopic to giant multicelled kelps.

bryophyte Nonvascular land plant requiring free water to complete fertilization. Haploid dominance in life cycle. Cuticle and stomata present in some species. A moss (e.g., peat moss), liverwort, or hornwort.

bud Undeveloped shoot, mainly meristematic tissue. Small, protective scales often cover it.

buffer system A weak acid and the base that forms when it dissolves in water. The two work as a pair to counter slight shifts in pH.

bulk flow In response to a pressure gradient, a movement of more than one kind of molecule in the same direction in the same medium, as in blood, sap, or air.

C3 plant Plant that uses three-carbon PGA as the first intermediate for carbon fixation during the second stage of photosynthesis.

C4 plant Plant that uses oxaloacetate (a four-carbon compound) as the first intermediate for carbon fixation during the second stage of photosynthesis. CO_2 is fixed twice, in two cell types. Carbon dioxide accumulates in leaf and helps counter photorespiration.

Calvin–Benson cycle Cyclic, light-independent reactions; "synthesis" part of photosynthesis. Uses ATP and NADPH from light-dependent reactions. RuBP or some compound to which carbon has been affixed is rearranged and regenerated, and a sugar phosphate forms.

CAM plant Type of plant that conserves water by opening stomata only at night, when it fixes carbon dioxide by means of a C4 pathway.

camouflage Adaptation in coloration, form, patterning, or behavior that helps predators or prey hide in the open (blend with their surroundings) and escape detection.

cancer Malignant tumor; mass of cells that have grossly altered plasma membrane and cytoplasm, grow and divide abnormally, and adhere weakly to home tissue (which leads to metastasis). Lethal unless eradicated.

capillary, blood [L. *capillus*, hair] Blood vessel with thin endothelial wall and small diameter; functions in the exchange of carbon dioxide, oxygen, and other solutes between blood and interstitial fluid.

capillary bed Diffusion zone, consisting of great numbers of capillaries, where blood and interstitial fluid exchange substances.

carbohydrate [L. *carbo*, charcoal, + *hydro*, water] Molecule of carbon, hydrogen, and oxygen mostly in a 1:2:1 ratio. Carbohydrates are structural materials, energy stores, and transportable energy forms. Monosaccharide, oligosaccharide, or polysaccharide.

carbon cycle An atmospheric cycle. Carbon moves from its largest reservoirs (sediments, rocks, and the ocean), through the atmosphere (mostly as CO_2), through food webs, and back to the reservoirs.

carbon dioxide fixation First step of light-independent reactions. Enzyme action affixes carbon (from CO_2) to RuBP or to some other compound for entry into the Calvin–Benson cycle.

carcinogen (kar-SIN-uh-jen) Any substance or agent that can trigger cancer.

cardiac cycle (KAR-dee-ak) [Gk. *kardia*, heart, + *kyklos*, circle] Sequence of muscle contraction and relaxation in one heartbeat.

cardiac pacemaker Sinoatrial (SA) node; basis of normal rate of heartbeat. Self-excitatory cardiac muscle cells spontaneously generate rhythmic waves of excitation over heart.

cardiovascular system Organ system that has blood, one or more hearts, and blood vessels.

carnivore [L. *caro, carnis*, flesh, + *vovare*, to devour] Animal that eats other animals.

carotenoid (kare-OTT-en-oyd) An accessory pigment. Different kinds absorb blue-violet and blue-green wavelengths, the energy of which is transferred to chlorophylls. They reflect yellow, orange, and red wavelengths.

carpel (KAR-pul) Female reproductive part of a flower. One or more carpels make up the ovary and a stigma, and often a style.

carrying capacity The maximum number of individuals in a population (or species) that a given environment can sustain indefinitely.

cartilage Connective tissue with solid, pliable intercellular material that resists compression.

Casparian strip Narrow, waxy, impermeable band between walls of abutting cells making up root endodermis and exodermis.

cDNA DNA molecule copied from a mature mRNA transcript by reverse transcription.

cell [L. *cella*, small room] Smallest living unit; organized unit with a capacity to survive and reproduce on its own, given DNA instructions, energy sources, and raw materials.

cell count The number of cells of a given type in one microliter of blood.

cell cycle Events by which a cell increases in mass, roughly doubles its cytoplasmic components, duplicates its DNA, then divides its nucleus and cytoplasm. Extends from the time a cell forms until it completes division.

cell differentiation Key development process. Different cell lineages become specialized in their composition, structure, and function by activating and suppressing some fraction of the genome in different ways.

cell junction Site that joins cells physically, functionally, or both (e.g., tight junction in animals; plasmodesma in plants).

cell plate Disklike structure that forms in a plant cell after nuclear division; becomes a crosswall, with new plasma membrane on both surfaces, that divides the cytoplasm.

cell theory Theory stating that all organisms consist of one or more cells, the cell is the smallest unit with a capacity for independent life, and all cells arise from preexisting cells.

cell wall Of most bacteria, many protistan and fungal cells, and plant cells, the outermost, semirigid, permeable structure that helps the cell retain its shape and resist rupturing when the internal fluid pressure increases.

central nervous system Brain and spinal cord.

central vacuole Large, fluid-filled organelle of living, mature plant cell. Stores amino acids, sugars, ions, and toxic wastes. As it enlarges during growth, it forces the primary cell wall to expand and cell surface area to increase.

centriole (SEN-tree-ohl) Structure that gives rise to microtubules of cilia and flagella.

centromere (SEN-troh-meer) [Gk. *kentron*, center, + *meros*, a part] Constricted portion of each chromosome; location of a kinetochore to which spindle microtubules become attached.

cephalization (sef-ah-lah-ZAY-shun) [Gk. *kephalikos*, head] The concentration of sensory structures and nerve cells in a head; occurred during evolution of bilateral animals.

cerebellum (ser-ah-BELL-um) [L. diminutive of *cerebrum*, brain] Hindbrain region with reflex centers for maintaining posture and smoothing out limb movements.

cerebral cortex Thin surface layer of cerebral hemispheres; receives, integrates, and stores sensory information; coordinates responses.

cerebrospinal fluid Clear extracellular fluid, enclosed in a continuous system of canals and chambers, that bathes and protects the brain and spinal cord.

cerebrum (suh-REE-bruhm) Part of forebrain that integrates olfactory input, selects motor responses. In mammals, it evolved into the most complex integrating center.

channel protein Transport protein that serves as an open or gated channel where ions and other solutes move across a cell membrane.

chemical bond A union between the electron structures of two or more atoms or ions.

chemical energy Potential energy of molecules.

chemical synapse (SIN-aps) [Gk. *synapsis*, union] Narrow cleft between a presynaptic neuron and a postsynaptic cell. Molecules of neurotransmitter diffuse across it.

chemiosmotic theory (KIM-ee-oz-MOT-ik) Idea that an electrochemical gradient drives ATP formation. H+ accumulates in a membranous compartment, then flows out through proteins (ATP synthases) in response to concentration and electric gradients. The flow drives the attachment of inorganic phosphate to ADP.

chemoautotroph (KEE-moe-AH-toe-trofe) Type of bacterium that can synthesize its own organic compounds using only carbon dioxide as the carbon source and an inorganic substance as the energy source.

chemoreceptor Sensory receptor that detects chemical energy (ions or molecules dissolved in the fluid bathing it).

chlorofluorocarbon (KLORE-oh-FLOOR-oh-car-bun) Compound of chlorine, fluorine, and carbon that has been contributing to ozone thinning.

chlorophyll (KLOR-uh-fill) [Gk. *chloros*, green, + *phyllon*, leaf] Main photosynthetic pigment; absorbs violet-to-blue and red wavelengths but transmits green.

chloroplast (KLOR-uh-plast) The organelle of photosynthesis in plants and many protistans.

chordate Animal with a notochord, dorsal hollow nerve cord, pharynx, and gill slits in pharynx wall during at least part of life cycle.

chorion (CORE-ee-on) Type of extraembryonic membrane that becomes part of placenta. Villi (absorptive structures) form at its surface and facilitate exchanges of substances between the embryo and mother.

chromatid (CROW-mah-tid) Of a duplicated eukaryotic chromosome, one of two DNA molecules (with associated proteins) attached at centromere. Mitosis or meiosis separates them; each becomes a separate chromosome.

chromatin A cell's collection of DNA and all of the proteins associated with it.

chromosome (CROW-moe-some) [Gk. *chroma*, color, + *soma*, body] Of eukaryotic cells, a DNA molecule, duplicated or unduplicated, with a profusion of associated proteins. Of prokaryotic cells (bacteria), a circular DNA molecule with few, if any, proteins attached.

chromosome number All chromosomes in a given type of cell. *See* haploidy; diploidy.

cilium (SILL-ee-um), plural **cilia** Short motile or sensory structure projecting from surface of certain eukaryotic cells; its core is a 9 + 2 array of microtubules.

circadian rhythm (ser-KAYD-ee-un) [L. *circa*, about, + *dies*, day] Cycle of physiological events completed every twenty-four hours or so independently of environmental change.

circulatory system Organ system that moves substances to and from cells, and often helps stabilize body temperature and pH. Typically consists of a heart, blood vessels, and blood.

cladogram [Gk. *clad-*, branch] Evolutionary tree diagram with groups arranged by branch points to show relative relationships. Groups closer together share a more recent common ancestor than those farther apart.

classification scheme A way of organizing and retrieving information about species.

cleavage Early stage of animal development. Mitotic cell divisions divide a fertilized egg into many smaller, nucleated cells; original volume of egg cytoplasm does not increase.

cleavage furrow Ringlike depression defining cleavage plane for dividing animal cells.

cleavage reaction Enzyme action that splits a molecule in two or more parts; hydrolysis is an example.

climate For a specified region, the prevailing weather conditions (e.g., temperature, cloud cover, wind speed, rainfall, and humidity).

climax community Array of species that has stabilized under prevailing habitat conditions.

climax pattern model Idea that physical conditions and other environmental factors often vary in their influence over a large region, so that stable communities other than the climax stage may also persist in that region.

cloaca (kloe-AY-kuh) Chamber or duct in last portion of gut of some animals that also serves in excretion, reproduction, and sometimes respiration.

cloning vector Plasmid that has been modified in the laboratory to accept foreign DNA.

club fungus Fungus with club-shaped cells that produce and bear spores.

cnidarian Radial invertebrate at tissue level of organization; the only nematocyst producer. Medusae and polyps are typical body forms.

coal Nonrenewable energy source that formed over 280 million years ago from submerged, undecayed, and compacted plant remains.

codominance In heterozygotes, simultaneous expression of a pair of nonidentical alleles that specify different phenotypes.

codon mRNA base triplet; its linear sequence corresponds to a linear sequence of amino acids in a polypeptide chain. Of 64 codons, 61 code for amino acids; 3 of these are also START signals for translation, 1 is a STOP signal for translation.

coelom (SEE-lum) [Gk. *koilos*, hollow] Cavity lined with peritoneum between the gut and body wall of most animals.

coenzyme Nucleotide that acts as an enzyme helper; it accepts electrons and hydrogen atoms stripped from substrates at a reaction site and transfers them to another reaction site.

coevolution Joint evolution of two species that interacting closely. When one evolves, the change affects selection pressures that are operating between the two, so that the other also evolves.

cofactor Metal ion or coenzyme; it helps an enzyme catalyze a reaction or it transfers electrons, atoms, or functional groups to a different substrate.

cohesion Capacity to resist rupturing when placed under tension (stretched).

cohesion theory of water transport Theory that the collective cohesive strength of their hydrogen bonds allows water molecules to be pulled up through a plant's xylem in response to transpiration from leaves.

collenchyma (coll-ENG-kih-mah) A simple plant tissue (one cell type only). Gives flexible support for primary growth, as in stems that are lengthening.

colon (CO-lun) Large intestine.

commensalism [L. *com*, together, + *mensa*, table] Ecological interaction between two (or more) species in which one benefits directly and the other is affected little, if at all.

communication display Behavior pattern that is a social signal. Often ritualized (has intended changes in function of common patterns).

communication signal Social cue encoded in stimuli (e.g., specific body coloration, body patterning, odors, sounds, and postures).

community All populations in a habitat. Also, a group of organisms with similar life-styles in a habitat, such as a bird community.

companion cell Living parenchyma cell that is specialized to help load organic compounds into adjacent conducting cells of phloem.

comparative morphology [Gk. *morph*, form] Scientific study of comparable body parts of adults or embryonic stages of major lineages.

competitive exclusion Theory that two or more species that require identical resources cannot coexist indefinitely.

complement system Set of proteins circulating in inactive form in vertebrate blood. Different kinds promote inflammation, induce lysis of pathogens, and stimulate phagocytes during nonspecific defenses and immune responses.

compound Substance consisting of two or more elements in unvarying proportions.

concentration gradient Between two adjoining regions, a difference in the number of molecules (or ions) of a substance. All molecules are in constant motion. As they collide, they careen outward to an adjoining region where they are less concentrated. Barring other forces, all substances diffuse down such a gradient.

condensation reaction Two molecules become covalently bonded into a larger molecule, and water often forms as a by-product.

cone cell In vertebrate eye, a photoreceptor that responds to intense light and contributes to sharp daytime vision and color perception.

conifer Type of gymnosperm; generally, an evergreen woody tree or shrub with needle-like or scale-like leaves.

connective tissue proper Animal tissue with a characteristic proportion of fibroblasts and other cells, fibers (e.g., collagen, elastin), and a ground substance consisting of modified polysaccharides.

conservation biology International efforts to identify all species (particularly in marine and land ecosystems most vulnerable to habitat destruction and high extinction rates) and to design long-term management programs based on ecological and evolutionary principles.

consumer [L. *consumere*, to take completely] A heterotroph that feeds on cells or tissues of other organisms for carbon and energy (e.g., herbivores, carnivores, and parasites).

continuous variation Of a population, a more or less continuous range of small differences in a given trait among its individuals.

contractile vacuole (kun-TRAK-till VAK-you-ohl) [L. *contractus*, to draw together] Organelle that takes up excess water in cell body and contracts to expel water through a pore.

control group A group used as a standard for comparison with an experimental group. Ideally, a control group is identical with the experimental group in all respects except for the one variable being studied.

cork cambium Lateral meristem that replaces epidermis with cork on woody plant parts.

corpus callosum (CORE-pus ka-LOW-sum) A band of numerous axons linking two cerebral hemispheres.

corpus luteum (CORE-pus LOO-tee-um) Of female mammals, a glandular structure that forms from cells of a ruptured ovarian follicle; it secretes progesterone and estrogen.

cortex [L. *cortex*, bark] Generally, a rindlike layer such as kidney cortex or cell cortex. In vascular plants, a ground tissue that supports plant parts and stores food.

cotyledon (KOT-uhl-EE-dun) Seed leaf. One develops in monocot seed and has digestive roles; two develop in dicot seed and store food for germination and early growth.

courtship display Pattern of ritualized social behavior between potential mates. Commonly incorporates frozen postures, exaggerated yet simplified movements, and visual signals.

covalent bond (koe-VAY-lunt) [L. *con*, together, + *valere*, to be strong] Sharing of one or more electrons between atoms or groups of atoms. If electrons shared equally, bond is nonpolar. If shared unequally, it is polar (slightly positive at one end, slightly negative at the other).

cross-bridge formation Reversible, ATP-driven interaction between a sarcomere's actin and myosin filaments; results in a short power stroke that is the basis of muscle contraction.

crossing over At prophase I of meiosis, an interaction in which nonsister chromatids (of a pair of homologous chromosomes) break at corresponding sites and exchange segments; genetic recombination is the result.

culture Sum of behavior patterns of a social group, passed between generations by way of learning and symbolic behavior, especially language.

cuticle (KEW-tih-kull) Body covering. Of plants, a transparent covering of waxes and cutin on outer epidermal cell walls. Of annelids, a thin, flexible coat. Of arthropods, a lightweight exoskeleton hardened with protein and chitin.

cyclic AMP (SIK-lik) Short for cyclic adenosine monophosphate. A nucleotide having roles in intercellular communication.

cyclic pathway of ATP formation Most ancient photosynthetic pathway, at plasma membrane of some bacteria and at chloroplast's thylakoid membrane. Photosystems in the membrane give up electrons to transport systems, which return them to photosystems. Electron flow across the membrane sets up H^+ gradients that drive ATP formation at nearby sites.

cyst Of many microorganisms, a resting stage with thick outer layers that typically forms under adverse conditions. Of skin, abnormal, fluid-filled sac without an external opening.

cytochrome (SIGH-toe-krome) [Gk. *kytos*, hollow vessel, + *chrōma*, color] Iron-containing protein molecule of electron transport systems, as used in photosynthesis, aerobic respiration.

cytokinesis (SIGH-toe-kih-NEE-sis) [Gk. *kinesis*, motion] Cytoplasmic division; splitting of a parent cell into daughter cells.

cytokinin (SIGH-toe-KYE-nin) Type of plant hormone; stimulates cell division, promotes leaf expansion, and retards leaf aging.

cytology Study of cell structure and function.

cytomembrane system System of organelles that modify, package, and distribute newly formed proteins and lipids. Endoplasmic reticulum, Golgi bodies, lysosomes, and various vesicles are its components.

cytoplasm (SIGH-toe-plaz-um) All cell parts, particles, and semifluid substances between the plasma membrane and the nucleus (or nucleoid, in bacteria).

cytoplasmic localization Parceling of a portion of maternal messages in the egg cytoplasm to each blastomere that forms during cleavage.

cytosine (SIGH-toe-seen) Pyrimidine; one of the nitrogen-containing bases in nucleotides.

cytoskeleton Dynamic internal framework of eukaryotic cells. Its microtubules and other components structurally support the cell and organize and move its internal parts. It helps free-living cells move in their environment.

cytotoxic T cell T lymphocyte that touch-kills body cells that are infected or altered, as by cancerous transformation.

decomposer [partly L. *dis-*, to pieces, + *companere*, arrange] Fungal or bacterial heterotroph. Obtains carbon and energy from remains, products, or wastes of organisms. Collectively, decomposers help cycle nutrients to producers in ecosystems.

deductive logic Pattern of thinking by which an individual makes inferences about specific consequences or specific predictions that must follow from a hypothesis.

deforestation Removal of all the trees from a large tract of land (e.g., Amazon River Basin).

degradative pathway A stepwise series of metabolic reactions that break down organic compounds to products of lower energy.

deletion At the cytological level, loss of a segment from a chromosome. At the molecular level, loss of one to several base pairs from a DNA molecule.

denaturation (deh-NAY-chur-AY-shun) Loss of a molecule's three-dimensional shape as weak bonds (e.g., hydrogen bonds) are disrupted.

dendrite (DEN-drite) [Gk. *dendron*, tree] Short, slender extension from cell body of a neuron; commonly a signal input zone.

denitrification (DEE-nite-rih-fih-KAY-shun) Conversion of nitrate or nitrite by certain soil bacteria to gaseous nitrogen (N_2) and a small amount of nitrous oxide (N_2O).

density-dependent control Factor that limits population growth by reducing the birth rate, increasing death and dispersal rates, or all of these. Predation, parasitism, disease, and competition for resources are examples.

density-independent factor Factor that causes a population's death rate to rise independently of population density. Storms and floods are examples.

dentition (den-TIH-shun) Collectively, the type, size, and number of an animal's teeth.

dermal tissue system Tissues that cover and protect all exposed surfaces of a plant.

dermis Skin layer beneath the epidermis that consists primarily of dense connective tissue.

desert Biome that forms where the potential for evaporation greatly exceeds rainfall, and where soil is thin and vegetation sparse.

desertification (dez-urt-ih-fih-KAY-shun) The conversion of grassland or irrigated or rain-fed cropland to a desertlike condition, with a drop in productivity of 10 percent or more.

detrital food web Network of food chains in which energy flows mainly from plants through arrays of detritivores and decomposers.

detritivore (dih-TRY-tih-vore) [L. *detritus*; after *deterere*, to wear down]. Heterotroph that ingests decomposing particles of organic matter (e.g., crab, earthworm, roundworm).

deuterostome (DUE-ter-oh-stome) [Gk. *deuteros*, second, + *stoma*, mouth] Category of bilateral animals in which the first indentation to form in the early embryo becomes an anus (e.g., an echinoderm or chordate).

development Of multicelled species, emergence of specialized, morphologically distinct body parts according to a genetic program.

diaphragm (DIE-uh-fram) [Gk. *diaphragma*, to partition] Muscular partition between thoracic and abdominal cavities with role in breathing. Also a contraceptive device inserted into the vagina to prevent sperm from entering uterus.

dicot (DIE-kot) [Gk. *di*, two, + *kotylēdōn*, cup-shaped vessel] Dicotyledon. Flowering plant generally characterized by embryos with two cotyledons; net-veined leaves; and floral parts arranged in fours, fives, or multiples of these.

diffusion Net movement of like molecules or ions down their concentration gradient. In the absence of other forces, they collide constantly and randomly owing to their inherent energy. It occurs most frequently where they are most crowded, so the net movement is outward from regions of higher to lower concentrations.

digestive system Body sac or tube having one or two openings and often specialized regions for ingesting, digesting, and absorbing food, then eliminating undigested residues.

dihybrid cross Experimental cross between true-breeding parents that differ in two traits. The hybrid F_1 offspring inherit two gene pairs, each with two nonidentical alleles.

diploidy (DIP-loyd-ee) Presence of two of each type of chromosome (pairs of homologous chromosomes) in a cell nucleus at interphase. *Compare* haploidy.

directional selection Mode of natural selection by which allele frequencies underlying a range of phenotypic variation shift in a consistent direction, in response to directional change or to new conditions in the environment.

disaccharide (die-SAK-uh-ride) [Gk. *di*, two, + *sakcharon*, sugar] A common oligosaccharide; two covalently bonded sugar monomers.

disease Outcome of infection when defenses against a pathogen cannot be mobilized fast enough, and the pathogen's activities interfere with normal body functions.

disruptive selection Mode of natural selection by which forms of a trait at both ends of range of variation are favored and the intermediate forms are selected against.

distal tubule Tubular portion of nephron that selectively reabsorbs water and sodium.

distance effect Idea that only species adapted for long-distance dispersal can be potential colonists of islands far from their home range.

diversity of life Sum total of all variations in form, function, and behavior that accumulated in different lineages. Such variations generally are adaptive to prevailing conditions or were adaptive to conditions in the past.

DNA Deoxyribonucleic acid (dee-OX-ee-RYE-bow-new-CLAY-ik). For cells and many viruses, a nucleic acid that is the molecule of inheritance. Hydrogen bonds hold its two helically twisted nucleotide strands together. DNA's nucleotide sequence encodes instructions for synthesizing proteins, hence new individuals of a species.

DNA clone Many identical copies of foreign DNA that was inserted into plasmids and later replicated repeatedly after being taken up by population of host cells (typically, bacteria).

DNA–DNA hybridization *See* nucleic acid hybridization.

DNA fingerprint DNA fragments, inherited in a Mendelian pattern from each parent, that give each individual a unique identity. For humans, the most informative fragments are from tandem repeats (short regions of repeated DNA) that differ greatly among individuals.

DNA ligase (LYE-gaze) Enzyme that seals new base-pairings during DNA replication; also used in recombinant DNA technology.

DNA polymerase (poe-LIM-uh-raze) Enzyme of replication and repair that assembles a new strand of DNA on a parent DNA strand; it uses exposed nucleotide bases as the template.

DNA repair Process that restores the original base sequence when part of a DNA molecule gets altered. DNA polymerases, DNA ligases, and other enzymes execute the repair.

DNA replication Process by which molecules of DNA are duplicated for later distribution to daughter nuclei. Completed before mitosis and meiosis in eukaryotic cells and during prokaryotic fission in bacterial cells.

dominance hierarchy Social organization in which some individuals of the group have adopted a subordinate status to others.

dominant allele Of diploid cells, an allele that masks phenotypic effect of any recessive allele paired with it.

dormancy [L. *dormire*, to sleep] A predictable time of metabolic inactivity for many spores, cysts, seeds, perennials, and some animals.

double fertilization Of flowering plants only, fusion of a sperm nucleus with an egg nucleus (a zygote forms), and fusion of another sperm nucleus with nuclei of endosperm mother cell, which gives rise to a nutritive tissue.

doubling time Time it takes for a population to double in size.

drug addiction Chemical dependence on a drug, which assumes an "essential" biochemical role in the body following habituation and tolerance.

dry shrubland Biome that forms when annual rainfall is less than 25 to 60 centimeters; short, multibranched woody shrubs dominate.

dry woodland Biome that forms when annual rainfall is about 40 to 100 centimeters; it may have tall trees, but no dense canopy.

duplication Gene sequence repeated several to many hundreds or thousands of times. Even normal chromosomes have such sequences.

ecdysone Hormone that has major influence over early development of many insects.

echinoderm Type of invertebrate with calcified spines, needles, or plates on body wall. Radially symmetrical, but with some bilateral features. Sea stars and sea urchins are examples.

ecology [Gk. *oikos*, home, + *logos*, reason] Scientific study of how organisms interact with one another and with their physical and chemical environment.

ecoregion Large area that is representative of a globally important biome or water province.

ecosystem Array of organisms and their physical environment, all interacting by a flow of energy and a cycling of materials.

ecosystem modeling An analytical method, based on computer programs and models, of predicting unforeseen effects of specific disturbances to an ecosystem.

ectoderm [Gk. *ecto*, outside, + *derma*, skin] The first-formed, outermost primary tissue layer of animal embryos; gives rise to nervous system tissues and integument's outer layer.

effector Muscle (or gland) that responds to signals from an integrator (e.g., a brain) by producing movement (or chemical change) to adjust the body to changing conditions.

effector cell Differentiated cell of a lymphocyte subpopulation that immediately engages and destroys an antigen-bearing agent during an immune response.

egg Mature female gamete; an ovum.

El Niño Massive eastward movement of warm surface waters of the western equatorial Pacific that displaces cooler water off coast of South America. A recurring event that disrupts the climate and ecosystems throughout the world.

electromagnetic spectrum All wavelengths from radiant energy less than 10^{-5} nm long to radio waves more than 10 km long.

electron Negatively charged unit of matter, with particulate and wavelike properties, that occupies one of the orbitals around the atomic nucleus. Atoms gain, lose, or share electrons.

electron transfer A molecule donates one or more electrons to another molecule.

electron transport phosphorylation (FOSS-for-ih-LAY-shun) Last stage of aerobic respiration, when electrons from reaction intermediates flow through a membrane transport system that gives them up to oxygen. The flow sets up an electrochemical gradient that drives ATP formation at other sites in the membrane.

electron transport system Organized array of membrane-bound enzymes and cofactors that accept and donate electrons in series. It sets up an electrochemical gradient that makes H^+ flow across the membrane. The flow energy drives ATP formation at other reaction sites.

element Substance that cannot be degraded by ordinary means into a substance having different properties.

embryo (EM-bree-oh) [Gk. *en*, in, + *bryein*, to swell] Of animals, a multicelled body formed by way of cleavage, gastrulation, and other early developmental events. Of plants, a young sporophyte, from the time of the earliest cell divisions after fertilization until germination.

embryo sac Female gametophyte of flowering plants.

embryonic induction In a growing embryo, release of a gene product from one tissue that affects the developmental fate of an adjacent tissue.

emerging pathogen Deadly pathogen, either a newly mutated strain of an existing type or one that evolved long ago and is now exploiting an increased presence of human hosts.

emulsification In chyme, a suspension of fat droplets coated with bile salts.

end product Substance present at the end of a metabolic pathway.

endangered species A species that is highly vulnerable to extinction by natural events (e.g., genetic drift after a severe bottleneck) or by human activities (e.g., accidental introduction of competitive, exotic species).

endergonic reaction (en-dur-GONE-ik) Chemical reaction having a net gain in energy.

endocrine gland Ductless gland that secretes hormones, which later enter the bloodstream.

endocrine system Integrative system of cells, tissues, and organs, functionally linked to the nervous system, that exerts control by way of its hormones and other chemical secretions.

endocytosis (EN-doe-sigh-TOE-sis) Cell uptake of substances when part of plasma membrane forms a vesicle around them. Three routes are receptor-mediated endocytosis, bulk transport of extracellular fluid, and phagocytosis.

endoderm [Gk. *endon*, within, + *derma*, skin] Inner primary tissue layer of animal embryos; source of inner gut lining and derived organs.

endodermis Sheetlike wrapping of single cells around root vascular cylinder; helps control uptake of water and dissolved nutrients.

endometrium (EN-doe-MEET-ree-um) [Gk. *metrios*, of the womb] Inner lining of uterus.

endoplasmic reticulum or **ER** (EN-doe-PLAZ-mik reh-TIK-yoo-lum) Organelle that starts at nucleus and curves through cytoplasm. New polypeptide chains acquire side chains inside rough ER (with ribosomes on its cytoplasmic side); smooth ER (with no ribosomes) is a site of lipid synthesis.

endoskeleton [Gk. *endon*, within, + *sklēros*, hard] Of chordates, an internal framework of cartilage, bone, or both that works with skeletal muscle to support and move body, and to maintain posture.

endosperm (EN-doe-sperm) Nutritive tissue that surrounds an embryo sporophyte inside the seed of a flowering plant.

endospore Resting structure formed by some bacteria; encloses a duplicate of the bacterial chromosome and a portion of cytoplasm.

endosymbiosis Continuing physical contact between one species and another species that lives in its body.

energy Capacity to do work.

energy carrier Molecule that delivers energy from one metabolic reaction site to another. ATP is the most common energy carrier.

energy flow pyramid Pyramidal diagram of an ecosystem's trophic structure. It depicts the energy losses at each transfer to another trophic level.

enhancer A short DNA base sequence that is a binding site for an activator protein.

entropy (EN-trow-pee) Measure of the degree of disorder in a system (how much energy has become so disorganized, usually as dissipating heat, that it is no longer available to do work). A system must receive and use energy to stay organized; it tends toward entropy when its energy outputs exceed energy inputs.

enzyme (EN-zime) A protein or one of a few RNAs that greatly speed (catalyze) reactions between substances, most often at functional groups.

eosinophil Fast-acting white blood cell; its enzyme secretions digest holes in parasitic worms during an inflammatory response.

epidermis Outermost tissue layer of plants and all animals above sponge level of organization.

epiglottis Flaplike structure between pharynx and larynx; its controlled positional changes direct air into trachea or food into esophagus.

epinephrine (ep-ih-NEF-rin) Adrenal hormone that raises blood levels of sugar and fatty acids; increases heart's rate and force of contraction.

epistasis (eh-PISS-tah-sis) Interaction among the products of two or more gene pairs.

epithelium (EP-ih-THEE-lee-um) Tissue that covers the animal body's external surfaces and lines its internal cavities and tubes. It has one free surface and one resting on a basement membrane that is next to a connective tissue.

equilibrium, dynamic [Gk. *aequus*, equal, + *libra*, balance] Point at which a reaction runs forward as fast as in reverse, so no net change in reactant or product concentrations.

erosion Movement of land under the force of wind, running water, and ice.

erythrocyte (eh-RITH-row-site) [Gk. *erythros*, red, + *kytos*, vessel] Red blood cell.

esophagus (ee-SOF-uh-gus) A muscular tube just after the pharynx in vertebrate and many invertebrate digestive tracts.

essential amino acid Any amino acid that an organism cannot synthesize for itself and must obtain from a food source.

essential fatty acid Any fatty acid that an organism cannot synthesize for itself and must obtain from a food source.

estrogen (ESS-trow-jen) Female sex hormone that helps oocytes mature, induces changes in the uterine lining during menstrual cycle and pregnancy, helps maintain secondary sexual traits, and affects growth and development.

estrus (ESS-truss) [Gk. *oistrus*, frenzy] Of most mammals, a cyclic period during which the female is sexually receptive to the male.

estuary (EST-you-ehr-ee) A partially enclosed coastal region where seawater mixes with fresh water and runoff from the land, as by streams and rivers.

ethylene (ETH-il-een) Plant hormone that stimulates fruit ripening and abscission.

Eubacteria Kingdom of all prokaryotic cells except archaebacteria.

eukaryotic cell (yoo-CARE-EE-oh-tic) [Gk. *eu*, good, + *karyon*, kernel] Cell having a nucleus and other membrane-bound organelles.

eutrophication The enrichment of a body of water with nutrients; typically leads to reduced transparency and a phytoplankton-dominated community.

evaporation [L. *e-*, out, + *vapor*, steam] The conversion of a substance from liquid state to gaseous state under input of heat energy.

evolution, biological [L. *evolutio*, unrolling] Genetic change in a line of descent over time, brought about by microevolutionary events (gene mutation, natural selection, genetic drift, and gene flow).

evolutionary tree Treelike diagram in which each branch signifies a separate line of descent from a common ancestor and each branch point signifies a time of divergence.

excitatory postsynaptic potential (EPSP) A graded potential that arises at an input zone of an excitable cell and that drives the cell membrane closer to threshold.

excretion Removal of excess water, solutes, and wastes, and some harmful substances from body by way of a urinary system or glands.

exergonic reaction (EX-ur-GONE-ik) Chemical reaction that shows a net loss in energy.

exocrine gland (EK-suh-krin) [Gk. *es*, out of, + *krinein*, to separate] Glandular structure that secretes products, usually through ducts or tubes, to a free epithelial surface.

exocytosis (EK-so-sigh-TOE-sis) Release of the contents of a vesicle at the cell surface, where the vesicle's membrane fuses with and becomes part of the plasma membrane.

exodermis Cylindrical sheet of cells just inside the root epidermis of most flowering plants; helps control uptake of water and solutes.

exon One of the base sequences of an mRNA transcript that will become translated.

exoskeleton [Gk. *skléros*, hard, stiff] External skeleton, as in arthropods.

exotic species Species that left its established home range, deliberately or accidentally, and successfully took up residence elsewhere.

experiment Test that simplifies observation in nature or in the laboratory by manipulating and controlling conditions under which the observations are made.

exponential growth (EX-po-NEN-shul) Pattern of population growth in which population size expands by ever increasing increments during successive intervals, because its reproductive base becomes ever larger. Increases in size against time plot out as a J-shaped curve.

extinction Irrevocable loss of a species.

extracellular fluid Of most animals, all fluid not inside cells; plasma (the liquid portion of blood) plus interstitial fluid (the liquid that occupies spaces between cells and tissues).

extracellular matrix The ground substance, fibrous proteins, and other materials between cells of animal tissues (e.g., cartilage).

FAD Flavin adenine dinucleotide; a type of nucleotide coenzyme that transfers electrons and unbound protons (H^+) from one reaction site to another. At such times it is abbreviated $FADH_2$.

fall overturn Vertical mixing of a body of water in autumn as its upper layer cools. Cool water is more dense, and sinks. The oxygen dissolved in surface water moves down and nutrients from bottom sediments move up.

family pedigree Chart of genetic relationship of family individuals through the generations.

fat Lipid with a glycerol head and one, two, or three fatty acid tails. Tryglycerides have three. The carbon backbone of unsaturated tails has single covalent bonds; that of saturated tails has one or more double bonds.

fate map Surface diagram of certain early embryos (e.g., of *Drosophila*) showing where the differentiated cells of the adult originate.

fatty acid Molecule with a backbone of as many as thirty-six carbon atoms, a carboxyl group ($-COO^-$) at one end, and hydrogen atoms at most or all of the other bonding sites.

feedback inhibition Mechanism by which a cellular change resulting from some activity shuts down the activity that brought it about.

fermentation [L. *fermentum*, yeast] Anaerobic pathway of ATP formation that starts with glycolysis and ends with transfer of electrons to a breakdown product or an intermediate. NAD^+ is regenerated. Net energy yield: two ATP per glucose molecule.

fertilization [L. *fertilis*, to carry, to bear] The Fusion of a sperm nucleus with the nucleus of an egg, which thus becomes a zygote.

fever Any core temperature higher than the set point in the hypothalamic region that functions as the body's thermostat.

fibrous root system All lateral branchings of adventitious roots that arose from a young stem.

filter feeder Animal that filters food from a current of water directed through a body part (e.g, through a sea squirt's pharynx).

filtration First step in urine formation; the pressure of heart contractions filters blood by forcing water and all solutes except proteins from glomerular capillaries into Bowman's capsule of nephrons.

fin Of fishes generally, appendage that helps stabilize, propel, and guide body in water.

first law of thermodynamics [Gk. *therme*, heat, + *dynamikos*, powerful] Law of nature that the total amount of energy in the universe remains constant. Energy cannot be created from nothing and existing energy cannot be destroyed.

fish Aquatic animal of the most ancient, diverse vertebrate lineage; a jawless, cartilaginous, or bony fish.

fitness Increase in adaptation to environment brought about by genetic change.

fixation Loss of all but one kind of allele at a gene locus; all individuals in a population are homozygous for it.

fixed action pattern Program of coordinated, stereotyped muscle activity that is completed independently of feedback from environment.

flagellum (fluh-JELL-um), plural **flagella** Motile structure of many free-living eukaryotic cells. Its core has a 9 + 2 array of microtubules.

flower Of angiosperms only, a reproductive structure with nonfertile parts (sepals, petals) and fertile parts (stamens, carpels) attached to a receptacle (modified base of a floral shoot).

fluid mosaic model Idea that cell membranes consist of a lipid bilayer and proteins. The lipids impart basic structure, impermeability to water-soluble molecules, and (by packing variations and movements) fluidity. Diverse proteins span the bilayer or associate with one of its surfaces and perform most membrane functions (e.g., transport, signal reception).

follicle (FOLL-ih-kul) Small sac, pit, or cavity, as around a hair; also a mammalian oocyte with its surrounding layer of cells.

food chain Straight-line sequence of who eats whom in an ecosystem.

food web Cross-connecting, interlinked food chains consisting of producers, consumers, and decomposers, detritivores, or both.

forebrain Most complex portion of vertebrate brain; includes cerebrum (and cerebral cortex), olfactory lobes, and hypothalamus.

forest Biome where tall trees grow close enough to form a fairly continuous canopy.

fossil Recognizable, physical evidence of an organism that lived in the distant past.

fossil fuel Coal, petroleum, or natural gas; a nonrenewable energy source that formed long ago from remains of swamp forests.

fossilization How fossils form. An organism or traces of it become buried in sediments or volcanic ash. Water and dissolved inorganic compounds infiltrate remains. Accumulating sediments exert pressure above the burial site. Over time, the pressure and chemical changes transform the remains to stony hardness.

founder effect A form of bottlenecking. By chance, allele frequencies of a few individuals that establish a new population are not the same as those of original population. With no further gene flow, natural selection affects allele frequencies in drastically different ways owing to its interactions with genetic drift.

free radical Any highly reactive molecular fragment having an unpaired electron.

fruit [L. after *frui*, to enjoy] Flowering plant's mature ovary, often with accessory structures.

FSH Follicle-stimulating hormone produced and secreted by the anterior lobe of pituitary gland; has reproductive roles in both sexes.

functional group An atom or a group of atoms that is covalently bonded to the carbon backbone of an organic compound and that influences its chemical behavior.

functional-group transfer Enzyme-mediated event in which a molecule donates one or more functional groups to another molecule.

Fungi Kingdom of fungi which, as a group, are major decomposers. Also includes diverse pathogens and parasites.

fungus Eukaryotic heterotroph that secretes digestive enzymes to break down food outside the body into molecules that its cells absorb (e.g., extracellular digestion and absorption). Saprobes feed on nonliving organic matter, and parasites feed on cells or tissues of living organisms.

gall bladder Organ that stores bile secreted from liver; its duct connects to small intestine.

gamete (GAM-eet) [Gk. *gametēs*, husband, and *gametē*, wife] Haploid cell, formed by meiotic cell division of a germ cell; required for sexual reproduction. Eggs and sperm are examples.

gamete formation Formation of sex cells (e.g., sperm and eggs); occurs in reproductive tissues or organs in most eukaryotic species.

gametophyte (gam-EET-oh-fite) [Gk. *phyton*, plant] Haploid gamete-producing body that forms during plant life cycles.

ganglion (GANG-lee-on), plural **ganglia** Distinct cluster of cell bodies of neurons.

gastrulation (gas-tru-LAY-shun) Stage of animal development; major reorganization of new cells into two or three primary tissue layers.

gel electrophoresis Laboratory technique used to distinguish different molecules in a given sample. Application of an electric field forces molecules to migrate through a viscous gel and distance themselves from one another on the basis of length, size, or electric charge.

gene [short for German *pangan*, after Gk. *pan*, all, + *genes*, to be born] Unit of information about a heritable trait, passed from parents to offspring. Each gene has a specific location on a chromosome (e.g., its locus).

gene flow Microevolutionary process; alleles enter and leave a population as an outcome of immigration and emigration, respectively.

gene frequency Abundance of a given allele relative to other alleles at same locus in a population.

gene library Mixed collection of bacteria that house many different cloned DNA fragments.

gene locus A gene's chromosomal location.

gene pair Two alleles at the same gene locus on a pair of homologous chromosomes.

gene pool All genotypes in a population.

gene therapy Generally, the transfer of one or more normal genes into an organism to correct or lessen adverse effects of a genetic disorder.

genetic code [After L. *genesis*, to be born] The correspondence between nucleotide triplets in DNA (then mRNA) and specific sequences of amino acids in resulting polypeptide chains; the basic language of protein synthesis in cells.

genetic disease Illness in which expression of one or more genes increases susceptibility to an infection or weakens immune response to it.

genetic disorder Inherited condition that causes mild to severe medical problems.

genetic divergence Gradual accumulation of differences in gene pools of populations or subpopulations of a species after a geographic barrier arises and separates them; mutation, natural selection, and genetic drift thereafter are operating independently in each one.

genetic drift Change in allele frequencies over the generations, as brought about by chance alone. Population size influences its effect on genetic and phenotypic diversity, because small populations are more vulnerable to losing alleles entirely.

genetic engineering Altering the information content of DNa molecules with recombinant DNA technology.

genetic equilibrium In theory, a state in which a population is not evolving. These conditions must be met: no mutation, the population very large in size and isolated from others of same species, and no natural selection (all members reproduce equally by random mating).

genetic recombination Result of any process that can incorporate new genetic information into a chromosome or DNA fragment. As one example, allelic combinations in chromosomes emerging from meiosis usually differ from the parental combinations (say, *Ab* and *aB* compared to parental types *AB* and *ab*). Also, nonreciprocal gene transfers in nature or the laboratory result in genetic recombination.

genome All of the DNA in a haploid number of chromosomes for a given species.

genotype (JEEN-oh-type) Genetic constitution of an individual; a single gene pair or the sum total of an individual's genes.

genus, plural **genera** (JEEN-US, JEN-er-ah) [L. *genus*, race or origin] A grouping of all species perceived to be more closely related to one another in their morphology, ecology, and evolutionary history than to other species at the same taxonomic level.

geographic dispersal Directional movement in which residents of an established community leave their home range and successfully settle in a new location, where they are exotic species.

geologic time scale Time scale for the Earth's history. Its major subdivisions correspond to mass extinctions. Its absolute dates have been refined by radiometric dating.

germ cell Animal cell of a lineage set aside for sexual reproduction; gives rise to gametes.

germination (jur-min-AY-shun) Of seeds and spores, resumption of growth after dispersal, dormancy, or both.

gibberellin (JIB-er-ELL-un) Type of plant hormone that promotes elongation of stems, helps seeds and buds break dormancy, and contributes to flowering.

gill Organ of respiration. Most have a thin, moist, vascularized layer for gas exchange.

gland Secretory cell or structure derived from epithelium and often connected to it.

glomerular capillary One of a set of blood capillaries in Bowman's capsule of a nephron.

glomerulus (glow-MARE-you-luss) [L. *glomus*, ball] First portion of the nephron, where water and solutes are filtered from blood.

glucagon (GLUE-kuh-gone) Hormone secreted by pancreas; stimulates cells to convert their stores of glycogen and amino acids to glucose.

glyceride (GLISS-er-eyed) Molecule with one, two, or three fatty acid tails attached to a glycerol backbone; one of the fats or oils.

glycerol (GLISS-er-ohl) [Gk. *glykys*, sweet, + L. *oleum*, oil] Three-carbon compound with three hydroxyl groups; component of fats and oils.

glycocalyx Sticky mesh of polysaccharides, polypeptides, or both around the cell wall of many bacteria.

glycogen (GLY-kuh-jen) A highly branched polysaccharide made of glucose monomers; the main storage carbohydrate in animals.

glycolysis (gly-CALL-ih-sis) [Gk. *glykys*, sweet, + *lysis*, breaking apart] Breakdown of glucose or another organic compound to two pyruvate molecules. First stage of aerobic respiration, fermentation, and anaerobic electron transport. Oxygen has no role in glycolysis, which occurs in the cytoplasm of all cells. Two NADH form. Net yield: two ATP per glucose molecule.

glycoprotein Protein with linear or branched oligosaccharides covalently bonded to it. Nearly all surface proteins of animal cells and many proteins circulating in blood are examples.

gnetophyte Woody gymnosperm, unusual in having vessels in xylem; vine or shrub.

Golgi body (GOHL-gee) Organelle of lipid assembly, polypeptide chain modification, and packaging of both in vesicles for export or for transport to locations in cytoplasm.

gonad (GO-nad) Primary reproductive organ in which animal gametes are produced.

graded potential Of a neuron and other excitable cells, a local signal that can slightly alter the resting membrane potential and can vary in magnitude, depending on the stimulus. With prolonged or intense stimulation, such signals may spread to a trigger zone of the membrane and initiate an action potential.

granum, plural **grana** In many chloroplasts, a stack of flattened, membranous compartments that have chlorophyll and other light-trapping pigments and reaction sites for ATP formation.

grassland Biome that has flat or rolling land, 25–100 centimeters of annual rainfall, warm summers, distinct array of grazing animals, and recurring fires that regenerate dominant plant species.

gravitropism (GRAV-ih-TROPE-izm) [L. *gravis*, heavy, + Gk. *trepein*, to turn] Tendency of a plant to grow directionally in response to the Earth's gravitational force.

gray matter Unmyelinated axons, dendrites, and cell bodies of neurons, plus neuroglial cells, in the brain and spinal cord.

grazing food web Network of food chains in which energy flows from plants to an array of herbivores, then to carnivores.

green alga Type of protistan evolutionarily, structurally, and biochemically most similar to plants (e.g., nearly all are photoautotrophs with starch grains and chlorophylls *a* and *b* in chloroplasts; and some have cell walls of cellulose, pectin, and other polysaccharides typical of plants).

green revolution In developing countries, the use of improved crop varieties, modern agricultural equipment and practices (e.g., heavy fertilizer and pesticide application) to increase crop yields.

greenhouse effect Overall warming of lower atmosphere. Gaseous molecules (e.g., carbon dioxide and methane) impede the escape of infrared wavelengths (heat) from the Earth's sunlight-warmed surface. The gases continually absorb those wavelengths and radiate much of their energy back toward Earth.

ground meristem (MARE-ih-stem) [Gk. *meristos*, divisible] Primary meristem that gives rise to the ground tissue system).

ground substance Intercellular material in some animal tissues; made of cell secretions and other noncellular components.

ground tissue system Tissues (parenchyma, especially) making up most of the plant body.

growth Of multicelled species, increases in the number, size, and volume of cells. Of bacteria, increase in the number of cells in a population.

guanine Nitrogen-containing base in one of the four nucleotide monomers in DNA or RNA.

guard cell Either of two adjoining cells that influence movement of carbon dioxide, oxygen, and water vapor across leaf or stem epidermis. When both guard cells swell with water and move apart, an opening (stoma) forms; when they lose water and collapse into each other, the stoma closes.

gut Generally, a sac or tube from which food is absorbed into internal environment. Also, a gastrointestinal tract (from stomach onward).

gymnosperm (JIM-noe-sperm) [Gk. *gymnos*, naked, + *sperma*, seed] Vascular plant that bears seeds at exposed surfaces of its reproductive structures (e.g., on cone scales).

habitat [L. *habitare*, to live in] Type of place where an organism or species normally lives; characterized by physical and chemical features and by its array of species.

hair cell Mechanoreceptor that may give rise to action potentials when bent or tilted.

half-life The time it takes for half of a given quantity of any radioisotope to decay into a different, and less unstable, daughter isotope.

halophile Archaebacterium of saline habitats.

haploidy (HAP-loyd-ee) Presence of only half of the parental number of chromosomes in a spore or gamete, as brought about by meiosis.

Hardy–Weinberg rule Statement that allele frequencies stay the same over the generations when there is no mutation, the population is infinitely large and is isolated from other populations of the same species, mating is random, and all individuals are reproducing equally and randomly (no natural selection).

HCG Short for human chorionic gonadotropin. Hormone that helps maintains endometrium during the menstrual cycle and first trimester of pregnancy.

heart Muscular pump; its contractions keep blood circulating through the animal body.

heat Thermal energy; a form of kinetic energy.

helper T cell T lymphocyte with central roles in both antibody-mediated and cell-mediated immune responses. When activated, it makes and secretes chemicals that induce responsive T and B cells to undergo rapid divisions into populations of effector and memory cells.

hemoglobin (HEEM-oh-glow-bin) [Gk. *haima*, blood, + L. *globus*, ball] Iron-containing, oxygen-transporting protein of red blood cells.

hemostasis (HEE-mow-STAY-sis) [Gk. *haima*, blood, + *stasis*, standing] Process that stops blood loss from damaged blood vessel; involves coagulation, blood vessel spasm, platelet plug formation, and other mechanisms.

herbivore [L. *herba*, grass, + *vovare*, to devour] Plant-eating animal (e.g., snail, deer, manatee).

hermaphrodite Individual having both male and female gonads.

heterocyst (HET-er-oh-sist) Cyanobacterial cell that modifies itself and makes a nitrogen-fixing enzyme when nitrogen supplies dwindle.

heterotroph (HET-er-oh-trofe) [Gk. *heteros*, other, + *trophos*, feeder] Organism unable to make its own organic compounds; feeds on autotrophs, other heterotrophs, organic wastes.

heterozygous condition (HET-er-oh-ZYE-guss) [Gk. *zygoun*, join together] For a given trait, having a pair of nonidentical alleles at a gene locus (that is, on a pair of homologous chromosomes).

higher taxon (plural, **taxa**) One of ever more inclusive groupings that reflect relationships among species. Family, order, class, phylum, and kingdom are examples.

hindbrain Medulla oblongata, cerebellum, and pons of vertebrate brain. Includes reflex centers for respiration, blood circulation, and other basic functions; also helps coordinate motor responses and many complex reflexes.

histone Type of protein intimately associated with eukaryotic DNA and largely responsible for organization of eukaryotic chromosomes.

homeostasis (HOE-me-oh-STAY-sis) [Gk. *homo*, same, + *stasis*, standing] State in which physical and chemical aspects of internal environment (blood, interstitial fluid) are being maintained within ranges suitable for cell activities.

homeotic gene A master gene governing the development of specific body parts.

hominid [L. *homo*, man] All species on or near evolutionary road leading to modern humans.

hominoid Apes, humans, and recent ancestors.

homologous chromosome (huh-MOLL-uh-gus) [Gk. *homologia*, correspondence] Of cells with a diploid chromosome number, one of a pair of chromosomes that are identical in size, shape, and gene sequence, and that interact at meiosis. Nonidentical sex chromosomes (e.g., X and Y) also interact as homologues during meiosis.

homologous structures The same body parts that became modified differently, in different lines of descent from a common ancestor.

homology Similarity in one or more body parts in different species; attributable to descent from a common ancestor.

homozygous condition (HOE-moe-ZYE-guss) For a specified trait, having a pair of identical alleles at a gene locus (on a pair of homologous chromosomes).

homozygous dominant condition Having a pair of dominant alleles at a gene locus (on a pair of homologous chromosomes).

homozygous recessive condition Having a pair of recessive alleles at a gene locus (on a pair of homologous chromosomes).

hormone [Gk. *hormon*, stir up, set in motion] Signaling molecule secreted by one cell that stimulates or inhibits activities of any other cell having receptors for it. Animal hormones are picked up by bloodstream, which delivers them to cells some distance away. In plants, hormones do not travel far from source cells.

horsetail Seedless vascular plant with ancient, tree-size ancestors. The sporophytes of the one existing genus have rhizomes, scale-shaped leaves, and hollow photosynthetic stems with silica-reinforced ribs.

human genome project Worldwide research project to sequence all 3.2 billion nucleotides in the DNA of human chromosomes. Now scheduled for completion within a few years.

humus Decomposing organic matter in soil. Amount varies in soils of different types.

hybrid offspring Of a genetic cross, offspring with a pair of nonidentical alleles for a trait.

hydrogen bond Weak interaction between a small, highly electronegative atom of a molecule and a neighboring hydrogen atom already taking part in a polar covalent bond.

hydrogen ion Free (unbound) proton; that is, a hydrogen atom that lost its electron and now bears a positive charge (H^+).

hydrologic cycle A biogeochemical cycle, driven by solar energy, in which water moves through the atmosphere, on or through land, to the ocean, and back to the atmosphere.

hydrolysis (high-DRAWL-ih-sis) [L. *hydro*, water, + Gk. *lysis*, loosening] Cleavage reaction that breaks covalent bonds and splits a molecule into two or more parts. Commonly, H^+ and OH^- (derived from a water molecule) become attached to the exposed bonding sites.

hydrophilic substance [Gk. *philos*, loving] A polar substance that dissolves easily in water. Sugars are examples.

hydrophobic substance [Gk. *phobos*, dreading] A nonpolar substance; it strongly resists being dissolved in water. Oil is an example.

hydrosphere Collectively, all of the Earth's liquid or frozen water.

hydrostatic pressure Pressure exerted by a volume of fluid against a wall, membrane, or some other structure that encloses the fluid.

hydrothermal vent ecosystem Ecosystem near a steaming fissure in the ocean floor. Chemoautotrophic bacteria are the basis of its food webs.

hypertonic solution A fluid having a greater concentration of solutes relative to another fluid.

hypha (HIGH-fuh), plural **hyphae** [Gk. *hyphe*, web] Fungal filament with chitin-reinforced walls; component of a mycelium.

hypodermis Subcutaneous layer with stored fat that helps insulate the body; anchors skin but still allows it to move somewhat.

hypothalamus [Gk. *hypo*, under, + *thalamos*, inner chamber, or *tholos*, rotunda] Forebrain region that controls visceral activities (e.g., salt–water balance, core temperature, and reproduction); influences related behaviors (e.g., hunger, thirst, and sex) and emotional states (e.g., sweating with fear).

hypothesis In science, a possible explanation of a phenomenon, one that has the potential to be proved false by experimental tests.

hypotonic solution A fluid that has a lower concentration of solutes relative to another fluid.

immune response Events by which B cells and T cells recognize antigen and give rise to antigen-sensitized populations of effector cells and memory cells.

immunoglobulin (Ig) One of five classes of antibodies, each with antigen-binding sites as well as other sites with specialized functions.

implantation Event in pregnancy. A blastocyst burrows into endometrium and establishes connections by which a mother will exchange substances with the embryo (and fetus) that develops from the blastocyst's inner cell mass.

imprinting Time-dependent form of learning, usually during a sensitive period for a young animal, triggered by exposure to sign stimuli.

in vitro fertilization Conception outside the body ("in glass" petri dishes or test tubes).

inbreeding Nonrandom mating among close relatives that share many identical alleles; a form of genetic drift in a small group of relatives that are preferentially interbreeding.

incomplete dominance Condition in which one allele of a pair is not fully dominant over the other; a heterozygous phenotype in between both homozygous phenotypes emerges.

independent assortment theory Mendelian theory that by the end of meiosis, each pair of homologous chromosomes (and linked genes on each one) are sorted out for shipment to gametes independently of how all the other pairs were sorted. Later modified to account for the disruptive effect of crossing over on linkages.

indirect selection theory Idea that altruistic individuals can pass on their genes indirectly by helping relatives survive and reproduce.

induced-fit model Idea that a substrate alters the shape of an enzyme's active site when bound to it, causing a more precise molecular fit between the two that promotes reactivity.

inductive logic Pattern of thinking by which an individual derives a general statement from specific observations.

infection Invasion and multiplication of a pathogen in a host. Disease follows if defenses are not mobilized fast enough; the pathogen's activities interfere with normal body functions.

inflammation, acute Rapid response to tissue injury by phagocytes and diverse proteins (e.g., histamine, complement, clotting factors). Signs include localized redness, heat, swelling, pain.

inheritance The transmission, from parents to offspring, of genes that specify structures and functions characteristic of the species.

inhibiting hormone Hypothalmic signaling molecule that suppresses a particular secretion by the anterior lobe of the pituitary gland.

inhibitor Substance able to bind with a specific molecule and interfere with its functioning.

inhibitory postsynaptic potential (IPSP) Graded potential at an input zone of an excitable cell that drives its membrane away from threshold.

instinctive behavior A behavior performed without having been learned by experience.

insulin Pancreatic hormone that lowers level of glucose in blood by causing cells to take up glucose; also promotes protein and fat synthesis and inhibits protein conversion to glucose.

integration, neural [L. *integrare*, coordinate] Moment-by-moment summation of excitatory and inhibitory synapses acting on a neuron.

integrator A control center (e.g., brain) that receives, processes, and stores sensory input, then puts together and issues commands for coordinated responses to it.

integument Of animals, protective body cover (e.g., skin). Of seed-bearing plants, one or more layers around an ovule; becomes a seed coat.

integumentary exchange (in-teg-you-MEN-tuh-ree) Respiration across a thin, moist, and often vascularized surface layer of animal body.

interleukin Type of protein that mediates the signaling between a variety of cell types (e.g., cells of immune system and of hypothalamus).

intermediate Substance that forms between the start and end of a metabolic pathway.

intermediate filament One of the ropelike cytoskeletal elements that impart mechanical strength to animal cells and tissues.

interneuron Neuron of brain or spinal cord.

internode In vascular plants, the stem region between two successive nodes.

interphase Of a cell cycle, interval between nuclear divisions when a cell increases in mass and roughly doubles the number of its cytoplasmic components. It also duplicates its chromosomes (replicates its DNA) during interphase, but *not* between meiosis I and II.

interspecific competition Ecological interaction in which individuals that belong to different species compete for a share of resources in the same habitat.

interstitial fluid (IN-ter-STISH-ul) [L. *interstitus*, to stand in the middle of something] The portion of extracellular fluid that occupies the spaces between animal cells and tissues.

intertidal zone The region above the low water mark and below the high water mark of a rocky or sandy shore; tides alternately submerge and expose its inhabitants.

intervertebral disk Disk-shaped, cartilaginous structure that is a shock absorber and flex point between a backbone's bony segments.

intraspecific competition Ecological interaction in which individuals that belong to the same population compete for a share of resources in their habitat.

intron One of the noncoding portions of a pre-mRNA transcript. All introns are excised before translation.

inversion A linear stretch of DNA within a chromosome that has become oriented in the reverse direction, with no molecular loss.

invertebrate Any animal without a backbone. Of the 2 million named species in the animal kingdom, all but 50,000 are invertebrates.

ion, negatively charged (EYE-on) An atom or a molecule that acquired an overall negative charge by gaining one or more electrons.

ion, positively charged An atom or a molecule that acquired an overall positive charge by losing one or more electrons.

ionic bond Two ions being held together by the attraction of their opposite charge.

isotonic solution A fluid having the same solute concentration as a fluid against which it is being compared.

isotope (EYE-so-tope) Of an element, an atom with more or fewer neutrons than the atoms having the most common number.

joint Area of contact or near-contact between bones.

juvenile Of some animals, a post-embryonic stage that changes only in size and proportion to become the adult (no metamorphosis).

J-shaped curve Type of diagrammatic curve that emerges when unrestricted exponential growth of a population is plotted against time.

karyotype (CARE-ee-oh-type) For an individual or a species, a preparation of metaphase chromosomes sorted by length, centromere location, and other defining features.

keratin Tough, water-insoluble protein made by vertebrate epidermal cells, which die and accumulate as keratinized bags at the surface of skin to form a barrier against dehydration, bacteria, and many toxins.

key innovation A structural or functional modification to the body that, by chance, gives a lineage the opportunity to exploit the environment in more efficient or novel ways.

keystone species A species that dominates a community and dictates its structure.

kidney One of a pair of vertebrate organs that filter ions and other substances from blood; it controls the amounts returned and so helps maintain the solute concentrations and volume of extracellular fluid.

kilocalorie 1,000 calories of heat energy (the amount required to raise the temperature of 1 kilogram of water by 1°C). Used as the unit of measure for the caloric content of foods.

kinase Enzyme that catalyzes phosphate-group transfers (e.g., a protein kinase).

kinetic energy Energy of motion.

kinetochore Cluster of proteins and DNA at the centromere region of a chromosome; spindle microtubules become attached to it at mitosis or meiosis. One is present on each chromatid of a duplicated chromosome.

Krebs cycle Cyclic pathway in mitochondria only; together with a few preparatory steps, the stage of aerobic respiration in which pyruvate is broken down to carbon dioxide and water. Coenzymes accept electrons and unbound protons (H^+) from intermediates and deliver them to next stage; two ATP form.

lactate fermentation An anaerobic pathway of ATP formation. Pyruvate from glycolysis is converted to three-carbon lactate, and NAD^+ is regenerated. Net energy yield: two ATP.

lactation Production and secretion of milk by hormone-primed mammary glands.

large intestine Colon; gut region that receives unabsorbed residues from small intestine, and concentrates and stores feces until expulsion.

larva, plural **larvae** An immature stage that develops between the embryo and adult in many animal life cycles.

larynx (LARE-inks) Tubular airway leading to lungs. Contains vocal cords in some animals.

lateral meristem Meristem in plants that show secondary growth; vascular or cork cambium.

lateral root Outward branching from the first (primary) root of a taproot system.

leaching Removal of some nutrients from soil as water percolates through it.

leaf Chlorophyll-rich plant part adapted for sunlight interception and photosynthesis.

learned behavior Lasting modification in behavior as a result of experience or practice.

lek Communal display ground for courtship behavior by some animals, including birds.

lethal mutation Mutation with drastic effects on phenotype; usually causes death.

LH Luteinizing hormone. Anterior pituitary hormone that has reproductive roles in both males and females.

lichen (LY-kun) Symbiotic interaction between a fungus and a photoautotroph.

life cycle Recurring pattern of genetically programmed events from the time individuals are produced until they themselves reproduce.

life table Tabulation of age-specific patterns of birth and death for a population.

ligament A strap of dense connective tissue that bridges a joint.

light-dependent reactions The first stage of photosynthesis. Sunlight energy is trapped and converted to chemical energy of ATP, NADPH, or both, depending on the pathway.

light-independent reactions Second stage of photosynthesis. ATP makes phosphate-group transfers required to build sugar phosphates. NADPH delivers electrons and hydrogen atoms for the synthesis reactions, which also require carbon from carbon dioxide. Sugar phosphates enter other reactions by which starch, cellulose, and other end products are assembled.

lignification Deposition of lignin in secondary wall of many plant cells. Lignin imparts strength and rigidity, stabilizes and protects other wall components, and forms a waterproof barrier. Important in vascular plant evolution.

limbic system A system of centers in cerebral hemisphere that governs emotions and affects memory. Distantly related to olfactory lobes; it still deals with the sense of smell.

limiting factor Any essential resource that, in short supply, limits population growth.

lineage (LIN-ee-age) Line of descent.

linkage group All the genes on a chromosome.

lipid Mainly a greasy or oily hydrocarbon. Lipid molecules strongly resist dissolving in water but quickly dissolve in nonpolar substances. Some types serve as the main reservoirs of stored energy in all cells; others are structural materials (as in membranes) and cell products (e.g., surface coatings).

lipid bilayer Structural basis of cell membranes. Two layers of mostly phospholipid molecules. Hydrophobic tails of the lipids are sandwiched between the hydrophilic heads, and heads are dissolved in intracellular or extracellular fluid.

lipoprotein Molecule that forms when proteins circulating in blood combine with cholesterol, triglycerides, and phospholipids absorbed from the small intestine.

liver In vertebrates and many invertebrates, a large gland that stores, converts, and helps maintain blood levels of organic compounds; inactivates most hormone molecules that have completed their tasks; inactivates compounds that can be toxic at high concentrations.

local signaling molecule One of the secretions by many cell types that alter chemical conditions only in localized tissue regions. Prostaglandin is an example.

logic Thought patterns by which an individual draws a conclusion that does not contradict evidence used to support that conclusion.

logistic population growth (low-JISS-tik) Pattern of population growth. A low-density population slowly increases in size, enters a phase of rapid growth, then levels off in size once the carrying capacity has been reached.

loop of Henle Hairpin-shaped, tubular part of a nephron that reabsorbs water and solutes.

lung Internal sac-shaped respiratory surface that evolved in oxygen-poor habitats as an adaptation that increases the surface area for gas exchange. A pair occur in a few fishes and in amphibians, birds, reptiles, and mammals.

lycophyte Seedless vascular plant with tree-size ancestors of ancient swamp forests. They require free water to complete life cycle. Most have leaves, branching rhizome, vascularized roots and stems (e.g., club mosses).

lymph (LIMF) [L. *lympha*, water] Tissue fluid that drained into vessels of lymphatic system.

lymph capillary Small-diameter vessel where tissue fluid moves into lymph vascular system.

lymph node Lymphoid organ that serves as a battleground for immune responses. Organized arrays of lymphocytes packed inside it cleanse lymph before it can reach the bloodstream.

lymph vascular system [L. *lympha*, water, + *vasculum*, a small vessel] Parts of lymphatic system that take up excess tissue fluid, absorbed fats, and reclaimable solutes for delivery to the bloodstream.

lymphatic system Supplement to vertebrate circulatory system. Its vessels deliver fluid and solutes from interstitial fluid to blood; and its lymphoid organs have roles in body defenses.

lymphocyte A T cell or B cell.

lysis [Gk. *lysis*, a loosening] Gross damage to a plasma membrane, cell wall, or both that lets the cytoplasm to leak out; causes cell death.

lysogenic pathway Latent period that extends many viral replication cycles. Viral genes get integrated into host chromosome and may stay inactivated through many host cell divisions but eventually are replicated in host progeny.

lysosome (LYE-so-sohm) Important organelle of intracellular digestion.

lysozyme Infection-fighting enzyme present in mucous membranes (e.g., of mouth, vagina).

lytic pathway Of viruses, a rapid replication pathway that ends with lysis of host cell.

macroevolution Large-scale patterns, trends, and rates of change among higher taxa.

macrophage Phagocytic white blood cell; roles in nonspecific defenses and immune responses. One of the key antigen-presenting cells.

mammal Only vertebrate whose females nourish offspring with milk from mammary glands.

mass extinction Catastrophic event or phase in geologic time when entire families or other major groups are irrevocably lost.

mass number Sum of all protons and neutrons in an atom's nucleus.

mast cell A basophil-like cell that releases histamine during tissue inflammation.

mechanoreceptor Sensory cell or nearby cell that detects mechanical energy (changes in pressure, position, or acceleration).

medulla oblongata Hindbrain region with reflex centers for basic tasks (e.g., respiration); coordinates motor responses with complex reflexes (e.g., coughing); also influences brain centers concerned with sleep and arousal.

medusa (meh-DOO-sah) [Gk. *Medousa*, one of three sisters in Greek mythology having snake-entwined hair] Of cnidarian life cycles, a free-swimming, bell-shaped stage, often with oral lobes and tentacles extending below the bell.

megaspore Haploid spore that forms by way of meiosis in the ovary of seed-bearing plants; one of its cellular descendants develops into an egg.

meiosis (my-OH-sis) [Gk. *meioun*, to diminish] Two-stage nuclear division process that halves the chromosome number of a parental germ cell (to haploid number). Each daughter nucleus receives one of each type of chromosome. Basis of gamete formation. Also, basis of formation of spores that give rise to gamete-producing bodies (gametophytes).

memory The capacity to store and retrieve information about past sensory experience.

memory cell B or T cell that forms during an immune response but that does not act at once; it enters a resting phase, from which it is released for a secondary immune response.

menopause (MEN-uh-pozz) [L. *mensis*, month, + *pausa*, stop] The time when the reproductive potential of human females draws to a close.

menstrual cycle A recurring cycle, lasting twenty-eight days on average in adult human females. A secondary oocyte is released from one of a pair of ovaries, and the lining of the uterus is primed for pregnancy. The hormones estrogen, progesterone, FSH, and LH control the cyclic activity.

menstruation Sloughing of a blood-enriched endometrium when pregnancy does not occur.

mesoderm (MEH-zoe-derm) [Gk. *mesos*, middle, + *derm*, skin] Primary tissue layer important in evolution of all large, complex animals; gives rise to many internal organs and part of the integument.

mesophyll (MEH-zoe-fill) A photosynthetic parenchyma with abundant air spaces for gas exchange between its cells.

messenger RNA (mRNA) A single strand of ribonucleotides transcribed from DNA, then translated into a polypeptide chain. The only RNA encoding protein-building instructions.

metabolic pathway (MEH-tuh-BALL-ik) Orderly sequence of enzyme-mediated reactions by which cells maintain, increase, or decrease the concentrations of particular substances.

metabolism (meh-TAB-oh-lizm) [Gk. *meta*, change] All the controlled, enzyme-mediated chemical reactions by which cells acquire and use energy to synthesize, store, degrade, and eliminate substances in ways that contribute to growth, survival, and reproduction.

metamorphosis (me-tuh-MOR-foe-sis) [Gk. *meta*, change, + *morphe*, form] Major changes in body form during the transition from the embryo to the adult; involves hormonally controlled size increases, reorganization of tissues, and remodeling of body parts.

metaphase Of meiosis I, a stage when all pairs of homologous chromosomes have become positioned at the spindle equator. Of mitosis or meiosis II, a stage when all the duplicated chromosomes have become positioned at the equator of the microtubular spindle.

metazoan Any multicelled animal.

methanogen Anaerobic archaebacterium that produces methane gas as by-product.

MHC marker Self-marker protein. Some are on all body cells of the individual; others are unique to macrophages and lymphocytes.

micelle (my-SELL) Of fat digestion, tiny droplet of bile salts, fatty acids, and monoglycerides; role in fat absorption from small intestine.

microevolution Of a population, any change in allele frequencies resulting from mutation, genetic drift, gene flow, natural selection, or some combination of these.

microfilament [Gk. *mikros*, small, + L. *filum*, thread] Cytoskeletal element. Each consists of two thin, twisted polypeptide chains; it has roles in cell movement and in producing and maintaining cell shapes.

micrograph Photograph of an image that came into view with the aid of a microscope.

microorganism Organism, usually single celled, too small to be observed without a microscope.

microspore Walled haploid spore; becomes a pollen grain in gymnosperms and angiosperms.

microtubular spindle Bipolar array of many microtubules; forms during nuclear division and moves chromosomes apart in controlled ways.

microtubule (my-crow-TUBE-yool) Cylindrical, hollow cytoskeletal element that consists of tubulin subunits; roles in cell shape, growth, and motion (e.g., key skeletal element of cilia, flagella, spindle apparatus).

microtubule organizing center MTOC; mass of substances in cytoplasm of eukaryotic cells; number, type, and location dictate orientation and organization of cell's microtubules.

microvillus (MY-crow-VILL-us) [L. *villus*, shaggy hair] Slender extension from free surface of certain cells; arrays of many microvilli greatly increase absorptive or secretory surface area.

midbrain Part of vertebrate brain with centers for coordinating reflex responses to visual and auditory input; also relays signals to forebrain.

migration Recurring pattern of movement between two or more locations in response to environmental rhythms; e. g., circadian rhythms and seasonal changes in daylength. It requires activation or suppression of internal timing mechanisms that govern physiological and behavioral functions of migratory animals.

mimicry (MIM-ik-ree) Close resemblance in form, behavior, or both between one species (the mimic) and another (its model). Serves in deception, as when an orchid mimics a female insect and so attracts males that pollinate it.

mineral Any element or inorganic compound that formed by natural geologic processes and is required for normal cell functioning.

mitochondrion (MY-toe-KON-dree-on) Double-membrane organelle of ATP formation. Only site of aerobic respiration's second and third stages. May have endosymbiotic origins.

mitosis (my-TOE-sis) [Gk. *mitos*, thread] Type of nuclear division that maintains the parental chromosome number for daughter cells. The basis of growth in size, tissue repair, and often asexual reproduction for eukaryotes.

mixture Two or more elements intermingled in proportions that can and usually do vary.

model Theoretical, detailed description or analogy that helps people visualize something that has not yet been directly observed.

molar Tooth with a platform having cusps (surface bumps) that help crush, grind, and shear food; one of the cheek teeth.

molecular clock Model used to calculate the time of origin of one lineage or species relative to others. The underlying assumption is that neutral mutations accumulate in a lineage at predictable rates that can be measured as a series of ticks back through time.

molecule A unit of matter in which chemical bonds hold together two or more atoms of the same or different elements.

mollusk Only invertebrate with a tissue fold (mantle) draped over a soft, fleshy body; most have an external or internal shell. Enormous diversity in body plans and sizes, as in chitons, gastropods (e.g, snails), bivalves (e.g., clams), and cephalopods (e.g., octopuses, squids).

molting Periodic shedding of body structures that are too small, worn out, or both. Permits certain animals to grow in size or renew some parts (e.g., exoskeletons, shells, hairs, feathers, and horns). Especially characteristic of insects and other arthropods.

Monera In earlier classification schemes, a prokaryotic kingdom that encompasses both archaebacteria and eubacteria.

monocot (MON-oh-kot) Monocotyledon; a flowering plant with one cotyledon in seeds, floral parts generally in threes (or multiples of three), and often parallel-veined leaves.

monohybrid cross Experimental cross between two parents that are homozygous for different versions of the same trait (e.g., *AA* and *aa*). F_1 offspring are heterozygous; each inherits a pair of nonidentical alleles (*Aa*) for the trait.

monomer Small molecule used as a subunit of polymers, such as sugar monomers of starch.

monosaccharide (MON-oh-SAK-ah-ride) [Gk. *monos*, alone, single, + *sakcharon*, sugar] One of the simple carbohydrates; a single sugar monomer. Glucose is an example.

monosomy Presence of a chromosome that has no homologue in a diploid cell.

morphogenesis (MORE-foe-JEN-ih-sis) [Gk. *morphe*, form, + *genesis*, origin] Inherited program of orderly changes in size, shape, and proportions of an animal embryo, leading to specialized tissues and early organs.

morphological convergence Macroevolutionary pattern. In response to similar environmental pressures over time, evolutionarily distant lineages evolve in similar ways and end up being alike in appearance, functions, or both.

morphological divergence Macroevolutionary pattern. Genetically diverging lineages slowly undergo change from the body form of their common ancestor.

motor neuron Type of neuron specialized to swiftly relay commands from the brain or spinal cord to muscle cells, gland cells, or both.

multicelled organism Organism composed of many cells with coordinated metabolic activity; most show extensive cell differentiation into tissues, organs, and organ systems.

multiple allele system Three or more slightly different molecular forms of a gene that occur among individuals of a population.

muscle fatigue Decline in tension of a muscle kept in a state of tetanic contraction as a result of continuous, high-frequency stimulation.

muscle tension Mechanical force exerted by a contracting muscle; resists opposing forces (e.g., gravity or weight of an object being lifted).

muscle tissue Tissue with arrays of cells able to contract under stimulation, then passively lengthen and return to their resting position.

mutagen (MEW-tuh-jen) Any environmental agent, such as a virus or ultraviolet radiation, that can alter DNA's molecular structure.

mutation [L. *mutatus*, a change, + *-ion*, an act, a result, or a process] Heritable change in the molecular structure of DNA. Original source of all new alleles and, ultimately, the diversity of life.

mutation frequency Of a population, the number of times that a mutation at a particular locus has arisen.

mutation rate Of a gene locus, the probability that a spontaneous mutation will occur during or between DNA replication cycles.

mutualism [L. *mutuus*, reciprocal] Symbiotic interaction that benefits both participants.

mycelium (my-SEE-lee-um), plural **mycelia** [Gk. *mykes*, fungus, mushroom, + *helos*, callus] Mesh of tiny, branching filaments (hyphae); the food-absorbing portion of most fungi.

mycorrhiza (MY-coe-RIZE-uh) "Fungus-root." A form of mutualism between fungal hyphae and young plant roots. The plant gives up some carbohydrates and the fungus gives up some of its absorbed mineral ions.

myelin sheath Axonal sheath around many sensory and motor neurons; enhances the long-distance propagation of action potentials.

myofibril (MY-oh-FY-brill) One of the many internal threadlike structures of a muscle cell, each divided into sarcomeres.

myosin (MY-uh-sin) Motor protein, often bound to microtubules; key roles in cell movements.

NAD+ Nicotinamide adenine dinucleotide. A nucleotide coenzyme; abbreviated NADH when carrying electrons and H+ to a reaction site.

NADP+ Nicotinamide adenine dinucleotide phosphate. A phosphorylated nucleotide coenzyme; abbreviated NADPH$_2$ when it is carrying electrons and H+ to a reaction site.

natural killer cell Cytotoxic lymphocyte that reconnoiters for tumor cells and virus-infected cells, then touch-kills them.

natural selection Microevolutionary process; the outcome of differences in survival and reproduction among individuals that vary in details of heritable traits. Over generations, it typically leads to increased fitness.

necrosis (neh-CROW-sis) Passive death of many cells that results from severe tissue damage.

negative feedback mechanism A homeostatic mechanism by which a condition that has changed as a result of some activity triggers a response that reverses the change.

nematocyst (NEM-at-uh-sist) [Gk. *nema*, thread, + *kystis*, pouch] Cnidarian capsule that has a dischargeable, tube-shaped thread, sometimes barbed; releases a toxin or sticky substance.

nephridium (neh-FRID-ee-um), plural **nephridia** Unit that controls composition and volume of fluid in some invertebrates (e.g., earthworms).

nephron (NEFF-ron) [Gk. *nephros*, kidney] One of the urine-forming tubules in a kidney; it filters water and solutes from blood, then selectively reabsorbs adjusted amounts of both in ways that help maintain the volume and composition of extracellular fluid.

nerve Cordlike bundle of the axons of sensory neurons, motor neurons, or both sheathed in connective tissue.

nerve cord A prominent longitudinal nerve. Most animals have one, two, or three. The nervous system of chordate embryos develops from a tubular, dorsal nerve cord.

nerve impulse *See* action potential.

nerve net Simple nervous system in epidermis of cnidarians and some other invertebrates; a diffuse mesh of simple, branching nerve cells interacts with contractile cells, sensory cells, or both.

nervous system Integrative organ system with nerve cells interacting in signal-conducting and information-processing pathways. Detects and processes stimuli, and elicits responses from effectors (e.g., muscles and glands).

nervous tissue Connective tissue composed of neurons and often neuroglia.

net population growth rate per individual (*r*) For population growth equations, a single variable combining birth and death rates; the assumption is that both remain constant.

neural tube Embryonic and evolutionary forerunner of brain and spinal cord.

neuroglia (NUR-oh-GLEE-uh) Collectively, cells that structurally and metabolically support neurons. They make up about half the volume of nervous tissue in vertebrates.

neuromodulator Any of a variety of signaling molecules that magnify or reduce the effects of a neurotransmitter.

neuromuscular junction A chemical synapse between axonal endings of a motor neuron and a muscle cell.

neuron (NUR-on) Type of nerve cell; basic communication unit in most nervous systems.

neurotransmitter Any of a class of signaling molecules secreted by neurons. It acts on cell next to it, then is rapidly degraded or recycled.

neutral mutation Mutation that has little or no effect on phenotype. Natural selection cannot change its frequency in a population because it does not affect survival or reproduction.

neutron Unit of matter, one or more of which occupies the atomic nucleus and has mass but no electric charge.

neutrophil The most abundant, fastest acting white blood cell; it phagocytizes bacteria.

niche (NITCH) [L. *nidas*, nest] Sum total of all activities and relationships in which individuals of a species engage as they secure and use the resources required to survive and reproduce.

nitrification (nye-trih-fih-KAY-shun) Process by which certain bacteria in soil break down ammonia or ammonium to nitrite, then other bacteria break down nitrite to nitrate (which is a form that plants can take up). Key part of the nitrogen cycle.

nitrogen cycle Atmospheric cycle. Nitrogen moves from its largest reservoir (atmosphere), through the ocean, ocean sediments, soils, and food webs, then back to the atmosphere.

nitrogen fixation Process by which certain bacterial species convert gaseous nitrogen to ammonia, which swiftly dissolves in their cytoplasm to form ammonium. Ammonium can be used in biosynthesis.

nociceptor (NO-SEE-sep-tur) Pain receptor that detects tissue damage, as by burns.

node Stem site where one or more leaves form.

noncyclic pathway of ATP formation (non-SIK-lik) [L. *non*, not, + Gk. *kylos*, circle] Light-dependent reactions of photosynthesis that requires photolysis, two photosystems, and two transport chains. Water molecules are split. They release electrons and hydrogen that are used in ATP and NADPH formation, plus oxygen as a by-product (the basis of Earth's oxygen-rich atmosphere).

nondisjunction Failure of sister chromatids or failure of a pair of homologous chromosomes to separate at meiosis or mitosis. As a result, daughter cells end up with too many or too few chromosomes.

notochord (KNOW-toe-kord) Of chordates, a rod of stiffened tissue (not cartilage or bone) that is a supporting structure for the body.

nuclear envelope Outermost portion of a cell nucleus; composed of a double membrane (two lipid bilayers and associated proteins).

nucleic acid (new-CLAY-ik) Single- or double-stranded chain of four kinds of nucleotides joined at phosphate groups. Nucleic acids differ in their base sequences. DNA and RNA are examples.

nucleic acid hybridization Any base-pairing between sequences of DNA or RNA from different sources.

nucleoid (NEW-KLEE-oid) Portion of bacterial cell interior in which the DNA is physically organized but not enclosed by a membrane.

nucleolus (new-KLEE-oh-lus) [L. *nucleolus*, little kernel] In the nucleus of a nondividing cell, an assembly site for the protein and RNA subunits that will later form ribosomes in the cytoplasm.

nucleosome (NEW-klee-oh-sohm) A stretch of eukaryotic DNA looped twice around a spool of histone molecules; one of many units that give condensed chromosomes their structure.

nucleotide (NEW-klee-oh-tide) Small organic compound consisting of a five-carbon sugar (deoxyribose), a nitrogen-containing base, and a phosphate group. The structural unit of adenosine phosphates, nucleotide coenzymes, and nucleic acids.

nucleotide coenzyme Protein that assists an enzyme by delivering electrons and hydrogen atoms released at a reaction site to another reaction site.

nucleus (NEW-klee-us) [L. *nucleus*, a kernel] Of atoms, a central core of one or more protons and (in all but hydrogen atoms) neutrons. In a eukaryotic cell, the organelle that physically separates DNA from cytoplasmic machinery.

numerical taxonomy Study of the degree of relatedness between an unidentified organism and a known group through comparisons of traits. Used to classify prokaryotes, which are poorly represented in the fossil record.

nutrient Element with a direct or indirect role in metabolism that no other element fulfills.

nutrition Processes by which food is selectively ingested, digested, absorbed, and converted to the body's own organic compounds.

obesity Excess of fat in adipose tissue; caloric intake has exceeded the body's energy output.

oligosaccharide (oh-LIG-oh-SAC-uh-RID) Short-chain carbohydrate of two or more covalently bonded sugar monomers. Disaccharides (two monomers) are examples.

omnivore [L. *omnis*, all, + *vovare*, to devour] An animal that feeds at more than one trophic level.

oncogene (ON-koe-jeen) Any gene having the potential to induce cancerous transformation.

oocyte Immature egg of all animals and some protistans.

oogenesis (oo-oh-JEN-uh-sis) Process by which a germ cell develops into a mature oocyte.

operator Very short base sequence between a promoter and bacterial genes; a binding site for a repressor that can block transcription.

operon Promoter–operator sequence that services more than one bacterial gene; part of a control mechanism that adjusts transcription rates upward or downward.

orbital One of the volumes of space around the atomic nucleus in which one or at most two electrons are likely to be at any instant.

organ Body structure having definite form and function that consists of more than one tissue.

organ formation Developmental stage in which primary tissue layers give rise to cell lineages unique in structure and function. Descendants of those lineages give rise to all the different tissues and organs of the adult.

organ system Two or more organs that are interacting chemically, physically, or both in a common task.

organelle (or-GUN-ell) Membrane-bound sac or compartment in the cytoplasm having one or more specialized metabolic functions. Most eukaryotic cells have a profusion of them.

organic compound Molecule of one or more elements covalently bonded to some number of carbon atoms.

osmoreceptor Sensory receptor that detects changes in water volume (solute concentration) in the fluid bathing it.

osmosis (oss-MOE-sis) [Gk. *osmos*, pushing] The diffusion of water in response to water concentration gradient between two regions that are separated by a selectively permeable membrane. The greater the number of ions and molecules dissolved in a solution, the lower its water concentration.

osmotic pressure Force that operates after hydrostatic pressure develops in a cell or in an enclosed body region; the amount of force that stops further increases in fluid volume by countering the inward diffusion of water.

ovary (OH-vuh-ree) In most animals, a female gonad. In flowering plants, the enlarged base of one or more carpels. A fruit is a mature ovary often combined with other floral parts.

oviduct (OH-vih-dukt) One of a pair of ducts through which eggs travel from an ovary to the uterus. Formerly called a fallopian tube.

ovulation (OHV-you-LAY-shun) Release of a secondary oocyte from an ovary during one menstrual cycle.

ovule (OHV-youl) [L. *ovum*, egg] Tissue mass in a plant ovary that develops into a seed. Consists of female gametophyte with egg cell, nutrient-rich tissue, and a jacket (cell layers) that will become a seed coat.

ovum (OH-vum) Mature secondary oocyte.

oxaloacetate (ox-AL-oh-ASS-ih-tate) A four-carbon compound with roles in metabolism (e.g., the point of entry into the Krebs cycle).

oxidation–reduction reaction An electron transfer between atoms or molecules. Often an unbound proton (H^+) is transferred at the same time.

ozone thinning Pronounced seasonal thinning of the atmosphere's ozone layer, especially above the Earth's polar regions.

pancreas (PAN-cree-us) Gland with roles in digestion and organic metabolism. Secretes enzymes and bicarbonate into small intestine; also secretes insulin and glucagon, hormones that travel the bloodstream to target cells.

pancreatic islet Any of the 2 million or so clusters of endocrine cells in the pancreas.

parapatric speciation Idea that neighboring populations can become distinct species while maintaining contact along a common border.

parasite [Gk. *para*, alongside, + *sitos*, food] Organism that lives in or on a host organism for at least part of its life cycle. It feeds on specific tissues and usually does not kill its host outright.

parasitism Symbiotic interaction in which one species (a parasite) benefits and the other (its host) is harmed. The parasite lives inside or on a host and feeds on its cells or tissues.

parasitoid Type of insect larva that grows and develops in a host organism (usually another insect), consumes its soft tissues, and kills it.

parasympathetic nerve (PARE-uh-SIM-pu-THET-ik) An autonomic nerve. Signals carried by such nerves tend to slow overall body activities and divert energy to basic tasks, and to help make small adustments in internal organ activity by acting continually in opposition to sympathetic nerve signals.

parenchyma (par-ENG-kih-mah) Simple tissue that makes up the bulk of a plant; has roles in photosynthesis, storage, secretion, other tasks.

parthenogenesis (par-THEN-oh-GEN-uh-sis) An unfertilized egg giving rise to an embryo.

passive transport Process by which a transport protein that spans a cell membrane passively permits a solute to diffuse through its interior. Also called facilitated diffusion.

pathogen (PATH-oh-jen) [Gk. *pathos*, suffering, + *genēs*, origin] Any virus, bacterium, fungus, protistan, or parasitic worm that can infect an organism, multiply in it, and cause disease.

pattern formation theory Explanation of the orderly, sequential sculpting of embryonic cells into specialized animal tissues and organs. Cytoplasmic localization and, later, inductive interactions among master genes are responsible. Gene products map out the basic body plan and create chemical gradients that dictate how specific body parts develop.

PCR Polymerase chain reaction. A method of enormously amplifying the quantity of DNA fragments cut by restriction enzymes.

peat bog Compressed, soggy, highly acidic mat of accumulated remains of peat mosses.

pedigree Diagram of the genetic connections among related individuals through successive generations; uses standardized symbols.

pelagic province (peh-LAD-jik) Total volume of water in the world ocean.

penis Male copulatory organ by which sperm is deposited in a female reproductive tract.

peptide hormone Amino acid hormone that, when bound to a membrane receptor, activates enzyme systems that trigger a response. Often a second messenger in the cell relays its signal.

perennial [L. *per-*, throughout, + *annus*, year] Flowering plant that lives for three or more growing seasons.

pericycle (PARE-ih-sigh-kul) [Gk. *peri-*, around, + *kyklos*, circle] One or more cell layers just inside the endodermis that give rise to lateral roots and contribute to secondary growth.

periderm Protective cover that replaces plant epidermis during extensive secondary growth.

peripheral nervous system (per-IF-ur-uhl) [Gk. *peripherein*, to carry around] All nerves leading into and out from the spinal cord and brain. Includes ganglia of those nerves.

peristalsis (pare-ih-STAL-sis) Recurring waves of contraction and relaxation of muscles in the wall of a tubular or saclike organ.

peritoneum (pare-ih-tuh-NEE-um) Membrane that lines the coelom and helps maintain the positions of soft organs inside it.

peritubular capillary One of the set of blood capillaries around tubular parts of a nephron. Reabsorbs water and solutes; secretes excess H^+ and other substances to be excreted later.

permafrost Permanently frozen layer beneath the soil surface in arctic tundra. Water cannot penetrate it even in summer.

peroxisome Vesicle in which fatty acids and amino acids are first digested to hydrogen peroxide, then converted to harmless products.

pest resurgence Directional selection for an insecticide-resistant strain of a pest species.

PGA Phosphoglycerate (FOSS-foe-GLISS-er-ate) Intermediate of glycolysis and of the Calvin–Benson cycle.

PGAL Phosphoglyceraldehyde. Intermediate of glycolysis and of the Calvin-Benson cycle.

pH scale Measure of the concentration of free hydrogen ions (H^+) in blood, water, and other solutions. pH 0 is the most acidic, 14 the most basic, and 7, neutral.

phagocyte (FAG-uh-sight) [Gk. *phagein*, to eat, + *kytos*, hollow vessel] Cell that captures prey by phagocytosis (e.g., amoebas); also cells that use same process for defense and day-to-day tissue housekeeping (e.g., macrophages).

phagocytosis (FAG-uh-sigh-TOE-sis) [Gk. *phagein*, to eat, + *kytos*, hollow vessel] Engulfment of foreign cells or particles by way of pseudopod formation and endocytosis.

pharynx (FARE-inks) Among invertebrates, a muscular tube to the gut. In some chordates, a gas exchange organ. In land vertebrates, a dual entrance to the esophagus and trachea.

phenotype (FEE-no-type) [Gk. *phainein*, to show, + *typos*, image] Observable trait or traits of an individual that arise from gene interactions and gene–environment interactions.

pheromone (FARE-oh-moan) [Gk. *phero*, to carry, + -*mone*, as in hormone] Hormone-like, nearly odorless exocrine gland secretion. A signaling molecule between individuals of the same species that integrates social behavior.

phloem (FLOW-um) Plant vascular tissue that conducts sugars and other solutes. Includes living cells (sieve tubes) that connect to form conducting tubes, and adjoining companion cells that help load solutes into the tubes.

phospholipid Organic compound that has a glycerol backbone, two fatty acid tails, and a hydrophilic head of two polar groups (one being phosphate). Phospholipids are the main structural material of cell membranes.

phosphorus cycle Movement of phosphorus (mainly phosphate ions) from land, through food webs, to ocean sediments, then back to land. As for other minerals, Earth's crust is the largest reservoir in this sedimentary cycle.

phosphorylation (FOSS-for-ih-LAY-shun) A common means of activating molecules for a reaction. An enzyme either attaches inorganic phosphate to a molecule or mediates a transfer of a phosphate group from one molecule to another (as when ATP phosphorylates glucose).

photoautotroph Photosynthetic autotroph; any organism that synthesizes its own organic compounds using carbon dioxide as the source of carbon atoms and sunlight as the energy source. Nearly all plants, some protistans, and a few bacteria are photoautotrophs.

photolysis (foe-TALL-ih-sis) [Gk. *photos*, light, +-*lysis*, breaking apart] Reaction sequence in which photon energy splits water molecules. The released electrons and hydrogen take part in the noncyclic pathway of photosynthesis, and the oxygen is released as a by-product.

photoperiodism Biological response to change in relative lengths of daylight and darkness.

photoreceptor Light-sensitive sensory cell.

photosynthesis Trapping of sunlight energy, followed by its conversion to chemical energy (ATP, NADPH, or both) and then synthesis of sugar phosphates, which become converted into sucrose, cellulose, starch, and other end products. The main pathway by which energy and carbon enter the web of life.

photosystem One of many clusters of light-trapping pigment molecules embedded in photosynthetic membranes. A chlorophyll of the system gives up electrons necessary for the light-dependent reactions of photosynthesis.

phototropism [Gk. *photos*, light, + *trope*, a turning, direction] Change in the direction of cell movement or growth in response to light (e.g., as when differences in cell elongation cause a stem to bend toward light).

photovoltaic cell Device that converts energy of sunlight into electricity.

phycobilin (FIE-koe-BY-lin) Type of accessory pigment that extends the functional range of chlorophyll in photosynthesis. Abundant in red algae and in cyanobacteria especially.

phylogeny Evolutionary relationships among species, starting with an ancestral form and including branches leading to descendants.

phytochrome A light-sensitive pigment. Its controlled activation and inactivation take part in plant hormone activities that govern leaf expansion, stem branching, stem lengthening and often seed germination and flowering.

phytoplankton (FIE-toe-PLANK-tun) [Gk. *phyton*, plant, + *planktos*, wandering] Aquatic community of floating or weakly swimming photoautotrophs (e.g., "pastures of the seas").

pigment Any light-absorbing molecule.

pioneer species Any opportunistic colonizer of barren or disturbed habitats. Adapted for rapid growth and dispersal.

pituitary gland Endocrine gland that, with the hypothalamus, controls many physiological functions, including activity of many other endocrine glands. Its posterior lobe stores and secretes hypothalamic hormones. Its anterior lobe produces and secretes its own hormones.

placenta (play-SEN-tuh) Blood-engorged organ of pregnant female placental mammals; made of some endometrial tissue and extraembryonic membranes. Allows exchanges between the mother and fetus without an intermingling of their bloodstreams, thus sustaining the new individual and allowing its blood vessels to develop apart from the mother's.

plankton [Gk. *planktos*, wandering] Of aquatic habitats, a community of suspended or weakly swimming organisms, mostly microscopic.

plant Generally, a multicelled photoautotroph with well-developed root and shoot systems; photosynthetic cells that include starch grains as well as chlorophylls *a* and *b*; and cellulose, pectin, and other polysaccharides in cell walls.

Plantae Kingdom of plants.

plasma (PLAZ-muh) Liquid portion of blood; mainly water in which ions, proteins, sugars, gases, and other substances are dissolved.

plasma membrane Outermost cell membrane; structural and functional boundary between cytoplasm and the fluid outside the cell.

plasmid Of many bacteria, a small, circular molecule of extra DNA that carries only a few genes and that is replicated independently of the bacterial chromosome.

plasmodesma, plural **plasmodesmata** (PLAZ-moe-DEZ-muh) Plant cell junction; membrane-lined channel that crosses both walls of two adjacent cells and connects their cytoplasm.

plasmolysis Osmotically induced shrinkage of a cell's cytoplasm.

plate tectonics Theory that great slabs (plates) of the Earth's outer layer (lithosphere) float on the hot, plastic, underlying mantle. All plates are in motion and have rafted continents to new positions over time. The geologic changes have profoundly affected the evolution of life.

platelet (PLAYT-let) Megakaryocyte fragment that releases substances used in clot formation.

pleiotropy (PLEE-oh-troe-pee) [Gk. *pleon*, more, + *trope*, direction] Positive or negative effects on two or more traits owing to expression of alleles at a single gene locus. Effects may or may not emerge at the same time.

polar body One of four cells that form by the meiotic cell division of an oocyte but that does not become the ovum.

pollen grain [L. *pollen*, fine dust] Immature or mature, sperm-bearing male gametophyte of gymnosperms and angiosperms.

pollen sac Chamber inside an anther in which pollen grains develop.

pollen tube Sperm-carrying tube that grows from a germinated pollen grain, through carpel tissues to the egg inside an ovule.

pollination Arrival of a pollen grain on the landing platform (stigma) of a carpel.

pollutant Natural or synthetic substance with which an ecosystem has no prior evolutionary experience, in terms of kinds or amounts; it accumulates to disruptive or harmful levels.

polymer (POH-lih-mur) [Gk. *polus*, many, + *meris*, part] Large molecule consisting of three to millions of monomers of the same or different kinds.

polymerase (puh-LIM-ur-aze) Enzyme that catalyzes a polymerization reaction (e.g., the DNA polymerase of DNA replication/repair).

polymorphism (poly-MORE-fizz-um) [Gk. *polus*, many, + *morphe*, form] The persistence of two or more qualitatively different forms of a trait (morphs) in a population.

polyp (POH-lip) Vase-shaped, sedentary stage of cnidarian life cycles.

polypeptide chain Organic compound with a sequence of three or more amino acids. Peptide bonds between them result in a regular pattern of nitrogen atoms in the carbon backbone: —N—C—C—N—C—C— . Every protein consists of one or more polypeptide chains.

polyploidy (POL-ee-PLOYD-ee) Having three or more of each type of chromosome in the nucleus of cells at interphase.

polysaccharide [Gk. *polus*, many, + *sakcharon*, sugar] Straight or branched chain of many covalently linked sugar units of the same or different kinds. In nature, the most common polysaccharides are cellulose, starch, and glycogen.

polysome A number of ribosomes translating the same mRNA molecule at the same time, one after the other, during protein synthesis.

pons Hindbrain region; traffic center for signals between cerebellum and forebrain centers.

population All individuals of the same species occupying the same area.

population density Count of individuals of a population occupying a specified area or specified volume of a habitat.

population distribution Dispersal pattern for individuals of a population through a habitat.

population size The number of individuals that make up the gene pool of a population.

positive feedback mechanism A homeostatic control mechanism. It sets in motion a chain of events that intensifies change from an original condition; after a limited time, intensification reverses the change.

potential energy Capacity of any stationary object to do work owing to its position in space or to the arrangement of its parts (e.g., a cat in a frozen posture, about to spring at a mouse).

predation Ecological interaction in which a predator feeds on a prey organism.

predator [L. *prehendere*, to grasp, seize] A heterotroph that feeds on other living organisms (its prey), that lives neither in or nor on them (as parasites do), and that may or may not end up killing them.

prediction Statement about what you should observe in nature if you were to go looking for a particular phenomenon; the if–then process.

pressure flow theory Explanation of how organic compounds move through phloem of vascular plant. The compounds follow solute concentration and pressure gradients between sources (e.g., photosynthetically active leaves where they form) and sinks (e.g., growing parts where they are being used or stored).

primary growth Plant growth originating at root tips and shoot tips.

primary immune response Defensive actions by white blood cells and their secretions, as elicited by first-time recognition of antigen. Includes antibody- and cell-mediated responses.

primary productivity, gross Of ecosystems, the rate at which primary producers capture and store a given amount of energy in their cells and tissues during a specified interval.

primary productivity, net Of ecosystems, the rate of energy storage in primary producer cells and tissues in excess of rate of aerobic respiration during a specified interval.

primary wall A wall of polysaccharides, glycoproteins, and cellulose that is flexible and thin enough to allow new plant cells to divide or change shape during growth and development.

primate Mammalian lineage dating from the Eocene; includes prosimians, tarsioids, and anthropoids (monkeys, apes, and humans).

primer Short nucleotide sequence designed to base-pair with any complementary DNA sequence; later, DNA polymerases recognize it as a START tag for replication.

prion Small infectious protein that causes rare, fatal degenerative diseases of nervous system.

probability The chance that each outcome of a given event will occur is proportional to the number of ways the outcome can be reached.

probe Very short stretch of DNA designed to base-pair with part of a gene being studied and labeled with an isotope to distinguish it from DNA in the sample being investigated.

procambium (pro-KAM-bee-um) A meristem that gives rise to primary vascular tissues.

producer Autotroph (self-feeder); it nourishes itself using sources of energy and carbon from its physical environment. Photoautotrophs and chemoautotrophs are examples.

progesterone (pro-JESS-tuh-rown) A sex hormone secreted by ovaries and the corpus luteum of female mammals.

prokaryotic cell (pro-CARE-EE-oh-tic) [L. *pro*, before, + Gk. *karyon*, kernel] Archaebacterium or eubacterium; single-celled organism, most often walled; lacks the profusion of membrane-bound organelles observed in eukaryotic cells.

prokaryotic fission Cell division mechanism by which a bacterial cell reproduces.

promoter Short stretch of DNA to which RNA polymerase can bind and start transcription.

prophase Of mitosis, a stage when duplicated chromosomes start to condense, microtubules form a spindle, and the nuclear envelope starts to break up. Duplicated pairs of centrioles (if present) are moved to opposite spindle poles.

prophase I The first stage of meiosis I. Each duplicated chromosome starts to condense. It pairs with its homologue; nonsister chromatids usually undergo crossing over. Each becomes attached to microtubular spindle. One of the duplicated pairs of centrioles (if present) is moved to opposite spindle pole.

prophase II First stage of meiosis II. In each daughter cell, spindle microtubules attach to kinetochores of each chromosome and move them toward spindle's equator. One centriole pair (if present) is already at each spindle pole.

protein Organic compound composed of one or more polypeptide chains.

Protista Kingdom of protistans. Chytrids; water molds; slime molds; protozoans; sporozoans; euglenoids; chrysophytes; dinoflagellates; and red, brown, and green algae are major groups.

protistan (pro-TISS-tun) [Gk. *prōtistos*, primal, very first] Diverse species, ranging from single cells to giant kelps, that are photoautotrophs, heterotrophs, or both. Some are thought to be most like the earliest eukaryotic cells. All are unlike bacteria in having a nucleus, large ribosomes, mitochondria, ER, Golgi bodies, chromosomes with many proteins attached, and cytoskeletal microtubules.

proton Positively charged particle; one or more reside in nucleus of each atom. An unbound (free) proton is called a hydrogen ion (H+).

proto-oncogene Gene sequence similar to an oncogene but coding for a protein that is used in normal cell functions. When mutated, it may trigger cancerous transformation.

protostome (PRO-toe-stome) [Gk. *proto*, first, + *stoma*, mouth] Lineage of coelomate, bilateral animals that includes mollusks, annelids, and arthropods. The first indentation to form in protostome embryos becomes the mouth.

protozoan Type of protistan that may resemble the single-celled heterotrophs that gave rise to animals. Amoeboid, animal-like, and ciliated protozoans are major categories.

proximal tubule Nephron's tubular portion into which water and solutes enter after being filtered from blood at Bowman's capsule.

pulmonary circuit Vertebrate cardiovascular route in which oxygen-poor blood flows from the heart to the lungs, where it is oxygenated before flowing back to the heart.

Punnett-square method Construction of a diagram of a genetic cross that is a simple way to predict the probable outcomes.

purine Nucleotide base having a double ring structure (e.g., adenine or guanine).

pyrimidine (pih-RIM-ih-deen) Nucleotide base having a single ring structure (e.g., cytosine or thymine).

pyruvate (PIE-roo-vate) A small organic compound with a backbone of three carbon atoms. Two molecules form as end products of glycolysis.

r Variable in population growth equations that signifies net population growth rate. Birth and death rates are assumed to remain constant and are combined into this one variable.

radial symmetry Animal body plan having four or more roughly equivalent parts around a central axis (e.g., sea anemone).

radioisotope Unstable atom (uneven number of protons and neutrons). It spontaneously emits particles and energy; over a predictable time span, it decays into a different atom.

radiometric dating Method of measuring the proportions of (1) a radioisotope in a mineral trapped long ago in newly formed rock and (2) a daughter isotope that formed from it by radioactive decay in the same rock. Used to assign absolute dates to fossil-containing rocks and to the geologic time scale.

rain shadow A reduction in rainfall on the leeward side of a high mountain range that results in arid or semiarid conditions.

reabsorption In a kidney, diffusion or active transport of water and reclaimable solutes from a nephron into peritubular capillaries; under control of ADH and aldosterone. At a capillary bed, osmotic movement of interstitial fluid into a capillary when water concentration between plasma and interstitial fluid differ.

rearrangement, molecular Conversion of one organic compound to another through changes in its internal bonds.

receptor, molecular Type of membrane protein that binds an extracellular substance (e.g., hormone).

receptor, sensory Sensory cell or specialized cell adjacent to it that can detect a stimulus.

recessive allele [L. *recedere*, to recede] In heterozygotes, an allele whose expression is fully or partially masked by expression of its partner. Fully expressed only in homozygous recessives.

recombinant chromosome Of eukaryotes, a chromosome that emerges from meiosis with a combination of alleles that differs from a parental combination of alleles.

recombinant DNA Any molecule of DNA that incorporates one or more nonparental nucleotide sequences. Outcome of microbial gene transfer in nature or recombinant DNA technology.

recombinant DNA technology Procedures by which DNA molecules from different species are isolated, cut up, and spliced together. A population of rapidly dividing bacterial cells or PCR is then used to amplify the recombinant molecules to useful quantities.

recombination Any enzyme-mediated reaction that inserts one DNA sequence into another. "Generalized" recombination uses any pair of homologous sequences between chromosomes as substrates, as during crossing over. Site-specific recombination uses only a short stretch of homology between viral and bacterial DNA. A different reaction can insert transposable elements at new, random sites in bacterial or eukaryotic genomes; no homology is required.

red alga Type of protistan. Most are multicelled, aquatic photoautotrophs with an abundance of phycobilins that mask chlorophyll *a*.

red blood cell Erythrocyte. Cell that serves in rapid transport of oxygen in blood.

red marrow Site of blood cell formation in the spongy tissue of many bones.

reflex [L. *reflectere*, to bend back] Stereotyped, simple movement in response to stimuli. In simple reflex arcs, sensory neurons synapse directly on motor neurons.

refractory period Brief interval following an action potential when a small patch of neural membrane is insensitive to stimulation.

regulatory protein Component of mechanisms that control transcription, translation, and gene products by interacting with DNA, RNA, new polypeptide chains, or proteins (e.g., enzymes).

releasing hormone Hypothalamic signaling molecule that enhances or slows secretions from target cells in anterior lobe of pituitary gland.

repressor Protein that binds with an operator on bacterial DNA to block transcription.

reproduction Any process by which a parental cell or organism produces offspring. Among eukaryotes, asexual modes (e.g., binary fission, budding, vegetative propagation) and sexual modes. Bacteria employ prokaryotic fission.

Viruses do not reproduce themselves; host organisms execute their replication cycle.

reproductive isolating mechanism A heritable feature of body form, functioning, or behavior that prevents interbreeding between two or more genetically divergent populations.

reproductive success Production of viable offspring by the individual.

reptile Carnivorous species belonging to the first vertebrate lineage to escape dependency on free water, by way of internal fertilization, efficient kidneys, amniote eggs, and other adaptations. Examples are dinosaurs (extinct), crocodilians, snakes, lizards, and tuataras.

resource partitioning Of two or more species that compete for the same resource, a sharing of the resource in different ways or at different times, which allows them to coexist.

respiration [L. *respirare*, to breathe] Of all animals, exchange of environmental oxygen with carbon dioxide from cells (e.g., through integumentary exchange or a respiratory system).

respiratory surface Thin, moist epithelium that functions in gas exchange in animals.

resting membrane potential Of a neuron and other excitable cells, a steady voltage difference across the plasma membrane in the absence of outside stimulation.

restoration ecology Attempt to reestablish biodiversity in ecosystems that have become altered as a result of agriculture, mining, and other severe disturbances.

restriction enzyme One of a class of bacterial enzymes that cut apart foreign DNA injected into the cell body, as by viruses. Important tool of recombinant DNA technology.

reticular formation Low-level pathway of information flow through vertebrate nervous system. Mesh of interneurons that extends from the upper spinal cord, through the brain stem, and into the cerebral cortex.

reverse transcription Synthesis of DNA on an RNA template by using reverse transcriptase, a viral enzyme. Basis of RNA virus replication cycle and of cDNA synthesis in laboratory.

RFLP Short for restriction fragment length polymorphism. DNA fragments of different sizes, cleaved by restriction enzymes, that reveal genetic differences among individuals.

Rh blood typing Method of characterizing red blood cells according to a self-marker protein at their surface. Rh$^+$ cells have it; Rh$^-$ cells do not.

rhizoid Simple rootlike absorptive structure of some fungi and nonvascular plants.

ribosomal RNA (rRNA) Type of RNA that combines with proteins to form ribosomes, on which polypeptide chains are assembled.

ribosome Structure composed of two subunits of rRNA and proteins. Has binding sites for mRNA and tRNAs, which interact to produce a polypeptide chain in translation stage of protein synthesis.

RNA Ribonucleic acid. Any of a class of single-stranded nucleic acids that function in transcribing and translating the genetic instructions encoded in DNA into proteins. A molecule of mRNA, rRNA, or tRNA.

rod cell Vertebrate photoreceptor sensitive to very dim light. Contributes to the coarse perception of movement across visual field.

root Plant part, typically belowground, that absorbs water and dissolved minerals, anchors aboveground parts, and often stores food.

root hair Threadlike extension of a specialized epidermal cell of a young root. Increases root surface area for absorbing water and minerals.

root nodule Localized swelling on a root of certain legumes and other plants. Develops when nitrogen-fixing bacteria infect the plant, multiply, and interact symbiotically with it.

rubisco RuBP carboxylase; an enzyme that catalyzes attachment of the carbon atom from CO_2 to RuBP and so starts the Calvin–Benson cycle of the light-independent reactions.

RuBP Short for ribulose bisphosphate. Organic compound that has a backbone of five carbon atoms that serves in carbon fixation and that is regenerated in the Calvin–Benson cycle in C3 plants.

salinization Salt buildup in soil by evaporation, poor drainage, and heavy irrigation.

saliva Glandular secretion that is mixed with food and starts starch breakdown in the mouth.

salt Compound that releases ions other than H$^+$ and OH$^-$ in solution.

saltatory conduction Of myelinated neurons, a rapid form of action potential propagation. Excitation hops to nodes between jellyrolled membranes of cells making up myelin sheath.

sampling error Use of a sample (or subset) of a population, an event, or some other aspect of nature for an experimental group that is not large enough to be representative of the whole.

saprobe Heterotroph that obtains energy and carbon from nonliving organic matter and so causes its decay (e.g., many fungal species).

sarcomere (SAR-koe-meer) One of many basic units of contraction, defined by Z lines, that subdivide a myofibril (muscle cell). Every sarcomere shortens in response to ATP-driven interactions between its parallel arrays of actin and myosin components.

sarcoplasmic reticulum (sar-koe-PLAZ-mik reh-TIK-you-lum) System of membranous chambers threading around myofibrils that take up, store, and release the calcium ions required for cross-bridge formation.

Schwann cell Type of neuroglial cell that wraps around certain axons like a jellyroll. A series of these cells, separated only by small nodes, form a myelin sheath.

sclerenchyma (skler-ENG-kih-mah) Simple plant tissue that supports mature plant parts and commonly protects seeds. Most of its cells have thick, lignin-impregnated walls.

sea-floor spreading Ongoing event in which molten rock erupts from immense, continuous ridges on the ocean floor, flows laterally in both directions, and hardens to form new crust. Elsewhere, it forces older crust down into vast trenches in the seafloor.

second law of thermodynamics A law of nature stating that the spontaneous direction of energy flow is from forms organized to less organized forms. The total amount of energy in the universe is spontaneously flowing from forms of higher to lower quality; with each conversion, some energy becomes randomly dispersed in a form (heat, most often) not as readily available to do work.

second messenger Molecule within a cell that mediates a hormonal signal by initiating the cellular response to it.

secondary immune response Immune action against previously encountered antigen, more rapid and prolonged than a primary response owing to swift participation of memory cells.

secondary sexual trait A trait associated with maleness or femaleness but with no direct role in reproduction (e.g., distribution of body hair and body fat). The primary sexual trait is the presence of male or female gonads.

secondary wall A wall on the inner surface of the primary wall of an older plant cell that stopped growing but needs structural support. Contains lignin in older cells of woody plants.

secretion A cell acting on its own or as part of glandular tissue releases a substance across its plasma membrane, to the surroundings.

sedimentary cycle Biogeochemical cycle. An element having no gaseous phase moves from land, through food webs, to the seafloor, then returns to land through long-term uplifting.

seed Mature ovule with an embryo sporophyte inside and integuments that form a seed coat.

segmentation Of animal body plans, a series of units that may or may not be similar to one another in appearance. Of tubular organs, an oscillating movement produced by rings of circular muscle in the tube wall.

segregation, theory of [L. *se-*, apart, + *grex*, herd] Mendelian theory. Sexually reproducing organisms inherit pairs of genes (on pairs of homologous chromosomes), the two genes of each pair are separated from each other at meiosis, and they end up in separate gametes.

selective gene expression Control of which gene products a cell makes or activates during a specified interval. Depends on the type of cell, its adjustments to changing chemical conditions, which external signals it is receiving, and its built-in control systems.

selective permeability Of a cell membrane, a capacity to let some substances but not others cross it at certain sites, at certain times. The capacity arises as an outcome of its lipid bilayer structure and its transport proteins.

selfish behavior An individual protects or increases its own chance to produce offspring regardless of consequences to its social group.

selfish herd Social group held together simply by reproductive self-interest.

semen (SEE-mun) Sperm-bearing fluid expelled from a penis during male orgasm.

semiconservative replication [Gk. *hēmi*, half, + L. *conservare*, to keep] Mechanism of DNA duplication. The DNA double helix unzips, and a complementary strand is assembled on exposed bases of each strand. Each conserved strand and its new partner wind up together to form a double helix, thus being a half-old, half-new molecule.

senescence (sen-ESS-cents) [L. *senescere*, to grow old] Processes leading to the natural death of an organism or to parts of it (e.g., leaves).

sensation Conscious awareness of a stimulus.

sensory neuron Type of neuron that detects a stimulus and relays information about it toward an integrating center (e.g., a brain).

sensory system The "front door" of a nervous system; it detects external and internal stimuli and relays information to integrating centers that issue commands for responses.

sessile animal (SESS-ihl) Animal that remains attached to a substrate during some stage (often the adult stage) of its life cycle.

sex chromosome A chromosome with genes that influence primary sex determination (whether male or female gonads will develop in the new individual). Depending on the species, somatic cells have one or two sex chromosomes, of the same or different type. In mammals, females are XX and males XY.

sexual dimorphism Occurrence of female and male phenotypes among the individuals of a sexually reproducing species.

sexual reproduction Production of offspring by meiosis, gamete formation, and fertilization.

sexual selection A microevolutionary process. Natural selection favors a trait that gives the individual a competitive edge in attracting or keeping a mate, hence in reproductive success.

shell model Model of electron distribution in which all orbitals available to electrons of atoms occupy a nested series of shells.

shifting cultivation Cutting and burning trees in a plot of land, followed by tilling ashes into soil. Once called slash-and-burn agriculture.

shoot system Aboveground plant parts (e.g., stems, leaves, and flowers).

sieve-tube member One of the cells that join together as phloem's sugar-conducting tubes.

sign stimulus Simple environmental cue that triggers a response to a stimulus that the nervous system is prewired to recognize.

sink Any region of a plant where cells are storing or using food (e.g., roots).

sister chromatid Of a duplicated chromosome, one of two DNA molecules (and associated proteins) attached at the centromere until they are separated from each other during mitosis or meiosis. After separation, each is then called a chromosome in its own right.

six-kingdom classification scheme A recent phylogenetic scheme that groups all organisms into the kingdoms Eubacteria, Archaebacteria, Protista, Fungi, Plantae, and Animalia.

skeletal muscle An organ with hundreds to many thousands of muscle cells bundled inside a sheath of connective tissue, which extends past the muscle as tendons.

sliding filament model Explanation of how muscles contract. Myosin heads projecting from microtubules at the center of a sarcomere repeatedly bind to neighboring actin filaments (which project inward from the sarcomere's sides), and with short, ATP-driven power strokes make them ratchet toward the center of the sarcomere. The sarcomere shortens as the actin filaments slide toward its center.

small intestine Part of the vertebrate digestive system in which digestion is completed and most dietary nutrients are absorbed.

smog, industrial Polluted, gray-colored air that forms above industrialized cities during cold, wet winters.

smog, photochemical Polluted, brown, smelly air that forms above large cities with many gas-burning vehicles during warm weather.

social behavior Diverse interactions among individuals of a species, which display, send, and respond to shared forms of communication that have genetic and learned components.

social parasite Animal that exploits the social behavior of another species to assure its own survival and reproduction.

sodium–potassium pump Type of membrane transport protein that, when activated by ATP, selectively transports potassium ions across a membrane, against its concentration gradient, and passively allows sodium ions to cross it in the opposite direction.

soil Mixture of mineral particles of variable sizes and decomposing organic material; air and water occupy spaces between particles.

solute (SOL-yoot) [L. *solvere*, to loosen] Any substance dissolved in a solution. Spheres of hydration around charged parts of its ions and molecules keep them dispersed.

solvent Any fluid (e.g., water) in which one or more substances are dissolved.

somatic cell (SO-MAT-ik) [Gk. *somā*, body] Any body cell that is not a germ cell. (Germ cells are the forerunners of gametes.)

somatic nervous system Nerves leading from a central nervous system to skeletal muscles.

somite One of many paired segments in a vertebrate embryo that give rise to most bones, skeletal muscles of head and trunk, and dermis.

source Any plant part where photosynthetic cells are making organic compounds.

speciation (spee-see-AY-shun) The formation of a daughter species from a population or subpopulation of a parent species by way of microevolutionary processes. Routes vary in their details and in length of time before the required reproductive isolation is completed.

species (SPEE-sheez) [L. *species*, a kind] One kind of organism. Of sexually reproducing organisms, one or more natural populations in which individuals are interbreeding and are reproductively isolated from other such groups.

sperm [Gk. *sperma*, seed] Mature male gamete.

spermatogenesis (sper-MAT-oh-JEN-ih-sis) Formation of mature sperm from a germ cell.

sphere of hydration A clustering of water molecules around individual molecules or ions of a substance placed in water owing to positive and negative interactions among them.

spinal cord The part of the central nervous system in a canal inside the vertebral column; site of direct reflex connections between sensory and motor neurons; also has tracts to and from the brain.

spleen A lymphoid organ that is a filtering station for blood, a reservoir of red blood cells, and a reservoir of macrophages.

spore Reproductive or resting structure of one or a few cells, often walled or coated and adapted for resisting adverse conditions, for dispersal, or both. May be nonsexual or sexual (formed by way of meiosis). Sporozoans, fungi, plants, and some bacteria form spores.

sporophyte [Gk. *phyton*, plant] A vegetative body that grows, by mitotic cell divisions, from a plant zygote and that produces spore-bearing structures.

spring overturn Of many bodies of water (e.g., large lakes in temperate zones), a springtime downward movement of dissolved oxygen near the surface, and an upward movement of nutrients from bottom sediments to the surface that fans primary productivity.

S-shaped curve Type of diagrammatic curve that emerges when logistic population growth is plotted against time.

stabilizing selection Mode of natural selection by which intermediate phenotypes in the range of variation are favored and extremes at both ends are eliminated.

stamen (STAY-mun) A male reproductive part of a flower; usually a pollen-bearing structure (anther) on a single stalk (filament).

start codon Base triplet in mRNA that serves as the START signal for translation.

stem cell Self-perpetuating animal cell that stays unspecialized. Some of its daughter cells also are self-perpetuating; others differentiate into specialized cells (e.g., red blood cells that arise from stem cells in bone marrow).

steroid hormone Lipid-soluble hormone made from cholesterol that acts on a target cell's DNA by entering the nucleus alone or bound to intracellular receptor. Some act by binding to a receptor on a target's plasma membrane.

sterol (STAIR-all) Lipid with a rigid backbone of four fused carbon rings. Sterols differ in the number, position, and type of their functional groups. Cholesterol is one; it is a precursor of steroid hormones and occurs in animal cell membranes.

stigma Sticky or hairy surface tissue on upper part of a carpel (or fused carpels) that captures pollen grains and favors their germination.

stimulus [L. *stimulus*, goad] A specific form of energy (e.g., pressure, light, and heat) that activates a sensory receptor able to detect it.

stoma (STOW-muh), plural **stomata** [Gk. *stoma*, mouth] One of many gaps between two guard cells in leaf and stem epidermis. Opens and closes to control inward movement of carbon dioxide and outward movement of water vapor and oxygen, depending on whether conditions in the environment call for water conservation.

stomach A muscular, stretchable sac that mixes and stores ingested food, helps break it apart mechanically and chemically, and controls its expulsion (e.g., into the small intestine).

stop codon Base triplet in a strand of mRNA that serves as a STOP signal during translation; it blocks further additions of amino acids to a newly forming polypeptide chain.

strain One of two organisms with differences that are too minor to classify it as a separate species (e.g., *Escherichia coli* strain 018:K1:H).

stratification Stacked layers of sedimentary rock that resulted from a gradual deposition of volcanic ash, silt, and other materials over time.

stream A flowing-water ecosystem that starts out as a freshwater spring or seep.

stroma [Gk. *strōma*, bed] A semifluid matrix between the thylakoid membrane system and the two outer membranes of a chloroplast; a zone where sucrose, starch, cellulose, and other end products of photosynthesis are assembled.

stromatolite Fossilized mats of shallow-water, microbial communities (mainly cyanobacteria) of Archean to Precambrian times. Their tacky gel secretions blocked out ultraviolet radiation but trapped sediments; so new mats had to grow over older ones, like cake layers. Stromatolites are found on all continents; some are over a half mile thick and hundreds of miles across.

substrate Reactant or precursor for a specific enzyme-mediated metabolic reaction.

substrate-level phosphorylation The direct, enzyme-mediated transfer of a phosphate group from a substrate to a molecule, as when an intermediate of glycolysis gives up a phosphate group to ADP to form ATP.

succession, primary (suk-SESH-un) [L. *succedere*, to follow after] Ecological pattern by which a community develops in orderly progression, from the time pioneer species colonize a barren habitat to the climax community.

succession, secondary Ecological pattern by which a disturbed area of a community recovers and moves back toward the climax state. It is typical of abandoned croplands, forest burns, and storm-battered intertidal zones.

surface-to-volume ratio Mathematical relation in which volume increases with the cube of the diameter, but surface area increases only with the square. If a growing cell were simply to expand in diameter, its volume of cytoplasm would increase faster than the surface area of the plasma membrane required to service it. In general, this constraint keeps cells small, elongated, or with infoldings or outfoldings of its plasma membrane.

survivorship curve Plot of age-specific survival of a group of individuals in the environment, from the time of birth until the last one dies.

swim bladder Adjustable flotation device that changes in volume as it exchanges gases with blood; it helps many fishes maintain neutral buoyancy in water.

symbiosis (sim-by-OH-sis) [Gk. *sym*, together, + *bios*, life, mode of life] Individuals of one species live near, in, or on those of another species for at least part of life cycle (e.g., in commensalism, mutualism, and parasitism).

sympathetic nerve An autonomic nerve that deals mainly with increasing overall body activities at times of heightened awareness, excitement, or danger; also works continually in opposition with parasympathetic nerves to make minor adjustments in internal organ activities.

sympatric speciation [Gk. *sym*, together, + *patria*, native land] A speciation event within the home range of an existing species, in the absence of a physical barrier. Such species may form instantaneously, as by polyploidy.

synaptic integration (sin-AP-tik) Moment-by-moment combining of all excitatory and inhibitory signals arriving at the trigger zone of a neuron or some other excitable cell.

syndrome A set of symptoms that may not individually be a telling clue but collectively characterize a genetic disorder or disease.

systemic circuit (sis-TEM-ik) Of vertebrates, a cardiovascular route in which oxygen-enriched blood flows from the heart through the rest of the body (where it gives up oxygen and takes up carbon dioxide), then back to the heart.

T cell T lymphocyte; a type of white blood cell vital to immune responses (e.g., helper T cells and cytotoxic T cells).

taproot system A primary root together with all of its lateral branchings.

target cell Any cell with molecular receptors that can bind with a particular hormone or some other signaling molecule.

taxonomy Field of biology that deals with identifying, naming, and classifying species.

tectum Midbrain's roof. Coordinating center for most sensory inputs and initiating motor responses in fishes and amphibians. In most vertebrates (not mammals), a reflex center that relays sensory input to forebrain.

telophase (TEE-low-faze) Of meiosis I, the stage when one member of each pair of homologous chromosomes has arrived at a spindle pole. Of mitosis and of meiosis II, the stage when chromosomes decondense into threadlike structures and two daughter nuclei form.

temperature A measure of the kinetic energy of ions or molecules in a specified region.

tendon A cord or strap of dense connective tissue that attaches a muscle to bone.

territory An area that an animal is defending against competitors for mates, food, water, living space, other resources.

test A means to determine the accuracy of a prediction, as by conducting experimental or observational tests and by developing models. Scientific tests are made under controlled conditions in nature or the laboratory.

testcross Experimental cross to determine whether an individual of unknown genotype that shows dominance for a trait is either homozygous dominant or heterozygous.

testis, plural **testes** A primary reproductive organ (gonad) of some male animals; it produces male gametes and sex hormones.

testosterone (tess-TOSS-tuh-rown) A type of sex hormone necessary for the development and functioning of the male reproductive system of vertebrates.

tetanus (TET-uh-nuss) Of a muscle, a large contraction in which repeated stimulation of a motor unit causes muscle twitches to mechanically run together. In a disease by the same name, a toxin prevents muscles from being released from contraction.

thalamus (THAL-uh-muss) A forebrain region that is a coordinating center for sensory input and a relay station for signals to the cerebrum.

theory, scientific A testable explanation of a broad range of related phenomena, one that has been subjected to extensive experimental testing and can be used with a high degree of confidence. A scientific theory remains open to tests, revision, and tentative acceptance or rejection.

thermal inversion The trapping of a layer of dense, cool air beneath a layer of warm air. The inversion can result in an accumulation of air pollutants close to the ground.

thermophile A type of archaebacterium that is adapted to unusually hot aquatic habitats, such as hot springs and hydrothermal vents.

thermoreceptor Sensory cell or specialized cell next to it that detects radiant energy (heat).

thigmotropism (thig-MOE-truh-pizm) [Gk. *thigm*, touch] An orientation of the direction of growth in response to physical contact with a solid object, as when a vine curls around a fencepost.

threshold Of an excitable cell (e.g., a neuron or muscle cell), the minimum amount of change in the resting membrane potential that will trigger an action potential.

thylakoid Of chloroplasts, part of an internal membrane system folded repeatedly into a stack of disks. Such stacks (grana) have light-absorbing pigments and enzymes required to form ATP, NADPH, or both in photosynthesis. The stacks connect (by membranous channels) as a single functional compartment.

thymine Nitrogen-containing base; one of the nucleotides in DNA.

thymus gland A lymphoid organ that has endocrine functions. Lymphocytes of the immune system multiply, differentiate, and mature in its tissues; its hormone secretions affect their functioning.

thyroid gland Endocrine gland located in front of the trachea; its hormones have widespread effects on growth and development, and on the overall metabolic rates of warm-blooded animals.

tissue Of multicelled organisms, a group of cells and intercellular substances that function together in one or more specialized tasks.

tonicity (TOE-niss-ih-TEE) Relative solute concentrations of two fluids (e.g., cytoplasmic fluid relative to extracellular fluid).

touch-killing Mechanism by which cytotoxic T cells directly release perforins and toxins onto a target cell and cause its destruction.

toxin A normal metabolic product of one species with chemical effects that can hurt or kill individuals of a different species.

trace element Any element that represents less than 0.01 percent of body weight.

tracer Substance with a radioisotope attached, like a shipping label, that researchers can track after delivering it into a cell, body, ecosystem, or some other system. Laboratory devices detect emissions from the tracer as it moves through a pathway or reaches a destination.

trachea (TRAY-kee-uh), plural **tracheae** An air-conducting tube of respiratory systems. Of land vertebrates, the windpipe through which air passes between the larynx and bronchi.

tracheal respiration Of certain invertebrates (e.g., insects), respiration by way of finely branching tracheae that start at openings in the integument and dead-end in body tissues.

tracheid (TRAY-kid) One of two types of cells in xylem that conduct water and minerals.

tract A cordlike bundle of axons of sensory neurons, motor neurons, or both inside the brain or spinal cord. Comparable to a nerve.

transcription [L. *trans*, across, + *scribere*, to write] First stage of protein synthesis. An RNA strand is assembled on exposed bases of one unwound strand of a DNA double helix. The transcript's base sequence is complementary to that of the DNA template.

transfer RNA (tRNA) An RNA that binds with and delivers amino acids to a ribosome and that pairs with an mRNA codon during the translation stage of protein synthesis.

translation Stage of protein synthesis when an mRNA's base sequence becomes converted to a sequence of particular amino acids in a new polypeptide chain. rRNA, tRNA, and mRNA interact to bring this about.

translocation Of cells, a stretch of DNA that moved to a new location in a chromosome or in a different chromosome, with no molecular loss. Of vascular plants, a process by which organic compounds are distributed through the phloem.

transpiration Evaporative water loss from a plant's aboveground parts, leaves especially.

transport protein One of many kinds of membrane proteins involved in active or passive transport of water-soluble substances across the lipid bilayer of a cell membrane. Solutes on one side of the membrane pass through the protein's interior to the other side.

transposable element A stretch of DNA that can move at random from one location to another in the individual's genome. Often it inactivates the genes into which it becomes inserted and causes changes in phenotype.

triglyceride (neutral fat) A type of lipid that has three fatty acid tails attached to a glycerol backbone. Triglycerides are the body's most abundant lipids and its richest energy source.

trisomy (TRY-so-mee) The presence of three chromosomes of a given type in a cell rather than the two characteristic of a parental diploid chromosome number.

trophic level (TROE-fik) [Gk. *trophos*, feeder] Of an ecosystem, all organisms that are the same number of transfer steps away from the energy input into the system.

tropical rain forest A biome characterized by regular, heavy rainfall, an annual mean temperature of 25°C, humidity of 80 percent or more, and stunning biodiversity. Presently being obliterated in regions with fast-growing human populations but limited food, fuel, and lumber; projections are that most will disappear by 2035.

tropism (TROE-pizm) Of plants, a directional growth response to an environmental factor (e.g., growth toward light).

true breeding Of a sexually reproducing species, a lineage in which one version only of a trait shows up through the generations in all parents and their offspring.

tumor Tissue mass composed of cells that are dividing at an abnormally high rate. If benign, its cells remain in their home tissue; if malignant, they have metastasized.

turgor pressure (TUR-gore) [L. *turgere*, to swell] Internal fluid pressure on a cell wall when water moves into the cell by osmosis.

ultrafiltration Bulk flow of a small amount of protein-free plasma from a blood capillary when the outward-directed force of blood pressure is greater than the inward-directed osmotic force of interstitial fluid.

uniformity theory Early theory that the Earth's surface changes in gradual, uniformly repetitive ways (major floods, earthquakes, and other infrequent annual catastrophes were not considered unusual). Helped change Darwin's view of evolution. Has since been replaced by plate tectonics theory.

upwelling An upward movement of deep, nutrient-rich water along coasts; it replaces surface waters that move away from shore when the prevailing wind direction shifts.

uracil (YUR-uh-sill) Nitrogen-containing base of a nucleotide in RNA but not DNA. Like thymine, uracil can base-pair with adenine.

ureter One of a pair of tubes that conduct urine from the kidneys to the urinary bladder.

urethra Tube that conducts urine from the urinary bladder to an opening at the body's surface.

urinary bladder The distensible sac in which urine is stored before being excreted.

urinary excretion Mechanism by which excess water and solutes are removed from the body through the urinary system.

urinary system Organ system that adjusts the volume and composition of blood, and thereby helps maintain extracellular fluid.

urine Fluid consisting of any excess water, wastes, and solutes; it forms in kidneys by filtration, reabsorption, and secretion.

uterus (YOU-tur-us) [L. *uterus*, womb] Of a female placental mammal, a muscular, pear-shaped organ in which embryos are contained and nurtured during pregnancy.

vaccination Immunization procedure against a specific pathogen.

vaccine An antigen-containing preparation, swallowed or injected, designed to increase immunity to certain diseases by inducing formation of armies of effector and memory B and T cells.

vagina The part of the reproductive system of mammalian females that receives sperm, forms part of the birth canal, and channels menstrual flow to the exterior.

variable Of an experimental test, a specific aspect of an object or event that may differ over time and among individuals. A single variable is directly manipulated in an attempt to support or disprove a prediction; any other variables that might influence the results are identical (ideally) in both the experimental group and one or more control groups.

vascular bundle An array of primary xylem and phloem in multistranded, sheathed cords that thread lengthwise throughout the ground tissue system.

vascular cambium A lateral meristem that increases stem or root diameter.

vascular cylinder Arrangement of vascular tissues as a central cylinder in roots.

vascular plant Plant with xylem and phloem, and usually with well-developed roots, stems, and leaves.

vascular tissue system Xylem and phloem, the conducting tissues that distribute water and solutes through a vascular plant.

vein Of a cardiovascular system, any of the large-diameter vessels that lead back to the heart. Of leaves, one of the vascular bundles that thread through photosynthetic tissues.

venule A small blood vessel that serves as a transitional conducting tube between a small-diameter capillary and a larger diameter vein.

vernalization Environmental stimulation of flowering, by exposure to low temperatures.

vertebra, plural **vertebrae** One of a series of hard bones, with cushioning intervertebral disks stacked between them, that serve as a backbone and that protect the spinal cord.

vertebrate Animal with a backbone.

vesicle (VESS-ih-kul) [L. *vesicula*, little bladder] One of a variety of small, membrane-bound sacs in the cytoplasm that function in the transport, storage, or digestion of substances.

vessel member Type of cell in xylem; dead at maturity, but its wall becomes part of a water-conducting pipeline (a vessel).

vestigial (ves-TIDJ-ul) Applies to a small body part, tissue, or organ abnormally developed or degenerated and unable to function like its normal counterpart (e.g., vestigial wings of mutant fruit flies; "tail bones" of humans).

villus (VIL-us), plural **villi** Any of several fingerlike absorptive structures projecting from the free surface of an epithelium.

viroid An infectious particle consisting only of very short, tightly folded strands or circles of RNA. Viroids might have evolved from introns, which they resemble.

virus A noncellular infectious agent that is composed of DNA or RNA and a protein coat; it can become replicated only after its genetic material enters a host cell and subverts the host's metabolic machinery.

viscera All soft organs inside an animal body (e.g., heart, lungs, and stomach).

vision Perception of visual stimuli. Requires a focusing of light precisely onto a layer of photoreceptive cells that is dense enough to sample details of a light stimulus, followed by image formation in a brain.

visual signal An observable action or cue that functions as a communication signal.

vitamin Any of more than a dozen organic substances that an organism requires in small amounts for metabolism but that it generally cannot synthesize for itself.

vocal cord Of certain animals, one of the thickened, muscular folds of the larynx that help produce sound waves for vocalization.

water potential Sum of two opposing forces (osmosis and turgor pressure) that can cause a directional movement of water into or out of an enclosed volume (e.g., a walled cell).

water table Upper limit at which the ground in a specified land region is fully saturated with water.

watershed Any specified region in which all precipitation drains into one stream or river.

wavelength A wavelike form of energy in motion. The horizontal distance between the crests of every two successive waves.

wax Molecule having long-chain fatty acids packed together and linked to long-chain alcohols or to carbon rings. Waxes have a firm consistency and repel water.

white blood cell An eosinophil, neutrophil, macrophage, T or B cell, or other leukocyte which, with chemical mediators, counters tissue invasion and tissue damage. Different kinds take part in nonspecific and specific defense responses. Some (e.g., eosinophils) are fast-acting. Some (e.g., macrophages) take part in sustained immune responses.

white matter The portion of the brain and spinal cord with axons that have glistening white sheaths and that specialize in rapid signal transmission.

wild-type allele Allele that occurs normally or with the greatest frequency at a given gene locus among individuals of a population.

wing A body part that serves in flight, as among birds, bats, and many insects. A bird wing is a forelimb having feathers, strong muscles, and extremely lightweight bones. A bat wing is a modified forelimb; four thin, elongated digits are the framework for a thin integumentary membrane. An insect wing develops as a lateral fold of the exoskeleton.

X chromosome A type of sex chromosome. In mammals, an XX pairing causes an embryo to develop into a female; an XY pairing causes it to develop into a male.

X-linked gene Any gene that is located on an X chromosome.

X-linked recessive inheritance Recessive condition in which the responsible, mutated gene is located on the X chromosome.

xylem (ZYE-lum) [Gk. *xylon*, wood] A tissue having pipelines that conduct water and solutes through vascular plants. The pipelines are the interconnecting walls of cells that are dead at maturity.

Y chromosome Distinctive chromosome in males or females of many species, but not both (e.g., human males are XY and human females, XX).

yellow marrow A fatty tissue in the cavities of most mature bones that produces red blood cells when blood loss from the body is severe.

Y-linked gene Gene on a Y chromosome.

yolk sac An extraembryonic membrane. In most shelled eggs, it holds nutritive yolk; in humans, part of the yolk sac becomes a site of blood cell formation and some cells give rise to forerunners of gametes.

zero population growth A state in which the number of births in a population is balanced by the number of deaths over a specified period, assuming that immigration and emigration are balanced, also.

zooplankton A community of suspended or weak-swimming heterotrophs of freshwater or marine habitats. Most of its species are microscopic; commonly, rotifers and copepods are among the most abundant.

zygote (ZYE-goat) The first cell of a new individual, formed by fusion of a sperm nucleus with egg nucleus at fertilization; a fertilized egg.

CREDITS AND ACKNOWLEDGMENTS

11.22 Carolina Biological Supply Company / **11.23** Bonnie Kamin/Stuart Kenter Associates /

CHAPTER 12 **12.1** A. C. Barrington Brown © 1968 J. D. Watson / **12.2** A. Lesk/SPL/Photo Researchers / **12.3** Art by Raychel Ciemma / **12.4** (b) Micrograph Lee D. Simon/Science Source/ Photo Researchers / **12.6** Micrograph Biophoto Associates/SPL/Photo Researchers / **12.7** Art by Precision Graphics / **12.8** Courtesy of the Cold Springs Harbor Laboratory Archives / **12.11** (b) PA News Photo Library /

Chapter 13 **13.1** (above) Dennis Hallinan/FPG; (below) © Bob Evans/Peter Arnold, Inc. / **13.4** From Stephen L. Wolfe, *Molecular and Cellular Biology*, Wadsworth, 1993 / **13.8** (a) 3-D model of tRNA by David B. Goodin, Ph.D; Art by Lisa Starr / **13.10** Art by Lisa Starr / **13.13** Nik Kleinberg / **13.14** Art by Lisa Starr / **13.15** Photograph courtesy of the National Neurofibromatosis Foundation /

CHAPTER 14 **14.1** (a) Ken Greer/Visuals Unlimited; (b) Biophoto Associates/Science Source/Photo Researchers; (c) James Stevenson/SPL/Photo Researchers; (d) Photograph Gary Head / **14.2** Lennart Nilsson © Boehringer Ingelheim International GmbH / **14.3** Brian Matthews, University of Oregon / **14.4** Art by Palay/Beaubois and Hans & Cassady / **14.5** (b) Jack Carey / **14.6** Frank B. Salisbury / **14.7** (a) Slim Films; (b) Dr. Brian V. Harmon, Queensland University of Technology. From *Methods in Cell Biology*, 46, 1995, "Anatomical Methods in Cell Death," Academic Press. Reprinted by permission / **14.8** Art by Betsy Palay/Artemis /

CHAPTER 15 **15.1** Lewis L. Lainey / **15.2** Science VU/Visuals Unlimited / **15.3** (a) Dr. Huntington Potter and Dr. David Dressler; (b) Stanley N. Cohen/Science Source/Photo Researchers / **15.4, 15.5** Art by Lisa Starr / **15.7** (right) Cellmark Diagnostics, Abingdon, U.K. / **15.8, 15.9** Art by Lisa Starr / **15.10** (a) From *Molecular and Cellular Biology*, Stephen L. Wolfe / **15.11** Photograph Hervé Chaumeton/Agence Nature/Art by Lisa Starr / **15.13** Photograph courtesy Calgene LLC / **15.14** R. Brinster and R. E. Hammer, School of Veterinary Medicine, University of Pennsylvania / **Page 237** S. Stammers/SPL/Photo Researchers /

CHAPTER 16 **16.1** Elliot Erwitt/Magnum Photos, Inc. / **16.2** (a) Jen and Des Bartlett/Bruce Coleman Ltd.; (b) Kenneth W. Fink/Photo Researchers; (c) Dave Watts/A.N.T. Photo Library / **16.3** Art by Raychel Ciemma / **16.4** (a, left) Courtesy George P. Darwin, Darwin Museum, Down House; (a, right) Heather Angel; (b) Christopher Ralling; (c) Photograph Dieter and Mary Plage/Survival Anglia; Art by Leonard Morgan / **16.5** (a) Field Museum of Natural History, Chicago, and the artist, Charles R. Knight (Neg. No. CK21T); (b) © Philip Boyer/ Photo Researchers / **16.6** (a) Heather Angel; (b) David Cavagnaro/Bruce Coleman, Inc.; (c) Dr. P. Evans/Bruce Coleman, Inc.; (d) Alan Root/Bruce Coleman Ltd. / **16.7** Down House and The Royal College of Surgeons of England / **16.8** P. Morris/Ardea London; Art by Raychel Ciemma / **Page 246** Photos courtesy Derrell Fowler, Tecumseh, OK / **16.9** Alan Solem / **16.12** J. A. Bishop and L. M. Cook / **16.14** (a) (b) Warren Abrahamson; (c) Forest W. Buchanan/Visuals Unlimited; (d) Kenneth McCrea and Warren Abrahamson / **16.16** (a) Thomas Bates Smith / **16.18** Bruce Beehler / **16.19** (above) David Neal Parks; (below) W. Carter Johnson / **16.20** Courtesy of Jerry Coyne / **16.21** Photograph David Cavagnaro / **16.22** Kjell Sandved/Visuals Unlimited /

CHAPTER 17 **17.1** (a) Photograph Gary Head; Snail courtesy of Larry Reed; (b) Adapted from R. K. Selander and D. W. Kaufman, *Evolution*, 29, 3, December 31, 1975 / **17.3** After F. Ayala and J. Valentine, *Evolving*, Benjamin-Cummings, 1979 / **17.4** (a) G. Ziesler/ZEFA; (b) Alvin E. Staffan/ Photo Researchers /**17.5** (a) Fred McConnaughey/ Photo Researchers; (b) Patrice Geisel/Visuals Unlimited; (right) Tom Van Sant/The Geosphere Project, Santa Monica, CA / **17.7** Art by Raychel Ciemma / **17.8** (left) Art by Precision Graphics based on U. Schliewen et al., *Nature*, 368:629-632, April 14, 1994. Used by permission of Macmillan Magazines, Ltd.; (right) Tom Van Sant/The Geosphere Project, Santa Monica, CA / **17.10** (left) H. Clarke, © VIREO/Academy of Natural Sciences; (center) Photograph Robert C. Simpson/Nature Stock / **17.12** (above) Jack Dermid; (below) Leonard Lee Rue III/FPG; (right) After P. Dodson, *Evolution: Process and Product*, Third edition, Prindle, Weber & Schmidt / **17.13** Jen and Des Bartlett/Bruce Coleman Ltd.

CHAPTER 18 **18.1** (left) Vatican Museums; (right) Martin Dohrn/SPL/Photo Researchers / **18.2** (a) N. Simmons and J. Geisler/ © American Museum of Natural History; (b) A. Feduccia, *The Age of Birds*, Harvard University Press, 1980; (c) H. P. Banks; (d) Jonathan Blair / **18.4** Art by Raychel Ciemma / **18.5** (top) Douglas P. Wilson/Eric & David Hosking; (center) Superstock, Inc.; (bottom) E. R. Degginger / **18.6** (a) Photograph Gary Head; (a) (b) (c) Art by Precision Graphics after E. Guerrant, *Evolution*, 36:699-712 / **18.7** (b) From T. Storer et al., *General Zoology*, Sixth edition, McGraw-Hill, 1979, reproduced by permission of McGraw-Hill, Inc. / **18.8** Art by Victor Royer / **18.10** (left) Kjell B. Sandved/ Visuals Unlimited; (center) Thomas D. Mangelsen/Images of Nature; (right) Jeffrey Sylvester/FPG / **18.11** (left to right) Larry Lefever/Grant Heilman Inc.; R.I.M. Campbell/Bruce Coleman Ltd.; Runk and Schoenberger/Grant Heilman Inc.; Bruce Coleman Ltd / **18.14** © Dan Peha/The Picture Cube, Inc. / **18.16** Art by Leonard Morgan / **18.17** (right) Martin Land/Photo Researchers; (left) Maps by Lloyd K. Townsend after A. M. Ziegler, C. R. Scotese, and S. F. Barrett, "Mesozoic and Cenozoic Paleogeographic Maps" and J. Krohn and J. Sündermann, "Paleotides Before the Permian" in F. Brosche and J. Sündermann (Eds.), *Tidal Friction and the Earth's Rotation II*, Springer-Verlag, 1983; Photograph Martin Land/SPL/ Photo Researchers / **18.18** (a) Kingsley R. Stern / **Page 289** © 1990 Arthur M. Greene

CHAPTER 19 **19.1** Jeff Hester and Paul Scowen, Arizona State University, and NASA / **19.2** Painting by William K. Hartmann / **19.3** (a) Painting by Chesley Bonestell / **19.4** Art by Precision Graphics / **19.5** (a) Sidney W. Fox; (b) W. Hargreaves and D. Deamer / **19.7** Art by Precision Graphics / **19.8** Bill Bachman/Photo Researchers / **19.9** (a) Stanley W. Awramik; (b) (c) (d) (e) (f) Andrew H. Knoll, Harvard University / **19.10** P. L. Walne and J. H. Arnott, *Planta*, 77: 325-354, 1967 / **19.11** Art by Raychel Ciemma / **19.12** Robert K. Trench / **19.13** (a) (b) Neville Pledge/South Australian Museum; (a) (b) Chip Clark / **19.14** (a) (b) (d) Art by Raychel Ciemma; (c) Patricia G. Gensel / **19.15** (above) Painting by Megan Rohn courtesy of David Dilcher; (below) Art by Precision Graphics / **19.16** © John Gurche 1989 / **19.17** (a) NASA Galileo Imaging Team / **19.18** Art by Raychel Ciemma / **19.19** (a) (b) Paintings © Ely Kish; (c) Field Museum of Natural History, Chicago, and the artist Charles R. Knight (Neg. No.CK8T) / **19.20** (left) Maps by Lloyd K. Townsend after A. M. Ziegler, C. R. Scotese, and S. F. Barrett, "Mesozoic and Cenozoic Paleogeographic Maps" and J. Krohn and J. Sündermann, "Paleotides Before the Permian" in F. Brosche and J. Sündermann (Eds.), *Tidal Friction and the Earth's Rotation II*, Springer-Verlag, 1983 /

CHAPTER 20 **20.1** (a) (b) (c) Tony Brain and David Parker/SPL/Photo Researchers; (d) Lee D. Simon/ Photo Researchers; (e) Gary W. Grimes and Steven L'Hernault / **20.2** Art by Lisa Starr / **20.3** (a) Stanley Flegler/Visuals Unlimited; (b) P. Hawtin, University of Southampton/SPL/Photo Researchers / **20.4** L. J. LeBeau, University of Illinois Hospital/BPS / **20.5** (a) Barry Rokeach; (b) Photograph courtesy Jack Jones. *Archives of Microbiology*, 1983, 136:254–261, reprinted by permission of Springer-Verlag / **20.6** (a) John D. Cunningham/Visuals Unlimited; (b) Tony Brain/SPL/Photo Researchers; (c) P. W. Johnson and J. McN. Sieburth, University of Rhode Island/BPS / **20.7** (above) Centers for Disease Control; (below) Photograph Edward S. Ross / **20.8** Art by Raychel Ciemma; Micrograph L. Santo / **20.9** (a) K. G. Murti/Visuals Unlimited; (b) Kenneth M. Corbett / **20.11** Art by Palay/Beaubois and Precision Graphics / **20.12** Art by Lisa Starr / **20.13** (a) Heather Angel; (b) W. Merrill / **20.14** (a) Art by Leonard Morgan; (b) M. Claviez, G. Gerish, and R. Guggenheim; (c) London Scientific Films; (d) (e) (f) Carolina Biological Supply Company; (g) Photograph courtesy Robert R. Kay from R. R. Kay et al., *Development*, 1989 Supplement, pp. 81-90, © The Company of Biologists Ltd. 1989 / **20.15** Edward S. Ross / **20.16** (a) M. Abbey/Visuals Unlimited; (b) Dr. Howard Spero; (c) G. Shih and R. Kessel/Visuals Unlimited / **20.17** (a) Sidney L. Tamm; (b) Art by Raychel Ciemma redrawn from V. & J. Pearse and M. & R. Buchsbaum, *Living Invertebrates*, The Boxwood Press, 1987. Used by permission. (c) Art by Raychel Ciemma; (d) Sidney L. Tamm; (e) Photograph by Ralph Buchsbaum and sketch from V. & J. Pearse and M. & R. Buchsbaum, *Living Invertebrates*, The Boxwood Press, 1987. Used by permission / **20.18** (a) Jerome Paulin/ Visuals Unlimited; (b) David M. Phillips/Visuals Unlimited; (c) John D. Cunningham/Visuals Unlimited / **20.19** Art by Leonard Morgan; Micrograph Steven L'Hernault / **20.20** R. Regnery, Centers for Disease Control and Prevention, Atlanta / **20.21** Art by Palay/Beaubois / **20.22** (a) Greta Fryxell, University of Texas, Austin; (b) (c) Ronald W. Hoham, Dept. of Biology, Colgate University / **20.23** (a) C. C. Lockwood; (b) © S. Berry/Visuals Unlimited; (c) Florida Department of Environmental Protection, Florida Marine Research Institute, St. Petersburg / **20.24** Art by Raychel Ciemma / **20.25** (above) from Tom Garrison, *Oceanography: An Invitation to Marine Science*, Wadsworth, 1993; (below) Lewis Trusty/Animals Animals / **20.26** (a) Manfred Kage/Peter Arnold, Inc.; (b) Ronald W. Hoham, Dept. of Biology, Colgate University; (c) © Linda Sims/Visuals Unlimited; (d) Brian Parker/Tom Stack and Associates / **20.27** Photograph D. J. Patterson/ Seaphot Limited: Planet Earth Pictures; Art by Raychel Ciemma / **20.28** Carolina Biological Supply Company /

CHAPTER 21 **21.1** © 1997 Sherry K. Pittam / **21.2** (a–e) Robert C. Simpson/Nature Stock / **21.3** (a) (b) Robert C. Simpson/Nature Stock / **21.4** Art by Raychel Ciemma; Micrograph Garry T. Cole, University of Texas, Austin/BPS / **21.5** (a) Jane Burton/Bruce Coleman Ltd.; (b) Thomas J. Duffy / **Page 336** Micrograph J. D. Cunningham/Visuals Unlimited / **21.6** Micrographs Ed Reschke; Art by Raychel Ciemma / **21.7** (a) (b) Robert C. Simpson/ Nature Stock; (c) G. L. Barron, University of Guelph; (d) Garry T. Cole, University of Texas, Austin/BPS / **21.8** (a) Dr. P. Marazzi/SPL/Photo Researchers; (b) Eric Crichton/Bruce Coleman Ltd. / **21.9** (a) After Raven, Evert, and Eichhorn, *Biology of Plants*, Fourth edition, Worth Publishers, New York, 1986; (b) (c) Photographs Gary Head; (d) Mark Mattock/Planet Earth Pictures; (e) Edward S. Ross / **21.10** (a) © 1990 Gary Braasch; (b) F. B. Reeves / **21.11** John Hodgin /

CHAPTER 22 **22.1** (a) Pat and Tom Leeson/Photo Researchers; (b) (c) Edward S. Ross / **22.2** Art by Raychel Ciemma after E.O. Dodson and Peter

Dodson, *Evolution Process and Product,* 3rd Edition, p. 401. Prindle Weber and Schmidt / **22.4** (a) Craig Wood/Visuals Unlimited; (b) Jane Burton/Bruce Coleman Ltd.; Art by Raychel Ciemma / **22.5** (a) Fred Bavendam/Peter Arnold; (b) John D. Cunningham/Visuals Unlimited / **22.6** Brian Parker/Tom Stack & Associates; (above) Field Museum of Natural History, Chicago (Neg. #7500C) / **22.7** (a) Kingsley R. Stern; (c) Ed Reschke/Peter Arnold; (d) William Ferguson; (e) W. H. Hodge; (f) Kratz/ZEFA / **22.8** Art by Raychel Ciemma; Photograph (inset) A. & E. Bomford/Ardea, London; Photograph Lee Casebere / **22.9** (left) Photograph Edward S. Ross; (inset below) R. J. Erwin/Photo Researchers; (inset above) Robert and Linda Mitchell; Art by Raychel Ciemma / **22.10** © 1994 Robert Glenn Ketchum / **22.11** (a) Jeff Gnass Photography; (b) Ed Reschke / **22.12** (a) Joyce Photographics/Photo Researchers; (b) Kingsley R. Stern; (c) Runk/Schoenberger/Grant Heilman, Inc.; (d) Cath Ellis, University of Hull/SPL/Photo Researchers; (e) William E. Ferguson / **22.13** (a) Art by Jennifer Wardip; (b) M.P.L. Fogden/Ardea, London; (c) Heather Angel; (d) L. Mellichamp/ Visuals Unlimited; (e) Peter F. Zika/Visuals Unlimited / **22.14** Art by Raychel Ciemma / **22.15** (a) George Loun/Visuals Unlimited; (b) Earl Roberge/Photo Researchers; (c) John Mason/Ardea London; (d) Dick Davis/Photo Researchers; (e) ZEFA-Rein / **22.16** Bob Cerasoli / **22.17** (left, center) © 1989, 1991 Clinton Webb; (right) Photograph Gary Head /

Chapter 23 **23.1** (a) Courtesy of Department of Library Services, American Museum of Natural History (Neg. # K10273) / **23.2** Jack Carey and Lisa Starr / **23.3** Art by D. and V. Hennings / **23.4** Art by Raychel Ciemma / **23.5** (left) After Laszlo Meszoly in L. Margulis, *Early Life,* Jones and Bartlett Publishers / **23.6** Bruce Hall / **23.7** (a) (b) (c) Art by Raychel Ciemma; (d) (left) Art by Raychel Ciemma, after Bayer and Owre, *The Free-Living Lower Invertebrates,* © 1968 Macmillan; (right) Don W. Fawcett/Visuals Unlimited; (e) Marty Snyderman/Planet Earth Pictures / **23.8** Art by Raychel Ciemma; (a) Photograph F. Schensky; (b) Kim Taylor/Bruce Coleman Ltd. / **23.9** Art by Raychel Ciemma / **23.10** (a) Art by Precision Graphics after T. Storer et al., *General Zoology,* Sixth edition, © 1979 McGraw-Hill; (b) Douglas Faulkner/Photo Researchers; (c) Christian DellaCorte; (d) Andrew Mounter/Seaphot Limited: Planet Earth Pictures / **23.11** Photograph Kim Taylor/Bruce Coleman Ltd.; Art by Raychel Ciemma / **23.12** (a) Cath Ellis, University of Hull/SPL/Photo Researchers; (b) Robert and Linda Mitchell / **23.13, 23.14** Art by Raychel Ciemma / **23.15** Art by Raychel Ciemma; (d) Photograph Carolina Biological Supply Company / **23.16** (a) Lorus J. and Margery Milne; (b) Dianora Niccolini / **23.17** (b) Photograph Gary Head; (c) Alex Kerstitch; (d) Jeff Foott/Tom Stack & Associates; (e) Frank Park/A.N.T. Photo Library; (f) Bob Cranston / **23.19** J.A.L. Cooke/Oxford Scientific Films / **23.20** (a) © Cabisco/Visuals Unlimited; (b) Jon Kenfield/Bruce Coleman Ltd.; (c) From Eugene N. Kozloff, *Invertebrates,* © 1990 by Saunders College Publishing, reproduced by permission of the publisher; (d) Adapted from Rasmussen, *Ophelia,* Vol. 11 in Eugene N. Kozloff, *Invertebrates,* 1990 / **23.21** Art by Raychel Ciemma / **23.22** Jane Burton/Bruce Coleman Ltd. / **23.23** (a) Jane Burton/Bruce Coleman Ltd.; (b) Angelo Giampiccolo/FPG; (c) John H. Gerard; (d) Redrawn from *Living Invertebrates,* V. & J. Pearse/ M. & R. Buchsbaum, The Boxwood Press, 1987. Used by permission; (e) P. J. Bryant, University of California, Irvine/BPS; (f) Ken Lucas/Seaphot Limited: Planet Earth Pictures / **23.24** (a) Franz Lanting/Bruce Coleman Ltd.; (b) Hervé Chaumeton; (c) Fred Bavendam/Peter Arnold,

Inc.; (d) Agence Nature; (above) Art by Raychel Ciemma / **23.25** After David H. Milne, *Marine Life and the Sea,* Wadsworth 1995 / **23.26** (a) Steve Martin/Tom Stack & Associates; (b) Z. Leszczynski/Animals Animals / **23.27** Art by D. & V. Hennings / **23.28** From Georges Pasteur, "Jean Henri Fabre," *Scientific American,* July 1994. Copyright © 1994 by Scientific American, Inc. All rights reserved / **23.29** (a) (b) (c) Edward S. Ross; (d) David Maitland/Seaphot Limited: Planet Earth Pictures; (e–i) C. P. Hickman, Jr.; (j) Edward S. Ross / **23.30** (a) Chris Huss/The Wildlife Collection; (b) John Mason/Ardea, London; (c) Kjell B. Sandved; (d) Ian Took/Biofotos / **23.31** (a) Art by L. Calver; (b) (c) Hervé Chaumeton/ Agence Nature / **Page 379** Jane Burton/Bruce Coleman Ltd. / **23.32** Art by Raychel Ciemma / **23.33** (a) J. Solliday/BPS; (b) Hervé Chaumeton/ Agence Nature /

CHAPTER 24 **24.1** (a) Jean Phillipe Varin/Jacana/ Photo Researchers; (b) Tom McHugh/Photo Researchers / **24.3** (a) Photograph Rick M. Harbo; (b) (c) Sketches redrawn from *Living Invertebrates,* V. & J. Pearse and M. & R. Buchsbaum, The Boxwood Press, 1987. Used by permission / **24.4** (a) Art by Raychel Ciemma; (b) Runk & Schoenberger/Grant Heilman, Inc. / **24.6** Art by Raychel Ciemma adapted from A. S. Romer and T. S. Parsons, *The Vertebrate Body,* Sixth edition, Saunders College Publishing, 1986; Photograph Al Giddings/Images Unlimited / **24.7** (a) (b) After David H. Milne, *Marine Life and the Sea,* Wadsworth, 1995; (c) Heather Angel / **24.8** (a) Erwin Christian/ZEFA; (b) Allan Power/Bruce Coleman Ltd.; (c) Tom McHugh/Photo Researchers / **24.9** (a) Bill Wood/Bruce Coleman Ltd.; (b) Art by Raychel Ciemma; (c) Robert and Linda Mitchell; (d) Patrice Ceisel/ © 1986 John G. Shedd Aquarium; (e) © Norbert Wu/Peter Arnold, Inc.; (f) From Tom Garrison, *Oceanography: An Invitation to Marine Science,* Wadsworth, 1993 / **24.10** After C. P. Hickman et al., *Integrated Principles of Zoology,* Sixth edition, St. Louis: C. V. Mosby Co., 1979 / **24.11** (a) © Marianne Collins; (b) (c) Art by Laszlo Meszoly and D. & V. Hennings / **24.12** (a) Jerry W. Nagel; (b) Stephen Dalton/Photo Researchers; (c) John Serraro/ Visuals Unlimited; (d) Juan M. Renjifo/Animals Animals / **Page 391** Art by Leonard Morgan adapted from A. S. Romer and T. S. Parsons, *The Vertebrate Body,* Sixth edition, Saunders College Publishing, 1986, and others / **24.13** Art by Raychel Ciemma / **24.14** (a) © 1989 D. Braginetz; (b) Zig Leszczynski/Animals Animals / **24.15** (a) D. Kaleth/Image Bank; (b) Andrew Dennis/A.N.T. Photo Library; (c) Stephen Dalton/Photo Researchers; Art by Raychel Ciemma; (d) Heather Angel / **24.16** (a) Art by Lisa Starr; (b) Gerard Lacz/A.N.T. Photo Library; (c) Rajesh Bedi; (d) Thomas D. Mangelsen/Images of Nature; (e) J. L. G. Grande/Bruce Coleman Ltd./ **24.17** (b) Art by D. & V. Hennings / **24.18** (a, left) Sandy Roessler/FPG; (a, right) Leonard Lee Rue III/FPG; (b) Art by Raychel Ciemma after M. Weiss and A. Mann, *Human Biology and Behavior,* Fifth edition, HarperCollins, 1990 / **24.19** Art by Raychel Ciemma / **24.20** (a) Tom McHugh/Photo Researchers; (b) D. & V. Blagden/A.N.T. Photo Library; (c) J. Scott Altenbach, University of New Mexico; (d) Clem Haagner/Ardea, London; (e) Douglas Faulkner/ Photo Researchers; (f) Christopher Crowley / **24.21** (a) Bruce Coleman Ltd.; (b) Tom McHugh/ Photo Researchers; (c) Larry Burrows/Aspect Picture Library / **24.22** © Time Inc. 1965/Larry Burrows Collection / **24.23** Art by D. & V. Hennings / **24.24** (a) Dr. Donald Johanson, Institute of Human Origins; (b) Louise M. Robbins; (c) (d) Kenneth Garrett/National Geographic Image Collection / **24.25** (a) Jean Paul Tibbles; (b) (c) Photographs by John Reader

© 1981 / **24.27** Art by Raychel Ciemma / **Page 405** Photograph Gary Head

CHAPTER 25 **25.1** (a) Roger Werth; (b) (c) © 1980 Gary Braasch; (d) Don Johnson/Photo Nats, Inc. / **25.2, 25.4** Art by Raychel Ciemma / **25.5** Micrograph James D. Mauseth, *Plant Anatomy,* Benjamin-Cummings, 1988 / **25.6** (a) (b) Biophoto Associates / **25.7** (a) Jan Robert Factor/ Photo Researchers; (b) D. E. Akin and I. L. Risgby, Richard B. Russel Agricultural Research Center, Agricultural Research Service, US, Department of Agriculture, Athens, GA; (c) Kingsley R. Stern / **25.9** George S. Ellmore / **25.10** Art by D. & V. Hennings / **25.11** Art by Raychel Ciemma / **25.12** (a) Robert and Linda Mitchell; (b) Roland R. Dute; (c) Photograph Gary Head / **25.13** Art by D. & V. Hennings; (a, left) Ray F. Evert; (right) James W. Perry; (b, left) Carolina Biological Supply Company; (right) James W. Perry / **25.14** Art by D. & V. Hennings / **25.15** (a) Heather Angel; (b) Photograph Gary Head / **25.16** (a) Art by Raychel Ciemma; (b) C. E. Jeffree et al., *Planta,* 172(1):20-37, 1987, reprinted by permission of C. E. Jeffree and Springer-Verlag; (c) Jeremy Burgess/SPL/ Photo Researchers / **25.17** Art by Raychel Ciemma / **25.18** (a) Art by Raychel Ciemma; (b) John Limbaugh/Ripon Microslides, Inc. / **25.19** (a) Micrographs Chuck Brown; (b) Carolina Biological Supply Company / **25.20** John Limbaugh/Ripon Microslides; Sketch after T. Rost et al., *Botany: A Brief Introduction to Plant Biology,* Second edition, © 1984, John Wiley & Sons / **25.21** (a) (b) Art by Raychel Ciemma / **25.23** H. A. Core, W. A. Coté, and A. C. Day, *Wood Structure and Identification,* Second edition, Syracuse University Press, 1979 / **25.24** (b) (c) Art by Raychel Ciemma / **25.25** Edward S. Ross / **Page 421** © Stephen J. Krasemann/The National Audubon Society Collection/Photo Researchers /

CHAPTER 26 **26.1** (a) (b) Robert and Linda Mitchell; (c) John N.A. Lott, *Scanning Electron Microscope Study of Green Plants,* St. Louis: C. V. Mosby Company, 1976; (d) Robert C. Simpson/ Nature Stock / **26.2** David Lavagnaero/Peter Arnold / **26.3** William Ferguson / **26.4** U.S. Department of Agriculture / **26.5** (a) Micrograph Chuck Brown; (b) (c) Art by Leonard Morgan / **26.6** Micrograph Jean Paul Revel / **26.7** (b) Mark E. Dudley and Sharon R. Long; (c) Adrian P. Davies/Bruce Coleman Ltd.; (d) NifTAL Project, University of Hawaii, Maui; Art by Jennifer Wardrip / **26.8** (a, left) Micrograph W.A. Cote of N.C. Brown Center for Ultra-Structure Studies; (a) (b) (c) Micrographs H. A. Core, W. A. Coté, and A. C. Day, *Wood Structure and Identification,* Second edition, Syracuse University Press, 1979 / **26.9** Art by Raychel Ciemma / **26.10** Photographs Frank B. Salisbury / **26.11** George S. Ellmore / **26.12** W. Thomson, *American Journal of Botany,* 57(3): 316, 1970 / **26.14** T. A. Mansfield / **26.15** (a) (b) Jeremy Burgess/SPL/Photo Researchers; Photograph John Lawlor/FPG / **26.16** Art by Hans & Cassady, Inc. / **26.17** Martin Zimmermann, *Science,* 133:73-79, © AAAS 1961 / **26.18** (a) (b) (c) Art by Palay/Beaubois / **26.20** James T. Brock /

CHAPTER 27 **27.1** (a) Merlin D. Tuttle, Bat Conservation International / **27.2** (a) Robert A. Tyrrell; (b) Thomas Eisner, Cornell University / **27.3** (a) Photograph Gary Head / **27.4** John Shaw/Bruce Coleman Ltd. / **27.5** (a) David M. Phillips/Visuals Unlimited; (c) David Scharf/ Peter Arnold, Inc. / **27.6** Art by Raychel Ciemma / **27.7** (a) (b) Patricia Schulz; (c) (d) Ray F. Evert; (e) (f) John Limbaugh/Ripon Microslides, Inc. / **27.8** (b) B. J. Miller, Fairfax, VA/BPS; (c) Richard H. Gross; (d) Janet Jones; (e) R. Carr / **27.9** Russell Kaye / © 1993 The Walt Disney Co. Reprinted with permission of Discover Magazine / **27.10** (a) Runk/Schoenberger/Grant Heilman, Inc.; (b) Kingsley R. Stern / **27.11** B. Bracegirdle and P.

Miles, *An Atlas of Plant Structure*, Heinemann Educational Books, 1977 / **27.12** (a) (b) Art by Raychel Ciemma; (c) Hervé Chaumeton/Agence Nature / **27.13** (d) (e) Art by Raychel Ciemma; (c) Barry L. Runk/Grant Heilman, Inc.; (d) Mauseth / **27.14** (a) (b) Art by Raychel Ciemma; (c) Biophot / **27.15** (a) Kingsley R. Stern / **27.16** Sylvan H. Wittwer/Visuals Unlimited / **27.17** (a) (b) Micrographs courtesy of Randy Moore from "How Roots Respond to Gravity," M. L. Evans, R. Moore, and K. Hasenstein, *Scientific American*, December 1986; (c) John Digby and Richard Firn / **27.18** (c) Frank B. Salisbury / **27.19** Photograph Gary Head / **27.20** Cary Mitchell / **27.22** Frank B. Salisbury / **27.23** Diana Starr / **27.25** (a) Jan Zeevart; (b) Photograph Gary Head / **27.26** N. R. Lersten / **27.27** R. J. Downs / **p. 457** © Kevin Schafer

CHAPTER 28 **28.1** David Macdonald / **28.2** (a) Focus on Sports; (inset) Manfred Kage/Bruce Coleman Ltd.; Art by Palay/Beaubois; Photographs (b, left) Lennart Nilsson from *Behold Man*, © 1974 by Albert Bonniers Forlag and Little, Brown and Company, Boston; (center) Manfred Kage/Bruce Coleman Ltd.; (right) Ed Reschke/Peter Arnold, Inc. / **28.3** Art by Raychel Ciemma / **28.4** Gregory Dimijian/Photo Researchers; Art by Raychel Ciemma adapted from C. P. Hickman, Jr., L. S. Roberts, and A. Larson, *Integrated Principles of Zoology*, Ninth edition, Wm. C. Brown, 1995 / **28.5** Micrographs (a) (b) (c) (e) (f) Ed Reschke; (d) Fred Hossler/Visuals Unlimited; (f) (above) © P. Motta/Dept. of Anatomy/University "La Sapienza," Rome/SPL/Photo Researchers / **28.6** Roger K. Burnard; (left) Art by Joel Ito; (right) Art by L. Calver / **28.7, 28.8** Micrographs Ed Reschke / **28.9** Art by Robert Demarest / **28.10** (a) Lennart Nilsson from *Behold Man*, © 1974 Albert Bonniers Forlag and Little, Brown and Company, Boston; (b) Kim Taylor/Bruce Coleman Ltd. / **28.11** Art by L. Calver / **28.14** Fred Bruemmer /

CHAPTER 29 **29.1** Comstock, Inc. / **29.3** (d) Micrograph Manfred Kage/Peter Arnold, Inc. / **29.4, 29.6** Art by Raychel Ciemma / **29.8** Art by Kevin Somerville; Micrograph Ed Reschke / **29.9** (b) Micrograph Dr. Constantino Sotelo from *International Cell Biology*, p. 83, 1977. Used by copyright permission of the Rockefeller University Press / **29.11** Micrograph R. G. Kessel and R. H. Kardon from *Tissues and Organs: A Text-Atlas of Scanning Electron Microscopy*, W. H. Freeman and Company, 1979. Used by permission of R. G. Kessel; Art by Robert Demarest / **29.13** (b) Art by Robert Demarest / **29.14** Painting by Sir Charles Bell, 1809, courtesy of Royal College of Surgeons, Edinburgh / **29.15** Art by Raychel Ciemma / **29.17** (b) Art by Raychel Ciemma / **29.18** Art by Kevin Somerville / **29.21** (a) Art by Robert Demarest; (b) Micrograph Manfred Kage/Peter Arnold, Inc. / **29.22** (a) Colin Chumbley/Science Source/Photo Researchers; (b) Photograph C. Yokochi and J. Rohen, *Photographic Anatomy of the Human Body*, Second edition, Igaku-Shoin Ltd., 1979 / **29.23** (b) Marcus Raichle, Washington University School of Medicine / **29.25** (a) (b) From Edythe D. London et al., *Archives of General Psychiatry*, 47:567-574, 1990; (right) photograph Ogden Gigli/Photo Researchers /

CHAPTER 30 **30.1** (a) Eric A. Newman; (b) Merlin D. Tuttle, Bat Conservation International / **30.2** (above) Art by Kevin Somerville; (below) Art by Lisa Starr / **30.3** From Hensel and Bowman, *Journal of Physiology*, 23:564-568, 1960 / **30.4** (left) Art by Palay/Beaubois after Penfield and Rasmussen, *The Cerebral Cortex of Man*, © 1950 Macmillan Publishing Company, Inc. Renewed 1978 by Theodore Rasmussen; (right) Photograph Colin Chumbley/Science Source/Photo Researchers / **30.5** Art by Raychel Ciemma / **30.7** Art by Robert Demarest; Micrograph Omikron/SPL/Photo Researchers / **30.6** Art by Robert Demarest / **30.8** (a) (b) Robert E. Preston, courtesy Joseph E. Hawkins, Kresge Hearing Research Institute, University of Michigan Medical School / **30.9** Art by Raychel Ciemma; (d) Keith Gillett/Tom Stack & Associates / **30.10** Photograph E. R. Degginger; Sketch after M. Gardiner, *The Biology of Vertebrates*, McGraw-Hill, 1972 / **30.11** Art by Robert Demarest / **30.12** Chase Swift / **30.11, 30.13** Art by Kevin Somerville / **30.14** Micrograph Lennart Nilsson © Boehringer Ingelheim International GmbH / **30.17** (a) (b) Photographs Gerry Ellis/The Wildlife Collection /

CHAPTER 31 **31.1** Hugo van Lawick / **31.2** Art by Kevin Somerville / **31.5, 31.6** Art by Robert Demarest / **31.7** (a) Mitchell Layton; (b) Syndication International (1986) Ltd.; (c) Photographs courtesy of Dr. William H. Daughaday, Washington University School of Medicine, from A. I. Mendelhoff and D. E. Smith, eds., *American Journal of Medicine*, 20:133, 1956 / **31.8** Art by Leonard Morgan / **31.9** (a) (b) Art by Raychel Ciemma; (c) Corbis-Bettmann / **31.11** Biophoto Associates/SPL/Photo Researchers / **31.12** Art by Leonard Morgan / **31.13** (a) John S. Dunning/Ardea, London; (b) Evan Cerasoli / **31.14** (a) (b) From R. C. Brusca and G. J. Brusca, *Invertebrates*, © 1990 Sinauer Associates. Used by permission; (c) Frans Lanting/Bruce Coleman Ltd.; (d) Robert and Linda Mitchell / **31.15** Photograph Roger K. Burnard /

CHAPTER 32 **32.1** John Brandenberg/Minden Pictures / **32.2** Art by L. Calver / **32.3** (left) Ed Reschke; (right) CNRI/SPL/Photo Researchers; Art by Robert Demarest / **32.4** Michael Keller/FPG / **32.5** Linda Pitkin/Planet Earth Pictures / **32.6** (a) Stephen Dalton/Photo Researchers; (b) Art by Raychel Ciemma / **32.7** Art by Raychel Ciemma / **32.8** (a, left) Art by Raychel Ciemma; (a) (b) Art by Joel Ito; Micrograph Ed Reschke / **32.9** National Osteoporosis Foundation / **32.10** C. Yokochi and J. Rohen, *Photographic Anatomy of the Human Body*, Second edition, Igaku-Shoin Ltd., 1979 / **32.11** Art by Raychel Ciemma / **32.12** (a) NHPA/A.N.T. Photo Library / **32.13** Art by Robert Demarest / **32.14** Art by Kevin Somerville / **32.15** (b) Micrograph John D. Cunningham; (c) Micrograph D. W. Fawcett, *The Cell*, Philadelphia: W. B. Saunders Co., 1966; Art by Robert Demarest / **32.16** Art by Nadine Sokol / **32.18** Art by Robert Demarest / **32.22** After Stephen L. Wolfe, *Molecular and Cellular Biology*, Wadsworth, 1993 / **32.20** Photograph Michael Neveux /

CHAPTER 33 **33.1** (a) From A. D. Waller, *Physiology, The Servant of Medicine*, Hitchcock Lectures, University of London Press, 1910; (b) Photograph courtesy of The New York Academy of Medicine Library / **33.3** (c) After M. Labarbera and S. Vogel, *American Scientist*, 70:54-60, 1982 / **33.4** Art by Precision Graphics / **33.5** Art by Palay/Beaubois / **33.6** CNRI/SPL/Photo Researchers / **33.7** Art by Raychel Ciemma / **33.8** (a) (b) Lester V. Bergman & Associates, Inc.; (c) After F. Ayala and J. Kiger, *Modern Genetics*, © 1980 Benjamin-Cummings / **33.9** Art by Nadine Sokol after Gerard J. Tortora and Nicholas P. Anagnostakos, *Principles of Anatomy and Physiology*, Sixth edition, Copyright © 1990 by Biological Sciences Textbooks, Inc., A & P Textbooks, Inc. and Elia-Sparta, Inc., reprinted by permission of HarperCollins Publishers / **33.11** Art by Kevin Somerville / **33.12** (a) C. Yokochi and J. Rohen, *Photographic Anatomy of the Human Body*, Second edition, Igaku-Shoin Ltd, 1979; (b) (c) Art by Raychel Ciemma / **33.14** Micrograph Michael Abbey/Photo Researchers; Art by Raychel Ciemma / **33.15** Art by Raychel Ciemma / **33.16** Art by Robert Demarest based on A. Spence, *Basic Human Anatomy*, Benjamin-Cummings, 1982 / **33.18** Sheila Terry/SPL/Photo Researchers / **33.19** Art by Lisa Starr / **33.20** Art by Kevin Somerville / **33.21** (a) Ed Reschke; (b) F. Sloop and W. Ober/Visuals Unlimited / **33.24** Lennart Nilsson © Boehringer Ingelheim International GmbH / **33.25, 33.26** Art by Raychel Ciemma / **33.27** Lennart Nilsson from *Behold Man*, © 1974 by Albert Bonniers Forlag and Little, Brown and Company, Boston /

CHAPTER 34 **34.1** (a) The Granger Collection, New York; (b) Lennart Nilsson © Boehringer Ingelheim International GmbH / **34.2** Micrograph Robert R. Dourmashkin, courtesy of Clinical Research Centre, Harrow, England / **34.3** Art by Lisa Starr / **34.4** (a) NSIBC/SPL/Photo Researchers, Inc.; (b) Biology Media/Photo Researchers, Inc. / **34.5** Art by Raychel Ciemma / **34.6** Art by Lisa Starr / **34.8** Art by Raychel Ciemma / **34.9** Art by Lisa Starr; Micrograph Dr. A. Liepins/SPL/Photo Researchers, Inc. / **34.10** Art by Lisa Starr; Photograph courtesy Don C. Wiley, Harvard University / **34.11** Art by Hans & Cassady, Inc. / **34.13, 34.14** Art by Palay/Beaubois after B. Alberts et al., *Molecular Biology of the Cell*, Garland Publishing Company, 1983 / **34.16** (above) Lowell Georgia/Science Source/Photo Researchers; (below) Matt Meadows/Peter Arnold, Inc. / **34.17** (left) David Scharf/Peter Arnold, Inc.; (right) Kent Wood/Photo Researchers / **34.18** Ted Thai/Time Magazine / **34.19** © Zeva Olbaum/Peter Arnold, Inc. / **34.20** (a) After Stephen L. Wolfe, *Molecular Biology of the Cell*, Wadsworth, 1993; (b) Micrographs Z. Salahuddin, National Institutes of Health / **34.21** Courtesy of Dr. Wayne A. Hendrickson and Dr. Peter Kwong /

CHAPTER 35 **35.1** Galen Rowell/Peter Arnold, Inc. / **35.4** (a) Peter Parks/Oxford Scientific Films; (b) Hervé Chaumeton/Agence Nature / **35.5** Micrograph Ed Reschke / **35.6** (a) (d) (e) Art by Raychel Ciemma; (b) (c) Art by Lisa Starr / **35.8** Art by Lisa Starr / **35.9** Micrograph H. R. Duncker, Justus-Liebig University, Giessen, Germany / **35.10** Art by Kevin Somerville / **35.12** Art by K. Kasnot; Photographs SIU/Visuals Unlimited / **35.11** Modified from A. Spence and E. Mason, *Human Anatomy and Physiology*, Fourth edition, 1992, West Publishing Company / **35.13** Giorgio Gualco/Bruce Coleman Ltd. / **35.15** Art by Leonard Morgan / **35.16** CNRI/SPL/Photo Researchers / **35.17** O. Auerbach/Visuals Unlimited / **35.18** Micrograph Lennart Nilsson from *Behold Man*, © 1974 by Albert Bonniers Forlag and Little, Brown and Company, Boston /

CHAPTER 36 **36.1** Hulton Getty Collection/Tony Stone Images / **36.3** Art by Raychel Ciemma / **36.4** (a) D. Robert Franz/Planet Earth Pictures; (c) Adapted from A. Romer and T. Parsons, *The Vertebrate Body*, Sixth edition, Saunders College Publishing, 1986 / **36.5** Gunter Ziesler/Bruce Coleman Inc. / **36.6** Art by Kevin Somerville / **36.8** Art by Robert Demarest; Micrograph Omikron/SPL/Photo Researchers / **36.9** After A. Vander et al., *Human Physiology: Mechanisms of Body Function*, Fifth edition, McGraw-Hill, 1990. Used by permission; (c) Redrawn from page 763 of *Human Anatomy and Physiology*, Fourth edition, by A. Spence and E. Mason. © 1992 by West Publishing Company. All rights reserved / **36.10** Art by Lisa Starr; (b, above) Microslide courtesy Mark Nielsen, University of Utah; (below) © D.W. Fawcett/PhotoResearchers, Inc. / **36.11** Art by Raychel Ciemma / **Page 607** (in-text art) After A. Vander et al., *Human Physiology: Mechanisms of Body Function*, Fifth edition, McGraw-Hill, 1990. Used by permission / **36.12** Photograph Ralph Pleasant/FPG / **36.14** Elizabeth Hathon/The Stock Market / **36.16** Photograph Gary Head / **36.17** Photographs Dr. Douglas Coleman, The Jackson Laboratory /

CHAPTER 37 **37.1** (above) David Noble/FPG; (below) Claude Steelman/Tom Stack & Associates; (right) Photograph Gary Head / **37.3, 37.4** Art by Robert Demarest / **37.5** (above) Art by Robert Demarest; (below) Art by Precision Graphics / **37.6** Photograph Gary Head / **37.7** (a) (b) From Tom Garrison, *Oceanography: An Invitation to Marine Science*, Wadsworth, 1993; (c) Thomas D. Mangelsen/Images of Nature / **37.8** Bob McKeever/Tom Stack & Associates / **37.9** Colin

Monteath, Hedgehog House, New Zealand / **37.10** Corbis-Bettmann /

CHAPTER 38 **38.1** (a) Hans Pfletschinger; (b) Art by Raychel Ciemma; (c) John H. Gerard; (d) (e) © David M. Dennis/Tom Stack & Associates; (f) John Shaw/Tom Stack & Associates / **38.2** (a) Frieder Sauer/Bruce Coleman Ltd.; (b) Wisniewski/ZEFA; (c) L.L. Rue III; (d) Carolina Biological Supply Company; (e) FredMcKinney/FPG; (f) Evan Cerasoli / **38.4** Carolina Biological Supply Company / **38.7** Photograph Gary Head / **38.8** (a) Art by Raychel Ciemma after V. E. Foe and B. M. Alberts, *Journal of Cell Science*, 61:32, © The Company of Biologists, 1983; (b) J. B. Morrill / **38.9** (left) Photographs Carolina Biological Supply Company; (far right) Peter Parks/Oxford Scientific Films/Animals Animals / **38.10** Art by Lisa Starr / **38.11** Carolina Biological Supply Company / **38.12** (a) Art by Palay/Beaubois after Robert F. Weaver and Philip W. Hedrick, *Genetics*, copyright © 1989 Wm. C. Brown Publishers; (b) Art by Raychel Ciemma after Scott Gilbert, *Developmental Biology, 4/E.* Sinauer / **38.13** Art by Raychel Ciemma after Scott Gilbert, *Developmental Biology, 4/E.* Sinauer / **38.14** Art by Raychel Ciemma / **38.15** (a) Art by Raychel Ciemma; (b) Ed Reschke / **38.16, 38.17, 38.18** Art by Raychel Ciemma / **38.19** (above) Art by Robert Demarest; (below) Art by Raychel Ciemma / **38.20** Photograph Lennart Nilsson from *A Child Is Born*, © 1966, 1977 Dell Publishing Company, Inc. / **38.21** Art by Robert Demarest / **38.22, 38.23, 38.24, 38.25** Art by Raychel Ciemma / **38.26** Photographs Lennart Nilsson, *A Child Is Born*, © 1966, 1977 Dell Publishing Company, Inc.; Art by Raychel Ciemma / **38.27** (a) Art by Raychel Ciemma modified from Keith L. Moore, *The Developing Human: Clinically Oriented Embryology*, Fourth edition, Philadelphia: W. B. Saunders Co., 1988 / **38.28** (a) From Lennart Nilsson, *A Child Is Born*, © 1966, 1977 Dell Publishing Company, Inc.; (b) James W. Hanson, M.D. / **38.29** Art by Robert Demarest / **38.30** Art by Raychel Ciemma / **38.31** Art by Raychel Ciemma adapted from L. B. Arey, *Developmental Anatomy*, Philadelphia, W. B. Saunders Co., 1965 / **38.33** CNRI/SPL/Photo Researchers / **38.34** (a) John D. Cunningham/Visuals Unlimited; (b) David M. Phillips/Visuals Unlimited / **Page 669** Alan and Sandy Carey /

CHAPTER 39 **39.1** (a) Photograph Gary Head; (b) Antoinette Jongen/FPG / **39.3** (a) E. R. Degginger; (a, inset) Jeff Fott Productions/Bruce Coleman, Ltd.; (b) © Paul Lally/Stock Boston / **39.4** (c) Stanley Flegler/Visuals Unlimited / **39.7** Photograph E. Vetter/ZEFA / **39.8** (a) Jonathan Scott/Planet Earth Pictures; (b) Wisniewski/ZEFA; (c) Fred Bavendam/Peter Arnold, Inc. / **Page 679** Photographs John A. Endler / **39.9** Photograph NASA / **39.10** After G. T. Miller, Jr., *Environmental Science*, Sixth edition, Wadsworth, 1997 / **39.11** Photograph United Nations / **39.12** Data from Population Reference Bureau after G. T. Miller, Jr., *Living in the Environment*, Eighth edition, Wadsworth, 1993 / **39.13** Photograph United Nations / **39.14, 39.15** After G. T. Miller, Jr., *Environmental Science*, Sixth edition, Wadsworth, 1997 /

CHAPTER 40 **40.1** (left) Donna Hutchins; (right) Edward S. Ross / **40.2** (a) (c) Harlo H. Hadow; (b) Bob and Miriam Francis/Tom Stack & Associates / **40.3** Clara Calhoun/Bruce Coleman Ltd. / **40.4** After G. Gause, 1934 / **40.5** (a) (c) Jane Burton/Bruce Coleman Ltd.; (b) Heather Angel; (d) (e) Jane Lubchenco, *American Naturalist*, 112:23-29, © 1978 by The University of Chicago Press / **40.6** After N. Weland and F. Bazazz, *Ecology*, 56:681-188, © 1975 Ecological Society of America / **40.8** Photograph Ed Cesar/Photo Researchers / **40.9** © C. James Webb, Phototake, NY / **40.10** (a) James H. Carmichael; (b) Edward

S. Ross; (c) W. M. Laetsch / **40.11** Edward S. Ross / **40.12** (a) (b) Thomas Eisner, Cornell University; (c) Douglas Faulkner/Sally Faulkner Collection; (d) Edward S. Ross / **40.13** (a) (b) (c) (d) Roger K. Burnard; (e) Roger A. Powell/Visuals Unlimited / **40.14** Glenn M. Oliver/Visuals Unlimited / **40.15** (a) John Carnemolla/Australian Picture Library/Westlight; (b) Peter Bird/Australian Picture Library/Westlight / **40.16** Angelina Lax/Photo Researchers / **40.17** After W. Dansgaard et al., *Nature*, 364:218-220, July, 15 1993; D. Raymond et al., *Science*, 259:926-933, February 1993; W. Post, *American Scientist*, 78:310-326, July-August 1990 / **40.18** (above) Photograph Dr. Harold Simon/Tom Stack & Associates; (below) After S. Fridriksson, *Evolution of Life on a Volcanic Island*, Butterworth: London, 1975 / **40.19** (a) (b) After J. M. Diamond, *Proceedings of the National Academy of Sciences*, 69:3199-3201, 1972; (c) David Cavagnaro / **40.20** (a) (b) Edward S. Ross / **40.21** Heather Angel/Biofotos /

CHAPTER 41 **41.1** (a) (b) Wolfgang Kaehler / **41.3** Art by Gary Head; Photograph Bruce Coleman, Ltd. / **41.4** Gene C. Feldman and Compton J. Tucker/NASA, Goddard Space Flight Center / **41.8** Gerry Ellis/The Wildlife Collection / **41.9** Art by Lisa Starr / **41.10** (a) Photograph by Gene E. Likens from G. E. Likens and F. H. Bormann, *Proceedings First International Congress of Ecology*, pp. 330-335, September 1974, Centre Agric. Publ. Doc. Wagenigen, The Hague, The Netherlands; (b) Photograph by Gene E. Likens from G. E. Likens et al., *Ecology Monograph*, 40(1):23-47, 1970; (c) After G. E. Likens and F. H. Bormann, "An Experimental Approach to New England Landscapes" in A. D. Hasler (ed.), *Coupling of Land and Water Systems*, Chapman & Hall, 1975 / **41.11** Art by Lisa Starr / **41.12** D. W. Schindler, *Science*, 184:897-899 / **41.13** (a) Art by Lisa Starr and Paul Hertz; (c) Photograph John Lawlor/FPG / **41.14** Art by Lisa Starr / **41.16** After W. Dansgaard et al., *Nature*, 364:218-220, 15 July 1993; D. Raymond et al., *Science*, 259:926-933, February 1993; W. Post, *American Scientist*, 78:310-326, July-August 1990 / **41.17** Art by Lisa Starr / **41.18** Art by Lisa Starr and Paul Hertz / **41.21** Photograph Ted Scambos, National Snow and Ice Data Center /

CHAPTER 42 **42.1** (a, above) Edward S. Ross; (below) David Noble/FPG; (b, above) Edward S. Ross; (below) Richard Coomber/Planet Earth Pictures / **42.2** Edward S. Ross / **42.3** Art by Lisa Starr; Photograph NASA / **42.4** (b) Art by L. Calver / **42.6, 42.7, 42.8** Art by Lisa Starr / **42.10** Art by Lisa Starr; Map courtesy of the World Wildlife Fund / **42.11** Art by Raychel Ciemma / **42.12** Art by D. & V. Hennings after Whittaker; Bland; and Tilman / **42.13** Harlo H. Hadow / **42.14** (above) John D. Cunningham/Visuals Unlimited; (inset) AP/Wide World Photos; (below) Jack Wilburn/Animals Animals / **42.15** (a) Kenneth W. Fink/Ardea, London; (b) Ray Wagner/Save the Tall Grass Prairie, Inc.; (c) Jonathan Scott/Planet Earth Pictures / **42.16** (left) © 1991 Gary Braasch; (right) Thase Daniel; (below, left) Adolf Schmideker/FPG; (right) Edward S. Ross / **42.17** Thomas E. Hemmerly / **42.18** (a) Dennis Brokaw; (b) Jack Carey; (c) Nigel Cook/Dayton Beach News Journal/Sygma / **42.19** (a) Fred Bruemmer; (b) Doug Sokell/Visuals Unlimited / **42.20** D. W. MacManiman / **42.21, 42.22** Modified after Edward S. Deevy, Jr., *Scientific American*, October 1951 / **42.23** (a–e) E. F. Benfield, Virginia Tech. / **42.24** Art by Lloyd K. Townsend / **42.25** (a) Dennis Brokaw; (b) McCutcheon/ZEFA; (c) Robert Hessler; Woods Hole Institution of Oceanography; (d) J. Frederick Grassle, Woods Hole Institution of Oceanography / **42.27** (top) Jim Doran; (center inset) Sea Studios/Peter Arnold, Inc.; (below) Douglas Faulkner/Sally Faulkner Collection; **page 745**, (moray eel) Jeff Rotman; (chambered nautilus) Alex Kerstitch; All other photographs Douglas Faulkner/Sally

Faulkner Collection / **42.28** E. R. Degginger; Art by D. & V. Hennings / **42.29** (a) Courtesy of J. L. Sumich, *Biology of Marine Life*, Fifth edition, William C. Brown, 1992; (b) © 1991 Gary Braasch / **42.30** Art by Lisa Starr. Adapted from Tom Garrison, *Essentials of Oceanography*, Wadsworth, 1995 / **41.31, 42.32** Art by Lisa Starr / **42.35** (a) Jack Carey / **42.33** NOAA/Laboratory for Satellite Altimerty / **42.34** (a) © Dennis Kunkel/Phototake NYC; (b) © Roland Birke/OKAPIA/Photo Researchers, Inc. (c) (d) Courtesy Rita Colwell and A. Huq/UMBI and Byron Wood and Brad Lotitz/NASA; (e) Raghu Rai/Magnum Photos /

CHAPTER 43 **43.1** (above) Photograph Gary Head; (below) U. S. Geological Survey / **43.2** Photograph United Nations / **43.3** (a) USDA Forest Service; (b) Heather Angel; Art after G. T. Miller, Jr., *Environmental Science: An Introduction*, Wadsworth, 1986, and the Environmental Protection Agency / **43.4** Center for Air Pollution Impact and Trend Analysis (CAPITA), Washington University, St. Louis, MO / **43.5** (a) (c) NASA; photograph National Science Foundation / **43.6** Data from G. T. Miller, Jr. / **43.7** (above) R. Bieregaard/Photo Researchers; (below) After G. T. Miller, Jr. *Living in the Environment*, Eighth edition, Wadsworth, 1993 / **43.8** NASA / **43.9** William Campbell/ *Time* magazine / **43.10** Gerry Ellis/The Wildlife Collection / **43.11** Thomas G. Meier/USDA Soil Conservation Service / **43.12** Agency for International Development / **43.13** Dr. Charles Henneghien/Bruce Coleman Ltd. / **43.14** From Water Resources Council / **43.15** Ocean Arks International / **43.16** After G. T. Miller, Jr. *Living in the Environment*, Tenth edition, Wadsworth / **43.17** Photograph AP/Wide World; Art by Precision Graphics after Marvin H. Dickerson, "ARAC: Modeling an Ill Wind" in *Energy and Technology Review*, August 1987. Used by permission of University of California, Lawrence Livermore National Laboratory and U.S. Department of Energy / **43.18** Alex MacLean/Landslides / **Page 767** J. McLoughlin/FPG / **43.19** © 1983 Billy Grimes /

Chapter 44 **44.1** (left) John Bova/Photo Researchers; (right) Robert Maier/Animals Animals / **44.2** (a) From L. Clark, *Parasitology Today*, 6(11)/1990, Elsevier Trends Journals, Cambridge, U.K.; (b) Jack Clark/Comstock, Inc / **44.3** (a) Eugene Kozloff; (b) (c) Stevan Arnold / **44.4** Photograph Hans Reinhard/Bruce Coleman Ltd.; (a) (b) From G. Pohl-Apel and R. Sussinka, *Journal for Ornithologie*, 123:211-214 / **44.5** (a) Evan Cerasoli; (b) From A. N. Meltzoff and M. K. Moore, "Imitation of Facial and Manual Gestures by Human Neonate," *Science*, 198:75-78. Copyright 1977 by the AAAS / **44.6** Eric Hosking / **44.7** Nina Leen in *Animal Behavior*, Life Nature Library / **44.8** Mark Bekoff / **44.9** (a) Edward S. Ross; (b) E. Mickleburgh/Ardea, London / **44.10** Art by D. & V. Hennings / **44.11** John Alcock / **44.12** (left) David Fritts/Animals Animals; (right) Ray Richardson/Animals Animals / **44.13** Michael Francis/The Wildlife Collection / **44.14** Frank Lane Agency/Bruce Coleman Inc. / **44.15** Fred Bruemmer / **44.16** John Alcock / **44.17** Timothy Ransom / **44.18** A. E. Zuckerman/Tom Stack & Associates / **44.19** Patricia Caulfield / **44.20** John Alcock / **44.21** Kenneth Lorenzen / **44.22** Gregory D. Dimijian/Photo Researchers / **44.23** Yvon Le Maho / **44.26** F. Schutz / **44.24** Lincoln P. Brower / **44.25** Photograph John Dominis, *Life* Magazine, © Time Inc. /